D1754303

From Alexandria, Through Baghdad

Nathan Sidoli
Glen Van Brummelen
Editors

From Alexandria, Through Baghdad

Surveys and Studies in the Ancient Greek
and Medieval Islamic Mathematical Sciences
in Honor of J.L. Berggren

Springer

Editors
Nathan Sidoli
School for International Liberal Studies
Waseda University
Tokyo, Japan

Glen Van Brummelen
Quest University
Squamish
British Columbia, Canada

Picture on Page VI: Copyright 2000, Simon Fraser University Media Center.

ISBN 978-3-642-36735-9 ISBN 978-3-642-36736-6 (eBook)
DOI 10.1007/978-3-642-36736-6
Springer Heidelberg New York Dordrecht London

Mathematics Subject Classification (2010): 01A20, 01A30, 01A35, 01A40, 01A85, 01A90

© Springer-Verlag Berlin Heidelberg 2014
This work is subject to copyright. All rights are reserved by the Publisher, whether the whole or part of the material is concerned, specifically the rights of translation, reprinting, reuse of illustrations, recitation, broadcasting, reproduction on microfilms or in any other physical way, and transmission or information storage and retrieval, electronic adaptation, computer software, or by similar or dissimilar methodology now known or hereafter developed. Exempted from this legal reservation are brief excerpts in connection with reviews or scholarly analysis or material supplied specifically for the purpose of being entered and executed on a computer system, for exclusive use by the purchaser of the work. Duplication of this publication or parts thereof is permitted only under the provisions of the Copyright Law of the Publisher's location, in its current version, and permission for use must always be obtained from Springer. Permissions for use may be obtained through RightsLink at the Copyright Clearance Center. Violations are liable to prosecution under the respective Copyright Law.
The use of general descriptive names, registered names, trademarks, service marks, etc. in this publication does not imply, even in the absence of a specific statement, that such names are exempt from the relevant protective laws and regulations and therefore free for general use.
While the advice and information in this book are believed to be true and accurate at the date of publication, neither the authors nor the editors nor the publisher can accept any legal responsibility for any errors or omissions that may be made. The publisher makes no warranty, express or implied, with respect to the material contained herein.

Cover Illustration: Leiden University Library. Ms. Or. 680, f. 43r.

Printed on acid-free paper

Springer is part of Springer Science+Business Media (www.springer.com)

For Len

An Appreciation

After a full lifetime in academia, one learns to appreciate colleagues who have a great sense of humor, who do not take themselves too seriously, and who are enthusiastic about life both within and away from the academy. One especially learns to appreciate these characteristics since they have become so rare. With the rise of over-specialization, the ever-increasing market drive in colleges and universities, and the regrettable incursion of competitive pressures for career advancement, academicians are becoming more inward focused and less committed to their colleagues and their institutions. Fortunately, there are still "throw-backs" to an earlier tradition and Len Berggren personifies all that is positive in that tradition.

My first acquaintance with Len was in the spring of 1983, when I had organized the first meeting of what became the Columbia History of Science Group. As an organizational neophyte, I had invited people to attend the meeting, but had neglected to require pre-registration or any other form of financial commitment. Thus, after I had reserved the meeting space, the accommodations, and the catering requirements, I found myself on the financial hook for the entire bill. I was expecting twenty-five attendees, but soon the cancellations began to trickle in at the last moment, cancellations that pointed toward personal financial ruin. When the numbers had shrunk to twenty, I knew my assistant professor's salary was in real trouble. Then Professor Berggren from Simon Fraser drove up in his Volvo, not alone, but with his lovely wife Tasoula and their two young boys. With twenty-three attendees, I broke even!

Although I had not previously met Professor Berggren, I was immediately drawn to Len and his family, but not just because they saved my bank account. First of all, his last name and choice of automobile both indicated that Len and I shared the same nationality (Sweden). Then, the manner in which the family interacted gave a clear indication that here was an academic who cared more for his family than his career. Indeed, the four Berggrens seemed genuinely excited to be together and also seemed to enjoy each other's company. Finally, when Len presented his paper on some arcane subject of ancient mathematics—indeed, probably ancient Islamic mathematics!—his entire family was in the audience. I was soon to learn that Tasoula, with her roots in the eastern Mediterranean basin, was an active collaborator with Len in his work and that the sons, too, knew of their parents' academic interests.

But I have not commented on Len's wonderful sense of humor and, above all, his personal humility. Since our auspicious meeting in 1983, I have had the pleasure of spending a long weekend at Friday Harbor with Len almost every year. One of the traditions we have had at our meeting is that the annual Milosian speaker never materializes (indeed, we always invite someone who cannot, for many reasons, attend the meeting). But Len was not in on the joke and, one year, volunteered to give the evening talk. With slides and a prepared talk, no one could bear to tell Len that the talk was designed *not* to be given. But in a way, the joke was on all of us, since Len gave a very engaging talk, all the time poking fun at the academic enterprise and providing us all with a heightened sense of enlightenment and levity.

Len and Tasoula are also gracious hosts and superb companions. When I moved to Vancouver, they invited me to spend a weekend in their mountain condominium at Whistler. It was a real pleasure to be with Len in such a relaxed setting. Walking, fishing, visiting, and laughing were all high on the agenda. And since I am not an historian of mathematics, nor an ancient scholar, nor an expert on Islamic science, our conversations were hardly on the professional level. But that is what makes Len such a wonderful person. He is a remarkable scholar, but above all, he is a wonderful human being and one person for whom I am enormously grateful to have known.

Vancouver, October 2010 *Keith R. Benson*

Preface

The practice of the history of mathematics is in flux. This statement may seem ironic or even paradoxical, for a discipline that relies seemingly on logic and precision. However, trends in the scholarly practice of history are gradually causing substantial changes in the questions raised by practitioners of the discipline, and the methods used to try to arrive at answers. Fifty years ago the discipline was populated mostly by mathematicians, emphasizing logical reconstructions and explanations; today, the concerns of the historian are being heard. Surprisingly, nowhere are these changes more true than in the study of ancient Greek mathematics, which has seen precious little genuinely new source material come to light in the past decades. Debates concerning, for instance, how to consider the Greek notion of geometric algebra led to an increasing realization that even mathematical structures can be shaped by cultural perspective; therefore, modern reconstructions of ancient concepts contain the hidden danger of saying as much about the historian as about the history.

With respect to medieval Islamic mathematics, the manuscript situation is entirely different: hundreds or even thousands of texts remain in libraries, unedited and not yet easily available to the scholar. Nevertheless similar trends have been felt. Categorizing subdisciplines according to our modern mathematical perspective is rightly viewed with suspicion. The border between mathematics and science was entirely different, and subjects like astrology need to be placed in a context appropriate for its time, not ours. The assumption of a single Islamic mathematical culture has been criticized, and nuanced evaluations of different mathematical communities are starting to arise. Even the traditional focus on the best theoretical mathematics of the time period has been questioned, raising the need to clarify the purpose of studying the history of mathematics in the first place. Is our primary task to provide a plausible logical path that led to the present mathematics, or should we rather be concerned with how mathematics clarifies the human condition within and between different cultures?

The literature accompanying these shifts in perspective has been vast, and negotiating the terrain can be difficult even for a scholar in these areas, let alone a new graduate student or an interested outsider. We are thus truly fortunate that a small tradition of surveys of the current state of the art, both in Greek and in Islamic mathematics, was initiated in the 1980s by J. Lennart Berggren. His 1984 survey of Greek mathematics (the first article in this volume), followed by a 1998 survey by Ken Saito (the second article), vividly tell the story of changing foci and perspectives. On the Islamic side, Len's 1985 and 1997 surveys (the fourth and fifth articles) tell the story of a vast literature coming to light, and the interaction of different points of view in coming to grips with it.

These four articles, so helpful to us as developing scholars, have not received updates since then (other than a couple of surveys in particular subdisciplines). As a fitting tribute to a man so important to us, and as a vital service to new scholars, graduate students, and interested readers, we have brought these four papers together in this volume, and written two new survey papers (one Greek, and one Islamic) to take the narrative to the present day. Through this we hope to provide a handbook or guide to developments in the field over the past forty years. In addition, we hope that the combined bibliographies of these articles, some 900 entries altogether, can form a kind of paper database to provide a guide to the recent practice of the discipline. They also bear witness to the shifts in perspective that

Len has witnessed, reported, and influenced since he began his historical career almost forty years ago.

The survey papers were just one aspect of the many contributions that Len has made to the history of both Greek and Islamic mathematics over this period. Dozens of journal articles in both areas reveal Len's role as a leader in shaping new and more nuanced approaches to the history of mathematics, while preserving the great achievements of our predecessors. The diversity of topics within these papers, including geometry, arithmetic, astronomy, and geography, is an impressive record of following the historical actors where they went, rather than using the material to one's own purposes — the art of the true historian. We are particularly grateful for Len's commitment to the primary literature as the key to grounding one's perspective. This is witnessed in many of his papers as well as his two translations of ancient texts (Euclid's *Phaenomena* with Robert Thomas, and Ptolemy's *Geography* with Alexander Jones).

But Len is much more than just a scholar. We find among his publications a devotion to taking the results of our labors to the broader community. Len worked tirelessly producing and editing encyclopedia articles, and wrote a variety of articles for pedagogical and popular publications. Most important among these is his *Episodes in the Mathematics of Medieval Islam* (1986), a popular book that continues to have a deep impact in increasing public awareness of Islamic contributions to mathematics. Recently it has been translated into German, and will soon appear in a second edition. Some of Len's most valued works combine his dedication to the original sources with public outreach: he edited the Islamic section of Victor Katz's *The Mathematics of Egypt, Mesopotamia, China, India and Islam* and is preparing a section for a supplementary volume; and he edited (with Jonathan and Peter Borwein) three editions of *Pi: A Sourcebook*. It is out of this latter project that we are fortunate to have from one of Len's non-historical colleagues, Jonathan Borwein, a survey of the story of π in this volume, as a tribute to Len's commitment to connecting the academic discipline of the history of mathematics with the broader community. Finally, Len is an active devotee of sundials and their many varieties; he has contributed articles to *The Compendium*, the official publication of the North American Sundial Society. One of the editors' fondest memories is of helping Len and his sundial colleague Brian Albinson to paint a split analemmatic sundial on the parking lot of Simon Fraser University, Len's home institution. A picture of Len serving as the gnomon for this dial is available on the Internet.

It is a truly a mark of the great respect that Len's colleagues hold for him that so many of them wished to join in this project in his honor. The second part of this book is a collection of seventeen studies by his friends that contribute to Len's interests in various ways. Each of these studies exemplifies Len's approach, respecting the original texts and following the historical trail where it leads. We offer this volume to Len as a tribute to his powerful scholarship and gentle leadership, and with our best wishes for a healthy and productive future.

San Diego, January 2013 *Glen Van Brummelen* and *Nathan Sidoli*

Acknowledgements

We wish to express our gratitude to the editors at Springer, Clemens Heine and Mario Aigner, for encouraging us in this project and accommodating our needs for this somewhat involved book. We are grateful to Tony Lévy and Kim Plofker for reviewing two papers, the contents of which were outside the areas we felt competent to judge. We are grateful to our research assistants for clerical and TEX related work: Jonathan von Ofenheim, Shauna Bryce and Nicholas Hanpeter. A special thanks goes to Nicholas Hanpeter for doing much of the work involved in producing the final documents for processing with XeTeX.

This book was typeset using XeTeX. We received support and advice from a number people on the XeTeX listserve and at http://tex.stackexchange.com. Among them, a special thanks to Vafa Khalighi for his work accommodating the bidi package to Springer's style. The font for the Roman script is Peter Baker's Junicode. Those for the Arabic and Greek scripts are Scheherazade and Gentium Plus, both maintained by SIL International. All three of these fonts are available under the Open Font License. For the Hebrew script we have licensed John Hudson's SBL Hebrew from the Society of Biblical Literature.

Nathan Sidoli and *Glen Van Brummelen*

Contents

Part I Surveys

History of Greek Mathematics: A Survey of Recent Research [1984] 3
J. Lennart Berggren
References 12

Mathematical Reconstructions Out, Textual Studies In: 30 Years in the Historiography of Greek Mathematics [1998] 17
Ken Saito
References 22

Research on Ancient Greek Mathematical Sciences, 1998–2012 25
Nathan Sidoli
References 44

History of Mathematics in the Islamic World: The Present State of the Art [1985] 51
J. Lennart Berggren
References 66

Mathematics and Her Sisters in Medieval Islam: A Selective Review of Work Done from 1985 to 1995 [1997] 73
J. Lennart Berggren
References 91

A Survey of Research in the Mathematical Sciences in Medieval Islam from 1996 to 2011 ... 101
Glen Van Brummelen
References 124

Index of Subjects (Survey Papers) 139

Part II Studies

Mechanical Astronomy: A Route to the Ancient Discovery of Epicycles and Eccentrics 145
James Evans and Christián Carlos Carman
References 171

Some Greek Sundial Meridians 175
Alexander Jones
References 187

An Archimedean Proof of Heron's Formula for the Area of a Triangle: Heuristics Reconstructed ... 189
Christian Marinus Taisbak
References ... 198

Reading the Lost Folia of the Archimedean Palimpsest: The Last Proposition of the *Method* . 199
Ken Saito and Pier Daniele Napolitani
References ... 225

Acts of Geometrical Construction in the *Spherics* of Theodosios ... 227
Robert Thomas
References ... 236

Archimedes Among the Ottomans: An Updated Survey ... 239
İhsan Fazlıoğlu and F. Jamil Ragep
References ... 246

The "Second" Arabic Translation of Theodosius' *Sphaerica* ... 255
Richard Lorch
References ... 258

More Archimedean than Archimedes: A New Trace of Abū Sahl al-Kūhī's Work in Latin ... 259
Jan P. Hogendijk
References ... 273

Les mathématiques en Occident musulman (IXe–XVIIIe s.) : Panorama des travaux réalisés entre 1999 et 2011 ... 275
Ahmed Djebbar
Littérature ... 290

Ibn al-Raqqām's *al-Zīj al-Mustawfī* in MS Rabat National Library 2461 ... 297
Julio Samsó
References ... 325

An Ottoman Astrolabe Full of Surprises ... 329
David A. King
References ... 341

Un algébriste arabe : Abū Kāmil Šuǧā^c ibn Aslam ... 343
Adel Anbouba (avec les commentaires de Jacques Sesiano)

Abū Kāmil's *Book on Mensuration* ... 359
Jacques Sesiano
References ... 408

Hebrew Texts on the Regular Polyhedra ... 409
Tzvi Langermann
References ... 466

A Treatise by al-Bīrūnī on the Rule of Three and its Variations ... 469
Takanori Kusuba
References ... 485

Safavid Art, Science, and Courtly Education in the Seventeenth Century ... 487
Sonja Brentjes
References ... 501

Part I
Surveys

History of Greek Mathematics: A Survey of Recent Research [1984]

J. Lennart Berggren

Abstract This survey reviews research in four areas of the history of Greek mathematics: (1) methods in Greek mathematics (the axiomatic method, the method of analysis, and geometric algebra); (2) proportion and the theory of irrationals (controversies over the origins of the theory of incommensurables); (3) Archimedes (aspects of controversies over his life and works); and (4) Greek mathematical methods (including discussion of Ptolemy's work, connections between Greek and Indian mathematics, the significance of Greek mathematical papyri, Arabic texts, and even archaeological investigations of scientific instruments).

When I first became interested in the history of mathematics, around the year 1970, I wrote to Ken May to ask his advice on which areas would be most fruitful for research, and I listed a variety of areas, asking for his opinion on each one. As was Ken's custom, he replied by returning my letter, annotated with his *marginalia*. I do not remember all the areas or his answers, but two do stand out in my memory. The one was "modern mathematics" and the other was "Greek mathematics." Next to the first he had penned an enthusiastic "Yes, badly needs work;" next to the second there was only the laconic "Over-researched." No one who has given advice to another will be surprised to learn I decided to do the history of Greek mathematics.

If one takes the remarks of Ken that I quoted above in the sense of saying that the history of ancient
Greek mathematics has been much more thoroughly re|searched than that of modern mathematics, 395
then of course they are quite true. However, far from awaiting only the filling-in of minor details, the history of Greek mathematics is an area where there is still controversy over some of the main features and where issues of considerable historical importance are still unsettled.

To give some focus (and boundaries) to this survey I have limited it to a review of four areas: (1) methods in Greek mathematics, (2) proportion and the theory of irrationals, (3) Archimedes, and (4) Greek mathematical methods.

Editors: Originally published as Berggren, J.L., 1984, History of Greek mathematics: A survey of recent research, Historia Mathematica 11, 394–410. We thank J.L. Berggren and Elsevier for permission to republish this paper.
To extend the bibliography even earlier for Greek mathematics, see Len's Greek mathematics: An annotated bibliography, in Dauben, J.W. (ed.), History of Mathematics from Antiquity to the Present: A Selective Bibliography, Garland, New York, 1984, pp. 25–64.

N. Sidoli and G. Van Brummelen (eds.), *From Alexandria, Through Baghdad: Surveys and Studies in the Ancient Greek and Medieval Islamic Mathematical Sciences in Honor of J.L. Berggren*, DOI 10.1007/978-3-642-36736-6_1,

© Springer-Verlag Berlin Heidelberg 2014

Methods in Greek Mathematics

Anyone who has studied the mathematics of any period realizes that the key to understanding it lies as much in a deep understanding of its methods as in a comprehensive knowledge of its results. Thus I want to begin with an account of the recent studies of three very general methods in Greek mathematics: the axiomatic method, the method of analysis, and geometric algebra.

The Axiomatic Method

There has been considerable debate concerning the origins of the axiomatic method in Greek mathematics. In [Szabó 1968] there appeared the thesis that it was the influence of Eleatic philosophy (in particular Zeno) that shaped the "theoretical foundations of deductive mathematics." Szabó believes that these foundations were made necessary by the discovery of incommensurables within such problems in music theory as that of dividing the octave into two equal intervals. The ratio that determines the octave is $2 : 1$, and to divide it into equal intervals means to find a number x so that $2 : x = x : 1$. This is but one example of the general problem of finding a mean proportional between two numbers in the ratio of $(n + 1) : n$, and thus it was within music theory, according to Szabó, that the theory of proportions, and indeed much of Euclidean arithmetic, originated. Completing Szabó's break from much of the current view of the history of Greek mathematics is a consequence of his belief that the theory of incommensurables was worked out in the fifth century BC, viz., that Theodoros and his pupil Theaetetos played no very important role in the history of Greek mathematics, since the achievements that Plato's dialogue *Theaetetos* seems to attribute to them must have been well-known parts of mathematics considerably before their time. (However, the interpretation of the "*Theaetetos* Passage" has practically become a separate branch of the history of Greek mathematics, and I shall deal later with its extensive literature.)

The point I am making here is that Szabó — like [Mugler 1948] — sees an important part of Greek mathematics as having originated in response to stimulus from philosophers. Objections of a general sort to Szabó's arguments on the Eleatic background have been raised in the review by [Knorr 1981] on the grounds that known history is not consistent with consequences of Szabó's thesis on the Eleatic origins of axiomatic mathematics. A more extensive critical discussion of Szabó's basic thesis on the
396 priority of logic over mathematics may be found in [Berka 1980]. |

I must say that it is not clear to me how far Szabó means to press his thesis, whether he is saying no Greek had ever phrased a mathematical argument prior to Parmenides, or if he is claiming only that the creation of an axiomatic foundation (as distinct from a loose deductive structure) for mathematics was a response to Eleatic criticism. On the other hand, on the early relationship between mathematics and music, I know of no serious effort to dispute Szabó's philological arguments on the musical origins of the vocabulary and concepts of proportion theory.

The Method of Analysis

The basic paper here is [Mahoney 1968], which goes beyond the past discussions of Greek geometrical analysis, which attempt to show how the ancient accounts could be squared with modern views, and studies, instead, the historical development of geometrical analysis as a body of techniques that are useful as much for solving problems as for proving theorems. Mahoney has shown that analysis not only led to solutions of problems such as duplicating the cube or trisecting the angle but also led to the recognition of general problems — such as verging constructions, in which one is asked to place a segment of fixed length in such a way that its endpoints lie on two given curves and it verges toward a given point. As general problems these arose out of specific cases occurring in the analysis of other problems.[1]

An account of ancient analysis that is, to some extent, an alternative to that offered in [Mahoney 1968] is found in [Hintikka and Remes 1974], where it is argued that a "directional" account of analysis as a search for consequences is misleading since Pappos' analysis is largely a search for "concomitants" in geometrical figures rather than consequences in proofs. From the point of view of historical method the work is an attempt to use the philosophy of mathematics to shed light on historical questions, an approach that is more common elsewhere in the history of science.

With regard to the results of analysis, it is a hypothesis in [Zeuthen 1896] that the origin of the loci that were to become known as conic sections is to be found in Menaechmos' analysis, early in the fourth century BC, of the problem of duplicating the cube.[2] [Knorr 1982] again advances Zeuthen's hypothesis on the origin of the curves, but goes on to argue that it was not until several decades after Menaechmos' statement of their analytic definition that they were realized to be plane sections of cones. To argue thus, Knorr must maintain that several generations of historians have erred in reading Eratosthenes' famous injunction "Do not cut the cone in the triads of Menaechmos" as implying that Menaechmos realized his triad of curves (hyperbola, parabola, and ellipse) were sections of a cone. Of course, Greek *testimonia* and *scholia* are often vague, but, since this has always seemed one of the less vexing ones, there are no adequate grounds for reading it other than as it has been read.

Knorr also supports Zeuthen's view that the somewhat special initial geometrical definition of the conic sections, as sections of an isosceles cone (of varying vertex angle) by a plane perpendicular to a generator of that cone, arose out of a | desire to show that the locus of points, described in analytic 397
terms as one in which certain squares were equal to certain rectangles, did exist as a continuous curve. To give this proof it was most convenient to use the above method of sectioning the cone, and thus it came about that what appears to be an awkward requirement was deliberately adopted for technical reasons.

The origins of analysis are another topic that has tempted historians, and some recent writings on analysis have attempted to find these origins in philosophical thought — in the school of Plato if we are to believe Proklos. Thus [Mugler 1948] argues that Plato was inspired to discover analysis by analogy with his idea of the upward movement of the mind in the dialectic, from a proposition to the hypothesis that implies it, from that hypothesis to a higher one, and so on until the mind reaches the ultimate source, which is ahypothetic. However, a convincing refutation of this view of the origins of analysis is given in [Cherniss 1951]. Cherniss argues that it is highly unlikely that the dialectic, an "upward" movement from proposition to hypothesis, until one reaches the ahypothetic, would have inspired analysis, a passage from a proposition to consequence. On the other hand [Lafrance 1980] argues that, far from inventing geometrical analysis, Plato was inspired by the use of it he saw in the practice of contemporary mathematics to form his philosophy of the dialectic. However, Cherniss' objections to any account that would have analysis derive from the dialectic would seem to apply equally well to accounts that would have the mathematician's practice of analysis inspire the Platonic dialectic.

Geometric Algebra

The scholarly point at issue here is whether it is historically justified to interpret parts of Greek mathematics, typified by Book II of Euclid's *Elements*, as translations of Babylonian algebraic identities and procedures into geometric language. The view that this represents what happened historically arose from conjectures of Zeuthen and Tannery and the subsequent publication by Neugebauer and others of the Babylonian mathematical cuneiform texts. The *locus classicus* is [Neugebauer 1936]. So far as I know

---

[1] This passage from particular cases of a problem to the recognition of the existence of a general area of investigation reminds one of a point made in [Waterhouse 1972] that the discovery of "regular solids" goes beyond the discovery of cube, pyramid, etc., to a conscious recognition of the idea of regular solids as a class of elements having structural features in common.

[2] For an alternative hypothesis on the origin of the conic sections see [Neugebauer 1948].

this theory is widely accepted,[3] but an early dissenting voice was that of [Szabó 1968], who suggested that "geometric algebra" ought to be replaced by the term *geometry of areas* in order to emphasize that the theorems are geometrical theorems, used to prove other theorems in geometry, and that there is no concrete evidence that pre-Euclidean Greeks took over Babylonian algebra and recast it in geometric form.

This point was raised again, rather more polemically, by [Unguru 1975], who argued that modern accounts of Greek mathematics have been so strongly affected by the concept of geometric algebra that it is now necessary to rewrite the whole subject. Unguru's attack provoked many vigorous responses [Van der Waerden 1976a; Freudenthal 1977; Weil 1978]. (The last word in that round of the controversy seems to have gone to the offense in [Unguru 1979].)

The point at issue between Neugebauer and Van der Waerden on the one hand and Szabó and Unguru
398 on the other has been fairly analyzed by [Mueller 1981], | who reaches the conclusion, based in large part on an analysis of how various propositions in *The Elements* are used, that "a strictly geometric reading of *The Elements* is ... sufficiently plausible to render the importation of algebraic ideas unnecessary" [p. 44]. However, when Mueller, in the context of this debate, writes of the propositions in *The Elements*, Book II, that "a final judgement on their meaning depends upon an examination of these applications in other parts of the work" one wonders what he is suggesting by the word "meaning." If this is to be a synonym for "use" or even "significance" then, unquestionably, a study of the applications of Book II in the *Elements* will shed light on this question, but a study of the use of the theorems of Book II cannot give an convincing answer to the question of the origins of the method of geometrical algebra.

My own view, therefore, is that to establish geometrized algebra as a historical fact still requires that considerable research be done on the time and method of transmission of Babylonian mathematical knowledge to the Greek world. Some Babylonian ideas — for example, the gnomon — seem to have been transmitted at an early date, whereas other notions — that of degree measurement of angles and the sexagesimal system, even in the modified form in which the Greeks used it, for example — seem to have arrived after Euclid wrote. When, in this interval, one is to date the importation and geometrization of Babylonian algebra is a historical question to be settled not by conjecture but by research. If the event cannot be located historically one must recognize the possibility that it may not have occurred.

The Theory of Proportion

The traditional form of the history of this theory relates how the early Pythagoreans, influenced by their belief that "all is number," held to the conviction that any ratio is the ratio of two whole numbers. This account was developed further in [Becker 1933], where it is argued, on the basis of testimony of Aristotle and other ancient sources, that there was a theory of proportion that replaced the old numerical one but was not yet the theory supplied by Eudoxos. This intermediate theory was based on the repeated subtraction of the lesser of two magnitudes to be compared from the greater, then the remainder from the lesser, etc., and went under the name of *anthyphairesis* or *antanairesis*. This procedure is precisely equivalent to the present-day procedure of finding a continued fraction expansion, so that four terms are in proportion ($A : B :: a : b$) if (in modern language) the terms of the continued fraction expansion of $A : B$ are the same as those of $a : b$. (The proof of proportionality in the case when A and B are incommensurable obviously presents great difficulties and can be executed in general only when some periodicity can be established). Then, the account continues, Eudoxos established the general theory of proportion that a *scholium* to Book V credits him with, one which applies equally well to any two magnitudes of a given kind, whether commensurable or incommensurable — and all was light.

Eudoxos enters into the history of science, also, as the discoverer of the model of homocentric
399 spheres, and [Riddell 1979] shows how, by working with se|quences of numbers arising out of these

[3] See [Knorr 1975] for a reconstruction of late fifth- to early fourth-century mathematics that makes important use of the hypothesis of a geometrized algebra (e.g., [Knorr 1975, p. 91]) and [Van der Waerden 1954] for a general history where geometric algebra figures prominently.

investigations, Eudoxos could have been led to his famous definition of equality of ratios. Riddell's paper is one of the more interesting examples of the method of historical investigation that seeks the origin of a mathematical theory credited to a particular person in other work of that person. Another example of this *genre* is the argument in [Von Fritz 1945] that it was Hippasos of Metapontum who discovered the irrational in the course of his investigations of the pentagonal faces of the dodecahedron, but this is a subject we shall return to.[4]

Within the past five years, however, there have been two challenges to the above history of the theory of proportion. The first is in [Knorr 1978c], which argues that, although Eudoxos invented a theory of proportion, the Eudoxan theory is not that found in Euclid's *Book V* but rather the following: If A and B and c and d are two pairs of commensurable magnitudes and if A and B and c and d have common measures E and f resp. so that $A = nE$ and $c = nf$ while $B = mE$ and $d = mf$ then $A : B :: c : d$. (Note that this definition would be a corollary of a version of *Elements* VII, 10, stated for commensurable magnitudes rather than for numbers.) Knorr finds evidence of this theory in Archimedes' treatment of the proof of the law of the lever in the *Equilibrium of Planes*, Book I (Propositions 6 and 7) and in a theorem on the areas of circular sectors reported by Heron and Pappos and ascribed by Heron to Archimedes. Characteristic of this theory is a separation of proofs into commensurable and incommensurable cases, a use of the above definition in the commensurable case, and an indirect argument, based on a lemma found as a *scholium* to Theodosios' *Spherica* III, 9, to handle the incommensurable case.

Quite independently of whether or not Archimedes used the approach to ratios suggested by Knorr (and his work raises the whole question of "the young Archimedes"), it is certainly true that the pieces Knorr cites are genuine pieces of Greek mathematics and their approach to proving statements involving proportions does invite investigation.

The present state of affairs seems to be that Knorr has drawn attention to an interesting method for proving that four magnitudes are proportional, but the question of whether his method is based on a pre-Eudoxan theory may be regarded as being still open. It is, after all, a hazardous business to reconstruct a theory on the basis of arguments that, being entirely inferential, are unsupported by extant *testimonia*.

Very shortly after the above speculations on a third ancient proportion theory appeared, [Fowler 1979] reconsidered the character of the anthyphairetic theory and argued that it was the theoretical logistic Plato referred to and constituted, in fact, a theory of ratio and not of proportion; i.e. it assigned a precise meaning to the ratio $A : B$ in terms of the sequence of whole numbers characterizing the continued fraction expansion of A/B. [Fowler 1980, 1982] goes on to argue that, in terms of this theory, much of Book II of *The Elements* can be understood as a treatise incorporating theorems useful in carrying out the proof of the periodicity of the sequence of numbers obtained when the ratio $n : m$ is investigated by the | anthyphairetic method. Fowler's principal arguments for the correctness of his 400
view are that:

(1) It gives an area of mathematical investigation that could reasonably be called theoretical logistic and provides a sensible interpretation of passages in Plato that are otherwise obscure.

(2) It explains why many pre-Euclidean fragments (e.g., Hippocrates of Chios) speak of "ratio" rather than proportion and why the term "ratio" even occurs in places in Euclid, despite the fact that "the *Elements* does not contain a precise definition of ratio."

As for the second point, the operative word is surely "precise." Fowler himself admits that Book V of the *Elements* does contain a definition of ratio, and whether a modern reader considers it precise or not is less important than whether Euclid considered it sufficient. The evidence in the *Elements* suggests that Euclid felt he had defined it as much as he had defined any other term and therefore had no hesitation about using it. One feels that [Mueller 1981] makes a valuable point when he speaks of Euclid's "conception of definitions as characterizations of independently understood notions," and, although Mueller applies this to explain the presence of two unreconciled theories of proportion in *The Elements*, it seems to me equally applicable to Euclid's definition of ratio.

---

[4] Mugler's attempt to find the inspiration for the method of analysis in Plato's dialectic is yet another example. That none of these searches for origins have convinced most scholars indicates that the genre, handled carefully, may be useful in suggesting leads for further research but cannot, alone, be convincing.

As for Fowler's first point, that an anthyphairetic theory of ratios gives us a means of interpreting what seem to be cryptic references in Plato, one can agree with him that it does provide an interpretation, while still wondering about the validity of interpretations that we see in the *Charmides* 116A "Thus reckoning (*logistikē*), I suppose, is concerned with the even and odd in their numerical relations to themselves and one another ..." a reference to the fact that "the ordering by size of two ratios (expressed as continued fractions) does not follow a simple lexicographical rule ... but has this rule reversed in the odd-numbered places of the anthyphairetic sequence" [Fowler 1979, 830-831].

Theory of Incommensurables

Fowler's study of the theory of ratios raises the problems faced by any study of ancient mathematics that relies for part of its evidence on interpretations of passages in Plato, in this case the reference to theoretical logistic. Plato's writings are, of course, a major witness to early mathematical activity in Greece, and great energy has been expended to extract information on the state of mathematics from these writings, whose primary purpose was philosophical. One's judgment of the success of such efforts must depend in large measure on his sympathy with the conjectural reconstructions of "the whole story" that are so necessary in attempts to reconstruct a century of activity from the fragments found in the writings of Plato.

Another case in point is [Knorr 1975], in which a key element is a minute examination of a short passage in the dialogue *Theaetetos* that relates to the early history of irrationals. Knorr uses the resources of philology, comparisons with passages elsewhere in Plato, and a study of the ancient commentators
401 and modern | historians to put forward a reconstruction of the history of Greek mathematics up to the time of Eudoxos that takes issue with many widely accepted theories. For example:

(1) [Van der Waerden 1947/1949] develops a suggestion of [Zeuthen 1910] that Theodoros of Cyrene used anthyphairesis to prove the irrationality of square roots of non-squares from 3 to 17. Knorr, however, argues that the assumption of an anthyphairetic approach contradicts such features of the text as its implication that Theodoros got "tangled up" in difficulties at 17 (for the case of 17, as opposed to 13 or 19, is particularly easy). Thus Knorr concludes that the approach was not anthyphairesis.

(2) [Van der Waerden 1947/1949] carries further an argument in [Becker 1936] to support his claim that a deductive formulation of Euclidean number theory by Pythagoreans early in the fifth century included not only the theory of even and odd, identified by Becker at the end of the *Elements*, Book IX, but all of Book VII as well, a book that [Van der Waerden 1961] argues is a Pythagorean transformation of the ordinary arithmetic of fractions into the number theory of integers. Knorr is much more restrictive in the number theory he ascribes to the early Pythagoreans (or, indeed, to any of the mathematicians prior to Theaetetos) and argues that Theaetetos was the one who developed number theory from the rudimentary state that Theodoros had to make use of.

(3) Finally Knorr holds that Theaetetos worked out only part of the theory found in Book X, notably a theory of "three classes of irrational lines, corresponding to the Euclidean medial, binomial and apotome" and that Eudoxos of Knidos played a large part in the development of the rest. This goes against the generally held view (e.g., [Van der Waerden 1949]) that Book X is almost entirely the work of Theaetetos.

Many of Knorr's arguments have gained a positive response, and [Van der Waerden 1976b] appears to be convinced of Knorr's interpretation of the *Theaetetos* passage, in general, as well as his reconstruction of Theodoros' proofs; in particular [Van der Waerden 1976b] also appears to have modified his conviction in 1947 that Book VII goes back to the early Pythagoreans, at least to the extent that he now regards the issue as "open."

However, shortly before Knorr published his book, there appeared [Neuenschwander 1973], which argues, on the basis of an exhaustive examination of the logical structure of the first four books of the *Elements*, that the propositions of Book II, although brought up to date in their terminology at Euclid's time, revert to the Pythagoreans. This view of the origins of Book II (or at least its Propositions 5, 6,

11, and 14) is shared by [Szabó 1974], but for a word of caution see [Fischler 1979].

The recent controversy between [Knorr 1979] and [Burnyeat 1978, 1979] is as much concerned with Plato's intent in writing the section of *Theaetetos* dealing with Theodoros' geometry lesson as it is with philological matters. Knorr argues the passage must be interpreted in a way that makes mathematical sense. Burnyeat argues that, on the contrary, just as context is important in settling
philolog|ical questions, context is equally important in settling the question of the intent of the passage. 402
According to Burnyeat, the context is a discussion of the philosophical point that a definition is more than a list of specific instances of what is to be defined. Plato had no intention, says Burnyeat, of writing a history of incommensurables.

Thus the debate, even on a work as thoroughly researched as [Knorr 1975], continues, as it must whenever there are sufficient documents to support a variety of reconstructions but an insufficient number to narrow the list of contending theories to one.

Archimedes

Here, too, there are also a number of controversies, this time relative to the life and work of Archimedes, although here we have an incomparably greater fraction of known works extant than is the case with pre-Euclidean mathematics. This may be, of course, an illustration of what I call Goldstein's law, after my colleague, B. Goldstein, who related it to me as follows: "The discovery of manuscripts does not solve problems. It creates them."

On the other hand, the controversies on Archimedes may only illustrate the fact that genius is more easily recognized than understood and, for this reason, Archimedes' works have challenged the best efforts of generations of scholars.

However, [Knorr 1978d] puts the argument that earlier studies of Archimedes and his works display an "ahistorical attitude" in assuming "for the whole corpus of Archimedes' works, apart from the *Method*, an ever-present and uniformly rigorous formal technique." To counter this, Knorr proposes a new chronology of Archimedes' works in which the key to the new classification is "the greater application of standard Euclidean techniques in the earlier works," and one result of the study is the identification of fragments from Heron and Pappus as excerpts from the work of a young Archimedes. Knorr argues that, contrary to Heath, the *Method* is the last of Archimedes' extant works, and he also urges the view that *Measurement of the Circle* is one of his earliest works. (In regard to the latter work [Sato 1979] tries to restore the Greek text of Proposition I of that work on the basis of, among other sources, Euclid's phrasing of theorems on quadrature, for the author believes that Archimedes was under the influence of Euclidean terminology when he wrote that work.)

Much of Knorr's argument rests on his analysis of convergence techniques, where the Axiom of Archimedes plays a crucial role, and here he evidently views the import and function of that axiom in a way quite different from that of [Dijksterhuis 1956] and [Hjelmslev 1950]. Both of these scholars made a penetrating study of the function of the axiom in Archimedes' works and distinguished its import from that of Euclid V, Definition 4, and Euclid X, 1, although they differ on the motivation for Archimedes' introducing it.

Unfortunately these discussions seem not to have had the impact on modern scholarship that they ought to have had, and in a work as recent as the general survey of Archimedes studies by [Schneider
1979], one reads [p. 48] that Archi|medes' Axiom and Euclid X, 1, are equivalent. Although there is 403
much that is valuable in [Schneider 1979] it retains such misconceptions as the above, as well as the idea propounded in [Drachmann 1967] that Archimedes discovered some of his quadratures by actually making models and weighing them.

This myth has been further developed in [Sato 1981], an attempt to use philological investigations to restore an early Archimedean corpus of three works, generally cited by Archimedes as Τὰ μηχαγικά. The author comes to the odd conclusion that what Archimedes found nonrigorous in his *Method* was its foundation on actual physical experiment, whereas (we are assured) Archimedes had no doubts about the acceptability of using indivisibles.

On the subject of Archimedes' mechanical works [Drachmann 1963] performs a very valuable service in (to my mind correctly) identifying many fragments of Archimedes' lost works on statics and so provides additional material for study for one who is interested in this side of Archimedes' work. On the other hand [Berggren 1976] has the opposite effect, urging the removal of trivial and logically isolated parts of *Equilibrium of Planes*, I (E.P.I.) from the Archimedean corpus and suggesting that even the proof of the Law of the Lever in that work may be spurious. This work has been criticized by [Knorr 1978d] but supported by Souffrin [1980], who also argues that Archimedes' treatise E.P.I. contains a virtual definition of the concept of center of gravity. The relative importance of Archimedes' mechanical investigations for understanding the development of his mathematical work has been stressed by practically all recent writers, but one consequence of [Knorr 1978d] would be that there was a group of early studies having nothing to do with mechanics. Later, in his 30s, Archimedes became interested in mechanics, and his discovery of his mechanical method of investigation led to his publication, when he was into his 40s, of the great works we associate with his name, many of them revising purported, earlier studies.

The attempt to reconstruct the contents of Archimedes' early writings on spirals is the subject of [Knorr 1978b, a]. In the former, the argument is presented that the heuristic background to Archimedes' mature studies on areas and tangents of spirals is to be found in Book IV of Pappos' *Mathematical Collection* and that the heuristic considerations involved rather elegant arguments concerning spirals on cones, cylinders, and spheres. In [1978a] the author shows that weaker constructions, not requiring the vergings of Propositions 7 and 8 of *On Spirals*, will suffice for Proposition 18 and concludes that, since Archimedes chose to use the stronger verging construction when it was not necessary, he must have regarded these as quite legitimate and in need of no further explanation.

A different approach to the treatise *On Spiral Lines* is found in [Bashmakova 1956, 1964]. In the second, more extensive, study the argument is put that Archimedes possessed methods which, although not algorithmic, were equivalent in conception to the 17th- and 18th-century notions of differentials. Through a close study of the mathematical arguments in *On Spiral Lines*, Bashmakova has identified as a key part of Archimedes' method of tangents the perception that a certain infinitely small triangle
404 is similar to a finite triangle. A similar analysis of Archi|medes' solution to the problem of cutting a sphere by a plane so that the two segments have to each other a given ratio leads her to conclude that Archimedes possessed a method for finding extrema that was used by M. Ricci and E. Torricelli in the 16th and 17th centuries.

Mathematical Methods

There has been an increasing recognition that the history of mathematical methods is as much a part of the history of mathematics as is the history of number theory or algebra. Whether judged in terms of breadth of coverage or depth of analysis, first among the recent studies of Greek mathematical astronomy is [Neugebauer 1975], *A History of Ancient Mathematical Astronomy*, a three-part work that is a comprehensive history of mathematical astronomy in both the Babylon and the Greco-Roman world, which also offers careful analyses of mathematical methods in other fields as well.[5]

Ptolemy's work is, of course, central to the study of mathematical methods.[6] For example, his work on refraction, and especially the tables accompanying it, furnishes [Schramm 1965] with one of its chief pieces of evidence for a discussion of the genesis of the idea of a function in Western science. One of

[5] Neugebauer's original papers on many of the topics dealt with in [Neugebauer 1975] can now conveniently be consulted in [Neugebauer 1983], a collection from his papers that were published from 1932 to 1980.

[6] The reader should also consult [Pedersen 1974a] for a good introduction not only to Ptolemy's mathematical methods but also to his planetary models: however, he will want to consult [Toomer 1977] to correct Pedersen on certain points of detail.

the main points of this paper is that it was the mathematical methods that Greek astronomers used in dealing with tabular data that provide the key to understanding the methods they, and the later Muslim astronomers, employed in dealing with optical data. These mathematical methods, as they appear in the *Almagest*, form the object of study of [Pedersen 1974b], an illuminating account of Ptolemy's treatment of tabulated functions of one-, two-, and three-variables. Both Schramm and Pedersen draw attention to the importance of Greek geometric modeling in providing a basis for the computational treatment of continuous phenomena, where, prior to the Greeks, the Babylonians had tabulated the appearances of discrete phenomena.

One area of study of the history of mathematical methods in Greek astronomy that has received some careful consideration is the study, via Sanskrit texts found largely in India, of the history of methods in astronomy predating those of Ptolemy. Although such work goes far beyond the scope of this survey, in that it is of primary significance for history of Indian mathematics, I must mention [Pingree 1970], *Census of the Exact Sciences in Sanskrit*, a series of volumes that will doubtless open up to researchers in many areas paths that lead to real treasures. Examples of the retrieval of pre-Ptolemaic methods from Sanskrit are found in [Pingree 1971, 1976].

That much may also be learned from a study of Greek mathematical papyri has been shown in [Neugebauer 1972], which discusses an epicyclic theory of planetary motion in which the motion of the planet on the epicycle is the wrong way. Neugebauer conjectures that the theory is pre-Apollonian (the work of this astronomer and writer on conics being taken as the starting point of the tradition in Greek mathematical astronomy that found its full development in Ptolemy's *Almagest*), so that the device of an epicycle — if not its correct use — predates Apollonios. | 405

The figure of Ptolemy looms so large in the history of mathematical methods that it is difficult to see behind him, a fact that gives added importance to the above-mentioned researches. Another area in which Ptolemy looms large is in the history of trigonometry. Theon tells of a work by Hipparchos on chords, and [Toomer 1973] shows that, with no theorem deeper than the Pythagorean, Hipparchos could have written out a table of chords for a sequence of arcs that increase by steps of $7\frac{1}{2}^\circ$ from 0° to 90°. Toomer also adduces evidence from the *Almagest* that suggests this was Hipparchos' procedure. The net effect of the paper is to sketch out a plausible chapter in early Greek trigonometry.

Toomer argues, against Tropfke, that Archimedes did not have trigonometric methods at his disposal,[7] and this view is also adopted by [Shapiro 1975]. Shapiro gives an ingenious mathematical argument, which uses only methods known from Archimedes' other works, that accounts for the bounds on the angle subtended by the sun that Archimedes gives in his *Sandreckoner*.

One of the refreshing aspects of the study of mathematical methods in antiquity is that one occasionally encounters texts that have been little studied or even untouched by modern scholarship, and so the debate is fed by new data. This is much rarer in the history of the areas traditionally regarded as belonging to mathematics. However, the increased activity in the study of Arabic mathematical texts has produced some unexpected finds. For example, [Toomer 1976] is an edition of a small collection of works by the Greek mathematician Diocles, including his treatise *On Burning Mirrors*, and this text, together with Toomer's accompanying historical commentary, casts considerable light on the history of the conic sections at the time of Diocles and his better-known contemporary Apollonios. It has also set Diocles himself on much firmer historical ground than he was previously. Another Greek work we have obtained through an Arabic translation is a section consisting of four books from Diophantos' *Arithmetica*. The recent discovery of a unique Arabic manuscript led to an initial edition and study in [Rashed 1974, 1975a, b]. This was followed by [Sesiano 1982], an edition, with translation and extensive commentary, which represented an expanded version of Sesiano's thesis of 1975. A third example is the Archimedean treatise *On Tangent Circles*, which was translated into Russian by B. A. Rosenfeld and then into German, from the unique surviving Arabic text, in [Dold-Samplonius et al. 1972].

Another work, which is certainly an Arabic version of a Greek text, and possibly an Archimedean text, is in [Dold-Samplonius 1977], the recently translated *Book of Assumptions* by Aqāṭun. The text is extremely interesting in content, but work still remains to be done in putting this text into the history not only of Greek mathematics but of the translations of Greek into Arabic.

---

[7] In light of Toomer's arguments here, much of what [Sezgin 1974, 122–123] writes needs to be modified.

Another entry on the list of works recently translated from Arabic, and arguably of Archimedean origin, is [Hill 1976], "On the Construction of Water-Clocks." According to Hill, who has made a detailed study of the relevant evidence, the treatise is probably an Arabic translation of a Byzantine elaboration on an Archimedean treatise. If so, this would open up new areas of research to scholars who have, up to now, accepted Plutarch's characterization of Archimedes' spirit as so lofty he disdained
406 to write on any mechanical subject but sphere making. |

An important field was [Goldstein 1967], the discovery, in an Arabic version, of a lost section of Ptolemy's derivation of the sizes and distances of the planets. The mathematical argument uses data on the apparent sizes of the planets as well as the relative sizes of their orbits, according to Ptolemy's epicyclic models, and the size of the earth.

Also significant, for the history of non-Euclidean geometry, is [Sabra 1969], which contains a version of a purported proof of the parallel postulate by a writer who is, arguably, Simplicius.

Finally, [De Solla Price 1975] is a reminder that historians of mathematics can learn much from an archaeological investigation of scientific instruments. Professor Price expresses the opinion that the geared calendar computer, whose design he exposed with such care, is part of a clockwork tradition that goes back to Archimedes' planetarium and that some "rather elegant number manipulation ... is necessary to get a set of correct ratios for turning the various planetary markers" [p. 58].

Conclusions

The three decades that have just ended have been decades of considerable change in the history of ancient Greek mathematics. The fact that no major new texts, such as the Archimedes palimpsest or the mathematical cuneiform texts, have come to light has given scholars a chance to reflect on the previous material and perhaps to begin to modify conclusions that once seemed fairly secure. The absence of spectacular new discoveries has sent all of us back to the old texts, and has led some to the compilation of texts, whose existence is conjectured, out of old materials. Clearly, in order to do this, one must be sensitive to the editorial methods of Heron, Pappos, and others, and some recent research has considered this issue.

Thus, the history of Greek mathematics is alive and well. It draws on civilizations to the east — notably India and the Islamic world — for new blood, the old texts are read in new ways by a new generation of scholars, and the old puzzles are as intriguing as ever. The challenge of understanding the development of the system of thought that became a paradigm for mathematics continues to inspire scholars' best efforts and so the history of Greek mathematics remains an important, and growing, part
407 of the history of science. |

References

Bashmakova, I.G., 1956. Differential methods in Archimedes' works. Actes du VIIIe Congres Internationale d'Histoire des Sciences (Florence), 120–122.

—— 1964. Les méthodes différentielles d'Archimède. Archive for History of Exact Sciences 2, 87–107.

Becker, O., 1933. Voreudoxische Proportionenlehre. Quellen und Studien zur Geschichte der Mathematik, Astronomie und Physik, Abt. B2, 311–333.

—— 1936. Lehre vom Geraden und Ungeraden im Neunten Buch der Euklidischen Elemente. Quellen und Studien zur Geschichte der Mathematik, Astronomie und Physik, Abt. B3, 533–553.

—— 1954. Grundlagen der Mathematik in geschichtlicher Entwicklung. 2nd, enlarged ed., Alber, Freiburg.

Berggren, J.L., 1976. Spurious theorems in Archimedes' Equilibrium of Planes, Book I. Archive for History of Exact Sciences 16, 87–103.

Berka, K., 1980. Was there an Eleatic background to pre-Euclidean Mathematics? In: Pisa Conference Proceedings, vol. I, 125–131. Reidel, Dordrecht.

Burnyeat, M.F., 1978. The philosophical sense of Theaetetus' mathematics. Isis 69, 489–513.

Cherniss, H., 1951. Plato as mathematician. Review of Metaphysics 4, 396–425.

Dijksterhuis, E.J., 1956. Archimedes. Munksgaard, Copenhagen.

Dold-Samplonius, Y., Hermelink, H., Schramm, M. (trans.), 1972. Archimedes, Uber einander beruhrende Kreise. Vol. 4 of Archimedis Opera mathematica. Teubner, Stuttgart.

Dold-Samplonius, Y., 1977. Book of Assumptions by Aqāṭun. Ph.D. thesis, Amsterdam, 1976.

Drachmann, A.G., 1963. Fragments from Archimedes in Heron's Mechanics. Centaurus 8, 91–146.

——— 1967/1968. Archimedes and the science of physics. Centaurus 12, 1–11.

Fischler, R., 1979. A remark on Euclid II, 11. Historia Mathematica 6, 418–422.

Fowler, D.H., 1979. Ratio in early Greek mathematics. Bulletin of the American Mathematical Society (N.S.) 1, 807–846.

——— 1980. Book II of Euclid's Elements and a pre-Eudoxan theory of ratio. Archive for History of Exact Sciences 22, 5–36.

——— 1982. Book II of Euclid's Elements and a pre-Eudoxan theory of ratio, Part 2. Sides and diameters. Archive for History of Exact Sciences 26, 193–209.

Freudenthal, H., 1977. What is algebra and what has it been in history? Archive for History of Exact Sciences 16, 189–200.

Von Fritz, K., 1945. The discovery of incommensurability by Hippasus of Metapontum. Annals of Mathematics 46, 242–264. | 408

Goldstein, B.R., 1967. The Arabic version of Ptolemy's Planetary Hypotheses. Transactions of the American Philosophical Society (N.S.) 57, Part 4.

Hill, D.R., (ed. and trans.), 1976. On the construction of water clocks (K. Arshimidas fi amal al-binkamat). Turner & Devereux, London.

Hintikka, J., Remes, U., 1974. The Method of Analysis. Boston Studies in Philosophy of Science, Vol. XXV. Reidel, Dordrecht and Boston.

Hjelmslev, J., 1950. Eudoxos' axiom and Archimedes' lemma. Centaurus 1, 2–11.

Knorr, W.R., 1975. The Evolution of the Euclidean Elements. Reidel, Dordrecht.

——— 1977. Incommensurability and irrationality: A new historical interpretation. History of Science 15, 216–227.

——— 1978a. Archimedes' neusis-constructions in spiral lines. Centaurus 22, 77–98.

——— 1978b. Archimedes and the spirals: The heuristic background. Historia Mathematica 5, 43–75.

——— 1978c. Archimedes and the pre-Euclidean proportion theory. Archives Internationales d'Histoire des Sciences 28, 183–244.

——— 1978d. Archimedes and the Elements: Proposal for a revised chronological ordering of the Archimedean corpus. Archive for History of Exact Sciences 19, 211–290.

——— 1979. Methodology, philology, and philosophy, With a reply by M. F. Burnyeat. Isis 70, 565–570.

——— 1981. Review of The Beginnings of Greek Mathematics by Árpád Szabó. Isis 72, 135–136.

——— 1982. Observations on the early history of the conics. Centaurus 26, 1–24.

Lafrance, Y., 1980. Platon et la geometrie : La methode dialectique en République 509d–511e. Dialogue 19, 46–93.

Mahoney, M.S., 1968. Another look at Greek geometrical analysis. Archive for History of Exact Sciences 5, 318–348.

Mueller, I., 1981. Philosophy of Mathematics and Deductive Structure in Euclid's Elements. MIT Press, Cambridge, Mass.

Mugler, C., 1948. Platon et la recherche mathématique de son époque. P. H. Heitz, Strasbourg-Zurich.

Neuenschwander, E., 1973. Die erster vier Bücher der Elemente Euklids. Archive for History of Exact Sciences 9, 325–380.

Neugebauer, O., 1936. Zur geometrischen Algebra. Quellen und Studien zur Geschichte der Mathematik, Astronomie und Physik, Abt. B3, 245–259.
——— 1948. The astronomical origin of the theory of conic sections. Proceedings of the American Philosophical Society 92, 136–138.
——— 1959. The equivalence of eccentric and epicyclic motion according to Apollonius. Scripta Mathematica 24, 5–21.
——— 1972. Planetary motion in P. Mich. 149. Bulletin of the American Society of Papyrology 9, 19–22.
——— 1975. A History of Ancient Mathematical Astronomy, Parts 1–3. Springer-Verlag, New York.
——— 1983. Astronomy and History: Selected Essays. Springer-Verlag, New York.
Pedersen, O., 1974a. Survey of the Almagest. Odense University Press, Odense.
——— 1974b. Logistics and the theory of functions: An essay in the history of Greek mathematics. Archives Internationales d'Histoire des Sciences 24, 29–50.
Pingree, D., 1970–1979. Census of the Exact Sciences in Sanskrit. Series A, Memoirs of the American Philosophical Society, Philadelphia. vol. 81 (1970) (vol. I of Census), vol. 86 (1971) (vol. 2 of Census),
409 vol. 111 (1976) (vol. 3 of Census), vol. 146 (1979) (vol. 4 of Census). |
——— 1971. On the Greek origin of the Indian planetary model employing a double epicycle. Journal for the IIistory of Astronomy 2, 80–85.
——— 1976. The recovery of early Greek astronomy from India. Journal for the History of Astronomy 7, 109–123.
Price, D.J.d.S., 1975. Gears from the Greeks. Science History Publications, New York.
Rashed, R., 1974. Les travaux perdus de Diophante (I). Revue d'Histoire des Sciences 27, 97–122.
——— 1975a. Les travaux perdus de Diophante (II). Revue d'Histoire des Sciences 28, 3–30.
——— 1975b. L'art de l'algèbre de Diophante. Matābiʿ al-hai'a al-misrīyya al-ʿāmma al-kitāb, Cairo.
Riddell, R.C., 1979. Eudoxan mathematics and the Eudoxan spheres. Archive for History of Exact Sciences 20, 1–19.
Sabra, A.I., 1969. Simplicius's proof of Euclid's parallels postulate. Journal of the Warburg and Courtauld Institutes 32, 1–24.
Sato, T., 1979. Archimedes' On the Measurement of a Circle, Proposition I: An attempt at reconstruction. Japanese Studies in the History of Science 18, 33–99.
——— 1981. Archimedes' lost works on the centers of gravity of solids, plane figures and magnitudes. Historia Scientiarum 20, 1–41.
Schneider, I., 1979. Archimedes. Wissenschaftliche Buchgesellschaft, Darmstadt.
Schramm, M., 1965. Steps toward the ideas of function. History of Science 4, 70–103.
Sesiano, J., 1982. Books IV to VII of Diophantus' Arithmetica: In the Arabic Translation Attributed to Qustā ibn Lūqā. Springer-Verlag, New York.
Sezgin, F., 1974. Geschichte des arabischen Schrifttums, vol. IV, Mathematik. Brill, Leiden.
——— 1978. Geschichte des arabischen Schrifttums, vol. VI, Astronomie, Brill, Leiden.
Shapiro, A.E., 1975. Archimedes' measurement of the sun's apparent diameter. Journal for the History of Astronomy 6, 75–83.
Souffrin, P., 1980. Trois études sur l'oeuvre d'Archimède, Cahiers d'histoire et de philosophie des Sciences vol. 14. Centre de documentation sciences humaines, Paris.
Szabó, Á., 1968. The Beginnings of Greek Mathematics. Reidel, Dordrecht.
——— 1974. Die Muse der Pythagoreer: Zur Frühgeschichte der Geometrie. Historia Mathematica 1, 291–316.
Toomer, G.J., 1973. The chord table of Hipparchus and the early history of Greek trigonometry. Centaurus 18, 6–28.
——— (ed.), 1976. Diocles on Burning Mirrors. Springer-Verlag, New York.
——— 1977. Review of Pedersen [1974a]. Archives Internationales d'Histoire des Sciences 100, 137–150.
Unguru, S., 1975. On the need to rewrite the history of Greek mathematics. Archive for History of Exact Sciences 15, 67–114.

——— 1977. Review of Knorr [1975]. History of Science 15, 216–227.
——— 1979. History of ancient mathematics: Some reflections on the present state of the art. Isis 70, 555–565.
Van der Waerden, B.L., 1947/1949. Die Arithmetik der Pythagoreer. Mathematische Annalen 120, 127–153, 676–700.
——— 1954. Science Awakening. P. Noordhoff, Groningen.
——— 1976a. Defence of a "shocking" point of view. Archive for History of Exact Sciences 15, 199–210.
——— 1976b. Review of Knorr [1975]. Historia Mathematica 3, 497–499.
Waterhouse, W.C., 1972. The discovery of the regular solids. Archive for History of Exact Sciences 9, 212–221.
Weil, A., 1978. Who betrayed Euclid? (Extract from a letter to the editor.) Archive for History of Exact Sciences 19, 91–93.
Zeuthen, H.G., 1896. Die geometrische Konstruction als "Existenzbeweis" in der antiken Geometrie. Mathematische Annalen 47, 222–228.
——— 1910. Sur la constitution des livres arithmétiques des Elements d'Euclide et leur rapport a la question de l'irrationalité. Oversigt over det Kongelige Danske Videnskabernes Selskabs Forhandlinger, 395–435.

Mathematical Reconstructions Out, Textual Studies In: 30 Years in the Historiography of Greek Mathematics [1998]

Ken Saito

History of Greek Mathematics Before and After 1970

Thirty years ago, at the end of the sixties, the history of Greek mathematics was considered an almost closed subject, just like physics was at the turn of the twentieth century. People felt that they had constructed a definitive picture of the essence of Greek mathematics, even though some details remained unclear due to irrecoverable document losses. Critical editions had been established, mainly by Heiberg, while two of the great scholars of the history of Greek mathematics, Tannery and Zeuthen, had built on this material. Then, the standard book [Heath 1921] brought much of this material together. Through his discoveries in Mesopotamian mathematics, Neugebauer was led to think that he had given substance to legends about the Oriental origin of Greek mathematics. Originally published in Dutch in 1950, the book [van der Waerden 1954] reflected scholars' self-confidence in this period.

One may well compare what happened after 1970 in the historiography of Greek mathematics to the developments of physics in the first decades of this century. In some sense the change in the history of Greek mathematics was even more dramatic, because no new important material was discovered since 1906, at which time the *Method* was brought to light by Heiberg. This great interpretative change was mainly due to a shift in scholars' attitudes. | 132

In the following, the historiography of Greek mathematics before 1970 will be briefly contrasted, with no pretense of being exhaustive, with that which followed.[1]

The Origins: Who Was the First Mathematician?

A tradition reaching as far back as Eudemus (late 4th century BC), via citations found in Proclus' *Commentary on the First Book of Euclid's Elements*, considers Thales (ca. 585 BC) to have been the founder of the Greek mathematical sciences. However, if by the phrase *"the origins of Greek mathematics"* we mean an embryo of the rigorous deductive structure found in the *Elements*, Thales had little to do with it. Eudemus may well have constructed a story of a mathematician from fragmentary sources at his disposal, which described the practical knowledge of a wise man (see [Dicks 1959], and also [Vitrac 1996]).

Editors: Originally published as Saito, K., 1998, Mathematical reconstructions out, textual studies in: 30 years in the historiography of Greek mathematics, Revue d'histoire des mathématiques 4, 131–142. We thank K. Saito and the editors of Revue d'histoire des mathématiques for permission to republish this paper.

[1] Accounts given below are a personal view, even if I tried to be as impartial as possible in the bibliography. I restricted myself to studies of mathematics in the classical period, that is, mathematics before Apollonius, although developments of research in late antiquity, including the rediscovery of part of the lost books of Diophantus' *Arithmetica* in Arabic, should not be underestimated.

N. Sidoli and G. Van Brummelen (eds.), *From Alexandria, Through Baghdad: Surveys and Studies in the Ancient Greek and Medieval Islamic Mathematical Sciences in Honor of J.L. Berggren*, DOI 10.1007/978-3-642-36736-6_2,

© Springer-Verlag Berlin Heidelberg 2014

Dismantling the myth of origins became the subject of hot debates centered around the figure of Pythagoras (ca. 572–ca. 494 BC), who enjoys no less enthusiastic advocates today than in the ancient world. However, considerable if not decisive, is the damage done to *"Pythagoras the mathematician"* by the blow of the epoch-making study [Burkert 1972].[2] Now we see Pythagoras as the founder of a prevalently (but not exclusively) religious community, established on doctrines of reincarnation and metempsychosis. To be a Pythagorean meant choosing a certain way of life based on these doctrines, without being necessarily involved in philosophical or scientific inquiries.

Thus we are rather concerned, now, with the role of the Pythagoreans in the development of Greek mathematics after the middle of the fifth century. The picture once prevailed that the discovery of
133 incommensurability was a scandal for the Pythagoreans and provoked a *crisis*. This | belief was deduced from 1) the alleged Pythagorean monopoly on the mathematical sciences in the fifth century; 2) their central dogma "all is number;" and 3) Iamblichus' testimony. However, 1) has no good evidence to support it; 2) is very likely an Aristotelian summary deduced from Philolaus' (ca. 470–ca. 390 BC) book; and 3) is so confused that it is hardly reliable (which means that we have no authentic document to credit the Pythagoreans with the discovery of incommensurability). The scandal, or foundations-crisis thesis has thus turned out to be scarcely plausible (see [Freudenthal 1966], [Knorr 1975, 21–61], [Fowler 1987, 294–308]). More recently [Fowler 1994] has even suggested that this discovery itself may have been no more than an incidental event. After all, the above thesis may have been a retroprojection of early twentieth-century interests in the foundations of mathematics.

Therefore, the roles traditionally ascribed to Pythagoreans are also to be reconsidered and greatly modified, a point to which we shall later return. For the moment let us examine modern studies devoted to the theory of proportions.

Mathematical Reconstructions

If a foundation-crisis theory was soon dismissed, the assumption that incommensurability constituted a turning point in Greek mathematics enjoyed better support. In fact, it seems natural to us, today, to suppose that the discovery of incommensurability called for a new definition of proportions (sameness of ratios) applicable to incommensurable magnitudes. This assumption gave birth to the most influential historical approach in this century: mathematical reconstruction.

[Becker 1933] pointed out that a passage of Aristotle's *Topics* can be construed as evidence for the existence of a definition of proportions based on *anthyphairesis* (Euclidean algorithm), which can be dated to a period between the discovery of incommensurability (probably second half of the fifth century) and Eudoxus' time (ca. 390–ca. 337). This paper not only called attention to the technique of *anthyphairesis*, but also encouraged scholars to use mathematical reconstructions in order to venture new conjectures and hypotheses. One eminent example of this technique is [Fritz 1945], which proposed, with no direct textual evidence, that incommensurability had first been found in a study of the relation between the
134 side and diameter of regular pentagons by the method of | *anthyphairesis*. Even [Knorr 1975], the most critical and thoroughgoing study of the development of incommensurability theory to date, remained highly speculative, and in a sense, this book marked a culmination in the tradition of the reconstruction approach opened by Becker.[3]

Although the significance of this kind of study cannot be denied, its danger also is obvious: one has no general criterion to judge whether the reconstructed argument ever existed in antiquity. Moreover, while most reconstructions deal with the period around and before 400 BC, sources come from later

[2] Reasonable doubts over whether Pythagoras actually was the first mathematician and philosopher go back at least to [Vogt 1908/09], and perhaps even to [Zeller 1844–1852]. Burkert's central thesis, as well as scholarly developments after 1972, are very skillfully, and with extraordinary clarity and concision, described in [Centrone 1996]. For opinions sympathetic to the view of Pythagoras and the Pythagoreans as scientists, see [van der Waerden 1979] and [Zhmud 1997].

[3] I exclude from this tradition the book [Fowler 1987] whose reconstructions undoubtedly are more sophisticated: see below.

periods.[4]

From Mathematical Reconstructions to Textual Studies

However, since any significant interpretation of ancient mathematics is bound to involve some kind of conscious or unconscious reconstruction, one may well ask whether it is actually possible to distinguish recent research from previous reconstructive approaches. Let me try to justify this distinction, granted that, here, my account inevitably is more personal than other parts of this paper.

Previous scholars (say, from Tannery and Zeuthen to van der Waerden) were, I believe, confident in the power of something like universal reason, and took it for granted that a careful mathematico-logical reasoning was able to restore the essence of ancient mathematics. Today scholars are more skeptical: the type of reasoning that once played an essential role tends to be regarded as a mere rationalizing conjecture. They are even convinced that the modern mind will always err when it tries, without the guide of ancient texts, to think as the ancients did (I personally think | that this opinion can be attributed 135
to an indirect influence of Thomas Kuhn).

Thus, texts are read in a different manner by recent historians of mathematicians, as well as by historians of science in general. For example, apparently redundant or roundabout passages call for more attention, because these might reveal some of the ancients' particular thoughts of which modem minds are unaware.[5] This is one of the attitudes typical of what I call "textual studies" in a broad sense (I do not restrict them to textual criticism), an attitude based on reasonable doubts as to the validity of logical conjectures.

Happily for the French-reading public, the spirit and results of this new textual approach are best embodied in the French translation of the *Elements* now in progress (see [Euclide 1990-2001]),[*] but a review of other studies will also help us understand the new historiography. Renewed interest in text led to careful examinations of the extant mathematical documents and their logical structure. Since most of these documents are series of propositions, the logical interdependence of propositions, or the "deductive structure," is one of the important subjects of recent studies. Most of this work has been limited to Euclid's *Elements*: see [Beckmann 1967], [Neuenschwander 1972], [Neuenschwander 1973], and the comprehensive and influential book [Mueller 1981].[6] Lately, [Netz 1999][**] has proposed brand-new, insightful approaches to texts.[7]

---

[4] Anomalies and idiosyncrasies in the logical structure of the *Elements* have been used by many scholars (including myself) in order to reconstruct the earlier phase of Greek mathematics. For example, the first four books of the *Elements* contain several demonstrations more easily proved with the theory of proportions. These demonstrations have been either located in the period when no adequate theory of proportions was available or attributed to some mathematician who compiled earlier versions of the *Elements*. [Artmann 1985] and [Artmann 1991] are a remarkable outcome of this approach. However, this approach relies on the assumption that Euclid's editorial intervention was minimal and the extent to which this assumption can be justified is unknown to us. With a bit of irony, Vitrac called this kind of approach an *"enquête archéologique"* [Vitrac 1993, xi]. See also [Gardies 1998], which developed very specific reconstructions based on logical analyses, and [Caveing 1994–1998].

[5] Here, one cannot but recall the attractive work of Árpád Szabó (I am thinking of [Szabó 1969] and less known [Szabó 1964]), who, using philosophical arguments, was the first seriously to criticize the trend of mathematico-logical reconstructions. His approach predated the present research trend. He was however concerned with finding traces of the earliest developments of Greek mathematics, and his arguments inevitably remain no less speculative than the theses he challenges.

[*] Editors: The original paper reads "[Euclide 1990–]." This project has been completed and we have updated the references.

[6] Concerning the proposition used by later authors, indices devoted to Pappus, Apollonius and Archimedes are available on my web page, where one will find how propositions of the *Elements* were used (or not used) in other mathematicians' works. My paper [Saito 1994] indicates the reasons that prompted me to assemble such indices.

[**] Editors: The original paper reads "[Netz (Forthcoming)]," however, this book has been published and we have updated the references.

[7] In this book, Netz examines the form and style of Greek mathematical texts, which, like Homer's epics, largely depended on "formulae" — fixed expressions regularly used to denote certain mathematical objects or relations. He illustrates the

136 The shift from reconstructions to textual studies can also be illustrated | by the works of Wilbur Knorr (1945–1997), the greatest historian of Greek mathematics in the second half of the twentieth century (and the restriction "second half" may be unnecessary). After the book [Knorr 1975], based on his Ph.D. thesis, he proposed a reconstruction of another theory of proportion in [Knorr 1978], an outcome of a thorough reading of Archimedes, attesting Knorr's shift toward a more substantial study of documents. Then, after producing [Knorr 1986], which is important for its emphasis on autonomous developments of problem-solving techniques independent of alleged philosophical interests, he arrived at textual studies in his monumental work [Knorr 1989].

A shift to textual studies entails a change in the subject of investigations. Even if one does not always embark on studies of Arabic and medieval Latin documents as Knorr did,[8] the weight necessarily moves from the fifth century (where the scarcity of documents provides ample room for conjectures and reconstructions) to the fourth century (where Plato and Aristotle are contemporary witnesses and more documents available: for example see [Mueller 1991]) and to the third century (where Archimedes' and Apollonius' works are at our disposal). When we calmly consider the status of extant documents, we can surely observe that it is extremely hazardous to speak of the history of Greek mathematics before 399 BC, the dramatic date of Plato's *Theaetetus*.

But reconstructions have not been dismissed. Rather, this approach also benefited from the same skepticism concerning rationalising conjectures. For example, [Fowler 1987], an outstanding work among recent attempts, is characterized by thorough investigations of extant documents. Its basic attitudes have more in common with recent research trends than with the anthyphairetic reconstruction tradition.

Algebraic vs. Geometrical Interpretation

A word is in order concerning the decline of the algebraic interpretation that served to combine such ingredients as Babylonian mathematics, Pythagorean interests, incommensurable magnitudes, and the
137 so-called geometric algebra. Following the bitter conflict caused by the provocative | paper [Unguru 1975], which strenuously argued against the prevailing algebraic interpretation, and the reactions to it (for a survey of this polemic, see [Berggren 1984]), scholars were reluctant to discuss this problem for some time. With the changes in research approach described above, the belief that algebraic interpretation could provide a sufficient description of the treatment of magnitudes in Greek geometry has become less convincing, and scholars can now speak as calmly about this as in [Grattan-Guinness 1996].

Pythagoreans, Are They Out?

Now, let us return to Pythagoreans, and try to replace them in the history of Greek mathematics. Two outstanding figures have drawn scholars' attention: Philolaus of Croton (ca. 470–ca. 390) and Archytas of Tarentum (fl. ca. 400–350).

In the study of these figures, the difficulty consists, above all, in the evaluation of documents. Since it already was very common among Plato's disciples in the Academy to attribute their own ideas to Pythagoreans, any such attribution is in itself suspicious. Of course, this hardly means that all

---

way in which mathematical deduction is constructed and how formulae work. He moreover analyses the relation between text and diagram, and elucidates their interdependence, showing the indispensable role played by diagrams.

[8] [Knorr 1989] thoroughly investigated Arabic and Latin traditions of Archimedes' *Dimensions of the Circle*. [Knorr 1996] convincingly showed that an Arabo-Latin tradition of Book XII of Euclid's *Elements* preserved a text preferable to Heiberg's Greek edition.

attributions are wrong, and the first task of scholars is to distinguish genuine fragments from spurious ones.

As for Philolaus, a contemporary of Socrates, [Huffman 1993] marked a great step ahead with a thoroughgoing analysis of extant fragments and the attempts to identify genuine ones. He described Philolaus as a natural philosopher facing epistemological problems with the help of the model given by the rigorous mathematical sciences. What is striking here is that Philolaus himself does not seem to have made any original contributions to mathematics, yet it was mathematics that provided him with a model for science. Philolaus' place is not yet settled and needs further research.

Archytas, Plato's contemporary and friend, was no doubt the most brilliant mathematician among the Pythagoreans. He was the first to solve the problem of finding two mean proportionals; and if his fragment n°1 is genuine,[9] as recent scholars are inclined to believe, his importance in music theory would be well established, although music theory was not exclusively a Pythagorean interest (see [Bowen 1982] and [Bowen 1991]). | 138

The important question, in my own opinion, is how much of Archytas' achievements owed to the Pythagorean tradition. In other words, can we assume that Archytas was a great mathematician *because* he was a Pythagorean? No document proves this directly and incontestably. And if the answer turned out to be negative, what sense would it make to speak of Pythagorean mathematics?

Let me give an example. In the *Elements* (Book VII to IX), arithmetic consists of two levels of propositions. Proposition VII-20 serves as a breakthrough to the higher level, and to several propositions concerning the non-existence of mean proportional numbers (for example, that no integer x satisfies $n : x :: x : 2n$ can easily be deduced from *Elements* VIII-8). These propositions are not likely to have been proved by pebble (*psephoi*) arithmetic ascribed to the Pythagoreans [Burkert 1972, 433–436]. Moreover, they are, in a sense, extensions of the recognition that the side and diagonal of squares are incommensurable (although we do not know whether the former were *historical* extensions of the latter). Our attention therefore is focused on whether or not these "propositions of a higher level" are Pythagorean achievements.

Boethius credits Archytas with the proof of the non-existence of mean proportional numbers for an epimoric ratio,[10] a special case of a group of propositions to be found in Book VIII of the *Elements*. If the Pythagoreans had actually discovered incommensurability (and it must have been discovered sometime before Archytas), and if some continuous Pythagorean tradition had enabled Archytas to prove his theorem, then we would have every right to speak of "Pythagorean mathematics." If, on the contrary, Archytas simply applied a theorem already known outside of the Pythagorean community to his interests in harmonics (perhaps even because of the Pythagorean tradition), then Pythagorean mathematics would look somewhat faded. Many other assumptions are possible between these two extremes, and we should try to determine how "Pythagorean" Archytas' achievements were. In this sense, Pythagoreans are by no means out, and they will continue to be the subject of studies and discussions.[11] | 139

---

[9] Once doubted by [Burkert 1972, 379–380 n.46], the authenticity of this fragment (DK 47B1) is effectively defended by [Bowen 1982, 83–85] and [Huffman 1985]. However, [Centrone 1996, 69–70, n. 21] expressed some reservations.

[10] The same proposition is included in Euclid's *Sectio Canonis* (prop. 3). The epimoric ratio is the ratio of two magnitudes whose greater term's excess over the lesser is a part (divisor) of both: it is therefore expressed in the form $(n + 1) : n$.

[11] A warning concerning astronomy seems appropriate here. Even though Archytas may have said that music and astronomy are sister sciences (DK 47B1), his astronomy was at best the very beginning of the Greek mathematical astronomy tradition culminating in Ptolemy. Note also that the word *sphairikas* in this fragment is found only in some of Nichomachus' manuscripts, not in the parallel passage in Porphyry, and [Bowen 1982, 80] omits this word. Greek geometrical astronomy began with Archytas' contemporary Euxodus (see [Berggren and Thomas 1996, 6 ff.]). Then, from the third century onwards, it developed more and more elaborate spherical models, and incorporated the purely arithmetical Babylonian data into this geometric framework (see [Jones 1991] and [Jones 1996]).

Conclusion

With some oversimplifications, I would sum up this discussion as such: Pythagoras out, Pythagoreans in (but without attributing to them a monopoly over the mathematical sciences). Fifth century out, fourth and third centuries in. Mathematical reconstructions out, textual studies in. What the Greeks could and should have done out, but what they actually did in. Today the history of Greek mathematics (and probably the history of mathematics in general) has become a branch of the history of ideas more than a branch of mathematics, as it used to be.

References

Artmann, B., 1985. Über voreuklidische 'Elemente,' deren Autor Proportionen vermied. Archive for History of Exact Sciences 33, 291–306.

—— 1991. Euclid's Elements and its prehistory. In: Mueller, I. (ed.), Peri Tōn Mathēmatōn (Apeiron 24), Academic Printing & Publishing, Alberta, pp. 1–47.

Becker, O., 1933. Voreudoxische Proportionenlehre. Quellen und Studien zur Geschichte der Mathematik, Astronomie und Physik, Abt. B2, 311–333.

Beckmann, F., 1967. Neue Gesichtspunkte zum 5. Buch Euklids. Archive for History of Exact Sciences 4, 1–144.

Beggren, J.L., 1984. History of Greek mathematics: A survey of recent research. Historia Mathematica 11, 394–410.

Berggren, J.L., Thomas, R.S.D., 1996. Euclid's Phaenomena: A Translation and Study of a Hellenistic Treatise in Spherical Astronomy. Garland, New York.

Bowen, A.C., 1982. The foundations of early Pythagorean harmonic science: Archytas, Fragment 1.
140 Ancient Philosophy 1, 79–104. |

—— 1991. Euclid's Sectio canonis and the history of Pythagoreanism. In: Bowen, A.C., Rochberg-Halton, F. (eds.), Science and Philosophy in Classical Greece. Garland, New York, pp. 164–187.

Burkert, W., 1972. Lore and Science in Ancient Pythagoreanism (Minar, E.L., trans.). Harvard University Press, Cambridge, Mass..

Caveing, M., 1994–1998, Essai sur le savoir mathématique dans la Mésopotamia et l'Egypte anciennes, Lille, Presses universitaires de Lille, 1994 (= vol. 1); La figure et le nombre. Recherches sur les premières mathématiques des Grecs, Lille, Presses universitaires du Septentrion, 1997 (= vol. 2); L'irrationalité dans les mathématiques grecques jusqu' á Euclide, Lille, Presses universitaires du Septentrion, 1998 (= vol. 3).

Centrone, B., 1996. Introduzione ai Pitagorici. Laterza, Roma-Bari.

Dicks, D.R., 1959. Thales. Classical Quarterly NS 9, 294–300.

Euclide, 1990-2001. Les Eléments (Vitrac, B., trad. et com.). Presses universitaires de France, Paris.

Fowler, D.H., 1987. The Mathematica of Plato's Academy. Clarendon Press, Oxford. Reprinted with corrections, 1990.

—— 1994. The story of discovery of incommensurability, revisited. In: Gavroglu, K., Christianidis, J., Nicolaïdis, E. (eds.), Trends in the Historiography of Science. Kluwer Academic Publishers, Dordrecht, pp. 221–235.

Freudenthal, H., 1966. Y avait-il une crise des fondements des mathématiques dans l'antiquité? Bulletin de la Société mathématique de Belgique 18, 43–55.

von Fritz, K., 1945. Discovery of incommensurability by Hippasus of Metapontum. Annals of Mathematics 46, 242–264.

Gardies, J.-L., 1988. L'héritage épistémologique d'Euxode de Cnide, Un essai de reconstitution. J. Vrin, Paris.

Grattan-Guiness, I., 1996. Numbers, magnitudes, ratios and proportions in Euclid's Elements: How did he handle them? Historia Mathematica 23, 355–375.

Heath, T.L., 1921. A History of Greek Mathematics (2 vols.). Clarendon Press, Oxford. Reprinted by Dover Publications, New York, 1981.
Huffman, C.A., 1985, The authenticity of Archytas Fr. 1. Classical Quarterly 35, 344–348.
——— 1993. Philolaus of Croton: Pythagorean and Presocratic. Cambridge University Press, Cambridge. | 141
Jones, A., 1991. The adaptation of Babylonian methods in Greek numerical astronomy. Isis 82, 441–453.
——— 1996. On Babylonian astronomy and its Greek metamorphoses. In: Jamil R.F., Ragep, S.P. (eds.), Tradition, Transmission, Transformation. Brill, Leiden, pp. 139–155.
Knorr, W.R., 1975. The Evolution of the Euclidean Elements. Reidel, Dordrecht.
——— 1978. Archimedes and the pre-Euclidean proportion theory. Archive for History of Exact Sciences 19, 211–290.
——— 1986. The Ancient Tradition of Geometric Problems. Birkhäuser, Boston. Reprinted by Dover Publications, New York, 1993.
——— 1989. Textual Studies in Ancient and Medieval Geometry. Birkhäuser, Boston.
——— 1996. The wrong text of Euclid: On Heiberg's text and its alternatives. Centaurus 38, 208–276.
Mueller, I., 1981. Philosophy of Mathematics and Deductive Structure in Euclid's Elements. The MIT Press, Cambridge, MA.
——— 1991. On the notion of a mathematical starting point in Plato, Aristotle, and Euclid. In: Bowen, A.C., Rochberg-Halton, F. (eds.), Science and Philosophy in Classical Greece. Garland, New York, pp. 59–97.
Netz, R., 1999, The Shaping of Deduction in Greek Mathematics. Cambridge University Press, Cambridge.
Neuenschwander, E., 1972/1973, Die ersten vier Bücher der Elemente Euklids. Archive for History of Exact Sciences 9, 325–380.
——— 1974/1975. Die stereometrischen Bücher der Elemente Euklids: Untersuchungen über den mathematischen Aufbau und die Entstehungsgeschichte. Archive for History of Exact Sciences 14, 91–125.
Saito, K., 1985. Book II of Euclid's Elements in the light of the theory of conic sections. Historia Scientiarum 28, 31–60.
——— 1986. Compounded ratio in Euclid and Apollonius. Historia Scientiarum 3, 25–59.
——— 1994. Proposition 14 of Book V of the Elements – A proposition that remained a local lemma. Revue d'histoire des sciences 47, 273–284.
Szabö, A., 1964. Ein Beleg für die voreudoxische Proportionenlehre? Aristoteles: Topik Θ.3, p. 158b29–35. Archiv für Begriffsgeschichte 9, 151–171.
——— 1969. Die Anfänge der griechischen Mathematik. Oldenbourg, München.
Unguru, S., 1975. On the need to rewrite the history of Greek mathematics. Archive for History of Exact Sciences 15, 67–114.
van der Waerden, B.L., 1954. Science Awakening (Dresden, A., trans.). Oxford University Press, Oxford.
——— 1979. Die Pythagoreer. Artemis Verlag, Zürich.
Vitrac, B., 1993. De quelques questions touchant au traitement de la proportionnallité dans les *Éléments* d'Euclide, Thèse de doctorat de l'École des hautes études en sciences sociale, Paris, sous la dir. de Dhombres (J.), soutenue le 17/12/1993.
——— 1996. Mythes (et réalités?) dans l'histoire des mathématiques grecques anciennes. In: Goldstein, C., Gray, J., and Ritter, J. (eds.), L'Europe Mathématique. Éditions de la Maison des Sciences de l'Homme, Paris, pp. 33-51.
Vogt, H., 1908/1909. Die Geometrie des Pythagoras. Bibliotheca Mathematica, III. Folge, 9, 15-54.
Zeller, E., 1844-1852. Die Philosophie der Griechen in ihrer geschichtlichen Entwicklung, 5th ed. Leipzig, 1923; reprint Hildesheim, 1990.
Zhmud, L. Wissenschaft, Philosophie und Religion im frühen Pythagoreismus. Akademie Verlag, Berlin.

Research on Ancient Greek Mathematical Sciences, 1998–2012

Nathan Sidoli

Introduction

This survey deals with research in ancient Greek mathematics and, to a lesser extent, mathematical sciences in the long first decade of the 21st century. It is modeled on the two foregoing surveys by J.L. Berggren and K. Saito and gives my personal appraisal of the most important work and trends of the last years, with no attempt to be exhaustive.

My overview of the work in this period is divided into two main sections: the "Methods," which researchers have applied, and the "Topics," which have garnered significant research interest. Many of the methods and topics that have been addressed in this period are not new, but the emphasis has often changed in response to a changing historiography. Of course, many of the individual works I discuss utilize more than one method and address more than one topic. Hence, the placement in my paper of any particular study is necessarily somewhat arbitrary.

I should mention some restrictions that must necessarily be made in such a survey. I deal only with mathematics, and some of the exact sciences insofar as they relate to mathematics, mathematical practice and mathematicians. This is a fairly arbitrary division. Ptolemy and Heron regarded themselves as mathematicians; astrologers of the Hellenistic periods were called mathematicians. A similar survey for astronomy, however, would be as long as that for mathematics, and those for mechanics and astrology would be nearly as long again. I only consider work in major European languages. In general, I restrict myself to the first decade of this century, give or take a few years on either side. I do, however, make a few exceptions to this to mention works published in the previous periods that were not discussed in either of the foregoing surveys by Berggren or Saito.

It may be useful if I explain my general views on the changing historiography of the period, which led me to organize this paper as I have. The majority of scholars working in the field now share certain methodological and historiographical commitments. There have been no polemics along disciplinary lines such as erupted in the 1970s (see pages 5–6 and 20, above). Of course, there have been disputes about historiographic methods, but these have largely focused around questions about the best way to rewrite the history of Greek mathematics, not about whether or not this needs to be done at all. The majority of scholars in the field now appear to accept the idea that Greek mathematics is *not* our mathematics, and that it needs to be understood in the context of its own historical setting. It appears that the old historiography has been overcome. There are very few who still believe in such historiographical artifacts as Pythagoras' deductive mathematics, geometric algebra, or a crisis precipitated by the discovery of incommensurability.

In general, the historiography of Greek mathematics has responded to changing trends in the historiography of science and of mathematics, but only slowly and tentatively. This is partly because the methodologies and approaches of these various fields are still necessarily different. Historians of Greek mathematics cannot directly employ many of the methodological approaches, such as studies of institutional practice or instrumentation, used by historians of later periods; and they are compelled to

N. Sidoli and G. Van Brummelen (eds.), *From Alexandria, Through Baghdad: Surveys and Studies in the Ancient Greek and Medieval Islamic Mathematical Sciences in Honor of J.L. Berggren*, DOI 10.1007/978-3-642-36736-6_3,
© Springer-Verlag Berlin Heidelberg 2014

use methods, such as philology and reconstruction, which historians of later periods have no interest in, or find distasteful. Nevertheless, in the period under consideration, historians of Greek mathematics have followed earlier trends in the history of science, such as focusing on local topics for which there is good evidence, exploring the implications of material culture, and fleshing out the influence of cultural, religious and intellectual contexts on mathematical activity. Most importantly, historians of Greek mathematics, like other historians of science, are now no longer primarily interested chronicling a series of results, which are arbitrarily assumed to be important, but are rather focused on practices of knowledge production and discourse in the mathematical sciences and how these were related to such practices in other areas of ancient life.

With the passage of time, there have been changes in the group of researchers focusing on ancient Greek mathematical sciences. A number of senior scholars have moved on to other interests, or reduced their output; but the most significant change, in this regard, has been the untimely loss of the late W.R. Knorr, the most prolific and prominent historian of Greek mathematics in the second half of the twentieth century [Fowler 1998]. At the same time, a number of new scholars have entered the field, to the extent that it now appears to be more active than it had been for decades. Of the new scholars, I will mention only the two most active and influential, R. Netz and F. Acerbi.

A snapshot of changing trends in the field can be captured by considering three books that have been published during the period covered by this survey. The most influential book during this period has been R. Netz's *The Shaping of Deduction in Greek Mathematics* [1999a], which is perhaps the only book on Greek mathematics to have received a wide audience and recognition among non-specialists. This book has done much to open new approaches to deduction in Greek mathematics by focusing on cognitive, linguistic and diagrammatic practices. S. Cuomo's *Greek Mathematics* [2001] takes a broad scope of mathematical activity in the ancient Mediterranean world and gives as much attention to the historical evidence as to the topics treated. Although this book provides almost no coverage of the technical accomplishments of the Greek mathematical sciences, it situates mathematical practices in the broader social and political context and addresses the fact that mathematics was used by many professional practitioners in the course of their work. F. Acerbi's *Il silenzio delle sirene* [2010a] is a compact, technical work that will probably not attract much attention outside of the circle of specialists. But for them it will be essential. The book deals with historiographical issues concerning methods and topics, treats the techniques that Greek mathematicians used to approach their subject matter, and ends with a full appendix on the Greek manuscript traditions of the major sources. Although the arguments are often brief, this work shows how the old historiography has been left behind and argues for a reevaluation of many suggestions and assumptions of previous historians that have often been accepted as fact.

Methods

Many of the methods employed by scholars during this period have been used for decades, even centuries — such as manuscript studies and mathematical formalization — while others are techniques emphasized by historians of science working on more modern periods — such as studies of material culture, or contextualization in various social and religious milieus.

Textual Studies

Those who study the history of the ancient mathematical sciences face a conspicuous scarcity of evidence. For Greek work, although there is some papyrological evidence, there is almost no other material evidence and the descriptions of mathematical activity in literary and philosophical texts are often difficult to reconcile and interpret. Hence, for the history of Greek mathematics, our principal sources are the mathematical texts preserved in the medieval manuscript tradition.

For these reasons, the foundation of scholarship in the Greek mathematical sciences has been, is, and will remain, textual studies, in which all serious scholars of Greek mathematics engage to one degree or another. Textual studies, however, encompass a wide range of activity including codicological studies of the papyri and manuscripts themselves, producing critical editions on the basis of these, grammatical and philological analysis and description of the text, translation, and mathematical and scientific commentary.

Ancient Texts

Our most direct textual evidence for Greek mathematical activity comes from papyri and other written sources, of which there are very few. Arithmetical tables, and publications and studies of these, have been catalogued by Fowler [1988, 1995]. A couple of papyri dealing with "algebraic" type problems have been published and studied by Sesiano [1999]. Bülow-Jacosen and Taisbak [2003] have edited, and studied, a papyrus containing "practical" geometrical problems of the kind that we find in the Heronian corpus, and Jones [2012], relying on new readings of certain abbreviations, has reedited it, and produced a new technical study.

For the history of Greek astronomy, the largest new source of documents has been published and studied by Jones [1999], in his *Astronomical Papyri from Oxyrhynchus*. This work shows that the tabular methods of Babylonian mathematical astronomy were circulating in Greek by at least the early centuries of the common era and that these were in use at the same time as tables based on pre-Ptolemaic geometric modeling, and Ptolemy's *Handy Tables*.

There is still no comprehensive study of mathematics and mathematical astronomy in Greek papyri, but Friberg [2005, 193–267] has made a study of the mathematical methods contained in Greek papyri from Egypt, and shown that the technical methods used in these texts go back to Old Babylonian methods.

Medieval Manuscripts

The vast majority of work on Greek mathematics deals, not with the relatively few ancient texts that have survived in various forms, but with the much greater quantity of texts that are preserved in medieval manuscripts. There have been a number of new studies of the medieval manuscript traditions of Greek mathematical works. Along with older work by Mogenet [1950] and Noack [1992] on Autolycus and Aristarchus, respectively, a new study of the manuscripts of Theodosius' *Spherics*, by Czinczenheim [2000], gives us a good sense for the overall manuscript tradition of the so-called *Little Astronomy*. Decorps-Foulquier [1999a] has produced a thorough study of the manuscripts of the scholarship on Apollonius' *Conics* carried out by Eutocius.

There is no work that covers all of the manuscript evidence for Greek mathematics, but Acerbi [2010a, 269–375] gives a treatment of the Greek traditions of a number of major works. This will be invaluable for experts and can help others get a sense for the real basis of our current editions and appreciate how unreliable this sometimes is. This means that we now have overview coverage of the major manuscript sources in both Greek and Arabic.

The Direct and Indirect Traditions

Historians of Greek mathematics divide the manuscript sources into direct and indirect traditions. The direct tradition is constituted by source texts, in Greek, found in manuscripts, or papyri, while the indirect traditions are commentaries and summaries of these in Greek along with translations, and their commentaries and summaries, largely in Arabic and Latin. A useful overview of the medieval transmission of mathematical texts is provided by Lorch [2001].

Although it might seem that only manuscripts in the direct tradition should be treated by historians as primary sources, even in the direct tradition the mathematical texts were subjected to numerous revisions over the centuries. The texts were often edited by scholars who were themselves practitioners, or teachers, of the fields that the texts transmitted, and who took the scope of their role to include a correction of the words of text based on their own understanding of the ideas that these words conveyed. Hence, in order to understand the nature of our source documents, we are often in a position of having to reconstruct a lost context of mathematical activity on the basis of both the direct and indirect traditions.

The question of the relative importance of the direct and indirect traditions has been reopened by Knorr [1996, 2001], who revisited a late nineteenth century debate between the classical philologist J.L. Heiberg and the Arabist M. Klamroth. The debate concerned the importance of the indirect tradition for determining the original form of that most canonical of Greek mathematical works, Euclid's *Elements*. At the time, Heiberg had argued that the Arabic sources were not reliable and could be ignored, but Knorr looked at propositions from Book XII in Greek, Arabic and Latin versions and came to the conclusion that certain Arabic traditions were more valuable than the direct tradition. This question was taken up again by Rommevaux, Djebbar and Vitrac [2001], who enlarged the scope by examining Book X, and concluded that the Greek sources were considerably more varied than Heiberg acknowledged, but that the relationships between the various versions were too complex to simply assert that the Arabic witnesses were more valuable. Acerbi [2003a] has studied the importance of the Arabo-Latin tradition for studying the contexts of Book V. The manuscript tradition has also been studied by looking at the scholia in Greek and Arabic manuscripts [Vitrac 2003; Djebbar 2003]. This work has culminated in a detailed study by Vitrac [2012] of the issues involved, including an introduction to the entire textual history of the *Elements*, which will be of great use to both experts and non-experts alike. He concludes that the indirect tradition is also very diverse, that differences between the various traditions are rarely global, and that, in our present state of knowledge, it would more useful to produce editions of the various versions of the indirect tradition than a new edition from the Greek manuscripts.

Because of the fidelity of William of Moerbeke's Latin translations of Archimedes, and the loss of a number of the medieval Greek manuscripts, the indirect Latin tradition has long been an important resource for studying his work. Recently, d'Alessandro and Napolitani [2012] have edited the fourteenth century Latin translations of Archimedes' works by Jacob of San Cassiano. They conclude that Jacob had access to a Greek manuscript which is not otherwise known to us, and which sometimes contains better readings than those which are still extant. This reconfirms the importance of the Latin tradition for assessing Archimedes' mathematics.

For some topics and authors, the indirect tradition is essential. For example, Hogendijk [1999/2000] has found traces of a lost book by Menelaus on the foundations of geometry in works by al-Sijzī, and I have argued that an examination of various Arabic and Latin versions of Menelaus' *Spherics* indicates that the so-called Menelaus Theorem was not his discovery but was adopted by him from previous work [Sidoli 2006]. Details of scholarship on the indirect tradition in Arabic are discussed by Van Brummelen in his contribution to this book (see below, pages 102–107).

New Editions

There have been a number of new critical editions, in both the direct and indirect traditions. Some of these are based on a reassessment of the manuscript evidence, but many of them make a critical edition available for the first time. The bulk of these are editions of versions in the indirect traditions. Arabic editions are treated in detail by Van Brummelen (pages 102–107), and I will only mention a few of them here.

Czinczenheim [2000] has produced a new edition of Theodosius' *Spherics*, along with a French translation. Tihon [1999] has completed the project, begun by Mogenet, of editing Theon of Alexandria's larger commentary on Ptolemy's *Handy Tables*. Tihon [2011] and Mercier [2011] have also begun the project of editing Ptolemy's *Handy Tables*, along with an English translation and technical commentaries. Stückelberger and Grasshoff [2006] have produced a new edition and German translation of Ptolemy's *Geography*. This includes extensive commentaries and a CD-ROM containing all the location data.

In Arabic sources, Berggren and I have produced an edition of the Arabic text, along with English translation and commentary, of Ptolemy's *Planisphere*, previously only published in a later Latin version [Sidoli and Berggren 2007]. Rashed and Bellosta [2010] have produced the first critical edition of Apollonius' *Cutting off a Ratio*, previously only available in Halley's Latin translation.

In Latin sources, Jones [2001] has reedited, along with an English translation and commentary, the Latin of William of Moerbeke's translation of a Greek work on catoptrics, erroneously ascribed to Ptolemy and usually thought to be due to Heron. Takahashi [2001] has provided the first edition, along with English translation and commentary, of a medieval Latin translation of Euclid's *Catoptrics*.

A welcome, and important, new trend has been the publication of editions that include multiple versions of a text. Kunitzsch and Lorch [2010, 2011] have edited Arabic versions of Theodosius' *Spherics* and *Habitations* together with Latin translations based on these. Vitrac and Djebbar [2011] have produced a new Greek edition of the so-called Book XIV of the *Elements*, actually due to Hypsicles of Alexandria, along with the first edition of an Arabic translation of this text. Rashed, Decorps-Foulquier and Federspiel [2008–2010] have carried out the massive task of providing a full edition of the Greek and Arabic texts for Apollonius' *Conics*, along with a French translation and commentaries. This is the first time that the first four books of the Arabic text have been edited.

F. Acerbi and B. Vitrac have started a new series of editions, including translation and commentary, of Greek works, focusing especially on less well-known treatises. Acerbi [2012] has produced the first volume of the series, a new edition of Diophantus' *Polygonal Numbers*, with Italian translation and technical commentary. The anonymous introduction to Ptolemy's *Almagest*, the first three parts of which have already appeared, will be published in this series [Acerbi, Vinel and Vitrac 2010].

The Language of Greek Mathematics

Following previous work by Mugler [1959] and Aujac [1984], Federspiel has carried out a series of studies on the technical language of Greek mathematical works. Of these, I will mention a paper on the usage of the definite and indefinite article, which introduces the concept of the "anaphor" as a part of a proposition in which the mathematician explicitly draws inferences that result from the construction [Federspiel 1995], and a series of papers on the language of Apollonius' *Conics*, including one on the problems, which is relevant to the way we understand constructions in Greek geometry [Federspiel 1994, 1999–2000, 2002, 2008].

Chapters 4 and 5 of Netz's *Shaping of Deduction in Greek Mathematics*, on "Formulae" and "The shaping of necessity," deal with the linguistic structure of Greek mathematical prose [Netz 1999a, 127–239]. Netz shows how Greek mathematical prose is structured from the bottom up — from simple objects, such as points and lines, to clusters of assertions, including previous theorems or aspects of the diagram, into arguments. Acerbi [2011a], in his monograph study *La sintassi logica della matematica greca*, gives extensive coverage of the linguistic choices that Greek mathematicians used to express the logical development of their arguments.

There have been a number of other papers that treat the language of Greek mathematics, of which I will mention only two. Vitrac [2002] gives a careful analysis of the language of a locus theorem in Aristotle's *Meteorology* III.5, which he argues, against the majority opinion, may have been due to Aristotle himself. Acerbi [2011b] provides a careful study of the "givens" terminology of Greek mathematicians, arguing that they used the concept of "given" to deal with the determinate existence of mathematical objects and as a way to give a deductive framework to constructions and calculations.

New Translations

There have been many new translations of source texts during this period. Most of the new editions are accompanied with modern language translations, and a number of the translations are discussed in the next section, on canonical authors. Here, I will mention only a few works that do not fit into these other categories.

Solomon [2000] has made a new translation, with commentary, of Ptolemy's *Harmonics*, and Berggren and Jones [2000] have translated the theoretical chapters of Ptolemy's *Geography*, adding an introduction and detailed annotations. Bowen and Todd [2004] have produced an English translation of the short treatise by Cleomedes, the *Heavens*, which they argue was part of a general exposition of Stoicism. The translation is accompanied by critical notes and many diagrams. Evans and Berggren [2006] have translated Geminus' *Introduction to the Phenomena*, and included a thorough introduction, commentary and translations of a number of related texts. Winter [2007] has made a new translation of the Aristotelian *Mechanical Problems*, which he argues was written by Archytas, based on the rather dubious assumption of a linear progression in the history of early mechanics. Sefrin-Weis [2010] has made a new translation, with detailed commentary, of Pappus' *Collection* IV, which although providing the first English translation includes a number of peculiar translation choices. This book also includes an edition of the Greek text, which was made by comparing two previous critical editions with the only manuscript source. The diagrams, however, are not an accurate reflection of what we find in the manuscript, but have been reconstructed from the mathematical argument.

Textual Studies on Canonical Authors

For Greek mathematics, the canonical authors have always been the three great geometers of the Hellenistic period: Euclid, Archimedes and Apollonius. Since our understanding of ancient Greek mathematical sciences centers around these three figures, they have been the focus of considerable textual work.

Euclid

Euclid is probably the most canonical author in the history of mathematics, and most studies of Greek mathematics begin, and unfortunately often end, with an investigation of his texts.

There have been a number of new translations of his works. Vitrac [1990–2001] has finished his project of making a French translation of the *Elements* with commentaries, and Acerbi [2007] has translated the complete works of Euclid into Italian, along with a fair number of related texts and provided the whole with commentaries. Acerbi's translation has Greek on the facing pages. Taisbak [2003] has made a detailed study of the much neglected *Data*, including commentaries and an English translation facing a reprinting of the Greek text.

The work referred to above on the direct and indirect traditions of ancient Greek works has highlighted the importance of the medieval versions of Euclid's *Elements*, not just for understanding the medieval period in its own right, but also for our understanding of Euclid's own work, with the core text being the *Elements*. There has been considerable research on the medieval tradition of Euclid's works, especially by S. Brentjes, G. De Young, A. Djebbar, E. Kheirandish, and A. Lo Bello. The details of this work are discussed by Van Brummelen, below (see pages 102–106). This work constitutes the greatest recent addition to our knowledge of the textual sources for Euclid's work, but it should be stressed that although we have a Greek edition and a number of Latin editions of his *Elements*, there is still no complete edition of any of the Arabic versions of this essential text.

Archimedes and the Archimedes Palimpsest

Arguably the greatest mathematician of the Hellenistic period, Archimedes had a different approach from Euclid. In place of systematic, foundational treatises, he focused on the measuration of curvilinear figures, and the mathematization of mechanics. Netz has begun a project of producing the first English translation of his works, with linguistic and literary commentaries, of which the first volume has appeared [Netz 2004a].

The most important development in textual studies of the Archimedes corpus has been the rediscovery, and sale at auction, of the so-called Archimedes Palimpsest. This manuscript, which contains some unique treatises by Archimedes, was read by Heiberg in the first decade of the twentieth century and then subsequently lost. Wilson [1999] gives a description of the codex, and the fascinating story of this object is told by Netz and Wilson [2009]. Netz, Noel, Wilson and Tchernetska [2011] have now produced a transcription of the text with facing pages of various images of the palimpsest,[1] accompanied with papers on the history, conservation and imaging of the manuscript and The Archimedes Palimpsest Project.

Since Heiberg was able to read the majority of the Archimedes text, including some parts that have since been lost, this work on the palimpsest has not revolutionized our knowledge of Archimedes, as some researchers had initially anticipated. Nevertheless, this textual work has lead to some new readings, such as a new interpretations of *Method* 14 [Netz, Saito and Tchernetska 2001–2002], and a new reconstruction of the final propositions, based on considerations of the physical conditions of the original Archimedes codex [Saito and Napolitani, below]. It seems certain that further study of this important source will continue to yield insight into Archimedes' mathematics.

Apollonius

The textual work on Apollonius during this period has been extensive. As well as Decorps-Foulquier's work on the Greek Apollonius manuscript tradition, and Federspiel's work on prose usage in the Greek text, discussed above, they have collaborated with R. Rashed in producing a new edition of the Greek and Arabic sources of Apollonius *Conics*, along with a French translation [Rashed, Decorps-Foulquier and Federspiel 2008–2010]. Taking this work together with the new edition of the Arabic, and oldest surviving, edition of Apollonius' *Cutting off a Ratio* [Rashed and Bellosta 2010], we now have a solid textual basis on which to form an assessment of Apollonius' contribution.

At the end of their lengthy study of Apollonius' *Conics*, Fried and Unguru [2001] have appended the first English translation, by Fried, of *Conics* IV. It has also appeared in a separate printing, with an introduction discussing the contents of the book [Fried 2002].

Diagrams

Another major area of research interest related to textual studies has been on the diagrams preserved in the manuscript sources, treated as both material object and text. The recent interest in this area was largely motivated by Chapter 1 of Netz's *Shaping of Deduction in Greek Mathematics*, "The lettered diagram" [Netz 1999a, 12-67]. Saito [2009] has given an overall discussion of the importance of textual studies for understanding Greek mathematics, which situates diagram studies within the scope of textual studies.

With the exception of the Belgian school, earlier editors, particularly Heiberg, paid little attention to manuscript diagrams, so that the critical editions of the canonical works do not contain the diagrams that were transmitted in the ancient or medieval periods. This situation is now beginning to be rectified. Netz's translation of Archimedes' *Sphere and Cylinder* includes a critical edition of the manuscript figures

[1] The palimpsest has been imaged using different spectra of light, including ultraviolet and x-ray fluorescence, to try to bring out the erased text.

[Netz 2004a]. K. Saito has undertaken a major project of examining and editing the manuscript images, focusing first on the work of Euclid. The diagrams of *Elements* I–IV and VI have been published, along with those of a number of Euclid's other works, such as the *Phenomena* and *Optics* [Saito 2006b, 2008].[2] The diagrams in the Euclidean tradition in Arabic have been investigated by De Young [2005].

The overall characteristics of the manuscript diagrams have been described by Saito [2012], with respect to diagrams in Euclid, and by Saito and myself, with respect to diagrams in Euclid, Archimedes, Apollonius, Theodosius, and others [Saito and Sidoli 2012]. The main results of this work shows that diagrams in ancient Greek works functioned as a kind of schematic, characterized by oversimplification and indifference to visual accuracy. In many cases, especially in solid geometry, the diagrams were meant to be redrawn on the basis of the text and the manuscript diagrams. This implies that it is the construction, rather than the diagram, that acts as a new source of necessity in the argument.

There have already been a number of results from this interest in diagrams. Decorps-Foulquier [1999b] has used the diagrams in Apollonius' *Conics* to explore the editorial work that Eutocius performed in preparing the Greek edition of the *Conics* that has come down to us. I have used the variant diagrams of a proposition from Aristarchus' *On the Sizes and Distances of the Sun and Moon* to show that the text was read in different ways by different readers in the medieval period, and to point out some technical issues that had previously gone unnoticed [Sidoli 2007]. Malpangotto [2010] has used the diagrams in ancient, medieval and early modern version of Theodosius' *Spherics* to show that different authors and editors used diagrams as a way of organizing their approach to spherical geometry and their conceptualization of the objects on the sphere.

During this period, there have also been a number of philosophical and logical studies on diagrams and diagrammatic reasoning. These are discussed below.

Material Culture

The material culture of science became a major area of interest and research for historians of science in the 1970s and 80s. Probably due to a scarcity of evidence, and the difficulties involved in interpreting the little evidence that there is, it took a long time for this trend to influence historians of the ancient exact sciences. During the period under consideration, however, a number of scholars have attempted to bridge this divide through various techniques.

In the first place, there have been a number of new discussions of the material evidence we do have. In her general overview, *Greek Mathematics*, Cuomo [2001] discusses some of the material evidence for the use of mathematics by various types of professionals, such as builders and acountants. Lewis [2001] has provided a textual, archeological and reconstructive study of the surveying instruments that were used by engineers and builders. This study includes translations of a number of related texts. Evans [1999] gives an important discussion of the material evidence for astronomy in the Greco-Roman world. In this regard, Jones has studied the meridian lines on a group of three sundials used to indicate the time of year (see below). Evans [2004] also discusses the material objects that would have made up the practice of professional astrologers, who were often called mathematicians in the ancient period.

There have been a number of treatments of astronomical works that were presented in the form of monuments, or public installations. Jones [2005b] gives a new edition, with translation and commentaries, of Ptolemy's *Canobic Inscription*. The original installation is not known, but a transcription of it is preserved in the manuscript tradition. Jones [2006] has also made a study of the Keskintos Astronomical Inscription, based on the actual artifact. This includes a text, translation and a reconstruction of an epicycle model that could produce the values found on the inscription. Lehoux [2007a] has made a complete study of the Greco-Roman material, both artifacts and text, that relate to parapegmata — monuments that were constructed in public spaces to keep track of a local position in some cyclical, usually astronomical, phenomena.

---

[2] See www.greekmath.org/diagrams/diagrams_index.html for the latest results of this project.

An important area of prolific activity has been research on the so-called Antikythera Mechanism — that is, the remains of a system of gears used to produce analog calculations of astronomical positions. There has been so much work done in this area that I cannot hope to cover it all in this context. I will mention only the main projects and individuals.

Since the mid 90s, M.T. Wright, and his collaborators, have been working on the Antikythera Mechanism, using new x-ray imaging and physical reconstruction. Wright [2007] has summarized the fruits of this work, and also included a bibliography of over ten of his papers that have resulted from this project. In the early years of this century, a new research project, The Antikythera Mechanism Research Project, has brought together researchers from a variety of backgrounds to study the device. This project has produced computed tomographic data of the internal structure of the device as well as new surface images. The project has led to a number of new results, of which I will mention only a few. Freeth, et al. [2006] set out the preliminary results of the project and proposed a new structure for the internal gearing. Freeth, Jones, Steele, and Bitsakis [2008] describe much more detail of the chronological and eclipse prediction dials on the surface of the object, and Freeth and Jones [2012] give an overview of their current view of the workings of the device and its relationship to ancient astronomy and cosmology. J. Evans, C.C. Carman and A.S. Thorndike have also been working on the device using the new images from the Antikythera Mechanism Project, comparison with ancient geometric models, and physical reconstruction. They have made a number of interesting findings, and offer a somewhat different interpretation of the gearing [Evans, Carman and Thorndike 2010; Carman, Thorndike and Evans 2012]. In their contribution to this volume, Evans and Carman argue that the Antikythera Mechanism can be used to help us reconsider the material culture of mathematical astronomy in the ancient world [Evans and Carman, below].

There have also been a number of studies which attempt to get a picture of the material conditions of Greek mathematicians by seeing what we can learn about this from their written works. For example, Tybjerg [2004] has argued that Heron makes mechanical devices tools of demonstration, and applies geometry to the mechanical objects. K. Saito and I have argued that the presentation of Theodosius' *Spherics* indicates that mathematicians were drawing diagrams on real globes in the course of studying spherical geometry [Sidoli and Saito 2009]. In harmonics, Creese [2010] has given a study of the texts related to the monochord as a scientific instrument, and Barker [2000, 192–229] has analyzed Ptolemy's treatment of scientific instruments in his *Harmonics*. This work makes it clear that Heron and Ptolemy were working in a similar tradition of using mechanical devices in their work, and applying mathematics to the study of instruments.

Social and Institutional Context

Another research methodology which has slowly affected the history of the ancient exact sciences from history of science has been research into the social and intellectual contexts in which ancient practitioners worked. Again, the primary reason why this trend has so weakly effected research on the ancient period is the almost total lack of evidence. For even the most famous of the Greek mathematicians, we often know with certainty nothing about the actual circumstances of their lives.

Nevertheless, there have been some studies that try to bridge this evidential gap in roundabout ways. Netz [2002a] has used loose demographic techniques to argue that there were very few mathematicians throughout the whole period of Greco-Roman antiquity, and a broad survey of the textual evidence to assert that those whose lives we know anything about came from the upper classes and did not generally earn a living through their mathematical activities. Asper [2003] has argued that theoretical mathematics arose out of a background of the practical mathematics practiced by professionals, and that mathematicians developed an interest in impersonalization, proof, and diagrams as way to distinguish themselves from other elite intellectuals, such as philosophers and sophists. Asper [2009] has also argued that theoretical mathematics, at least in the early period, was characterized by having no institutional setting, whereas the practical mathematics of professionals was loosely institutionalized in the educational settings of oral instruction where professionals learned their trades. Cuomo [2000, 9–56] has provided a discussion of the textual evidence we have for the social role of professionals who used mathematics, such builders, accountants and astrologers, and for the general, albeit vague, praise that intellectuals of all types had for a mathematical education.

If we take a strict sense of context, we still know little of the life circumstances of most mathematicians — how they made their living, where they learned mathematics, if they had any students, or how many people they might expect to read and understand their work. In fact, however, if we take a slightly broader concept of context, there has been a fair bit of work done to contextualize Greek mathematicians. In later sections, I will discuss work that broadens our understanding of the philosophical and intellectual contexts in which mathematicians worked, and the context of mathematical practice in which they produced their results.

Mathematical and Logical Studies

Historians of Greek mathematics now put less focus on producing the kinds of formal descriptions — in the logical or mathematical sense — of Greek mathematical works that were common in the past. Nevertheless, philosophers, logicians and mathematicians continue to have an interest in Greek mathematics, particularly Euclid's *Elements*. Moreover, there have been a number of developments along these lines which are important for our understanding of Greek geometry. In particular, it is now widely recognized that the predicate logic at the basis of Hilbert's program of critically reevaluating the foundations of geometry is insufficient to grapple with constructive practices and diagrammatic inferences. Hence, the constructive nature of Greek geometry was never properly understood under this program. Recently, a number of researchers have been working from various perspectives to try to develop the logical apparatus to handle the constructive, diagrammatic aspects of Greek geometry.

The authors of the following works are largely concerned with aspects of their projects that will be of interest to working mathematicians and logicians. In the following summaries, however, I only highlight aspects of their work to which I think historians of Greek mathematics should pay heed.

Using Martin-Löf's intuitionistic type theory, Mäenpää and von Plato [1990] give a formal treatment of Euclid's construction postulates (Posts. 1–3) and provide deduction rules that show how theorems and problems can be described in a single logical structure. Mäenpää [1997] gives a type-theoretic reading of Greek mathematical analysis which shows that predicate logic does not give a natural treatment of the fact that geometrical analysis treats a configuration of objects, and cannot explain the fundamental role of auxiliary constructions. He also discusses the overall logical form of an analysis-synthesis pair, which gives a new reading of the sense in which analysis is the reverse of synthesis.

There has been some work put into formalizing Euclidean geometry. Avigad, Dean and Mumma [2009] have developed a formal system that gives a model of Euclid's geometric practice. This includes attention to diagrammatic reasoning, which is provided with its own system of axioms — such as axioms of betweenness or intersection. In this model, constructions are performed by rules that guarantee the existence of certain objects, given certain conditions. This is not, however, fully constructive in the sense that construction rules are "built-in" theorems which assert that given some conditions certain objects exist that have various properties. Beeson [2010], however, provides a formal model of Euclidean constructive geometry as a intuitionistic geometry, and shows that if the geometry proves an existence theorem, the object can be constructed using the constructive methods provided by the theory. He argues that Euclid's constructions are meant to be well-formed algorithms in the strict sense. As well as developing formal models, specific questions can also be addressed using such methods. For example, Alvarez [2003] has given a logical analysis, in terms of construction, of the proof structure of those propositions in the *Elements* that rely on superposition, I.4, I.8 and III.24.

Pambuccian [2008] gives an overall treatment — historical, conceptual and technical — of work devoted to axiomatizing geometric constructions throughout the twentieth century. He is only briefly concerned with Greek geometry, but points out that constructive axiomatics preserves the distinction attributed to Geminus by Proclus between postulates, which ask for the production of something not yet given, and axioms, which refer to relationships that obtain between givens. He defines constructive approaches as quantifier-free, and points out that the intuitionistic approach mentioned above fails to meet this criterion.

Logical and Philosophical Studies of Diagram-Based Reasoning

Closely related to this material are a number of philosophical and logical studies that deal with diagrammatic practice, and the use of diagrams to draw inferences in Euclidean geometry.

K. Manders has published a work on Euclidean diagrammatic practice that has circulating widely among philosophers of mathematics in preprint form since 1995, introduced by a shorter overview of his ideas and how they have changed [Manders 2008a,b]. In these papers, he asserts that diagrams introduce concepts and starting points into the Euclidean argument and claims that Euclidean geometry is a practice of making inferences on diagrams and statements, but that it does not depend on geometric objects. Miller [2007] has created a formal system for diagrammatic inference and proof. He argues that the idea that diagrammatic methods are inherently informal is incorrect, and that the Euclidean practice can be understood as an informal implementation of a potentially rigorous approach. Panza [2012] provides a new study of the diagrammatic practices of Euclid's plane geometry, incorporating the new approaches that have been developed by logicians and philosophers, but arguing that Euclidean geometry is fundamentally about geometric objects and that the diagrams represent these. He argues that diagrams allow Euclid to deal with issues of identity, construction, existence and being "given," and that the geometric objects under study in the text inherent some properties from their diagrams.

Unfortunately, all this work focuses strictly on Euclid's *Elements*, and usually only the first book. Hence, a number of the conclusions reached are incompatible with the broader context of constructive and diagrammatic practices in which the *Elements* should be situated.

Balanced Reconstruction

For good reasons, elaborate reconstructions of ancient mathematical results and theories have fallen out of favor. In particular, historians of ancient mathematics no longer try to recreate the historical development of periods for which the texts have been completely lost, or never existed. The primary reason for this is that our reconstructions must always be plausible and linear whereas we know that the history of mathematics and science, for periods in which we have ample evidence, does not always proceed in such an idealized way. Moreover, the number of plausible, reconstructed histories is often equal to the number of scholars working along such lines.

Nevertheless, since so many of the Greek texts have been lost, historians still engage in a more balanced form of reconstruction, in which they try to show that certain results were possible, that certain methods could have been used, or how a certain result may have been shown, relying strictly on methods that are consistent with the available textual evidence. The scope for such reconstructions is more limited than the reconstruction of historical developments. For example, Masià-Fornos [2010] has shown that a putative lacuna in Archimedes' *Sphere and Cylinder* can be supplied rather naturally using concepts and methods readily available to Archimedes. In a similar vein, I have shown that mathematical methods evidenced in Ptolemy can be used to reconstruct the full argument for an obscure chapter in Heron's *Dioptra* [Sidoli 2005]. Acerbi [2011c] has given a reconstruction of the proof for the final, incomplete proposition of Diophantus' *Polygonal Numbers*. He has also given a proof in five theorems, using ancient methods, of the ancient claim that there are only thee homeomeric lines [Acerbi 2010b]. Such an argument does not prove that this proof was actually made in antiquity, but does show that a proof of its kind could have been produced, so that there is no reason to doubt the ancient claim that there was a proof of this fact.

New Readings

As the study of Greek mathematical sciences has come more strongly under the influence of the history of science, it has also come under the general tendency of scholars in the humanities to produce new readings of canonical texts. Of course, in some sense all historical work on this period involves the production of new readings of familiar texts; nevertheless a number of influential tendencies can be discerned.

Ancient Mathematical Contexts

In a sense, the most pervasive tendency of the new historiography has been the continuing practice of reading Greek mathematical texts in the context of ancient mathematical methods: that is, using the often terse technical sources to try to understand how Greek mathematicians actually thought about, and did, mathematics. Hence, many studies in the period under consideration could be placed in this section, but I present here only a selection of significant examples.

Probably the most complete project in this vein is a monograph study of Apollonius' *Conics* by Fried and Unguru [2001]. This book presents a new reading of the complete extant text and argues that Apollonius' mathematics should be understood and explained in the geometrical form in which it is presented. It also takes a polemical attitude towards the old historiographical methods and presents a case study for the success of the new historiography advocated by Unguru in the 1970s. Fried [2002] has studied the role of analogy in Apollonius' introduction and the use, in *Conics* VII, of the so-called *homologue*, a line which is defined by cutting the transverse axis in the same ratio that the transverse axis has to the latus rectum. He argues that proportion, as *analogia*, was an organizing principle in Apollonius' theory.

There have been new readings of the theory of ratio developed in *Elements* V. Acerbi [2003a] gives a new reading of both the Greek and Arabo-Latin traditions of *Elements* V, in which he argues that by the time the theory was composed, the concept of generality and the linguistic tools used to express it were fully developed. Looking at the same material, but reading it in a different context, Saito [2003] argues that the pre-Eudoxean theory of proportion reconstructed by Becker is unnecessary to explain the usage of proportion theory in the ancient mathematical context. The proportion theory put forward in *Elements* V can be understood better as a ratio theory that is useful for doing elementary geometry than as a complete, and general, treatment of ratio. Hence, any pre-Eudoxean theory should also be understood as having been some loose set of theorems and techniques for ratios that are useful in geometry.

There are also a number of specific results, or new readings of texts that have been produced in this way. By showing that Heron's *Dioptra* 35 can be read in the context of mathematical methods found in Ptolemy's works, I have argued that ancient gnomonics can be applied to the solution of problems in spherical astronomy [Sidoli 2005]. Saito [2006a] examines the treatment of quadrature in a range of Archimedean works and shows that the texts can be read more naturally when we realize that Archimedes did not work with a general concept of quantity but rather investigated well-defined figures and compared their various attributes with one another.

This approach has also been used to help us understand the goal and approach of lost works. For example, Acerbi [2008] provides a background for understanding the goal of Euclid's *Fallacies* by discussing the role of false proof in a wide range of philosophical and mathematical sources, and identifying a number of false proofs as possibly having an origin in this lost text. While we still do not know about the detailed contexts of this work, we can now develop a better sense of the questions it may have addressed and the approaches it may have followed.

Intellectual and Philosophical Contexts

Another type of scholarship which is not new, but which has certainly be influenced by tendencies in history of science and other areas of humanities scholarship, has been the practice of reading ancient mathematical authors in the contexts of other philosophical and intellectual fields with which they were in contact and with which they competed.

Mansfeld [1998] has examined the introductions to mathematical texts from Apollonius' *Conics* to Theon's commentaries. This examination of the preliminary material, in contrast to the austere and often strictly technical core texts, shows that Greek mathematicians situated their work in the intellectual and philosophical currents of their time. Cuomo [2000], in her book on Pappus, spends considerable effort situating his work in the general intellectual context of his time.

Bernard [2003b] gives a programmatic argument that categories of ancient rhetoric can be used to analyze mathematical texts. He also provides a detailed example of this program, arguing that an obscure passage in Pappus's *Collection* can be understood in terms of the techniques developed by orators and rhetoricians in late antiquity [Bernard 2003a].

Feke and Jones [2010] give a general overview of Ptolemy's philosophical approach, which shows both that he conceived of his project in the terms of the philosophy of his time and that his works had an internal philosophical consistency. Bernard [2010] makes a reading of the ethical and practical aspects of Ptolemy's mathematical project and argues that many aspects of its presentation may have been addressed towards the discourse of professional astrologers. Feke [2012] shows that Ptolemy situated his work in the philosophical currents of his time, while maintaining that the best way to pursue true philosophy is through mathematical investigations of the natural world.

Literary Readings

A relatively new trend has been a handful of studies that introduce non-mathematical, almost literary readings of Greek mathematical texts.

Netz [1999a], in passages of his *Shaping of Deduction in Greek Mathematics*, argues that there is a similarity between the use of formulaic language in Greek mathematical texts and in Homeric epic poetry. In his *Ludic Proof* [2009], he argues that the Hellenistic geometers can be read as using literary techniques that are comparable to what we find in the literature of this period — such as compositional variation between themes and methods, an element of surprise in the narrative structure, and experimentation with various genres. Wagner [2009] has given a Deleuzian reading of his own speculative reconstruction of Greek mathematical practice, and argues that Deleuze's analysis of Francis Bacon's approach to painting may be of use in understanding diagrammatic practice in Greek mathematics.

Unguru and Fried [2007] have gently ribbed the idea of highly postmodern readings by producing a fictional, purely literary, reading of Apollonius's *Conics* as a work on sexual politics and comparing this with a H.G. Zeuthen's highly mathematical and their own more historical readings.

Topics

In this section, I discuss a number of topics that have been addressed by historians over the long first decade of this century. Some of these topics have been perennial favorites — such as the origins of Greek mathematics, or geometrical analysis — while others are fairly new, such as late ancient mathematicians and commentators, or combinatorics.

Origins and Early Evidence

The origins of Greek mathematics, although poorly documented and highly speculative, was once a favorite topic for historians of the subject. Despite the fact that most historians have grown wary of such origin stories, it still receives some attention.

The relationship between Greek mathematics and Mesopotamian, that is Old Babylonian, mathematics can now be reappraised in the light of the many new studies of the latter that have been made since the early 1990s. Friberg [2007], in a study of various similarities between the two, has tried to revise the old view of a strong and direct influence of Babylonian mathematics on the development of theoretical Greek mathematics. It should be noted, however, that these similarities are only found in the subjects studied — which are relatively obvious objects of elementary mathematical investigation — not in the concepts or methods. Robson [2005], on the other hand, expresses the opposite opinion, namely, that given the rather stark differences of conception and approach, and the fact that Old Babylonian mathematics had already died out over 1,000 years before Greek mathematics began to be practiced more than 1,000 miles away, it is unlikely that the one greatly influenced the other. Given the current state of evidence, it seems unlikely that consensus can be reached on this issue. Although there certainly seems to be some continuity from Old Babylonian and Selucid mathematics to the "practical" material found in Greek papyri and the Heronian corpus, as shown by Friberg [2005, 193–267], the more advanced material of the Old Babylonian period shows such divergence from the approaches and methods of the Greek geometers that it is possible to argue that the later is independent of the former.

Although most historians of Greek mathematics are now highly doubtful that Pythagoras and the early Pythagoreans played a significant or unique role in the origins of Greek mathematical science, Zhmud [2006] has presented an updated version of his views on this in the new English translation of his book on the early Pythagoreans. In particular, he has strengthened his arguments by references to his work on Eudemus' histories of the mathematical sciences [Zhmud 2012]. It should be pointed out, however, that for the earliest history he must still rely on reconstructions, which are based on the assumption that the mathematical sciences must develop in a linear, cumulative way. Also on the subject of Pythagorean science, Borzacchini [2007] takes what he calls a cognitive approach to reconstructing an origin for the idea of incommensurability in Pythagorean music theory. Most of his concrete evidence, however, is drawn from the work of Archytas, which is rather late for an origins story. Mueller [2003] reconstructs a more intuitive and less rigorous proof of *Elements* I.32, that the sum of the angles of a triangle are two right angles, and argues that this could have been written as early as the fifth century, possibly by Pythagoreans. The majority opinion, however, now seems to be that the early Pythagoreans did not play a unique role in the development of the mathematical sciences. For example, Asper [2003] argues that theoretical mathematics, which arose first among the sixth century Ionians, crystalized in the Athenian milieu in the attempts of mathematicians to differentiate themselves from other elite intellectuals, such as philosophers and sophists. Netz [2004b], however, associates the origins of theoretical mathematical work with the production of mathematical texts, and hence studies the earliest source we have of for written proofs, Simplicius' account of Hippocrates' quadrature of lunes, which was taken from Eudemus' history.

There have also been a number of studies of the mathematics of Plato's time. Zhmud [1998] has made a skeptical study of the common idea that Plato was an "architect of science" by showing that this idea originated in the early Academy, which suggests that Plato was more influenced by mathematicians than they by him. Huffman [2005] has produced a monograph study of Archytas, a late Pythagorean contemporary of Plato's, who everyone agrees did important work in theoretical arithmetic and harmonics. Acerbi [2000, 2005] has done some work on the often cryptic references to mathematics in the writings of Plato himself.

Late Ancient Mathematicians and Commentators

Most of what we think we know about the origins of Greek mathematics comes from texts written many centuries later, in the late ancient period. This material used to be primarily of interest to historians of Greek mathematics because of its usefulness for studying earlier periods. A welcome change in this recent period of historical scholarship is that historians have developed an interest in the mathematics and scholarship of late antiquity for its own sake, and not merely for the historical material it contains.

Cuomo [2000] has produced a monograph study of Pappus, the most creative mathematician of late antiquity, and shown that his treatment of the mathematics of his predecessors was both an essential part of his mathematical project and explicable in the context of other late-ancient cultural and intellectual practices. Bernard [2003a] has explored the rhetorical aspects of Pappus' work, and linked it with the rhetoric of the period. He has also shown that this can be used as an example to argue for the general utility of sophistical and rhetorical categories as a context for understanding mathematical work [Bernard 2003b].

There has also been new work on Eutocius' commentaries on Archimedes and Apollonius. Netz [1999c] has argued that Eutocius' treatment of a problem due to Archimedes actually transforms the mathematical concepts in important ways, and he later situated this study between a study of Archimedes' approach and that taken by the medieval mathematician ʿUmar Khayyām [Netz 2004c]. Decorps-Foulquier [1998, 1999a,b] has made a number of studies of Eutocius as an editor of, and commentator on, Apollonius' *Conics*.

Netz [1999b] has studied the famous divisions of a proposition set out by Proclus in his commentary on Euclid's *Elements* I. He shows that this division was probably a description produced by Proclus that affected the later transmission of the Euclidean text. Hence, it should not be used as a universal model for all Greek mathematical texts.

Netz [1998] has written a general description of the commentary tradition and its effect on our image of Greek mathematics. Chemla [1999] has responded to this by pointing out the difficulties often involved in distinguishing between source text and commentary, and Bernard [2003c] has responded by highlighting the didactic aspects of this tradition.

Numeracy

Although there has been much study of literacy in the Greco-Roman world, our understanding of the state of numeracy in this period is still very slight. Netz [2002b] introduced a program of studying Greek numeracy, defined as counting with physical counters. More recently, Cuomo [2012] has announced a project to study Greco-Roman numeracy more broadly by focusing on textual and archeological evidence. This work in still in a preliminary stage.

Greek Combinatorics

One of the most interesting developments in the period covered by this paper has been some work that demonstrates that combinatorics, in a fairly developed sense, was practiced by Greek mathematicians. This refutes the previously prevailing view that the Greeks did no substantial work in combinatorics.

The first significant step in this work was made by mathematicians who recognized the relationship between two numbers attributed to Hipparchus by Plutarch and Schröder numbers in modern combinatorics [Stanley 1997; Habsieger, Kazarian and Lando 1998]. This was then built upon by Acerbi [2003b], in a historical paper in which he shows how Hipparchus' work in combinatorics could have been related to technical features in Stoic logic, and gives a discussion of many of the passages by Greek mathematicians and philosophers that have a combinatorial significance. He argues that this work on combinatorics probably took place in the context of technical studies of logic, the source texts of which were lost as interest in the technical study of logic declined.

Another argument for the use of combinatorics in Greek works concerns Archimedes' *Stomachion*, which only survives in two separate Greek and Arabic fragments. Netz, Acerbi and Wilson [2005] argue that Archimedes used combinatorial reasoning in his analysis of the stomachion puzzle-game. There is no direct evidence of combinatorics in the fragments that survive, however, and Morelli [2009] has argued that a combinatorial interpretation is not supported by any of the other ancient evidence for the game.

Computation and Algebra

There are still relatively few studies of the practical mathematics that was used by professionals, such as builders, accountants and astrologers. J. Sesiano has made editions and new studies of a number of texts related to these problem solving techniques, which clearly had relations to the mathematical practice of other Mediterranean and Near Eastern cultures [Sesiano 1998, 1999]. The last chapter of J. Friberg's book on the mathematics of Babylon and Egypt is also a key study of this material [Friberg 2005, 193–268].

Diophantus

A significant achievement of the last ten years has been a reevaluation of the work on Diophantus from the perspective of the new historiography. Although there has been some recent work on Diophantus along the lines of the old historiography [Meskens 2009, 2010], most work during this period has moved away from an algebraic interpretation of his work. That is, it seeks to understand the positive characteristics of his mathematical practice, without simply situating him as a precursor to the development of algebra.

A number of recent studies have shown that, although Diophantus is not doing algebra in the sense of exploring methods of solving certain types of equations, he does present general and systematic methods for solving arithmetic problems, that is, producing rational numbers that satisfy various conditions. For example, Thomaidis [2005] has shown that Diophantus transforms arithmetical problems into the special terms of his arithmetic theory in such a way as to finally produce an equation that can be solved. In a later paper, he identifies two types of equations in Diophantus' approach, what he calls equalities, which are statements of equality that are not to be solved, and the equation proper, which is the end result of the transformation into the arithmetic theory and which is then solved using the standard techniques of ancient and medieval algebra [Thomaidis 2011]. Christianidis [2007] has helped elucidate the general approach (*hodos*) by showing that the introduction can be read as an explanation of how to transform a problem into the arithmetical theory, and then solve the resulting equation. He then applies this analysis to a famous problem, *Arithmetica* II.8, to divide a proposed square into two squares. Bernard and Christianidis [2012] present a new framework for the first three books of *Arithmetica*, in which they attempt to catalog all the types of methods that are used to transform the conditions of the problem into an equation, and to explain the enigmatic role of "positions" (*hypostaseis*). Acerbi [2009] also supports an essentially non-algebraic reading by arguing that the technical term *plasmatikon* is a late interpolation and cannot be used to support an algebraic reading of the text, as had been done by some scholars.

The result of this research is that we will probably soon be in a position to characterize Diophantus' work — which has so far seemed idiosyncratic at best and cryptic at worst — using the terms and methods in which he expresses himself, and, in this way, develop a historically sound reading of his text.

Analysis

Discussions of geometrical analysis remain a regular pastime of historians of Greek mathematics. As well as the articles discussed above, in relation to logical studies, there have been a number of studies of the goals, language and techniques of analysis.

In a collection of essays in memory of W.R. Knorr, Berggren and Van Brummelen [2000] provide a clear overview of the structure of an analysis-synthesis pair and give examples from both Greek and Arabic texts. In the same collection, Netz [2000] argues that analysis was not heuristic in a meaningful sense, but that a Greek mathematician would publish an analysis in order to reveal the key idea behind the solution to a particular problem.

There has been new research into the terminology of "givens," which is fundamental to Greek mathematics. A given object is one that (1) is present at the beginning of a piece of mathematical discourse, (2) assumed by the mathematician, or (3) determined on the basis of either of these. Fournarakis and Christianidis [2006] make a linguistic study of the terminology in geometrical analysis and conclude that different grammatical forms of the base verb *didōmi* are used to make philosophical distinctions regarding the direction of the inference, which they relate to Pappus' discussion of analysis. It is not clear, however, that these distinctions have any meaning in mathematical practice. Acerbi [2011b] has studied the use of this terminology in all its forms and in the full range of mathematical texts and shows that the concept of "given" was used to discuss the concepts of uniqueness and existence, and to create a general framework within which mathematicians could treat different types of argumentative steps, such as constructions, calculations and operations.

K. Saito and I have made two studies of geometrical analysis that seek to revive the idea that analysis was a heuristic technique that could be used to help solve problems and work out theorems. We have argued that the *diorisms* that are found in problematic analysis are remnants of a practice of both exploring the possibility of solutions and enumerating all possible solutions [Saito and Sidoli 2010]. We have also explored a form of theoretic analysis not previously noticed by historians, and argued that this can be taken as evidence that Greek mathematicians actually did do theoretic analysis, contrary to the claims of some historians [Sidoli and Saito 2012].

Greek Foundations of Mathematics

There has always been interest in the foundations of Greek mathematics, but the influence of the new historiography has produced a focus on what Greek mathematicians themselves thought about the logical foundations of their own activity. Going back to the work of Stenius [1978] and others, there has been a recognition that Greek mathematicians approached foundations by doing mathematics, not meta-mathematics.

Acerbi [2010c] has studied what remains of the foundational work of Apollonius and Geminus and shown that for Greek mathematicians, foundations were approached in directly mathematical ways by reworking the elements — that is, invoking new definitions and construction postulates, justifying constructions by superpositions, rewriting the proofs for accepted theorems and trying to prove assumptions such as common notions and postulates. In this sense, we can see that most Greek mathematicians were involved in foundational issues. For example, Borzacchini [2006] argues that Archimedes introduced the axiom that was subsequently named after him in order to deal with the quadrature of curvilinear figures, to which the Eudoxian and Euclidean approach may not have been readily adaptable.

One mathematician of the Imperial period who was especially interested in foundations was Menelaus. He was proud of the fact that he had used no indirect arguments in his *Spherics* and that he had been able to avoid constructing lines inside the sphere, for both of which he criticized Theodosius. He also wrote an *Elements*, which has been lost, but which was almost certainly meant to be read as work on the foundations of mathematics. Hogendijk [1999/2000] has recovered a number of fragments of this text from al-Sijzī's geometry.

Another way in which Greek mathematicians did foundational work was by paying attention to the structure of their arguments, both at the local and global level. Into this category fall debates about the relative importance and position of problems and theorems, the pros and cons of indirect argument, the use of double indirect arguments to show equalities that were established by various means, the suppression or multiplication of cases, and so on. K. Saito has argued that Greek mathematicians used such structures to address patterns of argument, such that an individual proof often served as a paradigm, in the sense of a model [Chemla 2012, 30–31]. Mendell [2007] has shown that a paradigmatic type of two-step argument, involving showing that a proportion holds first for commensurable and next for incommensurable magnitudes, can be found already in Aristotle's writings.

Constructions

Greek mathematicians seem to have regarded constructions and problems as closely bound up with foundations, as evidenced by the work of Euclid, Apollonius and Menelaus. For much of the twentieth century Zeuthen's thesis that problems served as proofs of existence dominated thinking about Greek constructive practice [Thiel 2005]. Especially, following Knorr [1983], however, scholars came to doubt this view.

Harari [2003] argues that constructions in the *Elements* serve to exhibit spacial relations between objects and that, furthermore, Aristotle's ontology is not compatible with construing constructions as existence proofs. K. Saito and I have used a study of the constructions in Theodosius' *Spherics* to show that ancient geometers distinguished between constructions for the sake of a proof and constructions that can be used to solve problems [Sidoli and Saito 2009]. The former can be counterfactual, while the later are abstractions of operations that can be carried out in practice. This makes it clear that Greek geometers clearly distinguished between their diagrams and the objects the diagrams modeled. This is made clear in their use of the word "to imagine" (*noein*), which is used to mark situations when the diagram may not fully model the objects under consideration. Netz [2010] has studied all the occurrences of this word in the mathematical literature and determined that it primarily denotes three types of situations: three dimensional figures, models of physical objects, and disparities between the diagram and the objects it models.

The Exact Sciences

The exact sciences were regarded by Greek practitioners, not as applications, but as essential branches, of mathematics. From Archytas to Ptolemy we read of the division of mathematics into geometry, arithmetic, astronomy and harmonics, with each individual practitioner setting the significance and relationships of these fields according to his own research interests. In authors like Geminus, we find more elaborate divisions of the mathematical sciences.

There has been considerable work in each area of the exact sciences and I will not be able to do justice to many of them here. I have already discussed a number of editions and translations of works in the exact sciences, but here I will describe a number of important studies in the main fields of the exact sciences.

For the early history of mathematics, the most important science was always astronomy. There has been considerable work on Greek astronomy during the period under consideration, of which I will mention only a few examples. J. Evans' *History and Practice of Ancient Astronomy* [1998] gives a wonderful introduction to many aspects of ancient Mediterranean, and to a lesser extent, Near Eastern, astronomy, with an emphasis on practical methods and concepts. Although this book was designed as a textbook, it is also of use to scholars. Jones [2005a] has made a study of the relationship between mathematical models and the physical reality that they represented in Ptolemy's work. This paper explores the function of mathematical modeling in all of Ptolemy's major writings. Berggren and I have studied Aristarchus' *On the Sizes and Distances of the Sun and the Moon* as a work of mathematics and shown that physical hypotheses are used in this work simply to show how mathematics can be applied to them to produce new knowledge [Berggren and Sidoli 2007]. Van Brummelen [2009], in his book on the early history of trigonometry, has devoted two chapters to Greek trigonometry and its relationship to astronomy (Chaps. 1 and 2).

Spherics, which was the combined study of spherical geometry and spherical astronomy, was a science that bridged our modern conceptual distinction between mathematics and astronomy. I have studied the mathematics involved in Heron's *Dioptra* 35, which shows how to use simultaneous lunar eclipse observations to determine the distance between two terrestrial locations, and shown that this can be interpreted as allowing an exact calculation using the so-called analemma methods [Sidoli 2005]. Malpangotto [2003] has studied Pappus' treatment of, and commentary to, Theodosius' *Spherics*. K. Saito and I have shown that Theodosius' *Spherics* provides evidence that ancient spherics was often practiced by producing drawings on actual globes [Sidoli and Saito 2009].

Another essential branch of the exact sciences was harmonics, which became a field of the quadrivium and was developed by mathematicians from Archytas to Ptolemy. A number of recent studies have fully embraced harmonics as an ancient science. Barker [2000] has investigated Ptolemy's scientific method as it functions in his *Harmonics*, with due attention to conceptual, mathematical and instrumental approaches. He has also produced a monograph study of harmonics in the period from the late Pythagoreans to Theophrastus [Barker 2007]. He treats harmonics as a science and sets out the details of the various methodologies, with their different approaches to mathematization and quantification. Finally, Creese [2010] has produced a thorough study of the monochord, which became the key scientific instrument of ancient harmonics. This work helps us understand the relationship between mathematization and instrumentation in harmonics.

Optics was another area developed by a number of Greek mathematicians. Smith [1999] has made a sort of textbook for studying Greek optics based on original sources in English translation, with an emphasis on Ptolemy's *Optics*. Lehoux [2007b] discusses Galen and Ptolemy's accounts of vision and argues that Ptolemy was deeply concerned to construct a sound epistemological grounding for vision, which he regarded as the most mathematical of the senses.

Another important area of mathematical work was mechanics. Tybjerg [2004] has studied Heron's mathematical approach to mechanics and argues that he integrates mechanics and geometry and presents mechanics as a theoretical discipline based on mathematical demonstration. Berryman [2009] studies the history of Greek mechanics and argues that mechanical conceptions of processes and objects were a vital part of ancient natural philosophy. J. Evans and C. Carman have argued, using the Antikythera Mechanism, that mechanics and the mechanical hypothesis may have been important in the development of astronomical modeling in the Hellenistic period [Evans and Carman, below].

Conclusion

I hope it is clear from this often brief survey that the first decade of the 21st century has seen considerable work done on the history of the ancient Greek mathematical sciences. More researchers have entered the field than have left, and it has been invigorated by new approaches and ideas.

It is clear that the old historiography has been overcome. The new historiographic approach that was so hotly debated in the 1970s has become mainstream. There are now almost no serious scholars of the subject trying to determine how Greek mathematics must have originated based on what seems likely from some mathematical or logical perspective, or trying to understand the motivation for methods found in Apollonius or Diophantus using mathematical theories and concepts developed many centuries after these mathematicians lived. At the same time, Greek mathematics is *mathematics* and attempts to read these works using literary or postmodern approaches have not gained much traction.

There is still no book that does for the new historiography what T.L. Heath's monumental *A History of Greek Mathematics* [1921] did for the old historiography. Hence, scholars of the subject will still make use of Heath's work, and they will still suggest that their students read it. So, while the old historiography has been overcome, it has not been left behind. There is still value in making a mathematical reading of an ancient text, as a number of the studies surveyed above have shown.

The wealth of new studies on both familiar and novel topics means that we will soon be in a position to produce a new synthesis, which will describe the practices of Greek mathematicians as mathematical activity, which can be related to other intellectual and cultural activities of the period, and compared with the mathematical activities of other ancient cultures.

References

Acerbi, F., 2000. Plato: Parmenides 149a7–c3. A proof by complete induction? Archive for History of Exact Sciences 55, 57–76.

——— 2003a. Drowning by multiples. Remarks on the fifth book of Euclid's Elements, with special emphasis on prop. 8. Archive for History of Exact Sciences 57, 175–242.

——— 2003b. On the shoulders of Hipparchus: A reappraisal of ancient Greek combinatorics. Archive for History of Exact Sciences 57, 465–502.

——— 2005. A reference to perfect numbers in Plato's Theaetetus. Archive for History of Exact Sciences 59, 319–348.

——— 2007. Euclide: Tutte le opere. Bompiani, Milan.

——— 2008. Euclid's Pseudaria. Archive for History of Exact Sciences 62, 511–551.

——— 2009. The meaning of πλασματικόν in Diophantus' Arithmetica. Archive for History of Exact Sciences 63, 5–31.

——— 2010a. Il silenzio delle sirene. Carocci Editore, Rome.

——— 2010b. Homeomeric lines in Greek mathematics. Science in Context 23, 1–37.

——— 2010c. Two approaches to foundations in Greek mathematics: Apollonius and Geminus. Science in Context 23, 151–186.

——— 2011a. La sintassi logica della matematica greca. Archives-ouvertes.fr, Sciences de l'Homme et de la Société, Histoire, Philosophie et Sociologie des sciences. hal.archives-ouvertes.fr:hal-00727063.

——— 2011b. The language of the "givens": Its form and its use as a deductive tool in Greek mathematics. Archive for History of Exact Sciences 65, 119–153.

——— 2011c. Completing Diophantus, De polygonis numeris, prop. 5. Historia Mathematica 38, 548–560.

——— 2012. Diofanto: De polygonis numberis. Fabrizio Serra Editore, Pisa.

Acerbi, F., Vinel, N., Vitrac, B., 2010. Les Prolégomènes á l'Almageste. Une édition á partir des manuscrits les plus anciens : Introduction générale — Parties I–III. SCIAMVS 11, 53–210.

Alvarez, C., 2003. Two ways of reasoning and two ways of arguing in geometry. Some remarks concerning the application of figures in Euclidean geometry. Synthese 134, 289–323.

Asper, M. 2003. Mathematik, Milieu, Text. Sudhoffs Archiv 87, 1–31.

——— 2009. The two cultures of mathematics in ancient Greece. In: Robson, E., Stedall, J. (eds.), The Oxford Handbook of the History of Mathematics, Oxford University Press, Oxford, pp. 107–132.

Aujac, G., 1984. Le langage formulaire dans la géométrie grecque. Revue d'histoire des sciences 37, 97–109.

Avigad, J., Dean, E., Mumma, J., 2009. A formal system for Euclid's Elements. The Review of Symbolic Logic 2, 700–768.

Barker, A., 2000. Scientific Method in Ptolemy's Harmonics. Cambridge University Press, Cambridge.

——— 2007. The Science of Harmonics in Classical Greece. Cambridge University Press, Cambridge.

Beeson, M., 2010. Constructive geometry. In: Arai, T., Brendle, J., Kikyo, H., Chong, C.T., Downey, R., Feng, Q., Ono, H. (eds.), Proceedings of the 10th Asian Logic Conference. World Scientific, Hackensack, New Jersey, pp. 19–84.

Berggren, J.L., Jones, A., 2000. Ptolemy's Geography: An Annotated Translation of the Theoretical Chapters. Princeton University Press, Princeton.

Berggren, J.L., Sidoli, N., 2007. Aristarchus's On the Sizes and Distances of the Sun and the Moon: Greek and Arabic texts. Archive for History of Exact Sciences 61, 213–254.

Berggren, J.L., Van Brummelen, G., 2000. The role and development of geometric analysis and synthesis in ancient Greece and medieval Islam. In: Suppes, P., Moravcsik, J., Mendell, H. (eds.), Ancient & Medieval Traditions in the Exact Sciences. CSLI Publications, Stanford, pp. 1–31.

Bernard, A., 2003a. Sophistic aspects of Pappus's Collection. Archive for History of Exact Sciences 57, 93–150.

——— 2003b. Ancient rhetoric and Greek mathematics: A response to a modern historiographical dilemma. Science in Context 16, 391–412.

——— 2003c. Comment définir la nature des textes mathématiques de l'antiquité grecque tardive? Proposition de réforme de la notion de 'texts deutéronomiques.' Revue d'histoire des mathématiques 9, 131–173.

——— 2010. The significance of Ptolemy's *Almagest* for its early readers. Revue de synthèse 131, 495–521.

Bernard, A., Christianidis, J., 2012. A new analytical framework for the understanding of Diophantus's Arithmetica I-III. Archive for History of Exact Sciences 66, 1–69.

Berryman, S., 2009. The Mechanical Hypothesis in Ancient Greek Natural Philosophy. Cambridge University Press, Cambridge.

Borzacchini, L., 2006. Why rob Archimedes of his lemma? Mediterranean Journal of Mathematics 3, 433–448.

——— 2007. Incommensurability, music and continuum: a cognitive approach. Archive for History of Exact Sciences 61, 273–302.

Bowen, A.C., Todd, R.B., 2004. Cleomedes' Lectures on Astronomy. University of California Press, Berkeley.

Bülow-Jacosen, A., Taisbak, C.M., 2003. P.Cornell inv. 69. Fragment of a handbook in geometry. In: Pilz, A., et al. (eds.), For Particular Reasons: Studies in Honour of Jerker Blomqvist. Nordic Academic Press, Lund, pp. 55–70.

Carman, C.C., Thorndike, A.S., Evans, J., 2012. On the pin-and-slot device of the Antikythera Mechanism, with a new application to the superior planets. Journal for the History of Astronomy 43, 93–116.

Chemla, K., 1999. Commentaires, éditions et autres textes seconds: quel enjeu pour l'histoire des mathématiques? Réflections inspirées par la note de Reviel Netz. Revue d'histoire des mathématiques 5, 127–148.

——— (ed.), 2012. The History of Mathematical Proof in Ancient Traditions. Cambridge University Press, Cambridge.

Christianidis, J., 2007. The way of Diophantus: Some clarifications on Diophantus' method of solution. Historia Mathematica 34, 289–305.

Creese, D., 2010. The Monochord in Ancient Greek Harmonic Science. Cambridge University Press, Cambridge.

Cuomo, S., 2000. Pappus of Alexandria and the Mathematics of Late Antiquity. Cambridge University Press, Cambridge.

——— 2001. Ancient Mathematics. Routledge, London.

——— 2012. Exploring ancient Greek and Roman numeracy. BSHM Bulletin 27, 1–12.

Czinczenheim, C., 2000. Edition, traduction et commentaire des Sphériques de Théodose. These de docteur de l'Universite Paris IV. Printed: Atelier national de reproduction des thèses, Lille.

d'Alessandro, P., Napolitani, P.D., 2012. Archimede Latino, Iacopo da San Casiano e il Corpvs Archimedeo alla metà del quatrtrocento. Les Belles Lettres, Paris.

De Young, G., 2005. Diagrams in the Arabic Euclidean tradition: A preliminary assessment. Historia Mathematica 32, 129–179.

Decorps-Foulquier, M., 1998. Eutocius d'Ascalon éditeur du traité des Coniques d'Apollonius de Pergé et l'exigence de clarté: un exemple des pratiques exégétiques et critiques des héritiers de la science alexandrine. Dans: Argoud, G., Guillaumin, J.Y. (réd.), Sciences exactes et appliquées à Alexandrie. Publications de l'université de Saint-Étienne, Saint-Étienne, pp. 87–101.

——— 1999a. Recherches sur les Coniques d'Apollonios de Pergé et leurs commentateurs grecs: Histoire de la transmission des Livres I–IV. Klincksieck, Paris.

——— 1999b. Sur les figures du traité des Coniques d'Apollonios de Pergé édité par Eutocius d'Ascalon. Revue d'histoire des mathématiques 5, 61–82.

Djebbar, A., 2003. Quelques exemples de scholies dans la tradition arabe des Éléments d'Euclide. Revue d'histoire des sciences 56, 293–321.

Evans, J., 1998. The History and Practice of Ancient Astronomy. Oxford University Press, New York.

——— 1999. The material culture of Greek astronomy. Journal for the History of Astronomy 30, 237–307.

——— 2004. The astrologer's apparatus: A picture of professional practice in Greco-Roman Egypt. Journal for the History of Astronomy 35, 1-44.

Evans, J., Berggren, J.L., 2006. Geminos's Introduction to the Phenomena. Princeton University Press, Princeton.

Evans, J., Carman, C.C., Thorndike, A.S., 2010. Solar anomaly and planetary displays in the Antikythera Mechanism. Journal for History of Astronomy 41, 1–39.

Federspiel, M., 1994. Notes critiques sur le livre I des Coniques d'Apollonius de Pergè. Revue de études greques 107, 203–218.

——— 1995. Sur l'opposition défini/indéfini dans la langue des mathématiques grecques. Les études classiques 63, 249–293

——— 1999–2000. Notes critiques sur le livre II des Coniques d'Apollonius de Pergè. (Première partie, Seconde partie.) Revue de études greques 112, 409–443; 113, 359–391.

——— 2002. Notes critiques sur le livre III des Coniques d'Apollonius de Pergè. Première partie. Revue de études greques 115, 110–148.

——— 2008. Les Problèmes des livres Grecs des Coniques d'Apollonius de Pergè. Les études classiques 76, 321–360.

Feke, J. 2012. Ptolemy's defense of theoretical philosophy. Apeiron 45, 61–90.

Feke, J., Jones, A. 2010. Ptolemy. In: Gerson, L.P. (ed.), The Cambridge History of Philosophy in Late Antiquity. Cambridge University Press, Cambridge, vol. 1, pp. 197–209.

Fournarakis, P., Christianidis, J., 2007. Greek geometrical analysis: A new interpretation through the "givens"-terminology. Bollettino di Storia delle Scienze Matematiche 26, 3–56.

Fowler, D.H., 1998. In memoriam: Wilbur Richard Knorr (1945–1997): An appreciation. Historia Mathematica 25, 123–132.

——— 1988. A catalogue of tables. Zeitschrift für Papyrologie und Epigraphik 75, 273-280.

——— 1995. Further arithmetical tables. Zeitschrift für Papyrologie und Epigraphik 105, 225-228.

Freeth, T., Bitsakis, Y., Moussas, X., Seiradakis, J.H., Tselikas, A., Mangou, H., Zafeiropoulou, M., Hadland, R., Bate, D., Ramsey, A., Allen, M., Crawley, A., Hockley., P., Malzbender., T., Gelb, D., Ambrisco, W., Edmunds, M.G., 2006. Decoding the ancient Greek astronomical calculator known as the Antikythera Mechanism. Nature 444, 587–591.

Freeth, T., Jones, A., 2012. The Cosmos in the Antikythera Mechanism. ISAW Papers 4. http://dlib.nyu.edu/awdl/isaw/isaw-papers/4.

Freeth, T., Jones, A., Steele, J., Bitsakis, Y., 2008. Calendars with Olympiad display and eclipse prediction on the Antikythera Mechanism. Nature 454, 614–617.

Friberg, J., 2005. Unexpected Links between Egyptian and Babylonian Mathematics. World Scientific: Singapore.

Friberg, J., 2007. Amazing Traces of a Babylonian Origin in Greek Mathematics. World Scientific: Singapore.

Fried, M., 2002. Apollonius of Perga: Conics Book IV. Green Lion Press, Santa Fe.

——— 2003. The use of analogy in Book VII of Apollonius' Conica. Science in Context 16, 349–365.

Fried, M., Unguru, S., 2001. Apollonius of Perga's Conica: Text, Context, Subtext. Brill, Leiden.
Habsieger, L., Kazarian, M., Lando, S., 1998. On the second number of Plutarch. The American Mathematical Monthly 105, 446.
Harari, O., 2003. The concept of existence and the role of constructions in Euclid's Elements. Archive for History of Exact Sciences 57, 1–23.
Heath, T., 1921. A History of Greek Mathematics. Clarendon Press, Oxford.
Hogendijk, J.P., 1999/2000. Traces of the lost Geometrical Elements of Menelaus in two texts of al-Sijzī. Zeitschrift für Geschichte der Arabisch-Islamischen Wissenschaften 13, 129–164.
Huffman, C.A., 2005. Archytas of Tarentum: Pythagorean, Philosopher and Mathematician King. Cambridge University Press, Cambridge.
Jones, A., 1999. Astronomical Papyri from Oxyrhynchus. P. Oxy. 4133–4300a. Memoirs of the American Philosophical Society, 233. American Philosophical Society, Philadelphia, PA.
——— 2001. Pseudo-Ptolemy, De speculis. SCIAMVS 2, 145–186.
——— 2005a. Ptolemy's mathematical models and their meaning. In: Kinyon, M., Van Brummelen, G. (eds.), Mathematics and the Historian's Craft. Springer, New York, pp. 23–42.
——— 2005b. Ptolemy's Canobic Inscription and Heliodorus' observation reports. SCIAMVS 6, 53–98.
——— 2006. The Keskintos Astronomical Inscription, text and interpretations. SCIAMVS 7, 3–41.
——— 2012. P.Cornell inv. 69 revisited: A collection of geometrical problems. In: Ast, R., Cuvigny, H., Hickey, T.M., Lougovaya, J. (eds.), Papyrological Texts in Honor of Roger S. Bagnall. The American Society of Papyrologists, Durham, pp. 159–175.
Kheirandish, E., 1999. The Arabic Version of Euclid's Optics. Vol. I, II. (Kitāb Uqlīdis fī Ikhtilāf al-manāẓir). Springer, New York.
Knorr, W.R., 1983. Construction as existence proof in ancient geometry. Ancient Philosophy 3, 125–148.
——— 1996. The wrong text of Euclid: On Heiberg's text and its alternatives, Centaurus 38, 208–276.
——— 2001. On Heiberg's Euclid. Intercultural transmission of scientific knowledge in the middle ages: Graeco-Arabic-Latin (Berlin, 1996). Science in Context 14, 133–143.
Kunitzsch, P., Lorch, R., 2010. Theodosius Sphaerica: Arabic and Medieval Latin Translations. Franz Steiner Verlag, Stuttgart.
——— 2011. Theodosius, De habitationibus: Arabic and Medieval Latin Translations. Verlag der Bayerischen Akademie der Wissenschaften, Munich.
Lehoux, D., 2007a. Astronomy, Weather, and Calendars: Parapegmata and Related Texts in Classical and Near-Eastern Societies. Cambridge University Press, Cambridge.
——— 2007b. Observers, objects, and the embedded eye; or, seeing and knowing in Ptolemy and Galen. Isis 98, 447–467.
Lewis, M.J.T., 2001. Surveying Instruments of Greece and Rome. Cambridge University Press, Cambridge.
Lorch, R.P., 2001. Greek-Arabic-Latin: The transmission of mathematical texts in the Middle Ages. Science in Context 14, 313–331.
Mäenpää, P., 1997. From backwards reduction to configurational analysis. In: Otte, M., Panza, M. (eds.), Analysis and Synthesis in Mathematics. Kluwer, Netherlands, pp. 201–226.
Mäenpää, P., von Plato, J., 1990. The logic of Euclidean construction procedures. Acta Philosophica Fennica 49, 275–293.
Malpangotto, M., 2003. Sul commento di Pappo d'Alessandria alle «Sferiche» di Teodosio. Bollettino di Storia delle Scienze Matematiche 23, 121–148.
Malpangotto, M., 2010. Graphical choices and geometrical thought in the transmission of Theodosius' Spherics from antiquity to the Renaissance. Archive for History of Exact Sciences 64, 75–112.
Manders, K., 2008a. Diagram-based geometric practice. In: Mancosu, P. (ed.), The Philosophy of Mathematical Practice. Oxford University Press, Oxford, pp. 65–79.
——— 2008b. The Euclidean diagram (1995). In: Mancosu, P. (ed.), The Philosophy of Mathematical Practice. Oxford University Press, Oxford, pp. 80–133.

Mansfeld, J., 1998. Prolegomena Mathematica: From Apollonius of Perga to Late Neoplatonism. Brill, Leiden.

Masià-Fornos, R., 2010. A "lacuna" in Proposition 9 of Archimedes' On the Sphere and the Cylinder, Book I. Historia Mathematica 37, 568–578.

Mendell, H., 2007. Two traces of two-step Eudoxan proportion theory in Aristotle: A tale of definitions in Aristotle, with a moral. Archive for History of Exact Sciences 61, 3–37.

Mercier, R., 2011. Ptolemy's Handy Tables, Volume 1b, Tables A1-A2. Institute Orientaliste, Louvain.

Meskens, A., 2009. Reading Diophantos. In: Van Kerkhove, B. (ed.), New Perspectives on Mathematical Practices. World Scientific, Hackensack, pp. 28–46.

——— 2010. Travelling Mathematics: The Fate of Diophantos' Arithmetic. Berkhäuser, Basel.

Miller, N., 2007. Euclid and His Twentieth Century Rivals: Diagrams in the Logic of Euclidean Geometry. CSLI Publications, Stanford, CA.

Mogenet, J., 1950. Autolycus de Pitane : Histoire du texte suivie de l'édition critique des traités de La sphére en mouvement et Des levers et couchers. Publications universitaires de Louvain, Louvain.

Morelli, G., 2009. Lo Stomachion di Archimede nelle testimonianze antiche. Bollettino di Storia delle Scienze Matematiche 29, 181–206.

Mueller, I., 2003. Remarks on Euclid's Elements I,32 and the parallel postulate. Science in Context 16, 287–297.

Mugler, C., 1959. Dictonnaire historique de la terminologie géomeétrique des grecs. C. Klinckseick, Paris.

Netz, R., 1998. Deuteronomic texts: Late antiquity and the history of mathematics. Revue d'Histoire des Mathématiques 42, 261–288.

——— 1999a. The Shaping of Deduction in Greek Mathematics. Cambridge University Press, Cambridge.

——— 1999b. Proclus' division of the mathematical proposition into parts: How and why was it formulated? Classical Quarterly (N. S.) 49, 282–303.

——— 1999c. Archimedes transformed: The case of a result stating a maximum for a cubic equation. Archive for History of Exact Sciences 54, 1–47.

——— 2000. Why did Greek mathematicians publish their analyses? In: Suppes, P., Moravcsik, J., Mendell, H. (eds.), Ancient & Medieval Traditions in the Exact Sciences. CSLI Publications, Stanford, pp. 139–157.

——— 2002a. Greek mathematicians: A group picture. In: Wolpert, L., Tuplin, C.J., Rihll, T. E. (eds.), Science and Mathematics in Ancient Greek Culture. Oxford University Press, Oxford, pp. 196–216.

——— 2002b. Counter culture: Towards a history of Greek numeracy. History of Science 40, 312–352.

——— 2004a. The works of Archimedes, Vol. I: The Two Books on the Sphere and the Cylinder, Translated into English, Together with Eutocius' Commentaries, with Commentary, and Critical Edition of the Diagrams. Cambridge University Press, Cambridge.

——— 2004b. Eudemus of Rhodes, Hippocrates of Chios and the earliest form of a Greek mathematical text. Centaurus 46, 243–286.

——— 2004c. The Transformation of Mathematics in the Early Mediterranean World: From Problems to Equations. Cambridge University Press, Cambridge.

——— 2009. Ludic Proof: Greek Mathematics and the Alexandrian Aesthetic. Cambridge University Press, Cambridge.

——— 2010. Imagination and Layered Ontology in Greek Mathematics. Configurations 17, 19–50.

Netz, R., Acerbi, F., Wilson, N., 2005. Towards a Reconstruction of Archimedes' Stomachion. SCIAMVS 5, 67–99.

Netz, R., Noel, W., Wilson, N., Tchernetska, N. (eds.), 2011. The Archimedes Palimpsest. 2 vols. Cambridge University Press, Cambridge.

Netz, R., Saito, K., Tchernetska, N., 2001–2002. A new reading of Method Proposition 14: Preliminary evidence from the Archimedes palimpsest. (Parts I and II.) SCIAMVS 2, 9–29; 3, 109–125.

Netz, R., Wilson, N., 2010. The Archimedes Codex: Revealing the Secrets of the World's Greatest Palimpsest. Phoenix, London.

Noack, B., 1992. Aristarch von Samos: Untersuchungen zur Überlieferungsgeschichte der Schrift περὶ μεγεθῶν καὶ ἀποστημάτων ἡλίου καὶ σελήνης. Ludwig Reichert Verlag, Wiesbaden.
Pambuccian, V., 2008. Axiomatizing geometric constructions. Journal of Applied Logic 6, 24–46.
Panza, M., 2012. The twofold role of diagrams in Euclid's plane geometry. Synthese 186, 55–102.
Rashed, R., Bellosta, H., 2010. Apollonius de Perge, La section des droites selon des rapports. Walter de Gruyter, Berlin.
Rashed, R., Decorps-Foulquier, M., Federspiel, M., 2008–2010. Apollonius de Perge, Coniques : Text grec et arabe établi, traduit et commenté. Walter de Gruyter, Berlin.
Robson, E., 2005. Influence, ignorance, or indifference? Rethinking the relationship between Babylonian and Greek mathematics. BSHM Bulletin 4, 1–17.
Rommevaux, S., Djebbar, A., Vitrac, B., 2001. Remarques sur l'histoire du texte des Éléments d'Euclide. Archive for History of Exact Sciences 55, 221–295.
Saito, K., 2003. Phantom theories of pre-Eudoxean proportion. Science in Context 16, 331–347.
——— 2006a. Between magnitude and quantity: Another look at Archimedes' quadrature (trans. of Sūgaku 55, 166–179). Sugaku Expositions 19, 35–52.
——— 2006b. A preliminary study in the critical assessment of diagrams in Greek mathematical works. SCIAMVS 7, 81–144.
——— 2008. Diagrams in Greek Mathematical Texts: Report of Research Grant 17300287 of the Japan Society for the Promotion of Science. Sakai, Osaka. (See http://www.hs.osakafu-u.ac.jp/~ken.saito/diagram/ for further updates.)
——— 2009. Reading ancient Greek mathematics. In: Robson, E., Stedall, J. (eds.), The Oxford Handbook of the History of Mathematics. Oxford University Press, Oxford, pp. 801–826.
——— 2012. Traditions of the diagram, traditions of the text: A case study. Synthese 186, 7–20.
Saito, K., Sidoli, N., 2010. The function of diorism in ancient Greek analysis. Historia Mathematica 37, 579–614.
——— 2012. Diagrams and arguments in ancient Greek mathematics: Lessons drawn from comparisons of the manuscript diagrams with those in modern critical editions. In: Chemla, K. (ed.), History of Mathematical Proof in Ancient Traditions. Cambridge University Press, Cambridge, pp. 135–162.
Sefrin-Weis, H., 2010. Pappus of Alexandria: Book 4 of the Collection. Springer, London.
Sesiano, J., 1998. An early form of Greek algebra. Centaurus 40, 276–302.
——— 1999. Sur le Papyrus graecus genevensis 259. Museum Helveticum 56, 26–32.
Sidoli, N., 2005. Heron's Dioptra 35 and analemma methods: An astronomical determination of the distance between two cities. Centaurus 47, 236–258.
——— 2006. The sector theorem attributed to Menelaus. SCIAMVS 7, 43–79.
——— 2007. What we can learn from a diagram: The case of Aristarchus's On the Sizes and Distances of the Sun and the Moon. Annals of Science 64, 525–547.
Sidoli, N., Berggren, J.L., 2007. The Arabic version of Ptolemy's Planisphere or Flattening the Surface of the Sphere: Text, translation, commentary. SCIAMVS 8, 37–139.
Sidoli, N., Saito, K., 2009. The role of geometrical construction in Theodosius's Spherics. Archive for History of Exact Sciences 63, 581–609.
——— 2012. Comparative analysis in Greek geometry. Historia Mathematica 39, 1–33.
Smith, A.M., 1999. Ptolemy and the Foundations of Ancient Mathematical Optics. A Source Based Guided Study. Transactions of the American Philosophical Society, 89, Pt. 3. American Philosophical Society, Philadelphia, PA.
Solomon, J., 2000. Ptolemy's Harmonics. Brill, Leiden.
Stanley, R.P., 1997. Hipparchus, Plutarch, Schröder, and Hough. The American Mathematical Monthly 104, 344–350.
Stenius, E., 1978. Foundations of Mathematics: Ancient Greek and Modern. Dialectica 32, 255–290.
Stückelberger, A., Grasshoff, G., 2006. Ptolemaios: Handbuch der Geographie (2 Bände). Schwabe, Basel.
Taisbak, C.M., 2003a. Euclid's Data. Or the Importance of Being Given. Museum Tusculanum Press, Copenhagen.

Takahashi, K., 2001. A manuscript of Euclid's De speculis: a Latin text of MS 98.22 of the Archivo y Biblioteca Capitulares de la Catedral, Toledo. SCIAMVS 2, 75–143.
Thiel, C., 2005. Becker und die Zeuthensche These zum Existenzbegriff in der antiken Mathematik. In: Peckhaus, V. (ed.), Oskar Becker und die Philosophie der Mathematik. Wilhelm Fink Verlag, Paderborn, pp. 32–45.
Thomaidis, Y., 2005. A framework for defining the generality of Diophantos' methods in Arithmetica. Archive for History of Exact Sciences 59, 591–640.
——— 2011. Some remarks on the meaning of equality in Diophantos' Arithmetica. Historia Mathematica 38, 28–41.
Tihon, A., 1999. Le Grand commentaire de Théon d'Alexandrie aux Tables faciles de Ptolémée. Livre IV. Biblioteca Apostolica Vaticana, Vatican City.
——— 2011. Les Tables faciles de Ptolémée, Volume 1a, Tables A1-A2. Institute Orientaliste, Louvain.
Tybjerg, K., 2004. Hero of Alexandria's mechanical geometry. Apeiron 37, 29–56.
Unguru, S., Fried, M., 2007. Apollonius, Davidoff, Rorty, and Zeuthen: From A to Z, or, What else is there? Sudhoffs Archiv 91, 1–19.
Van Brummelen, G., 2009. The Mathematics of the Heavens and the Earth: The Early History of Trigonometry. Princeton University Press, Princeton.
Vitrac, B., 1990–2001. Euclide : Les Élements. Presses Universitaires de France, Paris.
——— 2002. Note textuelle sur un (problème de) lieu géométrique dans les Météorologiques d'Aristote (III.5, 375 b 16–276 b 22). Archive for History of Exact Sciences 56, 239–283.
——— 2003. Les scholies grecques aux Éléments d'Euclide. Revue d'histoire des sciences 56, 27–292.
——— 2004. À propos des démonstrations alternatives et autres substitutions de preuves dans les Éléments d'Euclide. Archive for History of Exact Sciences 59, 1–44.
——— 2012. The Euclidean ideal of proof in the Elements and philological uncertainties of Heiberg's edition of the text. In: Chemla, K. (ed.), History of Mathematical Proof in Ancient Traditions. Cambridge University Press, Cambridge, pp. 69–134.
Vitrac, B., Djebbar, A., 2011–2012. Le Livre XIV des Élements d'Euclid : versions greques et arabe (première partie, seconde partie). SCIAMVS 12, 29–158; 13, 3–158.
Wagner, R., 2009. For some histories of Greek mathematics. Science in Context 22, 535–565.
Wilson, N.G., 1999. Archimedes: The palimpsest and the tradition. Byzantinische Zeitschrift 92, 89–101.
Winter, T.N., 2007. The Mechanical Problems in the corpus of Aristotle. Faculty Publications, Classics and Religious Studies, University of Nebraska.
Wright, M.T., 2007. The Antikythera Mechanism reconsidered. Interdisciplinary Science Reviews 32, 27–43.
Zhmud, L., 1998. Plato as "architect of science." Phronesis 43, 211–244.
——— 2006. The Origin of the History of Science in Classical Antiquity. Chernoglazov, A. (trans.), Walter de Gruyter, Berlin.
——— 2012. Pythagoras and the Early Pythagoreans. Windle, K., Ireland, R. (trans.), Oxford University Press, Oxford.

History of Mathematics in the Islamic World: The Present State of the Art [1985]

J. Lennart Berggren

In recent years, many discoveries in the history of Islamic mathematics have not been reported outside the specialist literature, even though they raise issues of interest to a larger audience. Thus, our aim in writing this survey is to provide to scholars of Islamic culture an account of the major themes and discoveries of the last decade of research on the history of mathematics in the Islamic world.[1] However, the subject of mathematics comprised much more than what a modern mathematician might think of as belonging to mathematics, so our survey is an overview of what may best be called the "mathematical sciences" in Islam; that is, in addition to such topics as arithmetic, algebra, and geometry we will also be interested in mechanics, optics, and mathematical instruments.

We must, however, mention two limitations on our survey. The first is that for astronomy we simply refer the reader to the recent surveys in King 1980, a general survey published in this journal, and King 1983a, restricted to the Mamlūk period. The other is that our survey is limited to those works published in English, French, and German, so that the reader must look elsewhere for an account of the large number of sources and studies published in Russian.[2]

General Studies

A useful survey of the mathematical sciences in Islam is Kennedy 1970 and, although restricted to the Mongol and Seljuk periods, Kennedy 1968. An account organized around historical periods, but ending with the death of al-Bīrūnī (ca. A.D. 1050), is found in the introduction to the bio-bibliographical survey Sezgin 1974, 1978. (To supplement and correct Sezgin's important contribution, consult the essay reviews King 1979a | and Rosenfeld 1978.) Finally, the excellent history Youschkevitch 1976 (translated 10
and up-dated from the Russian original) recounts the history of pure mathematics in Islam to the death of al-Kāshī (ca. A.D. 1430). Indeed, our aim is to provide an overview of historical writing on Islamic mathematics since Youschkevitch's book appeared, as well as to supplement his account by references

Editors: Originally published as Berggren, J.L., 1985, History of mathematics in the Islamic world: The present state of the art, Middle East Studies Association Bulletin 19, 9–33. We thank J.L. Berggren and the Middle East Studies Association for permission to republish this paper, along with all of the images.

[1] The author wishes to thank A. Djebbar and J. Sesiano for supplying photographs of Figures 3 and 4 respectively, as well as the directors of the libraries mentioned in the captions of the individual photographs for permission to publish the photographs. He also acknowledges the generous help provided in support of his research by the Natural Sciences and Engineering Research Council of Canada in the form of Grant #3486.

[2] Mathematical Reviews, published by the American Mathematical Society, provides some account of the many contributions made in the Russian literature.

N. Sidoli and G. Van Brummelen (eds.), *From Alexandria, Through Baghdad: Surveys and Studies in the Ancient Greek and Medieval Islamic Mathematical Sciences in Honor of J.L. Berggren*, DOI 10.1007/978-3-642-36736-6_4,

© Springer-Verlag Berlin Heidelberg 2014

to material relevant to all the mathematical sciences. We may also mention our book *Mathematics in Medieval Islam*, to be published this year by Springer-Verlag.

Finally, there have recently appeared four publications that will do much to facilitate further research in mathematics in the Islamic world. Three of these are King 1981, 1987,* an Arabic catalogue of the astronomical and mathematical manuscripts in the Egyptian National Library, and King 1984, an English-language survey of the catalogue. The fourth is Mach 1977, a catalogue of the Yahuda collection of Arabic manuscripts at Princeton. These are, of course, only beginnings, and one hopes that there will one day be a catalogue of the vast holdings in Istanbul. We cannot go beyond our sources and, for most researchers, access to these is only through the catalogues.

Number Theory

Greek authors, considerably before the time of Euclid, were interested in special kinds of whole numbers (even, odd, squares, etc.), and this interest continued to the end of the ancient period. For example, Euclid (Book 7, Def. 22) speaks of perfect numbers, those (such as 6) which are the sum of their proper divisors ($6 = 3 + 2 + 1$). A generalization of this idea is that of "amicable numbers," where two whole numbers are said to be amicable if each is the sum of the proper divisors of the other. Perhaps Iamblichos's attribution of the discovery of this idea to Pythagoras may tell us more about the Greek love of attributing discoveries to great men of the remote past than about the history of amicable numbers; however, there is no reason to doubt the idea was one with a long history in the ancient world before Thābit b. Qurra (d. A.D. 901) stated a theorem in his *Kitāb al-a'dād al-mutahābbat* (Saidan 1977a) that provides a method of discovering amicable numbers by finding certain kinds of prime numbers;[3] but despite this, for several centuries after Thābit, the sole known pair of amicable numbers was the
11 ancient example of 220, 284.[4] Recently, however, | both Naini (1982) and Rashed (1983) showed that the thirteenth-century Persian mathematician Kamāl al-Dīn al-Fārisī, who gave a correct theory of the rainbow, discovered two amicable numbers (17,296 and 18,416) and then, in the early 1600s, his compatriot Muḥammad Bāqir Yazdī discovered the pair 9,363,584 and 9,437,056. Both mathematicians obtained their results by using Thābit's theorem, and Yazdī's treatise contains a generalization of the idea of amicability.[5] For both the Maghreb (Djebbar 1981) and the East (Rashed 1980, 1983) recent researches have shown that the study of number theory formed a continuous tradition and led to the discovery of theorems or problems usually ascribed to Western mathematicians several centuries later — for example, the appearance of Wilson's Theorem[6] in the work of Ibn al-Haytham, Bachet's problem

* Editors: The original paper reads "1985(?)." This paper was actually published in 1987, and we have changed the reference in the bibliography.

[3] Thābit's theorem says that if p, q, and r are prime numbers such that $p = 3 \cdot 2^n - 1$, $q = 3 \cdot 2^{2n-1}$, and $r = 9 \cdot 2^{2n-1} - 1$, then $2^n pq$ and $2^n r$ are amicable numbers.

[4] The divisors of 220 are 1, 2, 4, 5, 10, 11, 20, 22, 44, 55, 110, and their sum is 284. The divisors of 284 are 1, 2, 4, 71, 142, and their sum is 220. The paper Hogendijk 1985 points out that Thābit carried through his proof of his celebrated theorem for the case when the basic parameter $n = 7$ (referring here to the "n" in the previous note), and Hogendijk hypothesizes that, since this case produces the amicable pair 17,296 and 18,416, Thābit knew that this pair is amicable.

[5] The generalization is the idea of equiponderant numbers, that is, any pair such as 25 and 6 of which both members have the same sum for their proper divisors. Further details on Yazdī's study of these numbers are in Naini 1982. Rashed 1983 is a careful study of the development of a theory of proper divisors and the introduction of arithmetic functions in Islamic mathematics, and, incidentally, shows that equiponderant numbers appear as early as the treatise *Al-Takmila fī al-ḥisāb* of Abū Mansūr al-Baghdādī, who died in 1037. Sample problems from this treatise are given in Saidan 1987,* where a forthcoming edition of the work is promised. Rashed 1982 contains editions of Arabic texts important for the history of combinatorics and number theory.

[6] This theorem says that if p is a prime number, then p divides the expression $(p - 1) \cdot (p - 2) \ldots 2 \cdot 1 + 1$. For example, 5 divides $4 \cdot 3 \cdot 2 \cdot 1 + 1 = 25$. Rashed 1980 contains an edited Arabic text and French translation of a short work by Ibn al-Haytham, in which the theorem is used to solve a system of congruences, and, as Rashed observes, Ibn al-Haytham's description of the method based on the theorem as being "canonical," and his omission of a proof, strongly suggest that this is not the first appearance of the theorem.

of the weights in al-Khāzinī,[7] or the summation of the fourth powers of the integers 1, 2, ..., *n* in the work of the tenth-century mathematician Abū Ṣaqr al-Qabīṣī.[8] Indeed, the judgement in Youschkevitch 1976 (p. 69) that "the works of mathematicians of the Islamic countries on number theory present very little originality ..." can no longer be sustained.

Arithmetic

A good, general survey of the whole area is Sabra 1971, but it is indicative of how rapidly the history of Islamic mathematics has developed that this account must now be supplemented in several areas, one of them being the treatment of fractions in Islamic mathematics. One of the most exciting recent discoveries in the history of Islamic mathematics appeared in Saidan 1966. It was already known, from
Luckey 1951, that al-Kāshī, astronomer | royal for the Timurid prince Ulugh Beg at Samarkand, used 12
decimal fractions with facility, and several of his remarkable feats of calculation were carried out in both the decimal and sexagesimal systems. However, in 1966 Saidan summarized a text he had discovered that showed, in his words, "not al-Kāshī ... but al-Uqlīdisī, who flourished five centuries earlier, is the first Muslim mathematician so far known to write about decimal fractions." This statement has been widely accepted by historians of mathematics, but Rashed 1978, a study of root extractions and the invention of decimal fractions, contains a serious challenge to the view. Rashed argues that al-Uqlīdisī's treatment of decimal fractions shows that he realized neither their importance nor their proper mathematical context, and that in fact al-Uqlīdisī's work is not to be taken seriously as marking the invention of decimal fractions. Rather, argues Rashed, decimal fractions were invented in the school of al-Karajī, and to substantiate this claim Rashed published a text by the twelfth-century algebraist and commentator on al-Karajī, al-Samawʾal, where decimal fractions are treated systematically in a serious mathematical context and as part of general theory.

The larger point that Rashed is arguing, both here and in other publications (see Rashed 1975b for a statement of the main thesis), is that there was, in the eleventh and twelfth centuries, a "dialectic" between arithmetic and algebra, which had two effects. The first was an arithmetization of the algebra of al-Khwārizmī and his successors, that is, the deliberate application of arithmetic algorithms to algebraic expressions, and the other is the reaction of the "renewed algebra" (to use Rashed's phrase) on the arithmetic which gave rise to it. The importance of decimal fractions for Rashed's thesis is that their invention in the school of al-Karajī is one of the principal examples of this dialectic, which would be impossible had decimal fractions existed before al-Karajī.

Decimal fractions, however, are part of just one of the arithmetic systems used within *dār al-Islām*, where there were at least three different kinds of arithmetic: the base-60 system of the astonomers, the decimal arithmetic that originated in India, and the finger arithmetic of the treasury officials. Far from being a simple finger counting, this last-named system was one of mental arithmetic in which intermediate results were stored on the fingers of the hands, to be used at a later stage of the calculation. Saidan 1971 is the publication of a major text in this area, the *Arithmetic for Government Officials* by the renowned Abū al-Wafāʾ al-Būzjānī, a short summary of which is contained in Saidan 1974. Further information on Abū al-Wafāʾ's treatment of arithmetic used by tax officials is contained in two studies by Ehrenkreutz (1962, 1964). Another useful source for material on the treatment of fractions is al-Karajī's *Al-Kāfī fī al-ḥisāb*. The German translation Hochheim 1877–80 is very rare, and one can only
wish the recent publication of part of the Arabic text in Saidan 1971 were more complete. | 13

[7] This problem is about the least number of weights needed to weigh any load whose weight is a whole number that does not exceed a given number of unit weights, and the early twelfth-century writer al-Khāzinī treated it in his *Book of the Balance of Wisdom* (Rozhanskaya 1987). (Editors: The original paper reads "1985." This paper was actually published in 1987, and we have changed the reference in the bibliography.)

[8] The *Treatise on Different Kinds of Numbers* by the writer of the early Būyid period Abū Ṣaqr al-Qabīṣī contains this result. See Anbouba 1982 for the Arabic text and a French summary, and Sesiano 1987 for an English translation and commentary. (Editors: The original paper reads "1985." This paper was actually published in 1987, and we have changed the reference in the bibliography.)

As regards the base-60 system employed by the astronomers, the two papers King 1974a and 1979c are devoted to the sexagesimal multiplication tables that appeared in Islamic times. Figure 1 shows two facing pages (for the first sixty multiples of 41–50) from such a table, found in BN ar. 2531, which may have been used by the fifteenth-century Egyptian astronomer Ibn al-Majdī. Although its 3600 entries may make it seem large, tables with 60 times this number of entries were compiled to aid astronomers in multiplying two-digit sexagesimal numbers.

A recent study of the history of arithmetic in another of the languages of the Islamic world is that in Hermelink 1975 on Persian arithmetic texts. The two papers Yadegari 1978 and 1980 record the appearance of a special case of the binomial theorem in the thirteenth-century Persian author al-Zanjānī, but neither succeeds in convincing one that the text cited betrays any knowledge of the general theorem.

Algebra

The survey Hartner 1965 is now out-of-date and must be supplemented by Anbouba 1978a (devoted to the ninth and tenth centuries, the period roughly from al-Khwārizmī to al-Karajī). The book Rashed 1984a is a collection of Rashed's previous papers on topics related to the history of algebra in Islam through the eleventh and twelfth centuries. Finally, Saliba 1973 is an examination of the usages of the terms *al-jabr* and *al-muqābala* in the algebraic works of al-Khwārizmī, al-Karajī, al-Samawʾal, and others.

An important part of the Muslim algebraic tradition is the algebra of inheritance (*ʿilm al-farāʾiḍ*), whose practitioners were called the *faraḍī*. (D. A. King, who discusses the literature on this topic in King 1980, informs us that there are a large number of tables, for use by the *faraḍī*, extant in the literature from the fourteenth and fifteenth centuries.) In the (virtually unique) study Gandz 1938 the distinction is drawn between the *ʿilm al-farāʾiḍ*, "the science of the legal shares of the natural heirs," to quote Gandz, where the problems are largely the arithmetic problem of calculating sometimes complicated fractional parts, and the *ʿilm al-waṣāya*. This latter subject, which treats of inheritance problems where there is a legacy to a stranger, is where algebraic methods enter. (Al-Baghdādī's *Takmila*, Saidan 1987, contains an example of arithmetical problems arising from Muslim religious requirements, namely, the calculation of the community's share of one's wealth: the *zakāt*.) The surveys in the *Encyclopedia of Islam* under the headings "Mirāth" and "Farāʾid" contain the basic stipulations of Islamic law on the topic.

The effect of much of the recent research on the history of algebra in Islam has been to reveal the major role played by al-Karajī and al-Samawʾal in the developments in that subject from the late tenth
15 century onward. Large parts of al-Karajī's major extant mathematical works have been published, | and we have also obtained major sections from his lost works through the quotations in the algebra of al-Samawʾal. Figure 2 shows one major surprise discovered in this way, namely, a table of what are now known as the binomial coefficients. Such a table is of great mathematical importance because, with it, one can extend methods for root extraction, which date back to the seventh-century Indian writer Brahmagupta, to obtain algorithms for calculating fourth and higher roots. It is also of great historical importance to find such a table in the writings of a mathematician who flourished early in the eleventh century, because, since the table of binomial coefficients had been previously found only in such late works as those of al-Kāshī (fifteenth century) and Naṣīr al-Dīn al-Ṭūsī (thirteenth century), some had suggested that the table was a Chinese import. However, the use of binomial coefficients by Islamic mathematicians of the eleventh century, in a context which had deep roots in Islamic mathematics, suggests strongly the table was a local discovery — most probably of al-Karajī.

The publication Ahmad and Rashed 1972 also showed that al-Karajī had begun to develop algorithms, which were completed by al-Samawʾal, to treat the arithmetic of expressions which we would write today as

$$6x^7 + \frac{1}{2}x^4 + 8x^3 + 2x + 101,$$

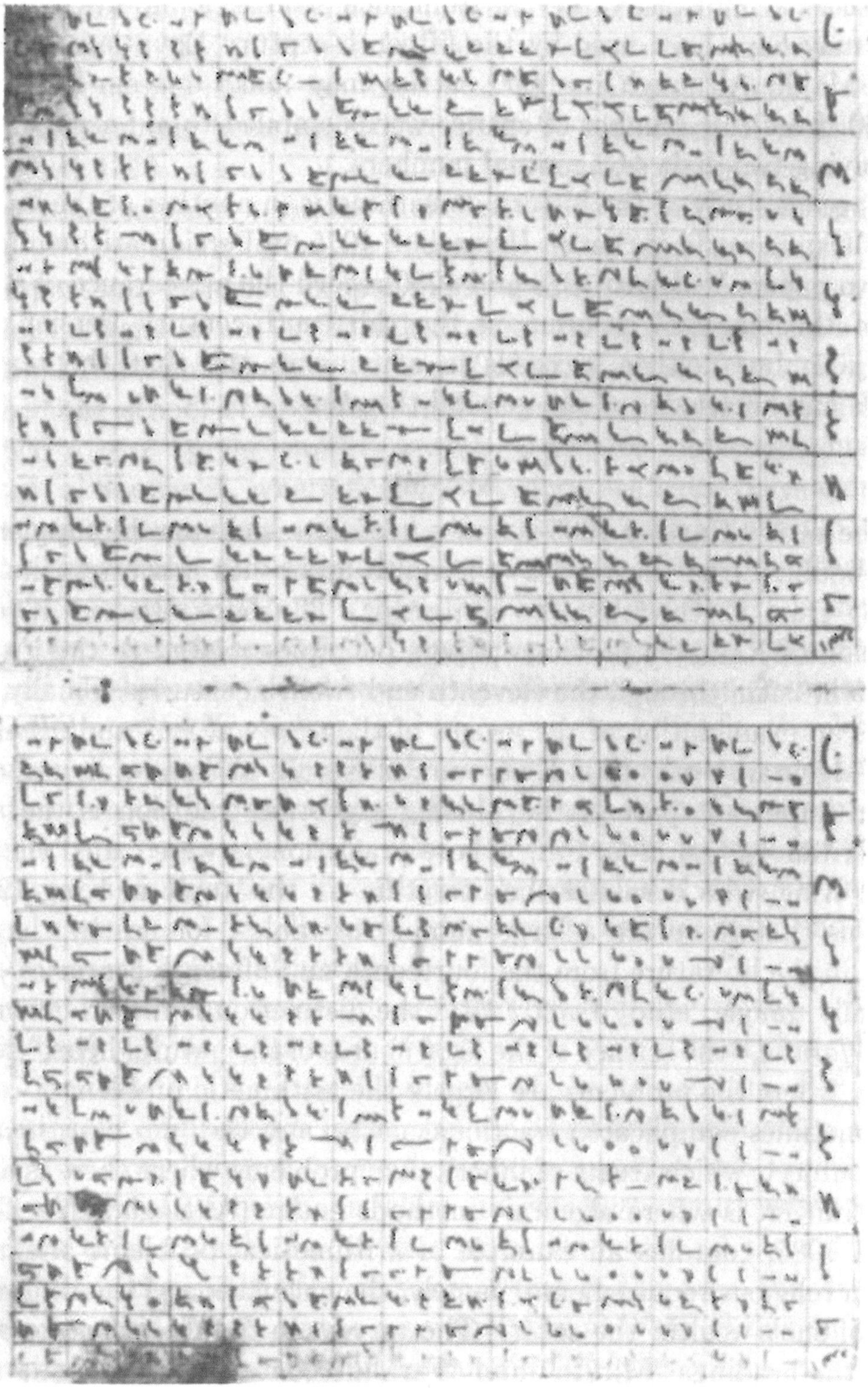

Figure 1: Part of a multiplication table expressed in the base-60 system using *abjad* numerals. The table is from the Paris ms. BN ar. 2531 and shows the products of the integers from 41 to 50 by 1, 2, ..., 59, 60.

that is, of polynomials with positive rational coefficients. These algorithms use only the coefficients, arranged in definite order, and manipulate them in a grid. In this way, one can add, subtract, multiply, divide, and even extract square roots of such expressions. One of al-Samawʾal's contributions was to extend these procedures to expressions with negative coefficients.

The solution of equations formed an important part of algebraic investigations, but, apart from those of first and second degree, there is no evidence that Islamic mathematicians were able to solve such equations by algebraic means. Rather, geometric methods, employing conic sections, were used by ʿUmar Khayyām to provide a systematic discussion of all possible equations of the third degree, and what is likely to be the definitive Arabic text and French translation of that work is Rashed and Djebbar 1981. The edition also contains the other known work of ʿUmar on algebra, an opuscule concerned with a geometric problem that leads to a particular cubic equation from among the complete list of types that ʿUmar solved in his larger treatise.

Geometric methods such as these may be appropriate for the solution of cubic equations rising from problems of geometrical construction, such as finding the side of a regular heptagon or nonagon in a given circle. When, however, the problem is one of finding the length of the chord in a given circle that subtends a central angle of 1°, and the result is needed for the construction of trigonometric tables, then
17 purely geometric constructions | are inadequate, and methods are required that will produce numerical answers. Thanks to the work of nineteenth-century scholars such as L. P. Sedillot and Fr. Woepcke, as well as the more recent work Luckey 1951, scholars have long known that the fifteenth-century astronomer al-Kāshī possessed algorithms for approximating the roots of certain higher equations to arbitrary accuracy. The methods were iterative, in the sense that there was a basic procedure which one followed, using the given data and an approximation to the answer, and then did again, this time using the given data and the result obtained from the last calculation. Successive iterations would produce results increasingly close to the true value of the root of the equation. Here again, the hypothesis that these methods entered Islam from China no longer seems necessary in light of Rashed 1974, which is a study (with a promise of a forthcoming edition of the text) of the *Algebra* of Sharaf al-Dīn al-Ṭūsī, who taught in Damascus in the twelfth century. The material contained in the al-Ṭūsī text shows that al-Kāshī's methods were the result of work undertaken by the algebraists of the eleventh and twelfth centuries.

Indeterminate Equations

The problem of finding solutions, either in whole numbers or fractions, to equations involving several unknown quantities was one that challenged Greek, Hindu, and Islamic mathematicians. Since such equations as $x^2 + y^2 = z^2$, to cite only one example, could have many solutions, they were known to the Arabic authors as *sayyāla* (indeterminates), and among the Islamic scholars who made important contributions to this field were Abū Kāmil, al-Karajī, and Abū Jaʿfar al-Khāzin, whose work spanned the period from the late ninth to the early eleventh century.

As for the first two, Sesiano 1977a and 1977b are studies of methods for finding the solutions of indeterminate equations found in the *Book on Algebra* by the early tenth-century Egyptian mathematician Abū Kāmil and in the *Badīʿ fī al-ḥisāb* of al-Karajī.

As for Abū Jaʿfar, he was one of the tenth-century mathematicians who interested themselves in the equation $x^3 + y^3 = z^3$, and his work criticizing al-Khūjandī's attempt to show that the equation has no whole-number solutions was studied by Woepcke over 120 years ago. (Whether any Islamic mathematician succeeded in giving a correct proof of this fact is not known.)[9]

Recently there has been considerable study of a second mémoire of Abū Jaʿfar, his *Treatise on Rational-Sided Right Triangles*, where the main problem is to find, for a given number n, a square
18 (a^2) so that both the sum of the square and the given number ($a^2 + n$) as well as their difference |

[9] Rosenfeld's claim (*Mathematical Reviews* #81k:01012) that Abū Jaʿfar gave a proof of this is based on a misunderstanding of the contents of Rashed 1979a.

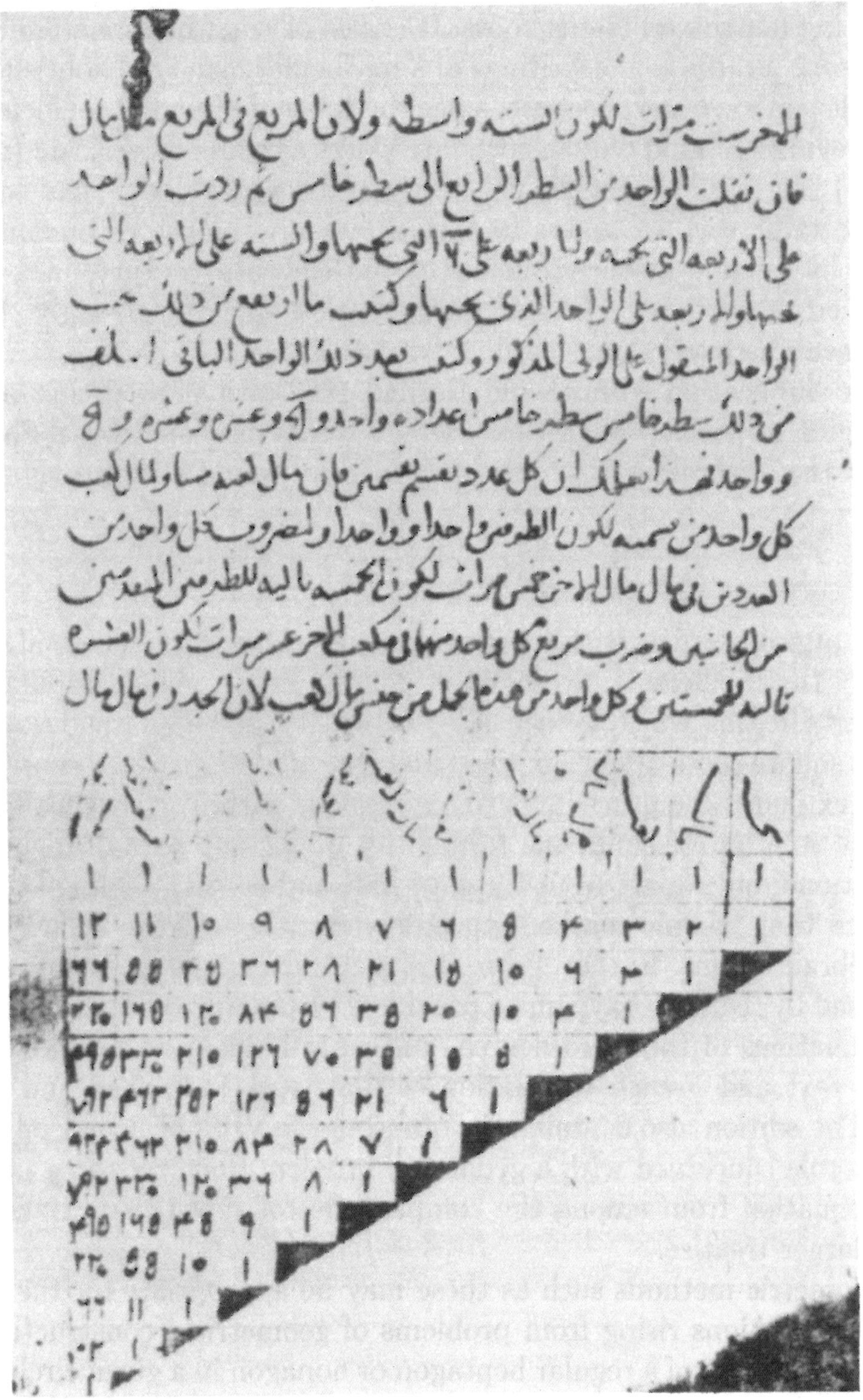

Figure 2: A table of binomial coefficients from a copy of *Al-Kitāb al-Bāhir* by al-Samawʾal, found in the Istanbul ms. AS 2118.

$(a^2 - n)$ are squares.[10] Saidan 1978b is an Arabic text of this treatise, together with an English summary of its contents, intended to form part of a series of texts illustrating the history of arithmetic in Islam. (One part of arithmetic was, at least in the view of some, number theory, which included aspects of the solution of indeterminate equations. However, this involved algebra as well, so it is not always clear where arithmetic ends and algebra begins.) A subsequent, independent publication was Anbouba 1979, another edition of the Arabic text, together with a complete French translation and a variety of notes — one of them commenting that Saidan erred in saying the the treatise of Abū Jaʾfar on *Rational-Sided Right Triangles* was the same as the treatise presented in Woepcke 1861. Finally, Rashed 1979a is a French summary of Abū Jaʾfar's treatise in the context of a study of the solutions of indeterminate equations during the tenth century, and the relation between this subject and algebra and arithmetic.

Recent publications have also contributed to our knowledge of the sources for indeterminate equations that Islamic mathematicians had to draw on. Sesiano 1982 is an edition of the Arabic text of four hitherto lost books of Diophantos's *Arithmetica* together with an English translation and a commentary that sheds considerable light on the mathematical, historical, and linguistic questions the text raises. Rashed 1984b is an independent edition of the Arabic text, together with a French translation and commentary. (Since we have only just received a copy of this work, we cannot comment on its relation to the earlier publication of the text in Rashed 1975a.)

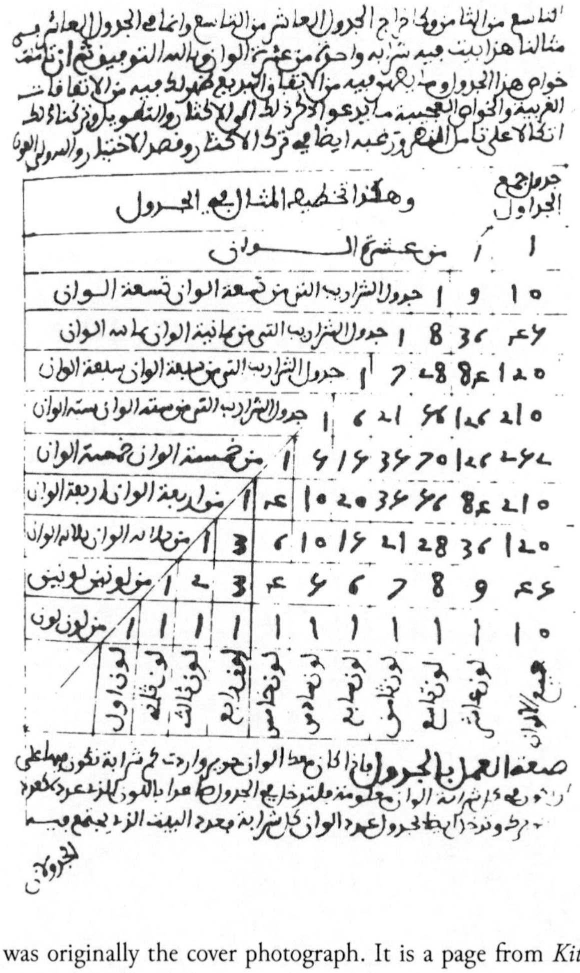

(Editors: This image was originally the cover photograph. It is a page from *Kitāb fiqh al-ḥisāb* by Ibn Munʾim.)

[10] It appears that al-Baghdādī's *al-Takmila* also contains material on this topic (Saidan 1987).

Combinatorics

Closely related, both historically and mathematically, to the previous topic we have discussed is the subject of combinatorial mathematics, one to which the advent of electronic computers has given great impetus. Perhaps it is more than a coincidence that, at the same time mathematicians have been taking increased interest in the subject, historians of mathematics have begun to investigate all stages of its history.

The cover photograph is from the work *Kitāb fiqh al-ḥisāb* by Ibn Munʿim, published in Djebbar 1985. Ibn Munʿim wrote this work probably under the patronage of the Almohad caliph Abū ʿAbdallāh Muḥ. b. Yaʿqūb al-Nāṣir in the first decade of the thirteenth century. The chart shows the number of pom-poms one can make from at most 10 colors of silk, when no color is repeated. For example, "6" in the column of the fifth color and the row for 3 colors means that in making 3-colored pom-poms out of Colors 1–5, and using Color 5. one can also use all pairs ($3 + 2 + 1 = 6$ in all) | of the other 19
four colors. For a given row then, say that for three colors, the sum of its entries equals the number of pom-poms using 3 colors of silk.

This table and others like it are quite unlike astronomical tables in that whereas, in the case of most astronomical tables, each entry must be computed *ab initio*, in the combinatorial tables simple additions usually suffice to fill in the whole table from a trivial row — *once the principle of the table is understood.*

The author of this table, Ibn Munʿim, then proceeds through a series of questions, such as the number of permutations of the letters of a word of which the number of letters is known and of which one or more letters are repeated a known number of times, or "to enumerate the number of words such that a human being cannot express himself except by means of one of them." For his purposes, Ibn Munʿim says the Arabic alphabet has 28 letters, that the longest word has 10 letters (e.g., أرسطاطاليس), that any of three vowels or a sukūn may vocalize a letter, that one cannot begin with a sukūn, and two sukūns cannot follow each other. (Philologists would find interesting Ibn Munʿim's discussion of possible objections that may be made to these conventions, and mathematicians will find interesting his reply that such objections are beside the point, since his methods, once understood, allow the reader to solve the problem for any set of conventions.)

From the standpoint of the progress of combinatorics, whose roots Djebbar correctly notes lie in such areas as linguistics, music, and astrology (see also Rashed 1973), the importance of Ibn Munʿim's work is that it was part of the mathematization of the subject (e.g., its foundation on general proofs) and the entrance of the subject into manuals designed for instruction. One important development for framing general proofs was the explicit use of various forms of mathematical induction, a technique widely used in combinatorial arguments, but which also appeared in number theory and algebra as well. A topic for further research would be the extent to which the impressive combinatorial studies in the twelfth and thirteenth centuries in the Maghreb had counterparts, or forerunners, in the East. (Rashed 1972 has shown that a form of mathematical induction, having strong similarities, as regards attempted generality, to the work of Pascal, may be found in the reasoning of al-Karajī and al-Samawʾal on the binomial theorem and formulas in number theory.)

Recreational Mathematics

Closely allied to the previous fields we have discussed is recreational mathematics, consisting of a group of problems that are usually easy to state, tempting to work on and, sometimes, annoyingly difficult to solve. Hermelink 1978 is an attempt to create a typology of such problems and to point out the utility of their study as an approach to the study of transmission of knowledge from one culture to another. | 21

One part of recreational mathematics that has enjoyed considerable popularity in East and West is the construction of magic squares. The square illustrated in Figure 3 is one taken from the twelfth-century writer al-Būnī, and the patient reader may verify that the sum of the entries in any row equals that in any column as well as the sum of any diagonal, or he may read either of the two studies Sesiano 1980

١١	١٥	٣٠	٣٤	٣٥	١
١٤	١٨	٢١	٤٤	١١	٢٢
٢٧	٣٤	١٢	١٧	٢٢	١٠
١٩	١٣	١٦	١٩	١٦	٢٨
٢٩	٣٠	١٥	٤	١٥	١٨
٣٦	٣٤	١٧	٣	١٢	٣١

Figure 3: A six-by-six magic square from the treatise *al-Durr al-manẓūm fī ʿilm al-awfaq wa'l-nujūm* by al-Būnī who died ca. 1225.

and 1981, where medieval Islamic discussions of methods for constructing magic squares are given.[11] No doubt, considerable light will be shed on the early history of these squares in Islam when Sesiano publishes his account of the major treatise on magic squares by the tenth-century mathematician whom we have already referred to, Abū al-Wafā᾽, whose work is so important for understanding the history of mathematics in Iraq during the time of the Būyids.

Another source of problems in recreational mathematics is the game of chess, of which a study is found in Wieber 1972. A section in Rozhanskaya and Rosenfeld 1987[*] concerns al-Khāzinī's discussion of the total number of dirhams necessary in order to put one on square one of a chessboard and, on each following square, double that on the previous one. The same problem was discussed almost 170 years earlier by al-Qabīṣī, together with an unusual variation, in a treatise that is also of relevance to the history of geodesy (Anbouba 1982; Sesiano 1987).

Geometry

Figure 4 is taken from a manuscript that is also a product of Būyid patronage, namely, a treatise by the mathematician and astronomer Abū Sahl al-Kūhī. Its subject is the construction of a figure with seven equal sides (the so-called "regular heptagon") in a given circle, and the fact that the problem is not elementary made it a favorite of mathematicians from the late ninth to the early eleventh century.

The history of this particular problem has recently attracted almost as much attention among historians of Islamic mathematics as the original problem did among the mathematicians themselves. For example, there was the Arabic survey article Anbouba 1977 on work on this problem during the fourth hijra century and a (regrettably) very much shorter French summary Anbouba 1978b. In the following year appeared Rashed 1979b, an edition, translation, and study of two treatises of Ibn al-Haytham on the regular heptagon, and Hogendijk 1984b is an extensive study of Greek and Arabic constructions of the regular heptagon to the end of the twelfth century. Hogendijk's researches show that the problem of the heptagon stimulated Islamic geometers to do original research on conic | sections and raised the 23
question of what it meant to solve a problem involving geometrical constructions; for during the tenth century there was a decided tightening of rigor in Islamic geometry, with the result that methods that had seemed perfectly acceptable to Archimedes were seen as needing further explanation, and what was called "moving geometry" was to be reduced to "fixed geometry." This was part of an Islamic tradition with Greek roots, and belonged to the general set of problems concerning what constructions one could do with straight-edge and compass only.

Other problems were such constructions as the regular nonagon, trisecting the angle, and constructing a cube whose volume is double that of a given cube. Work on these problems was one of the stimuli to the geometric treatment of cubic equations that one finds in ʿUmar Khayyām. On the first of these, Berggren 1981b is a translation of an anonymous treatise which seems to have been part of the "moving geometry" tradition mentioned above.

Constructing a regular nonagon, however, is a special case of trisecting the angle, and the transmission from the Greek to the Islamic world of methods to do this has been the subject of the recent studies Hogendijk 1981 and Knorr 1983, which arrive at different conclusions regarding the relation of some of the Islamic material to the Greek material. Finally, there is the study Tekeli 1968 illustrating the continuation of the classical solutions to the problem of the duplication of the cube late into Islamic times, as well as the recent edition Muwafi and Philippou 1981 of an Arabic translation of Eratosthenes on duplicating the cube.

Although certain problems whose origins lie in Greek mathematics, notably that of trisecting the angle and constructing polygons with equal sides and angles, have been relatively well studied, the same

[11] Among the contents of these treatises is a procedure for constructing bordered magic squares, that is, those in which the squares obtained by consecutively removing the borders are also magic.

[*] Editors: The original paper reads "1985." This paper was actually published in 1987, and we have changed the reference in the bibliography.

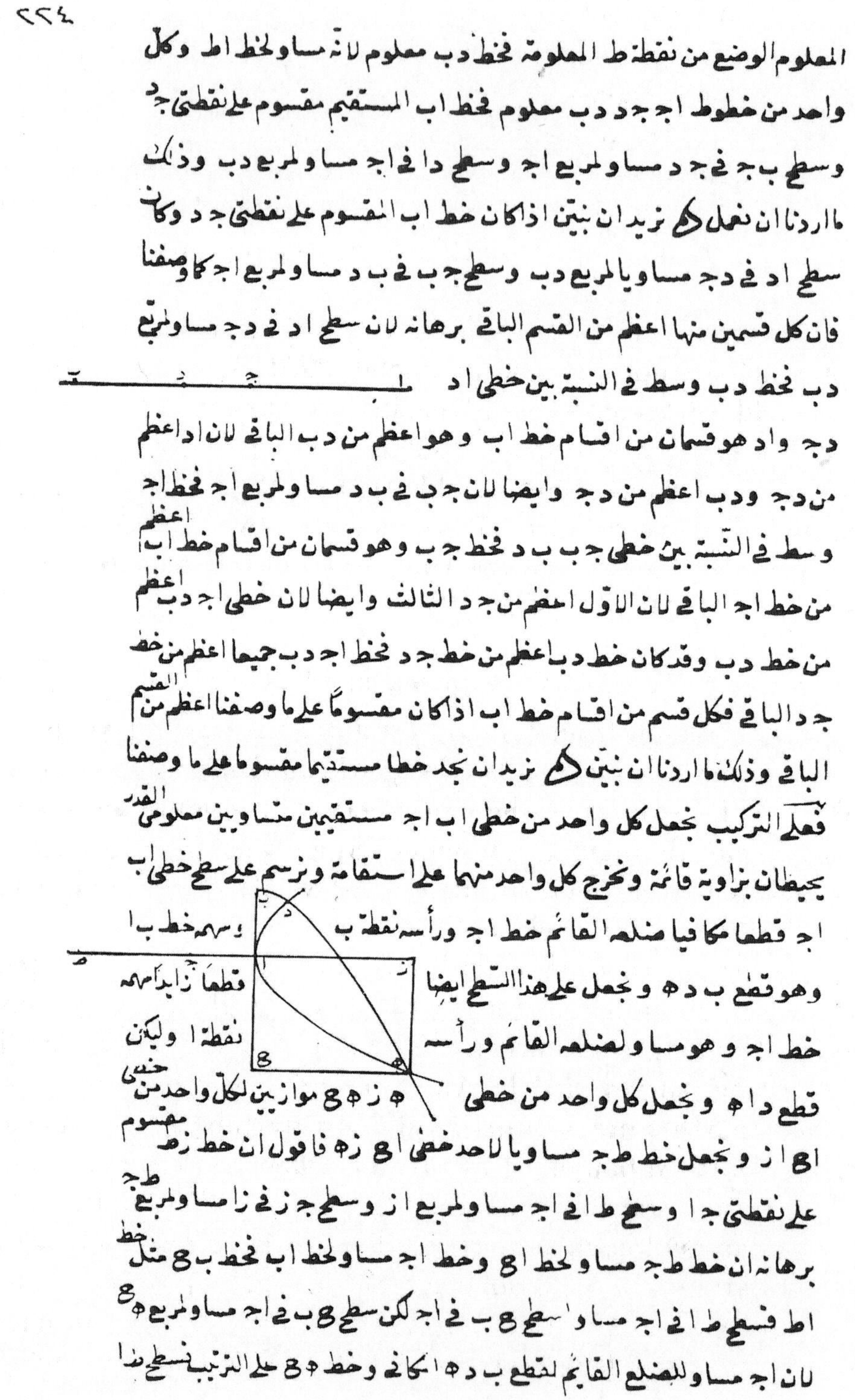

٢٢٤

المعلوم الوضع من نقطة ط المعلومة فخط د ب معلوم لانه مساو لخط ا ط وكل
واحد من خطوط ا ج ج د د ب معلوم فخط ا ب المستقيم مقسوم على نقطتي ج د
وسطح ب ج في ج د مساو لمربع ا ج وسطح د ا في ا ج مساو لمربع د ب وذلك
ما اردنا ان نعمل ﻫ نريد ان نبين اذا كان خط ا ب المقسوم على نقطتي ج د وكان
سطح ا د في د ج مساويا لمربع د ب وسطح ج ب في ب د مساو لمربع ا ج كما وصفنا
فان كل قسمين منها اعظم من القسم الباقي برهانه لان سطح ا د في د ج مساو لمربع
د ب فخط د ب وسط في النسبة بين خطي ا د
د ج وا د هو قسمان من اقسام خط ا ب وهو اعظم من د ب الباقي لان ا د اعظم
من د ج ود ب اعظم من د ج وايضا لان ج ب في ب د مساو لمربع ا ج فخط ا ج
وسط في النسبة بين خطي ج ب ب د فخط ج ب وهو قسمان من اقسام خط ا ب اعظم
من خط ا ج الباقي لان الاول اعظم من ج د الثالث وايضا لان خطي ا ج د ب اعظم
من خط د ب وقد كان خط د ب اعظم من خط ج د فخط ا ج د ب جميعا اعظم من خط
ج د الباقي فكل قسم من اقسام خط ا ب اذا كان مقسوما على ما وصفنا اعظم من القسم
الباقي وذلك ما اردنا ان نبين ﻫ نريد ان نجد خطا مستقيما مقسوما على ما وصفنا
فعلى التركيب نجعل كل واحد من خطي ا ب ا ج مستقيمين متساويين معلومي القدر
يحيطان بزاوية قائمة ونخرج كل واحد منهما على استقامة ونرسم على سطح خطي ا ب
ا ج قطعا مكافيا ضلعه القائم خط ا ج ورأسه نقطة ب وسهمه خط ب ا
وهو قطع ب د ه ونجعل على هذا السطح ايضا قطعا زائدا رأسه
خط ا ج وهو مساو لضلعه القائم ورأسه نقطة ا ولكن
قطع د ا ه ونجعل كل واحد من خطي ه ز ه ع موازيين لكل واحد من خطي
ا ع ا ز ونجعل خط ط ج مساويا لاحد خطي ا ع ز ه فاقول ان خط ز ط مقسوم
على نقطتي ج ا وسطح ط ا في ا ج مساو لمربع ا ز وسطح ج ز في ز ا مساو لمربع ط ج
برهانه ان خط ط ج مساو لخط ا ع وخط ا ج مساو لخط ا ب فخط ب ع مثل خط
ا ط فسطح ط ا في ا ج مساو لسطح ع ب في ا ج لكن سطح ع ب في ا ج مساو لمربع ه ع
لان ا ج مساو للضلع القائم لقطع ب د ه المكافي وخط ه ع على الترتيب فسطح ط ا

Figure 4: Diagram showing a construction with conic sections from a manuscript of a work by Abū Sahl al-Kūhī on constructing a regular heptagon (*risāla fī istikhrāj dilʿ al-musabbaʿ*…), from Cairo ms. Dar al-kutub, riyād. 40m, fol. 224a.

cannot be said generally of the Greek geometric tradition in Islam. For example, nothing like Clagett's *Archimedes in the Middle Ages*, whose focus is on medieval Latin texts dealing with Archimedes, has been published for Arabic texts. As for Apollonios and *The Conics*, the only recent contribution here is Hogendijk 1984a, an edition and study of Ibn al-Haytham's *Completion of "The Conics,"* and no edited versions of Arabic translations of Apollonios's works have appeared, although we are informed that G. J. Toomer is preparing an edition of the Arabic text of three books lost in Greek. With Euclid the situation improves somewhat, and the work of De Young and Engroff should do much to provide the scholar with information on what was a major tradition of study of one of the most important mathematical texts ever written.[12] Studies of the Arabic commentaries on Euclid are | Plooij 1950 (on 24
Book 5) and Busard and Konigsveld 1973 (on Book 3).[13] In addition, Sabra 1968 and 1969 are studies of the history of Euclid's parallel postulate, and soon to appear is an English translation of a book by B. A. Rosenfeld on the history of non-Euclidean geometry, which contains a considerable study of Islamic contributions to the subject.

Again, Arabic versions of the so-called "middle" books (Greek texts intermediate between the study of Euclid and that of Ptolemy) exist, but the Arabic tradition of study of these is virtually untouched. An examination of Sezgin 1974, 1978 shows that on texts as basic as those of Autolykos and Theodosios virtually no work has been done on the Arabic tradition.

Trigonometry

A good general orientation to the study of the history of trigonometry in Islam is Kennedy 1969, while the numerous studies published by the same author over the past several decades illustrate particular episodes in the history of the subject. (Many of these studies are now conveniently available in Kennedy et al. 1983, but a recent contribution, studying Abū al-Wafā'’s calculation of the distance from Baghdad to Mecca, is Kennedy 1984.) However, in addition to these short studies, several major sources have become available over the last decade. One of these is Kennedy 1976, a translation and commentary of al-Bīrūnī's *Treatise on Shadows*, which contains al-Bīrūnī's account of his own contribution, as well as those of others, to the study of what we would call the tangent and cotangent functions. Another is al-Bīrūnī's *Keys of Astronomy* (Debarnot 1980), a valuable collection of mathematical and historical material on the development of spherical trigonometry. (Debarnot 1978 is another interesting account of an early use of the idea of polar triangles in the Islamic world by al-Bīrūnī's teacher, the Prince Abū Naṣr Manṣur b. ʿIrāq.) Further, Villuendas 1979 furnishes an edition, translation, and study of spherical trigonometry as found in the *Kitāb majhūlāt qusiy al-kura* of the eleventh-century Andalusian qaḍī and astronomer Ibn Muʿādh. Less extensive is Berggren 1987,[*] which makes available to English readers Ibn Labbān's version of the elements of spherical trigonometry. Finally, one should note that King 1972 is a case study of spherical trigonometry as it appears in an important astronomical handbook (*zīj*) by the eleventh-century Egyptian astronomer Ibn Yūnus, who was a contemporary of Ibn Muʿādh.

A standard application of trigonometric methods in Islamic culture was in the calculation of the times and direction (*qibla*) of prayer. A review | of the literature here is found in King 1979b, but 25
we may add to the references there the studies Berggren 1981a and 1984 (the history of a particular method for finding the qibla as it developed over five centuries from a simple procedure of limited validity into an example of what King 1981 has called a "universal method" in Islamic mathematics)

[12] This work is still available only in the form of the Harvard theses Engroff 1980 and De Young 1981; however, an account of some of the problems this work raises for the standard history of Euclid's *Elements* in the Islamic world is given in De Young 1984.

[13] A recent, major contribution to the study of Arabic commentaries on Euclid is Matvievskaya 1987,[*] a survey of nine commentaries on Book 10 of the *Elements*, in which particular attention is paid to the commentaries of al-Māhānī, al-Ahwāzī, al-Khāzin, and Ibn al-Baghdādī. All of these not only treated Book 10 arithmetically but, in their attempts to justify such a treatment, approached the concept of real numbers.

[*] Editors: The original paper reads "1985." This paper was actually published in 1987, and we have changed the reference in the bibliography.

as well as King 1985b and 1983b. This latter is particularly important as an attempt to retrieve early Abbasid trigonometric methods.[14]

Much of the actual use of trigonometry depended on the tables of various trigonometric functions. A short recent account of the tables of the celebrated fifteenth-century Samarkand astronomer al-Kāshī is found in Hamadanizadeh 1980, but still wanting is an account of the methods employed by the astronomer-mathematicians in computing these tables.

A series of studies has shown that some Islamic astronomers also compiled one- and two-argument auxiliary tables. These tables, constructed from those of the basic trigonometric functions, were intended to simplify calculating functions of spherical trigonometry. Irani (1956) drew attention to such tables in the *zīj* of the ninth-century Damascene astronomer Ḥabash al-Ḥāsib, and Jensen (1971) argues that Abū Naṣr Manṣūr constructed his *Table of Minutes* to show the increased ease of using auxiliary tables when the trigonometric functions are computed (as they are today) for radius 1 instead of the usual 60. Similar tables appear in the *zīj* of Abū Naṣr's Egyptian contemporary Ibn Yūnus (King 1972, 29–33), and King (1973) shows how double-argument tables of the late fourteenth-century astronomer al-Khalīlī bear on the history of trigonometry.

Finally, there are methods which, although non-trigonometric, developed side by side with trigonometric methods and interacted with them in a way whose history is yet to be written. One such method, involving rotation and orthogonal projection of arcs on a sphere into one plane, is that of the analemma, and aspects of its history as applied to the problem of finding the direction of prayer are discussed in Berggren 1980.

Numerical Mathematics

Despite the above-mentioned lack of study of methods used to calculate trigonometric tables, there have been studies of methods used to calculate the values of trigonometric, and other, functions for values of the basic variables that are intermediate between two tabulated values. The thesis Hamadanizadeh 1976 and the supplementary paper Hamadanizadeh 1978 are both devoted to accounts of various methods used in the Islamic world for interpolating in tables.

Further, methods of calculating astronomical tables from rules expressing the dependence of one
26 quantity on another is the subject of the Arabic | article Saliba 1977a, while Tichenor 1967 and Saliba 1977b are devoted to the study of tables designed to simplify the application of astronomical theories to finding the position of a celestial body on a given day. (See also the supplementary mathematical remarks in Tee 1977 on the mathematical theory relevant to the construction of these tables.)

Optics

Al-Fārābī lists optics among the mathematical sciences as one which should hardly be distinguished from geometry, and, though this does not tell the whole story of optics in the Islamic world, it is true that some of the most interesting applications of geometry are to be found in problems arising in optics. Fundamental to the study of the geometrical tradition in this science is the recently published edition of the Arabic text of Ibn al-Haytham's *Kitāb al-manāẓir* (Optics), Sabra 1983. (A detailed study of Ibn al-Haytham's solution of what has become known as Alhazen's problem may be found in Sabra 1982.)

Although Ibn al-Haytham's *Optics* is far from a treatise on pure geometry thinly disguised as optics, a body of literature that more closely fits this description did arise in antiquity and gave rise to a lively tradition which continued up to the time of Ibn al-Haytham, and that tradition is the study of mirrors that cause burning. A recent publication relevant to the Greek and Islamic traditions of these "burning

[14] See also D. King, "Some Early Islamic Mathematical Methods and Tables for Finding the Direction of Mecca," forthcoming.

mirrors" is Toomer 1976, and the present author is preparing for publication an Arabic work on burning mirrors that incorporates material from a Greek treatise quite different from that studied by Toomer.

Mathematical Geography

The fundamental Arabic source for methods of determining latitudes and longitudes of cities, the distances between them, and the azimuth of one relative to another is al-Bīrūnī's *Kitāb taḥdīd nihāyāt al-amākin*, of which we have a well-edited Arabic text (Bulgakov 1962), English translation (Ali 1967), and commentary (Kennedy 1973). Although this work is silent on the subject of cartography, there is another writing of al-Bīrūnī, entitled *Kitāb tasṭīḥ al-kuwar* (Arabic text in Saidan 1977b), devoted to a discussion of map projections. On the basis of this, Berggren 1982 provides an English translation, with a commentary that is considerably more detailed than that accompanying the German translation in Suter 1922.[15] In this treatise al-Bīrūnī mentions al-Ṣāghānī's generalization of the astro|labic projection 27
(stereographic projection), which is studied in Lorch 1987.[*] Map and astrolabe projection belonged, in the Islamic classification of the the sciences, to the science of instruments, and the publications King 1974b and Lorch 1984 contain expositions of the theory and history of several instruments designed to produce solutions to a large number of problems arising in applications of mathematics.

Mechanics

A favorite topic of discussion in Arabic literature on mechanics was the instrument known as the *qarasṭūn* (or *qabbān*), the word signifying an unequal-arm balance. This instrument inspired a variety of treatises, one of the most original being that of Thābit b. Qurra (*Kitāb al-qarasṭūn*), now published as Jaouiche 1976. Another example of the depth of the mathematical tradition in mechanics in Islam is to be found in the correspondence between Abū Sahl al-Kūhī and Abū Isḥāq al-Ṣābī, part of which is summarized and studied in Sesiano 1979. (This correspondence is an example of serious work in Arabic on the determination of centers of gravity, which goes considerably beyond the work of Archimedes 1200 years earlier, and contains discussions of interest for geometry and the philosophy of mathematics. It has been edited and translated by the present author.)[16] An interesting discovery in this correspondence is Abū Sahl's location of the center of gravity of an arc of a circle.

A major source of information on mechanics in the Muslim world is *The Book of the Balance of Wisdom* by al-Khāzinī. We have already mentioned material it contains relative to the history of number theory and recreational mathematics, but it is also important for obtaining information on the contents of treatises that are lost or have reached us only in abridged form. A recent example is the restoration of the contents of a treatise by al-Bīrūnī on the specific gravity of gems (Rozhanskaya and Rosenfeld 1987).

---

[15] Other contributions to the study of this treatise are Richter-Bernburg 1982 (a translation of al-Bīrūnī's introduction to the treatise, not translated in Berggren 1982) and Kennedy and Debarnot 1984(?) (pointing out that the globular projection al-Bīrūnī describes may have been intended as a simplification of the azimuthal equidistant projection).

[*] Editors: The original paper reads "1985." This paper was actually published in 1987, and we have changed the reference in the bibliography.

[16] To appear in *Journal for History of the Arabic Sciences* under the title "The Correspondence of Abū Sahl al-Kūhī and Abū Isḥāq al-Ṣābī: A Translation with Commentaries."

Conclusions

Certain impressions emerge fairly clearly from recent research on the history of mathematics in Islam. The first is that of the mathematical autonomy and originality of this civilization, where innovations in arithmetic and algebra that once seemed due to outside influence have emerged as integral parts of the corpus of Islamic mathematics. Another is that the period from the mid-tenth to the mid-eleventh century was a highly creative period for many of the mathematical disciplines in Islam, one that saw significant advances in arithmetic and algebra, the development of spherical trigonometry, and brilliant
28 contributions to mechanics, optics, | and cartography. At the same time, one is impressed to see that, as late as the seventeenth century, Islamic scholars were still obtaining new results in number theory, and further investigations of the later period may show that Islamic scholars did not stop thinking about mathematics when al-Kāshī died. Finally, since past researches now allow us to identify with some certainty the major figures in the history of Islamic mathematics, it is important that scholars begin to publish series of works by these men or their students. Only on the basis of reliably edited texts can the history of Islamic mathematics move beyond what has often been a random development dependent on chance discoveries, and so explore the many inviting avenues that now appear to be open to further research.

References

Ahmad, S., Rashed, R. (eds.), 1972. Al-Bāhir en algèbre d'as-Samawʾal. (In French and Arabic.) University Press of Damascus, Damascus.

Ali, J. (trans.), 1967. The Determination of the Coordinates of Cities. American University of Beirut Press, Beirut.

Anbouba, A., 1977. Construction de l'heptagone régulier par les Arabes au 4e siècle de l'hégire. Journal for the History of Arabic Science 1, 352–84.

——— 1978a. L'algèbre arabe aux IXe et X^{e} siécles, Aperçu général. Journal for the History of Arabic Science 2, 66–100.

——— 1978b. Construction de l'heptagone régulier par les Arabes au 4e siècle de l'hégire. Journal for the History of Arabic Science 2, 264–69. (This is a summary of Anbouba 1977.)

——— 1979. Un traité d'Abū Jaʿfar (al-Khāzin) sur les triangles rectangles numériques. Journal for the History of Arabic Science 3, 134–78.

——— 1982. Un mémoire d'al-Qabīṣī (4e siècle H.) sur certaines sommations numériques. Journal for the History of Arabic Science 6, 208–81.

Berggren, J.L., 1980. A comparison of four analemmas for determining the azimuth of the Qibla. Journal for the History of Arabic Science 4, 69–80.

——— 1981a. On al-Biruni's 'Method of the Zijes' for the Qibla. In: Proceedings of the 16th International Congress of the History of Science, Section C, 237–45. Academy of the Socialist Republic of Romania, Bucharest.

——— 1981b. An anonymous treatise on the regular nonagon. Journal for the History of Arabic Science 5, 37–41.

——— 1982. Al-Bīrūnī on plane maps of the sphere. Journal for the History of Arabic Science 6, 47–112.

——— 1985. On the origins of al-Bīrūnī's 'Method of the Zījes' in the theory of sundials. Centaurus 28, 1–16. (Editors: In the original paper the details were given as "to appear." We have supplied the full reference.)

——— 1987. Spherical trigonometry in the zīj of Kūshyār ibn Labbān. In: King, D.A., Saliba, G. (eds.), From Deferent to Equant: A Volume of Studies in the History of Science in the Ancient and Medieval Near East in Honor of E.S. Kennedy, New York Academy of Sciences, New York, pp. 15–33. (Editors: We have supplied the full reference.)

Bulgakov, P.G. (ed.), 1962. Taḥdīd nihāyāt al-amākin li-taṣḥīḥ masāfāt al-masākin. Published in Majallat maʿhad al-makhṭūṭāt al-ʿarabiyya. The Arab League, Cairo.

Busard, H.L.L., van Konigsveld, P.S., 1973. Der Liber de arcibus similibus des Ahmed ibn Jusuf. Annals of Science 30, 381–406. | 29

De Young, G., 1981. The Arithmetic Books of Euclid's Elements in the Arabic Tradition. Ph.D. dissertation, Harvard University.

—— 1984. The Arabic textual traditions of Euclid's Elements. Historia Mathematica 11, 147–60.

Debarnot, M.-Th., 1978. Introduction du triangle polaire par Abū Naṣr b. ʿIrāq. Journal for the History of Arabic Science 2, 126–36.

—— 1980. Les Clefs d'astronomie d'Abū al-Rayḥān ... al-Bīrūnī : La trigonometrie sphérique chez les arabes de l'est à la fin du Xe siècle. Thèse du 3eme cycle. Paris.

Djebbar, A., 1981. Enseignement et recherche mathématiques dans le Maghreb des XIIIe– XIVe siècles. Publications mathématiques d'Orsay. Université de Paris-Sud, Paris.

—— 1985. L'analyse combinatoire au Maghreb : l'example d'ibn Munʿim (XIIe– XIIIe siècles). Publications mathématiques d'Orsay. Université de Paris-Sud, Paris.

Ehrenkreutz, A.S., 1962. The kurr system in medieval Iraq. Journal of the Economic and Social History of the Orient 5, 309–314.

—— 1964. The taṣrīf and Tasʾīr calculations in medieval Mesopotamian fiscal operations. Journal of the Economic and Social History of the Orient 7, 46–56.

Engroff, J.W., 1980. The Arabic tradition of Euclid's Elements, Book V. Ph.D. dissertation, Harvard University.

Gandz, S., 1938. The algebra of inheritance. Osiris 5, 319–91.

Hamadanizadeh, J., 1976. Medieval interpolation theory. Ph.D. dissertation, Teacher's College, Columbia University.

—— 1978. Interpolation schemes in Dastūr al-Munajjimīn. Centaurus 2,44–52.

—— 1980. The trigonometric tables of al-Kāshī in his Zīj-i Khāqānī. Historia Mathematica 7, 38–45.

Hartner, W., 1965. Al-Djabr wa'l-muḳābala. In: Encyclopedia of Islam, ed. 2, vol. 2. E.J. Brill, Leiden, pp. 360–62.

Hermelink, H., 1975. The earliest reckoning books existing in the Persian language. Historia Mathematica 2, 299–303.

—— 1978. Arabische Unterhaltungsmathematik als Spiegel jahrtausendealter Kulterbeziehungen zwischen Ost und West. Janus 65, 105–17.

Hochheim, A. (trans.), 1877–1880. Al-Kāfī fī ḥisāb. 3 pts. Halle/Saale.

Hogendijk, J.P., 1981. How trisections of the angle were transmitted from Greek to Islamic geometry. Historia Mathematica 8, 417–38.

—— 1984a. Ibn al-Haytham's Completion of the Conics. Springer-Verlag, New York.

—— 1984b. Greek and Arabic constructions of the regular heptagon. Archive for History of Exact Sciences 30, 197–330.

—— 1985. Thābit ibn Qurra and the pair of amicable numbers 17296, 18416. Historia Mathematica 12, 269–273. (Editors: We have completed the reference.)

Irani, R.A.K., 1956. The Jadwal al-taqwīm of Ḥabash al-Ḥāsib. Unpublished M.A. thesis, American University of Beirut.

Jaouiche, K. (ed. and trans.), 1976. Le livre du qarasṭūn de Ṯābit ibn Qurra. E. J. Brill, Leiden.

Jensen, C., 1971. Abū Naṣr Manṣūr's approach to spherical astronomy as developed in his treatise "The table of minutes." Centaurus 16, 1–19.

Kennedy, E.S., 1968. The exact sciences in Iran under the Saljuqs and Mongols. In: Boyle, J.A. (ed.), Cambridge History of Iran, Volume 5. Cambridge University Press, Cambridge, 659–679. | 30

—— 1969. An overview of the history of trigonometry. In: Baumgart, J.K. (ed.), Historical Topics for the Mathematics Classroom. National Council of Teachers in Mathematics, Washington, D.C., pp. 333–359. Reprinted in Kennedy et al., 1983.

—— 1970. The Arabic heritage in the exact sciences. Al-Abhath 23, 327–344. Reprinted in Kennedy et al., 1983.

——— 1973. A commentary upon Biruni's Kitāb taḥdīd al-amākin. University of Beirut Press, Beirut.

——— 1976. The Exhaustive Treatise on Shadows by Abū al-Rayḥān al-Bīrūnī. vol. 1, trans., vol. 2, commentary. Institute for History of Arabic Science, Aleppo.

——— 1984. Applied mathematics in the tenth century: Abu'l-Wafā' calculates the distance Baghdad-Mecca. Historia Mathematica 11, 193–206.

Kennedy, E.S., et al., 1983. Studies in the Islamic exact sciences. American University of Beirut Press, Beirut.

Kennedy, E.S., Debarnot, M.-Th., 1984. Two mappings proposed by Bīrūnī. Zeitschrift für Geschichte der arabisch-islamischen Wissenschaften 1, 145–147. (Editors: We have completed the reference.)

King, D.A., 1972. The astronomical works of Ibn Yūnus. Ph.D. dissertation, Yale University.

——— 1973. Al-Khalīlī's auxiliary tables for solving problems of spherical astronomy. Journal for the History of Astronomy 4, 99–110.

——— 1974a. On medieval multiplication tables. Historia Mathematica 1, 317–23. Reprinted in King 1985a.

——— 1974b. An analog computer for solving problems of spherical astronomy. Archives internationales d'histoire des sciences 24, 219–242. Reprinted in King 1986.

——— 1979a. Notes on the sources for the history of early Islamic mathematics. Journal of the American Oriental Society 99, 450–59.

——— 1979b. Ḳibla. In: Encyclopedia of Islam, ed. 2, vol. 5, pp. 83–88. E.J. Brill, Leiden.

——— 1979c. Supplementary notes on medieval Islamic multiplication tables. Historia Mathematica 6, 405–417. Reprinted in King 1985a.

——— 1980. The exact sciences in medieval Islam: Some remarks on the present state of research. Middle East Studies Association Bulletin 14, 10–26.

——— 1981–1986. A catalog of the scientific manuscripts in the Egyptian National Library. 2 vols. American Research Center in Egypt, Cairo.

——— 1981. Universal solutions in medieval Islamic astronomy. Abstract of talk, in Proceedings of the 16th International Congress of the History of Science, Part A., 144. Academy of the Socialist Republic of Romania, Bucharest.

——— 1983a. The astronomy of the Mamluks. Isis 74, 531–555. Reprinted in King 1985a.

——— 1983b. Al-Khwārizmī and new trends in mathematical astronomy in the ninth century. Hagop Kevorkian Center for Near Eastern Studies, New York University, New York.

——— 1984. A Survey of the Scientific Manuscripts in the Egyptian National Library. Undena Publication, Malibu, California.

——— 1985a. Islamic Mathematical Astronomy. London, Variorum Reprints.

——- 1985b. Some early Islamic approximate methods for determining the Qibla. To appear in King and Saliba 1985. (Editors: This paper was, apparently, not published in this form. King's contribution to this collection was: Some early Islamic tables for determining lunar crescent visibility. In: King, D.A., Saliba, G. (eds.), 1987, From Deferent to Equant: A Volume of Studies in the History of Science in the Ancient and Medieval Near East in Honor of E.S. Kennedy, New York Academy of Sciences, New York, pp. 185-225.)

——— 1986. Islamic Astronomical Instruments. Variorum Reprints, London.

King, D.A., Saliba, G.A. (eds.), 1987. From Deferent to Equant: Studies in the History of Science in the Ancient and Medieval Near East in Honor of E.S. Kennedy. New York Academy of Sciences, New York. (Editors: In the original article, this book was noted as "to appear." We have supplied the published details.)

Knorr, W., 1983. On the transmission of geometry from Greek into Arabic. Historia Mathematica 10, 71–78.

Lorch, R., 1984. Qibla diagrams and associated instruments. To appear. (Editors: We have not found a
31 published version of this paper.) |

——— 1987. Al-Ṣaghānī's treatise on projecting the sphere. In: King, D.A., Saliba, G. (eds.), From Deferent to Equant: A Volume of Studies in the History of Science in the Ancient and Medieval

Near East in Honor of E.S. Kennedy, New York Academy of Sciences, New York, pp. 237–252. (Editors: We have supplied the published details.)

Luckey, P., 1951. Die Rechenkunst bei Ǧamsīd b. Masʿūd al-Kāshī. Abhandlungen fur die Kunde des Morgenlandes, Deutsche Morgenlandische Gesellschaft 31, Wiesbaden.

Mach, R., 1977. Catalogue of Arabic manuscripts (Yahuda Collection) in the Garrett Collection, Princeton University. Indexed by R.D. McChesney. Princeton University Press, Princeton.

Matvievskaya, G., 1987. The theory of quadratic irrationals in medieval oriental mathematics. In: King, D.A., Saliba, G. (eds.), From Deferent to Equant: A Volume of Studies in the History of Science in the Ancient and Medieval Near East in Honor of E.S. Kennedy, New York Academy of Sciences, New York, pp. 419–426. (Editors: We have supplied the published details.)

Muwafi, A., Philippou, A.N., 1981. An Arabic version of Eratosthenes on mean proportionals. Journal for the History of Arabic Science 5, 147–74.

Naini, A.D., 1982. Geschichte der Zahlentheorie im Orient. Verlag Klose and Co., Braunschweig.

Plooij, E.B., 1950. Euclid's conception of ratio ... as criticized by Arabian commentators. Ph.D. dissertation, Rijksuniversiteit te Leiden.

Rashed, R., 1972. L'induction mathématique : al-Karajī, as-Samawʾal. Archive for History of Exact Sciences 9, 1–21.

——— 1973. Algèbre et linguistique : l'analyse combinatoire dans la science arabe. In: Cohen, R. (ed.), Boston Studies in the Philosophy of Sciences. D. Reidel, Dordrecht, pp. 383–99.

——— 1974. Résolution des equations numériques et algèbre : Šaraf-al-Din al-Ṭūsī, Viète. Archive for History of Exact Sciences 12, 244–90.

——— 1975a. Ed. L'art de l'algèbre de Diophante. Maṭabiʿ al-haiʾa al-miṣriyya al-ʿāmma al-kitāb, Cairo.

——— 1975b. Récommencements de l'algèbre au XIe et XIIe siècles, In: Murdoch, J.E., Sylla, E.D. (eds.), The Cultural Context of Medieval Learning. D. Reidel, Dordrecht, pp. 33–60.

——— 1978. L'extraction de la racine n^{ième} et l'invention des fractions decimales (XIe–XIIe siècles). Archive for History of Exact Sciences 18, 191–243. Reprinted in Rashed 1984a.

——— 1979a. L'analyse diophantienne au Xe siècle : l'exemple d'al-Khāzin, Revue d'histoire des sciences 32, 193–222. Reprinted in Rashed 1984a.

——— 1979b. La construction de l'heptagone régulier par Ibn al-Haytham. Journal for the History of Arabic Science 3, 309–87.

——— 1980. Ibn al-Haytham et le théorème de Wilson, Archive for History of Exact Sciences 22, 305–21. Reprinted in Rashed 1984a.

——— 1983. Nombres amiables, parties aliquotes et nombres figurés au XIIIeme– XIVeme siècles. Archive for History of Exact Sciences 28, 107–47. Reprinted in Rashed 1984a.

——— 1984a. Entre arithmétique et algèbre : recherches sur l'histoire des mathématiques arabes. Les Belles Lettres, Paris.

——— (ed. and trans.) 1984b. Diophante : les arithmétiques. Tome III (livre IV), tome IV (livres V, VI, VII). Les Belles Lettres, Paris.

Rashed, R., Djebbar, A., 1981. L'Œuvre algébrique d'al-Khayyām. Institute for the History of Arabic Science, Aleppo.

Richter-Bernburg, L., 1982. Al-Bīrūnī's Maqāla fī tasṭīḥ al-ṣuwar wa-tabṭīkh al-kuwar. Journal for the History of Arabic Science 6, 113–22. | 32

Rosenfeld, B.A., 1978. Review of Fuat Sezgin's Geschichte des arabischen Schrifttums, Bd. 5. Archives internationales d'histoire des sciences 28, 325–29.

Rozhanskaya, M., 1987. On a mathematical problem in al-Khāzinī's Book of the Balance of Wisdom. In: King, D.A., Saliba, G. (eds.), From Deferent to Equant: A Volume of Studies in the History of Science in the Ancient and Medieval Near East in Honor of E.S. Kennedy, New York Academy of Sciences, New York, pp. 427–435. (Editors: We have supplied the full reference.)

Rozhanskaya, M., Rosenfeld, B.A., 1987. On al-Bīrūnī's Densimetry. In: King, D.A., Saliba, G. (eds.), From Deferent to Equant: A Volume of Studies in the History of Science in the Ancient and Medieval Near East in Honor of E.S. Kennedy, New York Academy of Sciences, New York, pp. 404–417. (Editors: We have supplied the full reference.)

Sabra, A.I., 1968. Thābit ibn Qurra on Euclid's parallels postulate. Journal of the Warburg and Courtauld Institutes 31, 12–32.
—— 1969. Simplicius's proof of Euclid's parallels postulate. Journal of the Warburg and Courtauld Institutes 32, 1–24.
—— 1971. Ilm al-ḥisāb. In Encyclopedia of Islam, ed. 2, vol. 3, 1138–1141.
—— 1982. Ibn al-Haytham's lemmas for solving 'Alhazen's problem.' Archive for History of Exact Sciences 26, 299–324.
—— 1983. Ed. Kitāb al-manāẓir (The Optics of Ibn al-Haytham). Arabic text of books 1, 2, and 3 on direct vision. National Council for Culture, Arts, and Letters, Arabic Heritage Department, Kuwait.
Saidan, A.S., 1966. The earliest extant Arabic arithmetic. Isis 57, 475–90. See also Saidan 1978a.
—— (ed.) 1971. ʿIlm al-ḥisāb al-ʿArabī: ḥisāb al-yad (Arabic arithmetic: The Arithmetic of Abū al-Wafā' al-Būzjānī). Jamʿiat ʿummāl al-maṭābiʿ al-taʿā-wunīyat, Amman.
—— 1974. The arithmetic of Abu'l-Wafā. Isis 65, 367–75.
—— (ed.) 1977a. Kitāb al-a'dād al-mutaḥābbat li-Thābit ibn Qurra (Amicable numbers, by Thābit ibn Qurra). The Jordanian University. Amman.
—— (ed.) 1977b. Kitāb tasṭīḥ al-ṣuwar wa tabṭīh al-kuwar li-Abi all-Rayḥān al-Bīrūnī. Dirāsāt. The Jordanian University (Amman) 4, 7–22.
—— 1978a. The arithmetic of al-Uqlīdisī. The story of Hindu-Arabic arithmetic as told in Kitāb al-fuṣūl fī al-ḥisāb al-hindī by Abū al-Ḥasan Aḥmad ibn Ibrāhīm al-Uqlīdisī. D. Reidel, Dordrecht.
—— 1978b. Ḥawl khawāṣṣ al-aʿdād li-Abī Jaʿfar, Muḥammad ibn al-Ḥusain (Theorems in number theory, by Abū Jaʿfar Muḥammad ibn al-Ḥusain), Dirāsāt (December), 7–49.
—— 1987. The Takmila fī al-ḥisāb by al-Baghdādī. In: King, D.A., Saliba, G. (eds.), From Deferent to Equant: A Volume of Studies in the History of Science in the Ancient and Medieval Near East in Honor of E.S. Kennedy, New York Academy of Sciences, New York, pp. 437–443. (Editors: We have supplied the full reference.)
—— 1973. The meaning of al-jabr wa'l-muqābalah. Centaurus 17, 189–204. Reprinted in Kennedy et al. 1983.
—— 1976. The double-argument tables of Cyriacus. Journal for the History of Astronomy 7, 41–46.
—— 1977a. Asālib ḥisābat al-jadāwal al-falakīyat al-islāmīyat. Proceedings of the First International Symposium for the History of Arabic Science, vol. 1 (papers in Arabic). University of Aleppo Press, Aleppo, pp. 275–94.
—— 1977b. Computational techniques in a set of late medieval astronomical tables. Journal for the History of Arabic Science, 1, 24–32.
Sesiano, J., 1977a. Le traitement des equations indéterminées dans le Badīʿ fī'l-ḥisāb d'Abū Bakr al-Karajī. Archive for History of Exact Sciences 17, 297–379.
—— 1977b. Les méthodes d'analyse indéterminée chez Abū Kāmil. Centaurus 21, 89–105.
—— 1979. Note sur trois théorèmes de mécanique d'al-Quhi et leur conséquence. Centaurus 22, 281–97.
—— 1980. Herstellungsverfahren magischer Quadrate aus islamischer Zeit (I). Sudhoffs Archiv 64, 187–96.
—— 1981. Herstellungsverfahren magischer Quadrate aus islamischer Zeit (II). Sudhoffs Archiv 65,
33 251–65. |
—— 1982. Books IV to VII of Diophantus' Arithmetica in the Arabic translation attributed to Qusṭā ibn Lūqā. Springer Verlag, New York.
—— 1987. A treatise by al-Qabīṣī (Alchabitius) on arithmetical series. In: King, D.A., Saliba, G. (eds.), From Deferent to Equant: A Volume of Studies in the History of Science in the Ancient and Medieval Near East in Honor of E.S. Kennedy, New York Academy of Sciences, New York, pp. 483–500. (Editors: We have supplied the full reference.)
Sezgin, F., 1974, 1978. Geschichte des arabischen Schrifttums. 5. Mathematik; 6. Astronomie. E.J. Brill, Leiden.
Suter, H., 1922. Über die Projektion der Sternbilder und der Länder von al-Bīrūnī. Abhandlungen zur Geschichte der Naturwissenschaften und der Medizin, Erlangen 4, 79–93.

Tee, G., 1977. Letter to the editor: On computational techniques. Journal for the History of Arabic Science 1, 323–24.

Tekeli, S., 1968. 'The duplication of the cube' Zail-i Tahrir al Uqlidas, Majmuaʿ and Sidra al Muntahâ, In Proceedings of the 12th International Congress of the History of Science, Paris, pp. 137–40.

Tichenor, M.J., 1967. Late medieval two-argument tables for planetary longitudes. Journal of Near Eastern Studies 26, 126–28. Reprinted in Kennedy et al. 1983.

Toomer, G.J., 1976. Diocles on Burning Mirrors: An Arabic Translation of the Lost Greek Original. Springer-Verlag, New York.

Villuendas, M.V., 1979. La trigonometria europea en el siglo XI: Estudio de la obra de ibn Muʿād El kitāb mayhūlāt. Instituto de Historia de la Ciencia de la Real Academia de Buenas Letras, Barcelona.

Wieber, R., 1972. Das Schachspiel in der arabischen Literatur von den Anfangen bis zur zweiten Halfte des 16 Jahrhunderts. Ph.D. dissertation, University of Bonn.

Woepcke, F., 1861. Recherches sur plusieurs ouvrages de Léonard de Pise, Atti dell'Academia Pontificia dei nuovi Lincei 14, 211–27, 241–69, 301–24, 343–56.

Yadegari, M., 1978. The binomial theorem described by Amir Kalan al-Bukhari circa 1297 A.D. Islamic Quarterly, 20–22, 36–39.

——— 1980. The binomial theorem: A widespread concept in medieval mathematics. Historia Mathematica 7, 401–406.

Youschevitch, A.P., 1976. Les mathématiques arabes (VIIIe–XVe siècles), Cazenave, M., Jaouiche, K. (trans.). J. Vrin, Paris.

Mathematics and Her Sisters in Medieval Islam: A Selective Review of Work Done from 1985 to 1995 [1997]

J. Lennart Berggren

Abstract This paper surveys work done over the past decade, largely in Western Europe and North America, in the history of the mathematical sciences as practiced in medieval Islam from central Asia to Spain. Among the major topics covered, in addition to the usual branches of mathematics, are mathematical geography, astronomy, and optics. We have also given accounts of some current debates on the interpretation of important texts and, in addition, we have surveyed some of the literature dealing with the interrelation of mathematics and society in medieval Islam.

Ten years ago we surveyed recent work in the history of the mathematical sciences in medieval Is- 408
lam [10], including mathematics itself and the sciences of astronomy, astrology, cartography, optics, and music, where mathematics played an essential role. Since that time much has been published in these areas, and it seems fitting once again to survey what has been learned about medieval Islamic achievements in the mathematical sciences during the past decade.[1]

At the outset, however, two comments are called for. The first is that the present survey is largely restricted to works published in North America or Europe and, in particular, refers neither to works published in the former Soviet Union, where a long tradition of significant work exists, nor to works entirely in Arabic. Moreover, although the author has tried to reference most of the contributions of those working actively within the selected areas, it is inevitable that some important work will have been left out, for which sins of omission he can only ask forgiveness.

Second, although some cases have been noted in which a particular work has occasioned some critical comment this does not mean that all other work has been, or should be, accepted as established truth. All results in an area such as this are more or less tentative, if only because the mass of unexplored material is so large in relation to what has been studied, and the decision to report certain controversies means only that the issues raised by the controversy appear to be of particular interest.

The words "mathematical sciences" emphasize that the medieval Islamic scientists worked on such a variety of topics that to focus only on what we might call mathematics is to ignore areas in which the medievals exercised their creativity to great effect and which many of them saw as the *raison d'être* of the whole. Thus, one of the earliest (and possibly the best known) of the medieval Islamic scientists,

Editors: Originally published as Berggren, J.L., 1997, Mathematics and her sisters in medieval Islam: A selective review of work done from 1985 to 1995, Historia Mathematica 24, 407–440. We thank J.L. Berggren and Elsevier for permission to republish this paper.

[1] The author has made a serious search for all published papers from the period 1985–1995 not cited in his earlier survey [10] and has referenced such as he felt might appeal to a nonspecialist audience. As of this writing the year 1996 has not finished so a thorough survey of that year is impossible. The author has therefore referred to such papers from 1996 as have come to his attention and which add something significant to a major theme of this paper.

N. Sidoli and G. Van Brummelen (eds.), *From Alexandria, Through Baghdad: Surveys and Studies in the Ancient Greek and Medieval Islamic Mathematical Sciences in Honor of J.L. Berggren*, DOI 10.1007/978-3-642-36736-6_5,
© Springer-Verlag Berlin Heidelberg 2014

Muḥammad ibn Mūsā al-Khwārizmī, contributed significantly not only to arithmetic and algebra but also to astronomy and geography. To ignore these is to impoverish the history of mathematics.[2]

Al-Khwārizmī and His Times

Of course the role of al-Khwārizmī's treatise on Hindu arithmetic in introducing that arithmetic to the Latin West in the 12th century is well known, and André Allard's study [171] of the four main relevant Latin texts seemed to take the subject just about as far as it could go. For that very reason special interest is given to Menso Folkerts's discovery [51] of a new manuscript — found in New York! —
409 of | al-Khwārizmī's introduction to Hindu arithmetic. This new manuscript not only is more detailed than the previously known Cambridge exemplar, of which an English translation [32] has recently been published, but also is more carefully written and has fewer nontrivial errors. Additionally, it has the virtue of being complete, whereas the Cambridge manuscript breaks off midway through the discussion of division of fractions and before the discussion of square roots. With Folkerts's discovery we are now closer to a knowledge of the original form of the text that introduced arithmetic to Western Europe.

As far as basic numeration itself, David King is investigating (see, for example, [94] and [95]) the history of a forgotten system of numeration. Although it has been studied by other scholars (e.g., Jacques Sesiano [142]) King has done a thorough study of its history, from its origins in Greek stenography down to the Nazi party in Germany. Never intended for use in calculation, this simple system in its final form allows one to write any numeral from 1 to 9999 as a single cipher. King has documented the use of the system in medieval astrolabes and has found in Arabic treatises[3] on cryptography characters very close to those in the European manuscripts.

Whether al-Khwārizmī, who wrote in the first half of the ninth century, was also the first Islamic writer on algebra continues to be a subject of some controversy. Further ammunition for the doubters comes from an Arabic fragment from the Yemen [89] which names the seventh-century Caliph ʿUmar as a patron of certain "algebraists" who came from the province of Fars and taught orally. It also mentions both the future Caliph, ʿAlī, as an evidently able pupil, and al-Khwārizmī as one who, much later, under the reign of al-Maʾmūn, first wrote down this knowledge. King thinks the report is dubious, and this may well be true of the part about ʿAlī, but the suggestion that algebra was initially an oral tradition, which came early to the Islamic world from Persia, fits well with arguments of Jens Høyrup [71], who has argued that medieval Islam fell heir to two ancient "sub-mathematical" traditions, and employed ancient algebraic techniques and a cut-and-paste geometry of the sort one finds systematized in *Elements*, II, to justify these techniques.[4]

Methods for solving quadratic equations were known to the Babylonians more than two millennia before al-Khwārizmī's *Algebra*. However, al-Khwārizmī's work transforms what had been with the Babylonians a systematized method for solving quadratics into the science of algebra, i.e., a method derived from explicitly stated principles which made it clear to the learner why the procedures worked. It also became a step to the solution of higher degree equations, starting with the pure equation $x^n = a$. However, since the neighboring Chinese civilization had also discovered methods for solving higher degree equations, it is inevitable that questions of dependence have arisen. In a close analysis of the
410 algorithms employed | for root extraction, Chemla [29] has used a method of Allard [4] to compare the Chinese and Arabic algorithms and to reach conclusions that upset some previously held beliefs.

Sharaf al-Dīn al-Ṭūsī's monograph on algebra [152] is by a mathematician who worked in the late

[2] King [91, 125–126], for example, stresses that "our [astronomical] material is based on mathematical procedures" and goes on to itemize the mathematical riches to be found in the astronomical texts, e.g., tabulation and graphical representations of double-argument functions, double-order interpolation, and computational devices.

[3] The author thanks Professor King for sending him relevant material from MS Leiden Or. 14, 121 and another MS in a private collection in Frankfurt, to be discussed along with other material on Arabic numerical codes in his book, *The Ciphers of the Monks: A Forgotten Number-Notation of the Middle Ages* currently in preparation.

[4] This idea of proofs is, according to Høyrup [71, 475], where the Greek element enters.

12th century and carried forward the work of Omar Khayyām's *Algebra* in two respects. First, Sharaf al-Dīn gave complete and correct discussions of the possibilities of solving each cubic equation by means of intersecting conics, whereas Omar had not been able to do so.[5] Second, he employed what is essentially the Ruffini-Horner algorithm for solving the possible cases numerically, a topic Omar does not discuss.

A debate has arisen over the ideas behind Sharaf's procedures. In discussing the possibilities of solution of a cubic equation of the form $f(x) = x^3 + ax^2 + bx = c$, Sharaf al-Dīn exhibits a quadratic polynomial equivalent to $f'(x)$ and shows that if a is the root of $f'(x) = 0$ that yields the larger value of $f(a)$ then $f(x) = c$ has no root for $c > f(a)$ and one root if $c = f(a)$.[6] Roshdi Rashed feels that this is evidence that Sharaf al-Dīn had recognized the derivative, but Jan Hogendijk [64][7] argues that this is not the case and that one can obtain Sharaf al-Dīn's conditions by arguments that do not go beyond the mathematics of *Elements*, II. The continuing efforts to understand Sharaf al-Dīn's motivation in more modern terms, found in Nicholas Farès [50] and Christian Houzel [73], show that this controversy has not yet been settled.[8]

Further hints of algebra prior to al-Khwārizmī come from Barnabas Hughes [74] in an extract from the work of Ayyūb al-Baṣrī, an estate divider, found in a treatise compiled in the 11th century by a certain Abraham or Ibrāhīm. Ayyūb gives a threefold rule (called in Latin the *regula infusa*) for solving linear equations by reducing the coefficient to one. Hughes suggests that this may be evidence of earlier methods which al-Khwārizmī replaced by a single approach.

Whatever al-Khwārizmī's priority (or lack thereof) in Arabic algebra, his theory of quadratic equations was widely influential, and a study of this influence forms part of a survey of the history of quadratic equations in medieval Islam and Europe in Yvonne Dold-Samplonius [45]. (An account with less mathematical detail may be found in Dold-Samplonius [46].)

Al-Khwārizmī is as well a seminal figure in the history of astronomy, to which he contributed a treatise on the astrolabe, one on astrology (possibly), and an | astronomical handbook, called *zīj* in 411
Arabic, consisting of tables and rules for their use. His *Zīj al-Sindhind*, one version of which is extant in a Latin translation of an edition put together ca. A.D. 1000 in Spain, is unique among surviving *zījes* in being based largely on Indian material.

Al-Khwārizmī's astronomy shows that he learned more from the Hindus than simply the decimal arithmetic of whole numbers and sexagesimal fractions. Recent evidence of this comes from a reconstruction of the Sine[9] table underlying a table of an astronomical function compiled by al-Khwārizmī. Hogendijk [66] has shown that the underlying radius for the Sine table is 150, a known Indian parameter, and that the values for the Sines of multiples of 15° are also Indian. To interpolate between these values al-Khwārizmī used a modified version of linear interpolation that takes account of the fact that the values of the Sine increase most rapidly near 0° and less so as the argument approaches 90°. Adolphe Rome discovered this sensible, if crude, method of interpolation in the works of Ptolemy (ca. A.D. 150), and it has recently been found in the works of al-Battānī, Kūshyār ibn Labbān, and al-Khalīlī (Van Brummelen [154; 156]).

More sophisticated methods of interpolation were also used. Glen Van Brummelen [154] studies those of al-Khalīlī, and a general (though very concise) survey may be found in [53]. A recent addition to the list has been studied by Rashed [123], who has found in a work by a 12th-century scientist,

---

[5] It is enlightening to compare Omar's and Sharaf's treatments of the equation $x^3 + c = a^2$.

[6] Readers not familiar with medieval Arabic texts should be told that symbolic expressions such as $f(x), f(a), f'(x)$ and symbolism such as $=$ and $>$ do not appear in these texts, although a limited use of symbolism does occur in certain late texts from northwest Africa (al-Maghrib). Given the lack of symbolism some question may arise about our reference to what are, in the texts, purely verbal expressions (such as "cubes and squares and things") as "polynomials." However, since medieval Islamic mathematicians had developed tabular methods even for extracting roots of such expressions when the number of cubes, squares, etc., were specified (i.e., the coefficients were known) we feel the description of such expressions as "polynomials" is justified.

[7] This paper also contains a useful and concise summary of the contents of Sharaf's treatise.

[8] It is regrettable that Farès makes no reference to Hogendijk [64].

[9] We capitalize the names of the trigonometric functions to remind the reader that the medieval trigonometric functions denoted the lengths of certain lines, not their ratios.

al-Samaw'al, against the astrologers an extract from a work of al-Bīrūnī that shows that the latter knew of Brahmagupta's method of interpolation as well as one other Indian method and tried out both on the Cotangent function.

A problem that challenged the ingenuity of Islamic scientists is that of calculating the aspects between two planets in an astrological prediction. The solution that Hogendijk [63] studied in two versions of material from al-Khwārizmī's *zīj* is calculated in a set of tables based on a complex geometrical projection from the ecliptic to the equator and back again.

Another problem that astrology suggested to the astronomer-mathematicians is that of dividing the ecliptic into 12 astrological "houses."[10] Two papers by Edward S. Kennedy [81; 82] give accounts of two popular methods for accomplishing this division, the one using the prime vertical and the other using the equator.[11] An interesting problem that was solved in this connection is to find two angles, given their sum or difference and the ratio of their Sines.

Hogendijk [62] has found evidence of considerable ingenuity in the methods used in astronomical investigations as early as A.D. 780 when Yaʿqūb ibn Ṭāriq, in order to calculate the brightness of the
412 moon in terms of its position relative to the sun, | needed to calculate the values Sin(d), where d is the great circle arc between the sun and the moon. He had no spherical trigonometry, not even Menelaus's theorem, so he approximated Sin(d) as $\sqrt{(\mathrm{Sin}^2(\beta) + \mathrm{Sin}^2(\eta))}$, where β is the lunar latitude and η is the elongation between the sun and the moon. The idea of approximating a spherical triangle by a plane triangle and applying the Pythagorean theorem works because β does not exceed 5° and at the time when one is interested in computing the brightness the elongation η is not very great either, so approximating arcs by their Sines is reasonable. Although the Sine table Yaʿqūb used has $R = 3438'$, another Indian parameter for the radius of the reference circle, his use of d to compute lunar brightness does not seem to be Indian.

A valuable collection of sources and studies of early Islamic astronomy is Regis Morelon's collection of the astronomical works by or attributed to Thābit ibn Qurra [75], the great scholar from ninth-century Ḥarrān, who was employed as translator by the Banū Mūsā, one of whom recommended his appointment as astrologer to the Calif al-Muʿtaḍid. A summary of the main problems attacked in these treatises and some of the mathematical problems these involve — from the analysis of apparent solar motion to the study of conic sections — has been published by Morelon [114].

Finally, in regards to al-Khwārizmī's geography, King [87] has published a study of material, whose discovery he announced in the early 1980s, which he felt was arguably due to al-Khwārizmī but, in any case, was written in the early 9th century. In this material we have the earliest exact method for computing the direction of Mecca (the *qibla*) relative to one's locality, the direction in which Muslims are to pray. The problem of computing this direction can be described in a variety of ways, most directly perhaps as the geographical problem of finding the angle between two great circles: that joining one's locality to Mecca and the local meridian.

This solution employs methods using given points and arcs on the surface of a sphere to obtain lines and triangles inside the sphere, which one in turn uses to calculate or construct other angles and arcs on the surface. These methods, known from Indian sources — where they are used particularly for time reckoning — involve three basic tools (the Pythagorean theorem, the theory of similar triangles, and the Sine and Cosine) and differ considerably from the more elegant and popular solution known as the "method of the *zīj*es." However, there are intriguing similarities between the two methods, and there seems to be some agreement that this early method was modified by the great 10th-century astronomer Ḥabash al-Ḥāsib, on the basis of his researches on sundials, to create the latter method.[12]

[10] These are not the same as the zodiacal signs. In particular they depend on both the locality and the time for which they are calculated.

[11] In the West, these methods were ascribed to Campanus and Regiomontanus, respectively, but Kennedy quotes recent work showing that al-Bīrūnī claimed credit for the method using the prime vertical and that the equatorial method was known to Ibn Muʿādh al-Jayyānī of Jaen in 11th-century Spain.

[12] See Berggren [9] for an argument that this method originated in gnomonics. See also Lorch and Kunitzsch [107] for the publication of Ḥabash's treatise on the uses of a celestial sphere set in graduated horizontal and vertical rings. (A survey of surviving celestial spheres in medieval Islam is Savage-Smith [141].)

Two other studies on the derivation of the *qibla* are those of Juan Carandell [27] and Ahmad Dallal [36]. The former deals with a method for the *qibla* which breaks down into two analemmas by Ibn al-Raqqām, who died in Granada in A.D. 1315. (The discovery of this method in a treatise on the construction of sundials for | religious purposes shows how the historian of mathematics must be 413
willing to expand the search for sources of material beyond the obvious ones.) Ibn al-Raqqām's method is interesting in that not only is it an analemma but it is one that can be derived from the method of al-Bīrūnī, though the error in al-Bīrūnī's exposition has been corrected. Dallal's study [36] shows that the working out of the details of a solution to the *qibla* problem valid for all localities must be attributed not to al-Kāshī, as the present author once thought, but to Alhazen (Ibn al-Haytham), who lived some four centuries earlier.

The Islamic Study of Euclid's Works

One of the Islamic mathematical sciences whose theoretical traditions were just being formed when al-Khwārizmī wrote was geometry, for al-Khwārizmī's lifetime coincided with the period when Euclid's *Elements* was first being translated into Arabic. A knowledge of the various Arabic translations of this key work is essential not only to an understanding of an important genre of Arabic mathematical literature, the commentaries on the *Elements*, but perhaps to a more secure knowledge of the Greek text of that work.

Yet, the complicated, and often obscure, story of how medieval Islam came to master the contents and mathematical spirit of this seminal work has not yet been written, and only recently have some essential elements of the history become available. For example, historians have long taken at face value al-Nairīzī's claim that his commented edition of the first six books was based on the second translation of the *Elements* into Arabic by al-Ḥajjāj ibn Yūsuf, one done for the Calif Harūn al-Rashīd.[13] However, Sonja Brentjes, in her studies of the text of *Elements*, I and II [22] found in al-Nairīzī's work, has strengthened doubts that this is the case raised by Engroff in his earlier study of *Elements*, V. We now realize that although al-Nairīzī's text does represent one of al-Ḥajjāj's two versions it is one with many interventions from a later translation, one by Isḥāq ibn Ḥunain and revised by Thābit ibn Qurra, whose astronomical works we mentioned earlier. It cannot be used uncritically as a guide either to al-Ḥajjāj's methods of translation or to his vocabulary.[14]

Vocabulary is a key element in two recent studies aimed at uncovering which features of existing translations of the *Elements* are due to al-Ḥajjāj. One of these studies is that of Gregg De Young [161], who shows that earlier material is mixed into manuscripts of the *Elements* from Northwest Africa (al-Maghrib) and Spain (al-Andalus) as: (1) marginalia giving al-Ḥajjāj's version of the enunciations of the propositions for *Elements*, II, where one encounters, e.g., *ḍarb* (product) instead of the usual *saṭḥ* (area); (2) the addition of further cases to *Elements*, III, — including an elegant variation of the construction in III, 32; (3) alternate proofs of *Elements*, VIII, 20–21; and (4) three condensed and altered proofs following *Elements*, X, 67. | De Young raises, without proposing any definite answer, the question of 414
why al-Ḥajjāj was closer to the Greek in *Elements* III, IV, and VIII than in II and X.

In a recent study Ahmed Djebbar [40] calls attention to other linguistic peculiarities of the al-Ḥajjāj material in the above-mentioned sources and elsewhere and suggests that perhaps the first of al-Ḥajjāj's translations was done under the influence of the Arabic mathematical vocabulary that developed furthest by the early ninth century, that of treatises on reckoning.

Brentjes [21] identifies material in a Paris manuscript as the first connected fragment we have either of al-Ḥajjāj's work or of something very closely connected with it. She argues, on the basis of this study, that al-Ḥajjāj's second so-called "translation" was, rather, a reworking of his first, that he did not work from a Syriac but a Greek text, and that selections from the *Elements* in secondary sources, such as the

[13] Al-Ḥajjāj's two translations were the first translations of the *Elements* into Arabic. The first of the two was done for the Calif Harūn al-Rashīd and the second for the Calif al-Ma'mūn.

[14] On an addition to *Elements*, I, ascribed to Thābit, see Brentjes [20].

Epistles of the Brethren of Purity,[15] have more significance for the study of Islamic acquisition of the *Elements* than has heretofore been thought. Her work again rests on, among other methods, a detailed study of the vocabulary.[16]

Roger Herz-Fischler [56] investigates the Arabic tradition of a proposition on mean and extreme ratio that is implicit in the proofs of Theorems 2 and 7 of the Supplement to the *Elements* known as Book XIV, with a view to finding out whether the proposition itself was originally in the Greek text. He concludes, on the basis of evidence from Pappus, Arabic editions of the *Elements*, and the Arabo-Latin tradition that the Greek text did originally contain such a proposition. This is but one example of a larger point made by Wilbur Knorr [98] that Heiberg erred in his negative judgment of Klamroth's position that the Arabic text of the *Elements* is an essential witness to the original Greek text.

In addition to the history of the text of the *Elements* in Arabic, there is also the history of individual topics found in Euclid's work. For example, there is a rich tradition of Arabic commentaries on the theories of proportion found in Euclid, the standard study of which is the thesis by Edward Plooij [116]. On this topic, Bijan Vahabzadeh [153] draws attention to the close similarities between two attempts to use the notion of "part" to justify Euclid's rather nonintuitive definition of proportion by means of multiples — one of them by the 11th-century, Spanish scientist al-Jayyānī and the other by the British mathematician Nicolas Saunderson (1682–1739).

Another piece of Euclideana is, of course, the parallel postulate, and recently many Arabic texts dealing with the problem of the status of Postulate 5 of *Elements*, I, have been collected, translated into French, and published by Khalil Jaouiche [76], a work which will be of use not only to historians of mathematics but to those interested in the philosophy of mathematics as well.[17] Another important
415 survey | of the Arabic theory of parallels is by Boris Rosenfeld [128]. Whereas Jaouiche emphasizes how the Islamic mathematicians broke with most of the ancient commentators on Euclid, to create a true theory of parallels, Rosenfeld emphasizes the continuity of the whole endeavor and places it within the context of the history of non-Euclidean geometry.

Questions of parallel lines raise the problem of infinity in mathematics, but the problem arose in other ways as well. A discussion of some medieval Islamic thought on infinity and questions of atomism is found in Rachid Bebbouchi's work [6], and a late testimony to mathematical atomism is found in Rosenfeld [129, 97–101]. (The earlier cited work by the same author [128, 193–195] also contains a discussion of atomism in medieval Islam.) For the very important religious overtones of atomism, see Alnoor Dhanani's work [38].

Of the Euclidean works lost in the Greek, Hogendijk [60] identifies fragments of the *Porisms* in the works of two 10th-century geometers on the strength of Pappus's description of this work, and his auxiliary lemmas thereto, in his *Collection*, VII. In another study [67], he presents al-Sijzī's version of the problems in Euclid's *On Divisions*. Because he found them quite simple, al-Sijzī omitted the solutions and proofs for all but four of the problems from his source, which Hogendijk argues was a revised translation by Thābit ibn Qurra.

A witness to the active approach the medieval mathematicians took toward the ancient texts is Abū Sahl al-Kūhī's reworking of *Elements*, II. Al-Kūhī was one of the leading geometers of the late 10th century, and De Young's study [162] shows how he rewrote *Elements*, II, 1–10, to give the work greater unity and then added 17 propositions of his own.

A recent contribution by Sesiano [145] illuminates the motives of the early translators and shows how the best of them read the ancient sources actively, pen in hand. It presents a short piece by the above-mentioned Thābit solving a problem that Euclid suggested in his *On Divisions*. Thābit showed that if the side of a regular hexagon in a circle is contained in the smaller segment of that circle cut off by a side of an equilateral triangle, then the area between the two sides is one-sixth the area of the

---

[15] This was a group of Muslim scholars of the late 10th century who composed an encyclopedic treatment of the sciences in support of a particular religious or political cause.

[16] A striking example is the use in the Paris manuscript of "talbīn" for "rectangle" or "square." Brentjes is able to point to the ascription of the use of the word to al-Ḥajjāj by a writer of the first half of the 12th century, Ibn al-Sarī.

[17] Remarks of both historical and didactic interest on the parallel postulate may be found in Hogendijk's review [61] of this work.

circle.

Number Theory and Recreational Mathematics

The above "miniature" of Thābit at work on a text fits well with Hogendijk's argument [58] that Thābit not only discovered a rule for generating amicable numbers but used it to find a new pair of amicable numbers (17296 and 18416). Thābit's theorem on amicable numbers inspired the Persian mathematician Kamāl al-Dīn al-Fārisī (d. ca. 1320) to give a new proof. On the basis of al-Fārisī's work, however, the question arose in the early 1970s of whether medieval Muslim mathematicians proved the Fundamental Theorem of Arithmetic. Ahmet Ağargün and Colin Fletcher [3] argue convincingly that, despite a recent claim, al-Fārisī's treatise does not contain a proof of that theorem but something quite different, an explicit construction of the divisors of a number in terms of its prime factorization. | 416

Another topic in number theory with a long history in medieval Islam is that of magic squares, and Sesiano's [149] is an exposition of an 18th-century text by al-Kishnāwī containing a clear exposition of much earlier methods for the construction of simple magic squares of odd order, bordered magic squares,[18] partitioned magic squares of order $r \cdot n$, consisting of r^2 magic squares of order n, and squares in which all odd numbers appear in the central part and all even numbers appear near the corners. There is also a treatment of magic squares of odd order in which the center square is empty. Sesiano's most recent publication on this topic is [174], an edition of a work which he argues dates from the 11th century, since the obviously very thorough (but anonymous) author knows the 10th-century discoveries but is unaware of the full solution of constructing magic squares for numbers of the form $4k + 2$, an achievement of the 12th century.

Magic squares stand at the border of number theory and recreational mathematics, but a work part of which is solidly in this latter tradition is the *Liber mahameleth*, written in Latin in 12th-century Spain on the basis of Arabic sources. Sesiano [144] surveys a group of problems from this source, such as ladder and tree problems,[19] most of whose solutions depend on the Pythagorean theorem, and points out that the three approaches to the most common of these problems (solution by formula, proof by geometry, and solution by algebra) correspond to a traditional periodization of the history of mathematics: ancient Mesopotamia, Greece, and medieval Islam.

Other Aspects of the Hellenistic Tradition in Medieval Islam

An older contemporary of al-Kūhī (whose work on *Elements*, II, we discussed earlier) was Thābit's grandson, Ibrāhīm ibn Sinān, a talented mathematician of the first half of the 10th century. His treatise *On Analysis and Synthesis* is one of several written by prominent Islamic mathematicians on these two classical geometrical methods. Hélène Bellosta [7] shows that Ibrāhīm had two concerns in this work: to lay out an elaborate classification of geometrical problems according to the number and type of solutions they had, and to show how to do analyses and syntheses so that there is a complete correspondence between the two procedures. Ibrāhīm says he does this because of those who criticize analyses done in the usual manner, where one is always being surprised by the appearance of lines, etc., in the synthesis not found in the analysis.

The seriousness with which the classification of problems and the study of analysis and synthesis

---

[18] Sesiano [143] has edited the Arabic text of an earlier work, that of the 13th-century scholar al-Zanjānī on bordered-magic squares, a summary of which he had published earlier. For a recent publication in English on this topic, see Sesiano [147]. Sesiano [150] has published the Arabic text and a German paraphrase of a treatise on magic squares by Jamāl al-Zamān al-Kharaqī, who died in 1138/9.

[19] Such denomination of problems corresponds to medieval usage. For example, Sesiano [144, n. 17] cites al-Bīrūnī putting "ladder problems" into the domain of recreational mathematics.

417 were taken in the 10th century is shown by the fact that not only did | Ibrāhīm in the early part of the century write on the subject, but late in the century Alhazen, whose universal solution to the problem of finding the *qibla* we mentioned above, also wrote on it, in a work which has been studied by Jaouiche [77]. Jaouiche points out that the Arabic writers saw analysis as not being confined to geometry, viewing algebra too as a kind of analysis, and Alhazen showed how analysis could be applied to all four classical branches of mathematics: geometry, arithmetic, astronomy, and music. In addition he stressed the need for creativity in analysis, that it was not a simple algorithmic procedure. (A subsequent study of Alhazen on analysis and synthesis is Rashed [181]; a critical edition of the text, with translation and commentary by A. Djebbar and Kh. Jaouiche, is to appear.)

Much of the best of the medieval Arabic work, however, was not inspired by Euclid but by Archimedes, and in our previous survey [10] we called attention to the need for a study of the Archimedean tradition in the Islamic middle ages. Happily this gap in the literature has begun to be filled, and two studies of the Archimedean tradition in medieval Islam are (1) a philological study, by Richard Lorch [109], of Arabic, Hebrew, and Latin material germane to the problem of the direct transmission of the text of *On the Sphere and Cylinder* and (2) a study, also by Lorch [108], of the indirect transmission of the same work based on the theorems and proofs of a treatise on isoperimetry by the 10th-century astronomer Abū Ja'far al-Khāzin, who, according to Omar Khayyām, used conics to solve the cubic equation to which al-Māhānī had reduced the problem Archimedes posed in *On the Sphere and Cylinder*, II, 4.

A curious feature of the Archimedean tradition in medieval Islam is that only two works of the present Archimedean canon were transmitted to medieval Islam, namely both parts of *On the Sphere and Cylinder* and *On the Measurement of the Circle*. If, however, one accepts the view, for which a case can be made,[20] that the medieval period in Islamic science extended into the 19th century, it is relevant to note here the recension of Archimedes' *On Spiral Lines* done by al-Yanyawī* ca. 1700 (Berggren [16]).

The other Archimedean works known in medieval Islam are works which, in their entirety, are spurious but may well contain parts of authentic Archimedean works. Thus it is entirely consistent with the general character of the Archimedean tradition in medieval Islam that Sesiano [146] should have found, among the theorems in an Arabic manuscript in Teheran, part of a lost work in which Archimedes established the formula for the area of a triangle in terms of its sides.

Other recent explorations of the Archimedean tradition in medieval Islam are in Rashed [124; 125]. The former, on the tradition in planimetry and stereometry, shows how Islamic mathematicians from the 9th to the 11th centuries modified Archimedes' proofs and extended his methods to solve what were for them new problems. The latter makes available the commentary of the Arab savant al-Kindī
418 on Prop. 3 (on the circumference and diameter of a circle) of Archimedes' *Measurement of the Circle.* |

An extended, close examination of the medieval (Arabic and Latin) textual tradition of Archimedes' *Measurement of the Circle*, together with Arabic treatments of the other two famous classical problems of cube duplication and angle trisection, is found in Knorr [97].

Also in the Archimedean tradition is al-Kūhī's solution of a problem inspired by Archimedes' *On the Sphere and Cylinder*, II, 5–6: to construct a spherical segment whose (curved) surface is equal to that of one spherical segment and whose volume is equal to that of another, recently published by Berggren [19]. In modern terms this problem demands the solution of two cubic equations, but al-Kūhī did not phrase the problem in algebraic terms. Rather, he viewed it as a geometrical problem of constructing line segments satisfying certain relations, which he constructed by intersecting conic sections — in this case a hyperbola and a parabola, the same two types of conics that Omar Khayyām used 70 years later to solve the type of cubic equations arising from al-Kūhī's problem.

One of al-Kūhī's contemporaries was al-Sijzī, whose writings have been an important source for Hogendijk's successful searches for material from lost works of Euclid and Apollonius [60] and [59].[21]

---

[20] Certainly there was a tradition of studying and copying medieval works until that time.

* Editors: In the original paper the name is spelled al-Yānināwī; however, al-Yanyawī is the more common spelling.

[21] Kunitzsch and Lorch [106] give convincing codicological evidence that a Paris manuscript long supposed to have been copied by al-Sijzī was in fact copied by him.

(The latter work [59] contains traces of the *Neuses*, the *Plane Loci*, and *On Tangencies*, which must therefore have been available in Arabic.) Al-Sijzī's works have also been sources for study in their own right. Two examples are Rashed [121], devoted to an influential treatise by al-Sijzī on the asymptotes of the hyperbola, and Crozet [33], which is devoted to a treatise in which al-Sijzī attempts to calculate the volumes between, for example, three spheres, a pair of which are disjoint but contained in the third. Crozet is certainly right in saying that the point of the work is not easy to see, but his conjecture that it is a semi-intuitive exploration of ideas of "dimension" going beyond those of al-Sijzī's time into the idea of four-dimensional spheres is unwarranted.

Much of the medieval work in the Archimedean tradition, to say nothing of that on trisection, cube duplication, and related problems, required considerable knowledge of conic sections, and a major new source for the study of the Islamic tradition of conics is the publication of an Arabic translation, by the Banū Mūsā and Thābit ibn Qurra, of Apollonius's *Conics*, V–VII, the Greek text of which is lost, available in Gerald Toomer's edition, translation, and study [151].

In view of the theoretical importance ellipses assumed in astronomy as a result of the work of Kepler it is interesting to note that the 11th-century, Spanish astronomer Azarquiel (Ibn al-Zarqālluh) used them to approximate the path of the center of Mercury's epicycle in his design of an equatorium. Mercè Comes [31] has published the treatise in which this occurs, along with another by Ibn al-Samḥ.[22]
As Julio Samsò and Honorino Mielgo [140, 292] point out, in using the ellipse Azarquiel | had no 419
"theoretical pretentions," and indeed they establish that none of the writers of Spanish or northwest African *zij*es used the elliptical approximation as a basis for the difficult task of computing the positions of Mercury.

Another medieval application of conic sections, which has been published in Rashed [126],[23] is Abū Saʿd al-ʿAlā ibn Sahl's application of hyperboloids of revolution in the 10th century to design (1) a plano-convex lens that would focus an incoming bundle of parallel light rays to cause burning at a given distance and (2) a biconvex lens that would focus a pencil of rays emanating from a point source at a finite distance.

Other evidence of medieval Islamic investigations of the hyperbola is al-Kūhī's discovery of the focus-directrix property of that curve [57].[24] He used this property for the more difficult of the two cases of inscribing an equilateral pentagon in a given square, which is equivalent to constructing a solution of a fourth-degree equation.

A more recent publication of one of al-Kūhī's works that makes essential use of conic sections is Philippe Abgrall's publication [2] of a treatise in which al-Kūhī uses the method of analysis and conic sections to find on a line given in position the center of a circle tangent to two objects, each of which may be a point,[25] a line, or a circle.

A fourth Hellenistic writer (other than Euclid, Archimedes, and Apollonius) whose works are of first importance for the history of the mathematical sciences is Ptolemy, and of his works the most influential in Islamic science was the *Almagest*. During the past decade Paul Kunitzsch has followed up his history of the Arabic–Latin tradition of the *Almagest* in medieval Islam [101] with a study [117] of the fate of the catalog of 1025 stars, which forms VII, 5, and VIII, 1, of that work, during medieval times. On a more popular level Kunitzsch and Smart [102] provide reliable accounts of the derivations of the names of 254 of the brightest stars.

[22] Both treatises concern the equatorium, a medieval instrument for analogue computation of the position of the planets. The fact that the only known application of conics prior to the time of Kepler occurs in a treatise on an astronomical instrument is just one instance of many that could be adduced of the importance of the history of instruments for the history of mathematics.

[23] This was discussed earlier in Rashed [122].

[24] This property was known to Pappus and, doubtless, to other Greek geometers, but there is no evidence that the Islamic mathematicians learned this from the Greeks.

[25] A circle is, of course, tangent to a point when it passes through the point.

Projections of the Sphere in Medieval Islam

One of the most ancient treatments of the projection of a sphere on a plane was that of Ptolemy in *The Analemma*, where he laid out the mathematical apparatus for a theoretical consideration of the design of sundials. In the horizontal sundial with a vertical gnomon the celestial sphere is projected onto the plane of the horizon, and in the medieval Islamic world the question arose among specialists in gnomonics whether the hour lines are straight or not. In a brief but informative paper, Hogendijk [68] summarizes the history of the problem and the contribution of Alhazen, who was the first to show that all hour lines except the noon line on the sundial differ from straight lines.

Another work of considerable consequence for Islamic civilization was Ptolemy's *Planisphaerium*,
420 and the notes on this treatise by an 11th-century, Spanish astrono|mer, Maslama of Madrid, together with some related texts have been published and studied by Kunitzsch and Lorch [105]. One year later, Lorch [111] contributed an account of material found in Maslama's notes that one may interpret as being a proof of the circle-preserving property of stereographic projection, a proof which he suggests may antedate what has been considered as the oldest proof of that result, that of al-Farghānī.

The instrument derived from Ptolemy's *Planisphaerium* is the astrolabe,[26] and al-Kūhī's work *On the Construction of the Astrolabe with Proof* has been published both by Berggren [17] and by Rashed [126]. Al-Kūhī begins with a comprehensive, theoretical discussion of the possible projections of the sphere onto the plane and the kinds of curves one obtains by such projections. After giving a general proof that stereographic projection maps circles onto circles or straight lines, he devotes the bulk of the work to solving problems in which certain parts of an astrolabe have been, as it were, effaced and one is required to reconstruct the whole from what remains.[27]

This treatise shows the importance of studying a person's work in its entirety before coming to any firm conclusions on it. For example, in his work *On the Division of a Line According to the Ratio of Areas* al-Kūhī poses the following problem: Given four points A, G, D, and B on a given line segment AB, divide GD at E so that $AE \cdot ED/GE \cdot EB$ is equal to a known ratio. Had we not known of his treatise on the astrolabe we would hardly have suspected that a problem forming the subject of Apollonius's *On the Determinate Section* would play a role in the problem of constructing an astrolabe given a circle parallel to a horizon of known latitude and the image of a point of known declination.

One reason that astrolabes are important in the history of mathematics is that they provided a ready challenge to geometers to find new methods of representing spherical arcs and circles on the plane. One of these circles, of course, is the ecliptic which, as the annual course of the sun, is fundamental for the astrolabe, and Christopher Anagnostakis [5] has studied some methods of representing it. Among the most difficult of the curves to draw on the astrolabe, however, are ones that medieval Islam contributed to that ancient instrument, the azimuth circles. Berggren [13] has given a survey and mathematical analysis of 10 medieval Islamic geometric methods for drawing these curves on the astrolabe. The study shows, among other results, that al-Kūhī's method for drawing azimuth circles may well have been discovered by him. In a subsequent study, Berggren [14] argues that Ḥabash al-Ḥāsib's method for drawing the azimuth circles was originally based on a geometric method, popular in both Greek and
421 Islamic mathematics, known as an analemma. |

The Muslim geometers realized that one could exploit the symmetry resulting from stereographic projection to replace redundant parts from projection from the north pole with necessary parts for projection from the south pole. This gave rise to a variety of "mixed" astrolabes, which the geometers were as ingenious in naming as they were in designing. One wanting a guide through the maze of terms ("the myrtle-form," "the crab-form," etc.) can consult the nicely illustrated, short piece by Lorch [110].

Another mapping of the sphere that stimulated considerable discussion in the 10th century is one that al-Bīrūnī in his *Treatise on Projection of the Constellations* [8, 52] referred to as producing a "melon-

[26] Several papers not bearing directly on mathematical issues, but highly informative about the history of the astrolabe both in medieval Islam and the West, may be found in a collection of some of Kunitzsch's papers [103].

[27] The tradition of such problems stems from Ptolemy's *Planisphaerium*, where he solves the problem of finding the radius of the equator given the greatest circle in the plane (typically that representing the Tropic of Capricorn) and the distance from the south pole of the circle corresponding to it on the sphere.

shaped" astrolabe, and of which he said that eminent astronomers were counted both among the supporters and the detractors. The name "melon-shaped" seemed so strange that some doubted the validity of the reading of the Arabic, but Kennedy and Lorch [83] describe the mathematics behind this instrument as it is found in a treatise by Ḥabash, from which it emerges that the melon-shaped astrolabe is based on the polar azimuthal projection, which preserves distances between circles parallel to the equator and distances along the meridians. The great difficulty with the projection, which Ḥabash solved, is drawing the horizon and its almucantars and azimuths. Evidently, the astrolabe was called "melon-shaped" because the image of the horizon resembles a melon.

A third publication of an Arabic treatise on projections is Dallal's work [35] devoted to al-Bīrūnī's treatise, *Book of Pearls on the Projection of Spheres*. This work is devoted to the construction and use of the planispheric astrolabe for astrology and concludes with a discussion of such unusual types of astrolabes as "the boat astrolabe" and "the spiral astrolabe."

A more direct method of modeling the heavens — for purposes of visualization, instruction, and solving problems — is by means of a sphere. Such a sphere is marked with the principal celestial circles and (usually) some of the principal stars and/or constellations, and the surviving examples of such instruments have been studied carefully by Emilie Savage-Smith [141]. Such a sphere could rotate about its poles within a frame of two perpendicular circles, which represent the horizon and meridian of an observer. Lorch and Kunitzsch [107] have published both the text and translation of Ḥabash al-Ḥāsib's treatise on this instrument.

Important work on the Spanish and Northwest African tradition of astrolabes and other instruments has been done by the Barcelona school, under the direction of Samsò, whose students have published a series of works on the subject. Although much of their work has been published in Catalan, for which reason we do not cite it here, two recent English pieces are by Emilia Calvo [25; 26], on the construction and use of the universal plate of Ibn Bāṣo, who died in Granada in 1316. (The epithet "universal" refers to the fact that, unlike an ordinary astrolabe, Ibn Bāṣo's instrument can be used at all latitudes.) This ingenious device combines stereographic projections from two different points and gives the same set of curves different interpretations according to the problem to be solved. (Calvo has pointed out that the plate was rediscovered by Latin astronomers.) One also notes two | papers by Roser Puig [118; 119], 422
in the first of which she argues that the plate of the shakkāziyya astrolabe, one of the two universal astrolabes known from 11th-century Spain, was conceived as a simplification of the universal astrolabe invented by the Spanish astronomer Azarquiel, whose equatorium for Mercury we mentioned earlier. In the other, she studies the mathematics of a device found on the back of Azarquiel's astrolabe, where one applies an Indian technique to determine lunar parallax according to Ptolemy's theory.

Cartography and Geometry

In addition to their astronomical uses, projections were of cartographic interest, and one of the joys of the history of cartographic projections in Islam is that new material keeps turning up. A recent example is a *qibla*-finder which was sold at Sotheby's in London in 1989 and which consists of a projection of the medieval Islamic world onto a disk centered at Mecca. It is equipped with a rotatable ruler, and when the ruler is positioned on the square corresponding to a given city the *qibla* for that city may be read from the rim of the instrument and the distance to Mecca from the scale on the ruler.

The instrument, which was made in Isfahan, probably in the early 18th century, incorporates a projection which was thought to have been invented by the German historian of Islamic science, Carl Schoy, in the 1920s. But the instrument dates from 1710, and specialists believe that it is a descendant of a medieval prototype. Indeed, King [96, 8] argues that "al-Bīrūnī himself produced a Mecca-centered map something like the Isfahan world-map." Very much in the medieval tradition, the instrument maker represented the slightly curved meridians on either side of Mecca by straight lines. Had such an instrument not been found, no one would have suspected the existence of this ingenious cartographic solution to the problem of finding the *qibla*.

Most of the coordinates of localities on the *qibla*-finder come ultimately from a *Book of Longitudes and Latitudes of the Persians* (via the *zījes* of Naṣīr al-Dīn al-Ṭūsī and Ulugh Beg). A valuable source for these and other coordinates is E. S. and M. H. Kennedy's publication [79] of thousands of coordinates of localities as recorded in medieval sources. For years it existed only as computer printouts which one could inspect in Beirut, Providence, or New Haven, so its publication in any form would have been welcome. However, E. S. Kennedy's long experience with using the catalog prompted four different sortings of the data (by locality, by source, by increasing longitude, and by increasing latitude), so the wait has resulted in a volume of much-increased usefulness.

An interesting mathematical problem arising from the ancient geographical doctrine of climata is the subject of Dallal [34]. The climata of classical geography were belts of the Earth's surface parallel to the equator determined by the maximum length of daylight, and in V, 9 of his *zīj*, the *Masʿūdīc Canon*, al-Bīrūnī calculates their areas. In his study of this treatise, Dallal argues that al-Bīrūnī used Archimedes' exact result for the calculation of areas of spherical segments — which leads to the expression $\pi \cdot$
423 $[(180 \cdot 56\ 2/3)/\pi]^2(\sin\varphi' - \sin\varphi)$ for the area of the belt bounded | by latitudes φ and φ' — and the
approximation of 22/7 for π. This same problem appears in a different context, with a different solution, as we shall see below.

Optics in Medieval Islam

In addition to cartography, another science of considerable interest to medieval Muslim scientists was optics, and several recent publications have considerably advanced our knowledge of medieval Islamic achievements in this area. Foremost among these is Abdulhamid Sabra's [133], an English translation and study of Alhazen's *Optics*, I–III, which treats the subject of direct vision.[28] (Sabra's translation is based on his earlier publication [55] of the Arabic text.) As an attempt "to examine afresh, and in a systematic manner, the entire science of vision and to place it on new foundations" Alhazen's work synthesizes physics and mathematics in a way that was radically new.[29]

An earlier study of Alhazen's optical writings is Sabra's discussion [132], which deals with his treatise, *On Seeing the Stars*, in the context of other writings by that author and Ptolemy. In this treatise Alhazen attempts to reconcile Ptolemy's treatment of what has come to be called the moon illusion[30] in his *Almagest* with that presented in his *Optics*. Alhazen shows mathematically that, on the basis of the principle of refraction and the assumption that there is a layer of air above the denser, vaporous atmosphere surrounding us but below the rarer aether, it is possible that the celestial luminaries will appear larger to our eyes than they ought.[31] Sabra has recently carried further his study of Alhazen's optical writings concerned with astronomical phenomena with [173], an edition, translation, and study of some of the questions raised in three sections (as Sabra calls them) of a collection of Alhazen's writings called *Solution of Doubts in the Book of the Almagest, Which a Certain Scholar Has Raised*.

Sabra's views that (1) Ptolemy's *Optics* was of limited influence (because it was largely unavailable) in medieval Islam and (2) Alhazen's "science of optics" was a science of vision and not a science of instruments designed to focus light, such as lenses or burning mirrors, have both been contested by, among others, Rashed, whose publication [126] makes available treatises dealing with, among other topics, mirrors and lenses that cause burning and optical magnification. One of the most interesting of these is a treatise by Abū Saʿd al-ʿAlā ibn Sahl,[32] which, according to Rashed, forces a radical revision of our view of medieval optics because it contains the first statement of the Law of Refraction (Snell's
424 Law). Sabra [134], among others, has dissented from this conclusion, but there is no doubt that it is |

[28] See the reviews by Berggren [18] and Kheirandish [84] (who raises some interesting historiographic issues).

[29] Sabra [133, II: liv].

[30] This refers to the apparently greater size of the full moon near the horizon than when it is high in the heavens.

[31] A critical edition of the text and an English translation of Ibn al-Haytham's *On Seeing the Stars* appears in Sabra and Heinen [131].

[32] See Rashed's earlier study of this treatise in [122].

an important text for a variety of reasons, including its mathematical treatment of the focal properties of lenses.[33]

Another point at issue between Sabra and Rashed is the strength of the Ptolemaic optical tradition in medieval Islam. But, whatever medieval Islam's acquaintance with Ptolemy's *Optics*, there is no doubt that Euclid's work of the same name had a strong Islamic tradition. In her study [85] of the problems of transmission of Euclidean optics, Elaheh Kheirandish has suggested that linguistic transformations, rather than conceptual innovations, either in the late antique versions of the *Optics* or in the Arabic translation, led to variants of the Euclidean optical theory such as that found in al-Kindī's *De aspectibus*, one which we find developed in Alhazen's *Optics*. She argues that the several possible meanings for many of the Arabic translations of terms that had come to have precise technical meanings in Greek may have provided parts of what one had come to think of as conceptual innovations in later medieval optics (in Arabic and Latin).

Kheirandish's forthcoming monograph[34] on the Arabic version of Euclid's *Optics* will, with Lorch's forthcoming study of the Arabic tradition of Theodosius's *Spherics*, contribute valuable material to the study of the history of what the Arabic writers called "the middle books" in medieval Islam, i.e., those books that were to be studied after one had mastered the *Elements* and before one began a study of the *Almagest*.[35]

Islamic Spain and Northwest Africa

Although Alhazen's *Optics* appears to have attracted little attention in the eastern part of the Islamic world prior to the work of al-Fārisī (whose work on number theory we discussed above), it appears that it was read in Islamic Spain (al-Andalus) soon after it was written, a fact which challenges the oft-repeated view that the western part of the Islamic world lagged scientifically behind the eastern part.[36] Indeed, one of the most interesting documents to come to light in the past decade has been that discovered by Djebbar and Hogendijk, a substantial part of the mathematical compendium, *Book of Perfection* (*Istikmāl*), written in the 11th century by Yūsuf al-Mu'taman ibn Hūd, a savant who ruled the Spanish kingdom of Saragossa from 1081 until his assassination in 1085. A striking feature of the structure of the treatise is its philosophically based classification of theorems according to the Aristotelian concepts of genera and species. From Hogendijk's study [65] of the geometrical parts, we learn that Islamic mathematicians were aware of Ceva's | theorem as well as what we would now call 425
the projective invariance of the cross ratio. Djebbar's study of the arithmetical parts will appear soon.[37]

In his earlier study of algebra in the western part of the Islamic world, Djebbar [42] makes clear the considerable influence of the school of the algebraist of the eastern part, al-Karajī, who wrote in the latter part of the 10th century. He also makes some interesting conjectures about the sociopolitical reasons for the almost total lack of evidence of research in algebra in the western part of Islam prior to the 12th century.

The first part of another large mathematical treatise from the Islamic west has recently come to light. Aballagh and Djebbar [1] announced the discovery of 117 folios of the book by the (probably) 12th-century mathematician al-Ḥaṣṣār titled *The Complete [Book] on the Art of Number*.

Al-Ḥaṣṣār, Ibn al-Yāsamīn, Ibn Mun'im, and Ibn al-Bannā' are the mathematicians whose work is

---

33 A clear account of what Ibn Sahl accomplished, along with some interesting speculations on how he might have discovered his result (and other matters of interest), may be found in Hogendijk's forthcoming review in *Physis*.

34 This work is based on her Harvard thesis, *The Medieval Arabic Tradition of Euclid's Optika*, and will be published in [86].

35 An important work in this corpus, Euclid's *Phaenomena*, has recently appeared. See [166].

36 For this last point and for knowledge of the *Optics* in the eastern part of the Islamic world, see Sabra [133 II: lxiv]. For knowledge of at least some mathematical parts of the *Optics* in Islamic Spain, see Hogendijk [65, 220–222].

37 See Djebbar [44]. A complete text of al-Mu'taman's work has recently been found in a commentary by the mathematician Ibn Sartāq, his *K. al-Ikmāl*. Hogendijk and Djebbar plan a joint study of it.

highlighted in Djebbar's survey [43] of mathematical activity in northwest Africa from the 8th to the 16th centuries. The survey touches on arithmetic, combinatorics, and algebra as well as mathematics in the life of the medieval Muslim city; it emphasizes the impact of the teaching activity in mathematics, and it speculates on sociopolitical factors that may have influenced the growth and decline both of mathematical activities and of the reputations of certain mathematicians. The survey also provides a bibliographic entry to the work on the history of mathematics in northwest Africa by Djebbar's students over the past decade. The astronomical literature of that region is surveyed by King [91].

Djebbar's survey brings out the important role Islamic Spain played for the above region as a source of both texts and scholars. Further evidence of the important role Spain played in the history of mathematics comes from Lorch's publication [113] of the contents of those sections of the Sevillian astronomer Jābir ibn Aflaḥ's critique of the *Almagest* dealing with plane and spherical trigonometry. Particularly interesting are Jābir's different approaches to the exposition of plane and spherical trigonometry, the mystery of the sources of his work, and its considerable influence on the Latin West.

The Problem of Decline in The Islamic Mathematical Sciences

With al-Ḥaṣṣār and the other writers mentioned above we have entered what was at one time supposed to be the period of decline in Islamic science. However, recent research in the history of astronomy has shown the extent to which such a periodization of Islamic history does violence to historical reality. For over 40 years scholars have known that highly sophisticated alternatives to the Ptolemaic models for planetary motion were developed by the Damascus astronomer Ibn al-Shāṭir. More recent research has shown, however, that his work is simply one development in a tradition of reform of Ptolemaic
426 astronomy that began with Alhazen in the | 11th century, one whose ultimate goal was to produce planetary models that yielded predictions at least as good as Ptolemy's but, unlike his, did not require physical impossibilities, such as uniform motion of a sphere around a point other than its center. This reform movement developed principally at Marāgha in Iran under the patronage of Hūlāgū Khān, and the first serious models, which were proposed by al-ʿUrḍī in the mid-13th century, were developed later in that century by Naṣīr al-Dīn al-Ṭūsī and his student Quṭb al-Dīn al-Shirāzī. These non-Ptolemaic models continued to be explored through the 17th century. A primary source for these developments is al-Ṭūsī's exposition of cosmology, called the *Tadhkira*, in which work, recently translated and published by Ragep [120], al-Ṭūsī gives a detailed explanation of his modifications of Ptolemy's models.

The basic device that allowed this successful reform of astronomy was one that Kennedy has called the Ṭūsī couple: a linkage of two segments of equal and constant length, initially pointing in the same direction and rotating with constant angular velocities in the same plane. The angular velocity of the second segment relative to the first is twice that of the first around its center, and the effect is that the end of the second segment produces rectilinear motion. George Saliba and Kennedy [80] give an account of an extension of the device, also by al-Ṭūsī, which allowed it to operate not in a plane but on the surface of a sphere in order to model the planet's change in latitude during its course around the ecliptic.

Al-Ṭūsī confidently predicted that the spherical device worked just as the planar one did in changing rotating motion to rectilinear motion, but al-Shirāzī recognized that this was only approximately the case, and another of Naṣīr al-Dīn's students, al-Nīsābūrī, proved, using Menelaus's *Sphærica*, I, 11, that in fact one has what is now called hippopede motion (evidently first used in astronomy by Eudoxus of Knidus (400 B.C.) and recently studied by John North [115]). However, the width of the hippopede is very slight and all who wrote on it after al-Ṭūsī, right up through the early 16th century, assert that it is a good approximation to rectilinear motion.

Saliba has published a collection of his papers [137] treating many aspects of the whole reform movement, both the technical details of the theories and the wider issues of periodization in the history of science, the relation between theory and observation and the motivation for the reform. Of particular interest is the essential role that two pieces of mathematics, the Ṭūsī couple and ʿUrḍī's Lemma, played in the whole development, as well as the discussion of the mathematical equivalence of models developed by this school to those of Copernicus.

Further evidence of important later work in medieval Islamic science comes from one of the most brilliant periods of the whole enterprise in the late 14th and early 15th centuries. During this time the Timurid prince, Ulugh Beg in Samarqand, patronized a group of scholars, chief among them being the human microchip, Jamshīd al-Kāshī.

One of al-Kāshī's great achievements is the *Khāqānī Zīj*, based on an earlier *zīj*, the *Ilkhānī*, of Naṣīr al-Dīn al-Ṭūsī, and it is regrettable that this has not yet been published. Yet there have been a number of studies of various parts of it, among them Kennedy's study of its spherical trigonometry [78]. By
al-Kāshī's time, over | four centuries had passed since the contemporaries of al-Bīrūnī had worked out 427
the basic theorems of spherical trigonometry, such as the Sine Law and the spherical analog of the Pythagorean theorem: $\cos a \cdot \cos b = \cos c$. Despite this, however, al-Kāshī uses not these theorems but the theorem of the complete quadrilateral, known to Menelaus ca. A.D. 100. Nor was he alone in this. Al-Bīrūnī's older contemporary, al-Kūhī, wrote a separate treatise on the calculation of rising times by Menelaus's theorem, which he recognized as being "old-fashioned" but which he defended on other grounds, and Dallal [36, 151] calls attention to this feature of Alhazen's treatment of the problem of determining the *qibla* as well.

Perhaps one has here an example of different conceptions of what is elegant: ours might be to use the more recent theory whereas al-Kāshī, al-Kūhī, and Alhazen evidently preferred a unified approach involving repeated applications of a single theorem.

Dold-Samplonius [48] studies al-Kāshī's work on the measurement of *qubba*'s (domes) in IV, Ch. 9, of his *Calculator's Key*. According to Dold-Samplonius [49, 94], earlier writers on the subject, such as al-Būzjānī, had also treated the topic of *qubba*'s, but architecture had advanced sufficiently by al-Kāshī's time that the earlier treatments no longer sufficed, for the vertical cross sections were no longer simple segments of spheres or cones. They were, however, still constructed by rotating an arc of a circle about an axis, which allowed al-Kāshī to approximate their volumes and surface areas by cones and their frusta. For the surface areas he uses results from Archimedes' *Sphere and Cylinder*, I, to calculate a factor (1.775) by which one should multiply the square of the diameter of the base to obtain the surface area. In fact, Dold-Samplonius points out that modern methods produce a value of 1.784, so al-Kāshī's value is only about 0.5% too small.

In the case of the volumes, al-Kāshī recommends multiplying the cube of the diameter by 0.306, whereas modern methods yield the factor 0.313. Once again, and for the same reasons, al-Kāshī has an underestimate, but this time the error is on the order of 2.3%, still a very moderate error, as Dold-Samplonius points out, compared with earlier manuals on the subject which, for the much easier case of the hemispheric dome, erred by almost 18% in the volume.

Dold-Samplonius's [47] is a study of al-Kāshī's directions for measuring the surface of the *muqarnas*, a honeycomb of squinches that gave an aesthetic solution to the problem of putting a round dome on a square base that so strikes the visitor to Islamic shrines and other buildings.[38] Al-Kāshī approaches this seemingly complex structure by analyzing it into simple polygonal shapes and providing factors that may be used to calculate the area of those shapes from dimensions that one can measure.

Another study showing al-Kāshī's virtuosity as a calculator is Kennedy's study [178] of two methods
for calculating the equation of time,[39] those of Kūshyār ibn | Labbān and al-Kāshī. Although the part 428
dealing with Kūshyār must be modified according to the study of van Dalen [158], it is the study of al-Kāshī's method (in the Khāqānī *zīj*) that concerns us here. Al-Kāshī calculated for each integral value of the solar longitude λ the value of $\bar{\lambda}$, where $\bar{\lambda}$ satisfies the equation $\lambda = \bar{\lambda} + eq(\bar{\lambda} - \lambda_a)$, where λ_a is a constant and *eq* is a term correcting for the fact that the earth is not the center of the sun's orbit. He solves the above equation 360 times by an iterative method, and then calculates the value of the equation of time in each case. In every case but one, al-Kāshī's value is within one second of that calculated by a

[38] For a study of the construction of this feature of Islamic architecture by Mohammad al-Asad as well as a geometric analysis of many features of Islamic architectural decoration, see the recent work by Gülru Necipoğlu [172].

[39] This is the number of minutes that, depending on the time of the year, one must add or subtract to the time shown on an accurate sundial in order to obtain mean local time.

computer, and Kennedy supposes that the one exception is a scribal error rather than a computational slip.

Science and Society in Medieval Islam

A topic that cuts across the various branches of the mathematical sciences in medieval Islam is that of the interaction of these sciences with medieval Islamic society, an interaction that began with the formation of Islamic science. As we indicated earlier, an important part of this formation was what has been traditionally called the "reception" of foreign science — in this case principally Greek, but also Indian and Sassanid Persian — which several recent papers treat in different ways. Sabra [130] argues that the process must be seen not as a passive one of simple "reception" of ancient knowledge but, rather, as an active process of appropriation, in which Islamic scholars played a crucial role. Following on the appropriation, Sabra claims, came assimilation and naturalization of the formerly "foreign" sciences, and it is in this process, he argues, that we must search for answers to questions about the decline of science in medieval Islam. Berggren [15] adopts Sabra's concept of active recipients and explores how the Islamic features of the society affected both the acquisition and the subsequent development of the foreign sciences, including even such a basic science as arithmetic. Høyrup [72] emphasizes the unique character of the end result of this process and writes of the Islamic "miracle" of integrating "subscientific" mathematical traditions with the scientific traditions of Greek and Indian mathematics, an integration whose roots, he says, lay in the close connection of the religious and the secular in early Islamic fundamentalism.

The attitude of Muslim society, especially that of its religiously orthodox intellectuals, toward this foreign learning has been variously assessed since Goldziher's fundamental study of 1915 (translation in [167]), and Sabra discusses it at some length in his paper cited above. It also figures importantly in Sabra's study [135] of the development and role of an influential school of theology in the intellectual life of medieval Islam as reflected in the writings of, among others, Ibn Khaldūn and the 14th-century theologian al-Ījī. Particularly germane to the present essay are this school's views of scientific (in particular, astronomical) knowledge, for example, the question of the reality of the entities considered in mathematical astronomy. Consistently with his theological position, al-Ījī takes a fictionalist view of these entities and assures his readers [135, 37] that "... once you see these things [ecliptic, solsticial points, etc.] as mere imaginings which are more tenuous than a spider's web you will cease to be fright-
429 ened by the clanking noise of these words." | Brentjes [23] urges a re-examination of the question of the status of the foreign sciences in medieval Islam and offers, as a beginning, a study of al-Nuʿaimī's history of institutions of higher learning (*madrasas*) in Damascus between the 12th and 15th centuries. On the basis of this she argues that both the traditional picture of orthodox Islam's hostility to the ancient sciences and a belief in the exclusion of these sciences from the regular curriculum at the *madrasas* are mistaken.[40]

Another aspect of science in Islamic society is the existence of parallel traditions, the one "Islamic" and the other "foreign," of methods for attacking a range of problems with scientific overtones. Over the past 25 years King has explored the coexistence in medieval Islam of a strong tradition of a folk astronomy which employed no advanced mathematics and a mathematical astronomy stemming ultimately from foreign sources. As King points out in his introduction to a collection of his papers [180], the two traditions were virtually independent, and both offered methods to support three of the "five pillars of Islam" (prayer, fasting during the month of Ramadan, and the *ḥajj*). A broadly based survey of an important aspect of the folk tradition is in King [90], devoted to the use of shadows in time-keeping. Still to appear is King's *Survey of Astronomical Timekeeping in Medieval Islam*, I & II.[41]

In reckoning, too, a tradition of mental computation indigenous to many of the Islamic lands[42]

[40] On these points, see also Berggren [182, 314–318].

[41] To be published by Springer-Verlag. (Editors: The books were eventually published by Brill.)

[42] Because intermediate results in the mental calculations are stored by holding the fingers in certain positions, the system

coexisted with foreign systems stemming ultimately from ancient Mesopotamia and India. The book by Ulrich Rebstock [127] is a study of arithmetic in the life of Islamic society — as it was used by the merchants, functionaries, builders, and legal scholars — and it surveys the various ramifications of arithmetic in daily practice of a Muslim society on the basis of an extraordinary range of texts. Two other studies of the various arithmetic traditions in the Islamic world are Djebbar's study [41] of fractions in the Western part of the Islamic world, and Chemla's and Djebbar's comparisons [30] of Arabic, Chinese, and Indian treatments of fractions.

An interesting case of a science whose roots lay within Islamic society (e.g., in Arabic lexicography and poetry) is that of what we now call combinatorics but which, in medieval Islam, was considered simply one more branch of "reckoning" (*ḥisāb*). Djebbar [39] provides details on the many counting arguments, the progress from tabular enumeration to manipulation of rules, and the widened field of application of combinatorial methods one encounters in works such as those of the 13th-century writer Ibn Munʿim and of Ibn al-Bannāʾ in the 14th century. Despite this, however, he points out that combinatorics never became recognized as a branch of mathematics in its own right and asks, for example, whether the complete lack of mention of the work of Ibn Munʿim was in any way the result of the displeasure of an orthodox ideology which had come to power at a mathematician whose nonmathematical writings showed him to be a nonconformist. | 430

Another topic that Djebbar mentions as a field for combinatorial calculation was the theorem on the well-known figure from *Almagest*, I, 11, called "the secant figure," which asserted that for certain lines a, b, c, d, e, and f the ratio $a : b$ was equal to $c : d$ composed with $e : f$. The problem was to state and prove all 18 possible cases of this, and Sabine Koelblen [99; 100] discusses a variety of treatments of the problem in Arabic (and Latin) literature, from the separate geometric proofs of each case by Aḥmad ibn Yūsuf through the proof based on the idea of permutations by Thābit ibn Qurra to the elegant proof by Naṣīr al-Dīn al-Ṭūsī in his *Treatise on the Complete Quadrilateral.* In [99] Koelblen deals principally with the history of the problem from the point of view of combinatorics and finds in the texts she studies two different approaches to counting the number of cases. In [100] she studies the different modes of demonstration in the texts discussed in [99] and concludes with some well-taken warnings against seeing medieval work on composition of ratios as a step on the road to the conception of ratios as real numbers.

Mathematical Methods in the Study of Ancient and Medieval Sciences

In recent years a variety of techniques, combining mathematical, historical, and textual methods, have been developed for analyzing tables found in the *zīj*es. A recent example of the fruits these methods, long known, can produce is the study by José Chabás and Bernard Goldstein [28] of tables for the solar equation and planetary latitudes in a Spanish *zīj*. An ingenious suggestion the authors make is that the reference in the *zīj* to "Ebi Iusufi byn Tarach" is to Yaʿqūb ibn Ṭāriq, on the basis (among other matters) that the Biblical Yaʿqūb (Jacob) was indeed the father of Joseph (Ebi Iusufi)!

Within the recent past, two studies of ancient astronomical tables using newly developed statistical methods have been published. The first to appear was that of Benno van Dalen who, in his thesis at Utrecht [157], used four statistical estimators and ingenious *ad hoc* methods to shed new light on the parameters underlying the many tables found in ancient astronomical handbooks. Such handbooks are notoriously difficult to understand fully, not least because often the values of the parameters are not given in the text or, when they are, they may not be the ones used in the computation of the tables.

Van Dalen [158] shows the power of such methods when he determines not only the tabulated function and the underlying parameter values but even the author of a table about which no textual information is available. He shows that one of the tables appended to a version of Kūshyār's *Jāmiʿ Zīj* found in Berlin was most likely computed according to a mathematical formula known as "the method

is also called "finger reckoning" in Islamic sources.

of declinations" by Yaḥyā ibn Abī Manṣūr, a contemporary of al-Khwārizmī and the author of the solar
431 longitude table found in the Ashrafī *zīj*.[43] |

Finally, Van Brummelen [155] addresses the problem of how a medieval author computed large (on the order of 5000 entries) tables of auxiliary functions — specifically, what rounding procedures were used, on what table a given table is based, and, when interpolation is used (as it often is), how does one identify both the grid of values that are computed according to the rule given and the interpolation method used for the others? Van Brummelen was able to discover a grid as well as a possible interpolation scheme for one of the tables and to show that a second table depends on it.

General Surveys

In addition to the specialized literature referred to above there are a number of publications, whose intended audience ranges from the interested layperson to the historian of science with a specialty far from medieval Islamic mathematics but who is interested in a general picture of the field.

One such book is Berggren [11], whose focus is the history of arithmetic, algebra, geometry, and trigonometry in that part of the medieval Islamic world between Cairo and Samarqand. It also takes into account Islamic dimensions of these developments–including inheritance problems, the payment of tax, astronomical timekeeping, and the direction of prayer.

Samsò's book [139] provides coverage of the western part of the Islamic world with a history of the "ancient sciences"[44] in Islamic Spain.[45] A work in the same general area is by Samsò and Juan Vernet [138], valuable both for its text and for its beautifully reproduced high-quality color illustrations. In this work the reader will also find pieces by Mercè Viladrich (for astrolabes), Roser Puig (for the so-called universal instruments),[46] Mercè Comes (for equatoria), and David King (for sundials).[47] This work provides, in fact, a general overview of science in Islamic Spain and, to mention only the parts most relevant to this paper, contains essays on Spanish mathematics (by Ahmed Djebbar), astronomy (by Julio Samsò), navigation (by Juan Vernet), and the *anwā'* literature (by Miquel Forcada).

Berggren [165] provides a brief overview of the Islamic acquisition and development of the sciences discussed in this paper, as well as their transmission to the Latin West. The author stresses the technical depth of Islamic achievements because "it is too often emphasized that the primary interest of medieval
432 Islamic society in mathematics was due to the subject's utility."[48] |

An extensive survey of Islamic mathematics during its first three centuries is found in Sesiano's account [148], noteworthy for the number of specimens of actual mathematics included and for its discussion of recreational mathematics and magic squares. (In the same volume is Kunitzsch's survey of astronomy from the 8th to the 10th centuries [104], aimed at a general scholarly readership.) Another brief survey of pure mathematics in medieval Islam is that of Hogendijk [69].

More specialized is the volume of studies [163] and the papers therein on astronomy (King [92]), astrology (Pingree), geography (Hopkins), and al-Bīrūnī (Saliba). Mathematics proper is treated rather too briefly, with a bare five pages in a paper (Hill) whose focus is mechanical technology.

[43] In his study [70] of the *qibla* table of this *zīj*, Hogendijk shows that, despite what had been previously thought, the table was not carelessly computed from an approximate formula but that parts of it were computed according to an exact formula, and there is evidence that the table was copied from an earlier source.

[44] By this term the Arabic writers denoted the sciences they had first learned from earlier civilizations, principally the Greek.

[45] Those who have no Spanish will find the leading ideas in this work in the paper, "Andalusian Astronomy: Its Main Characteristics and Influence in the Latin West," in Samsò [179, 1–23].

[46] The term "universal instrument" refers to an instrument that solves a certain set of problems for all latitudes.

[47] King is currently directing a survey of all known Islamic astrolabes and sundials. A recent report on this project is his [93].

[48] Berggren [165, 141]. Indeed, elsewhere [164] he gives examples to show that many medieval Islamic mathematicians were concerned not only with highly technical mathematics but with the subtle questions of what constituted a proof and how the demands for rigor squared with those for clarity.

An expert history of premodern Islamic cartography occupies about half of John Harley and David Woodward's volume [54] which contains informative and authoritative treatments of, among much else of interest, Emilie Savage-Smith on celestial mapping [54, 12–70], Gerald Tibbetts on geographical mapping [54, 90–155], S. Maqbul Ahmad on Idrīsī [54, 156–174], Raymond Mercier on geodesy [54, 175–188], and David King and Richard Lorch on *qibla* charts, maps, and related instruments [54, 189–205].

Some Conclusions

The history of medieval Islamic mathematics and her sister sciences continues to be an active area of contemporary scholarship, one that regularly produces both significant new discoveries — be they new map projections, numeration systems, or planetary models — and lively debate over their interpretation. Moreover, these discoveries are as likely to come from the study of scientific instruments and astronomical or geographical tables as from treatises on number theory or geometry. Finally, one trusts that "the best is yet to be." Al-Kūhī was regarded by his contemporaries as the best geometer of his age, but less than half of his treatises have been translated or properly studied. His successor, al-Bīrūnī, was one of the great scientists of any age, but neither his *Al-Qānūn al-Mas'ūdī* nor his great treatise on the astrolabe has been translated into a European language other than Russian. And, to give only one other author among those we have mentioned, the same may be said of al-Kāshī's most famous works — his *Calculator's Key* and his Khāqānī *Zīj* (of which no translation has been published). Clearly the field still offers an abundance of material for further investigation, and there is much work yet to be done.

Acknowledgements This paper is an expanded version of a talk delivered at a conference at York University, April 24, 1995 in honor of Professor Hardy Grant on the occasion of his retirement. The author thanks the organizers of the conference, and in particular Professor Israel Kleiner, for inviting him to participate. He also thanks Sonja Brentjes and Ahmed Djebbar for comments on an earlier version of the paper, Ms. Nathalie Sinclair for her valuable assistance in locating and checking bibliographic citations and in preparing the final draft for submission, and the referees of the paper for their careful reading. He also thanks Jan Hogendijk both for comments on an earlier version and for his careful editorial work on the present version. | 433

References

1. Aballagh, M., Djebbar, A., 1987. Découverte d'un écrit mathématique d'al-Ḥaṣṣār (XIIe S.) : Le livre I du Kāmil. Historia Mathematica 14, 147–158.
2. Abgrall, P., 1995. Les cercles tangents d'al-Qūhī. Arabic Sciences and Philosophy 5, 263–295.
3. Ağargün, A.G., Fletcher, C.R., 1994. Al-Fārisī and the Fundamental Theorem of Arithmetic. Historia Mathematica 21, 162–173.
4. Allard, A., 1991. The Arabic origins and development of Latin algorithms in the twelfth century. Arabic Sciences and Philosophy 1, 233–284.
5. Anagnostakis, C., 1987. How to divide the ecliptic on an astrolabe. In: [12, pp. 133–144].
6. Bebbouchi, R., 1990. L'infini et les mathématiciens arabes. Deuxième colloque maghrébin sur l'histoire des mathématiques arabes : Tunis, les 1-2-3 Décembre 1988, Actes du colloque. University of Tunis, Tunis, pp. 20–26.
7. Bellosta, H., 1991. Ibrāhīm ibn Sinān: On Analysis and Synthesis. Arabic Sciences and Philosophy 1, 211–232.
8. Berggren, J.L., 1982. Al-Bīrūnī on plane maps of the sphere. Journal for the History of Arabic Science 6, 47–96.

9. ——— 1985. The origin of al-Bīrūnī's "method of the zījes" in the theory of sundials. Centaurus 28, 1–16.
10. ——— 1985. History of mathematics in the Islamic world: The present state of the art. Middle East Studies Association Bulletin 19, 9–33.
11. ——— 1986. Episodes in the Mathematics of Medieval Islam. Springer-Verlag, New York.
12. Berggren, J.L., Goldstein, B.R. (eds.), 1987. From Ancient Omens to Statistical Mechanics: Essays in the Exact Sciences Presented to Asger Aaboe. Munksgaard, Copenhagen.
13. Berggren, J.L., 1991. Medieval Islamic methods for drawing azimuth circles on the astrolabe. Centaurus 34, 309–344.
14. ——— 1991–1992. Ḥabash's analemma for representing azimuth circles on the Astrolabe. Zeitschrift für Geschichte der arabisch-islamischen Wissenschaften 7, 23–30.
15. ——— 1992. Greek and Islamic elements in Arabic mathematics. In: Mueller, I. (ed.), ΠΕΡΙ ΤΩΝ ΜΑΘΗΜΑΤΩΝ. Academic Print Publications, Edmonton, Alberta, pp. 195–217.
16. ——— 1987. Archimedes among the Ottomans. In: [12, pp. 101–109].
17. ——— 1994. Abū Sahl Al-Kūhī's Treatise on the Construction of the Astrolabe with Proof: Text, translation and commentary. Physis 31, 141–252.
18. ——— 1995. Review of The Optics of Ibn al-Haytham, Books I–III: On Direct Vision. Physis 32, 143–153.
19. ——— 1996. Al-Kūhī's "Filling a Lacuna in Book II of Archimedes" in the version of Naṣīr al-Dīn al-Ṭūsī. Centaurus 38, 140–207.
20. Brentjes, S., 1992. Der Thābit ibn Qurra zugeschriebene Zusatz I, 46 zu Euklid I, 46 in MS Leiden 399, 1. In: [37, pp. 91–120].
21. ——— 1993. Varianten einer-Ḥaǧǧāǧ-Version von Buch II der Elemente. In: [52, pp. 47–67].
22. ——— 1994. Textzeugen und Hypothesen zum arabischen Euklid in der Überlieferung von al-Ḥaǧǧāǧ ibn Yūsuf ibn Maṭar (zwischen 786 und 833). Archive for History of Exact Sciences 47, 53–92.
23. ——— 2007. Reflections on the role of the exact sciences in Islamic culture and education between the twelfth and the fifteenth centuries. In: Abattouy, M. (ed.), Études d'histoire des sciences arabes: textes réunis et présentés. Fondation du roi Andul-Aziz Al Saoud, Casablanca, pp. 15-33. (Editors: In the original paper, the details of this reference are given as "unpublished preprint." We have suppled the details.)
24. Butzer, P.L, Lohrmann, D. (eds.), 1993. Science in Western Civilization in Carolingian Times.
434 Basel, Birkhäuser. |
25. Calvo, E., 1992. Ibn Bāṣo's universal plate and its influence on European astronomy. Scientiarum Historia 18, 61–69.
26. ——— 1992–1994. On the construction of Ibn Bāṣo's universal astrolabe (14th c.) according to a Moroccan astronomer of the 18th century. Journal for the History of Arabic Science 10, 53–67.
27. Carandell, J., 1984. An analemma for the determination of the azimuth of the qibla in the Risālā fī ʿilm al-ẓilāl of Ibn al-Raqqām. Zeitschrift für Geschichte der arabisch-islamischen Wissenschaften 1, 61–72.
28. Chabás, J., Goldstein, B.R., 1994. Andalusian astronomy: al-Zīj al-Muqtabis of Ibn al-Kammād. Archive for History of Exact Sciences 48, 1–41.
29. Chemla, K., 1994. Similarities between Chinese and Arabic mathematical writings: (I) Root extraction. Arabic Sciences and Philosophy 4, 207–266.
30. Chemla, K., Djebbar, A., Mazars, G., 1992. Mondes arabe, chinois, indien : Quelques points communs dans le traitement des nombres fractionnaires. In: [175, pp. 263–276].
31. Comes, M., 1991. Ecuatorios andalusies : Ibn al-Samḥ, al-Zarqālluh y Abū-l-Ṣalt. Universidad de Barcelona, Barcelona.
32. Crossley, J.N., Henry, A.S., 1990. Thus spake al-Khwārizmī: A translation of the Cambridge University Library ms. Ii.vi.5. Historia Mathematica 17, 103–131.
33. Crozet, P., 1993. L'idée de dimension chez al-Sijzī. Arabic Sciences and Philosophy 3, 251–286.
34. Dallal, A., 1984. Al-Bīrūnī on climates. Archives internationales d'histoire des sciences 34, 3–18.

35. ——— 1988. Bīrūnī's Book of Pearls Concerning the Projection of Spheres. Zeitschrift für Geschichte der arabisch-islamischen Wissenschaften 4, 81–138.
36. ——— 1995. Ibn al-Haytham's universal solution for finding the direction of the qibla by calculation. Arabic Sciences and Philosophy 5, 145–194.
37. Demidov, S.S., Folkerts, M., Rowe, D.E., Scriba, C.J. (eds.), 1992. Amphora, Festschrift for Hans Wussing on the Occasion of his 65th Birthday. Birkhäuser, Basel.
38. Dhanani, A., 1994. The physical theory of kalam: Atoms, space, and void in Basrian Muctazili Cosmology. Brill, Leiden and New York.
39. Djebbar, A., 1981. L'analyse combinatoire dans l'enseignement d'Ibn Munʿim, XIIe– XIIIe siècles. Université de Paris-Sud, Département de mathématiques.
40. ——— 1996. Quelques commentaires sur les versions arabes des Eléments d'Euclide et sur leur transmission à l'Occident musulman. In: [176, 91–114].
41. ——— 1995. Le traitement des fractions dans la tradition mathématique arabe du Maghreb. In: [175, pp. 263–276].
42. ——— 1988. Quelques aspects de l'algèbre dans la tradition mathématique arabe de l'Occident musulman. Actes du premier colloque Maghrébin d'Alger sur les mathématiques arabes. Maison du Livre, Alger, pp. 99–123.
43. ——— 1995. Mathematics in Medieval Maghreb. AMUCHMA Newsletter 15, 3–42. [Published by the African Mathematical Union]
44. ——— 1998. La tradition arithmétique euclidienne dans le *Kitāb Al-Istikmāl* d'Al-Mu'taman et son prolongement en Andalus. Veme Colloque Maghrebim sur l'histoire des mathématiques arabes : Actes du colloque, Hammamet 1-2-3 Décembre 1994. L'Association Tunisienne des Sciences Mathématiques, Tunis, pp. 62–84. (Editors: In the original paper, the date of this reference is given as "to appear." We have suppled the final details.)
45. Dold-Samplonius, Y., 1987. Developments in the solution to the equation $cx^2 + bx = a$ from al-Khwārizmī to Fibonacci. In: [88, pp. 71–87].
46. ——— 1990. Quadratic equations in Arab mathematics. In: Balmer, H., Glaus, B. (eds.), Die Blütezeit der arabischen Wissenschaft. Verlag der Fachvereine, Zürich, pp. 67–78.
47. ——— 1992. Practical Arabic mathematics: Measuring the muqarnas by al-Kāshī. Centaurus 35, 193–242. | 435
48. ——— 1992. The XVth century Timurid mathematician Ghiyāth al-Dīn Jamshīd al-Kāshī and his computation of the qubba. In: [37, pp. 171–181].
49. ——— 1993. The volume of domes in Arabic mathematics. In: [52, pp. 93–106].
50. Farès, N., 1995. Le calcul du maximum et la 'dérivée' selon Sharaf al-Dīn al-Ṭūsī. Arabic Sciences and Philosophy 5, 219–238.
51. Folkerts, M., 1997. Die älteste lateinische Schrift über das indische Rechnen nach al-Hwārizmī: Edition, Übersetzung und Kommentar. Bayerische Akademie der Wissenschaften, Philosophisch-historische Klasse, Abhandlungen (Neue Folge) Heft 113, Verlag der Bayerischen Akademie der Wissenschaften, München.
52. Folkerts, M., Hogendijk, J.P. (eds.), 1993. Vestigia Mathematica: Studies in Medieval and Early Modern Mathematics in Honour of H.L.L. Busard. Rodopi, Amsterdam.
53. Hamadanizadeh, J., 1987. A survey of medieval Islamic interpolation schemes. In: [88, pp. 143–152].
54. Harley, J.B., Woodward, D. (eds.), 1992. The History of Cartography, vol. 2, book 1: Cartography in the Traditional Islamic South Asian Societies. University of Chicago Press, Chicago.
55. Ibn al-Haytham, Abū ʿAlī al-Ḥasan ibn al-Ḥasan, 1983. Kitāb al-Manāẓir, Books I–III On Direct Vision, Sabra, A. I. (ed.). National Council for Culture, Arts and Letters, Kuwait.
56. Herz-Fischler, R., 1988. Theorem xiv,** of the First "Supplement" to The Elements. Archives internationales d'histoire des sciences 38, 3–66.
57. Hogendijk, J.P., 1985. Al-Kūhī's construction of an equilateral pentagon in a given square. Zeitschrift für Geschichte der arabisch-islamischen Wissenschaften 1, 100–144.
58. ——— 1985. Thābit ibn Qurra and the pair of amicable numbers 17296, 18416. Historia Mathe-

matica 12, 269–273.
59. —— 1986. Arabic traces of lost works of Apollonius. Archive for History of Exact Sciences 35, 187–253.
60. —— 1987. On Euclid's lost Porisms and its arabic traces. Bollettino di storia delle scienze matematiche 7, 93–115.
61. —— 1987. Review of La théorie des parallèles en pays d'Islam. Centaurus 30, 293–294.
62. —— 1987. New light on the lunar crescent visibility table of Yaʿqūb Ibn Ṭāriq. Journal of Near Eastern Studies 47, 95–103.
63. —— 1989. The mathematical structure of two Islamic astronomical tables for "Casting the Rays." Centaurus 32, 171–202.
64. —— 1989. Sharaf al-Dīn al-Ṭūsī on the number of positive roots of cubic equations. Historia Mathematica 16, 69–85.
65. —— 1991. The geometrical parts of the Istikmāl of Yūsuf Al-Muʾtaman Ibn Hūd (11th century): An analytical table of contents. Archives internationales d'histoire des sciences 41, 207–281.
66. —— 1991. Al-Khwārizmī's table of the "sine of the hours" and the underlying sine table. Historia Scientiarum 42, 1–12.
67. —— 1993. The Arabic version of Euclid's On Division. In: [52, pp. 143–162].
68. —— 1994. Le traité d'Ibn al-Haytham sur les lignes horaires. Cahier du seminaire Ibn al-Haytham 4, 5–7.
69. —— 1994. Pure mathematics in Islamic civilization. In: Grattan-Guinness, I., (ed.), Companion Encyclopaedia of the History and Philosophy of the Mathematical Sciences. Routledge, London and New York, pp. 70–79.
70. —— 1994. The qibla table in the Ashrafī Zīj. In: [159, pp. 81–96].
71. Høyrup, J., 1986. Al-Khwārizmī, Ibn Turk, and the Liber Mensurationum: On the origins of
436 Islamic algebra. Erdem 5, 445–484. |
72. —— 1987. The formation of "Islamic mathematics": Sources and conditions. Science in Context 1, 281–329.
73. Houzel, C., 1995. Sharaf al-Dīn al-Ṭūsī et le polygone de Newton. Arabic Sciences and Philosophy 5, 239–262.
74. Hughes, B., 1994. Problem-solving by Ajjūb al-Baṣrī: An early algebraist. Journal for the History of Arabic Sciences 10, 31–40.
75. Thābit Ibn Qurrā, 1987. Oeuvres astronomiques. (Régis, M., éd., trad. et comm.) Les Belles Lettres, Paris.
76. Jaouiche, K., 1986. La théorie des parallèles en pays d'Islam. Vrin, Paris.
77. —— 1988. L'analyse et la synthèse dans les mathématiques arabo-islamiques : Le livre d'Ibn al-Haytham. Histoire des mathématiques arabes. Maison des Livres, Algiers, pp. 106–124.
78. Kennedy, E.S., 1985. Spherical astronomy in Kāshī's Khāqānī Zīj. Zeitschrift für Geschichte der arabisch-islamischen Wissenschaften 2, 1–46.
79. Kennedy, E.S., Kennedy, M.H., 1987. Geographical Coordinates of Localities from Islamic Sources. Institut für Geschichte der arabisch-islamischen Wissenschaften, Frankfurt am Main.
80. Kennedy E.S., Saliba, G., 1991. The spherical case of the Ṭūsī couple. Arabic Sciences and Philosophy 1, 285–291.
81. Kennedy, E.S., 1994. The prime vertical method for the astrological houses as presented in Kāshī's Khāqānī Zīj. In: [159, pp. 95–97].
82. —— 1994. Ibn Muʿādh on the astrological houses. Zeitschrift für Geschichte der arabisch-islamischen Wissenschaften 9, 153–160.
83. Kennedy, E.S., Lorch, R.P., 1997. Ḥabash al-Ḥāsib on the melon astrolabe. In: Anellis, I.H., Demidov, S.S., Drucker, T., Hogendijk, J.P. (eds.), Mathematics, Past Tense Present: Festschrift on the History of Mathematics in Honor of Boris Rosenfeld's 80th Birthday. Modern Logic Publishing, Ames, IA.
84. Kheirandish, E., 1994. Review of The Optics of Ibn al-Haytham, Books I–III: On Direct Vision. Harvard Middle Eastern and Islamic Review 1, 188–194.

85. —— 1996. The Arabic "version" of Euclidean Optics: Transformations as linguistic problems. In: [177, pp. 227–246].
86. —— 1998. The Arabic Tradition of Euclid's Optics (Kitāb Uqlīdis fī Ikhtilāf al-manāẓir). Springer-Verlag. New York. (Editors: In the original paper this item was "to appear." We have modified the reference to agree with the published book.)
87. King, D.A., 1986. The earliest Islamic mathematical methods and tables for finding the direction of Mecca. Zeitschrift für Geschichte der arabisch-islamischen Wissenschaften 3, 82–149.
88. King, D.A., Saliba, G. (eds.), 1987. From Deferent to Equant: A Volume of Studies in the History of Science in the Ancient and Medieval Near East in Honor of E. S. Kennedy. New York Academy of Sciences, New York.
89. King, D.A., 1988. A medieval Arabic report on algebra before al-Khwārizmī. Al-Masāq 1, 25–32.
90. —— 1990. A survey of medieval Islamic shadow schemes for simple time-reckoning. Oriens 32, 191–249.
91. —— 1990. An overview of the sources for the history of astronomy in the medieval Maghrib. Deuxième colloque maghrébin sur l'histoire des mathématiques arabes : Tunis, les 1-2-3 Décembre 1988, Actes du colloque. University of Tunis, Tunis, pp. 125–157.
92. —— 1990. Astronomy. In [163, pp. 274–289].
93. —— 1994. Astronomical Instruments between East and West, Kommunikation zwischen Orient und Okzident, Alltag und Sachkultur. Sitzungsberichte der Osterreichische Akademie der Wissenshaften, Philosophisch-Historische Klasse, 619. Band, Verlag der Akademie der Wisssenschaften, Wien, 143–198. | 437
94. —— 1992. The ciphers of the monks and the astrolabe of Berselius reconsidered. In: [37, pp. 375–388].
95. —— 1994. Ein vergessenes Zahlensystem des mittelalterlichen Mönchtums. In: [159, pp. 405–420].
96. —— 1994. World-maps for finding the direction and distance of Mecca: A brief account of recent research. Preprint of a paper presented at the Symposium on Science and Technology in the Turkish and Islamic World, Istanbul, June 3–5.
97. Knorr, W. R., 1989. Textual Studies in Ancient and Medieval Geometry. Birkhäuser, Boston.
98. —— 1996. The wrong text of Euclid: On Heiberg's text and its alternatives. Centaurus 38, 208–276.
99. Koelblen, S., 1993. Un exercice de combinatoire : Les relations issues de la figure sécante de Ptolémée des six quantités en proportion. Un parcours en histoire des mathématiques : Travaux et recherches, Sciences et Techniques en Perspective, 26, 1–21. Université de Nantes, Nantes.
100. —— 1994. Une pratique de la composition des raisons dans un exercice de combinatoire. Revue d'histoire des sciences 2, 209–247.
101. Kunitzsch, P., 1974. Der Almagest: Die Syntaxis Mathematica des Claudius Ptolemäus in arabisch-lateinischer Überlieferung. O. Harrassowitz, Wiesbaden.
102. Kunitzsch, P., Smart, T., 1986. Short Guide to Modern Star Names and Their Derivations. O. Harrassowitz, Wiesbaden.
103. Kunitzsch, P., 1989. The Arabs and the Stars: Texts and Traditions on the Fixed Stars, and Their Influence in Medieval Europe. Variorum, Northampton.
104. —— 1993. Arabische Astronomie im 8. bis 10. Jahrhundert, in [24, pp. 205–220].
105. Kunitzsch, P., Lorch, R., 1994. Maslama's Notes on Ptolemy's Planisphaerium and Related Texts. Sitzungsberichte der Bayerischen Akademie der Wissenschaften, Philosophisch-historische Klasse, Heft 2.
106. —— 1993. A note on codex Paris BN ar. 2457. Zeitschrift für Geschichte der arabisch-islamischen Wissenschaften 8, 235–240.
107. Lorch, R., Kunitzsch, P., 1985. Ḥabash al-Ḥāsib's Book on the Sphere and Its Use. Zeitschrift für Geschichte der arabisch-islamischen Wissenschaften 2, 68–98.
108. Lorch, R., 1986. Abū Jaʿfar al-Khāzin on isoperimetry and the Archimedean tradition. Zeitschrift für Geschichte der arabisch-islamischen Wissenschaften 3, 150–229.

109. ——— 1989. The Arabic transmission of Archimedes' Sphere and Cylinder and Eutocius' Commentary. Zeitschrift für Geschichte der arabisch-islamischen Wissenschaften 5, 94–114.
110. ——— 1994. Mischastrolabien im arabisch-islamischen Kulturgebiet. In: [159, pp. 231–236].
111. ——— 1995. Ptolemy and Maslama on the transformation of circles into circles in stereographic projection. Archive for History of Exact Sciences 49, 271–284.
112. ——— 1995. Arabic Mathematical Sciences: Instruments, Texts, Transmission. Variorum, Northampton.
113. ——— 1995. Jābir ibn Aflaḥ and the establishment of trigonometry in the west. Published as No. VIII in [112].
114. Morelon, R., 1994. Thābit ibn Qurra and Arab astronomy in the 9th century. Arabic Sciences and Philosophy 4, 111–139.
115. North, J.D., 1994. The Hippopede. In: [159, pp. 143–154].
116. Plooij, E.B., 1950. Euclid's Conception of Ratio and his Definition of Proportional Magnitudes as Criticized by Arabian Commentators. Ph.D. Thesis, University of Leiden. [Includes the text in facsimile with translation of the commentary on ratio of Abū ʿAbd Allāh Muḥammad ibn Muʿādh
438 al-Jayyānī]. |
117. Ptolemaeus, Claudius, 1986–1991. Der Sternkatalog des Almagest, die arabisch-mittelalterliche Tradition (Kunitzsch, P., hsg. und bearb.), Teil I: Die arabischen Übersetzungen; Teil II: Die lateinische Übersetzung Gerhards von Cremona; Teil III: Gesamtkonkordanz der Sternkoordinaten. O. Harrasowitz, Wiesbaden.
118. Puig, R., 1985. Concerning the Ṣafīḥa Shakkāziyya. Zeitschrift für Geschichte der arabisch-islamischen Wissenschaften 2, 123–139.
119. ——— 1989. Al-Zarqālluh's graphical method for finding lunar distances. Centaurus 32, 294–309.
120. Ragep, J., 1993. Naṣīr al-Dīn al-Ṭūsī's Memoir on Astronomy. Springer-Verlag, New York and Berlin.
121. Rashed, R., 1987. Al-Sijzī et Maimonide : Commentaire mathématique et philosophique de la proposition ii, 14 des Coniques d'Apollonius. Archives internationales d'histoire des sciences 37, 263–296.
122. ——— 1990. A Pioneer in anaclastics: Ibn Sahl on burning mirrors and lenses. Isis 81, 464–491.
123. ——— 1991. Al-Samawʾal, al-Bīrūnī et Brahmagupta : Les méthodes d'interpolation. Arabic Sciences and Philosophy 1, 101–160.
124. ——— 1992. Archimède dans les mathématiques arabes. In: Mueller, I. (ed.), ΠΕΡΙ ΤΩΝ ΜΑΘΗΜΑΤΩΝ. Academic Print Publications, Edmonton, Alberta, pp. 173–193.
125. ——— 1993. Al-Kindī's Commentary on Archimedes' The Measurement of a Circle. Arabic Sciences and Philosophy 3, 7–53.
126. ——— 1993. Géométrie et dioptrique au Xe siècle : Ibn Sahl, al-Qūhī et Ibn al-Haytham. Les Belles Lettres, Paris.
127. Rebstock, U., 1992. Rechnen im islamischen Orient: die literarischen Spuren der praktischen Rechenkunst. Wissenschaftliche Buchgesellschaft, Darmstadt.
128. Rosenfeld, B.A., 1988. A History of Non-Euclidean Geometry: Evolution of the Concept of a Geometric Space. Springer-Verlag, New York. Translated from the original Russian edition Istoriya Neevklidovoĭ Geometriĭ. Nauk, Moscow, 1976.
129. ——— 1993. Tashkent manuscripts on mathematical atomism. Studies in the History of Medicine and Science 12, 97–101.
130. Sabra, A.I., 1987. The appropriation and subsequent naturalization of Greek science in medieval Islam: A preliminary report. History of Science 25, 223–243.
131. Sabra, A.I., Heinen, A., 1987. On Seeing the Stars: Edition and translation of Ibn al-Haytham's Risāla fī Ru'yat al-kawākib. Zeitschrift für Geschichte der arabisch-islamischen Wissenschaften 7, 31–72.
132. Sabra, A.I., 1987. Psychology versus mathematics: Ptolemy and Alhazen on the moon illusion. In: Grant, E., Murdoch, J. (eds.), Mathematics and Its Applications to Science and Natural Philosophy in the Middle Ages: Essays in Honor of Marshall Clagett. Cambridge University Press,

Cambridge, pp. 217–247.
133. —— 1989. The Optics of Ibn al-Haytham, Books I–III: On Direct Vision, 2 Parts. The Warburg Institute, London.
134. —— 1994. Review of Géométrie et dioptrique au Xe siècle : Ibn Sahl, al-Qūhī et Ibn al-Haytham. Isis 85, 685–686.
135. —— 1994. Science and philosophy in medieval Islamic theology: The evidence of the fourteenth century. Zeitschrift für Geschichte der arabisch-islamischen Wissenschaften 9, 1–42.
136. Saliba, G., 1991. The astronomical tradition of Maragha: A historical survey and prospects for future research. Arabic Sciences and Philosophy 1, 67–100.
137. —— 1994. A History of Arabic Astronomy: Planetary Theories during the Golden Age of Islam. New York University Press, New York. | 439
138. Samsó, J., Vernet, J. (eds.), 1992. El legado cientifico andalusí. Museo Arqueológico Nacional, Madrid.
139. —— 1992. Las ciencias de los antiguos en al-Andalus. Mapfre, Madrid.
140. Samsó, J., Mielgo, H., 1994. Ibn al-Zarqālluh on Mercury. Journal for the History of Astronomy 25, 289–296.
141. Savage-Smith, E., 1985. Islamicate Celestial Globes: Their History, Construction, and Use. The Smithsonian Institution Press, Washington, DC.
142. Sesiano, J., 1985. Un système artificiel de numération au moyen ăge. In: Folkerts, M., Lindgren, U. (eds.), Mathemata: Festschrift für Helmuth Gericke. Franz Steiner Verlag Wiesbaden, Stuttgart, pp. 165–196.
143. —— 1987. Herstellungsverfahren magischer Quadrate aus islamischer Zeit (II). Sudhoffs Archiv 71, 78–87.
144. —— 1987. Survivance médiévale en Hispanie d'un problème né en Mésopotamie. Centaurus 30, 18–61.
145. —— 1987/88. Un complément de Ṯābit Ibn Qurra au Peri Diaireseon d'Euclide. Zeitschrift für Geschichte der arabisch-islamischen Wissenschaften 4, 149–159.
146. —— 1991. Un fragment attribué à Archimède. Museum Helveticum 48, 21–32.
147. —— 1991. An Arabic treatise on the construction of bordered magic squares. Historia scientiarum 42, 13–31.
148. —— 1993. Arabische Mathematik im 8.–10. Jahrhundert. In: [24, pp. 399–442].
149. —— 1994. Quelques méthodes arabes de construction des carrés magiques impairs. Bulletin de la Société vaudoise des sciences naturelles 83, 51–76.
150. —— 1995. Herstellungsverfahren magischer Quadrate aus islamischer Zeit (III). Sudhoffs Archiv 79, 193–226.
151. Toomer, G.J. (ed. and trans.), 1990. Apollonius Conics, Books V–VII, The Arabic Translation of the Lost Greek Original in the Version of the Banū Mūsā, 2 vols. Springer-Verlag, New York.
152. Sharaf al-Dīn al-Ṭūsī, 1986. Oeuvres mathématiques : Algèbre et géométrie au XIIe siècle. Rashed, R., (ed. and trans.). Les Belles Lettres, Paris.
153. Vahabzadeh, B., 1994. Two commentaries on Euclid's definition of proportional magnitudes. Arabic Sciences and Philosophy 4, 181–198.
154. Van Brummelen, G., 1991. The numerical structure of al-Khalīlī's auxiliary tables. Physis 28, 667–687.
155. —— 1994. A survey of the mathematical tables in Ptolemy's Almagest. In: [159, pp. 155–170].
156. —— 1994. From drudgery to invention: Astronomical computation in medieval Islam. Proceedings of the Canadian Society for the History and Philosophy of Mathematics 7, 178–189. (Editors: In the original paper, the details of this reference was given as "unpublished manuscript." We have provided the published details.)
157. van Dalen, B., 1993. Ancient and Mediaeval Astronomical Tables: Mathematical Structure and Parameter Values. Ph.D. Dissertation, Universiteit Utrecht.
158. —— 1994. A table for the true solar longitude in the Jāmiʿ Zīj. In: [159, 171–190].
159. von Gotstedter, A. (ed.), 1994. Ad Radices: Festband zum fünfzigjahrigen Bestehen des Instituts

für Geschichte der Naturwissenschaften der Johann Wolfgang Goethe-Universität Frankfurt am Main. Franz Steiner Verlag, Stuttgart.
160. De Young, G., 1984. The Arabic textual traditions of Euclid's Elements. Historia Mathematica 11, 147–160.
161. —— 1991. New traces of the lost al-Ḥajjāj Arabic translations of Euclid's Elements. Physis 38, 647–666.
162. —— 1992. Abū Sahl's additions to Book II of Euclid's Elements. Zeitschrift für Geschichte der
440 arabisch-islamischen Wissenschaften 7, 73–135. |
163. Young, M.J.L., Latham, J.D., Robert B.S. (eds.), 1990. Religion, Learning and Science in the ʿAbbasid Period. Cambridge University Press, Cambridge.

Supplementary Bibliography

Since submitting the paper we have become aware of other papers relevant to this article which ought to be listed in the bibliography. The size of the bibliography made renumbering impractical so we have listed them here and numbered them consecutively with the last entry of the above bibliography.

164. Berggren, J.L., 1990. Proof and pedagogy in medieval Islamic mathematics. Interchange 2, 36–48. [A special issue titled Creativity, Thought and Mathematical Proof, Hanna, G., Winchester, I. (eds.)]
165. —— 1994. Historical reflections on scientific knowledge: The case of medieval Islam. In: Hayhoe, R. (ed.), Knowledge Across Cultures: Universities East and West. OISE Press, Toronto, pp. 137–153.
166. Berggren, J.L., Thomas, R.S.D., 1996. Euclid's Phaenomena: A Translation and Study of a Hellenistic Treatise in Spherical Astronomy. Garland, New York.
167. Goldziher, I., 1981. The attitude of orthodox Islam toward the ancient sciences. In: Swartz, M., (trans. and ed.) Studies on Islam. Oxford University Press, New York and Oxford, pp. 185–215.
168. Hogendijk, J.P., 1988. Three Islamic lunar crescent visibility tables. Journal of the History of Astronomy 19, 29–44.
169. —— 1994. An Arabic text on the comparison of the five regular polyhedra, "Book XV" of the Revision of the Elements by Muḥyī al-Dīn al-Maghribī. Zeitschrift für Geschichte der arabisch-islamischen Wissenschaften 8, 133–233.
170. —— 1995. Mathematics in medieval Islamic Spain. In: Chatterji, S.D., (ed.), Proceedings of the International Congress of Mathematicians, August 3–11, Zurich. Birkhäuser, Basel, vol. 2, pp. 1568–1580.
171. Muḥammad ibn Mūsā al-Khwārizmī, 1992. Le calcul indien (algorismus), André Allard (ed. and trans.). Albert Blanchard, Paris.
172. Necipoğlu, G., 1995. The Topkapi Scroll: Geometry and Ornament in Islamic Architecture. The Getty Center for the History of Art and the Humanities, Santa Monica, CA.
173. Sabra, A.I., 1995/6. On Seeing the Stars, II: Ibn al-Haytham's "answers" to the "doubts" raised by Ibn Maʿdān. Zeitschrift für Geschichte der arabisch-islamischen Wissenschaften 10, 1–59.
174. Sesiano, J., 1996. Un traité médiéval sur les carrés magiques : De l'arrangement harmonieux des nombres. Presses Polytechniques et Universitaires Romandes, Lausanne.
175. Benoït, P., Chemla, K., Ritter, J., (eds.), 1992. Histoire de fractions, fractions d'histoire. Birkhäuser, Basel.
176. Folkerts, M., (ed.), 1996. Mathematische Probleme in Mittelalter: der lateinische und arabische Sprachbereich. Harrassowitz, Wiesbaden.
177. Ragep, F.J., Ragep, S.P., (eds.), 1996. Tradition, Transmission, Transformation, Proceedings of Two Conferences on Premodern Science Held at the University of Oklahoma. Brill, Leiden and New York.
178. Kennedy, E.S., 1988. Two medieval approaches to the equation of time. Centaurus 31, 1–8.

179. Samsó, J., 1994. Islamic Astronomy and Medieval Spain. Variorum, Aldershot.
180. King, D.A., 1993. Astronomy in the Service of Islam. Variorum, Aldershot.
181. Rashed, R., 1991. L'Analyse et la synthèse selon Ibn al-Haytham. In: Rashed, R. (ed.), Mathématiques et philosophie de l'antiquité à l'âge classique : Hommage à Jules Vuillemin. Editions du CNRS, Paris, pp. 131–162.
182. Berggren, J.L., 1992. Islamic acquisition of the foreign sciences: A cultural perspective. American Journal of Islamic Social Sciences 9, 310–324. Reprinted in [177, 263–284].

A Survey of Research in the Mathematical Sciences in Medieval Islam from 1996 to 2011

Glen Van Brummelen

Introduction

Len Berggren's two surveys of the mathematical sciences in medieval Islam, the two preceding articles in this volume, synthesized a vast body of literature. Their scope extended well beyond mathematics itself to include overlapping disciplines such as optics, geography and astronomy. In addition, the surveys recognized questions in the literature related to transmission, regional differences, and the issue of the perception of decline in later periods. In the fifteen years since the last survey in 1995, many of the same disciplines and debates continue to provoke interest, but much has changed. The field has been increasingly professionalized, partly through the birth of several new avenues of publication. Two new journals dedicated specifically to the history of mathematics and science in medieval Islam, *Suhayl* (operated at the University of Barcelona) and the Iranian *Tarikh-e Elm*, provide regular forums for the dissemination of research. *SCIAMVS (Sources and Commentaries in Exact Sciences)*, dedicated to all pre-modern cultures, has included within its pages a number of Arabic texts. Add to this the already-existing specialized journals (*Arabic Sciences and Philosophy* and *Zeitschrift für Geschichte der arabisch-islamischen Wissenschaften*), along with several other periodicals with broader interests, and we see that there has never been greater opportunity to explore issues related to Islamic science. As we shall see, attention to certain questions has waned, while new debates have arisen that question at a fundamental level why we do what we do, and what we should write. Given the increased volume of the literature and shifting academic ground since 1995, it is clear that a new survey is in order.

The mass of recent scholarly production that falls within our purview necessitates a few restrictions. Firstly, following Berggren's two surveys, we limit ourselves to research published at least partly in European languages, and do not include works published entirely in Russian, Arabic, or Persian. Secondly, we shall concentrate mostly on original research, referring only occasionally to survey papers and not to the many encyclopedia articles that have appeared since 1995. Four sources excluded by this restriction nevertheless should be brought to the reader's attention. These are (i) the massive 120-volume series *Islamic Mathematics and Astronomy* published by the Institut für Geschichte der arabisch-islamischen Wissenschaften in Frankfurt, which makes available again virtually all the literature in the field prior to 1960; (ii) the three-volume *Encyclopedia of the History of Arabic Science* [Rashed and Morelon 1996], whose 30 survey articles include many related to the mathematical sciences; (iii) Rosenfeld and İhsanoğlu's massive bio-bibliographic reference work *Mathematicians, Astronomers and Other Scholars of Islamic Civilisation and their Works (7th - 19th c.)* (Rosenfeld and İhsanoğlu [2003] and supplements in Rosenfeld and İhsanoğlu [2004] and Rosenfeld [2006]); and (iv) Berggren's collection of Arabic texts in Victor Katz's *The Mathematics of Egypt, Mesopotamia, China, India, and Islam: A Sourcebook* [Katz 2007]. Finally, in a survey of this magnitude it is inevitable that some sources will be omitted accidentally, for which the author can only apologize and ask for understanding. A supplement to this article will be maintained at https://pub.questu.ca/-gvb/islamsci.html, including all texts dating between Berggren's 1995 survey and the end of 2011, brought to my attention after this article's publication.

N. Sidoli and G. Van Brummelen (eds.), *From Alexandria, Through Baghdad: Surveys and Studies in the Ancient Greek and Medieval Islamic Mathematical Sciences in Honor of J.L. Berggren*, DOI 10.1007/978-3-642-36736-6_6,
© Springer-Verlag Berlin Heidelberg 2014

Over the past decades we have become more and more aware that Islamic mathematics should not be separated from surrounding subjects; mathematical creativity, then as now, occurred only occasionally within disciplinary silos. Topics such as geography, optics, astronomy and even astrology have been arenas for many mathematical innovations. However, if we were to cover thoroughly all these fields, this survey might easily reach the length of a monograph. Therefore, we shall attempt to restrict ourselves to episodes that produced mathematical innovations. In astronomy the dividing line is especially difficult to draw; we crave the reader's indulgence in our choices. Finally, we shall deal with issues of transmission, both into medieval Islam and within it (especially between east and west). But generally we exclude questions of transmission from Islam to Europe. As interesting and important as these questions are, we prefer here to deal with medieval Islamic scientific culture on its own terms rather than on those of a receiving culture.

The Foreign Legacy: Reception, Appropriation, Transformation

Perhaps the single greatest point of contention among scholars of the history of the Islamic mathematical sciences in recent decades has been the portrayal of its interactions with its Greek and Indian predecessors. The extreme positions in the debate, taken by almost no one, are (a) that Islamic scientists simply inherited and extended the work of their ancestors without introducing anything new of significance; and (b) that they created out of their inheritance an entirely new science, bearing little resemblance to anything that went before and establishing paradigms for the West to follow. Although both Greece and India made significant impacts on Islamic science (see Plofker [2002] for one of many examples of this, tracing Indian influence on Islamic mathematical astronomy through iterative approximations), the past fifteen years of research have emphasized a careful tracing of the reception of classical Greek texts. The choice of words used to describe the interactions can color the debate dramatically. Sabra ([Sabra 1996a], originally published 1987) argues that one should speak not of a passive "reception" of Greek science, but rather its "appropriation" and "naturalization" into an Islamic context that inevitably restructures its content and goals. In an article reprinted in the same volume, Berggren [1996b] extends the appropriation thesis by identifying four factors affecting assimilation of foreign knowledge: religious requirements, Arabic as a nearly universal *lingua franca*, tolerance of other religious groups, and the division of larger kingdoms into smaller ones. Gutas's *Greek Thought, Arabic Culture* [Gutas 1997] explores the nuances of this merger in detail, considering especially the movement to translate Greek texts to Arabic in the early ʿAbbāsid period. He emphasizes the symbiotic relationship between the translation efforts and social movements (particularly Persian elements in the courts of the early ʿAbbāsid caliphs), and denies that religious forces were meaningfully opposed to the Greek sciences. In *Islamic Science and the Making of the European Renaissance* [Saliba 2007], Saliba attributes the rise of Islamic science in general, and the translation movement in particular, to the advent of Arabic administrative structures in government and culture in the eighth century. As Dallal notes [Dallal 2010], Gutas's and Saliba's approaches in parallel help to fill a gap in the literature, proposing nuanced answers to the question of the reasons behind the seemingly sudden emergence of Islamic science. Dallal emphasizes that the absorption of foreign sciences was almost always in a context of critique and creativity.

Euclid's Elements manuscript studies: The project of reading Greek mathematics usually starts with Euclid's *Elements*, so we begin our survey of transmission there. Translation is a subtle business; even minor shadings of meaning in translated technical terms can alter the meaning of the text substantially for the receiving culture. The translation history of the *Elements* in Islam is rather complicated, although ongoing research continues to clarify the picture. Two major manuscript traditions dominate the field. The first is a pair of translations by al-Ḥajjāj ibn Yūsuf ibn Maṭar in the early ninth century, one predating the other by about 25 years. The latter, produced under caliph al-Maʾmūn, may have been a substantial revision of the original translation rather than an entirely new one; since both are lost, we must infer what we can from documents that relied upon them. An *Elements* commentary by al-Nayrīzī (translated in Lo Bello [2003a], Lo Bello [2003c], and Lo Bello [2009]) accompanies a Euclidean text that may be a representative of al-Ḥajjāj's second edition, but al-Nayrīzī's scholarly fingerprints might obscure al-Ḥajjāj's work. The second tradition begins about fifty years later with a new translation by Isḥāq ibn Ḥunayn, edited by Thābit ibn Qurra. This new work adhered to the original Greek syntax more closely than had al-Ḥajjāj's edition, although it is likely that Isḥāq was more fastidious on this point than Thābit. The situation after the composition of these two editions is complicated, involving about fifty Arabic scholars' revisions and commentaries.

The reconstruction of the al-Ḥajjāj tradition is particularly fascinating, since it is the earliest substantial record of the *Elements* in Islam. The choices Ḥajjāj made in translating technical terms (at least, what we can infer) help us see how Euclid was understood in the early days. For instance, a particular textual fragment that Djebbar [1996] considers to derive from the al-Ḥajjāj tradition uses the word *talbīn*, the making of bricks, to describe rectangles. This word represents a shift away from the Greek geometric conception of rectangle areas and toward a more arithmetic conception in terms of multiplication of length and width. However, a manuscript recently described in Brentjes [2006] contains remarks about the shift in technical language that suggests a more philosophical basis for the choice of words like *talbīn*. Brentjes considers this manuscript to be the closest available witness to al-Ḥajjāj's second translation, and points out that this evidence suggests that al-Ḥajjāj was working from a simpler Greek text than the underlying text of the Isḥāq-Thābit tradition. A study by De Young [2002a] of an anonymous commentary related to the al-Ḥajjāj tradition sheds more light on the history of al-Ḥajjāj's translation. De Young suggests (but does not assert) that al-Ḥajjāj's translation may have been more popular than Isḥāq-Thābit initially, possibly until as late as the 13th century, when Naṣīr al-Dīn al-Ṭūsī's redaction of the *Elements* (which preferred Isḥāq-Thābit) more or less ended the al-Ḥajjāj tradition. In another study, De Young [2005] discovers a manuscript of al-Ṭūsī's edition that contains diagrams in the margins directly ascribed to al-Ḥajjāj. Finally, Brentjes [1996a] suggests that non-primary sources might be helpful in disentangling this complicated history, using a discussion of the *Elements* by Ibn al-Haytham as an example to gain some insight into al-Ḥajjāj's second translation.

Al-Ṭūsī's volume, the *Taḥrīr Kitāb Uṣūl Uqlīdis*, had a wide medieval circulation but not as much modern coverage. A step is taken to rectify this lack of attention in De Young [2008c], where De Young finds that the *Taḥrīr*'s source is mostly the "Group A" manuscripts of Isḥāq-Thābit, with some terminology borrowed from al-Ḥajjāj's first translation. Many of the *Taḥrīr*'s alternate demonstrations appear to derive from Ibn al-Haytham's *Solution of the Difficulties in Euclid's Elements*. The Persian journey of the *Taḥrīr* has been the subject of a couple of recent studies. Brentjes [1998a] considers four Persian versions of the *Elements* and finds distinctions between them as well as differences with al-Ṭūsī, casting into doubt the hypothesis of a single Persian ancestor. These texts' intended readers impacted how the editions were composed: for instance, al-Āmulī's summary of Book I (14th-century Iran) was intended for a wider audience; but Abū al-Ḥayr (18th century Delhi) needed a more technical edition for his work in the Mogul court, and used al-Ṭūsī as a point of departure. De Young [2007] follows up Brentjes' study of the Persian sources by considering Qutb al-Dīn al-Shīrāzī's translation of the *Taḥrīr*. Al-Shīrāzī remains close to the Arabic terminology, but he introduces new material, presumably for pedagogical purposes. De Young points out the curious fact that Persian renditions of Euclid tend to turn to the *Taḥrīr* rather than to the original translations of Euclid; perhaps they were attracted by the *Taḥrīr*'s creative mathematical insights.

Historical data may also be generated by considering the interaction between Arabic and Latin translations of the *Elements*. The relation between the *Elements* commentary attributed to al-Nayrīzī (ca. AD 900), presumably based on al-Ḥajjāj, and the Latin text ascribed to Anaritius that is thought to be a 12th-century translation by Gerard of Cremona, is unclear. Busard [1996] argues that the Arabic commentary is not based on a pure al-Ḥajjāj text, that the two works are not independent, and that the Latin translation relies on both al-Ḥajjāj and Isḥāq-Thābit. Brentjes [2001b] disagrees, asserting that the Latin text derives mostly from al-Nayrīzī. De Young [2004] identifies a mixture of the two Arabic traditions in Gerard of Cremona, with influences differing from one Book of the *Elements* to the other; while Adelard of Bath's translation seems more consistently related to the al-Ḥajjāj tradition. Meanwhile, Brentjes's study of Book I of Hermann of Carinthia's translation of the *Elements* (also from the 12th century) locates several mixed influences; it is based primarily on an Arabic edition from the al-Ḥajjāj tradition but contains contaminations from the Isḥāq-Thābit tradition and other sources [Brentjes 2001a]. Another 12th-century Latin manuscript of the *Elements* combines the Greek-Latin tradition with the Arabic-Latin tradition [Brentjes 1996b]. The proofs in Book I appear to derive from a hitherto unknown Arabic tradition, with lost texts by al-Māhānī, Ibn Samḥ, and Jābir ibn Aflaḥ as possible source candidates.

Medieval Arabic texts have also been studied for many decades not just for their own sake, but also for their aid in restoring Euclid's original Greek work. The great 19th-century classicist J. L. Heiberg took the position that the surviving Greek texts are better representatives of the original Euclid than the Arabic texts; he was opposed by Klamroth in 1881. Knorr [1996] revives the century-old debate with a re-examination of Gerard of Cremona's and Adelard of Bath's Latin translations of Arabic editions of the *Elements* in comparison with a Greek manuscript of part of Books XI and XII. Knorr concludes from this evidence that Klamroth may have been correct in championing the value of the Arabic translations. Rommevaux, Djebbar and Vitrac [2001] study Book X on a textual level. Although the authors note that the situation is complicated, generally they support Klamroth's position: at least one of the Arabic traditions is closer to Euclid than the Greek tradition on which Heiberg relied. Arabic sources are also helpful in retrieving information on Greek commentaries on the *Elements*. Djebbar [2003b] studies scholia from two Arabic manuscripts, suggesting tentatively that the differences between them might be ascribed to underlying differences between the commentaries of Proclus and Simplicius, although other explanations are possible.

Additions to the Elements: The *Elements* was much more than a historical record for its commentators and translators, but a living document — a basis on which to build and extend mathematics of interest to practitioners. Additions were sometimes made within the text itself, but there was also a thriving tradition of separate addenda. These variations on Euclid's themes are becoming more accessible to scholars than they once were. Lo Bello's ongoing translation of al-Nayrīzī's extensive commentary is a leading example; the series now includes Book I in Lo Bello [2003a], Gerard of Cremona's translation of it in Lo Bello [2003c], and Books II-IV in Lo Bello [2009]. The latter contains all of Simplicius's commentary on Euclid, and fills lacunae from Book I using the Arabic text of a new manuscript discovered by Brentjes and edited in Arnzen [2002]. (Lo Bello's series also includes Albertus Magnus's commentary on Book I [Lo Bello 2003b].) Some commentators kept closer to Euclid's text than others. Crozet [1997] notes that al-Sijzī's tenth-century commentary was more faithful to the original structure than most; he tended to add alternate proofs of existing theorems rather than re-invent entire theories within the *Elements*. In a study of Aḥmad al-Karābīsī's commentary, Brentjes [2000] observes that al-Karābīsī adapts some of the Greek commentaries (especially Simplicius's and Heron's) into his work more than might appear from a direct reading of the text.

Book I: Faithfulness to Euclid was far from universal, and breaks from the Euclidean tradition would be among Islam's most important contributions to mathematics. Many of these new directions are at a fundamental level, which draws attention especially to Arabic revisions of *Elements* Book I. Brentjes's study of a variety of such texts identifies Greek sources for a number of the additions made to Book I, including sources beyond the standard Greek commentaries (Simplicius, Heron, and Pappus) [Brentjes 1997]. However, other Arabic variants of Book I reveal differences original to the Arabic authors. Bouzoubaâ Fennane [2003] studies Ibn al-Haytham's commentary on Proposition 7 in *Solution of the Difficulties in Euclid's Elements*, and argues that it represents a new and important stage in the explicit formulation of the principle of continuity. Tenth-century geometer Abū Sahl al-Kūhī's revision of Book I (edited and translated in Berggren and Van Brummelen [2005]) removes all its geometric constructions, leaving only the theorems behind. Al-Kūhī also shows none of Euclid's discomfort with the parallel postulate, moving it forward into a more natural logical position.

The parallel postulate: A number of mathematicians attempted to prove the parallel postulate over the centuries. Two such proofs have been published or discussed recently: one by al-Nayrīzī likely based on the otherwise unknown Greek author Aghānis, edited and translated into English in Hogendijk [2000b]; and another, discussed in Guergour [2009], in the late 11th-century *Book of Perfection* by al-Muʾtaman ibn Hūd, king of the Islamic kingdom of Zaragoza in Spain. (We shall have more to say about the *Book of Perfection* later.) Thābit ibn Qurra's two earlier proofs are already well known; both depend on a substitution of Euclid's definition of parallels with a definition based on the equidistance of one line from another. Rashed and Houzel [2005] (reprinted in Rashed [2009e]) provide editions and French translations of both of Thābit's works, and describe his attempts to prove the existence of a straight line equidistant from another. The most famous of all "proofs" is that of ʿUmar al-Khayyāmī in his *Commentary on the Problems Posed by Certain Postulates of Euclid's Treatise*, recently edited and translated in Rashed and Vahabzadeh [1999], translated to French in Djebbar [2002b], and explained in Kanani [2000]. Khayyāmī's approach replaces the parallel postulate with two statements: lines that converge must intersect, and two lines that converge can never diverge in the direction of convergence.

The theory of ratios and irrational numbers: More controversial, and eventually more productive, were Muslim responses to Euclid's theory of ratios, which have attracted modern researchers over the past fifteen years. When the quantities to be placed in ratio (usually line segments) are incommensurable, Euclid follows Eudoxus: $A/B = C/D$ when, for any whole numbers x, y, the magnitudes xA and xC are both (i) greater than, (ii) equal to, or (iii) less than the magnitudes yB and yD. From the very beginning Euclid's cumbersome definition was examined and questioned. Already in the early ninth century, al-ʿAbbās ibn Saʿīd al-Jawharī, a friend of caliph al-Maʾmūn (AD 813–833), made a rather crude attempt to prove or explain Euclid's definitions of equal and greater ratio (edited and translated in De Young [1997], and studied in Persian translation in De Young [2008b]).

It took little time after that for Muslim mathematicians to propose an alternate approach, the *anthyphairetic* definition: two ratios are deemed to be equal if the Euclidean algorithm is applied to both, and the same sequence of numbers emerges both times. (In the case of incommensurable magnitudes, these processes might go on forever, but that does not necessarily prevent a determination that the sequences are identical.) Depending on how the operations were interpreted by practitioners, geometric magnitudes could be represented almost arithmetically — a decidedly un-Euclidean move. Abū ʿAbd Allāh Muḥammad ibn Aḥmad al-Māhānī's *Treatise on the Difficulty Concerning Ratio* (AD 860), edited and translated to English by Vahabzadeh [2002], is a very early example of an anthyphairetic definition; al-Māhānī's treatise lays out the definition, but also attempts to demonstrate its equivalence to the Euclidean definitions. Vahabzadeh conjectures that the anthyphairetic definition may have been preferred because it can be configured to allow ratios to exist on their own, not only in equality to other ratios. This property shifts the basis of ratio toward a numerical understanding. Hogendijk [2002a] considers al-Māhānī's work to be superior to later efforts by al-Nayrīzī and ʿUmar al-Khayyāmī, and hypothesizes that it may have a Greek origin. Finally, Ben Miled [1999] argues that al-Māhānī's ninth-century commentary on Book X includes a very early algebraization of *Elements* material, including definitions of both negative and irrational numbers. Ben Miled [2004] extends this analysis, studying al-Karajī's tenth-century treatment of n-th roots in this context.

The composition of ratios within the Euclidean tradition is a subtle logical matter; in Book V he *defines* (not proves) the ratio composed of $a : b$ and $b : c$ to be equal to $a : c$. This caused difficulty from a surprising corner: Thābit ibn Qurra's study of the sector figure (or Menelaus's Theorem), the fundamental result of spherical trigonometry. This theorem asserts that a particular ratio of sines is equal to the composition of two other ratios of sines. In Thābit's *On the Sector Figure*, he demonstrates a number of propositions involving manipulations of these ratios that would seem obvious to any middle school student today, but require proof in Euclid's ratio theory. Later, in his related work *On the Composition of Ratios*, Thābit takes a more arithmetical approach. Lorch [2001] is an edition and English translation of both of Thābit's works. Bellosta [2004b] (reprinted in Rashed [2009e] with a French translation of the text) comments on the geometric aspects of *On the Sector Figure*. Crozet [2004b] (reprinted in Rashed [2009e] with edition and French translation) argues that *On the Composition of Ratios* planted seeds for the movement of ratio away from Euclid's conception and toward arithmetic, and provided a connection with the geometric tradition of Euclid's *Data*.

These issues continued to resonate later. Euclid's approach to ratios in book V, dealing with arbitrary magnitudes, is not perfectly compatible with his later work in Book VII, dealing specifically with numbers. In the early 11th century, Ibn al-Haytham attempted to recast some of Book V to harmonize it with Book VII. But De Young [1996] points out that he was not as successful as Ibn al-Sarī a century later, who instead altered Book VII to harmonize it with Book V. The most renowned Islamic work on ratios is within ʿUmar al-Khayyāmī's *Commentary on the Problems Posed by Certain Postulates of Euclid's Treatise*, edited and translated into French in Rashed and Vahabzadeh [1999]. Opinions on Khayyāmī's work focus on the extent to which the modern concept of real numbers is to be found within its pages. Vahabzadeh [1997, 2004] takes a more positive view (also arguing that Khayyāmī preferred the anthyphairetic definition because it allows a ratio to be defined on its own rather than in comparison to another ratio), while Vitrac [2000] is more guarded. Ben Miled [2008] links this topic to Khayyāmī's work on cubic equations, which Khayyāmī solved by converting them to geometric problems, choosing units of measure for lengths, surfaces and volumes. Ben Miled also observes that notions of ratio and number are found among the algebraists of the school of al-Karajī.

The arithmetic recasting of the *Elements* extended beyond ratio theory. Ahmed Djebbar's study of the first chapter of the 11th-century Spanish king al-Muʾtaman ibn Hūd's *Book of Perfection* reveals an extensive arithmetization of parts of the *Elements* (as well as some influence carried over from Nicomachus's *Arithmetica*), including the axioms and Books II, VII, and IX [Djebbar 1999]. De Young [2001] translates the popular and influential late 13th-century *Fundamental Theorems* by Shams al-Dīn al-Samarqandī, which transforms material mostly from Book I into a form amenable to algebra and mensuration. (See also De Young's treatment of Qāḍīzāde al-Rūmī's extensive early 15th-century commentary on al-Samarqandī's work, dealing specifically with questions regarding the role of a commentator [De Young 2002c].)

While the above topics have dominated recent research into the transformation of the Euclidean tradition in Islam, interest in and reaction to the *Elements* affected many (if not all) of its Books. Matvievskaya [1996] presents the commentary by Abū Naṣr Manṣūr ibn ʿIrāq (al-Bīrūnī's teacher) on Book XIII, focusing on the construction of the regular heptagon. Extending past the end of the *Elements* altogether, De Young [2008a] presents an anonymous addendum to Naṣīr al-Dīn al-Ṭūsī's redaction of the *Elements* entitled Book XVI, building on the late Greek Book XV, on the construction of polyhedra within spheres and within other polyhedra. De Young comments that this material is closer to the Archimedean tradition than to the Euclidean.

Euclid's Optics: Euclid wrote more than just the *Elements*; among these other books, the *Optics* was one of the most prominent in medieval Islam. Kheirandish's English translation of the Arabic version is accompanied by her study of the substantive changes caused by linguistic transformations of the text, especially the definitions [Kheirandish 1999, 1996]. Rashed has also re-edited and translated Diocles's text on burning mirrors (Rashed [1997c] in English, Rashed [2000a] in French; previously edited by Toomer in 1976), which survives only in Arabic. Rashed's edition and translation to French of al-Kindī's mid-ninth century commentary on the *Optics*, a starting point that aided the transition from Euclid eventually to Ibn al-Haytham's famous *Optics*, appears as one of the treatises in a volume devoted to al-Kindī's works on optics and catoptrics [Rashed 1997d]. Rashed discusses the significance of this manuscript for our understanding of optics both in Islam and in late antiquity in Rashed [1997b]. We shall discuss optics more fully later in this survey.

Other Greek authors: Although the appropriation and transformation of Euclid has been a major focus of recent research, the treatment received by other Greek authors has not been ignored. Menelaus of Alexandria (ca. AD 100) had a major impact on Islamic science, but the same may be said in reverse. Arabic editions of Menelaus's *Spherics* substantially alter the text, beginning with the replacement of the chord function with the sine. The Arabic manuscript history of this important geometric and trigonometric work is studied in Sidoli [2006]. Commentators such as Abū Naṣr Manṣūr and Naṣīr al-Dīn al-Ṭūsī did not feel the need to respect the boundary that the Greeks had erected between spherical geometry and its astronomical applications; their expositions are described in Nadal, Taha and

Pinel [2004]. (A more detailed study of al-Ṭūsī's commentary on Menelaus appeared in [Pinel and Taha 2003].)

Although Arabic transformation of Greek works was substantial, it is still possible to some extent to read backward through them to the Greeks. Menelaus's lost work, the *Geometrical Elements*, finds some expression in Hogendijk's reconstruction of fragments within two works by al-Sijzī (an older colleague of al-Bīrūnī) [Hogendijk 1999]. Hogendijk concludes, contrary to previous wisdom, that a trisection of the angle by conics is absent from this work. Sidoli and Berggren [2007] provide an edition and English translation of the Arabic translations of Ptolemy's *Planisphere*, revealing that Ptolemy assumed his readers were aware that stereographic projection preserves circles, but was apparently unaware of any general proof that it preserves angles. Brentjes [1998b] studies the Arabic transmission of Nicomachus's *Introduction to Arithmetic*, paying special attention to al-Kindī's inserted comments on whether 1 can be considered to be a number in the Euclidean sense, concluding that it cannot (on Nicomachus's text in Arabic, see also Freudenthal and Lévy [2004]).

Although there were many different degrees of respect toward the integrity of the original Greek texts, many Muslim authors treated their subjects more as living mathematical documents than as records of a bygone culture. Thus, for instance, Berggren and Sidoli [2007] show that Thābit ibn Qurra's version of Aristarchus's *On the Sizes and Distances of the Sun and Moon* contains remarks added by Thābit on the astronomical significance of certain passages, and revisions of the structure of some of the proofs. Berggren and Sidoli also show that Naṣīr al-Dīn al-Ṭūsī's 13th-century edition of Aristarchus's treatise is closer to Thābit's text than Thābit's is to Aristarchus's. With respect to Theodosius's *Spherics*, Sidoli and Kusuba [2008] note that al-Ṭūsī took the same sorts of mathematical liberties as Thābit had done earlier. (Complete editions of Theodosius in both Arabic and Latin were recently published in Kunitzsch and Lorch [2010].) Often, however, an Arabic text can be our best witness to the original Greek. Taha [1998] presents an edition and French translation of Thābit's edition of part of Archimedes's *Book of Lemmas*, and argues that Thābit's work is closer to the spirit of Archimedes than are several more modern translations. Sometimes Arabic is our only source, such as the edition of Ptolemy's *Planisphere* presented by Sidoli and Berggren, discussed above.

Modifying the ancient texts was only the beginning; a number of recent studies have shown the extent to which Muslim authors extended or even "completed" ancient works. Some of these extensions occurred at a small scale, such as al-Kūhī's filling in lacunae in Apollonius's *Conics* [Berggren and Van Brummelen 2002], al-Kūhī solving a problem posed by Archimedes in *On the Sphere and Cylinder* [Berggren 1996a], or Ibrāhīm ibn Sinān extending his techniques beyond Apollonius's geometry in his *Selected Problems* [Bellosta 1997]. Scholars judge the novelty and magnitude of these extensions in different ways. Rashed argues that some of the work of the Banū Mūsā, three brothers working on mathematics already in the mid-ninth century, can be read as employing geometric point transformations, thereby extending beyond Archimedean methods [Rashed 1996a; Rashed 1997a].

Other extensions of ancient texts could take place on a grand scale. Perhaps the most impressive example is Ibn al-Haytham's *Completion of the Conics* of Apollonius. Previously edited and translated to English by Hogendijk in 1985, the work was given a new treatment in Rashed's edition and translation into French within Rashed [2000e]. (Hogendijk responds to Rashed in Hogendijk [2002b] and laments duplications of effort such as this in Hogendijk [2003a], suggesting that our efforts be better channeled into presenting our findings to a wider audience.) The *Conics* itself is one of the most challenging of Greek texts, and it was a subject of fascination from some of the earliest Arabic sparks of interest in mathematics. In recent years Rashed has directed a massive project to edit and translate to French the entire seven books of the *Conics* that survive in Arabic [Rashed 2008a; Rashed 2010a; Rashed 2009b; Rashed 2008b; and Rashed 2009c], accompanied by editions and French translations by Decorps-Foulquier and Federspiel of the surviving Greek text (Books I-IV) [Decorps-Foulquier and Federspiel 2008; Decorps-Foulquier and Federspiel 2010]. The Arabic editions, supervised by the Banū Mūsā in the ninth century, reveal great detail, sophistication and mastery of its contents by the scholars involved in the project. Apollonius's other surviving work in Arabic, *Cutting Off of a Ratio*, has also been edited and translated to French by Rashed and Bellosta [2010].

Regional Influences

Locality versus essence: Medieval Islamic science is often considered implicitly to be a single entity; yet it spans over at least seven centuries, a vast geographical area, different interpretations of Islam and indeed different religions altogether. A. I. Sabra's influential paper "Situating Arabic science — locality versus essence" [Sabra 1996b] advocates for better sensitivity of local cultures and conditions, also emphasizing different loci of scientific activity (court, college, and mosque). His urge to respect the complexity of cultural, religious, and other contexts that shape the experiences and writings of historical scientists, a lesson for all historians of science, is only gradually being put into practice by historians of ancient and medieval mathematics.

This shift, which is happening across the discipline of the history of mathematics, is being pioneered in the Islamic exact sciences especially by Brentjes. Her survey of the mathematical sciences in Safavid Iran (around 1500 to 1700) advocates replacing a "vertical" with a "horizontal" approach [Brentjes 2010]. The former emphasizes the great works and searches for a narrative of intellectual progress; the latter considers social constructions, examinations of local history and comparisons with nearby and preceding cultures, paying attention to mathematics at social levels other than the elite. Some of Brentjes's recent efforts explore the role of patronage of the exact sciences in various Islamic contexts. Brentjes [2009] surveys patronage from the eighth century onward mostly in courts, and from the 12th century also in endowed teaching institutions. Brentjes [2008a] argues that patronage did not disappear after 1200, although it changed in various ways; hence the decrease in patronage cannot be used as an explanation for the so-called decline of the ancient sciences in Islamic cultures. Brentjes [2008b] considers the journey of Euclid's *Elements* in Iran through the courts, at the *madrasa*, and in the hands of collectors. In Brentjes [2007], she uses a treatise on the history of education by 15th-century Damascene scholar al-Nuʿaymī to examine the role of religious and legal scholars in the previous couple of centuries, arguing that the modern perception of their hostility to the ancient sciences must be re-evaluated. She has also studied the interplay between patronage of science and art in the 17th century Safavid court [Brentjes, below]. Dallal Dallal [2010] remarks that partly through Brentjes's work, there can now be little doubt that the rational sciences played a substantial role in courts and institutions of learning.

It can be much harder to write a proper social history of mathematics than a conventional history due to paucity of data, as Samsó points out in his evaluation of the evidence for al-Andalus [Samsó 2002]. Occasionally unique sources of evidence surprise us, such as the discovery of a second letter from Jamshīd al-Kāshī to his father [Bagheri 1997b] describing social and scientific life in the court of Ulugh Beg. Even though the sources may be scarce, it is incumbent upon the community to revisit the deeper questions raised by this new approach: what are the ultimate goals of our study? In light of our answer to this question, in what ways should we approach our subjects to achieve these goals?

Regional surveys: Nowhere in medieval Islamic science is the issue of locality more acute than in mathematics and astronomy in Spain and North Africa. As Samsó states in his study of the evidence for al-Bīrūnī's work in western Islam [Samsó 1996], political and cultural differences between east and west led to barriers discouraging transmission of scientific works after the tenth century. (See Djebbar [2002a] for a more general study of the circulation of ideas across Muslim cultures.) Thus Djebbar's surveys of mathematical activity in these regions reveal unique milieus. Djebbar [2003a] is a survey of research on mathematics in al-Andalus and the Maghreb from the ninth to the 16th centuries, raising several questions for future scholars to address. The Maghreb is the focus of Djebbar [1998], which notes the early influence of al-Andalus and the increasing role played by mosque schools from the 14th century onward. Two different schools of thought emerged, characterized by different attitudes toward the importance of demonstration in mathematics. A study examining the role of demonstration in various exact sciences in eastern Islamic lands may be found in Kheirandish [2008]. Djebbar [2000a] deals with the Ottoman period in the Maghreb (16th to 19th centuries), arguing that the change in political culture may not have impacted science as much as one might expect.

Many studies of specific topics in the mathematical sciences in certain regions, especially al-Andalus (many of these written at the University of Barcelona), have been composed in recent years. Since the literature still tends to organize most of the following topics more by their intellectual connections than by regional origin, these topics are discussed as they arise topically in the remainder of this survey.

Geometry

One of the most well-developed areas of Islamic mathematics, geometry was inspired initially by the Greek tradition. However, as Berggren [2002b] points out, the acceptance of geometry as a foreign science sometimes encountered religious barriers. Religious concerns, such as determining the direction of Mecca and the times of prayer, also spurred interest in geometry. Often, intellectual fascination proved to be stimulus enough. Hogendijk [1996b] contends that the Greek tradition was assimilated and extended in various ways, especially in trigonometry and the linking of ratio more closely with the concept of number. But no transformation of geometry occurred at the "paradigm" level, at the scale of the birth of analytic geometry in early modern Europe. This argument remains unresolved in the scholarly community; the studies in this section of our survey provide rich data to help decide the issue.

Heuristics; analysis and synthesis: Before the medieval period, works on heuristic strategies to solve geometric problems were almost unheard of (save for Archimedes's lost book *The Method*). In the tenth century al-Sijzī compiled a medieval equivalent to George Polya's *How to Solve It*, containing seven different approaches to geometric problems; this work is now available in Farsi translation by Bagheri and English translation by Hogendijk [Bagheri and Hogendijk 1996], and in edition and French translation in Rashed [2002b]. One of the methods discussed by al-Sijzī is analysis and synthesis, the ancient Greek technique that took on a robust and transformed life of its own in medieval Islam. Berggren and Van Brummelen [2001d] explore the ways in which this procedure was altered: in particular, analysis moved to the forefront, and it shifted its emphasis somewhat from demonstrating constructibility toward determining the unique existence of the sought object. Often the argument omitted not only the entire synthesis, but also the first half of a Greek analysis (the "transformation"), preserving only the second half of the analysis (the "resolution"). A typical work of "analysis by knowns" (where some part of a diagram is found to be "known" if others are assumed to be "known" in the style of Euclid's *Data*), Thābit ibn Qurra's *Book of Assumptions*, has received treatment from two authors. Dold-Samplonius [1996a] summarizes the contents of the work and notes the many differences between Thābit's original work and al-Ṭūsī's later popular edition; Dold-Samplonius [1996c] discusses the problem of the two towers that appears within its pages and other historical texts. Bellosta [2002a] examines some of the theorems and considers Thābit's application of the word "known" in the book. Rashed [2002b] edits and translates to French two large works by Ibn al-Haytham on the subject; the first deals with analysis itself, and the second with the meaning of the word "known." Mawaldi [2000] describes Kamāl al-Dīn al-Fārisī's 13th-century text *Fundamentals of Rules on the Principles of Useful Things*, in particular its application of analysis and synthesis to the solution of an algebraic equation. Analysis by knowns was prevalent in many Islamic geometrical works; it appears also in a number of studies described elsewhere in this section of the survey.

Conic sections: Islamic geometers' fascination with conic sections was inspired in part by their ability to solve the otherwise unsolvable. Hogendijk [1998] notes that conics were appealing for a variety of reasons, including practical applications and uses in other sciences such as astronomy and optics, but the primary motive may have been simple intellectual fascination. Understanding of the subject was superior to contemporaneous cultures, even to early Europe; conics were used to arrive at new solutions to a number of problems, including the construction of the regular heptagon and the trisection of the angle. Recent research has concentrated on 10th and early 11th century texts. Luther [2002] studies Ibrāhīm ibn Sinān's treatment of Apollonius's tangent circles problem using methods from Apollonius's *Conics* (again, mostly analysis without synthesis), and Jaouiche [1998] compares Ibn Sinān's approach with those of Ibn al-Haytham and François Viète. Also working with tangent circles was Abū Sahl al-Kūhī, in a treatise described in Abgrall [1997] and edited in Abgrall [2004b]. This text outlines al-Kūhī's invention of an instrument known as the "perfect compass" capable of drawing conic sections. This new device received attention in a treatise by al-Sijzī, who describes how to use the perfect compass to trace a conic section similar to one that is given (edited and translated into French in Rashed [2003a]; see also Rashed [2004], the first of a promised series of volumes on al-Sijzī's works written with Crozet). Abū al-Jūd Muḥammad ibn al-Layth used conics in the tenth century to solve cubic and quartic equations (edited and translated into French in Rashed [2010c]), anticipating Khayyāmī's more famous work a century later. We have already discussed the appearance of Rashed's new edition and French translation of Ibn al-Haytham's *Completion of the Conics* [Rashed 2000e]; this volume also contains a number of Ibn al-Haytham's other works, including an exposition of the use of conics to construct the regular heptagon and his *Correction to a Lemma by the Banū Mūsā*, relevant to Book VI of Apollonius's *Conics*.

Abū Sahl al-Kūhī: Two mathematical figures have received sustained attention in recent years, both dating since before the starting date of this survey: Abū Sahl al-Kūhī and al-Muʾtamān ibn Hūd. The former, who worked mostly in Baghdad under the Būyid kings, was both a great innovator and a great appreciator of the Greek heritage. Berggren [2003] notes in his survey of al-Kūhī's work that he was both "master of his age in the art of geometry" and "the last mathematician to look on mathematics with the eyes of the great Hellenistic geometers." We have already seen his respect for the Greek heritage in several works: solving a problem left behind by Archimedes' *Sphere and Cylinder* (edited and translated to English in Berggren [1996a] and discussed in Abgrall [2004a]), and adding propositions to Book II of the *Elements*, which we now know were intended to fill in logical gaps in Apollonius's *Conics* [Berggren and Van Brummelen 2002]. His spirited defense of Menelaus's Theorem of spherical trigonometry in the face of a revolution that would discard it in favor of simpler alternatives is a particularly striking example of his appreciation of the ancients [Berggren and Van Brummelen 2001a]. But his revision of *Elements* Book I reveals an ability to critique and even revise substantially the Greek classics: he dispenses with the *Elements*'s geometric constructions altogether and re-organizes the subject matter by shifting the parallel postulate much earlier in the book [Berggren and Van Brummelen 2005].

As a practicing geometer al-Kūhī attacked a wide variety of problems, working with conics in his treatises on the perfect compass and tangent circles (discussed in Abgrall [1997] and Rashed [2003a]) and on the volume of a rotated parabolic segment (edited and translated by Rashed [1996b] and discussed in Abgrall [2004a]). Several of his more Euclidean geometrical works have now received treatment. These include (a) *On Drawing Two Lines from a Point at a Known Angle*, which employs the Islamic version of analysis discussed above (edited and translated by Berggren and Van Brummelen [2001c] and Abgrall [2002]); (b) *On the Ratio of the Segments of a Line that Falls on Three Lines*, which employs both analysis and synthesis (edited and translated by Berggren and Van Brummelen [2000]); (c) a very short untitled work with only two unrelated propositions, one of which might be interpreted as dealing with homotheties of circles (edited and translated by Berggren and Van Brummelen [1999]). Al-Kūhī's geometric writings are being located in other sources as well, including fragments in al-Sijzī's works [Berggren and Hogendijk 2004], and (of all places) in a 17th-century Dutch translation [Hogendijk 2008c]. Al-Kūhī did venture beyond the discipline of pure geometry in several texts on practical applications, but relied on Euclidean geometry to achieve his goal. Examples of this include his treatise describing a method for finding the distance to the shooting stars [Berggren and Van Brummelen 2001b; Rashed 2001], and for finding the dip angle to the horizon when on an elevated platform (see Rashed [2001] for an English edition and translation, and Berggren and Van Brummelen [2003] for a critical response by the 12th-century Iranian mathematican al-Samawʾal). Al-Kūhī's work on the astrolabe and stereographic projection, edited and translated before the period of this survey (Rashed 1993 and Berggren 1994), is discussed and compared with Ibn Sahl's commentary and a similar treatise by al-Ṣāghānī in Abgrall [2000]. In his correspondence with Abū Isḥāq al-Ṣābī he deals with centers of gravity of certain geometric solids (originally edited and translated to English by Berggren in 1983; edited and translated to French in Abgrall [2004b] and discussed in Abgrall [2004a] and Bancel [2001], the latter in conjunction with several related texts). Al-Kūhī even uses geometry to venture into philosophy, taking on Aristotle with respect to the possibility of moving an infinite distance in finite time, by considering the projection of a semicircle onto a complete branch of a hyperbola [Rashed 1998; Rashed 1999a].

Ibn Hūd and the Istikmāl: The second mathematician to attract research attention is Yūsuf al-Muʾtaman ibn Hūd, king of Zaragoza, Spain in the late 11th century. His mostly geometrical *magnum opus*, the *Book of Perfection* (title notwithstanding, the book was never finished), is known to have made an impact due to a number of references to it in other treatises. But it was thought to be lost until Hogendijk announced the discovery of most of its contents in 1986. In 1997 Djebbar presented a 14th-century manuscript by Ibn Sartāq containing a redaction of the entire text [Djebbar 1997], including sections missing from the other manuscripts. Hogendijk [2003c] provides an analytical table of contents of the geometrical parts of the work that have been restored with the discovery of Ibn Sartāq's work, supplementing his 1991 table of contents published in *Archives Internationales d'Histoire des Sciences.*

Hogendijk had already begun to analyze the contents of the *Book of Perfection* in a series of studies prior to 1995. Since then, a number of scholars have joined the investigation. Djebbar [1999], whose interest in this text predates its discovery, published a study of the first chapter on arithmetic, noting that al-Muʾtaman brings together material for this chapter from *Elements* Book II (recast in arithmetic rather than its original geometric form), Book VII, and Book IX, as well as some material from Nicomachus's *Introduction to Arithmetic.* Within Rashed [1996b] is an edition and French translation of several geometric propositions, on topics such as conics and the areas of parabolic segments. Rashed [2002b] contains a set of theorems from the *Book of Perfection* relating to Ibn al-Haytham's works on analysis and synthesis, especially analysis by knowns. We find original material on spherics in the book as well; Rashed and al-Houjairi [2010] consider al-Muʾtaman's generalization of a result originally due to Theodosius, and Hogendijk [1996c] identifies the manuscript of Menelaus's *Spherics* that al-Muʾtaman must have used. Guergour [2005] studies al-Muʾtaman's use of the Pythagorean Theorem, and Guergour [2009] considers his "proof" of the parallel postulate. And, finally, al-Muʾtaman simplifies the approach taken by al-Haytham in solving Alhazen's problem in his *Optics* [Hogendijk 1996a]. Al-Muʾtaman seems to have written an *Optics* himself, but the work is lost and Hogendijk does not think it would have been a major step forward from al-Haytham's famous work.

Assorted geometric problems: Geometry remained an active domain throughout medieval Islam, but research attention lately has focused on earlier periods. Several ninth century treatises are of interest due to their appearance at a formative stage in the subject's development:

- Qusṭā ibn Lūqā's *Introduction to Geometry*, mostly borrowed from Greek sources, has been made available in Arabic by Guergour in the Arabic section of *Suhayl* 6 (2006), and translated to English and discussed in Hogendijk [2008a].
- Nuʿaim ibn Muḥammad ibn Mūsā, likely a son of one of the Banū Mūsā, composed a collection of assorted geometric problems and their solutions in the late ninth century. It was edited and translated to English in Hogendijk [2003b], and edited and translated into French in Rashed and Houzel [2004]. Panza [2008] argues that Nuʿaim's treatise uses a form of inference that might be described as algebraic, by reducing geometric problems to equations of the type found in al-Khwārizmī's *Algebra.*
- Several geometric works by Thābit ibn Qurra, including various lemmas and the construction of a semi-regular polyhedron with 14 faces, are presented and translated to French in Rashed [2009e].

Many of the works currently being published date from the tenth and 11th centuries. We are approaching a critical mass of literature that will soon allow a reliable account of this crucial period to be written. Treatises as yet uncited here include the following:

- Rashed [2000e] provides editions of texts on the construction of the regular heptagon, one of the more enduring geometric problems in medieval mathematics. Authors include Abu'l-Jūd, al-Sijzī, al-Kūhī, al-Shannī, and Kamāl al-Dīn ibn Yūnus.

- Rashed and Bellosta [2000] have compiled editions of five of Ibrāhīm ibn Sinān's works in logic and geometry, including his autobiography, *The Quadrature of the Parabola* (commentary only; the text is in Rashed [1996b]), *Tracing the Conic Sections*, *On Sundials*, and *Anthology of Problems*.
- Crozet has engaged in several studies of al-Sijzī's geometrical work, including Crozet [2004a] where al-Sijzī tackles problems relating to the division of a polygon into regions according to various conditions, and Crozet [2010] where al-Sijzī works with geometric transformations and invariants. We have already noted al-Sijzī's commentary on the *Elements* in Crozet [1997]; see also Crozet [1999], where he uses al-Sijzī's work as an example of the need for scholars to pay more attention to diagrams in Arabic manuscripts. Bellosta [1999] studies the manuscript BN 2457, which contains a number of geometrical treatises copied by al-Sijzī himself; this document allows us to witness what happened when a mathematician served as a copyist.
- Abū Saʿīd al-Ḍarīr al-Jurjānī's late tenth-century *Geometric Problems* and *The Extraction of the Meridian Lines* are presented in Hogendijk [2001b]. Hogendijk shows the influence that the former work had on al-Bīrūnī and Ibn al-Haytham.
- Rashed [2002b] compiles Arabic editions and French translations of six texts on plane geometry by Ibn al-Haytham, including two works related to analysis by knowns and a philosophical work arguing against Aristotle's definition of space. (In an earlier volume in the same series published before the chronological starting point of this survey, Rashed hypothesized that there were two Ibn al-Haythams, one a philosopher and another a mathematician; Sabra [1998, 2002b] disagrees.)

Disciplines Allied with Geometry

Architecture: Architecture was among the most prominent clients of the geometrical art. For years Dold-Samplonius has been producing valuable studies concerning the role of mathematics in constructing various types of buildings, especially those with religious significance. Her efforts are summarized by Dold-Samplonius [2003], where she describes the constructions and accompanying calculations needed for arches, vaults, *qubbas* (mausoleums), and *muqarnas*. Some of the biggest names in mathematics contributed, including Ibn al-Haytham and al-Karajī, as did some with less mathematical reputation (Aḥmad ibn Thabāt and Muḥammad Bahāʾ al-Dīn al-ʿĀmulī, for example). One of the most active figures in mathematics and architecture, the renowned 15th century Persian scientist Jamshīd al-Kāshī, calculated arches and domes in his *Key of Arithmetic* [Dold-Samplonius 2000]. The decorative patterns on the walls of religious buildings also attracted the attention of geometers. Sakkal [1995] surveys the geometry of ribbed domes, particularly in Spain and North Africa. Özdural [1996, 2000, 2002] identifies and describes what he calls "conversazioni" between artisans and mathematicians on the generation of these patterns. Özdural focuses on Abū 'l-Wafā''s tenth-century *On What the Artisan Requires of Geometrical Constructions* and a 13th-century anonymous work *On Interlocking Similar and Corresponding Figures*. The techniques often relied upon cut-and-paste methods. In more complicated situations when the equivalent of a solution to a cubic was required, the so-called "verging" constructions were preferred over solutions by conics, presumably due to their ease of use in practice (for this see especially Özdural [2002]). Sarhangi [2008] examines the constructions of an icosahedron and dodecahedron in a sphere in Abū 'l-Wafā''s work, pointing out that the former is not correct.

Mensuration: Applications in mensuration overlap between geometry proper and practical calculation. This understudied area could prove to be an extremely interesting mixture of existing traditions, relating to the influx of Euclidean geometry and algebra. Two works in this domain have been discussed recently. The first, *Book of Mensuration* by late ninth-century Egyptian mathematician Abū Kāmil [Sesiano 1996b], contains instructions for finding the areas and volumes of a number of shapes and solids, and includes some surveyors' approximations among the precise formulas (see Sesiano's contribution to this volume). The second, Ibn ʿAbdūn's *Epistle on Surface Measuring* (Djebbar [2006], with the Arabic text included in the 2006 issue and the previous issue of *Suhayl*), covers much the same ground. The particular interest of this latter work is that it is the earliest Andalusian attestation of an eastern tradition that existed before either Euclidean geometry or algebra arrived on the scene.

π: Continuing with interactions between geometry and calculation but returning to purer mathematical ground, we turn to various studies related to 15th-century Persian mathematician Jamshīd al-Kāshī's calculation of π. His *Treatise on the Circumference*, in which he computes π to a then unprecedented 16 decimal places, has previously been translated to German and Russian, but never to English. Azarian [2004] has made part of the text more accessible to English-speaking readers with a translation of the fundamental theorem in al-Kāshī's work, a translation of the introduction (including al-Kāshī's criticisms of previous work by Abū 'l-Wafā' and al-Bīrūnī) in Azarian [2009], and a summary of the work's mathematical contents in Azarian [2010]. Hogendijk [2008b] provides a facsimile of a manuscript of the entire treatise and transcribes al-Kāshī's arguments into modern mathematical notation. He compares al-Kāshī's method with related work by late 17th-century scholars Adriaan van Roomen and Ludolph van Ceulen, speculating that the striking similarity between van Ceulen and al-Kāshī may be explained by a joint debt to Archimedes's work. Hogendijk [2007] explores this similarity in more detail.

Philosophical issues: Mathematical arguments, usually but not exclusively geometrical, were often called upon to resolve philosophical dilemmas, with Aristotle generally serving as the starting point. El-Bizri [2007] discusses early 13th-century philosopher ʿAbd al-Laṭīf al-Baghdādī's defense of Aristotle's definition of place in the *Physics*, reacting against al-Ḥasan ibn al-Haytham's use of geometry to refute Aristotle and asserting the "sovereignty of philosophy." (El-Bizri accepts Rashed's hypothesis that there were two al-Haythams.) Naṣīr al-Dīn al-Ṭūsī uses combinatorial methods to demonstrate how a multiplicity can emanate from the One in emanationist neo-Platonic philosophy (Rashed [1999b] in French, and see also Rashed [2000b]; and Rashed [2000d] in German). Rashed [2003b] pursues this historical trail further into the work of Ibrāhim al-Ḥalabī, who took the combinatorial connection further and produced a treatise that "is the first we know of to be entirely devoted to combinatory analysis."

Notions of the infinitely large and infinitely small often provoke reflections that bridge the gap between philosophy and mathematics. We have already noted al-Kūhī's treatise responding to Aristotle on the possibility of infinite motion in finite time [Rashed 1998; Rashed 1999a]. McGinnis [2006] argues that Avicenna's analysis of motion at an instant contains a concept of limit, at least in an Aristotelian sense if not in a purely mathematical one. A collection of correspondence between Thābit ibn Qurra and one of his students has him asserting, against Aristotle, the existence of actual infinity [Sabra 1997].

Whether or not truly infinitesimal methods existed in medieval Arabic mathematical or philosophical treatises is a point of contention, between those who seek innovation in the texts and those who seek roots in the Greek tradition. Rashed's point of view is made clear by the title of his five-volume series of editions and French translations of texts, *Les Mathématiques Infinitésimales du IXe au XIe Siècle* ([Rashed 1996b; Rashed 2000e; Rashed 2002b; Rashed 2006] — vol. 2 is dated 1993, outside the purview of this survey). Although many of the works Rashed discusses in later volumes of the series do not relate directly to this theme, some deal with finding areas and volumes of curved figures. Rashed's respondents find Greek antecedents to Arabic texts in procedures such as the method of exhaustion. (Two treatises by Ibn al-Haytham edited in Rashed [2002b] have philosophical relevance more generally: one on the meaning of "known" in the context of analysis and synthesis, and one discussing Aristotle's definition of "place.") Rozhanskaya [1997] takes the debate into mathematical astronomy, claiming (against some earlier commentators) that both Thābit ibn Qurra and al-Bīrūnī used infinitesimal methods when dealing with the varying motion of the Sun around the ecliptic. Rashed [2010b] also follows the Arabic trail of Apollonius's proof that a hyperbola and its asymptotes approach each other indefinitely in the work of al-Qummī, a successor of al-Sijzī.

Finally, for a general overview of philosophy and mathematics in Islam see Endress [2003]. Among other topics, this article deals with Islamic reactions to the Platonic tradition (especially the writings of al-Kindī), Aristotle on philosophy and Ibn al-Haytham on mathematics as their respective demonstrative sciences, and the interplay between theology and science.

Optics: The early science of optics in Islam was influenced by the appropriation of Euclid's *Optics*. This process was rather complicated, but the story has been clarified markedly in recent years. Kheirandish [1999] combines the five known Arabic manuscripts into a single text that most closely approximates the original Greek as it might have been when it was translated around AD 800. As she remarks in Kheirandish [1996], the flexible Arabic language (especially with respect to the definitions) allowed for the growth of the discipline in ways that the more rigid Greek terminology did not allow. Kheirandish's volume includes a critical edition of al-Kindī's *Correction of the Optics*, and deals with al-Ṭūsī's commentary as well. The existence of al-Kindī's commentary was announced in Rashed [1997b], where he argued that the two known forms of the Greek *Optics* and the two Arabic forms are all independent. This commentary and several other of al-Kindī's optical writings (including the earliest known Arabic text on burning mirrors) are collected and edited together in Rashed [1997d]. Rashed argues that al-Kindī's conception of optics is closer to a physical reality than is Euclid's geometric expression.

Arabic optics beyond Euclid is dominated by the early 11th-century *Optics* by Ibn al-Haytham, but there were important developments before him. The state of affairs prior to AD 950 is summarized in a section of al-Fārābī's *Catalogue of the Sciences*, which locates optics as an established discipline within the mathematical sciences. Kheirandish's account of this text warns that the tradition is more complex than a reading of al-Fārābī's catalogue might suggest [Kheirandish 2003]. The best mathematicians did practice optics prior to al-Haytham: for instance, we have Abū 'l-Wafā''s presentation of two methods (but no proofs) for constructing a parabolic mirror in the late tenth century. Rashed presents a previously unpublished explanation of Abū 'l-Wafā''s methods by Neugebauer in terms of descriptive geometry [Neugebauer and Rashed 1999]; in contrast to Neugebauer, Rashed's explanation uses "conical projection methods." See also Bellosta [2002b] on Ibn Sahl and the application of conic sections in optics, and a discussion by Rashed [2002c] of transmission and innovation in this area dealing especially with Ibn Sahl and Ibn Ṣāliḥ. Rashed [2005], an English translation and considerable expansion of his 1993 *Géometrie et Dioptriques au Xe Siècle*, contains an edition and translation of Ibn Sahl's text.

The first three books of the seven in al-Haytham's *Optics* appeared with translation and commentary by A. I. Sabra in 1983 and 1989; in 2002, Books IV and V (*On Reflection, and Images Seen by Reflection*, Arabic text only) were published [Sabra 2002a]. Smith provides an edition and English translation of Books I-III of the altered Latin version of the work in Smith [2001] and Books IV-V in Smith [2006]; Hamedani [1999] discusses manuscript issues relating to another of Ibn al-Haytham's optical treatises, *On the Light of the Moon*. Sabra [2003] gives a physical account of what made al-Haytham's *Optics* innovative: it was the first Arabic work to be inspired directly by Ptolemy's *Optics*, employs for the first time a theory of light radiation to the exclusion of visual radiation (rays emanating from the eye to the seen object), and includes an element of the psychology of vision. In Book V, al-Haytham presents what came to be known as "Alhazen's problem": given a spherical mirror, and a point of sight and a point of radiation both facing that mirror, find the point(s) of reflection on its surface. Smith [2008] argues that al-Haytham's solution, while deficient compared with Huygens' 17th century effort (reproduced in an appendix), represents a *tour de force* of mathematical reasoning. This part of al-Haytham's work found its way to al-Mu'taman ibn Hūd, the late 11th-century king of Zaragoza, Spain, who simplified the mathematics [Hogendijk 1996a]. But generally, as Sabra [2007] points out, Ibn al-Haytham's *magnum opus* seems to have received little Arabic response until a 13th-century commentary by al-Fārisī; most reaction after that was filtered through al-Fārisī's commentary.

Mathematical Methods in Astronomy and Geography

Astronomy in medieval Islam was at least in part a mathematical discipline. However, the sheer magnitude of research in astronomy (both then and now) necessitates that we restrict our focus here, since there is enough astronomical material to write another survey article. We emphasize here those aspects of medieval Islamic astronomy that specifically involve mathematical innovations or methods. This boundary is problematic, but choices must be made.

Trigonometry: The quantitative study of arcs, angles, and lengths in geometric figures occurred mostly in, and was the foundation of, mathematical astronomy. Indeed, it is difficult to find works before late in the medieval period that deal with trigonometry separate from astronomy. (An account of the history of trigonometric methods in medieval Islamic astronomy and geography may be found in *The Mathematics of the Heavens and the Earth: The Early History of Trigonometry* [Van Brummelen 2009, 135–222].) The quantification of the motions of celestial bodies required a sine table as a starting point (sometimes, the Greek chord function; see Moussa [2010] for a discussion of conceptions of trigonometric functions in Islamic astronomy), and the key parameter underlying the sine table was sin 1°. A number of geometrically-inspired methods were proposed to compute an accurate approximation to this quantity, but the problem was finally dispatched by Persian scientist Jamshīd al-Kāshī in the early 15th century using an algebraic method, namely, a series of successive approximations to the solution of a particular cubic equation with root sin 1°. The *Treatise on the Determination of the Sine of 1 Degree*, inspired by al-Kāshī, is considered in Ahmedov and Rosenfeld [2000]. Although the treatise is attributed to Qāḍīzāde al-Rūmī in several manuscripts, Ahmedov and Rosenfeld argue that al-Kāshī's patron Ulugh Beg is the actual author. The debate is continued in Rosenfeld and Hogendijk [2002], where Rosenfeld maintains his attribution and Hogendijk sides with Qāḍīzāde al-Rūmī. The latter article includes an English translation and is supplemented with a facsimile reproduction of the entire treatise.

Spherical trigonometry, particularly important to mathematical astronomers, was to undergo a dramatic transformation in the late 10th and early 11th centuries; but prior to that the Greek legacy was strong. The fundamental result was Menelaus's Theorem, called the Sector (or Transversal) Figure in Islam. Menelaus's Theorem is actually two different propositions, each involving ratios of the sines of six arcs in a spherical quadrilateral. Thābit ibn Qurra composed two treatises on the subject in the late ninth century, *On the Sector Figure* and *On the Composition of Ratios*. Both were published in critical edition and English translation in Lorch [2001] (see also Lorch's description of a short commentary on the former work by Maslama al-Majrīṭī in Lorch [1996a] and Knobloch's study of its Latin translation in Rashed [2009e]); Bellosta [2004b] discusses the former work (especially the different forms in which the ratios appear) and the demonstrations within its pages. *On the Composition of Ratios*, discussed in Crozet [2004b], is noteworthy for the combinatorial issues that arise as the ratios are manipulated in various ways, and for the replacement of the Euclidean definition of ratio with an arithmetical definition. Crozet suggests that Thābit was anticipating new ways of understanding ratio concepts inherited from the Greeks that would arise in the following centuries.

The story of the theorems that replaced Menelaus, such as the Rule of Four Quantities and the Law of Sines, is told by al-Bīrūnī himself in *Keys of the Science of Astronomy*, published by Debarnot in 1985 and summarized by Saidan [2000] along with another of al-Bīrūnī's works on plane trigonometry. Although the theorems appear to have originated in eastern Islam (see a brief summary in Boulahia [2007]), they seem to have found their way to al-Andalus (Spain) quite quickly, in the work of Ibn Muʿādh. The new results were seemingly not known to al-Muʾtaman ibn Hūd, the scholar king of Zaragoza, but his *Book of Perfection* does include some innovations in spherical geometry, including a generalization of a proposition from Theodosius's *Spherics* and an integration of three propositions from Menelaus's *Spherics* [Rashed and al-Houjairi 2010].

Graphical methods, sometimes inscribed as curves on brass instruments, were sometimes used to solve trigonometric problems. Lorch [1998] discusses works by ninth-century authors Ḥabash al-Ḥāsib and al-Māhānī using these methods to solve problems in spherical astronomy. Some of these solutions are analemmas; others do not even have a geometric origin. Lorch [2000b] discusses two tenth-century texts on the sine quadrant; the solutions found in these treatises bear distinct resemblances to the graphical methods of Ḥabash and al-Māhānī. See also Charette [1999] for a description of a "winged" sine quadrant from the 14th century.

Mathematics in planetary theory: Models for the motions of the planets were the original applications of trigonometry, and mathematical methods continued to play a fundamental role in the construction of new and more effective models throughout the medieval Islamic period. However, Morelon's discussion of the interaction of physical and mathematical astronomy argues that the mathematical side has drawn attention from modern scholars at the cost of the physical side, and that the trail of improved astronomical models that occurred in Islam was spurred by both physical and mathematical concerns [Morelon 1999].

A tradition of critiques of Ptolemy's astronomical models started at least as early as Ibn al-Haytham. The goal was not so much to achieve a better fit with astronomical data, but rather to achieve a better fit with philosophical and cosmological principles, namely, uniform circular motion. Proposals for improved models began in earnest in the 13th century with the proposal of Mu'ayyad al-Dīn al-ʿUrḍī to replace Ptolemy's equant point, using a result now known as ʿUrḍī's Lemma. The fundamental work on al-ʿUrḍī was originally published by Saliba in 1990 and has been released in a third corrected edition in 2001 [Saliba 2001]. ʿUrḍī's contemporary Naṣīr al-Dīn al-Ṭūsī was the inventor of the Ṭūsī couple, a pair of circles that travel in uniform circular motion and yet cause a point to move along a straight line. This device allowed al-Ṭūsī to move parts of a planetary model closer to or further away from the center without violating uniform circular motion. Ragep [2000] discusses the origin and development of the couple in Ṭūsī's astronomical works, identifying some partial inspiration in Ibn al-Haytham's earlier model. All these developments raised the issue of the extent to which philosophical concerns should interact with astronomy; see Dallal [2010] (pp. 54-89) for a survey.

Although the new models were usually not used to compute planetary positions, an exception may be found in Jamshīd al-Kāshī's *Khāqānī Zīj*, which contains a complete calculation of a position of Venus. Al-Kāshī's calculation uses a model inspired by Ibn al-Haytham and al-Ṭūsī that replaces the epicycle with a sphere and applies spherical trigonometry [Van Brummelen 2006]. But the most marked change in the role of mathematics in planetary models took place very late, in the work of 16th-century astronomer Shams al-Dīn al-Khafrī. Al-Khafrī applied multiple models to the same planetary phenomenon, reconceiving the role of mathematics in astronomy as a modeling tool for physical astronomy, as opposed to an intellectual pursuit of the truth (originally described in a paper by Saliba in *Journal for the History of Astronomy* in 1994; see also Saliba [1997] and Saliba [2000]). An account of the development of all these models may be found in Saliba's *Islamic Science and the Making of the European Renaissance* [Saliba 2007], which (as we have seen) includes other themes such as a reconsideration of the emergence of science in early Islamic society.

One curious mathematical theory of the motion of the planets that turned out to have no basis was known as trepidation, or accession and recession. To explain a perceived gradual decrease in the obliquity of the ecliptic and changes in the velocity of the precession of the equinoxes, Theon of Alexandria and Indian and Islamic successors altered the Ptolemaic geometric models to include a backward and forward motion of the equinoctial points through an arc of 8°. A survey of the early history of trepidation from Theon to early tenth-century eastern authors Ibrāhīm ibn Sinān and al-Battānī may be found in Ragep [1996]. Trepidation later became a characteristic of Andalusian and Maghribi astronomy, starting with late 11th-century Toledan astronomer Ibn al-Zarqālluh and splitting into two distinct schools through the 14th century [Comes 1996; Comes 2007]. A Castilian textual fragment likely from the lost *zīj* of Ibn al-Kammād (an immediate successor of al-Zarqālluh) seems to fit the third of al-Zarqālluh's three trepidation models [Mancha 1998]; Chabás and Goldstein in 1994 had hypothesized al-Zarqālluh's second model as the source. Ibn al-Kammād is criticized in the early 13th-century *zīj* (astronomical handbook) of Ibn al-Hāʾim for apparently abandoning the Zarqāllian models, as described in Calvo [2002] (which also deals with the solar year and timekeeping). Comes [2001] is an edition of and commentary on Ibn al-Hāʾim's *zīj*; unusually for western *zījes*, it contains no tables to calculate trepidation. The work is based on the third of Zarqālluh's models, and intriguingly uses the newer spherical trigonometric techniques that had been invented to replace Menelaus's Theorem. Mercier [1996] studies trepidation tables in *De motu octave sphere* that appear in some copies of the *Toledan Tables*, and denies the claim by Gerard of Cremona that Thābit ibn Qurra was the source. Since these tables are designed to work with the *Toledan Tables*, the source must be more recent than Thābit;

Mercier conjectures Zarqālluh and Ibn al-Kammād. Comes [2002] surveys Maghribi texts from the 13th century onward, noting the application of trepidation to astrological predictions, and pointing out that these texts criticize but do not rework the existing models. Finally, Díaz-Fajardo [2001] studies a text by 15th-century Moroccan astronomer Abū ʿAbd Allāh al-Baqqār, who surveys the work of his Andalusian predecessors on trepidation and expresses misgivings about the entire theory.

Projections of the sphere: The mathematical study of projections of the sphere was motivated by the construction of the astrolabe, a representation of the celestial sphere on a flat surface using stereographic projection. The fundamental Greek text on the subject is Ptolemy's *Planisphere*, which we discussed earlier and recently has been edited and translated to English from Arabic manuscripts by Sidoli and Berggren [2007]. Stereographic projection was of use to medieval astronomers especially because of two mathematical properties: circles on the sphere are preserved through the mapping (known to Ptolemy), as are angles (not known, at least explicitly). Thus astrolabe makers needed only to draw circles and lines on the plate.

The astrolabe was a lively topic throughout the medieval Islamic period. The earliest complete extant treatise on the subject by mid-ninth century Baghdad scientist al-Farghānī, recently edited and translated to English [Lorch 2005], is noteworthy for its inclusion of numerical tables to help instrument makers. In the tenth century a number of detailed studies of the astrolabe appeared. Abgrall [2000] is a mathematical account of treatises by al-Ṣaghānī, Abū Sahl al-Kūhī, and a commentary by Ibn Sahl on al-Kūhī's work (which had been edited and translated by Rashed [1993] and Berggren [1994]; see also Rashed [2000c] for two fragments by Ibn Sahl, one on stereographic projection). Berggren [1998] presents two tenth-century methods for drawing azimuth circles on an astrolabe (useful for finding the direction of Mecca), the first by al-Kūhī and the second by Abū Maḥmūd al-Nasafī. Almost contemporaneous with these texts is Abū Naṣr Manṣūr ibn ʿIrāq's letter on the subject to al-Bīrūnī which contains a proof that seasonal hour lines on sundials are not straight and that hour curves on astrolabes are generally not circles — an improvement on al-Ṣaghānī's approach [Hogendijk 2001a]. Finally, Lorch [2000a] presents a treatise by Ibn al-Ṣalāḥ from the 12th century, which appears to contain a trace of Maslama's commentary on Ptolemy's *Planisphere*.

Stereographic projection was not the only means of representing a sphere on a flat surface. Kennedy, Kunitzsch and Lorch [1999] present a collection of texts on the "melon-shaped astrolabe." This device uses the "azimuthal equidistant" mapping, which is more intuitive than stereographic projection but results in melon-shaped curves on the instrument's plate rather than circles — more difficult for the instrument maker to produce on a brass plate (although not impossible, as we shall see later). Another astonishing discovery is a pair of 17th-century Iranian world maps, announced in King [1997] and published in a large illustrated volume in King [1999b]. These devices place Mecca at the center of the map, and the projection of the known world onto the device preserves accurately the direction and distance to Mecca for all locations. King hypothesizes that the theory underlying the map has its origin in the ninth century, in the work of Ḥabash al-Ḥāsib.

Geography and mathematics had many fruitful interactions, described in Sezgin's three volumes on Islamic geography, published as volumes X, XI, and XII of his series *Geschichte des Arabischen Schrifttums* [Sezgin 2000a,b,c]. The first volume in particular interacts with mathematical topics; it includes an argument that ninth-century mapmakers already used a stereographic projection of the terrestrial sphere, later corrupted by less competent successors. Sezgin also argues that Islamic influence on European geography persisted through the 18th century.

Religious requirements: Islam may be the only religion that may use trigonometry and spherical astronomy to perform some of its rituals. This manifests itself especially in the determination of the direction of Mecca (the *qibla*), the times of prayer (determined by the Sun's altitude), and the beginning of the sacred month of Ramadan (defined by the relative positions of the Sun and Moon). We have already described the remarkable world maps discovered by King that allow the user to find the *qibla* quickly, but most attempts to calculate the *qibla* use spherical trigonometry. Hogendijk [2000a] is an edition, translation to English and commentary of al-Nayrīzī's determination of the *qibla* of Baghdad using a predecessor of the eventually popular "method of the *zījes*." The text is dated around AD 900, which makes it the earliest mathematically precise calculation we have. Unfortunately numerical instabilities in the method render it so inaccurate in practice as to be be useless. Moussa [2011] provides a detailed mathematical analysis of the *qibla* determination in Abū 'l-Wafā''s *Almagest* about a century later, including the construction of the underlying trigonometric tables and a description of the observational instruments used. Rius [2000] has produced a volume surveying a variety of manuscripts on the *qibla* from al-Andalus and the Maghrib, discussing both the mathematical solutions and the more practical concerns of implementing the *qibla* in daily life (see also Rius [1996] and Rius and Comes [2006]). She covers treatises from the 12th century through the 18th, and edits and translates to Spanish the *Book of the Qibla* by 14th-century Berber scholar al-Maṣmūdī. The reaction of legal scholars to mathematical approaches to the *qibla*, not always positive, is examined in Dallal [2010] (pp. 3–9).

The *qibla* was only the beginning of the relation between astronomy and ritual. The study of timekeeping using the Sun and the stars, needed to determine the daily times of prayer, produced a rich and multi-faceted history beginning as early as the eighth century. This discipline has been brought to modern attention by King, whose past studies on the subject are included and extended in two volumes totalling almost 2000 pages under the title *In Synchrony with the Heavens: Studies in Astronomical Timekeeping and Instrumentation in Medieval Islam*. The first volume (*The Call of the Muezzin* [King 2004]), emphasizes texts and tables related to prayer times, extending beyond the mathematics to topics such as architecture and the role of the *muezzin* and the *muwaqqit* (timekeeper). The studies in the second volume (*Instruments of Mass Calculation* [King 2005]) deal specifically with mathematical and astronomical instruments constructed for the purpose of timekeeping.

Our final application of mathematical astronomy to religious practice is calendrical in nature. The emergence of the moon's crescent after new moon signified the beginning of the month in the lunar calendar, which was especially important for the month of Ramadan. Yazdi [2009] provides a detailed study of al-Khāzinī's mathematical tables to determine lunar crescent visibility.

Mathematical astronomical tables: The primary literature on mathematically-defined astronomical tables is almost overwhelming. About 250 *zījes* (astronomical handbooks containing tables allowing the reader to compute astronomical quantities) are known in the literature, and each *zīj* contains between a couple of dozen and a few hundred tables. A thriving tradition of arithmetic, trigonometric, and astronomical tables also existed outside of the *zīj* tradition. An early attempt to map out this vast terrain was Kennedy's 1956 classic *Survey of Astronomical Tables*. In 2001 King and Samsò, with a contribution from Goldstein, published a supplement to Kennedy's original survey [King and Samsó 2001]. Longer than the original, this article contains sections on the non-Ptolemaic planetary model-building tradition and on tables outside of the *zīj* tradition. Van Dalen is working on a comprehensive *zīj* survey that is anticipated to replace both of these works. Part of his research has revealed new and unusual astronomical works, including a new manuscript of the early *Mumtaḥan Zīj* [van Dalen 2004a], and the *Zīj-i Nāsirī* with Indian connections [van Dalen 2004b].

The *Khāqānī Zīj* by early 15th-century Persian scientist Jamshīd al-Kāshī is one of the most interesting *zījes*, for its inclusion of extensive theoretical passages explaining his innovations along with the tables. In 1998 Kennedy published a slim volume [Kennedy 1998] containing a detailed table of contents of this *zīj*. This book should help open the document to future researchers. Two articles on topics within this *zīj* have appeared recently: Kennedy [1995] describes al-Kāshī's determination of the ascendent in astrology (discussed below), while Van Brummelen [2006] gives an account of al-Kāshī's use of a spherical epicycle to calculate positions of the inferior planets (discussed above).

The entries in astronomical tables can often be analyzed to gain historical information about the table's origin. One example is van Dalen [1996], which analyzes the equation of time in the Latin translation of al-Majrīṭī's recension of al-Khwārizmī's *Sindhind Zīj*. Van Dalen uses a general analytical method he developed in 1989 to determine that the parameters embedded within the table's entries are mostly Ptolemaic in origin, implying that either the table itself or its parameters found their way somehow from eastern to western Islam. For other computational methods that have been used to gain historical information from tables, see Mielgo [1996] on mean motion tables and Van Brummelen

[1997] on dependences between tables.

Charette [1998] takes on the task of analyzing what might be the largest table ever constructed by hand: the 400,000-entry table by Cairene astronomer Najm al-Dīn al-Miṣrī around AD 1300. Although the original purposes of this monumental table were the calculation of altitudes of celestial objects and solar timekeeping, this table may be used to solve almost any problem of spherical astronomy by recasting the definitions of the table's arguments — provided that the user has the patience to deal with such a massive table.

The prediction of planetary longitudes was at the heart of most *zījes*, and there was some variety in methods of their tabulation. Longitudes were computed first by determining a "mean motion" (an average position arrived at by assuming that the planet travels at a constant speed). Next, two "equations" were added or subtracted to account for variations in speed due to the geometry of the Ptolemaic model. Samsó and Millás [1998] present the work of Andalusian astronomer Ibn al-Bannā᾽ al-Marrākushī around AD 1300, who adopts the practice of "displacement" in the tabulation of equations — adding to the equation a constant sufficiently large so that its value is always positive, and subtracting that constant from the mean motion. This calculation saved the user from deciding whether to add or subtract the equation, removing a major potential source of errors. Late tenth-century astronomer Kūshyār ibn Labbān uses the same procedure in his *Jāmiʿ Zīj*, but also tabulates one of the equations in an innovative way that decreases the computational burden compared to the usual Ptolemaic method [Van Brummelen 1998]. Finally, Dorce [2002] presents the *Crown of Astronomical Handbooks* of the 13th-century Andalusian and Maghribi astronomer Muḥyī al-Dīn al-Maghribī, noting that some of his planetary equation tables are "homothetic" — that is, they are computed simply by taking constant multiples of the entries in other tables. As for the tabulation of planetary latitudes, van Dalen has compiled a very helpful survey of Islamic mathematical approaches [van Dalen 1999] as part of his survey of a Chinese astronomical document.

Instruments: The construction of mathematical instruments to solve astronomical problems goes back to the beginning of Islamic science. This is illustrated by a set of eleven short treatises in two manuscripts on the astrolabe and other instruments from the early ninth century (two of which are attributed to al-Khwārizmī), edited and translated to English in Charette and Schmidl [2004]. Charette and Schmidl [2001] also study a text describing a later and rather complex ninth-century instrument, complete with several sine quadrants and trigonometric scales, which they attribute to Ḥabash al-Ḥāsib; they suggest that the instrument was used to find the time from the Sun's altitude. Although circles and straight lines were the geometric building blocks of most instruments, this is not true of Nastulus's AD 900 solar timekeeping instrument [King 2008]. The mathematically complex curves of this unique device are rendered with mathematical precision on its plate. A treatise by Ibrāhīm ibn Sinān on sundials was the subject of Luckey's 1941 doctoral dissertation; it has been published for the first time along with Hogendijk's edition of the Arabic text in Luckey [1999]. Sinān's work seems to have contained a proof that hour lines on a sundial are not straight, but the manuscript ends before the proof; and Ibn al-Haytham comments in another text (edited and translated to French in Rashed [2006]) that the demonstration was incomplete. (See also a French translation of Sinān's text, published in Rashed and Bellosta [2000].) Finally, the author of the gigantic 400,000-entry astronomical table we discussed earlier, Najm al-Dīn al-Miṣrī, also wrote a *Treatise on Instruments*. This book describes the construction of over 100 devices, including astrolabes, quadrants, sundials, and trigonometric instruments. The volume was edited and translated to English in Charette [2003].

Mathematics and astrology: The blurring of the boundary between astronomy and astrology in medieval Islam is highlighted by the fact that all of the medieval authors we shall mention here have already appeared elsewhere in this survey. It would be a mistake to assume that astrology was mathematically sparse; many of the quantities needed for its practice gave rise to some of the fiercest problems in spherical trigonometry. The lamented passing of E. S. Kennedy has impacted this area, but several recent studies of particular texts are helping to expose scholars to what is available. Kennedy [1996] is a good starting point to study the problem of dividing the ecliptic into the twelve astrological "houses", which vary depending on time and locality. Kennedy discusses nine mathematical solutions in his survey of 28 manuscripts, including two methods that were not in North's *Horoscopes in History*. North's reply to Kennedy's article in the same volume [North 1996] speculates that the problem arose fairly late in astrological history, around the fifth century AD. 11th-century Andalusian astronomer Ibn Muʿādh al-Jayyānī, known for bringing the revolution in spherical trigonometry to Spain from eastern Islam, discusses both the astrological houses and the astrological aspects (dividing the ecliptic according to "rays" cast by the planets) in his *On the Projection of Rays*. Hogendijk [2005] provides an account, edition and English translation of the relevant parts of this work as well as the related Latin treatise *Tabulae Jahen*, speculating an influence on Regiomontanus; Casulleras [2010] has since published and translated to Spanish Ibn Muʿādh's entire treatise. (See Casulleras [2008] for a general survey of mathematical astrology in the medieval Islamic west.) [Kennedy 1995] describes Jamshīd al-Kāshī's account of the determination of the ascendent (the point of the ecliptic on the eastern horizon at a given moment) in the *Khāqānī Zīj*. Al-Kāshī gives fifteen methods, some of which apply somewhat complicated spherical trigonometry. Complete and comprehensive studies of mathematical astrological texts are difficult to find; one notable exception is Kennedy [2009] on al-Battānī's astrological history, completed after Kennedy's passing by colleagues and former students.

Other problems: Astronomical problems with substantial mathematical content not captured in the above categories include the following:

- Rashed [2006] is an edition and translation into French of five treatises in mathematical astronomy by Ibn al-Haytham. One of these is the work mentioned above that completes Ibrāhīm ibn Sinān's proof that hour lines on a sundial are not straight. By far the largest work of the five (on which see also Rashed's commentary in [Rashed 2007b]) is a mathematical treatise on the (extremely small) effects of the planets' motions with respect to the background of the fixed stars on their daily rotations on the celestial sphere. Other works include *On the Compass for Drawing Great Circles*, and two short treatises on the altitude of a fixed star above the horizon and on horizontal sundials.
- A small tradition seems to have existed on the problem of finding the dip angle to the horizon when on an elevated platform. Rashed presents an Arabic edition, English translation [Rashed 2001] and French translation [Rashed 2002a] of tenth-century geometer Abū Sahl al-Kūhī's *What is Seen of Sky and Sea*, and Berggren and Van Brummelen [2003] study al-Samawʾal's dubious criticism of al-Kūhī's method in his *Exposure of the Errors of the Astronomers*. A later treatise on the same topic, attributed to 13th-century poet Bābā Afḍal al-Dīn al-Kāshī, is ascribed by Bagheri to the famous 15th-century astronomer Jamshīd al-Kāshī [Bagheri 2002].
- Al-Kūhī also wrote a treatise on a method to find the distance to a shooting star. Kūhī applies the geometric technique of analysis by knowns, but he seems not to have performed the actual calculation. See Arabic editions and English translations in Berggren and Van Brummelen [2001b] and Rashed [2001], and a French translation in Rashed [2002a].

Arithmetic

At first glance a rather banal subject, arithmetic provides the historian of medieval mathematics with a rich source of information, into both the transmission of mathematical concepts and the surprisingly sophisticated procedures developed by historical authors.

Use and transmission of numeration systems: The number system we use today originated in India, and was transmitted to us via a rather complicated route that passed through Islam. The system and its accompanying methods of calculation found their way first to eastern Islam in the eighth century, with al-Khwārizmī's *Arithmetic* playing a major role. Study of its transmission into western Islam and the new forms the numerals took in this journey suffers from a lack of manuscript evidence, particularly between the 10th and 13th centuries. The transmission to Latin sources is aided by the discovery by Folkerts of a new manuscript of a 12th-century Spanish Latin reworking of al-Khwārizmī's text (the first manuscript to contain the complete text; announced in Folkerts [1998], outlined in Folkerts [2001], and edited and translated into German in Folkerts [1997]). Two articles by Paul Kunitzsch (Kunitzsch [2003b] in English; Kunitzsch [2005] in German) provide detailed analyses of the state of research on the entire process of transmission of Hindu-Arabic numeration. See also Berggren [2002a], which surveys the arrival of Arabic numerals in Europe, including the role of astronomy and the use of sexagesimal (base 60) fractions. Multiple numeration systems in concurrent use were common; see Guergour [2000] for discussion of a treatise by Ibn al-Bannā᾽ on *rūmī* numbers, one of five systems circulating in the Maghreb at the time of the Ottoman Empire.

Numbers in practice: A number of texts explaining how to perform arithmetic calculations may be found in the literature, aimed at several audiences. They were composed by a wide variety of authors. Some are not well known otherwise, such as the 12th century text by Abū Bakr al-Ḥaṣṣār; a new manuscript of this treatise is announced in Kunitzsch [2003a]. Some are known especially for their arithmetic texts, such as 15th century Andalusian and Moroccan author al-Qalaṣādī, whose early intellectual journey is described in Marin [2004]. Others are among the leading lights in Islamic mathematics, including the great 15th-century Persian calculator Jamshīd al-Kāshī, whose *Key to Arithmetic* describes (among other topics) finding the greatest common divisor and least common multiple of a set of numbers and a method of computing the n-th root of a given number. This work is summarized in Azarian [2000]. Johansson [2011] considers various procedures for calculating cube roots in medieval cultures, confirming a hypothesized link between Persia and China. Berggren [1995] explains "finger reckoning" using a work by al-Baghdādī, in which fractions whose denominators exceed 10 were avoided. Sometimes genuinely new mathematics can be discovered in these works; a 14th-century commentary on a tenth-century Yemeni treatise on practical computation by al-Ṣardafī contains a new iterative method for determining square roots [Rebstock 1999].

However, much of the arithmetic art was geared toward practical (rather than scholarly) purposes, and most of the Islamic arithmetic works under recent scholarly scrutiny have emphasized applications. These especially included mensuration, particularly with respect to weights and measures. (On this topic, see also the section on mensuration in "Geometry," above.) Rebstock [1995] is an edition of *On Business Arithmetic*, a treatise by Ibn al-Haytham geared specifically to uses in commerce. Other arithmetic works dealing with applications include the following:

- *Book Containing the Operations Connected with the Royal Court and Descriptions of the Arithmetic of Secretaries*, a 12th century work written by a student of one al-Shaqqāq [Rebstock 2002/03];
- 13th-century Iranian poet Nāṣer-e Khosrow's *Curiosities of Arithmetic and Wonders of Arithmeticians* (written in Persian), which contains currency exchange problems requiring the solution of indeterminate linear equations [Bagheri 1997a];
- another Persian work from the 16th century by Qāsim ibn Yūsuf Abū Naṣr, which includes the determination of land areas (see the summary of the section on multiplication in Kennedy [2000]).

A recurring use of arithmetic was the correct allocation of an estate among claimants according to inheritance law. Rebstock [2008/09] traces the history of this topic, presenting both a logical and an arithmetical tradition. Arithmetical works containing inheritance problems that have been treated recently include al-Hubūbī's tenth century text from Khwārizm, which uses methods that delve into algebra and even geometry [Laabid 1998]; and a work by 11th-century Damascene al-Qurashī (described in Rebstock [2002a] and translated to German in Rebstock [2001]).

Recreational mathematics; magic squares: Every mathematical culture designs problems for recreational purposes, such as those found in the *Key to Transactions* by 11th-century Persian mathematician Muḥammad ibn Ayyūb Ṭabarī (surveyed in Bagheri [1999]). Research in Islamic recreational mathematics in the past fifteen years (and earlier) has been dominated by Jacques Sesiano's continuing studies of magic squares. Although magic squares may have existed earlier, serious study of their properties and methods of generation began in the early Islamic period. We know that interest began as early as the ninth century; our earliest treatises are from the tenth. The volume Sesiano [2004a] is a survey that begins with two of these early texts and illustrates an increase in sophistication over the centuries, including eventually the construction of magic squares of any size and subject to a number of conditions. One particular magic square in a tenth-century work by al-Anṭākī [Sesiano 2003c] deserves the name *quadratus mirabilis*; it is a "bordered" square (if one removes the cells on the outside border, a magic square remains; and again on the remaining square, and so on) with special conditions on the positions of the even and odd numbers. The other early work is *Treatise on the Magic Disposition of Numbers in Squares* by Abū 'l-Wafā', also dealing with bordered squares and other methods of construction (discussed in Sesiano [2003c] and edited in Sesiano [1998b]). Sesiano discusses several anonymous treatises on magic squares: two texts probably from the 11th century, *On the Harmonious Arrangement of Numbers* [Sesiano 1996c] and *Primer on the Harmonious Arrangement of Numbers* [Sesiano 1996a]; and a late 12th-century treatise that also contains some material on perfect and amicable numbers [Sesiano 2003a]. Constructions of particular types of magic square are surveyed in Sesiano [1998a] (on simple magic squares) and Sesiano [2003b] (on squares constructed using the moves of chess pieces, especially the knight). Although magic squares were of interest to mathematicians, after the 12th century they were also used as talismans, as exemplified in a Latin (but originally Arabic) text given and translated in Sesiano [2004b]. Two regional studies also have appeared: Sesiano [2004c] on magic squares in Iran, and Comes and Comes [2009] on magic squares in al-Andalus (especially a treatise by Azarquiel).

Algebra

The language of algebra: As powerful as modern algebraic notation is, it contains hidden assumptions and styles of thought that can make it something of a corrupting influence when presenting historical mathematics. We are starting to realize how careful one must be when considering the possible meanings of algebraic words and symbols, especially when studying pre-modern cultures. A number of studies (led by Oaks) are drawing attention to these matters in medieval Islam. It has been assumed commonly that words like "*al-jabr*" ("restoration") were technical and had specific mathematical meanings, but Oaks and Alkhateeb [2007] note that ninth-century algebraic manuscripts still used these words in non-technical senses. The word "*māl*", for instance, could take on different meanings at different stages of a problem — "quantity" if it was used in the problem's statement, and "square" in its solution [Oaks and Alkhateeb 2005]. When we read polynomial expressions like $3x + 4$ we think of the terms $3x$ and 4 as added, but in Arabic it was interpreted as an aggregate of seven objects [Oaks 2009]. Subtler distinctions, like different flavors of algebraic meaning that we subsume under the same $=$ sign, were also made [Oaks 2009]. In fact, algebra itself had a different relation to geometry than it has today: as an artificial language, algebra was restricted to problem solving while geometry was the language of theory expression [Oaks 2007]. For the most part Arabic algebra was rhetorical (expressed verbally), but an abbreviated algebraic notation arose in the 12th century. This notation is surveyed in Abdeljaouad [2002], concentrating on an 18th-century Tunisian manuscript of Ibn al-Hā'im. Oaks [2009] also explains this notation and applies it to other treatises; in Oaks [2007] he argues that this notation qualifies as "symbolic", not merely "syncopated".

Solving equations: The art of solving equations, so influenced today by our symbolic algebra, was conceived differently in the medieval period. Not only its language, but also its practice, differed from author to author. The classic early work (from which "algebra" derives its name) is al-Khwārizmī's early ninth-century *Algebra*, edited and translated to French in Rashed [2007a] and to English in Rashed [2009a]. Rashed argues that the style of al-Khwārizmī's work was distinctly Arabic, different from the earlier Greek practice sometimes called "geometric algebra," and not influenced by Indian texts. Certainly later authors were more directly influenced by the Greek tradition, such as Thābit ibn Qurra in his treatise showing the correspondence between the solutions of second-degree equations by algebraists and those by geometers (edited and translated to French in Rashed [2009d]). The next major algebraic work is Abū Kāmil's *Algebra*, around AD 900. It relies partly on al-Khwārizmī, but also on authors such as Heron and Euclid. Chalhoub [2004] provides the first German translation of the main part of Abū Kāmil's work, and Oaks and Alkhateeb use both authors' works in their study Oaks and Alkhateeb [2005] of the various uses of the word "*māl*".

Later authors also employed differing approaches. A thirteenth-century Iranian treatise by al-Fārisī, for example, a commentary on an earlier work by al-Baghdādī, describes the use of the Greek geometric techniques of analysis and synthesis to solve algebraic equations [Mawaldi 2000]. One of the most famous treatises in algebra is ʿUmar al-Khayyāmī's treatise on solving cubic equations, edited and translated into French in Rashed and Vahabzadeh [1999] (along with a related smaller treatise). In this work Khayyāmī solved 14 different types of cubic equation by means of the intersection of conic sections. (See Djebbar [2000c] for a survey of Khayyāmī's work, context, and influence.) The 14 types emerge due to the lack of negative numbers; without them, $ax^3 + bx^2 = c$ needs to be approached differently from $ax^3 = bx^2 + c$, for instance. Khayyāmī's use of the restricted Greek notion of number did not prevent him from supplying a rigorous foundation for algebra by regarding real numbers as "the homogeneous measures of continuous magnitudes" [Oaks 2011]. However, the emergence of a concept of negative numbers may be seen to begin earlier, with algebraic treatises such as those by al-Karajī and al-Samawʾal, as argued in Bellosta [2004a].

Readers looking for a broad overview of the development of algebra from al-Khwārizmī through the 12th century may be interested in Djebbar [2005], intended for a more general audience. It discusses various cultural influences and transmissions, dealing with western Islam in some detail.

Number theory: Although medieval Islamic number theory perhaps has not been on the scholarly agenda recently as much as it should have, several studies have shed some light on the subject. Amicable numbers (two numbers so that the divisors of each number sum to the other one) were a recurring interest, as witnessed by two works separated by five centuries. Thābit ibn Qurra's ninth-century *On the Extraction of Amicable Numbers*, edited and translated to French in Rashed and Houzel [2009], provided a method to calculate such pairs. Al-Fārisī's 14th-century *Memorandum to Colleagues Explaining the Proof of Amicability* contains a sequence of propositions that might be interpreted as part of a proof of the Fundamental Theorem of Arithmetic (described in Ağargün [2000]). Figurate numbers also attracted attention, including in al-Andalus and the Maghrib. Djebbar [2000b] studies an early 13th-century work by Ibn Munʿim that discusses the history of these numbers in these regions beginning in the 11th century; its mathematical discussion involves some use of polynomials.

Concluding Remarks

The magnitude of research reported here is a clear sign that the study of medieval Islamic mathematical sciences is healthy and developing rapidly. Thanks to a flood of new editions and translations, access to original sources is much easier than it was fifteen years ago. We are also closer to developing a more rounded picture: rather than concentrating on sources that capture our interest, scholars today are representing more effectively the diversity of topics and interests found in the primary literature. The ground on which we can build an understanding of the field is fertile.

However, substantial questions face us on how to proceed. As we have noted, larger debates concerning how to engage in the history of mathematics are starting to impact the Islamic exact sciences. When writing a mathematical culture's history, should one do as most have done in the past — concentrate on the greatest intellectual achievements and trace a narrative from one great man to the next? Should we emphasize instead the daily practice of mathematical practitioners, thereby giving a better picture of the intellectual life of a typical mathematical scholar of the time? Should we go even further, and emphasize the social context of the mathematical ideas, the institutions, and the mathematical education of all people? Although currently most work continues to focus on the historical leading edge of mathematical accomplishments, these alternative approaches hold great potential to enrich our efforts. They allow the history of Islamic exact sciences to address a broader and deeper range of historical issues, thereby making the field more relevant to society at large — without needing to exclude or reject our traditional pathways.

Finally, the increased access to primary sources might allow us to develop meaningful perspectives on the unique character of medieval Islamic mathematics. We need to find appropriate methods and vocabulary to characterize what makes this period similar to and different from its antecedent, contemporaneous and successor scientific cultures. Indeed, as Sabra has pointed out, we must find the words to describe the different characters within the many and diverse subcultures within medieval Islam. Certainly a shared (although not universal) religion and language have great influence on scientific character, although the nature of those influences are not always obvious. But are there also other ways of identifying styles of reasoning, types of mathematical and scientific interest, and shifting borders of thought? The community at large does not yet have an agreed-upon collection of issues, or even a shared grammar, to come to consensus on these thorny matters. Even so, the fruitfulness of our research and increasing awareness of the deep questions lead inevitably to confidence that well-supported and fascinating approaches will continue to emerge in the next fifteen years.

Acknowledgements Without doubt, inter-library loan librarians are miracle workers. As the reader may well imagine paging through the following bibliography, Shauna Bryce at Quest University provided extraordinary service and support far beyond the usual call of duty. My thanks go also to my student Jonathan von Ofenheim for TEX support and data entry, and to Ariel Van Brummelen for help in editing both the paper and the bibliography.

References

Abdeljaouad, M., 2002. Le manuscrit mathématique de Jerba : Une pratique des symboles algébriques maghrébins en pleine maturité. Quaderni di Ricerci in Didattica 11, 63. pp. 110–173. http://math.unipa.it/~grim/quaderno11.htm.

——— 2003. Commentary on the poem of al-Yāsamīn. Association Tunisienne des Sciences Mathématiques, Le Bardo.

——— 2005. 12th century algebra in an Arabic poem: Ibn Al-Yāsamīn's Urjūza fi'l-Jabr wa'l-Muqābala. LLULL 28, 181–194.

Abgrall, P., 1997. Al-Qūhī et les courbes coniques. In: Hasnawi, A., Elaramni-Jamal, A., Aouad, M. (eds.), Perspectives arabes et médiévales sur la tradition scientifique et philosophique grecque. Peeters, Leuven, pp. 21–29.

——— 2000. La géométrie de l'astrolabe au Xe siècle. Arabic Sciences and Philosophy 10, 7–77.

——— 2002. Une contribution d'al-Qūhī à l'analyse géométrique. Arabic Sciences and Philosophy 12, 53–89.

——— 2004a. Al-Qūhī, archimédien. In: Morelon, R., Hasnawi, A. (eds.), De Zénon d'Élée à Poincaré : Recueil d'études en hommage à Roshdi Rashed. Peeters, Leuven, pp. 85–188.

——— 2004b. Le Développement de la Géométrie aux IXe–XIe Siècles—Abū Sahl al-Qūhī. Librairie Scientifique et Technique Albert Blanchard, Paris.

Ağargün, A., 2000. Kamal al-Din al-Farisi and the Fundamental Theorem of Arithmetic. In: İhsanoğlu, E., Günergun, F. (eds.), Science in Islamic Civilisation. IRCICA, Istanbul, pp. 185–192.

Ahmedov, A., Rosenfeld, B.A., 2000. The mathematical treatise of Ulugh Beg. In: İhsanoğlu, E., Günergun, F. (eds.), Science in Islamic Civilisation. IRCICA, Istanbul, pp. 143–150.

Arnzen, R., 2002. Abū l-ʿAbbās an-Nayrīzīs Exzerpte aus (Ps.-?) Simplicius' Kommantar zu den Definitionen, Postulaten und Axiomen in Euclids Elementa I. Rüdiger Arnzen, Köln.

Azarian, M.K., 2000. Meftah al-Hesab: A summary. Missouri Journal of Mathematical Sciences 12, 75–95.

——— 2004. Al-Kāshī's fundamental theorem. International Journal of Pure and Applied Mathematics 14, 499–509.

——— 2009. The introduction of al-Risāla al-Muhītīyya: An English translation. International Journal of Pure and Applied Mathematics 57, 903–914.

——— 2010. Al-Risāla al-Muhītīyya: A summary. Missouri Journal of Mathematical Sciences 22, 64–85.

Bagheri, M., 1997a. Mathematical problems of the famous Iranian poet Nāṣer-e Khosrow. Historia Mathematica 24, 193–196.

——— 1997b. A newly found letter of al-Kāshī on scientific life in Samarkand. Historia Mathematica 24, 241–256.

——— 1999. Recreational problems from Ḥāsib Ṭabarī's Miftāḥ al-muʿāmalāt. Gaṇita-Bhāratī 21, 1–9.

——— 2002. A new treatise by al-Kāshī on the depression of the visible horizon. In: Dold-Samplonius, Y., Dauben, J., Folkerts, M., van Dalen, B. (eds.), From China to Paris: 2000 Years Transmission of Mathematical Ideas. Steiner, Stuttgart, pp. 357–368.

Bagheri, M., Hogendijk, J., 1996. Al-Sijzī's Treatise on Geometrical Problem Solving. Fatemi, Tehran.

Bancel, F., 2001. Les centres de gravité d'Abū Sahl al-Qūhī. Arabic Sciences and Philosophy 11, 45–78.

Bellosta, H., 1997. Ibrāhīm ibn Sinān : Apollonius arabicus. In: Hasnawi, A., Elaramni-Jamal, A., Aouad, M. (eds.), Perspectives Arabes et Médiévales sur la Tradition Scientifique et Philosophique Grecque. Peeters, Leuven, pp. 31–48.

——— 1999. The specific case of geometrical manuscripts using the example of manuscript B.N. 2457. In: Yusuf, I., (ed.), Editing Islamic Manuscripts on Science. Al-Furqān Islamic Heritage Foundation, London, pp. 181–191.

——— 2002a. Un complément arabe aux Données d'Euclide : Le Kitab al-Mafrudat de Tabit ibn Qurra. Ansari, S.M.R. (ed.), Science and Technology in the Islamic World: Proceedings of the XXth International Congress of History of Science. Brepols, Turnhout, pp. 71–82.

——— 2002b. Les instruments ardents dans la tradition arabe. Matapli 67, 73-88.

——— 2004a. L'émergence du négatif. In: Morelon, R., Hasnawi, A. (eds.), De Zénon d'Élée à Poincaré : Recueil d'études en hommage à Roshdi Rashed. Peeters, Leuven, pp. 65–83.

——— 2004b. Le traité de Thābit ibn Qurra sur la figure secteur. Arabic Sciences and Philosophy 14, 145–168.

Ben Miled, M., 1999. Les commentaires d'al-Māhānī et d'un anonyme du Livre X des Éléments d'Euclide. Arabic Sciences and Philosophy 9, 89–156.

——— 2004. Les quantités irrationnelles dans l'œuvre d'al-Karajī. In: Morelon, R., Hasnawi, A. (eds.), De Zénon d'Élée à Poincaré : Recueil d'études en hommage à Roshdi Rashed. Peeters, Leuven, pp. 27–54.

——— 2008. Mesurer le continu, dans la tradition arabe des Livres V et X des Éléments. Arabic Sciences and Philosophy 18, 1–18.

Berggren, J.L., 1995. Numbers at work in medieval Islam. Journal for History of Arabic Science 11, 45–51.

——— 1996a. Al-Kūhī's "Filling a lacuna in Book II of Archimedes" in the version of Naṣīr al-Dīn al-Ṭūsī. Centaurus 38, 140–207.

——— 1996b. Islamic acquisition of the foreign sciences: A cultural perspective. In: Ragep, J., Ragep, S., Livesey, S. (eds.), Tradition, Transmission, Transformation. Brill, Leiden, 1996, pp. 263–283.

——— 1998. Geometric methods in medieval Islam: The case of the azimuth circles. In: Actes du troisième colloque maghrébin sur l'histoire des mathématiques arabes, vol. 1. Office des Publications Universitaires, Algiers, pp. 13–21.

——— 2002a. Medieval arithmetic: Arabic texts and European motivations. In: Contreni, J.J., Casciani, S. (eds.), Word, Image, Number: Communication in the Middle Ages. Edizioni del Galluzzo, Florence, pp. 351–365.

——— 2002b. The transmission of Greek geometry to medieval Islam. Cubo Matemática Educacional 4, 1–13.

——— 2003. Tenth-century mathematics through the eyes of Abū Sahl al-Kūhī. In: Hogendijk, J.P., Sabra, A.I., (eds.), The Enterprise of Science in Islam: New Perspectives. Dibner Institute Studies in the History of Science and Technology, MIT Press, Cambridge, MA, pp. 177–196.

Berggren, J.L., Hogendijk, J., 2004. The fragments of Abū Sahl al-Kūhī's lost geometrical works in the writings of al-Sijzī. In: Burnett, C., Hogendijk, J., Plofker, K., Yano, M. (eds.), Studies in the History of the Exact Sciences in Honour of David Pingree. Brill, Leiden, pp. 609–655.

Berggren, J.L., Sidoli, N., 2007. Aristarchus's On the Sizes and Distances of the Sun and the Moon: Greek and Arabic texts. Archive for History of Exact Sciences 61, 213–254.

Berggren, J.L., Van Brummelen, G., 1999. Abū Sahl al-Kūhī on two geometrical questions. Zeitschrift für Geschichte der arabisch-islamischen Wissenschaften 13, 165–187.

——— 2000. Abū Sahl al-Kūhī's On the Ratio of the Segments of a Single Line that Falls on Three Lines. Suhayl 1, 11–56.

——— 2001a. Abū Sahl al-Kūhī on rising times. SCIAMVS 2, 31–46.

——— 2001b. Abū Sahl al-Kūhī on the distance to the shooting stars. Journal for the History of Astronomy 32, 137–151.

——— 2001c. Abū Sahl al-Kūhī's On Drawing Two Lines from a Point at a Known Angle. Suhayl 2, 161–198.

——— 2001d. The role and development of geometric analysis and synthesis in ancient Greece and medieval Islam. In: Suppes, P., Moravcsik, J.M., Mendell, H. (eds.), Ancient & Medieval Traditions in the Exact Sciences. Essays in Memory of Wilbur Knorr, CSLI Publications, Stanford, pp. 1–31.

——— 2002. From Euclid to Apollonius: Al-Kūhī's lemmas to the Conics. Zeitschrift für Geschichte der arabisch-islamischen Wissenschaften 15. 165–174.

——— 2003. Al-Samawʾal versus al-Kūhī on the depression of the horizon. Centaurus 45, 116–129.

——— 2005. Al-Kūhī's revision of Book I of Euclid's Elements. Historia Mathematica 32, 426–452.

Boulahia, N., 2007. Quelques contributions arabes en trigonométrie sphérique. In: Abattouy, M. (ed.), Études d'histoire des sciences arabes. Fondation du roi Abdul-Aziz Al Saoud pour les études Islamiques et les sciences humaines, Casablanca, pp. 35–41.

Bouzoubaâ Fennane, K., 2003. Réflexions sur le principe de continuité à partir du commentaire d'Ibn al-Haytham sur la Proposition I.7 des Éléments d'Euclide. Arabic Sciences and Philosophy 13, 101–136.

Brentjes, S., 1996a. The relevance of non-primary sources for the recovery of the primary transmission of Euclid's Elements into Arabic. In: Ragep, J., Ragep, S., Livesey, S. (eds.), Tradition, Transmission, Transformation. Brill, Leiden, pp. 201–225.

——— 1996b. Remarks about the proof sketches in Euclid's Elements, Book I as transmitted by MS Paris, B.N. Fonds Latin 10257. In: Folkerts, M. (ed.), Mathematische Probleme im Mittelalter: Der Lateinische und Arabische Sprachereich. Harrassowitz, Wiesbaden, pp. 115–138.

——— 1997. Additions to Book I in the Arabic traditions of Euclid's Elements. Studies in History of Medicine and Science 15, 55–117.

——— 1998a. On the Persian transmission of Euclid's Elements. In: Vesel, Ž., Beikbaghban, H., Thierry de Crussol des Epesse, B. (eds.), La science dans le monde iranien à l'époque islamique. Institut français de recherche en Iran, Tehran, pp. 73–94.

——— 1998b. La transmission arabe de Introductio arithmetica dans des travaux non-mathématiques au cours du IXe siècle. In: Actes du troisième colloque maghrébin sur l'histoire des mathématiques arabes, vol. 1. Office des Publications Universitaires, Algiers, pp. 23–29.

——— 2000. Aḥmad al-Karābīsī's commentary on Euclid's Elements. In: Folkerts, M., Lorch, R. (eds.), Sic Itur ad Astra. Harrassowitz, Wiesbaden, pp. 31–75.

——— 2001a. Observations on Hermann of Carinthia's version of the Elements and its relation to the Arabic transmission. Science in Context 14, 39–84.

——— 2001b. Two comments on Euclid's Elements? On the relation between the Arabic text attributed to al-Nayrīzī and the Latin text ascribed to Anaritius. Centaurus 43, 17–55.

——— 2006. An exciting new Arabic version of Euclid's Elements: MS Mumbai, Mullā Fīrūz R.I.6. Revue d'histoire des mathématiques 12, 169–197.

——— 2007. Reflections on the role of the exact sciences in Islamic culture and education between the twelfth and the fifteenth centuries. In: Abattouy, M. (ed.), Études d'histoire des sciences arabes. Fondation du Roi Abdul-Aziz al Saoud pour les Études Islamiques et les Sciences Humaines, Casablanca, pp. 15–33.

——— 2008a. Courtly patronage of the ancient sciences in post-classical Islamic societies. Al-Qanṭara 29, 403–436.

——— 2008b. Euclid's Elements, courtly patronage and princely education. Iranian Studies 41, 441–463.

——— 2009. Patronage of the mathematical sciences in Islamic societies. In: Robson, E., Stedall, J. (eds.), Oxford Handbook of the History of Mathematics. Oxford University Press, Oxford, pp. 301–327.

——— 2010. The mathematical sciences in Safavid Iran: Questions and perspectives. In: Hermann, D., Speziale, F. (eds.), Muslim Cultures in the Indo-Iranian World during the Early-Modern and Modern Period. Klaus Schwarz Verlag, Berlin, pp. 325–372.

Busard, H.L.L., 1996. Einiges über die Handschrift Leiden 399,1 und die arabisch-lateinische Übersetzung von Gerard von Cremona. In: Dauben, J.W., Folkerts, M., Knobloch, E., Wussing, H. (eds.), History of Mathematics: States of the Art. Academic Press, San Diego, pp. 173–205.

Calvo, E., 2002. Ibn al-Kammād's astronomical work in Ibn al-Hā'im's Al-Zīj al-Kāmil fī-l-Ta'ālīm. I. Solar year, trepidation, and timekeeping. In: Ansari, S.M.R. (ed.), Science and Technology in the Islamic World: Proceedings of the XXth International Congress of History of Science. Brepols, Turnhout, pp. 109–120.

Casulleras, J., 2008/09. Mathematical astrology in the medieval Islamic west. Zeitschrift für Geschichte der arabisch-islamischen Wissenschaften 18, 241–268.

——— 2010. La Astrología de los Matemáticos. La Matemática Aplicada a la Astrología a Través de la Obra de Ibn Mu'āḏ de Jaén. Universitat de Barcelona, Barcelona.

Chalhoub, S., 2004. Die Algebra: The Kitab al-Gabr wal-muqabala of Abu Kamil Soga ibn Aslam. Institute for the History of Arabic Science, University of Aleppo, Aleppo.

Charette, F., 1998. A monumental medieval table for solving the problems of spherical astronomy for all latitudes. Archives internationales d'histoire des sciences 48, 11–64.

——— 1999. Der geflügelte quadrant: Ein ungewöhnliches sinusinstrument aus dem 14. jh. In: Eisenhardt, P., Linhard, F., Petanides, K. (eds.), In Der Weg der Wahrheit: Aufsätze sur Einheit der Wissenschaftsgeschichte. Olms, Hildesheim, pp. 25–36.

——— 2003. Mathematical Instrumentation in Fourteenth-Century Egypt and Syria: The Illustrated Treatise of Najm al-Dīn al-Mīṣrī. Brill, Leiden.

Charette, F., Schmidl, P., 2001. A universal plate for timekeeping by the stars by Ḥabash al-Ḥāsib: Text, translation and preliminary commentary. Suhayl 2, 107–159.

——— 2004. Al-Khwārizmī and practical astronomy in ninth-century Baghdad. The earliest extant corpus of texts in Arabic on the astrolabe and other portable instruments. SCIAMVS 5, 101–198.

Comes, M., 1996. The accession and recession theory in al-Andalus and the north of Africa. In: Casulleras, J., Samsó, J. (eds.), From Baghdad to Barcelona. University of Barcelona, Barcelona, pp. 349–364.

——— 2001. Ibn al-Hā'im's trepidation model. Suhayl 2, 291–408.

——— 2002. Some new Maghribi sources dealing with trepidation. In: Ansari, S.M.R. (ed.), Science and Technology in the Islamic World: Proceedings of the XXth International Congress of History of Science. Brepols, Turnhout, pp. 121–141.

——— 2007. La trépidation dans les tables astronomiques d'al-Andalus et d'Afrique du Nord. In: Abattouy, M. (ed.), Études d'histoire des sciences arabes. Fondation du Roi Abdul-Aziz Al Saoud pour les Études Islamiques et les Sciences Humaines, Casablanca, pp. 99–120.

Comes, M., Comes, R. Los cuadrados mágicos matemáticos en al-Andalus. El tratado de Azarquiel. Al-Qanṭara 30, 137–169.

Crozet, P., 1997. Al-Sijzī et les Éléments d'Euclide : Commentaires et autres démonstrations des propositions. In: Hasnawi, A., Elaramni-Jamal, A., Aouad, M. (eds.), Perspectives Arabes et médiévales sur la tradition scientifique et philosophique grecque. Peeters, Leuven, pp. 61–77.

——— 1999. À propos des figures dans les manuscrits arabes de géométrie : L'exemple de Siğzī. In: Ibish, Y. (ed.), Editing Islamic Manuscripts on Science. Al-Furqān Islamic Heritage Foundation, London, pp. 131–163.
——— 2004a. Al-Siğzī et la tradition des problèmes de division des figures. In: Morelon, R., Hasnawi, A. (eds.), De Zénon d'Élée à Poincaré : Recueil d'études en hommage à Roshdi Rashed. Peeters, Leuven, pp. 119–159.
——— 2004b. Thābit ibn Qurra et la composition des rapports. Arabic Sciences and Philosophy 14, 175–211.
——— 2010. De l'usage des transformations géométriques à la notion d'invariant : La contribution d'al-Sijzī. Arabic Sciences and Philosophy 20, 53–91.
van Dalen, B., 1996. Al-Khwārizmī's astronomical tables revisited: Analysis of the equation of time. In: Casulleras, J., Samsó, J. (eds.), From Baghdad to Barcelona, University of Barcelona, Barcelona, pp. 195–252.
——— 1999. Tables of planetary latitude in the Huihui li (II). In: Kim, Y.S., Bray, F. (eds.), Current Perspectives in the History of Science in East Asia. Seoul National University Press, Seoul, pp. 316–329.
——— 2004a. A second manuscript of the Mumtaḥan Zīj. Suhayl 4, 9–44.
——— 2004b. The Zīj-i Nāsirī by Maḥmūd ibn ʾUmar: The earliest Indian-Islamic astronomical handbook with tables and its relation to the ʿAlāʾī Zīj. In: Burnett, C., Hogendijk, J., Plofker, K., Yano, M. (eds.), Studies in the History of the Exact Sciences in Honour of David Pingree. Brill, Leiden, pp. 825-862.
Dallal, A., 2010. Islam, Science, and the Challenge of History. Yale University Press, New Haven.
Decorps-Foulquier, M., Federspiel, M., 2008. Apollonius de Perge, Coniques, Tome 1.2: Livre I. De Gruyter, Berlin.
——— 2008. Apollonius de Perge, Coniques, Tome 2.3: Livres II-IV. De Gruyter, Berlin.
De Young, G., 1996. Ex aequali ratios in the Greek and Arabic Euclidean traditions. Arabic Sciences and Philosophy 6, 167–213.
——— 1997. Al-Jawharī's additions to Book V of Euclid's Elements. Zeitschrift für Geschichte der arabisch-islamischen Wissenschaften 11, 153–178.
——— 2001. The Ashkāl al-Tasis of al-Samarqandī: A translation and study. Zeitschrift für Geschichte der arabisch-islamischen Wissenschaften 14, 57–117.
——— 2002a. The Arabic version of Euclid's Elements by al-Ḥajjāj ibn Yūsuf Ibn Maṭar. New light on a submerged tradition. Zeitschrift für Geschichte der arabisch-islamischen Wissenschaften 15, 125–164.
——— 2002b. Euclidean geometry in two medieval Islamic encyclopaedias. al-Masāq 14, 47–60.
——— 2002c. Kadizade al-Rumi on Samarkandi's Ashkāl al-Tasis: A mathematical commentary. In: Ansari, S.M.R. (ed.), Science and Technology in the Islamic World: Proceedings of the XXth International Congress of History of Science. Brepols, Turnhout, pp. 83–90.
——— 2004. The Latin translation of Euclid's Elements attributed to Gerard of Cremona in relation to the Arabic transmission. Suhayl 4, 311–383.
——— 2005. Diagrams in the Arabic Euclidean tradition: A preliminary assessment. Historia Mathematica 32, 129–179.
——— 2007. Qutb al-Dīn al-Shirazī and his Persian translation of Naṣīr al-Dīn al-Ṭūsī's Taḥrīr Kitāb Uṣūl Uqlīdis. Farhang 20, 17–75.
——— 2008a. Book XVI: A mediaeval Arabic addendum to Euclid's Elements. SCIAMVS 9, 133–209.
——— 2008b. A Persian translation of al-Jawharī's additions to the Elements, Book V. Tarikh-e Elm 7, 1–24.
——— 2008c. The Taḥrīr Kitāb Uṣūl Uqlīdis of Naṣīr al-Dīn al-Ṭūsī: Its sources. Zeitschrift für Geschichte der arabisch-islamischen Wissenschaften 18, 1–71.
Díaz-Fajardo, M., 2001. La Teoría de la Trepidación en un Astrónomo Marroquí del Siglo XV. University of Barcelona, Barcelona.

Djebbar, A., 1996. Quelques commentaires sur les versions arabes des Éléments d'Euclide et sur leur transmission à l'occident musulman. In: Folkerts, M. (ed.), Mathematische Probleme im Mittelalter: Der Lateinische und Arabische Sprachereich. Harrassowitz, Wiesbaden, pp. 91–114.

——— 1997. La rédaction de l'Istikmāl d'al-Mu'taman (XIe s.) par Ibn Sartāq, un mathématicien des XIIIe–XIVe siècles. Historia Mathematica 24, 185–192.

——— 1998. Les activités mathématiques dans les villes du Maghreb central (XIe–XIXe s.). In: Actes du troisième colloque maghrébin sur l'histoire des mathématiques arabes, vol. 1. Office des Publications Universitaires, Algiers, pp. 73–115.

——— 1999. Les livres arithmétiques des Éléments d'Euclide dans le traite d'al-Mu'taman du XIe siècle. LLULL 22, 589–653.

——— 2000a. Les activités mathématiques au Maghreb à l'époque Ottomane (XVIe–XIXe siècles). In: Günergun, F., İhsanoğlu, E., Djebbar, A. (eds.), Science, Technology and Industry in the Ottoman World. Brepols, Turnhout, pp. 49–66.

——— 2000b. Figurate numbers in the mathematical tradition of al-Andalus and the Maghrib. Suhayl 1, 57–70.

——— 2000c. ʿOmar Khayyām et les activités mathématiques en pays d'Islam aux XIe–XIIe siècles. Farhang 12, 1–31.

——— 2000. La production scientifique arabe, sa diffusion et sa réception au temps des croisades : L'exemple des mathématiques. In: van den Abeele, D., Tihon, A., Draelants, I. (eds.), Occident et Proche-Orient: Contacts scientifiques au temps des croisades, réminiscences. Brepols, Turnhout, pp. 343–368.

——— 2002. La circulation des mathématiques entre l'Orient et l'Occident musulmans : Interrogations anciennes et éléments nouveaux. In: Dold-Samplonius, Y., Dauben, J., Folkerts, M., van Dalen, B. (eds.), From China to Paris: 2000 Years Transmission of Mathematical Ideas. Steiner, Stuttgart, pp. 213–235.

——— 2002b. Épître d'Omar Khayyam Sur l'explication des prémisses problématiques du livre d'Euclide. Farhang 14, 79–136.

——— 2003a. A panorama of research on the history of mathematics in al-Andalus and the Maghrib between the ninth and sixteenth centuries. In: Hogendijk, J.P., Sabra, A.I. (eds.), The Enterprise of Science in Islam: New Perspectives. MIT Press, Cambridge, MA, pp. 309–350.

——— 2003b. Quelques exemples de scholies dans la tradition arabe des Éléments d'Euclide. Revue d'Histoire des Sciences 56, 293–321.

——— 2005. L'Algèbre Arabe: Genèse d'un Art. Vuibert, Paris.

——— 2006. Ibn Abdun's epistle on surface measuring — a witness to the pre-algebraical tradition. Suhayl 6, 255–257.

Dold-Samplonius, Y., 1996a. *The Book of Assumptions*, by Thābit ibn Qurra (836–901). In: Dauben, J.W., Folkerts, M., Knobloch, E., Wussing, H. (eds.), History of Mathematics: States of the Art. Academic Press, San Diego, pp. 207–222.

——— 1996b. How al-Kāshī measures the muqarnas: A second look. In: Folkerts, M. (ed.), Mathematische Probleme im Mittelalter: Der Lateinische und Arabische Sprachbereich. Harrassowitz, Wiesbaden, pp. 56–90.

——— 1996c. Problem of the two towers. In: Franci, R., Pagli, P., Toti Rigatelli, L. (eds.), Itinera Mathematica: Studi in Onore di Gino Arrighi per il Suo 90° Compleanno. Università di Siena, Siena, pp. 44–69.

——— 2000. Calculation of arches and domes in 15th century Samarkand. In: Williams, K. (ed.), Nexus III: Architecture and Mathematics. Pacini Editore, Ospedaletto, pp. 45-55.

——— 2003. Calculating surface areas and volumes in Islamic architecture. In: Hogendijk, J.P., Sabra, A.I. (eds.), The Enterprise of Science in Islam: New Perspectives. MIT Press, Cambridge, MA, pp. 235–265.

Dorce, C., 2002. The Tāj al-Azyāj of Muḥyī al-Dīn al-Maghribī (d. 1283): Methods of computation. Suhayl 3, 193–212.

El-Bizri, N., 2007. In defence of the sovereignty of philosophy: Al-Baghdādī's critique of Ibn al-Haytham's geometrisation of place. Arabic Sciences and Philosophy 17, 57–80.

Endress, G., 2003. Mathematics and philosophy in medieval Islam. In: Hogendijk, J.P., Sabra, A.I. (eds.), The Enterprise of Science in Islam: New Perspectives. MIT Press, Cambridge, MA, pp. 121–176.

Folkerts, M., 1997. Die älteste lateinische Schrift über das indische Rechnen nach al-Ḫwārizmī. Verlag der Bayerischen Akademie der Wissenschaften, Munich.

——— 1998. Remarks on al-Khwārizmī's Arithmetic. In: Actes du troisième colloque maghrébin sur l'histoire des mathématiques arabes, vol. 1. Office des Publications Universitaires, Algiers, pp. 117–123.

——— 2001. Early texts on Hindu-Arabic calculation. Science in Context 14, 13–38.

Freudenthal, G., Lévy, T., 2004. De Gérase à Bagdad : Ibn Bahrīz, al-Kindī, et leur recension arabe de l' Introduction arithmétique de Nicomaque, d'après la version Hébraïque de Qalonymos ben Qalonymos d'Arles. In: Morelon, R., Hasnawi, A. (eds.), De Zénon d'Élée à Poincaré : Recueil d'études en hommage à Roshdi Rashed. Peeters, Leuven, pp. 479–544.

Guergour, Y., 2000. Les différents systèmes de numérotation au Maghreb à l'époque Ottomane : L'exemple des chiffres rūmī. In: Günergun, F., İhsanoğlu, E., Djebbar, A. (eds.), Science, Technology and Industry in the Ottoman World. Brepols, Turnhout, pp. 67–74.

——— 2005. Le roi de Saragosse al-Mu'taman Ibn Hūd (m. 1085) et le théorème de Pythagore: Ses sources et ses prolongements. LLULL 28, 415–434.

Guergour, Y., 2009. Le cinquième postulat des parallèles chez al-Mutaman ibn Hūd, roi de Saragosse (1081–1085). LLULL 32, 59–72.

Gutas, D., 1997. Greek Thought, Arabic Culture. Routledge, London/New York.

Hamedani, H.M., 1999. Remarks on the manuscript tradition of some optical works of Ibn al-Haytham. In: Ibish, Y. (ed.), Editing Islamic Manuscripts on Science. Al-Furqān Islamic Heritage Foundation, London, pp. 165–180.

Hogendijk, J.P., 1996a. Al-Muʾtaman's simplified lemmas for solving "Alhazen's problem." In: Casulleras, J., Samsó, J. (eds.), From Baghdad to Barcelona. University of Barcelona, Barcelona, pp. 59–101.

——— 1996b. Transmission, transformation, and originality: The relation of Arabic to Greek geometry. In: Ragep, J., Ragep, S., Livesey, S., (eds.), Tradition, Transmission, Transformation. Brill, Leiden, pp. 31–64.

——— 1996c. Which version of Menelaus' Spherics was used by al-Muʾtaman ibn Hud in his Istikmal? In: Folkerts, M. (ed.), Mathematische Probleme im Mittelalter — der Lateinische und Arabische Sprachbereich. Harrassowitz, Wiesbaden, pp. 17–44.

——— 1998. L'étude des sections coniques dans la tradition arabe. In: Actes du troisième colloque maghrébin sur l'histoire des mathématiques arabes, vol. 1. Office des Publications Universitaires, Algiers, pp. 147–158.

——— 1999. Traces of the lost Geometrical Elements of Menelaus in two texts of al-Sijzī. Zeitschrift für Geschichte der arabisch-islamischen Wissenschaften 13, 129–164.

——— 2000a. Al-Nayrīzī's mysterious determination of the azimuth of the qibla at Baghdād. SCIAMVS 1, 49–70.

——— 2000b. Al-Nayrīzī's own proof of Euclid's parallel postulate. In: Folkerts, M., Lorch, R. (eds.), Sic Itur ad Astra. Harrassowitz, Wiesbaden, pp. 252–265.

——— 2001a. The contributions by Abū Naṣr ibn Irāq and al-Ṣaghānī to the theory of seasonal hour lines on astrolabes and sundials. Zeitschrift für Geschichte der arabisch-islamischen Wissenschaften 14, 1–30.

——— 2001b. The geometrical works of Abū Saʿīd al-Ḍarīr al-Jurjānī. SCIAMVS 2, 47–74.

——— 2002a. Anthyphairetic ratio theory in medieval Islamic mathematics. In: Dold-Samplonius, Y., Dauben, J., Folkerts, M., van Dalen, B. (eds.), From China to Paris: 2000 Years Transmission of Mathematical Ideas. Steiner, Stuttgart, pp. 187–202.

——— 2002b. Two editions of Ibn-al-Haytham's Completion of the Conics. Historia Mathematica 29, 247–265.

——— 2003a. The Burning Mirrors of Diocles: Reflections on the methodology and purpose of the history of pre-modern science. Early Science and Medicine 7, 181–197.

——— 2003b. The geometrical problems of Nuʿaim ibn Muḥammad ibn Mūsā (ninth century). SCIAMVS 4, 59–136.

——— 2003c. The lost geometrical parts of the Istikmāl of Yūsuf al-Muʾtaman ibn Hūd (11th century) in the redaction of Ibn Sartāq (14th century): An analytical table of contents. Archives internationales d'histoire des sciences 53, 19–34.

——— 2005. Applied mathematics in eleventh century al-Andalus: Ibn Muādh al-Jayyānī and his computation of astrological houses and aspects. Centaurus 47, 87–114.

——— 2007. Similar mathematics in different cultures: Jamshīd al-Kāshī and Ludolph Van Ceulen on the determination of π. In: Daelemans, F., Duvosquel, J.-M., Halleux, R., Just, D. (eds.), Mélanges offerts à Hossam Elkhadem par ses amis et élèves. Archives et Bibliothèques de Belgique, Brussels, pp. 189–203.

——— 2008a. The Introduction to Geometry by Qusṭā ibn Lūqā: Translation and commentary. Suhayl 8, 163–221.

——— 2008b. Al-Kāshī's determination of π to 16 decimals in an old manuscript. Zeitschrift für Geschichte der arabisch-islamischen Wissenschaften 18, 73–153.

——— 2008c. Two beautiful geometrical theorems by Abū Sahl Kūhī in a 17th century Dutch translation. Tarikh-e Elm 6, 1–36.

Jaouiche, K., 1998. Aperçu sur le problème des cercles tangents chez Ibrāhīm ibn Sinān, Ibn al-Haytham et Viète. In: Actes du troisième colloque maghrébin sur l'histoire des mathématiques arabes, vol. 1. Office des Publications Universitaires, Algiers, pp. 179–193.

Johansson, B.G., 2011. Cube root extraction in medieval mathematics. Historia Mathematica 38, 338–367.

Kanani, N., 2000. Omar Khayyām and the parallel postulate. Farhang 12, 107–124.

Katz, V., ed., 2007. The Mathematics of Egypt, Mesopotamia, China, India, and Islam: A Sourcebook. Princeton University Press, Princeton.

Kennedy, E.S., 1995. Treatise V of Kāshī's *Khāqānī Zīj*: Determination of the ascendent. Zeitschrift für Geschichte der arabisch-islamischen Wissenschaften 10, 123–145.

——— 1996. The astrological houses as defined by medieval Islamic astronomers. In: Casulleras, J., Samsó, J. (eds.), From Baghdad to Barcelona. University of Barcelona, Barcelona, pp. 535–578.

——— 1998. On the Contents and Significance of the Khāqānī Zīj by Jamshīd Ghiyāth al-Dīn al-Kāshī. Institute for the History of Arabic-Islamic Science, Frankfurt am Main.

——— 2000. The operation of multiplication in a sixteenth century Persian treatise. In: Folkerts, M., Lorch, R. (eds.), Sic Itur ad Astra. Harrassowitz, Wiesbaden, pp. 304–306.

Kennedy, E.S., colleagues and former students, 2009–2010. Al-Battānī's astrological history of the Prophet and the early caliphate. Suhayl 9, 13–148.

Kennedy, E.S., Kunitzsch, P., Lorch, R., 1999. The Melon-Shaped Astrolabe in Arabic Astronomy. Steiner, Wiesbaden/Stuttgart, 1999.

Kheirandish, E., 1996. The "Arabic version" of Euclidean optics: Transformations as linguistic problems in transmission. In: Ragep, J., Ragep, S., Livesey, S., (eds.), Tradition, Transmission, Transformation. Brill, Leiden, pp. 227–245.

——— 1999. The Arabic Version of Euclid's Optics. Springer, New York.

——— 2003. The many aspects of "appearances": Arabic optics to 950 AD. In: Hogendijk, J.P., Sabra, A.I. (eds.), The Enterprise of Science in Islam: New Perspectives. MIT Press, Cambridge, MA, pp. 55–83.

——— 2008. Science and mithāl: Demonstrations in Arabic and Persian scientific traditions. Iranian Studies 41, 465–489.

King, D.A., 1997. Two Iranian world maps for finding the direction and distance to Mecca. Imago Mundi 49, 62–82.

——— 1999a. Aspects of Fatimid astronomy: From hard-core mathematical astronomy to architectural orientations in Cairo. In: Barrucand, M. (ed.), L'Égypte Fatimide : Son art et son histoire. Presses de l'Université de Paris-Sorbonne, Paris, pp. 497–517.
——— 1999b. World-Maps for Finding the Direction and Distance to Mecca. Al-Furqān Islamic Heritage Foundation, London.
——— 2000. Mathematical astronomy in Islamic civilisation. In: Selin, H., (ed.), Astronomy Across Cultures. Kluwer, Dordrecht, pp. 585–613.
——— 2004. In Synchrony with the Heavens. Studies in Astronomical Timekeeping and Instrumentation in Medieval Islamic Civilization, vol. 1, The Call of the Muezzin. Brill, Leiden.
——— 2005. In Synchrony with the Heavens. Studies in Astronomical Timekeeping and Instrumentation in Medieval Islamic Civilization, vol. 2, Instruments of Mass Calculation. Brill, Leiden.
——— 2008. An instrument of mass calculation made by Nastulus in Baghdad. Suhayl 8, 93–119.
King, D.A., Samsó, J., 2001. Astronomical handbooks and tables from the Islamic world (750–1900): An interim report. Suhayl 2, 9–105.
Knorr, W., 1996. The wrong text of Euclid: On Heiberg's text and its alternatives. Centaurus 38, 208–276.
Kunitzsch, P., 2003a. A new manuscript of Abū Bakr al-Ḥaṣṣar's Kitāb al-Bayān. Suhayl 3, 187–192.
——— 2003b. The transmission of Hindu-Arabic numerals reconsidered. In: Hogendijk, J.P., Sabra, A.I. (eds.), The Enterprise of Science in Islam: New Perspectives. MIT Press, Cambridge, MA, pp. 3–21.
——— 2005. Zur geschichte der arabischen Ziffern. Bayerische Akademie der Wissenschaften, Philosophisch-Historische Klasse, Sitzungsberichte 3, 1–39.
Kunitzsch, P., Lorch, R., 2010. Theodosius Sphaerica: Arabic and Medieval Latin Translations. Steiner, Stuttgart.
Laabid, E., 1998. Les donations dans les mathématiques médiévales : L'exemple d'al-Ḥubūbī. In: Actes du troisième colloque maghrébin sur l'histoire des mathématiques arabes, vol. 1. Office des Publications Universitaires, Algiers, pp. 207–220.
Lo Bello, A. (ed.), 2003a. The Commentary of al-Nayrīzī on Book I of Euclid's Elements of Geometry. Brill, Boston.
——— 2003b. The Commentary of Albertus Magnus on Book I of Euclid's Elements of Geometry. Brill, Boston.
——— 2003c. Gerard of Cremona's Translation of the Commentary of al-Nayrīzī on Book I of Euclid's Elements of Geometry. Brill, Boston.
——— 2009. The Commentary of al-Nayrīzī on Books II-IV of Euclid's Elements of Geometry. Brill, Boston.
Lorch, R., 1996a. Maslama al-Majrīṭī and Thābit ibn Qurra's al-Shakl al-Qaṭṭaʿ. In: Casulleras, J., Samsó, J. (eds.), From Baghdad to Barcelona. University of Barcelona, Barcelona, pp. 49–57.
——— 1996b. The transmission of Theodosius' Sphaerica. In: Folkerts, M. (ed.), Mathematische Probleme im Mittelalter: Der Lateinische und Arabische Sprachereich. Harrassowitz, Wiesbaden, pp. 159–184.
——— 1998. Graphical methods in spherical astronomy in treatises by Ḥabash al-Ḥāsib and al-Mahānī. In: Actes du troisième colloque maghrébin sur l'histoire des mathématiques arabes, vol. 1. Office des Publications Universitaires, Algiers, pp. 221–226.
——— 2000a. Ibn al-Ṣalāḥ's treatise on projection: A preliminary survey. In: Folkerts, M., Lorch, R. (eds.), Sic Itur ad Astra. Harrassowitz, Wiesbaden, pp. 401–408.
——— 2000b. Some early applications of the sine quadrant. Suhayl 1, 251–272.
——— 2001. Thābit ibn Qurra: On the Sector-Figure and Related Texts. Institut für Geschichte der arabisch-islamischen Wissenschaften, Frankfurt am Main. Reissued 2008, Erwin Rauner Verlag, Augsberg.
——— 2005. Al-Farghānī on the Astrolabe. Franz Steiner Verlag, Wiesbaden/Stuttgart.
Luckey, P., Hogendijk, J.P., 1999. Die Schrift des Ibrāhīm b. Sinān b. Tābit über die Schatteninstrumente. Institut für Geschichte der arabisch-islamischen Wissenschaften, Frankfurt am Main.

Luther, I., 2002. The solution of Apollonius' problem in the medieval Arab East. In: Ansari, S.M.R. (ed.), Science and Technology in the Islamic World: Proceedings of the XXth International Congress of History of Science. Brepols, Turnhout, pp. 91–99.

Mancha, J.L., 1998. On Ibn al-Kammād's table for trepidation. Archive for History of Exact Sciences 52, 1–11.

Marin, M., 2004. The making of a mathematician: Al-Qalaṣādī (d. 891/1486) and his Riḥla. Suhayl 4, 295–310.

Matvievskaya, G.P., 1996. New information on the mathematical creativity of Abū Naṣr Manṣūr ibn ʿIrāq (treatise on regular polyhedra). Uzbekskiĭ Matematicheskiĭ Zhurnal 2, 59–68.

Mawaldi, M., 2000. Méthode de l'analyse et de la synthèse de Kamal al-Dīn al-Fārisī. In: İhsanoğlu, E., Günergun, F. (eds.), Science in Islamic Civilisation. IRCICA, Istanbul, pp. 193–199.

McCarthy, D.P., Byrne, J.G., 2003. Al-Khwārizmī's sine tables and a western table with the Hindu norm of $R = 150$. Archive for History of Exact Sciences 57, 243–266.

McGinnis, J., 2006. A medieval Arabic analysis of motion at an instant: The Avicennan sources to the forma fluens/fluxus formae debate. British Journal for the History of Science 39, 189–205.

Mercier, R., 1996. Accession and recession: Reconstruction of the parameters. In: Casulleras, J., Samsó, J. (eds.), From Baghdad to Barcelona. University of Barcelona, Barcelona, pp. 299–347.

Mielgo, H., 1996. A method of analysis for mean motion astronomical tables. In: Casulleras, J., Samsó, J. (eds.), From Baghdad to Barcelona. University of Barcelona, Barcelona, pp. 159–179.

Morelon, R., 1999. Astronomie physique et astronomie mathématique dans l'astronomie précopernicienne. In: Rashed, R., Biard, J. (eds.), Les doctrines de la science de l'antiquité à l'âge classique. Peeters, Leuven, pp. 105–129.

Moussa, A., 2010. The trigonometric functions, as they were in the Arabic-Islamic civilization. Arabic Sciences and Philosophy 20, 93–104.

—— 2011. Mathematical methods in Abū al-Wafāʾ's Almagest and the qibla determinations. Arabic Sciences and Philosophy 21, 1–56.

Nadal, R., Taha, A., Pinel, P., 2004. Le contenu astronomique des Sphériques de Ménélaos. Archive for History of Exact Sciences 58, 381–436.

Neugebauer, O., Rashed, R., 1999. Sur une construction du miroir parabolique par Abū al-Wafāʾ al-Būzjānī. Arabic Sciences and Philosophy 9, 261–277.

North, J.D., 1996. A reply to Prof. E. S. Kennedy. Comment on: The astrological houses as defined by medieval Islamic astronomers. In: Casulleras, J., Samsó, J. (eds.), From Baghdad to Barcelona. University of Barcelona, Barcelona, pp. 579–582.

Oaks, J.A., 2007. Medieval Arabic algebra as an artificial language. Journal of Indian Philosophy 35, 543–575.

—— 2009. Polynomials and equations in Arabic algebra. Archive for History of Exact Sciences 63, 169–203.

—— 2011. Al-Khayyām's scientific revision of algebra. Suhayl 10, 47–75.

Oaks, J.A., Alkhateeb, H.M., 2005. Māl, enunciations, and the prehistory of Arabic algebra. Historia Mathematica 32, 400–425.

—— 2007. Simplifying equations in Arabic algebra. Historia Mathematica 34, 45–61.

Özdural, A., 1996. On interlocking similar or corresponding figures and ornamental patterns of cubic equations. Muqarnas 13, 191–211.

—— 2000. Mathematics and arts: Connections between theory and practice in the medieval Islamic world. Historia Mathematica 27, 171–201.

—— 2002. The use of cubic equations in Islamic art and architecture. Nexus Network Journal 4 (electronic).

Panza, M., 2008. The role of algebraic inferences in Naīm ibn Mūsā's collection of geometrical propositions. Arabic Sciences and Philosophy 18, 165–191.

Pinel, P., Taha, A., 2003. Le travail d'al-Ṭūsī sur les Sphériques de Ménélaus : Établissement critique du texte, apport mathématique, interpretation astronomique. Farhang 15–16, 33–109.

Plofker, K., 2002. Use and transmissions of iterative approximations in India and the Islamic world. In: Dold-Samplonius, Y., Dauben, J., Folkerts, M., van Dalen, B. (eds.), From China to Paris: 2000 Years Transmission of Mathematical Ideas. Steiner, Stuttgart, pp. 167–186.

Ragep, F.J., 1996. Al-Battānī, cosmology, and the early history of trepidation in Islam. In: Casulleras, J., Samsó, J. (eds.), From Baghdad to Barcelona. University of Barcelona, Barcelona, pp. 267–298.

——— 2000. The Persian context of the Ṭūsī couple. In: Poutjavady, N., Vesel, Ž. (eds.), Naṣīr al-Dīn Ṭūsī, Philosophe et Savant du XIIIe Siècle. Institut Français de Recherche en Iran/Presses Universitaires d'Iran, Tehran, pp. 113–130.

Rashed, R., 1996a. Archimedean learning in the middle ages: The Banū Mūsā. Historia Scientiarum 6, 1–16.

——— 1996b. Les Mathématiques Infinitésimales du IXe au XIe Siècle, vol. I, Fondateurs et Commentateurs. Al-Furqān Islamic Heritage Foundation, London.

——— 1997a. Les commencements des mathématiques archimédiennes en arabe : Banū Mūsā. In: Hasnawi, A., Elamrani-Jamal, A., Aouad, M. (eds.), Perspectives arabes et médiévales sur la tradition scientifique et philosophique grecque. Peeters, Leuven, pp. 1–19.

——— 1997b. Le commentaire par al-Kindī de l' Optique d'Euclide : Un traité jusqu'ici inconnu. Arabic Sciences and Philosophy 7, 9–56.

——— 1997c. Dioclès et Dtrūms : deux traités sur les miroirs ardents. Mélanges Institut Dominicain d'Études Otientales du Caire 23, 1–155.

——— 1997d. Œuvres philosophiques et scientifiques d'al-Kindī, vol. I, L'Optique et la Catoptrique. Brill, Leiden.

——— 1998. Al-Qūhī contre Aristote : Sur le mouvement. Oriens Occidens 2, 95–117.

——— 1999a. Al-Qūhī vs. Aristotle: On motion. Arabic Sciences and Philosophy 9, 7–24.

——— 1999b. Combinatoire et métaphysique : Ibn Sinā, al-Ṭūsī et al-Ḥalabī. In: Rashed, R., Biard, J. (eds.), Les doctrines de la science de l'antiquité à l'âge classique. Peeters, Leuven, pp. 61–86.

——— 2000a. Les catoptriciens grecs, vol I. Société d'Édition Les Belles Lettres, Paris.

——— 2000b. Histoire de l'analyse combinatoire. In: İhsanoğlu, E., Günergun, F. (eds.), Science in Islamic Civilisation. IRCICA, Istanbul, pp. 151–166.

——— 2000c. Ibn Sahl and al-Qūhī : The projections. Addenda and corrigenda to: Géométrie et Dioptrique au Xe Siècle. Arabic Sciences and Philosophy 10, 79–100.

——— 2000d. Kombinatorik und Metaphysik. Ibn Sinā, aṭ-Ṭūsī und al-Ḥalabī. In: Thiele, R. (ed.), Mathesis, Festschrift zum siebzigsten Geburtstag von Matthias Schramm. Verlag für Geschichte der Naturwissenschaften und der Technik, Berlin, pp. 37–54.

——— 2000e. Les mathématiques infinitésimales du IXe au XIe siècle, vol. III, Ibn al-Haytham: Théorie des coniques, constructions géométriques et géométrie pratique. Al-Furqān Islamic Heritage Foundation, London.

——— 2001. Al-Qūhī: From meteorology to astronomy. Arabic Sciences and Philosophy 11, 157–204.

——— 2002a. Al-Qūhī : De la météorologie à l'astronomie. Oriens Occidens 4, 1–57.

——— 2002b. Les mathématiques infinitésimales du IXe au XIe siècle, vol. IV, Ibn al-Haytham : méthodes géométriques, transformations ponctuelles et philosophie des mathématiques. Al-Furqān Islamic Heritage Foundation, London.

——— 2002c. Transmission et innovation : L'exemple du miroir parabolique. In: Ansari, S.M.R. (ed.), Science and Technology in the Islamic World: Proceedings of the XXth International Congress of History of Science. Brepols, Turnhout, pp. 101–108.

——— 2003a. Al-Qūhī et al-Sijzī : Sur le compas parfait et le tracé continu des sections coniques. Arabic Sciences and Philosophy 13, 9–43.

——— 2003b. Metaphysics and mathematics in classical Islamic culture: Avicenna and his followers. In Peters, T., Iqbal, M., Haq, S.N. (eds.), God, Life, and the Cosmos: Christian and Islamic Perspectives. Ashgate, Surrey, pp. 173–193.

——— 2004. Œuvre mathématique d'al-Sijzī, vol. I, Géométrie des coniques et théorie des nombres au Xe Siècle. Peeters, Leuven.

——— 2005. Geometry and Dioptrics in Classical Islam. Al-Furqan Islamic Heritage Foundation, London.
——— 2006. Les mathématiques infinitésimales du IXe au XIe siècle, vol. V, Ibn al-Haytham : Astronomie, géométrie sphérique et trigonométrie. Al-Furqān Islamic Heritage Foundation, London.
——— 2007a. Al-Khwārizmī: Le commencement de l'algèbre. Librairie Scientifique et Technique Albert Blanchard, Paris.
——— 2007b. The celestial kinematics of Ibn al-Haytham. Arabic Sciences and Philosophy 17, 7–55.
——— 2008a. Apollonius de Perge, Coniques. Tome 1.1: Livre I. De Gruyter, Berlin.
——— 2008b. Apollonius de Perge, Coniques. Tome 3: Livre V. De Gruyter, Berlin.
——— 2009a. Al-Khwārizmī: The Beginnings of Algebra. SAQI, London.
——— 2009b. Apollonius de Perge, Coniques. Tome 2.2: Livre IV. De Gruyter, Berlin.
——— 2009c. Apollonius de Perge, Coniques. Tome 4: Livre VI, VII. De Gruyter, Berlin.
——— 2009d. Résolution géométrique des équations du second degré. In: Rashed, R. (ed.), Thābit ibn Qurra: Science and Philosophy in Ninth-Century Baghdad. Walter de Gruyter, Berlin, pp. 153–169.
——— 2009e. Thābit ibn Qurra: Science and Philosophy in Ninth-Century Baghdad. Walter de Gruyter, Berlin/New York.
——— 2010a. Apollonius de Perge, Coniques. Tome 2.1: Livres II, III. De Gruyter, Berlin.
——— 2010b. L'asymptote : Apollonius et ses lecteurs. Bollettino di Storia delle Scienze Matematiche 30, 223–254.
——— 2010c. Les constructions géométriques entre géométrie et algèbre : L'épître d'Abū al-Jūd à al-Bīrūnī. Arabic Sciences and Philosophy 20, 1–51.
Rashed, R., Bellosta, H., 2000. Ibrāhīm ibn Sinān : Logique et géométrie au Xe siècle. Brill, Leiden.
——— 2010. Apollonius de Perge : La Section des Droits selon des Rapports. De Gruyter, Berlin/New York.
Rashed, R., al-Houjairi, M., 2010. Sur un théorème de géométrie sphérique : Théodose, Ménélaüs, ibn Irāq et ibn Hūd. Arabic Sciences and Philosophy 20, 207–253.
Rashed, R., Houzel, C., 2004. Recherche et enseignement des mathématiques au IXe Siècle. Le Recueil de Propositions Géométriques de Naīm ibn Mūsā. Peeters, Leuven.
——— 2005. Thābit ibn Qurra et la théorie des parallèles. Arabic Sciences and Philosophy 15, 9–55.
——— 2005. Théorie des nombres amiables. In: Rashed, R. (ed.), Thābit ibn Qurra: Science and Philosophy in Ninth-Century Baghdad. Walter de Gruyter, Berlin, pp. 77–151.
Rashed, R., Morelon, R., 1996. Encyclopedia of the History of Arabic Science. 3 vols. Routledge, London.
Rashed, R., Vahabzadeh, B., 1999. Al-Khayyām Mathématicien. Librairie Scientifique et Technique Albert Blanchard, Paris.
Rebstock, U., 1995/96. Der Muʿāmalāt-Traktat des Ibn al-Haiṯam. Zeitschrift für Geschichte der arabisch-islamischen Wissenschaften 10, 61–121.
——— 1998. If numbers are right: On the use of reckoning in the Islamic middle age. In: Actes du troisième colloque maghrébin sur l'histoire des mathématiques arabes, vol. 1. Office des Publications Universitaires, Algiers, pp. 239–249.
——— 1999. The Kitāb al-Kāfī fī Mukhtaṣar (al-Ḥisāb) al-Hindī of al-Ṣardafī. Zeitschrift für Geschichte der arabisch-islamischen Wissenschaften 13, 189–204.
——— 2001. Al-Qurašī, ʿAlī ibn Ḫiḍr. At-Taḏkira bi-uṣūl al-ḥisāb wa l-Farāʾiḍ (Buch über die Grundlagen der Arithmetik und der Erbteilung). Institut für Geschichte der arabisch-islamischen Wissenschaften, Frankfurt am Main.
——— 2002a. An early link of the Arabic tradition of practical arithmetic: The Kitāb al-Tadhkira bi-uṣūl al-ḥisāb wa 'l-farāʾiḍ wa-ʿawlihā wa-taṣḥīḥihā. In: Dold-Samplonius, Y., Dauben, J., Folkerts, M., van Dalen, B. (eds.), From China to Paris: 2000 Years Transmission of Mathematical Ideas. Steiner, Stuttgart, pp. 203–212.
——— 2002/03. Réhabilitation d'un texte mathématique malconnu. Zeitschrift für Geschichte der arabisch-islamischen Wissenschaften 15, 175–184.

——— 2008/09. Arithmetik (ḥisāb) und Erbteilungslehre (ʿilm al-farāʾiḍ): Symbiose einer islamischen Wissenschaftsdisziplin. Zeitschrift für Geschichte der arabisch-islamischen Wissenschaften 18, 269–285.

Rius, M., 1996. La orientatión de las mezquitas según el Kitāb dalāʾil al-Qibla de al-Mattīyī. In: Casulleras, J., Samsó, J. (eds.), From Baghdad to Barcelona. University of Barcelona, Barcelona, pp. 781–830.

——— 2000. La Alquibla en al-Andalus y al-Magrib al-Aqsà. University of Barcelona, Barcelona.

Rius, M., Comes, M., 2006. Finding the qibla in the Islamic Mediterranean milieu. In: Actes du Huitième Colloque Maghrebin sur l'Histoire des Mathématiques Arabes. Association Tunisienne des Sciences Mathématiques, Tunis, pp. 131–138.

Rommevaux, S., Djebbar, A., Vitrac, B., 2001. Remarques sur l'histoire du texte des Éléments d'Euclide. Archive for History of Exact Sciences 55, 221–295.

Rosenfeld, B.A., 2006. A second supplement to: Mathematicians, Astronomers, and other Scholars of Islamic Civilisation and their Works (7th–19th c.). Suhayl 6, 9–79.

Rosenfeld, B.A., Hogendijk, J.P., 2002. A mathematical treatise written in the Samarqand observatory of Ulugh Beg. Zeitschrift für Geschichte der arabisch-islamischen Wissenschaften 15, 25–65.

Rosenfeld, B.A., İhsanoğlu, E., 2003. Mathematicians, Astronomers, and other Scholars of Islamic Civilisation and their Works (7th–19th c.). IRCICA, Istanbul.

——— 2004. A supplement to: Mathematicians, Astronomers, and other Scholars of Islamic Civilisation and their Works (7th–19th c.). Suhayl 4, 87–139.

Rozhanskaya, M.M., 1997. Les méthodes infinitésimales dans la mécanique arabe. Archives Internationales d'Histoire des Sciences 47, 255–270.

Sabra, A.I., 1996a. The appropriation and subsequent naturalization of Greek science in medieval Islam: A preliminary statement. In: Ragep, J., Ragep, S., Livesey, S. (eds.), Tradition, Transmission, Transformation. Brill, Leiden, pp. 3–27.

——— 1996b. Situating Arabic science: Locality versus essence. Isis 87, 654–670.

——— 1997. Thābit ibn Qurra on the infinite and other puzzles: Edition and translation of his discussions with Ibn Usayyid. Zeitschrift für Geschichte der arabisch-islamischen Wissenschaften 11, 1–33.

——— 1998. One Ibn al-Haytham or two? An exercise in reading the bio-bibliographical sources. Zeitschrift für Geschichte der arabisch-islamischen Wissenschaften 12, 1–50.

——— 2002a. Kitāb al-Manāẓir of Al-Ḥasan ibn al-Haytham. Books IV-V. On Reflection, and Images Seen by Reflection. National Council for Culture, Arts and Letters, Safat.

——— 2002b. One Ibn al-Haytham or two? Conclusion. Zeitschrift für Geschichte der arabisch-islamischen Wissenschaften 15, 95–108.

——— 2003. Ibn al-Haytham's revolutionary project in optics: The achievement and the obstacle. In: Hogendijk, J.P., Sabra, A.I. (eds.), The Enterprise of Science in Islam: New Perspectives. MIT Press, Cambridge, MA, pp. 85–118.

——— 2007. The commentary that saved the text. The hazardous journey of Ibn al-Haytham's Arabic Optics. Early Science and Medicine 12, 117–133.

Saidan, A.S., 2000. Al-Bīrūnī on trigonometry. In: İhsanoğlu, E., Günergun, F., (eds.), Science in Islamic Civilisation. IRCICA, Istanbul, pp. 167–178.

Sakkal, M., 1995. Geometry of ribbed domes in Spain and North Africa. Journal for the History of Arabic Science 11, 53–73.

Saliba, G., 1997. A redeployment of mathematics in a sixteenth-century Arabic critique of Ptolemaic astronomy. In: Hasnawi, A., Elaramni-Jamal, A., Aouad, M. (eds.), Perspectives arabes et médiévales sur la tradition scientifique et philosophique grecque. Peeters, Leuven, pp. 105–122.

——— 2000. The ultimate challenge to Greek astronomy: Ḥall mā lā Yanḥall of Shams al-Dīn al-Khafrī (d. 1550). In: Folkerts, M., Lorch, R., (eds.), Sic Itur ad Astra. Harrassowitz, Wiesbaden, pp. 490–505.

——— 2001. The Astronomical Work of Muʾayyad al-Dīn al-ʿUrḍī (d. 1266): A Thirteenth Century Reform of Ptolemaic Astronomy, ʿUrḍī's Kitāb al-Hayʾa, Third corrected edition. Centre for Arab Unity Studies, Beirut.

——— 2007. Islamic Science and the Making of the European Renaissance. MIT Press, Cambridge, MA.

Samsó, J., 1996. "Al-Bīrūnī" in al-Andalus. In: Casulleras, J., Samsó, J. (eds.), From Baghdad to Barcelona, University of Barcelona, Barcelona, pp. 583–612.

——— 2002. Is a social history of Andalusī exact sciences possible? Early Science and Medicine 7, 296-299.

Samsó, J., Millás, E., 1998. The computation of planetary longitudes in the zīj of Ibn al-Bannāʾ. Arabic Sciences and Philosophy 8, 259–286.

Sarhangi, R., 2008. Illustrating Abu al-Wafāʾ Būzjānī: Flat images, spherical constructions. Iranian Studies 41, 511–523.

Sesiano, J., 1996a. L'abrégé enseignant la disposition harmonieuse des nombres, un manuscrit arabe anonyme sur la construction des carrés magiques. In: Casulleras, J., Samsó J. (eds.), From Baghdad to Barcelona, University of Barcelona, Barcelona, pp. 103–157.

——— 1996b. Le Kitāb al-Misāḥa d'Abū Kāmil. Centaurus 38, 1–21.

——— 1996c. Un traité médiéval sur les carrés magiques. Presses Polytechniques et Universitaires Romandes, Lausanne.

——— 1998a. Quelques constructions de carrés à magie simple dans les textes arabes. In: Actes du troisième colloque maghrébin sur l'histoire des mathématiques arabes, vol. 1. Office des Publications Universitaires, Algiers, pp. 251–262.

——— 1998b. Le traité d'Abū'l-Wafā sur les carrés magiques. Zeitschrift für Geschichte der arabisch-islamischen Wissenschaften 12, 121–244.

——— 2003a. Une compilation arabe du XIIe siècle sur quelques propriétés des nombres naturels. SCIAMVS 4, 137–189.

——— 2003b. Construction of magic squares using the knight's move in Islamic mathematics. Archive for History of Exact Sciences 58, 1–20.

——— 2003c. Quadratus mirabilis. In: Hogendijk, J.P., Sabra, A.I. (eds.), The Enterprise of Science in Islam: New Perspectives. MIT Press, Cambridge, MA, pp. 199–233.

——— 2004a. Les carrés magiques dans les pays islamiques. Presses Polytechniques et Universitaires Romandes, Lausanne.

——— 2004b. Magic squares for daily life. In: Burnett, C., Hogendijk, J., Plofker, K., Yano, M. (eds.), Studies in the History of the Exact Sciences in Honour of David Pingree. Brill, Leiden, pp. 715–734.

——— 2004c. La science des carrés magiques en Iran. In: Pourjavady, D.N., Vesel, Ž., Sciences, techniques et instruments dans le monde iranien (X^e^-XIXe siècle). IFRI/Tehran University Press, Tehran, pp. 165-181.

Sezgin, F., 2000a. Geschichte des arabischen Schrifttums. Band X. Mathematische Geographie und Kartographie im Islam und ihr Fortleben im Abendland. Kartenband. Institut für Geschichte der arabisch-islamischen Wissenschaften, Frankfurt am Main.

——— 2000b. Geschichte des arabischen Schrifttums. Band XI. Mathematische Geographic und Kartographie im Islam und ihr Fortleben im Abendland. Kartenband. Institut für Geschichte der arabisch-islamischen Wissenschaften, Frankfurt am Main.

——— 2000c. Geschichte des Arabischen Schrifttums. Band XII. Mathematische Geographie und Kartographie im Islam und ihr Fortleben im Abendland. Kartenband. Institut für Geschichte der arabisch-islamischen Wissenschaften, Frankfurt am Main.

Sidoli, N., 2006. The sector theorem attributed to Menelaus. SCIAMVS 7, 43–79.

Sidoli, N., Berggren, J.L., 2007. The Arabic version of Ptolemy's Planisphere or Flattening the Surface of a Sphere: Text, translation, commentary. SCIAMVS 8, 37–139.

Sidoli, N., Kusuba, T., 2008. Nasīr al-Dīn al-Tūsī's revision of Theodosius's Spherics. Suhayl 8, 9–46.

Smith, A.M., 2001. Alhacen's Theory of Visual Perception. American Philosophical Society, Philadelphia.

——— 2006. Alhacen on the Principles of Reflection. American Philosophical Society, Philadelphia.
——— 2008. Alhacen's approach to "Alhazen's problem." Arabic Sciences and Philosophy 18, 143–163.
Taha, A., 1998. La version arabe des lemmes d'Archimède. In: Actes du troisième colloque maghrébin sur l'histoire des mathématiques arabes, vol. 1. Office des Publications Universitaires, Algiers, pp. 263–275.
Vahabzadeh, B., 1997. Al-Khayyām's conception of ratio and proportionality. Arabic Sciences and Philosophy 7, 247–263.
——— 2002. Al-Māhānī's commentary on the concept of ratio. Arabic Sciences and Philosophy 12, 9–52.
——— 2004. Umar al-Khayyām and the concept of irrational number. In: Morelon, R., Hasnawi, A. (eds.), De Zénon d'Élée à Poincaré : Recueil d'études en hommage à Roshdi Rashed. Peeters, Leuven, pp. 55–63.
Van Brummelen, G., 1997. Determining the interdependence of historical astronomical tables. Journal of the American Statistical Association 92, 41–48.
——— 1998. Mathematical methods in the tables of planetary motion in Kūshyār ibn Labbān's Jāmiʿ Zīj. Historia Mathematica 25, 265–280.
——— 2006. Taking latitude with Ptolemy: Jamshīd al-Kāshī's novel geometric model of the motions of the inferior planets. Archive for History of Exact Sciences 60, 353–377.
——— 2009. The Mathematics of the Heavens and the Earth: The Early History of Trigonometry. Princeton University Press, Princeton.
Vitrac, B., 2000. Omar Khayyām et Eutocius : Les antécédents grecs du troisième chapitre du commentaire Sur certaines prémisses problématiques du Livre d'Euclide. Farhang 12, 51–105.
Yazdi, H.-R.G., 2009–2010. Al-Khāzinī's complex tables for determining lunar crescent visibility. Suhayl 9, 149–184.

Index of Subjects (Survey Papers)

ʿAbbāsid period, 64, 102
accession and recession, *see* trepidation of the equinoxes
algebra, 40, 54–56, 58, 59, 66, 74–75, 85, 86, 90, 121–123
  geometric, 5–6, 20, 25, 122
Alhazen's Problem, 64, 111, 114
amicable numbers, 52, 79, 122, 123
analemma, 43, 64, 77, 82, 115
analogia, 36
analysis and synthesis, 4–5, 30, 34, 41, 79, 81, 109–113, 120, 123
anaphor, 29
al-Andalus, 77, 81, 83, 85–86, 90, 108–109, 112–113, 115–123
angle measurement, 6
antanairesis, *see* anthyphairetic definition of ratio
anthyphairetic definition of ratio, 6–8, 18, 20, 105–106
Antikythera Mechanism, 12, 33, 43
anwāʾ literature, 90
appropriation of science, 88, 102
arches, 112
architecture, 87, 112, 118
arithmetic, 4, 11, 21, 38, 40, 53–54, 58, 66, 74, 80, 86, 89, 90, 118, 120–122
  pebble, 21
ascendent, 118, 120
aspects (astrology), 76, 120
astrolabe, 74, 75, 82–83, 90, 91, 110, 117, 119
  almucantar, 83
  azimuth circles, 82, 83, 117
  boat, 83
  melon-shaped, *see* azimuthal equidistant projection
  spiral, 83
  universal, 83
astrological history, 120
astrological houses, 76, 119
astrology, 25, 32, 34, 37, 40, 59, 73, 75–76, 90, 117–120
astronomy, 10–12, 21, 25, 27, 32–33, 36, 42–43, 73–76, 80, 81, 86–88, 90, 102, 107, 109, 113–118, 121
  folk, 88
atomism, 78
Axiom of Archimedes, 9, 41
axiomatic method, 4, 34–35
azimuthal equidistant projection, 65, 83, 117

Babylon, 5–6, 10, 11, 17, 20, 21, 27, 38, 40, 74
Bachet's problem of the weights, 53
binomial coefficients, 54
binomial theorem, 54, 59
Brethren of Purity, 78
burning mirrors, 11, 65, 81, 84, 106, 114
Būyid period, 53, 61, 110

cartography, 65, 66, 73, 83–84, 91
catoptrics, 29, 106
celestial sphere, 76, 83
centers of gravity, 10, 65, 110
Ceva's Theorem, 85
chess, 61, 122
China, 54, 56, 74, 89, 119, 121
chord, 11, 106, 115
climata, 84
combinatorics, 39–40, 52, 59, 86, 89, 113, 115
conic sections, 5, 11, 27, 29, 31, 32, 36, 37, 39, 56, 61, 63, 75, 76, 81, 107, 109–112, 114, 123
constructions, 42
continued fractions, 6–8
continuity, 104

N. Sidoli and G. Van Brummelen (eds.), *From Alexandria, Through Baghdad: Surveys and Studies in the Ancient Greek and Medieval Islamic Mathematical Sciences in Honor of J.L. Berggren*, DOI 10.1007/978-3-642-36736-6,
© Springer-Verlag Berlin Heidelberg 2014

cosmology, 33, 86, 116
cotangent, 63, 76
cubic equation, 56, 61, 75, 80, 106, 110, 112, 115, 123

ḍarb, 77
decimal system, 53, 75, 120
derivative, 1, 75
diagrams, 31–33, 35, 42
dialectic, 5, 7
didōmi, *see* given, mathematical objects
differential, *see* derivative
diorisms, 41
dip angle to the horizon, 110, 120
displacement, 119
dodecahedron, 7, 112
domes, 112
doubling the cube, 61, 80, 81
duplicating the cube, 4, 5

Egypt, 27, 40
Eleatic philosophy, 4
ellipse, 5, 81
epicycle, 11, 12, 81, 118
equation of time, 87, 118
equations, planetary, 119
equatorium, 81, 83, 90
equiponderant numbers, 52
Euclidean algorithm, 18, 105
exhaustion, method of, 113

figurate numbers, 123
finger arithmetic, 53, 89, 121
foundations of mathematics, 41–42
fractions, 8, 53, 74, 89
Fundamental Theorem of Arithmetic, 79, 123

geodesy, 61, 91
geography, 29, 30, 65, 74, 76, 83–84, 101, 114–118
geometry, 61–63, 80, 90, 109–114
  fixed, 61
  intuitionistic, 34
  moving, 61
given, mathematical objects, 29, 35, 41, 109
gnomonics, 6, 36, 76, 82
greatest common divisor, 121

harmonics, 21, 30, 33, 38, 42–43
heptagon, 56, 61, 106, 109–111
hexagon, 78
Hindu-Arabic numeration, 120–121
hippopede, *see* homocentric spheres
homeomeric lines, 35
homocentric spheres, 6, 86
homologue, 36
homothety, 110, 119
hyperbola, 5, 80, 81, 111, 113
hyperboloid, 81
hypostaseis, 40

icosahedron, 112
ʿilm al-farāʾiḍ, 54, 90, 121
ʿilm al-waṣāya, 54
incommensurability, 4, 6–9, 18–21, 25, 38, 42, 105
indeterminate equations, 56–58, 121
India, 11, 12, 53, 54, 75–76, 83, 88, 89, 102, 116, 118, 120, 122
infinity, 78, 111, 113
inheritance, *see* *ʿilm al-farāʾiḍ*
instruments, 119
interpolation, 64, 74, 75, 90
isoperimetry, 80
iteration, 56, 87, 102, 121

al-jabr, 54, 122

Keskintos Astronomical Inscription, 32
known, mathematical objects, *see* given, mathematical objects

latitude, planetary, 86, 89, 119
Law of Sines, *see* Sine Law
law of the lever, 7, 10
least common multiple, 121
lexicography, 89
limit, 113
linear equation, 75, 121
linguistics, 59
logistikē, 8
longitude, planetary, 119
longitude, solar, 87, 90

madrasas, 88, 108
Maghreb, 52, 59, 75, 77, 81, 83, 85–86, 108–123
magic squares, 59, 79, 90, 121–122
māl, 122
Mamlūk period, 51
mathematical induction, 59
mean and extreme ratio, *see* mean proportional
mean motion, 118, 119
mean proportional, 4, 21, 78
mechanics, 10, 12, 25, 30, 31, 43, 65–66
Menelaus's Theorem, 28, 76, 87, 89, 105, 110, 115, 116

mensuration, 106, 112–113, 121
method of declinations, 90
Mirāth, 54
Mongol period, 51
monochord, 33, 43
moon illusion, 84
muezzin, 118
multiplication, 54, 121
al-muqābala, 54
muqarnas, 87, 112
music, 4, 21, 38, 59, 73, 80
muwaqqit, 118

navigation, 90
negative numbers, 123
non-Euclidean geometry, 12, 63, 78
nonagon, 56, 61
number theory, 8, 52–53, 58, 59, 66, 79, 85, 123
numeracy, 39
numeration systems, 74
numerical mathematics, 64

optics, 11, 32, 43, 64–66, 73, 81, 84–85, 101, 106, 109, 111, 114
Ottoman period (Maghreb), 108

palimpsest, Archimedes, 12, 31
parabola, 5, 80, 110–112, 114
parallax, lunar, 83
parallel postulate, 12, 63, 78, 104–105, 110, 111
parapegma, 32
patronage, 108
pentagon, 18, 81
perfect compass, 110
perfect numbers, 52, 122
π, 84, 113
planetary theory, 10, 11, 116–117
planimetry, 80
plasmatikon, 40
poetry, 37, 89
polar triangle, 63
polygon, 112
polyhedron, 5, 106, 111, 112
polynomial, 56, 75, 81, 122, 123
prayer times, 109, 117, 118
prime numbers, 52, 79
projections of the sphere, 82–83, 117
proof, 19–20, 26, 29, 33–35, 42, 90
proportion, *see* ratio theory
Pythagorean Theorem, spherical, 87
Pythagoreans, 6, 8, 18, 20–21, 38, 43

qabbān, 65
qarasṭūn, 65
qibla, 63, 76–77, 83, 87, 90, 91, 109, 117–118
quadrant, 119
quadratic equation, 74, 75, 122
quadrature, 9, 36, 41
  of lunes, 38
quartic equation, 110
qubba, 87, 112

rainbow, 52
Ramadan, 117, 118
ratio theory, 4, 6–8, 18–20, 36, 78, 89, 105–106, 109, 115
rays, *see* aspects (astrology)
real numbers, 106, 123
reception of science, 88, 102
recreational mathematics, 59, 79, 90, 121–122
refraction, 10, 84
regular infusa, 75
regular solids, *see* polyhedron
rhetoric, 37, 39
rising times, 87
root extraction, 53, 54, 56, 74, 105, 121
Ruffini-Horner Equation, 75
Rule of Four Quantities, 115
rūmī numbers, 121

Safavid Iran, 108
Samarqand, 53, 64, 87
Saragossa, *see* Zaragoza
saṭḥ, 77
sayyāla, 56
Schröder Numbers, 39
Sector Theorem, *see* Menelaus's Theorem
Seleucid period, 38
Seljuk period, 51
Seville, 86
sexagesimal system, 6, 53–54, 75, 121
shooting stars, 110, 120
sine, 75, 76, 106, 115
Sine Law, 87, 115
sine quadrant, 115, 119
Snell's Law, 84
solar equation, 89
spherical quadrilateral, 115
spherics, 27, 28, 32, 33, 41–43, 106, 107, 111, 115
spiral, 10, 80
squaring the circle, 80
statics, 10
statistics, 89
stereographic projection, 65, 82, 107, 110, 117

stereometry, 80
Stoicism, 30, 39
sundial, 32, 76, 77, 82, 87, 90, 112, 117, 119, 120
surveying instruments, 32

tables, numerical, 11, 27, 29, 54, 56, 59, 64, 75, 76, 89–90, 115–119
  auxiliary, 64, 90
  double argument, 64, 74
tangent, 63
tangent circles, 11, 109, 110
timekeeping, 76, 88, 90, 116, 118–119
Timurids, 53, 87
Transversal Figure, *see* Menelaus's Theorem
trepidation of the equinoxes, 116–117
trigonometry, 11, 43, 63–64, 75, 86, 90, 109, 115, 119
  spherical, 63, 66, 76, 86, 87, 106, 110, 115–117, 119, 120
trisecting the angle, 4, 61, 81, 107, 109
Ṭūsī Couple, 86, 116

universal instrument, 90
ʿUrḍī's Lemma, 86, 116

verging constructions, 4, 10, 112

water clock, 12
Wilson's Theorem, 52

zakāt, 54, 90
Zaragoza, 85, 105, 111, 114, 115
zīj, 63, 64, 75, 81, 87, 89, 90, 116, 118–119

Part II
Studies

Mechanical Astronomy: A Route to the Ancient Discovery of Epicycles and Eccentrics

James Evans and Christián Carlos Carman

Abstract The ancient Greek art of *sphairopoiïa* was devoted to the building of models of the universe such as celestial globes and armillary spheres. But it also included the construction of geared mechanisms that replicated the motions of the Sun, Moon, and planets, such as the famous orrery that Cicero attributed to Archimedes or the spectacular Antikythera mechanism, found in an ancient shipwreck of about 60 BC. Was *sphairopoiïa* merely an imitative art, in which the modelers followed the precepts of the theoretical astronomers? Or could theoretical astronomy also learn something from the art of mechanics? In this paper, we examine the relation of astronomy to mechanics in the ancient Greek world, and argue that we should imagine astronomy and mechanics in conversation with one another, rather than in a simple, one-way transmission of influence.

Introduction

The emergence of deferent and epicycle theory in Greek planetary astronomy is shrouded in mystery. The homocentric spherical models associated with Eudoxus, Callippus and Aristotle were abandoned in planetary theory sometime in the century after the death of Aristotle (322 BC), though the idea that the universe consists of nested spherical orbs continued to dominate cosmological thinking until the Renaissance. The next theoretical, geometrical tools known to us are epicycles and eccentrics, which probably appeared within a few decades one way or the other of 200 BC. We do not know what may have motivated them. One key development was Greek absorption and adaptation of Babylonian astronomy, which was already well under way in the third century BC, though of course a decisive episode was centered around the work of Hipparchus in the second. It is possible that Greek epicycle-and-eccentric theories arose, in part, as an effort to model Babylonian "phenomena" — i.e., the phenomena predicted by the Babylonian theories, which provided a much more convenient and comprehensive account of the planetary motions than mere observation ever could.

At least from the time of Archimedes, in the late third century, Greek astronomers and mechanics (*mechanikoi*) also constructed models to imitate the workings of the heavens. Such a model was called, in Greek, a *sphairopoiïa* (spherical construction), or often simply a *sphaira*. (The corresponding Latin term was *sphaera*.) *Sphairopoiïa* was also the name for the branch of mechanics devoted to this art.[1] The art of *sphairopoiïa* included the art of building simple teaching tools such as celestial globes and armillary spheres. But it also included the construction of more elaborate machines intended to replicate the motions of the Sun, Moon, and planets. We know from the remains of the Antikythera mechanism that the operative principle of these more elaborate planetarium-style *sphairopoiïai* was the concrete realization of astronomical period relations by means of gear trains. That gears emerged in Greek

[1] For an introduction to *sphairopoiïa*, see Evans and Berggren [2006, 47, 52–53, 246–249]. For a study stressing *sphairopoiïa* as a tool of discovery rather than merely of representation, see Aujac [1970].

N. Sidoli and G. Van Brummelen (eds.), *From Alexandria, Through Baghdad: Surveys and Studies in the Ancient Greek and Medieval Islamic Mathematical Sciences in Honor of J.L. Berggren*, DOI 10.1007/978-3-642-36736-6_7,
© Springer-Verlag Berlin Heidelberg 2014

mechanics and epicycles appeared in Greek astronomy at roughly the same time (almost certainly within a century of one another), is certainly suggestive. And it is all the more remarkable that the gears are probably the older of the two. So the question arises: Were gearwork *sphairopoiïai* simply imitative models meant to illustrate the theories of the geometrical astronomers? This view, which has been almost universally held, we might describe as "theory first, mechanical model later." Or was theoretical astronomy also able to borrow inspiration from mechanics? In this paper we shall explore the possibility that mechanical invention played a role in the development of Greek theoretical astronomy. Gearwork mechanisms may have provided the insight that led to the invention of epicycles and eccentrics.

Chronology

The oldest mention known to us of something like gears appears in the pseudo-Aristotelian *Mechanics* (sometimes also called *Mechanical Problems*).[2] In later antiquity, it was generally believed to have been written by Aristotle himself. Diogenes Laertius [v, 26] mentions a *Mechanikon* in his list of books by Aristotle. And Athenaeus the Mechanic (perhaps first century BC) provides an earlier attestation, mentioning Aristotle in his list of authors on mechanics who can provide a reader with a theoretical introduction but not with instruction in anything practical [Whitehead and Blythe 2004, 44–45]. Since the middle of the nineteenth century, most (though not all) authorities have held that it is not really by Aristotle, but comes from the Peripatetic school of the late fourth or early third century BC. There is a voluminous literature on the question, spanning the whole period from the early nineteenth century to our own day. Summaries of the debate up to the recent past have been made by Berryman [2008, 107–109] and, in much greater detail, by Bottecchia Dehò [2000, 27–51]. Here we shall only mention what is essential for our argument.

The text attempts to derive the principle of the lever from that of the circle, and those of other simple powers from that of the lever. Because of its unsophisticated approach to the lever itself, it is reasonable to suppose that it was written some time before Archimedes' *On the Equilibrium of Planes*. Of course, by itself this argument would not be enough, as we have plenty of instances of "crude" treatises being written later than more sophisticated ones. But in any case, the text makes no use of Archimedes' results, nor of the concept of a center of gravity. Moreover, as Berryman [2008, 108] points out, in its enumeration of various mechanical powers (lever, wheel, pulley, wedge) with their many applications (to rudders, forceps, nutcrackers, rollers, etc.) the text makes no mention of the screw, which, again, suggests a pre-Archimedean date. Historians of mathematics place the composition in the Peripatetic school on the basis of the similarity of some of its demonstrations to demonstrations found in genuine works of Aristotle, such as *On the Heavens* and the *Physics* [Heath 1921, 1:344–346]. A strong argument for placing it rather early in the history of the school is based on its mathematical terminology — a case developed by Heiberg, based on a detailed examination of the technical vocabulary. Thus, according to Heiberg, the *Mechanical Problems* could have been written either before Euclid had made mathematical terminology more consistent and convenient, or perhaps a while after Euclid but in circles that were still dominated by the older, Aristotelian terminology [Heiberg 1904, 30–32]. An effort has occasionally been made to ascribe it to Strato of Lampsacus, who was head of the school c. 288–269 BC, based partly on the ascription of a work on mechanics to Strato [Diogenes Laertius v, 59], but this has not won wide support.[3] Marshall Clagett [1959, 4–9] revived interest in the *Mechanical Problems* when he argued for the significance of its dynamic approach to problems of statics (unlike the approach of the later Archimedean treatises), which was to become so important in the Middle Ages. Since then, the original ascription to Aristotle has been again defended, notably by Krafft [1970, 13–20].[4] Clearly, the authorship of the text remains an open question; but the authorship issue is not important for our

[2] Pseudo-Aristotle, *Mechanical Problems* 848a25–38 [Hett 1936, 334–337].

[3] The attribution to Strato has been most recently reassessed (unfavorably) by Bodnár [2011].

[4] Arguments against Krafft's conclusion were given by Knorr [1982, 100–101, n27], whose own view was for a date early in the third century.

purposes. In view of the history and present state of scholarship, we shall be adopting a conservative position if we take the *Mechanical Problems* as no later than the middle of the third century BC. Indeed, there seems to be no recent authority who would make it later than Strato.[5]

It is true that the text does not mention teeth, but only circles in contact, so some scholars have been reluctant to credit it with a discussion of real gears, but only "friction wheels."[6] Of course, the best way to improve a friction wheel is to eliminate the possibility of slipping by adding teeth. And, as far as we know, no one has pointed to an actual ancient artifact that involves a pair of interacting "friction wheels." By contrast, gears are not merely theoretical entities postulated by modern historians, but things that really did exist. Moreover, the writer says that the objects he is discussing are sometimes seen in temples, where they have been dedicated as offerings. And they are arranged so that, from one motion, many circles move at the same time. Although a *pair* of "friction wheels" may be workable, the reliable movement of many wheels by one driver seems plausible only for toothed wheels. The writer says that, using the principle of the circle, the craftsmen construct an instrument in which the first cause (τὴν ἀρχήν, perhaps referring to the first wheel in the system) is concealed, "so that only the wonder of the machine is apparent, while the cause is unseen" (ὅπως ᾖ τοῦ μηχανήματος φανερὸν μόνον τὸ θαυμαστόν, τὸ δ' αἴτιον ἄδηλον). It appears, then, that the writer is describing actual machines that he has seen. The use of θαυμαστόν also suggests that we are dealing with an early stage of the wonder-working art, in which the mere fact of multiple circular motions produced from one input would have been enough to amaze. Finally, it is noteworthy that Book I of the *Mechanica* of Hero of Alexandria begins (after the discussion of a winch that is almost certainly misplaced or interpolated) with a general discussion of the theory of gears and here the discussion makes no use or mention of teeth — simply ratios of circumferences, etc.[7] Teeth are introduced later on. So it seems that an introductory discussion of the mathematics of gear systems that makes no mention of the details of teeth would not be out of the ordinary.

Let us turn now to the other evidence for early gearing. The dates of the Alexandrian mechanic Ctesibius have been contested. But Drachmann places his *floruit* around 270 BC on the basis of an epigram by Hedylos, quoted by Athenaeus of Naucratis (11.497a–e), which tells of a musical cornucopia that he made for the statue of Arsinoë, the sister and wife of Ptolemy II Philadelphus (reigned 285–247 BC) [Drachmann 2008]. Now, Vitruvius discusses a water clock that he claims was made by Ctesibius, in which a rack engaged a toothed wheel.[8] So here is an argument (not a proof, certainly) for situating the first gears by the middle of the third century. Unfortunately, Athenaeus (4.174d) makes the situation a bit murky by saying elsewhere in the same work that Ctesibius, the inventor of a hydraulic organ, lived in the reign of "the second Euergetes" (Ptolemy VIII Euergetes, who reigned jointly with Ptolemy VI and Cleopatra II, in 170–164 BC, and on his own, 146–116 BC). So either Athenaeus has made a slip or there was a second Ctesibius. This issue, which has a long history, has been discussed in detail by Drachmann [1951], who argues that the second Ctsebius is unlikely. In any case, the placement of *a* Ctsebius in the reign of Ptolemy II, based as it is on the detail added by Hedylos's account of the statue of Arsinoë, seems reasonably secure.

Archimedes, who died in 212 BC, is said by Diodorus Siculus (twice) as well as by Athenaeus of Naucratis to have invented the water pump called the *cochlias*, now often known as the Archimedean screw.[9] According to Athenaeus, a water screw was used to pump out the bilge of the ship *Syracosia*, which Hieron II of Syracuse (reigned c. 271–216 BC) built and sent to Egypt as a gift for King Ptolemy. Athenaeus's source was a certain Moschion, who wrote a book in which the construction of this ship was treated in considerable detail. The mention of the water-screw bilge pump is embedded in the course of the longer description of the ship, which lends this detail more credibility. A water screw, of course, is not a gear; but its central element, the helical screw, is similar in form to one of the two

[5] An outlying position is that of Winter [2007], who argues that the text is even older and was written by Archytas.

[6] Drachmann [1963, 13] opts for friction wheels. Berryman [2009, 113] comes down on the side of gears.

[7] Carra de Vaux [1894, 42–45]. The *Mechanica* of Hero survives only in Arabic.

[8] Vitruvius, *On Architecture* ix, 8.5.

[9] Diodorus Siculus, *Bibliotheca historica* i, 34.2 and v, 37.3–4 [Oldfather 1933, 1:112–115 and 3:199]. Athenaeus of Naucratis, *Deipnosophistae* v, 208f.

key elements of an endless screw. And the endless screw, in which a helical worm gear engages a plane gear, can be regarded as a natural development of a rack and pinion. These three technologies, then, are closely allied. It is, of course, irrelevant for our purposes whether Archimedes really invented the water screw or drew upon an already existing technology.[10]

For the endless screw or screw-windlass itself, the picture is less clear. Athenaeus of Naucratis [v207b], again, tells us that Archimedes was the discoverer of the screw-windlass (ἕλιξ), which he reportedly used to launch a ship. However, other writers say that he used pulleys for this task. The various testimonia have been collected and discussed by Drachmann [1958]. We will not indulge in legends of Archimedes' ship-launching, nor speculate on just which power he was thinking of when he boasted that if he had a place to stand he could move the Earth. That the rack and pinion, the water-screw, and the screw-windless should have emerged around the same time is inherently plausible and the key point is that we have attestations of all three for the third century BC.

Several Arabic manuscripts preserve a work on a water clock attributed to Archimedes. This device involves a crown gear engaging a lantern pinion, which are illustrated in diagrams. A key feature of the clock is its bird's head that disgorges a ball once each hour [Hill 1976]. Three nearly complete manuscripts exist in Paris, London, and New York and a fragment is preserved at Oxford. All four present Archimedes as the inventor. The work is also mentioned in the *Fibrist* of Ibn al-Nadīm, an Arabic bio-bibliographical work of the tenth century, in which it is also attributed to Archimedes [Dodge 1970, 636]. Some scholars have regarded the work as Byzantine or Muslim in origin (though drawing on Hellenistic ideas), Carra de Vaux going so far as to characterize the use of the name of Archimedes as most likely the "banal ruse of an author desirous of being read."[11] Moreover, the Oxford fragment is dedicated to one "Māristūn." Since Philo of Byzantium dedicated his works to "Ariston," Drachmann [1948, 38] remarks, "One might almost regard the dedication to Ariston the hall-mark of a work by Philon." And he concludes that the "whole thing is the work of a Moslem inventor, who has put together details from several sources, one of them doubtless Philon, another probably Heron...." The fullest discussion of its possible origins is given by Hill [1976, 6–9], who notes that the problem is complex. But, as Hill points out, the Arabic writers are unanimous in ascribing the first section (involving the water machinery and the release of the balls) to Archimedes. He concludes that the treatise is at least based on Hellenistic models and that it shows signs of having been translated into Arabic from Greek. Hill's own view is that the roots of the treatise are indeed Archimedean, but that it was reworked by Philo, and that the later sections with their Eastern motifs are later additions. For Philo, it is difficult to arrive at a secure date, but usually his *floruit* is placed in the latter part of the third century BC.[12]

Archimedes is also said to have devised a machine that represented the movements of the Sun, the Moon, and the planets. This instrument (along with a simpler celestial globe) was reportedly taken to Rome by the general Marcellus after the sack of Syracuse. Our chief source is Cicero, who describes Archimedes' device in a philosophical dialogue, the *Republic*, which was modeled on Plato's.[13] Now, Cicero wrote his *Republic* around 54 BC, but its dramatic date is set around 129. In the course of the dialogue one of the speakers recounts an episode in the life of Gaius Sulpicius Gallus, around 166, when Gallus saw and explained the *sphaera* that had been brought back to Rome in 212. Whether this device still survived in Cicero's day we have no way to know. Such a wonderful machine would certainly have been a "keeper" and its location in a wealthy household may possibly have helped to preserve it. However, Cicero does not say that he himself had seen it and perhaps he was supplying details on the

---

[10] Stephanie Dalley has argued that the water screw appeared at Nineveh in the 8th century BC; see Dalley and Oleson [2003]. A more commonly encountered proposal is that Archimedes drew upon an already-existing Egyptian technology, for a discussion (and refutation) of which see Oleson [1984, 291–294]. Oleson also discusses the iconographical evidence for the water screw, none of which pre-dates the Roman period. Terracotta reliefs in London and Cairo show a man or boy treading a water screw (Figure 71 and 86) and a wall painting in Pompeii shows the same (Figure 101).

[11] Carra de Vaux [1891, 296]. A German translation of the text with commentary was published by Wiedemann and Hauser in 1918, reprinted in Wiedemann [1970].

[12] Drachmann [2008b] has him flourishing c. 250 BC; Toomer, c. 200 BC in his entry "Philon of Byzantium" in the Oxford Classical Dictionary, 3rd ed.

[13] Cicero, *Republic* i, 21–22 [Keyes 1994, 40–43].

basis of what he had seen of the *sphaera* of Posidonius of Rhodes, his teacher and friend. But that Archimedes built some sort of device seems certain, and he is said by Pappus of Alexandria, on the authority of Carpos of Antioch, to have written a treatise on *sphairopoiïa* [Ver Eecke 1933, 2:813–814]. This may plausibly have included a description of his machine or of its principles. If we accept that Archimedes did build such a machine — even while admitting that we can say nothing with certainty about its features — it is difficult to see what its fundamental principle might have been if it was not the gear. And there must have been some line of development before anything as complex as a *sphairopoiïa* could have been built. (We may, for example, imagine the first gearwork mechanisms as simple displays of an amazing principle, such as the machine mentioned in the *Mechanical Problems*).

Here, then, are half a dozen strands of evidence placing gears in the third century. While no single episode in the early history of Greek mechanics is decisive, and any one of them may certainly be subject to reservations, together they allow us, with reasonable confidence, to situate the appearance of gears by about the middle of the third century BC.

As for epicycles and eccentrics, the tradition has been to place their origin around the time of Apollonius of Perge, if not with Apollonius himself, based on some remarks of Ptolemy.[14] Of course, it is possible that someone might have imagined epicycles and eccentrics before Apollonius. An epicycle for an inferior planet seems such an obvious way of explaining the planet's limited elongations from the Sun that it could conceivably have been invented more than once.[15] But Apollonius is the first figure for whom we have any evidence for an interest in epicycles or eccentrics as mathematical objects for which theorems can be proven. For this reason, it is necessary to say a little about how Apollonius's lifetime is best established.

Eutocius, in his *Commentary* (early sixth century AD) on the *Conics* of Apollonius, says that Apollonius lived (or, perhaps, was born) in the reign of Ptolemy III Euergetes (247–222 BC) [Heiberg 1891–93, 168]. The verb used is γέγονε, a perfect of γίγνομαι, so the basic meaning is usually (though not always) "was in existence" rather than "came into being."[16] Heiberg translated it by *vixit* ("lived") and Heath [1921, 2:126] by "flourished." Photius, the ninth-century patriarch of Constantinople, quoting a more doubtful source, the second-century grammarian Ptolemaeus Chennus of Alexandria, says that an Apollonius was famous for astronomy in the reign of Ptolemy IV Philopater (222–205 BC), and it is usually supposed that this must be Apollonius of Perge.[17] Heath, following Hultsch, discussed this evidence and concluded that Apollonius "was probably born about 262 BC, or 25 years after Archimedes," a conclusion that is still widely quoted.[18]

---

[14] Ptolemy, *Almagest* xii, 1 [Toomer 1984, 555 & 558].

[15] However, it should be pointed out that the common attribution of circumsolar orbits for Venus and Mercury to Heraclides of Pontus has been thoroughly refuted. See Eastwood [1992] and Toomer [2008b].

[16] Rhode [1878] studied 129 instances of the use of γέγονε in the *Suda* (a Byzantine historical compilation of about the tenth century). He found that the meaning was

"certainly the time of flourishing in	88 cases
probably the time of flourishing in	17
certainly the time of birth in	6
perhaps the time of birth in	4
no obstacle to meaning ἤκμαζεν ["he was in his prime"] in	9
wholly undecidable in	5."

[17] Bekker [1824–1825, Codex 190, 1:151b18]. Henry and Schamp [1959–1991, 3:66]. This passage is the source of the well-known story that Apollonius was given the nickname ε, on account of the resemblance of the letter to the shape of the Moon, which he had investigated most thoroughly. It occurs in a short list of people who had letters of the alphabet as nicknames.

[18] Apollonius addressed the first three books of his *Conics* to Eudemus, but the fourth and following books to Attalus. Heath supposed that this is King Attalus I of Pergamum (reigned 241–197 BC), which supported his dating. But Toomer [2008a] has argued that Attalus was a common name among those of Macedonian descent and that "it is highly unlikely that Apollonius would have neglected current etiquette so grossly as to omit the title of 'King' (βασιλεύς) when addressing the monarch." We set the Attalus association aside as too insecure. Similarly, the mention by Pappus in Book 7 of the *Mathematical Collection* [Hultsch 1876–1878, 2: 678–679; Ver Eecke 1933, 507] that Apollonius studied with the pupils of Euclid is set aside, as it is not clear whether Pappus's source meant pupils that Euclid himself had taught, or the intellectual descendants in the "school."

But Apollonius himself provides some autobiographical detail that may contradict so early a dating. In the introduction to Book II of the *Conics*, Apollonius says that he is sending the work to Eudemus (of Pergamum) by way of his own son Apollonius and he requests that Eudemus should also give a copy of it to "Philonides the geometer, whom I introduced to you in Ephesus," if the latter ever be in the vicinity of Pergamum. Now an Epicurean philosopher named Philonides is known from an anonymous papyrus biography found at Herculaneum (*P. Herc.* 1044). The text was published in 1900 by Crönert, who pointed to its possible utility for dating Apollonius.[19] (This text was available when Heath published his *History of Greek Mathematics*, but he did not use it in his discussion of the date of Apollonius.)[20] Philonides is known also from inscriptions found at Athens and Delphi.[21] From the biography, we see that Philonides the Epicurean was well-connected at the Seleucid court of Antiochus IV Epiphanes (reigned 175–c. 164 BC) and his nephew Demetrius I Soter (162–150), which provides key evidence for dating him. Moreover, we learn that his first teacher was one Eudemus and that he then followed lectures by "Dionysodorus the son of Dionysodorus of Caunus" [Fragment 25; Gallo 1980, 82].[22] Philonides wrote an exegesis "of Book 8 of [Epicurus's] *On Nature* and many others of various kinds concerning his doctrines, many [of these exegeses] geometrical concerning the Minimum" (ἐλάχιστον).[23] *Elachiston* here perhaps refers to the Epicurean theoretical minimum magnitude. As Sedley [1976, 24] has suggested, resolving the apparent conflict between the existence of a minimum magnitude and the ordinary practices of Greek geometry (which assumed continuously divisible lines) could have been a significant issue for an Epicurean geometer. Of course, Epicurus's animosity towards geometry is well known, but Sedley [1976] and Mueller [1982, 95] argue that the later Epicureans were not all anti-geometers. Still, the papyrus life has little to say of Philonides' geometrical accomplishments — we have no actual geometrical discovery attributed to him. Rather the emphasis is on his role as an Epicurean philosopher, his conversion of Demetrius to Epicureanism, and his role in saving his home town of Laodicea-by-the-Sea (Syria) from destruction. Philonides also composed epitomes of the letters of Epicurus, Metrodorus, Polyaenus and Hermarchus that could be "useful to lazy young people." Thus Philonides seems, at least later in life, to have been more a philosopher than a geometer. On the other hand, the papyrus life gives us the names of his geometry teachers and also mentions a Zenodorus with whom Philonides was on friendly terms [Fragments 31, 34, Gallo 1980, 88–89], who is perhaps, though not certainly, the geometer mentioned by Pappus and Theon of Alexandria as the author of a treatise on isometric figures.[24] Finally, the papyrus seems also to associate Philonides with Ephesus at one stage of his life[25] — where, of course, we know that Apollonius introduced Eudemus and Philonides. In all, the identification of Apollonius's Philonides the geometer with Philonides the Epicurean seems reasonably secure, based as it is on the latter's demonstrated interest in geometry, his association with a certain Eudemus and his probable connection to Ephesus. But having Apollonius overlap with Philonides the Epicurean obviously requires taking Euctocius's γέγονε as indicating the birth of Apollonius and not his flourishing. This is somewhat unusual but not terribly rare.

The birth date of Philonides the Epicurean is often taken as about 200 BC, on the basis of his

[19] Crönert [1900]. A new edition of the text, with Italian translation and copious notes, is provided by Gallo [1980 2:23–166]. Much of the evidence for Apollonius's date is discussed in Huxley [1963, 100–103] as well as Fraser [1972, 1: 415–418]. Fraser [1972, 2:600–604] prints most of the relevant Greek texts. Toomer [2008a] provides a concise and cogent discussion.

[20] Heath did discuss this papyrus in the section of his book dealing with the identity and dating of the geometer Dionysodorus [Heath 1921, 2:218].

[21] Köhler [1900]. The inscription from Delphi has been published in Plassart [1921]; Philonides and his brother appear at IV 78–80 (p. 24).

[22] A mathematician named Dionysodorus is said to have solved a cubic and to have studied the torus [Heath 1921, 2:218–219]. Vitruvius, *On Architecture* ix, 8, also mentions someone of this name as the inventor of a conical sundial. But there was more than one mathematician named Dionysodorus, as is clear from Strabo, *Geography* xii, 548c. See Knorr [1986, 263–276].

[23] Fragments 13 inf.–14. See Gallo [1980, 67–68], which corrects the text of Crönert [1900, 947] by removing the first line to a greater distance on the papyrus.

[24] See Heath [1921, 2:207], Knorr [1986, 233–234, 272–274] and Toomer [1972].

[25] Fragment 37 [Gallo 1982, 91]. The reading of Ephesus is not certain, however, due to damage to the papyrus.

association with Antiochus and Demetrius, as well as the evidence of the inscriptions.[26] However, Gera [1999] has argued that Philonides' association with the Seleucid court began, not in the reign of Antiochus Epiphanes, as is usually supposed, but in the reign of his predecessor on the throne, his brother Seleucus IV. If this is correct, we should probably back up Philonides' career by a decade or so. So Apollonius's introduction of the young Philonides to Eudemus could have happened as early as about 190, but not much earlier. Some years later, when Apollonius sent Book II to Eudemus, Apollonius had a grown son but was not yet very far along in the revision of his *Conics*; so he was then perhaps 40 years old. If he did his astronomy early in life, we might suppose that the work on epicycles and eccentrics was as early as about 210 BC, but not much earlier. Thus it is clear that Apollonius' mathematical work on planetary theory came well after (by one or two generations) the introduction of gears into Greek mechanics. Gears would pre-date Apollonius's astronomical investigations even with Heath's birth date for Apollonius; but the Philonides connection makes it all the more certain.[27]

Mechanical Geometry

In pure geometry, there was early discussion of mechanical contrivances for solving problems. A striking example involves the problem of finding the two mean proportionals in a continued proportion [Heath 1921, 1:255–258]. That is, given a and b, to find x and y such that $b/y = y/x = x/a$. This problem is closely related to the duplication of the cube. For, suppose that we are given a cube of side a and volume a^3 and we seek b, such that $b^3 = 2a^3$. If we can find two mean proportionals x and y, where $b/y = y/x = x/a = r$, say, we have

$$\left(\frac{b}{a}\right)^3 = \left(\frac{b}{y}\frac{y}{x}\frac{x}{a}\right)^3 = 2,$$

or $r^3 = 2$.

Thus the duplication of the cube is solved, for the required side length is $b = ra$. Hippocrates of Chios is said to have been the first to show that the duplication of the cube could be reduced to finding two mean proportionals[28] and from then on the problem was usually approached in the latter form.

As Pappus was later to say, geometrical problems could be divided into three categories. Plane problems (ἐπίπεδα προβλήματα) could be solved with straightedge and compass alone.[29] Solid problems (στερεά προβλήματα) required the use of conic sections. They are called solid, says Pappus, because they make use of solid figures for their construction. Finally, the most complex problems required the use of other special curves, such as the quadratrix, or spirals, or conchoids. These problems are called *grammika* (γραμμικά προβλήματα). There is no satisfactory English translation of *grammika* ("linear" won't do, but we might approximate it by "making use of curves").[30] Now, the duplication of the cube (and, equivalently, the finding of two mean proportionals) does not belong to the class of plane problems, but to the class of solid problems. Ancient geometers did not, therefore, abandon all hope when

[26] Philippson [1941] gives 200–130 BC for Philonides' lifespan, on the basis that he was in his prime in the reign of Demetrius. Gallo [1980, 36] holds that Philonides' birth date is unlikely to have been after 200 BC, and rather likely to have been before this date, though not by much. Knorr [1986, 276] prefers a birth date of 220 BC. Fortunately, none of this fine-tuning is consequential for our argument.

[27] We know, e.g., from the inscription at Delphi that Philonides the Epicurean had a father who also was named Philonides; see Köhler [1900] and Plassart [1921, 24]. This elder Philonides was a man of some distinction, who played a role in facilitating diplomacy with the Seleucid court. On balance, it does seem probable that it is indeed the son, Philonides the Epicurean, who was Apollonius's "Philonides the geometer." But if it were the father who was Apollonius's Philonides, this would back Apollonius up by perhaps twenty years.

[28] Eutocius, in his Commentary on Archimedes' *On the Sphere and Cylinder II* [Mugler 1972, 64; Netz 2004, 294].

[29] Pappus, *Mathematical Collection* iii, 7 [Ver Eecke 1933, 1:38].

[30] Alexander Jones translated γραμμικά with "curvilinear" in his translation of Pappus's *Mathematical Collection* vii, 27 [Jones 1976, 2:112–113].

faced by a problem that could not be solved by ruler and compass alone, but had recourse to more complex methods.

Eutocius, in his commentary on Archimedes' *On the Sphere and Cylinder*, describes a mechanical solution to the problem of finding two mean proportionals, which he (probably wrongly) attributes to Plato, and which makes use of a special sliding instrument.[31] Eutocius attributes a second mechanical solution to Eratosthenes, and this involves an instrument called a *mesolabon* (a "mean-taker"). Pappus also discusses a number of solutions, including the *mesolabon*.[32]

According to Plutarch, Plato condemned such methods as the corruption of the good of geometry, since they involved a descent from the incorporeal things of pure thought to the realm of the perceptible.[33] (And this is a good reason for doubting Eutocius' attribution of a mechanical solution to Plato. A second reason is that Pappus does not mention a solution by Plato and he surely would have, had he known about it.) If Plutarch's story is not apocryphal, Plato must have been reacting to what he regarded as an unfortunate trend in geometry, so it is possible that geometers were engaged with mechanical solutions already in the early Academy. In any case, we have good evidence for mechanical approaches all through the Hellenistic period. The mechanical methods added to the quiver of available techniques.

And, of course, we have Archimedes' famous discussion, in the *Method*, of the use of mechanical methods as an *aid in discovery* of new theorems, which must then be proven in more conventional ways.[34] Archimedes finds it easier to divine the areas or volumes of figures if he imagines slicing them up and weighing them against slices of other figures by means of a balance. Here, then is a case in which mechanics helps guide speculation in pure geometry.

But it would be wrong to think of mechanics simply as an assistant to geometry. Rather, as Sidoli and Saito [2009, 605–607] have stressed, Greek geometry emerged in the context of instruments and this instrumental context helped shape the methods of geometry. Thus it is no accident that Euclid's rules of construction admitted of compass and straightedge only — it was by playing about with straightedge and compass that the early geometers imagined new problems and clarified their thinking about them. Sidoli and Saito point out that also in the *Spherics* of Theodosius, the great majority of constructions mentioned could actually be carried out on the surface of a real globe (leaving aside those that mandate a slicing of the globe): mechanical thinking — thinking about instruments — played a role in the development of the treatise. Perhaps most importantly, in the case of plane geometry, mechanical approaches served to broaden the scope of geometry, and to extend its range beyond the field of play imagined by the early geometers.

Mechanical Astronomy

If even pure geometry could benefit from mechanics, is it possible that astronomy benefitted from its relation to the art of *sphairopoiïa*? Ordinarily, we are disposed to think of *sphairopoiïa* as merely representational, as involving models meant to inspire contemplation or wonder, or tools that could be used in teaching. But a well-made celestial globe could also be used to solve problems of spherical trigonometry without tedious calculation — that is, a globe could serve as a specialized analogue computer. The use of a globe has often been claimed for Ptolemy's treatment of the heliacal risings and settings of the fixed stars (in his *Phaseis*) for climes outside of Alexandria, as well as for Hipparchus's sidereal phenomena in his *Commentary on the Phenomena of Aratus and Eudoxus* [Neugebauer 1975, 930–931].

---

[31] The mechanical proofs are discussed in Heath [1921, 1:255–260]. Also relevant is the construction of a chonchoid curve by Nicomedes, using a special mechanical instrument [Heath 1921, 1:238–240].

[32] Pappus, *Mathematical Collection* iii, 7 [Ver Eecke 1933, 40–41].

[33] Plutarch, *Marcellus* xiv 5 [Perrin 1917, 471–473]. Plutarch, *Symposia* viii 2.1 (718 E–F) [Minar et al. 1961, 121–123].

[34] In this volume, see the chapter contributed by Ken Saito and Pier Daniele Napolitani, "Reading the Lost Folia of the Archimedean Palimpsest: The Last Proposition of the *Method*."

But we wish to suggest something well beyond the simple use of a globe as an instrument of calculation. For *sphairopoiïai* could also be used as tools of discovery. In the early period of Greek astronomy, when mathematicians were still mastering the theory of the celestial sphere, it is likely that propositions were sometimes discovered with the aid of a celestial globe or an armillary sphere before being proven geometrically. Many propositions of early spherics, of the kind that appear in Autolycus of Pitane's *On the Moving Sphere* and *On Risings and Settings* (c. 320 BC) may have been discovered and demonstrated on actual models, before being reduced to geometrical proof. For example, Autolycus writes that if two points of the celestial sphere rise at the same time, the one which is further north will set later.[35] This proposition can be seen to be true with a mere glance at a celestial globe; but the proof using methods available in Autolycus's day runs to two pages. This sort of mechanical astronomy — using an instrument to discover possible theorems and then subjecting them to geometrical proof — would be analogous to what Archimedes claims he did in the *Method*.

To take an example from much later, in the *Sphere* of Sacrobosco (13th century), widely used for teaching elementary astronomy in the medieval universities, we read that in far northern latitudes some of the zodiac signs may rise or set *prepostere*, that is, in the reverse of the usual order. "They rise backwards, as Taurus before Aries, Aries before Pisces, Pisces before Aquarius. Yet the signs opposite these rise in the right order. They set backwards, as Scorpio before Libra, Libra before Virgo. Yet the signs opposite these set in the direct order."[36] It seems that Sacrobosco or one of his sources has simply noticed this odd phenomenon on a celestial globe; in a non-mathematical introduction such as the *Sphere*, a detailed proof was not required.

It is not obvious whether these examples of the usefulness of mechanics to spherics are applicable to Greek planetary astronomy, cosmology, or philosophy of nature. For, a common metaphor for expressing the ancient Greek attitude toward the cosmos was to consider the universe, not as a machine, but as a living animal. In Plato's *Timaeus*, the demiurge creates not only a body, but also a soul for the world. In the first century AD, Pliny could refer to the Sun as the soul and mind of the world.[37] And in the second century, Ptolemy still ascribed souls to the planets. However, mechanical metaphors and models are not unknown in ancient natural philosophy, a point recently emphasized by Sylvia Berryman, who points to Aristotle's explanation of the action of limbs (in *On the Motion of Animals*) by analogy to that of a rudder [Berryman 2009, 67]. Berryman adduces a good deal of evidence to demonstrate that mechanical hypotheses sometimes supplied analogies for the functioning of organisms, and played a role in medical theory.

But let us turn to evidence for mechanical thinking in Greek astronomy. In Geminus's work (first century BC), we see "*sphairopoiïa*" used with a range of meanings. It can mean, of course, the branch of the mechanical art devoted to building models of the heavens.[38] But, in his *Introduction to the Phenomena*, Geminus sometimes uses the word to mean a theoretical picture of the world that can be said to be after or according to nature. For example, Geminus criticizes Krates the grammarian for readjusting Homer to make his verses appear to agree with contemporary astronomy. Homer mentions Aethiopians living near the rising of the Sun, and others living near the setting of the Sun, both equally burned by the Sun. Krates attempted to make sense of this by claiming Homer meant there must be Aethiopians living around the winter tropic as well as around the summer tropic. But Krates' interpretation is nonsense, says Geminus. For Homer and the other ancient poets believed that the Earth is flat and that it extends all the way to the sphere of the cosmos, with Ocean ranged all around, and that the risings are out of the Ocean and the settings are into the Ocean. Naturally enough, Aethiopians living at the extreme east and west could both be burned. This notion was consistent with their idea of the world, "but alien to the spherical construction (*sphairopoiïa*) in accord with nature," for, says Geminus, the Earth lies at the middle of the whole cosmos, and "the risings and settings of the Sun

[35] Autolycus of Pitane, *On the Moving Sphere*, proposition 9 [Aujac 1979, 60].

[36] Thorndike [1949, 138], slightly modified.

[37] Pliny, *Natural History* ii, 13 [Rackham 1947, 178–179].

[38] Geminus discussed the branches of mathematics, including *sphairopoiïa*, and their relations to one another in his *Philokalia*, which was cited at length by Proclus in his *Commentary on the First Book of Euclid's* Elements [Evans and Berggren 2006, 243–249].

are from the ether and into the ether, since the Sun is always equally distant from the Earth."[39]

Yet again, Geminus uses *sphairopoiïa* for a spherical arrangement that actually exists in nature. Geminus says, for example, that "there is a certain spherical construction (*sphairopoiïa*) proper for each [planet], in accordance with which they pass sometimes toward the following [signs], sometimes toward the preceding, and they sometimes stand still."[40] Or, again (at xvi 19), Geminus invokes the spherical construction (*sphairopoiïa*) to prove that there exists a second temperate zone in the southern hemisphere of the Earth. In these passages, there is no question of a "model" to save the phenomena; rather, Geminus is speaking of the *sphairopoiïa* of the world itself. If the world is considered in terms appropriate to a mechanical construction, then perhaps understanding can proceed in either direction — from world to model, or from model to world.

In Theon of Smyrna (early second century AD), we have a nice example of mechanical imagination leading from a gearwork machine to the world. Theon is in the course of discussing the nested spheres postulated by Eudoxus. He raises the question of how it can be that in the universe some spheres turn eastward and some turn westward, when it might be more natural to suppose that they all turn in one direction. Here he is referring to the fact that in Eudoxus's system, for each planet, the outermost sphere turns toward the west and is responsible for the daily revolution, while the next sphere interior to it turns eastward and is responsible for producing the planet's zodiacal motion. But, says Theon, maybe there are gears between these spheres, which could reverse the motion, just as in the case of a *mechanosphairopoiïa*.[41] Theon is pondering a machine and reasoning from the machine to answer to a question about the natural world. And he apparently coins a word, ***mechano**sphairopoiïa* (which seems not to be used by any other author), to make it clear that he means a man-made machine, and not the *sphairopoiïa* of the cosmos itself.

And when Ptolemy begins to describe his theory of latitudes for the planets, he makes a famous plea that no one should complain about the difficulty of his hypotheses. For it is not appropriate to compare human contrivances with the divine, nor to form beliefs about celestial things on the basis of very dissimilar analogies. And he goes on: "We see that in the models constructed on Earth the fitting together of these [elements] to represent the different motions is laborious, and difficult to achieve in such a way that the motions do not hinder each other, while in the heavens no obstruction whatever is caused by such combinations."[42] We should not, says Ptolemy, judge simplicity in celestial things from what might appear to be simple on the Earth. This is perhaps a sign that in Ptolemy's day there were people trying to argue about celestial reality on the basis of mechanical models and that this is what motivated his criticism.

The Case of the Antikythera Mechanism

This gearwork astronomical computing machine was discovered in an ancient shipwreck at the beginning of the twentieth century. The date of the shipwreck is most securely established from the coins found in association with the wreck, from which it appears that the ship sank in the decades just after 60 BC.[43] This is well supported by the dating of everyday objects (such as the crew's pottery dishes) that were carried on board [Davidson Weinberg et al. 1965, 4]. But the date for the mechanism itself is not as tightly constrained. Analysis of the letter forms in the Greek inscriptions has been said to imply a most likely date in the range 150–100 BC,[44] but some epigraphers believe that one should allow a century in either direction of 125 BC.[45]

[39] Geminus, *Introduction to the Phenomena* xvi, 28–29. See also the discussion in Evans and Berggren [2006, 1–53].

[40] Geminus, *Introduction to the Phenomena* xii, 23.

[41] Theon of Smyrna iii, 30 [Dupuis 1892, 290].

[42] Ptolemy, *Almagest* xiii, 2 [Toomer 1984, 601].

[43] See Panogiotis Tselekas, "The Coins" [Kaltsas 2012, 216–219].

[44] Freeth et al. [2006, Supplementary Information, p. 7].

[45] We thank Alexander Jones for sharing this view.

Most of the moving parts of the mechanism were actuated by gear wheels driven by a single input. However, one part had to be moved by hand. This is the Egyptian calendar ring, which was divided into the 12 months (30 days each) and five additional days of the Egyptian year. Because the Egyptian calendar year was always 365 days long, with no leap days, the calendar ring had to be displaced "by hand" by one day every four years. Beneath the Egyptian calendar ring is a circle of closely spaced holes drilled into the underlying plate. There was probably a little post (or posts) on the back of the calendar ring. The ring could therefore be pulled off, turned to the appropriate orientation for the year under consideration, and then plugged back in.

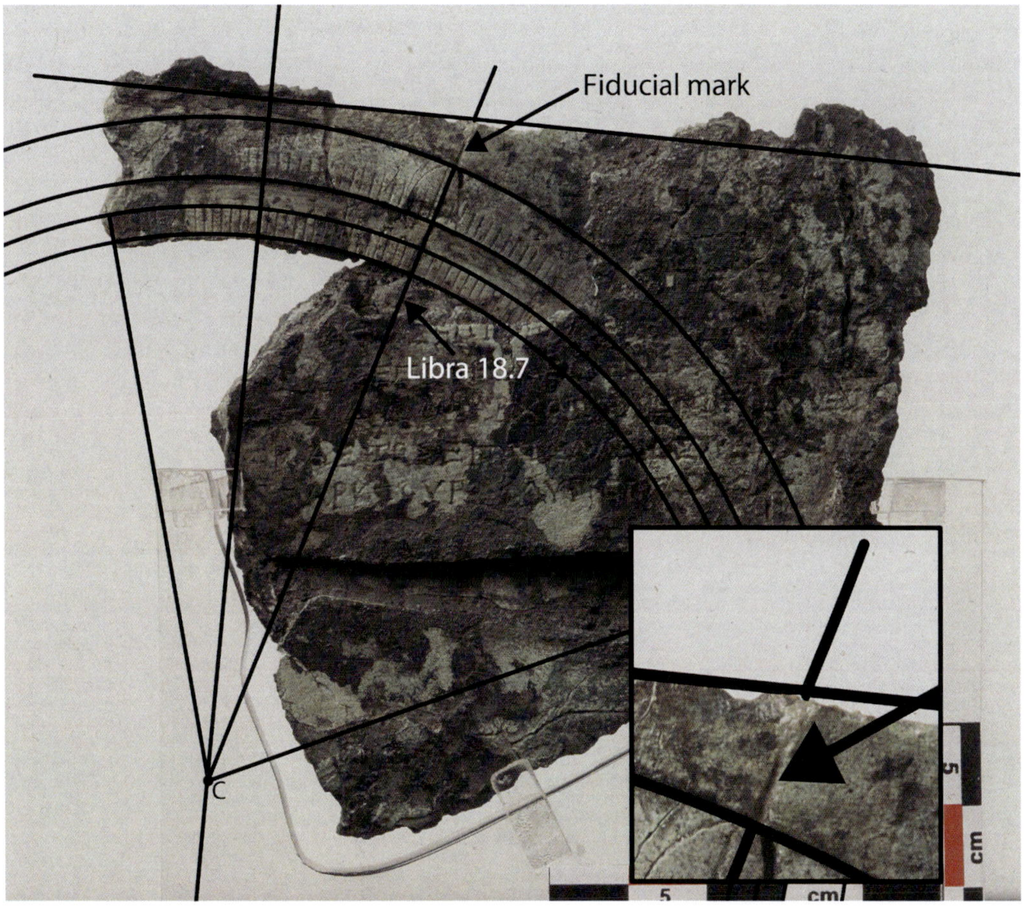

Figure 1: Fragment C of the Antikythera mechanism, carrying the remains of the zodiac and Egyptian calendar scales. National Archaeological Museum, Athens. © Hellenic Ministry of Education and Religious Affairs, Culture and Sports / Archaeological Receipts Fund. (Photograph by Kostas Xenikakis.)

On the plate just outside the Egyptian calendar scale, at about the beginning of the month of Payni, is a clear, uniformly made mark, shown in Figure 1. Price [1974, 19–20] drew attention to this and argued that it was a fiducial mark for setting the Egyptian calendar ring for some initial date. But in his analysis Price assumed that the calendar ring is still in its original position and, when this led to impossible dates, that it was set at the correct day of the month, but the wrong month of the year. However, as is known, the Egyptian calendar ring is out of its proper position by several months for the epoch of the Antikythera mechanism, so no inference can be drawn from the day of the year that now happens to lie against the fiducial mark. But, as we shall see, something interesting can be said about the zodiac degree corresponding to the mark, as the mark is inscribed on the same plate as the zodiac. To be sure, the x-ray CT (computed tomography) scans show that the region of the plate in the vicinity of the mark has cracks under the surface (and one crack is even visible in the surface images), so one could wonder whether this is a deliberately made mark or some sort of damage — a break in the plate, for example. However, to Price, who had the advantage of examining it directly, it seemed a deliberately made mark. The authors, separately on two different occasions, had the chance to view the fragment in its glass case at the National Archaeological Museum in Athens. To us, it also seems deliberately made, and all the more convincingly so when the mark is viewed in person (as opposed to in photographs or x-rays). But we must admit that this is not certain.[46]

We point out that the fiducial mark is nearly perfectly radial, that is, directed toward the center of the circular scales. In Figure 1, C is the geometrical center of the system of scales, found by simultaneous fitting of the four circles shown. The radial direction of the mark supports the view that it is indeed associated with the scales. Like Price, we ask whether it might have been intended as the "$t = 0$" setting mark for the Egyptian calendar ring. That is, we imagine that, someplace in the inscriptions, there would have been a line that read, "For such and such a year, set the first of Thoth (i.e., the first day of the Egyptian year) at the mark." (Alternatively, it would be conceivable to prescribe the setting of the calendar ring without the use of a separate fiducial mark, if, for example, there were an inscription that said: for such and such a year, place 1 Thoth against a certain degree of the zodiac.)

We are lucky in the portion of the zodiac that is preserved (approximately a quadrant). For the Sun's position on the first of Thoth fell in the extant portion of the zodiac between the years 425 BC and 72 BC. This encompasses practically the whole range of possible dates for the construction of the Antikythera mechanism, except perhaps for a very few years at the most recent end of the interval, immediately before the shipwreck. Thus, if there were a calibration mark for the first of Thoth, it would almost certainly have to fall in the preserved portion of the zodiac. But, there is one, and only one, such mark visible in the CT. As there is only one, it must in all likelihood be the setting mark for the calendar scale.

Year	1 Thoth	Sun
198	Oct 12	195.2
202	Oct 13	196.2
206	**Oct 14**	**197.2**
210	**Oct 15**	**198.2**
214	**Oct 16**	**199.1**
218	Oct 17	200.1
222	Oct 18	201.1

Table 1: Longitudes of the Sun (calculated from modern theory) at noon on the first day of Thoth, for geographical longitude 23° E.

Let us enquire for just which year the beginning of Thoth would be aligned with the fiducial mark. In Table 1, for each year in column 1, column 2 indicates the Julian calendar date corresponding to 1 Thoth [Bickerman 1980, 115–112]. Column 3 gives the longitude of the Sun calculated from modern theory for noon of 1 Thoth in the given year, at 23° east longitude (roughly the longitude of Antikythera itself, but, more importantly, in the middle of the Greek cultural zone from which the mechanism likely originated).[47]

---

[46] Other scholars of the Antikythera mechanism whom we have consulted (each of whom has had ample opportunity to view it in person) are of divided opinion: two were of the opinion that the mark was most likely purposely made, one was convinced that it is due to accidental damage, and one was noncommittal.

[47] The National Renewable Energy Laboratory maintains a solar calculator at http://www.nrel.gov/midc/solpos/spa.html

The fiducial mark lies at about Libra 17.7°, i.e., longitude 197.7°, according to the modern convention, which assigns to the first mark of Libra the value 0°. However, it is probable that the ancient mechanic would have considered the long mark at the beginning of Libra to represent Libra 1°, which means the fiducial mark lies at 198.7°. To allow for either possibility, we look for years in which the Sun's noon longitude on the first of Thoth falls in the range from about 197.2° to 199.2° (thus allowing half a day one way or another about noon for either possibility). As can be seen, the result is the range 214–206 BC (shown in bold print in table). But we do not know just how the ancient mechanic would have calculated solar longitudes for this calendrical problem. Would he simply have used mean longitudes, for example? Moreover, how accurate was the equinox or solstice date that was used to tie the Sun to the calendar? If we allow a total of 2½° (roughly the size of the maximum solar equation) above 198.7° and below 197.7°, we look for years for which the Sun's longitude at local noon on the first of Thoth fell in the broader range 195.2°–201.2°. This gives us the more conservative estimate of 222–198 BC.[48] Of course, we have no way to know whether or not a single epoch date characterized the entire mechanism — the lunar and eclipse gear trains, for example, along with the Egyptian calendar. A single epoch is a plausible assumption, but no more than that.

Even if the mechanism should have an epoch date in the range 222–198 BC, it does not necessarily follow that it was built in this period. For example, (1) the extant machine could be a later copy or an elaboration of a mechanism built in this period. Or (2) perhaps a later designer drew upon a body of knowledge (a list of epoch positions of the Sun, Moon, planets, eclipse cycle, and calendar, etc.) from the late third century BC. Or (3) the designer, for some reason of convenience, could have adopted an epoch date that was substantially earlier than the date of construction. Ptolemy, in the planetary tables of the *Almagest* that he constructed in the second century AD adopted as his epoch the beginning of the reign of Nabonassar (747 BC), to ensure that neither he nor his readers would ever have to calculate a planetary position for a date before the epoch, which would have required a separate set of precepts. However, it is not so clear that the example of planetary tables applies very well to a gearwork mechanism: for a mechanical device, there would be no need for separate precepts for dates before "$t = 0$" — one would simply turn the input knob backwards. Also, while in the case of tables a far-distant epoch would pose no significant extra labor for practical calculation, the same cannot be said of a mechanical device. Here an epoch in the remote past would be inconvenient, as it would require lots of manual cranking to bring the machine up to the user's own date. Thus, if the fiducial mark is genuine, there are grounds to consider the possibility that it reflects a date not too remote from the date of construction (at least of a prototype or ancestor, if not of the extant machine).

The Lunar Anomaly in the Antikythera Mechanism

One of the most remarkable aspects of the Antikythera mechanism is its incorporation of a device to represent the lunar anomaly — the speeding up and slowing down of the Moon as it moves around the zodiac. The central idea is that one gear is mounted on another one of the same size and tooth count, but with the two axles slightly eccentric to one another. The driving wheel engages the follower by means of a pin that fits in a slot of the follower.[49] Because the wheels rotate about different centers, the uniform

that was used for these calculations. However, this calculator does not accept values of ΔT greater than 8000s, so adaptations have to be made. For ΔT (the excess of atomically defined Terrestrial Time over Universal Time, which arises from the "clock error" of the Earth as its rotation decelerates), we used the value $3\frac{1}{2}^h$, which is appropriate for the years around −200 (201 BC). See Morrison and Stephenson [2004].

48 In the course of our work, we discussed our conclusions with Alexander Jones, and found that he had independently arrived at a rather similar view: "(1) that if the mark is deliberately engraved, it must indicate the epoch position of Thoth 1 and thus an epoch date in the latish 3rd century BC, and (2) that the mark does indeed look deliberate though the coincident break makes certainty impossible." (Private communication.)

49 Price [1975, 35] noticed this slot, but conjectured that it might be the result of an attempt to repair a broken gear. Wright [2002] described the pin-and-slot device and observed that it would suitable for modeling an anomaly, and perhaps a lunar anomaly. But because of problems with the tooth counts in the lunar gear trains, he could not settle on

motion of the first produces *nonuniform* motion of the second. Let us see how this pin-and-slot device works.

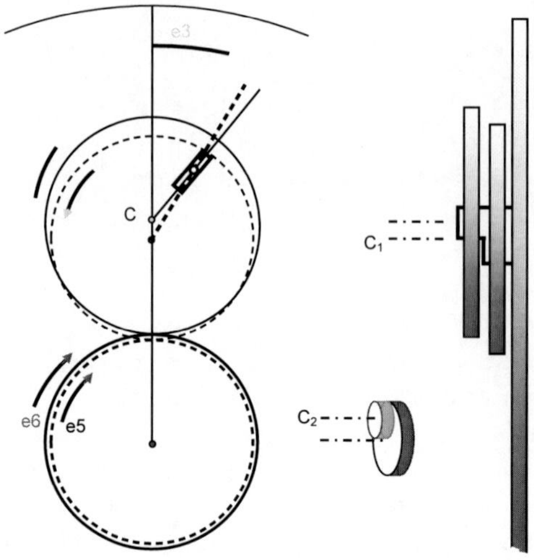

(A) (B)

Figure 2: The device for producing the non-uniform motion of the Moon in the Antikythera mechanism. Reproduced from Evans, Carman and Thorndike [2010].

The mechanism for the lunar inequality involves four gears of identical tooth number (50), called e5, e6, k1, and k2, illustrated in Figure 2A. This figure is based on the reconstruction by Freeth et al. [2006]. The input motion is from a hollow pipe at E (perpendicular to the plane of the diagram) that turns e5 at the rate of the Moon's mean sidereal frequency, ω_{si}. Concentric with e5, but turning freely from it, is a large wheel e3, which turns at the rate of the Moon's line of apsides. From a modern point of view, the orientation of the Moon's major axis does not stay invariable. Rather, the Moon's elliptical orbit itself turns in its own plane, so that the perigee advances in the same direction as the Moon moves, taking about 9 years to go all the way around the zodiac. The ancient astronomers were aware that the position of fastest speed in the Moon's orbit itself advances around the zodiac, the Greeks modeling the motion geometrically and the Babylonians by means of arithmetical period relations. In the Antikythera mechanism the advance of the Moon's perigee and apogee is modeled by letting e3 turn with a frequency we shall denote ω_A.

Riding on e3 at center C_1 is gear k1, which is driven by e5. A second gear, k2, turns about an axis, C_2, also attached to e3 but slightly offset from C_1. (In Figure 2, we have drawn wheels e5 and k1 slightly smaller than e6 and k2, in order to show the relationships among the wheels more clearly. But these wheels should all be thought of as having the same size.) The offset is achieved by using a stepped stud, with its larger diameter centered at C_1 and its smaller diameter centered at C_2, as shown in Figure 2B and in the perspective view. Wheel k1 has a small pin that engages a radial slot in k2. Thus k1, turning at a uniform angular speed, drives k2, producing a quasi-sinusoidal oscillation in the angular speed of k2. The motion of k2 is transferred to e6, rotating freely about axis E, which communicates the nonuniform motion of the Moon to the other parts of the mechanism. Uniform motion in (at e5) is transformed into non-uniform motion out (at e6) around the same axis.[50]

this interpretation of the pin and slot and left it as an unexplained mystery. The first complete demonstration of the pin and slot as a device for the lunar anomaly is in Freeth et al. [2006].

[50] The output is by means of a central shaft attached to e6; and this shaft runs inside the hollow pipe to which e5 is attached.

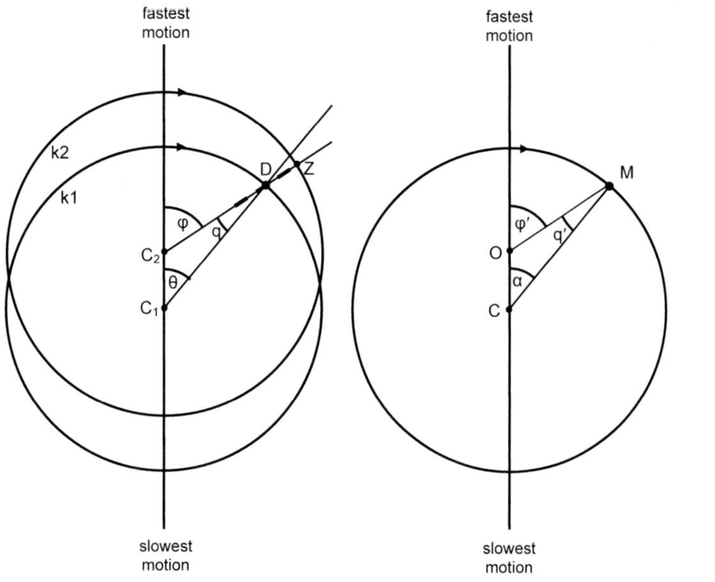

Figure 3: The pin-and-slot device of the Antikythera mechanism (A) compared with a standard eccentric-circle model (B).

In Figure 3A we see k1 and k2 in isolation. Wheel k1, carrying the pin D, turns uniformly about center C_1. On wheel k2, which rotates about center C_2, the radial slot is represented by the heavy dashed line.[51] Suppose k1 rotates uniformly, so that θ increases uniformly with time. Then a point Z on the perimeter of k2 will be seen from C_2 to rotate at a non-uniform angular speed. Angle φ increases more rapidly than θ when D is up in the diagram, and more slowly when D is down (though, of course, both wheels complete one period in the same time). At any instant, $\varphi = \theta + q$, where q functions as an equation of center. It is easy to show that

$$\sin q = \frac{e \sin \theta}{\sqrt{1 + e^2 - 2e \cos \theta}} \tag{1}$$

where $e = C_1C_2/C_1D$. This is the same equation of center as one gets with an ordinary eccentric circle.

To see the equivalence to an eccentric circle in terms of simple geometry (rather than trigonometry) let us examine Figure 3B, which represents the standard eccentric-circle model. (For the time being we suppose that the eccentric is fixed — i.e., we temporarily ignore the advance of the line of apsides. We will take up the question of a moving line of apsides below.) O is the Earth, and C is the center of the eccentric circle, around which the Moon M moves uniformly, so that angle α (the mean anomaly) increases uniformly with time. Then, as viewed from the Earth O, the angular position of the Moon at any time is $\varphi' = \alpha + q'$, where q' is the equation of center in the standard eccentric-circle model. It is immediately obvious that the pin-and-slot mechanism will reproduce the angular position of the Moon provided we put $\theta = \alpha$ (as we obviously need the mean motion to be the same in both models), and we require that $C_1C_2/C_1D = OC/CM$. Then the two equations of center will always be equal, i.e., $q = q'$. To put it another way, the direction C_2D (defined by φ) in the pin-and-slot mechanism is the same as the direction OM (defined by φ') of the Moon as viewed from the Earth in the eccentric-circle

[51] In Figure 3A, we have for the sake of simplicity shown k1 and k2 turning in the same direction as the Moon goes in Figure 3B. The key point is that the pin and slot reproduce the angular motion around an eccentric circle; the reversals of direction at the e5/k1 and k2/e6 interfaces need not concern us at this stage.

model. Now, the clever thing about the mechanism is that the non-uniform rotation of k2 is then transferred to e6, which rotates about the geometrical center of the zodiac scale. So a Moon marker driven by e6 will travel around a circle that is centered on E in Figure 2, but it will speed up and slow down on this circle.[52]

As far as is known, there is no extant ancient mention of the quasi-equivalence of the pin-and-slot mechanism to the eccentric circle model. This is a quasi-equivalence because the pin-and-slot mechanism produces the same motion in angle, but not the same physical motion in space as the eccentric-circle model. The output of the pin-and-slot device is a point moving at non-uniform speed on circle k2 — and, ultimately, after the motion is transferred back to e6, nonuniform motion on a circle concentric with the Earth. But the output of the eccentric-circle model is a point moving uniformly around a circle that is eccentric to the Earth, O.

An ancient Greek astronomer trained in the philosophical-geometrical tradition of Hipparchus and Ptolemy would not have regarded the pin-and-slot mechanism as a realistic representation of the lunar theory, for the pin-and-slot mechanism suppresses the motion in depth, though it does give a motion in longitude that agrees with what the eccentric theory prescribes. In the *Planetary Hypotheses*, Ptolemy criticized *sphairopoiïa* as traditionally practiced, saying that it "presents the phenomenon only, and not the underlying [reality], so that the craftsmanship, and not the models, becomes the exhibit."[53] It is possible that Ptolemy is merely complaining about a closed box, on the exterior of which the phenomena are displayed, but whose inner workings are kept sealed out of sight. But it seems to us likely that he is complaining just as much about the nature of the inner workings themselves. Suppose one took the lid off the box and saw inside, not epicycles and eccentrics, but pin-and-slot mechanisms, whose motions are a far cry from the real motions of the planets. *Quelle horreur*! For Ptolemy, the best *sphairopoiïa* would be one that offered a faithful display of the phenomena on the outside but that, when opened, revealed the true nature of planetary motion. Neither would Aristotle have approved of the pin and slot, as he maintains that each simple body (e.g., a celestial orb) should be animated by a single simple motion.[54] And here, the final output motion is the rotation of e6, which consists in a steady rotation with a superimposed oscillation. Did the ancient mechanic who designed the Antikythera mechanism realize the equivalence in angle of the pin-and-slot mechanism to the eccentric-circle theory? Or was this mechanism considered a rough-and-ready approximation to the behavior of the Moon — good for giving the final output angle, but not necessarily considered exact? In any case, no proof of the equivalence survives.

Perhaps the contrast between applied mechanics and accepted celestial physics should not surprise us, for there is a well-documented example of a similar contrast. Greek astronomers grounded in the philosophical-geometrical tradition (e.g., Theon of Smyrna, early second century AD) wrote treatises on deferent and epicycle theory while their contemporaries were busy mastering and adapting the non-geometrical planetary theory of the Babylonians [Jones 1999]. The philosophically-based astronomy of the high road explicitly endorsed uniform circular motion as the only motion proper to celestial bodies, while the numerically-minded astronomers (who needed quick and reasonably reliable results for astrology) made free and easy with nonuniformity of motion. In a similar way, it is possible that mechanical tricks of the trade such as the pin-and-slot mechanism were used in a craft tradition of model-building, quite apart from the practices of the "serious" (i.e., geometrically-minded) astronomers. On the other hand, Figure 3 shows that a proof of the equivalence in angle would have been well within the reach of Greek geometry. But the first historical accounts of ancient Greek astronomy were written by travelers of the high road (e.g., Ptolemy's historical remarks in the *Almagest* and Proclus's account in his *Sketch of Astronomical Hypotheses*). We should not be surprised that their accounts left no trace of the influence of mechanics on theoretical astronomy. Their silence on the issue should not be taken as evidence.

One last detail: We gave our equivalence proof above for a stationary line of apsides. Now we must

[52] The equivalence (in angular motion) of the pin-and-slot mechanism to a standard epicycle model was demonstrated by Freeth et al. [2006]. For a simpler demonstration of this equivalence, see Carman, Thorndike and Evans [2012].

[53] Ptolemy, *Planetary Hypotheses* i, 1 [Hamm 2011, 45].

[54] Aristotle, *On the Heavens* 268b28–269a2.

show that the pin-and-slot mechanism remains equivalent (in angular motion) to the eccentric-circle model, even when the pin-and-slot device is mounted on the turning wheel e3. The generalization is very simple. The trick is to view the motions in the frame of reference of wheel e3. This approach, involving the change of a frame of reference, was well within the scope of Greek mathematics. Ptolemy, for example, often subtracts angular velocities to effect a change of reference frame.

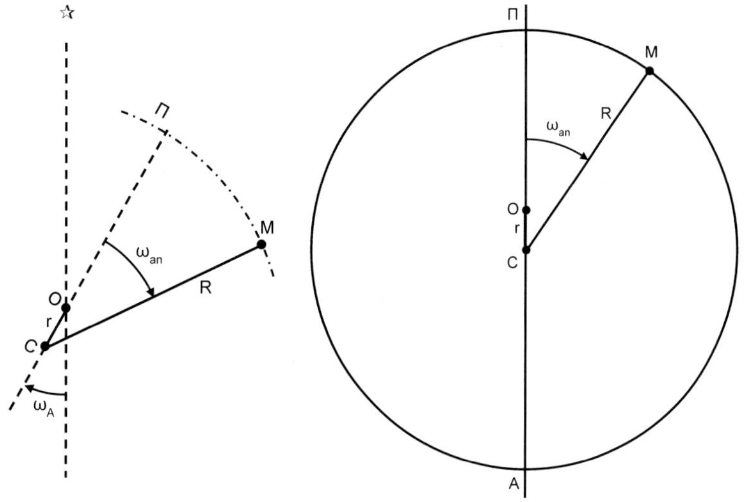

Figure 4: (A) The standard eccentric-circle model for the motion of the Moon, incorporating an advancing line of apsides. (B) The lunar theory of (A) as viewed in the frame of reference in which the line of aspides ACOΠ is stationary.

In Figure 4A, we see the standard eccentric-circle lunar theory, often associated with Hipparchus. The Earth O and the vertical reference line are fixed with respect to the stars. C is the center of the Moon's eccentric, and it moves in a circle of radius r around the Earth at a frequency ω_A. Thus the line of apsides CO slowly rotates. At the moment shown in the figure, the instantaneous position of the perigee is Π, and the instantaneous position of the Moon's circle (radius R) is shown by the dot-dash arc. Meanwhile, the Moon M moves in such a way that the mean anomaly increases uniformly with time, at the anomalistic frequency ω_{an}. (And $\omega_{an} = \omega_{si} - \omega_A$, where ω_{si} is the sidereal frequency.) In Figure 4B, we see the same lunar theory, but as viewed in a rotating frame of reference that is at rest with respect to the line of apsides ACOΠ.

Let us turn now to the Antikythera mechanism. In Figure 5A, we see the front face of the Antikythera mechanism as viewed in "absolute space." Wheel e5 turns at the Moon's mean sidereal frequency ω_{si} (corresponding to the sidereal month). Wheel e3, on which wheels k1 and k2 are mounted, turns at the frequency ω_A at which the line of apsides advances. Thus it is clear that the rate at which e5 turns with respect to the line of apsides is $\omega_{si} - \omega_A = \omega_{an}$, the anomalistic frequency. In Figure 5B, we see the same system, as viewed by an observer standing on e3. The line of apsides is fixed, and e5 rotates at angular speed ω_{an}. Thus it is clear that k1 also rotates at angular speed ω_{an} in this frame.

Comparing Figures 5B and 4B, we see that we have in each case a stationary line of apsides, and a mean angular speed of the Moon with respect to the line of aspides that is equal to ω_{an}. Thus the equivalence proof (in the rotating frame of reference) may proceed exactly as we outlined above for the case of a stationary line of apsides. And if the angular motion of k2 in Figure 5B is the same as the angular motion of the M about O in Figure 4B, as viewed in the rotating frame of reference, these two motions will remain equivalent to one another as viewed in any other frame of reference, including the space frame.

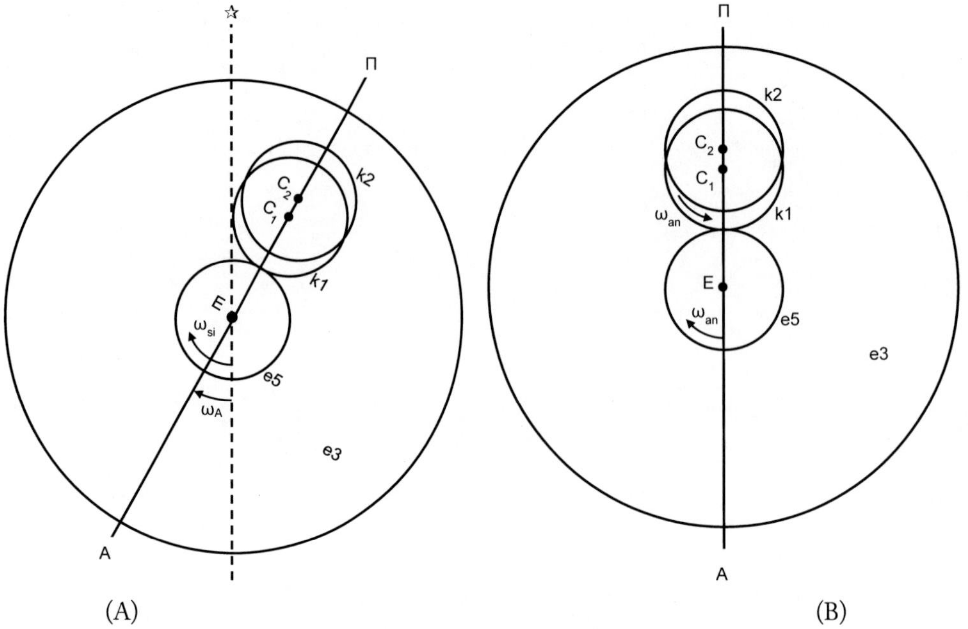

Figure 5: (A) The lunar device of the Antiykthera mechanism, as viewed in the frame of reference at rest with respect to the box that contains the machine. (B) The mechanism of (A) as viewed in the frame of reference of wheel e3, such that the line of apsides AOCΠ is stationary.

Outer Planets

Recently, it has been shown that a pin-and-slot mechanism could also be used to represent the synodic cycle of a superior planet, including retrograde motion.[55] Again the pin-and-slot mechanism turns out to be exactly equivalent, in terms of angular motion (but ignores or suppresses the variation in distance), to the output of a simple concentric-deferent-plus-epicycle model. In the case of the lunar inequality, the pin-and-slot mechanism must be mounted on wheel e3 (which is fixed with respect to the lunar line of apsides). In the case of the outer planets, which retrograde when they are in opposition to the Sun, the main solar gear b1 plays the role of e3. That is, the pin-and-slot mechanisms for the outer planets must be mounted on the main solar wheel.

One simple way to accomplish this is shown in Figure 6. The input frequency of wheel u is the planet's sidereal frequency ω_{si}. (Thus we begin with a simple construction similar to that for the Moon shown in Figure 2: for the moment we do not worry about how the input frequency ω_{si} is obtained.)[56] The main solar wheel b1 turns at the Sun's sidereal frequency $\omega_{\odot}$. Thus, in the frame of reference of b1, u turns at frequency $\omega_{si} - \omega_{\odot}$. For a superior planet we must have $\omega_{\odot} = \omega_{si} + \omega_{an}$, where ω_{an} is the anomalistic (or synodic) frequency. Thus, in the frame of b1, the rotation frequency of wheel u is $\omega_{si} - \omega_{\odot} = -\omega_{an}$. Therefore, in the frame of b1, wheel u is turning counterclockwise at the correct angular speed to produce an anomaly with respect to the Sun. A pin-and-slot mechanism can be mounted on b1, as shown. In the frame of b1, the rotation frequency of x is ω_{an} and the mechanism introduces a nonuniformity into the rotary motion of wheel y in the usual way. The nonuniform motion of y would be transferred to a final wheel z (not shown), concentric with point E. Note that one need

[55] See Carman, Thorndike and Evans [2012] as well as Freeth and Jones [2012]. A film of a metal pin-and-slot mechanism producing retrograde motion may be seen at http://www2.ups.edu/faculty/jcevans/.

[56] As we shall see below, this is probably not the way the motions of the superior planets were modeled on the Antikythera mechanism. A more efficient design would allow us to produce the input frequency without using more gears. But the simple conceptual model shown in Figure 6 displays all that is essential for understanding the key points.

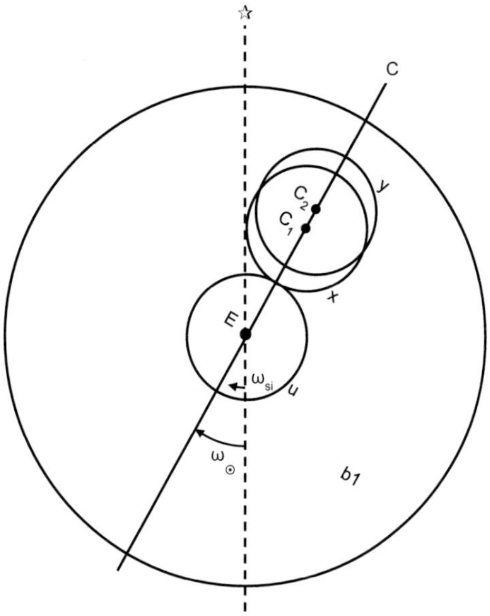

Figure 6: A pin-and-slot pair (wheels x and y) for a superior planet, riding on the main solar gear b1. b1 turns at the Sun's mean frequency $\omega_\odot$. In this hypothetical model the input wheel u turns at the planet's sidereal frequency ω_{si}, which is less than $\omega_\odot$.

not know anything about epicycles or eccentrics to arrive at such a mechanical solution. It is enough to know that the planet's retrograde motion is an anomaly with respect the Sun and that the pin and slot can be used to introduce a suitable wiggle into the steady motion.

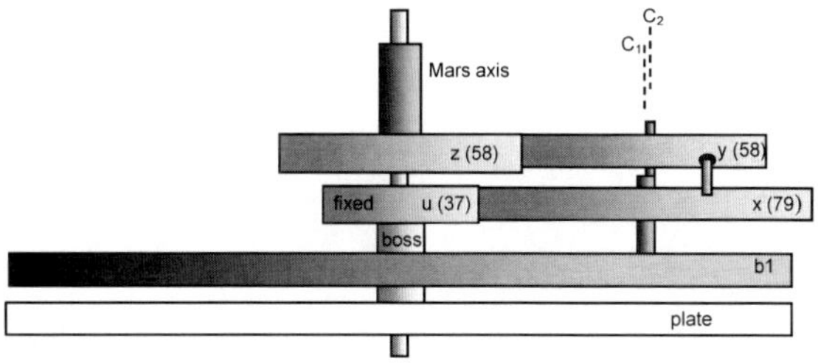

Figure 7: A more realistic pin-and-slot mechanism for Mars, as it might possibly have been realized on the Antikythera mechanism. The "input wheel" u is fixed, and x is driven from b1. From Carman et al. [2012].

Although the planetary gearing has not survived, most researchers believe that the Antikythera mechanism did also model the motions of the planets. The evidence for this comes from the inscriptions, which mention most of the planets by name [Freeth and Jones 2012, 8–10], as well as planetary phenomena, such as "stations" [Freeth at al. 2006, 587]. Additional evidence comes from the remnants of mounting hardware on b1: it looks as if *something* substantially complicated were originally mounted on b1, though it is now nearly all lost [Wright 2002; Freeth and Jones 2012]. While we have no way to know whether the pin-and-slot mechanism was used to model the motions of the planets, we do know that such a mechanism was used to represent the lunar anomaly. It seems plausible to suppose that a similar mechanism was used for the planets. The output of such a planetary device is motion in a circle concentric with the point representing the Earth, but motion that speeds up and slows down and occasionally reverses direction.

But the actual machinery for the superior planets built into the Antikythera mechanism probably more resembled Figure 7. Rather than supplying an initial drive of wheel u at frequency ω_{si}, one could make use of the freedom to vary the number of teeth on wheels u and x, and let wheel x be driven directly from b1. So, in Figure 7, drawn for Mars, wheel u is fixed to a stationary boss attached to the underlying plate. Wheel b1, carrying axle C_1, turns around once in a year. Because x engages u, x will be forced to rotate. If the gear ratio u/x is properly chosen, then x can be made to rotate at the anomalistic frequency, as seen in the frame of reference of b1. If x has 79 teeth and u has 37, for example, we will get for the rotation frequency of x with respect to b1 $(37/79)\omega_\odot$, which is a good match to ω_{an}, the anomalistic frequency of Mars. (79 years = 37 anomalistic periods is a preserved Babylonian period relation.) As far as the operation of the pin-and-slot mechanism goes, it makes no difference whether we use a solution like Figure 6 or one like Figure 7. The key thing is to mount the pin-and-slot device on the main solar wheel and to have x turn at the anomalistic frequency in the frame of reference of b1. Of course, economy of construction might well have an influence on just how we choose to get x turning at the right frequency.

Inner Planets

The inner planets may also be modeled with a pin and slot. Consider Figure 8, which shows the view in the frame of reference of "absolute space". The question is: At what frequency ω_* must the input wheel u turn? (The output wheel z, not shown, is concentric with u.) When we transform to the frame of reference of b1, by subtracting $\omega_\odot$, the motion of wheel u must be at the anomalistic frequency ω_{an} of the inferior planet. Thus it must be the case that $\omega_* = \omega_{an} + \omega_\odot$. In the frame of reference of b1, the frequency of wheel u is therefore just ω_{an}, as is appropriate for generating an anomaly with respect to the Sun.

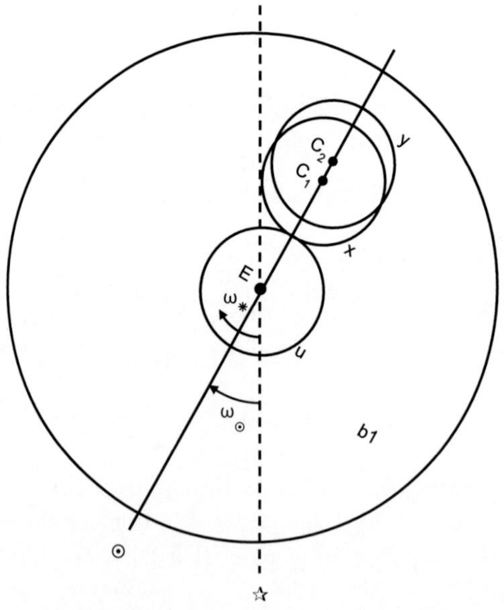

Figure 8: A hypothetical pin-and-slot representation of an inner planet, as viewed in the frame of reference of the box.

However, for an inferior planet, in frame b1 there can be no progressive motion of the planet — merely an oscillation, with no net forward progress. Thus the pin-and-slot mechanism must be modified, in conformity with Figure 9. The key thing is that the distance C_1C_2 between the centers of wheels x and y must now be *greater* than the distance C_1D of the pin from the center of wheel x. As before, the heavy dashed line shows the radial slot in wheel y. Wheel x turns at the constant anomalistic frequency, so θ increases uniformly with time. The resulting motion of wheel y is then a quasi-sinusoidal oscillation, with no net forward progress. It is easy to show that the equation of center q is also given by Equation 1, where $e = C_1D/C_1C_2$. Thus, a point Z on the perimeter of wheel y simply oscillates back and forth in the frame of b1. This motion is transferred to wheel z (concentric with u). Finally, when we return to the frame of reference of the box, the steady forward motion at frequency $\omega_\odot$ is added to the oscillation. The result is exactly what we need — a back and forth oscillation superimposed on a steady forward motion that keeps pace with the mean Sun.

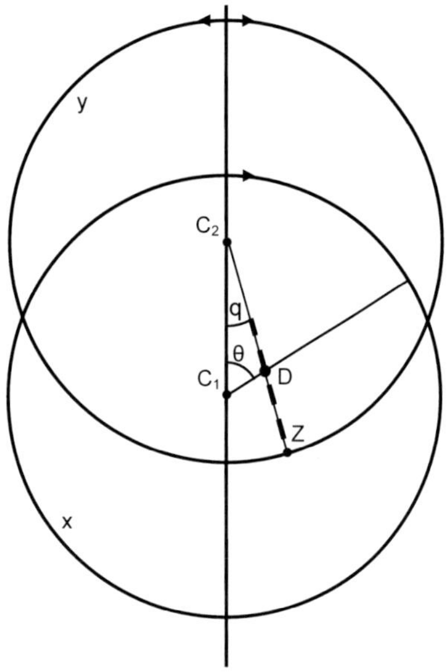

Figure 9: The arrangement of the pin-and-slot mechanism required for an inner planet: $C_1D<C_1C_2$. This effectively destroys the steady forward motion and results in a simple oscillation of wheel y in the frame of b1.

And here we may establish a connection with Ptolemy's *Planetary Hypotheses*. For the input frequency that we have denoted $\omega_*(=\omega_{an}+\omega_\odot)$ is precisely the angular speed of the planet on its epicycle, reckoned from the direction of the vernal equinox. (See Figure 10.) Now, in giving planetary parameters in forms that he felt would be helpful for those who wish to build models, Ptolemy formed his so-called compound or complex periods. For the inferior planets, the period in question is precisely that of the planet on its epicycle, but as reckoned from the vernal equinox.[57] It has never been quite clear why Ptolemy introduced this period. But if Ptolemy is thinking that builders are going to be using pins and slots, we can see why. This suggests a continuous tradition of working with pin-and-slot mechanisms at least from the time of the Antikythera mechanism (second century BC ?) to Ptolemy's own day (second century AD).

It should be noted that in seeking a pin-and-slot mechanism to use for the inner planets, we had no need of, and did not use, anything from epicycle or eccentric theory. It was enough to know that ω_* must necessarily be equal to $\omega_{an}+\omega_\odot$ and that, in the frame of b1, wheel y can make no net forward

[57] Ptolemy, *Planetary Hypotheses* i, 2 [Hamm 2011, 172–174]. These complex periods are discussed in Duke [2009].

progress.[58] Also, in the particular case of the Antikythera mechanism, if the motions of the inferior planets were represented kinematically by means of pins and slots, it is more likely that a construction like that in Figure 7 was used, since it allows one to obtain the correct rotation frequency of wheel x more economically.

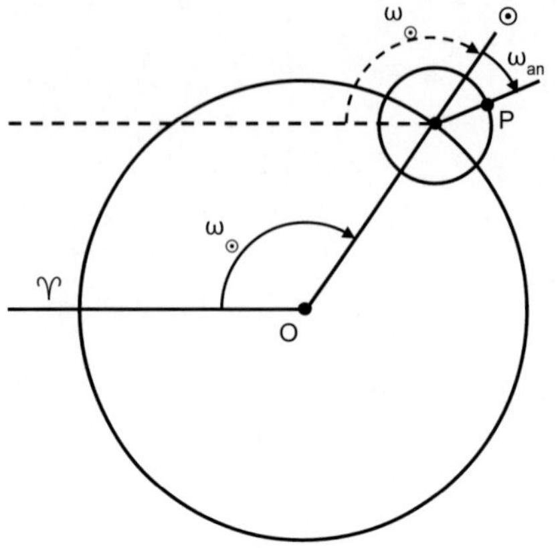

Figure 10: Standard epicycle model for an inner planet. $\omega_\odot + \omega_{an}$ is equal to the angular frequency of the planet P on its epicycle, as measured from a line parallel to the direction to the vernal equinox ♈.

Possible Origins

How might the pin-and-slot mechanism have originated? One possibility, of course, is that it is an after-the-fact adaptation of eccentrics or epicycles. In such a case, an astronomer steeped in geometrical planetary theory, or the mechanic with whom he was collaborating, dreamed up a clever way to separate out the angular part of the motion and display it alone. In this case, in Figure 3, the predominant influence went from right to left, that is, from pure theoretical astronomy to mechanical simulation. This is probably the majority view among historians of ancient astronomy. It is perfectly possible. And we do know of two mechanisms from times suitably later than the invention of epicycles and eccentrics. According to Cicero, Posidonius built such a model.[59] (Cicero invokes the *sphaera* of Posidonius in an early instance of the "watchmaker" version of the argument from design for the existence of god: if we came across this *sphaera* we would hardly doubt that it was built by a rational being; but what then of the cosmos that it imitates?) And Strabo mentions that when the Roman general Lucullus took Sinope (on the north coast of Asia Minor) in 70 BC, one of the objects carried off was the "sphere" (*sphaira*) of Billarus.[60] As there is no detail, we cannot be sure whether this was a planetarium-style "sphere" like those of Archimedes and Posidonius or a simple celestial globe; but the former seems somewhat more likely, as an ordinary globe would have been rather a commonplace by 70 BC. So here are one or two devices, known to have existed in the early first century BC, which comfortably post-date the invention of epicycles and eccentrics. The Antikythera mechanism could well be a third.

[58] We should note the existence of an alternative solution. Above, we argued that ω_* for an inferior planet must be chosen so that the rotation frequency of u in frame b1 is equal to ω_{an}. But it would also be possible to have the rotation frequency of u in frame b1 equal to $-\omega_{an}$. Then we would have $\omega_* = -\omega_{an} + \omega_\odot$. In epicycle theory, this would correspond to the planet revolving in the backwards direction on its epicycle.

[59] Cicero, *On the Nature of the Gods* ii, 88 [Rackham 1952, 206–208].

[60] Strabo, *Geography* xii, 3.11. For a proposal that the Antikythera mechanism may be the lost sphere of Billarus, see Mastrocinque [2009].

A second possibility is that the pin-and-slot mechanism owes its inspiration to the system of Eudoxus and Callippus. We don't mean to suggest that the pin and slot are a literal representation of Eudoxian planetary theory. But a mechanic, in an attempt to adapt the spherical models to a plane in building a machine for display, may have borrowed an idea. Here the connection is most easily seen in the case of planetary motion. Let us recall that for replicating retrograde motion, Eudoxus imagines two spheres that turn in opposite directions, with their axes offset at a small angle, and the axle of one sphere set into the surface of the other. The outermost sphere of this pair has its axis inserted into the "equator" of a sphere that rotates with the planet's zodiacal motion. If we imagine flattening the system out — taking say just the northern hemisphere this might readily suggest something close to Figure 3A. Against this is the fact that in Eudoxus's system, the two spheres we have mentioned turn in opposite directions, rather than in the same direction. But the general idea of an off-axis wobble might have been transferable. In addition, Eudoxus's idea of *homocentric* spheres might well be modeled by pin-and-slot devices, as they do effectively produce suitably non-uniform motion on a circle concentric with the Earth.

A third possibility is that the pin and slot arose from an effort to model Babylonian phenomena (and this need not be inconsistent with Eudoxian influence on the choice of mechanism). If this were the case, we need not suppose that the mechanic had a detailed understanding of the equivalence of the pin-and-slot mechanism to epicycles or eccentrics, or of epicycles and eccentrics to one another. It would be enough to provide a back-and-forth motion superimposed on a mean motion, without any worry about the geometrical details. Here the simplest case to consider is that of the Moon. Geminus, in his *Introduction to the Phenomena* (first century BC) describes for Greek readers the essential features of the Babylonian lunar theory now known as System B.[61] The Moon's daily motion changes from day to day according to a simple saw-tooth-pattern, with uniform daily changes of 0;18° between maximum and minimum daily displacements of 15;14,35° and 11;6,35°. (We use the Neugebauer notation, in which whole degrees stand to the left of the semi-colon, and successive sexagesimal parts stand to the right and are themselves separated by commas.) This leads to a lunar "equation of center" of quadratic form:

$$q = DT\left[\frac{1}{2}\frac{t}{T} - \left(\frac{t}{T}\right)^2\right], \quad \text{for any time } t \text{ between 0 and } T/2, \text{ and}$$
$$q = DT\left[-\frac{1}{2}\left(\frac{t}{T} - \frac{1}{2}\right) + \left(\frac{t}{T} - \frac{1}{2}\right)^2\right], \quad \text{for } t \text{ between } T/2 \text{ and } T, \tag{2}$$

where q is the difference between the Moon's longitude according to System B and the longitude it would have if it moved uniformly. The time t is reckoned from the moment of fastest motion, T is the length (in days) of the anomalistic month, and D is the difference between the greatest and least daily motion (in degrees per day). The associated length of the anomalistic month is T = 27;33,20 days. The curves for q are segments of parabolas (given by Eq. 2), alternately concave upward and concave downward. See Figure 11.

The graph in Figure 11 also shows the equation of center given by Equation 1 and using the dimensions of the pin-and-slot device in the Antikythera mechanism. Freeth [2006, 590] reported distance C_1C_2 (in Figure 3A) as 1.1 mm and distance C_1D as 9.6 mm. Clearly, the pin-and-slot device of the Antikythera mechanism does a fine job of modeling Babylonian lunar theory. The maximum equation of center is larger than one would expect for a Greek theory based on an epicycle or eccentric.[62] Of course, this could be an accidental result of construction — a few tenths of a millimeter in C_1C_2 would make a big difference. For example, if C_1C_2 were 0.8 mm (instead of the measured 1.1), the maximum equation would be in the 5° zone that Ptolemy adopted.

For comparison's sake, we also show the equation of center for Ptolemy's first lunar model (a simple epicycle model with epicycle radius 5;15 and deferent radius 60), as well as for the two versions of Hipparchus's lunar theory. According to Ptolemy, Hipparchus found two different values for the lunar

[61] Geminus, *Introduction to the Phenomena* xviii, 4-19. See Evans and Berggren [2006, 228–230 and 96–98].

[62] We are grateful to Dennis Duke for pointing this out.

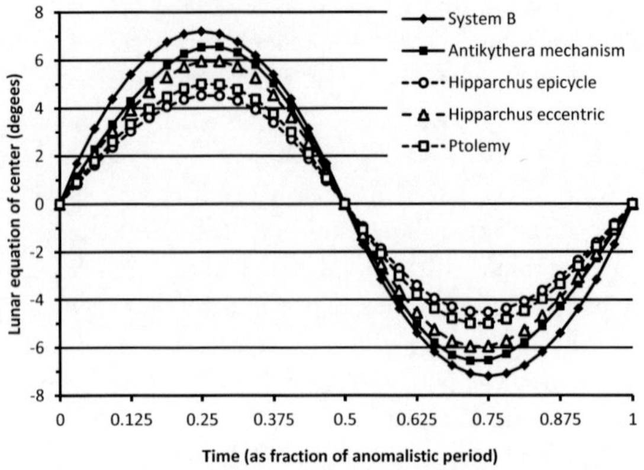

Figure 11: The equation of center of the Moon, according to various ancient models.

eccentricity.[63] Using the eccentric-circle model he found 327.67/3144 for the ratio OC/CM (Figure 3B). But using the epicycle model, he found the ratio 247.5/3122.5 for the ratio of the epicycle's radius to the deferent's radius. According to Ptolemy, Hipparchus used different data for the two determinations, but the results were also marred by faulty computation. But, Ptolemy says, some people have wrongly thought that the difference in the results must be due to some difference between the two hypotheses. This interesting remark shows us that, in the period between Hipparchus and Ptolemy, the equivalence of epicycle and eccentric may not have been thoroughly understood by all astronomers. Toomer [1967] presented a good argument that Hipparchus eventually adopted the smaller of his two results for the epicycle radius.

However, we would like to stress that we are not making an argument about whether it is Greek or Babylonian lunar theory that is represented on the Antikythera mechanism. We just wish to suggest that the Babylonian planetary theory could conceivably have served as model for a gearwork lunar mechanism that incorporated a moving line of apsides, that the Babylonian theory is a reasonably close match to the behavior of the later Greek epicycle theory, and that no knowledge of epicycles and deferents would have been required to incorporate the Babylonian theory into a gearwork mechanism.

In the lunar theory of System B, each parabolic segment is symmetric about its maximum or minimum. But the equation of center for an epicycle or an eccentric has a slight asymmetry, due to the form of the equation printed above. The pin-and-slot device, being equivalent in angle to either an eccentric or an epicycle, shows a similar asymmetry. However this assumes that the pin leads the slot — i.e., that the input motion is to the wheel carrying the pin (as is the case on the Antikythera mechanism). If the slot leads the pin, it turns out that the functional form of the equation of center is different [Carman, Thorndike and Evans 2012, 99], and one has instead:

$$\sin q = e \sin \theta. \qquad (3)$$

In this case, the bumps of the equation of center curve are symmetric about their maxima and minima. At the level of precision of astronomy in the second century BC, there would be no way to choose empirically between these two forms. Is the fact that the pin leads the slot in the Antikythera mechanism a sign that the mechanic understood the equivalence to an eccentric-circle model? This could be the case, of course, but there are two reasons why the evidence is ambiguous. First, the mechanic would likely have regarded the pin as the active element, and the slot as passive, so it would have been intuitive to place the pin on the wheel with the input motion. And second, the lesson of the planets could have been determinative.

[63] Ptolemy, *Almagest* iv, 11 [Toomer 1984, 211]. See the discussion by Neugebauer [1975, 317–319].

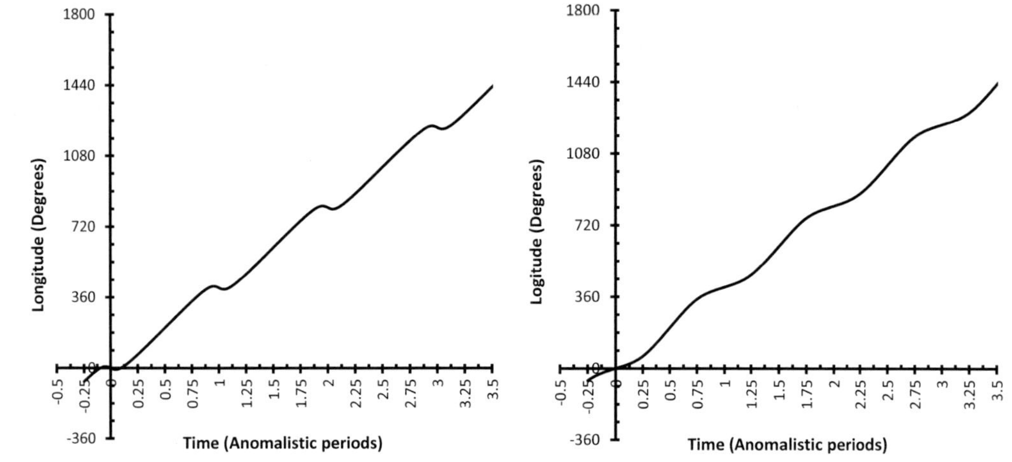

(A) (B)

Figure 12: (A) Motion in longitude produced by a pin-and-slot device scaled appropriately for Mars, with the pin leading the slot. The equation of center is given by Equation 1. (B) Motion in longitude produced by a pin-and-slot device scaled appropriately for Mars, but with the slot leading the pin. The equation of center is given by Equation 3.

For it happens that in the case of a planet with a large epicycle and with an anomalistic period longer than the sidereal period (such as Mars), the version of the mechanism with the slot leading the pin will not actually produce retrograde motion at all [Carman, Thorndike and Evans 2012, 113]. In Figure 12A we see a graph of longitude versus time for a pin and slot device with the synodic and sidereal frequencies and the eccentricity e chosen appropriately for Mars, and with the pin leading the slot: retrogradation appropriately occurs once every synodic cycle. But in Figure 12B we see the outcome with the same frequencies and the same e, but with the slot leading the pin. So, once the mechanism was applied to the planets, the "correct" relation between pin and slot would be settled on almost automatically. Our research group discovered this behavior of the pin and slot *empirically* — the fact that it makes an important difference whether the pin leads the slot or the slot leads the pin. Due to a miscommunication between the designer of our pin-and-slot device for Mars and the machinist who made it, our first model was built with the slot leading the pin, and it displayed no retrograde motion. This is an example of the way that mechanics and theoretical astronomy, in contact with one another, may have helped lead to refinements in planetary theory in its early, formative period.

If there were pins and slots before there were epicycles and eccentrics, the predominant direction of historical transmission in Figure 3 would be from left to right. Epicycles would have emerged through some mathematical astronomer steeped in natural philosophy taking seriously the possibility of one turning wheel riding on another. Did epicycles emerge, then, through an effort to model Babylonian phenomena with a mechanical invention? While we cannot know, there are some arguments that can be offered in support of this possibility. First, there is the fact that the gear trains in the Antikythera mechanism are based on Babylonian period relations. Moreover, the gearwork shows its designer's preference for directly relating synodic months to years and to anomalistic months, without using the day as a fundamental unit.[64] This is characteristic of Babylonian lunar theory.

Second, the representation of the solar anomaly on the Antikythera mechanism appears to be based simply on a nonuniform (but piecewise-uniform) division of the zodiac.[65] The whole of the extant portion of the zodiac happens to lie in the fast zone of the Babylonian solar theory of System A. In this

---

[64] We thank Alexander Jones for his insight on this point.

[65] Evans, Carman Thorndike [2010]. It is not quite possible to exclude an underlying geometrical theory (eccentric circle), but the statistics favor System A.

theory, the Sun travels at a uniform speed of 30° per synodic month in the fast zone (and at a speed of 28⅛° per synodic month in the slow zone). Simply by matching degrees on the zodiac against days on the Egyptian calendar scale, one can see that over the useable 69° of the scales the correspondence is indeed consistent with 30° per synodic month. This result does not depend on being able to find accurate centers of the extant portions of the scales; it comes from simply looking to see which degree mark is against which day mark. And it is hard to see how such a large effect could result from a sloppily performed division. One would imagine that in inscribing a zodiac, for example, the mechanic would begin by dividing the circle into quadrants. So it is difficult see how an effective equation of center could rise steadily from 0° to more than 2° over the course of 69° of longitude simply by chance error.[66]

And third, there are some complexities in the application of the pin-and-slot mechanism that would have to be mastered to go from epicycles and eccentrics to pins and slots. Notably, there is the fact that in standard theory, the Moon goes around backwards on its epicycle, but the planets go forward on theirs. It turns out, however, that it all works out fine with pins and slots. The key thing is that for the superior planets, the solar wheel turns faster than the sidereal frequency of the planets, but in the case of the Moon, wheel e3 turns more slowly than the sidereal frequency of the Moon. This leads to a very nice sort of geometrical reversal that allows both the outer planets and the Moon to be modeled in the same way.[67] And to transfer the design from the superior planets to the inferior, there is also the reversal in the relative sizes of C_2C_1 and C_1D (compare Figures 3 and 9) that was discussed above. Everything considered, we believe it would have been easier to arrive at a mechanical representation of Moon and inner and outer planets based on the pin-and-slot mechanism simply by starting from the phenomena than by starting from epicycle-and-deferent theory.

We know from the preface to Book I that Apollonius composed his *Conics* while living in Alexandria, which was the capital city of mechanical modelers. Apollonius himself may have been involved with mechanics and wonder-working, for several Arabic manuscripts preserve all or part of a description of an automaton flute-player attributed to a certain Apollonius.[68] This apparatus involves a water tank, valves, and gears. In the manuscripts the title reads: "Apollonius (a-b-l-n-y-w-s) the carpenter [and] the geometer. The art of the flute player." Now, "the Carpenter" is an epithet often attached to the name of Apollonius of Perge in medieval Arabic literature. For example, Ṣāʿid al-Andalusī in his *Book of the Categories of Nations* (11th century) wrote, "Among the Greek mathematicians, we have Ablūniūs the Carpenter, who wrote the book on *Makhrūṭāt* [Conics], which discusses bent lines that are neither straight lines nor arc segments ..." [Salem and Kumar 1991, 27].[69] The treatises in the Paris manuscript containing the "Apollonius" treatise are:[70] 1st, a "revision" or "improvement" of Theodosius on the sphere by Muḥyi al-Dīn al-Maghribī. 2nd (from fol. 29v), the treatise on a water-clock attributed to Archimedes, discussed above. 3rd (fol. 39v-42v), the treatise of Apollonius the Carpenter on the flute player. So, whether the compiler of this manuscript had the attribution right or not, it does seem likely, in view of the presence of Theodosius and Archimedes, that he had an ancient (and not a later) Apollonius in mind. Moreover, the "revision" of Theodosius does indeed begin with a selection of theorems from the first book Theodosius's *Spherics*; the second and third parts show more originality.[71]

---

[66] New work, still to be published, shows that the equation of center effect extends over the entire preserved 88° of the zodiac.

[67] This is discussed in detail in Carman et. al. [2012, 101-103].

[68] Bibliothèque Nationale, Paris, ms. Arabe 2468 (which may be viewed at http://gallica.bnf.fr/ark:/12148/btv1b52000453w/f90.image), British Library, Add. 23391, New York Public Library, Spencer, Indo-Persian ms. 2. For a description of this ms. see Schmitz [1992, 165–168]. An Arabic text based on all three manuscripts, with English translation and discussion, may be found in Shehadeh, Hill and Lorch [1994]. There is a German translation and discussion, based on the Paris and London mss., first published in 1914, in Wiedemann [1970, 2:50–56]. An additional manuscript at the Université St. Joseph, Beirut, was closely related to the London ms; this has disappeared, but photographs of it survive (see Shehadeh, Hill and Lorch [1994], who also mention a fragment of the treatise at Damascus).

[69] Perhaps, as Len Berggren has suggested (personal communication), an early Arabic writer thought that "carpenter" was a suitable thing to call a man who occupied himself with taking sections of cones.

[70] For details, see the on-line BN catalogue description of Arabe 2468 at http://archivesetmanuscrits.bnf.fr/ead.html?id=FRBNFEAD000030385.

[71] We have compared Carra de Vaux's [1891, 291-294] summary of the contents of the Arabic "revision" with Ver Eecke's

4th (from fol. 36), a hodgepodge on various topics, including regular polygons, the comparative weights of various minerals, and leveling. 5th and final (from fol. 52), a description of a "perfect compass," by Abū Sahl al-Kūhī and presented to the sultan Saladin, by means of which one can draw all the conic sections.[72] As Shehadeh, Hill and Lorch [1998, 355–356] remark, the authorship of the "Apollonius" flute-player treatise may never be known, and it could conceivably be of Hellenistic, Byzantine or even Islamic origin. However, Hill [1976, 9] previously suggested that the first part of the "Archimedes" treatise, including the arrangement of the gears, may possibly be the work of Archimedes. This, they argue, somewhat strengthens the case that the "Apollonius" treatise was composed by Apollonius of Perga. However, they note that the Arabic of the "Apollonius" treatise differs from that of the "Archimedes," which suggests different translators, so one has no way of knowing whether these two treatises were grouped together in Hellenistic times or long afterward. Finally, the authors point out that the mention by the "Apollonius" treatise of the vertical water wheel as a recent invention supports a Hellenistic date of composition.[73]

We do have the earlier example of Archimedes as a geometer who was also interested in mechanics and astronomy. And much later, according to the Suda, Ptolemy wrote a work on mechanics in three books.[74] That Apollonius might have had an interest in mechanics is not implausible. Certainly, as a mathematician living in Alexandria he could not have been unaware of gears and their uses. Much of what we have proposed in this paper must remain speculative. We have used the Antikythera mechanism frequently in our argument, as it offers us the only real insight we have into the design of early astronomical gearing. But our goal has been to explore a new approach to the early, formative period of Greek planetary theory, rather than the history of this particular machine, which could well turn out to be considerably later. We are left with a fascinating possibility to consider. Early Greek mechanics may have contributed in a significant way to the development of Greek theoretical astronomy. We should imagine planetary astronomy in conversation with mechanics, rather than a one-way transmission.

Acknowledgements We would like to express our thanks to the following friends and colleagues: to Brett Rodgers for discussion of Greek dialects and the Delphi inscription mentioned in the text, as well as of the papyrus life of Philonides; to Alexander Jones and Nathan Sidoli for their careful reading and comments on a draft of this paper, which helped us improve it materially; to Len Berggren for a discussion of the treatise on the flute player by Apollonius the Carpenter; to Mike Edmunds and Tony Freeth of the Antikythera Mechanism Research Project for their generosity in providing access to data; and to the National Archaeological Museum, Athens, for permission to reproduce the photograph in Figure 1.

References

Aujac, G., 1970. La sphéropée, ou la mécanique au service de la découverte du monde. Revue d'histoire des sciences et de leurs applications 23, 93–107.

——— 1979. Autolycos de Pitane : La sphère en mouvement; Levers et couchers héliaques; Testimonia. Les Belles Lettres, Paris.

Bekker, I., 1824–1825. Photii Bibliotheca. G. Reimer, Berlin. Accessible at http://gallica.bnf.fr/ark:/12148/bpt6k5489656d.

Berryman, S., 2009. The Mechanical Hypothesis in Ancient Greek Natural Philosophy. Cambridge University Press, Cambridge.

Bickerman, E.J., 1980. Chronology of the Ancient World. Cornell University Press, Ithaca.

[1927] translation of the Greek text and with Kunitzsch and Lorch's [2010] account of the medieval Arabic and Latin tradition of Theodosius.

72 The text of this treatise has been published with a French translation by Woepcke [1874].

73 On the other hand, Wilson [2008, 338] does not hesitate to ascribe the treatise to Apollonius of Perga.

74 See *Suda On Line: Byzantine Lexicography* at http://www.stoa.org/sol/.

Bodnár, I., 2011. The Pseudo-Aristotelian "Mechanics": The Attribution to Strato. In: Desclos, M.L., Fortenbaugh, W.W., (eds.), Strato of Lampsacus: Text, Translation and Discussion. Transaction Publishers, New Brunswick, p. 171–204.

Bottecchia Dehò, M.E., 2000. Aristotele. Problemi Meccanice: Introduzione, Testo Greco, Traduzione Italiana, Note. Rubbettino Editore, Soveria Mannelli (Catanzaro).

Carman, C.C., Thorndike, A.S., Evans, J., 2012. On the pin-and-slot device of the Antikythera Mechanism, with a new application to the superior planets. Journal for the History of Astronomy 43, 93–116.

Carra de Vaux, B., 1891. Notice sur deux manuscrits arabes. Journal asiatique, 8e série, 17, 287-322.

——— 1894. Les mécaniques ou l'élévateur de Héron d'Alexandrie, publiées pour la première fois sur la version arabe de Qostâ ibn Lûqâ. Imprimerie Nationale, Paris.

Clagget, M., 1959. The Science of Mechanics in the Middle Ages. University of Wisconsin Press, Madison.

Crönert, W., 1900. Der Epikureer Philonides. Sitzungsbericht der Königlich Preussischen Akademie der Wissenschaften zu Berlin. 2nd half-volume, 942–959.

Dalley, S., Oleson, J.P., 2003. Sennacherib, Archimedes, and the water screw: The context of invention in the ancient world. Technology and Culture 44, 1–26.

Davidson Weinberg, G., Grace, V.R., Edwards, G.R., Robinson, H.S., Throckmorton, P., Ralph E.K., 1965. The Antikythera shipwreck reconsidered. Transactions of the American Philosophical Society, New Series, 55, 3–48.

Decorps-Foulquier, M., Federspiel, M., 2010. Apollonius de Perge : Coniques. Walter de Gruyter, Berlin.

Dodge, B., 1970. The Fihrist of al-Nadīm: A Tenth-Century Survey of Muslim Culture. Columbia University Press, New York.

Drachmann, A.G., 1948. Ktesibios, Philon and Heron: A Study in Ancient Pneumatics. Munksgaard, Copenhagen.

——— 1951. On the alleged second Ktsebios. Centaurus 2, 1–10.

——— 1958. How Archimedes expected to move the Earth. Centaurus 5, 278–282.

——— 1963. The Mechanical Technology of Greek and Roman Antiquity: A Study of the Literary Sources. International Booksellers and Publishers, Munksgaard; University of Wisconsin Press, Madison.

——— 2008a. Ctesibius (Ktesibios). Complete Dictionary of Scientific Biography, 2: 491–492. Charles Scribner's Sons, Detroit.

——— 2008b. Philo of Byzantium. Complete Dictionary of Scientific Biography, 10: 586–589. Charles Scribner's Sons, Detroit.

Duke, D., 2009. Mean motions in Ptolemy's 'Planetary Hypotheses.' Archive for History of Exact Sciences 63, 635–654.

Dupuis, J., 1892. Théon de Smyrne, philosophe platonicien. Exposition des connaissances mathématiques utiles pour la lecture de Platon. Paris. Reprinted: Culture et Civilisation, Bruxelles, 1966.

Eastwood, B., 1992. Heraclides and heliocentrism: Texts, diagrams, and interpretations. Journal for the History of Astronomy 23, 233–260.

Evans, J., Berggren, J.L., 2006. Geminos's Introduction to the Phenomena: A Translation and Study of a Hellenistic Survey of Astronomy. Princeton University Press, Princeton.

Evans, J., Carman, C.C., Thorndike, A.S., 2010. Solar anomaly and planetary displays in the Antikythera Mechanism. Journal for the History of Astronomy 41, 1–39.

Fraser, P.M., 1972. Ptolemaic Alexandria. Clarendon Press, Oxford.

Freeth, T., Bitsakis, Y., Moussas, X., Seiradakis, J.H., Tselikas, A., Mangou, H., Zafeiropoulou, M., Hadland, R., Bate, D., Ramsey, A., Allen, M., Crawley, A., Hockley, P., Malzbender, T., Gelb, D., Ambrisco, W., Edmunds, M.G., 2006. Decoding the ancient Greek astronomical calculator known as the Antikythera Mechanism. Nature 444, 587–591. There is substantial "Supplementary Information" at http://www.nature.com/nature.

Freeth, T., Jones, A., 2012. The cosmos in the Antikythera Mechanism. ISAW Papers 4. http://dlib.nyu.edu/awdl/isaw/isaw-papers/4.

Gallo, I., 1980. Frammenti biografici da papyri. Edizioni dell'Ateneo & Bizzarri, Roma.

Gera, D., 1999. Philonides the Epicurean at court: Early connections. Zeitschrift für Papyrologie und Epigraphik 125, 77–83.

Hamm, E.A., 2011. Ptolemy's Planetary Theory: An English Translation of Book One, Part A of the Planetary Hypotheses with Introduction and Commentary. Ph.D. Dissertation, University of Toronto.

Heath, T.L., 1921. A History of Greek Mathematics. Clarendon Press, Oxford.

Heiberg, J.L., 1891–1893. Apollonii Pergai quae graece exstant opera cum commentariis antiquis. Teubner, Leipzig.

——— 1904. Mathematisches zu Aristoteles. In Cantor M. (ed.), Abhandlungen zur Geschichte der mathematischen Wissenschaften, 18. Hft., p. 3–49. B.G. Teubner, Leipzig.

——— 1910–1915. Archimedis opera omnia cum commentariis Eutocii, 3 vols. Teubner, Leipzig. Reprinted: Stuttgart, 1972.

Henry, R., Schamp, J., 1959-1991. Photius, Bibliothèque. 9 vols. Les Belles Lettres, Paris.

Hett, W.S., 1936. Aristotle: Minor Works. Harvard University Press, Cambridge, MA.

Hill, D.R., 1976. On the Construction of Water-Clocks. Kitāb Arshimīdas fī ʿamal al-binkamāt. Turner & Devereux, London.

Hultsch, F., 1876–1878. Pappi Alexandrini Collectionis quae supersunt. 3 vols. Weidmann, Berlin.

Huxley, G., 1963. Studies in the Greek astronomers. Greek, Roman and Byzantine Studies 4, 83–105.

Jones, A., 1976. Pappus of Alexandria: Book 7 of the Collection. Springer, New York.

——— 1999. Astronomical Papyri from Oxyrhynchus. Memoirs of the American Philosophical Society 233. American Philosophical Society, Philadelphia.

Kaltsas, N., 2012. The Antikythera Shipwreck: The Ship, the Treasures, the Mechanism. Kapon, Athens.

Keyes, C.W., 1994. Cicero. De re publica. De legibus. Harvard University Press, Cambridge, MA, and London (first published, 1928).

Knorr, W.R., 1982. Ancient Sources of the Medieval Tradition of Mechanics: Greek, Arabic and Latin Studies of the Balance. Istituto e museo di storia della scienza, Firenze.

——— 1986. The Ancient Tradition of Geometric Problems. Birkhäuser, Boston.

Köhler, U., 1900. Ein Nachtrag zum Lebenslauf des Epikureers Philonides. Sitzungsbericht der Königlich Preussischen Akademie der Wissenschaften zu Berlin, 2nd half-volume, 999–1001.

Krafft, F., 1970. Dynamische und statische Betragtungsweise in der antiken Mechanik. Franz Steiner Verlag, Wiesbaden.

Kunitzsch, P., Lorch, R., 2010. Theodosius Sphaerica. Arabic and Medieval Latin Translations. Franz Steiner Verlag, Stuttgart.

Mastrocinque, A., 2009. The Antikythera shipwreck and Sinope's culture during the Mithridatic wars. In: Højte, J.M. (ed.), Black Sea Studies 9: Mithridates VI and the Pontic Kingdom. Aarhus University Press, Aarhus. Available at http://www.pontos.dk/publications/books/BSS 9

Minar, E.L., Sandbach, F.H., Helmbold, W.C., 1961. Plutarch. Moralia, Vol. IX. Harvard University Press, Cambridge.

Morrison, L.V., Stephenson, F.R., 2004. Historical values of the Earth's clock error ΔT and the calculation of eclipses. Journal for the History of Astronomy 35, 327–336.

Mueller, I., 1982. Geometry and Scepticism. In: Barnes, J., Brunschwig, J., Burnyeat, M., Schofield, M. (eds.), Science and Speculation: Studies in Hellenistic Theory and Practice. Cambridge, Cambridge University Press.

Mugler, C., 1972. Archimède. Oeuvres. Tome IV, Commentaires d'Eutocius. Fragments. Les Belles Lettres, Paris.

Netz, R., 2004. The Works of Archimedes. Vol. I, The Two Books On the Sphere and the Cylinder. Cambridge University Press, Cambridge.

Neugebauer, O., 1975. A History of Ancient Mathematical Astronomy. Springer, Berlin and New York.

Oldfather, C.H., 1933. Diodorus Siculus, Library of History, Vol. I. Harvard University Press, Cambridge.

Oleson, J.P., 1984. Greek and Roman Mechanical Water-Lifting Devices: The History of a Technology. University of Toronto Press, Toronto.

Perrin, B., 1917. Plutarch. Lives, Vol. V. Harvard University Press, Cambridge.

Philippson, R., 1941. Philonides (5). Paulys Realencyclopädie der classischen Altertumswissenschaft. Neue Bearbeitung, 39:63–73. Alfred Druckenmüller, München.

Plassart, A., 1921. Inscriptions de Delphes. La liste des théorodoques. Bulletin de correspondance hellénique 45, 1–85.

Price, D.d.S., 1974. Gears from the Greeks: The Antikythera Mechanism — A Calendar Computer from ca. 80 B.C. Transactions of the American Philosophical Society, New Series, vol. 64, part 7.

Rackham, H., 1947–1963. Pliny: Natural History. Harvard University Press, Cambridge, MA, W. Heinemann, London.

——— 1952. Cicero: De natura deorum. Academica. Harvard University Press Cambridge, MA, London.

Rhode, E., 1878. Γέγονε in den Biographica des Suidas. Rheinisches Museum für Philologie, N. F. 33, 161–220.

Salem, S.I., Kumar, A., 1991. Science in the Medieval World: "Book of the Categories of Nations" by Ṣāʿid al-Andalusī. University of Texas Press, Austin.

Sedley, D., 1976. Epicurus and the Mathematicians of Cyzicus. Cronache Ercolanesi 6, 23–54.

Shehadeh, K., Hill, D.R., Lorch, R., 1994. Construction of a Fluting Machine by Apollonius the Carpenter. Zeitschrift für Geschichte der arabisch-islamischen Wissenschaften 9, 326–356. Reprinted in Hill, D.R., Studies in Medieval Islamic Technology. Ashgate Publishing, Aldershot (Hampshire) and Brookfield (Vermont), 1998.

Sidoli, N., Saito, K., 2009. The role of geometrical construction in Theodosius's Spherics. Archive for History of Exact Sciences 63, 581–609.

Thorndike, L., 1949. The Sphere of Sacrobosco and its Commentators. University of Chicago Press, Chicago.

Toomer, G. J., 1967. The size of the lunar epicycle according to Hipparchus. Centaurus 12, 145–150.

——— 1972. The mathematician Zenodorus. Greek, Roman and Byzantine Studies 13 177–192.

——— 1984. Ptolemy's Almagest. Duckworth, London.

——— 2008a. Apollonius of Perga. Complete Dictionary of Scientific Biography, 1: 179–193. Charles Scribner's Sons, Detroit.

——— 2008b. Heraclides Ponticus. Complete Dictionary of Scientific Biography, 15: 202–205. Charles Scribner's Sons, Detroit.

Ver Eecke, P., 1927. Les Sphériques de Théodose de Tripoli : oeuvres traduites pour la première fois du grec en français. Desclée de Brouwer et Cie, Bruges. Nouveau tirage: Albert Blanchard, Paris, 1959.

——— 1933. Pappus d'Alexandrie. La Collection mathématique. Albert Blanchard, Paris.

Whitehead, D., Blythe, P.H., 2004. Athenaeus Mechanicus, On Machines (Περὶ μεχανημάτων). Franz Steiner Verlag, Stuttgart.

Wiedemann, E., 1970. Aufsätze zur arabischen Wissenschaftsgeschichte, 2 vols. Georg Olms Verlag, Hildesheim and New York.

Wilson, A.I., 2008. Machines in Greek and Roman technology. In: Oleson, J.P. (ed.), The Oxford Handbook of Engineering and Technology in the Classical World. Oxford University Press, Oxford, 2008, p. 337–366.

Winter, T.N., 2007. The Mechanical Problems in the Corpus of Aristotle. Faculty Publications, Classics and Religious Studies, University of Nebraska-Lincoln. Available at http://digitalcommons.unl.edu/classicsfacpub/68/.

Woepcke, F., 1874. Trois traités arabes sur le compas parfait. Notices et extraits des manuscrits de la Bibliothèque imperiale et autres bibliothèques. Paris, Imprimerie Nationale, t. 22, 1e partie, p. 1–176.

Wright, M.T., 2002. A planetarium display for the Antikythera Mechanism. Horological Journal 144, 169–73 and 193.

Some Greek Sundial Meridians

Alexander Jones

Abstract In Greco-Roman sundials, the indication of time of year often has a prominence comparable to the indication of time of day. The meridian hour-line offered a convenient scale for time of year, and we have instances of sundials reduced to just the meridian, so that the only time of day when a reading was possible was noon. The present paper discusses three interesting ancient meridian scales. The first, *I. Milet. inv. 46*, was found during the German excavations at Miletos in the early 20th century. It has not been previously published or discussed in print, and its present location is not known, but a fairly precise drawing with measurements survives. This was a mere meridian inscribed on a south-facing vertical surface, with graduations marking the dates of the Sun's entries into the zodiacal signs. *Breccia Alexandria Mus. No. 185* is an inscription explaining the use of a sundial that itself no longer exists. The surviving part of the text concerns how to read the stage of the year off the sundial; there were indications not only of the Sun's entries into the zodiacal signs but also of dates of stellar visibility phenomena, indexed by letters of the Greek alphabet. *National History Museum of Romania inv.* 18.757 is a sundial with a meridian marked at irregular intervals with letters of the Greek alphabet that again probably represent stellar visibility phenomena, though in this case the accompanying inscription identifying the phenomena is lost.

I. Milet. inv. 46.

In 1900, the German excavations directed by Theodor Wiegand at Miletos turned up a marble fragment of a column inscribed with a long vertical line running from near the present top of the fragment to about halfway down, crossed by several shorter horizontal lines, with Greek letters in various places that evidently did not make sense to the archeologist who made a pencilled copy ("*Schede*") for the planned publication of inscriptions from Miletos (Figure 1).[1] According to this sheet, the fragment (assigned inventory number 46) was of bluish marble, with dimensions 68 cm (height) by 47 cm (diameter). The findspot was "in the field by the gate of the *Heilige Strasse*," that is, the city gate named "Heiliges Tor" by the excavators leading out to the ancient road running south from Miletos to Didyma.

In 1905, Albert Rehm, who had joined the excavation as an epigrapher, examined the inscription, and annotated the *Schede* with an improved transcription of the text, measurements of the intervals between the transverse lines, and the remark that the date of the inscription was "surely the Roman imperial period." He noted that it was "jetzt vor dem Inschr.-Mus." (at the archeological site), and below this he

I am delighted to be able to offer this little batch of gnomonic curiosities to Len Berggren as a token of homage and of nearly three decades of friendship.

[1] Bayerische Staatsbibliothek (Munich), Rehmiana Suppl., Box 1–300. I thank the Bayerische Staatsbibliothek for providing a scan and permission to reproduce it, and the Deutsches Archäologisches Institut and particularly Norbert Ehrhardt for information about the inscription and permission to publish it.

N. Sidoli and G. Van Brummelen (eds.), *From Alexandria, Through Baghdad: Surveys and Studies in the Ancient Greek and Medieval Islamic Mathematical Sciences in Honor of J.L. Berggren*, DOI 10.1007/978-3-642-36736-6_8,

© Springer-Verlag Berlin Heidelberg 2014

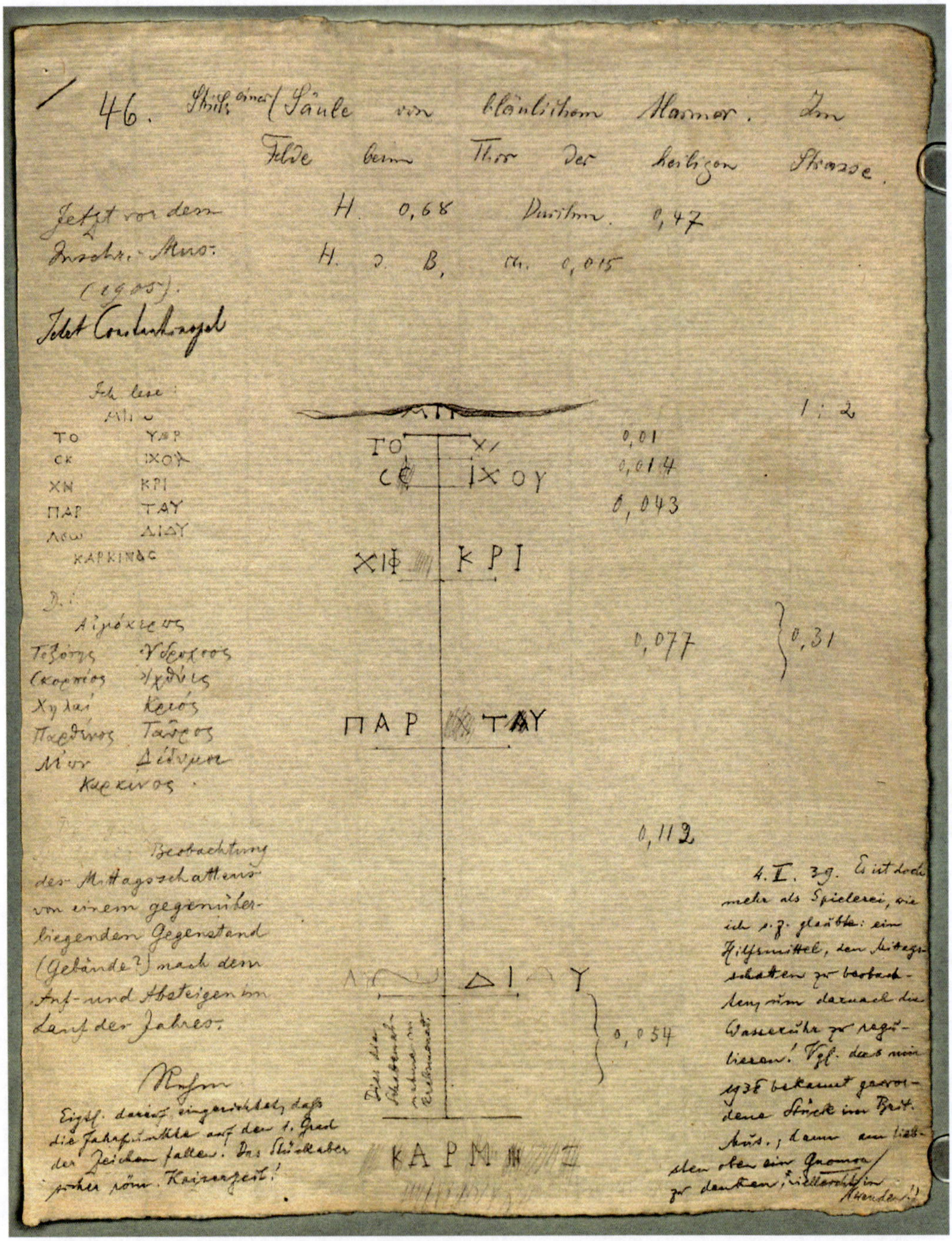

Figure 1: "Schede" of I. Milet. inv. 46. Photo courtesy of Bayerische Staatsbibliothek.

subsequently wrote, "Jetzt Constantinopel." Despite inquiries, I have been unable to confirm whether it is now in Istanbul, and the *Schede* is for the time being the primary evidence for the inscription. It was not included in the published series *Inschriften von Milet*, and to my knowledge the present article is the first publication of it.

A transcription based on a tracing from the *Schede*, with Rehm's measurements (in meters), is given in Figure 2.

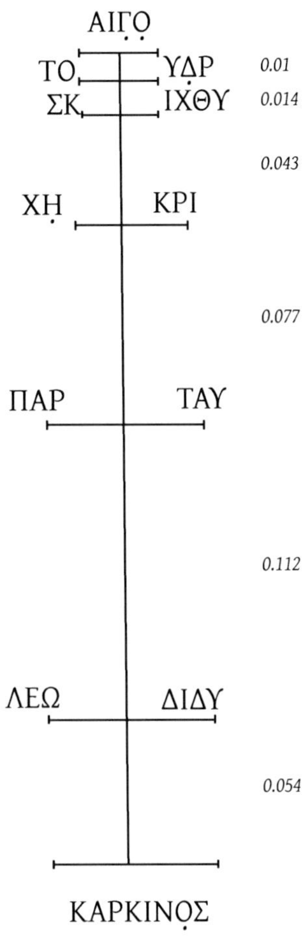

Figure 2: Transcription of I.Milet. inv. 46, with Rehm's measurements (meters).

As Rehm realized, this is a meridian line upon which the shadow of a gnomon fell at noon, showing the zodiacal sign currently occupied by the Sun. The names of the signs of the zodiac, all of them abbreviated except for Cancer, are inscribed clockwise, with Cancer at the bottom and Capricorn at the top; the transverse lines must represent the Sun's entry into the relevant signs, and their symmetry shows that the solstitial and equinoctial points were assumed to be at the sign boundaries. This was not a conventional sundial displaying the time of day (except at noon) but one merely showing the current stage of a "natural" solar year, independent of the civil calendar.

According to Rehm's measurements, the placement of the transverse lines is only moderately accurate for the latitude of Miletos (approximately 37°32′ N). Assuming that the tip of the gnomon was approximately 9 cm in perpendicular distance from the column's surface (which gives a near optimal fit to the data) and approximately 5 cm above the top mark, we may calculate how far below the top mark the other marks ought to be and compare with those resulting from summing Rehm's measurements of the individual intervals:[2]

Mark	*Theoretical (cm)*	*Measured (cm)*	*Difference (cm)*
Capricorn	0.0	0.0	
Sagitt./Aquarius	0.7	1.0	+0.3
Scorpio/Pisces	2.8	2.4	−0.4
Libra/Aries	6.8	6.7	−0.1
Virgo/Taurus	14.6	14.4	−0.2
Leo/Gemini	24.2	25.6	+1.4
Cancer	31.6	31.0	−0.6

The largest discrepancy, in the Leo/Gemini line, would correspond to more than three days' change in the Sun's declination. Of course, even an accurately constructed meridian would not reveal the exact dates of the solstices since the rate of change of the Sun's declination is then close to zero.

In an unpublished monograph on ancient meteorology that Rehm submitted (unsuccessfully) for a prize competition in 1905, he speculated that the Miletos meridian was intended to display the stage of the solar year as a basis for predictions of weather, such as one finds in the *parapegmata*.[3] Much later, in 1939, he wrote on the *Schede* a different notion, that the purpose of the meridian was to allow calibration of a water clock so that it would display seasonal hours of appropriate length for the current time of year. In the absence of knowledge of the kind of site where the column was originally erected, it is impossible to say whether either of Rehm's suggestions is likely to be correct.

I am aware of only one close parallel to the Miletos meridian, namely a Hellenistic vertical meridian, similarly inscribed with the beginnings of zodiacal signs, first described by D.W.S. Hunt in 1947; when he saw it, the block had been built into the coping of a well in a remote village on Chios [Hunt 1940–1945, 41–42].[4] Recently the block was extracted and transferred to the Archeological Museum of Chios, and as K. Schaldach has shown, it turns out to bear planar sundials on its east and west faces as well as the meridian on the south face [Schaldach 2011].[5] The column on which the Miletos meridian was inscribed does not appear to have borne any other sundials.

The so-called *Solarium Augusti* partially excavated in Rome in 1979–1980 is now widely held to have been an immense stand-alone horizontal meridian, rather than a complete sundial, with an obelisk as its gnomon [Buchner 1982].[6] The scale of the monument allowed it to be marked with cross-strokes not only for the beginnings of the zodiacal signs but also for subdivisions presumably representing single days.

Breccia, Alexandria Mus. No 185

In his meteorological manuscript, Rehm offers an excursus on another inscription relating to a meridian divided into segments corresponding to the zodiacal signs, Breccia Alexandria Mus. No. 185, discovered in 1901 at Lake Maryut near Alexandria [Rehm 1906, 6–11; Breccia 1911, text 185 on pp. 105–106 and plate XXV]. The block had been recut in antiquity to make an open water channel ending in a lion's head, with the inscribed surface on the bottom (Figure 3). The text as read by Breccia is given below (omitting the mostly illegible final lines), with a literal translation:

---

[2] "Theoretical" values were computed assuming latitude 37°30′ for Miletos and an obliquity of 23°40′. The criterion adopted for estimating the gnomon's length (more precisely, 8.9 cm) was a least-squares fit for the differences between theoretical and measured values.

[3] Rehm [1906]. (The remainder of the manuscript of the monograph is Bayerische Staatsbibliothek, Rehmiana III/10.) The Miletos meridian is discussed on pp. 4–6. The famous parapegma inscription fragments from Miletos were found in or near the Theater, far from the findspot of the meridian: Diels and Rehm [1904, 92].

[4] On this meridian the zodiacal sign names run counterclockwise.

[5] I am grateful to Karlheinz Schaldach for sharing this article with me before its publication.

[6] See Heslin [2007] for a review of more recent discussions and the arguments for the identification as a meridian.

Figure 3: Breccia, Alexandria Mus. No. 185 [Breccia 1911, plate XXV].

δ̣ι̣ὰ̣ π̣εριφ̣ερ̣ε̣ι̣ῶ̣ν̣ τῶν ἐφε[ξ]ῆς
τῶν διατεινουσῶν ἀπ' ἀνατολῶν ἐπὶ δύσεις,
ἀπὸ τῆς ἑτέρας ἐπὶ τὴν ἑτέραν
μεθίσταται τὸ ἄκρον τῆς σκιᾶς
ἐν ἡμέραις τριάκοντα. ἀπὸ χειμερινῶν δὲ τροπῶν
[ἐ]πὶ θερινὰς τροπὰς μεθιστάμενο[ν]
[τ]ὸ ἄκρον τῆς σκιᾶς, δι'
οὗ ἂμ φέρητα[ι]
[τ]ῶμ πρὸ μεσημβρίας Ζωιδίων,
[ἐν] τούτωι τῶι Ζωιδίωι
[σ]ημαίνει τὸν ἥλιον εἶναι
[κα]ὶ ἐπὶ τῶν ἰῶτα φερόμενον
[σ]ημαίνει ζεφύρου πνοήν·
[ἀπ]ὸ θερινῶν δὲ τροπῶν
[ἐπ]ὶ χειμερινὰς τροπᾶς μεθιστάμεν[ον]
[τὸ] ἄκρον τῆς σκιᾶς δι' οὗ ἂμ φέρητα[ι]
[τῶ]ν ἐγ μεσημβρίας Ζωιδίων
ἐν [τ]ουτῶι τῶι Ζωιδίωι
[ση]μαίνει τὸν ἥλιον εἶναι
[καὶ ἐ]πὶ τῶν ἰῶτα φερόμενον
[σημ]αίνει Πλειάδος δύσιν
[ἑκά]στης δὲ ἡμέρας
[.]ν τὸ ἄκρον τῆς σκιᾶς
[..] τῆς πέρας γραμμῆς α

Through arcs in sequence,
extending from east to west,
from one to another
the shadow's tip shifts
in 30 days. While shifting from winter solstice
to summer solstice,
the shadow's tip,
through whichever one it is moving
of the zodiacal signs before noon,
in this zodiacal sign
it indicates that the Sun is,
and moving to the iotas,
it indicates the blowing of the Zephyros wind;
from the summer solstice
while shifting to winter solstice,
the shadow's tip through whichever it moves
of the zodiacal signs after noon,
in this sign
it indicates that the Sun is,
and moving to the iotas,
it indicates the setting of the Pleiades,
every day
... the shadow's tip
... the limiting line ...

Rehm was surely correct in interpreting this text as part of the description of a sundial (perhaps, but not necessarily, spherical) having not only solstitial and equinoctial circles but also a series of other day circles marking the Sun's entry into each zodiacal sign (labelled in clockwise order), essentially combining the principle of the Miletos and Chios meridians with a conventional sundial. Some examples of extant sundials of this type are Gibbs nos. 1068G (hemispherical, Rome), 4007 (horizontal plane, Pompeii), 4010 (horizontal plane, Rome), 7002G (globe, Prosymna), and the spherical north face of 7001G (multiple, Tinos, without surviving labels) [Gibbs 1976]. The beginnings of the zodiacal signs are regularly aligned with the solstices and equinoxes.[7]

A more uncommon feature of this lost Egyptian sundial was the presence of a mark along the meridian indicating (during the interval between summer and winter solstice) the annual onset of the Zephyros wind, and (during the interval between winter and summer solstice) the morning setting of the Pleiades. These events were in fact believed to be approximately equidistant from the winter solstice, about the middle of the Sun's passage through Aquarius and Scorpio respectively, and conventionally marked the beginnings respectively of spring and winter.[8] Stellar and meteorological phenomena are rarely marked on Greco-Roman sundials. Whereas the zodiacal sign boundaries define a fixed and symmetrical set of eight events (in addition to the solstices and equinoxes) and four meridian marks, parapegma phenomena are fundamentally disorderly: there is no pattern to their distribution through the year, their dates are not very precisely determined by astronomical conditions that in any case vary with terrestrial latitude, and it is an arbitrary decision how many phenomena and which ones should be recorded. In a "transposed" hemispherical sundial from Kea (Gibbs 1073) a day circle inscribed near the summer solstitial circle is labelled κύων ἐκφανής, "Sirius conspicuous," that is, Sirius' morning rising. A fragment of a spherical sundial from Delos (Gibbs 1001) has a day circle in approximately the same location, labelled] ἑῶια ἐπι[τολή, "morning rising" — undoubtedly of Sirius again. The spherical south face of a multiple sundial from Tinos (Gibbs 7001G) has parts of day circles, terminating

[7] An apparent exception is a fragment (now lost) of a planar (presumably horizontal) sundial allegedly found at Rome "in the Mausoleum of Augustus in the Campus Martius," of which an engraving was published in Gruterus [1707, cxxv]. Six day lines are shown, of which the three furthest from the summer solstitial line are labelled, respectively, TAUR, AEQ. VER, and ARIES. But the copy on which the engraving was based was evidently incompetent [Arnaldi 1999].

[8] In the parapegma appended to Geminus' *Isagoge* (translated by Evans and Berggren [2006, 231–240]) the onset of Zephyros is assigned to the 14th, 16th, and 17th day in Aquarius according respectively to Eudoxus, Democritus, and Euctemon and Callippus. The morning setting of the Pleiades is assigned to the 15th, 16th, and 19th days in Scorpio according respectively to Euctemon, Callippus, and Eudoxus. Thus according to this parapegma, Euctemon placed the winter solstice 43 days after the morning setting of the Pleiades, and the Zephyr 43 days after the solstice.

at the fifth and seventh hour circles, labelled πλειάδων δύσις/χειμῶνος ἀρχή ("[morning] setting of the Pleiades/beginning of winter," above the equinoctial circle by roughly two-fifths of the interval to the winter solstitial circle), πλείας ἐκφανής/[θέ]ρους ἀρχή ("Pleias conspicuous [i.e. morning rising]/beginning of summer," below the equinoctial circle about three-fifths of the interval to the summer solstitial circle), and κύων ἐκφανής ("Sirius conspicuous," a little above the summer solstitial circle). The excavated sector of the *Solarium Augusti*, mentioned above, has a marking at the first day-division of Virgo labelled ἐτήσιαι παύονται ("the Etesian winds cease"). This exhausts the extant examples known to me.

Unlike these extant sundials, the lost one described in Alexandria Mus. No. 185 seems not to have had the names of the astronomical and meteorological events inscribed next to the mark on the meridian, but rather just the letter iota (probably on each side of the line, since the text speaks of "iota" in the plural, τῶν ἰῶτα). This raises the question, why iota, which is neither the initial of either of the phenomena nor the first letter of the Greek alphabet? One possible answer is that this was not the only marking along the meridian, but one of a series labelled by the letters of the alphabet in order starting with alpha. That would mean that a large number of annual phenomena were marked, at least ten (counting the two specified in the inscription). Since the beginnings of the zodiacal signs, because of their symmetry, only require four day lines in addition to the conventional solstitial and equinoctial lines, other kinds of phenomena, most likely stellar risings and settings, would have had to be included. Alphabetic labels would imply the presence elsewhere of an indexed list of the phenomena. The extant inscription singles out the "iota" phenomena for special mention because of the special status of these events as defining the beginnings of seasons.

National History Museum of Romania inv. 18.757.

One sundial with such an indexed meridian has come to light (Figs. 4–5). This marble fragment (height 210 mm, length 174 mm, width from 95.5 to 122 mm) was formerly catalogued as inventory number L.2023 in the National Museum of Antiquities, Bucharest,[9] and now as inventory number 18.757 in the National History Museum of Romania, Bucharest. It was found in 1950 in the sanctuary area of ancient Histria (Istros), near the west coast of the Black Sea. It had been reused as filler rubble in late antiquity so that no precise archeological context exists for its original use. A dating to the late fourth or early third century B.C. was proposed by E. Popescu on the basis of the inscribed letter forms [Ionescu-Cârligel 1970, 124], but since these consist of no more than an imperfectly preserved sequence of isolated alphabetic letters (from beta through mu) and five letters from two words (ΟΣ ΧΡΟ), we do not have a very satisfactory basis for paleographical dating. The most I would say is that the form of the sigma, with its horizontal hastae diverging towards the right rather than parallel, indicates a Hellenistic date, but I see nothing that would rule out a date as late as the first century B.C.

The fragment preserves the central part of the bowl of a spherical sundial, including parts of the meridian, the fifth hour circle, the equator, and the summer solstitial circle. The engraved lines, which have been executed with care, are about 1 mm thick. Ionescu-Cârligel measured the arc between the fifth hour circle and the meridian along the equator as 41.2 mm, and the corresponding arc along the solstitial circle as 48.5 mm; assuming 24° for the obliquity of the ecliptic, these arcs imply that the sundial was constructed for a locality with a longest day of 14.126 hours and a latitude of approximately 44°31′ [Ionescu-Cârligel 1970, 124].[10] This appears remarkably close to the actual latitude of Histria, 44°33′, but the precision of the match is probably accidental, since changing either measurement by as little as 0.2 mm affects the latitude by more than half a degree, and such exactitude can scarcely be

[9] I am deeply grateful to the National History Museum of Romania, and in particular to Dr. Adela Bâltâc of their Archeology Department, for providing new photographs of the fragment with the permission to publish them, and for the measurements reported here. The sundial was previously discussed by Ionescu-Cârligel [1969 and 1970] and Feraru [2008]. It is no. 1044 in Gibbs [1976].

[10] Dr. Bâltâc has confirmed that the measurements are correct.

Figure 4: National History Museum of Romania 18.757, facing view. Photo courtesy of Adela Bâltâc.
© 2010, Adela Bâltâc.

claimed for either the ancient engraving or a modern measurement.[11] If, for example, the sundial was constructed exactly for the parallel where the longest day is exactly 15 1/2 hours (which Ptolemy situates at latitude 45°1′ and calls the "parallel through the middle of Pontus"), it would require an increase in the solstitial arc or a decrease in the equatorial arc of only 0.1 mm relative to Ionescu-Cârligel's measurements.

The sundial shows a comparatively uncommon feature. An inscribed line crosses the fifth hour circle at an angle such that it evidently met the point, now broken away, where the meridian intersected the winter solstitial circle. This was part of a "daylight triangle," comprising two lines extending from the winter solstitial noon point to two points equidistant from the summer solstitial noon point along the summer solstitial circle, such that the arcs of the summer solstitial circle outside the daylight triangle correspond to the same length of time as the entire arc of the winter solstitial circle. The triangle thus displays an approximation of the increase in the length of daylight relative to the winter solstice through the course of the year. Since the sundial was constructed for a locality near the latitude along which the ratio of longest to shortest day is 15.5 : 8.5 (Ptolemy's "parallel through the middle of Pontus"), the preserved line ought to have met the summer solstitial line at about the point corresponding to $6\times(8.5/15.5)$ hours, thus a little to the left of the third hour circle. In fact, the extrapolation of the small length of the line that is preserved, if it was constructed as a great circle, would have intersected the summer line very near the third hour circle.[12] The fragmentary inscription along the line was probably something along the lines of "extent of the time belonging to every day."[13] Daylight triangles are attested in only a handful of ancient sundials (Gibbs lists three besides the present one).[14]

But the feature that makes this sundial particularly interesting is a series of inscribed horizontal lines extending from the meridian to the left or right, and labelled with Greek letters in alphabetical order, starting with alpha (broken off) at the stroke nearest to the summer solstitial circle and continuing to at least nu (also broken off) about half way between the equator and the winter solstitial circle. The strokes, as Ionescu-Cârligel realized, must mark certain events tied to the solar year, i.e. at noon on the day when a particular event is reached, the gnomon's shadow will cross the meridian at the corresponding stroke. Since every point along the meridian between the solstice points corresponds to two dates, one in the half of the year from summer to winter solstice and the other in the remaining half, one would expect some indication of which half each event belongs to; clearly the distinction between strokes extending leftwards and rightwards from the meridian serves this purpose. The nature of each event must have been given in a separate list indexed by the alphabetic letters, perhaps on the base of the sundial. The index letters follow the order determined by the placement of the crossing points along the meridian from bottom to top, not the order that they occur in the solar year, so that the indexed list would not have been in chronological sequence.

The use of the alphabet as an indexing device, a kind of system of proto-footnotes, is itself noteworthy in the history of organization of data. The motivation was obviously the desire to present more information to the spectator than could be legibly inscribed in the appropriate places on the sundial. Another astronomical artifact using alphabetic indexing is the Antikythera Mechanism (second or early first century B.C.). On the face of the Mechanism conventionally referred to as the "front," a partly preserved circular dial is graduated into the twelve zodiacal signs and their degrees, with index letters inscribed at irregular intervals along the degree marks. These letters refer to lettered lines of an inscription on a bronze plate, also partially extant, listing risings and settings of stars and constellations. When a revolving pointer representing the longitude of the Sun lined up with one of the index letters, the corresponding stellar event was predicted.[15] On the Mechanism's "back," a spiral dial that was traced by

[11] The calculation is much less sensitive to the value assumed for the obliquity.

[12] The day triangle line crosses the fifth hour circle 40 mm above the equator circle, and the meridian (presumably) 68.5 mm above the equator circle.

[13] Ionescu-Cârligel [1970, 136–137, n. 22] draws attention to the corresponding inscriptions on the day triangle of the planar sundial from Delos (Gibbs 4001G), ποῦ χρόνος πάσης ἡμέρας παρήκει/λοιπός, "where the time belonging to every day reaches/remaining."

[14] Gibbs nos. 1068G (hemispherical, Rome), 3046 (conical, Samos), and 4001G (horizontal plane, Delos).

[15] Price [1974, 46–51] published a transcription, partly indebted to Rehm's unpublished work, of the parts of the "para-

Figure 5: National History Museum of Romania 18.757, lateral view. Photo courtesy of Adela Bâltâc. © 2010, Adela Bâltâc.

a revolving pointer is divided into the 223 lunar months of a "Saros" eclipse cycle, with brief inscriptions in the cells belonging to months within which a lunar or solar eclipse was possible. The inscribed cells have index letters that apparently referred to a separate inscription giving further details of the eclipse predictions [Freeth, Jones, Steele, and Bitsakis 2008].

I am not aware of other instances of this kind of alphabetic indexing in Greco-Roman texts or artifacts. (Alphabetic sorting of words is a different principle.) It would be interesting to know whether it was a method that originated in or that was especially characteristic of astronomy.

Let us now see what is the pattern of meridian marks on the Histria sundial. In Table 1, the marks pointing to the left are separated from those pointing to the right, on the hypothesis that the direction is meant to indicate whether the associated event precedes or follows the summer solstice. The columns of the table contain (i) the index letter, (ii) the mark's signed distance in millimeters from the equinoctial circle, such that positive is towards the summer solstice circle,[16] (iii) the mark's signed length in millimeters, such that right is positive, (iv) the distance of the mark from the equinoctial circle divided by the measured distance between the equinoctial and summer solstice circles, i.e. the ratio r of the Sun's declination to the obliquity ε,[17] (v) the unsigned elongation of the Sun's longitude λ from the summer solstitial point, where λ is estimated as follows:

$$\sin(\lambda) = \sin(r\varepsilon)/\sin(\varepsilon), \text{ assuming } \varepsilon = 23°40',$$

(vi) the approximate change in this elongation corresponding to a change of 0.5 mm in the measured interval on the sundial's meridian, (vii) the Sun's longitude, assuming that the event precedes the summer solstice, (viii) the Sun's longitude, assuming that the event follows the summer solstice, and (ix) the differences in longitude corresponding to consecutive marks on the same side of the meridian.

We can dismiss the hypothesis that zodiacal sign entries account for most or all of the markings. There is no sign of the symmetries expected if the sign entries were inscribed according to the normal convention that the equinoxes and solstices coincide with sign entries. Nor can one find more than one or, at the most generous, two pairs of marks on the same side of the meridian for which the solar longitudes are separated by close to 30°.[18]

Can we identify any stellar phenomena associated with the marks? The undertaking is necessarily speculative, because the ancient Greek tradition of stellar visibility dates, as preserved in *parapegmata* and other sources, involved a large number of stars and constellations, with potentially four phenomena recorded for each (morning and evening risings = MR and ER, morning and evening settings = MS and ES, with occasional distinction between the dates of partial and complete visibility of a constellation); the sundial could only have registered a small selection. Moreover, the ancient sources and the authorities that they cite assign diverging dates, sometimes widely diverging, to the same phenomena. Modern astronomical theory is of limited usefulness as a control on the ancient data, because the predictions of modern models for stellar visibility do not agree very closely with any particular ancient set of visibility

pegma" inscription that were visible in Rehm's and in Price's time. A new, augmented edition and study of the parapegma and dial inscriptions by A. Jones, T. Freeth, and Y. Bitsakis is in preparation.

[16] These measurements were made along the curved surface from the upper edge of the summer solstice circle to the lower edge of the mark, and subtracted from the measured distance (67.5 mm) from the upper edge of the solstice circle to the equinoctial circle; they are thus effectively the distance from the lower edge of the mark to the lower edge of the equinoctial circle.

[17] The center-to-center distance from the solstice to the equinoctial circle is assumed to be 68.5 mm, allowing 1 mm for the width of the inscribed lines.

[18] Ionescu-Cârligel [1970, 134–135] attempts to establish that nine of the marks are sign entries (not aligned with the solstices and equinoxes), with three of the remainder indicating the beginnings or middle of seasons, and one event unexplained. His values for the Sun's elongation from the summer solstice point, obtained by an approximative graphical method, show discrepancies greater than 2° — in one case nearly 14°! — with respect to the numbers in my column v for all but five of the marks. Comparison of his diagram with a photograph shows that he has placed several marks inaccurately. Moreover, while in most instances he assumes that a mark to the right of the meridian means an event preceding the solstice and a mark to the left means after the solstice, contrary to this principle he has the event for Z before the solstice. Even so, the intervals between the supposed sign entries, which ought all to be approximately 30°, vary by his calculations between 26° and 38°.

i	ii	iii	iv	v	vi	vii	viii	ix
summer solstice	+68.5		1.000	0				
[A]	+62.5	+14.5	0.912	23.5	±1.0	66.5	113.5	
Γ	+53	+14	0.774	38.5	±0.7	51.5	128.5	15.0
Δ	+38	+21.5	0.555	55.5	±0.5	34.5	145.5	17.0
I	−7	+15.5	−0.102	96.0	±0.4	354.0	186.0	40.5
Λ	−17	+20	−0.248	104.8	±0.4	345.2	194.8	8.7
[N]	−37.5	+19.5	−0.547	124.0	±0.5	326.0	214.0	19.2
B	+58.5	−16	0.854	30.6	±0.8	59.4	120.6	
E	+36	−14	0.526	57.6	±0.5	32.4	147.6	26.9
Z	+22.5	−14.5	0.328	70.3	±0.5	19.7	160.3	12.8
H	+12	−14.5	0.175	79.6	±0.4	10.4	169.6	9.3
Θ	−2	−10.5	−0.029	91.7	±0.4	358.3	181.7	12.1
K	−8	−17	−0.117	96.9	±0.4	353.1	186.9	5.2
M	−21.5	−28.5	−0.314	108.8	±0.5	341.2	198.8	11.9
limit of uninscribed space above N	−59.5		−0.869	151.0	±0.8	299.0	299.0	

Table 1: Analysis of meridian marks on the Histria sundial.

dates, and it is unclear how one should apply them to entire constellations or to such an object as the Pleiades. Furthermore, the ancient sources seldom show awareness that visibility dates are dependent on geographical latitude.

Probably the best hope is to limit consideration to the phenomena that were regarded as the most fundamental markers of stages of the solar year, and that are attested in nonspecialist sources such as Hesiod and the Hippocratic writers, in particular the risings and settings of the Pleiades, Arcturus, and Sirius. We may begin with the morning rising of Sirius, the only one of this star's phenomena that was commonly invoked in lay sources. The "Geminus" parapegma, which dates the phenomena by the day number within the variable-length interval that the Sun spends in each zodiacal sign, gives the following (I give the zodiacal sign, day number, and length of the zodiacal "month" from the parapegma, and an approximate solar longitude derived from these data):

Authority	*Day*	*Total days*	*Solar longitude for sign*
Dositheos (Egypt)	Cancer 23	31	112°
Meton	Cancer 25	31	114°
Euctemon	Cancer 27	31	116°
Eudoxos	Cancer 27	31	116°
Callippos	Cancer 30	31	119°
Euctemon (again)	Leo 1	32	121°

According to modern theory (which ought at least to be a meaningful predictor of relative dates of visibility in different localities), Sirius' morning rising in the 2nd century B.C. would have occurred about seven days later at latitude 44° than at latitude 38°, which is a reasonable "average" northern Mediterranean latitude against which to calibrate the ancient sources;[19] thus for our sundial we would expect a mark corresponding to a solar longitude around 120° for this phenomenon. From Table 1 it would seem that A, B, or Γ could mark Sirius' morning rising, depending on whether the chronological sequence of the phenomena is to be read counterclockwise (A or Γ) or clockwise (B).

[19] I have used Alcyone Software's "Planetary, Stellar and Lunar Visibility" for these calculations (http://www.alcyone.de/).

We can carry out the same procedure for the three visibility phenomena of Arcturus that fall within the part of the solar year preserved on the sundial fragment, and for the four phenomena of the Pleiades, sometimes finding reasonably good matches on the hypothesis of a clockwise order, sometimes on the hypothesis of a counterclockwise order, sometimes on both hypotheses. The following are two conceivable reconstructions; for each phenomenon I give just the average estimated solar longitude derived from the data in the "Geminus" parapegma, and the adjustment predicted by modern theory:

I. Clockwise order (Pleiades ES and MS and Arcturus ES omitted).

Phenomenon	*Estimated longitude from parapegma*	*Latitude correction*	*Index letter*	*Estimated longitude from sundial*
Pleiades MR	47°	+6°	Γ	52°
Arcturus MS	66°	+21°	A	66°
Sirius MR	116°	+7°	B	121°
Arcturus MR	167°	+5°	H	170°
Pleiades ER	187°	−9°	K	187°

II. Counterclockwise order (Pleiades MS and Arcturus MR omitted).

Phenomenon	*Estimated longitude from parapegma*	*Latitude correction*	*Index letter*	*Estimated longitude from sundial*
Pleiades ES	12°	−1°	H	10°
Arcturus MS	66°	+21°	B	59°
Sirius MR	116°	+7°	A	114°
Pleiades ER	187°	−9°	I	186°
Arcturus ES	217°	+25°	N	214°

On balance, I think the clockwise reconstruction is preferable since it gives a more regular pattern by including all risings while excluding most settings. One might experiment with including other bright stars (such as Vega and Capella) that are frequently included in *parapegmata*, though I doubt whether one would get more conclusive results since the exercise becomes increasingly arbitrary as one goes beyond the most "essential" stars. It is noteworthy that, following either the clockwise or the counterclockwise reconstruction, applying the latitude corrections makes the agreement between the *parapegma* data and the sundial's marks worse, suggesting that the sundial was probably based on traditional data compiled in more southerly latitudes rather than from local observations.[20]

References

Arnaldi, M., 1999. Alla redazione de Gnomonica. Gnomonica, Arte, Storia, Cultura e Tecniche degli orologi solari 3, Maggio 1999, 7–8. www.nicolaseverino.it/Rivista%20Gnomonica/gnomonica3.pdf.

Breccia, E., 1911. Iscrizioni grechi e latine = Service des antiquités de l'Égypte. Catalogue géneral des antiquités égyptiennes du Musée d'Alexandrie N°s 1-568, Cairo.

Buchner, E., 1982. Die Sonnenuhr des Augustus: Nachdruck aus RM 1976 und 1980 und Nachtrag über die Ausgrabung 1980/1981. Philipp von Zabern, Mainz.

Diels, H., Rehm, A., 1904. Parapegmenfragmente aus Milet. Sitzungsberichte der königlich preussischen Akademie der Wissenschaften, philosophisch-historische Klasse 23, 92–111.

Evans, J., Berggren, J.L., 2006. Geminos's Introduction to the Phenomena. Princeton University Press, Princeton.

[20] I wish to thank an anonymous reader of this article for detecting systematic errors in my calculations relating to the Histria sundial.

Feraru, R.M., 2008. Nouvelles contributions à l'étude des cadrans solaires découverts dans les cités grecques de Dobroudja. Dialogues d'histoire ancienne 34/2, 65–80.

Freeth, T., Jones, A., Steele, J.M., Bitsakis, Y., 2008. Calendars with olympiad display and eclipse prediction on the Antikythera Mechanism. Nature 454, 614–617.

Gibbs, S., 1976. Greek and Roman Sundials. Yale University Press, New Haven.

Gruterus, J., 1707. Inscriptiones antiquae totius orbis Romani. F. Halma, Amsterdam.

Heslin, P., 2007. Augustus, Domitian and the so-called Horologium Augusti. Journal of Roman Studies 97, 1–20.

Hunt, D.W.S., 1940–1945. An archaeological survey of the classical antiquities of the Island of Chios carried out between the months of March and July, 1938. Annual of the British School at Athens 41, 29–52.

Ionescu-Cârligel, C., 1969. Cadrane solare grecesti si romane din Dobrogea. Pontica 2, 199–208.

——— 1970. Contributions à l'étude des cadrans solaires antiques. Dacia : revue d'archéologie et d'histoire ancienne N.S. 14, 119–137.

Price, D.J.d.S., 1974. Gears from the Greeks. Transactions of the American Philosophical Society N.S. 64.7, Philadelphia.

Rehm, A., 1906. Meteorologische Instrumente der Alten. Unpublished manuscript written as chapter I of Untersuchungen über die meteorologischen Theorien der griechischen Altertums auf Grund der literarischen und monumentalen Überlieferung, Bayerische Staatsbibliothek, Rehmiana III/7.

Schaldach, K., 2011. Eine seltene Form antiker Sonnenuhren: Der Meridian von Chios. Archäologisches Korrespondenzblatt 41, 73–83.

An Archimedean Proof of Heron's Formula for the Area of a Triangle: Heuristics Reconstructed

Christian Marinus Taisbak

Abstract I believe, as did al-Bīrūnī, that Archimedes invented and proved Heron's formula for the area of a triangle. But I also believe that Archimedes would not multiply one rectangle by another, so he must have had a another way of stating and proving the theorem. It is possible to "save" Heron's received text by inventing a geometrical counterpart to the un-Archimedean passage and inserting that before it, and to consider the troubling passage as Archimedes' own translation into terms of measurement. My invention is based on a reconstruction of the heuristics that led to the proof.

I prove a crucial lemma: If there are five magnitudes of the same kind, a, b, c, d, m, and m is the mean proportional between a and b, and $a : c = d : b$, then m is also the mean proportional between c and d.

Introduction

A triangle is the mean proportional between two rectangles, *one* of which is contained by the semiperimeter of the triangle and the semiperimeter diminished by one of the sides, whereas *the other* is contained by the semiperimeter diminished by either of the remaining sides.

The statement above is a reconstruction, seen nowhere in the received sources. In Heron's text[1] we learn that the area of the triangle is the side of a square equal to one of the said rectangles multiplied by the other one. Obviously, this kind of statement is alien to standard Greek geometry: a side, a line segment, cannot be equal to an area, and multiplication of rectangles cannot be represented by a square — so it must needs be understood arithmetically, and no wonder, since mensuration is what it was meant for. The invention of the proof is attributed to Archimedes by al-Bīrūnī, but E.J. Dijksterhuis [1956, 412] had "some doubt whether the proof in the form in which it is quoted by Heron, can really originate from Archimedes." That doubt is quite legitimate as to the form, but then it is also legitimate to guess an answer to the question: Since the theorem is proved by sound propositions from the *Elements*, how would Archimedes state the theorem geometrically?

An answer to that depends, I am sure, on a reconstruction of the heuristics that led to the proof. Below I venture such a reconstruction, and I propose a "missing" geometrical passage which will "save" the peculiar arithmetical statement. I find it quite tenable that Archimedes, after the geometrical part, himself "translated" it into arithmetic to serve its purpose of mensuration. For all we know, Archimedes did not hesitate to put (approximate) numbers to lines' lengths, e.g. in his *Mensuration of a Circle* 3.

After some typographical conventions, I present my analysis and heuristics, followed by commented translations of *Metrika* I.7 and I.8. You might want to read the translations first to form your own

A revised and extended version of my paper on this subject that appeared in *Centaurus* (which, besides being marred by a glaring erratum, contained no heuristics) [Taisbak 1980].

[1] *Metrika* I 8 [Schöne 1903, 22, ll. 15–19; 24, ll. 10–21].

N. Sidoli and G. Van Brummelen (eds.), *From Alexandria, Through Baghdad: Surveys and Studies in the Ancient Greek and Medieval Islamic Mathematical Sciences in Honor of J.L. Berggren*, DOI 10.1007/978-3-642-36736-6_9,
© Springer-Verlag Berlin Heidelberg 2014

opinion about a possible analysis leading to his synthesis.

Typographical Conventions

A *triangle ABC* is denoted *ABC*. Its *angle B* is denoted $\angle B$ or $\angle ABC$.

A rectangle with sides $AB = l$ and $BC = w$ is denoted $(AB \cdot BC)$ or $(l \cdot w)$. If AB and BC are numbers (i.e. lengths), $AB.BC$ denotes their product. The context will guide us.

The geometrical square on (i.e. with side) $AB = q$ is denoted $AB^{\square}$ or $q^{\square}$. The arithmetical square of (the number) AB is denoted AB^2.

The *ratio* of two homogeneous magnitudes[2] A and B is written $A : B$. A *proportion* "A is to B as C to D" is written $A : B = C : D$.

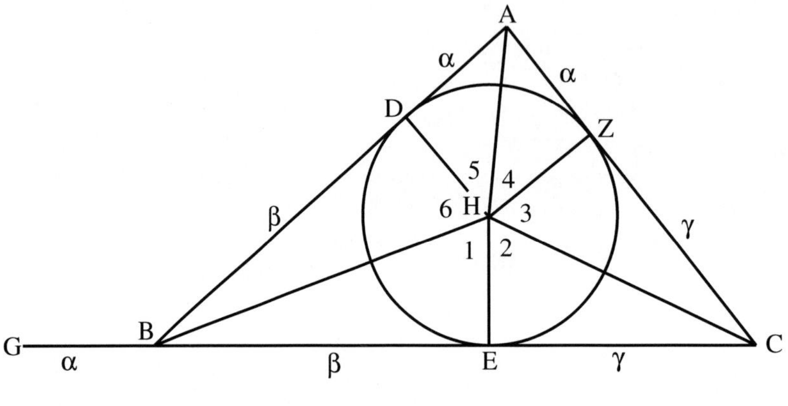

Figure 1: Relations of the semiperimeter.

The *semiperimeter* of a triangle *ABC* with sides a, b, c, is the sum $(a + b + c)/2 = s$. In Figure 1, $\beta + \gamma = a$, $\gamma + \alpha = b$, $\alpha + \beta = c$. Adding these equations we have $2\alpha + 2\beta + 2\gamma = a + b + c = 2s$, so that $\alpha + \beta + \gamma = s$, and

$$\alpha = s - a, \beta = s - b, \gamma = s - c.$$

To visualize s, CB is prolonged to G with $BG = AD$, so we get $BG = AD = \alpha$, $BE = \beta$, $CE = \gamma$, and therefore $CG = s$.

Historical warning: The lower-case letters a, b, c, l, s, w, α, β, γ should be understood as names of line segments, not (real) numbers. The Greeks would use BC, CA, AB, etc. In certain parts of his propositions, Heron will think of them as (approximate) lengths. I prefer lower-case letters for the sake of readability, running the risk of misinterpretation.

Analysis and Heuristics

With these conventions the opening statement — a triangle *ABC* is the mean proportional between two rectangles — can now be written as

$$(s \cdot \alpha) : ABC = ABC : (\beta \cdot \gamma),$$

[2] The homogeneous magnitudes involved are straight line segments, triangles, or rectangles (squares included).

and it is time to disclose that, arithmetically, this is equivalent to the formula

$$\text{Area of } ABC = \sqrt{s(s-a)(s-b)(s-c)}.$$

Lemma 1: If the radius of the triangle's incircle is r, the triangle is equal to (i.e. has the same area as) $(r \cdot s)$. This is well known, and was proved by Heron [Schöne 1903, 22, ll. 2–10].

Lemma 2: A rectangle is the mean proportional between the squares on its sides. This can be inferred from *Elements* X.53, lemma, which states that the mean proportional between two squares is the rectangle contained by their sides. *Lemma* 2 can also be proved by *Elements* VI.1 (see Figure 2):

$$l : w = l^{\square} : (l \cdot w) = (l \cdot w) : w^{\square}.$$

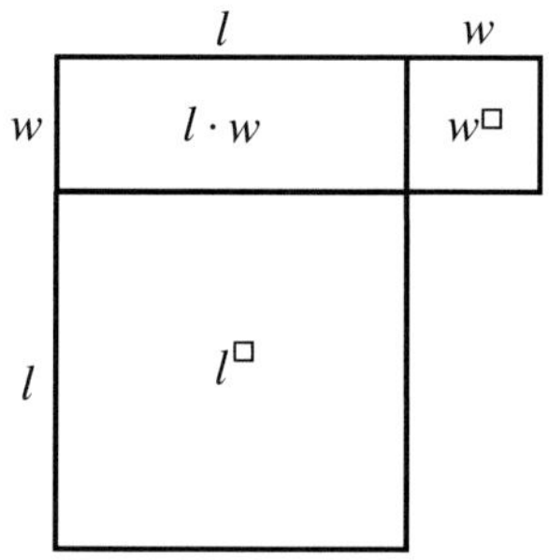

Figure 2: *Lemma 2.*

Lemma 3.1: If m is the mean proportional between a and b, there exist (infinitely many) magnitudes c and d, such that m is also the mean proportional between c and d. For line segments and rectangles this can be proved by *Elements* VI.12, which shows how to find a fourth proportional to three given line segments. Hence, we need not bother about the existence of a fourth proportional; the Greeks never did.[3]

Lemma 3.2: If m is the mean proportional between a and b, and if m is also the mean proportional between c and d, then $a : c = d : b$ (inverse proportion).

Suppose that

$$a : m = m : b, \tag{1}$$

and that

$$c : m = m : d,$$

then *enallax*[4] [*Elements* V def. 12],

$$m : c = d : m,$$

[3] While discussing "The Distinctive Assumptions of Book V" (of the *Elements*), Ian Mueller wrote [1981, 139, n. 24], "This explanation is put forward and developed by Becker in 'Warum haben die Griechen die Existenz der vierten Proportionale angenommen?'"

It seems clear that no Greek ever questioned this "assumption of the existence of a fourth proportional," perhaps because the use was not noticed, but more probably because the existence of such a proportional to three given geometrical objects was considered obvious on the basis of intuitive ideas about continuity.

[4] Heath [1926] refers to this operation with the expression "alternately."

and by *perturbed analogy* [*Elements* V.23]

$$a : c = d : b. \tag{2}$$

And conversely (the crucial lemma in the heuristics of *Metrika* I.8): If there are five magnitudes of the same kind, a, b, c, d, m, and m is the mean proportional between a and b, and $a : c = d : b$, then m is also the mean proportional between c and d. This is proved by taking (2) alternatively with (1) and using *Elements* V.23.[5]

Lemma 4: In Figure 1, $\angle BHC + \angle AHD = 2$ right angles. Since $\angle 1 = \angle 6$, $\angle 2 = \angle 3$, and $\angle 5 = \angle 4$, thus,

$$\angle 1 + \angle 2 + \angle 5 = \angle 6 + \angle 3 + \angle 4.$$

But since the six are equal to 4 right angles,

$$\angle 1 + \angle 2 + \angle 5 = \angle BHC + \angle AHD = 2 \text{ right angles}.$$

This lemma is proved by Heron [Schöne 1903, 22, ll. 23–28].

Lemma 5: [*Elements* VI.8, corollary] If, in a right triangle, a perpendicular is drawn from the right angle to the base (hypotenuse), the perpendicular is the mean proportional between the segments of the base. That is, the square on the perpendicular is equal to the rectangle contained by the segments of the base.

In Figure 3, let $\angle CHK$ be a right angle. Then $HE^{\square} = (KE \cdot EC)$, which we rename (for readability)

$$r^{\square} = (\varepsilon \cdot \gamma).$$

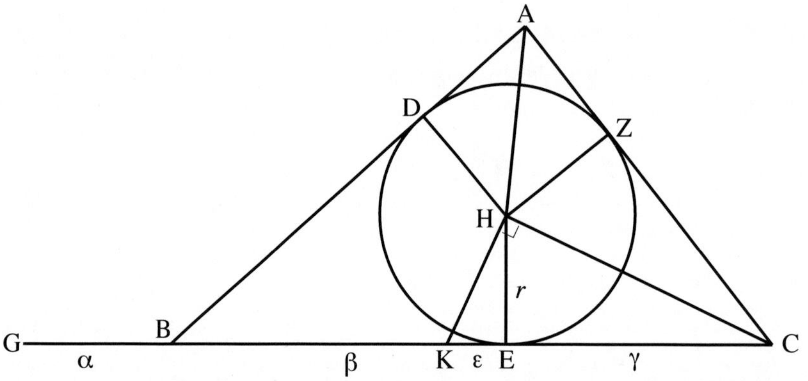

Figure 3: *Metrika* I 7 & 8, preliminary figure.

With these lemmas in mind (and let me emphasize that *Lemma* 3, as far as I know, is not known from any received text, but inspired by Book X of the *Elements*) we may turn to the heuristic proper: We learned by *Lemma* 1 that the triangle ABC is equal to $(r \cdot s)$, and therefore, by *Lemma* 2, that ABC is the mean proportional between $s^{\square}$ and $r^{\square}$. That is,

$$s^{\square} : ABC = ABC : r^{\square},$$

so that, by *Lemma* 5,

[5] When real numbers are invented, *Lemma* 3.2 becomes trivially obvious.

$$s^{\square} : ABC = ABC : (\varepsilon \cdot \gamma). \tag{1.0}$$

This first result of our heuristics shows a rectangle involving $\gamma = CE = s - c$. We might want to involve $\beta = BE = s - b$, so we use *Lemma* 3 to see what will happen to $s^{\square}$ if, in (1.0), we substitute $(\beta \cdot \gamma)$ for $(\varepsilon \cdot \gamma)$, to get the proportion

$$(s \cdot ?) : ABC = ABC : (\beta \cdot \gamma).$$

Now, since $(\beta \cdot \gamma) > (\varepsilon \cdot \gamma)$, the rectangle $(s \cdot ?) < s^{\square}$, according to *Lemma* 3.2 (inverse proportion). Let us consider a rectangle $(s \cdot z)$ with $z < s$, so that

$$(s \cdot z) : ABC = ABC : (\beta \cdot \gamma). \tag{1.1}$$

Is z given by this proportion if the sides of ABC are given?

According to *Lemma* 3.2

$$s^{\square} : (s \cdot z) = (\beta \cdot \gamma) : (\varepsilon \cdot \gamma), \tag{1.2}$$

and by "cancelling" [*Elements* VI.1],

$$s : z = \beta : \varepsilon. \tag{1.3}$$

To a Greek experienced in handling proportions, this is very inviting because ε is part of β (KE is part of BE). Therefore, *diairesis logou* [*Elements* V def. 15], subtraction within the ratio, will render a new proportion:

$$s - z : z = \beta - \varepsilon : \varepsilon = BK : EK. \tag{1.4}$$

I imagine that here the analyser gets the (b)right idea: to make BK the side of a right triangle similar to KEH by prolonging HK to meet the perpendicular to CG from B in L (see Figure 4). Then, $BL : EH = BK : EK$, which we rename

$$\delta : r = \beta - \varepsilon : \varepsilon. \tag{1.5}$$

The geometer will see immediately that CL, if joined, subtends two right angles, $\angle LBC$ and $\angle LHC$, and thus is a diameter in a circle passing through H and B, inviting the following arguments about angles and similar triangles. Since the quadrilateral $CHBL$ is inscriptible in a circle, the opposite angles $\angle BHC$ and $\angle BLC$ are together 2 right angles [*Elements* III.22], but so are also $\angle BHC + \angle AHD$, by *Lemma* 4 [Schöne 1903, 22, 22–28].

Therefore $\angle AHD = \angle CLB$, and the triangle AHD is similar to CLB. Among other properties, this renders

$$BL : DH = BC : DA,$$

that is

$$\delta : r = \beta + \gamma : \alpha. \tag{1.6}$$

Proportions (1.4), (1.5), and (1.6) and the transitivity of ratio [*Elements* V.11] ensure that

$$s - z : z = \beta + \gamma : \alpha,$$

inviting *synthesis logou* [*Elements* V def. 14], to get a new proportion,

$$s : z = \alpha + \beta + \gamma : \alpha.$$

But $\alpha + \beta + \gamma = s$, and therefore the unknown $z = \alpha$, which was probably what the analyser hoped for, to be able to rewrite (1.1), in our terms,

$$(s \cdot \alpha) : ABC = ABC : (\beta \cdot \gamma).$$

That is, the triangle ABC is the mean proportional between two rectangles:

$$R_1 = (s \cdot \alpha) \text{ and } R_2 = (\beta \cdot \gamma).$$

We are now ready to read *Metrika* I.7 and I.8, and to put them into a form that both respects geometry and is useful for mensuration — as we are entitled to believe was Archimedes' (or Whoever's) intention.

Comment: *Metrika* I.7 is a parenthesis in a series of theorems about how to find the area of triangles. It is meant to explain a surprising passage in *Metrika* I.8. At the same time, it gives us an idea of how Heron thought of, and handled, numbers. It is well-known that some of our numerical terminology was born in geometry: numerical multiplication is *thought of*, but not *illustrated* geometrically; the numbers are lengths of straight line segments, their product is thought of as the rectangle "contained by" the straight lines, as defined in *Elements* II def. 1. Particularly, the square *on* a line represents the square *of* the number;[6] "square root" translates πλευρά, literally "side of the square." But the crux in this proposition, and the next, is that the operations transcend the geometrical representation. How can the product of squares be represented, since geometry has no fourth dimension? That is what this proposition is about: behind these arithmetical statements lurks our *Lemma* 2, that any rectangle is the mean proportional between two squares, the squares on its sides. It is worth mentioning that Euclid, in his number-theoretical books *Elements* VII–IX never illustrates products by rectangles, but always by line segments — in that way, using lines as we use the alphabet to denote random numbers.

Heron's *Metrika* I.7

We turn now to a commented translation of Heron's text [Schöne 1903, 16–24].

> If there be two numbers AB and BC, then the square root of the-square-of-AB-multiplied-by-the-square-of-BC will be the product $(AB \cdot BC)$.[7] For, since [S 18] as AB is to BC, so is the square of AB to the product $(AB \cdot BC)$, and also the product $(AB \cdot BC)$ to the square of BC, therefore also as the square of AB is to the product $(AB \cdot BC)$, so will the product $(AB \cdot BC)$ be to the square of BC.
>
> But if three numbers are in proportion (ἀνάλογον), the product of the extremes will be equal to the square of the mean. Therefore the square of AB multiplied by the square of BC will be equal to the number $(AB \cdot BC)$ multiplied by itself. Therefore the square root of the product of the-square-of-AB-and-the-square-of-BC will be the number $(AB \cdot BC)$.

Assertion: $\sqrt{AB^2 \cdot BC^2} = (AB \cdot BC)$. (Heron tells us that the square root of a square number n^2 is n.)

Proof: Since $AB : BC = AB^2 : (AB \cdot BC)$ and $AB : BC = (AB \cdot BC) : BC^2$, therefore $AB^2 : (AB \cdot BC) = (AB \cdot BC) : BC^2$ (transitivity of ratio [*Elements* V.11]).

Definition: If $p^2 : pq = pq : q^2$, then the three numbers p^2, pq, q^2 are said to be *analogon*, in (continuous) proportion. That is, pq is the mean proportional between p^2 and q^2, and $(pq)^2 = p^2 \cdot q^2$.

By *Elements* VI.17 and VII.19,

$$AB^2 \cdot BC^2 = (AB \cdot BC) \cdot (AB \cdot BC)$$

That is,

$$\sqrt{AB^2 \cdot BC^2} = (AB \cdot BC).$$

[6] If, in Heron's text, AB is understood to be a number, I translate with the "square *of* AB."

[7] Literally, "the number contained by ABC" (τὸν ὑπὸ ΑΒΓ περιεχόμενον ἀριθμόν). In the next sentence, by a fairly standard practice of ellipsis, this becomes "the by ABC" (τὸν ὑπὸ ΑΒΓ).

Heron's *Metrika* I.8

There is, however, a general method[8] to find the area of any triangle without [knowing] a height if [the] three sides are given. An example: let the [lengths of the] sides of the triangle be 7, 8, and 9 units. Add together 7 + 8 + 9, that is 24. Take half of them: 12. Subtract 7 units, 5 left; then, subtract 8 from 12, 4 left. And also [subtract] 9, 3 left. Multiply 12 by 5, result 60; and those by 4, result 240; and those by 3, total 720. Extract the square root, which will be the area of the triangle. Since, however, 720 has no rational root, we will take its root with the least difference in the following way: Because the nearest square to 720 is 729 and has root 27, divide 720 by 27, that is 26 and two thirds; add the 27, that is 53 two thirds. Take half of that, 26 $\frac{1}{2}$ $\frac{1}{3}$. Therefore the square root of 720 is approximately 26 $\frac{1}{2}$ $\frac{1}{3}$, for 26 $\frac{1}{2}$ $\frac{1}{3}$ multiplied by itself makes 720 $\frac{1}{36}$, such that the difference [S 20] is $\frac{1}{36}$. But if we want the difference expressed in a lesser part than $\frac{1}{36}$, we may use the value just found, 720 $\frac{1}{36}$ instead of 729; and by so doing we will find that the difference becomes much less than $\frac{1}{36}$.[9] The geometrical proof for that is the following:

To find the area of a triangle, given its sides.

It is of course possible to draw one height and calculate its length and find the area of the triangle, but now we must calculate the area without [knowing] the height. [S 22] Let the given triangle be *ABC*, and let each of [the sides] *AB*, *BC*, *CA* be given; to find the area. [See Figure 4.]

Let the incircle *DEZ* with centre *H* be inscribed in the triangle, and let *AH*, *BH*, *CH*, *DH*, *EH*, *ZH* be joined.

Now the rectangle ($BC \cdot EH$) is double the triangle *BHC* [*Elements* I 41], the rectangle ($CA \cdot ZH$) is double the triangle *CHA*, and the rectangle ($AB \cdot DH$) is double the triangle *AHB*. Therefore the rectangle contained by the perimeter of the triangle *ABC* and *EH*, viz. the radius of circle *DEZ*, is double the triangle *ABC*.

Let *CB* be produced [to *G*] and let *BG* be made equal to *AD*; thus [the straight line] *CBG* is half the perimeter of the triangle *ABC* because $AD = AZ$, $BD = BE$, and $CZ = CE$, and so the rectangle ($CG \cdot EH$) is equal to the triangle *ABC*.

The next passage[10] is the one that troubled Dijksterhuis [1956, 412], among others, although we should be warned by *Metrika* I.7, above. Dijksterhuis rightly commented that the "squares on *CG* and *EH*" have lost their direct geometrical meaning and are looked upon as dimensionless magnitudes (or numbers) which can be squared in their turn. I think that the passage can be understood by inventing its geometrical counterpart (marked < ... >), inserting that before it, and considering the troubling passage as Archimedes' own translation into terms of measurement, in accordance with what we learned in *Metrika* I.7.

But < since any rectangle is the mean proportional between the squares on its sides, the rectangle ($CG \cdot EH$) is the mean proportional to the square on *CG* and the square on *EH*. Thus the triangle *ABC* is the mean proportional to the square on *CG* and the square on *EH*. Therefore > the rectangle ($CG \cdot EH$) is the side [i.e. square root] of the square of *CG* multiplied by the square on *EH*; thus the area of the triangle *ABC* multiplied by itself is equal to the square of *CG* multiplied by the square on *EH*.

As a matter of fact, it is safe to invent a geometrical counterpart because the following reasonings are perfectly geometric and in accordance with the theory of magnitudes and proportion in the *Elements*. In most texts in Greek geometry, analysis and heuristics are suppressed and only a synthesis is presented; such is also the case here: Heron now starts a construction at random, it seems, conjuring up a very informative diagram (Figure 4) that sequentially proves the whole thing. It is, however, instructive (and often very entertaining) to try to reconstruct the heuristics by turning the synthesis upside down. I hope to have done so above in the introduction.

Let *HL* have been drawn at right angles with *CH*, and *BL* with *CB*, and let *CL* be joined. Since either of the angles *CHL*, *CBL* is right, [*CL* is a diameter and] the quadrilateral *CHBL* is [inscriptible] in a circle, and so the angles *CHB* and *CLB* are [together] equal to two right angles [*Elements* III.22].

---

[8] That is, besides the various methods shown in the previous theorems.

[9] Heron's method can be understood as follows: If q^2 is the square nearest to n, we have $n = q^2 \pm r$, and $\sqrt{n} = q \pm f$, $(f < 1)$. So $n = q^2 \pm 2qf + f^2 = q^2 \pm r$. If we ignore f^2, we have $r \approx 2qf$, and so $f \approx r/2q$. This method was known from Babylonian mathematics and was probably used at all times.

Heron uses the Egyptian concept of unit fractions instead of 26 $\frac{5}{6}$, a normal practice in Hellenistic arithmetic.

[10] Schöne [1903, 22, ll. 15–19].

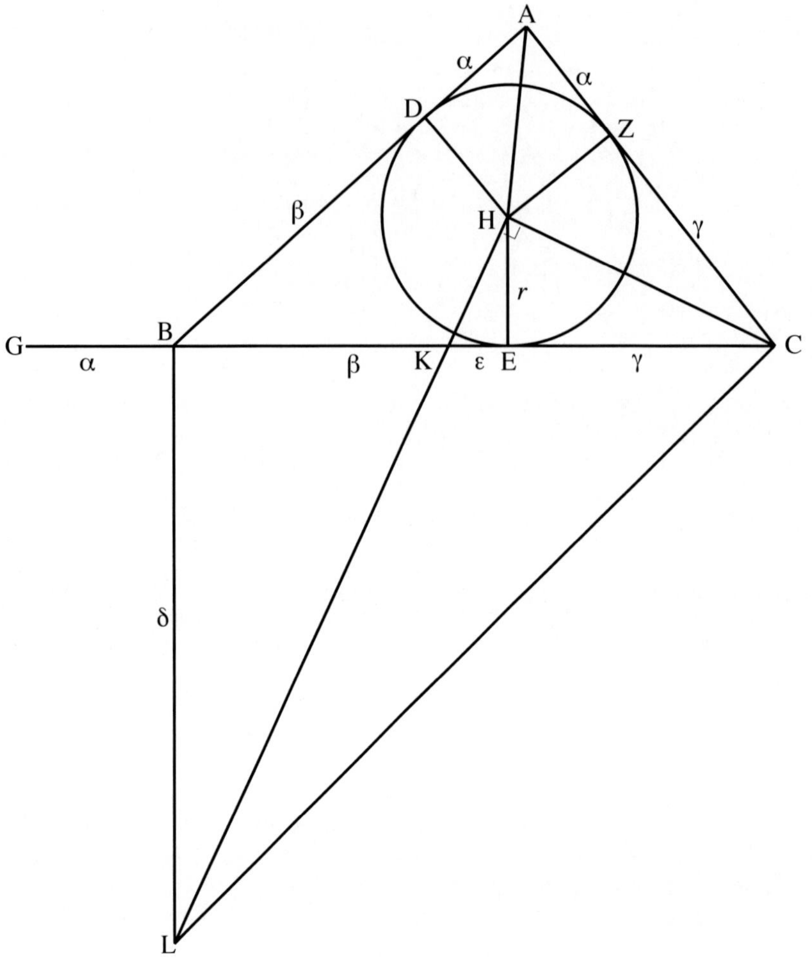

Figure 4: *Metrika* I.7 and I.8.

Why does he construct two right angles, $\angle CHL$ and $\angle CBL$? Oh, yes, to ensure an inscriptible quadrilateral, this is a good idea because the similar triangles below, *AHD* and *CLB*, seem to drop out — if not out of the blue, then out of the quadrilateral at any rate. If we only knew what they are good for. This reticence about what we are heading for is one of the most charming and irritating features in Hellenistic mathematics.[11]

> But also the angles *CHB* and *AHD* are [together] equal to two right angles; for the angles at the centre *H* are halved by *AH*, *BH*, *CH*, and the angles *CHB* and *AHD* are [together] equal to the angles *AHC* and *DHB* [together], and the sum of all of them equals four right angles. Therefore the angle *AHD* is equal to *CLB*. And the right angle *ADH* is equal to the right angle *CBL*; [S 24] thus the triangle *AHD* is similar to the triangle *CBL*. Therefore, as *CB* is to *BL*, so is *AD* to *DH*, that is as *BG* to *EH*, and *enallax* as CB is to BG, so is BL to EH [*Elements* V def. 12], that is *BK* to *KE*, because *BL* is parallel to *EH* [*Elements* VI.4]. And *synthenti*, as *CG* is to *BG*, so is *BE* to *EK* [*Elements* V.18].

That is,

$$\beta + \gamma : \delta = \alpha : r$$

and *enallax*,

[11] The reticence of the ancient mathematicians has been much discussed by early modern mathematicians and modern scholars, but Netz [2009] has recently discussed it in some detail.

$$\beta + \gamma : \alpha = \delta : r = \beta - \varepsilon : \varepsilon,$$

and *synthenti*,

$$\alpha + \beta + \gamma : \alpha = \beta : \varepsilon.$$

But, since $\alpha + \beta + \gamma = s$,

$$s : \alpha = \beta : \varepsilon.$$

> Therefore also, as the square on *CG* is to the rectangle $(CG \cdot GB)$, so is the rectangle $(BE \cdot EC)$ to [the rectangle] $(CE \cdot EK)$ [*Elements* VI.1], that is to the square on *EH*, for *EH* is drawn in a right triangle perpendicular from the right angle to the base [i.e. hypotenuse, *Elements* VI.8, corollary].

That is,

$$s^{\square} : (s \cdot \alpha) = (\beta \cdot \gamma) : (\varepsilon \cdot \gamma) = (\beta \cdot \gamma) : r^{\square}.$$

> Therefore the square of *CG* multiplied by the square of *EH*, the square root of which [product] is the area of the triangle *ABC* [because *ABC* is the mean proportional between those squares], is equal to the rectangle $(CG \cdot GB)$ multiplied by the rectangle $(CE \cdot EB)$.

That is,

$$(ABC \cdot ABC) = s^2 \cdot r^2 = (s \cdot \alpha) \cdot (\beta \cdot \gamma).$$

> And each of [the segments] *CG*, *GB*, *BE*, *CE* is given, for *CG* is half the perimeter of the triangle *ABC*, *BG* is the difference between half the perimeter and *CB*, *BE* is the difference between half the perimeter and *AC*, and *EC* is the difference between half the perimeter and *AB*, because $EC = CZ$, $BG = AZ = AD$. Thus the area of the triangle *ABC* is given.

As often, Heron ends with a *synthesis*, in geometry meant as a constructive demonstration of the validity of the proposition. In this case, however, the *synthesis* is simply a numerical example, which does not prove any validity unless one calculates the area of the said triangle by another method. He may have thought of that, however, when choosing the lengths of the sides: I suspect that he knew how to find triangles with sides of integer length, by first finding two right triangles with one side of equal length; *in casu* 5, 12, 13 and 9, 12, 15. A method to find such triangles (of which there are infinitely many, even with prime lengths) was well known in Hellenistic mathematics. The length 12 is the height of the triangle on the base 14.

> It is calculated in the following way: Let *AB* be 13 units, *BC* 14 units, and *AC* 15 units. Add 13, 14 and 15, and 42 results; of which half becomes 21. Subtract 13, 8 remain; the same with 14, 7 remain; and lastly 15, 6 remain. 21 by 8, and the product by 7, and yet again the product by 6, 7056 results. The square root thereof is 84; so big will the area of the triangle be.

Epilogue

I have no doubt that this theorem was meant, stated and proved as a genuine geometric proposition, and then — when applied in mensuration, which of course was its *raison d'être* — summed up in an arithmetical style. Obvious relatives are the propositions in Archimedes' *Mensuration of a Circle*, and like them it is more than probable that the text underwent several "emendations" on its way to classrooms. However, Heron seems very painstaking, in *Metrika* I.7, in preparing our minds for the obnoxius concept of multiplying a square by a square and finding the "side" of such a monster-square. It remains (to me, at least) a wonder when looking into Hellenistic mathematics why millenia had to pass before arithmetic got a footing as solid as, or more than, Euclid's geometry. Why didn't it trouble them? But then, what do I know about the troubles they've seen?

References

Becker, O., 1933. Eudoxos-Studien II. Warum haben die Griechen die Existenz der vierten Proportionale angenommen? Quellen und Studien zur Geschichte der Mathematik, Astronomie, und Physik, Abteilung B, 2, 369–387.

Dijksterhuis, E.J., 1956. Archimedes. Ejnar Munksgaard, Copenhagen.

Heath, T.L., 1926. The Thirteen Books of Euclid's Elements. Cambridge University Press, Cambridge.

Mueller, I., 1981. Philosophy of Mathematics and Deductive Structure in Euclid's Elements. MIT Press, Cambridge.

Netz, R., 2009. Ludic Proof. Cambridge University Press, Cambridge.

Schöne, H., 1903. Heron von Alexandria: Vermessungslehre und Dioptra, Heronis Alexandrini opera quae supersunt omnia, vol. 3. Teubner, Leipzig.

Taisbak, C.M., 1980. An Archimedean proof of Heron's Formula for the area of a triangle; Reconstructed. Centaurus 24, 110–116.

Reading the Lost Folia of the Archimedean Palimpsest: The Last Proposition of the *Method*

Ken Saito and Pier Daniele Napolitani

Abstract We examine the determination of the volume of solids by means of a virtual balance, in Archimedes' *Method*, a work preserved only in the Archimedes palimpsest.

We concentrate on the intersection of two cylinders, whose volume is correctly stated in the preface, although the folia containing the proof are lost. All the reconstructions of Archimedes' argument by previous scholars concerning the volumes of the intersection of cylinders are unsatisfactory, for the description in the preface states that the argument will contain a demonstration (not only an heuristic argument by the virtual balance), while the reconstruction of the configuration of the quires of the palimpsest shows that the space for this argument was no more than three pages — too short for a rigorous demonstration.

We take up Reinach's remark that the intersection of cylinders can be divided into eight equal hoofs whose volume has been determined in previous propositions, and we argue that Archimedes' demonstration consisted of decomposing the intersection of cylinders into hoofs.

This means that Archimedes was not aware of the fact that the hoof, the sphere and the intersection of cylinders can be treated in much the same way by virtual balance, although the process of the determination of their volumes is similar. Archimedes did not try to extract some quantitative property common to these three solids and to treat them in a similar way. Thus, Archimedes was a much more "classical" geometer than we tend to assume.

In the final part, we suggest that Archimedes' invention of the problem of determination of the intersection of cylinders may have been inspired by some building existing at his time. Indeed, excavations in Morgantina, Sicily, have revealed two barrel vaults arranged at a right angle, which can be dated to Archimedes' time. This construction may well have given him the idea to consider the intersection of two cylinders.

Appendix 1 shows the outline of propositions in the *Method* concerning the sphere and the hoof, with suggestions of how Archimedes may have found certain arguments. It also contains reconstructions by previous scholars as to the volume of intersection of cylinders.

Appendix 2 explains how one can infer, with pretty high certainty, the number of lost folia by the reconstruction of the quires of parchment.

Appendix 3 shows a schematic presentation of the three quires of the palimpsest (with the reconstruction of lost folia) which contains the *Method*.

Introduction

The *Method* is the work in which Archimedes sets out his way of finding the areas and volumes of various figures. It can be divided into three parts. The first part is the preface addressed to Eratosthenes, in

The authors thank Paolo d'Alessandro for his valuable advices on codicological issues treated in this article.

N. Sidoli and G. Van Brummelen (eds.), *From Alexandria, Through Baghdad: Surveys and Studies in the Ancient Greek and Medieval Islamic Mathematical Sciences in Honor of J.L. Berggren*, DOI 10.1007/978-3-642-36736-6_10,
© Springer-Verlag Berlin Heidelberg 2014

which Archimedes explains his motivation for writing the work. We find that he was sending demonstrations of results that he had communicated before — the volume of two novel solids, which we call "hoof" and "vault" in this article.[1]

As Archimedes thought that it was a good occasion to reveal his way of finding results that he had previously published with rigorous demonstration, he decided to include an exposition of this "way" (*tropos* in Greek, not method, as is usually assumed in modern accounts.)[2]

Thus, the first eleven propositions show how the results in his previous works (*Quadrature of the Parabola*, *Sphere and Cylinder* and *Conoids and Spheroids*) were found. We call this group of propositions the second part of the work.

The third and last part, beginning with Prop. 12, treats the two novel solids and gives a demonstration of their volumes. Unfortunately, the end of the *Method* is lost. As is well known, the *Method* is known only through the palimpsest found in 1906, and some pages had already been lost. The text of the *Method* breaks off definitively near the end of the demonstration of the volume of the hoof, the first of the two novel solids announced in the preface. We have no testimony concerning how Archimedes demonstrated the volume of the vault, the second novel solid.

In this article, we try to reconstruct this lost demonstration, based on recent studies made after the reappearance of the palimpsest in 1998.

The Archimedean "Way" of Finding Results

First, let us briefly look at the "way" that Archimedes presents in this work. In this section, we will see the simplest case of the paraboloid, and an application to the sphere.

The Simplest Example: Paraboloid (Prop. 4)

The simplest example can be found in Prop. 4, where Archimedes compares a paraboloid to the cylinder circumscribed about it[3] The paraboloid *BAG* having axis *AD*, is cut by a plane *MN*, perpendicular to its axis.[4] Archimedes shows that the segment *BAG* is half the cylinder circumscribed about it. By the property of the parabola, the following proportion holds:

$$\text{circle } CO : \text{ circle } MN = \text{sq}(CS) : \text{sq}(MS) = SA : AD.$$

Let us imagine a cylinder *BGHE*, circumscribed about the segment of the paraboloid *BAG*. Prolong axis *DA* to *Q*, so that *QA* is equal to *DA*, and imagine a balance *DAQ* whose fulcrum is the point *A*. Then, the above proportion means that if the circle *CO* (section of the segment of parabola) is moved to point *Q*, it is in equilibrium on the balance with the circle *MN* (section of the cylinder) remaining in place.

[1] We use the words "area" and "volume" only for the sake of convenience. Archimedes did not use these words, and his results were stated as comparisons between figures. For example he says, "the surface," not the area of the surface, "of any sphere is four times its greatest circle." This is a characteristic feature of Greek theoretical geometry, common at least to Euclid, Archimedes and Apollonius. The use of words like length, area and volume is based on the possibility of expressing geometrical magnitudes by real, positive numbers, taking any magnitude of the same dimension as unit. However, this was not possible for Greek geometers who did not have the concept of real number.

[2] The manuscript gives the title *ephodos*, not *methodos*, to this work which is usually called the *Method*, following Heiberg. Moreover, neither *ephodos* nor *methodos* appears in the preface or the text of this work. Archimedes always uses the word *tropos* to refer to his "method" of virtual balance which is discussed in the present paper. See Knobloch [2000, 83].

[3] The manuscript does not have proposition numbers. We use the propositions numbers in Heiberg [1910–1915].

[4] The diagrams of solid figures found in the manuscript are always planar, like Figure 1 (A), and we have often provided perspective drawings like Figure 1 (B).

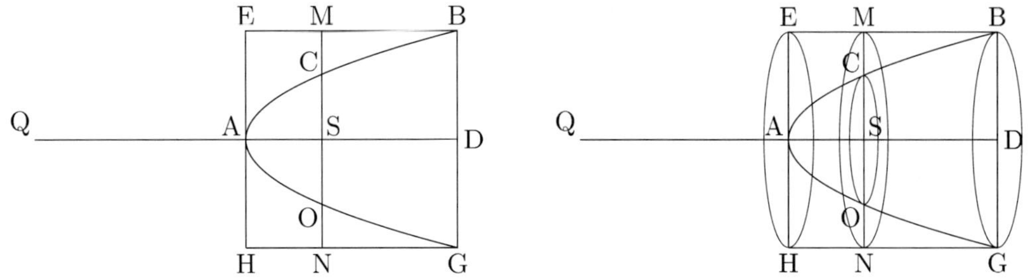

Figure 1: (A) *Method* Prop. 4. (B) Perspective diagram.

If all the sections of the paraboloid are thus moved and balanced, then the whole paraboloid, moved to point *Q*, is in equilibrium with the cylinder remaining in place.[5] As the barycenter of the cylinder is the midpoint of the axis *AD*, it follows that the (volume of the) paraboloid is half the cylinder.

This argument works because all the sections of the paraboloid are moved to one and the same point *Q*, while all the sections of the cylinder remain in place. This is possible because the circle sections of the paraboloid, such as the circle *CO*, increase in direct proportion with the distance *AS* from the vertex *A*, which is the fulcrum of the balance.[6]

Sphere: Invention of an Auxiliary Solid (Prop. 2)

Let us look at another, slightly more complicated proposition. Prop. 2 determines the volume of the sphere. In the following, we present the outline of Archimedes' argument, which is described in more in detail in Appendix 1, Prop. 2: Sphere.

Let the sphere *AG* be cut by a plane *MN*, perpendicular to the diameter *AG*. The section of the sphere is circle *CO*. This section is by no means in proportion to the distance from the point *A*.

Archimedes then adds a cone, *AEZ*, having as height the diameter of the sphere, *AG*, and as base a circle *EZ*, whose diameter is twice the diameter of the sphere. Then, the sum of the sections of the sphere and the cone, that is the circle *CO* together with the circle *PR*, is in direct proportion to the distance from *AS*, and these two circles moved to the other end of the balance, *Q*, are in equilibrium with the circle *MN*, remaining in place. The barycenter of the cylinder is the point *K*, the midpoint of its axis *AG*, and *AK* is half *AQ*. Therefore, by virtue of the law of the lever, the cylinder is twice the sphere and the cone taken together. The rest of the proposition is quite simple (for details, see Appendix 1, Prop. 2: Sphere).

By adding an auxiliary solid (cone *AEZ* in this case), Archimedes succeeds in extending the use of the virtual balance to the sphere and other solids.[7]

[5] Archimedes, carefully enough, does not say that the sections of the segment of the paraboloid moved to point *Q* makes up again the segment itself. He says that the segment of paraboloid is "filled" by its sections.

[6] The expression "increase in direct proportion" is modern, with an algebraic background. This is never found in Greek geometry and we use it for the sake of convenience.

[7] The volume of the hyperboloid, for which Archimedes does not describe the details of the argument in Prop. 11, can be determined by the same technique of adding an auxiliary solid. For a reconstruction of this argument, see Hayashi [1994].

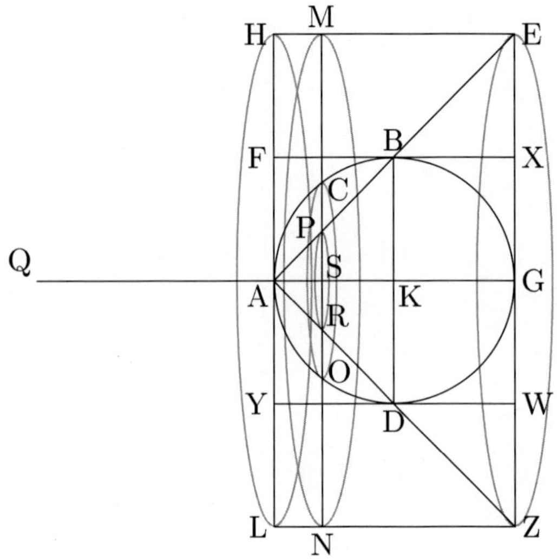

Figure 2: *Method* Prop. 2: Perspective diagram.

The First Novel Solid: Hoof

The hoof is one of the two novel solids that induced Archimedes to write the letter to Eratosthenes, now known as the *Method*. The hoof is generated by cutting a cylinder with an oblique plane passing through the diameter of the base circle.

Let there be a prism with a square base, and let a cylinder be inscribed in it. And let a diameter of the base circle of the cylinder be drawn, parallel to a side of the square, and let the cylinder be cut by an oblique plane passing through this diameter and one of the sides of the square opposite to the base of the prism. The hoof is the solid contained by the semicircle in the base of the cylinder, the semi-ellipse in the cutting plane, and the surface of the cylinder.

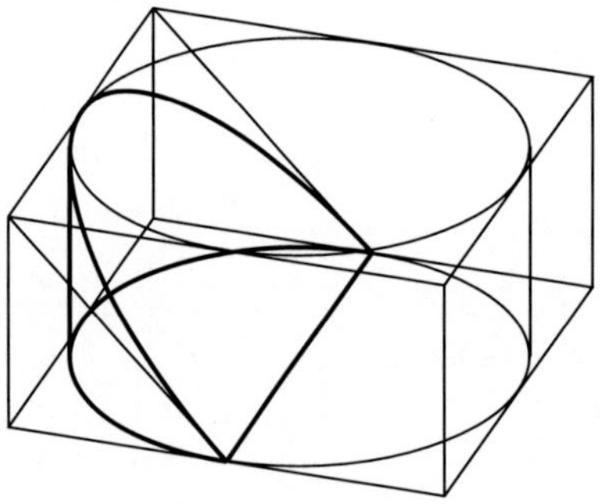

Figure 3: A hoof.

Archimedes found that the hoof is one sixth of the prism. This is the first solid found to be equal to some other solid contained by planes only. Although he had obtained several results concerning solids contained by various curved surfaces of different solids, as sphere, paraboloid, etc., they were always compared with other solids contained by at least one curved surface, as cone and cylinder (see note 1, above). The fact that Archimedes was very proud of this novel result can be seen from what he says about it in the preface of the *Method*.

He gives no less than three arguments for the volume of this solid:

(1) First, in Prop. 12+13, by using the virtual balance, which was already familiar to him (for details, see Appendix 1, Volume of the Hoof: Part 2).[8] What is strange to us in this argument is that Archimedes does not see that the virtual balance could be used in much the same way as in Prop. 2, where he determined the volume of the sphere; and the whole argument would have been much simpler. We have reconstructed this argument in Appendix 1, Volume of the Hoof: Part 1.

(2) Then, Archimedes gives another argument using plane sections without breadth (like Cavalieri's indivisibles) in Prop. 14.

(3) In the following Prop. 15, this argument is transformed into a rigorous demonstration using *reductio ad absurdum* (Appendix 1, Volume of the Hoof: Part 3) twice. We shall call this kind of argument, often called the "method of exhaustion," simply double *reductio ad absurdum.*[9]

A large portion of Prop. 15, the last extant proposition of the *Method*, is lost, and there is no further folium which contains text from this work. So we have no direct textual witness for the reconstruction of Archimedes' arguments for the other solid, the vault.

Possible Propositions for the Vault

We now proceed to the second novel solid in the *Method* which we call "vault." It is the solid bounded by the surfaces of two cylinders having equal bases whose axes meet at a right angle each other. In short, it is the intersection of two equal cylinders.

Let AA' and BB' be axes of cylinders, meeting at point K. Their base circles are $EZHZ'$ and $EYHY'$ respectively. (Archimedes describes the solid within the cube to which the intersection is inscribed, but we have prolonged both cylinders in our figure to make the intersection clearly visible. Note that we have only Archimedes' verbal presentation and no diagram for this solid is extant in the manuscript.) In the figure, only the half of the intersection is shown; the other half of the solid, behind the plane of $YZY'Z'$, is symmetrical to the part shown in the figure.

Archimedes states, in the preface, that this solid is two-thirds of the cube circumscribed about the intersection.

For us, the most important property of the vault is a square section formed by passing a plane parallel to the axes of the two cylinders (hatched in the figure). Indeed, all of the reconstructions hitherto proposed for Archimedes' lost arguments of the vault make use of this square section.

Our conclusion in the present paper, however, is that Archimedes cannot have argued in this way. Let us first look at the mathematically plausible reconstructions hitherto proposed.

Scholars have unanimously claimed that there were at least two different arguments: first a mechanical and heuristic one, then a geometric and rigorous one. This assumption seems to be natural, for there were three arguments for the hoof: besides the mechanical arguments in Prop. 12+13, there were two geometrical arguments, one using "indivisibles" (Prop. 14) and another by a rigorous *reductio ad absurdum* (Prop. 15).

Let us now consider the reconstructed arguments or demonstrations, bearing in mind that they are all mathematical reconstructions, with no direct textual evidence, as Ver Eecke rightly observed.[10]

---

[8] Heiberg divided this argument into two propositions, probably because there are diagrams at the end of what he named Prop. 12. We follow his numbering, and write Prop. 12+13 when we refer to the whole argument.

[9] On the so-called method of exhaustion, and its appearance as a real method in Western mathematics, see Napolitani and Saito [2004].

[10] Ver Eecke, referring to the reconstructions in Heiberg and Zeuthen [1907], Reinach [1907] and Heath [1912], expresses his doubts about their significance as historical research:

> Ces reconstitutions, qui pourraient du reste être étendues à un grand nombre d'autres propositions, n'intéressent que comme applications de la méthode mécanique d'Archimède, ou comme exercices d'archéologie mathématique [Ver Eecke 1921, vol. 2, 519].

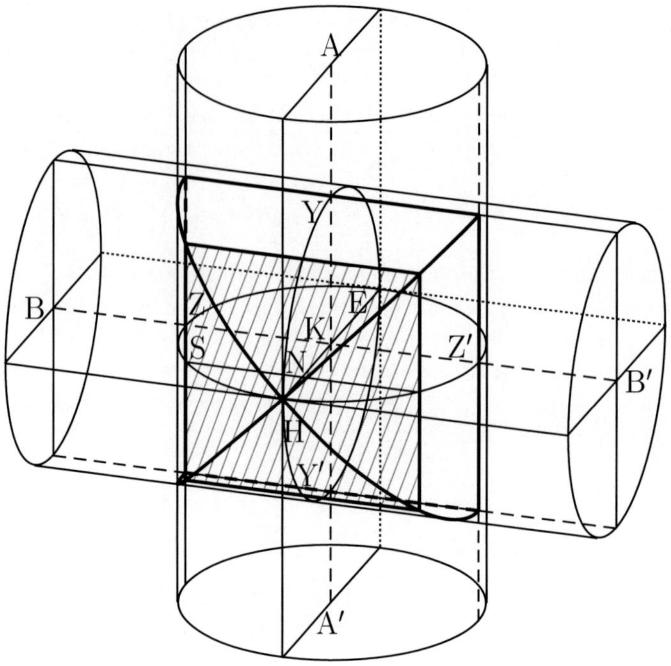

Figure 4: A "vault" (intersection of cylinders).

Reconstruction by Virtual Balance

As for the use of virtual balance, the argument for the sphere (Prop. 2) is valid also for the vault. This mathematical fact was pointed out as early as 1907, only one year after the discovery of the *Method* [Heiberg and Zeuthen 1907; Reinach 1907]. Here we give the basic idea of the argument (for more details see Appendix 1, Volume of the Vault: Part 1).

In the preceding Figure 4, imagine a sphere of which $EZHZ'$ and $EYHY'$ are great circles. Then, its section by the plane which cuts the hatched square from the vault, is the circle inscribed in the square (or the square section of the intersection is circumscribed about the section of the sphere). So, if one substitutes the sphere, the cylinder and the cone appearing in the argument of the volume of the sphere (Appendix 1, Prop. 2) by the vault, the prism and the pyramid respectively, then the rest of the argument is practically the same and the vault turns out to be two-thirds of the cube.

Rufini [1926] refers to this argument as proposition 16. So we call this hypothetical proposition "Rufini 16."

Note that the arguments in Archimedes' Prop. 12+13 for the hoof are completely different from this reconstruction for the vault. We will return to this point later.

Reconstruction by Indivisibles, and by Double Reductio ad Absurdum

For the vault, an argument by indivisibles, like Prop. 14 for the hoof is, of course, also possible. This argument is based on the fact that the square section of the vault is eight times the triangular section of a particular hoof (see Appendix 1, Volume of the Vault: Part 2 and Figure 13 for details).

Thus, if one compares the vault with the circumscribed cube, just as Archimedes compared the hoof with the triangular prism circumscribed about it in Prop. 14, the rest of the proposition is so similar to

Prop. 14 that Sato [1986–1987] even tried to reconstruct the Greek text of this hypothetical argument, which he named proposition 17, depending heavily on the extant text of Prop. 14.[11] We call this Sato 17.

Once a proposition by indivisibles — similar to the extant Prop. 14 — has been reconstructed, it is no more than routine (though tedious) work to convert this argument by indivisibles, to a demonstration by double *reductio ad absurdum*. Rufini numbered this hypothetical proposition 17, (different from Sato 17), and described its outline [Rufini 1926, 174–178].

Thus, one can reconstruct three arguments for the lost pages at the end of the *Method*: (1) an heuristic argument for finding the volume of the vault by way of the virtual balance, modeled after Prop. 2 (Rufini 16), (2) then an argument by indivisibles like Prop. 14 (Sato 17), (3) and a rigorous demonstration like Prop. 15 (Rufini 17). In the preface, Archimedes promises only the last one, the rigorous demonstration. However, at least one of the former two arguments can also be expected, as in the case of the hoof. This has been the consensus of the scholars up to now.

	balance	indivisible	*reductio ad absurdum*
Hoof	Prop. 12+13 (Appen. 1, Hoof 2)	Prop. 14 (Appen. 1, Hoof 3)	Prop. 15
Vault	Rufini 16 (Appen. 1, Vault 1)	Sato 17 (Appen. 1, Vault 2)	Rufini 17

Table 1: Extant propositions for the hoof, and reconstructed propositions for the vault.

The Problem with the Current Reconstruction

In Table 1, above, three approaches are shown (balance, indivisibles and *reductio ad absurdum*) for each of the two novel solids, namely, the hoof and the vault. The arguments for the hoof are extant in the palimpsest either partially or fully, while those for the vault are completely lost, and are reconstructions. Among these, the indivisible argument (Sato 17) and the demonstration by *reductio ad absurdum* (Rufini 17) are simple adaptations of the extant propositions for the hoof (Prop. 14 and 15, respectively). This was made possible by the fact that the square section of the vault is always eight times the triangular section of the hoof.[12]

However, the arguments by virtual balance for the two solids are completely different. Archimedes applies the virtual balance to the hoof in Prop. 12+13, and his argument depends on a particular property of the hoof, that its height is in direct proportion to the distance from the diameter of the base.

So what would happen if we accepted the reconstructions for the vault? If at least one of the two reconstructions that do not use the virtual balance (i.e., Sato 17 or Rufini 17 in Table 1) corresponded to what Archimedes really wrote, then the parallelism between the arguments between the hoof and vault would have been obvious to any careful reader, to say nothing of Archimedes himself, for the square section of the latter is eight times the triangular section of the former, and the structure of the arguments is the same.

And if, in addition, the manuscript had also contained the argument for the vault by means of a virtual balance (like Rufini 16 which uses the same square section as in Sato 17 or Rufini 17), then

[11] He assumed another proposition, 16 (equivalent to Rufini 16), before it, which would have had recourse to the virtual balance.

[12] This relation between the sections of the two solids is visually represented in Figure 13 in Appendix 1. Compare this figure with Figure 4.

it would have been rather difficult not to wonder if an argument by virtual balance, similar to Rufini 16, would not be possible for the hoof, too. This is mathematically possible, indeed, as is shown in Appendix 1, Volume of the Hoof: Part 1.

However, the extant text of Prop. 12+13 for the hoof, which is much more complicated than this reconstruction, does not show awareness of this fact on Archimedes' part. So if one accepts the current reconstructions treating the vault, one has difficulty in explaining the structure of the argument of Prop. 12+13.

Anticipating the conclusion of the present article, we reply that none of the reconstructed arguments for the vault existed in the palimpsest, and that Archimedes' approach to this solid was completely different.

The Space for the Lost Propositions: Mathematical Estimates vs. Codicological Arguments

We have pointed out a problem in accepting the reconstructions concerning the vault, which are mathematically fully acceptable (and have been accepted), consisting only of techniques used by Archimedes himself.

Now let us look at the problem from another point of view: how many pages of the manuscript were occupied by the lost proposition(s) for the vault?

Before entering into codicological arguments, let us estimate the length of the three hypothetical propositions in Table 1. The argument by virtual balance (Rufini 16) would have been approximately of the same length as Prop. 2, of which it is an adaptation. Prop. 2 occupies a little more than 2 pages.[13] The 'indivisible' argument for the hoof (Prop. 14) has 2 pages and some lines, while the rigorous proof by double *reductio* (Prop. 15) occupies about 6 pages.[14] The corresponding propositions to each of these (Sato 17 and Rufini 17, respectively) would have been more or less of the same length. So we would expect about ten pages in total for three propositions concerning this solid, and at the least six pages, because this would correspond to the rigorous demonstration that Archimedes promised in the preface.

Codicological arguments, however, show that there cannot have been even six pages at the end of the *Method* for these proposed propositions on the volume of the vault. The space is only about three pages, against any mathematical expectations — ten pages for all the three propositions, and the demonstration alone would require six pages!

Indeed, the folium which contains the last extant word of the text of the *Method* is followed (not immediately) by the folium containing the initial part of the *Spiral Lines*, and four folia or eight pages are lost between them — this is what the codicological argument shows.

Of the lost eight pages, the first half page, that is, one column, should be occupied by the concluding arguments of Prop. 15, which is not complete in the extant folium, and the last four pages and a half are necessary to accommodate the beginning of the *Spiral Lines* (the part preceding the text in the extant folium 168), so that only three pages are left for the last proposition of the *Method*, which is completely lost.[15]

---

[13] The 'page' is that of the codex, and consists of 2 columns, 34–36 lines, each line containing about 25 characters. One page of the codex corresponds to about three pages of the Greek text in Heiberg's edition.

[14] Only a part of this proposition is extant, and this estimate depends on the reconstruction of the quires of the codex, which we will discuss later.

[15] See the reconstruction of the quires proposed by Abigail Quandt in Netz, Noel, Tchernetska and Wilson [2011, 41–49]. The quire at issue is quire 5 (p. 41). Since this reconstruction is vital to our argument, we explain it in detail in Appendix 2.

What Demonstration Would Fit in Only Three Pages?

This space is surprisingly short. As we have argued, the lost text must contain a rigorous demonstration for the volume of the vault, and such a space is too small for the usual lengthy Archimedean arguments by double *reductio ad absurdum*, as the demonstration of the volume of a solid contained by curved surfaces (in this case, like Rufini 17). Judging from the extant Prop. 15, this proposition would be as long as six pages.

The lost pages at the end of the *Method* could not contain such a demonstration, let alone a set of three propositions as those for the hoof.

Then, in this short space, what kind of argument can we imagine for the vault that would be consistent with Archimedes' words in the preface where he promised to give its *demonstration*?

We seem to be at an impasse, but there is a very simple solution.[16] The vault can be divided into eight hoofs, all equal to each other. Figure 5 shows one of the eight such hoofs cut from the vault. One only has to divide the vault by two planes passing through the border lines of the surfaces of the two cylinders (shown with dotted lines in the figure), then by two planes, each through one axis of a cylinder and perpendicular to the other axis. We have argued above that the square section of the vault is eight times the triangular section of the hoof, but similar relations also hold between the entire solids. This fact was already pointed out by Heiberg and Zeuthen [1907, 357] and Reinach [1907, 960–61], but only *en passant*, after showing a reconstruction of a proposition by virtual balance (Rufini 16).

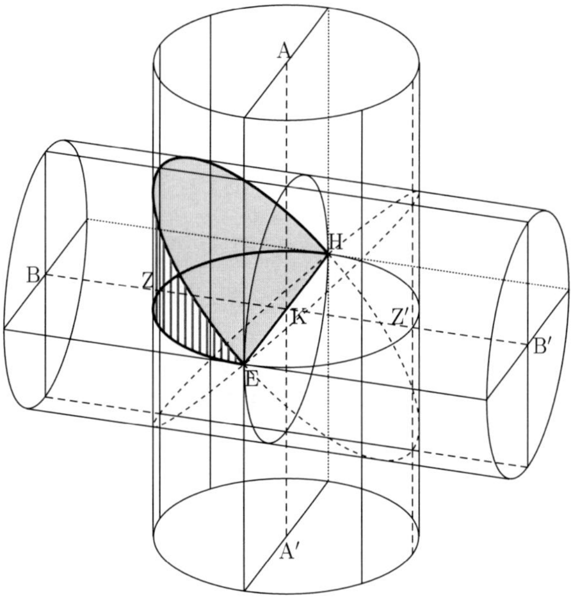

Figure 5: Decomposition of the vault into eight hoofs.

There may be some doubt about whether a proof by decomposition of the vault into hoofs would be too simple and straightforward to fill three manuscript pages.

However, the vault is not a simple solid like the sphere, and mere description of the solid requires some space. In the preface of the *Method*, Archimedes states the volume of the vault as follows:

[16] Our conclusion is also suggested in Netz, Noel, Tchernetska and Wilson [2011, vol. 1, 230]. Most of our arguments come from Hayashi and Saito [2009], from which we have borrowed the figures (except Figures 6, 15 and the plate of appendix 3).

> If in a cube a cylinder be inscribed which has its bases in the opposite parallelograms and touches with its surface the remaining four planes (faces), and if there also be inscribed in the same cube another cylinder which has its bases in other parallelograms and touches with its surface the remaining four planes (faces), then the figure bounded by the surfaces of the cylinders, which is within both cylinders, is two-thirds of the whole cube. [Heath 1912, suppl. p. 12]

This enunciation occupies sixteen lines in the manuscript, almost one fourth of a page (a page consists of two columns, which has usually 36 lines). The lost proposition must have begun with a description like this, then there must have been the exposition (*ekthesis*) referring to the diagram by the names of the points. To describe the solid, it is necessary to identify which of the cylinder surfaces appear as the surface of the intersection. Since Archimedes does not use perspective drawing in the *Method*, probably he drew a plane diagram like Figure 6, and developed some argument purporting to establish that the lines *EG* and *FH* are the borders of the two cylindrical surfaces constituting the surface of the solid of intersection, and that the areas *EKF* and *GKH* are the surface of the cylinder having the axis *AB*, while the areas *FKG* and *HKE* represent the surface of the cylinder having the axis *CD*, and so on. Such an affirmation must have been accompanied by some justificative statements. Only after such descriptions and arguments, is it possible to assert that the vault is decomposable into eight hoofs which are equal to one another. He may well have used another diagram to show the hoof obtained by decomposition.

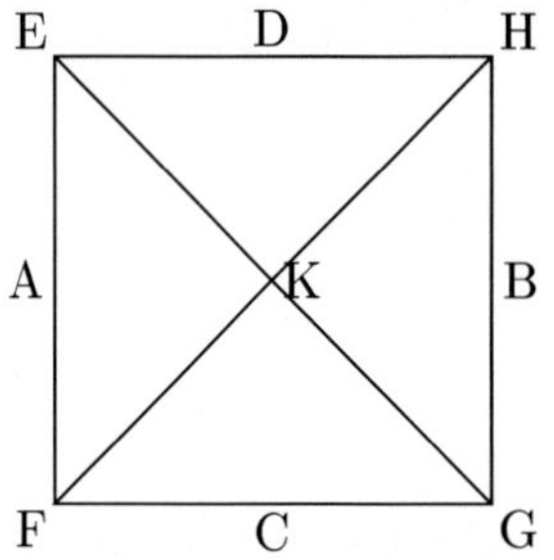

Figure 6: A possible planar diagram for the vault.

All these arguments and diagrams seem sufficient to fill most of the space of three pages. Moreover, the concluding part of Prop. 15, which we assumed to occupy just one column, may have been longer, and some concluding remarks pertaining to the whole work may have existed after the demonstration of the volume of the vault. Thus three pages seem to be just enough to contain the proof we propose.

Archimedes as Ancient Geometer: A Revised Portrait

We have argued that the lost demonstration of the volume of the vault was probably its decomposition into eight hoofs, whose volume had already been determined in Prop. 12–15.

In this section, we argue that this interpretation suggests a considerable change of the image of Archimedes as mathematician, which has been excessively modernized.

Difficulty Resolved: An Interpretation of the Method

We have pointed out the difficulty with the current reconstruction of the determination of the volume of the vault, above. If we assume that Archimedes cut the solid in such a way to obtain square sections and used the virtual balance in much the same way as in Prop. 2 for the sphere (Rufini 16; see Appendix

1, Volume of the Vault: Part 1), it is difficult to explain why he did not adopt a similar argument for the hoof, for which he developed a series of very complicated arguments in Prop. 12+13.[17]

In our interpretation, Archimedes did not cut the vault by such planes. He first observed that its surface consists of two parts — the surface of one of the intersecting cylinders, or that of the other — and cut the solid of intersection according to the border line of the two parts. Thus he gets four segments, each of which is one fourth of the whole solid. If one looks at this segment along the line KZ (see Figure 5 above), it is easy to see that its "height" is proportional to the distance from K. This is the convenient condition for an approach by the virtual balance. At this point, or earlier, he may have realized that he could divide the segment into two symmetrical parts cutting it by the circle $EZHZ'$, obtaining the hoof.

Thus the question of the volume of the vault is reduced to that of the hoof, and the natural approach is to introduce the virtual balance whose arm is KZ with fulcrum K (see Appendix 1, Volume of the Hoof: Part 2). This is Prop. 12 of the *Method*, and although this approach led to the apparently no less difficult problem of determining the barycenter of a semicircle, Archimedes somehow circumvented it in Prop. 13, and obtained the result. In the latter proposition, he cut the solids by planes parallel to the arm of the balance, and this new way of cutting the solid, through which he obtained the result he was looking for, probably suggested the "indivisible" solution (Prop. 14), which could easily be transformed into a rigorous demonstration by double *reductio ad absurdum* (Prop. 15).[18]

With this interpretation, the difficulty with Prop. 12+13 disappears. Archimedes did not cut the vault in the manner of generating square sections, for he first divided it into hoofs. We should add that his approach was rather natural. For us moderns, equipped with the diabolic effectiveness of integral calculus, the volume of a solid has little to do with its shape or appearance. One only has to find a set of parallel planes which generates "simple" sections (or more precisely, the sections whose areas can be expressed by integrable functions). And since Archimedes cuts the conoids and spheroids (paraboloids, hyperboloids and ellipsoids in our terms) always by planes perpendicular to their axis, we tacitly assume that Archimedes shared our idea, that is, to find the volume of a solid is to find appropriate parallel sections.

In short, we have thus overestimated the "modern" ingredients in Archimedes' works. If he treated the paraboloid, spheroid and the hyperboloid in the same manner in his preceding work *Conoids and Spheroids*, this was because they were all generated by rotation, and the same approach was valid for all of them. However, the vault is not a solid of rotation, and he observed its shape and appearance to find an appropriate approach. According to our interpretation, he first cut this solid by the planes through the border lines of the surfaces of two cylinders, so that the segments are part of one cylinder, not some entangled mixture of two cylinders. For him, this simplifies the situation. Should one ask why he did not cut the solid by planes that would generate square sections, the answer is now clear. First, he did not share our concept that determining volume implies finding appropriate parallel sections. There was no reason to cut the intersection of two cylinders in such a way as to mix up the two cylinders, while it can obviously be divided into segments, each of which consists of "one" cylinder, not of "two."[19]

---

[17] R. Netz suggests that Archimedes was "playful" and "sly" (Netz and Noel [2007, 37], though not in this context). Such an interpretation would resolve this difficulty, for Archimedes might well have written confusing and unnecessarily complicated arguments anywhere on purpose. Our arguments, however, try to defend an honest Archimedes.

[18] It seems that Prop. 14 offers a new, powerful approach for the determination areas and volumes, though we know nothing about its application to other figures. Probably, Archimedes did not have time to develop its potentiality after he wrote *Method*, which was very probably written after all his other major works had been sent to Alexandria, from *Quadrature of Parabola* to *Conoids and Spheroids*.

[19] It should be remembered, that eighteen centuries later, Piero della Francesca treated the vault, and he cut this solid and the circumscribed cube by a plane through the straight line passing through the intersection of the axes of cylinders, and perpendicular to both axes. (Then the cutting plane can be rotated around this straight line.) The section of the solid is always that of one cylinder, and is an ellipse. Then he compared this section with the circumscribed rectangle, which is the section of the cube produced by the same cutting plane. By ingenious, but not very rigorous inferences, he concluded correctly that the vault is two-thirds of the circumscribed cube. For details, see Gamba, Montbelli and Piccinetti [2006].

This is historical evidence that cutting the vault by planes which generate square sections is not a universally obvious approach.

Moreover, cutting the solid through the curve of the borders of the cylindrical surfaces, Archimedes obtains a segment whose height is proportional to the distance from the center of original solid as we have seen above. Then there is no reason to make other trials other than to introduce the usual tool of virtual balance, unless this approach happens to prove impracticable. This approach led him to a very difficult problem as we have seen, but fortunately, his genius found a solution to it in Prop. 13.

Our interpretation, then, suggests the figure of a mathematician much less modern than we are used to imagine. He did not recognize the general approach of cutting the solid by appropriate planes to determine its volume. His approach was much less general, and the appearance of a solid was a non-negligible factor in his investigation. Losing much of Archimedes' "modernity," we have instead recovered his honesty and sincerity at least in the *Method*, for we no longer have to ascribe to him a playful or sly character when he develops the complicated arguments in Prop. 12+13 of the *Method*. If this conclusion seems strange, it is because of what has been said about the *Method* since its discovery. Modern scholars have been misled by the title of the work "*Method*", invented by Heiberg (see note 2, above), and by his remarks that Archimedes' method in this work was equivalent to the integral calculus.[20]

Archimedes as Ancient Geometer

We should rather look at Archimedes in the context of Greek geometry, of which at least a basic part is still taught at schools. Indeed, our school geometry — with its theorems on congruence of triangles, similar figures, and so on — is an adaptation of Euclid's *Elements* which were directly used in the classrooms until the 19th century. The objects are figures which are described by words and shown in the diagrams. The demonstration is directly referred to the objects shown in the diagrams or at least connected to them by labels (e.g., the expression "square on *AB*," where the square is not always really drawn in the diagram). Geometry and arithmetic were clearly separated; there is no symbolic language similar to our symbolic algebra.

The objects (figures) are formalizations of either concrete objects, or of effective solution procedures. Let us illustrate this last point. For us, an ellipse is the locus of zeros of a polynomial of second degree with two variables:

$$Ax^2 + Bxy + Cy^2 + Dx + Ey + F = 0; \text{ where } B^2 - 4AC < 0.$$

In other words, a curve is defined by an abstract property of an algebraic nature, which precedes the object itself. For the Greeks, however, the ellipse is the curve obtained by cutting a cone with a plane that intersects all of its generatrices. A curve is defined by a specific procedure, then its properties are derived thereof. Greek mathematics is thus a mathematics of individual objects, each generated from a suitable constructive process.[21]

So every argument is referred to some figure shown in the diagram, and this means that there was no way of describing a general method of solution. If we have discussed the double *reductio*, not the *method* of exhaustion in the present article, this is because neither Euclid, nor Archimedes, nor any Greek mathematician has ever spoken of this demonstration technique in general terms. We have several propositions in which we find similar sequences of particular arguments, and it is we that give a name to this pattern of arguments. The same is true for what we have called the virtual balance of Archimedes in the present article.

If we adopt this point of view, some consequences immediately follow:

- Greek mathematics is not a general mathematics, unlike post-Cartesian mathematics.
- There are no general objects, still less general methods.

[20] "Die neue Methode des Archimedes ist tatsächlich mit der Integralrechnung identisch" [Heiberg 1907, 302].

[21] This is the view on the objects of Greek mathematics given by Giusti [1999] (esp. capitolo 5).

- The procedure of measuring an object is a formalization of some concrete process; indirect confrontations are applied only if the direct one is proved impossible.

Archimedes' works fits these general characteristics well. His extant works are divided into two groups according to two major themes: geometry of measure and mechanics. In the works of geometry (*Measurement of a Circle*, and the four works sent to Alexandria: *Quadrature of the Parabola*, *On the Sphere and the Cylinder*, *Spiral Lines*, *Conoids and Spheroids*), Archimedes deals with the problem of measuring, that is, determining the size of geometrical objects, through direct comparison between an "unknown" figure (e.g., a sphere or a paraboloid) and a better known one (e.g., a cylinder or a cone), and shows, for example, that the sphere is two-thirds of the circumscribed cylinder, or that the paraboloid is one and a half times the cone inscribed in it, and so on. This is why Archimedes was so proud to tell Eratosthenes, in the preface of the *Method*, that he had succeeded in demonstrating for the first time the equivalence between a solid curved figure and a "straight" one (a parallelepiped). Quadrature (or cubature in this case) of a figure was not the result of finding a formula, like $V = \frac{4}{3}\pi r^3$, but of finding the simplest known figure equal to it.

Our investigation in the present article confirms that Archimedes was working within the framework of Greek geometry, despite of his numerous and marvelous results.

Concluding Remarks: How Did Archimedes Come to Consider the Novel Solids?

Before concluding this article, we should mention a problem which is brought about by our interpretation of the last proposition of the *Method*.

According to the hitherto prevailing interpretation, Archimedes used a parallel argument for the hoof and the vault. Then, it was not so important to decide how he came to consider these particular solids. He might even have started from the method of determining the volume. A possibility was that he was perhaps looking for some solid for which the same argument for the volume of the sphere was valid, and found the vault which can be obtained by replacing the circles (section of the sphere by parallel planes) with squares. Then, replacing these square sections by similar triangles, a hoof can be obtained. Though one could only speak of a possibility, Archimedes' novel solids may have been invented "inversely" from the way of determining their volume.[22]

Our interpretation in the present article, however, has confirmed a "classical" interpretation of Archimedes, denying his recognition of the common method between the sphere and the two novel solids. But if Archimedes was not induced to consider the novel solids because of the common method used to determine their volume, how did he come to consider these novel solids? As we have proposed that Archimedes found the hoof during his investigation of the vault, the question is reduced to that of considering the vault.

Concerning this question, we have a very interesting piece of archaeological evidence. Recent excavations of a bath at Morgantina, in Sicily, have revealed the existence of two barrel vaults arranged at a right angle, although without intersecting.[23] This reminds us of the discovery of remains of a "hydraulic establishment" in Syracuse by the Italian archaeologist G. Cultrera in the 1930s, where the same technique was used [Cultrera 1938]. The existence of this type of construction at Morgantina — at that time part of the Syracusan kingdom of Hieron II — and the existence of at least one similar building in Syracuse itself, suggest the possibility that Archimedes was inspired by something to consider the volume of the vault. If we are allowed to put it dramatically, Archimedes, lying in the bath or having a massage, asked himself the question: what if those two vaults were to intersect? What kind of shape would result? It should be noted that the construction techniques used at Morgantina (and most likely at Syracuse) were not such that would easily have allowed the construction of a cross vault. However, it is not really a question of whether Archimedes knew this public bath directly or indirectly. What is

[22] One of the authors was once inclined to this position. See Saito [2006].

[23] For a more detailed description of the excavation, see Lucore [2009].

important is that there is a possibility that Archimedes may have found the inspiration of considering the vault from some real and existing objects like vaults, so that we do not have to assume that he started from some established method of determining the volume of a solid, and worked backwards to other solids for which the same method was valid.

In short, Archimedes was not so modern as we have been prone to imagine. He was an ancient. His arguments about the volume of solids always began with some concrete solid; he did not invent a solid from a method; the recognition, evident for us, that the volume of a solid is determined by its sections, was not necessarily evident to him.

Thus the inquiry into the number of lost pages at the end of the *Method* has revealed an Archimedes less modern but at the same time less sly and more honest and serious.

Appendix 1: Archimedes' Propositions and Reconstructions

Prop. 2: Sphere

The use of the virtual balance is based on the equilibrium between sections of figures whose volume (or area) is unknown, and corresponding sections of a known figure. To determine the volume of a solid (or the area of a plane figure in Prop. 1), it is necessary to carry its sections to the other end of the virtual balance, and find the section of another solid, which, in its place, is in equilibrium with it. Both the volume and the barycenter of the second solid must be known. In many cases, this second solid is a cylinder.

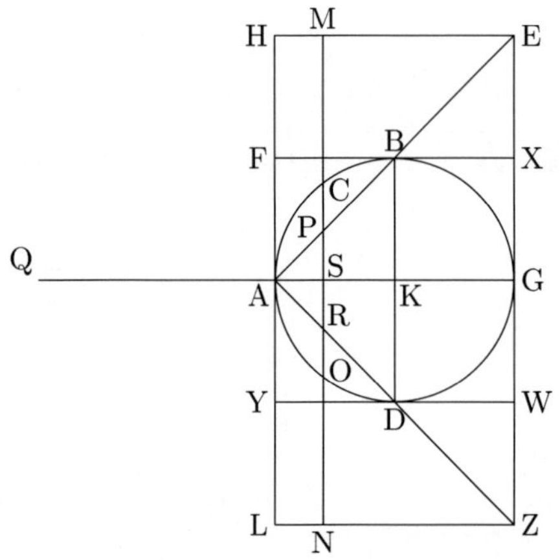

Figure 7: *Method* Prop. 2: Volume of the sphere.

In Figure 7, *ABGD* is a great circle of the sphere, *AG* and *BD* are two of its diameters, perpendicular to each other. Archimedes conceives another great circle in the sphere with diameter *BD* and perpendicular to the plane of *ABGD*, and constructs a cone having this circle as the base and the point *A* as vertex. This cone is extended to the plane through *G* and parallel to the base of the cone. This cone makes a circle whose diameter is *EZ*. Then a cylinder is constructed with the circle *EZ* as base, and the straight line *AG* as height. Finally, a balance *QG* is conceived, *QA* being equal to *AG*.

If one cuts the sphere, cone and cylinder by a plane *MN*, perpendicular to *AG*, then the sections of the sphere, the cone and the cylinder are all circles whose diameter are *CO*, *PR*, *MN*, respectively.

Since

$$\mathrm{sq}(CS) + \mathrm{sq}(PS) = \mathrm{sq}(CS) + \mathrm{sq}(AS) = \mathrm{sq}(AC),$$

and

$$\mathrm{sq}(GA) : \mathrm{sq}(AC) = GA : AS = QA : AS,$$

therefore

$$\mathrm{sq}(GA) : \mathrm{sq}(CS) + \mathrm{sq}(AS) = QA : AS.$$

The squares can be replaced by the circles having the sides of the square as radius. Therefore,

$$\text{circle } MN : \text{circle } CO + \text{circle } PR = QA : AS. \tag{1}$$

This means that the circle CO (section of the sphere) and the circle PR (section of the cone), taken together and carried to the point Q, are in equilibrium with the circle MN (section of the cylinder) remaining in place. Doing the same for other parallel planes, like MN, we have an equilibrium between the solids: the sphere and cone moved to point Q are in equilibrium with the cylinder remaining in place. From this equilibrium, the volume of the sphere is easily determined.

Volume of the Hoof

Part 1: A Possible Use of the Virtual Balance

The volume of the hoof can easily be determined in substantially the same way as that of the sphere (Prop. 2), though Archimedes did not take this approach.

Imagine a hoof, cut from a cylinder whose base is the circle EH, by the oblique plane through EH and FV.

Extend HZ and HW until they meet ED and EF, extended, at points D' and F', respectively. Imagine a pyramid having base $ED'F'$ and vertex H, and a triangular prism $D'EF' - G'HV'$. Their role corresponds to that of the cone AEZ (Figure 7) and cylinder $EZLH$ in Prop. 2, which treats the sphere.

Take the barycenter of the triangle $ED'F'$ and $HG'V'$, E_0 and H_0 respectively, and extend E_0H_0 to Q_0 so that $E_0H_0 = H_0Q_0$, and imagine the balance $Q_0H_0E_0$ with fulcrum H_0. The section of the hoof by any plane, $NM'Y'$, perpendicular to the arm of the balance is triangle NSX (shown by the shadowed triangle in the figure). In the case of the sphere, the section was the circle having center N and radius NS. Instead of the sections of the sphere, the cone and the cylinder in the case of the sphere, consider the sections of the hoof, the pyramid and the prism cut by the plane $NM'Y'$. The sections are triangles NSX, $NM'Y'$ and $NM''Y''$, respectively, and they are similar to each other (the section of the hoof is shadowed, and those of the pyramid and the cylinder are shown by dashed lines in the figure).

The rest of the argument is similar to that in Prop. 2. For the circle sections of the sphere, the cone and the cylinder, the proportion (1) was deduced; now between the similar triangles, which are sections of the hoof, the pyramid and the cylinder, one can deduce:

$$\text{triangle } NM'Y' : \text{triangle } NSX + \text{triangle } NM''Y'' = QH_0 : H_0N_0.$$

This proportion means an equilibrium on the virtual balance: the triangles NSX and $NM''Y''$, that is, the sections of the hoof and of the pyramid, taken together and carried to Q_0 are in equilibrium with the triangle $NM'Y'$, the section of the prism, remaining in place.[24] From this equilibrium of the sections follows the equilibrium between the solids — the hoof and the pyramid carried to Q_0 is in equilibrium with the prism remaining in place. This equilibrium means that the hoof and the pyramid taken together are half the prism, and it can be easily deduced that the hoof is one-sixth the prism $D'EF' - G'HV'$, or two-thirds the prism $DEF - GHV$.

[24] We have imagined a balance E_0Q_0 which passes the barycenter of each of the section of the prism (e.g., N_0 is the barycenter of the triangle $NM'Y'$).

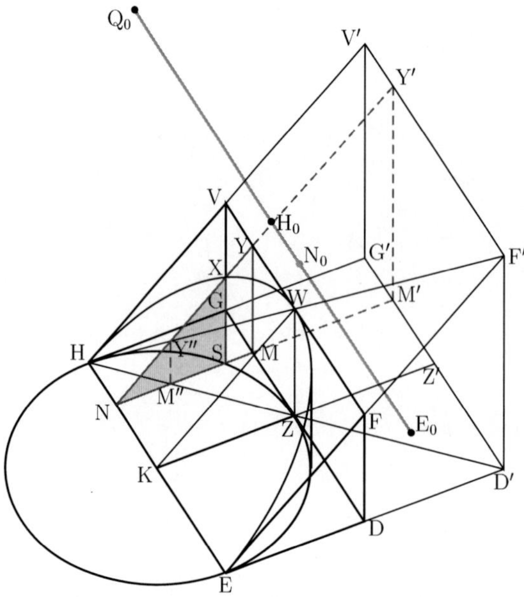

Figure 8: Volume of the hoof: A possible use of virtual balance.

Part 2: Archimedes' Use of the Virtual Balance, Prop. 12+13

Archimedes' approach to the hoof, however, was completely different from the above reconstruction. He imagines a balance *HJ* (Figure 9), perpendicular to the section of the half cylinder which contains the hoof, and passing through the center *O* of the section. Probably, Archimedes first saw that the "height" of the hoof is proportional to the distance from the diameter of the base *AC*. Indeed, if one cuts the hoof by a plane *LM*, perpendicular to the base and parallel to the diameter *AC* of the semicircle, the section is the parallelogram *MF*, whose height *PR* is proportional to *KR*. It is obvious that the section of the hoof (parallelogram *MF*), carried to point *H*, is in equilibrium with the section of the semicylinder (*ML*) remaining in its place. This is, indeed, the repeated pattern of the argument by virtual balance.

Considering all the sections by parallel planes, the hoof moved to point *H* is in equilibrium with the semicylinder having base *ABC* and height *BD*, left in its place. If one knew the barycenter of the semicylinder (this is, of course, equivalent to the barycenter of the semicircle) the volume of the hoof would be determined at once.

However, this was not the case, of course. Archimedes then finds another solid, whose volume and the barycenter is known, and in equilibrium with the semicylinder. The solid is a triangular prism (Figure 10A). The semicylinder and the prism are in equilibrium on the balance *CP* whose fulcrum is the point *Q*.

No attempt has been made, as far as the authors know, to explain how Archimedes discovered the triangular prism, but it is fairly easy to find a reasonable hypothesis. Obviously, the problem is reduced to finding a plane figure in equilibrium with a semicircle. In Figure 10B, *CP* is the arm of a virtual balance having the fulcrum at the point *Q*. It is required to find some figure on the left side of *RO*, which would be in equilibrium with the semicircle *OPR*.

Archimedes always cuts the figure by lines or planes perpendicular to the arm of the balance, but this approach was useless in this situation, for it would have taken him back to the hoof from which he started. Confronted with this difficulty, his genius invented another way of cutting the figure. Imagine that the semicircle is cut by a line *SK*, parallel to the arm *CP* of the balance, and look for the section *LX* which would be in equilibrium with *SK*, around the point *S*. If such a line *LX* is found for each section

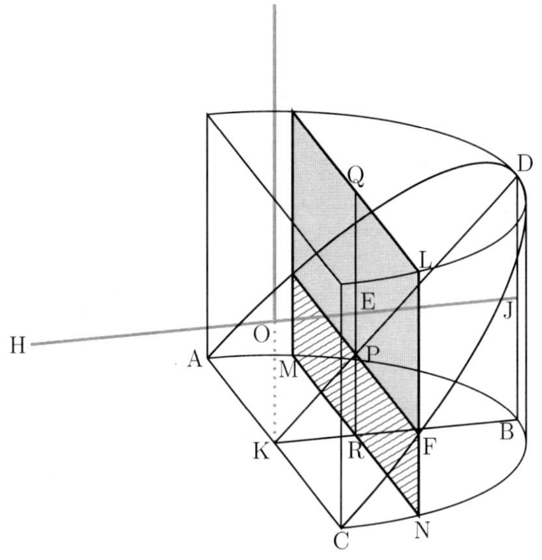

Figure 9: Prop. 12: Hoof and semicylinder.

SK, then the figure filled by all the lines *LX*, would be a figure in equilibrium with the semicircle.

Let the length of *LX* be *m* and the distance of the barycenter of *LX* (midpoint of *LX*) from *S* be *l*, Since this is in equilibrium with *SK*,

$$SK : m = l : \frac{SK}{2}.$$

That is,

$$\mathrm{rec}(l, m) = \frac{\mathrm{sq}(SK)}{2}.$$

Now, every Greek mathematician knew that $\mathrm{sq}(SK) = \mathrm{rec}(RS, SO)$ in a circle, so that

$$\mathrm{rec}(l, m) = \frac{\mathrm{sq}(SK)}{2} = \frac{\mathrm{rec}(RS, SO)}{2}. \tag{2}$$

Then, one might as well try assigning *RS*, *SO* and ½ to *l* and *m*, so that the equality (2) holds. There are not so many possibilities for such an assignment, and the assignment $l = SO/2$, $m = RS$ for sections between *R* and *Q* would create triangle *CHQ*. For the sections between *Q* and *O*, an assignment symmetrical to those between *R* and *Q* would create triangle *CMQ*. As a whole, triangle *HMQ* is found to be in equilibrium with the semicircle *OPR*. Then, considering the prism and the semicylinder having these plane figures as base, the semicylinder is in equilibrium with the triangular prism, so that the triangular prism is in equilibrium with the hoof, carried to the endpoint of the balance.

The whole argument of Prop. 12+13 is very long and complicated, but the first step of introducing the semicylinder in Prop. 12 is quite natural for someone who has become accustomed to the use of the virtual balance as seen in the other figures. The only impressive leap is found in Prop. 13, where Archimedes cuts the figure by planes parallel to the arm of the balance, while in the previous propositions he cut the figure by planes perpendicular to the arm of the balance.[25] Once this unusual way of cutting is found, then it must have been easy for any Greek geometer to find a section of some

[25] In our exposition above, we considered the semicircle which is the base of the semicylinder, so we cut the semicircle by lines parallel to the arm of the balance.

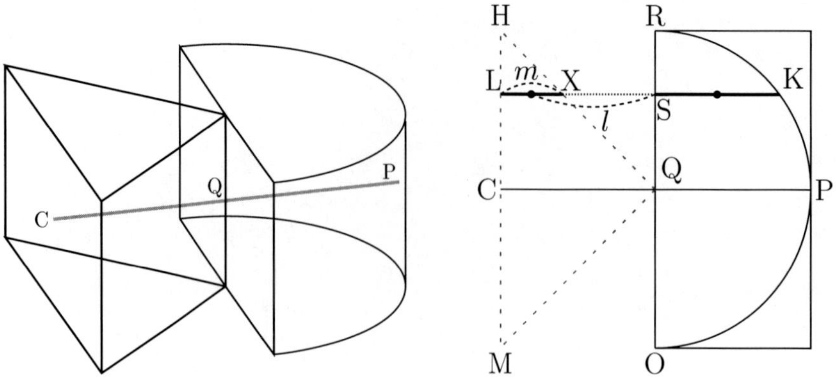

Figure 10: (A) Prop. 13: Semicylinder and triangular prism in equilibrium; (B) Look for a shape in equilibrium with a semicircle.

"manageable figure" in equilibrium with the section *SK* of the semicircle, and to find consequently the triangle *HMQ* (or some other appropriate figure), as we have shown.

This new way of cutting the figure by planes parallel to the arm of the balance probably opened the way for the argument without the balance in Prop. 14. Archimedes only had to try cutting the original hoof by the same plane with which he cut the semicylinder. The resulting sections of the hoof are triangles similar to each other and constructed on half chords of a circle (we will soon see it in Figure 11). At this point he could have realized that the determination of the volume of the hoof is identical with that of spheroids, but for some unknown reason he did not, and instead found another wonderful way of reducing the determination of the volume of the hoof to the quadrature of a parabola, one of his early findings. Rewriting the whole argument of Prop. 14 into a rigorous demonstration by double *reductio ad absurdum* must have been no more than routine work for Archimedes, who had already written the *Conoids and Spheroids*.

We moderns may find at once that cutting the figure by planes perpendicular to the diameter of the base gives the easiest solution, because the cubature of a solid involves finding some set of parallel planes which yield sections whose area are easily integrable. So our approach begins with cutting the solid by various parallel planes, and it is rather difficult not to find the "right" way of cutting the hoof. However, it was not possible for Archimedes to find this section before the usual and evident application of the virtual balance (Prop. 12), and the effort to resolve the difficulty he encountered (Prop. 13).

Part 3: By Indivisibles without the Balance, and Its "Exhaustion" Version (Prop. 14 and 15)

In this proposition, Archimedes cuts the hoof by a plane passing through N and perpendicular to the diameter of the semicircle EZH (Figure 11). This plane cuts, from the hoof, triangle NSX. If one considers the triangular prism $DEF - GHV$ circumscribed about the hoof, the same cutting plane cuts the triangle MNY from the prism. Archimedes compares the two sections MNY and NSX, which are two similar triangles, and shows that their ratio $MNY : NSX$ is reduced to a ratio of two line segments, $MN : NL$, where L is the point where the cutting plane meets the parabola with vertex Z, diameter ZK, and passing through E and H. That is,

$$MNY : NSX = MN : NL.$$

This proportion holds for any point N on the diameter EH, and gathering all the sections together, Archimedes concludes that[26]

[26] There are some twenty lines of text justifying this transition from the proportion of the sections to that of "all the

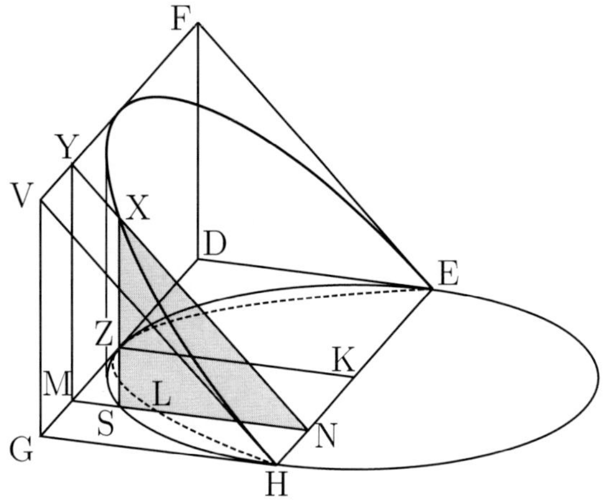

Figure 11: Prop. 14: "Indivisible" approach to the volume of the hoof.

$$\text{prism } DEF - GHV : \text{hoof} = \text{parallelogram } DH : \text{parabolic segment } EZH.$$

As the parabolic segment is two-thirds of the circumscribed parallelogram, the hoof is also two-thirds of the prism.

This argument, though not containing mechanical elements, was by no means at the level of rigor required in Greek geometry. However, it is only routine work to transform it into such a demonstration. One only has to divide the diameter *EH* into equal parts, and consider the triangular prism having as base the triangle *NSX*. Thus one can construct solids inscribed in, and circumscribed about, the hoof consisting of triangular prisms, which differ by a magnitude smaller than any assigned magnitude. Then the rest is the usual argument by *reductio ad absurdum*. This is exactly what Archimedes did in Prop. 15.

The Volume of the Vault

Part 1: By Virtual Balance (Rufini 16)

We present here an outline of the reconstruction of the argument by the virtual balance for the volume of the vault, which can be found in Heiberg and Zeuthen [1907, 357], Reinach [1907, 959–960], Heath [1912, suppl. p. 48–51], and Rufini [1926, 170–173].[27]

The outline of the argument by virtual balance is as follows. Using the diagram from Prop. 2, one only has to imagine that the circle *ABGD* is the section of the vault by the plane of this diagram (one of the axes of the intersecting cylinders is *BD*, the other axis is through *K* and perpendicular to the plane of the diagram), and that parallelogram *EL* and triangle *AEZ* are sections of a prism (or parallelopiped) and a pyramid respectively (both having square base). Then, the plane through *MN* cuts, from the

sections" (figures), which was illegible for Heiberg. Recent studies of the palimpsest has restored the text and shown that Archimedes was not developing a naïve argument by intuition, but was trying to provide a justification to this argument of "summing up" infinite sections applying a theorem valid for the proportion of the sum of a finite number of terms. See Netz, Saito and Tchernetska [2001–2002].

[27] These reconstructions are generally called "Prop. 15" except by Rufini [1926] who assigns number 16, because the current Prop. 8 did not appear in Heiberg's first report of the discovery of the palimpsest [Heiberg 1907], and the proposition numbers assigned to the subsequent propositions were less by one (see also Appendix 2). The proposition numbers we use are those in Heiberg [1910–1915].

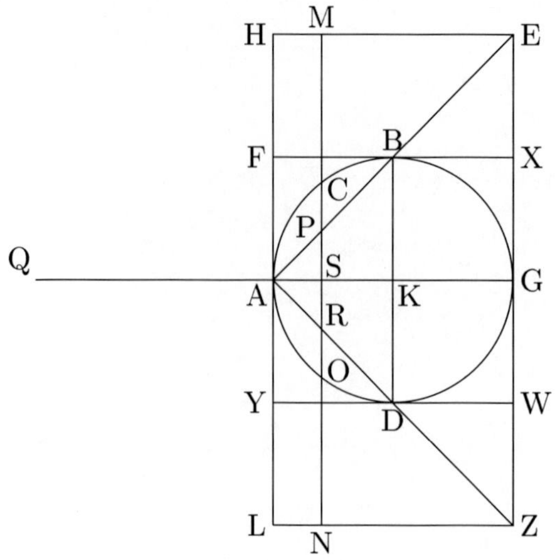

Figure 12: The diagram for the sphere, reused for the vault.

vault, the square on CO — more precisely, the line CO joins the midpoints of the opposite sides of the square — from the prism the square on MN, and from the pyramid the square on the PR. By the same argument for the sphere in Prop. 2, the square on CO and on PR, carried to the point Q are in equilibrium with the square on MN, remaining in its place. The same thing being done for other sections, it turns out that the vault and the pyramid together, carried to the other end of the balance so that their center of gravity is the point Q, are in equilibrium with the prism EH remaining in place. So (the vault) + (pyramid AEZ) is half the prism EH; and since the pyramid AEZ is one third the prism EH, the vault is one-sixth of the prism EH. And the cube FW is one-fourth the prism, so that the vault is two-thirds of the cube FW in which it is inscribed.

Part 2: By Indivisibles (Sato 17)

The volume of the vault can be determined in much the same way as Prop. 14, Archimedes' treatment of the hoof.

The reconstructed argument can best be understood by adding some lines to the hoof (Figures 11 and 13). Consider the hoof that is cut from a cylinder by a plane which makes half a right angle to the plane of the base circle, so that $ED = DF$ (Figure 11). Then construct a cube in which the vault is inscribed (Figure 13); one of the two cylinders is the cylinder of the hoof, the other (not drawn in the figure) has the axis ZKZ'. The section of the intersection by the same plane through N, which cuts the triangle NSX from the hoof, cuts a square from the vault. This square is hatched in Figure 13, and is obviously eight times the triangle NSX. The same cutting plane cuts from the circumscribed cube a square equal to the square of the surface of the cube. Then, just as in the case of the hoof, the following proportion holds (the point L is shown in Figure 11 only, not in Figure 13):

$$\text{(section of the cube)} : \text{(section of the intersection)} = MN : NL,$$

and it can be shown that the vault is two-thirds of the circumscribed cube.

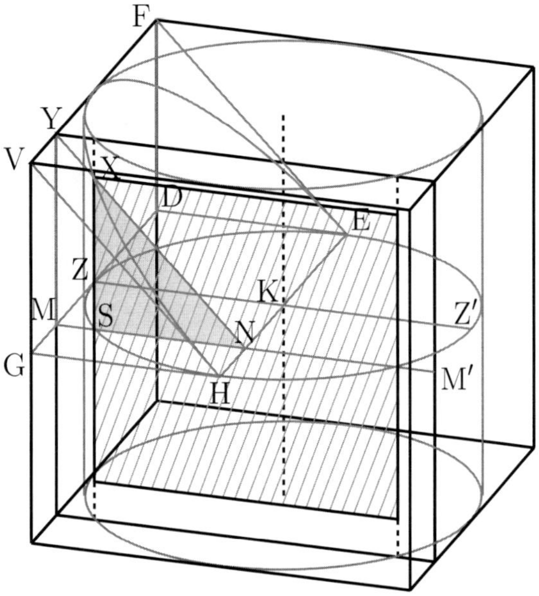

Figure 13: The volume of the vault by indivisibles.

Appendix 2: Reconstructing the Codex

The extant text of the *Method* breaks off before Prop. 15 finishes, and the remainder of this proposition and the proposition which followed it are found nowhere in the palimpsest. Nevertheless, we know for certain the number of the lost folia at the end of the *Method*, thanks to a successful reconstruction of the quires of the original Archimedean codex. In this appendix, we show how the reconstruction is made, and argue that its results are certain.

Reconstruction of the Quires of the Archimedean Codex

First, let us consider how the Archimedean codex was unbound, and its folia reused in the palimpsest.

Medieval manuscripts like the palimpsest, as well as the original Archimedean codex whose parchment was reused for the palimpsest, are materially compiled from quires. A quire consists of four parchment sheets (less often, three or five, or more), folded in half, placed one inside another other and sewn at the fold.

However, the Archimedean codex no longer exists in bound form, so the first challenge is to reconstruct its quires. When the palimpsest was made, the original codex was unbound and the folia were cut into two halves (that is, into single pages), then themselves folded in half and reused for the prayer book, which is thus half the size of the original Archimedean manuscript (Figure 14).[28] Of course, the page order of the original Archimedean manuscript is not preserved in the extant prayer book, and not all the folia that the original contained are present in it.

However, where the Archimedean text is readable, the text itself permits us to determine the order of the pages in the original codex, and now the folia of the Archimedean palimpsest are given double

[28] More precisely, the folia were trimmed in the process of making the palimpsest, so that the page of the palimpsest is smaller than the half of that of the original Archimedean codex. See Netz, Noel, Tchernetska and Wilson [2011, vol. 1, 144].

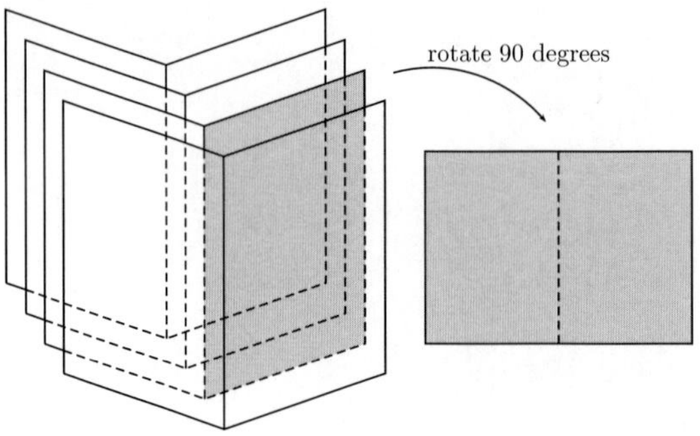

Figure 14: Recycling the parchment.

numberings. One is the folium number in the prayer book, the other is the folium number in the order of the Archimedean text. The former, like 46r-43v, is shown in the margin of Heiberg's edition of the *Method*, while the latter, like A15, can be found in the names of the digitized images of the pages of the palimpsest.[29]

Thus the order of the folia originating in the Archimedean codex is known, but the order of the text does not show where one quire began and how many folia were bound into one quire. However, a careful comparison of the folium number of the prayer book and that of the Archimedean codex can reveal, almost certainly, the composition of the quires in the original Archimedean codex. For example, we are sure that the eight folia from A14 to A21 constituted one quire, as is shown in Figure 15A.[30]

To illustrate how the construction of the quires is determined, let us take, for example, the two folia A15 and A20, which are separated by four intermediate folia, from A16 to A19. In the prayer book, they are bi-folia 46-43 and 45-44 respectively. This means that they are two consecutive folia in the prayer book. If this has not happened by chance, the only reasonable explanation is that A15 and A20 are two halves of one folium in the original Archimedean codex (see Figure 15A), and they were put one over the other to be reused in the prayer book.

This reasoning is supported by a similar examination of the arrangement of the eight folia from A14 to A21 in the prayer book (Figure 15A), and we are sure that this cannot have happened by chance. The table shows that the same thing has happened for other three folia of the same quire of the Archimedean codex, A14+A21, A16+A19 and A17+A18; although they ended up in different places in the prayer book, the two half folia obtained from one original folium are always consecutive in the prayer book. This approach turns out quite successful for all the folia of Archimedean manuscript, and the reconstruction is sometimes physically confirmed by the traces of binding [Netz, Noel, Tchernetska and Wilson 2011, vol. 1, 48].

A similar argument for the folia containing the following propositions of the *Method* (from A22) enables the reconstruction of the following quire, where one folium between A26 and A27 is missing, as is shown in Figure 15B.[31]

[29] The folia from 41 to 48 of the prayer book constitutes one quire, whose decomposition yields four sheets of parchments 41-48, 42-47, 43-46 and 44-45. The folium (or bi-folium) 43-46 is the 15th folium among the extant folia of Archimedean manuscript, and its *recto* side A15r is double pages 46r-43v of the prayer book. So A15=46r-43v. (See below.)

The page 46r is written before 43v, because, when this page is placed so that the Archimedean text is readable, the page 46r of the prayer book is the upper half of the page, and the 43v is the lower half. The images of this page on the web have the name beginning with "46r-43v Archi15r," and in the printed edition "Arch 15r 46r+43v" [Netz, Noel, Tchernetska and Wilson 2011, vol. 1, 41]. The images of all the pages of the palimpsest are available at http://www.archimedespalimpsest.org.

[30] This is based on the figure in Netz, Noel, Tchernetska and Wilson [2011, vol. 1, 41]. We have changed the folio number "Arch14r" to "A14r" etc.

[31] We have corrected typos in Netz, Noel, Tchernetska and Wilson [2011, vol. 1, 41], where 166r+167v and 166v+167r for A24r and A24v appear as "166r+166v" and "166v+166r," and we have described lost folia explicitly as "(lost)."

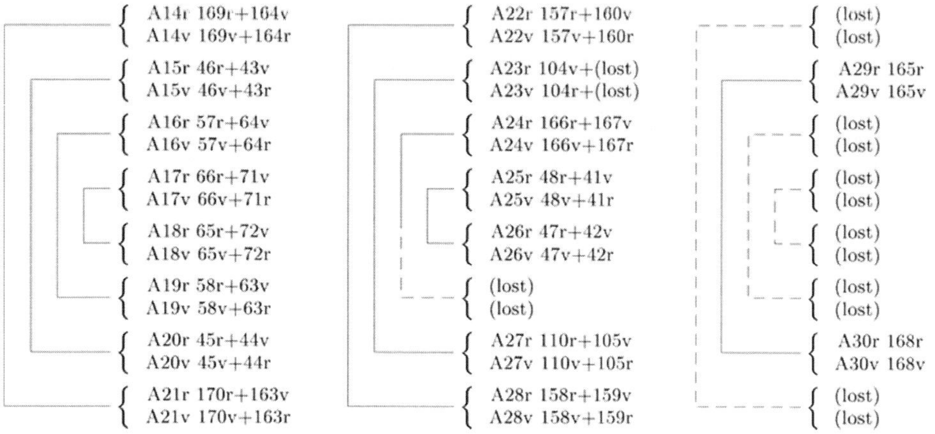

Figure 15: (A) Reconstruction of quire 3 of the Archimedes codex (the first containing the *Method*); (B) Reconstruction of quire 4 (the second containing the *Method*); (C) Reconstruction of quire 5 (the third containing the *Method*).

Much of the content of the lost folium between A26 and A27 can also be reconstructed. Prop. 13 begins in A26v, and the extant text (six lines and one whole column; see Heiberg [1910–1915, vol. 2, 492]) is just enough to infer the argument intended by Archimedes, which is fairly complicated and would have occupied most of the lost folium, and the following Prop. 14 begins exactly at the beginning of the following folium A27. So the possibility is excluded that there might have been another, unknown, proposition, between Prop. 13 and 14, except that there may have been some remarks by Archimedes like the ones between Prop. 1 and 2, or at the end of Prop. 2.

Prop. 14 continues to the next folium A28, which is the last in this quire, and ends in the middle of the first column of its recto page. Then comes Prop. 15, the last extant proposition of the *Method*, and continues into another quire of which only one folium is extant. We have thus reconstructed the first two quires containing the *Method*, and we can be sure no proposition has been completely lost up to this point.

Now let us examine the only extant folium of the following (third) quire of the *Method*, which has the folium number 165-168 in the prayer book. Figure 15C is the reconstruction of this quire. The readers may have two reasonable questions: first, why does the one folium 165-168 of the palimpsest appear in two Archimedean pages (this cannot occur if a folium of the palimpsest is one page of the original Archimedean codex), and then, how has it been possible to determine that this is the second folium from outside the quire while all other folia are lost? Now we respond to these questions.

Folium 165-168, the only one extant in this quire, is exceptional, for it is not one page of the Archimedes codex, as all other folia in the palimpsest, but spans two pages. It is, in fact, the central part of one original parchment sheet [Netz, Noel, Tchernetska and Wilson 2011, vol. 1, 45]. It was placed upside down when the text of the prayer book was written on it, apparently to minimize interference from the Archimedean text, which remains visible (Figure 16). On the left page (165v of the prayer book, then 165r), we read part of Prop. 15. Each column contains only 27 lines of the usual 36 lines in the Archimedean codex, so that about nine lines are completely lost.[32] On both pages, each line in the outer column is partly lost, either at the beginning or at the end, as is shown in Figure 16.

The opposite page (168v and 168r), contains text from the *Spiral Lines*. This means that a part of Prop. 15 of the *Method* and the beginning part of the *Spiral Lines* was in the same quire. How long was the length of the lost text between these two pages?

[32] According to Netz, Noel, Tchernetska and Wilson [2011, vol. 1, 34], six lines from the top and three lines from the bottom are lost.

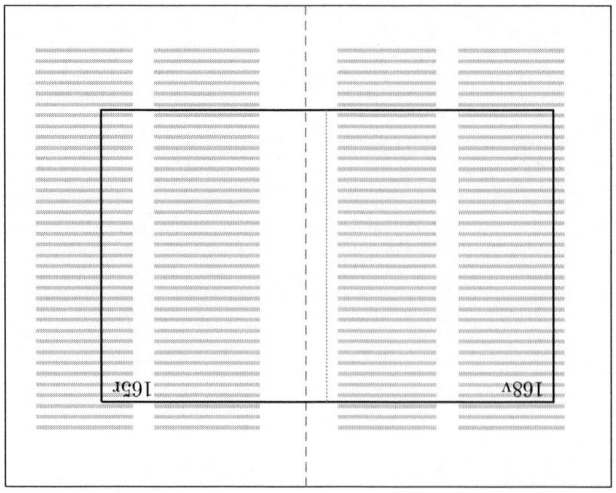

Figure 16: The last extant folium containing the *Method*.

Since most of the reconstructed quires of the Archimedean codex are quaternions, that is, quires of four folia (there are also a few ternions, quires of three folia), we may assume that this quire, of which we possess only (the central part of) one folium, was also a quaternion. We will see (note 35) that the possibility of a ternion is excluded.

Then, the number of intermediate pages depends on the position of the extant folium, 165-168, in the original quire. Fortunately, we know that this is the second folium from the outermost one. The subsequent text of the *Spiral Lines* is found in the palimpsest, after a lacuna corresponding to two pages, or one folium.[33] The part following just after this gap constitutes a ternion, then follows immediately a quaternion. All the (six plus eight) folia of these two quires are extant, and there is no gap in the text.

This means that the extant folium 165–168 is the second folium from the outside of the quire. Consequently, the lost folia of this quire are: (1) the first folium containing two pages of Prop. 15 of the *Method*, (2) four folia or eight pages between the extant text of *Method* Prop. 15 in folium 165 and the text of the *Spiral Lines* in folium 168, and (3) last folium (two pages) corresponding the lacuna in the text of the *Spiral Lines*.

We also know the length of the beginning part of the *Spiral Lines* before folium 168. In Heiberg's edition, there are about 8500 characters, which correspond to four pages and a column in the manuscript.[34] If we assume that the *Spiral Lines* begins at the top of a column as does the *Method*, the *Spiral Lines* very likely begins at the second column of the verso of the fourth folium of the quire (see Appendix 3).[35]

Therefore, there are only three pages and one column for the final part of the *Method*, of which at least one column was occupied by the concluding part of the Prop. 15. This leaves only three pages, perhaps less, for the whole set of the lost propositions on the vault.[36] In Appendix 3, we have shown in a somewhat schematic way, the content of each page of the three quires containing the text of the *Method*.

[33] We can precisely estimate the length of this lacuna, for the complete text of *Spiral Lines* is preserved in other manuscripts.

[34] In the following part of the *Spiral Lines*, where the text of the palimpsest is available, there is no discrepancy between the reading of the palimpsest and the other manuscripts that would affect the estimate of the length of the text. We assume that this is also the case in the beginning part, where the palimpsest is lost.

[35] The possibility of a ternion is excluded at this point, for if it had been a ternion there would not even have been enough space for the beginning part of the *Spiral Lines*.

[36] We have to admit that a slight possibility exists that the quire at issue was a quinternion, a quire of five folia, so that there were seven pages, instead of three, for the lost proposition(s) on the vault. However, there is no quinternion among the reconstructed quires of the Archimedean codex, and it seems arbitrary to assume here a quinternion which is not attested elsewhere in the codex.

Appendix 3: Page by Page Reconstruction of the Quires Containing the *Method*

The diagram of this appendix shows all the folia containing the *Method*, with the reconstruction of the original quires. The following are legenda and some comments:

1. The horizontal lines in each page show Archimedean text, while the vertical lines are text of the prayer book. However, these lines do not exactly correspond to the text of each page. The space occupied by the text and the number of lines in one page are different from one page to another. However, the same image (36 lines for Archimedes' text) is mechanically reproduced for all the pages in this diagram.
2. The lower half of A23 had already been lost when Heiberg consulted the palimpsest. One folium is lost between A26 and A27 in the second quire, and all the folia of the third quire are lost, except 165-168 (the central part of A29-A30). Very probably, they were simply not used in the palimpsest. In this diagram, they appear without the horizontal lines showing the text.
3. After Heiberg consulted the palimpsest, the recto page of A16 was covered by a forged illustration, and the upper half of A21 and A23 (recto and verso) were lost. These pages are indicated by different hatched lines. We have Heiberg's readings of these pages, which can no longer be examined.
4. The spaces occupied by each proposition are divided by straight lines, which are dashed when the border between propositions is not certain.
5. Prop. 8 consists only of a short enunciation without further argument, and belongs to a folium read by Heiberg but now lost. Judging from Heiberg's text, which shows the beginning of every line in the manuscript, Prop. 8 consists of eight full lines followed by seven very short lines. Very probably, the diagram of the Prop. 7 appeared beside the shortened lines of the text of Prop. 8.[37] This means that Prop. 8 was not meant as an independent proposition but was a mere corollary. Thus we have put the number of the Prop. 8 in parentheses.

[37] A similar arrangement of diagrams can be found at the ends of Prop. 3 and Prop. 4, for which the border lines between the following proposition goes through the middle of a column.

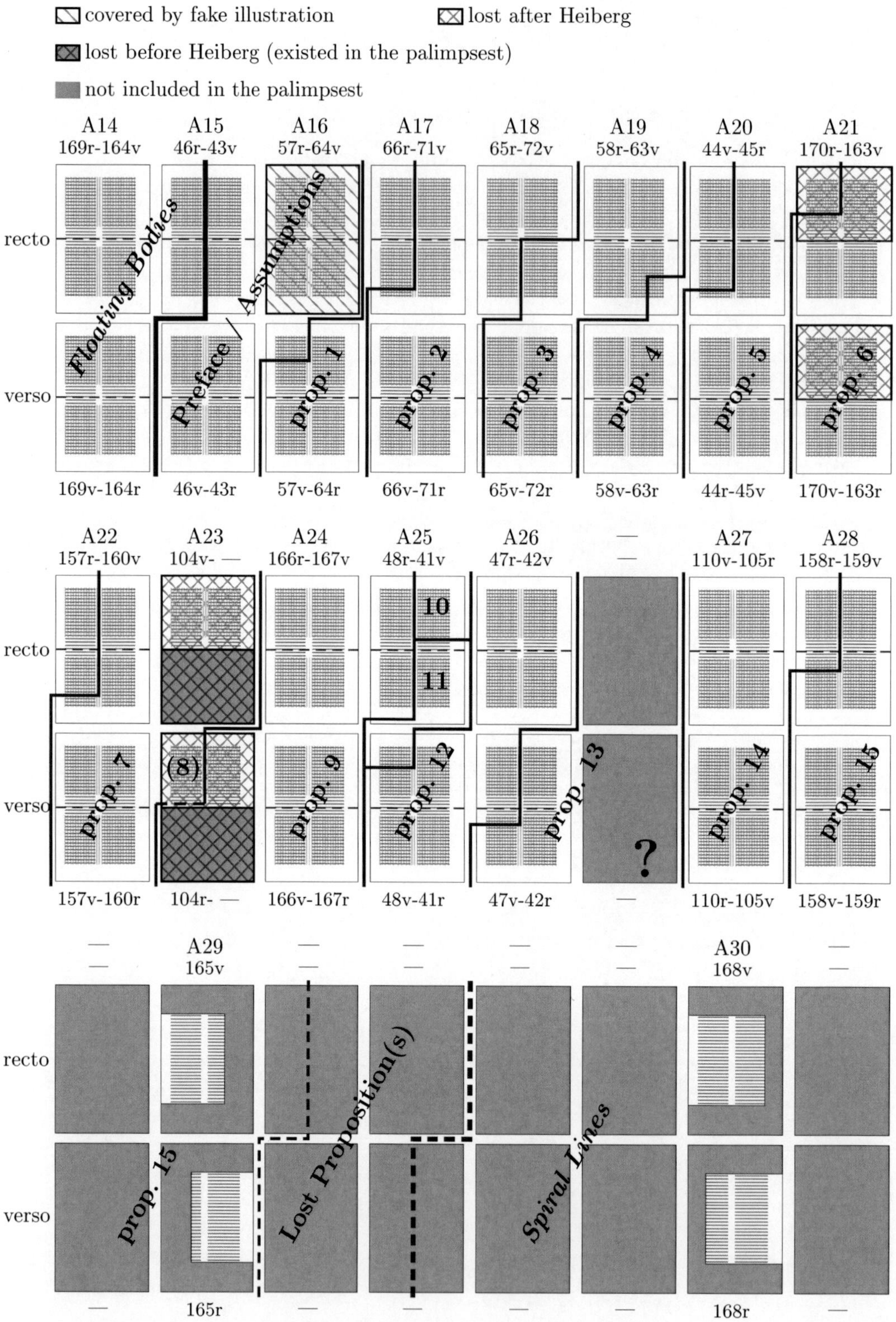
covered by fake illustration
lost after Heiberg
lost before Heiberg (existed in the palimpsest)
not included in the palimpsest
A14 169r-164v
A15 46r-43v
A16 57r-64v
A17 66r-71v
A18 65r-72v
A19 58r-63v
A20 44v-45r
A21 170r-163v
recto
verso
Floating Bodies
Preface / Assumptions
prop. 1
prop. 2
prop. 3
prop. 4
prop. 5
prop. 6
169v-164r
46v-43r
57v-64r
66v-71r
65v-72r
58v-63r
44r-45v
170v-163r
A22 157r-160v
A23 104v- —
A24 166r-167v
A25 48r-41v
A26 47r-42v
— —
A27 110v-105r
A28 158r-159v
10
11
prop. 7
(8)
prop. 9
prop. 12
prop. 13
?
prop. 14
prop. 15
157v-160r
104r- —
166v-167r
48v-41r
47v-42r
—
110r-105v
158v-159r
A29 165v
A30 168v
prop. 15
Lost Proposition(s)
Spiral Lines
165r
168r

References

Cultrera, G., 1938. Siracusa: Rovine di un antico stabilimento idraulico in contrada Zappalà. Atti della Reale Accademia Nazionale dei Lincei. Notizie degli scavi di antichità. 261–301.

Gamba, E., Montbelli, V., Piccinetti, P., 2006. La matematica di Piero della Francesca. Lettera matematica pristem 59, 49–59.

Giusti, E., 1999. Ipotesi sulla natura degli oggetti matematici. Bollati Boringhieri, Turin. (trans. La naissance des objets mathématiques. Ellipses, Paris.)

Hayashi, E., 1994. A reconstruction of the proof of Proposition 11 in Archimedes's Method: Proofs about the volume and center of gravity of any segment of an obtuse-angled conoid. Historia Scientiarum 3, 215–230.

Hayashi E., Saito K., 2009. Tenbin no Majutsushi: Arukimedesu no Sūgaku (Sorcerer of the Scales: Archimedes' Mathematics). Kyōritsu Shuppan, Tokyo. (Reviewed by Sidoli, N., 2012, Historia Mathematica 39, 222–224.)

Heath, T.L., 1912. The Works of Archimedes Edited in Modern Notation with Introductory Chapters by T.L. Heath, with a Supplement: The Method of Archimedes Recently Discovered by Heiberg. Cambridge University Press, Cambridge.

Heiberg, J.L., 1907. Eine neue Archimedeshandschrift. Hermes 42, 236–303.

——— 1910–1915. Archimedis opera omnia cum commentariis Eutocii, 3 vols. Teubner, Leipzig. Reprinted: Stuttgart, 1972.

Heiberg, J.L., Zeuthen, H.G., 1907. Eine neue Schrift des Archimedes. Bibliotheca Mathematica III[e] Folge, 321–363.

Knobloch, E., 2000. Archimedes, Kepler, and Guldin: The role of proof and analogy. In: Thiele, R. (ed.), Mathesis: Festschrift zum siebzigsten Geburtstag von Matthias Schram. Berlin, Diephloz, pp. 82–100.

Lucore, S., 2009. Archimedes, the North Baths at Morgantina and early developments in vaulted construction. In: Kosso, C., Scott A. (eds.), The Nature and Function of Water, Baths, Bathing and Hygiene from Antiquity through the Renaissance. Brill, Leiden, pp. 43–59.

Napolitani, P.D., Saito, K., 2004. Royal road or labyrinth? Luca Valerio's De centro gravitatis solidorum and the beginnings of modern mathematics. Bollettino di Storia delle Scienze Matematiche 24, 67–124.

Netz, R., Saito, K., Tchernetska, N., 2001–2002. A new reading of "Method" Proposition 14: Preliminary evidence from the Archimedes Palimpsest (Part 1 and Part 2). SCIAMVS 2, 9–29; 3, 109–129.

Netz, R., Noel, W., Tchernetska, N., Wilson, N. (eds.), 2011. The Archimedes Palimpsest. 2 vols. Cambridge University Press, Cambridge.

Netz, R., Noel, W., 2007. The Archimedean Codex: Revealing the Secrets of the World's Greatest Palimpsest. Orion Books, London.

Reinach, T., 1907. Un traité de géométrie inédit d'Archimède. Restitution d'après un manuscrit récemment découvert. Revue générale des sciences pures et appliquées 18, 911–928, 954–961.

Rufini, E. 1926. Il "Metodo" di Archimede e le origini del calcolo infinitesimale nell'Antichità. Stock, Bologna. Reprinted: Feltrinelli, Milano, 1961.

Saito, K., 2006. Between magnitude and quantity: Another look at Archimedes' quadrature. Sugaku Expositions 19, 35–52.

Sato, T., 1986–1987. A reconstruction of "The Method 17, and the development of Archimedes' thought on quadrature — Why did Archimedes not notice the internal connection in the problems dealt with in many of his works? Historia Scientiarum 31, 61–86; 32, 75–142.

Ver Eecke, 1921. Les œuvres complètes d'Archimède, 2 vols. Desclée, De Brouwer & cie, Bruges. Reprinted: Vaillant-Carmanne, Liège. 1960.

Acts of Geometrical Construction in the *Spherics* of Theodosios

Robert Thomas

Abstract Two ways of talking about mathematics, the ideal agents of Philip Kitcher and Brian Rotman, and David Wells's analogy to abstract games like chess and go, are brought together and put to the test on the Hellenistic mathematical text *Spherics* by Theodosios. The subject of agents or players, as in a game or play, is discussed.

Agents in Mathematics

In *The Wealth of Nations*, Adam Smith invented the metaphor of the invisible hand for the force that turns individually advantageous decisions into collective advantage. In economics, it is not clear who is doing what. One might think that mathematics would be free of such puzzles, but Brian Rotman's [1993] semiotic analysis of mathematical discourse identifies three characters playing roles in it as if it were a play. There is the *person* in his capacity as writer and unengaged reader of the text, totally free and in charge. If what is going on is a proof, the person takes on the task of judge, whereas if a complicated calculation is performed, then the person is reduced to a mere spectator on account of what Netz, making this point, calls "the thick texture of calculation" [Netz 2009, 40]. There is the *subject*, who obeys some of the commands in the text, for instance, "consider triangle *ABC*." While a reader can consider the triangle if she wants to enter into the spirit of the inquiry, the subject has no choice. But there is a third character in this drama, because persons such as ourselves are not able to carry out commands like letting *n* go to infinity or even joining *A* and *B*. Such obedience is the task of the *ideal agent*. Many actions in mathematical texts are not feasible for human agents, and that is why Brian Rotman [1993] and Philip Kitcher [1984] before him[1] adopted the notion of ideal agent to say who it is that does these things that we cannot do. And there is a lot of it. In geometry in particular, there is little beyond considering triangle *ABC* that the geometer *can* do. We cannot draw straight lines, and we cannot draw perfect circles. We can at best only approximate such actions; what we discuss are the lines and circles as idealized work of ideal agents. In his book *Mathematics Without Numbers* [1989], Geoffrey Hellman casts mathematics into modal terms, terms of possibility. It seems to me that talk of agents is another more concrete way of talking about possibilities, not so much logical possibilities as possibilities for someone. In this vein, I have put forward the notion that much mathematics could be regarded as basic strategic thinking for actions of ideal agents, which may be thought to be like game-playing.

I need to distance myself from two different uses of the notion of game in writing about mathematics. I am not trying to relate to the association of games and mathematics by Ludwig Wittgenstein in calling mathematics, like many other activities, a "language game." This was a rather idiosyncratic use of the

[1] And Moritz Pasch well before him (1920s); see Pollard [2010].

N. Sidoli and G. Van Brummelen (eds.), *From Alexandria, Through Baghdad: Surveys and Studies in the Ancient Greek and Medieval Islamic Mathematical Sciences in Honor of J.L. Berggren*, DOI 10.1007/978-3-642-36736-6_11,

© Springer-Verlag Berlin Heidelberg 2014

term "game" on his part, and he was well aware that the concept game is not an easy one, being at best a family-resemblance concept. There are no necessary and sufficient conditions for being a game.[2]

Another writer to associate games and mathematics is Reviel Netz with his recent book title *Ludic Proof* [2009]. I want to say a couple of things about it, especially since it is about Hellenistic mathematics. He attributes a certain *playfulness* to Hellenistic mathematics in common with Hellenistic poetry — playfulness not game-playing. This observation is more about form than content, and "ludic" does not indicate in his book the level of frivolity that the word could indicate. No implication, he says, is "intended by my reference, in the title, to the 'ludic' — as if the science, or the civilization that gave birth to it, were engaged in the playful *as against* the serious. As if the ludic should be seen as a kind of holiday from the central issues of either science or poetry" [Netz 2009, 211]. One of his many examples is after all Archimedes. Mathematicians' playfulness can be a way of being serious.

Since my game analogy is not in twenty-year-old books like Kitcher's and Rotman's, I elaborate a little on it. The idea is based on and began in reaction to the idea put forward by the English writer on mathematics David G. Wells [2010] that doing mathematics is like playing abstract games like chess and go.[3] I think that it is more like the strategic thinking that distinguishes expert play from the kind of chess that I play. I don't see it as like our *playing* a game at all: no opponent, no winner, no fixed rules, no equipment, no moves even. Our aim is understanding. But what is to be understood? Possibilities, that's what. We understand things like the impossibility of finding odd composite numbers less than nine and the implications for the sides of a triangle of two angles being equal. If an ideal agent were playing a game with positive integers, "where the composite ones are" could be just what it needed to know. If it were in the business of geometrical construction, then what you can and cannot do with ruler and compasses is just what it would need to know. When I said that mathematics can be regarded as "basic strategic thinking," I meant *really* basic. Before you formulate strategy you need to know your possibilities — *that* basic. Not even the choice of ends, just the lay of the land,[4] including what means and ends are to hand. Mathematical facts are those a mathematical robot would need to know if it were carrying out rather general instructions. In practice, the *person* keeps such information (facts and strategy) in his or her head, reminds the *subject* of bits of it as required, and instructs ideal agents in great detail, for example "join *AB*."

I have summarized this game analogy as follows:

1. The agent's mathematical activity (not playing a game) is analogous to the activity of playing a game like chess where it is clear what is possible and what is impossible — the same for every player — often superhuman but bound by rules...
2. Our mathematical activity is analogous to
   a. game invention and development,
   b. the reflection on the playing of a game like chess that distinguishes expert play from novice play, or
   c. consideration of matters of play for their intrinsic interest apart from playing any particular match — merely human but not bound by rules [Thomas 2009].

This analogy brought me face to face with a question I had not seen asked except in the most general terms: what are the powers of the ideal agent in a piece of mathematics?[5]

[2] Wittgenstein did, however, have something like the idea that mathematical conventions and deductions from them were norms for how mathematical concepts could be deployed in applications [Friederich 2011].

[3] Only *like*; no identification.

[4] Consider the example of chess. Corroboration of my amateur impression of the difference between amateur and expert perceptions in chess is given by research Chase and Simon [1973] into the different capacities of amateurs and masters in simply remembering game configurations. Such perceptions are of course fundamental to all strategic thinking. Amateurs were able to reproduce 30% and masters 70% of a configuration they had seen for five seconds, according to Sweller, Clark and Kirschner [2010]. They were equally unable to reproduce random non-game configurations.

[5] That the powers need to be superhuman has been explicitly recognized since Frege and I suppose has always been obvious in geometry if not arithmetic. What *exactly* the powers need to be has been a subject of logical/axiomatic explanation since Engeler [1967], according to Pambuccian [2008].

Introduction

It happened that with this question in the back of my mind I began work on translating the *Spherics* of Theodosios for a project on that work with Nathan Sidoli. As I read the first of the three books, I was struck by the difference between two sorts of thing that someone in it is expected to do or already to have done. It is a characteristic of the Hellenistic tradition in mathematics that many constructions are presumed to have been done. And other constructions are ordered. None of these constructions are things that humans can do, but they are not all of the same order of impossibility. It's a bit like numbers: among infinite numbers some are more infinite than others. Let me illustrate what I mean with the example of the very first proposition. On the one hand we have the situation of someone's having cut a sphere with a plane to produce a curve *ABC* in the surface of the sphere (to be proved a circle), and on the other hand we have the instruction to join the two points *A* and *B* to the foot of the perpendicular from the centre of the sphere to the plane at *D*. This latter instruction is the sort of thing to be done with a ruler in the first book of the *Elements*, whereas the slicing of a sphere with a plane to produce a circle requires a chef in Plato's heaven to wield a planar cleaver. With our compasses we can approximate a circle even on a sphere, but we have no instrument for slicing spheres. I hope that you see the distinction I am trying to draw and why, in the context of my reflection on who does what. I found this provocative enough to write about it early in 2009 when I had translated only Book One. I had just done so when I received a manuscript from Nathan in Japan called "The role of geometrical construction in Theodosius's *Spherics*," written with Ken Saito and since published [Sidoli and Saito 2009]. They had fastened on *exactly* the same difference between two sorts of thing done in the book.

Every schoolboy knows that the *Elements* of Euclid concerns lines and circles in the plane and ultimately in space — at least the better-educated schoolboys — since it ends with the construction of the Platonic solids. What many schoolboys think they know, incidentally, is that the *Elements* is a compendium of the known geometry at the time. One sees this written by adults that ought to know better. These schoolboys — and adults — are wrong. Euclid himself, if he wrote the book *Phenomena* that goes under his name, knew and used a substantial amount of geometry that is not in the *Elements*. And even if Euclid did not write that book, Autolykos used the same raw material in much the same way at about the same time. This extra mathematics considers, among other things but chiefly, plane sections of spheres. Conic sections for dummies, as it were. Yes, all the sections are circles; so what can be interesting? Well, to be honest it isn't as fascinating as Book One of the *Elements*, but it does form a gentle introduction to the topic — carefully organized.

The constructions in Book One of the *Spherics* are important, forming the apparent goal of the book. As Sidoli and Saito [2009] observe, the constructions at the end of Book One (problems 20 and 21) are used throughout the remainder of the work. They say 23 and 14 times in the remaining 23 plus 14 propositions of books two and three. The constructive goal is plainly vital to the remainder of the work. They also point out that the previous problem, 19, is called upon in the *Elements* in five of the six propositions XIII.13–18 without comment by either Euclid or Heath. The exception, 17, uses 15 and so is implicitly dependent on problem 19 — obviously not on Theodosios himself, who reworked this material after Euclid was dead. For the sake of brevity and sticking to the point, I'm going almost to ignore Book Two as the same things go on there as in Book One.

Book One of the *Spherics*

I paraphrase the definitions and propositions of Book One for the reader not already familiar with this work. First the definitions:

1. A *sphere* is a solid figure bounded by a single surface, all straight lines to which from a single point lying within the figure are equal to one another.
2. The point is the *centre* of the sphere.

3. An *axis* of the sphere is a straight line passing through the centre and bounded in each direction by the surface of the sphere and around which immobile straight line the sphere rotates.
4. The *poles* of the sphere are the endpoints of the axis.

Since I must interrupt here to say something about the next definition, I mention that the rotation of the sphere in definition 3 is never mentioned again.

The next definition takes a "circle in a sphere" as undefined. Its meaning was more obvious to its original audience than it is to us. A Greek circle is a circular disk,[6] so that a circular section does not lie on a sphere but lies in it. When they say a circle is in a sphere they mean two things, one that the disk is in the solid sphere (definition 1) but also that the circumference of the circle lies on the surface of the sphere. Any circle *in* a sphere is a plane section of a sphere; that any plane section of a sphere is a circle is the first proposition.

5. *Pole* of a circle in a sphere names a point on the surface of the sphere from which all straight lines meeting the circumference of the circle are equal to one another.

Now the theorems (for background, not essential to the argument):

1. If a spherical surface is cut by a plane, the curve produced in the surface of the sphere is the circumference of a circle.
   Corollary. If a circle is in a sphere, the perpendicular dropped from the centre of the sphere to it falls at its centre.
3. If a sphere Σ with centre B touches at A a plane Π not cutting Σ, then the line joining the point of contact A to the centre B is perpendicular to Π.
4. If a sphere Σ with centre B touches at A a plane Π not cutting Σ, then B will be on a perpendicular to Π erected into the sphere at A.
5. Circles through the centre of a sphere are great circles. Other circles in a sphere are equal to one another if equidistant from the centre of the sphere, and the farther away from the centre the smaller the circle.
6. If a circle C is in a sphere Σ, a straight line joining the centre of Σ to the centre of C is perpendicular to C.
7. If a perpendicular is dropped from the centre of a sphere to a circle in the sphere and produced in both directions, it will meet the sphere at the poles of the circle.
8. If a perpendicular is dropped to a circle in a sphere from one of its poles, it will fall at the centre of the circle, and produced it will meet the sphere at the other pole of the circle.
9. If a circle is in a sphere, the line joining its poles is perpendicular to the circle and will pass through the centers of the circle and of the sphere.
10. In a sphere, great circles bisect each other.
11. In a sphere, circles that bisect each other are great.
12. If a great circle in a sphere cuts a circle in the sphere at right angles, it will bisect it and pass through its poles.
13. If a great circle in a sphere bisects a small circle in the sphere, it will cut it at right angles and pass through its poles.
14. If a great circle in a sphere cuts a circle in the sphere through its poles, it will bisect it at right angles.
15. The secants to the circumference of a great circle in a sphere from its pole are equal to the side of the square inscribed in a great circle.
16. If the secants to the circumference of a circle C in a sphere from its pole are equal to the side of a square inscribed in a great circle, then C will be great.

There are as well construction problems, one at the start, and four at the end as the apparent goal of the book:

---

[6] The definition of a circle in Heath's *Elements* is "A circle is a plane figure contained by one line such that all the straight lines falling upon it from one point among those lying within the figure [the centre] are equal to one another." The wording of the first definition here appears to be modeled on this one.

2. To find the centre of a given sphere.
18. To set out a line equal to the diameter of a given circle in a sphere.
19. To set out a line equal to the diameter of a given sphere.
20. To draw a great circle through two given points on a spherical surface.
21. To find the pole of a given circle in a sphere.

You see that we find points, draw circles on the sphere, and set out — that is set out on a plane exterior to the sphere — lines of a specific length.

Constructions in Book One

With a pair of compasses, perhaps bowlegged, we can approximate a circle on an approximate sphere, and we know that the circle we approximate is the circumference of a plane section of the sphere. We are not bothered by the approximation here because we know that the ideal agent in the Platonic heaven has a perfectly flat plane that will cut the perfect sphere in the perfect circle that the agent can draw on its surface. These are the sort of operations that one carries out in the constructions of the *Elements*; we are used to them.[7] But in the *Spherics* there is another class of construction altogether that can be carried out also only by an ideal agent but that we are unable even to approximate. To find the centre of a given sphere, Theodosios begins by simply requiring that the sphere be cut by a plane. That it should have *been cut* by a plane, by someone unspecified somehow. In that plane we can carry out the construction of finding the centre of the circle, but getting the plane is a real difficulty. Nowadays one might obviate the difficulty by allowing some cutting plane to be given along with the sphere, as was the case in Theorem 1, but I think Theodosios would have regarded that as cheating. Theodosios seems to call upon an extra-ideal agent to perform this feat that is not just an idealization of what we can approximate but goes altogether beyond it in the manner of completed arithmetic infinities. Such a cut in a sphere is itself a completed geometric infinity of points, but so are circles drawn with compasses and lines drawn with straightedges.

Note that the culminating constructions of the book are trivial if one is prepared to use the full capabilities of such an extra-ideal agent. A great circle through two given points (20) is the plane section of the sphere through them and its centre, no challenge if one can cut the sphere to order. And the pole of a circle (21) is even easier with those capabilities, since the perpendicular erected at the centre of the circle cuts the sphere in its poles according to Theorem 8. But as humans outside the sphere we cannot approximate those operations. What the constructions of the book tell us is how to draw the great circle and find the pole using only operations that we can perform in our approximate way and that an agent merely idealizing us can carry out by doing perfectly what we can do imperfectly. A difficult question posed to me is why we subordinate or prefer the one to the other. I have no answer to that question apart from the origin of geometry in operations carried out by humans approximating the ideal actions that would, if possible, give exactly accurate determinations.[8] The actions of ideal agents are idealized from our human actions; the less far-fetched or magical the better.

---

[7] We are roughly used to them, but roughly is not necessarily good enough. Presumably dissatisfied with the extant versions of the Baghdad Arabic translation of the *Spherics*, al-Ṭūsī composed a version that was not a slavish copy but was meant to be read and understood. This has been published [al-Ṭūsī, 1940] and studied by, among others, Sidoli and Kusuba. He added three postulates.

1. That we make any point that happens to be on the surface of the sphere a pole and we draw about it with any distance, less than the diameter of the sphere, a circle on that surface; and
2. that we produce any arc that there is until it completes its circle; and
3. that we cut off what is equal to a known arc from an arc greater than it, when they belong to equal circles; ...[Sidoli and Kusuba 2008, 15 ff.; *cf.* al-Ṭūsī, 1940, 3]

[8] Such god-like idealization seems to be at work in the appropriation of measuring rod and rope as royal insignia in ancient Iraq [Robson 2008].

Problem 21 finds a pole of a small circle by drawing a great circle through diametrically opposite points *A* and *Z* on it and bisecting the arc *AZ* of the great circle inside the small circle.[9] So 21 depends on 20. Problem 20 is more complicated, even when the two points *A* and *B* are not the ends of a diameter so that the circle wanted is unique: The procedure is to draw two great circles with each of *A* and *B* as poles. This can be done with compasses, the proper polar distance to allow it being determined in Proposition 17 and available by Problem 19 and plane geometry. Each point of intersection *C* and *D* of the two great circles is the right distance from each of *A* and *B* because it is on those circles, and so it is a pole from which a great circle can be drawn through *A* and *B*. Either *C* or *D* will of course produce the same unique great circle through *A* and *B*. The aim of these constructions is to give approximating and ideal procedures for doing with real and ideal compasses what can in principle be easily done by magic.

Problem 2 is not at all like that. It is a construction entirely to be imagined, saying that to find the centre of a sphere you take the mid-point of a diameter of the sphere obtained by setting up a perpendicular at the centre of the circular disk obtained by cutting the sphere with some plane. These actions are those of an ideal agent of the higher order. You have to start somewhere, and you start conceptually. This problem is not used after the early theorems; it is a sort of scaffolding — needed at first but not kept.

The other two constructions, 18 and 19, are more like 20 and 21. Their object is to obtain, as line segments on a given plane, the diameter of a given circle in a sphere and the diameter of the sphere itself, that is, to extract from the surface of the sphere information that can be used for the construction in 20 of the great circle through two points. The method in 18 requires only knowing three points, *A*, *B*, and *C*, on the circle at issue. One takes the distances between the points of a triangle imagined inscribed in the embedded circle (there is no need to do anything but *imagine* this triangle inside the sphere) and uses them to construct a congruent triangle *DEF* in the plane (Figure 1). Perpendiculars at two vertices *E* and *F* of that triangle will meet at a point *H* that not only lies on the congruent circle through the vertices of the congruent triangle but is the opposite end of the diameter of that circle having the third vertex *D* as its other end. Presto, the diameter *DH* of the circle without even drawing the circle in the plane. The method in 19 also uses information obtainable on the surface of the sphere. One draws a circle with any pole and any radius. Using the method of 18 to obtain the diameter of that circle, one constructs a triangle in the plane with the diameter of the circle as one side and the polar distance as the other two sides. That triangle is congruent to a triangle inside the sphere with the pole and points at opposite ends of a diameter of the circle as its vertices. One knows by Theorem 13 that the great circle that cuts the circle at right angles passes through these three points. One then has a triangle in the plane congruent to an undrawn triangle inscribed in that undrawn great circle. The method of 18 is being used a second time and has proceeded as far as triangle *DEF*. If *H* is the intersection of perpendiculars erected at *E* and *F*, then again *DH* is the diameter of the circle, which being great has the diameter of the sphere as its diameter. The diameter of the sphere is available in a plane diagram; "presto" would be an exaggeration, but "efficiently" would not.

You see that Problems 18–21 are practical constructions for use with compasses and a solid globe. What needs to be done is idealized merely in exactness. This is not the case in the proofs of the theorems, where inside the sphere actions are performed to which we could not approximate even if we were inside an empty sphere. As with all actions of the extra-ideal agent, however magical, it suffices to imagine them — as in Problem 2. I see no existential commitment in imagining things, and for the most part even imagining things is optional, although the solution of Problem 19 begins (uniquely) with the instruction to imagine a sphere.

It is the process of Problem 19, setting out on the plane the diameter of a given sphere, that Euclid calls upon in the final six propositions of the *Elements*, where, in order to construct solids that could be inscribed in a given sphere, he says simply, "Let the diameter of the given sphere be set out."[10] Heath

[9] *Z* is found by bisecting an arc *DE* of the small circle, where the other arc *DE* is bisected by *A*. These bisections have attracted comment [Berggren 1991; Sidoli and Saito 2009] since Theodosios does not specify how they are to be done, and there are several possibilities.

[10] Netz misstates this as beginning with a given diameter [Netz 2009, 92]. He also says "It is as if Euclid has invented a

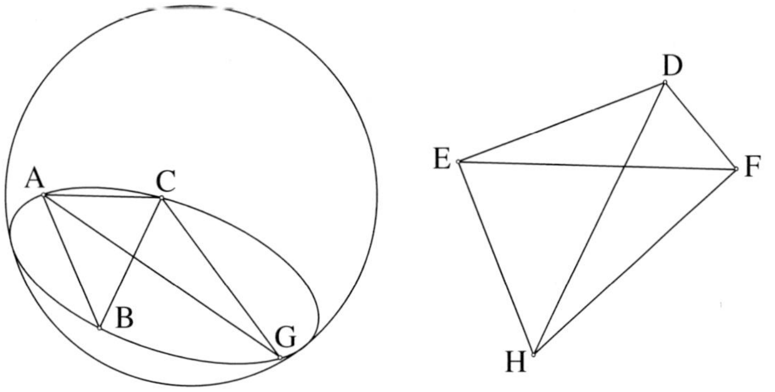

Figure 1: Congruent triangles *ABC* inscribed in a circle on the sphere and *DEF* in the plane with perpendiculars *EH* and *FH* and *DH*, diameter of the circle. Diagram due to N. Sidoli.

points out that the solution of Pappus to the problem of Proposition 13, to inscribe a tetrahedron in a given sphere, requires "a knowledge of some properties of a sphere which are of course not found in the *Elements* but belonged to treatises such as the *Sphaerica* of Theodosius." But he ignores that the method of setting out on the plane the diameter of a given sphere — required in the Euclidean text itself — is also included in such knowledge. This appears to be an instance of anti-spherics prejudice blinding an observer to the obvious.

Constructions in Book Two

Sidoli and Saito [2009] add to the above the two constructions from Book Two.

14. In a sphere, given a small circle and a certain point on its circumference, to draw a great circle through the given point touching the given circle.
15. Given a small circle in a sphere and a certain point on the surface of the sphere between it and the circle equal and parallel to it, to draw a great circle through the given point touching the given circle.

As they remark, the construction and proof of 15 are more complicated than those of 14, but in each case the aim is to enable the one sort of agent — our sort of agent — to draw the great circle with compasses as in I 20. While the constructions cannot require any constructions from Book Two, there being no other constructions in Book Two, the proofs use theorems of Book Two.

To return to the metaphor of actors in a play, the *dramatis personae* of the *Spherics* includes two characters with different responsibilities whom we may call the *construction agent*, that is, the lesser ideal agent that does exactly things we can't do exactly but can approximate, and the *imagination agent* that is the extra-ideal agent, whose doings beyond the construction agent's abilities we can imagine for the sake of learning what the construction agent can do. According to Proclus reporting Geminus, a corresponding distinction is that between postulates about what one can construct and axioms about what one can know. This is explained by Pambuccian thus:

> *postulates* ask for the production, the ποίησις of something not yet given, of a τι, whereas *axioms* refer to the γνῶσις of a given, to insight into the validity of certain relationships that hold between given notions... [Pambuccian 2008, 25]

---

parlor game 'fit solids inside spheres.'" Others try their hand at other solids — and polygons inside circles [Netz 2009, 92–94].

The construction agent is bound by the postulates,[11] whereas the imagination agent is bound only by our imaginations. It might be that someone with metaphysical inclinations would be tempted to argue that what the imagination agent does and creates need only be imagined and is therefore unreal or because created is therefore real. The agents, however, are unreal, imaginary, mere *façons de parler*.

Euclid

I wondered when I first thought of this topic whether the imagination agent is needed in the *Elements*. One hears about their use of ruler and compasses, and most of the books do in fact require no more. However, the dropping of a perpendicular from the centre of a sphere, as from any point not on a plane to the plane is not something that can be carried out with straightedge and compasses. This difficulty infects the whole of Books XI and XII of the *Elements* from Proposition 4 of Book XI where points on and off a plane must be "joined," as they are from the beginning in the *Spherics*. Construction agents at all like ourselves are not good enough for the late books of the *Elements*.[12] In Ian Mueller's massive examination of the logical structure of the *Elements*, he considers in great detail the constructions in Book XIII of Platonic solids to fit in a given sphere. In some ways this appears to be the culmination of the work as a whole. This is not altogether so, as there is more in the arithmetical Books VII–IX than is needed for X and XIII [Mueller 1981, 58] and more in the geometrical Books XI–XII than is needed for XIII, with the material in XII being ascribed to Eudoxos and in XIII to Theaetetus [Heath 1981, vol. 1, 212; Mueller, 1981, 207]. But one might say that geometrically that is where the work is headed. It is accordingly worth something to get as good an understanding of Book XIII as possible. I suggest that the present consideration can help with that understanding.

As Mueller notes [1981, 247, n. 1], Theodosios "uses a different definition of the sphere." Euclid's definition reads:

> A sphere is the figure comprehended when, the diameter of the semicircle remaining fixed, the semicircle is carried around and returned again to the same (position) from which it began to be moved. [XI.14. Mueller's translation, 1981, 361]

It is simply assumed that the solid so defined is the same as that of Theodosios. One can't prove it because the whole notion of motion — essential to Euclid's definition — is completely foreign to Theodosios's work, which is resolutely static in spite of being intended "for astronomical rather than geometrical purposes" [Mueller 1981, 247, n. 1]. The motion in the *Spherics* is just drawing circles with compasses and lines, as elsewhere in the *Elements*. (The astronomical purpose for which the *Spherics* is intended is the consideration of motion of a sphere as in Euclid's *Phenomena*. Sphere motion is astronomy; geometry is static.) Recall Definition 3 of "axis." The rotation mentioned there is not mentioned again.

In Propositions 13–17 of Book XIII, Euclid constructs the tetrahedron, octahedron, cube, isosahedron, and dodecahedron that a given sphere would circumscribe. He does three things given the sphere: he constructs the solid, he shows that it fits exactly into a sphere of the given size,[13] and he discusses the characteristic edge length of the solid. The earlier propositions in the book Heath calls "introductory" [1981, vol. 1, 212] and Mueller calls "lemmas" [1981, 251], criticizing their proofs rather severely. We are concerned with the constructive part, which Mueller describes thus:

> This procedure involves constructing the figure outside the [given] sphere, relative to a straight line equal to the diameter of the given sphere, and then arguing that the semicircle with the straight line as diameter will pass through all the vertices of the figure as it revolves around the diameter. There is no real mathematical difference between this procedure and inscribing the figure in the sphere. Apparently Euclid adopts the procedure and, hence, the generative definition of the sphere (XI, def. 14) as a means of avoiding a treatment of the sphere

[11] Note the postulates of the careful al-Ṭūsī; he cannot have overlooked the actions I attribute to the imagination agent.

[12] Straightedges at all like mine are not good enough for XI.4. Mine can help me to draw lines only on a plane already there, not between arbitrary points in space.

[13] This is also called "comprehending" the solid in the sphere and is explained by Heath as "the construction of the circumscribing sphere," which must be shown actually to circumscribe the solid.

> analogous to the treatment of the circle in Book III. From a foundational point of view the advantage gained is only apparent, since the procedure depends upon tacit assumptions about the properties of a semicircle revolving about its diameter. [Mueller 1981, 254]

More importantly from a foundational point of view, as we have seen, the whole procedure depends on the *Spherics*, which is precisely the required treatment of the sphere analogous to the treatment of the circle in Euclid's Book III. As Heath notes:

> It will be observed that, although in these cases Euclid's construction is equivalent to inscribing the particular regular solid in a given sphere, he does not actually construct the solid *in* the sphere but constructs a solid which a sphere *equal* to the given sphere will circumscribe. [Heath 1956, vol. 3, 472]

By the time of Pappus this subtlety was water under the bridge, and he constructs the circumscribed solids inside the given sphere [*ibid.*]. It is plainly true that the same solid results whether you construct it out of whole cloth on your working plane as you would do any other construction or you try to build it inside the given sphere. The difference, as it seems to me, is partly in what it means to be *given* a sphere. If you are given a sphere, then you have to determine its diameter from outside it, and that is the first thing Euclid does, "Let the diameter of the given sphere be set out" on the working plane. As there is nothing in the *Elements* about such an operation, it must be presumed to be done à la Theodosios I 19 by operations the construction agent can perform on the surface of the given sphere and on the plane. Operations then follow à la Euclid, including the erection of a perpendicular to the working plane to get the construction into space and joining points in space that do not lie on any given plane. So in Euclid the construction agent is more powerful than the construction agent in Theodosios. Whether it has the full power of Theodosios's imagination agent — or more — I do not know. It is able, however, to create a curved surface by the rotation about the diameter of a semicircle, something neither agent in Theodosios needs to do.

I see some gain for understanding in seeing that an agent of ideal-human powers can begin with the sphere and proceed to the construction. If one began by constructing a Euclidean sphere, one would already have the diameter, as that is the beginning of the construction of the sphere. You begin, in effect, with the diameter rather than with the sphere, as Netz [2009, 92] says that Euclid does.

The conjunction of Euclid and constructions raises — in some minds — the odd question, "What are such constructions for?" As I have been suggesting that they tell how things can be constructed in accordance with agreed-upon limitations, I am unsympathetic to what Orna Harari calls "the widely accepted contention that geometrical constructions serve in Greek mathematics as proofs of the existence of the constructed figures" [Harari 2009, 1]. Since I have encountered the contention by Zeuthen [1896] only where it has been opposed [Mueller 1981; Knorr 1986; Harari 2009] and Harari supplies no supporting citation, I think it is time — despite attempts to take it seriously and successfully to show that it is wrong — to dismiss it as an anachronism that arose in the nineteenth century because of that century's concerns and not Euclid's. I cannot doubt that the ancients assumed, for example, that there was a cube twice the size of any given cube; the famous problem was to construct it.[14] Moreover, the theme of this paper undercuts the notion that problems had any single purpose, the unlikelihood of which was brought to my attention by João Caramalho Domingues.[15]

Conclusion

The view of mathematics that I put forward at the beginning was one that expressed what we learn about mathematics in terms of what someone can do — the constraints on mathematical behavior. I first encountered this idea in a posthumously published paper of Leslie Tharp in which he wrote:

> We have claimed that the modal propositions of arithmetic are primarily about concepts, and are about ordinary objects and activities in the indirect sense that the concepts may be applied to ordinary objects arising from

[14] Where existence was a concern, the case of a square root of two, they were not shy about mentioning it.

[15] Personal communication, 2011 10 15.

> ordinary activities, such as an actually constructed inscription. In particular, existential assertions such as 'there is a number...' may go far beyond anything humanly feasible. The discomfort with modal treatments of mathematics is reminiscent of the everyday interchange of 'can' and 'may.' One sometimes says 'Herr Schmidt can drive 150 kph on the Autobahn' when he actually cannot (because, say, his Volkswagen won't go that fast). Obviously, what one means is that he *may,* that is, the relevant rules permit such speed. We interpret the mathematical modalities in such a 'may' sense: one may construct an inscription with 99^{99} strokes — the concepts undeniably permit it. [Tharp 1989, 187]

Some of the mathematical behavior that is permitted by our concepts is ours directly. If we are going to look for the subgroups of a finite group, we shall look among the divisors of the order of the group for their sizes. If the order is prime, the Autobahn is closed. When there are divisors, we can look for, find, and write down existent subgroups ourselves. But much of what we seek in mathematics can only be found by us nominally. The function $y = f(x) = x^2$ is one of which we can write a formula and approximately a graph over a finite interval, but the whole graph — even approximately — and an exact graph — even over a finite interval — are things that we cannot draw. The behavior that our knowledge of such a function governs is that of an ideal construction agent capable of actions we can perform only in principle. Only an ideal agent is licensed to drive on this particular Autobahn. When in contrast we come to some exercise of the axiom of choice or some more esoteric action involving the iterative hierarchy of sets in aid of proving some theorem, then we need the imagination agent whose powers vastly exceed those of our surrogate the construction agent. Whether there is a useful distinction to be drawn here between different ranks of agent, it is clear enough that such agents are required if we are to think of mathematical actions actually being performed. And since mathematical discourse is full of demands for such action, such agents are at least implicitly required. If one is going to make sense of mathematical language using modality, then one has implicit agents that one is making use of. Just what agents are presupposed in different times and parts of mathematics — even in different books of the *Elements* — varies and so is a subject for historical research.

References

Berggren, J.L., 1991. The relation of Greek spherics to early Greek astronomy. In: Bowen, A. (ed.), Science and Philosophy in Classical Greece. Garland, New York, pp. 227–248.

Berggren, J.L., Thomas R.S.D., 2006. Euclid's Phænomena: A Translation and Study of a Hellenistic Treatise in Spherical Astronomy. Garland, New York. Reprinted: American Mathematical Society and London Mathematical Society, Providence, 2006.

Chase, W.G., Simon, H.A., 1973. Perception in chess. Cognitive Psychology 4, 55–81.

Czinczenheim, C., 2000. Édition, traduction et commentaire des Sphériques de Théodose. Lille. (Thèse de docteur de l'Université Paris IV.)

Engeler, E., 1967. Algorithmic properties of structures. Mathematical Systems Theory 1, 183–195.

Friederich, S., 2011. Motivating Wittgenstein's perspective on mathematical sentences as norms. Philosophia Mathematica 19, 1–19.

Harari, O., 2003. The concept of existence and the role of constructions in Euclid's Elements. Archive for History of Exact Sciences 57, 1–23.

Hellman, G., 1989. Mathematics Without Numbers: Towards a Modal-Structural Interpretation. Oxford University Press, Oxford.

Heath, T.L., 1921. A History of Greek Mathematics, 2 vols. Clarendon Press, Oxford. Reprinted: Dover, New York, 1951.

——— 1926. The Thirteen Books of Euclid's Elements, 3 vols., 2nd edition. Cambridge University Press, Cambridge. Reprinted: Dover, New York, 1956.

Heiberg, J.L., 1927. Theodosius Tripolites [word deleted in *corrigenda*] Sphaerica. Abhandlungen der Gesellschaft der Wissenschaften zu Göttingen, Philologisch-historische Klasse (N.S.) 19, no. 3, i–xvi, 1–199.

Kitcher, P., 1984. The Nature of Mathematical Knowledge. Oxford University Press, Oxford.

Morrow, G.R., 1970. Proclus, A Commentary on the First Book of Euclid's Elements. Princeton University Press, Princeton.
Mueller, I., 1981. Philosophy of Mathematics and Deductive Structure in Euclid's Elements. MIT Press, Cambridge, MA.
Netz, R., 2009. Ludic Proof: Greek Mathematics and the Alexandrian Aesthetic. Cambridge University Press, Cambridge.
Pambuccian, V., 2008. Axiomatizing geometric constructions. Journal of Applied Logic 6, 24–46.
Pollard, S., 2010. "As if" reasoning in Vaihinger and Pasch. Erkenntnis 73, 83–95.
Robson, E., 2008. Mathematics in Ancient Iraq: A Social History. Oxford University Press, Oxford.
Rotman, B., 1993. Ad Infinitum: The Ghost in Turing's Machine. Stanford University Press, Palo Alto.
Sidoli, N., Kusuba, T., 2008. Naṣīr al-Dīn al-Ṭūsī's revision of Theodosius's Spherics. Suhayl 8, 9–46.
Sidoli, N., Saito, K., 2009. The role of geometrical construction in Theodosius's Spherics. Archive for History of Exact Sciences 63, 581–609.
Sweller, J., Clark, R., Kirschner, P., 2010. Teaching general problem-solving skills is not a substitute for, or a viable addition to, teaching mathematics. Notices of the American Mathematical Society 57, 1303–1304.
Tharp, L., 1989. Myth and mathematics: A conceptualistic philosophy of mathematics I. Synthese 81, 167–201.
Thomas, R., 2009. Mathematics is not a game but... The Mathematical Intelligencer 31, no. 1, 4–8.
al-Ṭūsī, Naṣīr al-Dīn, 1940. Kitāb al-Ukar li-Thā'ūdhūsiyūs, Osmania Oriental Publications Bureau, Hyderabad. Reprinted: Sezgin, F. (ed.), 1998. A Collection of Mathematical and Astronomical Treatises as Revised by Naṣīraddīn aṭ-Ṭūsī, 2 vols. Institute for the History of Arabic-Islamic Science at the Johann Wolfgang Goethe University, Frankfurt am Main.
Wells, D.G.., 2010. Philosophy and Abstract Games. Rain Press, London.
Zeuthen, H.G., 1896. Die geometrischen Construction als "Existenzbeweis" in der antiken Geometrie. Mathematische Annalen 47, 222–228.

Archimedes Among the Ottomans: An Updated Survey

İhsan Fazlıoğlu and F. Jamil Ragep

Abstract This paper provides a survey of Archimedean material that was produced and disseminated during the Ottoman period, mainly in the city of Istanbul. Moving from the founding figures of Ottoman science such as Dāwūd al-Qayṣarī and Muḥammad al-Fanārī in the fourteenth and fifteenth centuries to Muṣṭafā Ṣidqī in the eighteenth-century, the article discusses Ottoman work in several areas of Archimedean mathematics and science: 1) the number Pi; 2) Hydrostatics and Specific Gravity of Elements, for which an edition of a unique manuscript by Taqī al-Dīn al-Rāṣid is given in an appendix; 3) Geometry (e.g. the sphere and cylinder, squaring the circle, trisecting an acute angle, heptasecting a circle, and spiral lines). From an examination of the content of texts and extant manuscript witnesses, it is clear that Ottoman work on the Archimedean corpus owed a great debt to émigré scholars and was closely connected with major Islamic centers of learning such as the Marāgha Observatory and the Samarqand School; at the same time Ottoman scholars themselves made numerous contributions to the Archimedean heritage.

Introduction

A number of years ago, Len Berggren wrote an article that was quite audacious for its time, entitled "Archimedes Among the Ottomans."[1] Much more has been learned about Ottoman science in the intervening years, so in the following we provide an update of Berggren's pioneering article and also present a previously unedited text by Taqī al-Dīn Maʿrūf (1526–1585) on specific gravity that is in the Archimedean tradition.

By and large, the socio-political structure that we refer to as Ottoman was actually an outgrowth of earlier systems that had prevailed in Anatolia, namely those of the Seljuk Sultanate and the Turkish principalities. Well before the emergence of the Ottoman Empire, Anatolia had become part of a large cultural zone whose origins were in Central Asia and Iran. Particularly during the Īlkhānid period (1256–1335), this zone or milieu flourished politically and culturally, mainly due to contemporary scholars who shared its fundamental outlook and strove to maintain it despite the enormous changes and disruptions that were occurring. It was the outlook of this cultural milieu that would pave the way for Ottoman scientific and cultural progress. Significantly, this outlook itself changed and evolved, eventually itself being transformed.

This paper is based upon work supported by the Canada Foundation for Innovation, the Quebec Government, and McGill University under Grants no. 12587 (PL Robert Wisnovsky) and no. 203634 (PL F. Jamil Ragep). Any opinions, findings and conclusions or recommendations expressed in this material are those of the authors and do not necessarily reflect those of the Canada Foundation for Innovation, the Quebec Government, or McGill University. We thank Sally P. Ragep for many useful suggestions

[1] Berggren [1987].

N. Sidoli and G. Van Brummelen (eds.), *From Alexandria, Through Baghdad: Surveys and Studies in the Ancient Greek and Medieval Islamic Mathematical Sciences in Honor of J.L. Berggren*, DOI 10.1007/978-3-642-36736-6_12,

© Springer-Verlag Berlin Heidelberg 2014

In general, the term "Ottoman" denotes spatial and chronological contexts that are too extensive and too long to be amenable to generalization. To limit the scope of this paper, we concentrate on Ottoman scientific activities in Istanbul. We shall, though, have occasion to refer to scientific activities in other parts of the Islamic world insofar as they had an impact on developments in Istanbul. We should note some further limitations, in large part because of the lack of research in these areas. First, we will not be dealing with work by non-Muslim scientists working within the Ottoman realm except insofar as knowledge exchanges with Muslim scholars are involved. Second, we will not discuss, except in general terms, scientific exchanges between Ottoman and Western European domains. Given these limitations, our aim is to provide an overview of the status and significance of Archimedes for Ottoman science during its classical, pre-print manuscript period.[2]

Founding Figures

Dāwūd al-Qayṣarī (d. 751/1350), the head instructor of the first Ottoman madrasa, which was founded in Iznik in 1337, was educated in the cultural milieu mentioned above. Initially, he studied at the Niẓāmiyya madrasa in Nīksār, Toqat, under the supervision of Ibn Sartāq (Shams al-Dīn Muḥammad b. Sartāq b. Çoban b. Shīrkīr b. Muḥammad b. Sartāq al-Vararkīnī al-Marāghī, d. after 728/1327), part of the second-generation associated with the Marāgha Observatory.[3] Ibn Sartāq made a recension (*taḥrīr*) and taught the geometrical work *al-Ikmāl fī al-handasa*,[4] which was originally entitled *al-Istikmāl fī al-handasa* by its author, the sultan of Saragossa Abū ʿĀmir Yūsuf b. Aḥmad al-Muʾtaman b. Hūd (d. 478/1085). In the course of his education, Dāwūd al-Qayṣarī copied the works of his teacher and studied primary sources on topics related to geometrical mathematics (plane geometry, solid geometry, conic sections, plane and spherical trigonometry, geometrical number theory, and geometrical algebra). These sources included the contributions of Archimedes. One manuscript (Istanbul, Süleymaniye Library, Ayasofya MS 4830) indicates that it was Ibn Sartāq who introduced Dāwūd Qayṣarī to leading Greek figures of mathematics and philosophy such as Aristotle, Archimedes, Apollonius, Agathon, and Euclid, as well as other leading mathematicians from the classical Islamic period such as al-Kindī, al-Khwārizmī, Thābit ibn Qurra, Ibn al-Haytham, Ibn al-Ṣalāḥ, and al-Kūhī. As demonstrated from marginal study notes found within its two existing copies, *al-Ikmāl fī al-handasa* had a continuing impact on Ottoman science up until the nineteenth century.[5]

Ottoman science was greatly influenced by Muḥammad al-Fanārī (d. 834/1431), who lived at a time when the Ottoman state was transformed from an insignificant principality into a great empire. Those intellectuals interested in mathematical sciences who had gathered around Fanārī were evidently familiar with Archimedes. For instance, in his autobiography, entitled *Kitāb durrat tāj al-rasāʾil wa-ghurrat minhāj al-wasāʾil*, ʿAbd al-Raḥmān al-Bisṭāmī (d. 858/1454) refers to the mathematical work of Archimedes along with that of Pythagoras and Thābit ibn Qurra.[6] One notable work from this period is *Unmūdhaj al-ʿulūm* by Muḥammad Shāh al-Fanārī (d. 839/1435–6), the son of Muḥammad al-Fanārī.[7] The *Unmūdhaj* is a type of encyclopedic work collected according to the taste of the author. In the section on applied geometry (*ʿilm al-misāḥa*), Muḥammad Shāh presents the number pi, attributing it

[2] For a paper setting forth a framework for understanding Ottoman mathematical sciences, see Fazlıoğlu [2010].

[3] For more on Ibn Sartāq, see http://www.ihsanfazlioglu.net/yayinlar/makaleler/1.php?id=166 (accessed 6 October 2012). See also Rosenfeld and İhsanoğlu [2003, 221 (no. 612)] (a work to be used with caution).

[4] For copies of *al-Ikmāl*, see Istanbul, Askeri Müze [Military Museum] MS 64/2, folios 6b–290b and Cairo University MS 23209/2, folios 6b–225a. The copy available in Cairo was copied by Dāwūd al-Qayṣarī and includes correction notes of his teacher Ibn Sartāq along the margins.

[5] For more details, see Fazlıoğlu [1998]. Also, Ahmed Djebbar presents (in Arabic) the main content of the non-extant introduction of the *Kitāb al-Istikmāl* that is taken from Ibn Sartāq's redaction [Djebbar 1997]. Djebbar references (p. 185) the important marginal notes contained in Ayasofya MS 4830, ff. 108b, 165a, 169a, 171a, 178b.

[6] Istanbul, Nuruosmaniye Kütüphanesi [Library] MS 4905, ff. 1b–42b on f. 15a. For more, see Fazlıoğlu [1996].

[7] Istanbul, Süleymaniye Library, Hüsrev Paşa MS 482. For more on this work and its author, see Fazlıoğlu [2010]. Cf. Rosenfeld and İhsanoğlu [2003, 272 (no. 806)].

to Archimedes.[8] In another section, on the science of weights (*ʿilm al-athqāl*), he discusses the findings of Archimedes on hydrostatics and on specific gravity for seventeen substances.[9]

Muḥammad al-Fanārī and his intellectual circle restructured science in the Ottoman domains, thanks largely to a series of discoveries associated with the Marāgha Observatory that flourished in the second half of the thirteenth century.[10] These reached Anatolia during the time of the Seljuk Sultanate and the Turkish principalities. Afterwards Ottoman science continued to thrive not only due to its native scholars, such as those associated with the Fanārī-circle, but also as a result of contributions made by returning and emigrant scholars from Egypt, Syria, Iraq, Iran, and Central Asia. Especially after the Golden Horde went into a period of stagnation, a number of scholars migrated to Anatolia from the regions under its rule. Later, from the fifteenth century until the conquest of Crimea, successive waves of emigrant scholars also found their way to Ottoman lands.[11] Taken together with the conquest of Istanbul, which gave the Ottoman Empire leadership in the Islamic World, uninterrupted domestic stability persuaded even more foreign scholars to immigrate to Ottoman lands. We can mention here the scholars associated with the Ulugh Beg Observatory in Samarqand, who played an important role in Ottoman science. These included Qāḍīzāde al-Rūmī (d. ca. 847/1444), one of the founding members of this observatory who was originally from Bursa and learned astronomy in the circle of Muḥammad al-Fanārī. Later, such scholars as Fatḥallāh al-Shirwānī (d. 891/1486), ʿAlī Qūshjī (d. 879/1474), ʿAbd al-ʿAlī al-Birjandī (d. 932/1525–6) and their students helped Ottoman science flourish; these scholars transmitted works from the Ulugh Beg Observatory and put them into circulation and at the disposal of Ottoman scientific circles. The wave of migration from the regions of Central Asia and Iran into the Ottoman Empire gained momentum during the second half of the sixteenth century following the establishment of the Ṣafawid dynasty and the imposition of Shīʿism, but slowed down toward the end of the century.[12] Indeed, as we see in the case of Muṣliḥ al-Dīn al-Lārī (d. 979/1571), the Ottoman domain became the new home of scholars who had emigrated from as far away as the Indian Subcontinent. The fall of al-Andalus and persistent instability in North Africa also encouraged scholars from these lands, Muslim and non-Muslim, to resettle in the Ottoman Empire, particularly in Istanbul. Many other scholars did the same after the fall of the Mamluks, again heading to the Ottoman capital. As a result, significant contributions to Ottoman science were made by these numerous émigrés: studies on the mathematical sciences multiplied in the Ottoman Empire and the number of scholars dealing with these sciences increased. Archimedes and his works received their due share in this productive period of scientific and mathematical resurgence.

The Number Pi

One reason Archimedes is renowned as an exceptional mathematician is due to his research on the number pi. While some works use the number pi without reference to Archimedes, other works apply the number in connection with his name. As pointed out earlier, the oldest Ottoman source that uses the number pi with reference to Archimedes is Muḥammad Shāh al-Fanārī's *Unmūdhaj*. The next source known to us is *al-Iqnāʿ fī ʿilm al-misāḥa*, which an unknown author wrote in Arabic and presented to the reigning Sultan Mehmed the Conqueror (r. 848–50/1444–1446, 855–86/1451–1481). *Al-Iqnāʿ* mentions the number pi with reference to Archimedes and discusses whether a straight line (the diameter) can calculate a curved line (the circumference). In presenting the topic, the anonymous author states the following: "... and Archimedes came forth and studied the situation (*waḍʿiyya*)."[13]

Another source on the number pi is *Tadhkirat al-kuttāb fī ʿilm al-ḥisāb*, which was written by Ghars

[8] Hüsrev Paşa MS 482, ff. 170a–171a; Fazlıoğlu [2010, 144–145].

[9] Hüsrev Paşa MS 482, ff. 169a–170a; Fazlıoğlu [2010, 143–144].

[10] Ragep [2010].

[11] See Fazlıoğlu [2008a].

[12] For an elaborated discussion, see Fazlıoğlu [2008b].

[13] Istanbul, Süleymaniye Library, Ayasofya MS 715; see also Fazlıoğlu [2004, 72–73 (Turkish), 134–135 (Arabic)].

al-Dīn Aḥmad al-Ḥalabī, better known as Ibn Naqīb (d. 971/1563). An essential work in Ottoman mathematics and the mathematics of accounting, the *Tadhkira* was used extensively as shown by the number of copies found in manuscript libraries in Turkey. Due to its popularity in the field, Darwīsh b. Muḥammad b. Luṭfī (d. 979/1572), a student of Ibn Naqīb, translated the *Tadhkira* into Turkish at the request of Muḥammad Pāsha, the vizier of Sultan Salīm II. An important aspect of Ibn Naqīb's *Tadhkira* is that it applies Jamshīd al-Kāshī's (d. 832/1429) *al-Risāla al-Muḥīṭiyya* (Treatise on the Circumference) in studying the number pi and compares the approximation of al-Kāshī with that of Archimedes. The *Tadhkira* thus demonstrates that, at least during the sixteenth century, Ottoman mathematics, including the mathematics of accounting, was utilizing al-Kāshī's approximation of pi, which marked significant progress in the history of mathematics. We should also point out that the *Tadhkira* discussed decimal fractions when dealing with the number pi, which had also been a feature of al-Kāshī's work.[14]

Ottoman scholars were not merely content to mention Archimedes and his followers regarding the number pi but also undertook independent research on it. For instance, Jamāl al-Dīn Yūsuf b. Muḥammad al-Qurashī (10th/16th century) penned a treatise on the number pi entitled *Risāla fī maʿrifat kammiyyat muḥīṭ al-dāʾira*.[15] But it was Taqī al-Dīn al-Rāṣid (d. 993/1585) who conducted the most detailed research on the number pi. In his *Risāla fī taḥqīq mā qālahu ʿAllāma Ghiyāth al-Dīn Jamshīd fī bayān al-nisba bayn al-muḥīṭ wa-'l-quṭr*, Taqī al-Dīn reviews al-Kāshī's *al-Risāla al-Muḥīṭiyya* and elaborates on his ideas pertaining to operating with decimal fractions and the relationship between the circumference and diameter in a given circle.[16]

Furthermore, astronomical works produced in the Ottoman realm make frequent references to Archimedes as regards the number pi. One representative example is the *Khulāṣat al-hayʾa* by Sayyid ʿAlī Raʾīs (d. 970/1563), which is the first Turkish astronomy book in the field of planetary theory. Going beyond simply mentioning the ratio pi and attributing it to Archimedes, Sayyid ʿAlī Raʾīs provides a fairly extensive summary of the works of Archimedes on the circle.[17] As this case shows, future research on a number of different scientific genres will be needed to show to what degree Ottoman science was concerned with the works of Archimedes.

Hydrostatics and Specific Gravity of Elements

Archimedes's principle states "that a body immersed in a fluid is buoyed up by a force equal to the weight of the displaced fluid." Based on this principle, hydrostatics aims to determine the amount of each substance within a given composition made up of two different substances. That is, it studies the specific gravity of substances. We know of numerous works in hydrostatics that were written during the pre-Ottoman period. For instance, the specific gravity of elements and methods for calculation are provided in the section related to the science of measurement (*ʿilm al-misāḥa*) in Ibn al-Khawwām's (d. 724/1324) *al-Fawāʾid al-bahāʾiyya fī al-qawāʿid al-ḥisābiyya* and in two commentaries on it: ʿImād al-Dīn al-Kāshī's (d. ca. 740/1340) commentary entitled *Farāʾiḍ al-fawāʾid* and Kamāl al-Dīn al-Fārisī's (d. ca. 720/1320) *Īḍāḥ al-maqāṣid*. These two commentaries were available during the early years of the Ottoman Empire. The large number of extant copies suggests that all these works were studied later on as well. As was the case with many other arithmetic works that had a wide circulation in Ottoman lands, Jamshīd al-Kāshī's *Miftāḥ al-ḥisāb*, in a section under geometry, deals with existing findings in the field

[14] Istanbul, Köprülü Kütüphanesi MS 936, folio 120a–120b. For Ibn Naqīb, see İhsanoğlu et al. [1999, 1, 73–75] (henceforth OMALT); and Rosenfeld and İhsanoğlu [2003, 327 (no. 983)]. Also see Hogendijk [2008/2009].

[15] Istanbul, Süleymaniye Library, Laleli MS 2723/7, folios 47b–49a [autograph]. For Jamāl al-Dīn al-Qurashī, see OMALT [1, 100–101].

[16] Istanbul, Kandilli Rasathanesi [Observatory] MS 208/8, folios 94b–99b [autograph]. For Taqī al-Dīn, see OMALT [1, 83–87].

[17] Sayyid ʿAlī Raʾīs, *Khulāṣat al-hayʾa*, Istanbul, Süleymaniye Library, Ayasofya MS 2615, folios 52a–53a. For Sayyid ʿAlī Raʾīs, see İhsanoğlu et al. [1997, 1, 140–145].

and presents a table of specific gravity of minerals. Muḥammad Shāh al-Fanarı, in his *Unmūdhaj*, also presents information on the specific gravity of certain substances.[18]

Muṣliḥ al-Dīn ibn Sinān (fl. 906/1500) translated into Arabic, at the request of Sultan Mehmed II (the Conqueror), the study of an unknown Byzantine author on hydrostatics under the title *Sharḥ Kalima min kalimāt al-Aflāṭūniyya*. He finished the translation on 28 Rajab 905/≈28 February 1500 and presented it to Sultan Bāyazīd II.[19] As far as we know, this Muṣliḥ al-Dīn is the first compiler, as translator, of an independent work on hydrostatics in the Ottoman Empire up to this date. As should be clear, Muṣliḥ al-Dīn has misattributed this work to Plato; in a detailed analysis of the work, E. Wiedemann has shown that it relies on the works of al-Bīrūnī and follows al-Bīrūnī's methodology in the field of hydrostatics.[20]

It would, though, be left to the already mentioned Taqī al-Dīn al-Rāṣid to undertake the most comprehensive study of hydrostatics in the Ottoman realm. In his treatise *Risāla fī ʿamal al-mīzān al-ṭabīʿī* (Treatise on using the natural balance), Taqī al-Dīn begins by reviewing the ideas of Ibn al-Khawwām, Kamāl al-Dīn al-Fārisī, and Ghiyāth al-Dīn Jamshīd al-Kāshī before putting forward his own views on hydrostatics.[21] We present an edition of the unique manuscript witness below in an appendix to this article.

Geometry

As we have seen, a number of Ottoman works referred to Archimedes with regards to the number pi and the problem of squaring the circle (*tarbīʿ al-dāʾira*). Other problems associated with Archimedes included trisecting an acute angle (*tathlīth al-zāwiya*) and doubling the cube [or altar] (*taḍʿīf al-madhbaḥ*). Among Ottoman writers, we find these group of problems first being mentioned, though not necessarily in a strictly mathematical context, in various works of ʿAbd al-Raḥmān al-Bisṭāmī (d. 858/1454)[22] and the *Taḍʿīf al-madhbaḥ* of Luṭfallāh al-Tūqātī (Molla Luṭfī) (d. 899/1494).[23]

Scholarly interest in, and major research on, the works of Archimedes would come later in the eighteenth century. In this regard, we shall first mention Asʿad Efendī al-Yanyawī (d. 1143/1730).[24] Asʿad Efendī studied some geometric theorems of Archimedes that had not yet appeared in Arabic. Based on his studies of Archimedes's other works, he translated these theorems into Arabic under the title *Kitāb ʿamal al-murabbaʿ al-musāwī li-l-dāʾira*.[25] A careful study of the book indicates that Asʿad Efendī had mastered the relevant translations made under the early ʿAbbāsids. He had also acquainted himself with ancient Greek works of mathematics that European scholars had only recently rediscovered. In the introduction of his book, he describes Archimedes as a *muhandis*, which in classical Arabic meant geometrician but in more contemporary usage came to mean a technical-mechanical expert or engineer. He goes on to state that the main purpose of his book is to elaborate on certain geometrical premises of Archimedes. Asʿad notes that Archimedes's *Kitāb al-kura wa-'l-usṭuwāna* (Book on the sphere and cylinder) had already been translated into Arabic, whereas his *Kitāb khuṭūṭ lawlabiyya (?)* (Book on

[18] For more on the subject, see Rozhanskaya [1996], Mawaldi [1994, 434–439, 453–459], and Fazlıoğlu [2010, 131–163].

[19] Istanbul, Süleymaniye Library, Hasan Hüsnü Paşa MS 121/29, folios 144b–147a and Serez MS 3824/7, folios 72b–73b. For Muṣliḥ al-Dīn ibn Sinān, see İhsanoğlu et al. [2006, 1, 14–15]. Also see Adıvar [1943, 9].

[20] Wiedemann [1970, 1, 240–257 (=163–180, orig. pagination)]. This particular treatise is discussed in the third section ("Über eine dem Platon zugeschriebene Abhandlung über spezifische Gewichte," 1, 250–257 (=173–180)).

[21] Alexandria, Baladiyya (Municipal) Library MS 3762 jīm, folios 1a–4b.

[22] *Waṣf al-dawāʾ fī kashf āfāt al-wabāʾ*, Istanbul, Süleymaniye Library, Şehid Ali Paşa MS 2811/44, folios 260b–265b; *al-Adʿiya al-muntakhaba fī al-adwiya al-mujarraba*, Istanbul, Süleymaniye Library, Ayasofya MS 377/3, folios 51a–101b [autograph].

[23] Luṭfallāh al-Tūqātī [1940].

[24] His full name is Asʿad b. ʿAlī b. ʿUthmān b. Muṣṭafā al-Yanyawī. For his biography, see Salim [1315, 76–78], and OMALT [1, 175–176]. See also Berggren [1987, 102], who prefers to read the name as Yāninawī, which is a nisba referring to the city of Yannina, now in northwest Greece.

[25] Cairo, Dār al-kutub, Muṣṭafā Fāḍil Riyāḍa MS 41/22, ff. 144b–147a and Dār al-kutub Mīqāt MS 172/2, ff. 10b–16a.

spiral lines) and his *Kitāb qaṭʿ al-kura wa-'l-makhrūṭāt* (Book on the section of the sphere and the cone) had not yet appeared in Arabic.[26] Along with Archimedes's works, Asʿad Efendī also translated a Latin mathematical work as *Tarbīʿ al-dāʾira wa-nisba muʾallafa* (Quadrature of the circle and ratio of composition).[27]

The mathematician Badr al-Dīn Muḥammad (d. 1146/1733) continued the work of his father Asʿad Efendī and authored numerous works in the field of geometry. One of the two treatises that he wrote with reference to Archimedes is on heptasecting and polysecting the circle. His other treatise deals with trisecting the acute angle (*tathlīth al-zāwiya*) and the quadrature of the circle (*tarbīʿ al-dāʾira*), two major problems inherited from ancient geometry. In addition to Archimedes, Badr al-Dīn also dealt with the mathematical work of scholars such as Euclid, Apollonius, the Banū Mūsā, and Ibn al-Haytham. Interestingly, he presented his demonstrations in terms of kinematic geometry (*al-handasa al-mutaḥarrika*), meaning that he used motion (*ḥaraka*) in these demonstrations.[28] Badr al-Dīn Muḥammad generally published his works in volumes that included texts of previous authors with his annotations and studies.[29]

Another Ottoman mathematician interested in Archimedes's geometry works was Muṣṭafā Ṣidqī (d. 1183/1769).[30] As will be mentioned below, he not only reproduced relevant works in the field but in the year 1153/1740–41, he made a "correction" (*iṣlāḥ*) and "recension" (*taḥrīr*) of the *Kitāb ʿamal al-dāʾira al-maqsūma bi-sabʿa aqsām mutasāwiya* (Book on the construction of the circle divided into seven equal parts).[31] This work had originally been translated into Arabic by Thābit ibn Qurra and attributed to Archimedes.[32] As indicated in recent studies, the works of Muṣṭafā Ṣidqī made remarkable contributions to the field[33] and provide us with an indication of the advanced degree to which mathematics and geometry had reached in the pre-modern Ottoman period; they also show that someone raised and educated in the tradition of Ottoman science could have the competence and means to revise and edit sophisticated content found in ancient mathematical works.

Dissemination

Archimedes's own works circulated in the Ottoman realm just as widely as the studies mentioned above that referred to Archimedes's findings. Although determining the scope and extent of this circulation poses a daunting challenge, we believe the actuality of this circulation is beyond doubt based on the large number of extant manuscript witnesses. The reasons the works of Archimedes gained an important standing in the Ottoman realm is due in part to the legacy of the Seljuk Sultanate in Anatolia and in part to the immigration of scholars and knowledge referred to above. It is clear that the Marāgha Observatory must have played an important role: Quṭb al-Dīn al-Shīrāzī (d. 710/1311; from the observatory's first-generation of students) and Ibn Sartāq (from the observatory's second-generation) published and taught in Anatolia. Naṣīr al-Dīn al-Ṭūsī's (d. 672/1274) recensions (*taḥrīrāt*) also contributed greatly to the knowledge of Archimedes and the availability of his works in the Ottoman realm. Nevertheless, a more important source for knowledge of the Archimedian tradition arrived in Ottoman lands by way of the Samarqand mathematical-astronomical school (both before and after the death of Ulugh Beg [d. 853/1449]). Luminaries associated, whether directly or indirectly, with Ulugh Beg's Observatory, such as Qāḍīzāde, Fatḥallāh al-Shirwānī, ʿAlī al-Qūshjī, and ʿAbd al-ʿAlī al-Birjandī, brought their

---

[26] Berggren [1987, 102].

[27] Cairo, Dār al-kutub, Taymūr Riyāḍa MS 140/16.

[28] For another example of "moving geometry," see Berggren [1986, 82–84].

[29] Bir and Kaçar [2003].

[30] Fazlıoğlu [2006], and OMALT [1, 214–217].

[31] Cairo, Dār al-kutub, Muṣṭafā Fāḍil Riyāḍa MS 41/14.

[32] Whether this is an authentic work by Archimedes has been contested; see Hogendijk [1984]. The authenticity of the work is argued for by Knorr [1989].

[33] Abd al-Laṭīf [1993, 215, 246–256].

mathematical and geometrical books with them when they left Central Asia for Ottoman lands.[34] Once there, these works were reproduced and went into circulation. As an illustration of this process, we can point to the "middle books" (*Mutawassiṭāt*), the collection of ancient Greek and classical Islamic works on mathematics, geometry and astronomy that were supposed to be studied between Euclid's *Elements* and Ptolemy's *Almagest*. This collection, which contained two works by Archimedes, had been the subject of an important recension by Naṣīr al-Dīn al-Ṭūsī. The texts were collected and reissued in the Ottoman capital at the request of Sultan Mehmed II.[35]

Over time, Ottoman mathematicians would reproduce (*istinsākh*) these *Mutawassiṭāt* when deemed necessary. For instance, Muṣṭafā Ṣidqī — as we have seen a leading mathematician of the eighteenth century — copied the *Mutawassiṭāt* twice between 1144 and 1159 and taught them to students, including his favorite, Shukrzāde Fayḍallāh Sarmad.[36] He also produced a copy of *Taḥrīr kitāb al-ma'khūdhāt* (Book of Lemmas), which had been attributed to Archimedes, using both the translation of Thābit ibn Qurra and the recension by Naṣīr al-Dīn al-Ṭūsī. An examination of the mathematical corpus of Muṣṭafā Ṣidqī, as well as the manuscript copies he made, indicates that he had access to the following: texts written by Greek mathematicians and translated in Baghdad; works written by mathematicians during the classical Islamic period such as Thābit ibn Qurra, Ibrāhīm ibn Sinān, the Banū Mūsā, Abū al-Jūd ibn al-Layth, al-Bīrūnī, and Abū ʿAbdallāh al-Shannī; and Ottoman mathematicians such as Taqī al-Dīn al-Rāṣid, Asʿad al-Yanyawī and Badr al-Dīn Muḥammad.

It is beyond the scope of this article to present a complete list of all manuscript witnesses somehow associated with Archimedes that are extant in Turkish or even Istanbul manuscript libraries. The following partial list is meant to provide an indication of the manner in which they circulated and the extent to which Ottoman scholars were interested in them. A striking example is Üsküdar/Istanbul, Hacı Selim Ağa MS 743, which contains a copy of the *Taḥrīr kitāb al-kura wa-'l-usṭuwāna* (Recension of the book on the sphere and cylinder). There we discover that various scholars have penned comments: Abū al-Faraj (Barhebraeus, d. 1286) wrote his in Syriac, while Muḥammad b. Ibrāhīm al-Shirwānī, Mu'ayyad-zāde ʿAbd al-Raḥmān b. ʿAlī, a noted Ottoman scientist, and Asʿad Yanyawī, whom we have encountered above, added their notes in Arabic (f. 138a). Lastly we should note that a number of libraries holding copies of these works were madrasa libraries, thus indicating that works on the ancient sciences were held there and accessible to madrasa professors and students, contrary to what is sometimes alleged.

1. *Kitāb al-kura wa-'l-usṭuwāna* (Book on the sphere and cylinder): Istanbul, Süleymaniye Library, Hacı Beşir Ağa MS 440; Carullah MS 1475; Fatih MS 510.
2. *Kitāb misāḥat al-dā'ira* (Book on the area of the circle): Istanbul, Süleymaniye Library, Fatih MS 3414.
3. *Kitāb al-ma'khūdhāt* (Book of Lemmas): Istanbul, Süleymaniye Library, Hacı Beşir Ağa MS 440.
4. *Kitāb ʿamal sāʿat al-mā'* (Book on constructing a waterclock): Istanbul, Süleymaniye Library, Ayasofya MS 4861.[37]
5. *Maqālā Arshimīdis fī taksīr al-dā'ira* (Treatise by Archimedes on the area of the circle): Istanbul, Süleymaniye Library, Esad Efendi MS 2034; Askeri Müze nr. 73.[38]
6. *Taḥrīr kitāb al-kura wa-'l-usṭuwāna li-Arshimīdis* (Recension [by Naṣīr al-Dīn al-Ṭūsī] of the book on the sphere and cylinder of Archimedes): Istanbul, Köprülü Fazıl Ahmed Paşa MS 930; Köprülü

[34] For example, seventeen texts included in the *Mutawassiṭāt* are available in Askeri Müze [Military Museum] under MS 83. The copyist (*mustansikh*) of these texts is the famous mathematician and writer on optics Kamāl al-Dīn al-Fārisī. Another example is a collection that includes the geometry works of Ibn al-Haytham, which must have reached Istanbul around this time (Askeri Müze MS 3025).

[35] Askeri Müze [Military Museum] MS 82. The nineteen texts in this codex were reproduced in the years 882–883/1477–1479 at the request of Sultan Mehmed II (the Conqueror), in order that he be able to study them.

[36] Cairo, Dār al-kutub, Muṣṭafā Fāḍil Riyāḍa MS 40 (containing 22 works; 227 folios) and MS 41 (containing 31 works; 187 folios). For more details on Muṣṭafā Ṣidqī, see İzgi [1997, 1, 302–306]

[37] Şeşen [1997, 289–290].

[38] Şeşen [1997, 820].

Fazıl Ahmed Paşa MS 931; Süleymaniye Library, Ayasofya MS 2760; Süleymaniye Library, Ayasofya MS 2758; Üsküdar/Istanbul, Hacı Selim Ağa MS 743.

7. *Sharḥ tarjamat kitāb al-maʾkhūdhāt li-Arshimīdis* (Commentary [by ʿAlī ibn Aḥmad al-Nasawī] on the translation of the book of lemmas of Archimedes): Istanbul, Süleymaniye Library, Ayasofya MS 2760.
8. *Risāla fī al-burhān ʿalā al-muqaddima allatī ahmalahā Arshimīdis fī kitābihi tasbīʿ al-dāʾira wa-kayfiyyat ittikhādh dhālika* (Treatise on the proof of a premise that Archimedes omitted in his book on heptasecting the circle and the way to use it [by Abū al-Fatḥ Kamāl al-Dīn Mūsā ibn Yūnus al-Mawṣilī (d. 639/1242)]: Manisa, İl Halk Kütüphanesi [Provincial Public Library] MS 1706.[39]

References

ʿAbd al-Laṭīf, ʿAlī Isḥāq, 1993. Ibn al-Haytham: ʿālim al-handasa al-riyāḍiyya: dirāsa taḥlīliyya wa-taḥqīq. al-Jāmiʿa al-Urduniyyah, Amman.

Adıvar, A., 1943. Osmanlı Türklerinde İlim. Maarif matbaası, Istanbul.

Berggren, J.L., 1986. Episodes in the Mathematics of Medieval Islam. Springer-Verlag, New York.

——— 1987. Archimedes among the Ottomans. In: Berggren, J.L., Goldstein, B.R. (eds.), From Ancient Omens to Statistical Mechanics: Essays on the Exact Sciences Presented to Asger Aaboe. Acta Historica Scientiarum Naturalium et Medicinalium, vol. 39. University Library, Copenhagen, pp. 101–109.

Bir, A., Kaçar, M., 2003. Bedrettin Muhammed el-İstanbuli'nin Teslis-i Zaviye (Trisecting the angle) ve Tesbi'-i Da'ire (Heptagoning the circle) Risaleleri. Osmanlı Bilimi Araştırmaları 4/2, 1–20.

Djebbar, A., 1997. La rédaction de L'istikmāl d'al-Muʾtaman (XIe s.) par Ibn Sartāq, un mathématicien des XIIIe–XIVe siècles. Historia Mathematica 24, 185–192.

Fazlıoğlu, İ., 1996. İlk Dönem Osmanlı İlim ve Kültür Hayatında İhvânu's-Safâ ve Abdurrahmân Bistâmî. Dîvân İlmî Araştırmalar Dergisi 2, 229–240.

——— 1998. Osmanlı Coğrafyasında İlmî Hayatın Teşekkülü ve Davud el-Kayserî. In: Koç, T. (ed.), Uluslararası XIII. ve XIV. Yüzyıllarda Anadolu'da İslam Düşüncesi ve Davud el-Kayserî Sempozyumu. Kayseri Büyükşehir Belediyesi Kültür Müdürlüğü, Kayseri, pp. 25–30.

——— 2004. Uygulamalı Geometrinin Tarihine Giriş: el-İkna fi ilmi'l-misaha. Dergah Yayinlari, Istanbul.

——— 2006. Mustafa Sıdkı. In: Güran, K. (ed.), Türkiye Diyanet Vakfı İslam Ansiklopedisi, vol. 31. Türkiye Diyanet Vakfı, Istanbul, pp. 356–357.

——— 2008a. Altın-Orda Ülkesi'nde İlk Matematik Kitabı: Hesap Biliminde Şaheser [et-Tuhfe fî ilmi'l-hisâb]. In: Çakmak, C. (ed.), Teoman Duralı'ya Armağan: Bir Felsefe-Bilim Çağrısı. Dergâh Yayınları, Istanbul, pp. 224–259.

——— 2008b. The Samarqand mathematical-astronomical school: A basis for Ottoman philosophy and science. Journal for the History of Arabic Science 14, 3–68.

——— 2010. İthâf'tan Enmûzec'e Fetihten Önce Osmanlı Ülkesi'nde Matematik Bilimler. In: Yücedoğru, T., Koloğlu, O., Kılavuz, U.M., Gömbeyaz, K. (eds.), Uluslararası Molla Fenârî Sempozyumu (4–6 Aralık 2009 Bursa) (International symposium on Molla Fanārī, 4–6 December 2009 Bursa). Bursa Büyükşehir Belediyesi, Bursa, pp. 131–163.

——— 2012. İbn Sertâk [Muhammed Sertakoğlu]. http://www.ihsanfazlioglu.net/yayinlar/makaleler/1.php?id=166 (accessed 6 October 2012).

Hogendijk, J.P., 1984. Greek and Arabic constructions of the regular heptagon. Archive for History of Exact Sciences 30, 197–330.

——— 2008/2009. Al-Kāshī's determination of π to 16 decimals in an old manuscript. Zeitschrift für Geschichte der arabisch-islamischen Wissenschaften 18, 73–153.

[39] Şeşen [1997, 182].

İhsanoğlu, E. et al. (eds.), 1997. Osmanlı Astronomi Literatürü Tarihi (History of Astronomy Literature During the Ottoman Period). 2 vols. IRCICA, Istanbul.

——— 1999. Osmanlı Matematik Literatürü Tarihi (History of Mathematical Literature During the Ottoman Period). 2 vols. IRCICA, Istanbul.

——— 2006. Osmanlı Tabii ve Tatbiki İlimler Literatürü Tarihi (History of the Literature of Natural and Applied Sciences During The Ottoman Period). 2 vols. IRCICA, Istanbul.

İzgi, C., 1997. Osmanlı Medreselerinde İlim. 2 vols. İz, Istanbul.

Knorr, W.R., 1989. On Archimedes' construction of the regular heptagon. Centaurus 32, 257–271.

Luṭfallāh al-Tūqātī (Molla Luṭfī), 1940. La duplication de l'autel (Platon et le problème de Délos), texte arabe publié par Serefettin Yaltkaya, traduction française et introd. par Abdülhak Adnan Adıvar et Henry Corbin. E. de Boccard, Paris.

Mawaldi, M., 1994. Kamāl al-Dīn al-Fārisī's Asās al-qawāʿid fī uṣūl al-fawāʾid. Institute of Arabic Manuscripts, Cairo.

OMALT = İhsanoğlu, E. et al. (eds.), 1999.

Ragep, F.J., 2010. Astronomy in the Fanārī-Circle: The critical background for Qāḍīzāde al-Rūmī and the Samarqand school. In: Yücedoğru, T., Koloğlu, O., Kılavuz, U.M., Gömbeyaz, K. (eds.), Uluslararası Molla Fenârî Sempozyumu (4–6 Aralık 2009 Bursa) (International symposium on Molla Fanārī, 4–6 December 2009 Bursa). Bursa Büyükşehir Belediyesi, Bursa, pp. 165–176.

Rosenfeld, B.A., İhsanoğlu, E. (eds.), 2003. Mathematicians, Astronomers & Other Scholars of Islamic Civilisation and Their Works (7th–19th c.). IRCICA, Istanbul.

Rozhanskaya, M. (in collaboration with Levinova, I.S.), 1996. Statics. In: Rashed, R., Morelon, R. (eds.), Encyclopedia of the History of Arabic Science. 3 vols. Routledge, London and New York, vol. 2, pp. 614–642.

Salim, M.E., 1315 H. Tezkire-i Salim, Edited by Paşa Ahmet Cevdet. İkdam Matbaası, Der Saadet.

Şeşen, R., 1997. Mukhtārāt min al-makhṭūṭāt al-ʿArabiyya al-nādira fī maktabāt Turkiyya. Waqf al-abḥāth li-ʾl-tārīkh wa-ʾl-funūn wa-ʾl-thaqāfa al-islāmiyya (ĪSĀR), Istanbul.

Wiedemann, E., 1970. Über Bestimmung der spezifischen Gewichte (=Beiträge zur Geschichte der Naturwissenschaften. VIII.). In: Wiedemann, E., Aufsätze zur arabischen Wissenschaftsgeschichte, 2 vols. Georg Olms Verlag, Hildesheim and New York, vol. 1, pp. 240–257 (pp. 163–180 in the original pagination).

Appendix

Treatise on Constructing the Natural Balance

By Taqī al-Dīn al-Rāṣid

Alexandria, Public Library MS 3762 jīm

The unique manuscript has been mostly transcribed as is, which accounts for some inconsistencies such as in overlining; however, some corrections have been made (see notes), orthography of hamzas has been modernized, and shaddas (except for sun letters and feminine nisbas) have been supplied. The lemmas in the notes before the bracket indicate our corrections or readings; words after the bracket (with the exception of our comments in square brackets, [ ]) are what is in the manuscript. If there is more than one occurrence of a lemma in a line, the order of the noted expression is indicated with a superscript numeral. Additions and comments by the editors, in the text or the notes, are in square brackets, [ ].

We thank Ms. Sally P. Ragep for drafting the diagrams and Prof. Youssef Ziedan of the Bibliotheca Alexandrina for providing images of the manuscript.

[1a] رسالة في علم [!] الميزان الطبيعي

[1b] ... الطبيعى و ...

باسم نوءٍ اسم بواسم صبيب صلوات الرحمن على رسوله وحبيبه المشفع عند نصب الصراط والميزان وعلى آله المصطفين الأبرار، وأصحابه المحبّين الأخيار، ما تناسقت النسب الطبيعية وتعاقبت الأدوار وكوّر النهار على الليل والليل على النهار. وبعد:

فهذه رسالة لطيفة في عمل الميزان الطبيعي الذي يعلم به كمّية ما في الجسم المركّب من معدنين مختلفين من كلّ واحد منها من غير هدم لشكله المركّب عليه وفي معرفة العمل به وبيان النسبة الطبيعية بينها ألّفتها تحفة لحضرة آصف الدولة السليمانية ووزيرها ومدبّر أوامرها النافذة في الآفاق، ومشيرها ذو الذ[كر] الشديد الثاقب والفكر المجيد الصائب سمي ليث الله الغالب صفى خليفة الله ذي المناقب علي باشا، لا برحت موازين عدالة بالقسط قائمة على الدوام، مكلوؤة بعين عناية الملك العلاّم. ورتّبتها على مقدّمة وبابين وخاتمة، وبالله التوفيق والهداية إلى سواء الطريق.

المقدّمة في معرفة التفاوت بين الأجسام المتساوية الحجم وزناً

والكلام في هذا المقام يستدعي تمهيد قواعد نسبية يعلم منها ما بين كلّ جسمين من التفاوت المذكور وعكسه أي تفاوتهما في الحجم عند تساوي الوزن. فمن الحكماء من احتال لذلك بعمل مكعّبات متساوية من كلّ واحد من الفلزّات ثمّ وزنها فوقف على ذلك. ومنهم من عمل من كلّ منها شريطاً جذبه من ثقب واحد وقطعها على طول واحد وحرّر أوزانها فظهر لمدّ[؟] التفاوت.

2 ... الطبيعى و ...] [غير مقروء]. 3 الصراط] الطراط. 7 وبيان] وبياة. 8 آصف] ا آصف. ذو] ذوا. 13 قواعد] قوعد.

وعندي أنّ ذلك لا يفيد تقريباً قريباً فضلاً عن التحقيق. بيانه أنّه لا يمكن جعلها على مقادير محقّقة التساوي بالتطريق والمساواة بالمبرّد، وهو ظاهر. ولا بالسبك على قالب معيّن، لأنّ كلّ واحد منها له تخلخل إذا أذيب، وتكاثف إذا جمد؛ وذلك لا على التساوي بل على النسبة الوزنية فيه، وقوّة التركّب في معدنه وضعفها، وهو ظاهر أيضاً. وأمّا في الشريط المجذوب فلذلك إذ سعت الثقب الذي يجذب منه لا تستمرّ على حاله إلى آخر العمل، بل كلّما جذب منها واحداً اتّسعت، وأثر ذلك في المجذوب ثانياً ثخناً مّا. وأيضاً فالألين من الفلزّات يحصل له امتداد مّا بقوّة الجذب، ولا لذلك في الأيبس. ولم يزل ذلك يخالج فكري منذ طالعت كلام العماد الخوّام – رحمه الله تعالى – خصوصاً. وقد قال الحسن الفارسي "لا سبيل إلى تصحيح ذلك"؛ حتى وقفت من كلام الراصد المتقن جمشيد – تغمّد الله برحمته – في فنّ المساحة ما أسقى الغليل بواضح البرهان والتعليل، وها أنا ذاكر منها ما يحتاج إليه تبرّكاً بعبارته وبالله المستعان.

أقول: إذا كان جسمان متساويان في الحجم، مختلفان في الوزن، فإنّ نسبة وزن الأوّل إلى وزن الثاني عند تساوي حجمهما كنسبة حجم الثاني إلى حجم الأوّل عند تساوي وزنهما.

والحيلة إلى معرفة هذه النسبة أن نأخذ قمقم تكون أنبوبتها منحية مائلة الرأس إلى أسفل، ونملأها ماءً صافياً ونضع كفّة ميزان تحتها. فإذا أولجنا فيها شيئاً من الفلزّات أو الجواهر، وينبغي أن يكون مصمتاً لا مجوّفاً.

أقول: أمّا قوله وينبغي إن أراد به عدم التجويف مطلقاً سواء كان له منفذ أو لا فهو قيد وجوب إن أراد به بعض المجوّفات كالحقّة والطاسة، فقيد أولوية إذ يجوز إدخاله للماء بحافته بحيث لا يؤثّر [2a] تجويفه في كثرة إخراج الماء عن القدر المخصوص به.

وأقول: أيضاً إنّه ينبغي تقييد قوله ونملأها ماءً صافياً بحيثية سيلان الماء من الأنبوب، وانقطاعه لنفسه؛ وأن تبقى القمقمة على ذلك الوضع إلى آخر العمل؛ وأن يكون العمل من أوّله إلى آخره بماءٍ واحد، إذ للمياه ثقل وخفّة في الجملة.

عود إلى سوق كلامه رحمه الله قال: فخرج من الأنبوبة بقدر حجم ذلك الجسم ماءً؛ فإذا اسقطنا فيها جسماً آخر يكون وزنه مساوياً للجسم الأوّل فخرج منها مقدار آخر من الماء، فتكون نسبة وزن الماء الأوّل إلى وزن الماء الثاني كنسبة حجم الجسم الأوّل إلى حجم الجسم الثاني. وهذه صورة ذلك:

فإذا أولجنا القمقمة مائة مثقال مثلاً من كلّ واحد من الأجسام التي سنوردها في الجدول وعلمنا وزن كلّ واحد من المياه الخارجة، تحصل لنا نسبه حجم بعضها مع بعض عند تساوي الوزن، بل نسبة وزن بعضها مع بعض عند تساوي الحجم بالتكافؤ. ولاستخراج نسب المائعات، ينبغي أن يؤخذ إناء، ويعرف كم يسع ماء. وهكذا كم يسع من كلّ مائع ليعرف نسبة وزن الماء إلى وزن كلّ واحد منها عند تساوي الحجم.

---

[3] ظاهر] ظ. [10] وزن][1] ىالوزن. [11] حجمهما] حجمها. [13] وينبغي] وينبغ. [21] فتكون] فيكون. [23] الجدول] الجدوول. [25] بالتكافؤ] بالتكافى. ماء] ما.

أقول: وبذلك يعرف ما في الجسم المائع الحاصل من خلط مائعات متعدّدة معلومة الأسماء مجهولة المقادير كما لا يخفي. ولذلك لم أورد ذلك في هذه الرسالة، وهذا ما يحتاج إليه من جدوله الموعودة بذكره؛ أوردت ذلك فيه بالرقم الهندي، وإلى جانبه مجتنبها إلى الطساسيج ومرفوع ذلك، ومنحطّ إلى جانبه بالرقم الجملي، ضبطاً لصحّته عند تداول أيدي الكتاب.

الباب الأول
في معرفة عمل الميزان المذكور

وطريقه أن تتّخذ من الصفر عمود ميزان على الوضع المألوف إلاّ أنّ [2b] جناحي العمود يكونان على شكل مثلّث الثخن، زاويته العليا حادّة؛ وليبالغ في مساواة ثخن الجناحين، ووضعهما على خطّ مستقيم بقدر الإمكان. وليكون بعد الثقبين الّذين في طرفيه على بعد واحد من القطب، ويكون اللسان قائماً على العمود قياماً في نهاية الاعتدال، ومسقط حجره الحقيقي القطب. وليكن أحد الثقبين مخروقاً من الجهة العليا، وليأخذ من الجهة الأخرى من عند الثقب بقدر ذلك ليعتدلا. ثمّ يتّخذ كفّتين مسطّحتين من الصفر، وليكونان من صفر واحد غير مختلف وإن سبك؛ ثمّ عمل منه ذلك مبالغة في تساوي الماهية فيهما لكان أولى؛ وليكونان على قدر واحد ووزن واحد، له اختلاق فيه أصلاً، ليحترز غاية الاحتراز عن أن يقع في سبكها أو تطريقها مكان مجوّف أو مطبق بحيث يتخلّله الهواء، فلا يستقيم فيه عمل. ثمّ يعمل لكلّ منها من ذلك الصفر ثلاث علاقات من الشريط المتعادل وزناً وهيئة وطولاً، وتجمع رؤوس شريط كلّ كفّة بحلقة صغيرة، وتثبت إحداهما في الثقب الذي لم تخرقه، وتدع الأخرى لتعلّق في الثقب المخروق وقت العمل، وتكون سعة الحلقة بحيث إذا سيّرت على العمود سارت كثقالة الكرسطون. ثمّ يكمل هذا العمود بعلاقة كعادة الموازين مجتهداً في تحريره بحيث يتطفّف بأدنى ثقل يوضع في إحدى كفّتيه. وقد تمّ عمله وهذه صورته:

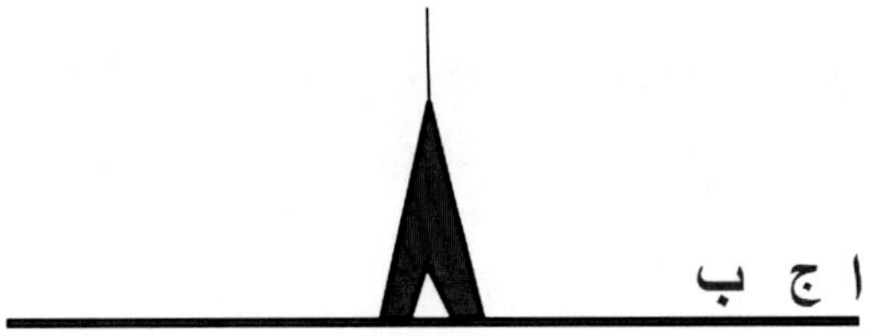

الباب الثاني
في
كيفية الوزن به ووضع المراكز على العمود

وطريقه أن يوضع في باطية ماء صاف. ثمّ تؤخذ قطعة مصمتة من الذهب الخالص، وتوضع في الكفّة السيّارة؛ ثمّ تقابل بوزنها في الهواء من الفضّة الخالصة بالتحرير التامّ؛ ثمّ تدلّى الكفّتان في الماء بحيث يغمر الماء كلاًّ من الجسمين. فمن الواضح أنّ كفّة الذهب ترجّح على أختها متردّ[؟] حلقتها عن مقرّها، وهو نقطة ا إلى جهة العمود حتّى يعتدلا في الماء، فتعلم على مقرّ الحلقة علامة بعلامة ب مثلاً، وهو مركز الذهب وح؛ فمتى أردنا معرفة ما في الجسم المركّب من فضّة وذهب مثلاً علمنا مركز الذهب كما تقدّم؛ ثمّ وضعنا الجسم المذكور في الكفّة الثابتة، ووزنّاه بثقله في الهواء من الذهب، أدليناهما في الماء، وحرّكنا الحلقة إلى أن يعتدلا. فمن الجلي أنّ الاعتدال يقع بين علامتي ا ب.

[3a] فننظر في نسبة ما بين نقطة ا وموقع الاعتدال كنقطة ج مثلاً إلى ما بين ا ب من المسافة، فما كان ففي الجسم الممتزج بمثله فضّة والباقي وهو نسبة ما بين جب إلى ا ب ذهب. والسبيل إلى تحرير هذه النسبة أن تجزّئ

---

16 وتكون] ويكون. 22 ماء] ماءً. تؤخذ] يوخد. 23 تدلّى] يدلى.

ما بين ا ب بأجزاء القيراط أو الأصابع، وتنظر في نقطة ج، أين وقعت منها، وعلى هذا فقس، وبالغ في تحرير القسمة مهما أمكن، وفي كثرة تجزئتها لتقف على التقريب المقارب للصواب إن شاء الله تعالى.

تنبيه

متى كان الجسم مصمتاً فذاك وإلّا فمتى كان مجوّفاً كمكحلة وما أشبه ذلك. فعند إنزاله الماء يجب المبالغة في إدخال الماء إلى جوفه بحيث لا يحتبس في جانبه شيء من الهواء أصلاً.

فإن قلت: إنّ ما في هذا الجدول قد علمت مقادير تفاوته بماءٍ مّا، والمياه تختلف خفّة وثقلاً كما تقدّم، فينبغي بل يجب استخراج ذلك الأصل في كلّ وقت من أوقات العمل؛ ثمّ يتمّم العمل بذلك الماء.

فأقول: إنّ ثقل الماء وخفّته لا أثر له في تغيير تلك الأصول متى كان امتحانها كلّها بماء واحد مرّةً ثمّ بأخرى مرّةً أخرى غاية ما في الباب أن يلزم من ذلك تغيّر أوزان مياهها، لكنّ على النسبة الطبيعية التي اعتبرت أوّلاً فالأصول المذكورة نسبتها باقية بالقياس إلى بعضها، وعلى هذا فلا يجوز في عمل واحد أن يستخرج بعضه بماءٍ وبعضه بماءٍ آخر مخالف للأوّل غير أنّه لو أتمّ العمل بماءٍ ثمّ استؤنف بماءٍ آخر مخالف له لم يوجب ذلك إلاّ تزحزح المراكز على نمط طبيعي لا تتغيّر به المقادير النسبية كما لو كانت بالثلث أو الربع لم تتغيّر أبداً وإن تزحزحت المراكز.

فإن قلت: إنّ تزحزح المراكز إنّما يلزم أن لو أدليت كلّ كفّة في إناء فيه ماءٌ مخالف لمقابله، وأمّا إذا وضعتا في أحدهما مرّة دفعة ثمّ في الآخر أخرى لم يلزم ذلك؛ وإن كان الماء الثاني مخالف لأنّ نسبة ما حسره أحد الجرمين من الماء الأوّل إلى ما حسره من الماء الثاني كنسبة ما حسره الجسم الآخر من الماء الأوّل إلى ما حسره من الماء الثاني، فلم يلزم تزحزح المراكز.

فأقول: إنّ النسبة كالنسبة مشابهة لا مساواة. وإذا كان الأمر كذلك فلا مساواة في المراكز باختلاف المياه بل فيها مشابهة على نمط واحد بتزحزح واحد متشابهة، كلّما اعتبر بمياه مختلفة، وذلك لأنّ قيام العمود بالفضّة وميله بالذهب في ماء عذب أضعف منه في ماء أجاج كما يشهد به العيان. وإذا تقرّر ذلك فنسبة القيام الأوّل في الماء الأوّل إلى القيام الثاني في الماء الثاني كنسبة بعد المركز الأوّل عن القطب إلى بعده عنه ثانياً وهو ظاهر.

تذنيب

متى أردنا معرفة [3b] ذلك بصنج من حديد فإنّنا نستخرج مركز الذهب مع الحديد كما تقدّم في مثاله مع الفضّة، وليكن على نقطة ب. ثمّ يستخرج الفضّة معه أيضاً، ويفرضه على نقطة ج. فمن الواضح أنّ المركز الأوّل يقع في جهة اللسان، والثاني في جهة كفّ الذهب. ثمّ نزن الجسم المركّب من الذهب والفضّة بمثله من الحديد ونُدليهما في الماء ونعادل بينهما بتسيير الحلقة كما مرّ؛ فبالضرورة يقع الاعتدال بينهما فيما بين المركزين، وليكن على نقطة د مثلاً، وننسب ما بين جد إلى جب. فما كان نفى الجسم بمثله ذهب والباقي فضّة.

وعلى هذا القياس لو اعتبرنا ذلك بصنج من النحاس أو القلعي، إذ هما أخفّ من الجسم المركّب المذكور، وبذلك يعتبر وزن خاتم من فضّة أو ذهب له فصّ من ياقوت أو غيره من الأحجار إمّا بالطريق الأوّل بأن يكون الصنج ذهباً إن كان الخاتم منه أو فضّة إن كان منها، وذلك بأن يستخرج مركزه مع الحجر ثمّ نكمل العمل وإمّا بالطريق الثاني وذلك بأن يستخرج مركز الفلزّ المركّب منه الخاتم مع ذلك الصنج الأخفّ من كلّ من جزء الخاتم، ثمّ مع

---

10 بعضه][1] [فوق السطر]. 19 ماء][2] ما. 20 الثاني][2] [أغلب الكلمة ناقصة من المخطوط]. ظاهر] ظ. 22 معرفة] [بعض الكلمة ناقصة من المخطوط]. 23 نقطة][1] نقط. ج] ح. 26 جب] حب. 30 جزء] جزوى.

الحجر ويكمل العمل كما مرّ. ولا يخفى أنّه لو اعتبر بأثقل من كلّ منهما بأن كان الخاتم مركّباً من فضّة وياقوت والصنج ذهباً لكان العمل بعكس ما مرّ، ولظهوره بالتأمّل لم انفرض له. أمّا لو كان الصنج أخفّ من أحدهما وأثقل من الآخر فلا يستقيم مطّرداً في هذا الميزان. وسيأتي الإيماء إلى ذلك إن شاء الله تعالى.

مهمّة هذا الباب له دخل كبير في معرفة ما في الذهب من الغشّ ومقداره؛ فلعلّ ذلك لا يظهر بالمحك نشرة علاج سطحه. فإنّ وضع المبرد أو القلم على كلّ ما يشترى من الحلي في كلّ وقت غير متيسّر وفيه تشويه، فمعرفته بهذا الطريق لا نظير لها.

خاتمة

أقول: إنّه يجوز تركّب جسم من جسمين بحيث يكون في ثقله وحجمه متساوياً لجسم آخر كجسم مركّب من ذهب وقلعي؛ نسبة ما بين الذهب والقلعي فيه كنسبة تفاوته مع الذهب إلى تفاوته مع القلعي في الأصول المذكورة بالإبدال في وزن الفلزّين. وليكن فضّة مثلاً: وإذا كان الأمر كذلك فزنته مع الفضة في الهواء معادل لزنته معها في الماء لاستوائهما في الحجم. وأنا أسمّي هذا الجسم المعتدل الفلزّي. وفيما مضى من الأصول قد تقرّر أنّ هذه النسبة في هذا المثال نسبة الثلث في إبدال ذلك [و]يكون ما في المعتدل الفلزّي المذكور من الذهب مثلان وزناً مساويان لمثل من فضّة حجماً ومثل من القلعي وزناً مساوٍ لثلثين منها حجماً فثلاثة أمثال من الفضّة مساوية لذلك حجماً [4ª] ووزناً. ويلزم من ذلك استواءهما في الميزان في الماء والهواء. فمتى وزنّا مركّباً من معدنين في طرفين بمعدن بينهما فتساويا في الماء والهواء، فإنّا نعلم من ذلك مقدار كلّ جزء من أجزائه بالنسبة المذكورة، ويكون ذلك معتدل فضّياً إن كان المقابل فضّة وحديدياً وإن كان حديداً وهلمّ جراً.

[التالي مشطوب في المخطوط]

الخاتمة في معرفة استخراج مقادير ما في إناء مركّب من ثلاثة أجزاء من كلّ واحد منهما. وليكن ذلك حقة من الفضة، مرصّعة بياقوت وعروق من ذهب. وطريقه أن يستخرج بصنج من الذهب مركز الذهب مع الفضّة وهو نقطة ج ، ثمّ مركزه من الياقوت، وهو نقطة ب ، ثمّ نضع الجسم المذكور في الكفّة اليسرى، ونعادله بمثله ذهباً خالصاً ونذليهما في الماء ونعادله بتسيير الحلقة كما مرّ. فمن الواضح أنّ ما في الحقة من الذهب نعادله مقداره من الذهب المعيّر به ويقع الاعتدال بين ذينك المركزين على نسبة ما فيها من الياقوت إلى ما فيها من الفضّة. وليكن على نقطة د ما بين نقطتي فضّة وما بين نقطتي ياقوت؛ وما بقي من وزنه ذهب.

[نهاية النصّ المشطوب]

لا يخفى على المتدبّر بعد ذلك كيف يتصرّف فيما لم يذكر من النسب. وفي هذا القدر كفاية، وبالله التوفيق والهداية إلى سواء الطريق. قال ذلك وكتبه فقير عفو ربّه الرؤوف محمّد تقي الدين بن القاضي معروف الحنفي الدمشقي لطف الله بهما، تمّ. [4ᵇ]

[13] فثلاثة] فثلثه. [15] جزء] جزو. [18] ثلاثة] ثلثه. [19] مرصّعة] +من[؟] [الكلمة الزائدة مشطوبة]. [23] نقطتي][1] [الحرفان ناقصان]. نقطتي][2] [الحرفان ناقصان]. [26] الرؤوف] الروف.

الأوزان / الأجسام	المثاقيل	دوانيقها	طساسيجها	قراريط الطسّوج		مجنّسها طساسيج	ضبطها بالجمل طساسيج مرفوع	[وحدات]	دقائق
الذهب	١٠٠	٠	٠	٠		٢٤٠٠	م	؛	؛
الزيبق	٧١	١	٠	١١		١٧٠٨	كح	كح	كح
الأسرب	٥٩	٢	٢	١٠		١٤٢٦	كج	مو	كه
الفضّة	٥٤	٠	٣	٢٠		١٢٩٩	كا	لط	نا
الصفر	٤٦	١	٣	١٨		١١١١	يح	لا	مو
النحاس	٤٥	٢	٣	١٧		١٠٩١	[يح]	يا	مب
الشبه	٤٥	٠	٠	٠		١٠٨٠	يح	؛	؛
الحديد	٤٠	٣	٣	١٢		٠٩٧٥[1]	يو	يه	كط
الرصاص القلعي	٣٨	٢	١	٢٣		٠٩٢١	يه	كا	نز
الياقوت الأحمر	٢٠	١	٠	١٥		٠۴٨٤	ح	د	لز
اللعل	١٨	٤	٣	٨		٠٤٥١	ز	لا	كا
الزمرّد	١٤	٢	٢	١٩		٠٣٤٦	٥	مو	لز
اللاجورد	١٤	٠	٣	٠		٠٣٣٩	٥	لط	ا
اللؤلؤ	١٣	٣	٣	٦		٠٣٢٦[2]	٥	كو	يو
العقيق	١٣	٢	٣	٢		٠٣٢٣	٥	كج	و
البرّ	١٣	٢	٢	٠		٠٣٢٢	٥	كب	ا
البلّور	١٣	٠	٣	٠		٠٣١٥	٥	يه	؛
الزجاج	١٣	٠	١	١٧		٠٣١٣	٥	يج	مب
الأنبوس	١١	١	١	١		٠٢٦٩[3]	د	كط	ب
العاج	٨	٣	٢	١٣		٠٢٠٦	ج	كو	لج

1 ٠٩٧٥] ٠٩٥٧.

2 ٠٣٢٦] ٠٣٢٧.

3 ٠٢٦٩] ٠٢٢٩.

The "Second" Arabic Translation of Theodosius' *Sphaerica*

Richard Lorch

Abstract From a short comparison of terminology it is suggested that the Arabic text of Theodosius' *Sphaerica* carried by two manuscripts in Hebrew script is a translation different from the one translated by Gerard of Cremona into Latin. An edition of a lemma for Proposition III.11 *inter alia* supports the hypothesis that this translation is the basis of Moses b. Tibbon's Hebrew version.

Like many other Greek mathematical and scientific works, Theodosius' *Sphaerica* was translated into Arabic in the ninth century. Of the Arabic version that Gerard of Cremona translated into Latin in the twelfth century at least three manuscripts are known:[1]

A: Istanbul, Seray, Ahmet III 3464, ff. 20v–53v
N: Lahore, private library, M. Nabī Khān, pp. 185–281
H: Paris, Bibliothèque nationale de France, heb. 1101, ff. 1r–53r, 86r–87r.

It will be referred to as **ANH**.

There was at least one other Arabic translation of the work: that in manuscripts

F: Florence, Laur. Med. 124
C: Cambridge, University Library, add. 1220, ff. 1r–50r.

Both manuscripts are in Hebrew script — and so may be dated, perhaps, to the fourteenth century and assigned to the western area of the Arabic tradition. A preliminary comparison with the text translated by Gerard may be made by taking as examples four short enunciations specifying construction (ex. 1–4), I 19, I 20, I 21 and I 22.[2]

1. Τοῦ δοθέντος ἐν σφαίρᾳ κύκλου τὴν διάμετρον ἐκθέσθαι.

ANH كيف نجد خطاً مساوياً لقطر دائرة معلومة فى كرة.
FC نريد أن نجد قطر دائرة مفروضة على كرة.

2. Τῆς δοθείσης σφαίρας τὴν διάμετρον ἐκθέσθαι.

ANH كيف نخط خطاً مثل قطر كرة معلومة.
FC نريد أن نجد قطر كرة مفروضة.

It is a pleasure to thank Paul Kunitzsch for considerable help in writing this paper. He is not responsible for the opinions expressed in it.

[1] The text was edited by Kunitzsch and Lorch [2010].

[2] The numeration of the edition (see previous note) is used here and throughout the article. It is taken from manuscript **A**. The corresponding propositions in Czinczenheim's Greek text are I 18, 19, 20 and 21 respectively [Czinczenheim 2000].

N. Sidoli and G. Van Brummelen (eds.), *From Alexandria, Through Baghdad: Surveys and Studies in the Ancient Greek and Medieval Islamic Mathematical Sciences in Honor of J.L. Berggren*, DOI 10.1007/978-3-642-36736-6_13,

© Springer-Verlag Berlin Heidelberg 2014

3. Διὰ δοθέντων σημείων, ἅ ἐστιν ἐπὶ σφαιρικῆς ἐπιφανείας, μέγιστον κύκλον γράψαι.

ANH كيف نرسم دائرة عظيمة تمر بنقطتين معلومتين فى بسيط كرة.

FC نريد أن نخط دائرة عظيمة على نقطتين معلومتين على بسيط الكرة.

4. Τοῦ δοθέντος ἐν σφαίρᾳ κύκλου τὸν πόλον εὑρεῖν.

ANH كيف نجد قطب دائرة معلومة فى كرة.

FC نريد أن نجد قطب دائرة معلومة الذى على الكرة.

In all four examples the Greek for "to find" or "to determine" is represented in **ANH** by كيف + an imperfect and in **FC** by نريد أن + an imperfect. **FC** favours مفروض ("assumed") to translate δοθείς ("given"), which it has in several places where **ANH** consistently has معلوم ("known").

There is some inconsistency in the terminology. For instance, ἐκθέσθαι is represented by وجد ("to find") in both translations, but in ex. 2 by خط ("to draw") in **ANH** and by وجد in **FC**. Again, in the first proposition τέμνω generally becomes قطع ("to cut") in **ANH** and فصل ("to cut off") in **FC**, but قطع is to be found in **FC**, about half-way through the proof.

In the definitions at the beginning of the work, the most striking difference between **ANH** and **FC** is the translation of ἄξων as محور ("axis") in **ANH** and as قطر ("diameter") in **FC**. But there are plenty more differences, e.g. in the definition of "sphere" **FC** has مستوية for ἴσαι ἀλλήλαις εἰσίν, where **ANH** has مساوٍ بعضها لبعض, a more accurate rendering. Similarly, in the definition of the pole of a circle **ANH** again has مساوٍ بعضها لبعض and **FC** has this time متساوية. In this definition, **FC** has, simply, وقطب الدائرة على الكرة هى نقطة ..., which agrees with some Greek manuscripts; the reading chosen for the edited Greek text of Czinczenheim has additionally λέγεται; this is represented in the fuller version of **ANH** by الشىء الذى يقال له فى الكرة قطب دائرة, which in Gerard's Latin becomes "Res que in spera polus circuli dicitur."

On the whole, the great differences in terminology and style indicate two translations. They are too numerous and not consistent enough to be the work of a redactor.

As an extended specimen of **FC**'s style (and to show further its independence from **ANH**), we give its version of the lemma to Proposition III 11. It will be noted that it corresponds to none of the forms of the lemma presented in manuscripts **ANH**; even the name of the point that carries the right angle is different (*A* in **FC**, *B* in the proofs in **A** and **H**). But it corresponds very well to the proof in the Hebrew translation by Moses b. Tibbon, as may be seen by comparing **FC** with Knorr's translation of a manuscript of the Moses b. Tibbon version in the Jewish Theological Seminary in New York [Knorr 1986, 235–237].[3] The following is a "mathematical translation:"[4]

When triangle *ABG* is right and the right angle is point *A*, draw *BD* to base *AG*. *I say:* $GA : AD > \angle ADB : \angle DGB$

Proof: Let $DE \parallel GB$

$\therefore DE > AD$ and $< DB$

Construct a circle about centre *D* and with radius *DE*, going beyond *A* and cutting *DB* at *Z*

Produce *DA* to meet the circle at *H*

$\therefore$ sect. $DEH > \triangle DAE$; and sect. $DEZ < \triangle DBE$

$\therefore \triangle DAE : \triangle DBE <$ sect. $DEH :$ sect. DEZ

But $\triangle DAE : \triangle DBE = \overline{AE} : \overline{EB} = AD : DG$

and sect. $DEH :$ sect. $DEZ = \angle ADE : \angle EDZ$

$\therefore \angle ADE : \angle EDB > \overline{AD} : \overline{DG}$

Componendo $AG : GD < \angle ADB : [\angle]BDE$

[3] The text by Jacob b. Machir (1290) is apparently an adaptation of the Moses b. Tibbon translation [Knorr 1986, 235–237; and private communication from Knorr].

[4] This is not an exact translation. It is intended to reproduce the mathematical reasoning. It is followed by the full Arabic text.

Convertendo [$\angle$]ADB : [$\angle$]ADE < AG : AD
And [$\angle$]ADE = [$\angle$]DGB
$\therefore AG : AD$ > [$\angle$]ADB : [$\angle$]DGB. Q.E.D.

The only difference from the Hebrew of any consequence is in the line

Componendo $AG : GD < \angle ADB$: [$\angle$]BDE.

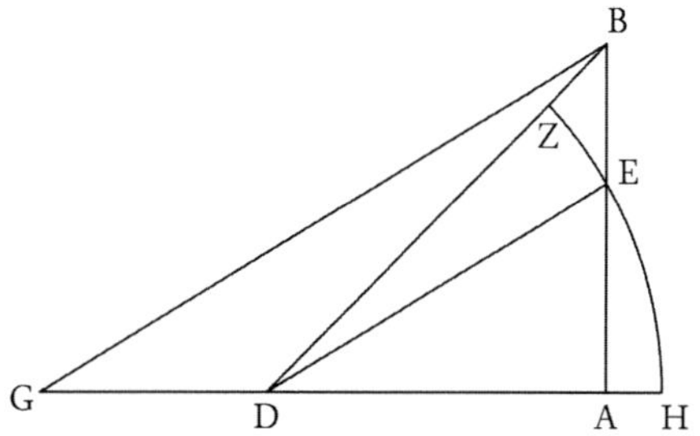

Figure 1: Lemma to Proposition III 11.

The term for *convertendo* in this Arabic text, إذا خلفنا,[5] that introduces the next line probably arose from a colloquial rendering of إذا قلبنا ("when we turn around"),[6] for قلب meant the conversion of a ratio $a : b$ into $a : a - b$.[7] It seems probable that the text is disturbed at this point. The Hebrew text translated by Knorr also has a deduction by *componendo*, but it is chosen so that it is the desired result (the argument forms the ratio $a + b : a$ from $a : b$, rather than **FC**'s $a + b : b$).

In conclusion, we may say that the text represented by **FC** was probably a translation, independent of **ANH**, and that it was the basis of Moses b. Tibbon's Hebrew.

إذا كان مثلث ا ب جـ قائم الزاوية وزاويته القائمة نقطة ا وخرج من نقطة ب إلى قاعدة ا جـ خط مستقيم كيف اتفق وهو خط ب د ، فأقول إن نسبة جـا إلى ا د أعظم من نسبة زاوية ا د ب إلى زاوية د جـ ب ، برهانه أنا نخرج من نقطة د خطاً موازياً لخط جـ ب وهو خط د ه فخط د ه أعظم من خط ا د وأصغر من خط د ب فلذلك إذا عملنا دائرة على مركز د ببعد د ه تجوز ا وتفصل خط د ب على نقطة ز ونخرج د ا إلى أن يلقى الدائرة على نقطة ح فقطع د ه ح أعظم من مثلث د ا ه وقطع د ه ز أصغر من مثلث د ب ه فنسبة مثلث د ا ه إلى مثلث د ب ه أصغر من نسبة قطع د ه ح إلى قطع د ه ز لكن نسبة مثلث د ا ه إلى مثلث د ه ب كنسبة خط ا ه إلى خط ه ب وهي كنسبة ا د إلى د جـ وقطع د ه ح إلى قطع د ه ز كزاوية ا د ه إلى زاوية ه د ز فإذاً نسبة زاوية ا د ه إلى زاوية ه د ب أعظم من نسبة خط ا د إلى خط د جـ وإذا ركبنا تكون نسبة ا جـ إلى جـ د أصغر من نسبة زاوية ا د ب إلى زاوية ب د ه وإذا خلفنا كانت نسبة ا د ب إلى ا د ه أصغر من نسبة ا جـ إلى ا د وزاوية ا د ه مساوية لزاوية د جـ ب فإذاً نسبة ا جـ إلى ا د أعظم من نسبة ا د ب إلى د جـ ب ، وذلك ما أردنا أن نبين .

[5] *sic* **C**, خالفنا **F**.

[6] This was suggested by Paul Kunitzsch (private communication).

[7] See the Euclid texts (Book V, definitions) in MSS Tehran, Malik 3586 (there is no visible foliation) and Leiden 399,1 [Besthorn and Heiberg 1932, 22] for the definition of قلب. The translation is by Isḥāq ibn Hunayn and revised by Thābit ibn Qurra.

[1] قائم] עלי .C *add. et del.* [4] يلقى] ילאק F، ילקא C. [5] وقطع د ه ز... د ا ه] .F *marg.* [7] وقطع] נסבה C *supra*. كزاوية] נסבה C *supra*. [9] خلفنا] כאלפנא F. أصغر] אעגם C. [10] أعظم] אצגר F، אעטם F *supra*. أن نبين] .F *om.*

References

Ancient Authors

Euclid: Besthorn and Heiberg [1932].
Theodosius: Czinczenheim [2000], Kunitzsch and Lorch [2010].

Modern Scholarship

Besthorn, R.O., Heiberg, J.L., 1932. Codex Leidensis 399,1. Euclidis elementa elementa ex interpretatione al-Hadschdschadschii cum commentariis al-Narizii, 3 vols. Finished by G. Junge, J. Raeder, W. Thomson. Gyldendaliana, Copenhagen.

Czinczenheim, C., 2000. Edition, traduction et commentaire des Sphériques de Théodose. Thèse de docteur de l'Universite Paris IV. Atelier National de Reproduction des Thèses, Lille.

Knorr, W.R., 1986. The medieval tradition of a Greek mathematical lemma. Zeitschrift für Geschichte der arabisch-islamischen Wissenschaften 3, 230–261.

Kunitzsch, P., Lorch, R., 2010. Theodosius, Sphaerica: Arabic and Medieval Latin Translations. Franz Steiner Verlag, Stuttgart.

Lorch, R., 1996. The transmission of Theodosius' Sphaerica. In: Folkerts, M. (ed.), Mathematische Probleme im Mittelalter. Der lateinische und arabische Sprachbereich, Harrassowitz, Wiesbaden, pp. 159–183.

More Archimedean than Archimedes: A New Trace of Abū Sahl al-Kūhī's Work in Latin

Jan P. Hogendijk

Abstract In 1661, Borelli and Ecchellensis published a Latin translation of a text which they called the *Lemmas* of Archimedes. The first fifteen propositions of this translation correspond to the contents of the Arabic *Book of Assumptions*, which the Arabic tradition attributes to Archimedes. The work is not found in Greek and the attribution is uncertain at best. Nevertheless, the Latin translation of the fifteen propositions was adopted as a work of Archimedes in the standard editions and translations by Heiberg, Heath, Ver Eecke and others. Our paper concerns the remaining two propositions, 16 and 17, in the Latin translation by Borelli and Ecchellensis, which are not found in the Arabic *Book of Assumptions*. Borelli and Ecchellensis believed that the Arabic *Book of Assumptions* is a mutilated version of a lost "old book" by Archimedes which is mentioned by Eutocius (ca. A.D. 500) in his commentary to Proposition 4 of Book 2 of Archimedes' *On the Sphere and Cylinder*. This proposition is about cutting a sphere by a plane in such a way that the volumes of the segments have a given ratio. Because the fifteen propositions in the Arabic *Book of Assumptions* have no connection whatsoever to this problem, Borelli and Ecchellensis "restored" two more propositions, their 16 and 17. Propositions 16 and 17 concern the problem of cutting a given line segment AG at a point X in such a way that the product $AX \cdot XG^2$ is equal to a given volume K. This problem is mentioned by Archimedes, and although he promised a solution, the solution is not found in *On the Sphere and Cylinder*. In his commentary, Eutocius presents a solution which he adapted from the "old book" of Archimedes which he had found. Proposition 17 is the synthesis of the problem by means of two conic sections, as adapted by Eutocius. Proposition 16 presents the diorismos: the problem can be solved only if $K \leqslant AB \cdot BG^2$, where point B is defined on AG such that $AB = \frac{1}{2}BG$. We will show that Borelli and Ecchellensis adapted their Proposition 16 not from the commentary by Eutocius but from the Arabic text *On Filling the Gaps in Archimedes' Sphere and Cylinder* which was written by Abū Sahl al-Kūhī in the tenth century, and which was published by Len Berggren. Borelli preferred al-Kūhī's diorismos (by elementary means) to the diorismos by means of conic sections in the commentary of Eutocius, even though Eutocius says that he had adapted it from the "old book." Just as some geometers in later Greek antiquity, Borelli and Ecchellensis believed that it is a "sin" to use conic sections in the solution of geometrical problems if elementary Euclidean means are possible. They (incorrectly) assumed that Archimedes also subscribed to this opinion, and thus they included their adaptation of al-Kūhī's proposition in their restoration of the "old book" of Archimedes.

Our paper includes the Latin text and an English translation of Propositions 16 and 17 of Borelli and Ecchellensis.

N. Sidoli and G. Van Brummelen (eds.), *From Alexandria, Through Baghdad: Surveys and Studies in the Ancient Greek and Medieval Islamic Mathematical Sciences in Honor of J.L. Berggren*, DOI 10.1007/978-3-642-36736-6_14,

© Springer-Verlag Berlin Heidelberg 2014

Introduction

Abū Sahl al-Kūhī was a geometer and astronomer of Iranian origin, who had an outstanding reputation among his contemporaries. He was even called "the master of his era in the art of geometry,"[1] and one can therefore understand why the study and translation of his works has been one of Len Berggren's passions. Just as in the case of other scientists in the Eastern Islamic world, none of his works were transmitted to Europe in the 12th and 13th centuries. Al-Kūhī's name occurs for the first time in Europe in the 17th century, in connection with the *Lemmas* of (pseudo?) Archimedes.

The *Lemmas* of Archimedes have not come down to us in Greek. A text entitled *Kitāb al-ma'khūdhāt* (Book of Assumptions) and attributed to Archimedes survives in an Arabic translation by Thābit bin Qurra (836–901).[2] All, or almost all, extant Arabic manuscripts of the work contain not the original translation by Thābit but a later edition by the famous Iranian scientist Naṣīr al-Dīn al-Ṭūsī, who died in 1274.[3] The *Book of Assumptions* was part of al-Ṭūsī's edition of a collection of mathematical texts which were called the *Middle Books*, and which were supposed to be studied between the *Elements* of Euclid and the *Almagest* of Ptolemy. Al-Ṭūsī presented his edition of the *Book of Assumptions* along with the commentary by Abu'l-Ḥasan ʿAlī ibn Aḥmad al-Nasawī (ca. 1010),[4] who had included two generalizations of Proposition 5 by Abū Sahl al-Kūhī.

The *Book of Assumptions* was first translated from Arabic into Latin by the English mathematician Greaves [1659].[5] Two years later, a much superior translation appeared in Florence [Borelli 1661]. This translation was the product of a collaboration between Abraham Ecchellensis (1605–1664), a Christian philosopher and Arabist from Northern Lebanon, and the Italian mathematician Giovanni Borelli (1608–1679) who did not know Arabic.[6] Both translations were based on the edition by al-Ṭūsī. Greaves and Ecchellensis and Borelli called the text which they translated the *Lemmas* of Archimedes, and the text has been known under this title ever since. The *Lemmas* are available in standard translations of the works of Archimedes by Heath [1912, 301–318] and Ver Eecke [1960, 523–542], although the attribution to Archimedes remains doubtful at best. Al-Kūhī's generalizations of Proposition 5 of the *Lemmas* occur in both Latin translations [Greaves 1659, 8–9; Borelli 1661, 393–395]. His name appears as "Abu Sohal Alkouhi" in Greaves [1659, 4], and as "Abusahal Alkuhi" and "Alkauhi" in Borelli [1661, 385, 393–4]. Al-Kūhī's generalizations also appear in Isaac Barrow's 1675 edition of the works of Archimedes [Barrow 1675, 269–270], where they are said to be "ad mentem Abi Sahl Cuhensis, percelebris Mathematici," and in Voogt's Dutch version of the *Elements* of Euclid of 1695. Since al-Kūhī's generalizations have been published and translated [Hogendijk 2008; Merrifeld 1866], there is no need to discuss the mathematics here.

The present paper concerns a hitherto unrecognized trace of al-Kūhī's work in 17th century Europe. The paper consists of this introduction, three sections on the Greek, Arabic and Latin traditions, and an appendix. The appendix contains the Latin text with English translations of the last two propositions, 16 and 17, of the *Lemmas* by Ecchellensis and Borelli [1661], which are not contained in the Arabic manuscripts of the *Book of Assumptions* that have been inspected hitherto, and which are therefore not considered as part of the *Lemmas* of Archimedes by modern historians. The aim of this paper is to show that Proposition 16 was adapted from al-Kūhī's work: *Filling a Lacuna in The Book by Archimedes on the Sphere and Cylinder* in the edition of al-Ṭūsī. This work was published in Arabic with English translation by Len Berggren [1996]. In the proposition in question, which we will quote below, al-Kūhī considers a straight segment ABG such that $AB = \frac{1}{2}BG$ [Berggren 1996, 201-203]. He proves that for all points $D \neq B$ on the segment AG, we have $AB \cdot BG^2 > AD \cdot DG^2$. In algebraic symbolism, putting

---

1 شيخ عصره في صناعة الهندسة [Berggren 2003, 178].

2 On Thābit bin Qurra and his mathematical works, see Sezgin [1974, 264–272].

3 Al-Ṭūsī's edition survives in several dozens of manuscripts, of which Sezgin [1974, 133] gives a non-exhaustive list. It is not clear whether the *Book of Assumptions* survives in a version which predates Naṣīr al-Dīn.

4 On al-Nasawī, see Sezgin [1974, 345–348].

5 On John Greaves (1602–1652), see Toomer [1996, 126-179].

6 On Borelli, see Gillispie [1973, 306–314].

$|AG| = a$, al-Kūhī's proposition implies that the maximum value of $x(a-x)^2$ for $0 < x < a$ is reached at $x = \frac{1}{3}a$.

The present paper continues with a brief section on the Greek tradition, which is about Proposition 4 of Book II of *On the Sphere and Cylinder* by Archimedes (ca. 250 BC), and the commentary by Eutocius of Ascalon (sixth century AD). In the commentary, Eutocius presents Archimedes' proof by means of conic sections that $AB \cdot BG^2 > AD \cdot DG^2$ (in al-Kūhī's notation). The next section, on the Arabic tradition, is about al-Kūhī's own proof of $AB \cdot BG^2 > AD \cdot DG^2$ without conic sections. In my final section on the Latin tradition, I argue that Proposition 16 of the *Lemmas* by Ecchellensis and Borell was adapted from al-Kūhī's proof, and I comment on the odd fact that Ecchellensis and Borelli attribute their Proposition 16 to Archimedes rather than to al-Kūhī.

The Greek Tradition: Archimedes, *On the Sphere and Cylinder* II.4, and the Commentary by Eutocius

In Proposition 4 of Book II of *On the Sphere and Cylinder*, Archimedes wants to cut a sphere into two parts by means of a plane in such a way that the volumes of the two parts are in a given ratio. He shows that this problem can be reduced to an auxiliary problem, namely the division of a given segment AB at a point O in such a way that the ratio of AO to a given length is equal to the ratio of a given area to the square of OB [Heath 1912, 62–64; Netz 2004, 202–204; Ver Eecke 1960, 101–103]; here I use lettering compatible with Proposition 17 by Ecchellensis and Borelli in the Appendix of this paper. In modern notation, the condition in the auxiliary problem boils down to $AO \cdot OB^2 = \ell \cdot \Delta$ where ℓ is the given length and Δ the given area. If we put $a = |AB|$, $x = |AO|$, $k = \ell \cdot \Delta$, the condition is equivalent to the equation $x(a-x)^2 = k$. For the special case of the sphere that has to be cut, the auxiliary problem always has a solution. Archimedes correctly states that the general auxiliary problem requires a diorism, that is, a necessary and sufficient condition for the existence of a solution. Archimedes gives no further details, but the diorism can be stated algebraically as $k \leqslant \frac{4}{27}a^3$; if $k > \frac{4}{27}a^3$ the equation has no root x with $0 < x < a$, so the geometric problem cannot be solved.

Archimedes promised to give the solution (analysis, synthesis, and diorism) of the auxiliary problem at the end, presumably at the end of Book II of *On the Sphere and Cylinder*. In the time of Eutocius, who lived seven centuries after Archimedes, the solution of the auxiliary problem was no longer found in the extant manuscripts of *On the Sphere and Cylinder*, so the Archimedean solution of the problem of cutting a sphere was left incomplete. In his commentary to *On the Sphere and Cylinder* II.4, Eutocius presents a solution of the auxiliary problem that he found in an old book, and which he attributes to Archimedes on plausible grounds [Heath 1912, 66–79; Netz 2004, 318–343; Ver Eecke 1960, 635–664]. In the solution, the required point O is constructed by means of the intersection of a parabola and a hyperbola. The solution consists successively of an analysis, a synthesis , and a diorism [Heath 1912, 67–72; Netz 2004, 319–328; Ver Eecke 1960, 636–646]. Borelli and Ecchellensis included the synthesis, with some changes in detail, in Proposition 17 of their edition of the *Lemmas*; see the appendix to this paper for the Latin text and an English translation.[7] In the diorism it is proved that the maximum of $AO \cdot OB^2$ occurs for $OB = 2AO$. The proof as presented by Eutocius uses a lengthy and ingenious argument involving the case where the parabola and a hyperbola in the synthesis are tangent. Eutocius then continues with solutions of the original problem of cutting the sphere by the Greek mathematicians Dionysodorus and Diocles. These solutions do not concern us here.

---

[7] In modern notation, the solution can be expressed as follows. Choose a rectangular coordinate system with origin A and positive x-axis AB. We have to construct point O on AB in such a way that $AO : \ell = \Delta : OB^2$ for a given segment ℓ and a given area Δ. Let $a = |AB|$ and describe the parabola $(a-x)^2 = \frac{\Delta}{a} \cdot (\ell - y)$ and the hyperbola $x(\ell - y) = a \cdot \ell$. Let $P(x_1, y_1)$ be a point where the hyperbola intersects the parabola. Then obviously $x_1(a-x_1)^2 = x_1 \cdot \frac{\Delta}{a} \cdot (\ell - y_1) = \frac{\Delta}{a} \cdot a \cdot \ell = \Delta \cdot \ell = k$, so if we choose point O on AB such that $|AO| = x_1$, O is a desired point. Compare Proposition 17 in the appendix to this paper, where $\ell = |AC|$, $\Delta = |BE^2| = \frac{4}{9}a^2$. $\Delta > 0$ can be chosen arbitrarily because only the product $\ell \cdot \Delta$ occurs in the equation.

The Arabic Tradition: Al-Kūhī's *Treatise on Filling the Gaps in* On the Sphere and Cylinder

On the Sphere and Cylinder of Archimedes was translated into Arabic at least twice [Sezgin 1974, 128–130]. One translation was revised by Thābit bin Qurra, who also translated the *Book of Assumptions*. The other translation was made by Isḥāq ibn Ḥunayn, who also produced an Arabic translation of the commentary by Eutocius. Naṣīr al-Dīn al-Ṭūsī had access to all these Arabic versions. He included the text by Archimedes and some (but not all) of the commentary of Eutocius in his edition of *On the Sphere and Cylinder*, which belonged to the *Middle Books* [al-Ṭūsī 1940b, 2–3]. Al-Ṭūsī presents the commentary by Eutocius immediately after the text by Archimedes to which it refers; thus Book II, Proposition 4 of *On the Sphere and Cylinder* [al-Ṭūsī 1940b, 86–89] is followed by the relevant commentary of Eutocius, including the analysis, synthesis, and diorism of the auxiliary problem [al-Ṭūsī 1940b, 89–96]. In the Arabic translation, the commentary to *Sphere and Cylinder* II.4 begins with a passage which I quote in detail for later use in the section on the Latin tradition below. The passage resembles the Greek text but is not completely equivalent to it because some of the nuances were lost in the Arabic translation. Al-Ṭūsī says [with my additions to the translation in square brackets]:

وقد ذكر أوطوقيوس العسقلابي في شرحه لهذا الكتاب أن أرشميدس وعد بيان ذلك في كتابه هذا ولم يوجد في شيء من النسخ ما وعده ولذلك سلك كل واحد من دينوسوذورس وديفليس بعده طريقاً غير الذي سلكه هو في هذا الكتاب إلى قسمة الكرة بقسمين على نسبة مفروضة.

قال وأنا وجدت في كتاب عتيق أشكالاً مستغلقة جداً لكثرة ما فيه من الخطأ وما في الأشكال من التحريف بسبب جهل الناسخين وكان فيه الفاظ من لغة ذريس التي كان أرشميدس يحب استعمالها واصطلاحات له خاصّة كما كان يعبر عن القطع المكافي والزائد بالقائم الزاوية والمنفرجة الزاوية فواظبت عليه إلى أن تقرر لي هذه المقدمة وهي هذه:

اذا كان خطان معلومان عليهما اب اج وسطح معلوم عليه د وأردنا أن نقسم اب على ه قسمة تكون نسبة سطح د إلى مربع به كنسبة اه إلى اج فلنجعل كأن ذلك قد كان ...

Eutocius of Ascalon mentioned in his commentary to this book that Archimedes promised the proof of that in this book of his, but what he promised was not found in any of the manuscripts. Therefore both Dīnūsūdhūrus [i.e., Dionysodorus] and Diyuflīs [i.e., Diocles] [who came] after him followed a method different from the method which he himself followed in this book, for the division of the sphere into two parts according to an assumed ratio.

He [= Eutocius] said: I found in an old book propositions which were very obscure because of the many errors in them and the corruptions in the figures caused by the ignorance of the scribes. But in it were words from the language of Dhurīs [i.e., the Doric dialect], which Archimedes liked to use, and technical terms which were special for him; thus he called the sufficient and exceeding section [i.e., the parabola and hyperbola] the right-angled and obtuse angled [sections]. So I worked assiduously on it until this preliminary was established for me, and it is as follows: If there are two known lines AB, AG, and a known area D, and we want to divide AB at E in such a way that the ratio of area D to the square of EB is as the ratio of AE to AG, so let us assume as if that exists ... [al-Ṭūsī 1940b, 89–90].

The Arabic text then continues with a translation of Eutocius' commentary, including the analysis, synthesis and diorism, which Eutocius attributes to Archimedes, and the two solutions by Dionysodorus and Diocles.

Al-Ṭūsī does not mention his own name anywhere in the text of his edition of *On the Sphere and Cylinder*, and he mentions the name of al-Kūhī for the first time towards the end, after the last proposition of Book II, Proposition 9. There, al-Ṭūsī says: "And I say: And by Abū Sahl Wayjan ibn Rustam al-Kūhī is a treatise, and he called it: *On Filling the Gaps in the Second Book of Archimedes* [i.e., *On the Sphere and Cylinder*]."[8] Al-Ṭūsī rendered only fragments of al-Kūhī's treatise, which have been edited and translated by Len Berggren [1996], and which are also available in an uncritical edition in [al-Ṭūsī 1940b, 115-127] and in a French summary by Woepcke [1851, 103-114]. Most of these

[8] أقول ولأبي سهل يحيا (يعني: ويجن) بن رستم القوهي رسالة وسمّها بسدّ الخلل الذي في المقالة الثانية من كتاب أرشميدس [al-Ṭūsī 1940b, 115:8–9].

fragments are related to al-Kūhī's solution to the following problem: to construct a spherical segment whose (spherical) surface area is equal to the surface area s of a given spherical segment and whose volume is equal to the volume v of a different given spherical segment [Berggren 1996, 140–141]. Al-Kūhī solves this problem by means of a parabola and a hyperbola, and he then turns to the diorism, that is the necessary and sufficient conditions for the existence of a solution. He shows that a certain ratio between two cones (which is determined only by the quantities s and v) is equal to the ratio $BD^3/BZ \cdot ZE^2$, where B, D, E, Z are points on a line segment such that $BD = 2DE$ and Z is another point on the segment. Thus the diorism boils down to the determination of the maximum value of $BZ \cdot ZE^2$. Al-Kūhī says: "But the solid of the line *BZ* by the square of *ZE* is the greatest possible when *BZ* is half of *ZE*, as was shown by (conic) sections in the account that we presented following (the method of) Eutocius. And we will give later a demonstration independent of conic sections" [Berggren 1996, 170:6–9, 196; al-Ṭūsī 1940b, 119:3–6]. Indeed, al-Kūhī first discusses the situation where the two conics in his construction are tangent, just like Archimedes (as presented by Eutocius) in the diorism of his auxiliary problem. At the end, al-Kūhī determines the maximum of $BZ \cdot ZE^2$ in a different way, without conic sections [Berggren 1996, 177:11–179:4, 202–203; al-Ṭūsī 1940b, 123:19–124:20; see also the commentary by Berggren 1996, 157–159; and the French summary by Woepcke 1851, 113–114].[9] Al-Kūhī's new determination of the maximum is of course much easier than the complicated argument by means of the tangent parabola and hyperbola, and Berggren remarks that al-Kūhī may have written his proposition "with perhaps a smile and a slight nudge in Archimedes' ribs" [Berggren 1996, 157].

I will now cite Berggren's translation because it will be essential in the next section of this paper. I have italicized the names of standard operations on ratios which are explained in Book V of the *Elements* of Euclid [Heath 1925, 2:112–137]. A summary of the proposition is to be found in the next section, where it will be compared to Proposition 16 in Borelli's edition of the *Lemmas* of Archimedes.

Al-Kūhī says (Figure 1):

> So, in order to prove it, let *AB* be half of *BG* and first let *D* be between *A* and *B*. I say that the solid of the line *AB* by the square of *BG* is greater than the solid of the line *AD* by the square of *DG*.
>
> We make *GE* equal to *GB*; hence, since the ratio of *AB* to *BG* is as the ratio of *BG* to *BE*, the surface of *AB* by *BE* equals the square of *BG*. But[10] the surface of *AB* by *BE* is larger than the surface of *AD* by *DE* because *B* is closer to the middle of *AE* than *D*; hence, the square of *BG* is greater than the surface of *AD* by *DE* and the ratio of the surface of *ED* by[11] *DB*, which is another magnitude, to the surface of *ED* by *AD*, i.e., the ratio of *BD* to *DA*, is greater than the ratio of the surface of *ED* by *DB* to the square of *BG*. So *by composition* the ratio of *BA* to *AD* is greater than the ratio of the surface of *ED* by *DB* together with the square of *BG*, i.e. the square of *DG*, to the square of *BG*. And so the solid of the line *BA* by the square of *BG* is greater than the solid of the line *AD* by the square of *DG*.
>
> Next, let *D* be between *B* and *G* and the rest as before. Then the surface of *AB* by *BE*, i.e. the square of *BG*, will be smaller than the surface of *AD* by *DE* because of *D* being nearer to the midpoint of *AE* than *B*. And the ratio of the surface of *BD* by *DE*, which is another magnitude, to the square of *BG* will be greater than its ratio to the surface of *AD* by *DE*, i.e. than the ratio of *BD* to *DA*. So, *by inversion*, the ratio of the square of *BG* to the surface of *BD* by *DE* is smaller than the ratio of *AD* to *DB*. Hence, *by separation*, the ratio of the square of *DG* to the surface of *BD* by *DE* is less than the ratio of *AB* to *BD*. And so, *by inversion*, the ratio of the surface of *BD* by *DE* to the square of *DG* is larger than the ratio of *DB* to *BA*. And *by composition* the ratio of the square on *BG* to the square of *DG* is greater than the ratio of *DA* to *AB*. And so the solid of *AB* by the square of *BG* is greater than the solid of *AD* by the square of *DG*, which is what we wanted. [Berggren 1996, 202–203]

[9] Note that the author of the demonstration must be al-Kūhī, not al-Ṭūsī, because the demonstration occurs in a long quotation from al-Kūhī that ends on the following page with "[a]nd the Shaykh Abū Sahl al-Kūhī solved the problem in another way which we shall not present" [Berggren 1996, 179:20-180:2, 203; al-Ṭūsī 1940b, 125:9]. Shortly afterwards, al-Ṭūsī concludes with the final sentence "[a]nd this is what Abū Sahl al-Kūhī presented" [Berggren 1996, 183:15, 204; al-Ṭūsī 1940b, 127:15].

[10] Berggren adds here "< by Premiss 9 >," and elsewhere in the translation similar references to other "premisses." Since these references occur only in the manuscript Leiden, Or. 14/25 and not in the other Arabic manuscripts which Berggren used, I assume that they were added by a later commentator to al-Ṭūsī's edition of al-Kūhī's work. I have therefore deleted the "premisses" from the translation.

[11] Berggren translates the same Arabic word *fī*, which indicates the multiplication of two quantities, as "by" but also as "in." I always use "by."

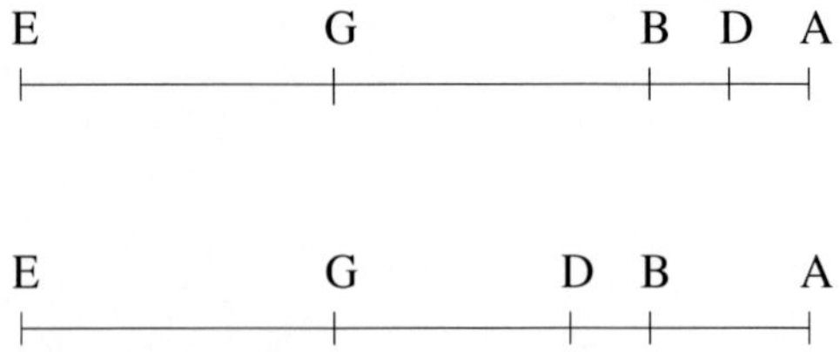

Figure 1: Al-Ṭūsī's version of al-Kūhī's lemma to *On the Sphere and Cylinder* II.4.

The Latin Tradition: Propositions 16 and 17 of the *Lemmas* of Archimedes in the Version of Borelli and Ecchellensis

Borelli and Ecchellensis begin their edition of the *Lemmas* of Archimedes with a long preface. They argue that the *Book of Assumptions*, which they translated from Arabic and which they call the *Lemmas*, is essentially the "old book" which Eutocius found and which he mentions in his commentary to *On the Sphere and Cylinder* II.4. They cite the introduction by Eutocius in which he talks about this old book, in a Latin translation based on the Greek text that had already been available since 1544.[12] Unfortunately for Borelli and Ecchellensis, the Arabic text of the *Book of Assumptions* consists of fifteen elementary geometrical theorems on straight lines and circles, which have no relationship whatsoever to *On the Sphere and Cylinder* II.4 and to Eutocius' commentary. Borelli and Ecchellensis argue that the end of the *Book of Assumptions* must have been lost in the course of time, just like the end of *On the Sphere and Cylinder*. They believe that this lost final part of the *Book of Assumptions* contained the propositions which Eutocius found in his old book, and they then decided to restore this lost final part. Thus Borelli and Ecchellensis say (the "I" who is speaking is Borelli):

"I have added at the end of this book two other propositions [i.e., 16 and 17] by Archimedes that were found by Eutocius. The last one [i.e., 17] is perhaps the same which is missing here, for Almochtasso [i.e., al-Nasawī][13] says in the preface that the propositions in this little work are sixteen, although the last one [in the extant text of the *Book of Assumptions*] is the fifteenth."[14] I note that in the Arabic manuscripts available to me, al-Nasawī says (in the edition of al-Ṭūsī) that the number of propositions is 15 and not 16, cf. [al-Ṭūsī 1940a, 2:4].

What exactly are these propositions 16 and 17 which Borelli and Ecchellensis added to the fifteen propositions of their Latin translation of the *Book of Assumptions*?

Proposition 17 is a construction of the auxiliary problem of Archimedes' *On the Sphere and Cylinder* II.4 by means of a parabola and hyperbola, The construction is essentially the same (with minor changes) as the synthesis in the commentary by Eutocius. This commentary was available to Ecchellensis and Borelli in Greek and in Arabic but a comparison between the figures shows that Proposition 17 is based on the (Latin translation of the) Greek version of *On the Sphere and Cylinder* in *Archimedis Syracusani philosophi* ... [1544] (see p. 34 of the Latin translation of Eutocius' commentary), because the labels of the points in the geometrical figures are exactly the same. In the Arabic, the labels are different [al-Ṭūsī

[12] In *Archimedis Syracusani philosophi* ... [1544], the Greek text is on p. 30 of *Eutokiou Askaloonitou Hypomnema*, at the end of the Greek section of the book, and a Latin translation is on p. 32 of *Commentarii Eutocii Ascalonitae* at the end of the book. The Latin translation in *Archimedis Syracusani philosophi* ... is different from the translation which Ecchellensis and Borelli present in their preface [Borelli 1661, 381–382], and which had been made for them by "learned friends;" see the text in the appendix of this paper.

[13] Almochtasso is derived from the Arabic *al-mukhtaṣṣ*, meaning "the distinguished [scholar]," which al-Ṭūsī used to refer to al-Nasawī.

[14] Addidi in fine huius libris duas alias Archimedis propositiones ab Eutocio repertas quarum altera fortasse illa eadem est que hic deficit, nam Almochtasso in proemio ait, propositiones huius Opusculi sexdecim esse, cum tamen postrema sit decimaquinta [Borelli 1661, 383].

1940b, 90–91]. For example, the point that corresponds to M in Figure 4 and 5 below by Borelli and Ecchellensis is labeled *nūn* in the Arabic, so if they had derived the figure from the Arabic, one would expect N here.

In their Proposition 16, Borelli and Ecchellensis show that if point C is on line segment AB such that $AB = 3AC$, and if $D \neq C$ is another point on line segment AB, then $AC \cdot CB^2 > AD \cdot DB^2$. This is the same result as in the proposition by al-Kūhī quoted in the previous section. I will now discuss the similar structure of the proofs by Borelli/Ecchellensis and al-Kūhī by summarizing the two propositions in a uniform notation. In Figure 2, I use the same labels of points in the figure as in the Latin text of Ecchellensis and Borelli and in Berggren's translation from the Arabic, with the exception of points P and Q. In the translation of al-Kūhī, one has to read $P = B, Q = G$, and in the version of Borelli and Ecchellensis $P = C, Q = B$. Note that the orientations of the two figures of Kūhī and Borelli/Ecchellensis are opposite, as one would expect because the directions of writing in Arabic and Latin are also opposite.

Figure 2: Borelli/Ecchellensis Proposition 16 compared with al-Kūhī's lemma.

We begin with a line segment AQ with a point P on it such that $AQ = 3AP$. Choose a point $D \neq P$ on segment AQ. Required to prove that $AP \cdot PQ^2 > AD \cdot DQ^2$.

Al-Kūhī and Borelli/Ecchellensis begin by extending AQ to point E in such a way that $QE = PQ$. They then remark that $AP \cdot PE = 4AP^2 = PQ^2$. Now D can be located in two different ways. If D is between A and P, we have $AP \cdot PE > AD \cdot DE$ because, as al-Kūhī and Borelli/Ecchellensis point out, point P is closer to the midpoint of AE than point D. Similarly, if D is between P and Q, we have $AD \cdot DE > AP \cdot PE$ because in this case, D is closer to the midpoint of AE than P. Al-Kūhī and Borelli/Ecchellensis base themselves here on Euclid's *Elements* II.5 [Heath 1925, 1:382–383], without saying so. If point M (not mentioned by the authors) is the midpoint of AE, then $AD \cdot DE = MA^2 - MD^2$ and $AP \cdot PE = MA^2 - MP^2$. Thus far the two proofs agree.

Al-Kūhī gives the two cases a separate treatment. If D is between A and P, we have

$$\frac{PD}{AD} = \frac{PD \cdot DE}{AD \cdot DE} > \frac{PD \cdot DE}{PQ^2},$$

since

$$AD \cdot DE < AP \cdot PE = PQ^2.$$

Therefore

$$\frac{AP}{AD} = 1 + \frac{PD}{AD} > 1 + \frac{PD \cdot DE}{PQ^2} = \frac{PD \cdot DE + PQ^2}{PQ^2} = \frac{DQ^2}{PQ^2};$$

hence, as required,

$$AP \cdot PQ^2 > AD \cdot DQ^2.$$

If D is between P and Q, al-Kūhī shows in a similar way that

$$\frac{PD \cdot DE}{PQ^2} > \frac{PD}{AD}.$$

He now argues, in a needlessly complicated way, via the three intermediate steps

$$\frac{PQ^2}{PD \cdot DE} < \frac{AD}{PD},$$

and

$$\frac{DQ^2}{PD \cdot DE} < \frac{AP}{PD},$$

or

$$\frac{PD \cdot DE}{DQ^2} > \frac{PD}{AP},$$

that

$$\frac{PQ^2}{DQ^2} > \frac{AD}{AP}.$$

Hence, as required,

$$AP \cdot PQ^2 > AD \cdot DQ^2.$$

Borelli and Ecchellensis shorten the argument by introducing an extra point O as follows. Let

$$\frac{PD}{DO} = \frac{PQ^2}{AD \cdot DE}.$$

Then, $DO < PD$ if D is between A and P, and $DO > PD$ if D is between P and Q. In both cases we have $AO < AP$. Now since

$$\frac{PD}{DO} = \frac{PQ^2}{AD \cdot DE},$$

also

$$\frac{AD}{DO} = \frac{PQ^2}{PD \cdot DE}.$$

If D is between A and P, then

$$\frac{AD}{AO} = \frac{AD}{DO + AD} = \frac{PQ^2}{PD \cdot DE + PQ^2} = \frac{PQ^2}{DQ^2}.$$

Also if D is between P and Q,

$$\frac{AD}{AO} = \frac{AD}{AD - DO} = \frac{PQ^2}{PQ^2 - PD \cdot DE} = \frac{PQ^2}{DQ^2}.$$

Hence, in both cases,

$$AO \cdot PQ^2 = AD \cdot DQ^2.$$

Because $AO < AP$ we conclude that $AD \cdot DQ^2 < AP \cdot PQ^2$, as required,

We note that the proofs by al-Kūhī and Ecchellensis/Borelli are not the only possible ones. The 12th-century mathematician Sharaf al-Dīn al-Ṭūsī (who is not the same as Naṣīr al-Dīn al-Ṭūsī mentioned above) proved the same theorem in a different way, but also without conic sections [Rashed 1986, vol. 2, 2-5].

In the introduction to Proposition 16 (cf. the appendix), Borelli says that the translation by Ecchellensis of the Arabic version of Eutocius' introduction was taken from "the edition of Abusahal Alkuhi." As pointed out in the previous section of this paper, al-Ṭūsī's name is nowhere mentioned in the text of the edition of *On the Sphere and Cylinder* so it is not surprising that his name does not occur in the Latin. But we also noticed that al-Kūhī's name is only mentioned (by al-Ṭūsī) after the end of *On the Sphere and Cylinder* itself, at the beginning of the treatise *On Filling the Gaps*. Thus Ecchellensis must have gone through al-Ṭūsī's entire edition of *On the Sphere and Cylinder* Book II, including al-Kūhī's work at the end. The close agreement between the first part of the proofs by al-Kūhī and Ecchellensis/Borelli now shows beyond any doubt that Proposition 16 of Ecchellensis and Borelli must be dependent on the corresponding proposition by al-Kūhī in the Arabic. As in the case of Proposition 17, some changes were made in the proof, but the basic structure remained the same. So Ecchellensis must have read and translated the proposition by al-Kūhī, and one wonders whether he also communicated to Borelli the other interesting parts of al-Kuhi's treatise.

Finally, I call attention to the end of the introduction to Proposition 16. There Borelli and Ecchellensis state that it is "not a small sin" in geometry to use conic sections in a proof if it can also be done by ruler and compass. Borelli and Ecchellensis conclude that the diorism by the tangent conic sections as presented in Eutocius' commentary cannot have been by Archimedes, but must have been mutilated, by Eutocius or by someone else. Accordingly, they present Proposition 16 as a reconstruction of a proposition by Archimedes. Thus they must have considered al-Kūhī as an editor, not as an author, and one may wonder what they thought about the authorship of al-Kūhī's construction of a spherical segment with a given spherical surface area s and a given volume v. There is no evidence that Archimedes had the same dogmatic preference for ruler-and-compass constructions and Euclidean methods as Borelli and Ecchellensis. One can find the first traces of such views in the work of the more bureaucratically minded Greek mathematician Apollonius of Perga (ca. 200 BC).[15] By attributing such opinions to Archimedes, Borelli and Ecchellensis made al-Kūhī's proposition look more Archimedean than Archimedes himself would have been. For there is no reason to deny Archimedes the authorship of the proof of the diorism by means of the tangent hyperbola and parabola that was preserved in Eutocius' commentary. Al-Kūhī might have been pleased and honored by the posthumous misattribution of his proposition to the greatest geometer of Greek antiquity.

Appendix: Texts and Translations from the Edition by Ecchellensis and Borelli [1661, 409–413].

All additions by me are in square brackets [ ]. Marginal remarks are inserted in the text for typographical reasons, and are included in pointed brackets < >. Parentheses also occur in the original Latin.

[Borelli 1661, 409] *In praefatione huius operis memini non esse omnino improbabile hunc libellum Archimedis non alium fuisse ab illo antiquo lemmatum libro ab Eutocio reperto, quod precipuè ex verbis eiusdem Eutocij in Comment. proposit. 4. lib. 2. de Sphaera & Cylindro comprobatum fuit: illa fidellisimè translata ex textu Graeco ab amicis doctissimis cum iam in praefatione excusa essent aliam translationem ex Arabico Manuscripto Serenissimi Magni Ducis misit Excell. Abrahamus Ecchellensis desumptam ex editione Abusahli Alkuhi qui pariter librum ordinationis [sic]*[16] *lemmatum Archimedis conscripsit, ut in proemio huius operis testatur Almochtasso. Verba eius sunt haec, quae paulò clarius propositum confirmare videntur;* & meminit Eutocius Ascalonita in Comment. huius libri, quod Archimedes promiserit demonstrationem huius in hoc suo libro, quod in nullo exemplari reperitur, quod promisit. Atque ita unusquisque tam Dyonisodorus, quàm Diocles post illum progressus est per aliam viam, quàm ille (scilicet Archimedes) in hoc libro in divisione Sphaerae in duas partes, quae datam habeant proportionem. Dixit, & ego reperi in [p. 410] Veteri Libro Theoremata satis obscura propter multitudinem errorum, qui in eo sunt, nec non menda, quae occurrunt in figuris propter ignorantiam amanuensium, erantque in eo Doricae dictiones, quarum usus Archimedi familiaris erat, & vocabula ipsi propria; hinc utebatur loca sectionum parabolae, & hyperbolae, rectanguli, & obtusanguli coni sectonibus quamobrem operam ipsi navavi, donec assecutus sum istam propositionem, & est ista, & c.

Modo quia in praedicto libro antiquo ab Eutocio reperto recensentur due propositiones, quarum unam promiserat se demonstraturum Archimedes, & utraque in nostro opusculo iniuria temporum deficit; earum altera forsan erit 16. illa propositio in proemio ab Almochtasso numerata ubi ait propositiones huius opusculi sexdecim esse, cum tamen postrema sit 15. quare inutile forsan non erit eas hic reponere, praecipuè quia Eutocius non rite eas restituit, nec omninò repurgavit a mendis, quibus scatebat exemplar antiquum ab ipso

[15] See *Conics* V.51–52, where Apollonius constructs two minimal straight lines from a given point to a conic section by intersecting it by a second conic section, but if there is only one minimum straight line and the conics would be tangent, he omits the second conic section and presents a more elementary proof, cf. [Toomer 1990, 144–172].

[16] "librum ornationis" would be a correct translation of the Arabic, but Ecchellensis and Borelli [1661, 385] also printed "ordinatio" (rather than "ornatio") in their translation of the preface to the *Lemmas*.

inventum. Et primo noto, quod Eutocius eas vocat theoremata, cum potius problemata sint, & sic etiam ad eodem Eutocio postmodum appellantur. Forsan hoc accidit, quia in libro illo antiquo in formam theorematum scripta erant, sed Eutocius ut ad propositionem Archimedis ea accomodaret [sic], forma problematica ea exposuit. Rursus Eutocius primum theorema se expositurum pollicetur, ut deinde analysi problematis Archimedei accomodetur [sic]. Unde conijcere licet alterum theorema additum, vel alteratum ab Eutocio, vel ab aliquo alio fuisse, in quo proponit, quod, si aliqua recta linea secta sit in duo segmenta, quorum unum duplum sit alterius, solidum parallelepipedum rectangulum contentum sub quadrato maioris, & sub minore segmento maximum erit omnium similium solidorum, quae ex divisione euisdem recta linea in quolibet alio eius puncto consurgunt. Et hoc quidem ostenditur per sectiones conicas, contra artis praecepta; peccatum enim est non parvum apud Geometras, problema planum per conicas sectiones resolvere cum via plana absolvi posse, hoc autem preclari nonnulli viri pariter adnotarunt, & praestiterunt, ut nuper accepi,

Translation:

In the preface of this book, I have mentioned that it is not altogether improbable that this booklet by Archimedes [i.e., the *Lemmas*] is the same as the old book of lemmas which was found by Eutocius. This is particularly confirmed by the words of the same Eutocius in the Commentary on Proposition 4 of Book 2 of *On the Sphere and Cylinder*. Since these [words] have already been presented in the preface in a most faithful translation from the Greek text by learned friends, the most excellent Abraham Ecchellensis made another translation from the Arabic Manuscript of the Most Exalted Great Duke,[17] taken from the edition of Abusahal Alkuhi who also wrote the book on the Arrangement[18] of the *Lemmas* of Archimedes, as Almochtasso [i.e., al-Nasawī] states in the preface of that work. His words are as follows, and they seem to make a little clearer what has been proposed [by me]:

And Eutocius of Ascalon mentioned in the Commentary to this work, that Archimedes promised a demonstration of this in this his book, [and] that what he promised was found in no copy. Then Dionysodorus as well as Diocles proceeded in another way than he (that is, Archimedes) in this book in [solving] the division of the sphere into two parts which have a given ratio. He [Eutocius] said: And I have found in an old book theorems which were quite unclear because of the multitude of errors in it, and the mistakes which occur in the figures because of the ignorance of the scribes. And in it were Doric expressions, which Archimedes commonly used, and terminology proper to him; thus instead of "sections of the parabola and hyperbola" he used "sections of a right-angled and obtuse-angled cone." For this reason, I zealously studied this work for myself until I grasped the proposition. It is as follows: etc.

But because in the above-mentioned old book which was found by Eutocius two propositions were recorded, and Archimedes had promised that he would demonstrate one of them, and each of them is missing in our little work [i.e. the *Lemmas*] because of the vicissitudes of time, [therefore] the second of them will perhaps be that 16th proposition which was enumerated by Almochtasso [= al-Nasawī] in the preface [of the *Lemmas*], where he says that the propositions of this work are sixteen, although the last one is nr. 15. This is why it will perhaps not be useless to place them here, especially because Eutocius did not restore them properly, and did not completely clean them from the errors which were abundant in the ancient book that he had found. First, I note that Eutocius calls them theorems, although they are better [called] problems, and this is how they are called by Eutocius later on. Perhaps this happened because they were written in the form of theorems in the old book, but in order to adapt them to the proposition of Archimedes, Eutocius presented them in the form of problems. Secondly, Eutocius declared that he would present the first theorem, in order to adapt it later to the analysis of the problem of Archimedes [i.e., the division of the sphere]. Here it is possible to conjecture that the other theorem that was added [i.e., the diorism], was changed either by Eutocius or by someone else; [this is

[17] The reference is to Fernando II de' Medici, 1610–1670.

[18] The Latin "ordinatio" is a mistake for "ornatio," meaning embellishment, which is the equivalent of the Arabic *tazyīn*.

the theorem] in which he stated the proposition that if some straight line is divided into two segments such that the one [segment] is twice the other, the rectangular parallelepipedal solid contained by the square of the greater [segment] and by the lesser [segment] is the maximum of all solids of the same kind, which emerge from the division of the same straight line in an arbitrary other point of it. And this is indeed shown [i.e., in Eutocius' text] by conic sections, against the rules of the art. For it is not a small sin among geometers to solve a plane problem by means of conic sections, since it can be solved in a plane way [i.e., by ruler and compass]. Several distinguished men have unanimously noted and expressed this [opinion], as I have learned recently.

Propositio XVI

Si recta linea *AB* sit tripla *AC*, non vero tripla ipsius *AD*; Dico parallelepipedum rectangulũ contentum sub quadrato *CB* in *AC* maius esse parallelepipedo sub quadrato *DB* in *AD*.

Producatur *AB* in *E*, ut sit *BE* aequalis *BC*. Quoniam *BC* dupla erat ipsius *AC*, erit *EC* quadrupla ipsius *AC*, & propterea rectangulum *ACE* aequale erit quadruplo quadrati *AC*, scilicet aequale erit quadrato *CB*: Est vero in primo casu, rectangulum *ADE* maius rectangulo *ACE*, in secundo vero minus, (eo quod punctum *D* in primo casu propinquius est semipartitioni totius *AE*, quàm *C*, in secundo verò remotius); igitur si fiat *CD* ad *DO*, ut quadratum *CB* ad rectangulum [p. 411] *ADE*, erit in primo casu *DO* maior, quàm *CD*, in secundo vero minor; & propterea *AO* minor erit, quam *AC* in utroque casu. Et quia quadratum *CB* ad rectangulum *ADE* est ut *CD* ad *DO*, igitur solida parallelepipeda reciproca erunt aequalia, scilicet solidum quadrato *CB* in *DO* ducto aequale erit solido, cuius basis rectangulum *ADE*, altitudo vero *CD*, seu potius aequale erit solido, cuius basis rectangulum *EDC*, altitudo vero *AD*, & propterea ut quadratum *BC* ad rectangulum *EDC*, ita erit reciproce *AD* ad *DO*, & comparando antecedentes ad terminorum differentias in primo casu, & ad eorundem summas in secundo casu, erit quadratum *BC* ad quadratum *DB* ut *AD* ad *AO*, & denuo solidum parallelepipedum rectangulum contentum sub quadrato *BC* in *AO* aequale erit ei, cuius basis quadratum *DB*, altitudo vera *AD*: Est vero *AO* ostensa minor, quàm *AC* in utroque casu, igitur parallelepipedum, cuius basis quadratum *BC*, altitudo *AC* maius est eo, cuius basis est idem quadratum *BC*, altitudo *AO*; ideoque parallelepipedum, cuius basis quadratum *BC*, altitudo *AC* maius est quolibet parallelepipedo, cuius basis quadratum *BD*, altitudo *AD*: quare patet propositum.

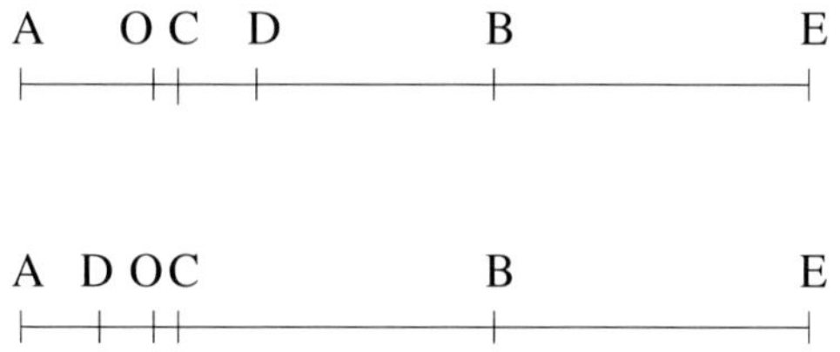

Figure 3: Borelli/Ecchellensis Proposition 16.

Translation:

Proposition 16

If the straight line *AB* is three times *AC*, but not three times *AD*, I say that the rectangular parallelepiped contained by the square of *CB* in *AC* is greater than the parallelepiped [contained] by the square of *DB* in *AD*.

Let *AB* be produced to *E* such that *BE* is equal to *BC*. Since *BC* was [assumed to be] twice *AC*, *EC* will be four times *AC*, and therefore the rectangle *ACE*[19] will be equal to four times the square of *AC*, that is to say, equal to the square of *CB*.

In the first case, the rectangle *ADE* is greater than the rectangle *ACE*, in the second case it is less (because point *D* is in the first case closer to the midpoint of the whole *AE* than *C*, but in the second case farther away). So if *CD* is made to *DO* as the square of *CB* to the rectangle *ADE*, *DO* will be greater than *CD* in the first case and less in the second case. And therefore *AO* will be less than *AC* in both cases. And since the square of *CB* is to the rectangle *ADE* as *CD* to *DO*, thus the reciprocal parallelepipedal solids will be equal, that is to say, the solid of the square of *CB* multiplied by *DO* will be equal to the solid whose base is the rectangle *ADE* and whose altitude is *CD*, and moreover, it will be equal to the solid whose base is the rectangle *EDC* and whose altitude is *AD*. And therefore, the square of *BC* will be to the rectangle *EDC*, reciprocally, as *AD* to *DO*. By comparing the antecedent terms to the differences of the terms in the first case, and to their sums in the second case, the square of *BC* will be to the square of *DB* as *AD* to *AO*, and again, the rectangular parallelepipedal solid contained by the square of *BC* and *AO* will be equal to the [solid], whose base is the square of *DB* and whose altitude is *AD*.

But *AO* was shown to be less than *AC* in both cases. Therefore the parallelepiped, whose base is the square of *BC* and whose altitude is *AC* is greater than the [solid] whose basis is the same square *BC* and whose altitude is *AO*; and therefore the parallelepiped whose basis is the square of *BC* and whose altitude is *AC* is greater than any parallelepiped whose base is the square of *BD* and whose altitude is *AD*. Thus the proposition is evident.

Propositio XVII

Sit *AB* tripla ipsius *AE*, maior vero quàm tripla alterius *CA*, secari debet eadem *AB* citra, & ultra *E*, in *O*, ita ut parallelepipedum, cuius basis quadratum *OB*, altitudo *OA* aequale sit parallelepipedo, cuius basis quadratum *EB*, altitudo *AC*.

Fiat rectangulum *ACBF*, & producantur latera *CA*, *FB* & fiat rectangulum *CFN* aequale quadrato *EB*, & ducta diametro *CEG* compleantur [p. 412] parallelogramma rectangula *AL*, *AK*, *LB*, *BK*, atque axe *FG*, latere recto *FN*[20] describatur parabole < Prop. 52. lib. 1. >[21] *FM* secans *HG* in *M*; erit igitur in parabola quadratum *MG* aequale rectangulo *GFN* sub abscissa, & latere recto contento < Prop. 11. lib. 1. > ideoque idem quadratum *FG* ad rectangulum *NFG*, atque ad quadratum *MG* eandem proportionem habebit: est vero quadratum *FG* ad rectangulum *NFG*, ut *FG* ad *FN*, cum *FG* sit illorum altitudo communis, nec non ut *CFG* ad *CFN* sumpta nimirum CF communi altitudine, ergo rectangulum *CFG* ad *CFN* eandem proportionem habebit, quàm quadratum *FG* ad quadratum *MG*, & permutando rectangulum *CFG* ad quadratum *FG* erit ut rectangulum *CFN* ad quadratum *GM*, sed ut rectangulum *CFG* ad quadratum *FG*, ita est *CF* ad *FG*, & *EA* ad *AC*, igitur *EA* ad *AC* erit ut rectangulum *CFN* ad quadratum *GM*, seu ut quadratum *EB*, vel *KG* ad quadratum *GM*: est vero *AC* minor, quàm *AE*, quae triens est totius *AB*, igitur *MG* minor est, quàm *GK*. Postea per *B* circa asymptotos *ACF* describatur hyperbole *BK*, quae transibit per punctum *K* < Prop. 4. & 12. lib. 2. >, cum parallelogramma *AF*, & *CK* aequalia sint propter diagonalem *CEG*, quare punctum *M* paraboles cadet intra hyperbolem *BK*, sed parabole *FM* occurrit asymptoto *CF* in vertice *F*, & occurrit etiam asymptoto *CA* in aliquo alio puncto < Prop. 26. lib. 1. >, cum *CA* sit parallela axi *FG* paraboles, & hyperbole semper intra asymptotos incedat < ex 1. & 2. lib. 2 >, igitur parabola *FM* bis hyperbole occurrit supra, & infra punctum *M*: sint occursus *X*, a quibus ductis parallellis ad asymptotos compleantur parallelogramma *RP*, & *AF*, quae erunt aequalia inter asymptotos, & hyperbolen constituta < Prop. 12. lib. 2. >, & propterea *COS* parallelogrammorum diameter erit, & una linea recta: & quia *OA* ad *AC* est ut *CF*

[19] By the rectangle *ACE*, Borelli means the rectangle contained by sides equal to *AC* and *CE*, that is in modern terms, $AC \cdot CE$.

[20] The printed text has *PN*.

[21] The references to the *Conics* of Apollonius, which Borelli put in the margin, appear in the text for typographical reasons.

ad *FS*, sive ut rectangulum *CFN* ad rectangulum *SFN*: erat autem quadratum *EB* aequale rectangulo *CFN* ex constructione, & quadratum [p. 413] *OB*, sive *XS* in parabola aequale est rectangulo *SFN*, ergo *AO* ad *AC* est ut quadratum *EB* ad quadratum *OB*, & propterea parallelepipedum, cuius basis quadratum *OB*, altitudo *OA* aequale erit parallelepipedo base quadrato *EB*, altitudine *AC* contento, quod erat propositum.

Note: Borelli adds some notes on the tangent *XI* in the figure, which do not concern us here.

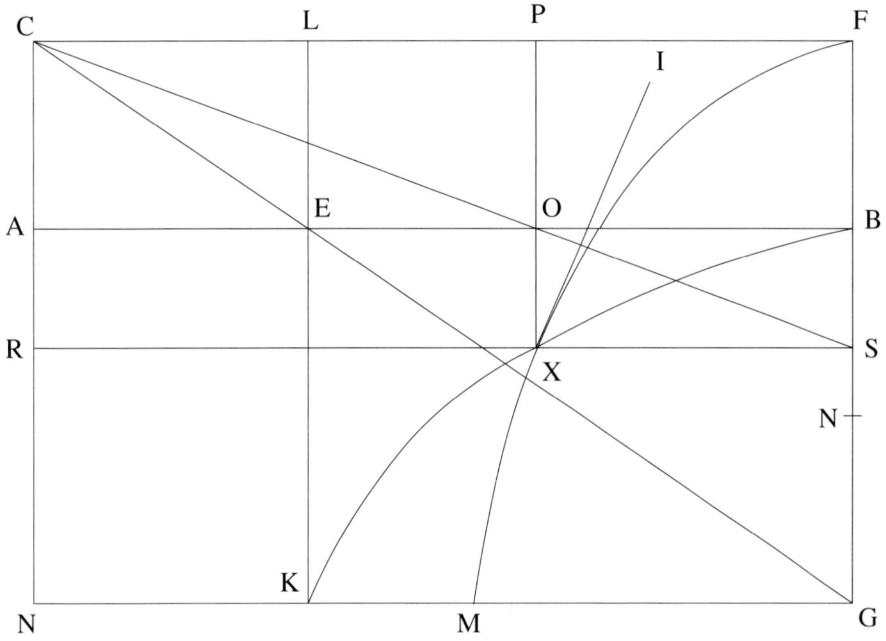

Figure 4: Borelli/Ecchellensis Proposition 17, part 1.

Translation:

Proposition 17

Let *AB* be three times *AE* but greater than three times another [segment] *AC*, then *AB* has to be divided on both sides of *E* at *O* in such a way that the parallelepiped whose base is the square of *OB* and whose altitude is *AO* is equal to the parallelepiped whose base is the square of *EB* and whose altitude is *AC*.

Let the rectangle *ACBF* be made, and let the sides *CA*, *FB* be extended, and let the rectangle *CFN* be made equal to the square of *EB*. After the diameter *CEG* has been drawn, let the right-angled parallelograms *AL*, *AK*, *LB*, *BK* be completed. With axis *FG* and latus rectum *FN*[22] let the parabola *FM* be described < Prop. 52 of book 1 >,[23] intersecting *HG* at *M*. Then in the parabola, the square of *MG* is equal to the rectangle *GFN* which is contained by the abscissa [*GF*] and by the latus rectum [*FN*] < Prop. 11 of book 1 >. Thus, the same square of *FG* has the same proportion to the rectangle *NFG* and to the square of *MG*. But the square of *FG* is to the rectangle *NFG* as *FG* to *FN*, since *FG* is a common altitude of them, and also as *CFG* to *CFN*, if *CF* is of course taken as common altitude, so the rectangle *CGF* has the same proportion to *CFN* as the square of *FG* to the square of *MG*.

[22] The printed text has *PN*.

[23] Borelli's references are to the *Conics* of Apollonius.

Permutando, the rectangle *CFG* will be to the square of *FG* as the rectangle *CFN* to the square of *GM*. But as the rectangle *CFG* is to the square of *FG*, so is *CF* to *FG*, and *EA* to *AC*. Therefore *EA* will be to *AC* as the rectangle *CFN* to the square of *GM*, or as the square of *EB*, that is *KG*, to the square of *GM*. But *AC* is less than *AE*, which is one-third of the whole *AB*. Therefore *MG* is less than *GK*.

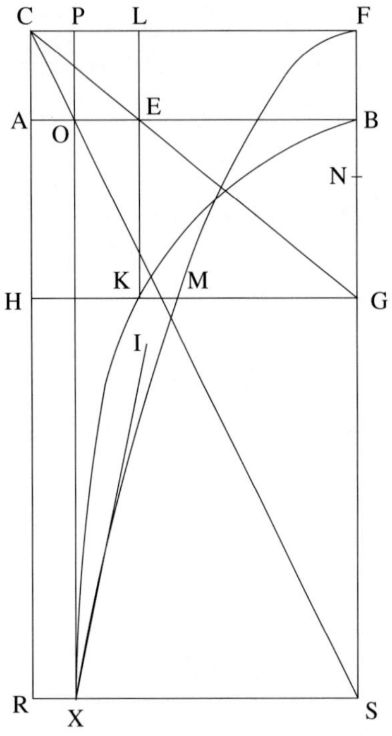

Figure 5: Borelli/Ecchellensis Proposition 17, part 2.

After this, let through *B* with asymptotes *ACF* the hyperbola *BK* be described, which will pass through point *K* < Prop. 4 and 12 of book 2 >, since the parallelograms *AF* and *CK* are equal because of the diagonal *CEG*, so point *M* of the parabola is located inside the hyperbola *BK*. But the parabola *FM* meets the asymptote *CF* in the vertex *F*, and it also meets the asymptote *CA* in some other point < Prop. 26 of book 1 >, since *CA* is parallel to the axis *FG* of the parabola, and the hyperbola is always inside its asymptotes < by Prop. 1 and 2 of book 2 >. Therefore, the parabola *FM* intersects the hyperbola twice, above and under point *M*. Let the intersections be point *X*, from which, after the parallels to the asymptotes have been drawn, parallograms *RP* and *AF* are completed, which [parallelograms] will be equal since they are constituted between the hyperbola and its asymptotes < Prop. 12 of book 2 >. Therefore *COS* will be the diameter of the [two] parallelograms, and it is one straight line. And since *OA* is to *AC* as *CF* to *FS*, that is, as the rectangle *CFN* to the rectangle *SFN*, and the square of *EB* was [assumed to be] equal to the rectangle *CFN* by construction, and the square of *OB*, that is *XS* in the parabola, is equal to the rectangle *SFN*, therefore *AO* is to *AC* as the square of *EB* to the square of *OB*. Therefore the parallelepiped whose base is the square of *OB* and whose altitude is *OA* will be equal to the parallelepiped contained by the square of *EB* as base and by the altitude *AC*, which is what was proposed.

References

Archimedes, 1544. Archimedis Syracusani philosophi ac geometrae excellentissimi Opera, quae quidem extant, omnia, ... nuncquam primum & Graece & Latine in lucem edita ... Adiecta quoque sunt Eutocii Ascalonitae, in eosdem Archimedis libros commentaria, item Graece & Latine, nunquam antea excusa. Joannes Hervagius, Basiliae.

Barrow, I. (ed.), 1675. Archimedis Opera; Apollonii Pergaei Conicorum libri IIII; Theodosii Sphaerica, methodo nova illustrata, et succincte demonstrata. Guil. Godbid, vœneunt apud Rob. Scott, London.

Berggren, J.L., 1996. Al-Kūhī's filling a lacuna in Book II of Archimedes in the version of Naṣīr al-Dīn al-Ṭūsī. Centaurus 38, 140–207.

—— 2003. Tenth-century mathematics through the eyes of Abū Sahl al-Kūhī. In: Hogendijk, J.P., Sabra, A.I. (eds.), The Enterprise of Science in Islam: New Perspectives. MIT Press, Cambridge, MA. pp. 177–196.

Borelli, G.A., 1661. Apollonii Pergaei Conicorum Lib. V. VI. VII. paraphraste Abalphato Asphahanensi nunc primum editi, Additus in calce Archimedis Assumptorum Liber ex codicibus arabicis mss ... Abrahamus Ecchellensis Maronita ... latinos reddidit Io.[hannes] Alfonsus Borellus in Geometricis versione contulit. Iosephus Cocchinus, Florentiae.

Gillispie, C.G. (ed.), 1973. Dictionary of Scientific Biography, vol. 2. Scribner's Sons, New York.

Graeves, J., 1659. Lemmata Archimedis apud graecos et latinos jam pridem desiderata, e vetuste codice M.S. arabico a Johanno Gravio traducta et nunc primum cum arabum scholis publicata, revisa et pluribus mendis expurgata a Samuele Foster. In: Foster, S., 1659. Miscellanea sive lucubrationes mathematicae. Ex Officina Leybouriana, Londini.

Heath, T.L., 1912. The Works of Archimedes, with the Method of Archimedes. Cambridge University Press, Cambridge.

—— 1925. The Thirteen Books of Euclid's Elements, 2nd edition, 3 vols. Cambridge University Press, Cambridge.

Hogendijk, J.P., 2008. Two beautiful geometrical theorems by Abu Sahl Kuhi in a 17th century Dutch translation. Ta'rikh-e Elm: Iranian Journal for the History of Science 6, 1–36.

Merrifield, C.W., 1866-1869. On a geometrical proposition, indicating that the property of the radical axis was probably discovered by the Arabs. Proceedings of the London Mathematical Society 2, 175–177. Reprint: Sezgin, F. (ed.), 1998. Islamic Mathematics and Astronomy. Institute for the History of Arabic-Islamic Sciences, Frankfurt, vol. 79, pp. 221–223.

Netz, R. (trans.), 2004. The Works of Archimedes: Translation and Commentary. Volume 1: The Two Books on the Sphere and Cylinder. Cambridge University Press, Cambridge.

Rashed, R., (ed.), 1986. Sharaf al-Dīn al-Ṭūsī, Oeuvres mathématiques, 2 vols. Les Belles Lettres, Paris.

Sezgin, F., 1974. Geschichte des arabischen Schrifttums, vol. 5. Mathematik bis ca. 430. H. Brill, Leiden.

Toomer, G.J., 1996. Eastern Wisedome and Learning: The Study of Arabic in Seventeenth-Century England. Oxford University Press, Oxford.

—— 1990. Apollonius' Conics: Books V to VII, the Arabic Translation of the Lost Greek Original in the Version of the Banū Mūsā, 2 vols. Springer, New York.

al-Ṭūsī, Naṣīr al-Dīn (ed.), 1940a (1359H). Kitāb al-Ma'khūdhāt li-Arshimīdis. Taḥrīr ... al-Ṭūsī. Osmania Oriental Publication Bureau, Hyderabad. Reprint: Sezgin, F. (ed.), 1998. Islamic Mathematics and Astronomy, vol. 48. Institute for the History of Arabic-Islamic Sciences, Frankfurt, pp. 99–131.

al-Ṭūsī, Naṣīr al-Dīn (ed.), 1940b (1359H). Kitāb al-kura wa l-usṭuwāna li-Arshimīdis. Taḥrīr ... al-Ṭūsī (in Arabic). Osmania Oriental Publication Bureau, Hyderabad. Reprint: Sezgin, F. (ed.), 1998. Islamic Mathematics and Astronomy, vol. 48. Institute for the History of Arabic-Islamic Sciences, Frankfurt, pp. 133–377.

Ver Eecke, P., 1960. Les oeuvres complètes d'Archimède, suivies des commentaires d'Eutocius d'Ascalon. Traduites du grec en français, 2 vols. Blanchard, Paris.

Voogt, C.S., 1695. Euclidis beginselen der Meet-Konst Waar bij t 16 Boeck Flussatis Candalla, door C.J. Voogt Geometra. Johannes van Keulen, Amsterdam.

Woepcke, F., 1851. L'Algèbre d'Omar Alkhayyâmî. Duprat, Paris. Reprint: Woepcke, F., 1986. Études sur les mathématiques arabo-islamiques, vol. 1. Institute for the History of Arabic-Islamic Sciences, Frankfurt, 1986, pp. 49–256.

Les mathématiques en Occident musulman (IXe–XVIIIe s.) : Panorama des travaux réalisés entre 1999 et 2011

Ahmed Djebbar

Abstract The first part of this paper presents an overview of research carried out up to 1998 on the history of mathematical activities in the Islamic West. The topics covered are algebra, the science of calculation, theory of numbers, geometry and combinatorics.

The second part covers different aspects of research carried out in the same field between 1999 and 2011: bibliographic work, the circulation of scientific writings between the Islamic East and West, the mathematical products of the Andalus and the Maghreb.

This article shows, in particular, the expansion of the scope of investigation into domains that had not been sufficiently studied before, such as mensuration, division of figures and tiling, and the science of dividing heritages.

Introduction

En 1998, à l'occasion du Colloque de Boston, intitulé « *New Perspectives on Science in Medieval Islam* », nous avons présenté un bilan des recherches qui avaient été effectuées, jusqu'à cette date, sur l'histoire des activités mathématiques en Andalus et au Maghreb [Djebbar 2003a, 309–350]. Ce travail s'inscrivait, modestement, dans le prolongement des publications de D.A. King puis de J.L. Berggren sur les recherches en histoire des mathématiques et de l'astronomie en pays d'Islam, réalisées entre 1970 et 1995 [King 1980, 10–26 ; Berggren 1985, 9–33 ; Berggren 1997, 407–440].

Au cours des dix dernières années, un certain nombre de travaux ont été réalisés sur différents aspects des mathématiques pratiquées en Occident musulman entre le IXe et le XIXe siècle, recherches universitaires (sous forme de magisters et de thèses le plus souvent non encore publiés), articles (parus dans des revues spécialisées ou dans des Actes de colloques), éditions critiques et ouvrages biobibliographiques. Il nous a donc semblé utile de faire le point sur la nature de cette production, sur son contenu et sur ses grandes orientations. Pour fournir le maximum d'informations au lecteur, nous avons signalé, pour chaque sujet de recherche évoqué, toutes les références le concernant, sans préjuger du niveau de chacune des études parce que l'expérience montre que, dans ce domaine, il est toujours possible de trouver des informations originales sur la tradition mathématique d'al-Andalus et du Maghreb dans des articles de vulgarisation ou des travaux universitaires de niveau modeste. Mais, avant cela, il nous paraît nécessaire de rappeler les éléments essentiels qui s'étaient dégagés du premier bilan qui avait été fait en 1998 sur la production de la période antérieure.

Avant les années 80, les publications sur la tradition mathématique de l'Occident musulman, réalisées par F. Woepcke, A. Marre, M. Steinschneider, H. Suter, M. A. Cherbonneau, G. Eneström et M. Souissi, ont concerné, essentiellement, deux disciplines : la science du calcul et l'algèbre. Et les écrits analysés avaient été produits par trois auteurs seulement, tous les trois postérieurs au XIe siècle, al-Ḥaṣṣār (XIIe s.), Ibn al-Bannā (m. 1321) et al-Qalaṣādī (m. 1486). Ces travaux avaient explicité cer-

N. Sidoli and G. Van Brummelen (eds.), *From Alexandria, Through Baghdad: Surveys and Studies in the Ancient Greek and Medieval Islamic Mathematical Sciences in Honor of J.L. Berggren*, DOI 10.1007/978-3-642-36736-6_15,
© Springer-Verlag Berlin Heidelberg 2014

tains aspects relatifs à la circulation, d'est en ouest, des écrits mathématiques d'Orient, ainsi que leur forte empreinte sur le contenu des écrits connus produits en Occident musulman depuis le IXe siècle. Ils avaient également révélé deux aspects alors inconnus des pratiques mathématiques en Andalus et au Maghreb. En premier lieu, l'élaboration d'un symbolisme arithmétique et algébrique et son introduction, vers la fin du XIIe siècle, dans une partie des manuels produits dans la région. En second lieu, la présence en Andalus, dès le milieu du XIe siècle, et sous une forme très élaborée, des outils trigonométriques produits en Orient et qui prolongeaient les traditions grecque et indienne dans ce domaine [Djebbar 1998, I, 33–60].

Les recherches réalisées à partir des années 80 du siècle précédent ont prolongé celles qui portaient sur l'algèbre et le calcul indien tout en s'intéressant à d'autres disciplines, comme la géométrie, la théorie des nombres, la science des héritages et l'analyse combinatoire.

L'algèbre

En algèbre, on a abouti à une meilleure connaissance de l'évolution de la classification des six équations canoniques du 1e et du 2e degré, accompagnée d'une extension du domaine des équations[1]. Cette évolution semble avoir été la conséquence du développement, en Orient, de la manipulation des monômes et des polynômes avec l'émergence de la notion de puissance. La découverte et l'analyse de nouveaux textes produits au Maghreb, en particulier le *Kitāb al-uṣūl wa l-muqaddimāt fi l-jabr wa l-muqābala* [Le livre des fondements et des préliminaires en algèbre] d'Ibn al-Bannā, le *Rashfat ar-ruḍāb min thughūr aᶜmāl al-ḥisāb* [Succion du nectar des bouches des opérations du calcul] d'al-Qaṭrawānī (XIVe s.) et quelques fragments du traité non encore retrouvé d'Abū l-Qāsim al-Qurashī (m. vers 1185), ont confirmé la circulation de notions, de techniques ou d'écrits algébriques orientaux postérieurs à la publication du livre d'al-Khwārizmī (m. vers 850). Mais, en dehors du traité d'Abū Kāmil (m. vers 930), aucun autre ouvrage de cette époque, ou d'une époque ultérieure, n'est explicitement cité par les praticiens de l'algèbre de l'Occident musulman. Ce qui laissait en suspens la question de la circulation, de l'Orient vers l'Occident, des contributions importantes d'al-Karajī (m. vers 1029), d'as-Samaw'al (m. 1175) et d'al-Khayyām (m. 1131). Ce silence des sources connues permettait aussi de supposer la publication, en Andalus, de contributions algébriques originales à partir de la seule connaissance des travaux d'al-Khwārizmī et d'Abū Kāmil.

Un second domaine de l'algèbre a bénéficié de recherches fructueuses, celui du symbolisme. Certains de ses aspects avaient été révélés, au cours de la période antérieure, par les travaux de Woepcke [Woepcke 1854]. Les nouvelles investigations, en particulier dans un ouvrage peu connu d'Ibn al-Yāsamīn (m. 1204), le *Talqīḥ al-afkār fī l-ᶜamal bi rushūm al-ghubār* [La fécondation des esprits sur l'utilisation des chiffres de poussière] ont permis d'avancer de plusieurs siècles la date d'utilisation des symboles connus [Zemouli 1993]. Elles ont également permis de conjecturer leur première apparition à Séville même si les recherches postérieures n'ont pas encore permis de confirmer cette hypothèse. Quoi qu'il en soit, les recherches de cette période ont établi que c'est au Maghreb, et dans des ouvrages écrits par des mathématiciens de cette région, aux XIVe-XVe siècles, qu'un symbolisme arithmétique et algébrique, relativement élaboré, a accompagné l'exposé rhétorique des problèmes et de leurs résolutions.

Un troisième constat concerne les différentes formes de circulation de l'algèbre après le XIIIe siècle. Les textes produits au cours des décennies 80 et 90 ont tous été produits au Maghreb. L'analyse de leurs contenus respectifs montre clairement que l'algèbre est présente essentiellement dans les commentaires aux écrits d'Ibn al-Bannā et dans les manuels de calcul. Tous ces ouvrages réservent un chapitre à l'exposé des outils de la discipline et des techniques de résolution des équations. Dans la première catégorie d'écrits, les démonstrations accompagnent, parfois, l'exposé des algorithmes de résolution.

[1] Il s'agit des équations suivantes (exprimées dans le symbolisme actuel et présentées dans l'ordre adopté par al-Khwārizmī), X étant le « bien », $\sqrt{X}$ la « racine du bien », a, b, c, des nombres entiers ou fractionnaires positifs :

(1) $aX = b\sqrt{X}$; (2) $aX = c$; (3) $b\sqrt{X} = c$;
(4) $aX + b\sqrt{X} = c$; (5) $aX + c = b\sqrt{X}$; (6) $b\sqrt{X} + c = aX$.

Dans la seconde, seuls les procédés sont présentés et illustrés.

La science du calcul

En science du calcul, une première orientation a concerné la description, sous forme d'articles, ou la publication sous forme de thèse ou d'ouvrages imprimés, d'un certain nombre d'écrits maghrébins des XIVe–XVe siècles. Une partie des travaux a révélé l'existence de copies de certains écrits considérés comme perdus, comme le *Fiqh al-ḥisāb* [La science du calcul] d'Ibn Mun^{c}im (m. 1228) et le *Kitāb al-kāmil fī ^{c}ilm al-ghubār* [Livre complet sur la science <des chiffres> de poussière] d'al-Ḥaṣṣār (XIIe s.). Une autre partie a permis de mettre à la disposition des chercheurs ou d'autres utilisateurs, des ouvrages qui étaient peu connus ou totalement inconnus, comme le *Raf^{c} al-ḥijāb ^{c}an wujūh a^{c}māl al-ḥisāb* [Soulèvement du voile sur les formes des opérations du calcul] d'Ibn al-Bannā, le *Ḥaṭṭ an-niqāb ^{c}an wujūh a^{c}māl al-ḥisāb* [Abaissement de la voilette sur les formes des opérations du calcul] d'Ibn Qunfudh (m. 1407), le *Talqīḥ al-afkār* d'Ibn al-Yāsamīn et le *Sharḥ at-Talkhīṣ* [Commentaire au Talkhīṣ] d'al-^{c}Uqbānī (m. 1408).

Dans le prolongement de ces contributions, quelques études se sont focalisées sur certains outils arithmétiques, comme la méthode de fausse position et les algorithmes de calcul, et sur des applications à un domaine juridique, la science des héritages. Pour la première fois, les aspects mathématiques de ce vaste champ de la législation musulmane ont fait l'objet d'analyses comparatives basées sur des sources de l'Occident musulman, le *Mukhtaṣar* [Abrégé] de l'andalou al-Ḥūfī (m. 1192) et le *Sharḥ Mukhtaṣar al-Ḥūfī* [Commentaire à l'Abrégé d'al-Ḥūfī] du maghrébin al-^{c}Uqbānī.

Il faut enfin signaler que l'ensemble des études faites sur les écrits que nous venons d'évoquer a apporté des éclairages significatifs sur deux aspects importants. Le premier concerne le phénomène de circulation de l'Orient vers l'Occident musulman de concepts et de procédures appartenant soit à la tradition du « *calcul indien* », soit à la tradition dite du « *ḥisāb maftūḥ* », c'est-à-dire celle du calcul digital et mental. Le second a trait aux spécificités de certaines pratiques calculatoires d'al-Andalus et du Maghreb, comme l'utilisation de la numération dite « *rūmī* » [byzantine] (appelée aussi « *chiffres de Fez* » et « *chiffre des registres* »)[2], ainsi que la manipulation des fractions, avec l'usage systématique d'un symbolisme relativement élaboré.

La théorie des nombres

En théorie des nombres, les recherches de cette période avaient révélé des préoccupations, des pratiques et des contributions originales. Elles ont également permis de constater que les contributions connues s'inscrivaient, essentiellement, dans le prolongement des trois grandes traditions grecques, c'est-à-dire celle des Livres VII–IX des *Eléments* d'Euclide (IIIe s. av. J.C.), celle de l'*Introduction arithmétique* de Nicomaque de Gérase (IIe s.) et, dans une moindre mesure, celle des *Arithmétiques* de Diophante (IIIe s.). Cet héritage est arrivé, partiellement, en Andalus, enrichi par les résultats de nouvelles recherches effectuées essentiellement à Bagdad, à partir du IXe siècle. On y trouve des preuves d'existence et des procédés permettant de déterminer des nombres premiers, parfaits, amiables et figurés ainsi que les sommes de certaines séries finies d'entiers. Au niveau des démonstrations, on remarque, à côté de différentes preuves par induction, l'introduction de l'analyse et de la synthèse dans l'établissement de résultats arithmétiques. Cette démarche, un peu artificielle pour de nombreuses propositions de théorie des nombres, nous renseigne en fait sur l'impact que semble avoir eu, par l'intermédiaire du *Kitāb al-istikmāl* [Livre du perfectionnement] d'al-Mu'taman (m. 1085), le *Kitāb at-taḥlīl wa t-tarkīb* [Livre

[2] Il s'agit d'un système de numération non positionnelle à 27 chiffres (9 pour les unités, 9 pour les dizaines et 9 pour les centaines), avec l'ajout, aux chiffres précédents, d'un même signe répété autant de fois qu'il y a de milliers dans chaque position).

de l'analyse et de la synthèse] d'Ibn al-Haytham (m. vers 1041).

La géométrie

En géométrie, et malgré les nombreux témoignages du biobibliographe andalou Ṣāᶜid al-Andalusī (m. après 1068), les écrits de l'Occident musulman qui étaient connus au début des années 1980 se limitaient, essentiellement, aux traductions latines ou hébraïques de quelques opuscules sur le mesurage et d'écrits sur les outils mathématiques de l'astronomie. Des investigations nouvelles ont permis d'exhumer puis d'analyser des documents de différentes formes : des épîtres, comme le *Mukhtasār fī l-misāḥa* [L'abrégé sur le mesurage] d'Ibn al-Bannā ; des résumés, comme la lettre d'Ibn Bājja (m. 1138) à son ami Ibn al-Imām, dans laquelle il synthétise les travaux originaux de son professeur Ibn Sayyid (XIe s.) ; des fragments, comme ceux de l'ouvrage d'Ibn as-Samḥ, qui traitent du cylindre et de ses sections planes ; des parties substantielles d'un ouvrage perdu, comme le *Kitāb al-istikmāl* d'al-Mu'taman.

Ces différents écrits ont confirmé une circulation partielle mais qualitativement importante, vers l'Andalus de versions arabes d'ouvrages grecs (les *Eléments* d'Euclide, les *Coniques* d'Apollonius (IIIe s. av. J.C.), les *Sphériques* de Ménélaüs (IIe s.) et de copies d'ouvrages réalisés par des mathématiciens d'Orient entre le IXe et le XIe siècle (comme *l'Epître sur l'aire de la parabole* d'Ibrāhīm Ibn Sinān (m. 940) et le *Livre de l'analyse et de la synthèse* d'Ibn al-Haytham) [Hogendijk 1991]. Ils ont également révélé certains aspects des pratiques géométriques de l'Occident musulman et certaines contributions, comme l'établissement du théorème dit de Ceva et l'étude de nouvelles courbes planes par projection d'une catégorie de courbes gauches, ayant pour but de réaliser la multisection d'un angle ou la détermination de *n* grandeurs proportionnelles entre deux grandeurs données.

L'analyse combinatoire

En combinatoire, la recherche a été plus féconde que pour les autres disciplines. Pour la première fois, des textes mathématiques arabes, produits au Maghreb, exposaient des résultats, des procédures et des applications de nature combinatoire. Il s'agit de deux écrits d'Ibn al-Bannā, le *Tanbīh al-albāb ᶜalā masā'il al-ḥisāb* [Avertissement aux gens intelligents sur les problèmes du calcul] et le *Rafᶜ al-ḥijāb*, ainsi qu'un ouvrage plus ancien, le *Fiqh al-ḥisāb* d'Ibn Munᶜim. Auparavant, les seuls témoignages connus attestant d'une pratique combinatoire en pays d'Islam se trouvaient, essentiellement, dans des ouvrages de lexicographie comme le *Kitāb al-ᶜayn* [Livre du ᶜayn] d'al-Khalīl Ibn Aḥmad (m. *ca.* 786), de grammaire comme la *Jamharat al-ᶜArab* [Le recueil des Arabes] d'Ibn Durayd (m. 933) ou de philosophie, comme le *Sharḥ al-ishārāt wa t-tanbīhāt* [Commentaires sur les 'Remarques et avertissements'] de Naṣīr ad-Dīn aṭ-Ṭūsī (m. 1274). D'autres dénombrements par énumération avaient été remarqués dans des ouvrages mathématiques, mais ils étaient isolés et aucun n'avait abouti à un développement théorique connu.

L'exhumation et l'analyse des trois textes maghrébins qui viennent d'être évoqués ont permis de révéler une véritable tradition, née à Marrakech à la fin du XIIe siècle, à partir de préoccupations lexicographiques, et entretenue dans cette même ville, et sous différentes formes, tout au long des XIIIe–XIVe siècles. Elles ont également permis de repérer des tentatives de théorisation de ces démarches combinatoires, en les accompagnant de démonstrations inductives, quand cela était possible, et en les rattachant à des résultats de l'arithmétique néopythagoricienne.

Nature et contenu des travaux realisés après 1998

Au cours des douze dernières années, on a assisté à un développement significatif de la production dans le domaine de l'histoire des activités mathématiques en Occident musulman. Ce développement a été stimulé par l'organisation de colloques internationaux ou régionaux et par un plus grand intérêt porté à l'histoire des sciences en général de la part de quelques institutions universitaires au Maghreb. Les thèmes qui ont été étudiés au cours de cette période peuvent être regroupés en trois catégories. La première englobe des travaux à caractère « *externaliste* » comprenant des présentations générales portant sur les grandes orientations de l'activité mathématique de la région, sur leurs auteurs, sur leur production et sur sa circulation partielle. La deuxième regroupe les études, de nature « *internaliste* », qui exposent le contenu connu de la production mathématique d'al-Andalus et du Maghreb, avec ses éventuelles caractéristiques. La troisième catégorie est constituée de toutes les recherches qui se sont intéressées à l'intervention d'outils et de techniques mathématiques dans d'autres disciplines et dans différentes activités de la cité.

Les grandes orientations des mathématiques en Andalus et au Maghreb

Les nouvelles recherches dans ce domaine se sont inscrites dans le prolongement des études, faites dans les décennies 80 et 90, sur les activités mathématiques dans les deux grands espaces culturels de l'Occident musulman et dans chacune des trois régions du Maghreb. Mais elles se sont intéressées, principalement, à certaines périodes sur lesquelles les sources désormais disponibles fournissent le plus d'informations. Il y a d'abord l'Andalus des XIe–XIIe siècles et plus précisément la production mathématique de cette période, qui s'est exprimée en hébreu. La découverte et l'exploitation de nouveaux documents ont permis de disposer d'informations supplémentaires sur les contributions de cette tradition scientifique, dans sa globalité et sur la production de certains de ses représentants les plus éminents, comme Ibn Ezra (m. *ca.* 1167) [Lévy 2000 ; 2001b] et Maïmonide (m. 1204) [Lévy 2004]. Au XIIIe et au XIVe siècle, les publications en hébreu, liées à la tradition arabe, se sont poursuivies et ont concerné essentiellement, au vu des sources qui ont été analysées, l'arithmétique et la géométrie [Lévy 2002].

Quant à la tradition arabe proprement dite, c'est la période postérieure au XIIe siècle qui a fait l'objet d'études. Elle correspond à trois grands moments de l'histoire de l'Occident musulman, la phase almohade, celle des trois royaumes magrébins et celle que l'on qualifie, abusivement, de période ottomane puisque le Maghreb extrême n'a jamais été sous la tutelle de la « *Sublime porte* ». Le premier moment a été celui de la prééminence de Séville relayée à partir du XIIIe siècle par Marrakech qui profite de l'installation dans la ville de quelques hommes de sciences d'al-Andalus [Djebbar 2005a ; Laabid 2005 ; Lamrabet 2008]. Le second moment correspond aux XIVe–XVe siècles. Avec la disparition des derniers centres intellectuels andalous (à l'exception de celui de Grenade), les mathématiciens les plus en vue travaillent désormais à Fez, à Tunis ou à Tlemcen. Leur production est, dans l'ensemble, une reprise des thèmes et des méthodes exposés dans les écrits antérieurs, accompagnés de commentaires, d'exemples et, parfois, de justifications [Lamrabet 2007 ; Laabid 2007].

La dernière période est celle où le Maghreb est régi par deux institutions politiques distinctes, à l'ouest, une royauté qui sera assumée par plusieurs dynasties (Wattassides, Saadiens et Alaouites) ; au centre et à l'est, des beylicats sous la tutelle du pouvoir ottoman (1515–1830). Mais, au niveau de l'activité scientifique, les situations dans les deux régions sont semblables dans la mesure où elles sont le résultat d'une dynamique générale dont les prémices étaient déjà présentes dans la production du XIVe siècle [Djebbar 2003b]. Dans le domaine mathématique, ce phénomène avait débuté par la réduction du champ de la recherche avec un ralentissement significatif et la marginalisation de certains chapitres qui étaient enseignés au cours des siècles précédents. Les nouvelles études ont montré que cette tendance a concerné toutes les régions du Maghreb, indépendamment de la nature de leurs systèmes politiques, et elle s'est accentuée après le XVe siècle, avec un arrêt complet de la recherche [Djebbar 2000b]. Au niveau des publications, elle s'est manifestée par la multiplication d'écrits sous forme de commentaires, d'abrégés et de gloses [Aballagh 2003].

Les travaux biobibliographiques

Ce constat global est confirmé par des études particulières concernant le profil, les activités ou le contenu de la production des mathématiciens de cette région. Dans le domaine biobibliographique, la plus importante est, incontestablement la version anglaise, de l'ouvrage de [Rosenfeld & Matvievskaya 1983] qui présente 1300 références. Dans la nouvelle version, qui regroupe 1700 références, le contenu a été actualisé et considérablement enrichi avec, en particulier, l'addition de nombreuses biographies de mathématiciens et d'astronomes de l'Occident musulman [Rosenfeld & İhsanoğlu 2003 ; Rosenfeld 2004 ; 2006].

D'autres contributions, sous forme de travaux universitaires ou de publications (articles de recherches et livres) ont été réalisées. Elles concernent, le plus souvent, des auteurs répertoriés dans l'ouvrage qui vient d'être évoqué, mais elles fournissent des informations plus détaillées sur leurs activités et sur leurs écrits mathématiques. Parmi ces scientifiques, certains ont bénéficié de publications consacrées soit à leur biographie soit à leur production mathématique soit à tous les aspects de leurs activités [Abdeljaouad 2005a ; Djebbar & Aballagh 2001 ; Lamrabet 2000 ; Marin 2004]. D'autres ont été présentés, ainsi que leurs contributions, à l'occasion de l'édition ou de l'exposé du contenu d'un de leurs ouvrages. C'est le cas d'al-Ghurbī (m. XIVᵉ s.), un mathématicien peu connu [Harbili 2006], et d'Ibn Haydūr, le plus important commentateur des ouvrages d'Ibn al-Bannā [Muṣliḥ 2006 ; Naghsh 2007].

La circulation des écrits mathématiques arabes d'Occident vers l'Orient et vers l'Afrique subsaharienne

Pendant longtemps l'une des préoccupations des chercheurs, dans le domaine de l'histoire des mathématiques en pays d'Islam, a été la recherche des informations permettant d'écrire l'histoire de la circulation des ouvrages grecs et arabes d'Orient vers l'Occident [Djebbar 2002]. Avec le développement quantitatif des recherches sur les activités mathématiques du Maghreb et d'al-Andalus, plusieurs éléments, puisés dans les ouvrages bibliographiques et, surtout, dans les textes mathématiques eux-mêmes, ont encouragé de nouvelles investigations. Ces dernières ont permis de rassembler suffisamment de matériaux pour entreprendre des études sur un phénomène peu connu auparavant et qui concerne la présence, dans des écrits scientifiques produits en Orient, de références explicites à des auteurs occidentaux et à certains de leurs ouvrages. L'illustration la plus frappante de ce phénomène est la circulation du *Kitāb al-istikmāl* d'al-Mu'taman : des copies de cet ouvrage (publié à Saragosse au XIᵉ siècle) sont d'abord signalées à Marrakech puis au Caire puis à Maragha où une nouvelle rédaction de son contenu est réalisée vers la fin du XIIIᵉ siècle ou au début du XIVᵉ [Djebbar 1997 ; Bouzari 2005].

On découvre aussi la diffusion, d'ouest en est, de pratiques et d'outils, caractéristiques de la tradition mathématique de l'Occident musulman, tels qu'ils nous ont été conservés dans les manuels des XIIᵉ-XIVᵉ siècles produits à Séville ou à Marrakech. Au niveau des publications, il y a, en particulier, le *Kitāb al-bayān wa t-tadhkār* [Livre de la démonstration et du rappel] d'al-Ḥaṣṣār, rédigé probablement à Séville et dont une copie a été réalisée à Bagdad [Kunitzch 2002–2003], quelques décennies à peine après la mort de l'auteur. Il y a aussi le *Talkhīṣ aᶜmāl al-ḥisāb* [L'abrégé des opérations du calcul] d'Ibn al-Bannā dont le contenu a dominé l'enseignement du calcul et de l'algèbre au Maghreb tout au long des XIVᵉ–XVᵉ siècles. Il a d'abord fait l'objet d'un résumé, à Jérusalem ou au Caire, par Ibn al-Hā'im (m. 1412) [Aballagh 2000a]. Puis il a été longuement commenté, dans cette dernière ville, par Ibn al-Majdī (m. 1447) dans un volumineux ouvrage, le *Ḥāwī l-lubāb fī sharḥ Talkhīṣ aᶜmāl al-ḥisāb* [Le recueil de la moelle qui commente l'Abrégé des opérations du calcul]. Dans cet écrit, l'auteur se réfère à deux autres ouvrages du même auteur, plus importants sur le plan mathématique que le *Talkhīṣ*, le

Raf^c al-ḥijāb et le *Kitāb al-uṣūl*. Il confirme aussi, par des citations, la circulation du poème algébrique d'Ibn al-Yāsamīn qui avait, quelques décennies auparavant, attiré l'attention d'Ibn al-Hā'im. Ce dernier lui a même consacré un commentaire développé [Abdeljaouad 2003]. Mais une édition critique et une analyse comparative du contenu très riche du *Ḥāwī* d'Ibn al-Majdī pourraient révéler d'autres liens avec la production mathématique de l'Occident musulman [Djebbar 2006].

On peut déjà affirmer l'existence de certains de ces liens grâce à la présence, dans l'ouvrage, du symbolisme arithmétique et algébrique qui est un élément caractéristique de la pratique mathématique du Maghreb des XIIIe–XVe siècles. Des recherches récentes ont d'ailleurs montré l'intervention de ce symbolisme dans d'autres écrits orientaux et leur utilisation jusqu'au XVIIIe siècle [Abdeljaouad 2007 ; 2011]. Il reste à identifier les manuels qui ont permis cette circulation. Parmi les sources possibles, il y a bien sûr les commentaires au *Talkhīṣ* d'Ibn al-Bannā dont nous connaissons le contenu grâce aux travaux de ces trois dernières décennies [al-Qalaṣadī 1999 ; Harbili 2006]. Mais il est possible que d'autres manuels d'enseignement, encore inconnus, aient bénéficié de cette circulation vers certains foyers scientifiques d'Orient, comme le Caire et Istanbul.

Il faut également signaler un second phénomène, peut-être plus tardif, et qui commence à faire l'objet de recherches, celui de la circulation des écrits mathématiques d'al-Andalus et du Maghreb vers les centres islamisés de l'Afrique subsaharienne. Depuis plusieurs décennies, des travaux bibliographiques ont révélé l'existence de nombreux manuscrits scientifiques dans les bibliothèques des pays de cette région. Une partie d'entre eux concerne les mathématiques. Il s'agit, au vu de ce qui est connu, de manuels d'enseignement et d'ouvrages utilisant le calcul et la géométrie pour résoudre des problèmes liés aux activités profanes ou cultuelles (mesurage, répartition des héritages, détermination des moments des prières, de la direction de la Mecque, etc.). Une première étude comparative, consacrée aux seuls écrits mathématiques et astronomiques, montre que les ouvrages maghrébins des XIVe–XVIe siècles tiennent une place importante dans la formation des lettrés subsahariens [Djebbar & Moyon 2011]. L'analyse, non encore faite, de nombreux manuscrits encore anonymes et l'exhumation d'autres documents appartenant à des bibliothèques ne possédant pas de catalogues, pourraient aider à expliciter les liens qui se sont tissés, dans certains domaines scientifiques, entre le nord de l'Afrique et sa partie subsaharienne.

Le contenu des mathematiques produites en occident musulman

La matière mathématique produite ou enseignée, en arabe, en Andalus et au Maghreb, entre le IXe et le XIXe siècle, a fait l'objet de nombreux travaux qui ont porté, en premier lieu, sur les quatre disciplines qui avaient déjà été étudiées dans la période antérieure. Mais de nouveaux thèmes ont également été abordés. Ils ont concerné des pratiques mathématiques liées aux activités quotidiennes de la cité islamique, comme le mesurage, le découpage et la répartition des héritages. Parallèlement, un certain nombre d'épîtres, d'ouvrages ou de chapitres de livre ont été édités, accompagnés parfois d'une traduction française. Il faut enfin signaler quelques études portant sur les aspects épistémologiques ou philosophiques de textes andalous ou maghrébins. Le contenu mathématique de ces écrits était connu, mais il n'avait pas encore été analysé sous cet angle et en tenant compte de l'environnement culturel dans lequel vivaient leurs auteurs [^cUthmānī 1999 ; Lamrabet 2001 ; Aballagh 2007].

L'analyse combinatoire

En analyse combinatoire, une nouvelle publication du chapitre 11 du *Fiqh al-ḥisāb* d'Ibn Mun^cim, qui est entièrement consacré à ce sujet, a été réalisée à l'occasion de l'édition de l'ouvrage [Ibn Mun^cim 2006, 201–236]. Mais, aucun écrit nouveau et original, produit en Occident musulman, n'a été exhumé au cours de la période concernée. Toutefois, des développements, faisant intervenir des dénombrements, sont repérés dans des textes qui reprennent des sources du XIIe et du XIVe siècle, comme celui d'Ibn al-

Bannā [Djebbar 2003c, 39–42], celui d'Ibn Haydūr (m. 1413) [Muṣliḥ 2006, 536–545, 746] et celui de Ṭfayyash (m. 1914) [Ms. Banī Yazgan, Bibliothèque privée, 469–471]. Le domaine des carrés magiques pourrait révéler des pratiques combinatoires élémentaires, comme le laisse penser une note anonyme où les carrés d'ordre trois sont obtenus, par permutation, à partir d'un carré donné [Ms. Rabat, Ḥasaniya n° 53, 161]. Les ouvrages bibliographiques et les catalogues des bibliothèques maghrébines mentionnent des écrits ayant été consacrés à la construction de ces carrés. Certains n'ont pas encore été retrouvés et pour ceux qui nous sont parvenus, l'étude de leurs contenus respectifs n'en est qu'à ses débuts [Sesiano 2004, 17, 126].

L'algèbre

En algèbre, les investigations de ces dernières décennies n'ont pas permis de révéler de nouveaux documents ou d'exhumer des copies d'écrits cités par des auteurs du XIVe siècle, comme le *Kitāb al-jabr* [Livre d'algèbre] d'al-Qurashī dont il nous est parvenu quelques fragments et des références précises. Mais les textes connus depuis longtemps, ou édités durant la période antérieure, ont alimenté des études à différents niveaux. Il y a eu d'abord une première présentation du contenu des pratiques algébriques d'expression arabe, en Andalus et au Maghreb, avec leurs prolongements en Europe à travers des traductions ou de nouvelles publications de textes en hébreu et en latin [Djebbar 2001b, 73–116 ; Moyon 2007 ; 2011]. Une seconde étude a concerné le plus ancien texte algébrique connu de l'Occident musulman, le fameux poème d'Ibn al-Yāsamīn. L'intérêt de ce texte est plus culturel et didactique que scientifique dans la mesure où son contenu est en deçà de ce que l'on connaissait en tant que savoir algébrique à l'époque de sa publication et même plus tard. Il est en fait l'illustration d'une tendance qui va s'accentuer au cours des siècles, et qui va accompagner le phénomène de réduction du champ des activités mathématiques au Maghreb et de leur extinction pure et simple en Andalus. Ce poème deviendra alors un outil indispensable dans l'enseignement de base de l'algèbre [Abdeljaouad 2005c].

Ce sont peut-être aussi des préoccupations liées à l'histoire de l'enseignement des mathématiques qui sont à l'origine d'une nouvelle étude sur le symbolisme algébrique pratiqué au Maghreb depuis la fin du XIIe siècle. Cette étude confirme et explicite le rôle de ce symbolisme dans l'enseignement de l'algèbre et dans la résolution des problèmes aboutissant à des équations du premier et du second degré [Abdeljaouad 2005b].

La théorie des nombres

En théorie des nombres, l'analyse des écrits disponibles a montré la forte présence de la tradition euclidienne dans le *Kitāb al-istikmāl* d'al-Mu'taman, sous la forme d'une nouvelle rédaction des Livres VII à IX des *Eléments*, avec l'abandon d'un certain nombre de propositions, considérées comme superflues dans la nouvelle présentation, et le regroupement des propositions restantes en fonction de leurs liens logiques. D'une manière plus précise, le chapitre consacré à ce thème est divisé en quatre sections traitant, respectivement, les propriétés intrinsèques des nombres, leurs propriétés en tant que grandeurs rapportées les unes aux autres, celles qui découlent de leur similitude aux trois grandeurs géométriques (lignes, surfaces, solides) et, enfin celles qui résultent du rapport des nombres à leurs parties. Dans le prolongement de cette tradition euclidienne, al-Mu'taman a inclus, dans la quatrième partie de son exposé, le contenu de la rédaction de la *Risāla fī l-a^c^dād al-mutaḥābba* [Epître sur les nombres amiables] de Thābit Ibn Qurra (m. 901), sans en modifier ni la structure ni les démonstrations [Djebbar 1999]. C'est peut-être à partir de cette initiative que l'intérêt pour les nombres amiables s'est développé chez des auteurs postérieurs. Quoi qu'il en soit, on a trouvé l'évocation plus ou moins détaillée de ce thème dans des écrits d'origine andalouse, comme le *Kitāb al-kāmil* d'al-Ḥaṣṣār et le *Fiqh al-ḥisāb* d'Ibn Mun^c^im [Ibn Mun^c^im 2006, 189–192], ou dans des manuels maghrébins, comme le commentaire d'al-Ghurbī au *Talkhīṣ* d'Ibn al-Bannā [Harbili 2006].

La tradition néopythagoricienne est absente du traité d'al-Mu'taman. Mais nous savons aujourd'hui qu'elle était bien connue en Andalus et qu'elle a inspiré de nouvelles études. La première est attribuée à Ibn Sayyid, le professeur du philosophe Ibn Bājja et la seconde à Ibn Ṭāhir (XIIe s.), probablement de Séville. Ces deux contributions ne nous sont pas parvenues. Mais nous sommes informés sur leur contenu grâce à la troisième contribution, celle d'Ibn Mun^c^im qu'il a exposée dans le neuvième chapitre de son *Fiqh al-ḥisāb*. L'analyse du contenu de ce chapitre a permis de lever le voile sur un aspect méconnu des pratiques arithmétiques andalouses. Il a aussi révélé le lien existant entre cette tradition et celle du Maghreb représentée par deux contributions, celle d'Ibn al-Bannā dans son *Raf^c al-ḥijāb*, et celle d'Ibn Haydūr dans son *Tuḥfat aṭ-ṭullāb fī sharḥ mā ashkala min Raf^c al-ḥijāb* [La parure des étudiants sur l'explication des difficultés du Lever du voile]. Ces auteurs se sont largement inspirés des écrits de leurs prédécesseurs, avec un élément original dans la démarche d'Ibn al-Bannā qui a tenté de faire le lien entre des propositions arithmétiques et des résultats de nature combinatoire [Aballagh 2000c ; Djebbar 2000a].

Quant à la tradition diophantienne, certains aspects avaient été révélés avec l'édition du *Kitāb al-uṣūl* d'Ibn al-Bannā [Djebbar 1990]. Mais, comme pour d'autres sujets traités au Maghreb, les problèmes de ce chapitre semblent trouver leur origine dans des écrits andalous antérieurs, comme le *Fiqh al-ḥisāb* d'Ibn Mun^c^im [Ibn Mun^c^im 2006, 193–200]. En attendant d'exhumer de nouvelles sources, il serait utile de réaliser une étude comparative du contenu des matériaux disponibles.

La science du calcul

Dans le domaine de la science du calcul, les travaux récents ont permis de mieux connaître à la fois les contours des pratiques de l'Occident musulman, leurs orientations essentielles et certains aspects de leur contenu technique. Des textes connus mais qui étaient encore inédits ont été publiés. C'est le cas du *Liber Mahameleth* d'un auteur latin anonyme du XIIe siècle. On était déjà informé de certaines particularités de cet important ouvrage, en particulier de ses liens avec la tradition andalouse du *Ḥisāb al-mu^c^āmalāt* [Calcul des transactions] [Sesiano 1988]. L'analyse de certains chapitres révèle aussi ses liens avec l'une des traditions orientales du calcul, celle dite du *calcul ouvert*, par opposition au *calcul indien* [Vlasschaert, 2010]. Le second ouvrage, édité et accompagné, lui aussi, d'une traduction française, est le commentaire d'al-Qalaṣādī au *Talkhīṣ* d'Ibn al-Bannā. C'est un écrit tardif, réalisé par l'un des derniers mathématiciens d'al-Andalus, et qui perpétue les caractéristiques et les éléments essentiels de la science du calcul tels qu'ils avaient été fixés par les professeurs de Marrakech au XIIIe siècle [al-Qalaṣādī 1999]. Le troisième ouvrage, dont une édition vient d'être publiée, est en hébreu mais il s'inscrit dans la tradition maghrébine qui vient d'être évoquée puisqu'il puise sa matière dans des écrits d'Ibn al-Bannā, le *Talkhīṣ* et le *Raf^c al-ḥijāb*, et de certains de ses commentateurs. Il s'agit de la *Iggeret ha-Mispar* [L'épître sur le nombre], réalisée à Palerme par Isaac Ibn al-Aḥdab (XIVe s.). On y trouve un exposé des objets et des outils du calcul dans une présentation qui suit le plan adopté par Ibn al-Bannā dans son *Talkhīṣ*, sans qu'il s'agisse d'une traduction dans la mesure où l'auteur a pris le soin de rédiger un texte nouveau et d'y insérer des éléments puisés dans d'autres écrits [Lévy 2003 ; Wartenberg 2007, 4–20].

Sur le plan de l'analyse mathématique des contenus, les nouvelles études se sont intéressées à un premier thème, celui de la numération et plus précisément celle qui utilise les chiffres « *rūmī* ». Les premières publications réalisées sur ce thème, tout au long du XXe siècle, avaient fourni de nombreuses informations sur les différents contextes de leur utilisation en Andalus et au Maghreb. Elles s'étaient également interrogées sur l'origine de ces chiffres et avaient proposé des réponses [Colin 1933 ; Labarta & Barceló 1988]. Les nouvelles études se situent dans le prolongement des précédentes en apportant de nouvelles informations sur l'utilisation de cette numération ou sur des écrits qui lui ont été consacrés au Maghreb [Guergour 2000 ; Guesdon 2002 ; Comes 2002–2003 ; Aballagh 2002 ; Lamrabet 2003]. A partir des résultats de toutes ces recherches et des sources maghrébines connues, il est devenu possible d'orienter l'étude de ce thème vers l'édition et l'analyse comparative des textes désormais disponibles et qui ont été rédigés par des mathématiciens du Maghreb ayant vécu entre le XIIe et le XIVe siècle.

Parmi les autres thèmes de la science du calcul, certains méritaient, au vu des sources désormais disponibles, des études particulières et des analyses comparatives reposant sur les pratiques connues dans les trois grandes régions de l'empire musulman (Asie centrale, Proche Orient et Occident musulman). C'est le cas des méthodes de simple et de double fausse position, des quatre opérations arithmétiques classiques et des procédés d'approximation. Le troisième sujet est le seul, à notre connaissance, à avoir fait l'objet, au cours de cette dernière décennie, d'études portant sur les écrits andalous et maghrébins disponibles. Ces études ont abouti à des travaux universitaires et à des publications. Les nouvelles investigations sont parties des résultats obtenus par des contributions antérieures sur les procédés d'approximation en Occident musulman [Lamrabet 1981 ; Djebbar 1986 ; Sesiano 1988 ; Aballagh 1988] et en Orient [Berggren 2002]. Elles ont ensuite profité de l'édition d'ouvrages révélés au cours des trois dernières décennies, en particulier le *Talqīḥ al-afkār* d'Ibn al-Yāsamīn, le *Fiqh al-ḥisāb* d'Ibn Mun^c^im, le *Raf^c^ al-ḥijāb* d'Ibn al-Bannā et la *Tuḥfat aṭ-ṭullāb* d'Ibn Haydūr. Elles ont également bénéficié de l'exploitation de sources encore inédites, comme le *Bayān wa t-tadhkār* d'al-Ḥaṣṣār, le commentaire au *Talkhīṣ* d'al-Ghurbī [Harbili 2006] et le *Ḥaṭṭ an-niqāb* d'Ibn Zakariyā' (XIVe s.). Grâce aux travaux issus de ces investigations, nous disposons aujourd'hui de l'édition de tous les fragments connus, produits ou utilisés au Maghreb, et dans lesquels sont exposées des techniques d'approximation des racines carrées et cubiques d'entiers, de fractions ou de combinaisons des deux, accompagnées parfois de démonstrations. Trois éléments essentiels se dégagent de toutes ces études : la reproduction de certains procédés d'approximation présents dans des manuels orientaux, publiés entre le IXe et le XIe siècle, l'élaboration de nouveaux procédés inspirés par les techniques de résolution des équations quadratiques et la justification de certains algorithmes à l'aide de preuves arithmétiques ou géométriques [Harbili 2005 ; 2011a ; 2011b].

La géométrie

En géométrie, les travaux de la dernière décennie ont concerné, en premier lieu, la circulation, vers l'Occident musulman, des versions arabes du corpus grec (*Eléments* et *Données* d'Euclide, *Coniques* d'Apollonius, *Sphériques* de Ménélaüs, *Mesure du cercle* et *Sphère et cylindre* d'Archimède) ainsi que les différents aspects de leur intervention dans les écrits andalous ou maghrébins [Guergour 2006 ; Bouzari 2009]. Certains travaux récents ont concerné le contenu de versions arabes, latines et hébraïques des *Eléments*. A partir de ce qui était connu au sujet de la circulation, vers l'Andalus, d'au moins deux versions arabes de cet ouvrage (l'une d'al-Ḥajjāj (VIIIe–IXe s.) et l'autre d'Isḥāq Ibn Ḥunayn (m. 910), révisée par Thābit Ibn Qurra [Djebbar 1996, 104–111]), les nouvelles recherches ont permis de mettre en lumière deux aspects importants, de la circulation des textes : le rôle de la tradition hébraïque dans la préservation du traité d'Euclide [Lévy 2005] et les liens existants entre le contenu des manuscrits arabes d'Occident et certaines versions latines réalisées à partir du XIIe siècle [Rommevaux, Djebbar & Vitrac 2001, 271–277].

La géométrie « *savante* » en Andalus

Parmi les écrits andalous qui prolongent la tradition grecque des *Eléments*, un seul texte nous est parvenu, celui d'al-Mu'taman. Il s'agit de plusieurs chapitres de son *Kitāb al-istikmāl.* C'est une rédaction abrégée de la partie proprement géométrique du traité d'Euclide, selon la démarche et le style que ce mathématicien avait adoptés pour la rédaction des Livres VII–IX [Hogendijk 1991]. Après la découverte de la partie manquante de l'ouvrage d'al-Mu'taman, dans une rédaction réalisée, au XIIIe siècle, par Ibn Sartāq, un mathématicien iranien [Djebbar 1997 ; Hogendijk 2004], le chapitre de l'ouvrage de l'*Istikmāl* correspondant aux Livres I–IV des *Eléments* a été édité et traduit en français. Son contenu est conforme à la démarche générale de l'auteur qui semble avoir conçu son ouvrage comme un recueil de résultats, de démonstrations et de constructions tirés du corpus géométrique grec mais exposés selon une architecture différente [Guergour 2005 ; 2006 ; 2009]. Quant au chapitre qui traite de la théorie des rapports, il se présente également comme une rédaction abrégée du Livre V des *Eléments*, avec le regroupement de certaines propositions et l'abandon d'autres [Djebbar 2011b].

On sait que l'ouvrage d'al-Mu'taman a circulé au Maghreb et il est possible que le contenu de ses chapitres géométriques ait alimenté de nouveaux écrits, comme le livre d'Ibn Mun^{c}im, *Tajrīd akhbār kutub al-handasa ^{c}alā ikhtilāf maqāṣidihā* [L'abstraction de matériaux des livres de géométrie sans distinction de leurs buts] et la *Risāla fī n-nisba* [Epître sur le rapport] d'Ibn Haydūr, deux textes qui n'ont pas encore été retrouvés.

La géométrie des coniques est également présente dans le livre d'al-Mu'taman qui lui a consacré deux chapitres. Le premier est intitulé « *Première section de la troisième espèce sur la détermination des sections <coniques> et sur leurs propriétés principales* ». Il reprend, partiellement, le contenu des Livres I–III du traité d'Apollonius. Il a fait l'objet d'une édition et d'une traduction française récente [Bouzari 2008]. Le second, qui n'a pas encore bénéficié d'une édition, est intitulé « *Seconde section de la troisième espèce sur les propriétés de lignes, des angles et des surfaces des sections <coniques> en combinaison* ». C'est un abrégé des Livres IV–VII du même traité.

Le second texte andalou existant, qui témoigne de l'étude des coniques, est un fragment d'un traité d'Ibn as-Samḥ (m. 1037) qui pourrait être celui qui a été décrit par Ṣā^{c}id al-Andalusī comme un « *grand livre en géométrie où il a épuisé <l'étude de> ses parties relatives aux lignes droite, arquée et courbe* ». L'auteur y étudie les ellipses engendrées comme sections de cylindre. Il nous est parvenu dans une version hébraïque dont le contenu a fait l'objet d'une traduction française et d'une analyse comparative [Lévy 1996, 927–973, 1080–1083 ; Rashed 1996, 885–927] qui devraient être complétées par une édition critique.

Nous n'avons pas encore d'éléments permettant de décrire la genèse des études andalouses sur les sections coniques puisque les deux ouvrages qui viennent d'être évoqués sont, pour le moment, les plus anciens qui témoignent de cette activité. Mais des recherches antérieures concernant les contributions d'Ibn Sayyid et d'Ibn Bājja avaient montré qu'à la fin du XIe siècle et au début du XIIe, les sections coniques étaient encore étudiées et utilisées à Valence et à Saragosse [Djebbar 1993] avec, peut-être, des prolongements à Marrakech à l'époque d'Ibn Mun^{c}im et d'Ibn al-Bannā puisque la présence du traité d'Apollonius y est signalée [Djebbar & Aballagh 2001, 61]. Ainsi, avec les résultats des dernières recherches, nous disposons d'un ensemble d'éléments textuels ou de témoignages qui contribueront à la réalisation d'un premier bilan sur la tradition des sections coniques en Occident musulman.

Quant à la géométrie archimédienne, elle est également présente dans les écrits andalous et, dans une moindre mesure, dans certains écrits maghrébins [Djebbar 2011a]. Il y a d'abord des références aux deux seuls écrits complets d'Archimède qui avaient été traduits en arabe en Orient, c'est-à-dire le *Traité sur la sphère et le cylindre* et l'*Epître sur la mesure du cercle*. Une partie de leurs contenus est traitée dans l'*Istikmāl* d'al-Mu'taman. Mais on trouve aussi dans cet ouvrage des éléments de la tradition arabe des IXe–Xe siècles qui sont parvenus en Andalus dans des écrits des frères Banū Mūsā et d'Ibrāhīm Ibn Sinān [Hogendijk 1991 ; 2004].

La géométrie du mesurage et du découpage

A côté de ce corpus géométrique savant, nourri par des œuvres de la tradition grecque et par de nouvelles contributions produites en Orient, on trouve, comme dans la science du calcul, un ensemble de textes qui témoignent de pratiques répondant à des besoins de la société. En effet, il ne s'agit pas de développements et de résultats théoriques mais de formules et de procédures permettant de trouver des solutions à des problèmes concrets. Pendant longtemps, ce corpus est resté en marge des préoccupations des chercheurs, malgré quelques publications qui en avaient révélé certains aspects. Pour l'Andalus, on peut citer le *Liber Embadorum*, version latine du *Chibbur ha-Meschicha we ha-Tischboreth* [La composition sur les mesures géométriques] d'Abraham Bar Ḥiyya (m. vers 1145) [Curtze 1902 ; Bar Ḥiyya 1912], suivie de sa traduction en catalan [Bar Ḥiyya 1931], l'étude concernant le *Liber mensurationum*, version latine d'une épître attribuée à un certain Abū Bakr [Busard 1968] et, enfin, l'édition d'une épître andalouse, plus tardive que la précédente, écrite par Ibn ar-Raqqām (m. 1315) et intitulée *at-Tanbīh wa t-tabṣīr fī qawānīn at-taksīr* [Rappel et éclaircissement au sujet des règles du mesurage] [al-Khaṭṭābī 1956, 39–42]. Pour le Maghreb, les années 60 du siècle dernier ont vu la publication de deux épîtres qui, comme la précédente, n'étaient pas accompagnées d'une analyse de leurs contenus. Il s'agit de la *Risāla fī l-ashkāl al-misāḥiya* [Epître sur les figures du mesurage] d'Ibn al-Bannā [al-Khaṭṭābī 1956, 43–47] et du *Sharḥ al-iksīr fī ᶜilm at-taksīr* [Commentaire sur l'Elixir de la science du mesurage] d'Ibn al-Qādī (m. 1630) [al-Khaṭṭābī 1957, 77–87]. Mais ces initiatives n'ont pas eu de suite au niveau de la recherche.

Le plus ancien écrit arabe de cette catégorie produit en Andalus est la *Risāla fī t-taksīr* [Epître sur le mesurage]. Elle a été publiée au Xe siècle par Ibn ᶜAbdūn (m. après 970), un enseignant de mathématique qui fit carrière en médecine. Son contenu traite, essentiellement, de questions de mesure. On y trouve l'exposé des formules donnant les aires et les volumes d'un ensemble de figures planes et solides. On y trouve aussi une famille de problèmes dans lesquels il s'agit de déterminer un élément d'une figure à partir de la connaissance de certaines données. Cette partie de l'épître révèle, au niveau des procédures et de la terminologie, des similitudes avec des textes mathématiques de la tradition babylonienne. En effet, la résolution de certains problèmes suit les mêmes étapes et utilise les mêmes opérations qui sont à l'œuvre dans certaines tablettes cunéiformes, mais sans utiliser la terminologie algébriques que l'on trouvera plus tard dans le traité d'al-Khwārizmī [Djebbar 2005–2006].

A partir des résultats de tous les travaux antérieurs, l'étude de la géométrie pratique en Occident musulman, avec ses prolongements dans l'espace latin, s'est poursuivie selon deux directions. La première a été favorisée par la découverte et l'analyse de nouveaux écrits arabes, publiés au XIIIe siècle en Andalus, en particulier le *Kitāb at-taqrīb wa t-taysīr li ifādat al-mubtadi' bi ṣināᶜat at-taksīr* [Livre qui vulgarise et facilite pour faire profiter le débutant de l'art du mesurage] d'Ibn al-Jayyāb et le *Kitāb al-qurb fī t-taksīr wa t-taqṭīᶜ* [Livre qui facilite le mesurage et le découpage] de ᶜAbdallāh al-Mursī. Leurs contenus nous renseignent sur certains aspects des pratiques géométriques locales au cours des deux siècles qui ont suivi la rédaction de l'épître d'Ibn ᶜAbdūn. On y retrouve les thèmes du mesurage de la tradition arabe d'Orient dont le plus ancien témoin connu est le chapitre géométrique du livre d'algèbre d'al-Khwārizmī [Rashed 2007, 202–231]. Mais on y trouve aussi des spécificités locales au niveau du traitement de certaines figures (planes ou solides) et au niveau de la terminologie, spécificités que ces écrits partagent avec ceux d'Ibn ᶜAbdūn et d'Abū Bakr [Djebbar 2007, 120–133].

A côté de ces thèmes, il y a l'exposé des différents aspects d'un chapitre consacré au découpage des figures planes, en tenant compte de différentes contraintes, comme la manière de faire le découpage, la forme géométrique des parties que l'on doit découper, ou bien le rapport, du point de vue de l'aire, entre la partie à découper et la figure donnée. En Orient, deux traditions de découpages ont vu le jour ou, pour être plus précis, ont été redynamisées dans le cadre de préoccupations et de besoins nouveaux. La première pourrait être qualifiée de savante parce qu'elle semble avoir pour origine un ouvrage d'Euclide, aujourd'hui perdu, le *Livre sur les divisions (des figures)*. Un certain nombre d'écrits arabes appartiennent à cette tradition, en particulier l'épître intitulée *Fī l-ḥujja al-mansūba ilā Suqrāṭ fī l-murabbaᶜ wa quṭrihī* [Sur la preuve attribuée à Socrate au sujet du carré et de sa diagonale] de Thābit Ibn Qurra et le *Kitāb fī mā yaḥtāju ilayhī aṣ-ṣāniᶜ min aᶜmāl al-handasa* [Livre de ce qui est nécessaire à l'artisan en constructions géométriques] d'Abū l-Wafā' (m. 997). La seconde tradition trouve son origine dans les problèmes d'arpentage qu'avaient à résoudre les fonctionnaires des administrations de l'empire musulman, et dans les problèmes de répartition des terres d'un héritage. Les problèmes et les activités de découpage qui en découlent sont attestés par la littérature juridique islamique depuis le VIIIe siècle.

Les deux livres andalous que nous avons évoqués mériteraient d'être édités parce qu'ils contiennent de précieuses informations sur différents domaines (techniques de résolution des problèmes, unités de mesure en usage à l'époque de leurs auteurs et dans différentes villes d'al-Andalus, statut des spécialistes du découpage, etc.). En attendant, l'analyse de leur contenu montre une perpétuation ou un prolongement de plusieurs traditions. La première est celle de la géométrie savante du mesurage d'origine orientale, avec des références à Euclide (même si les résolutions ne sont jamais accompagnées de démonstrations). La seconde est celle de la tradition locale du découpage dont on ignore les premiers pas en Andalus mais dont la plus ancienne manifestation connue dans un manuel de cette région se trouve dans le chapitre III du *Chibbur ha-Meschicha* d'Abraham Bar Ḥiyya [Lévy 2001a].

La seconde orientation prise par les recherches sur la tradition du mesurage a concerné les écrits latins du moyen âge qui ont exposé ce thème. Certains d'entre eux sont des traductions de textes arabes ou hébraïques, réalisées à partir du XIIe siècle, comme le *Liber Embadorum* d'Abraham Bar Ḥiyya, le *Liber mensurationum* [Livre sur le mesurage] d'Abū Bakr, le *Liber Saydi Abuothmi* [Livre de Sa^c^īd Abū ^c^Uthmān], le *Liber Aderameti* [Livre de ^c^Abd ar-Raḥmān] [Busard 1969] et le *De superficierum divisionibus liber* [Livre sur les divisions des surfaces]. D'autres sont des écrits originaux qui s'inscrivent dans le prolongement de la double tradition mathématique, celle qui se rattache, d'une manière ou d'une autre, à l'ouvrage d'Euclide et celle qui résout des problèmes concrets à la manière des auteurs d'al-Andalus. Les plus importants ouvrages de cette catégorie sont le *Liber Philotegni* [Livre de l'ami de l'art] de Jordanus de Nemore (XIIIe s.) et le *De arte mensurandi* [L'art du mesurage] de Jean de Murs (m. ca. 1351).

Dans ce domaine, les investigations ont été menées selon trois orientations complémentaires : en premier lieu, un bilan détaillé des pratiques antérieures à l'avènement des mathématiques des pays d'Islam, en particulier celles des traditions mésopotamienne, grecque et romaine. En second lieu, un exposé des éléments essentiels obtenus par la recherche au sujet des pratiques de mesurage et de découpage dans l'Orient musulman puis en Andalus. Enfin, une analyse comparative du contenu des textes arabes et latins disponibles. Il se dégage de cette étude approfondie un ensemble de conclusions qui permettent de prolonger les résultats antérieurs en les explicitant ou en les affinant. De plus, ce travail est complété par l'édition et la traduction française des versions latines des quatre écrits arabes sur le mesurage que nous avons déjà évoqués [Moyon 2008 ; 2010]. Ceci ouvre la voie à de nouvelles investigations, en particulier dans la partie du corpus latin qui n'a pas encore fait l'objet d'une étude comparative.

Les mathématiques au service d'autres activités

Les recherches déjà effectuées sur les écrits produits en Occident musulman montrent qu'en dehors de quelques interventions ponctuelles, comme celle de la notion de rapport en rhétorique [Aballagh 2000b], celle du dénombrement en lexicographie [Djebbar 2003c] et celle des procédés arithmétiques dans la résolution de problèmes ludiques [Djebbar 2004 ; Bouzari 2003], les outils mathématiques sont présents, essentiellement dans quatre domaines, l'astronomie, l'astrologie, les transactions commerciales et la science des héritages.

L'astronomie et l'astrologie

Pour les deux premières disciplines, les travaux récents ont exploité un certain nombre d'écrits andalous et maghrébins, inédits pour la plupart. Parmi les thèmes traités dans les articles publiés et qui concernent l'Andalus, on peut citer, pour le XIe siècle, les calculs d'Ibn Mu^c^ādh permettant de déterminer les maisons astrologiques et leurs aspects [Hogendijk 2005]. Pour le XIIe siècle, il y a le calcul de la grandeur des phases des éclipses solaires et lunaires par Jābir Ibn Aflaḥ (m. 1145) [Bellver 2008] et sa méthode des « *quatre éclipses* » pour déterminer la période d'anomalie de la Lune [Bellver 2006]. Pour le même siècle, deux contributions d'Ibn al-Hā'im (ca. 1205) ont fait l'objet de publications : le calcul de la longitude et de la latitude de la lune et l'étude de la théorie de la trépidation [Puig 2000 ; Comes 2001].

S'agissant du Maghreb, une première contribution a porté sur les outils mathématiques qui interviennent dans l'ouvrage d'al-Ḥasan al-Murrākushī (m. après 1260), *Kitāb al-mabādi' wa l-ghāyāt fī ᶜilm al-mīqāt* [Livre des principes et des buts sur la science du temps] [Assali 2000]. Mais, la richesse de cet ouvrage et ses liens avec la tradition antérieure de l'astronomie appliquée nécessitent une étude comparative du contenu de chacun de ses chapitres, comme cela s'est fait pour l'astrolabe universel [Puig 2005]. Il y a aussi les procédés arithmétiques exposés par deux astronomes maghrébins, Ibn al-Bannā et al-Jādirī (m. 1416), dans le but de résoudre des problèmes liés à la pratique religieuse (conversions entre calendriers lunaire et solaire, calcul des moments des prières et détermination de la direction de la Mecque) [Calvo 2004]. Il faut enfin signaler les études récentes sur les contributions d'Ibn ᶜAzzūz al-Qasanṭīnī (m. 1354), un astronome peu connu, originaire du Maghreb central. Ses *zīj*, qui ont été exhumés il y a moins de vingt-cinq ans, s'inscrivent, par leur contenu, dans le prolongement de la grande tradition andalouse des XIᵉ-XIIᵉ siècles, en particulier dans le traitement mathématique de certains problèmes astrologiques [Samsó 2007 ; Casulleras 2007].

Les travaux qui viennent d'être évoqués révèlent l'intervention d'outils appartenant à plusieurs disciplines ou chapitres mathématiques : science du calcul, géométrie du plan et de la sphère, trigonométrie, méthodes d'approximation ou d'interpolation, etc. Ainsi, en plus des informations qu'ils fournissent sur la place de ces outils dans les pratiques astronomiques et astrologiques d'al-Andalus et du Maghreb, ils ouvrent la voie à de nouvelles recherches sur leur circulation, d'Orient en Occident, et sur les supports de cette circulation.

Les transactions

Dans le vaste domaine du « *calcul des transactions* », l'unique ouvrage qui nous soit parvenu et qui, comme son titre l'indique, est consacré à ce thème, est le *Liber Mahamelet* [Livre des transactions] que nous avons déjà évoqué. C'est un volumineux traité dont le contenu suggère de nouvelles voies de recherche. D'abord, une première évaluation du contenu de la tradition andalouse des IXᵉ–XIIᵉ siècles, tant au niveau de la nature des problèmes résolus que des procédures utilisées pour ces résolutions. On sait en effet, que des ouvrages s'inscrivant dans cette problématique des « *mathématiques transactionnelles* » avaient été publiés, en Andalus, à la fin du Xᵉ siècle et au début du XIᵉ. Les plus connus d'entre eux sont le *Kitāb al-muᶜāmalāt* [Livre des transactions] d'Ibn as-Samḥ (m. 1035) et le *Kitāb al-arkān fī l-muᶜāmalāt* [Livre des fondements des transactions par la voie de la démonstration] et qui ne sont connus que par les références à leurs contenus respectifs, fournis par des auteurs du XIIᵉ siècle. En second lieu, la recherche, dans le corpus exposé, de ce qui est le résultat de la circulation des savoir-faire d'Orient vers l'Occident et de ce qui est l'expression mathématique de pratiques locales. En troisième lieu, des éléments de ce corpus pourraient aider à situer l'auteur et son apport personnel dans le cadre de la nouvelle dynamique scientifique d'expression latine.

A partir du XIIᵉ siècle, les mathématiciens d'al-Andalus et du Maghreb prennent l'habitude d'exposer les problèmes de transaction dans des ouvrages de calcul ou d'algèbre, après la présentation des opérations arithmétiques classiques et les procédures de résolution des problèmes (règle des quatre grandeurs proportionnelles, méthode de double fausse position et procédé algébrique). Quant aux thèmes traités, une grande partie est inspirée de la vie économique au sens large (vente, achat, salaires, conversion de monnaies, estimation de l'aumône légale, calcul des gains ou des pertes après plusieurs transactions, etc.). Mais on trouve aussi des problèmes « *pseudo-concrets* » dont le rôle est d'illustrer des méthodes de résolution et d'exercer les apprenants à maîtriser ces méthodes (problèmes de rencontre, problèmes de volatiles, etc.). Comme aucun auteur ne revendique la paternité de l'un ou l'autre de ces problèmes ni les procédés de résolution, nous avons là un ensemble de matériaux (énoncés, procédures, terminologie) qui suggère une étude approfondie de leur circulation entre les foyers scientifiques de l'empire musulman [Djebbar 2001a ; 2005b].

La science des héritages

L'autre grand domaine d'intervention des outils mathématiques est constitué par toutes les pratiques liées à la répartition des héritages. De nombreux ouvrages ont été consacrés à ce sujet, depuis le VIIIe siècle, à la fois en Andalus et au Maghreb, à la suite de la première production orientale dans ce domaine. La plus grande partie d'entre eux traite des aspects juridiques de la répartition d'un héritage entre les ayants droit et des conditions qui permettent de bénéficier d'une donation de la part du défunt. L'autre partie traite essentiellement des aspects techniques de la répartition. Et ce sont, le plus souvent, des mathématiciens qui en sont les auteurs. Parmi les ouvrages de cette seconde catégorie, les plus célèbres en Occident musulman jusqu'au XIVe siècle sont ceux d'al-Ḥūfī (m. 1192) et d'al-Qurashī. Ils se distinguent, essentiellement, par les méthodes utilisées pour résoudre les problèmes de répartition. L'ouvrage du premier est mentionné depuis longtemps dans les catalogues de certaines bibliothèques. Mais l'étude de son contenu n'a été réalisée que dernièrement [Laabid 2006]. Elle a été précédée par une autre étude sur l'un de ses commentaires par le mathématicien du Maghreb central Sa^{c}īd al-^{c}Uqbānī [Zerrouki 2000]. L'ouvrage d'al-Qurashī n'a pas encore été retrouvé mais on sait que son contenu a fait l'objet d'un commentaire rédigé par l'andalou Ibn Ṣafwān (m. 1371) et intitulé *Kifāyat al-fāriḍ al-murtāḍ fī at-tanbīh ^{c}alā mā aghfalahū jumhūr al-furrāḍ* [Le <livre> suffisant pour le répartiteur d'héritage satisfait sur le rappel de ce qu'ont négligé les spécialistes des héritages]. Ce dernier écrit a été retrouvé et son analyse a confirmé l'apport original d'al-Qurashī dans le calcul des parts des ayants droit [Laabid 2011].

L'analyse des textes aujourd'hui disponibles a permis de présenter, dans le détail, les différents types de problèmes que peut rencontrer un répartiteur d'héritage et les outils utilisés pour aboutir, dans chaque situation, à une répartition équitable. Pour nous limiter aux aspects techniques du sujet, on constate que deux démarches distinctes ont cohabité pendant des siècles. La première, que l'on pourrait qualifier de « *traditionnelle* », est purement arithmétique. Elle est basée sur la manipulation de fractions et sur la technique des tableaux. C'est la plus ancienne puisqu'elle remonte au premier siècle de l'Islam. Elle s'est diffusée comme un « *savoir-faire* » par un apprentissage direct dans le cadre des activités quotidiennes des juristes qui se sont spécialisés dans la répartition des héritages. La seconde fait intervenir des outils mathématiques qui ont été diffusés, par les traductions ou par l'enseignement direct, à partir de la fin du VIIIe siècle. Cette diffusion a été relativement rapide puisque, au Xe siècle, certains juristes d'Orient, comme al-Ḥubūbī, les ont utilisés dans leurs manuels consacrés aux problèmes d'héritage [Laabid 2001 ; 2003]. Parmi ces techniques, on trouve la méthode de fausse position et les procédés algébriques. Les recherches récentes ont montré que l'exposé d'al-Ḥūfī utilise toutes ces procédures, confirmant ainsi leur circulation en Occident musulman. Elles montrent aussi que la méthode d'al-Qurashī s'inscrit dans la tradition arithmétique ancienne mais en la rénovant par une manipulation des fractions qui fait intervenir le « *plus petit commun multiple* » et qui introduit dans les calculs plus de souplesse et de rapidité. C'est d'ailleurs ce qu'affirmaient certains mathématiciens maghrébins, en particulier al-^{c}Uqbānī et al-Qalaṣādī [Laabid 2005]. Ces derniers ont permis à cette méthode, grâce à leur enseignement, et malgré la résistance de nombreux juristes, de se diffuser tout au long du XIVe siècle et du XVe, en particulier à Tlemcen et à Tunis. Nous savons qu'al-Qalaṣādī a séjourné un certain temps au Caire et qu'il y a enseigné. Mais nous ne savons pas encore s'il a fait connaître cette méthode de calcul et nous ignorons la réaction des juristes égyptiens face à cette « *innovation* ».

Il faut enfin signaler un autre aspect de l'activité de répartition des héritages qui s'est révélé avec l'étude relativement récente d'un certain nombre d'écrits de calcul maghrébins. Le statut élevé de cette activité dans la cité islamique et son développement quantitatif au cours des siècles semblent être la cause d'une modification interne d'un certain nombre de manuels de calcul, avec le gonflement du chapitre des fractions au détriment des autres chapitres. C'est ce que l'on observe, en particulier dans le *Kitāb al-bayān* d'al-Ḥaṣṣār [Ms. Rabat, B. N. 917 Q, 23–55], dans le *Talqīḥ al-afkār* d'Ibn al-Yāsamīn et dans le *Fiqh al-ḥisāb* d'Ibn Mun^{c}im [Zemouli 1993, 147–187 ; Ibn Mun^{c}im 2006, 237–354]. La place qui y est réservée aux fractions représente, dans chacun d'eux, plus de 40% de la partie arithmétique de leurs contenus respectifs. Il est intéressant aussi de remarquer que, dans les manuels maghrébins de calcul postérieurs au XIIIe siècle, le volume du chapitre des fractions a été réduit d'une manière significative.

Il reste à savoir si cela est lié, directement ou indirectement, à l'enseignement de la méthode de calcul d'al-Qurashī.

Littérature

Aballagh, M., 1988. Raf[c] al-ḥijāb d'Ibn al-Bannā. Thèse de Doctorat. Université de Paris I-Pantheon-Sorbonne, Paris.

——— 2000a. Introduction à l'étude de l'influence d'Ibn al-Bannā sur les mathématiques en Egypte à l'époque ottomane. En : İhsanoğlu, E., Günergun, F., Djebbar, A. (eds.), Science, Technology and Industry in the Ottoman Word, vol. VI. Brepols, Turnhout, pp. 75–80.

——— 2000b. L'application du rapport mathématique dans la science de la rhétorique chez Ibn al-Bannā al-Murrākushī. En : Les outils de la démonstration en science. Publications de la Faculté des Lettres et Sciences Humaines, Rabat, pp. 63–80. (en arabe)

——— 2000c. Quelques aspects de la théorie des nombres dans les mathématiques grecques et ses prolongements arabo-islamiques. Publications de la Faculté des Lettres et sciences humaines, Rabat, pp. 63–80. (en arabe)

——— 2002. Découverte d'un nouvel ouvrage mathématique d'Ibn al-Bannā. Da[c]wat al-ḥaqq n° 363, 126–132. (en arabe)

——— 2003. Introduction à la lecture des gloses de Muḥammad Bannīs sur le "Désir des étudiants sur le commentaire du 'Souhait des calculateurs' d'Ibn Ghāzī." En : El Bouazzati, B. (ed.), La pensée scientifique dans le moyen âge tardif. Publications de la Faculté des Lettres et Sciences Humaines, Rabat, pp. 123–137. (en arabe)

——— 2007. Philosophie et histoire des mathématiques en Occident musulman, Essai de synthèse. Thèse de Doctorat d'Etat. Université Mohamed V, Rabat. (en arabe)

Abdeljaouad, M., 2003. Commentaire sur le poème al-Yāsamīniya de l'égyptien Ibn al-Hā'im. Publications de l'A.T.S.M, Tunis.

——— 2005a. The Eight Hundredth Anniversary of the Death of Ibn al-Yāsamīn, Bilaterality as part of his thinking and practice. Actes du 8e Colloque maghrébin sur l'histoire des mathématiques arabes (Tunis, 18–20 décembre 2004). Publications de l'A.T.S.M., Tunis, pp. 1–30.

——— 2005b. Le manuscrit mathématique de Jerba, une pratique des symboles algébriques maghrébins en pleine maturité. Actes du 7e Colloque maghrébin sur l'histoire des mathématiques arabes (Marrakech, 30 mai-1e juin 2002), vol. 2. al-Wataniya, Marrakech, pp. 9–98.

——— 2005c. Ibn al-Yāsamīn's Urjūza fī al-Jabr wa'l-muqābala. Llull 28, n° 61, pp. 181–194.

——— 2007. La circulation des symboles mathématiques maghrébins entre l'Occident et l'Orient musulmans. Actes du 9e Colloque maghrébin sur l'histoire des mathématiques arabes (Tipaza, 12-14 mai 2007). Imprimerie Fasciné, Alger, pp. 7–35.

——— 2011. Seker-Zade (m. 1787), le témoin le plus tardif faisant un usage vivant des symboles mathématiques maghrébins inventés au 12e siècle. Actes du 10e Colloque maghrébin sur l'histoire des mathématiques arabes (Tunis, 29–31 mai 2010). Publication de l'A.T.S.M., Tunis, pp. 7–32.

Assali, S.-A., 2000. Les outils mathématiques dans l'œuvre astronomique d'al-Ḥasan al-Murrākushī (XIIIe s.). Mémoire de Magister, E.N.S., Alger. (en arabe)

Bar Ḥiyya, A., 1912. Chibbur ha-Meschicha we ha-Tischboreth (M. Guttmann, ed.). Schriften des Vereins Mekize Nirdamin, Berlin.

——— 1931. Llibre de geometria, Hibbur hameixihà uehatixbòret, Millas, J. Vallicrosa (trad.). Editorial Alpha, Barcelone.

Bellver, J., 2006. Jābir b. Aflaḥ on the four-eclipse method for finding the lunar period in anomaly. Suhayl 6, 159–248.

——— 2008. Jābir Ibn Aflaḥ on lunar eclipses. Suhayl 8, 47–91.

Berggren, J. L., 1985. History of mathematics in the Islamic world: The present state of the art. Middle East Studies Association Bulletin 19, 9–33. (See pp. 51–71 of this volume.)

——— 1997. Mathematics and her sisters in medieval Islam: A selective review of work done from 1985 to 1995. Historia Mathematica 24, 407–440. (See pp. 72–99 of this volume.)
——— 2002. Some Ancient and Medieval Approximations to Irrational Numbers and Their Transmission. En : Dold-Samplonius, Y., Dauben, J.W., Folkerts, M., van Dalen, B. (eds.), From China to Paris, 2000 Years Transmission of Mathematical Ideas. Franz Steiner, Stuttgart, pp. 31–44.
Bouzari, A., 2003. Procédure et circulation des « nombres pensés » de l'Orient à l'Occident musulmans. Actes du Colloque « De la chine à l'Occitanie, Chemins entre arithmétique et algèbre » (Toulouse, 22-24 septembre 2000). Spiesser, M. et Guillemot, M. (eds.), C.I.H.S.O. - Université de Toulouse II, Toulouse, pp. 15–27.
——— 2005. Les coniques de l'Istikmāl d'al Mu'taman (m. 1085) dans la rédaction d'Ibn Sartāq (XIVe s.). Actes du 8ème Colloque sur l'Histoire des Mathématiques Arabes (Tunis, 20–23 décembre 2004). Association Tunisienne des Sciences Mathématiques (ed.), Tunis, pp. 83–92
——— 2008. La géométrie des coniques dans la tradition de l'Occident musulman à travers le Kitāb al-istikmāl [Livre de l'accomplissement] d'al-Mu'taman (m. 1085). Doctorat d'histoire des mathématiques. Université de Lille 1, Lille.
——— 2009. Les coniques en Occident Musulman entre le XIe et le XIVe siècle. Llull 32, 59–72.
Busard, H.L.L., 1968. L'algèbre au moyen âge. Le "Liber mensurationum" d'Abu Bekr. Journal des savants, Avril-Juin, 65–125.
——— 1969. Die Vermessungstraktate 'Liber Saydi Abuothmi' und 'Liber Aderameti.' Janus 56, 161–174.
Calvo, E., 2004. Two treatises on Mīqāt from the Maghrib (14th and 15th Centuries A. D.). Suhayl 4, 159–206.
Casulleras, J., 2007. Ibn ^c^Azzūz al-Qusanṭīnī's tables for computing planetary aspects. Suhayl 7, 47–114.
Colin, G.S., 1933. De l'origine grecque des 'chiffres de Fès' et de nos 'chiffres arabes.' Journal Asiatique 222, 193–215.
Comes, M., 2001. Ibn al-Hā'im's trepidation model. Suhayl 2, 291–408.
Comes, R., 2002–2003. Arabic, Rūmī, Coptic or merely Greek alphanumerical notation ? The case of a mozarabic 10th Century Andalusi Manuscript. Suhayl 3, 157–186.
Curtze, M., 1902, Der 'Liber Embadorum' des Savasorda in der Übersetzung des Plato von Tivoli. Urkunden zur Geschichte der Mathematik im Mittelalter und der Renaissance 1, 1–183.
Djebbar, A., 1984. Deux mathématiciens peu connus de l'Espagne du XIe siècle, al-Mu'taman et Ibn Sayyid. En : Folkerts, M., Hogendijk, J.P. (eds.), Vestigia Mathematica, Studies in medieval and early modern mathematics in honour of H.L.L. Busard. GA, Amsterdam, pp. 79–91.
——— 1986. Algorithmes et optimisation dans les mathématiques arabes, Premier Symposium International de l'ICOMIDC sur "Informatics and the teaching of mathematics in developing countries" (Monastir, 3–7 Février 1986). En : Amara, M., Boudriga, N., Harzallah, K. (eds.), Actes du Symposium. Tunis, Chap. 13.
——— 1990. Mathématiques et mathématiciens du Maghreb médiéval (IXe–XVIe siècles) : Contribution à l'étude des activités scientifiques de l'Occident musulman. Thèse de Doctorat. Université de Nantes, Nantes.
——— 1996. Quelques commentaires sur les versions arabes des Eléments d'Euclide et sur leur transmission à l'Occident musulman. En : Folkerts, M. (ed.), 1996. Mathematische Probleme im Mittelalter, der lateinische und arabische Sprachbereich. Harrassowitz Verlag, Wiesbaden, pp. 91–114.
——— 1997. La rédaction de l'Istikmāl d'al-Mu'taman (XIe s.) par Ibn Sartāq un mathématicien des XIIIe–XIVe siècles. Historia Mathematica, n° 24, 185–192.
——— 1998. Contribution à l'étude des activités mathématiques dans l'Occident musulman (IXe–XVIe siècles), Habilitation à diriger des recherches. Ecole des Hautes Etudes en Sciences Sociales, Paris.
——— 1999. Les livres arithmétiques des Eléments d'Euclide dans une rédaction du XIe siècle, le Kitāb al-istikmāl d'al-Mu'taman (m. 1085). Llull 22, n° 45, 589–653.
——— 2000a. Figurate Numbers in the Mathematical Tradition of Andalus and the Maghrib. Suhayl 1, 57–70.

——— 2000b. Les activités mathématiques au Maghreb à l'époque ottomane (XVIe–XIXe siècles). En : İhsanoğlu, E., Günergun, F., Djebbar, A. (eds.), Science, Technology and Industry in the Ottoman Word, vol. VI. Brepols, Turnhout, pp. 49–66.

——— 2001a. Les transactions dans les mathématiques arabes, classification, résolution et circulation. Actes du Colloque International « Commerce et mathématiques du Moyen âge à la Renaissance, autour de la Méditerranée » (Beaumont de Lomagne, 13-16 mai 1999). Editions du C.I.H.S.O, Toulouse, pp. 327–344.

——— 2001b. L'algèbre arabe, genèse d'un art. Editions Vuibert-Adapt, Paris.

——— 2002. La circulation des mathématiques entre l'Orient et l'Occident musulmans, interrogations anciennes et éléments nouveaux. En : Dold-Samplonius, Y., Dauben, J.W., Folkerts, M., van Dalen, B. (eds.), From China to Paris, 2000 Years Transmission of Mathematical Ideas. Franz Steiner, Stuttgart, pp. 213–236.

——— 2003a. A Panorama of Research on the History of Mathematics in al-Andalus and the Maghrib between the Ninth and Sixtenth Century. En : Hogendijk, J.P., Sabra, A.I. (eds.), The Enterprise of Science in Islam: New perspectives. The MIT Press, Londres, pp. 309–350.

——— 2003b. Les activités mathématiques au Maghreb à travers le témoignage d'Ibn Khaldūn, Actes des journées sur « Les sciences dans la phase de déclin » (Marrakech, 8-11 février 2001). Faculté des Lettres et Sciences Humaines, Rabat, pp. 7–22.

——— 2003c. Mathématiques et société à travers un écrit maghrébin du XIVe siècle. Actes du colloque international « De la Chine à l'Occitanie, chemins entre arithmétique et algèbre » (Toulouse, 22-24 septembre 2000). Editions du C.I.H.S.O., Toulouse, pp. 29–54.

——— 2004. Du nombre pensé à la pensée du nombre : quelques aspects de la pratique arithmétique arabe et de ses prolongements en Andalus et au Maghreb. En : Alvarez, C., Dhombres, J., Pont, J.-C. (eds.), Sciences et Techniques en Perspective, IIe série, vol. 8, fascicule 1, pp. 303–322.

——— 2005a. Les mathématiques dans le Maghreb impérial (XIIe–XIIIe s.), Actes du 7e Colloque maghrébin sur l'histoire des mathématiques arabes (Marrakech, 30 mai-2 juin 2002). al-Wataniya, Marrakech, pp. 97–132.

——— 2005b. Savoirs mathématiques et pratiques métrologiques arabes. Actes du colloque « La juste mesure, quantifier, évaluer, mesurer, entre Orient et Occident (VIII[e]–XVIII[e] siècle) ». Moulinier, L., Sullimann, Verna, L.C., Weill-Parot, N. (eds.), Presses Universitaires de Vincenne, Paris, pp. 59–78.

——— 2005–2006. L'épître sur le mesurage d'Ibn [c]Abdūn, un témoin des pratiques antérieures à la tradition algébrique arabe. Suhayl 5, partie arabe, 7–68 ; 6, partie arabe, 81–86.

——— 2006. Les traditions mathématiques d'al-Andalus et du Maghreb en Orient, l'exemple d'Ibn al-Majdī. Actes du 8[e] colloque maghrébin sur l'histoire des mathématiques arabes (Tunis, 18–20 décembre 2004). A.T.S.M., Graphimed, Tunis, pp. 155–184.

——— 2007. La géométrie du mesurage et du découpage dans les mathématiques d'Al-Andalus (X[e]–XIIIe s.). En : Radelet de Grave, P. (ed.), Liber Amicorum Jean Dhombres, Réminisciences 8. Brepols, Turnhout, pp. 113–147.

——— 2011a. Pratiques géométriques et géométrie savante au Maghreb, L'exemple d'Ibn Haydñr. Actes du 9e Colloque maghrébin sur l'histoire des mathématiques arabes (Tipaza, 12–14 mai 2007). Imprimerie Fasciné, Alger, pp. 53–79.

——— 2011b. La théorie des rapports entre Orient et Occident musulmans, L'exemple d'al-Mu'taman et d'Ibn Sartāq. Actes du 10e Colloque maghrébin sur l'histoire des mathématiques arabes (Tunis, 29–31 mai 2010). Publications de l'A.T.S.M., Tunis, pp. 104–152.

Djebbar, A., Aballagh, M., 2001. La vie et l'œuvre d'Ibn al-Bannā de Marrakech, Rabat. Publications de la Faculté de Lettres et Sciences Humaines. Université Mohammad V, Rabat. (en arabe)

Djebbar, A., Moyon, M., 2011. Les sciences arabes en Afrique, Mathématiques et astronomie (IX[e] au XIX[e] siècle), suivie de Nubdha fī [c]ilm al-ḥisāb d'Aḥmad Bābir al-Arawānī. Editions Grandvaux-Vecmas, Paris.

Guergour, Y., 2000. Les différents systèmes de numération au Maghreb à l'époque ottomane : l'exemple des chiffres rūmī. En : İhsanoğlu, E. Günergun, F., Djebbar, A. (eds.), Science, Technology and Industry in the Ottoman Word, vol. VI. Brepols, Turnhout, pp. 67–74.

——— 2005. Le roi de Saragosse Al-Mu'taman Ibn Hūd (m. 1085) et le théorème de Pythagore, ses sources et ses prolongements. Llull 28, n° 62, 415–434.

——— 2006. La géométrie euclidienne chez al-Mu'taman Ibn Hūd (m. 478/1085), Contribution à l'étude de la tradition géométrique arabe en Andalus et au Maghreb. Thèse de Doctorat. Université Badji Mokhtar, Annaba.

——— 2009. Le cinquième postulat des parallèles chez Al-Mu'taman Ibn Hūd, Roi de Saragosse (1081–1085). Llull 32, n° 69, 59–72.

Guesdon, M.-G., 2002. La numérotation des cahiers et la foliotation dans les manuscrits arabes datés jusqu'à 1450. Revue des Mondes Musulmans et de la Méditerranée 99–100, 101–115.

Harbili, A., 2005. Quelques procédés d'approximation dans les écrits mathématiques maghrébins des XIIe–XIVe siècles, vol. 1. Actes du 7e Colloque maghrébin sur l'histoire des mathématiques arabes (Marrakech, 30 mai–1e juin 2002). E.N.S., Marrakech, pp. 157–200.

——— 2006. Le Takhlīs d'al-Ghurbī, un commentaire inédit du Talkhīṣ d'Ibn al-Bannā. Actes du 8e Colloque maghrébin sur l'histoire des mathématiques arabes (Tunis, 18–20 décembre 2004). Publications de l'A.T.S.M., Tunis, pp. 199–216.

——— 2011a. Les procédés d'approximation dans les ouvrages mathématiques de l'Occident Musulman. Llull 34, n° 73, 39–60.

——— 2011b. Les procédés d'approximation dans les écrits mathématiques du Maghreb et d'al-Andalus (X^e^–XVe siècles), contribution à l'étude de la tradition mathématique de l'Occident musulman. Thèse de doctorat. Université Badji Mokhtar, Annaba.

Hogendijk, J.P., 1991. The geometrical parts of the Istikmal of Yusuf al-Mu'taman ibn Hud (11th century). An analytical table of contents. Archives Internationales d'Histoire des Sciences 41, 207–281.

——— 2004. The lost geometrical parts of the Istikmāl of Yūsuf al-Mu'taman ibn Hūd (11th century) in the redaction of Ibn Sartāq (14th century): An Analytical Table of Contents. Archives Internationales d'Histoire des Sciences 53, 19–34.

——— 2005. Applied mathematics in Eleventh Century al-Andalus, Ibn Mu^c^ādh al-Jayyānī and his computation of astrological houses and aspects. Centaurus 47, 87–114.

Ibn Mun^c^im, 2006. Fiqh al-ḥisāb [La science du calcul] (Lamrabet, D. Ed.). Dar al Amane, Rabat.

al-Khaṭṭābī, M.L., 1956. Deux épîtres sur la science du mesurage d'Ibn ar-Raqqām et Ibn al-Bannā. Da^c^wat al-ḥaqq n° 256, 39–47. (en arabe)

——— 1957. Commentaire de l'Elixir sur la science du mesurage. Da^c^wat al-ḥaqq n° 258, 77–87. (en arabe)

King, D.A., 1980. The exact sciences in medieval Islam, some remarks on the present state of research. Middle East Studies Association Bulletin 4, 10–26.

Kunitzsch, P., 2002–2003. A new manuscript of Abū Bakr al-Ḥaṣṣār's Kitāb al-bayān. Suhayl 3, 187–192.

Laabid, E., 2001. Le partage proportionnel dans la tradition mathématique maghrébine. Actes du Colloque International « Commerce et mathématiques du Moyen âge à la Renaissance, autour de la Méditerranée » (Beaumont de Lomagne, 13–16 mai 1999). Editions du C.I.H.S.O., Toulouse, pp. 315–326.

——— 2003. Procédés arithmétiques pouvant remplacer l'algèbre, exemples de la tradition des héritages au Maghreb médiéval. Actes du colloque international « De la Chine à l'Occitanie, chemins entre arithmétique et algèbre » (Toulouse, 22-24 septembre 2000). Editions du C.I.H.S.O., Toulouse, pp. 111–134.

——— 2005. Les problèmes d'héritage et de mathématiques au Maghreb des XIIe–XIVe siècles, essai de synthèse, vol 1. Actes du 7e Colloque maghrébin sur l'histoire des mathématiques arabes (Marrakech, 30 mai–2 juin 2002). al-Wataniya, Marrakech, pp. 241–261.

——— 2006. Les techniques mathématiques dans la résolution des problèmes de partages successoraux dans le Maghreb médiéval à travers le Mukhtaṣar d'al-Hūfī (m. 1192), sources et prolongements. Thèse de Doctorat d'Etat d'histoire des mathématiques. Université de Rabat, Faculté des Sciences de l'Education, Rabat.

——— 2007. Ibn Khaldūn et le ᶜilm al-farā'iḍ. Actes de la Table ronde sur « Les structures intellectuelles en occident musulman à l'époque d'Ibn Khaldūn » (Rabat, 23–26 février 2006). Publications de la Faculté des Lettres et des Sciences Humaines. Rabat ; Série « Colloques et Séminaires » n° 140, 15–26.

——— 2011. La contribution d'al-Qurashī dans la science des héritages, Entre Ibn Ṣafwān al-Mālaqī (m. 773/1361) et Saᶜīd al-ᶜUqbānī at-Tilimsānī (m. 811/1408). Actes du 2e Séminaire sur l'Histoire des Sciences (Alger, Université Bab Ezzouar, 7–9 juin 2011). A paraître.

Labarta, A., Barceló, C., 1988. Números y cifras en los documentos arábigohispanos. Gràfiques Canuda, Barcelone.

Lamrabet, D., 1981. Les mathématiques maghrébines au moyen âge, Traduction de manuscrits inédits, Motivations pédagogiques de leur étude actuelle, Mémoire de Post-Graduation en didactique des mathématiques. Université Libre de Bruxelles, Bruxelles.

——— 2000. Un mathématicien maghrébin méconnu, Ibn Haydūr at-Tādilī. Publications de la Faculté des Lettres et des Sciences Humaines. Rabat ; Série « Colloques et Séminaires » n° 83, 121–129.

——— 2001. Ibn Rashīq (XIIIe siècle) et la classification des sciences mathématiques. Publications de la Faculté des Lettres et des Sciences Humaines. Rabat ; Série « Colloques et Séminaires » n° 94, 43–56.

——— 2003. Aperçu sur les systèmes de numération en usage au Maghreb du XVIIe siècle. Publications de la Faculté des Lettres et des Sciences Humaines. Rabat ; Série « Colloques et Séminaires », n° 104, 23–37.

——— 2007. Ecrits mathématiques en circulation au Maghreb à l'époque d'Ibn Khaldūn (732–808H/1332–1406). Publications de la Faculté des Lettres et Sciences Humaines. Rabat, Série « Colloques et Séminaires » n° 140, 27–57.

——— 2008. Notes diverses sur l'enseignement des mathématiques au Maroc sous les Almohades (542-668h/1147-1269) et les Mérinides (668–870h/1269–1465). Publications de la Faculté des Lettres et Sciences Humaines. Rabat, Série « Colloques et Séminaires » n° 150, 19–48.

Lévy, T., 1996. Fragment d'Ibn al-Samḥ sur le cylindre et ses sections planes, conservé dans une version hébraïque. In : Rashed, R., Les mathématiques infinitésimales du IXe au XIe siècle, vol. 1, Fondateurs et commentateurs. Al-Furqān, Londres, pp. 927–973 and 1080–1083.

——— 2000. Abraham ibn Ezra et les mathématiques. Remarques bibliographiques et historiques. En : Tomson, P.J. (ed.), Abraham ibn Ezra, savant universel. Bruxelles, pp. 61–75.

——— 2001a. Les débuts de la littérature mathématique hébraïque, la géométrie d'Abrahām bar Ḥiyya (XIe–XIIe s.). Micrologus IX, pp. 35–64.

——— 2001b. Hebrew and Latin Versions of an Unknown Mathematical Text by Ibn Ezra, Aleph, vol. 1. pp. 295–305.

——— 2002. De l'arabe à l'hébreu, la constitution de la littérature mathématique hébraïque (XIIe–XVIe siècle). En : Dold-Samplonius, Y., Dauben, J.W., Folkerts, M., van Dalen, B. (eds.), From China to Paris, 2000 Years Transmission of Mathematical Ideas. Franz Steiner, Stuttgart, pp. 307–326.

——— 2003. L'algèbre arabe dans les textes hébraïques (I). Un ouvrage inédit d'Isaac ben Salomon al-Ahdab (XIVe siècle). Arabic sciences and philosophy 13–1, 269–301.

——— 2004. Maïmonide et les sciences mathématiques. En : Lévy, T., Rashed, R. (eds.), Maïmonide philosophe et savant (1138–1204). Peeters, Louvain, pp. 219–252.

——— 2005. Le manuscrit hébreu Munich 36 et ses marginalia : un témoin de l'histoire textuelle des Eléments d'Euclide au Moyen âge. En : Jacquart, D., Burnett, C. (eds.), Scientia in margine. Etudes sur les marginalia dans les manuscrits scientifiques du Moyen Age à la Renaissance. Droz, Genève, pp. 103–116.

Marin, M., 2004. The making of a mathematician, al-Qalaṣādī and his Riḥla. Suhayl 4, 295–310.

Moyon, M., 2007. La tradition algébrique arabe du traité d'al-Khwārizmī au Moyen âge latin et la place de la géométrie. En : Barbin, E., Bénard, D. (eds.), Histoire et Enseignement des mathématiques : rigueurs, erreurs, raisonnements. I.N.R.P., Lyon, pp. 289–318.

——— 2008. La géométrie pratique en Europe en relation avec la tradition arabe, l'exemple de mesurage et du découpage, Contribution à l'étude des mathématiques médiévales. Thèse de doctorat en épistémologie et histoire des sciences. Université des Sciences et des Technologies, Lille.

——— 2010. Le De Superficierum Divisionibus Liber d'al-Baghdādī et ses prolongements en Europe. Actes du 9e colloque maghrébin sur l'histoire des mathématiques arabes (Alger, 12–14 mai 2007). Imprimerie Fasciné, Alger, pp. 159–201.

——— 2011. Algèbre & Practica geometriae en Occident médiéval latin : Abū Bakr, Fibonacci et Jean de Murs. En : Rommevaux, S., Spiesser, M., Massa Estève, M.-R. (eds.), Pluralité de l'algèbre à la Renaissance. Editions Champion, Paris.

Muṣliḥ, A., 2006. La parure des étudiants et le souhait des calculateurs sur l'explication des difficultés du Rafc al-ḥijāb d'Ibn Haydūr at-Tādilī. Thèse de Doctorat en philosophie. Université Mohammed V, Faculté des Lettres et Sciences Humaines, Rabat. (en arabe)

Naghsh, I., 2007. L'approfondissement dans le commentaire sur le Talkhīṣ d'Ibn Haydūr at-Tādilī. Thèse de Doctorat en philosophie. Université Mohamed V, Faculté des Lettres et Sciences Humaines, Rabat. (en arabe)

Puig, R., 2000. The theory of the moon in the Al-Zīj al-Kāmil fī t-tacālīm of Ibn al-Hā'im (ca. 1205), Suhayl 1, 71–99.

——— 2005. La saphea (ṣafīḥa) d'al-Zarqālī dans le Kitāb Jāmic al-mabādi' wa l-ghāyāt fī cilm al-mīqāt d'Abū l-Ḥasan al-Marrākushī. Actes du 7e Colloque maghrébin sur l'histoire des mathématiques arabes (Marrakech, 30 mai–1e juin 2002). Ecole Normale Supérieure, Marrakech, pp. 271–280.

al-Qalaṣādī, 1999. Sharḥ Talkhīṣ acmāl al-ḥisāb [Commentaire sur l'Abrégé des opérations du calcul], F. Bentaleb (ed.). Dār al-Gharb al-islāmī, Beyrouth.

Rommevaux, S., Djebbar, A., Vitrac, B., 2001. Remarques sur l'histoire du texte des Eléments d'Euclide. Archives for the History of Sciences 55, 221–295.

Rashed, R., 1996. Les mathématiques infinitésimales du IXe au XIe siècle, vol. 1, Fondateurs et commentateurs. Al-Furqān, Londres.

——— 2007. Al-Khwārizmī, le commencement de l'algèbre, Blanchard, Paris.

Rosenfeld, B.A., 2004. A supplement to Mathematicians, Astronomers and Other Scholars of Islamic Civilization and their Works (7th–19th c.). Suhayl 4, 87–139.

——— 2006. A second supplement to Mathematicians, Astronomers and Other Scholars of Islamic Civilization and their Works (7th–19th c.). Suhayl 6, 9–79.

Rosenfeld, B.A., Matvievskaya, G., 1983. Matematiki i astronomi musulmanskogo srednevekovya i ikh trudi (VII–XVII vv). Nauk, Moscou.

Rosenfeld, B.A., İhsanoğlu, E., 2003. Mathematicians, Astronomers and other Scholars of Islamic Civilization and their Works (7th–19th c.). IRCICA, Istanbul.

Samsó, J., 2007. Andalusian Astronomy in 14th Century Fez, Al-Zīj al-Muwāfiq of Ibn cAzzūz al-Qusanṭīnī, n° IX. Variorum, Aldershot.

Sesiano, J., 1988. Le Liber Mahameleth, un traité mathématique latin composé au XIIe siècle en Espagne. Premier Colloque Maghrébin d'Histoire des Mathématiques Arabes (Alger, 1–3 décembre 1986). En : Actes du Colloque. Maison du Livre, Alger, pp. 69–98.

——— 2004. Les carrés magiques dans les pays islamiques. Presses Polytechniques et Universitaires Romandes, Lausanne.

cUthmānī, cA., 1999. La philosophie des mathématiques chez Ibn al-Bannā al-Murrākushī et ses commentateurs maghrébins, Diplîme d'Etudes Approfondies en Philosophie. Université Mohammed V, Faculté de Philosophie, de Sciences sociales et de Psychologie, Rabat. (en arabe)

Vlasschaert, A.-M., 2010. Le Liber mahameleth, édition critique et commentaires. Franz Steiner, Stuttgart.

Wartenberg, I., 2007 The Epistle of the Number by Isaac ben Solomon ben al-Aḥdab (Sicily, 14th century), An Episode of Hebrew Algebra. Thèse de Doctorat de philosophie. Université Paris VII, Paris.

Woepcke, F., 1854. Notice sur des notations algébriques employées par les Arabes. Journal Asiatique 5e série 4, 348–384.

Zemouli, T., 1993. Les écrits mathématiques d'Ibn al-Yāsamīn (m. 1204), Mémoire de Magister. E.N.S., Alger. (en arabe)

Zerrouki, M., 2000. Mathématiques et héritages à travers le commentaire par al-[c]Uqbānī du Mukhtaṣar d'al-Ḥūfī, Mémoire de Magister. E.N.S., Alger.

Ibn al-Raqqām's *al-Zīj al-Mustawfī* in MS Rabat National Library 2461

Julio Samsó

Abstract The paper contains a brief introduction on the figure of the astronomer and polymath Abū ʿAbd Allāh Muḥammad Ibn al-Raqqām (fl. Tunis, Bijāya and Granada and died in Granada on 25th May 1315), and on his three *zīj*es, based on the unfinished *zīj* compiled by Ibn Isḥāq al-Tūnisī (fl. Tunis and Marrākush ca. 1193–1222).

Two of the aforementioned *zīj*es (*Shāmil* and *Qawīm*) were carefully described by the late Professor E.S. Kennedy in 1997. Since then, a new *zīj* (*al-Mustawfī*) has been discovered and the purpose of this paper is to complete Kennedy's work by giving some information about this newly discovered text. I have also tried to establish a relative chronology of the three *zīj*es, which seem to have been compiled between 1280 and 1290, before Ibn al-Raqqām's arrival in Granada, although the *Qawīm* was the object of a revision during his stay in this Andalusian city.

The canons of the *Mustawfī Zīj* deal with chronology (Chapters 1–11); mean motions, with radices calculated for the beginning of the Hijra and for the longitude of Tunis (Chapters 12–13); equation of time and trepidation (Chapters 14–17); apogees and solar and lunar longitudes (Chapters 18–20); planetary longitudes and latitudes (Chapters 21–24); trigonometry and spherical astronomy (Chapters 25–55); luni-solar conjunctions and oppositions, lunar visibility, parallax and eclipses (Chapters 56–60); topics in mathematical astrology as projection of rays, *tasyīr*, year, nativity and month transfers; (Chapters 61–63); trigonometry and planetary visibility (Chapters 64–65).

The tables deal with 1) chronology; 2) mean motions (calculated with the same parameters as in Ibn al-Raqqām's two other *zīj*es, the anonymous Hyderabad recension of Ibn Isḥāq's *Zīj* and Ibn al-Bannā's *Minhāj*); 3) trepidation; 4) solar equation; 5) lunar equations; 6) planetary equations; 7) planetary latitudes; 8) solar declination, lunar latitude and obliquity of the ecliptic; 9) equation of time; 10) trigonometry and spherical astronomy; 11) solar and lunar velocity, parallax and eclipses; 12) auxiliary functions; 13) astrology ; 14) star table giving the tropical positions of fixed stars for year 680/1280–81; 15) longitudes and latitudes of 67 cities using the water meridian (27° west of Cordova) as the base meridian.

On the whole, this *zīj* contains further information of the influence that the Andalusian astronomer Ibn al-Zarqālluh (d. 1100) had in the development of astronomy in the Maghrib during the 13th and 14th centuries: sidereal positions which can be turned into tropical ones by using a model of trepidation, cyclical model of the obliquity of the ecliptic, motion of the solar and planetary apogees, solar model with variable eccentricity, corrections of Ptolemy's lunar model, etc., all follow the Zarqāllian tradition. On the other hand, Ibn al-Raqqām shows here, as well as in the *Shāmil Zīj*, the interest he felt in the solution of problems of spherical astronomy, to which he dedicates many pages of the *zīj*.

N. Sidoli and G. Van Brummelen (eds.), *From Alexandria, Through Baghdad: Surveys and Studies in the Ancient Greek and Medieval Islamic Mathematical Sciences in Honor of J.L. Berggren*, DOI 10.1007/978-3-642-36736-6_16,
© Springer-Verlag Berlin Heidelberg 2014

Ibn al-Raqqām and His *Zīj*es

Abū ʿAbd Allāh Muḥammad b. Ibrāhīm b. ʿAlī b. Aḥmad (Muḥammad?) b. Yūsuf al-Mursī al-Andalusī al-Tūnisī al-Awsī, known as Ibn al-Raqqām, belonged to "the people of Murcia" (*min ahl Mursiya*)[1] and died in Granada on 21st *ṣafar* 715 (26th May 1315) at an advanced age. He lived in Tunis and Bijāya and migrated from Bijāya to Granada during the reign of Muḥammad II (1273–1302), probably after 1288–89.

He compiled three "editions" of the *Zīj* of Ibn Isḥāq (fl. Tunis and Marrākush ca. 1193–1222), which had been left unfinished by its author. Two other "editions" were prepared by two contemporaries of Ibn al-Raqqām: the anonymous author of the compilation (ca. 1280-81) extant in MS Hyderabad Andra Pradesh State Library 298 and Ibn al-Bannāʾ (1256–1321), who prepared his *Minhāj al-ṭālib fī taʿdīl al-kawākib* [Mestres 1996 and 1999; Vernet 1952]. All these *zīj*es have the same numerical tables (especially those related to mean planetary motions and equations), copied from Ibn Isḥāq.

In 1997, Edward S. Kennedy published a detailed description of the two *zīj*es of Ibn al-Raqqām which were known at that time.[2] A few notes on these two *zīj*es follow:

1. *Al-Zīj al-Shāmil fī tahdhīb al-Kāmil*: extant in only one manuscript (Istanbul, Kandilli 249).[3] It was compiled in Bijāya in 679/1279–80, according to the introduction. The table of the lunar mansions, however, is dated 887 (*sic*, a mistake for 687/1288–89?) and it adds 16;50° to the longitudes of the *Almagest*. The maximum tabular value of the solar equation is 1;47,51°, which corresponds to year 689/1290, due to the fact that Ibn al-Raqqām uses a Zarqāllian model with variable solar eccentricity. It is possible to imagine that 679 (written in words in the introduction) could also be an error for 689. The *Kāmil* mentioned in the title is *al-Zīj al-Kāmil fī l-Taʿālīm*, compiled by Ibn al-Hāʾim (fl. 1204–05), an exceptional collection of astronomical canons which are not simple instructions for the use of a set of tables but develop all the underlying astronomical theory with the corresponding mathematical proofs [Samsó 2011, 321–326; Calvo 1998b; Puig 2000; Comes 2001]. Neither the only MS in which this work is preserved (Oxford, Bodleian Library Marsh 618) nor the copy available to Ibn al-Raqqām contain any tables, in spite of the fact that the canons mention a few of them. For that reason Ibn al-Raqqām copied, from the *Kāmil*, those canons that had a more practical character, removed the mathematical proofs, and completed them with the numerical tables of Ibn Isḥāq. To this he adds some sixty chapters on problems related to spherical astronomy and astrology, which seem to be independent from the *Kāmil*.
2. *Al-Zīj al-Qawīm fī funūn al-taʿdīl wa l-taqwīm*: The complete text is extant in MS Rabat National Library 260, and a fragmentary copy can be found in an unnumbered MS at the Museo Naval in Madrid, where we also find the *Minhāj* of Ibn al-Bannāʾ [Vernet 1980]. It was finished in Tunis (longitude 41;45°, latitude 36;37°).[4] It was obviously revised in Granada, after Ibn al-Raqqām's arrival in the city, for it contains a table for establishing the visibility of the crescent calculated for a latitude of 37;10°; this is precisely the modern value for the latitude of that city, with no precedents before Ibn al-Raqqām, who must have made a very careful determination of this value. What is more, Kennedy [1997, 38–41] has studied the procedures described by Ibn al-Raqqām, in the *Shāmil* and in the *Qawīm*, to determine crescent visibility and has shown that the method explained in this second *zīj* can only be applied to the latitude of Granada. It contains a table of stellar coordinates "observed" (*marṣūda*) in 680/1280–81 [MS Rabat 260, 100–103]. It must have been finished after 1290, because it contains the table of the solar equation that we find in the *Shāmil*. This *zīj* is,

---

[1] See Ibn al-Khaṭīb [1973–1978, vol. III, 69–70, 334]. A summary of what we know about this author can be found in Casulleras [2007c] and Samsó [2006].

[2] On Ibn al-Raqqām's methods for the division of the astrological houses, see Kennedy [1996, 557–568].

[3] A partial edition of the canons and tables was made by ʿAbd al-Raḥmān [1996], which we will use in this paper. Besides, ʿAbd al-Raḥmān prepared a provisional transcription of the whole set of canons which I have found particularly useful.

[4] In King's survey of tables for timekeeping calculated for Tunis the latitudes for this city are 36;40° and 37°, but not 36;37° [King 2004, 64–65, 110, 119, 157–158, 428–431].

apparently, the least ambitious of the three and the canons are the usual ones in works of this kind: mere instructions for the use of the tables, with very few theoretical materials.

In the same year, 1997, I spent a week at the Frankfurt Institut für Geschichte der arabisch-islamischen Wissenschaften exploring the collection of microfilms compiled by Prof. Fuat Sezgin, among which I found a microfilm of MS Rabat National Library 2461, which proved to be a late copy of a third *zīj* by Ibn al-Raqqām. Prof. Sezgin generously provided me with an enlargement of this microfilm and I, as well as two other colleagues of the University of Barcelona, have used it in three papers [Díaz-Fajardo 2005; Samsó 2008; Bellver and Samsó 2012]. A general description of this third *zīj* follows:

3. *Al-Zīj al-Mustawfī li-mā ḥāza min al-basṭ wa l-ḥaẓẓ al-awfar wa l-qisṭ al-awfā*[5] (*Zīj* with which everything acquired through donation and generous fortune in plentiful amount is fully returned): extant in MSS 534K (recently discovered by M. J. Parra) 2461 and 4157 of the National Library in Rabat, which also preserves MS 4156, containing a fragment of the same work related to the equation of time. Other MSS are Tunis National Library 11018 (Aḥmadiyya collection, old number 5584) and Cairo National Library DM 718,2 (only Chapters 50–51 of the canons). It was compiled in Tunis after 680/1280–81 (the date of the incomplete star table calculated with an increase of 16;46° compared with the longitudes of the *Almagest*). From the column of meridian altitudes of the stars, one can calculate that the latitude of the place of observation is 36;37° (Tunis). These agree with the coordinates given in the *Qawīm Zīj* and with the *radices* for mean motions, calculated for the meridian of Tunis for which a longitude of 41;45° is given, with the commentary that it was established by Ibn Isḥāq as a result of the observation of two lunar eclipses. The table of the solar equation is of little use for establishing a date: it was copied from Ibn Isḥāq's *Zīj*, and the original table was calculated for year 619/1222. Of the three *zīj*es by Ibn al-Raqqām, the *Mustawfī* is the one that achieved the greatest diffusion in the Maghrib. We know this from the number of extant manuscripts as well as from the fact that the *Mustawfī* was used by Maghribī *muwaqqits*. It is frequently quoted and used by two commentators of the well-known *urjūza* (written in Fez by al-Jādirī in 1391–92), *Rawḍat al-azhār fī ʿilm waqt al-layl wa l-nahār*: these are al-Habbāk, the author of the *Natāʾij al-afkār* (Tilimsān ca. 1514–15) and Ibn al-Muftī (fl. Marrākush, Fez and Tarudant, d. 1611).[6] This *zīj*, like the *Shāmil*, is clear proof of Ibn al-Raqqām's interest in spherical astronomy, a characteristic which is quite rare in Andalusian and Maghribī astronomy.

From the information available we might think that the three *zīj*es were compiled between ca. 1280 and ca. 1290, before Ibn al-Raqqām's departure for Granada, where he arrived before 1302, although additions were made to the *Qawīm* after the astronomer's arrival in the city. The constants of precession used in the star tables of the *Mustawfī*, the *Qawīm* (16;46°), and the *Shāmil* (16;50°) lead us to believe that the *Mustawfī* and the *Qawīm* were compiled before the *Shāmil*. On the other hand, the table of the solar equation in the *Mustawfī* (copied from Ibn Isḥāq), suggests that this *zīj* was written before the *Shāmil* and the *Qawīm* (table calculated for year 689/1290). As a result, we may suggest the following order: 1st *Mustawfī* (Tunis), 2nd *Qawīm* (Tunis, with a later revision in Granada) and 3rd *Shāmil* (Bijāya). If this suggestion proves to be true, it implies that Ibn al-Raqqām began his professional life in Tunis and, at a later date, moved to Bijāya.

Manuscript 2461 of the National Library of Rabat

As I have already said, this manuscript is very late. Some information is given in the colophon placed at the end of the canons (p. 226): the copyist was Muḥammad b. Aḥmad al-Hawārī who copied the

[5] This title contains a possible rhyme and this led me to believe that the first word was *Mustawfā* (rhyme with *al-awfā*) which both Díaz-Fajardo and I used in several publications. George Saliba suggested to me that *Mustawfī* was a better option and I follow his advice here.

[6] On these two works see Samsó [1998 and 2001] and Comes [2002].

manuscript for himself and, after him, for any of the sons of his country (*li-man shā'a Allāh ba'da-hu min abnā' jinsi-hi*). The copy was finished on 17th Ṣafar 1313/ 8th August 1895.

MS Rabat NL4157/D 2308, which contains the same *zīj*, is even later. It was copied by ʿAlī b. Muḥammad al-Ṭayyib b. ʿAbd Allāh b. Qāsim al-Ṣafāʿī al-Salāwī and is dated 11th Shawāl 1345/13th April 1927. This ʿAlī b. Muḥammad was a disciple of the aforementioned Muḥammad b. Aḥmad and his copy was made from the manuscript copied by Muḥammad b. Aḥmad al-Hawārī.

Only very recently and too late to use it in this paper, I have been able to obtain a copy of manuscript Tunis National Library 11018 (Aḥmadiyya collection, old number 5584) with the help of Dr Chedli Guesmi. Therefore, as Rabat NL4157/D 2308 does not add any information to the older manuscript, I will limit myself to a description of MS 2461, (pp. 153–289) with which I will try to complement the information given by Kennedy in 1997. My work in this field would have been impossible without the help of two unpublished Ph.D. theses presented at the University of Barcelona: ʿAbd al-Raḥmān [1996] and Mestres [1999].

A Table of Contents of Manuscript Rabat NL2461

The Canons

Pages 153–155: after the customary religious introduction, the complete title of the *zīj* appears: *Al-Zīj al-Mustawfī li-mā ḥāza min al-basṭ wa l-ḥaẓẓ al-awfar wa l-qisṭ al-awfā* (p. 153). This is followed by an index with the numbers and titles of the 65 chapters of the canons.

Chronology[7]

Chapter 1: *On the bases of calendars* (Dhikr uṣūl al-tawārīkh) (pp. 155–158).

The four calendars used are the Arabic, *Rūmī*, Persian and Coptic calendars (p. 155). The mean lunar month, according to the astronomers, has a length of 29;31,50,5 days, a value which, when multiplied by 12, gives a lunar year of 354;22,1 days. The leap (*kabā'is*) years correspond to numbers 2, 5, 7, 10, 13, 14, 16, 18, 21, 26, 29 of the 30-year cycle (p. 156).

The Byzantine solar calendar is called *al-Rūmī* or *al-ta'rīkh al-ʿajamī*. It uses the era of Alexander *Dhū l-Qarnayn*. A day is intercalated at the end of December of a bissextile year and at the end of February in the Syriac (*suryānī*) calendar. The Persians use the era of Yazdijird and a solar year of 365 days without fraction. The Coptic calendar uses the era of Philippos, who is called "King of the Copts" (*Malik al-Qibṭ*): a quotation from Abū Maʿshar al-Balkhī on the Coptic calendar is introduced here (p. 157).

Differences between eras (p. 158) [van Dalen 2000, 266; Mestres 1999, 184]:

- Philippos-Yazdijird: 955 [Persian] years and 3 months
- Alexander-Yazdijird: 942 Rūmī years and 259 days
- Hijra-Yazdijird: 3624 days

To these eras he adds

- Bukhtnaṣar-Yazdijird: 13[7]9 Persian years[8] and 3 months,
- Alexander – Bronze era (*ta'rīkh al-ṣufr*) : 273 Rūmī years and 3 months [−2 days]. This is the era used by the Christians.
- Alexander – Coptic era [Diocletian]: 594 Rūmī years and 332 days
- Alexander – Christian era: 311 Rūmī years, 5 months and 17 days [instead of 3 months and 2 days]

[7] On chronology see van Dalen [2000].

[8] 1399 years in the manuscript.

Interestingly, Ibn al-Raqqām is using the Bronze or Spanish era (1.1.–37) and the Christian era (1.1.1). Both appear in the *zīj* of al-Khwārizmī-Maslama (probably an interpolation due to Maslama) and in other 10th century Andalusian sources [Samsó 2001, 74–76].

Chapter 2: *Obtaining the Rūmī date from an Arabic date* (pp. 158–159).

The calendaric lunar year contains $354 + 3/10 + 2/3 \times 1/10$ days $= 354 + 11/30$ days $= 354;22$ days.

Chapter 3: *Obtaining the Persian date from an Arabic date* (p. 160).

Chapter 4: *Obtaining the Arabic date from a Rūmī date* (pp. 160–162).

Chapter 5: *Obtaining the Persian date from a Rūmī date* (p. 162).

Chapter 6: *Obtaining the Rūmī date from a Persian date* (pp. 162–163).

Chapter 7: *Obtaining the Arabic date from a Persian date* (p. 163).

Chapter 8: *Determining the leap* (kabīsa) *year in the Arabic and Rūmī calendars* (p. 163).

Chapter 9: *Determining the initial week days* (ʿalāmāt) *of Arabic years and months by computation and with tables* (pp. 163–164).

Chapter 10: *Determining the initial week days* (ʿalāmāt) *of Rūmī years and months by computation and with tables* (pp. 164–165).

Chapter 11: *Determining the initial week days* (ʿalāmāt) *of Persian years and months by computation and with tables* (p. 165).

Mean Motions

Chapter 12: *Radices of mean motions used in this* zīj (p. 165).

The radix date used is midday of Wednesday 1st of Muḥarram of the year in which the Prophet Muḥammad performed his migration (*Hijra*). The meridian for which this radix was calculated was the meridian of Tunis, whose longitude (41;45°) was established by Ibn Isḥāq by observing two lunar eclipses. The local latitude is 36;37°, obtained after multiple observations and with previous precise determinations of the values of trepidation, position of the apogees, altitude and depression of the pole and the variation of the obliquity of the ecliptic.[9]

Chapter 13: *Obtaining mean positions in longitude and anomaly* (pp. 165–166).

Equation of Time, Trepidation

Chapter 14: *Correcting [the time] in localities placed east or west of Tunis* (pp. 166–167).

Chapter 15: *How to calculate the hours used in the computation* (Fī maʿrifat taʿdīl sāʿāt al-taqwīm) (pp. 167–169).

The chapter deals with the length of day and night for different times of the solar year and for different latitudes, for both equal and unequal hours.

Chapter 16: *On the equation of time* (pp. 169–170).

This chapter is interesting and merits a detailed study. It begins by explaining the well-known causes of the existence of the equation of time (the sun moves on the ecliptic with unequal velocity; equal arcs of the ecliptic cross the horizon or the meridian simultaneously with unequal arcs of the equator). To this he adds that the motion of accession and recession (*al-iqbāl wa l-idbār*) affects the equation of time and, for that reason, in his tables for trepidation, Ibn al-Raqqām includes a table of the *iqbāl*

[9] The same geographical coordinates of Tunis are given in the *Qawīm Zīj*.

al-ʿamūd fī l-quṭr wa huwa tashrīq al-mabdaʾ (p. 245) with which one obtains the "second accession" (*al-iqbāl al-thānī*), the difference in right accession between two positions of the Head of Aries at two given moments. Tables of this kind appear in other Andalusī and Maghribī *zījes*[10] but I have not seen, so far, such a detailed commentary and explanation of its use as in the *Mustawfī Zīj*.

Chapter 17: *Computation of the motion of accession and recession* (pp. 170–171).

Most of this chapter is a careful description of the procedure to calculate the position of the movable Head of Aries (p. 229), with which one enters the table of the equation (p. 243) in order to obtain the amount of precession for a given date. At the end, he explains, not very clearly, another procedure for this computation using the method explained by Ibn al-Zarqālluh which takes into account the variable character of the obliquity of the ecliptic [Goldstein 1964].

Apogees, Solar and Lunar Longitudes

Chapter 18: *On the knowledge [of the position] of the apogees* (pp. 171–172).

This chapter is also interesting and it has already been edited and studied [Díaz Fajardo 2005]. The problem is the following one: Ibn al-Zarqālluh established that the solar apogee had a motion of 1° in 279 years [Toomer 1969]. The Andalusī and Maghribī school of Zarqāllian *zījes* extended this motion to the apogees of all planets (Ibn al-Kammād, Ibn al-Hāʾim, the anonymous Tunisian author of the Hyderabad recension and Ibn al-Raqqām in the *Shāmil Zīj*) or only to the apogees of Venus and Mercury (Ibn al-Bannāʾ, Ibn al-Raqqām in the *Qawīm Zīj*) [Samsó and Millás 1998, 268–270; Samsó 1997, 82–83, 102]. In this chapter, Ibn al-Raqqām adopts the same position as Ibn al-Bannāʾ and says that he is following the opinion of Ibn Isḥāq although he is not entirely convinced and expects future observations to confirm or disprove this idea.

Chapter 19: *Computing the solar longitude* (pp. 172–173).

This canon repeats that the mean motion solar radices are calculated for the longitude of Tunis. The solar longitude is computed with a "table of the partial equation" (*Jadwal taʿdīl al-shams al-juzʾī*, p. 246, with a maximum of 1;49,7°) copied from Ibn Isḥāq's *Zīj* (table calculated for 619/1222). Ibn al-Raqqām states that the result obtained is approximate because he uses a Zarqāllian solar model with variable eccentricity [Toomer 1969; Samsó and Millás 1994]. He also explains how to use the "table of the total solar equation" (*taʿdīl al-shams al-kullī,* pp. 247–248). He considers that a particular eccentricity can be used for a period slightly longer than fifty years, after which one should compute another table of the solar equation or use the "total" table. Obviously, here Ibn al-Raqqām's argument is not consistent, for he uses a partial table calculated by Ibn Isḥāq for 1222, and he is compiling his *zīj* ca. 1280, when, according to him, this table is no longer valid.

Chapter 20: *Computing the lunar longitude* (pp. 173–174).

Ibn al-Raqqām explains the standard method to calculate the lunar longitude and introduces two further corrections. The first is Zarqāllian, and considers that the centre of the lunar mean motion is not the centre of the earth but is displaced from it to another centre situated on the straight line which joins the centre of the earth to the solar apogee [Puig 2000]. The correction of the lunar mean longitude reaches a maximum value of 24′ and is tabulated in the *Jadwal taʿdīl al-qamar bi-buʿdi-hi min awj al-shams* (p. 249) which gives

$$\text{Sin}\, a \times 0;0,24°,$$

in which the argument (a) is the difference between the lunar mean longitude and the longitude of the solar apogee. This kind of correction is only needed when total exactitude is required, that is, in the computation of eclipses and for the determination of the visibility of the crescent. The second correction has been used in Eastern Islam at least since the period of Maʾmūnī observations (ca. 830). It is called "equation of the inclined sphere" (*taʿdīl al-falak al-māʾil*) and it takes into account that the

[10] In Ibn al-Zarqālluh's treatise on the motion of the fixed stars, for example [Millás 1943–50, 336].

moon does not move on the ecliptic but on a plane at an angle of 5°, approximately, from the plane of the ecliptic.

Planetary Longitudes and Latitudes

Chapter 21: *Computing the longitude of the five planets* (pp. 174–175).
Chapter 22: *Computing the position of the ascending and descending nodes* (p. 175).
Chapter 23: *Direct motion, retrogradation and stations of the planets* (pp. 175–176).
Chapter 24: *On the latitude of the planets* (pp. 176–178).
In this chapter he states that the maximum lunar latitude is 5°.

Trigonometry, Spherical astronomy

Chapter 25: *On the knowledge of the arc* (al-qaws), *sine* (al-jayb al-mustawī/al-jayb al-mabsūṭ), *cosine* (al-jayb al-mankūs/jayb al-tamām), *versed sine* (al-sahm/al-watar al-rājiʿ), *chord* (al-watar) *and how to calculate one of these [functions] from another one* (pp. 178–181).

Chapter 26: *Knowing the solar declination or the declination of each degree of the ecliptic* (p. 181).
For this purpose he uses a table called *Jadwal al-mayl al-juzʾī li falak al-burūj* (table of the partial declination) with a maximum value of 23;52,30° and a minimum 23;32,30° for the obliquity of the ecliptic (p. 271). Ibn al-Raqqām also includes a second table (*Jadwal al-mayl al-kullī wa daqāʾiq ikhtilāf al-muyūl*, table of the total declination and minutes of the variation of declination, p. 272). Both tables seem to be taken from Ibn Isḥāq's *Zīj* and the second one is based on a Zarqāllian model [Goldstein 1964; Samsó 1987; Comes 1992]. Once more, Ibn al-Raqqām is using a value of the obliquity (23;32,30°) calculated almost sixty years before and which corresponds to a minimum value. According to the model used, this value should have been increased for Ibn al-Raqqām's time.

Chapter 27: *On the knowledge of right ascensions and descensions* (pp. 181–182).
Two formulas are explained. The first is

$$\text{Cos}\ \alpha_0 = \frac{60\ \text{Cos}\ \lambda}{\text{Cos}\ \delta} \tag{1}$$

for α_0 = right ascension from Aries 0°, λ = longitude, and δ = declination, which is an application of the cosine law to a right spherical triangle in which the three sides are λ, δ and α_0. The same expression appears in Ibn al-Hāʾim's *Kāmil Zīj*,[11] and in Ibn al-Raqqām's *Shāmil Zīj*.[12]
The second formula is equivalent to

$$\text{Sin}\ \alpha_0 = \frac{60\ \text{Sin}\ \delta\ \text{Cos}\ \epsilon}{\text{Cos}\ \delta\ \text{Sin}\ \epsilon} = \frac{60\tan\delta}{\tan\epsilon} \tag{2}$$

for ϵ = obliquity of the ecliptic. It derives from the *Almagest* I.16 [Toomer 1984, 71–74; Neugebauer 1975, 31–32]. We find it in many sources well known in al-Andalus such as Khwārizmī, Chapter 25 [Suter 1914, 18–19; Neugebauer, 1962, 47–48], Ibn al-Muthannā [Millás 1963, 67–68; Goldstein 1967, 69–75, 202–204], al-Battānī [Nallino 1899–1907, vol. I, 13–14, 163], Ibn ʿEzra's *Fundamenta Tabularum* [Millás 1947, 147–152], and the *Toledan Tables* [Pedersen, 2002, 411–413, 612–615].

Chapter 28: *How to obtain the degree of longitude which corresponds to a degree of right ascension* (p. 182).

Chapter 29: *How to obtain the rising and setting amplitude of any degree of the ecliptic in any geographical location by means of the local latitude* (pp. 182–183).

---

[11] VII, 7: Bodleian Library MS pp. 266–268, fols. 84v–85v.

[12] Chapter 76: Kandilli MS, fol. 29r. See ʿAbd al-Raḥmān [1996, 115–116] and Kennedy [1997, 47].

The chapter describes two procedures to obtain the rising or setting amplitude of the sun, planet, or star when one knows its declination and the local latitude or the declination and half the day arc of the star or planet. I have not been able to find these two methods in Western Islamic sources except in Ibn al-Raqqām's *Shāmil*; the source is probably Chapter 7 of al-Battānī's *Zīj* [Nallino 1899–1907, vol. I, 20–21, 177–178].

Chapter 30: *How to obtain the ascensions of any degree of the ecliptic on an inclined horizon, which are called oblique ascensions* (al-maṭāliʿ al-ufuqiyya) (pp. 183–184).

This chapter explains four different procedures to calculate the ascensional difference ($\Delta\alpha$), equivalent to the equation of half daylight or any point of the ecliptic for a = rising amplitude, ϕ = local latitude, and $M/2$ = half the longest duration of daylight.

$$\text{Cos } \Delta\alpha = \frac{60 \text{ Cos } a}{\text{Cos } \delta}. \quad (1)$$

(In the text, this formula appears as Cos $\Delta\alpha$ = Cos a Cos $\delta/60$.) This is a simple application of the cosine law to a right spherical triangle, having as sides the rising amplitude, the declination and half of the equation of daylight.

$$\text{Sin } \Delta\alpha = 60 \tan \phi \tan \delta, \quad (2)$$

$$\text{Sin } \Delta\alpha = \frac{\text{Sin } \alpha_0 \text{ Cos } M/2}{60}. \quad (3)$$

(In the text, this formula appears as Sin $\Delta\alpha$ = (Sin α_0 Sin $M/2$)/60.)

Formula 3 is trivially equivalent to 2, because

$$-\cos M/2 = \tan \phi \tan \epsilon,$$

$$\sin \alpha_0 = \cot \epsilon \tan \delta.$$

One should take the right ascension of the ecliptic degree counted from Aries 0°. One then obtains the distance of the ecliptic degree from the nearest equinox and, using it as an argument, a, one enters the Table of the Sines of the Ascensional Differences (*jadwal juyūb fuḍūl al-maṭāliʿ*) and obtains a value F. The ascensional difference is

$$\text{Sin } \Delta\alpha = F \cdot 12 \cot (90° - \phi). \quad (4)$$

As the table in question (p. 286) calculates $5 \tan \delta (a)$, we have

$$5 \tan \delta \cdot 12 \cot (90° - \phi) = 60 \tan \phi \tan \delta = \text{ Sin } \Delta\alpha.$$

Expressions 1, 2 and 4 appear also in the *Shāmil Zīj*.[13]

Chapter 31: *On determining the day and night arcs of the sun, the moon, the rest of the planets and the fixed stars* (pp. 184–185).

In the case of the sun, Ibn al-Raqqām explains several procedures to calculate its day arc using tables and he refers to Chapter 30 in which he has explained how to compute $\Delta\alpha$, equivalent to the equation of half daylight. For the moon, planets and stars he repeats procedure 4 of Chapter 30.

Chapter 32: *On determining the time degrees of day and night unequal hours* (p. 185).

Chapter 33: *On equal hours of day and night* (pp. 185–186).

Chapter 34: *Transforming seasonal into equal hours and equal into seasonal hours* (p. 186).

Chapter 35: *On determining the digits of the extended* (mabsūṭ) *and the reversed* (mankūs) *shadows (cotangent and tangent) corresponding to a given altitude, determining the altitude by means of each one of the aforementioned shadows, and calculating one of the shadows from the other one* (pp. 186–187).

[13] Chapter 78 (fol. 29v of the Kandilli MS); see ʿAbd al-Raḥmān [1996, 116–118] and Kennedy [1997, 48].

In this chapter, Ibn al-Raqqām uses the cosecant function (*quṭr al-ẓill*).

Chapter 36: *On determining the azimuth by means of the altitude* (p. 187).

This chapter deals with the computation of the azimuth, *az*, of a star or planet when one knows its instantaneous altitude, *h*, as well as its declination, δ, and the local latitude, ϕ. Ibn al-Raqqām begins by calculating the rising amplitude, *a*, using a standard rule which can be found in the *Almagest*,[14]

$$\text{Sin } a = \frac{60 \text{ Sin } \delta}{\text{Cos } \phi}.$$

He then calculates *i*, which he calls *jayb ikhtilāf al-ufq*, with

$$i = \frac{\text{Sin } h \text{ Sin } \phi}{\text{Cos } \phi},$$

$$|\sin a| + |i| = s \text{ (}al\text{-}ḥāṣil\text{), and}$$

$$\text{Sin } az = \frac{60s}{\text{Cos } h}.$$

I have not been able to find this procedure in any earlier Western Islamic source or in the standard Eastern sources which were known in al-Andalus, although, slightly later, it reappears in the *Rawḍat al-azhār* by al-Jādirī (compiled in Fez in 1391–92) and and in Habbāk's commentary of the same work called *Natā'ij al-afkār fī sharḥ Rawḍat al-azhār* (written in Tlemcen in 1514-15): it is clear that both works use the *Mustawfī Zīj*.

Equivalent or identical methods can be found in authors from the Mashriq such as Abū'l-Wafā', al-Bīrūnī[15] or, much later, al-Kāshī.[16] The terminology used changes. Eastern sources call *i* the "share of the azimuth" (*ḥiṣṣat al-samt*) and *s* the "equation of the azimuth" (*ta'dīl al-samt*) defined by al-Kāshī as "the distance between the foot of the perpendicular dropped from the star to the horizon plane and the intersection of the horizon plane with the plane of the star's day circle."

Chapter 37: *On determining the latitude of any locality* (pp. 187–188).

Ibn al-Raqqām uses 1) the table of geographical coordinates; 2) the observation of the meridian solar altitude; 3) the maximum and minimum altitude of a circumpolar star. He considers the case of localities south of the equator, in spite of the fact that practically all the inhabited world lies north of the equator.

Chapter 38: *On determining the meridian solar altitude and the maximum northern or southern altitude of any star* (pp. 188–189).

Chapter 39: *On determining the time passed of a day by means of the observed solar altitude* (pp. 189–190).

Ibn al-Raqqām explains two procedures.

For $T =$ time before or after midday, $D/2 =$ half the day arc, $h =$ instantaneous solar altitude, and $h_m =$ meridian solar altitude, the first procedure is

$$\text{Vers } T = \frac{\text{Vers } D/2 \cdot \text{ Sin } h}{\text{Sin } h_m}, \tag{1a}$$

and

$$\frac{D}{2} \pm T = \textit{al-dā'ir min al-falak} \text{ (time since sunrise).} \tag{1b}$$

[14] *Almagest* II.3 [Toomer 1984, 78–79].

[15] In the *Taḥdīd* [Kennedy 1973, 61–62], *Ẓilāl* [Kennedy 1976, vol. II, 129–131], *Maqālīd* [Debarnot 1985, 244–247], *Qānūn* I, 438–441.

[16] Kennedy [1985, 19–20].

Expression (1a) appears in the two versions of the *zīj* of Ḥabash and it was demonstrated by Abū'l-Wafā' [Nadir 1960]. Equivalent formulas, using shadows instead of altitudes, can be found in Indian and Eastern Islamic astronomical sources [Davidian 1960].[17]

In the second procedure, Ibn al-Raqqām calculates

$$\text{Sin } h_m - \text{Sin } h = F\ (\textit{al-faḍla}). \tag{2a}$$

Then, he enters, with ϕ and δ as arguments, the "table of the proportion of parallels" (*jadwal nisbat al-madārāt*) and obtains the values N_ϕ and N_δ.

Finally,

$$\text{vers } T = F \cdot N_\phi \cdot N_\delta. \tag{2b}$$

This formula can be explained analysing the second column of the *Jadwal nisbat juyūb al-tafāḍul wa l-madārāt li l-shams wa l-kawākib* extant in p. 285 of the Rabat MS. A table of the same kind appears in Ibn al-Zarqālluh's *Almanac* and in the anonymous Hyderabad recension of the *Zīj* of Ibn Isḥāq [Millás 1943–50, 226; Mestres 1999, 275, 279–281]. Mestres has shown that the tabulated function is $1/\cos x$. This implies that formula (2b) is equivalent to

$$\text{Vers } T = \frac{\text{Sin } h_m - \text{Sin } h}{\cos\phi \cdot \cos\delta}.$$

The same rule is given in Ibn al-Zarqālluh's *Almanac* and in the Hyderabad recension [Millás 1943–50, 141–142; Mestres 1999, 76]. An explanation for it is given by al-Battānī [Nallino 1899–1907, vol. I, 191].

Chapter 40: *On determining the solar altitude from the number of hours passed of the day* (p. 190).

Inverse use of formula (1a) of Chapter 39.

Chapter 41: *On determining the northern and southern distances* (*abʿād* = declinations) *of the fixed stars, the planets and the moon from the equator* (p. 190).

The method explained by Ibn al-Raqqām is also used in many sources such as al-Battānī (incomplete),[18] Ibn al-Kammād's *Zīj*,[19] Ibn al-Bannā''s *Minhāj*[20] and the anonymous Hyderabad recension of the *Zīj* of Ibn Isḥāq;[21] as well as in Eastern Islamic sources such as Bīrūnī's *Maqālīd*.[22]

Chapter 42: *On determining the degree of the ecliptic which crosses the meridian at the same time as a star* (pp. 190–191).

This chapter describes a method to calculate the degree of mediation of a star or planet which derives from the *Almagest* VIII.5. See a detailed study by Samsó [2008].

Chapter 43: *On determining the degrees of the ecliptic which rise or set at the same time as a star on any inclined Northern horizon* (p. 191).

This chapter describes two different procedures, the first of which can also be found in the *Almagest* II.9 [Toomer 1984, 104], and is also explained by al-Battānī [Nallino 1899–1907, vol. III, 49–50]. The second method is to be found in al-Battānī, Ibn al-Kammād,[23] the Hyderabad recension of Ibn Isḥāq[24]

---

[17] See also al-Bīrūnī in *Shadows* [Kennedy 1976, vol. II, 123–124], *Taḥdīd* [Kennedy 1973, 113–115], *Mafātīḥ* [Debarnot 1985, 228–231].

[18] Nallino [1899–1907, vol. III, 47]. See also the commentary by Nallino [1899–1907, vol. I, 192–193].

[19] In an appendix to the Latin translation of the *Muqtabas Zīj* by Johannes de Dumpno, which contains passages from *al-Kawr ʿalā l-dawr*, a lost *zīj* of Ibn al-Kammād. See the Latin MS 10023 of the Biblioteca Nacional de Madrid, fol. 20 v [Chabás and Goldstein 1994].

[20] Vernet [1952, 44–45].

[21] Mestres [1999, 68].

[22] Debarnot [1985, 210–213]. See also Kennedy [1985, 9].

[23] Lat. MS 10023 of the Madrid NL, fol. 21r–21v.

[24] Mestres [1999, Ar. 118–119; Eng. 70].

and Ibn al-Bannāʾ.[25]

Chapter 44: *On determining the time of the night on the basis of the altitude of one of the fixed stars* (pp. 191–192).

The two procedures explained are the same as the ones applied to the sun in Chapter 39.

Chapter 45: *On determining the altitude of a star on the basis of the number of hours of the night which have passed* (pp. 192–193).

Inverse use of procedure (1a) of Chapters 39 and 44.

Chapter 46: *On determining the four cusps (ascendant, descendant, tenth and fourth houses) by means of the number of hours passed of day or night* (p. 193).

In this chapter, Ibn al-Raqqām explains the standard method for the determination of the tropical longitude of the ascendant and midheaven, using tables of right and oblique ascensions. The data of the problem are either the number of hours since sunrise or sunset or the arc, or the hour angle of a star measured from the meridian or from its rising point.

Chapter 47: *On determining the number of hours passed of day or night by means of the ascendant or one of the four cusps* (p. 193).

Inverse use of the method explained in Chapter 46.

Chapter 48: *On determining the division of the twelve houses by means of [the longitude] of the four cusps* (pp. 193–195).

This chapter deals with the division of the houses of the horoscope [North 1986; Kennedy 1996; Calvo 1998; Casulleras 2009], and explains two well-known methods: the standard one (in two variants) and the dual longitude method. In the Hyderabad recension of Ibn Isḥāq's *Zīj*, the second method is considered more suitable for everyday life, while the standard method should be used for important matters [Mestres 1999, 77]. Interestingly, Ibn al-Raqqām himself, in the *Shāmil Zīj*, only uses the standard method for the centre of the earth [Kennedy 1996, 558–559], while in the *Mustawfī*, he does not introduce this restriction.

Chapter 49: *On determining the distance of a star from the equator (= declination), and the degree of the ecliptic which crosses the meridian with the star by means of the degree of the ecliptic that rises or sets with it and its rising and setting amplitude* (p. 195).

Ibn al-Raqqām calculates the declination with a formula he has already used in Chapter 29. Then he obtains the right ascension of the degree of mediation by means of half the day arc (Chapter 31) and the oblique ascension of the rising or setting degree (see Chapters 42–43).

Chapter 50: *On determining the distance* (buʿd = declination) *of a star and the degree of its mediation with a procedure different from those explained in the previous Chapters 41 and 42, which does not use ascensions* (p. 195).

Ibn al-Raqqām explains a different method from the one used in Chapter 42. Its origins are not clear; they may be the result of his own research. A table of right ascensions is used in the last stage of the computation. See Samsó [2008].

Chapter 51: *On determining the degree of the ecliptic corresponding to a star, its latitude also measured from the ecliptic by means of its distance from the equator (declination) and the degree of the ecliptic which crosses the meridian together with the star* (pp. 195–196).

The method of computation is identical to the one in described in Chapter 25 of al-Battānī's *Zīj* and uses the same terminology [Nallino 1899–1907, vol. III, 54–57, vol. 1, 36–37, 197–200]. Besides, Schiapparelli remarked, in his commentary on Nallino's edition, that while the computation of latitude is correct, that of the longitude is erroneous, in spite of the fact that al-Battānī had successfully solved the inverse problem in Chapter 18 (to obtain the declination and mediation of a star or planet knowing its longitude and latitude: he tried to give a new solution but failed in his attempt) [Nallino 1899–1907, vol. I, 31–32]. Interestingly, al-Battānī's error appears in the only Arabic Escorial manuscript and in

[25] Vernet [1952, Ar. 48 ; Sp. 113].

Plato of Tivoli's Latin translation; it seems that Ibn al-Raqqām used an Arabic manuscript of the same family and did not check the details of the procedure.

Chapter 52: *On determining the distance between two stars or between two points of the sphere, expressed in degrees of a great circle passing through the centres of both stars by means of their longitudes and latitudes* (pp. 196–197).

Ibn al-Raqqām considers the cases in which the two celestial bodies have latitude 0° and in which only one body or point has latitude. The solution for this latter case is taken from al-Battānī's Chapter 26 [Nallino 1899–1907 vol. III, 57–61, vol. I, 37–40, 199–204]. For the general case (both stars have latitude) Ibn al-Raqqām explains a computation in five steps, for which I have not been able to find a source.

Chapter 53: *On determining the longitude of your location by means of the observation of an observer* (bi-raṣad rāṣid) (pp. 198–199).

The first procedure described by Ibn al-Raqqām is well known: simultaneous observation of the beginning or the end of a solar or lunar eclipse at two places, one of which corresponds to the location whose longitude we want to obtain. The second method is more interesting because it is rarely mentioned in the sources: the observation of an eclipse in a place whose longitude is unknown comparing the local hour with the hour calculated for a place whose longitude is known. A curious remark at the end of the chapter states that the method could, perhaps (*qad yumkin dhālika*), be applied to a simultaneous observation of the fall of a large [shooting?] star (*inqiḍāḍ kawkab ʿaẓīm*).

Chapter 54 (pp. 199–201): *On determining the azimuth of Mecca, may God the Exalted ennoble her, in a location of known longitude and latitude* (pp. 199–201).

Ibn al-Raqqām explains three methods for determining the *qibla*: the first one is almost a literal copy of Ibn al-Haytham's *Qawl fī samt al-qibla bi l-ḥisāb* [Dallal 1995], a variant of the "method of the *zīj*es," which I have not been able to find in any other Western Islamic source.

The second (called *talkhīṣ hādhā l-bāb*) is, again, a copy of Ibn Muʿādh's version of the "method of the *zīj*es" which obviously circulated among astronomers of this school, for Ibn Muʿādh's Arabic text was preserved in the anonymous Hyderabad recension of the *Zīj* of Ibn Isḥāq. It was also accessible to readers of Latin Europe, for it appears in the Latin translation of Ibn Muʿādh's *Tabulae Jahen* [Samsó and Mielgo 1994].

I have not been able to find any source for the third method, although it is obviously another variant of the *method of the zījes*. As the fifth step of the computation (in six steps) Ibn al-Raqqām obtains the distance between Mecca and the locality for which we are calculating the *qibla*, measured on a great circle of the earth's sphere: the whole procedure is almost the same as the one that appears in Chapter 52 (pp. 196–197). This latter chapter deals with an identical problem on the celestial sphere: computation of the distance between two stars or planets on a great circle passing through their centres. We have already seen that the second method explained in Chapter 52 could be an original contribution of Ibn al-Raqqām himself, who dealt with the same problem in his analemma construction to find the *qibla* in which there is a possible influence of al-Bīrūnī [Carandell 1984].

Chapter 55: *On determining the hour of the rise of dawn and extinction of twilight by means of the solar depression below the horizon and vice versa* (pp. 281–282).

The angle of solar depression used by Ibn al-Raqqām (and by Ibn Muʿādh [Goldstein 1976 and 1977; Smith and Goldstein 1993]) is $-19°$. The formula used by Ibn al-Raqqām corresponds to an application of al-Battānī's Chapters 21–22 (determination of the night hour if the altitude of a star is known and reciprocal) [Nallino 1899–1907, vol. III, 50–53; vol. I, 33–35, 195–196]. In both cases the half day arc of the Sun and its meridian altitude are used.

Luni-Solar Conjunctions and Oppositions, Lunar Visibility, Parallax, Eclipses

Chapter 56: *On determining [lunisolar] conjunctions and oppositions* (pp. 202–203).

In this chapter, Ibn al-Raqqām insists on the fact that this *zīj* has been computed for the longitude of Tunis. The lunar longitude should be calculated with the Zarqāllian correction which considers that the centre of the lunar mean motion depends on the position of the solar apogee (see above Chapter 20). At the end of the chapter, he promises to write a *maqāla* on planetary conjunctions (*qirānāt*), which, to the best of my knowledge, is not extant.

Chapter 57 (pp. 203–208): *On determining the lunar parallax* (pp. 203–208).

In this chapter, Ibn al-Raqqām begins with two different methods to calculate the geocentric distance of the Moon. The first one, for which he gives no source, is equivalent to the one explained in the *Almagest* V.6 [Toomer 1984, 233–234; Pedersen 1974, 193–195]. The second procedure derives from al-Battānī who calculates the distance of the Moon 1) for conjunctions and oppositions and 2) for the general case, using the functions of interpolation of the tables of lunar equations which introduce a correction due to the position of the centre of the epicycle from the apogee/perigee of the deferent [Nallino 1899–1907, vol. III, 116–118; vol. I, 78–79, 253–256].

He then continues with the computation of the horizontal parallax with two procedures which are almost identical, and with the calculation of its components in longitude and latitude, for which he offers two methods: the first one, for which he gives no source, is followed by a long passage from Chapter 39 of al-Battānī's *Zīj* [Nallino 1899–1907, vol. I, 79–83, vol. III, 118–125, 256–262]. Then, he follows with a set of instructions for the use of the tables and with an approximate method which seems to derive from Ibn al-Muthannā's commentary on al-Khwārizmī's astronomical tables [Goldstein 1967, 121–124]. The same rule is also to be found in canons Cc of the *Toledan Tables* [Pedersen 2002, 652–653]. This interesting chapter has been edited and analysed by Bellver and Samsó [2012].

Chapter 58: *On determining lunar crescent visibility in the evening and in the morning* (pp. 208–210).

The main part of Ibn al-Raqqām's chapter follows al-Battānī's Chapter 41, with refinements and corrections [Nallino 1899–1907, vol. III, 129–136, vol. I, 85–92, 265–272]. Thus, al-Battānī approximates the distance between the Sun and the Moon (*qaws al-buʿd*) by using Pythagoras' theorem, while Ibn al-Raqqām gives an exact solution which uses the cosine law. Moreover, as al-Battānī justifies the basic limit of visibility (12° according to the Indian tradition) on the mean difference between solar and lunar velocity per day, he uses a value of 12;11°. Ibn al-Raqqām uses Ibn Isḥāq's mean motion tables in which we find

$$\text{mean solar motion per day} = 0;59,8,11...^\circ,$$
$$\text{mean lunar motion per day} = 13;10,34,53...^\circ.$$

The difference between these two values is

$$12;11,26,42...^\circ \approx 12;11,27^\circ,$$

very near the basic limit of visibility used by Ibn al-Raqqām (12;11,28°).

Chapter 59: *On determining lunar eclipses* (pp. 210–212).

This chapter seems to be a copy (sometimes literal) of al-Battānī's Chapter 43 [Nallino 1899–1907, vol. III, 146–156, vol. I, 96–104, 275–276].

Chapter 60: *On determining solar eclipses either by computation or using the tables* (pp. 212–216).

This chapter derives also from al-Battānī's Chapter 44 [Nallino 1899–1907, vol. III, 157–171, vol. I, 104–113, 277–282]. Interestingly, Ibn al-Raqqām reproduces certain numerical values which have the same errors as the ones that appear in the Escorial manuscript and in the Latin translation of Plato of Tivoli. Thus, the limit of the visible latitude of the Moon for an eclipse is, according to Ibn al-Raqqām,

$$\beta_L < 0;30,35^\circ,$$

instead of 0;30,25°. The same value is used by Plato of Tivoli, while the Escorial MS of al-Battānī's *Zīj* has 0;34,36°.

Analogously, the difference between the apparent diameter of the Sun in its perigee and in its apogee should be

$$\text{diam. (perigee)} = 0;33,40°,$$
$$\text{diam. (apogee)} = 0;31,20°,$$

the difference being 0;2,20°. The Escorial MS, Plato and Ibn al-Raqqām give 0;2,15°.

Mathematical Astrology

Chapter 61: *On determining the projection*[26] *of the rays of the planets* (kawākib) *on the ecliptic without referring it to the horizon* (pp. 216–220).

This is one of the problems of mathematical astrology that make Ibn al-Raqqām's *Mustawfī Zīj* particularly interesting.[27] Eight different methods are explained.

1. The first method is concerned with the increase in ecliptical longitude (Δ_λ) one has to add to the longitude of a planet when it has latitude and projects its rays on the ecliptic in a right of left sextile. With some slight differences. this method is the same as the one applied in Chapter 54 of al-Battānī's *Zīj* [Nallino 1899–1907, vol. III, 197; vol. I, 129–131, 305–313]. One should add that both al-Battānī's and Ibn al-Raqqām's texts seem to be corrupt. An attempt to reconstruct them was made by Schiapparelli, based on the hypothesis that the method used was the same as the one that appears in al-Battānī's Chapter 26 (= Ibn al-Raqqām Chapter 52): to find the distance between two celestial bodies, only one of which has latitude while the other is on the ecliptic [Nallino 1899–1907, vol. I, 307–308]. In this particular case the distance is known (60°) and we have to find Δ_λ.
2. The same kind of problem is solved by the second method, which is a simple application of the cosine law: the same solution can be found in al-Bīrūnī's *Maqālīd* where it is attributed to al-Battānī and al-Ṣūfī [Debarnot 1985, 268–269].
3. Method attributed to the Persians: the planet which casts its rays is considered to be on the ecliptic and right ascensions are used. Other sources in which this method is mentioned are mainly Andalusī and Maghribī [Casulleras 2007a, 34–36], although it also seems to appear in a second manuscript of the *Mumtaḥan Zīj* described recently by Benno van Dalen [2004, 30]. Method 5 seems to be a variant of this one.
4. A method analogous to the previous one, using oblique ascensions [Casulleras 2007a, 36–37]. The planet is also considered to be on the ecliptic. It also appears in Chapter 114 of Ibn al-Raqqām's *Shāmil Zīj* [ʿAbd al-Raḥmān 1996, 159], as well as in other Western Islamic sources. According to Hogendijk [1998] we can find it in an appendix (a later addition?) to a treatise on the astrolabe by al-Khwārizmī.

6. & 8. Single hour-line method:[28] attributed to Ptolemy,[29] it can be found in many Islamic sources. It is one of the several attempts to combine right and oblique ascensions in such a way that the right ascension component increases when the planet approaches the meridian, while the oblique ascension component also increases when it approaches the horizon [Samsó 2009, 32–35 (referring to *tasyīr*)]. The method was studied by Kennedy and Krikorian-Preisler on the basis of a text extant in Bīrūnī's *Qānūn Masʿūdī* [Kennedy and Krikorian-Preisler 1972; Hogendijk 1998, Casulleras 2010]. It was known in al-Andalus where it reached Ibn al-Samḥ through Abū Maʿshar or another Eastern source. Ibn Muʿādh criticised it [Casulleras 2007a, 38–41]. Ibn ʿAzzūz al-Qusanṭīnī computed a set of tables for the latitude of Fez, using this method which could be applied for the calculation of both projection of rays and *tasyīr* [Casulleras 2007b]. In this method, the degree of longitude of the planet casting its rays is used as a starting point; Ibn al-Raqqām criticises this and proposes to use the degree of mediation of the planet. This is what he does in method 8, which he attributes to the school (*madhhab*) of Wālīs al-Miṣrī (= Vettius Valens) and, he adds, it is the method on which the

[26] The Rabat MS has here (p. 216) *maṭāliʿ* (ascensions) instead of *maṭāriḥ* (projection).

[27] On this topic see Kennedy and Krikorian-Preisler [1972] and Casulleras [2004, 2007a, 2007b, 2008–09, 2010].

[28] I am using the terminology coined by Hogendijk [1998].

[29] Not to be found in Ptolemy's works, although it derives from the single hour line method for the computation of *tasyīr* which actually appears in the *Tetrabiblos*.

community of astrologers (*jamāʿa min aṣḥāb al-aḥkām*) agree. This confirms Hogendijk's opinion that the single hour-line method is the standard procedure used in the Islamic world for casting the rays [Hogendijk 1998].

7. Method of the four position semicircles, in which the so-called *al-āfāq al-ḥāditha* (incident horizons)[30] are used, although Ibn al-Raqqām calls them *dawāʾir al-azmān* (time circles): these are artificial horizons for a given latitude ξ (arcs of a great circle passing through the north and south points of the local horizon and through the planet or through another significant point of the celestial sphere used in the procedure). Ibn al-Raqqām ascribes this method to "the moderns" (*al-mutaʾākhirūn*). The method is described in many Islamic sources. It was probably introduced in al-Andalus by Maslama al-Majrīṭī (d. 1007) who used it for the approximate computation of the additional tables on the projection of rays which he added to his revision of al-Khwārizmī's *Zīj* [Hogendijk 1989 and 1998]. Ibn Muʿādh gave two mathematical solutions for the application of this method: the first one gives an exact result and was studied by Hogendijk [2005]. The second is approximate and has been published by Casulleras [2004] who noted the existence of major similarities with the approximate method explained by al-Bīrūnī in his *al-Qānūn al-Masʿūdī*. The method explained here by Ibn al-Raqqām is also approximate and practically the same as Ibn Muʿādh's, although the terminology used is slightly different.

Chapter 62: *On determining the* tasyīr *(prorogation or progression) of the indicators* (pp. 220–222).

Three methods are explained, the first of which solves, again, the problem of determining the distance between two celestial bodies or two points of the celestial sphere measured on the great circle which passes by its centres or points (see Chapters 52 and 61). In his description of the first method, Ibn al-Raqqām considers three subcases: a) the two points are on the ecliptic; b) one of them has latitude and the other is on the ecliptic: he calculates the distance with the same procedure as in Chapter 52 (derived from al-Battānī, Chapter 26); c) both points have latitude: here he follows the different subcases also explained in al-Battānī, Chapter 26. Hogendijk [1998] remarks that he has only found this procedure for the computation of the *tasyīr*, based on the distance between two points on a great circle in Ibn Bāṣo's treatise on the universal astrolabe [Calvo 1993, Ar. 174, Sp. 199].

The second method is ascribed to Hermes and it uses two position circles (called *dāʾira zamāniyya*, time circle), apparently for latitudes ξ_1 and ξ_2. The arc of the *tasyīr* between the first, P_1, and the second, P_2, indicators will be

$$\alpha_{\xi 1}(P_1) - \alpha_{\xi 2}(P_2).$$

The third method corresponds to the "single hour-line method" and is attributed to Ptolemy for it appears in the *Tetrabiblos* III.10 [Robbins 1940, 286–307], as well as in many Islamic sources.[31]

Chapter 63 (pp. 222–225): *On determining year, nativity and month transfers* (pp. 222–225).

In this chapter there is another reference to the meridian of Tunis used in this *zīj*.

Trigonometry, Planetary Visibility

Chapter 64: *On determining the chords traced on a circle by computation* (p. 225).

This chapter calculates the chords of 60°, 120°, 90°, 36°, 72° etc. following *Almagest* I.10 [Toomer 1984, 48–60], although the source could also be al-Battānī's Chapter 3 [Nallino 1899–1907, Ar. 13–17; I, 9–11].

Chapter 65: *On determining the visibility and invisibility of the planets in every locality* (pp. 225–226).

The limits of visibility for each planet are the same as the ones that appear in the *Almagest* [Toomer 1984, 639; Neugebauer 1975, 235].

Page 226: Colophon of the manuscript, author and date of the copy.

[30] A term which seems to have been coined by Muḥyī al-Dīn al-Maghribī [Kennedy 1996, 555–556; Dorce 2002–03, 63–76].

[31] Ibn Abī l-Rijāl and Ibn Qunfudh, among many others [Samsó 2009, 32–35].

The Tables

Chronology (pp. 227–228)

Table for extracting dates from one calendar to another (p. 227).

The three calendars implied are the Arab, *rūmī* (Byzantine) and Persian.

Table of the initial weekdays of the Arab, Rūmī *and Persian years and months* (p. 228).

Mean Motions (pp. 229–242)

The mean motion tables in this *zīj* are sidereal, following the Zarqāllian tradition. Its arguments are standard: hours and days (except for very slow mean motions), months, single years (*mabsūṭa*) from 1 to 30 (the table of the motion of the apogees adds, after 30 years, 60 and 90), and collected years (*majmūʿa*) in multiples of 90 until 900 years. These values probably correspond to Ibn al-Raqqām's original tables. As the collected years have incorporated the value of the radix (with the sole exception of the table for the motion of the apogees), a late copyist has added a new radix for year 1260/1844: this is probably the manuscript from which Muḥammad b. Aḥmad al-Hawārī copied the present text in 1313/1895. It has been obtained by adding the table values corresponding to 900 and 360 years and subtracting the value of the radix from the result. In several cases, there is an error in this operation, resulting from the omission of the subtraction of the radix.

The mean motion parameters are the same as those of the Hyderabad recension, the *Minhāj* of Ibn al-Bannāʾ and the *Qawīm* and *Shāmil Zīj*es and they have been squeezed — to use the term of Otto Neugebauer — several times [Kennedy 1997, 36–37; Mestres 1996; Samsó and Millás 1994 (Sun) and 1998 (planets)]. This is why, in most cases, I omit them here. For slow motions (apogees, motion of the Head of Aries, centre of the solar eccentric) the radices are the same as in the rest of the aforementioned *zīj*es. In the other cases, the radices are the same as in the Hyderabad recension and the *Qawīm* (both calculated for the meridian of Tunis), while the others present a difference due to the longitude correction.

Table of the mean motion of the apogee (p. 229 right).

This table has been edited and commented, together with Chapter 18, by Díaz-Fajardo, 2005. The radix position of the solar apogee for the beginning of the Hijra is $2^s16;44,17°$, calculated for the longitude of Toledo.[32] A note above the table gives the position of the solar apogee for the end of 1314/1896–97 and the longitude of Fez.

Table of the mean motion of the Head of Aries (trepidation) (p. 229 left).

Radix (beginning of Hijra): 3;53,55°.[33] Mean motion per day: 0;0,0,54,57,17,48° (almost the same as the parameter established by Ibn al-Zarqālluh 0;0,0,54,57,17,38°, squeezed from the tables calculated for Julian and Persian years) [Samsó and Millás 1994, 10–12; Kennedy 1997, 55; Mielgo 1996, 164–178].

Table of the mean motion of the centre of the solar eccentric (p. 230).

This table allows the user to calculate the position of the centre of the solar eccentric in Ibn al-Zarqālluh's model of variable solar eccentricity (see Chapter 19). Radix: $2^s23;40,53°$. The copyist calculates a new radix for 1314/1896–97 and the longitude of Fez: $7^s11;0,4°$.

Table of the mean motion of the axis of the ecliptic (p. 231).

The pole of the ecliptic rotates on an epicycle whose centre moves on a circle around the pole of the equator in Ibn al-Zarqālluh's model that intends to justify the variation of the obliquity of the ecliptic (see Chapter 26). The radix position is $9^s11;14,34°$. The same radix appears in the *Shāmil Zīj* (Bijāya).

---

32 The same value appears in the Hyderabad recension and in Ibn al-Bannāʾ and the source is probably Ibn al-Zarqālluh himself [Samsó and Millás 1994, 7–9].

33 Ibn al-Zarqālluh: 3;51,11° [Millás 1943–50, 324].

Table of the solar mean motion (p. 232).

The radix position is $3^s23;18,58°$. A new radix for 1314/1896–97 and the longitude of Fez has been added.

Table of the lunar mean motion in longitude (p. 233).

Radix position: $4^s0;2,20°$.

Table of the lunar mean motion in anomaly (p. 234).

The radix in the MS is $3^s26, 38, 34°$. This value does not agree with the radices found in the related *zījes*, calculated for the meridian of Tunis, in which we find

$$\text{Hyderabad (Tunis)} : 3^s17;38,43°,$$
$$\textit{Qawīm} \text{ (Tunis)} : 3^s17;38,34°.$$

We can check the validity of this latter value for the *Mustawfī*, for the mean longitudes of the table for the collected years have the radix added to the corresponding mean anomalies. Thus, choosing the amount corresponding to 900 years gives

$$(900)\ 9^s12;37,35° - 3^s17;38,34° = 5^s24;59,1°.$$

This is precisely the value for 900 years that we find in the Hyderabad recension, Ibn al-Bannāʾ and the *Qawīm*, in which the radix has not been added to the mean anomaly. In the *Shāmil Zīj* we have $5^s24;59,0°$.

Table of the mean motion of Saturn (p. 235).

The radix ($11^s0;10,49°$) is, again, mistaken but this time it has been corrected by the copyist who writes, in the margin, $3^s26,17,9°$, which coincides with the Hyderabad recension and the *Qawīm* ($3^s26,17,10°$). We can check it with the value for the mean longitude corresponding to 900 years with

$$(900)\ 11^s18;18,7 - 3^s26;17,10° = 7^s22;0,57°,$$

which agrees with the rest of the sources.

It is obvious that the copyist is looking carefully at the values he finds in the manuscript from which he is copying. For 1260 years we read, in the table, $1^s23;23,40°$, which is the result of adding two values from the table of collected years,

$$11^s18;18,7° \text{ (900 yrs.)} + 2^s5;5,53° \text{ (360 yrs.)} = 1^s23;23,40°,$$

without remembering that the radix had been added to the two values. The copyist remarks that the correct value corresponding to 1260 years is

$$13^s23;23,40° - 3^s26;17,9° = 9^s27;6,31°.$$

Table of the mean motion of Jupiter (p. 236).

The radix value is $11^s0;10,49°$ which agrees with the *Qawīm*, while the Hyderabad recension has $11^s0;10°$. In the value corresponding to 1260 years ($10^s20;33,25°$) the copyist has found the same mistake as in the table for the mean motion of Saturn and replaces this value by $11^s20;22,36°$, which is correct.

Table of the mean motion of Mars (p. 237).

The radix value is $7^s0;36,13°$, which agrees with the *Qawīm* and the Hyderabad recension. The value for 1260 yrs ($1^s13;2,18°$) is, again, wrong; although in this case no correction given by the copyist.

Table of the mean anomaly of Venus (p. 238).

The radix value is $1^s14;28,34°$, which agrees with the *Qawīm* and the Hyderabad recension. The value for 1260 yrs ($9^s24;37,49°$) is, again, wrong, without correction.

Table of the mean anomaly of Mercury (p. 239).

The radix value is $2^s13;18,22°$, which agrees with the *Qawīm* and the Hyderabad recension. The value for 1260 yrs ($7^s5;25,58°$) is wrong, without correction.

Table of the mean motion of the lunar node (p. 240).

The radix value is $6^s24;19,30°$, which agrees with the *Qawīm* and the Hyderabad recension. The value ($7^s5,40,15°$) corresponding to 900 yrs should be replaced by $6^s5,40,15°$. This wrong value has been used to calculate the mean position of the node for 1260 yrs ($0^s10,32,3°$) in which we have, therefore, a double error.

Table of the mean motions of the planets and the lunar node for minutes of hours (pp. 241–242).

Arguments from 1 to 60 minutes for the mean motions in longitude of the sun and the moon, lunar mean motion in anomaly, mean motions in longitude of Saturn, Jupiter and Mars, mean motion in anomaly of Venus and Mercury, mean motion of the lunar node.

Same table in the *Qawīm Zīj*.

Trepidation (pp. 243–24)

Table of the equation of the motion of accession and recession (p. 243).

Arguments from 1° to 360°. The maximum tabular value is 10;24° for an argument of 90°. The table gives the value of precession (*P*) for a given position of the Head of Aries (*i*) in its motion around an equatorial epicycle (see Chapter 17 and the table of the mean motion of the Head of Aries). The table has been calculated with the expression

$$P = \text{arcSin } (\text{Sin } i \cdot \text{Sin } 10{:}24°).$$

The same table appears in the Hyderabad recension, Ibn al-Bannāʾ and the *Shāmil Zīj*. The corresponding table of the *Qawīm Zīj* has been calculated for a maximum value of 10;40,13°.

Table of the declination of the beginning point (al-mabdaʾ) *which is the declination of the first [point] of the sign of Aries* (p. 244).

Arguments from 1° to 360°. The maximum tabular value is 4;7,57 for an argument of 90°. 4;7,57 is the radius of the equatorial epicycle in the trepidation model and the table calculates the distance to the equator of each value of the argument (*i*, as in the previous table) using the expression

$$\delta(i) = 4;7,57 \text{ Sin } i.$$

The purpose of this table is to allow the user to obtain the value of precession, *P*, with the Zarqāllian procedure, which takes into account the variation of the obliquity of the ecliptic and uses the formula

$$P = \text{arcSin } \frac{\text{Sin } \delta(i)}{\text{Sin } \epsilon}.$$

The same table appears in the Hyderabad recension and in the *Shāmil*. It is not in the *Qawīm* or Ibn al-Bannāʾ.

Table of the accession of the perpendicular (iqbāl al-ʿamūd) *which is the "orientality" (motion towards the East?) of the beginning point* (tashrīq al-mabdaʾ) (p. 245).

Arguments from 1° to 360°. The maximum tabular value is 8;16 for an argument of 180°. 8;16 is, approximately, twice the radius of the equatorial epicycle (4;7,57). The table calculates

$$P_2 = 4;7,57 - 4;7,57 \cos i.$$

This table is used to calculate the equation of time (see Chapter 16) and it corresponds to the difference in right ascension between two positions of the Head of Aries in its motion on the equatorial epicycle. The same table appears in the Hyderabad recension, Ibn al-Bannāʾ and the *Shāmil Zīj*.

Solar Equation (pp. 246–248)

Table of the partial solar equation (taʿdīl al-shams al-juzʾī). The solar apogee was $2^s16;44,17°$ at the beginning of the Hijra (p. 246).

Arguments from 1° to 360°. The maximum tabular value is 1;49,7° for an argument of 92°. It is a table of the solar equation of a standard type calculated for a time in which the variable solar eccentricity (e) was

$$e = \text{Sin } 1;49,7° = 1;54,15^p.$$

An identical table can be found in the Hyderabad recension, and Mestres [1999, 43–44] has calculated that, using the tables extant in this recension to obtain the value of eccentricity for a given date, one gets year 619/1222 for an eccentricity of $1;54,15^p$. This is the year in which (according to a marginal note in the Escorial MS 909 which contains the *Minhāj* of Ibn al-Bannāʾ) Ibn Isḥāq made the observation of the planets upon which his *zīj* is based [Vernet 1952, 21]. The obvious consequence is that Ibn al-Raqqām is copying here a table authored by Ibn Isḥāq (see Chapter 19). The maximum solar equation in the corresponding tables of the *Shāmil* and the *Qawīm Zīj*es is 1;47,51° (e = 1;52,55°), calculated for 689/1290, probably by Ibn al-Raqqām himself.

The position of the solar apogee at the beginning of the Hijra (76;44,17°) is the same as the one we find in the Hyderabad recension, the *Minhāj* of Ibn al-Bannāʾ, the *Shāmil* and the *Qawīm Zīj*es.

Table of the adjusted (muḥarrar) *table of the total solar equation* (p. 247).[34]

Arguments from 1° to 360°. This is a standard table of the solar equation with a maximum value of 1;45,57° (for an argument 92°) which corresponds to an eccentricity of

$$e = \text{Sin } 1;45,57° = 1;50,56^p.$$

This value is the minimum solar eccentricity in the Hyderabad recension and the *Shāmil Zīj*: both *zīj*es contain an identical table which does not appear in the *Qawīm*. Ibn al-Kammād's *Muqtabas* contains a similar table with a maximum of 1;45,32° ($e = 1;50,30^p$). As for Ibn al-Bannāʾ, his *Minhāj* contains a table of the same kind, although displaced by 4° (in order to avoid negative values) which reaches a maximum of 5;46°, corresponding to 1;46° in a non-displaced equation. The values of this table seem to be the result of a rounding of the corresponding values in the Hyderabad recension or Ibn al-Raqqām's *Shāmil*: the original version was probably authored by Ibn Isḥāq.

Table of the minutes of the ratio (nisba) *of the variation* (ikhtilāf) *of the centre of the solar sphere* (p. 248 right).

This table calculates an interpolation function (m) designed to avoid calculating the large number of values required by Ibn al-Zarqālluh's solar model in which the equation of the centre depends on two variables: the mean longitude of the sun from its apogee and the position of the centre of the solar eccentric. Arguments from 1° (m = 0;59,59°) to 180° (m = 0;0°) and, symmetrically, from 180° to 360°. The same table appears in the Hyderabad recension and in Ibn al-Raqqām's *Shāmil*. Ibn al-Kammād's *Muqtabas* has tabulated a similar function to a precision of minutes.

Table of the solar adjusted increase (al-zāʾid al-muḥarrar li l-shams) (p. 248 left).

This table, with arguments from 1° to 360°, calculates the difference between a solar equation for a maximum eccentricity and one for a minimum eccentricity. The tabular values reach their maximum (0;36,51°) for arguments between 91° and 94°. The same table appears in Ibn al-Raqqām's *Shāmil* and in the *Minhāj* of Ibn al-Bannāʾ, although in this latter case, the function is displaced by 4°. A slightly different parameter is used in Ibn al-Kammād's equivalent table. The maximum solar equation in our *zīj* will be

$$1;45,57° + 0;36,51° = 2;22,48°,$$

and the maximum solar eccentricity will be

[34] On this table and the two following ones see Toomer [1969] and Samsó and Millás [1994]. See also Samsó [2011, 207–218, 491–492].

$$\text{Sin } 2;22,48° = 2;29,30^{p}.$$

Lunar Equations (pp. 249–251)

The first table in this collection has a clear Zarqāllian origin, while the second contains a correction which often appears in Islamic astronomy at least since Yaḥyā ibn Abī Manṣūr's *Mumtaḥan Zīj* (ca. 830). The rest constitute a standard Ptolemaic set of lunar equation tables (pp. 250–251), usually computed with Ptolemaic parameters. They are all to be found in the Hyderabad recension, the *Minhāj* of Ibn al-Bannāʾ (sometimes displaced), Ibn al-Raqqām's *Shāmil* and *Qawīm.*

Table of the lunar equation by means of its distance from the solar apogee (p. 249 right).

Arguments from 1° to 180°. As we have seen in Chapter 20 the argument (*a*) corresponds to the difference between the lunar mean longitude and the longitude of the solar apogee. The maximum tabular value is 0;24° for an argument 90° and the tabulated function is

$$\text{Sin } a \times 0;0,24°.$$

This correction appears in the canons of Ibn al-Kammād's *Muqtabas*, Ibn al-Hā'im's *Kāmil* and the Hyderabad recension. The same table appears in Ibn al-Bannāʾ.

Table of the equation of the lunar inclined sphere (p. 249 left).

Also mentioned in Chapter 20. Arguments from 1° to 180°, which correspond to the difference between the lunar position and the position of the lunar node. Maximum tabular values (0;6,39°) for arguments 45° through 46° and 134° through 135°. The table computes the difference between the lunar position in its inclined sphere and its position on the ecliptic. The same table appears in the Hyderabad recension (twice).

Table of the equation of the deviation (inḥirāf) *of the diameter of the lunar epicycle* (p. 250 right).

This table corresponds to the equation of the centre of the moon (c_3 in Neugebauer [1975, 93–94]).

Table of the equation of the ratio (nisba) *of the nearest distance* (al-buʿd al-aqrab) *of the lunar epicycle* (p. 250 left).

It corresponds to c_6, the interpolation function.

Table of the equation for the nearest distance of the lunar epicycle (p. 251 right).

It corresponds to c_5, the difference between the equation of the lunar anomaly at perigee (quadratures) and apogee (syzygies). The maximum tabular value is 2;40° (the same maximum as in the *Shāmil* and the *Qawīm*), not 2;39° as in the *Almagest*, the *Handy Tables* or the Hyderabad recension.

Table of the mufrad *lunar equation which is the radius of the epicycle* (p. 251 left).

It corresponds to c_4, the equation of anomaly at the syzygies, in which the centre of the epicycle is placed at the farthest distance from the earth. The maximum tabular value is 4;55,59°, not 5;1° as in the *Almagest* and the *Handy Tables*. 4;55,59° is also the value we find in the Hyderabad recension, while 4;56° is the maximum used in Ibn Muʿādh's *Tabulae Jahen*, the *Minhāj* of Ibn al-Bannāʾ and the *Alfonsine Tables*. It is an Indian parameter which appears in the *Khaṇḍakhādyaka*, the *Zīj al-Shāh* and many Eastern Islamic sources. The table has been calculated with Ptolemaic methods but with an epicycle radius of 5;9,41^{p} instead of 5;15^{p}, used for the other functions [Samsó and Millás 1998, 277–278].

Tables of Planetary Equations (pp. 252–264)

A set of five tables for the equations of each planet computed with Ptolemaic parameters, with the exception of the tables of the equation of the centre of the five planets. They are all to be found in the Hyderabad recension and in Ibn al-Raqqām's *Shāmil* and *Qawīm*. In the *Minhāj* of Ibn al-Bannāʾ the tables of the equations of the centre are displaced, while the tables used for the computation of the equation of anomaly of Saturn and Jupiter have an entirely different structure [Samsó and Millás 1998, 278–285].

The positions of the apogees are the following:

Saturn:	$7^s29;43,0°$	(p. 252 right),[35]
Jupiter:	$5^s9;43,0°$	(p. 254 left),[36]
Mars:	$4^s2;13,0°$	(p. 257 right),[37]
Venus:	$2^s16:44,17°$	(p. 259 left),[38]
Mercury:	$6^s18;24,17°$	(p. 262 right).[39]

It seems clear that the apogees of the superior planets in the three *zījes* of Ibn al-Raqqām seem to be the result of the rounding of the more precise values found in the Hyderabad recension and in Ibn al-Bannā'. I have argued elsewhere that the apogees of Saturn, Jupiter and Mars derive from al-Battānī and that they are sidereal, not tropical, calculated for a moment in which the value of precession was 0° (581 A.D. according to Ibn al-Zarqālluh) [Samsó and Millás 1998, 265–270]. The apogees of Venus and Mercury correspond to the beginning of the Hijra.

Equations of Saturn (pp. 252–254)

Table of the equation of the centre of Saturn (p. 252 right): c_3.

It reaches a maximum of 5;48° for arguments 90° through 94°, instead of 6;32° in the *Almagest*. This implies that the eccentricity used for the computation of the table is $3;2,30^p$ instead of $3;25^p$. I do not know the origin or the cause of this correction, which does not affect the rest of the tables.

Table of the minutes of the ratio of the minimum and maximum distances of Saturn (p. 252 left): c_8, the interpolation function.

Table of the equation of the deviation (inḥirāf) *of the maximum distance of Saturn's epicycle* (p. 253 right).

It corresponds to c_5, the difference between the equation of anomaly for the middle distance and the same equation when the centre of the epicycle is on the apogee. The maximum tabular value is 0;21°.

Table of the equation of the middle distance of Saturn's epicycle (p. 253 left).

It corresponds to c_6, the equation of the anomaly for the middle distance of the centre of the epicycle. The maximum tabular value is 6;13° for arguments 94° through 99°.

Table of the equation of the deviation (inḥirāf) *of the minimum distance of Saturn's epicycle* (p. 254 left).

It corresponds to c_7, which tabulates the difference between the equation of anomaly for the perigee and the equation of anomaly for the middle distance. The maximum tabular value is 0;25° for arguments 102° through 111°. This table contains a long set of erroneous values for arguments 71° through 88°.

Equations of Jupiter (pp. 254–256)

Table of the equation of the centre of Jupiter (p. 254 left)

The table calculates c_3. Its maximum is 5;41° for arguments 89° through 97°, a value which is not Ptolemaic (5;16°) and implies an eccentricity of $2;59^p$ whose origin is unknown to me and which is not used in the rest of the equation tables.

---

[35] The same positions as in the *Shāmil* and in the *Qawīm*. The Hyderabad recension and Ibn al-Bannā' have $7^s29;42,45°$.

[36] The same apogee as in the *Shāmil* and in the *Qawīm*, while in the Hyderabad recension and Ibn al-Bannā' it is $5^s9;42,45°$.

[37] The Hyderabad recension and Ibn al-Bannā' have $4^s2;12,45°$.

[38] The same position as the solar apogee. It is same in the rest of the sources.

[39] Placed at a fixed distance of 4;1,40° from the solar apogee. It is the same in the other sources. See Chapter 18, above.

Table of the minutes of the ratio between maximum and minimum distance of Jupiter (p. 255 right): c_8.

Table of the equation of the deviation of the maximum distance of Jupiter's epicycle (p. 255 left).
It corresponds to c_5. The maximum tabular value is 0;30° for arguments 107° through 118°.

Table of the equation of the minimum distance of Jupiter's epicycle (p. 256 right).
It corresponds to c_7. Its maximum value is 0;33° for arguments 109 through 120°.

Table of the equation of the mean distance of Jupiter's epicycle (p. 256 left)
It corresponds to c_6. Its maximum value is 11;3° for arguments 99° though 102°.

Equations of Mars (pp. 257–259)

Table of the equation of the centre (p. 257 right).
The table corresponds to c_3. Maximum value 11;25° for arguments 93° through 96°, which implies a small correction of the Ptolemaic value (11;32°). The origin of this correction is, probably, al-Battānī [Nallino 1899–1907, vol. II, 123].

Table of the minutes of the ratio between maximum and minimum distance of Mars (p. 257 right): c_8.

Table which determines the maximum distance of the epicycle of Mars (p. 258 right).
It corresponds to c_5. Its maximum is 5;38° for arguments 153° through 156°.

Table of the equation of the mean distance of the epicycle of Mars (p. 258 left).
It corresponds to c_6. Its maximum is 41;9° for arguments 130° through 132°.

Table of the equation of the minimum distance of the epicycle of Mars (p. 259 right)
It corresponds to c_7. Its maximum is 8;3° for 159°. This table seems to be corrupt in the Hyderabad recension, which reaches a maximum of 8;59°.

Equations of Venus (pp. 259–261)

Table of the equation of the centre of Venus (p. 259 left).
The table is c_3. Its maximum value is 1;51° for arguments 86° through 89°. Surprisingly, this value does not coincide with the maximum of the partial solar equation (1;49,7°, see p. 246) which, as we have seen, was calculated by Ibn Isḥāq for 1222. The corresponding eccentricity is 0;58,8° for Venus and twice that amount (1;56,16°) for the sun. If we calculate the corresponding date, using Ibn al-Zarqālluh's model of variable solar eccentricity, we obtain 1113, a date which cannot correspond to Ibn Isḥāq, but which fits the time of Ibn al-Kammād, whose *Muqtabas Zīj* uses a maximum solar equation of 1;52,44° [Samsó and Millás 1998, 273].

Table of the equation of the minutes of the ratio of the deviation of the minimum distance of Venus (p. 260 right): c_8.

Table of the equation of the maximum distance of Venus' epicycle (p. 260 left).
It corresponds to c_5. Its maximum is 1;42° for arguments 161° through 162°.

Table of the equation for the middle distance of the epicycle of Venus (p. 261 right).
It corresponds to c_6. Its maximum is 45;59° for arguments 135° through 136°.

Table of the equation of the deviation of the minimum distance of the epicycle of Venus (p. 261 left).
It corresponds to c_7. Its maximum is 1;55° for arguments 161° through 163°. The maximum in the *Almagest* is 1;52° as in the rest of our Maghribī sources.

Equations of Mercury (pp. 262–264)

Table of the equation of the centre of Mercury (p. 262 right).

The table corresponds to c_3. Its maximum is 3;2° for arguments 93° through 97°, which derives from al-Battānī; the maximum equation in the *Almagest* is 2;52°.

Table of the equation of the minutes of the ratio of the deviation of the minimum distance of Mercury (p. 262 left): c_8.

Table of the equation of the deviation of the maximum distance of Mercury's epicycle (p. 263 right).

It corresponds to c_5. Its maximum is 3;12° for arguments 129° through 131°.

Table of the equation for the middle distance of the epicycle of Mercury (p. 263 left).

It corresponds to c_6. Its maximum is 22;2° for arguments 111° through 112°.

Table of the equation of the deviation of the minimum distance of the epicycle of Mercury (p. 264 right).

It corresponds to c_7. Its maximum is 2;1° for arguments 130° through 136°.

Table of the stations before the direct planetary motion and its retrogradation (p. 264 left).

Tables of Planetary Latitudes (pp. 265–271)

With the exception of the two first tables (anomalies of the three superior planets and interpolation table) all the tables in this section are clearly Ptolemaic and follow the model of *Almagest* XIII.5 [Toomer 1984, 632–634]. Arguments from 1° to 360° with intervals of 1°. The same tables are also found in the *Shāmil* (except the first one) and in the *Qawīm*.

Table of the anomalies of the advance (ḥiṣaṣ al-masīr) *of the three superior planets* (p. 265).

This is a table of the mean motions in anomaly of Saturn, Jupiter and Mars. The arguments are given in days (1, 2, 4, 10, 20, 30, 40, 50, 60, 70, 80, 90, 100, 110, 120, 130) and hours (1, 2, 3, 4, 5, 10, 15, 20).

Table of the ratio of the anomalies (nisbat al-ḥiṣaṣ) *of the five planets in latitude* (p. 266 right).

Table of interpolation of the same kind as those calculated in the *Almagest* (XIII.5) for each planet, although here one table is considered to be valid for all of them. Arguments from 1° to 360° with an interval of 1°.

Table of the latitude of Saturn in the northern half of the ecliptic (p. 266 left).

Table of the latitude of Saturn in the southern half of the ecliptic (p. 267 right).

Table of the latitude of Jupiter in the northern half of the ecliptic (p. 267 left).

Table of the latitude of Jupiter in the southern half of the ecliptic (p. 268 right).

Table of the latitude of Mars in the northern half of the ecliptic (p. 268 left).

Table of the latitude of Mars in the southern half of the ecliptic (p. 269 right).

Table of the inclination (mayl) *of Venus' epicycle in the southern and northern sides* (p. 269 left).

Table of the slant (inḥirāf) *of the epicycle of Venus* (p. 270 right).

Table of the inclination (mayl) *of Mercury's epicycle in the southern and northern sides* (p. 270 left).

Table of the slant (inḥirāf) *of the epicycle of Mercury* (p. 271 right).

Solar Declination, Lunar Latitude and Obliquity of the Ecliptic (pp. 271–272)

Table of the partial declination (al-mayl al-juzʾī) *of the ecliptic* (p. 271 left).

Arguments from 1° to 360°. Standard table of declination with a maximum value: 23;32,30° (see Chapter 26). This table was probably calculated for the time of Ibn Isḥāq and is also preserved in the Hyderabad recension. The *Shāmil* and the *Qawīm* use an obliquity of 23;32,40°.

Table of the lunar latitude (p. 272 right).

Calculated for a maximum of 5°. The same table as in the *Shāmil* and in the *Qawīm.*

Table of the total declination and minutes of the variation (ikhtilāf) *of the declinations* (p. 272 left).

An equivalent table appears in the *Shāmil* [Kennedy 1997, 59], although it has a different structure and it lacks the column of the "minutes of the variation." In the *Mustawfī*, the table begins with the maximum value of the obliquity (23;52;30°) for argument 1° and ends with its minimum value (23;32,30°) for an argument 180°. This is precisely the value used by Ibn Isḥāq and Ibn al-Raqqām in their table of the partial equation. The arguments to be used in order to determine ϵ have to be calculated in the *Table of the mean motion of the axis of the ecliptic* (p. 231).[40] It is clear that, in the *Mustawfī*, the table begins with the position of the pole of the ecliptic at its furthest distance from the pole of the equator and ends at the nearest distance.[41] The table also includes a column with an interpolation function beginning with 60′ (argument 1°) and decreasing slowly until it reaches 0′ (argument 180°). The role of this function deserves further study, as there is no explanation in the canons.

Equation of Time (p. 273)

This page contains two different tables to calculate the equation of time.

Table of the equation of the days and their nights due to Abū l-ʿAbbās ibn Isḥāq.

A note above the table states: generated by the two motions of the sun and of trepidation (see Chapter 16). Arguments from 1° to 360°. The same table appears in the Hyderabad recension, and Mestres [1999, 266–270] has established that the tabular values are added to 24 hours in order to avoid negative results (-1 is expressed as 23;59^h). He has recalculated the whole table with the following parameters:

$$\begin{array}{ll} \text{Obliquity of the ecliptic:} & 23;32,30^\circ, \\ \text{Solar eccentricity:} & 1;56^p, \\ \text{Solar apogee:} & 88^\circ, \\ \text{Constant:} & 1;12, \\ \text{Displacement:} & 9;26^\circ. \end{array}$$

A second table for the equation of time with arguments from 1° to 360° and values in degrees. It seems to derive from al-Battānī's table [Nallino 1899–1907, vol. II, 61–64] (approximated to minutes) with the values truncated to degrees. Al-Battānī's table is also the source of several tables in the Toledan tradition [Pedersen 2002, 968–987]. There is no table for the equation of time in the *Shāmil* or the *Qawīm Zījes*.

Trigonometry and Spherical Astronomy (pp. 274–280)

Table of the arcs, sines (al-juyūb al-mustawiya) *and cosines* (al-juyūb al-mankūsa) (p. 274).

Table of right ascensions (p. 275).

---

[40] See also Chapter 26.

[41] On the procedure used for its computation see Goldstein [1964] and Samsó [1987].

Reckoned from the first degree of Capricorn. Entries approximated to minutes. The obliquity of the ecliptic used is 23;32,30°. Similar tables in the *Shāmil* and the *Qawīm* approximated to seconds.

Table of oblique ascensions for the latitude of Tunis (p. 276).

Entries approximated to minutes. Same table in the *Qawīm* in which a latitude of 36;37° for Tunis is mentioned.

Table of the rising times of the zodiacal signs (maṭāliʿ al-burūj) *for each degree of latitude* (p. 277).

Above the table, horizontally, the names of the signs using the corresponding symmetries (Aries/Pisces, Taurus/Aquarius etc.). On the right and left of the table, vertically, the degrees of latitude from 0° to 66°. Entries approximated to minutes. The table seems to be computed rather inaccurately.

Table of half the arc of daylight for a latitude 36;40° (p. 278).

36;40° corresponds to the latitude of Tunis (see the following table), for which Ibn al-Raqqām normally uses a latitude of 36;37°. Arguments from 1° to 360°. Entries calculated to the precision of minutes. The maximum tabular value (for a solar longitude of 90°) is 108;55°.

Table of the time degrees of the hours for a latitude 36;40°, which is the latitude of Tunis (p. 279 right).

Arguments from 1° to 360°. The table gives the time degrees corresponding to one seasonal hour for each degree of longitude. The maximum tabular length of 1 hour is 18;11°, an amount that gives 109;6° (not 108;55°, as in the previous table) for half the arc of daylight. Same table in the Hyderabad recension [Mestres 1999, 285].

Table for the equal hours of half a day for the latitude of Tunis (p. 279 left).

Arguments from 1 to 360°. The longest day has 7;16 equal hours which agrees slightly better with a half the arc of daylight of 109;6° than with 108;55°. Same table in the Hyderabad recension [Mestres 1999, 285].

Table of the altitude and the shadow [for a gnomon of 12 digits] *and* [for a gnomon of 6;30] *feet* (p. 280).

Arguments from 1° to 90°. The table calculates the cotangent (*al-ẓill al-mabsūṭ*) expressed in the two aforementioned units.

Solar and Lunar Velocity, Parallax, Eclipses (pp. 281–285)

Table of the solar motion per equal hour (p. 281 right).

Table of the solar velocity, the argument being the solar longitude in intervals of 1°. Its minimum and maximum values are 0;2,22° (arg. 1°) and 0;2,34° (arg. 180°). The table derives clearly from al-Khwārizmī's *Zīj* [Suter 1914, 175–180]. A table of the same kind with a different minimum (0;2,23°) and maximum (0;2,33°) can be found in the *Shāmil* and the *Qawīm*.

Table of the lunar motion per equal hour (p. 281 left).

Table of the lunar velocity, the argument being the lunar longitude in intervals of 1°. Its minimum and maximum values are 0,30,12° (arg. 1°) and 0;35,40° (arg. 180°). The table clearly derives from al-Khwārizmī's *Zīj* [Suter 1914, 175–180]. A table of the same kind with a different minimum (0;30,21°) and maximum (0;36,1°) can be found in the *Shāmil* and the *Qawīm*.

Table of the hours of the lunar parallax in longitude (p. 282).

The argument goes from 0;30° to 120°, the interval being 0;30°. The values of the table are expressed in hours and minutes and attain a maximum of 1;36^h for an argument 66°. This is the standard value for the maximum horizontal parallax in the Indo-Iranian tradition of al-Khwārizmī's *Zīj*, which contains the same table although the interval in the argument is 1°. E.S. Kennedy established the process of computation of al-Khwārizmī's table [Suter 1914, 28–29, 191–192; Neugebauer 1962, 69–71, 121, 123–126; Kennedy 1956, 49–52], which reappears in Ibn al-Kammād's *Muqtabas*, although the argument is expressed in time from 0;15^h to 9^h at intervals of 0;15^h.[42] The instructions for the use of the table in the

[42] MS Madrid BN 10023: canons in *Porta* 25, fols. 13v–14r; table in fol. 51v. See Chabás and Goldstein [1994, 22–23].

Hyderabad recension of Ibn Isḥāq's *Zīj* are the same as those we find in al-Khwārizmī and the *Mustawfī* but the only (incomplete) extant table derives from Ibn al-Kammād [Mestres 1999, 82]. Finally, Ibn al-Raqqām's *Qawīm Zīj* contains the same instructions for the use of the tables (Chapter 40, pp. 47–49 of the Rabat MS) and the same table as the *Mustawfī* (pp. 92–93).

Table of the lunar parallax in latitude and the ratio of the meridians (nisbat al-mamarrāt) (p. 283 right).
Arguments from 1° to 360° with intervals of 1°.

The first tabulated function corresponds to the parallax in latitude and the table is extant in the *Mustawfī* and in the *Qawīm* (p. 94). The arguments are given from 1° to 90° (which correspond to the distance between the lunar node and the nonagesimal) in steps of 1° and the computed values reach a maximum of 0;48,46°, which corresponds to the maximum horizontal parallax, P_H. The same table (with a maximum of 0;48,45°) appears in al-Khwārizmī's *Zīj* [Suter 1914, 191–192]. It is well known that this table has been computed with the expression $\sin a \times 0;48,45°$ (a being the argument). A similar table (calculated in the same way and reaching a maximum of 0;48,32°) is to be found in Ibn al-Kammād's *Muqtabas*[43] and in the Hyderabad recension of Ibn Isḥāq's *Zīj* [Mestres 1999, 82–85].

The second function, called *nisbat al-mamarrāt*, uses the same arguments as the previous table and reaches a maximum of 0;23,33° (arg. 90°). It looks like some kind of sinusoidal function but I have been unable to recalculate it or to discover its application.[44] The same table appears in the *Qawīm* (p. 94). A similar table can be found in Ibn al-Zarqālluh's *Almanach*, although, in this case, the maximum is 23;33°, and the interval in the arguments is 3° [Millás 1943–50, 227].

Table of the lunar eclipses for the maximum distance (p. 283 left).

The argument of the table is the argument of latitude. The limits used are:

10;48° – 0°,
169;52° – 180°,
180° – 190;48°,
349;52° – 360°.

I have not been able to find any source for this table, whose entries are different from those of the *Qawīm Zīj* and similar to those of the *Shāmil*. The maximum tabular values are

eclipsed digits of the diameter:	21;31^d,
minutes of immersion (*daqā'iq al-suqūṭ*):	50;31^m,
minutes of half totality (*daqā'iq al-makth*):	25;3^m.

Table of the lunar eclipses for the minimum distance (p. 284 right)

This table seems to derive from al-Battānī with only one important difference [Nallino 1899–1907, vol. II, 90]: the argument in al-Battānī is the latitude, while the *Mustawfī* uses the argument of latitude. The limits of the argument are

12° – 0°,
167:48° – 180°,
180° – 192°,
347;48° – 360°.

The maximum tabular values are

digits:	21;31^d	(Batt. 21;36^d),
minutes of immersion:	56;59^m	(Batt. id.),
minutes of half totality:	28;55^m	(Batt. 28;56^m).

[43] MS BN Madrid 10023 fol. 51r. See Chabás and Goldstein [1994, 19–20].

[44] The table has sectors which are clearly corrupt. It is neither a declination table calculated by means of the solar longitude or the solar right ascension nor a table of the "second declination."

Similar values can be found in the *Qawīm* but not in the *Shāmil*.

Table of the correction (taqwīm) *with arguments the distances* [from the earth] (p. 284 left).

The *Jadwal al-taqwīm bi-ḥiṣaṣ al-abʿād* (the *Shāmil* fol. 82v adds *min al-arḍ*) can also be found in the *Shāmil* (fol. 82v) and in the *Qawīm* (p. 97) and it has been carefully described by Kennedy [1997, 64–65]. It contains two interpolation functions called *taqwīm* and *ḥiṣaṣ al-abʿād*. The entries of the second function are the same as the ones that appear in the *Almagest* VI.8, although in the latter table the interval in the arguments is 6° [Toomer 1984, 308; see also Nallino 1899–1907, vol. II, 89]. The second function is used to correct the value of parallax taking into consideration the geocentric distance of the moon, as well as in the computation of eclipses for lunar positions between the apogee and the perigee.

Table of solar eclipses for maximum and minimum distances (p. 285 right).

This table derives clearly from *Almagest* V.8 [Toomer 1984, 306], although the argument (argument of latitude) is counted from the lunar node and not (as in the *Almagest*) from the northern point of the lunar inclined plane. The same table appears in the Hyderbad recension (derived from Ibn al-Kammād) and in the *Qawīm* (Rabat MS, p. 98). The table in the *Shāmil* (Kandilli MS, p. 84) is different.

Auxiliary Functions (pp. 285–286)[45]

Table of the ratio of the sines of the difference and of the parallels for the sun and the celestial bodies (Jadwal nisbat juyūb al-tafāḍul wa l-madārāt li l-shams wa l-kawākib) (p. 285 left).

Arguments, a, from 1° to 90° with intervals of 1°. It calculates two different functions

$$\textit{Nisbat al-tafāḍul}:\ 5 \tan a, \tag{1}$$

$$\textit{Nisbat al-madārāt}:\ \frac{1}{\cos a}. \tag{2}$$

This table appears twice in the Hyderabad recension [Mestres 1999, 275, 279–281], in the *Shāmil* and the *Qawīm*. We have seen an example of its use in Chapter 39 of the canons. The second function (*nisbat al-madārāt*) can also be found in Ibn al-Zarqālluh's *Almanach*, approximated to seconds and with an interval of 3° in the arguments [Millás 1943–50, 226].

Table of the ratio of the latitude for the visibility of the new moon (p. 286 right).

The arguments (latitude) of this table go from 1° to 90°, with an interval of 1°. The entries begin with 0;1 (the units are not indicated) and increase uniformly until they reach 1 for an argument 50°. From this value onwards until 90° the entry is always 1. It seems clear that the computer has stopped calculating after 50° because if, as it seems, the argument is the local latitude, it does not make much sense to bear in mind latitudes greater than 50°. I believe that this table computes $\tan \phi$ for a radius of the base circle 1. In fact, the table usually gives values with an error of $-0;1$ until entry 42°. From here onwards the errors increase until they reach $-0;12$ for entry 50°.

The use of this table is justified in an approximate procedure explained by Ibn al-Raqqām in the last lines of Chapter 58 (pp. 209–210). The text explains: "Enter with the latitude in the table of the ratio of the latitude and multiply the ratio by the lunar latitude at the moment of [the assumed] visibility. Add the result to the distance between the two luminaries expressed in degrees of their [oblique] descensions if the latitude is northern, or subtract it from [that distance] if it is southern. If the result of this subtraction or addition is less than 12°, the moon will not be seen. Otherwise, it will be seen." The same procedure is explained in the *Qawīm* [Kennedy 1997, 69].

The rationale of this method (obviously incomplete in the translated text) has been explained by King [1987, 186–187]. $\beta \tan \phi$ is a standard approximation, c, used to correct the lunar longitude, λ, in order to obtain λ', the longitude of the point of the ecliptic which sets simultaneously with the moon. Then:

[45] On auxiliary functions used in tables for timekeeping, see King [2004, 114–183].

$$\lambda' = \lambda + c$$

and, $D_{\phi}(\lambda')$ and $D_{\phi}(\lambda_s)$ being the oblique descensions of the corrected lunar longitude and the sun

$$s = D_{\phi}(\lambda') - D_{\phi}(\lambda_s),$$

s being the arc that should be greater than 12° if the new moon is going to be seen.

Table of the sines of the difference for obtaining oblique ascensions (p. 286 left).

The table computes $5 \tan \delta(x)$, x being the argument (1° through 90°), for ϵ = 23;33°. As seen in Chapter 30, this is a table that simplifies the computation of the ascensional difference, $\Delta\alpha$, which is

$$\sin \Delta\alpha = \tan \delta \tan \phi$$

when we use a shadow table calculated with a gnomon of 12^{d} and a sine table for R = 60.

On the origins of this kind of table, see Neugebauer and Schmidt [1952]. The same table can be found in Ibn al-Zarqālluh's *Almanach*, and in the Hyderabad recension [Millás 1943–50, 225; Mestres 1999, 282].

Astrological (p. 287)

Table of the transfer (taḥwīl) of the years and months (p. 287).

This table derives from table 174 of the Hyderabad recension which ascribes it to Ibn al-Hā'im, although the values are rounded to minutes or to seconds in the different sections. The Hyderabad table has been edited by M. ʿAbd al-Raḥmān and it is different from the table extant in the *Shāmil Zīj*, analysed by Kennedy [ʿAbd al-Raḥmān 1996b, 372–377; Kennedy 1997, 65–66].

Star Table (p. 288)

Table of the tropical positions of fixed stars for year 680[/1280–81] (p. 288).

This table gives the tropical longitude, latitude, mediation, declination, degree of the ecliptic which rises and sets together with the star, half day arc and meridian altitude of 30 stars. It is, in fact, an incomplete table and the complete version can be found in the *Qawīm Zīj* [MS Rabat 260, 100–103], where the same coordinates, calculated for the same date, are given for 88 stars. A third copy of the same table appears in MS Ambrosiana 338, fols. 145v–146r [Samsó 2002–03, 84–85], as an appendix to the Arabic translation, by al-Ḥajarī (beginning of the 17th century), of Abraham Zacut's *Almanach Perpetuum*. The longitudes are those of the star catalogue of the *Almagest* with an increase of 16;46° due to precession. The computed latitude is 36;37° (Tunis). The recomputation of all the columns of the complete table by María José Parra and Josep Casulleras[46] shows that most of the columns have been calculated within a reasonable degree of precision, with the exception of the two columns that give the rising and setting degree of the ecliptic, the only ones which depend on the use of a table of oblique ascensions, which show very important errors. The analysis by Casulleras tends to explain these errors as a result of copying mistakes, but one cannot dismiss the possibility that these two columns were copied from another source, which calculated them for a different latitude.

Geographical Table (p. 289)

Table of the longitudes and latitudes of the cities (p. 289).

Geographical coordinates of 67 cities of the Maghrib, Al-Andalus, Italy (Rome, Sardinia, Sicily), Egypt, Syria, Palestine, Lebanon. Among Maghribī cities we find Qusanṭīniya which is not Constantine

[46] In two papers published in the *Archives Internationales d'Histoire des Sciences* 62, 2012, 27–41 and 43–54.

but Constantinople (long. 49;30°, lat. 45;15°). Not a single city from Iraq or the Arabian Peninsula appears: the omission of Mecca is particularly significant but, at the end of the table, there are seven cities which I have not been able to identify at first sight, one of them being Yājūj (long. 172;45°, lat. 45;15°). The table seems incomplete in this manuscript: there is space left for seventeen more place names. Most of the cities appearing here can be found, with almost identical coordinates, in the corresponding tables of the *Shāmil* (97 cities) and the *Qawīm* (94 cities).[47]

The longitudes of the Western cities use the water meridian (27° west of Cordova), documented in al-Andalus since the 10th century, which makes it possible to make an important correction of the size of the Mediterranean [Comes, 1992, 1992b, 1993, 1997 and 2000]. Thus, if the longitude of Cordova is 27° and the longitude of Damascus is 66° (according to Ibn al-Raqqām's table), the difference of longitudes between the two cities amounts to 39°, fairly near the modern value (41;5°) when one compares it with the difference in Ptolemy's *Geography* (59;40°) or with the difference in al-Khwārizmī (50;40°). The latitude of Granada is 37;30° different from the very precise value 37;10° used in the *Qawīm*.

References

ʿAbd al-Raḥmān, M., 1996a. Ḥisāb aṭwāl al-kawākib fī l-Zīj al-Shāmil fī tahdhīb al-Kāmil li-Ibn al-Raqqām. Unpublished Ph.D. dissertation presented at the University of Barcelona.

—— 1996b. Wujūd jadāwil fī zīj Ibn al-Hāʾim. In: Casulleras, J., Samsó, J. (eds), From Baghdad to Barcelona: Studies in the Islamic Exact Sciences in Honour of Prof. Juan Vernet. Barcelona, pp. 365–381.

Bellver, J., Samsó, J., 2012. Ibn al-Raqqām on lunar parallax. Suhayl 11, 189–229.

Bīrūnī, 1954. al-Qānūn al-Masʿūdī, 3 vols. Hyderabad.

Calvo, E., 1993. Abū ʿAlī al-Ḥusayn ibn Bāṣo: Risālat al-ṣafīḥa al-ŷāmiʿa li-ŷamīʿ al-ʿurūḍ: Critical edition, Spanish translation and commentary. Madrid.

—— 1998a. La résolution graphique des questions astrologiques à al-Andalus. In: Histoire des Mathématiques Arabes : Actes du 3ème Colloque Maghrébin sur l'Histoire des Mathématiques Arabes. Alger. pp. 31–44.

—— 1998b. Astronomical theories related to the sun in Ibn al-Hāʾim's al-Zīŷ al-Kāmil fī l-Taʿālīm. Zeitschrift für Geschichte der arabisch-islamischen Wissenschaften 12, 51–111.

Carandell, J., 1984. An analemma for the determination of the azimuth of the qibla in the Risāla fī ʿilm al-ẓilāl of Ibn al-Raqqām. Zeitschrift für Geschichte der arabisch-islamischen Wissenschaften 1, 61–72.

Casulleras, J., 2004. Ibn Muʿādh on the astrological rays. Suhayl 4, 385–402.

—— 2007a. El cálculo de aspectos o la proyección de rayos en la astrología medieval árabe. Archives Internationales d'Histoire des Sciences 57, 25–46.

—— 2007b. Ibn ʿAzzūz al-Qusanṭīnī's tables for computing planetary aspects. Suhayl 7, 47–114.

—— 2007c. Ibn al-Raqqām. In: Hockey, T. (ed.), Biographical Encyclopedia of Astronomers, vol. I. Springer, New York, pp. 563–564.

—— 2009. Métodos para determinar las casas del horóscopo en la astrología medieval árabe. Al-Qanṭara 30, 41–67.

—— 2008–2009. Mathematical astrology in the Medieval Islamic West. Zeitschrift für Geschichte der arabisch-islamischen Wissenschaften 18, 241–268.

—— 2010. La astrología de los matemáticos: La matemática aplicada a la astrología en la obra de Ibn Muʿāḏ de Jaén. Barcelona.

Casulleras, J., Samsó, J., 1996. From Baghdad to Barcelona: Studies in the Islamic Exact Sciences in Honour of Prof. Juan Vernet. Barcelona.

[47] The geographical table of the *Shāmil* has been edited, comparing the values of the coordinates with those in the *Qawīm*, by ʿAbd al-Raḥmān [1996, 327–331].

Chabás, J., Goldstein, B.R., Andalusian Astronomy: al-Zîj al-Muqtabis of Ibn al-Kammād. Archive for History of Exact Sciences 48, 1–41.

Comes, M., 1992a. À propos de l'influence d'al-Zarqālluh en Afrique du Nord : l'apogée solaire et l'obliquité de l'écliptique dans le Zīdj d'Ibn Isḥāq. In: Actas del II Coloquio Hispano-Marroquí de Ciencias Históricas : Historia, Ciencia y Sociedad. Madrid, pp. 147–159.

——— 1992b. The Meridian of water in the tables of Geographical Coordinates of al-Andalus and North-Africa. Journal for the History of the Arabic Science 10, 41–51. Reprinted: Fierro, M., Samsó, J., (eds.), The Formation of al-Andalus, Part 2: Language, Religion, Culture and the Sciences. Ashgate Variorum, Aldershot, pp. 381–391.

——— 1993. Las tablas de coordenadas geográficas y el tamaño del Mediterraneo según los astrónomos andalusíes. Al-Andalus: El Legado Científico. Proyecto Sur de Ediciones, Barcelona, pp. 22–37.

——— 1997. al-Taḥdīd al-daqīq li-ṭūl al-baḥr al-abyaḍ al-mutawassiṭ allatī waṣala ilay- hi al-falakiyyūn al-ʿarab fī al-Andalus. Journal for the History of the Arabic Science 11, 19–26.

——— 2000. Islamic Geographical Coordinates: al-Andalus' contribution to the correct measurement of the size of the Mediterranean. Science in Islamic Civilization: Studies and Sources on the History of Science. Istanbul, pp. 123–138.

——— 2001. Ibn al-Hāʾim's trepidation model. Suhayl 2, 291–408.

——— 2002. Some new Maghribī sources dealing with trepidation. Science and Technology in the Medieval Islamic World. Turnhout, 121–141.

van Dalen, B., 2000. Taʾrīkh. Encyclopaedia of Islam, vol. X. Leiden, pp. 264–271.

——— 2004. A second manuscript of the Mumtaḥan Zīj. Suhayl 4, 9–44.

Dallal, A., 1995. Ibn al-Haytham's universal solution for finding the direction of the qibla by calculation. Arabic Sciences and Philosophy 5, 145–193.

Davidian, M.-L., 1960. Al-Bīrūnī on the time of day from shadow lengths. Journal of the American Oriental Society 80, 330–335. Reprint in: Kennedy, E.S., et al., 1983. pp. 274–279.

Debarnot, M.T., 1985. al-Bīrūnī, Kitāb maqālīd ʿilm al-hayʾa: La Trigonométrie sphérique chez les Arabes de l'Est à la fin du Xe siècle. Damas.

Díaz-Fajardo, M., 2005. Al-Zīŷ al-Mustawfà de Ibn al-Raqqām y los apogeos planetarios en la tradición andaluso-magrebí. Al-Qanṭara 26, 19–30.

Dorce, C., 2002–2003. El Tāŷ al-azyāŷ de Muḥyī al-Dīn al-Magribī. Barcelona.

Goldstein, B.R., 1964. On the theory of trepidation according to Thābit b. Qurra and al-Zarqālluh and its implications for homocentric planetary theory. Centaurus 10, 232–247.

——— 1967. Ibn al-Muthannā's Commentary on the Astronomical Tables of al-Khwārizmī. New Haven.

——— 1976. Refraction, twilight and the height of the atmosphere. Vistas in Astronomy 20, 105–107.

——— 1977. Ibn Muʿādh's Treatise on Twilight and the Height of the Atmosphere. Archive for the History of the Exact Sciences 17, 97–118.

Hogendijk, J.P., 1989. The mathematical structure of two Islamic astrological tables for casting the rays. Centaurus 32, 171–202.

——— 1998. Progressions, rays and houses in medieval Islamic astrology: A mathematical classification. Unpublished paper presented at the Dibner Institute Conference: New Perspectives on Science in Medieval Islam. Cambridge, MA. A revised version of this paper has been published with the same title: Casulleras, J., Hogendijk, J.P., 2012, Suhayl 11, 33–102.

——— 2005. Applied mathematics in eleventh century Spain: Ibn Muʿādh al-Jayyānī and his computation of astrological houses and aspects. Centaurus 47, 87–114.

Ibn al-Khaṭīb, 1973–1978. Al-Iḥāṭa fī akhbār Garnāṭa. Ed. ʿAbd Allāh ʿInān, M., 4 vols. Cairo.

Kennedy, E.S., 1956. Parallax theory in Islamic Astronomy. Isis 47, 33–53. Reprinted: Kennedy. E.S., et al., 1983, pp. 164–184.

——— 1973. A Commentary upon Bīrūnī's Kitāb Taḥdīd al-Amākin, Beirut.

——— 1976. The Exhaustive Treatise on Shadows by Abū al-Rayḥān Muḥammad b. Aḥmad al-Bīrūnī, 2 vols. Aleppo.

——— 1983. Colleagues and Former Students: Studies in the Islamic Exact Sciences. Beirut.

——— 1985. Spherical astronomy in Kāshī's Khāqānī Zīj. Zeitschrift für Geschichte der arabisch-islamischen Wissenschaften 2, pp. 1–46.
——— 1996. The astrological houses as defined by medieval Islamic astronomers. In: Casulleras, J., Samsó, J., (eds.), From Baghdad to Barcelona: Studies in the Islamic Exact Sciences in Honour of Prof. Juan Vernet. Barcelona, pp. 535–578.
——— 1997. The astronomical tables of Ibn al-Raqqām, a scientist of Granada. Zeitschrift für Geschichte der arabisch-islamischen Wissenschaften 11, 35–72.
Kennedy, E.S., Krikorian-Preisler, H., 1972. The astrological doctrine of projecting the rays. Al-Abhath 25, 3–15. Reprinted in: Kennedy, E.S., et al., 1983. 372–384.
King, D.A., 1987. Some early Islamic tables for determining lunar crescent visibility. In: King, D.A., Saliba, G., (eds.), From Deferent to Equant A Volume of Studies in the History of Science in the Ancient and Medieval Near East in Honor of E.S. Kennedy. The New York Academy of Sciences, New York.
——— 2004. In Synchrony with the Heavens: Studies in Astronomical Timekeeping and Instrumentation in Medieval Islamic Civilization (Studies I-IX), vol. I, The Call of the Muezzin. Brill, Leiden-Boston.
Mestres, A., 1996. Maghribī astronomy in the 13th century: A description of manuscript Hyderabad Andra Pradesh State Library 298. In: Casulleras, J., Samsó, J. (eds), From Baghdad to Barcelona: Studies in the Islamic Exact Sciences in Honour of Prof. Juan Vernet. Barcelona, pp. 383–443.
——— 1999. Materials andalusins en el Zīj de Ibn Isḥāq al-Tūnisī. Unpublished Ph.D. dissertation presented at the University of Barcelona.
Mielgo, H., 1996. A method of analysis for mean motion astronomical tables. n: Casulleras, J., Samsó, J., (eds.), From Baghdad to Barcelona: Studies in the Islamic Exact Sciences in Honour of Prof. Juan Vernet. Barcelona, pp. 159–179.
Millás Vallicrosa, J.M., 1947. El libro de los fundamentos de las tablas astronómicas de R. Abraham ibn ʿEzra. Barcelona.
——— 1943–1950. Estudios sobre Azarquiel. Madrid-Granada.
Millás Vendrell, E., 1963. El comentario de Ibn al-Muṯannà a las Tablas astronómicas de al-Juwārizmī. Estudio y edición crítica del texto latino en la versión de Hugo Sanctallensis. Madrid-Barcelona.
Nadir, N., 1960. Abū l-Wafāʾ on the solar altitude. The Mathematics Teacher 53, 460–463. Reprinted: Kennedy, E.S., et al., 1983. pp. 280–283.
Nallino, C.A., 1899–1907. Al-Battānī sive Albatenii: Opus astronomicum, 3 vols. Mediolani Insubrum.
Neugebauer, O., 1962. The Astronomical Tables of al-Khwārizmī. Translation with Commentaries of the Latin Version Edited by H. Suter, Supplemented by Corpus Christi College MS 283. Copenhagen.
——— 1975. A History of Ancient Mathematical Astronomy. Berlin.
Neugebauer, O., Schmidt, O., 1952. Hindu astronomy at Newminster in 1428. Annals of Science 8, 221–228. Reprinted: Neugebauer, O., 1983. Astronomy and History: Selected Essays. Springer Verlag, New York, pp. 425–432.
North, J.D., 1986. Horoscopes and History. London.
Pedersen, O., 1974. A Survey of the Almagest. Odense.
Pedersen, F.S., 2002. The Toledan Tables: A Review of the Manuscripts and the Textual Versions with an Edition. Copenhagen, 2002.
Puig, R., 2000. The theory of the moon in the al-Zīj al-Kāmil fī-l-Taʿālīm of Ibn al-Hāʾim (ca. 1205). Suhayl 1, 71–99.
Robbins, F.E., 1949. Ptolemy, Tetrabiblos. Loeb Classical Library, Cambridge, MA. Reprinted: 1980.
Samsó, J., 1987. Sobre el modelo de Azarquiel para determinar la oblicuidad de la eclíptica. In: Homenaje al Prof. Darío Cabanelas Rodríguez O.F.M. con motivo de su LXX aniversario, vol. II, Granada, pp. 367–377. Reprinted: Samsó, J., 1994, no. IX.
——— 1994. Islamic Astronomy and Medieval Spain. Variorum, Aldershot.

——— 1997. Andalusian astronomy in 14th century Fez: al-Zīj al-Muwāfiq of Ibn ʿAzzūz al-Qusanṭīnī. Zeitschrift für Geschichte der arabisch-islamischen Wissenschaften 11, 73–110. Reprinted: Samsó, J., 2007. no. IX.

——— 1998. An outline of the history of Maghribī zijes from the end of the thirteenth century. Journal for the History of Astronomy 29, 93–102. Reprinted: Samsó, J., 2007. no. XI.

——— 2001a. Astronomical observations in the Maghrib in the fourteenth and fifteenth centuries. Science in Context 14, 165–178. Reprinted: Samsó, J., 2007. no. XII.

——— 2001b. La medición del tiempo en al-Andalus en torno al año 1000. In: Ribot, L., Valdeón, J., Villares, R. (eds.), 2001. Año 1000, Año 2000: Dos milenios en la Historia de España. Madrid, pp. 71–92.

——— 2002–2003. In pursuit of Zacut's Almanach Perpetuum in the Eastern Islamic world. Zeitschrift für Geschichte der arabisch-islamischen Wissenschaften 15, 67–93. Reprinted: Samsó, J., 2007. no XVI.

——— 2006. Ibn al-Raqqām, Abū ʿAbd Allāh. In: Lirola, J. (ed.), 2006. Fundación Ibn Ṭufayl de Estudios Arabes, vol IV., Enciclopedia de la Cultura Andalusí: Biblioteca de al-Andalus: de Ibn al-Labbāna a Ibn al-Ruyūlī. Almería, pp. 440–444.

——— 2007. Astronomy and Astrology in al-Andalus and the Maghrib. Ashgate-Variorum, Aldershot.

——— 2008. The computation of the degree of mediation of a star or planet in the Andalusian and Maghribi tradition. In: Dauben, J.W., Kirschner, S., Kühne, A., Kunitzsch, P., Lorch, R.P., (eds.), Mathematics Celestial and Terrestrial: Festschrift für Menso Folkerts zum 65. Geburtstag. Deutsche Akademie der Naturforscher Leopoldina, Halle an der Saale, pp. 395–404.

——— 2009. La Urŷūza de Ibn Abī l-Riŷāl y su comentario por Ibn Qunfuḏ: Astrología e Historia en el Magrib en los siglos XI y XIV. Al-Qanṭara 30, 7–39, 321–360.

——— 2011. Las Ciencias de los Antiguos en al-Andalus, 2nd ed. Almería, 2011.

Samsó, J., Mielgo, H., 1994. Ibn Isḥāq al-Tūnisī and Ibn Muʿādh al-Jayyānī on the qibla. Reprinted: Samsó, J., 1994. no. VI.

Samsó, J., Millás, E., 1994. Ibn al-Bannāʾ, Ibn Isḥāq and Ibn al-Zarqālluh's solar theory. Reprinted: Samsó, J., 1994. no. X.

——— 1998. The computation of planetary longitudes in the Zīj of Ibn al-Bannā. Arabic Sciences and Philosophy 8, 259–286. Reprinted: Samsó, J., 2007. no. VIII.

Smith, A.M., Goldstein, B.R., 1993. The medieval Hebrew and Italian versions of Ibn Muʿādh's On Twilight and the Rising of the Clouds. Nuncius 8, 611–643.

Suter, H., 1914. Die astronomischen Tafeln des Muḥammed ibn Mūsā al-Khwārizmī in der Bearbeitung des Maslama ibn Aḥmed al-Madjrīṭī und der Latein: Uebersetzung des Athelhard von Bath. Copenhagen.

Toomer, G.J., 1969. The solar theory of al-Zarqāl: A history of errors. Centaurus 14, 306–336.

——— 1984. Ptolemy's Almagest. New York.

Vernet, J., 1952. Contribución al estudio de la labor astronómica de Ibn al-Bannāʾ. Tetuán.

——— 1980. La supervivencia de la astronomía de Ibn al-Bannā. Al-Qanṭara 1, 447–451.

An Ottoman Astrolabe Full of Surprises

David A. King

Abstract Recently a set of plates from an 11th-century Andalusi astrolabe has been discovered inside an Ottoman Turkish astrolabe from ca. 1700. Most of these plates conform to the tradition represented by some fourteen surviving Andalusi astrolabes from the 11th century. One of the plates, however, serves the latitude 16;30° south of the Equator, and this feature, not attested on any known astrolabe (Byzantine, Islamic or European), raises a number of intriguing questions.

Introduction

In 1987, Len Berggren's contribution to the *Festschrift* for Asger Aaboe was a paper entitled "Archimedes among the Ottomans" [Berggren 1987]. In this, he described Archimedean materials that he had discovered in a text by a 17th-century Ottoman scholar named al-Yanyawī (from the town that is now Ioannina or Jannena in Northern Greece). More recently, Ekmeleddin İhsanoğlu and his colleagues in Istanbul have published multiple volumes of bio-bibliographical materials on hundreds of scientists in the Ottoman world.[1]

There can be no doubt that Ottoman manuscript sources, not least when they contain earlier material that would otherwise be lost, have much to contribute to the history of Islamic science. In this paper I present an Ottoman 'document' of a different kind that merits our attention.

The astrolabe shown in Figure 1 was auctioned at Sotheby's in London in 1991 and passed into a private collection in Europe [Sotheby's 30.05.1991, 136 (lot no. 391)].[2] It is now in the Museum of Islamic Art in Doha. It is unsigned and undated, and it has been assigned the number #4040 in the International Instrument Checklist.[3] The front and the back are clearly Ottoman, doubtless made in Istanbul and probably from around 1700.[4] The rete was made by the same person, but shows Andalusi influence (see below).

Ottoman astrolabes are characterized by their relative simplicity.[5] A minimum of decoration and astronomical and trigonometric scales and diagrams, first encountered in earlier astrolabe traditions — Abbasid Iraqi and Iranian; early Andalusi;[6] later Iranian; later Andalusi and Maghribi; Mamluk Syrian

[1] See İhsanoğlu *et al.* [1997–2000], for the volumes relating to astronomy, mathematics and geography.

[2] The components are listed in King [2005, 1006 (1.3.11a), 1013 (2.3.10a)]. See also King [2005, 944–945, 956 (plates)].

[3] On this convention see King [2005, 360].

[4] Some of the productions of the Istanbul astrolabists Ibrāhīm ibn Muḥammad al-Balawī and Aḥmad Ayyūbī around 1700 include new components for older astrolabes [King 2005, 1013, nos. 2.3.5–6]. I have not been able to show that either of these men made the later additions in the astrolabe under study.

[5] See King [2005, 1013–1014], for a list of some of these, and King [2005, 774–796], for detailed descriptions of the earliest two examples.

[6] I use the convention of Prof. Julio Samsó of Barcelona and his colleagues that al-Andalus refers to that part of the

N. Sidoli and G. Van Brummelen (eds.), *From Alexandria, Through Baghdad: Surveys and Studies in the Ancient Greek and Medieval Islamic Mathematical Sciences in Honor of J.L. Berggren*, DOI 10.1007/978-3-642-36736-6_17,

© Springer-Verlag Berlin Heidelberg 2014

Figure 1: The front and back of the unsigned, undated Ottoman astrolabe. Photo courtesy of the former owner.

and Egyptian — are attested on Ottoman pieces. The inscriptions are in a mixture of Arabic and Arabic with Persian influence, and they are in *naskhī* script.

For me, the first surprise came in the early 1990s when I opened the Ottoman astrolabe to look at the plates; these were a set of seven plates from an 11th-century Andalusi astrolabe. Some 14 astrolabes survive from that milieu, though none earlier and, curiously, not one from the 12th century, although we do have some from the 13th century.[7] Two plates are shown in Figs. 3–4.

What has happened here is that our Ottoman astrolabist came into possession of this set of Andalusi plates and found them so imposing that he constructed a new mater to house them and a new rete to use with them. He even extended the range of latitudes represented with three new sets of markings. Perhaps, when he acquired the Andalusi plates, there was an original mater to house them, but this has not survived. Neither has the original rete, on which the star-positions would have been some 600 years out of date by his time.

One of the plates, in particular, attracted my attention, for it is unlike any other known plate, and it is this that occasioned the present study. The Ottoman mater and rete are of interest in their own right and so we look at the entire instrument in some detail.

Iberian peninsula under Muslim domination at a given time. Thus, the term does not relate to the medieval equivalent of the modern province of Andalucía.

[7] For a list, see King [2005, 1006]. Detailed descriptions of all of them have been prepared, and I hope to publish these eventually.

Some Preliminaries

Astrolabe plates from early Iraq and Iran (roughly, 800–1100) usually serve a series of latitudes appropriate for the Eastern realms: 33° (Baghdad), 36° (Rayy), and sometimes 21° (Mecca) and 24° (Medina) [King 2005, 439–544, 948–950]. Usually the length of longest daylight will be stated for those latitudes, invariably using the Ptolemaic value for the obliquity of the ecliptic, 23;51°. This tradition derives from the earliest Greek and earliest Islamic astrolabes (8th century) that had plates for each of the seven climates of Antiquity which are defined in terms of the length of longest daylight at those latitudes (see Figure 2) [King 2005, 421, 428–429, 948]. The idea was to produce instruments that were universal, serving all regions of the earth.

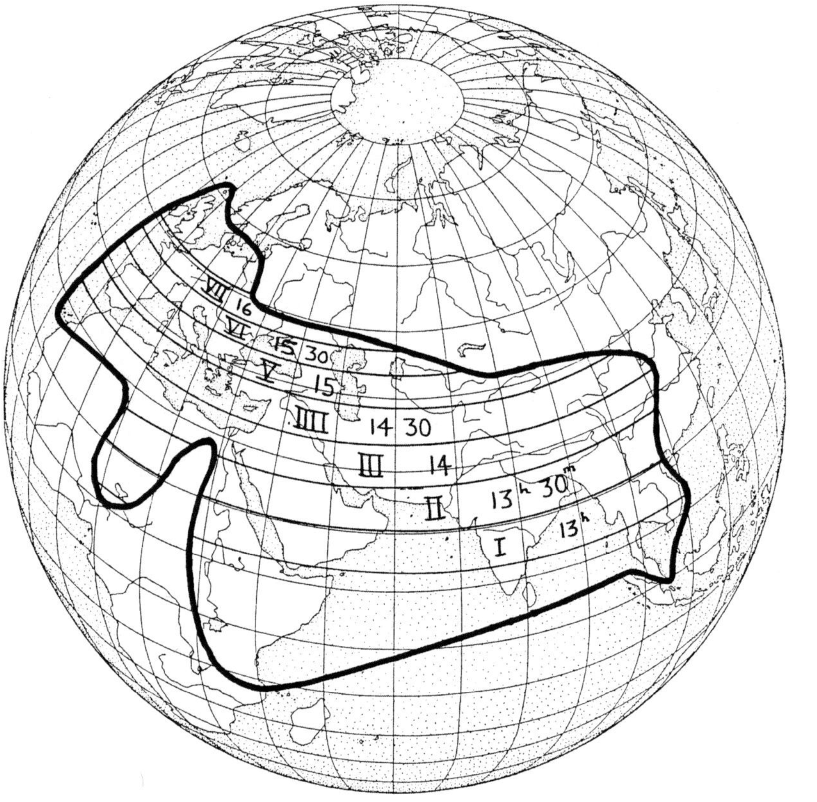

Figure 2: The seven climates of Antiquity with the world known to Ptolemy [King 2004, 689].

The earliest known Andalusi astrolabe, from the 10th century and surviving only in an illustration, has plates for the climates [King 2005, 383 928]. All 11th-century Andalusi astrolabes have plates for a wider range of latitudes than their Abbasid predecessors, usually with the associated length of half daylight and usually also with a series of names of localities mainly but not always in the Islamic West [King 2005, 951–957].

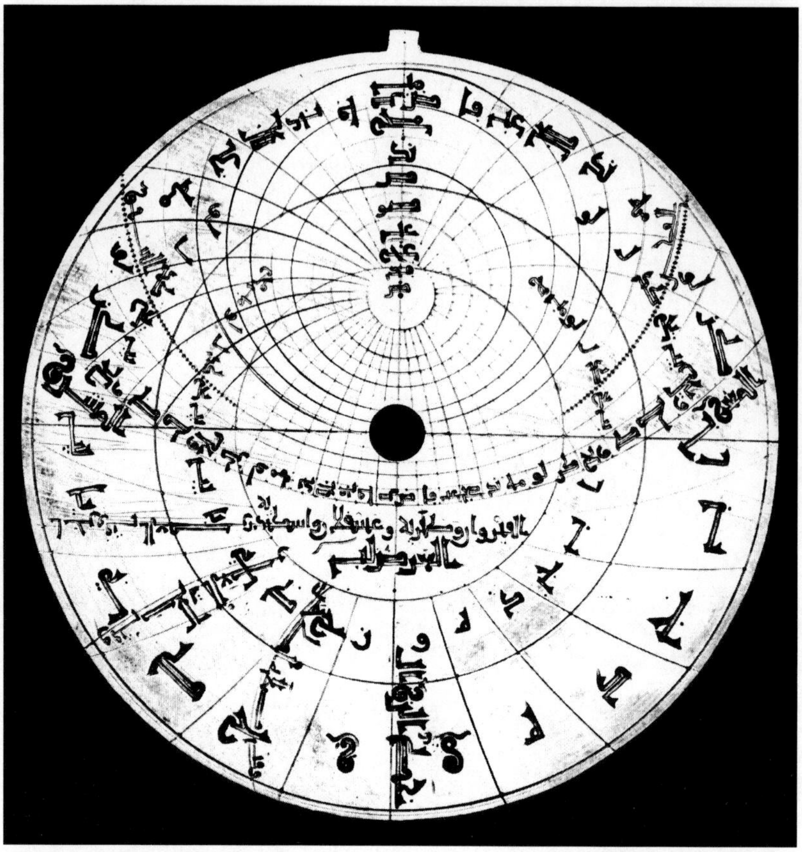

Figure 3: The plate for 32°. Photo courtesy of the former owner.

The First Surprise: The Andalusi Plates

One of the plates is shown in Figure 3. The inscriptions are in an elegant but heavy Andalusi *kūfī* script. A peculiarity, which could serve to identify the maker, is the 6 and 7 in the two sets of numbers for the seasonal hours: they are sometimes written as one or the other of each of "S" and a backwards "S." Alas, I have not been able to identify the maker by comparison with the engraving on other astrolabes from the same milieu (or with the localities chosen for the latitudes or with the unusual division of the azimuth curves — see below). The astronomical markings are all competently executed. Given what we know about 11th-century Andalusi astrolabes, it is not surprising that the main set of plates that we find inside the Ottoman astrolabe has markings for a wide range of latitudes and localities (this time without daylight lengths):

23°	Mecca, Jedda, Taif, Yamama, Ṣiraf, al-Manṣūra in China (!)
25°	Yathrib (= Medina), Hajar, Bahrein
30°	Miṣr (= Cairo-Fustat), Kirman, Siniz (near Ahwaz), ʿAyn Shams (= Heliopolis)
32°	Kairouan, Tiberias, Ascalon, Alexandria
33°	Baghdad, Hit, Damascus, Tunis (!), Salé
35°	Ceuta, Tangiers, Sicily, Mosul, Manbij, Qum
36°	Almería, Harran, Samarqand, Raʾs al-ʿAyn
37;30°	Seville, Málaga, Granada, Bukhara, Rayy
38;30°	Cordova, Murcia (written *m-r-s-y-l-h* !), Marwarrūdh, Balkh, Jurjan
45°	Constantinople, Burjān (??)

There are altitude circles for each 6° and azimuth circles for each 9°, as well as curves for the times of the prayers: the *ẓuhr* shortly after midday and the beginning and end of the *ʿaṣr* in the afternoon (*waqt ṣalāt al-ẓuhr*, *waqt ṣalāt al-ʿaṣr*, and *waqt ākhir al-ʿaṣr* or *ākhir waqt al-ʿaṣr*), as well as special markings for twilight at 18° above the horizon (*al-fajr*, daybreak, on the left, and *al-shafaq*, nightfall, on the right).[8] These are standard on Andalusi and Maghribi astrolabes.

Unusual is the marking of the azimuth circles *for each 9°*; this is attested only on two other 11th-century Andalusi astrolabes (#118, made by Ibrāhīm ibn Saʿīd al-Sahlī in Toledo in 460 H [= 1067/68], preserved in Oxford [Gunther 1932, vol. 1, 253–256]; and #1099, made by Aḥmad ibn Muḥammad al-Naqqāsh in Saragossa in 472 H [= 1079/80], preserved in Nuremberg [King 1992, vol. 2, 568–570]; see also below). This raises the question whether the altitude and azimuth circles were engraved using geometrical construction or using tables of polar coordinates, such as were compiled by al-Farghānī in Baghdad in the mid 9th century (although no manuscripts of his work on astrolabe construction are known from the Islamic West). With such tables, polar coordinates of the centres of the various circles with respect to the centre of the astrolabe are given for each degree of altitude and each degree of azimuth for each degree of latitude.[9] In any case, as we shall see, our Ottoman astrolabist was clearly quite taken by the idea of drawing the azimuth circles for each 9°.

Here there are more Eastern Islamic localities than is usual on early Andalusi plates, even though there are some astrolabes that give even more information.[10] Amongst the other pieces that have similar lists of localities are #118 and #1099, mentioned above. Frankly, such information is an *excès de délicatesse*; most of the places in the Islamic East would be unknown to the average Andalusi.

It is worth comment that medieval latitudes do not always reflect actual values. Here we find 23° used for Mecca, which is some 1½° too high. Alas, we do not find this value on any other contemporaneous astrolabe. (Muslim astronomers in the 9th century measured the latitude of Mecca as 21°, 21;30° and 21;40°.)[11]

Another good example is the 45° used for Constantinople, also not found in other contemporaneous astrolabes. This value, some 4° too high, is even attested in Byzantine sources, and results from putting the Rome of the East at the middle of the 6th climate rather than the middle of the 5th, where it belongs.[12] What is curious is that our Ottoman astrolabist did not see fit to delete the name Constantinople or to make a new set of markings for latitude 41°.

Notice also the error in the latitude of Tunis; this suggests that the maker was far from Tunis (that is, in al-Andalus and not in al-Maghrib). However, the same error is made on other 11th-century Andalusi astrolabes.

I suspect that at least one plate of markings for, say, latitudes 41° and 42°, is missing from the set. But it baffles me why our Ottoman astronomer would have dumped a plate that would have served Istanbul (and also why he did not prepare such a plate himself; see below).

Additional markings for the astrological houses and the casting of the rays (*taswiyat al-buyūt wa-maṭraḥ al-shuʿāʿ*)[13] for latitude 35° (see Figure 4) suggest at first sight that this latitude was favored by the maker, indeed either that this was the latitude of the locality where he made the astrolabe or of the destination for which the instrument was intended. Only the towns of Ceuta and Tangiers could come into consideration here, in spite of the fact that an Andalusi rather than Maghribi provenance for the plates is almost certain. We note that under the Ḥammūdids in the first part of the 11th century

8 On such markings, see King [2005, 46–50]. They are to be used with the point of the ecliptic opposite to the position of the sun.

9 See Lorch [2005, 111–293], for the tables, and King [2005, 39–41], for the context.

10 On the coordinates of localities in the written sources see Kennedy and Kennedy [1987]. For latitudes displayed on early Eastern and Western Islamic astrolabes (to ca. 1100), as well as the earliest European astrolabes, see King [2005, 915–962].

11 King [2012, 225–228 (paper IX)], on early measurements of the latitude of Mecca.

12 See King [1991] on this pathetic state of affairs.

13 For references to the rich literature on this subject see Casulleras [2004]. The article "Tasyīr" by O. Schirmer in EI_2 is still useful.

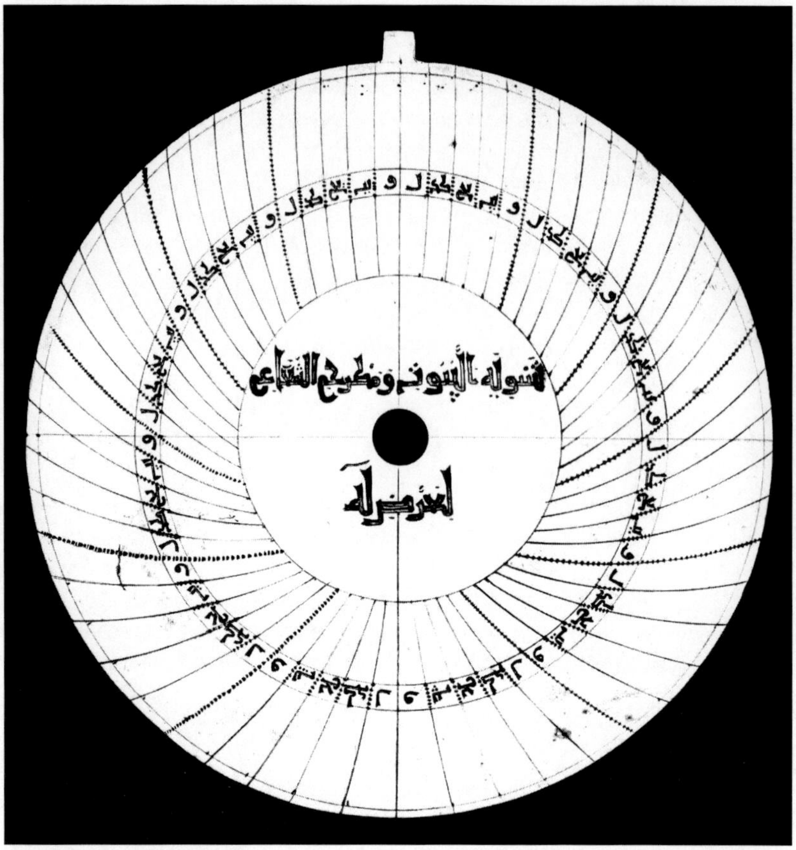

Figure 4: The astrological plate for latitude 35°. Photo courtesy of the former owner.

Ceuta, Tangiers, were united with Algeciras and Málaga in a single principality.[14]

Perhaps there was originally an additional plate with markings on both sides for the astrological houses for other latitudes, and our Ottoman astrolabist thought he could dispense with this. Such markings for latitudes further north in al-Andalus do occur on various three other 11th-century Andalusi astrolabes. The earliest attestation of such markings is on the spectacular astrolabe of al-Khujandī in Baghdad dated 984/5 [King [2005, 368, 508 (fig. 9f), 514].

There is also a set of markings for the horizons of 17 latitudes (Figure 5):

18° - 30° - 37° - 45° - 60° - 73° / 22° - 33° - 40° - 63° /
25° - 32° - 36° / 28° - 35° - 38° - 66°

This selection ensures that every few degrees are represented. The half horizon for latitude 66° is an arc of the circle of the ecliptic, but before we jump to any conclusions, we should look at the four radial declination scales, divided for each 6° and subdivided for each 1°. Noteworthy is the obliquity of the ecliptic engraved at the ends of each of these scales: it is 23;33°, a value associated with the observations conducted in Baghdad for the Caliph al-Ma'mūn in the early 9th century. This kind of markings for the horizons is associated with the 9th-century Baghdad astronomer Ḥabash al-Ḥāsib and is not attested on any of the other known 11th-century Andalusi astrolabes. Again, the first attestation of such markings is on the astrolabe of al-Khujandī [King 2005, 367, 509 (fig. 9g), 514].

The following is the arrangement of the markings on the fronts and backs (a/b) of the seven original plates:

1a: surprise coming up – 1b: houses for 35°

[14] See the articles "Sabta," "Ṭandja," and "Ḥammūdids" in EI_2.

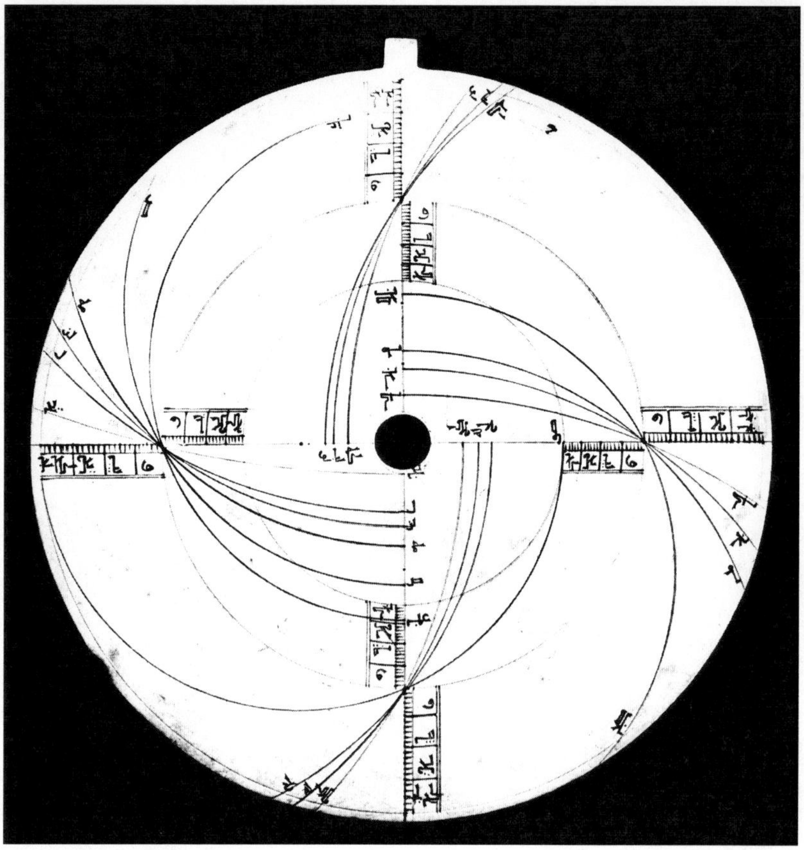

Figure 5: The plate of horizons. Photo courtesy of the former owner.

2a: 23° – 2b: 25° 3a: 30° – 3b: 32°
4a: 33° – 4b: 36° 5a: 35° – 5b: horizons
6a: 37;30° – 6b: 38;30° 7a: 45° – 7b: originally blank

The Biggest Surprise of All

There is a plate for 16;30° *South* (Figure 6).[15] This is labelled *khalf khaṭṭ al-istiwāʾ fī l-janūb*, literally "behind the equator in the South." The horizon is now concave downwards, whereas for northern latitudes it is concave upwards. The curve for altitude 18° is almost a straight line. The additional markings are the same as on the other plates, even as far as the prayers are concerned! It should nevertheless be understood that the presence of this plate is purely symbolic.

No other Islamic astrolabic plates for southern latitudes have survived from before the Moghul period in India (16th–17th century); from that milieu we even have two double astrolabes, each with two retes for the northern and southern skies and each plate with a northern astrolabic projection on one side and a southern one on the other.[16] The existence of this particular plate shows first of all that

[15] The importance of these markings is already signaled in King [2005, 944–945].

[16] Turner [1985, 74–83] and Pingree [2009, 82–87] feature the one formerly in the Time Museum, Rockford IL, and now in the Adler Museum in Chicago, in considerable detail. A second one by the same maker surfaced in 2011 in a private collection.

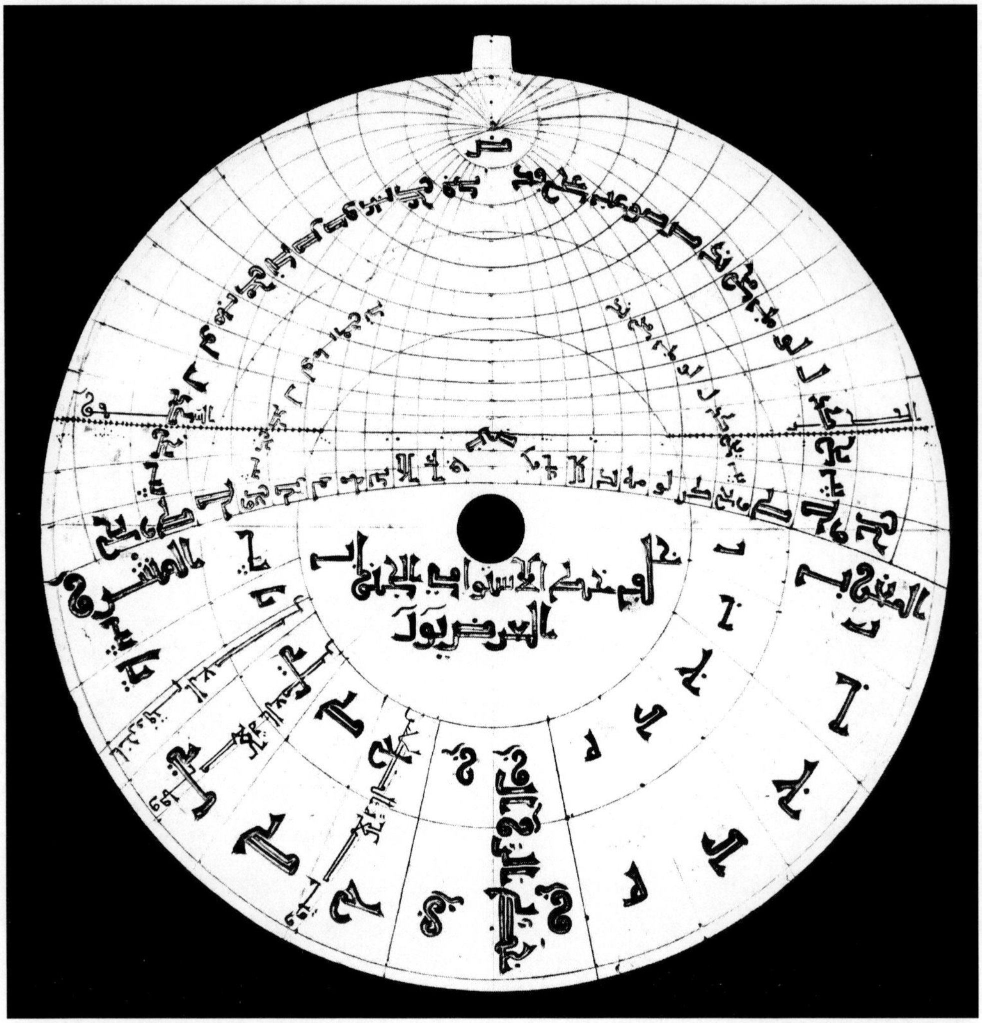

Figure 6: The plate for latitude 16;30° South. Photo courtesy of the former owner.

an Andalusi astrolabist in the 11th century would not shy away from constructing a plate for a southern latitude.

Already in the 9th and 10th centuries Muslim astronomers had developed mixed retes with parts of the ecliptic based on northern projections and others based on southern ones. The plates on these instruments were correspondingly complicated; alas, no instruments of this kind survive.[17]

The latitude 16;30° South is that of Anti-Meroë, corresponding to the middle of the first climate in the southern hemisphere with a length of longest day of 13 hours. It is the lower limit of the *oekumene* in the three cartographic grids associated with Ptolemy of Alexandria [Neugebauer 1969, 220–224; Neugebauer 1985, 934–940; Berggren and Jones 2000, 35–41]. Was our Andalusi astrolabist influenced by a tradition imported from the Islamic East, essentially Baghdad, or did he think of this himself or take the idea from some other Andalusi?

Now the upper limit of Ptolemy's world, the limit of the ocean surrounding the earth, was sometimes taken as $72° \pm 1°$. Notice this latitude (73°) is represented on the Andalusi plate of horizons. It is doubtless significant that two 11th-century Andalusi astrolabes — #116, made by Muḥammad ibn al-

[17] On mixed astrolabes see King [2005, 55–57, 376, 557–563].

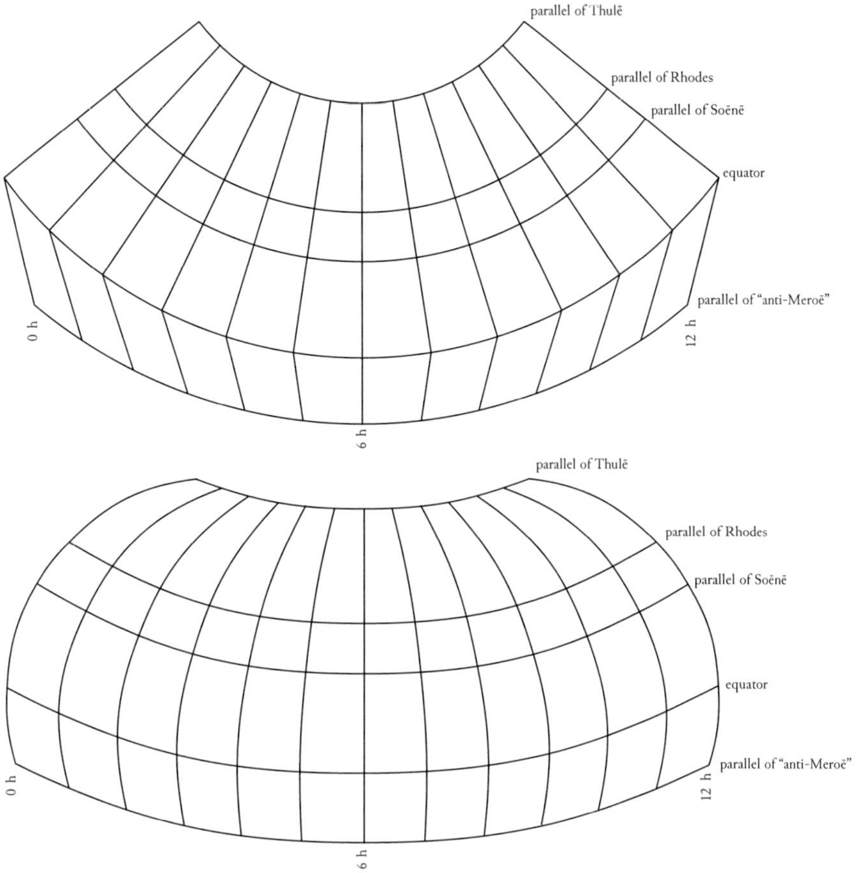

Figure 7: The extent of Ptolemy's first and second cartographical representations from 16½° South to 72° North. Graphics due to A. Jones.

Sahlī in Toledo in 420 H [= 1029/30], now in Berlin,[18] and #121, made by Ibrāhīm al-Sahlī in Valencia in 478 [= 1086], preserved in Kassel,[19] have a plate for latitude 72° [King 2005, 943–944]. These plates for 72° only make sense when one realizes that both of these astrolabes, as well as serving numerous latitudes, are also fitted with markings for latitude 0°; in other words, in a sense, they serve the whole world.

The plate for 16;30° South certainly makes no sense in relative isolation. I hypothesize that there must have been a plate for 72° as well, in which case the maker could have claimed that he had made an astrolabe for the whole of the known world. This would then be unique in the history of instrumentation. We know already of astrolabes with sets of plates for each of the seven climates or for a series of latitudes from 0° to 90°; here then we would be witness to an interesting twist at being universalistic.

One may wonder whether the European tradition of representing southern astrolabic projections on the face of clocks was inspired by Western Islamic practice. In this regard, we note that a geared astrolabe from northern France datable to ca. 1300 is based on a southern stereographic projection [Gunther 1932, vol. 1, 347, pls. LXXX and LXXXI (no. 198)].[20]

[18] See Woepcke [1858], for a detailed description of this piece, summarized by Gunther [1932, vol. 1, 251–252],

[19] See Gunther [1932, vol. 1, 263] and Schmidl [2007] for a summary of a detailed description prepared by the same author.

[20] Another more sophisticated device of this kind, dating from the same milieu, has recently come to light in a private collection.

The value 16;30° for the latitude of the middle of the first climate tells us something more, because for obliquity 23;33°, which our Andalusi astrolabist favored, it should be 16;40°. In fact, 16;30° corresponds more closely to Ptolemy's obliquity 23;51° (accurately the latitude is 16;27°, although Ptolemy himself used 16;25°). Ptolemy's obliquity was used, implicitly if not explicitly, on all known Islamic astrolabes, Eastern and Western, up to ca. 1100, with one exception.[21]

The Ottoman Additions

The Ottoman components are engraved in a plain and elegant *naskhī* script typical of the milieu from which they hail.

We note that there is a magnetic compass on the front of the throne, with no indication of the existence of magnetic declination. Already ca. 1300 various Islamic instruments were fitted with a compass [Schmidl 1997–1998], though not usually astrolabes, where they are superfluous anyway. Indeed, compasses are rarely found on late Islamic astrolabes. (They are also very rarely found on European astrolabes after ca. 1450.) On this particular compass there is no indication that the compass may not point due north. (In the Islamic world, magnetic declination was first measured by al-Wafāʾī in Cairo ca. 1450 but not widely discussed thereafter.)

On the front rim there are four altitude scales labelled for each 5° up to 90° and subdivided for each 1°.

On the back of the mater, we find precisely the standard markings that identify this piece as Ottoman. On the upper rim, there are two altitude scales for each degree, labelled for each 6°. In the upper left, there is a sine quadrant with equi-spaced horizontal and vertical lines for each 2 units. There is a declination quadrant of radius 24 and two axial semicircles for calculations of sines and cosines. On the upper right there is a universal horary quadrant with markings for each seasonal hour 1–6. The shadow squares below the horizontal diameter serve bases 7 on the left (*al-aqdām*) and 12 on the right (*al-aṣābiʿ*). These serve to find tangents on the horizontal scale (*ẓill al-mankūs*) and cotangents on the vertical ones (*ẓill al-mabsūṭ*). The circumferential scale on the lower left serves to find the altitude of the sun at the afternoon prayer (*ʿaṣr*) from the meridian altitude of the sun, and since this works for all latitudes it is labelled *āfāqī*, universal, literally, serving all horizons. The corresponding scale on the lower right serves the cotangent to base 12. Our astrolabist has misspelled the word for "digits" as *asābiʿ* when it should be *aṣābiʿ*; only a non-Arabic speaker would make such a mistake. The word *asābiʿ* does have a meaning in Arabic, namely, "seventh fractional parts." The same error occurs on another Ottoman astrolabe #4112, formerly preserved in Kandilli Observatory near Istanbul and now apparently in the Rehmi Koç Museum in that city, which appears to be by the same anonymous maker [Dizer 1986, 13 (no. 12, illustrations of the front and back)].

The rete is elegantly executed (see Figure 8). The style is not original but is inspired by an astrolabe of the Andalusi astronomer Ibn Bāṣo who worked in Granada ca. 1300. The most significant aspect of this design is the equatorial bar in the upper ecliptic. One of the three known astrolabes of Ibn Bāṣo — #144, dated 704 Hijra [= 1304/05], present location unknown, has Ottoman additions (see Figure 9) [Gunther 1932, vol. 1, 289 (confused); *Linton Catalogue*, 1980, 87–89 (no. 162)]. Maybe it was precisely this piece that our Ottoman had seen. Even the star-pointers on the Ottoman piece are influenced by those on Ibn Bāṣo's rete. (The other astrolabe by the same maker, #4112, has a standard Ottoman rete.)

The scale for the ecliptic is divided for the 12 signs of the zodiac, subdivided into 3° intervals. The name *al-asad* for Leo has been incorrectly engraved on the ecliptic ring as *a-l-s-d*. (Spelling mistakes on Islamic astrolabes are very rare.) The following 25 stars are named on their pointers, with one other unnamed, reading counter-clockwise from the vernal equinox on the left:

baṭn qayṭūs – ghūl – dabarān – ʿayyūq – rijl-i jawzā – mankib-i jawzā // ʿabūr – ghumayṣā – [*yad-i dubb*], written

[21] Inevitably al-Khujandī's spectacular astrolabe from Baghdad, 984/85. See King [2005, 513–517].

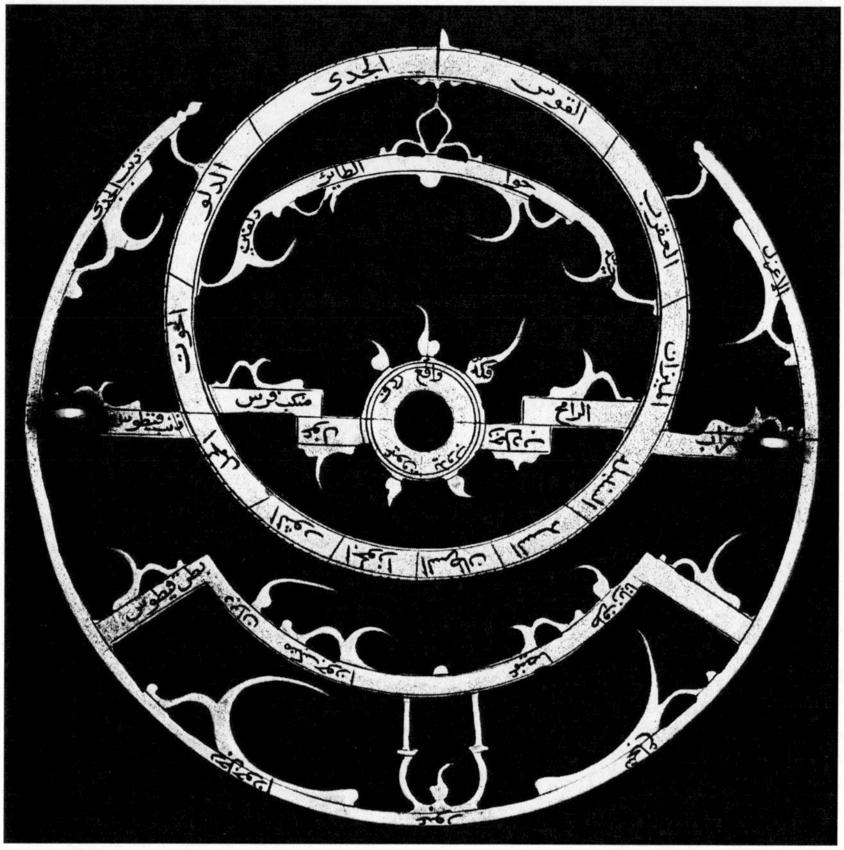

Figure 8: The Ottoman rete. Photo courtesy of the former owner.

y-d-w-b – ṭaraf-i zubānā – shujāʿ – rijl-i dubb – unnamed // *al-ghurāb – al-aʿzal – al-rāmiḥ – fakka – ḥayya – ḥawwā // wāqiʿ – al-ṭāʾir* (with a *hamza*!) *– ridf – dulfīn – dhanab al-jady – mankib-i faras – dhanab-i qayṭūs*

For identifications the reader should consult the list of astrolabe stars compiled by Paul Kunitzsch [Kunitzsch 1966, 59–69; Kunitzsch 1990]. Significant here is the use of the *Western Arabic* names of the stars Sirius and Procyon (α Canis maioris and α Canis minoris), namely, *ʿabūr* and *ghumayṣāʾ*. These are generally found *only* on Western Islamic astrolabes; on Eastern ones, which include Ottoman pieces, the standard names are the *Eastern Arabic* ones *al-shiʿrā al-yamāniya* and *al-shiʿrā al-shaʾamiya*. What has happened here is that our Ottoman astrolabist has copied these names from a Western Islamic rete, either the one that might have been on the original Andalusi astrolabe, or another, an astrolabe in the tradition of Ibn Bāṣo, with 28 pointers for a similar set of stars, that might have been available to him in Istanbul (#144, see above). I have not investigated the star-positions on these two astrolabes, though this would be worthwhile. On the other Ottoman instrument that may have been by the same maker (#4112, see above), two different names are given for Sirius: *ʿabūr* and *shiʿra-yi yamānī*; for Procyon we find *ghumayṣā* again.

We now turn to the three sets of latitude dependent markings that our Ottoman astrolabist included himself, two on a new plate. These serve the following latitudes, and the associated lengths of maximum daylight are accurately calculated for obliquity 23;35°:

8a: 27° – 13;43^{h} ; 7b: 43° – 15;12^{h} ; 8b: 48° – 15;52^{h}.

These latitudes complement those on the Andalusi plates, showing that our Ottoman astrolabist intended to render his new instrument serviceable and not simply aesthetic. The latitude 27° could serve Akhmim in Egypt and Qulzum on the Red Sea littoral. Latitude 43° would have served, say,

Figure 9: An example of the retes of Ibn Bāṣo. Photo courtesy of the former owner, and also used in Linton Catalogue [1980, 89].

Skopje and Sarajevo, but we can only speculate about his intentions with latitude 48°, close to the middle of the 7th climate, perhaps intended for Buda and Pest. On the other astrolabe by the same maker (#4112, see above) the latitudes served are

21;30° 24° 30° 32° 33;30° 36° 38° 39° 40° 41° 42° 43° 45°.

Our Ottoman astrolabist made no comment on the Andalusi plate for 45°, marked for Constantinople. He would have known perfectly well that the latitude of Istanbul was 41° or thereabouts, but for some reason he did not include a new plate for that latitude. This is curious indeed, since he went to the trouble of making new markings for 43° and 48°.

On the new markings altitude circles are engraved for each 6°, which is standard, and azimuth circles *for each 9°*, the only documented example apart from the Andalusi astrolabes mentioned above (common divisions are 6° or 10°). Clearly, our Ottoman astrolabist was, with good reason, fascinated by the Andalusi plates. He also drew special curves on each set of markings for the first and second afternoon prayers (*ʿaṣr-i awwal* and *ʿaṣr-i thānī*), and marked with fish-bones the altitude 18° above the horizon which serves for the determination of daybreak and nightfall.

The alidade is marked on one fiducial side with a scale divided for each 10 units subdivided for each 2 units. The sighting vanes have been flattened so that they are no longer serviceable. The pin seems to be by our Ottoman astrolabist, but the "horse" is missing.

Conclusions

My contribution to Asger Aaboe's *Festschrift* was a paper entitled "Universal solutions in Islamic astronomy," and in this I surveyed the numerous tables and instruments prepared by Muslim astronomers that served all terrestrial latitudes [Berggren and Goldstein, eds. 1987, 121–132; King 2004 (reprint), 679–709]. We may now perhaps add another example, two aspects of which were known already, namely, the preparation of astrolabe plates:

1. for each of the seven climates (on the earliest Eastern astrolabes and on the earliest Western astrolabes);
2. for latitudes between 0° and 90° (well attested over the centuries);

and now there is a possible third aspect (partly hypothetical):

3. for latitudes between the lower and upper limits of the Ptolemaic cartographic tradition (suspected on this one incomplete 11th-century Andalusi astrolabe).

I have argued elsewhere for medieval astronomical instruments to be treated with the same kind of respect as medieval manuscripts. They too are historical 'documents,' and of particular interest are pieces that display two or more layers of components or inscriptions [King 2011, study I (published 1994); King 2005, studies X and XIIIa]. May this one please Len, who has rediscovered medieval Islamic sundials and published several papers on them. Perhaps I can now persuade him in his retirement to have a look at some medieval Islamic astrolabes or related devices in between fishing trips. He will not have to look at many before he discovers some features that will take him by surprise.

References

Berggren, J.L., 1987. Archimedes amongst the Ottomans. In: Berggren, J.L., Goldstein, B.R. (eds.), From Ancient Omens to Statistical Analysis: Essays in the Exact Sciences Presented to Asger Aaboe. Copenhagen, pp. 101–109.

Berggren, J.L., Goldstein, B.R., (eds.), 1987. From Ancient Omens to Statistical Analysis: Essays in the Exact Sciences Presented to Asger Aaboe. Acta Historica Scientiarum Naturalium et Medicinalium 39, Copenhagen.

Berggren, J.L., Jones, A., 2000. Ptolemy's Geography: An Annotated Translation of the Theoretical Chapters. Princeton University Press: Princeton and Oxford.

Casulleras, J., 2004. Ibn Muʿādh on the astrological rays. Suhayl 4, 385–402.

Chicago Astrolabe Catalogue: see Pingree 2009.

Dizer, M., 1986. Astronomi hazineleri (Treasures of Astronomy preserved in Kandill Observatory, in Turkish). (Boğaziçi Üniversitesi Yayinlan No. 404), Boğaziçi Üniversitesi & Kandilli Rasathanesi, Istanbul.

EI_2 = The Encyclopedia of Islam, new edition., 13 vols. Brill, Leiden etc., 1960–2009.

Gunther, R.T., 1932. The Astrolabes of the World, 2 vols. Oxford. Reprinted: In 1 vol., Holland Press, London, 1967.

İhsanoğlu, E., et al., 1997, 1999, 2000. Ottoman Astronomical, Mathematical, and Geographical Literature, 2+2+2 vols. IRCICA, Istanbul.

Kennedy, E.S., Kennedy, M.H., 1987. Geographical Coordinates of Localities from Islamic Sources. IGAIW, Frankfurt am Main.

King, D.A., 1987. Islamic Astronomical Instruments. Variorum, London.

——— 1991. Notes on Byzantine Astronomy, a review of David Pingree, The Astronomical Works of Gregory Chioniades, I: The Zīj al-ʿAlāʾī, Amsterdam, 1985–1986, and Alexander Jones, An Eleventh-Century Manual of Arabo-Byzantine Astronomy, Amsterdam, 1987. Isis 82, 116–118.

——— 1992. Die Astrolabiensammlung des Germanischen Nationalmuseums. In: Bott, G. (ed.), Focus Behaim-Globus, 2 vols. Germanisches Nationalmuseum, Nuremberg. vol.1, 101–114, vol. 2, 568–602, 640–643.

——— 2004–2005. In Synchrony with the Heavens: Studies in Astronomical Timekeeping and Instrumentation in Islamic Civilization, 2 vols. Brill, Leiden and Boston.

——— 2011. Astrolabes from Medieval Europe. Ashgate - Variorum, Aldershot and Burlington.

——— 2012. Islamic Astronomy and Geography. Ashgate - Variorum, Aldershot and Burlington.

Kunitzsch, P., 1966. Typen von Sternverzeichnissen in astronomischen Handschriften des zehnten bis vierzehnten Jahrhunderts. Otto Harrassowitz, Wiesbaden.

——— 1990. Al-Ṣūfī and the astrolabe stars, Zeitschrift für Geschichte der arabisch-islamischen Wissenschaften 6, 151–166, Reprint: In Kunitzsch, P., Stars and Numbers: Astronomy and mathematics in the medieval Arab and Western worlds. Ashgate - Variorum, Aldershot and Burlington. XIII.

Linton Catalogue, 1980. Instruments scientifiques, livres anciens / Scientific instruments, rare books. Leonard Linton collection, Point Lookout, N.Y., Nouveau Drouot, Paris.

Lorch, R.P., 2005. Al-Farghānī on the Astrolabe: Arabic Text Edited with Translation and Commentary. Franz Steiner, Stuttgart.

Neugebauer, O., 1969. The Exact Sciences in Antiquity, second edition. Dover, New York.

——— 1985. A History of Ancient Mathematical Astronomy, 3 pts. in 3 vols. Springer, New York.

Nuremberg Astrolabe Catalogue: see King 1992.

Pingree, D., 2009. Eastern Astrolabes, vol. 2 of Historic Scientific Instruments of the Adler Planetarium and Astronomy Museum. Adler Planetarium and Astronomy Museum, Chicago.

Rockford Time Museum Astrolabe Catalogue: see Turner 1985.

Schmidl, P., 1997–1998. Two early Arabic sources on the magnetic compass. Journal of Arabic and Islamic Studies 1, 81–132. [Available on the Internet.]

——— 2007. 'Astrolabien' (with contributions by K. Gaulke, in Der Ptolemäus von Kassel. Landgraf Wilhelm IV. von Hessen-Kassel und die Astronomie (Kataloge der Museumslandschaft Hessen Kassel 38). Kassel, 218–231.

Sezgin, Fuat, et al., (eds.), 1991. Arabische Instrumente in orientalistischen Studien, 6 vols. IGAIW, Frankfurt am Main.

Sotheby's London 30.05.1991. Islamic Art Auction Catalogue.

Turner, A.J., 1985. The Time Museum - Time measuring instruments - Astrolabes / astrolabe related instruments. The Time Museum, Rockford.

Woepcke, F., 1958. Über ein in der Königlichen Bibliothek zu Berlin befindliches arabisches Astrolabium, Abhandlungen zur Königlichen Akademie der Wissenschaften zu Berlin, pp. 1–31 and plates (separatum), Reprinted: Sezgin et al., (eds.), 1991, 2, pp. 1–36.

Un algébriste arabe : Abū Kāmil Šuǧā[c] ibn Aslam

Adel Anbouba (avec les commentaires de Jacques Sesiano)

Abstract At the beginning of the 20th century, hardly anything was known about Abū Kāmil. By the end of it, he was known to have lived in the second half of the 9th century and to have held an important official position in Egypt. We owe this information to studies made by various German scholars in the first half of the 20th century. We are also indebted to Adel Anbouba, whose considerable knowledge of Arabic bibliographical works led him, independently, to the same conclusions. His study, though, appeared in a short-lived periodical. It thus deserves to be reprinted. A few additions have been made by the editor. (J. Sesiano)

Introduction

[L'article de M. Anbouba réimprimé ci-après nous a été envoyé voici plus d'une trentaine d'années par l'auteur (entre-temps décédé), avec des corrections et des additions manuscrites. En effet, ledit auteur n'avait pas reçu les épreuves de cet article, et il était désolé de voir la forme sous laquelle il était paru. De plus, la revue le contenant, les *Horizons techniques du Moyen-Orient*, dont il occupait les pp. 6 à 15 du n° 3 de 1963, n'a eu qu'une existence éphémère et une notoriété restreinte. Aussi l'auteur m'écrivait-il, dans une lettre datée du 26 janvier 1978 : *Mon article sur Abū Kāmil (est) paru il y a quinze ans, la direction de la revue —morte l'année même de sa parution— ne m'en a pas donné une seule copie, malgré la promesse de trente tirés à part. J'ai été obligé au bout d'un mois de racheter en librairie les quelques numéros qui en restaient. Plus encore, la Revue fit tirer l'article sans tenir compte de mes corrections et sans attendre le bon à tirer*. Cette étude méritait pourtant un meilleur traitement : l'auteur avait étudié quantité de sources bio- et bibliographiques, et c'est grâce à ses recherches minutieuses que quelques renseignements nous sont parvenus sur le deuxième algébriste de langue arabe, voire le premier pour l'influence, dont on conjecturait jusqu'à il n'y a pas si longtemps la date approximative de l'existence. H. Suter n'écrivait-il pas en 1892, dans son étude commentée des informations du *Fihrist* sur les mathématiciens, à l'article Abū Kāmil (*Abhandlungen zur Geschichte der Mathematik*, 6, pp. 37 & 69), « Ueber seine Lebenszeit habe ich keine Angaben gefunden » ? Certes, des renseignements avaient entre-temps été mis en lumière dans diverses études en allemand (par H. Suter, E. Wiedemann, Fr. Hauser), mais de manière éparse, et M. Anbouba, après avoir retrouvé ces renseignements (indépendamment), sut les réunir et leur donner l'importance qui convenait.

Notre échange épistolaire avec M. Anbouba a duré plusieurs années, et fut interrompu par la guerre civile du Liban. Il aimait à parler de ses projets et me parlait des difficultés qu'il éprouvait parfois, et qui n'avaient pas toutes un rapport avec les aléas de l'impression de ses articles. M. Anbouba a beaucoup travaillé sur la théorie des nombres chez les auteurs arabes. On lui doit l'édition du *Badī*[c] d'al-Karajī/al-Karaǧī, où ce sujet occupe une place considérable. On lui doit aussi la découverte de l'opuscule d'al-<u>Kh</u>āzin/al-<u>H</u>āzin, où plusieurs des sujets traités par Leonardo Fibonacci dans son *Liber*

N. Sidoli and G. Van Brummelen (eds.), *From Alexandria, Through Baghdad: Surveys and Studies in the Ancient Greek and Medieval Islamic Mathematical Sciences in Honor of J.L. Berggren*, DOI 10.1007/978-3-642-36736-6_18,
© Springer-Verlag Berlin Heidelberg 2014

quadratorum se retrouvent. Sa publication n'en fut toutefois pas facile, et eut lieu avec quelque retard, comme le dit pudiquement Anbouba au début de son article : *Cet article envoyé à l'édition aussi tôt que mai 1978 a subi, comme on le voit, un retard accidentel assez long* (voir la note au début de son article « Un Traité d'Abū Ja^cfar [al-Khāzin] sur les triangles rectangles numériques », *Journal for the history of Arabic science*, 3, 1 (1979), pp. 134–178). Le retard dans sa publication ne fut pas vraiment *accidentel*, comme M. Anbouba me le confia ultérieurement avec quelque amertume. Il avait envoyé son étude pour publication au *Journal for the history of Arabic science*, dans le comité d'édition duquel on trouvait MM. Roshdi Rashed et Edward Kennedy. Anbouba n'entendit plus parler de son article, mais il apprit un jour la prochaine publication d'une étude sur le même sujet par Roshdi Rashed. Il s'en ouvrit alors à E. Kennedy, qui, après recherches, retrouva son manuscrit, *oublié* dans un tiroir de l'Institut d'Alep où siégeait la direction du *Journal*. Kennedy fit alors hâter la publication de l'article d'Anbouba, qui put ainsi paraître peu avant celui dont il avait hâté, sinon inspiré, la publication (R. Rashed, « L'analyse diophantienne au Xe siècle : l'exemple d'al-Khāzin », *Revue d'histoire des sciences*, XXXII, 3 (1979), pp. 193–222).

Dans la réimpression de cet article, nous avons rétabli les signes précisant la valeur des lettres dans les transcriptions de l'arabe (ajoutés, avec plus ou moins de bonheur, à la main dans le texte édité) ; nous avons aussi vérifié et mis à jour les références, du moins pour les ouvrages qui nous étaient accessibles, en ajoutant des références aux éditions allemandes, d'une incomparable érudition mais ignorées de l'auteur qui n'y avait pas accès. En de rares endroits, là où les assertions de Anbouba nous paraissaient moins sûres qu'il lui semblait, nous nous sommes permis de le remarquer. Nous avons ajouté en appendice un sommaire de l'*Algèbre* d'Abū Kāmil, destiné à compléter la liste des sujets mathématiques qu'avait dressée Anbouba. Toutes les additions qui ne sont pas que de simples corrections au texte sont clairement visibles car elles sont enfermées dans des crochets. (J. Sesiano)]

Un algébriste arabe : Abū Kāmil

Lorsqu'en 1853 l'orientaliste allemand Franz Woepcke[1] révéla au monde savant l'algèbre arabe *al-Faḫrī*[2] d'al-Karağī[3], son attention fut retenue par deux problèmes absolument uniques dans l'ouvrage, résolus grâce à un système d'équations linéaires à plusieurs inconnues[4]. Il ne manqua pas alors de formuler l'opinion suivante : « Or c'est ici que je dois signaler un fait extrêmement curieux, à savoir qu'Alkarkhi[5], dans deux de ses problèmes, fait usage d'un terme spécial pour désigner une seconde inconnue, dont il se sert dans la résolution du problème, absolument comme nous calculons avec *x* et

[1] On doit à F. Woepcke, dont la vie aura été bien brève (1828–1864), de remarquables études sur l'histoire des sciences chez les Arabes, rédigées en français pour la plupart.

[2] *Al-Faḫrī* : du nom du grand vizir Faḫr al-Mulk, patron des lettres et des sciences, à qui ce livre fut dédié. Woepcke en publia des extraits en langue française : *Extrait du Fakhrî, traité d'algèbre (...), précédé d'un mémoire sur l'algèbre indéterminée chez les Arabes* (Paris 1853) [réimprimé dans : F. Woepcke, *Etudes sur les mathématiques arabo-islamiques* (2 vol., Francfort 1986), I, pp. 269–426].

[3] Al-Karağī : mathématicien d'origine persane, séjourna à Bagdad. C'est là qu'il publia, coup sur coup, ses trois oeuvres maîtresses : *al-Faḫrī*, *al-Badī^c* et *al-Kāfī* de 402 H / 1011-12 à 403 H / 1012–13 (ou 405 H ?). [Sur le premier, voir la note précédente ; édition du second par A. Anbouba, *L'algèbre al-Badī^c d'al-Karagī*, Beyrouth 1964 ; traduction du troisième par A. Hochheim, *Kâfî fîl Hisâb (Genügendes über Arithmetik)* (3 Hefte), Halle a. d. Saale 1878–1880, & la partie algébrique du texte arabe (Hochheim, III, pp. 4–27) en appendice à l'édition de l'Arithmétique d'Abū'l-Wafā' par A. Saidan (Amman 1971), pp. 368–407. Des éléments autobiographiques d'al-Karağī se trouvent dans son *Inbāṭ al-miyāh al-ḫafīya* (éd. Haydarabad 1359 H / 1940–41), voir l'édition du *Badī^c* par Anbouba (n. 3), pp. 12 & 18, ou notre étude sur ses équations indéterminées, *Archive for history of exact sciences*, 17 (1977), pp. 297–379.]

[4] Ce sont les problèmes 5 et 6 de la troisième section de problèmes. Voir *al-Faḫrī* d'al-Karağī, ms. [Dār al-kutub, riyāḍ. 23] de la Bibliothèque Nationale du Caire, fol. 40 [ou *Extrait du Fakhri*, p. 90].

[5] Al-Karkhi [al-Karḫī] : graphie erronée, autrefois suivie par plusieurs historiens (dont Cajori, Sarton, Smith, et les Encyclopédies américaine, britannique, italienne). Le manuscrit 2459 de la Bibliothèque Nationale de Paris, sur lequel travailla Woepcke, en est responsable.

y[6]. Cependant l'auteur ne se sert pas du même terme dans les deux problèmes pour désigner la seconde inconnue, et n'emploie ce procédé que cette seule fois. Cela nous prouve que nous avons ici affaire à un de ces premiers pas dans le chemin d'une découverte importante, que malheureusement il n'a pas été permis à la science arabe de poursuivre jusqu'au bout »[7].

A l'époque où Woepcke écrivait ces lignes, le dépouillement des œuvres mathématiques arabes ne faisait que commencer, et Woepcke était loin de soupçonner que *les deux problèmes en question figuraient dans une algèbre arabe antérieure d'au moins un siècle au Faḫrī, « al-ğabr wa'l-muqābala » de l'Egyptien Abū Kāmil Šuğāᶜ ibn Aslam*[8].

C'est à cet auteur et à l'inventaire de sa production que nous voudrions consacrer cet article ; nous espérons étudier ultérieurement son « al-ğabr wa'l-muqābala », ouvrage fondamental dans l'histoire de l'algèbre arabe. [L'Institut d'Alep lui en avait proposé la publication, ce qu'il avait accepté ; quoique son travail fût avancé, il y renonca ultérieurement. Une autre édition parut ensuite dans ledit Institut, voir n. 14.]

Abū Kāmil a laissé plusieurs écrits dont la sobre énumération faite en 377 H / 987–88 par Ibn al-Nadīm dans son *Fihrist* reste la plus valable[9]. Les historiens modernes en ont fait connaître quelques-uns. C'est ainsi qu'en 1896, Gustavo Sacerdote publia une traduction italienne du « Traité du pentagone et du décagone »[10], que suivit, en 1910, la traduction allemande de Heinrich Suter[11]. Puis, en 1911, parut la traduction allemande, par Suter également, des « Questions rares d'arithmétique »[12]. En 1912, L.Ch. Karpinski donna des extraits et une analyse de l'*Algèbre* d'Abū Kāmil, d'après une traduction latine du Moyen Âge, conservée à Paris[13]. Enfin, en 1935, Josef Weinberg publia à Munich une traduction allemande de l'*Algèbre* d'Abū Kāmil, d'après une traduction hébraïque ancienne[14].

Cependant, malgré l'attention dont bénéficia Abū Kāmil, peu de renseignements ont pu être réunis sur sa vie, dont les dates restent imprécises. On estime toutefois qu'il a fleuri entre 850 et 955 (236–345 H)[15]. Nous inclinons à croire, pour notre part, que sa vie active d'auteur et, en particulier, la publi-

---

[6] Al-Karağī utilise les mots *šay'* et *qisṭ* dans le problème 5, *šay'* et *qism* dans le n° 6.

[7] *Extrait du Fakhri*, p. 11. [On trouve déjà chez Diophante l'usage de πρῶτος et δεύτερος ἀριθμός, mais, il est vrai, avant que l'on ait des inconnues désignées en tant que telles.]

[8] *Kitāb al-ğabr wa'l-muqābala* de Abū Kāmil Šuğāᶜ ibn Aslam, ms. Kara Mustafa Paşa 379 [aujourd'hui : Beyazıt 19046]. Nous devons le microfilm de ce manuscrit et l'autorisation d'utiliser l'ouvrage dans nos publications à la bienveillance de la Direction des Bibliothèques à Istanbul. Nous sommes heureux de lui exprimer ici notre vive gratitude. [Le manuscrit est désormais disponible dans un fac-similé publié en 1986 par l'Institut für Geschichte der arabisch-islamischen Wissenschaften de Francfort comme volume C.24 (au fol. x^r du manuscrit correspond la page $2x - 1$ dans la reproduction).]

[9] Ibn al-Nadīm, *Fihrist*, Le Caire, non daté [1348 H / 1929–30], p. 406 [édition de G. Flügel, J. Roediger, A. Müller, *Kitâb al-Fihrist* (2 vol., Leipzig 1871–72), I, p. 281 ; le vol. I contient le texte, le vol. II des notes et l'index].

[10] « Il trattato del pentagono e del decagono di Abu Kamil Shogiaᶜ ben Aslam ben Muhammed », *Festschrift zum achzigsten Geburtstage Moritz Steinschneiders*, Leipzig 1896, pp. 169–194. Sacerdote utilisait une traduction hébraïque.

[11] « Die Abhandlung des Abū Kāmil Shoğāᶜ b. Aslam "über das Fünfeck und Zehneck" », *Bibliotheca mathematica*, 3. F., 10 (1909–10), pp. 15–42. Suter utilisait des traductions latine et hébraïque. [Suter utilisait la traduction de Sacerdote et la version latine conservée, publiée depuis par R. Lorch, *Vestigia mathematica* (Amsterdam 1993), pp. 215–252 (Livre IV de l'*Algèbre* d'Abū Kāmil, voir Appendice) ; le même ouvrage contient, aux pp. 315–452, le reste de la traduction effectuée par Guillelmus au XIVième siècle.]

[12] « Das Buch der Seltenheiten der Rechenkunst von Abū Kāmil el-Miṣrī », *Bibliotheca mathematica*, 3. F., 11 (1910–11), pp. 100–120. [Cette étude et la précédente ont été entre-temps (1986) réimprimées par l'Institut für Geschichte der arabisch-islamischen Wissenschaften de Francfort, dans les deux volumes des *Beiträge zur Geschichte der Mathematik und Astronomie im Islam* de Suter.]

[13] L.Ch. Karpinski, « The Algebra of Abu Kamil Shojaᶜ ben Aslam », *Bibliotheca mathematica*, 3. F., 12 (1911–12), pp. 40–55. [Publiée complètement depuis, voir l'ouvrage sus-mentionné (n. 11).]

[14] Josef Weinberg, *Die Algebra des Abū Kāmil Šoğāᶜ ben Aslam*. [Traduction jointe par Sami Chalhoub à son édition (partielle) du texte arabe, Alep 2004. Traduction anglaise, avec texte hébreu, par M. Levey, *The Algebra of Abū Kāmil, Kitāb fī al-jābr* (sic) *wa'l muqābala, in a Commentary by Mordecai Finzi*, Madison 1966. L'attribution à Finzi d'un commentaire est abusive. Les deux traductions, de Weinberg et de Levey, utilisent le ms. Munich hébr. 225, partiellement édité par Levey ; le fragment Paris, BNF hébr. 1029, fol. $296^r - 309^v$, n'est pas une copie du précédent, contrairement à l'assertion de Levey. Il correspond pour le contenu aux fol. $95^r - 114^r$ du texte arabe, mais l'expression en diffère parfois et il a des additions.]

[15] G. Sarton, *Introduction to the history of science*, vol. I (Washington 1927), pp. 630–631.

cation de son ouvrage fondamental « *al-ǧabr wa'l-muqābala* » se place au voisinage de 270 H / 883–84. Les raisons qui nous ont amené à cette conclusion sont exposées ci-après.

En 344 H / 955–56, comme nous l'apprend Ibn al-Nadīm dans son *Fihrist*[16], meurt Aḥmad al-ᶜImrānī de Mossoul, auteur d'un commentaire de l'*Algèbre* d'Abū Kāmil. Ce fait a fourni aux historiens une première date-limite pour la vie d'Abū Kāmil. Cependant, un autre commentateur du même ouvrage, al-Iṣṭaẖrī, est cité par Ibn al-Nadīm *dans une génération antérieure à celle d'al-ᶜImrānī* et comprenant notamment Muḥammad ibn Lurra[17]. De ce dernier, ingénieur et géomètre, nous trouvons mention dans *al-Aᶜlāq al-nafīsa* d'Ibn Rustah, livre écrit vers 290 H / 902–3[18]. Or, un commentaire se présente naturellement, dans l'histoire des sciences chez les Arabes, comme le fruit de plusieurs années d'enseignement. Si l'on tient compte de ce premier fait, puis du temps nécessaire à l'*Algèbre* d'Abū Kāmil pour s'imposer comme classique, et enfin de la génération à laquelle appartient al-Iṣṭaẖrī qu'Ibn al-Nadīm place, d'ailleurs, dans sa liste juste avant Ibn Lurra, on voit comme on se sent en droit de placer l'*Algèbre* d'Abū Kāmil avant 300 H / 912.

D'autre part, dans son « Livre sur les testaments » (*Kitāb al-waṣāyā*)[19], *postérieur à l'Algèbre*, Abū Kāmil s'élève vivement contre l'attribution, faite par Abū Barza, de la première algèbre arabe à son grand-père, le mathématicien turc ᶜAbd al-Ḥamīd[20]. La vivacité de la critique montre que l'affaire est encore chaude et que Abū Kāmil n'attaque pas un mort. Or, le mathématicien et faqīh [= juriste] Abū Barza est mort en 298 H / 910-11, d'après al-Qifṭī et Ibn al-Aṯīr[21], et en 292 H / 904–5 d'après Ibn al-Ǧauzī[22]. Sans doute sa réclamation en faveur de son grand-père a-t-elle trouvé place dans un de ses écrits mathématiques. Et nous voilà amenés, pour situer dans le temps l'*Algèbre* d'Abū Kāmil —antérieure, rappelons-le, à son *Livre des testaments*— à nous replier sur une date bien antérieure à 300 H, et qui pourrait être voisine de 270 H / 883[23].

Un argument différent en faveur de notre affirmation est fourni par l'*Algèbre* d'Abū Kāmil. *Ce livre qui traduit en maints endroits l'influence grecque et où ne manquent pas les « questions diophantiennes », ignore cependant l'Arithmétique de Diophante*. Or, vers cette époque —nous ne pouvons pas préciser davantage— l'*Arithmétique* est traduite par Qusṭā ibn Lūqā dont l'orientaliste G. Gabrieli place la mort vers 300 H / 912–13 et le séjour à Bagdad, au plus tard, jusqu'en 260 H / 873-74[24]. Si la traduction de l'*Arithmétique* —entreprise particulièrement difficile, donc œuvre de maturité plutôt— avait été antérieure à l'*Algèbre* d'Abū Kāmil, il est assez difficile d'admettre que ce dernier serait resté sans en prendre connaissance, tant les auteurs mettaient de soin à se documenter et tant les communications étaient bien assurées entre Bagdad et les différentes parties du royaume d'Islam. Abū Kāmil nous prévient

---

[16] *op. cit.* [n. 9], p. 408 [p. 283].

[17] *ibid.*, p. 407 [p. 282].

[18] *Kitâb al-Aᶜlâq al-nafīsa*, éd. M.J. de Goeje, *Bibliotheca geographorum Arabicorum*, VII [Leiden 1892], p. 160 (art. *Isfahan* [مدينة اصبهان]).

[19] Cité [comme كتاب الوصايا بالجذور] dans Ḥājjī H̱alīfa, *Kašf al-ẓunūn*, éd. Š. Yaltkaya & K. Bilge, 2 vol., Istanbul 1941–43 [réimpression Istanbul 1972], vol. 1, col. 664. [On pourra aussi se reporter à l'édition de G. Flügel (qui comprend une traduction latine), *Lexicon bibliographicum et encyclopaedicum a Haji Khalfa compositum*, 7 vol. (Leipzig et Londres 1835–58), vol. V, p. 168.]

[20] On trouvera des vues intéressantes sur la paternité de la première algèbre arabe dans Aydın Sayılı, *Logical necessities in mixed equations by ᶜAbd al-Ḥamīd ibn Türk and the algebra of his time*, Türk tarih kurumu yayınlarından, ser. 7, 41 (1962) [réimpression Ankara, 1985].

[21] Al-Qifṭī, *Iḫbār al-ᶜulamā'*, Le Caire 1326 H / 1908–9, p. 265 [*Ta'rīḫ al-ḥukamā'* (éd. par A. Müller et J. Lippert), Leipzig 1903, p. 406, ou M. Casiri, *Bibliotheca arabico-hispana Escurialensis* (Madrid 1760–70, 2 vol.), I, p. 408]. Ibn al-Aṯīr, *al-Kāmil fī'l-ta'rīḫ* (Le Caire 1929–1938, 9 vol.), rubrique de l'année 298 H / 910-11.

[22] Ibn al-Ǧauzī, *al-Muntaẓam* (Haydarabad 1357–1359 H / 1938–1941, 10 vol.), vol. 6, rubrique de l'année 292 H / 904–5.

[23] L'*Algèbre* d'Abū Kāmil contient à l'adresse d'al-H̱wārizmī des invocations de miséricorde, qui montrent qu'al-H̱wārizmī est mort (Ms. Kara Mustafa Paşa 379 [n. 8], fol. 109^v, 18. Voir aussi fol. 110^r, 10. La mort d'al-H̱wārizmī est placée par Nallino vers 233 H / 847–48, mais son *Algèbre* est bien antérieure (vers 205 H / 820–21). Voir l'article sur l'algèbre dans l'*Enciclopedia italiana* [II, 1929–VII, p. 423], ou l'article al-K̲h̲^{w}ārizmī dans l'*Encyclopédie de l'Islam* [II, 1927, pp. 965–966]. On voudra bien lire dans notre *Iḥyā al ǧabr*, Beyrouth 1955 [rééd. *ibid.* 1968], p. 3 ligne 12, 820 au lieu de 830.

[24] Voir l'article de Giuseppe Gabrieli sur Qusṭā ibn Lūqā dans les *Rendiconti della Reale Accademia dei Lincei* (Cl. di sc. mor., stor. e filolog.), s. V, XXI (1912), pp. 341–382.

d'ailleurs, dans sa préface, du soin qu'il a pris de lire les auteurs qui l'ont précédé. Plaçons donc vers 270 H / 883–84, et jusqu'à nouvel examen de la question, la publication de l'*Algèbre* d'Abū Kāmil et de la traduction de Qusṭā, en attendant que de nouveaux faits viennent nous éclairer.

[De nouveaux faits sont effectivement survenus entre-temps. D'abord, un manuscrit arabe de Diophante a été retrouvé en 1968 par F. Sezgin, contenant quatre livres autres que les six préservés en grec. Il en apparaît que la traduction de l'*Arithmétique* de Diophante en arabe était partielle : des 13 livres dont se composait originellement l'ouvrage, seuls les sept premiers furent traduits, et dans un commentaire de la fin de l'antiquité, probablement dû à Hypatia comme suggéré en 1978 déjà (*Archives internationales d'histoire des sciences*, 30 (1980), p. 183). Il en apparaît ensuite que les quatre livres découverts s'insèrent au milieu de ceux que nous connaissons en grec et qui, eux, n'appartiennent pas à la version commentée. Voir notre édition des quatre livres conservés, qui sont donc les livres IV à VII (New York 1982). Les équations diophantiennes de l'*Algèbre* d'Abū Kāmil auxquelles A. Anbouba fait allusion sont sûrement inspirées de travaux grecs *autres* que ceux de Diophante. Voir l'édition précitée, pp. 81–82 ; ou l'étude de ces problèmes dans *Centaurus*, 21 (1977), pp. 89–105 ; ou l'*Introduction à l'histoire de l'algèbre* [n. 49], pp. 46–49 (pp. 48-51 dans l'édition française).]

Les conclusions précédentes, fondées sur l'œuvre d'Abū Kāmil, seraient fortement consolidées si elles pouvaient s'appuyer sur un témoignage direct relatif à sa vie. Or ce témoignage existe. Dû à un familier de notre auteur, il trouve place dans un ouvrage littéraire, ce qui explique, jusqu'à un certain point, qu'il n'ait pas encore été utilisé par les historiens des sciences. Il s'agit du recueil d'histoires édifiantes *al-Mukāfa'a*[25] de l'écrivain, poète et mathématicien Abū Ǧaᶜfar Aḥmad ibn Yūsuf[26]. Ce dernier —comme son père d'ailleurs— servit la dynastie des Tulunides qui gouverna l'Egypte de 254 H à 292 H (868 à 905)[27]. Il en laissa une histoire dont il ne reste que quelques extraits, mais qui a passé en grande partie dans une histoire postérieure due à Abū Muḥammad al-Balawī[28]. Nous apprenons par cette dernière que *Abū Kāmil, d'abord emprisonné par Aḥmad ibn Ṭūlūn —qui accéda au pouvoir en 254 H / 868 et mourut en 270 H / 883-84—, fut relâché par lui et placé à la tête de l'arsenal*[29]. Ces arrestations arbitraires étaient assez fréquentes alors, et l'entourage d'Ibn Ṭūlūn eut souvent à souffrir de son caractère soupçonneux et dur, voire cruel[30]. Le père d'Abū Gaᶜfar, arrêté lui-même, on ne sait pour quel motif, ne dut son salut qu'aux supplications et aux pleurs d'une trentaine de malheureux dont il était le bienfaiteur et le soutien. En quelle année Abū Kāmil fut-il affecté à la direction de l'arsenal, l'histoire ne le dit pas et nous ne ferons pas là-dessus de conjecture ; mais qu'il nous suffise de dire qu'en 263 H / 876–77 Aḥmad ibn Ṭūlūn, qui redoutait une expédition punitive de la part du calife, avait fait fortifier l'île (située près du Caire) qui abritait son arsenal, et poussé fébrilement à la construction de bateaux de guerre[31]. Quoi qu'il en soit, nous savons maintenant que Abū Kāmil a rempli la charge d'ingénieur maritime dans l'administration tulunide, ce qui devait lui assurer des moyens de vie honorables —chose qu'il n'aurait pu obtenir à l'époque par sa science mathématique. Ainsi, vers 265 H / 878–79, et probablement avant, Abū Kāmil était déjà un homme de science reconnu.

Dans le recueil *al-Mukāfa'a*, cité plus haut, le mathématicien-écrivain Abū Gaᶜfar rapporte trois histoires, à lui contées par Abū Kāmil, lequel en tenait certainement deux, et probablement la troisième aussi, de l'astronome d'al-Ma'mūn, Sanad ibn ᶜAlī.

Dans l'une d'elles[32], fort savoureuse d'ailleurs, Sanad ibn ᶜAlī raconte comment, ayant terminé les

---

[25] Abū Ǧaᶜfar Aḥmad ibn Yūsuf, *al-Mukāfa'a*, édité par Aḥmad Amīn et ᶜAlī al-Ǧārim, Le Caire 1941.

[26] Voir Sarton, *op. cit.* [n. 15], vol. I, p. 598. Buṭrus al-Bustānī, *Dā'irat al-Maᶜārif*, tome III.

[27] Voir *al-Muntaẓam* [n. 22], vol. V, rubrique de l'année 270 H / 883–84. Voir Brockelmann, *Histoire des peuples et des états islamiques* (trad. française [de : *Geschichte der islamischen Völker und Staaten*, Munich 1939], Paris 1949), pp. 122–125.

[28] Abū Muḥammad al-Balawī, *Sīrat Aḥmad ibn Ṭūlūn* (éd. par Muḥ. Kurd ᶜAlī), Damas 1358 H / 1939–40.

[29] Al-Balawī, *op. cit.* [n. 28], p. 208. Le texte imprimé porte la mention Abū Kāmil Šuǧāʿ al-ḥāǧib (le chambellan), au lieu de al-ḥāsib (le calculateur). La rectification s'impose sans discussion.

[30] Voir la mise à mort du médecin d'Ibn Ṭūlūn dans Ibn abī Uṣaibiᶜa, *Kitāb ᶜuyūn al-anbā' fī ṭabaqāt al-aṭibbā'* [éd. Müller, 2 vol., Le Caire 1882], vol. 2, pp. 83–85. Voir aussi al-Balawī, *op. cit.* [n. 28], préface p. 28, pp. 113, 266–271, 319, 328–329. Voir *al-Mukāfa'a* [n. 25], p. 48.

[31] Al-Balawī, *op. cit.* [n. 28], pp. 85–88.

[32] *Al-Mukāfa'a* [n. 25], pp. 211–215.

Eléments d'Euclide à dix-sept ans, il vendit le mulet de son père, à l'insu de celui-ci, pour acheter l'*Almageste* de Ptolémée, au prix considérable de vingt dinars[33]. Il lui fallut trois ans de labeur acharné pour s'en rendre maître. C'est alors qu'il se présenta à une de ces réunions mathématiques qui se tenaient régulièrement chez al-ᶜAbbās ibn Saᶜīd al-Ǧauharī[34]. Le père de ce mathématicien, Saᶜīd, était à n'en pas douter un homme riche, cultivé et de bonnes manières. Le calife al-Rašīd l'avait chargé de veiller à l'éducation de son fils, le jeune al-Ma'mūn[35], et al-ᶜAbbās, d'ailleurs géomètre et astronome de valeur —probablement le premier mathématicien arabe à avoir tenté de réduire [déduire ? (*Réd.*)] le [cinquième] postulat d'Euclide[36]— reconnut la valeur du jeune Sanad et le présenta au calife al-Ma'mūn qui le patronna.

La deuxième anecdote raconte l'histoire touchante d'un géomètre tombé dans la misère, Ibrāhīm ibn al-ᶜAǧamī (dont nous ne connaissons d'ailleurs absolument rien), qui, dans son malheur, se tourna vers al-Ma'mūn, véritable providence des mathématiciens, et lui dédia une note, sans doute fort brève, sur une proposition de géométrie. Al-Ma'mūn interrogea les deux frères mathématiciens Muḥammad et Aḥmad, fils de Mūsā ibn Šākir, sur la valeur d'Ibrāhīm ibn al-ᶜAǧamī. Ils le taxèrent de médiocre. D'ailleurs celui-ci, introduit en leur présence devant al-Ma'mūn, se troubla et resta coi, ce qui ne les fâcha guère. Ibrāhīm fut cependant sauvé par l'intervention du ḥāǧib [chambellan] d'al-Ma'mūn, Sindī ibn Šāhak, qui fit valoir l'instruction géométrique qu'il avait reçue d'Ibrāhīm, comme il fut sauvé sans doute aussi par la bonté du calife. Sindī ibn Šāhak, interrogé par Sanad ibn ᶜAlī, lui avoua sans peine n'avoir jamais été l'élève d'Ibrāhīm, mais qu'il avait inventé cette fable, excédé par l'attitude des deux fils de Mūsā[37]. Nous ne nous attarderons pas sur le rôle de conseiller scientifique que les fils de Mūsā ont joué, peut-être accidentellement, auprès du calife, malgré l'intérêt qu'il y aurait de savoir si une telle charge existait. L'histoire a retenu de nombreuses interventions de conseillers littéraires auprès de califes ou de mécènes pour les éclairer sur la valeur des poèmes composés en leur louange et sur les gratifications à accorder[38]. Qu'il nous suffise de remarquer que cette anecdote permet de cerner d'assez près l'âge de Sanad ibn ᶜAlī. En effet, elle s'est déroulée « fī 'ayyām al-Ma'mūn » (au temps d'al-Ma'mūn), c'est-à-dire sous son califat. Or, al-Ma'mūn accéda à cette charge suprême en 198 H / 813–14 à Marw, en Ḫurāsān, et il ne quitta Marw que fin 203 H / 819 pour entrer à Bagdad au début de 204 H / 819[39]. *Quatre mois plus tard mourait Sindī ibn Šāhak, le 6. VII. 204 H*[40]. Ainsi, en l'an 200 H / 815–16, Sanad ibn ᶜAlī était au voisinage de ses vingt ans, al-Ma'mūn avait trente ans, al-ᶜAbbās ibn Saᶜīd al-Ǧauharī à peu près le même âge[41] qu'al-Ma'mūn, peut-être un peu plus.

La troisième anecdote[42] a pour personnages les deux mêmes fils de Mūsā, Sanad ibn ᶜAlī, le fa-

---

33 Sous Harūn al-Rašīd, le faqīh [juriste] Abū Yūsuf conseillait de verser dix dirhams par mois, soit un dinar, à chaque prisonnier, pour ses frais de nourriture (voir Abū Yūsuf Yaᶜqūb, *Kitāb al-Ḫarāǧ*, p. 88 —d'après Adam Mez, *Die Renaissance des Islam* [Heidelberg 1922], trad. arabe [*al-Ḥaḍāra al-islāmīya*, Le Caire 1940-1941, 2 vol.], vol. II, p. 145, n. 1). Dix dinars suffisaient pour la subsistance d'une famille durant un mois. Voir *al-Mukāfa'a* [n. 25], pp. 167–172. On trouvera dans les histoires bien des faits confirmant ce niveau de vie des classes modestes ou pauvres.

34 Voir le *Fihrist* [n. 9], p. 393 [p. 272], Ibn al-Qifṭī [n. 21], p. 143 [p. 255].

35 Aḥmad Farīd Rifāᶜī, *ᶜAṣr al-Ma'mūn*, Le Caire 1346 H / 1927–28 (3 vol.), vol. I, pp. 211–212. Voir le *T'arīḫ* d'al-Ṭabarī, rubrique de année 205 H, Abū Maryam, au service de Harūn al-Rašīd après avoir été à celui de Saᶜīd al-Ǧauharī.

36 Voir la démonstration du 5ᵉ postulat d'Euclide telle qu'elle est rapportée et réfutée par Naṣīr al-Dīn Ṭūsī dans sa *Risāla al-šafīya 'an al-šakk fī'l-ḫuṭūṭ al-mutawāzīya*, p. 4, l. 16 et pp. 17–26. (Recueil des *Rasā'il* de Naṣīr al-Dīn Ṭūsī (2 vol.), Haydarabad 1359 H / 1940–41). [Traduction par Kh. Jaouiche, *La théorie des parallèles en pays d'Islam* (Paris 1986), pp. 201–226.] On trouvera des résultats nouveaux sur les observations faites par Sanad ibn ᶜAlī, al-ᶜAbbās ibn Saᶜīd al-Ǧauharī et les autres astronomes d'al-Ma'mūn dans le remarquable ouvrage de Aydın Sayılı, *The observatory in Islam*, Ankara 1960, pp. 50–87.

37 *Al-Mukāfa'a* [n. 25], pp. 193–195.

38 Voir, par exemple, comment al-Ḫaṣīb, Wali d'Egypte, se fait conseiller par le grand poète Abū Nuwās (al-Ǧahšiyārī, *Kitāb al-wuzarā' wa'l-kuttāb*, Le Caire 1357 H / 1938, p. 205 [ou : E. Wagner, *Abū Nuwās* (Wiesbaden 1965), p. 70 et suiv.]).

39 Al-Ṭabarī [n. 35], rubrique année 204 H.

40 Ibn Taifūr, *Kitāb Baġdād*, Le Caire 1949, p. 187. Al-Ṣafadī, *'Umarā' Dimašq*, éd. Ṣalāḥ al-Dīn al-Munaǧǧid, Damas 1955, p. 39 (date de mort citée par l'éditeur d'après le *T'arīḫ* d'Ibn ᶜAsākir).

41 Renseignement donné par Sanad ibn ᶜAlī dans la première anecdote, *al-Mukāfa'a* [n. 25], p. 214.

42 *Al-Mukāfa'a*, pp. 195–198.

meux al-Kindī, le calife al-Mutawwakil et, en arrière-plan, l'ingénieur Aḥmad ibn Kaṯīr al-Farġānī qui construisit un nilomètre au Caire. L'action *se situe en 247 H / 861–62*, quelques mois avant la mort du calife, et *Sanad y tient le rôle d'expert-ingénieur*. Comme cette histoire a été citée par Ibn abī Uṣaibi^{c}a[43], elle a pu déjà être utilisée par l'historien turc Salih Zeki[44]. Muḥammad et Aḥmad, fils de Mūsā, y sont peints sous des traits assez noirs. Sanad ibn ^{c}Alī y joue, au contraire, le beau rôle. Bien que le texte ne précise pas que Abū Kāmil tienne le récit de Sanad lui-même, il ne semble pas qu'il y ait lieu d'en douter. En 247 H, donc, Sanad n'était pas loin de ses soixante-dix ans.

Ces trois histoires semblent nous avoir éloignés de notre personnage principal, Abū Kāmil. Mais ce n'est qu'en apparence. *En effet, nous tenons là une source importante par laquelle Abū Kāmil pouvait se documenter sur la vie scientifique à la cour d'al-Ma'mūn*. Par là, il a pu connaître les circonstances dans lesquelles l'*Algèbre* d'al-Ḫwārizmī fut élaborée. Sanad était d'ailleurs lui-même intéressé par l'algèbre et il composa un ouvrage sur cette matière, cité dans le *Fihrist* d'Ibn al-Nadīm[45].

Où Abū Kāmil a-t-il rencontré Sanad ? Est-ce au Caire, à Damas, à Bagdad même ? Les anecdotes rapportées ne donnent aucune indication là-dessus. Mais la stupéfiante facilité avec laquelle les hommes de ce temps-là entreprenaient de longs voyages laisse le champ libre à toutes les suppositions. Nous trouvons d'ailleurs, au-delà de 280 H / 893–94[46], le fils de Sanad ibn ^{c}Alī de séjour en Egypte. Retenons donc que, probablement vers 250 H / 864[47], Abū Kāmil, déjà versé dans les sciences mathématiques de son temps, interrogeait un illustre aîné sur les fastes scientifiques d'une époque glorieuse.

Dressons à présent un inventaire des œuvres d'Abū Kāmil. L'imprécision de la langue et des auteurs, la négligence des copistes, ont provoqué là de singulières confusions. On doit à Abū Kāmil :

(A I) Kitāb al-ǧabr wa'l-muqābala, mentionné dans le *Fihrist* (écrit, rappelons-le, en 377 H / 987–88) : (1) à l'article « Abū Kāmil » ; (2) à l'article « al-Iṣṭaḫrī », sous le nom déjà abrégé de *Kitāb al-ǧabr* ; (3) à l'article « al-^{c}Imrānī », sous le nom complet de *Kitāb al-ǧabr wa'l-muqābala*[48].

Ce livre est également cité par Ḥāǧǧī Ḫalīfa qui en rapporte la première page post-laudatoire et en expose les objectifs d'après la préface. Le manuscrit Kara Mustafa Paşa 379 [n. 8] répond en tout point au titre de l'ouvrage et à la description qui en est faite par Ḥāǧǧī Ḫalīfa. On reconnaît dans l'exposé de ce dernier les expressions mêmes de Abū Kāmil[49].

(A II) Kitāb kamāl al-ǧabr wa tamāmuh wa'l-ziyāda fī uṣūlih.

Cet ouvrage est cité par Ḥāǧǧī Ḫalīfa, qui s'exprime ainsi[50] : « Abū Kāmil Šuǧā^{c} ibn Aslam a dit dans (B I) *Kitāb al-waṣāyā bi'l-ǧabr wa'l-muqābala* : J'ai composé un livre connu sous le nom de *Kamāl*

---

43 *op. cit.* [n. 30], I, p. 207.

44 Qadrī Ṭūqān, *Turāṯ al-^{c}arab al-^{c}ilmī*, 2e éd. [Le Caire 1954], p. 139.

45 *Fihrist* [n. 9], p. 398 [p. 275]. Cependant disons que L.Ch. Karpinski met en doute cette information (cf. l'article al-Kh^{w}ārizmī dans l'*Encyclopédie de l'Islam*, II (Leyde 1927), p. 965).

46 Cela ressort des circonstances historiques de l'anecdote rapportée dans *al-Mukāfa'a*, p. 176.

47 Reporter cette date à 260 H / 873–74 serait porter l'âge de Sanad ibn ^{c}Alī à près de 80 ans. Nous ignorons à quel âge celui-ci mourut.

48 *Fihrist* [n. 9], pp. 406, 407, 408 [pp. 281, 282, 283 ; voir aussi Casiri [n. 21], I, p. 411].

49 Ḥāǧǧī Ḫalīfa. *op. cit.* [n. 19], col. 1407–08 [V, pp. 68–69]. [Il nous semble important de reproduire ce texte. Après avoir mentionné l'ouvrage d'Abū Kāmil sur l'algèbre testamentaire (voir ci-après), Ḥāǧǧī Ḫalīfa poursuit en évoquant son livre principal (les passages se retrouvant textuellement chez Abū Kāmil sont mis entre guillemets) : ولابى كامل المذكور كتاب الجبر والمقابلة مجلّد اوله « الحمد للّه اعدل من حكم واحكم من علم » الخ. ذكر انّه كان كثير النظر فى كتب العلمآء بالحساب فرأى « كتابَ محمّد بن موسى الخوارزمى المعروف بالجبر والمقابلة اصحّها اصلاً واصدقها قياساً وكان ممّا يجب له علينا (...) من التقدمة والاقرار له بالمعرفة والفضل اذ كان السابق الى كتاب الجبر والمقابلة والمبتدىء له والمخترع لما فيه من الاصول التى فتح اللّه لنا بها ما كان منغلقاً وقرّب بها ما كان متباعداً وسهل بها ما كان معسّراً (...) ورأيت فيه مسائل ترك شرحها وايضاحها (...) ففرّعتُ منها مسائل كثيرة يخرج اكثرها الى غير الضروب الستّة التى ذكرها الخوارزمى فى كتابه فدعانى الى كشف ذلك وتبيينه (...) فالّفتُ كتاباً فى الجبر والمقابلة ورسمتُ فيه بعض ما ذكره محمّد بن موسى (...) فى كتابه (...) وبيّنتُ شرحه واوضحتُ ما ترك الخوارزمى فى ايضاحه وشرحه » الخ. Sur cet éloge ambigu d'al-Ḫwārizmī par Abū Kāmil, voir pp. 63–64 dans notre *An introduction to the history of algebra*, Providence 2009 —ou p. 67 de l'édition française originale, Lausanne 1999.]

50 *ibid.* [n. 49, pp. 67–68].

al-ǧabr wa tamāmuh wa'l-ziyāda fī uṣūlih, et j'ai prouvé *dans mon deuxième livre* la supériorité et la priorité de Muḥammad ibn Mūsā en algèbre, et réfuté (la prétention) de l'imposteur (?)[51], connu sous le nom de Abū Barza, dans ce qu'il attribue à ᶜAbd al-Ḥamīd, qu'il nomme son grand-père. Quand j'eus montré son incapacité et son peu de savoir dans cette attribution, je jugeai bon de composer un livre sur les questions testamentaires par l'algèbre » [قال ابو كامل شجاع بن اسلم فى كتاب الوصايا بالجبر والمقابلة الفتُ كتاباً معروفاً بكمال الجبر وتمامه والزيادة فى اصوله واقمتُ الحجّة فى كتابى الثانى بالتقدمة والسبق للجبر والمقابلة لمحمّد بن موسى والردّ على المحرّق المعروف بابى بردة ممّا ينسب الى عبد الحميد الذى ذكر انّه جدّه وما بيّنتُ من تقصيره وقلّة معرفته فيما نسب الى جدّه رأيتُ ان اؤلّف كتاباً فى الوصايا بالجبر والمقابلة].

Le texte présente des difficultés. Voici comment nous l'entendons : Après avoir écrit un livre exhaustif sur l'algèbre, *qui est, selon nous,* (A I), *mais que ses contemporains désignaient entre eux sous le titre* (A II) —telle est, pensons-nous, la nuance de sens apportée par le mot « connu »— Abū Kāmil vint à savoir l'audace d'un certain Abū Barza qui, frustrant al-Hwārizmī de son mérite, proclamait ᶜAbd al-Ḥamīd l'auteur de la première algèbre arabe. Abū Kāmil réunit alors le faisceau de preuves qui montraient l'antériorité d'al-H̱wārizmī et la confusion de Abū Barza dans ses prétentions. C'est alors que, pour publier sa réfutation, il composa son (B I) *Kitāb al-waṣāyā bi'l-ǧabr wa'l-muqābala*, livre que tout algébriste arabe se devait, pour ainsi dire, d'écrire, et dont Abū Kāmil pouvait avoir certains éléments tout prêts. Ainsi nous estimons que l'expression *deuxième livre* désigne (B I), le second dans l'ordre chronologique de publication. Ni Ḥāǧǧī H̱alīfa ni Abū Kāmil n'éprouvent le besoin d'expliciter davantage, la phrase ne comportant pas d'autre mention de titre. Pour être déconcertante, cette relation des faits, dans l'ordre inverse de leur déroulement chronologique, n'est pas étrangère aux auteurs arabes.

Remarquons que Ḥāǧǧī H̱alīfa n'a pas pris connaissance de (A II). De plus, ce qui nous autorise, pensons-nous, à nous tenir à cette interprétation du texte cité plus haut de Ḥāǧǧī H̱alīfa, c'est *l'existence d'un passage analogue* sous la plume d'Abū Kāmil. Dans la préface de son livre (A I), *Kitāb al-ǧabr wa'l-muqābala*[52], il écrit : « Je trouvai que le livre de Muḥammad ibn Mūsā al-H̱wārizmī, *connu* sous le nom d'*al-ǧabr wa'l-muqābala* (...) » [فرأيتُ كتابَ محمّد بن موسى الخوارزمى المعروف بالجبر والمقابلة, voir note 49]. Or, il ressort, avec évidence, de l'examen du livre d'al-H̱wārizmī que ce dernier *ne lui a pas donné de titre.* Le mot « connu » traduit donc la nuance que nous soulignions plus haut de Ḥāǧǧī H̱alīfa : « nommé, désigné par les contemporains, non par l'auteur ». On aura, d'ailleurs, une idée de l'imprécision des écrivains quand on saura qu'un commentateur du livre d'al-H̱wārizmī, Sinān ibn al-Fatḥ, trouve naturel d'écrire, contre toute évidence : « Muḥammad ibn Mūsā al-H̱wārizmī a composé un livre *qu'il a nommé al-ǧabr wa'l-muqābala* » [وقد وضع محمّد ابن موسى الخوارزمى كتاباً سمّاه الجبر والمقابلة][53]. Ceci doit donc nous inciter à ne pas accorder à certaines expressions une précision qu'elles n'ont pas dans l'esprit de leur auteur.

Ailleurs, Ḥāǧǧī H̱alīfa écrit[54] : « (B II) *Kitāb al-waṣāyā bi'l-ǧuḏūr, de Abū Kāmil Šuǧāᶜ ibn Aslam.* Le début en est : Louange à Dieu qui parfait la grâce de ses créatures etc. Il (Abū Kāmil) y mentionne qu'il a composé un livre *connu* sous le nom de *Kamāl al-ǧabr wa-tamāmuh* et qu'il a prouvé dans son *deuxième livre* la supériorité et la priorité de Muḥammad ibn Mūsā dans l'algèbre [اوله الحمد لله المتمم نعمته على خلقه الخ. ذكر فيه انّه الّف كتاباً معروفاً بكمال الجبر وتمامه واقام الحجّة فى كتابه الثانى بالتقدمة والسبق للجبر والمقابلة لمحمّد بن موسى الخوارزمى]. Il jugea bon alors [فرأى] de composer un livre sur les testaments où il a mis, d'abord, les questions simples traitées par les fuqahā' [juristes] et une partie de l'ouvrage d'al-Ḥaǧǧāǧ ibn Yūsuf, connu sous le nom de *Kitāb al-waṣāyā*. Puis il a expliqué les questions dont on a besoin et exposé clairement et correctement par l'algèbre, par le dinar et le dirham [بالجبر والمقابلة والدينار والدرهم] les questions qu'on se doit d'exposer ». Ḥāǧǧī H̱alīfa ajoute : « Ce livre (B II)

[51] Le mot est douteux ; il est, suivant les éditions, *al-muḥtarif* ou *al-muḫtāriq*. Nous adoptons *al-muḫtāriq* pour les raisons de sens et de graphie : *iḫtaraqa* = inventer un mensonge [Flügel a محرّق ; mais une autre lecture serait مخرّف, « charlatan »].

[52] Ms. Kara Mustafa Paşa 379 [n. 8], fol. 2^r, 4.

[53] Sinān ibn al-Fatḥ, *Kitāb al-kaᶜb wa'l-māl wa'l-aᶜdād al-mutanāsiba*, Le Caire, Bibliothèque Nationale (*Dār al-kutub*), riyāḍ. 260, fol. 95^v.

[54] Ḥāǧǧī H̱alīfa, *op. cit.* [note 19], col. 1469–70 [V, p. 168].

est un livre agréable, de format moyen, en un volume ».

Le texte que nous venons de citer présente également une ambiguïté vers la fin. Quant au dinar et au dirham, nous pensons qu'ils désignent les deux inconnues x et y[55], de sorte que le livre (B II) aurait utilisé la résolution des équations à une inconnue du premier degré (*bi'l-ğuḏūr*) et aussi celle des systèmes de deux équations linéaires.

Parlant de (A II), Qadrī Ṭūqān, qui nous semble citer Salih Zeki, dit que (A II) était connu sous le nom d'*al-Kāmil*[56]. La remarque est précieuse, eu égard à l'érudition de Salih Zeki.

(A III) Al-Kāmil.

Dans sa petite encyclopédie *Iršād al-qāṣid* [n. 55], al-Akfānī, mort en Egypte en 749 H / 1348–49, cite, dans la rubrique *al-ğabr wa'l-muqābala*, parmi les algèbres développées, *al-Kāmil* de *Abū Šuğā^c ibn Aslam* (sic)[57]. Al-Akfānī avait des connaissances mathématiques valables, mais les renseignements qu'il donne sur les algèbres de ^cUmar al-Ḫayyām et d'al-Samaw'al rendent improbable qu'il ait lu les livres dont il rend compte. Utilisant le *Iršād al-qāṣid* d'al-Akfānī, l'Egyptien al-Qalqašandī (m. en 821 H / 1418–19) dans son *Ṣubḥ al-a^cšā* puis Ṭāšköprüzāde (m. en 962 H / 1554–55) dans son *Miftāḥ al-sa^cāda* reproduisent la citation, faute comprise : « al-Kāmil, de Abū Šuğā^c ibn Aslam ». Celle-ci se retrouve dans le *Kašf al-ẓunūn* de Ḥāğğī Ḫalīfa, empruntée au *Miftāḥ al-sa^cāda*[58].

(A IV) Al-Šāmil, mentionné par Ḥāğğī Ḫalīfa[59].

Disons qu'il existe à la Bibliothèque Nationale de Berlin une feuille manuscrite portant une liste de noms d'ouvrages d'arithmétique et d'algèbre. Il y est écrit : *al-Kāmil fi'l-ğabr wa'l-muqābala ou aš-Šāmil ou al-Kamāl*, de Šuğā^c ibn Aslam, sans qu'il soit possible de savoir si *ou* signifie l'équivalence ou l'exclusion[60].

Or nous sommes convaincu de l'identité des ouvrages (A I), (A II), (A III), (A IV), comme de celle de (B I) et (B II), et nous voudrions dire ici sur quelles raisons nous nous fondons. De toutes les sources d'information que nous possédons, le *Fihrist*, écrit en 377 H / 987–88, reste l'une des plus précieuses, vu son ancienneté, la probité reconnue de son auteur, et la richesse d'information que valait à ce dernier son métier de *warrāq*, « éditeur » de l'époque. Or, l'examen du *Fihrist* amène à la constatation [correction à la main par l'auteur du mot imprimé : construction (on attendrait plutôt (*Réd.*) : conclusion)] suivante, extrêmement importante : *les mathématiciens arabes cités par Ibn al-Nadīm ne donnent pas, en règle générale, de nom à leurs livres.*

Six mathématiciens sont cités par Ibn al-Nadīm comme auteurs d'une algèbre arabe : Sahl ibn Bišr, Sanad ibn ^cAlī, al-Ḫwārizmī (indirectement, dans les articles consacrés à ses commentateurs al-Ṣaidanānī et Sinān ibn al-Fatḥ), Abū Kāmil Šuğā^c ibn Aslam, al-Miṣṣīṣī et al-Dīnawarī[61]. Or, le titre de l'ouvrage dû aux six est invariablement *Kitāb al-ğabr wa'l-muqābala*, simple référence à la matière traitée, d'où nous concluons que, *selon toute vraisemblance, les auteurs ne donnaient pas de nom à leurs*

---

[55] Le mot *dīnār* est employé pour désigner une inconnue dans (A I), ms. Kara Mustafa Paşa 379 [n. 8], fol. 27^v, 17 [lat. 1399], 32^v, 19 [lat. 1689], l'autre inconnue étant désignée par *šay'*. On trouve *šay'* [lat. *res* ; hébr. דבר], *dīnār* [lat. *dinar* ; hébr. דינר], *fals* [lat. *bizantius*, *moneta*, *zoz* (= זוז) ; hébr. פלס] pour désigner trois inconnues aux fol. 38^r, 40^r, 41^r. On trouve aussi *šay'*, *dīnār*, *dirham* [דרהם], *fals* pour quatre inconnues au fol. 95^r [hébr. 181^r, 19]. L'essentiel aux yeux de Abū Kāmil est de distinguer les inconnues par des noms différents, mais peu importent ces noms (fol. 97^r, 14). [Néanmoins, il s'attribue le moyen de réduire le nombre des dénominations d'inconnues à trois par élimination successive des autres ; voir son *Algèbre*, fol. 96^v – 97^r ; ou le commentaire au problème B.381 dans l'édition du *Liber mahameleth*.] Ailleurs, il emploie *šay' kabīr*, *šay' ṣaġīr* (fol. 41^r [lat. *res magna*, *res parva* ; hébr. דבר גדול, דבר קטן], fol. 41^v). L'expression *ḥisāb al-dirham wa'l-dīnār* pour désigner les systèmes d'équations à plusieurs inconnues est utilisée par Ibn al-Akfānī, *Iršād al-qāṣid*, Beyrouth 1322 H / 1904–5, p. 126. Celui-ci est suivi par al-Qalqašandī, *Ṣubḥ al-a^cšā fī ṣinā^cat al-inšā'*, Le Caire 1914–28 (14 vol.), vol. 14, p. 220, et Ṭāšköprüzāde, *Miftāḥ al-sa^cāda*, Haydarabad 1329 H / 1911, vol. I, p. 328.

[56] Qadrī Ṭūqān, *op. cit.* [n. 44], p. 133.

[57] *op cit.* [n. 55], p. 125.

[58] Qalqašandī, *op. cit.* [n. 55], vol. I, p. 475 ; Ṭāšköprüzāde, *op. cit.* [n. 55], vol. 1, p. 327 ; Ḥāğğī Ḫalīfa, *op. cit.* [n. 19], vol. 1, col. 1381 [vol. V, p. 27 : كامل فى الجبر والمقابلة لابى شجاع ابن اسلم].

[59] *op. cit.* [n. 19], vol. 1, col. 579 [II, p. 585 (الشامل) ; IV, p. 10 (شامل فى الجبر والمقابلة)].

[60] W. Ahlwardt, *Verzeichniss der arabischen Handschriften der Königlichen Bibliothek zu Berlin* [Berlin 1887–1899 (10 vol.)], n° 6013.

[61] *Fihrist* [n. 9], pp. 397, 398, 404, 406, 122 [pp. 274, 275, 280, 281, 78].

ouvrages, de sorte que ceux-ci étaient désignés par le nom de la matière et de l'auteur[62]. Cette conclusion trouve sa confirmation dans l'examen des ouvrages qui nous sont conservés de cette époque.

En voici quelques exemples :

1°) Le fameux livre d'algèbre d'al-H̱wārizmī n'a pas reçu de titre de son auteur, comme il ressort clairement de la préface[63].

2°) On doit au même auteur une *maqāla* [= un opuscule] dont le titre, *Istiḫrāğ tārīḫ al-yahūd wa aᶜyādihim*, est visiblement l'œuvre du copiste ; le texte en témoigne[64].

3°) L'*Algèbre* d'Abū Kāmil, conservée dans le manuscrit Kara Mustafa Paşa 379 [n. 8], n'a pas reçu de titre de son auteur, comme le montrent la préface et le texte. S'exprimant de la même manière qu'al-H̱wārizmī, Abū Kāmil écrit dans la préface (fol. 2^v, 2) : « Je composai donc *un* livre [كتاباً] sur al-ğabr wa'l-muqābala » (entendez : le présent livre d'algèbre).

4°) Il nous reste d'Abū Kāmil un écrit mathématique de 17 pages, conservé à Leyde sous le nom de *Kitāb ṭarā'if al-ḥisāb de Abū Kāmil al-Miṣrī*[65], et à Paris sous le nom de *Risāla fī ᶜilm al-ḥisāb de Kāmil Šuğāᶜ ibn Aslam dit Abū Kāmil*. [Cette dernière indication est sans doute une addition postérieure, elle figure dans la table des matières sur la page de couverture du recueil. Dans le *Catalogue des manuscrits arabes des nouvelles acquisitions (1888–1924)* de E. Blochet (Paris 1925), on lit « problèmes d'arithmétique par Abou Kamil Shodjaᶜ ibn Aslam ». Le début du texte (fol. 3^v), lui, ne possède pas de titre et commence, après la *bismillah*, par قال شجاع بن اسلم المعروف بابى كامل][66]. L'examen de cet écrit montre que *l'auteur ne lui a pas donné de nom*, d'où l'embarras des copistes et des lecteurs. C'est cet écrit que les Anciens, comme Ibn al-Nadīm, ont désigné sous le titre de *Kitāb al-ṭair*[67] utilisant un détail caractéristique des énoncés. La preuve nous en est heureusement fournie par al-Samaw'al ibn Yaḥyā al-Mağribī, qui mentionne dans son algèbre *al-Bāhir*[68] la deuxième question du *Kitāb al-ṭair* de Abū Kāmil [comme exemple d'un problème ayant de nombreuses réponses : ومثال ما له اجوبة كثيرة متناهية نريد ان نشترى بمائة درهم مائة طائر من ثلاثة اصناف بطّ وحمام ودجاج وكلّ بطّة بدرهمين وكلّ ثلاث حمامات بدرهم وكلّ دجاجتين بدرهم, et on lit à la fin du traitement : وهذه المسألة هى الثانية من كتاب الطير لأبى كامل]. Elle est aussi la deuxième question des manuscrits de Leyde et Paris (texte original : فان دفع اليك مائة درهم فقيل لك ابتع بهذه مائة طائر من ثلاثة اصناف حمام وبطّ ودجاج البطّ بدرهمين درهمين والحمام كلّ ثلاثة بدرهم والدجاج كلّ دجاجتين بدرهم). Son énoncé permet de comprendre l'appellation de *Kitāb al-ṭair* : il s'agit d'acheter pour cent dirhams cent volailles de trois espèces différentes : des oies, des pigeons et des poules, sachant qu'une oie vaut deux dirhams, trois pigeons valent un dirham et deux poules également un dirham.

Nommons encore quelques écrits anciens édités ces dernières années et qui accusent cette absence de titre :

— Une *maqāla* appelée aussi *Istiḫrāğ t'ariḫ al-yahūd* due à Ibn Bāmšād al-Qā'inī ; le titre, absent du texte, est dû à un copiste[69].

---

[62] On s'en convaincra en dépouillant les noms des livres mathématiques depuis l'époque d'al-H̱wārizmī jusqu'à celle d'Ibn al-Nadīm dans le *Fibrist* (pp. 390–410 [pp. 271–284]). Ceci est une règle générale mais elle peut souffrir des exceptions.

[63] Al-H̱wārizmī, *Kitāb al-ğabr wa'l-muqābala*, éd. Le Caire 1939, d'après le ms. de la Bibliothèque d'Oxford [Hunt. 214 ; aussi utilisé pour son édition par Fr. Rosen, *The algebra of Mohammed ben Musa*, Londres 1831]. Le fait est confirmé également par le ms. 5955 de la Bibliothèque Nationale de Berlin, ouvrage anonyme portant le titre de *Kitāb fi'l-misāḥa wa'l-waṣāyā* que nous avons identifié avec le *Kitāb al-ğabr wa'l-muqābala*, d'al–H̱wārizmī.

[64] *Al-Rasā'il al-mutafarriqa fi'l-hai'a li'l-mutaqaddimīn wa muᶜāṣiray al-Bīrūnī*, Haydarabad 1948, n° 1. [Le (ou : un) titre *t'arīḫ al-yahūd* apparaît à la fin de l'opuscule.]

[65] Ms. Leyde Cod. Or. 199 [fol. $50^v – 58^v$]. Nous en devons la photocopie à l'obligeance du Dr. P. Voorheeve que nous sommes heureux de pouvoir remercier ici. [Entre-temps reproduit à partir d'un microfilm (en blanc sur noir) dans un article de A. Saidan sur cet opuscule, voir مجلّة معهد المخطوطات, 9 (1963), pp. 294–310].

[66] Ms. arabe 4946 de la Bibliothèque Nationale de Paris. Nous sommes heureux, à cette occasion, de remercier la Direction de la Bibliothèque Nationale de Paris pour les nombreux microfilms de manuscrits arabes que nous lui devons.

[67] *Fibrist*, p. 406 [p. 281].

[68] *Al-Bāhir*, ms. Istanbul, Aya Sofya 2718, fol. $102^r – 103^r$ [pp. 229–230 de l'édition entre-temps publiée par Salah Aḥmad et Roshdi Rashed, *Al-Bahir en algèbre d'As-Samaw'al*, Damas 1972].

[69] *Rasā'il* [n. 64], n° 3.

— Une *maqāla* intitulée *Fī istiḫrāǧ sāʿāt (...)* qui, après l'invocation du nom de Dieu très miséricordieux, commence sans transition : Abū'l-Ḥasan ʿAlī ibn ʿAbdallah ibn Muḥammad ibn Bāmšād al-Qā'inī dit : « On m'a demandé comment calculer les heures qui séparent l'aube du lever du Soleil »[70].

Ce dernier exemple suggère l'existence d'une vaste catégorie d'écrits que l'auteur devait laisser sans titre : celle des lettres-réponses aux questions posées par des correspondants.

Poursuivant notre examen du *Fihrist*, nous y relevons des livres d'arithmétique sous le nom uniforme de *Kitāb al-ǧāmʿ wa'l-tafrīq* dus à Sanad ibn ʿAlī, al-Ḫwārizmī (cité indirectement à l'article Ṣaidanānī), Abū Kāmil, Sinān ibn al-Fatḥ, al-Dīnawarī[71]. Le même fait se répète à l'occasion d'autres sciences, telles que l'usage de l'astrolabe (*al-ʿamal bi'l-asṭurlāb*), la mesure des surfaces et des volumes (*al-misāḥa*), les questions testamentaires où l'on peut relever les *Kitāb al-waṣāyā* de Sinān ibn al-Fatḥ, d'al-Miṣṣīṣī, d'al-Karābīsī et d'al-Dīnawarī[72].

Ce fait est donc caractéristique d'une époque. En général, les publications mathématiques sont reconnues grâce au nom de l'auteur et de la matière. Mais l'on conçoit que ce fait s'étende au domaine de la production littéraire. Ainsi, on doit à Abū ʿUbaida al-Taimī, mort en 209 H / 824–25, un livre sur les chevaux[73] ; à al-Ǧahšiyārī, mort en 331 H / 942–43, un livre sur les vizirs et les kuttāb [= secrétaires][74] ; à Abū Muḥammad al-Balawī un livre sur la vie d'Aḥmad ibn Ṭūlūn écrit en 330 H / 941–42[75]. Le texte montre d'une façon formelle que l'auteur a laissé son ouvrage sans titre. Cette coutume se perpétuera longtemps et on la retrouvera dans les livres connus sous le nom d'*al-Aḥkām al-sulṭāniya*, écrits l'un par Abū'l-Ḥasan al-Māwardī, mort en 450 H / 1058–59, l'autre par Abū Yaʿlā al-Ḥanbalī, mort en 458 H / 1065–66[76].

Dans les cinq premières années du Vième siècle hégirien, le mathématicien al-Karaǧī nommera formellement un de ses ouvrages *al-Badīʿ* (le Merveilleux)[77], mais il n'a pas donné de titre explicitement à son arithmétique désignée par la suite sous le nom d'*al-Kāfī* : le texte de l'ouvrage le montre (du moins tel qu'il nous a été conservé)[78].

Les observations que nous avons faites sur les noms des ouvrages dans le *Fihrist*, et qui sont confirmées par al-Qifṭī, Ibn abī Uṣaibiʿa, Yāqūt al-Ḥamawī et d'autres auteurs anciens, font donc éclater la singularité et, disons-le, l'invraisemblance de titres tels que *Kitāb kamāl al-ǧabr wa tamāmuh wa'l-ziyāda fī uṣūlih*, ou *al-Šāmil*, ou *al-Kāmil*. Nous pensons, quant à nous, que dans la bouche d'Abū Kāmil [n. 50] ce titre interminable *Kitāb kamāl al-ǧabr (...)* souligne l'estime de ses contemporains et sa satisfaction propre qui paraît en maints endroits de son livre (A I)[79].

[Cette opinion de l'auteur sur l'absence de titre dans les ouvrages est originale, mais n'est, selon nous, pas assurée —sauf pour une « lettre » (*risāla*), par nature plus courte. Il semble naturel, et il l'a semblé autrefois, que tout ouvrage devait avoir un titre et un nom d'auteur, ne serait-ce que pour la commodité du lecteur ou de l'acquéreur. Mais la multiplicité des titres identiques indiquant le sujet (à l'instar des livres d'aujourd'hui intitulés « Analyse mathématique », « Précis de littérature française » ou « La cuisine à la portée de tous ») a pu amener à modifier ou augmenter leur titre ultérieurement. De plus, la page

---

70 *Rasā'il* [n. 64], n° 4.

71 *Fihrist*, pp. 398, 404, 406 et 122 [pp. 275, 280, 281 et 78].

72 *Fihrist*, pp. 406 et 122 [pp. 281–282 et 78].

73 Abū ʿUbaida al-Taimī, *Kitāb al-Ḫail*, Haydarabad 1358 H / 1939–40.

74 Voir note 38.

75 Voir note 28.

76 Abū'l-Ḥasan al-Māwardī, *al-Aḥkām al-sulṭanīya*, Le Caire 1380 H / 1960–61 ; Muḥammad ibn Abī Yaʿlā al-Ḥanbalī, *al-Aḥkām al-sulṭanīya*, Le Caire 1356 H / 1937–38.

77 *Kitāb al-badīʿ fī'l-ḥisāb*, Bibl. apost. vatic., ms. Barb. Or. 36 [entre-temps publié par A. Anbouba lui-même, v. n. 3].

78 *Kāfī li'l-Karaǧī*, ms. Istanbul, Saray, Ahmet III, n° 3135 [n. 3]. Signalons que nous avons identifié avec le *Kāfī* l'ouvrage signalé, sous le nom différent de *Muḫtasar fī'l-ḥisāb wa'l-misāḥa*, cat. *Fibris al-funūn al-mutanawwiʿa*, Bibliothèque Municipale d'Alexandrie (1929, p. 80 [*Fihrist maḫṭūṭāt al-maktaba al-baladīya fī'l-Iskandarīya* (6 vol.), Alexandrie 1926–1929]) et mentionné par C. Brockelmann dans sa *Geschichte der arabischen Literatur* [I, p. 247 (éd. 1943)], art. « al-Karaǧī ».

79 Il est plaisant de voir Abū Kāmil ponctuer sa satisfaction, devant certaines de ses solutions, par l'expression *wa lā qūwata illa bi'llah*, ms. Kara Mustafa Paşa 379, fol. 96^v, 3–4 ; fol. 100^v, 14. [De fait, ceci indique seulement une transition, comme on le voit dans sa Géométrie pratique étudiée dans ce volume pour les diverses expressions eulogiques.]

de titre est souvent la plus maltraitée lors de la transmission, qu'elle soit gâtée, remplie d'annotations de lecteurs ou de noms de propriétaires du manuscrit, ou, simplement, perdue. C'est peut-être ce qui est arrivé au texte de géométrie pratique étudié dans ce volume. Ceci ne doit pourtant pas infirmer les conclusions d'Anbouba sur l'identité d'ouvrages traitant d'un même sujet et écrits par un même auteur mais mentionnés sous des noms différents, ses arguments étant parfaitement judicieux.]

L'inexistence des ouvrages (A II), (A III), (A IV), en tant qu'œuvres distinctes de (A I) ressort pour ainsi dire du *Fihrist*. Ibn al-Nadīm nomme neuf ouvrages d'Abū Kāmil, cinq de Sanad ibn ᶜAlī, trois d'al-Ṣaidanānī, deux d'Abū Barza, deux de ᶜAbd al-Ḥamīd, six de Sinān ibn al-Fatḥ, huit d'al-Miṣṣīṣī, cinq d'al-Karābīsī[80]. Ceci montre l'importance relative accordée à Abū Kāmil parmi les algébristes. Or, citant des écrits secondaires tels que le *Kitāb al-ṭair* ou le *Kitāb al-ᶜaṣīr* d'Abū Kāmil, Ibn al-Nadīm négligerait des œuvres fondamentales telles que (A II), (A III), (A IV) ? Alors qu'à l'article al-H̱wārizmī il parle du premier et du deuxième *Zīǧ* (Tables astronomiques) de celui-ci ; à l'article Euclide, de la première traduction des *Eléments* et de la deuxième[81] ; à l'article Ibn Amāǧūr du « Kitāb al-zīǧ dit [المعروف] al-Muzannar », du « Kitāb al-zīǧ dit al-Badīᶜ », du « Kitāb al-zīǧ dit al-H̱ālis »[82] ; à l'article al-Battānī de ses deux *zīǧ*, première et deuxième copie[83].

De même, citant les commentaires d'al-Iṣṭaḫrī et d'al-ᶜImrānī, Ibn al-Nadīm les appelle respectivement « Šarḥ kitāb al-ǧabr de Abū Kāmil », « Šarḥ kitāb al-ǧabr wa'l-muqābala de Abū Kāmil », sans craindre que cette légère altération du titre ne crée d'ambiguïté et sans sentir le besoin de souligner qu'il s'agit d'un seul et même livre d'Abū Kāmil, tant pour lui et pour ses lecteurs la chose allait de soi ; alors que parlant de deux Abū Yaḥyā al-Marwazī il signale qu'ils sont différents[84]. Objecte-t-on qu'Ibn al-Nadīm ignorait l'existence de (A II), (A III), (A IV) (remarquons qu'il ne nomme pas (B I), (B II)) ? Mais alors il est un fait absolument singulier. C'est que, au cours des siècles, tous les auteurs qui ont eu à citer l'*Algèbre* d'Abū Kāmil en aient connu une et ignoré trois. Ainsi al-Akfānī a connu le seul *al-Kāmil* ; al-Qurašī a commenté *al-Šāmil*[85] ; Ibn al-Nadīm cite le seul « al-ǧabr wa'l-muqābala », le seul également que Ḥāǧǧī H̱alīfa ait pu consulter. *Le fait sort des limites du probable et il est plus naturel d'admettre que les quatre ouvrages n'en font qu'un.* Mais il existe cependant un auteur qui nous a entretenu [lire plutôt : un auteur qui mentionne les noms] de deux algèbres sur quatre d'Abū Kāmil. C'est Ibn al-Hā'im, mort en 815 H / 1412–13, qui dans son commentaire d'*al-Yasmīnīya*, écrit : « Le premier à avoir fait connaître l'algèbre et à avoir composé là dessus est al-Ustāḏ [= Maître] Muḥammad ibn Mūsā al-H̱wārizmī, Dieu l'ait en sa miséricorde ; son mérite est inscrit dans les histoires et *son livre* est notoirement connu ; parmi les ouvrages développés (*mabsūṭa*), les plus précieux aux connaisseurs sont le livre appelé *al-Faḫrī* et *al-Kitāb al-šāmil al-kāmil* dû à *l'Imām Abū Kāmil* (والكتاب الشامل الكامل المنسوب للامام ابى كامل) ; et, parmi les livres intermédiaires (*mutawassiṭāt*), *al-Badīᶜ* de l'auteur du *Faḫrī* »[86]. Peut-on taxer Ibn al-Hā'im, auteur de plusieurs ouvrages d'arithmétique et d'algèbre, et érudit, d'ignorer la production d'Abū Kāmil ? Encouragé par le jeu de mots, il vient de nous révéler que le livre d'algèbre d'Abū Kāmil s'appelle *al-Šāmil* et *al-Kāmil*, épithètes qui soulignent son intégrité [lire plutôt (*Réd.*) : unicité]. En même temps, il nous dévoile le processus de prolifération des titres pour un même ouvrage. *Un titre peut être remplacé par un synonyme, un ouvrage peut être désigné par ses objectifs, sa matière, une qualité que l'on jugera caractéristique.*

80 *Fihrist*, pp. 397–406 [pp. 275, 280–282].

81 *Fihrist*, p. 385 [pp. 274 (al-H̱wārizmī), 265].

82 *Fihrist*, p. 404 [p. 280].

83 *Fihrist*, p. 404 [p. 279 (نسختان اولى وثانية)].

84 *Fihrist*, p. 382 [p. 263]. C'est d'ailleurs une coutume assez courante chez les historiens arabes, quand une confusion est possible, d'en prévenir le lecteur.

85 Ḥāǧǧī H̱alīfa, *op. cit.*, col. 579 [II, p. 585].

86 *Šarḥ ibn al-Hā'im ᶜalà'l-Yasmīnīya*, ms. (Le Caire IV, 189), p. 1 [p. 57 de l'édition par Mahdi Abdeljaouad, *Sharh al-Urjūza al-Yasmīnīya*, s.l. s.d. [Tunis 2004]]. Nous avons traduit par « dû à » l'expression arabe *al-mansūb ilà*, que le lecteur moderne croit mieux rendre par « attribué à ». Mais dans la pensée d'Ibn al-Hā'im il n'existe aucun doute sur l'appartenance de l'ouvrage à Abū Kāmil ; et l'expression *mansūb ilà* dans les textes anciens n'implique pas la nuance de doute qu'elle suggère aux modernes. Ainsi le géomètre Abū'l-Ǧūd Muḥammad ibn al-Laiṯ écrit : « Et j'y fis la *Risāla al-mansūba ilaiya* [qu'on devra donc ici nécessairement traduire par l' « opuscule *dû à* moi-même »] en 358 H [968–69] ».

Si l'on ne veut pas admettre l'identité de (A I), (A II), (A III), (A IV), il faudra prendre en considération que les traductions latines ou hébraïques faites au Moyen Âge, et que nous avons signalées en tête de cet article, appartiennent toutes soit au *Kitāb al-ṭair*, soit à (A I) ; en particulier, *le traité du pentagone et du décagone —ms. Paris lat. 7377A, n° 5— traduit par Suter et Sacerdote* [ce dernier de l'hébreu uniquement] *n'est qu'un chapitre* [le Livre IV] *de l'algèbre* (A I)*, ms. Kara Mustafa Paşa 379, fol.* 67^{r}*–*79^{v}. De même, le texte latin traduit en anglais par Karpinski, le texte hébraïque traduit par Weinberg [n. 14][87] sont des extraits de (A I). Tous les emprunts faits à Abū Kāmil par al-Karağī et Léonard de Pise [pour ce dernier la transmission *directe* n'est pas assurée] et signalés par les historiens des sciences sont puisés dans (A I).

[Le texte de l'auteur est ici quelque peu confus, mais il est lui-même la victime d'informations erronées. Son intention est de mentionner que toutes les traductions ou citations ou allusions à l'*Algèbre* d'Abū Kāmil existantes ne s'appliquent qu'à l'*Algèbre* d'Abū Kāmil tel que nous la connaissons (donc son A I). Quant à l'allusion qu'il fait au *Kitāb al-ṭair*, elle n'a pas lieu d'être. D'abord parce que c'est un tout autre sujet, ensuite parce qu'il n'en existe ni traduction latine ni hébraïque. Anbouba reprend ici une opinion qui remonte à une affirmation de Moritz Steinschneider aux pp. 408–409 de son « Zur Geschichte der Übersetzungen aus dem Indischen ins Arabische », *Zeitschrift der deutschen morgenländischen Gesellschaft*, 25 (1871), pp. 378–428. Mais cette allégation est dénuée de tout fondement, et repose seulement sur une interprétation d'un titre malencontreusement ajouté à la partie de l'*Algèbre* d'Abū Kāmil sur les équations indéterminées (תחבולות המספר, correspondant il est vrai assez bien à طرائف الحساب —quoique חשבון eût mieux convenu que מספר). Pourtant la lecture des premiers mots montre clairement que cet écrit n'a aucun point commun avec celui du manuscrit de Leyde ; la remarque de Steinschneider *die hebräische Übersetzung scheint vollständiger als die Leydener Hs* laisse donc pantois. Remarquons d'ailleurs que les titres dans la traduction en hébreu sont la conséquence d'un coup d'œil rapide sur le contenu ; ainsi, celui de l'algèbre proprement dite est חשבון השטחים, « calcul des surfaces », sans doute une impression laissée à la vue des figures illustrant l'enseignement de l'algèbre.]

Maintenant s'il existait quatre algèbres exhaustives d'Abū Kāmil, comment le seraient-elles toutes les quatre ? Ou bien elles sont un seul et même ouvrage —ce que nous prétendons— ou bien l'une d'entre elles représente un progrès par rapport aux autres et a dû finir par s'imposer et les faire tomber dans l'oubli, ce qui n'a pas eu lieu. Il est, de plus, inadmissible qu'un auteur donne à trois de ses ouvrages des titres synonymes, et nous avons vu que les noms épithètes *al-Kāmil*, *al-Šāmil* et autres sont contraires aux traditions de l'époque d'Abū Kāmil et doivent *a priori* soulever des doutes. Est-ce d'ailleurs une coïncidence si Ibn al-Nadīm ne cite qu'une *Algèbre* par auteur, O un *Kitāb al-waṣāyā*, un *Kitāb al-ǧam*^{c} *wa'l-tafrīq* ? Cette coïncidence se répète avec une telle fréquence qu'elle est ou bien le fruit d'un tri systématique de la part d'Ibn al-Nadīm —mais cette hypothèse est absurde pour qui connaît le *Fihrist* et les lexicographes arabes[88]— ou bien alors elle est l'image d'une réalité historique. La vérité qui s'impose est que, jusqu'à l'époque d'al-Karağī, un algébriste composait un ouvrage sur les principes de l'algèbre, un sur les waṣāyā, un sur le calcul indien, un sur l'arithmétique sans calcul indien. Il publiait, à côté, des écrits secondaires, sur des points particuliers de l'algèbre par exemple[89]. La raison en est dans la lenteur évidente avec laquelle la science se développait. Or, l'analyse de l'ouvrage (A I) d'Abū Kāmil va nous montrer que cet auteur a, vraiment, tout dit pour son époque.

Sur beaucoup de points, le livre d'Abū Kāmil reste supérieur au *Faḫrī* d'al-Karağī, paru plus d'un siècle plus tard. Il est impossible dans ces conditions d'admettre l'existence de (A II), (A III), (A IV) en tant que livres distincts de (A I).

[87] A en juger du moins [Anbouba n'avait pas pu consulter la traduction de Weinberg] par les nombreuses citations qui en sont faites dans J. Tropfke, *Geschichte der Elementar-Mathematik*, T. III (Berlin 1937), pp. 43, 75–76, 78, 80.

[88] Voir plus haut le dénombrement des *Zīğ* ; voir aussi le *Fihrist*, par exemple aux articles « Abū Ma^{c}šar », « Māšā'allāh », « al-Kindī », pp. 401, 396, 371–379 [pp. 277, 273, 255–261].

[89] On peut classer dans cette catégorie : Ṯābit ibn Qurra, *Qaul fī taṣḥīḥ masā'il al-ǧabr bi'l-barāhīn al-handasīya*, ms. Ayasofya 2457/3, fol. 39^{r} – 41^{v} [texte, traduction et commentaire par P. Luckey, « Ṯābit b. Qurra über den Richtigkeitsnachweis der Auflösung der quadratischen Gleichungen », *Berichte über die Verhandlungen der sächsischen Akademie der Wissenschaften, math.-phys. Kl.*, 93 (1941), pp. 93–114.] ; al-Karağī, ^{c}*Ilal al-ǧabr wa'l-muqābala*, Oxford, Seld. 3234. Ce dernier ouvrage tient en 24 pages et quelques lignes. Le texte montre que le titre a été ajouté par un copiste [opuscule édité en 1991 à Ankara par le Atatürk kültür merkezi].

Les considérations développées plus haut valent pour (B I) et (B II). En examinant de plus près les fragments que nous avons reproduits de Ḥāǧǧī Ḫalīfa[90], on s'aperçoit qu'en résumant les préfaces du
(B I) *Kitāb al-waṣāyā bi'l-ǧabr wa'l-muqābala*
et du
(B II) *Kitāb al-waṣāyā bi'l-ǧuḏūr*,
notre lexicographe rend compte, en réalité, d'un seul et même livre. Les titres de (B I) et (B II) sont, au degré de précision des copistes et auteurs arabes, synonymes. Le mot *ǧuḏūr* est le pluriel de *ǧiḏr*, « racine », « inconnue au premier degré ». [Ici l'auteur a biffé à la main la phrase, pourtant (*Réd.*) appropriée (et liée à la phrase qui suit, qui, elle, n'est pas biffée) : *Al-waṣāyā bi'l-ǧuḏūr* signifierait que seules les équations du premier degré interviennent dans la résolution de ces questions testamentaires.] Tel est, d'ailleurs le cas du *Kitāb al-waṣāyā* d'al-Ḫwārizmī[91]. Il est possible que le titre de (B I) ait été suggéré par celui d'al-Ḥaǧǧāǧ ibn Yūsuf : *Kitāb al-waṣāyā bi'l-ǧuḏūr*. Nous concluons pour notre part à l'identité de (B I) et (B II)[92].

Au terme de cette étude, il ne nous reste plus qu'à emprunter à Ibn al-Nadīm sa liste des ouvrages d'Abū Kāmil [par * nous (*Réd.*) notons les ouvrages conservés] :

Kitāb al-falāḥ, *Kitāb miftāḥ al-falāḥ*, *Kitāb al-ǧabr wa'l-muqābala**, *Kitāb al-ᶜaṣīr*, *Kitāb al-ṭair**, *Kitāb al-ǧamᶜ wa'l-tafrīq*, *Kitāb al-ḫaṭā'ain*, *Kitāb al-misāḥa wa'l-handasa**, *Kitāb al-kifāya*.

Nous considérons que les titres sus-nommés sont dûs aux usagers de ces livres, non à Abū Kāmil lui-même. De plus, des noms comme *al-falāḥ*, *al-kifāya*, ne permettent pas d'identifier la matière traitée dans l'ouvrage. Ajoutons à cette liste *kitāb al-waṣāyā* et ne soyons pas surpris que Abū Kāmil ait encore écrit un livre sur le calcul indien, qui pourrait bien être —simple conjecture de notre part— le premier ou le dernier de la liste d'Ibn al-Nadīm.

Voici le tableau des principales questions [les références aux fol. sont parfois approximatives] :

1) (ff. $3^r - 11^v$) La résolution, accompagnée de démonstration, des équations :

$$ax^2 = bx, \qquad ax^2 = c, \qquad bx = c$$

$$ax^2 + bx = c, \quad ax^2 + c = bx, \quad bx + c = ax^2.$$

2) (ff. $12^v - 16^v$) Règles de calcul de produits, avec démonstrations :

$$a \cdot x = ax, \qquad ax \cdot bx = abx^2$$

$$(a + x)\,x \cdot (a - x)\,x, \quad (a \pm x)^2, \quad (a + x)(a - x).$$

3) (fol. 19^r) Règles de calcul des radicaux, avec démonstrations :

$$\sqrt{a} \cdot \sqrt{b} = \sqrt{ab}, \qquad \frac{\sqrt{a}}{\sqrt{b}} = \sqrt{\frac{a}{b}}.$$

4) (ff. $26^r - 27^r$) Règles de calcul de fractions, avec démonstrations :

[90] *Kašf al-ẓunūn* [n. 19], col. 1407, col. 1469–1470 [V, pp. 68, 168–169].

[91] Bien que le *Kitāb al-waṣāyā* d'al-Ḫwārizmī semble faire partie de son *Algèbre*, nous pensons qu'il constitue un ouvrage séparé qui aura été mis à la suite de l'*Algèbre* par la faute d'un copiste. [Cette note a été biffée à la main par l'auteur, à juste titre : al-Ḫwārizmī fait allusion à cette partie dans l'introduction de son *Algèbre*.]

[92] Dans notre article « Al-Karaǧī », *al-Dirāsāt al-adabīya*, T. II et III, 1959, p. 84, nous avions distingué (B I) et (B II), suivant en cela l'opinion courante. Rappelons qu'il existait un manuscrit de (B II) à Mossoul, dans la bibliothèque de ᶜAlī al-Ṣā'iġ (cf. Dāwūd al-Čalabī, *Maḫṭūṭāt al-Mauṣil*, Bagdad 1346 H / 1927–28, p. 294). [Suit une note manuscrite de l'auteur :] Les démarches que nous avons faites à Mossoul pour obtenir le ms. n'ont pas pu aboutir, la bibliothèque de ᶜAlī al-Ṣā'iġ ayant été dispersée à sa mort. [Selon une lettre que nous (*Réd.*) avions reçue en 1970 du Conservateur de la Bibliothèque de l'Université à Mossoul, le manuscrit d'Abū Kāmil ne se trouvait pas dans sa Bibliothèque, quoique la famille de ᶜAlī al-Ṣā'iġ lui eût assuré que tant ses livres que ses manuscrits avaient été acquis par ladite Bibliothèque.]

$$\frac{a}{b} = \frac{a^2}{ab}, \qquad \frac{a}{b} + \frac{b}{a} = \frac{a^2+b^2}{ab}, \qquad \frac{a}{b} \cdot \frac{b}{a} = 1$$

$$\frac{a}{b} \cdot \frac{c}{d} = \frac{ac}{bd}. \qquad \text{(fol. } 66^r\text{)}$$

5) (ff. 27^v, 32^v, $38^r - 41^v$, $95^r - 100^v$) Systèmes d'équations linéaires à plusieurs inconnues.

6) (ff. $35^r - 36^r$, 48^r, $49^v - 56^v$) Equations irrationnelles à une inconnue.

7) (ff. $25^v - 34^v$, $36^v - 40^v$) Equations fractionnaires à une inconnue.

8) (ff. $67^v - 78^v$) Calcul du côté de polygones réguliers (pentagone, décagone, pentadécagone) en fonction du rayon du cercle (inscrit ou circonscrit) et inversement.

9) (ff. $101^r - 102^r$) Progressions arithmétiques et problèmes divers.

10) (ff. $103^r - 108^r$) Problèmes de courriers, de robinets, de mélanges, de salaires, et problèmes divers.

11) (ff. $108^v - 110^v$) Sommations de $1^2 + 2^2 + \cdots + n^2$, $1 + 2 + 2^2 + \cdots + 2^n$.

12) (ff. $21^v - 67^r$) Equations qui se ramènent au second degré (38 problèmes).

13) (ff. $79^r - 95^r$) Equations indéterminées à solutions rationnelles (38 problèmes). Exemples :

$$x^2 + y^2 = 5,$$

$$\begin{cases} x^2 - 2x = y^2 \\ x^2 - 3x = z^2, \end{cases} \qquad \begin{cases} x^2 - 5 = y^2 \\ x^2 - 5 + \sqrt{x^2 - 5} = z^2. \end{cases}$$

[Appendice]

[Nous (*Réd.*) ajoutons ci-après la division de l'ouvrage d'Abū Kāmil, en renvoyant à la foliotation et aux lignes, ou à la pagination et aux lignes, des textes manuscrits respectivement imprimés [nn. 8, 11, 14].

Livre I (Original arabe [n. 8] : fol. $1^r - 21^v$, 16 ; traduction latine [n. 11] : lignes 1 – 1057 ; traduction hébraïque [n. 14] : (éd. Levey) pp. 29 – 83, 14, Ms. Munich hébr. 225, $95^r - 111^r$). Résolution des six formes réduites des équations des deux premiers degrés ayant des coefficients positifs et au moins une solution positive ($bx = c$, $ax^2 = bx$, $ax^2 = c$, $ax^2 + bx = c$, $ax^2 = bx + c$, $ax^2 + c = bx$), et illustration géométrique de la formule de résolution. Calculs avec des expressions algébriques contenant une inconnue et des racines carrées.

Le contenu est à peu près le même que chez son prédécesseur al-H̲wārizmī, mais le niveau est plus élevé : son ouvrage est destiné non à des débutants, mais à des mathématiciens déjà formés, c'est-à-dire ayant pleine connaissance des *Eléments de géométrie* d'Euclide.

Livre II (fol. 21^v, 16 – 42^r, 8 ; latin : lignes 1058 – 2263 ; hébreu : (éd. Levey) pp. 83, 14 – 145, 19–21, Ms. Munich hébr. 225, $111^r - 130^v$ (lacune)). A nouveau comme chez son prédécesseur, Abū Kāmil présente six exemples de problèmes menant à chacune des équations, ainsi que la résolution de divers problèmes d'applications. Mais le niveau est ici encore plus élevé, et enrichi de démonstrations géométriques des formules utilisées ; Abū Kāmil fera école pour ce dernier point, comme on le voit dans le *Liber mahameleth* de Johannes Hispalensis (XIIième siècle).

Livre III (fol. 42^r, 9 – 67^r, 18 ; latin : lignes 2264 – 3559 ; hébreu : (éd. Levey) pp. 145, 22 – 215, Ms. Munich hébr. 225, (lacune) $130^v - 154^r$). Dans l'introduction de son ouvrage, Abū Kāmil mentionnait (fol. 2^r, 13–14) qu'il présenterait des problèmes menant à d'autres types d'équations que celles qu'avait expliquées al-H̲wārizmī dans son livre (غير الضروب الستّة التى ذكرها الخوارزمى فى كتابه ; voir ci-dessus note 49). Il répète ceci au début du livre III (وجدتُ مسائل كثيرة من حساب الجبر تخرجك الى غير الضروب الستّة التى ذكرتُها فى كتابى هذا ; notez l'élimination de la mention du livre d'al-H̲wārizmī !). De fait, ces nouvelles équations sont les anciennes, mais avec des coefficients, ainsi que les solutions, qui sont des quantités irrationnelles (racines carrées). La théorie n'est donc pas nouvelle, mais les calculs deviennent notablement plus compliqués ; surtout, c'est le premier usage systématique de tels irrationnels dans des problèmes (voir l'*Introduction à l'histoire de l'algèbre* précédemment mentionnée [n. 23], chap. 3, 3).

Livre IV (fol. 67^v,1 – 78^v ; latin (éd. Lorch) : lignes 1 – 448 ; hébreu : Ms. Munich hébr. 225, 155^r – 165^v). L'introduction des précédentes équations n'est pas fortuite : elles surviennent dans la résolution de problèmes algébriques sur les polygones réguliers inscrits ou circonscrits, où soit le rayon du cercle soit le côté du polygone est donné. Ces polygones (triangles, carrés, pentagones, décagones, pentadécagones) étant tous constructibles par la règle et le compas, et rentrant donc dans le domaine de la géométrie d'Euclide, les équations résultantes sont quadratiques ou réductibles à des équations quadratiques.

Livre V (fol. 79^r, 1 – 95^r, 14 ; latin (fragment initial seulement) : pp. 448 – 450 dans l'édition susmentionnée ; hébreu : Ms. Munich hébr. 225, 166^v – 181^r). Recueil de 38 équations indéterminées du type de celles qu'on trouve chez Diophante, mais provenant d'une source grecque *autre* que Diophante. On peut en inférer le type des problèmes qui devaient se trouver dans les trois livres perdus de l'*Arithmetica* de Diophante ; voir l'*Introduction à l'histoire de l'algèbre* précédemment mentionnée [n. 23], fin du chap. 2.

Livre VI (fol. 95^r,14 – 111^r ; hébreu : Ms. Munich hébr. 225, 181^r – 191^v (incomplet, se termine à 105^v, 13 de l'arabe). Problèmes divers, la majorité de nature plutôt récréative du fait des situations peu vraisemblables qu'ils décrivent, et largement présents dans les traités médiévaux en arabe ou en latin (leurs sources sont souvent antiques). On trouve ainsi : l'achat d'un cheval par plusieurs participants, l'échange de sommes d'argent entre partenaires, le vol d'argent mis en commun dans un coffre, des problèmes de courriers et de robinets, des paires d'équations linéaires déterminées ou non, des calculs de salaires d'ouvriers travaillant isolément ou en groupe, des calculs de prix ou de butins en progression arithmétique, la somme des carrés de nombres naturels successifs, la duplication sur les cases du jeu des échecs —laquelle s'appuye sur un texte (aujourd'hui perdu) d'al-Ḫwārizmī.

Remarque. Les problèmes récréatifs, souvent considérés comme de moindre importance, auront de fait un apport loin d'être négligeable pour le développement des mathématiques. Ils autoriseront le calcul et l'expression de grands nombres, comme le dernier problème susmentionné, lequel est l'extension d'un calcul antique des duplications jusqu'à 30. Quant aux problèmes relatifs à l'échange ou au partage de sommes d'argent, ils seront quelques siècles plus tard étroitement liés à l'apparition des nombres négatifs comme solutions dans des systèmes linéaires ; voir l'*Introduction* susmentionnée, chap. 4, ou l'étude, plus complète, dans les *Archive for history of exact sciences* 32 (1985), pp. 105–150. Cette source d'innovations n'est pas surprenante : l'introduction de conditions peu réalisables dans les hypothèses laisse aussi toute liberté pour l'usage de calculs ou de réponses inhabituels.]

Abū Kāmil's *Book on Mensuration*

Jacques Sesiano

Abstract The Egyptian Abū Kāmil (ca. 850 – ca. 930), said to be the second Arabic algebraist in time after al-Khwārizmī (ca. 820), was the first to write an algebraic treatise for trained mathematicians. Among the shorter treatises he authored, one is obviously written for a wider readership for, as Abū Kāmil himself says, it presents formulae but not, as a higher-level treatise would, demonstrations. This is the *Treatise on Mensuration*, which teaches how to calculate areas and volumes of the usual plane and solid figures. That is just what many ancient and mediaeval books also do. But, unlike most of them, Abū Kāmil's also teaches the relations between sides of regular polygons and radii of inscribed and circumscribed circles.

Introduction

The Egyptian Abū Kāmil (ca. 850 – ca. 930) is considered to be the second algebraist in time after al-Khwārizmī (ca. 820). As a matter of fact, he can be considered as the first true algebraist writing in Arabic; while al-Khwārizmī wrote elementary algebra for a general readership, Abū Kāmil wrote for mathematicians, that is to say, readers familiar with the fundamental work on mathematics, Euclid's *Elements*. His treatise was influential both in the eastern and in the western part of the Islamic world of the time, and thus served as a model for further text-books on the subject. Abū Kāmil was in particular followed in leaving out two topics which formed two sections of al-Khwārizmī's treatise, practical geometry and the application of algebra to legacies.

In addition to his *Algebra*, Abū Kāmil wrote other treatises. We know the subject of three of them. One teaches how to solve pairs of linear indeterminate equations with integral (and positive) solutions, so-called "bird problems" since they involve the purchasing with a given sum of various sorts of bird, knowing their total number and the individual prices.[1] Abū Kāmil was particularly interested in the widely varying number of acceptable solutions. Thus, in the six examples he presents, one has no solution whereas the last has 2676. To our present knowledge, he is the first to have attempted to enumerate the various solutions without being satisfied with just one. Another short treatise was entirely devoted to the science of legacies, that is, how to share out an estate — given that relatives are, by law, entitled to a fixed fraction of it, depending on their degree of kinship with the deceased (thus, the sum of these fractions is mostly not equal to 1).[2] Abū Kāmil found it made sense to separate this topic from the general study of algebra and its more common applications, since no more than a little specific knowledge of algebra was actually needed for dealing with legacies. The other topic treated separately was, as said, practical geometry. Like inheritance algebra, practical geometry, or, as commonly called,

---

[1] See the commented translation by H. Suter [1911].

[2] This treatise is mentioned twice in Ḥājjī Khalīfa's كشف الظنون (V, pp. 68 (كتاب الوصايا بالجبر والمقابلة) and 168–69 (كتاب الوصايا بالجذور) [Flügel 1835–1858]).

N. Sidoli and G. Van Brummelen (eds.), *From Alexandria, Through Baghdad: Surveys and Studies in the Ancient Greek and Medieval Islamic Mathematical Sciences in Honor of J.L. Berggren*, DOI 10.1007/978-3-642-36736-6_19,

© Springer-Verlag Berlin Heidelberg 2014

the science of mensuration, was then to become a separate field of knowledge, in particular in teaching computation formulae to be applied by land surveyors, that is, people who did not need to be trained in the use of algebraic reasonings or geometrical demonstrations. Such was the intended readership of Abū Kāmil's *Book on Mensuration.*[3]

The title given to this treatise in mediaeval Arabic bibliography is "Book on mensuration and geometry" (كتاب المساحة والهندسة) and in modern Arabic bibliography, "Mensuration of pieces of ground" (مساحة الارضين).[4] The latter title is easily explained: it is that of the first part, dealing with planimetry, which was thus taken, as in the manuscript source, to be the title of the whole work. The former might have been the original title, although it is more likely, as was usual for such works, to have been "Book on Mensuration." The third part of the book, which has less to do with mensuration proper, might account for the addition.

The subject-matter of this treatise is mostly well in keeping with that of other mensuration works found in antiquity and in Islamic times. The first part deals with elementary planimetry: finding the area, or elements, of plane figures, first rectilinear (squares, rectangles, triangles, trapezia) then circular (circles, segments of a circle). Then elementary stereometry is taught: finding area and volume of the usual solid bodies (spheres, parallelepipeds, cylinders, pyramids, cones). The third part, which deals with the first regular polygons, is not concerned with area, but with the relation between the side of a polygon and the diameter of an inscribed or circumscribed circle. In this, we may recognize one specific interest of Abū Kāmil, to which he devotes a whole chapter in his *Algebra* ("Book IV").

Everything is taught in the present treatise by solving one example of each type, from which the calculation formula may easily be inferred. There are no demonstrations whatsoever.[5] This absence corresponds to the author's purpose, as stated in his introduction: the book addresses beginners or those needing practical formulae, and adding proofs would have made the work too long and inaccessible for many such readers. Anyone wishing to study the subject from a theoretical point of view is referred to Euclid's *Elements*.

This work is preserved today in a single manuscript, in Tehran, bearing the shelf-mark 26 in the Parliament Library (كتابخانه مجلس). Until recent years, it was kept in the Senate Library (كتابخانه مجلس سنا), with shelf-mark 2672, but No 26 in that library's printed catalogue; thus this number became the new shelf-mark after the transfer. The manuscript is written in a very legible *naskhi*, with most diacritical points used correctly. Red ink is used for some headings and for marking new sections. The figures are fairly well drawn. The copy is dated: it was finished on Monday, 8 *dhū'l-qaʿda* 758, thus Monday 23 October 1357. Our treatise occupies the fol. 93^r to 107^r, following works written, or inspired, by the late 13th-century scholar Ibn al-Bannāʾ (his *Talkhīṣ*, a commentary on it, his *Algebra*) and preceding a work by ʿAlī ibn ʿAbdallah al-Tabrīzī (كتاب تذكرة الاخوان). Thus, all are notably later works than Abū Kāmil's.

The progenitor of the *Book on Mensuration*, or some ancestor of it, was, unlike the extant copy, read attentively: the text preserved today incorporates numerous former glosses or readers' remarks; see the passages in square brackets in Nos 16, 17, 25, 32, 34, 35, 40, 50, 52, 56, 57, 65, 66, 67. It also contained lacunae, as is seen from note 112, below. Although careful, and even taking the initiative to make some changes, our copyist may be responsible for some further lacunae, for he did not grasp much of the content. He is responsible, at least in part, for writing in symbols the numbers originally written in words: in the last problem (No 67), a mistakenly repeated sentence has once the verbal and once the symbolic expression; in No 62 we find a confusion between 50 (خَمسين) and $\frac{2}{5}$ (خُمسين), and a similar situation occurs in No 48; a wrong transcription is found in No 65. The writing of fractions is at first erroneous (inverting the terms of the fraction, also in the strange occurrence in No 18); but the transcription is correct from No 28 on. Thus, between copyists and readers, the form of the text has been altered in the course of time. But it may well not have been perfect to begin with.

Indeed, the foremost algebraist of the 9th century is not really recognizable in this short work, at

[3] A short account of it was published by us in 1996.

[4] Ibn al-Nadīm [1871–1872 I, 281] and Suter [1900, 43]. See also Sezgin [1974, 281].

[5] The form such proofs would take may be inferred from similar problems in Heron's *Metrica* and Leonardo Fibonacci's *Practica geometriae*.

least for the form. Whereas his *Algebra* is the masterpiece of a seasoned mathematician, presenting the subject-matter clearly, we find here a surprising lack of precision and didactic sense, as many instances show. Just after the enunciation, and before the calculation, we may already be given the answer (Nos 1, 4, 5, 16–18, 20, 23, 30, 35, 44–51, 56–67);[6] yet this is not done systematically. At times, a polygon is specified to be equilateral and equiangular (متساوى الاضلاع والزوايا, our "regular"), whereas sometimes we find only one of the two terms (Nos 44, 52, 55, 58, 59, 62) or even none (Nos 46, 53, 54, 63, 65); this cannot be wholly due to inattentive copyists. The figures, sometimes alluded to in the text, sometimes not, are absent for Nos 20–22, 59 and for all stereometric problems.[7] Such lack of rigor is particularly striking in a book intended for beginners who need a methodical and formally repetitive presentation of the matter.

In the introduction, Abū Kāmil refers to Euclid's *Elements* for proofs.[8] He does not refer to his own *Algebra*, though some of the formulae of the last section (our Nos 52–57, 66–67) are established there. This would suggest that the *Book on Mensuration* precedes the *Algebra*. Indeed, given the somewhat slipshod aspect of this work, we might imagine it to be the production of a mathematician at the beginning of his career.[9] On the other hand, in the introduction, Abū Kāmil alludes to the "many treatises" he has written. Thus, it is difficult to situate this work in time. But it is likely to have been written before or at the time of his official function at the Arsenal.[10]

Among the potential readers one would of course expect the official landsurveyors and the landowners. Indeed, as Abū Kāmil explains (No 24; see No 34), the formulae used by the former gave very rough approximations (of the kind used in ancient Egypt and in second-rate ancient treatises), and it would seem that local traditions remained uninfluenced by the works written in the major centre of ancient science, Alexandria.[11] As to the landowners (see No 25), they were apparently quite satisfied with the rough formulae, and were not inclined to change them for Abū Kāmil's exact methods. Which is not surprising, since the result of his vast erudition had no other effect than to diminish the surface areas of their domains. More surprising is Abū Kāmil's astonishment at their reaction, though this may well be no more than the ingenuousness of a young scholar, who still believes that scientific truth can overcome personal interests.

Summary of the Contents

A. Plane figures

Required the area (مساحة[12]) or elements of the figure considered: side (جانب, ضلع), diagonal (قطر = διάμετρος), height (عمود); diameter (قطر = διάμετρος), circumference (دور, تدوير and استدارة,[13] خط محيط, محيط), arc length (تدوير القوس), chord (وتر), arrow (سهم).

[6] Highlighting the result thus may indeed be useful for the reader, for in another similar problem he might infer the result by a rule of three.

[7] The one missing in No 63 is clearly a scribal omission since the corresponding text is present. Indeed, all extant figures have the numerical values written on their elements or, when this is too long, in a separate text; but all that has clearly been added later, for one figure is supposed to illustrate not just one, but the whole group of problems involving the same figure.

[8] An incidental reference to Ptolemy and al-Khwārizmī is found in No 43.

[9] The 'venerable' in the title is of course a later addition.

[10] On this activity, see Anbouba's study, above.

[11] Which is hardly surprising: Alexandria was a Greek city and the indigenous population had nothing to do with it (just as in Albert Camus' *La peste* [1947], Oran is seen as a French city).

[12] تكسير in No 20.

[13] "Perimeter" for a rectilinear area.

1. Quadrilateral figures

Square (مربّع متساوى الجوانب والقطرين): Given the side a, find the area S (No 1) and the diagonal D (No 2); given the diagonal, find the area (No 3) and the side (No 4).

Rectangle (مربّع متساوى الطولين والعرضين والقطرين): Given the sides a, b, find the area S (No 5) and the diagonal D (No 6); given the diagonal D and the difference between the sides $a - b$, find the area S (No 7); given the area S and the difference $a - b$, find the sides a, b (No 8).

Trapezium (مربّع مختلف الجوانب): Given the sides a (base), c (top), b and d (lateral sides), find the area S (No 23). First, find the vertical height h by means of the projections d_a, b_a of d and b on a.

General rectilinear figure: Decomposition into triangles (No 25).

2. Circular figures

Circle (مدوّر): Given the diameter D, find the area S (No 9) and the circumference P (No 10); given the diameter D and the circumference P, find the area S (No 11); given the circumference P, find the area S (No 12) and the diameter D (No 13); given the area S, find the diameter D (No 14) and the circumference P (No 15).

Segment of a circle (قوس): Given the arc length a, the lengths of the chord t and the arrow h, find the area S; first, determine the diameter D of the circle belonging to the segment (No 43).

3. Triangular figures

Right-angled triangle: Given the sides a, b, c with $a^2 + b^2 = c^2$, find the area S (No 16).

Equilateral triangle (مثلّث متساوى الاضلاع): Given the side a, find height h and area S (No 20); given the height h, find the side a (No 21).

Isosceles triangle (مثلّث متساوى الجانبين): Given the sides $a = b$, c, find the height h_c and the area S (No 22).

Scalene triangle (مثلّث مختلف الاضلاع): Given the sides a, b, c, find the heights and the area if the triangle is acute, thus $a^2 + b^2 > c^2$ (No 17), or obtuse, thus $a^2 + b^2 < c^2$ (No 18). The heights h_a, h_b, h_c are determined by means of the projections b_a, c_a; a_b, c_b; a_c, b_c.

General formula: Given the sides a, b, c and with σ designating the half perimeter, find the area S (No 19).

Formulae in use among (Egyptian) official land surveyors (عند مسّاح السلطان) for equilateral triangles, scalene triangles, trapezia and irregular quadrilaterals, circles and polygons (No 24).

B. Solid figures

Required the volume (مساحة, مساحة الجسم, جسم) or the lateral area (سطح, مساحة السطح) of regular bodies. New elements are the altitude (ارتفاع, سمك), or, for a well, the depth (عمق), the slant height (سمك العمود على السطح), the great circle (منطقة).

1. *Cylinder* (عمود مدوّر متساوى الطرفين): Given the diameter of the base D and the altitude h, find the volume V (No 28) or the lateral area S (No 29).

2. *Parallelepiped* (عمود مربّع): Given the sides a, b of the base and the altitude h, find the volume V (No 26) and the lateral area S (No 27).

3. Pyramid

Regular pyramid (مخروط لا رأس له, "pyramid without top," or رأسه مثل الزجّ, "with a top like a spear-head"): Given the altitude and the sides of the base, find the volume (Nos 35, 41) or, by means of the slant height, the lateral area (Nos 37, 42).

Frustum of a pyramid (عمود مربّع مخروط): Given the sides of the lower and upper bases and the altitude (for a well: depth), find the volume (No 30) or the lateral area (No 32).

4. Cone

Regular cone (عمود مدوّر, with specification as above): Given the altitude and the diameter of the base, find the volume (Nos 36, 41) or, by means of the slant height, the lateral area (No 38).

Frustum of a cone (عمود مدوّر مخروط): Given the diameter of the lower and upper bases and the altitude, find the volume (No 31), or the lateral area (No 33).

5. *Measuring Wood in Egypt* (مساحة الخشب عند اهل مصر): No 34.

6. *Sphere* (كرة): Given the diameter, find, by means of the great circle, the area (No 39) and the volume (No 40).

C. Regular polygons

We consider some regular polygons, either inscribed in (وقع داخل, وقع فى) or circumscribed about (وقع على, وقع خارج; احاط ب) a circle, and we look for the side when the diameter is known or for the diameter when the side is known.

1. *Square*: Given the diameter D of a circle, find the side a_4 of the square inscribed (No 44) or a_4' of the square circumscribed (No 45); given the side a_4 of a square, find the diameter D of the circle circumscribed (No 46) or D' of the circle inscribed (No 47).

2. *Triangle*: Given the diameter D of a circle, find the side a_3 of the triangle inscribed (No 48) or a_3' of the triangle circumscribed (No 49); given the side a_3 of a triangle, find the diameter D of the circle circumscribed (No 51) or D' of the circle inscribed (No 50).

3. *Pentagon*: Given the diameter D of a circle, find the side a_5 of the pentagon inscribed (No 52) or a_5' of the pentagon circumscribed (No 53); given the side a_5, find the diameter D of the circle circumscribed (No 56) or D' of the circle inscribed (No 57). These four cases are studied in detail in Book IV of the *Algebra*.

4. *Hexagon* : Given the diameter D of a circle, find the side a_6 of the hexagon inscribed (No 58) or the side a'_6 of the hexagon circumscribed (No 59); given the side a_6 of a hexagon, find the diameter D of the circle circumscribed (No 61) or D' of the circle inscribed (No 60).

5. *Octagon* : Given the diameter D of a circle, find the side a_8 of the octagon inscribed (No 62) or the side a'_8 of an octagon circumscribed (No 63); given the side a_8 of an octagon, find the diameter D of the circle circumscribed (No 65) or D' if it is inscribed (No 64).

6. *Decagon* : Given the diameter of a circle D, find the side a_{10} of the decagon inscribed (No 54) or the side a'_{10} of the decagon circumscribed (No 55); given the side of a decagon a_{10}, find the diameter D of the circle circumscribed (No 66) or D' of the circle inscribed (No 67). These four cases are studied in detail in Book IV of the *Algebra*.

D. Root approximations

Although he does not explain his method, it appears that Abū Kāmil, or the source of his estimations, has employed the usual ancient way of approximating square roots, namely, if a^2 is the largest integer below N,

$$\sqrt{N} \cong a + \frac{N - a^2}{2a}$$

(which produces a result larger than the true one); see Nos 2 (& 46, 63, 64), 20, (37), 44, 48 (& 49). In two cases (Nos 20, 51), the approximation may be connected with

$$\sqrt{N} \cong (a + 1) - \frac{(a + 1)^2 - N}{2(a + 1)}.$$

Editorial Procedures

In both the translated and the original text we shall use round brackets for our own additions, angular brackets where we add parts missing in the extant manuscript, and square brackets for some interpolations (namely those of some interest to the history of the text, the others being in the critical notes). In the Arabic text, we have kept such diacritical dots as are correct in the manuscript, and left unchanged the variant readings (ثلاثة, ثلثة; ثمانية, ثمنية; اربع مائة, اربعمائة). However, where the numerals are in symbols, we have adopted for their transliteration the *scriptio difficilior* since this is found several times in the manuscript (see also the copyist's correction of ثلثة عشر to ثلاثة عشر in No 23). In the translation, numbers are written in symbols, except in the enunciation of problems. For various suggestions and corrections I am indebted to my friend and colleague Ahmed Djebbar. This study of Abū Kāmil's manuscript was made possible by a grant from the Fonds national suisse de la recherche scientifique.

كتاب فى المساحة

للشيخ العالم شجاع بن اسلم المعروف بأبى كامل

رحمه اللّه بكرمه

آمين

(93v) بسم اللّه الرحمن الرحيم وصلّى اللّه على سيّدنا محمّد وآله

قال شجاع ابن اسلم المعروف بأبى كامل

باب مساحة الأرضين

امّا معرفة مساحة الأرضين فليس يتهيّأ ان يقيم البرهان على ما ارسم منه كما اقمتُه على كثير من كُتُبى ولو قدّمتُ البرهان على ما اذكره منها لاجتمعتُ على كثير ممّا فى ⟨كتاب⟩ اقليدس ولطال الكتاب وعسر على المتعلّمين اذ كنت انّما وضعتُ هذا الكتاب للمبتدئين للدخول فى هذه الصناعة والمحبّين لتعلّمها وانّما اروى ما ارسمه من المساحة روايةً بترتيب صحيح ونظم مستوٍ وتأليف حسن وقول برهانى فمَن تمسّك بالابواب التى ارويها وسلك الطريق الذى ابنيه صحّ له الجواب ونجا من الخطأ وامن الزلل وسهل عليه معرفة البرهان.

فان اراد البرهان على ذلك وعلى ما تولّد الحساب من مسائل الهندسة واالمساحة والعدد وغير ذلك فعليه بالنظر فى كتاب اقليدس فقد احكمه غاية الاحكام وبناه باوثق البنيان فاذا استوعب ما فيه وفهمه فهماً صحيحاً لا يدخله ريب ولا يختلجه شكّ ولا يشوبه دنس عرف اقامة البرهان على الهندسة والمساحة والعدد وغير ذلك من جميع الفنون. ولا قوّة الّا باللّه العلىّ العظيم وحسبُنا اللّه ونعم الوكيل.

(1) فان قيل لك أرض مربّعة متساوية الجوانب والقطرين كلّ جانب منها عشرة اذرع كم مساحتها.

امّا حسابه فانّ مساحتها مائة. وبابه ان تضرب الطول وهو عشرة اذرع وهو احد الجوانب فى العرض وهو الجانب الآخر وهو عشرة فتكون مائة (94r) وذلك مساحتها. وهذه صورتها.

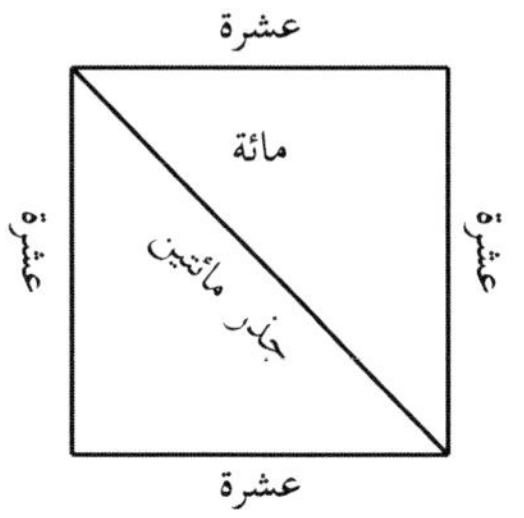

(2) فان قال كم قطرها.

فاضرب احد الجوانب وهو عشرة فى مثله فتكون مائة ثمّ اضرب الجانب الذى يليه وهو عشرة فى مثله فتكون مائة فتجمعهما فيكونان مائتين فخذ جذر مائتين فهو احد قطريها والقطر الآخر مثله.

---

1 كتاب فى المساحة] جزؤ فى مساحة الارضين (*rubro col.*). 8 مساحة] حساب. يقيم] نقيم. 9 على]2 الى. 11 بترتيب صحيح ... وقول برهانى] *infra in cod.* (فاذا استوعب *ante*). برهانى] برهان. 12 ابنيه] ابنته (ابنت *corr. ex*). 22 فيكونان] فتكون. فهو] هو.

وجذر العدد هو عدد تضربه فى مثله فيكون ما اجتمع منه مثل ذلك العدد الذى تريد جذره مثل جذر اربعة اثنان لانّك اذا ضربت الاثنين فى مثله كان اربعة وجذر تسعة ثلاثة لانّ ثلثة فى ثلاثة تسعة وجذر ستّة عشر اربعة لانّ اربعة فى اربعة ستّة عشر.

وجذر مائتين الذى قلنا انّه قطر المربّعة لا ننطق به ولا جذر لمائتين لانّك لا تجد عدداً تضربه فى مثله فيكون مائتين فان اردت جذر عدد يقرب من مائتين او يقرب المائتان منه فاربعة عشر وسُبع ⟨لانّك⟩ اذا ضربته فى مثله كان المائتان قريباً منه لانّك اذا ضربت الاربعة عشر وسُبعاً فى مثله كان مائتين وسُبع السُبع. فان سئلتَ عن مربّعة كلّ جانب منها عشرة اذرع كم قطرها فقل جذر مائتين فان قيل لك قربَه فقل اربعة عشر وسُبعاً.

(3) فان قيل لك قطرها عشرة اذرع كم مساحتها.

فاضرب القطر وهو عشرة فى مثله فيكون مائة فخذ نصفه فيكون خمسين فذلك مساحتها.

(4) فان قيل لك قطرها عشرة كم هى من كلّ جانب.

امّا حسابه فان كلّ جانب منها جذر خمسين. وبابه ان تضرب العشرة فى مثلها فتكون مائة فتأخذ نصفها فتكون خمسين فتقول جذر خمسين كلّ جانب منها.

(5) فان قيل لك (94^v) مربّعة متساوية الطولين متساوية العرضين متساوية القطرين طولها ثمانية وعرضها ستّة كم مساحتها.

امّا حسابه فان مساحتها ثمانية واربعون ذراعاً. وبابه ان تضرب احد الطولين وهو ثمانية فى احد العرضين وهو ستّة فتكون ثمانية واربعين وذلك مساحتها. وهذه صورتها.

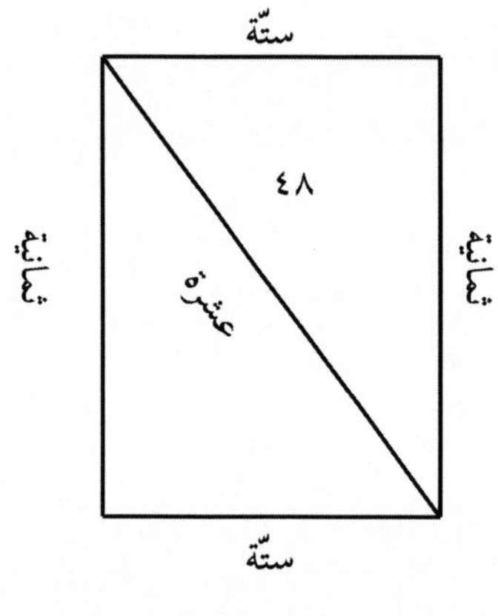

(6) فان قال كم قطرها.

فاضرب احد الطولين وهو ثمانية فى مثله فتكون اربعة وستّين واضرب احد العرضين وهو ستّة فى مثله فتكون ستّة وثلاثين فتجمعهما فيكونان مائة فتأخذ جذرها فيكون عشرة وهو قطرها.

(7) فان قال مربّعة متساوية الطولين والعرضين والقطرين قطرها عشرة وأحد الطولين اطول من احد العرضين باثنين كم مساحتها.

[1] تريد جذره] تريد ان شا الله خذ جذره. اثنان] اثنين. [2] ستّة عشر] ١٦. [3] ستّة عشر] ١٦. [5] يقرب المائتان] تقرب الماتين. [6] كان][1] كانت. قريباً] قريب. الاربعة عشر وسُبعاً] ١٤ وسبع. [7] اربعة عشر وسُبعاً] ١٤ وسبع. [12] خمسين][1] ٥٠. [13] ثمانية] ٨. ستّة] ٦. [15] ثمانية واربعون] ٤٨. ثمانية] ٨. [16] ستّة] ٦. ثمانية واربعين] ٤٨. [18] ثمانية] ٨. اربعة وستّين] ٦٤. ستّة] ٦. [19] ستّة وثلاثين] ٣٦. فيكونان] فتكون. مائة] ١٠٠. عشرة] ١٠.

بابه ان تضرب القطر وهو عشرة فى مثله فتكون مائة فتسقط منه ضرب فضل احد الطولين على احد العرضين فى مثله وهو اربعة فتبقى ستّة وتسعون فتأخذ نصفها فتكون ثمانية واربعين فذلك مساحتها.

(8) فان قال كم طولها وكم عرضها.

فنصّف الاثنين فيكون واحداً فاضربه فى مثله فيكون واحداً فزده على مساحتها وهى ثمانية واربعون فيكون تسعة واربعين فخذ جذرها فتكون سبعة فزد عليه الواحد فتكون ثمانية وهو احد الطولين وانقص الواحد من السبعة فتبقى ستّة وهو احد العرضين. واللّه اعلم.

باب حساب المدوّرات

(9) فان قال مدوّرة قطرها عشرة اذرع كم مساحتها.

بابه ان تضرب القطر وهو عشرة فى مثله فتكون مائة فتسقط من المائة سُبعها ونصف سُبعها وهو واحد وعشرون وثلاثة اسباع فيبقى ثمانية وسبعون واربعة اسباع (95^r) فذلك مساحتها. وهذه صورتها.

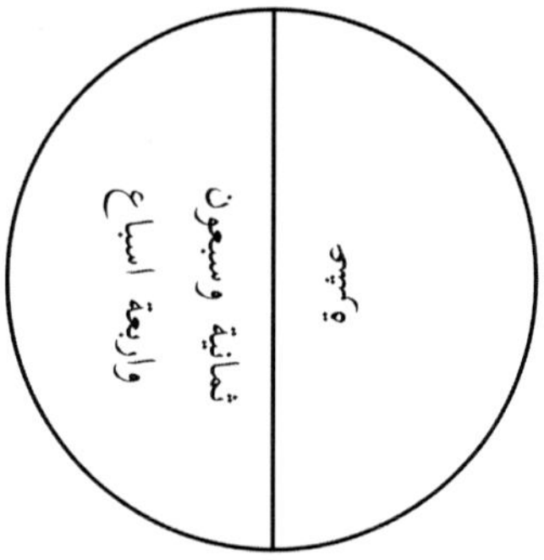

(10) فان قال مدوّرة قطرها عشرة اذرع كم دورها.

فاضرب القطر فى ثلاثة وسُبع ابداً فتكون واحداً وثلاثين وثلاثة اسباع فذلك دورها.

(11) فان قال كم مساحتها من قبل دورها وقطرها.

فاضرب نصف القطر وهو خمسة فى نصف الدور وهو خمسة عشر وخمسة اسباع فتكون ثمانية وسبعين واربعة اسباع فذلك مساحتها.

وكذلك تفعل بكلّ أرض مدوّرة اذا علمتَ كم قطرها.

وان شئت فاضرب القطر فى مثله واسقط ممّا اجتمع سُبعه ونصف سُبعه فما بقى فهو مساحتها.

وان شئت فاضرب القطر فى ثلاثة وسُبع ابداً فما كان فهو دورها ثمّ اضرب نصف القطر فى نصف الدور فما بلغ فهو مساحتها.

(12) فان قال دورها عشرة اذرع كم مساحتها.

فاضرب العشرة فى مثلها فتكون مائة ⟨ثمّ⟩ اسقط منها ثُمنها ابداً وهو فى هذه المسئلة اثنا عشر ونصف فيبقى سبعة وثمانون ونصف فاقسمها على احد عشر ابداً فتخرج سبعة وعشرة اجزاء ونصف جزء من احد عشر فذلك مساحتها.

---

2 ثمانية واربعين] ثمانية واربعون. 4 وهى] وهو. ثمانية واربعون] ٤٨. 4–5 تسعة واربعين] ٤٩. 5 سبعة] ٧. 9 واحد وعشرون] ٢١. 10 ثمانية وسبعون واربعة اسباع] ٧٨$\frac{٧}{٤}$. 14 خمسة عشر] ١٥. 21 اثنا عشر] ١٢. 22 سبعة وثمانون] ٨٧.

(13) وان قال دورها عشرة اذرع كم قطرها.

فاضرب العشرة فى سبعة ابداً فتكون سبعين فاقسمها على اثنين وعشرين ابداً فتخرج ثلاثة واربعة اجزاء من اثنين وعشرين فذلك قطرها.

(14) فان قال مساحتها مائة واربعة وخمسون كم قطرها.

فاضرب المائة واربعة وخمسين فى اربعة عشر ابداً فتكون الفين ومائة وستّة وخمسين واقسمها على احد عشر ابداً فيخرج مائة وستّة وتسعون فتأخذ جذرها فيكون اربعة عشر وذلك قطرها.

(15) فان قال مساحتها مائة واربعة وخمسون كم دورها.

فاضرب المائة واربعة وخمسين فى اثنى عشر واربعة اسباع ابداً فتكون الفاً وتسع مائة وستّة وثلاثين فتأخذ جذرها (95ᵛ) فتكون اربعة واربعين فذلك دورها. والله اعلم.

باب حساب المثلّثات

(16) فان قال مثلّث من جانب ثلاثة ومن جانب اربعة ومن جانب خمسة كم مساحته.

امّا حسابه فان مساحته ستّة. وكلّ مثلّث تضرب ⟨جانبيه⟩ الاقصرين كلّ واحد منهما فى مثله وتجمعهما فيكونان مثل ضرب الجانب الاطول فى مثله فان مساحته ان تضرب احد الجانبين الاقصرين فى الجانب الاقصر الآخر وتأخذ نصف ما اجتمع من الضرب فما كان فهو مساحته. فاضرب الثلاثة فى الاربعة فتكون اثنى عشر فخذ نصفها فتكون ستّة فذلك مساحة المثلّث وذلك بابه وحسابه.

وانّما ضربتَ الجانبين الاقصرين احدهما فى الآخر وهما ثلاثة واربعة وأخذت نصف ما اجتمع من الضرب لانّ ضرب ثلاثة فى مثلها واربعة فى مثلها مجموعين يكونان خمسة وعشرين وهو مثل ضرب الضلع الاطول وهو خمسة فى مثله. [وذلك بابه وحسابه] وهذه صورته.

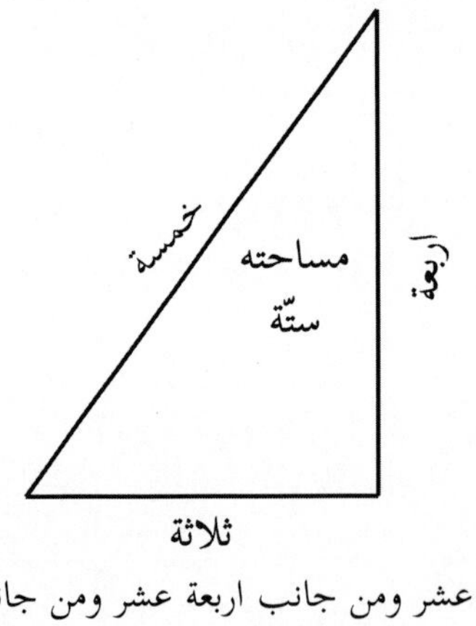

(17) فان قيل مثلّث من جانب خمسة عشر ومن جانب اربعة عشر ومن جانب ثلاثة عشر كم مساحته.

امّا حسابه فان مساحته اربعة وثمانون. وكلّ مثلّث تضرب جانبيه الاقصرين كلّ واحد منهما فى مثله وتجمعهما فيكون ذلك اكثر من ضرب الجانب الاطول فى مثله واردت مساحته من قبل عموده فاعلم ان لهذا المثلّث ثلاثة اعمدة كلّ عمود ⟨منها⟩ يقع على ضلع من اضلاع هذا المثلّث داخل المثلّث.

---

[2] اثنين وعشرين] ٢٢. [2–3] اثنين وعشرين] ٢٢. [4] وخمسون] وخمسين. [5] المائة واربعة وخمسين فى اربعة عشر] ١٥٤ في ١٤. الفين ومائة وستّة وخمسين] ٢١٥٦. احد عشر] ١١. [6] مائة وستّة وتسعون] ١٩٦. اربعة عشر] ١٤. [7] مائة واربعة وخمسون] ١٥٤. [8] المائة واربعة وخمسين فى اثنى عشر] ١٥٤ في ١٢. الفاً وتسع مائة وستّة وثلاثين] ١٩٣٦. [9] اربعة واربعين] ٤٤. [13] فيكونان] فيكون. [14] الثلاثة فى الاربعة] ٣ في ٤. اثنى عشر] ١٢. [16] ثلاثة واربعة] ٣ في ٤. [17] ثلاثة] ٣. يكونان] تكون. الضلع] الظلع. [19] مثلّث] المثلث. خمسة عشر] ١٥. اربعة عشر] ١٤. ثلاثة عشر] ١٣. [20] اربعة وثمانون] ٨٤.

فاذا اردت معرفة عمود واحد من هذه الثلاثة الاعمدة فاطلب العمود الذى يقع على جانب الاربعة عشر فانّه يقع صحيحاً لا كسر فيه فاذا اردت ذلك فاضرب الاربعة عشر فى مثلها فتكون مائة وستّة وتسعين [واضرب اىّ الجانبين الآخرين اردت فى مثله] فاضرب الثلاثة عشر (96^{r}) فى مثلها فتكون مائة وتسعة وستّين فزدها على المائة وستّة وتسعين فتكون ثلاث مائة وخمسة وستّين فاسقط منها الجانب الآخر وهو خمسة عشر فى مثله وذلك مائتان وخمسة وعشرون فيبقى مائة واربعون فتأخذ نصفها فتكون سبعين فتقسمها على الاربعة عشر فتخرج خمسة وهى مسقط الحجر على جانب اربعة عشر ممّا يلى الثلاثة عشر فتضرب الخمسة فى مثلها فتكون خمسة وعشرين وتسقطها من الثلاثة عشر فى مثلها وذلك مائة وتسعة وستّون فتبقى مائة واربعة واربعون فتأخذ جذرها فتكون اثنى عشر وهو العمود على جانب اربعة عشر فاضرب عمودها وهو اثنا عشر فى نصف الاربعة عشر وهو سبعة فتكون اربعة وثمانين فذلك مساحته.

واذا اردت ان تعلم مسقط الحجر من غير هذه الجهة على جانب اربعة عشر فاضرب الثلاثة عشر فى مثلها والخمسة عشر فى مثلها وخذ فضل ما بينهما فتجده ستّة وخمسين فتأخذ نصفها فيكون ثمانية وعشرين فتقسمه على الاربعة عشر فيخرج اثنان فتسقطهما من نصف الاربعة عشر فتبقى خمسة وهو مسقط الحجر على الاربعة عشر ممّا يلى الثلاثة عشر وان زدتَ الاثنين على نصف الاربعة عشر صار تسعة وهو مسقط الحجر على الاربعة عشر ممّا يلى الخمسة عشر فان ضربتَ الخمسة فى مثلها وأُسقط ما اجتمع من ضرب الثلاثة عشر فى مثلها بقى مائة واربعة واربعون فتأخذ جذرها فتكون اثنى عشر وهو العمود وان ضربتَ ايضاً التسعة فى مثلها واسقطتَ ما اجتمع من ضرب الخمسة عشر فى مثلها بقى مائة واربعة واربعون فتأخذ جذرها فتكون اثنى عشر وهو العمود.

فاذا القى عليك مثلّث يشبه هذا المثلّث وعملتَ كما وصفتُ لك فكان موقعا الحجر من الناحيتين جميعاً مثل طول الضلع وقع عليه الحجر و⟨اذا⟩ ضربتَ كلّ واحد من موقعى الحجر فى مثله واسقطتَ ما اجتمع من الضرب من ضرب ما اجتمع من كلّ واحد من جانبى المثلّث فى مثله وكلّ واحد ممّا يليه ⟨من⟩ الذى يليه وكان الذى بقى من احدهما مساوياً للذى بقى من الآخر فقد اصبتَ وان خالف فقد اخطأتَ (96^{v}) فاعد حسابك. فاذا عرفتَ العمود فاضربه فى نصف القاعدة فما كان فهو مساحة المثلّث. وفى استخراج العمود اعمال كثيرة وفيما رسمتُ لك منها كفاية. وعمود هذا المثلّث على الخمسة عشر احد عشر وخُمس وعلى الثلاثة عشر اثنا عشر واثنا عشر جزءاً من ثلاثة عشر.

(18) فان قال مثلّث من جانب اربعة ومن جانب ثلاثة عشر ومن جانب خمسة عشر كم مساحته.

امّا حسابه فان مساحته اربعة وعشرون ذراعاً. وكلّ مثلّث تضرب جانبيه الاقصرين كلّ واحد منهما فى مثله وتجمعهما فيكون ذلك اقلّ من ضرب الجانب الاطول فى مثله واردتَ مساحته من قبل عموده فاعلم ان لهذا المثلّث ثلاثة اعمدة عمود واحد منها يقع داخلَ المثلّث على الضلع الاطول والعمودين ⟨على الضلعين⟩ الاقصرين من خارج

2 صحيحاً] صحيح. الاربعة عشر] ١٤. مائة وستّة وتسعين] ١٩٦. 3 الثلاثة عشر] ١٣. مائة وتسعة وستّين] ١٦٩. 3–4 المائة وستّة وتسعين] ١٩٦. 4 ثلاث مائة وخمسة وستّين] ٣٦٥. خمسة عشر] ١٥. 4–5 مائتان وخمسة وعشرون] ٢٢٥. 5 مائة واربعون] ١٤٠. سبعين] ٧٠. الاربعة عشر] ١٤. خمسة] ٥. 6 اربعة عشر] ١٤. الثلاثة عشر] ١٣. الخمسة] ٥. خمسة وعشرين] ٢٥. 7 الثلاثة عشر] ١٣. مائة وتسعة وستّون] ١٦٩. مائة واربعة واربعون] ١٤٤. اثنى عشر] ١٢. 8 العمود] عمودها. اربعة عشر] ١٤. اثنا عشر] ١٢. الاربعة عشر] ١٤. سبعة] ٧. اربعة وثمانين] ٨٤. 9 مساحته] مساحتها. 10 اربعة عشر] ١٤. الثلاثة عشر] ١٣. 11 ستّة وخمسين] ٥٦. ثمانية وعشرين] ٢٨. 12 الاربعة عشر] ١٤. اثنان] ٢. فتسقطهما] فتسقطها. الاربعة عشر] ١٤. خمسة] ٥. الاربعة عشر] ١٤. 13 الثلاثة عشر] ١٣. الاثنين] ٢. الاربعة عشر] ١٤. تسعة] ٩. الاربعة عشر] ١٤. 14 الخمسة عشر] ١٥. الخمسة] ٥. ما] مما. الثلاثة عشر] ١٣. بقى] يبقي. 14–15 مائة واربعة واربعون] ١٤٤. 15 اثنى عشر] ١٢. 16 الخمسة عشر] ١٥. مائة واربعة واربعون] ١٤٤. اثنى عشر] ١٢. 17 فكان] وكان. موقعا] موقع. 18 وقع] ووقع. موقعى] موقع. 19 وكان] فكان. 22 الخمسة عشر] ١٥. الثلاثة عشر] ١٣. 23 ثلاثة عشر] ١٣. 24 اربعة] ٤. ثلاثة عشر] ١٣. خمسة عشر] ١٥. 25 اربعة وعشرون] ٢٤.
26 لهذا] لهذه *corr. ex*.

المثلّث احدهما يقع على استقامة احد الضلعين الاقصرين من خارج المثلّث ويقع الآخر خارج الضلع الآخر اذا اخرج على استقامته من خارج المثلّث.

فاذا اردت ان تعلم العمود الذى يقع من داخل المثلّث على الضلع الاطول فاعمل كما وصفتُ لك فى المسئلة التى قبل هذه وهو ان تضرب الاربعة فى مثلها والخمسة عشر فى مثلها وتجمعهما فيكونان مائتين وأحداً واربعين وتسقط منها الثلاثة عشر فى مثلها فتبقى اثنان وسبعون فتأخذ نصفها فتكون ستّة وثلاثين فتقسمها على الخمسة عشر فيخرج اثنان وخُمسان وهو موقع الحجر على الخمسة عشر ممّا يلى الاربعة.

فان اردت ان تعلم كم موقعه ممّا يلى الثلاثة عشر فانقص الاثنين وخُمسين من الخمسة عشر فيبقى اثنا عشر وثلاثة اخماس وهو موقع الحجر. وان اردته من الجهة التى اعلمتُك فاضرب الثلاثة عشر فى مثلها والخمسة عشر فى مثلها واجمعهما فيكونان ثلاث مائة واربعة وتسعين واسقط منها الاربعة فى مثلها فتبقى ثلاث مائة وثمانية وسبعون فخذ نصفها فتكون مائة وتسعة وثمانين فاقسمها على الخمسة عشر فيخرج اثنا عشر وثلاثة اخماس وهو مسقط الحجر على الخمسة عشر (97ʳ) ممّا يلى ثلاثة عشر.

فاذا اردت ان تعلم كم طول العمود فاضرب الاثنين وخُمسين فى مثله واسقطه من ضرب الاربعة فى مثلها فيبقى عشرة وخُمس وخُمس الخُمس فخذ جذرها وهو ثلاثة وخُمس وهو طول العمود فاضربه فى نصف الخمسة عشر وهو سبعة ونصف فتكون اربعة وعشرين وهو مساحته. وان ضربت مسقط الحجر من ناحية الثلاثة عشر وهو اثنا عشر وثلاثة اخماس فى مثله وأُسقط ما اجتمع من الضرب من الثلاثة عشر فى مثلها بقى عشرة وخُمس وخُمس الخُمس مثل ما بقى من الاوّل سواء فتأخذ جذرها فتكون ثلاثة وخُمساً وهو طول العمود. ولو خالف لكنتَ قد اخطأتَ فاعد الحساب.

وان اردت ان تعلم كم يخرج كلّ واحد من الخطّين الاقصرين على استقامته الى مسقط العمود على كلّ واحد منهما فابدأ بأحدهما وهو الاربعة واجعل مسقط العمود على استقامته الى موضع موقع العمود فانّه يخرج خمسة اذرع ويكون طول هذا الخطّ من موضع إلتقاءه مع طرف الخمسة عشر الى موضع مسقط العمود عليه تسعة اذرع. وبابه ان تضرب الاربعة فى مثلها وثلثة عشر فى مثلها وتجمعهما فيكونان مائة وخمسة وثمانين فتسقطها من الخمسة عشر فى مثلها وهو مائتان وخمسة وعشرون فتبقى اربعون فتأخذ نصفها فتكون عشرين فتقسمها على الاربعة فتكون خمسة وهو ما يخرج خطّ الاربعة على استقامته الى موضع مسقط العمود.

فاذا اردت ان تعلم كم العمود فاضرب الخمسة فى مثلها واسقط ما اجتمع من ضرب الثلاثة عشر فى مثلها فتبقى مائة واربعة واربعون فتأخذ جذرها فتكون اثنى عشر وهو طول العمود على طرف الخمسة الاذرع من خطّ الاربعة الذى صار جميعه تسعة. فاذا اردت ان تعلم كم مساحة هذا المثلّث من هذه الجهة فاضرب العمود وهو اثنا عشر فى نصف الاربعة وهو اثنان فتكون اربعة وعشرين وهو مساحة المثلّث.

---

2 على] فى *corr. ex*. استقامته] استقامه. 4 والخمسة عشر] و١٥. فيكونان] فتكون. مائتين وأحداً واربعين] ٢٤١. 5 منها] منهما. الثلاثة عشر] ١٣. اثنان وسبعون] ٧٢. ستّة وثلاثين] ٣٦. 5–6 الخمسة عشر] ١٥. 6 اثنان وخُمسان] ٢ $\frac{٥}{٢}$ (*sic*; *hic et infra*)؛ اثنين وخمسين *in figura:*. الخمسة عشر] ١٥. 7 الاثنين وخُمسين] ٢ $\frac{٥}{٢}$. الخمسة عشر] ١٥. 7–8 اثنا عشر وثلاثة اخماس] ١٢ $\frac{٥}{٣}$. 8 الثلاثة عشر] ١٣. 8–9 والخمسة عشر] و١٥. 9 فيكونان] فتكون. ثلاث مائة واربعة وتسعين] ٣٩٤. 9–10 ثلاث مائة وثمانية وسبعون] ٣٧٨. 10 مائة وتسعة وثمانين] ١٨٩. الخمسة عشر] ١٥. اثنا عشر وثلاثة اخماس] ١٢ $\frac{٥}{٣}$. 11 الخمسة عشر] ١٥. ثلاثة عشر] ١٣. 12 الاثنين وخُمسين] ٢ $\frac{٥}{٢}$. الاربعة] ٤. 13 عشرة] ١٠. الخمسة عشر] ١٥. 14 سبعة ونصف] ٧ $\frac{١}{١}$ (*sic*). اربعة وعشرين] ٢٤. مساحته] مساحتها. الثلاثة عشر] ١٣. 14–15 اثنا عشر وثلاثة اخماس] ١٢ $\frac{٥}{٣}$. 15 الثلاثة عشر] ١٣. 16 سواء] سوى. ثلاثة وخُمساً] ٣ $\frac{٥}{١}$ (*in figura:* ٣ $\frac{٥}{٢}$). 20 الخمسة عشر] ١٥. تسعة] ٩. 21 فيكونان] فتكون. مائة وخمسة وثمانين] ١٨٥. 21–22 الخمسة عشر] ١٥. 22 مائتان وخمسة وعشرون] ٢٢٥. اربعون] ٤٠. عشرين] ٢٠. 24 الثلاثة عشر] ١٣. 25 مائة واربعة واربعون] ١٤٤. اثنى عشر] ١٢. 26 الذى] التى. 27 اثنا عشر] ١٢. اثنان] ٢. اربعة وعشرين] ٢٤.

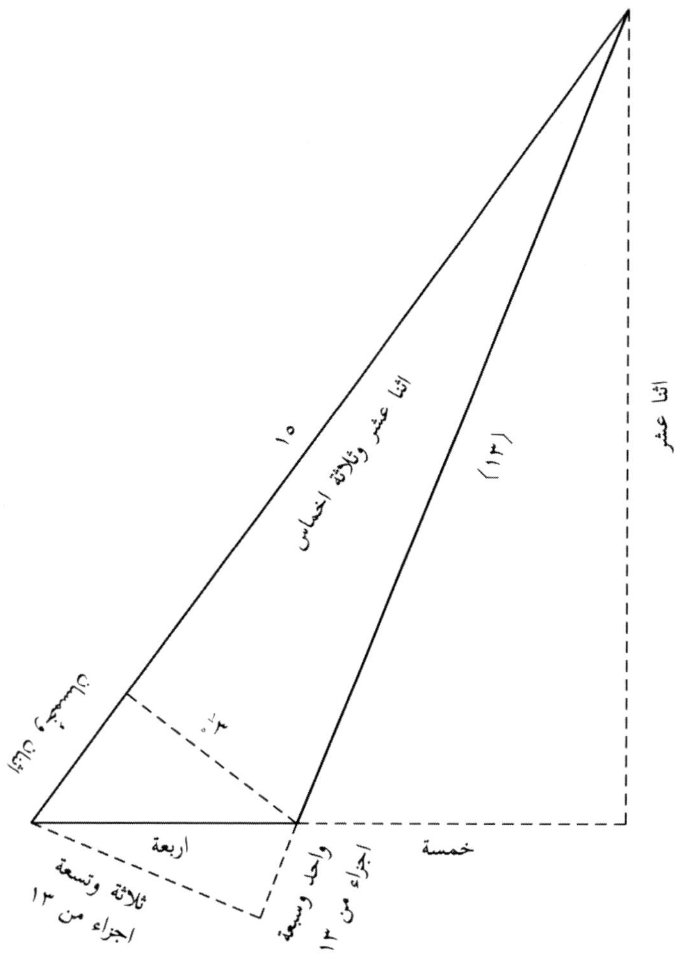

(97^v) فاذا اردت ان تعلم كم يخرج خطّ الثلاثة عشر على استقامته الى موضع موقع العمود فانّه يخرج واحداً وسبعة اجزاء من ثلاثة عشر ويكون طول هذا الخطّ من موضع إلتقاءه مع طرف خطّ الخمسة عشر الى موضع مسقط العمود عليه اربعة عشر وسبعة اجزاء من ثلاثة عشر. فاذا اردت ان تعلم كم العمود الذى يخرج من ملتقى خطّ الاربعة وخطّ الخمسة عشر ⟨الى موضع موقع العمود على خطّ الثلاثة عشر⟩ على استقامته وهو واحد وسبعة اجزاء من ثلاثة عشر فاضرب واحداً وسبعة اجزاء من ثلاثة عشر فى مثله واسقط ما اجتمع منه من ضرب الاربعة فى مثلها وخذ جذر ما بقى فيكون ثلثة وتسعة اجزاء من ثلاثة عشر وهو العمود. فاذا اردت ان تعلم مساحة المثلّث من هذه الجهة ايضاً فاضرب العمود وهو ثلثة وتسعة اجزاء من ثلاثة عشر ⟨فى نصف الثلاثة عشر⟩ وهو ستّة ونصف فتكون اربعة وعشرين وهو مساحة المثلّث.

(19) وفى مساحة المثلّث باب واحد يخرج به جميع المثلّثات وهو ان تجمع جوانب المثلّث الثلاثة ثمّ تأخذ نصف ما اجتمع فتحفظه ثمّ انظر كم فضل هذا النصف على كلّ جانب من جوانب المثلّث فتضرب الفضول بعضها فى بعض ثمّ تضرب ما اجتمع من ذلك فى الذى حفظتَ من نصف الاضلاع فما اجتمع أخذتَ جذره فما كان فهو

[1] الثلاثة عشر] ١٣. واحداً] واحد. [2] ثلاثة عشر] ١٣. الخمسة عشر] ١٥. [3] اربعة عشر] ١٤. ثلاثة عشر] ١٣. [4] الخمسة عشر] ١٥. [5] ثلاثة عشر] ١٣. ثلاثة عشر] ١٣. الاربعة] ٤. [6] ثلاثة عشر] ١٣. [7] ثلاثة عشر] ١٣. [8] اربعة وعشرين] ٢٤. [10] بعضها] يعضها.

مساحة المثلّث.

ومثل ذلك اذا قال مثلّث من جانب خمسة عشر ومن جانب اربعة عشر ومن جانب ثلاثة عشر. فاذا اردت ان تعلم كم مساحته فاجمع جوانبه الثلاثة وهى خمسة عشر واربعة عشر وثلاثة عشر فتكون اثنين واربعين فخذ نصفها فيكون واحداً وعشرين فاحفظه ثمّ انظر فضله على كلّ واحد (98ʳ) من جوانب المثلّث فتجد فضله على الخمسة عشر ستّة وعلى الاربعة عشر سبعة وعلى الثلاثة عشر ثمانية فتضرب ستّة فى سبعة فى ثمانية فتكون ثلاث مائة وستّة وثلاثين فتضربها فى نصف جوانب المثلّث الذى حفظتَ وهو واحد وعشرون فتكون سبعة آلاف وستّة وخمسين فتأخذ جذرها فتكون اربعة وثمانين وهى مساحة المثلّث. وهذا الباب يجرى فى كلّ مثلّث فاحتفظ به فانّه أصل صحيح.

(20) فان قال مثلّث متساوى الاضلاع كلّ جانب منه عشرة اذرع كم مساحته.

امّا حسابه فان مساحته جذر الف وثمانى مائة وخمسة وسبعين. فاذا اردت مساحته من قبل عموده فمعلوم ان عموده يقع على النصف من ايّ الاضلاع احببتَ ان يقع العمود عليه فتضرب نصف احد الاضلاع وهو خمسة فى مثله وتلقيه من ضرب ايّ الاضلاع شئت فى مثله فتبقى خمسة وسبعون فتقول العمود جذر خمسة وسبعين فتضربه فى نصف الضلع الذى وقع عليه العمود وهو خمسة فتضرب خمسة فى مثلها فتكون خمسة وعشرين فتضربها فى الخمسة وسبعين فتكون الفاً وثمانى مائة وخمسة وسبعين فجذرها هو التكسير وهو بالتقريب ثلاثة واربعون وثُلث.

وكلّ مثلّث متساوى الاضلاع فاذا اردت مساحته بالتقريب فاضرب احد اضلاعه فى مثله فما اجتمع فخذ ثُلثه وعُشره فما كان فهو مساحة المثلّث.

وان اردت ان تعلم مساحته بالتفاضل فاعمل كما وصفتُ لك فيخرج لك مساحته جذر الف وثمانى مائة وخمسة وسبعين.

وكلّ مثلّث متساوى الاضلاع فان معرفة عموده ان تضرب احد الاضلاع فى مثله ثمّ تسقط ممّا اجتمع رُبعه وتأخذ جذر ما بقى فما كان فهو العمود.

(21) فان قال مثلّث متساوى الاجناب عموده عشرة اذرع كم كلّ جانب منه.

فاضرب العمود فى مثله فتكون مائة فزد عليها ثُلثها ابداً وهو ثلاثة وثلاثون وثُلث فتكون مائة وثلاثة وثلاثين وثُلثاً فجذر مائة وثلاثة وثلاثين وثُلث كلّ جانب من المثلّث.

(22) فان قال مثلّث من جانبين (98ᵛ) عشرة عشرة ومن الجانب الآخر وهو القاعدة اثنا عشر كم مساحته.

فمعرفة مساحته من قبل عموده ان تخرج عموده على اثنى عشر فهو اسهل لانّه يقع على النصف. وكلّ مثلّث متساوى الجانبين واردت اخراج عموده على الجانب الآخر فانّه يقع على نصفه مساوياً كان لأحد الجانبين او مخالفاً الّا انّه لا بدّ ان يكون اقصر من الجانبين الآخرين مجموعين فان كان اطول منهما او مثلهما فانّه لا يكون من هذه الثلاثة الخطوط مثلّث لانّ كلّ مثلّث فلا بدّ ان يكون كلّ جانبين منه ⟨مجموعين⟩ اطول من الجانب الآخر. فان

---

[2] خمسة عشر] ١٥. اربعة عشر] ١٤. ثلاثة عشر] ١٣. [3] كم] *supra lin.* خمسة عشر واربعة عشر وثلاثة عشر] ١٥ و١٤ و١٣. اثنين واربعين] ٤٢. [4] واحداً وعشرين] ٢١. فتجد] فخد. [4–5] الخمسة عشر] ١٥. [5] الاربعة عشر] ١٤. الثلاثة عشر] ١٣. [5–6] ثلاث مائة وستّة وثلاثين] ٣٣٦. [6] واحد وعشرون] ٢١. سبعة آلاف وستّة وخمسين] ٧٠٥٦. [7] اربعة وثمانين] ٨٤. وهى] وسبع. [10] حسابه] مساحته. الف وثمانى مائة وخمسة وسبعين] ١٨٧٥. [11] يقع][2] تلقي. خمسة] ٥. [12] خمسة وسبعون] ٧٥. خمسة وسبعين] ٧٥. [13] خمسة] ٥. خمسة] ٥. خمسة وعشرين] ٢٥. [14] الخمسة وسبعين] ٧٥. الفاً وثمانى مائة وخمسة وسبعين] ١٨٧٥. ثلاثة واربعون وثُلث] ٤٣ ⅓. [17–18] الف وثمانى مائة وخمسة وسبعين] ١٨٧٥. [21] متساوى] مستوي. [22] فى مثله] العشرة فى عشرة ادرع. ثلاثة وثلاثون وثُلث] ٣٣ ⅓. مائة وثلاثة وثلاثين وثُلثاً] ١٣٣ ⅓. [23] مائة وثلاثة وثلاثين وثُلث] ١٣٣ ⅓. [24] الجانب] جانب. اثنا عشر] ١٢. [25] اثنى عشر] ١٢. على][2] *corr. ex* فى. [26] مساوياً] متساوي. مخالفاً] مخالف.

اردت اخراج عموده على جانب الاثنى عشر وقدّمنا لك انّه لا يقع الّا على نصفه فاضرب نصف الاثنى عشر وهو ستّة فى مثله واسقط ما اجتمع من ضرب احد الجانبين فى مثله فيبقى اربعة وستّون فتأخذ جذرها فتكون ثمانية وهو العمود فاضربه فى نصف القاعدة وهو ستّة فتكون ثمانية واربعين وهو مساحة المثلّث.

واذا اردت ⟨ان تعلم مساحته⟩ بالتفاضل فاعمل كما وصفتُ لك وهو ان تجمع جوانب المثلّث الثلاثة فتكون اثنين وثلاثين فتأخذ نصفه فتكون ستّة عشر فتنظر كم فضله على كلّ جانب من جوانب المثلّث فتجد فضله على الجانبين اللذين هما عشرة عشرة ستّة ستّة وعلى الجانب الذى ⟨هو⟩ اثنا عشر اربعة فتضرب ستّة فى ستّة فى اربعة فتكون مائة واربعة واربعين فتضربها فى ستّة عشر التى هى نصف جوانب ⟨المثلّث⟩ فتكون الفين وثلاث مائة واربعة فتأخذ جذرها فتكون ثمانية واربعين وهى مساحة المثلّث.

(20′) وكذلك ان اردت مساحة المثلّث المتساوى الجوانب الذى كلّ جانب منه عشرة بالتفاضل. فاجمع جوانبه الثلاثة فتكون ثلاثين فخذ نصفها فتكون خمسة عشر فانظر كم فضله على كلّ جانب من جوانب المثلّث فتجد فضله على كلّ جانب خمسة فتضرب خمسة فى خمسة ⟨فى خمسة⟩ فتكون مائة وخمسة وعشرين فتضربها فى نصف جوانب المثلّث الثلاثة وهو خمسة عشر فتكون الفاً وثمانى مائة وخمسة وسبعين فتأخذ جذرها فما كان فهو مساحة المثلّث فتقول (99r) مساحة المثلّث جذر الف وثمانمائة وخمسة وسبعين وهو بالتقريب ثلاثة واربعون وثُلث. وذلك بابه وحسابه.

(23) فان قال مربّع مختلف الجوانب من جانب تسعة عشر ويقابله ⟨خمسة والجانب الآخر خمسة عشر ويقابله⟩ ثلاثة عشر كم مساحته.

امّا حسابه فان مساحته مائة واربعة واربعون ذراعاً. فان اردت مساحته فاعرف عموده اوّلاً.

ومعرفة عموده اذا كان الجانب الذى قلنا انّه خمسة يوازى الجانب الذى قلنا انّه تسعة عشر ومعرفة موازاته ان العمودين اللذين يخرجان من طرفى خطّ الخمسة فيقعا على خطّ تسعة عشر هما متساويان فان كانا غير متساويين فانّ الجانبين غير متوازيين. فان اردت ان تعلم كم العمود الذى يخرج من طرف خطّ الخمسة حتّى يقع على خطّ تسعة عشر وأين يقع منه فاسقط الخمسة من التسعة عشر فيبقى اربعة عشر واسقط السطح الذى جانبان من جوانبه العمودان اللذان يخرجان من طرفى خطّ الخمسة ويقعان على خطّ التسعة عشر والجانبان الآخران خطّ الخمسة ومثلها يقابلها من خطّ التسعة عشر فاذا اخرجتَه فهم قطعتين من السطح مثلّثين احد جوانب احدهما ثلاثة عشر والجانب الآخر العمود والجانب الآخر قطعة من التسعة عشر وأحد جوانب المثلّث الآخر العمود الآخر الذى قلنا انّه مساوٍ للعمود الآخر والجانب الآخر ⟨خمسة عشر والجانب الآخر⟩ قطعة من خطّ التسعة عشر فيكون مع القطعة الاخرى من التسعة عشر التى فى المثلّث الآخر اربعة عشر فاذا قرنا المثلّثان احدهما الى الآخر والصقنا جانب العمود بجانب العمود صار مثلّث من جانب خمسة عشر ومن جانب ثلاثة عشر ومن جانب اربعة عشر.

فمعرفة عموده على حسب ما بيّنتُه لك وهو ان تضرب اذا اردت اخراج العمود (99v) على الاربعة عشر خمسة

---

[1] الاثنى عشر] ١٢. الاثنى عشر] ١٢. [2] اربعة وستّون] ٦٤. ثمانية] ٨. [3] ثمانية واربعين] ٤٨. [5] اثنين وثلاثين] ٣٢. ستّة عشر] ١٦. [6] الجانب] الجوانب. اثنا عشر] ١٢. [7] ستّة عشر] ١٦. الفين وثلاث مائة واربعة] ٢٣٠٤. [8] ثمانية واربعين] ٤٨. [9] بالتفاضل] التفاضل. [10] ثلاثين] ٣٠. خمسة عشر] ١٥. [11] خمسة] ٥. خمسة فى خمسة] ٥ في ٥. مائة وخمسة وعشرين] ١٢٥. [12] خمسة عشر] ١٥. الفاً وثمانى مائة وخمسة وسبعين] ١٨٧٥. [14] واربعون] واربعين. [15] تسعة عشر] ١٩. [16] ثلاثة عشر] ١٣. [17] مائة واربعة واربعون] ١٤٤. [18] تسعة عشر] ١٩. [19] اللذين] الدين. فيقعا] فيقعان. تسعة عشر] ١٩. [20] الخمسة] خمسه. [21] تسعة عشر] ١٩. التسعة عشر] ١٩. اربعة عشر] ١٤. جانبان] جانبين. [22] العمودان] العمودين. اللذان] الدين. يخرجان] يقعان في هدا السطح الدى قلنا انه يخرج. الخمسة] ٥. والجانبان الآخران] والجانبين الاخرين. [23] التسعة عشر] ١٩. [24] التسعة عشر] ١٩. [25] مساوٍ] مساوي. التسعة عشر] ١٩. [26] الاخرى] الاخرا. اربعة عشر] ١٤. قرنا] قدمنا. [27] خمسة عشر] ١٥. ثلاثة عشر] ١٣. اربعة عشر] ١٤. [28] الاربعة عشر] ١٤.

عشر فى مثلها وثلاثة عشر فى مثلها ثمّ تأخذ فضل ما بينهما فيكون ستّة وخمسين فتأخذ نصفها فتكون ثمانية وعشرين فتقسمها على الاربعة عشر فيخرج اثنان فتسقطهما من نصف الاربعة عشر وتزيدهما على نصف الاربعة عشر فيكون مسقط العمود على الاربعة عشر ممّا يلى الثلاثة عشر على خمسة وممّا يلى الخمسة ⟨عشر⟩ على التسعة. فمعرفة العمود ان شئت ان تضرب الخمسة فى مثلها وتسقط ما اجتمع من ضرب الثلاثة عشر فى مثلها فتبقى مائة واربعة واربعون وان شئت فاضرب التسعة فى مثلها واسقط ما اجتمع من ضرب الخمسة عشر فى مثلها فتبقى مائة واربعة واربعون ايضاً فتأخذ جذرها فتجده اثنى عشر وهو العمود.

ثمّ ترجع الى السطح الذى قلنا ⟨انّه⟩ احد جوانبه تسعة عشر ويقابله خمسة والجانب الآخر خمسة عشر ويقابله ثلاثة عشر وقد بيّنّا ان العمود الذى يخرج من احد طرفى خطّ الخمسة اثنا عشر فاجمع الخمسة والتسعة عشر فيكونان اربعة وعشرين فخذ نصفها فتكون اثنى عشر فاضربه فى العمود وهو اثنا عشر فتكون مائة واربعة واربعين فذلك مساحته. وذلك بابه وحسابه.

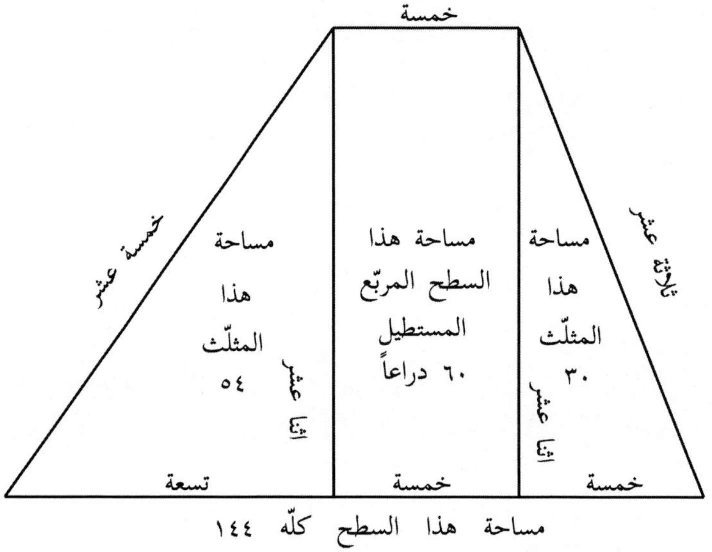

(24) وامّا مسّاح السلطان فانّهم اذا صادفوا مثل هذه الأرض وارادوا مساحتها جمعوا الخمسة والتسعة عشر ثمّ أخذوا نصف ذلك فوجدوه اثنى عشر ثمّ جمعوا الثلاثة عشر والخمسة عشر ثمّ أخذوا نصف ذلك فوجدوه اربعة عشر فضربوا الاربعة عشر فى الاثنى عشر فتكون مائة وثمانية وستّين فذلك مساحتها عندهم وهذا لا يصحّ وما اخبرناك اوّلاً فهو الصحيح.

وكذلك يعملون فى كلّ أرض مربّعة مختلفة الجوانب. يجمعون الجانبين اللذين يقابل احدهما الآخر ويأخذون نصف ما اجتمع ثمّ يجمعون الجانبين الآخرين ويأخذون نصف ما اجتمع ثمّ يضربون النصف فى النصف فما كان فهو مساحتها وقد اعلمناك ان هذا غير صحيح.

---

[1] وثلاثة عشر] وثلثة عشر *corr. ex*.   ستّة وخمسين] ٥٦.   [1–2] ثمانية وعشرين] ٢٨.   [2] الاربعة عشر] ١٤.   اثنان] اثنين. الاربعة عشر] ١٤.   [2–3] الاربعة عشر] ١٤.   [3] الاربعة عشر] ١٤.   الثلاثة عشر] ١٣.   على]³ بلى *corr. ex*.   [4] التسعة] السعه. الثلاثة عشر] ١٣.   [5] مائة واربعة واربعون] ١٤٤.   التسعة] السعه.   الخمسة عشر] ١٥.   [6] مائة واربعة واربعون] ١٤٤.   اثنى عشر] ١٢.   [7] تسعة عشر] ١٩.   ويقابله] وتقابله.   خمسة] ٥.   خمسة عشر] ١٥.   [8] ثلاثة عشر] ١٣.   الخمسة] السعه. اثنا عشر] ١٢.   [9] فيكونان] فتكون.   اربعة وعشرين] ٢٤.   اثنى عشر] ١٢.   اثنا عشر] ١٢.   مائة واربعة واربعين] ١٤٤. [10] مساحته] مساحتها.   [12] اثنى عشر] ١٢.   الثلاثة عشر والخمسة عشر] ١٣ و١٥.   اربعة عشر] ١٤.   [13] الاربعة عشر فى الاثنى عشر] ١٤ في ١٢.   مائة وثمانية وستّين] ١٦٨.   [15] اللذين] الدين.   [17] مساحتها] مساحته.

وكذلك يعملون فى المثلّث (100r) المتساوى الاضلاع. يأخذون احد جوانبه فيضربونه فى نصف جانب آخر فما اجتمع فهو مساحته وهذا ليس بصحيح والصحيح ما بيّنتُ لك.

وامّا المثلّث المختلف الاضلاع فيضربون اطول جانب فيه فى نصف اطول الجانبين فما كان فهو مساحته عندهم. فاذا ارادوا ان يعدّلون عند انفسمهم جمعوا جوانبه ثمّ أخذوا ثُلث ما اجتمع فضربوه فى نصف اطول جوانبه فاذا ارادوا ان يعدّلون العدل الصحيح ضربوا ثُلث جوانبه ⟨فى سُدس جوانبه⟩ فما كان فهو مساحته عندهم وليس فى هذا شيء صحيح والصحيح ما اعلمتُك به فتمسّك به ترشد. ولا حول ولا قوّة الّا باللّه.

واذا صار مسّاح السلطان الى أرض شبيهة بالمدوّرة أخذوا (100v) استدارتها ثمّ أخذوا ثُلثه فضربوه فى مثله فما كان فهو مساحتها عندهم فاذا ارادوا ان يعدّلوا ضربوا ثُلث استدارتها فى رُبع استدارتها فما كان فهو مساحتها عندهم.

وكذلك يفعلون فى أرض كثيرة الجوانب اعنى ⟨ان⟩ يكون لها خمسة جوانب وستّة واكثر من ذلك فانّهم يمسحونها على السبيل الذى اعلمتُك فى مساحة المدوّرة. وربّما قطّعوها فاخرجوا منها مربّعات ومثلّثات فيمسحون ما كان منها من المربّعات على النحو الذى اعلمتُك انّهم يمسحون مربّعاتهم وما كان منها ⟨من⟩ مثلّثات يمسحونه على السبيل الذى فسّرتُ لك فى مساحتهم للمثلّثات.

(25) والوجه الصحيح فى مساحة الأرض اذا كثرت جوانبها ان تقطع مثلّثات كلّها ثمّ تذرع كلّ مثلّث منها على ما فسّرتُ لك فى مساحة المثلّثات ثمّ تجمع ذلك كلّه فما كان فهو مساحة الأرض فان خرج فى تقطيعك مربّع متساوى الطولين والعرضين والقطرين فامسحه من غير ان تقطعه بمثلّثين [ومساحته ان تضرب طوله فى عرضه فما كان فهو مساحة المربّع] فاجمعه الى مساحة ما بقى من تقطيعك فما كان فهو مساحة الأرض.

واصحاب الأرض قد جروا فى مسايحهم على الجور عليهم فهو عندهم انّه قد عدل عليهم فليس يلتفتون الى غيره ولا يعرفون سواه واكثرهم ان مسحتُ عندهم بالعدل ونقصتُ عن مساحته الذى مسحها عليه مسّاح السلطان لم يرض به ولم يلتفت اليه ومال الى مسّاح السلطان وان جاروا عليه.

مساحة المجسّمات وسطوحها

(26) فان قال عمود مربّع اسفله اربعة فى اربعة ورأسه اربعة فى اربعة وارتفاعه عشرة كم مساحة جسمه.

فاذا كان المجسّم متساوى (101r) الطرفين فاضرب طول اسفله فى عرض اسفله فما بلغ فاضربه فى سمكه فما اجتمع فهو مساحة العمود. فاضرب اربعة فى اربعة فتكون ستّة عشر فاضربها فى السمك وهو عشرة فتكون مائة وستّين ذراعاً وهو مساحة العمود.

(26′) وكذلك ان كانت بئر مربّعة اسفلها اربعة فى اربعة واعلاها اربعة فى اربعة وسمكها الذى هو عمقها عشرة.

فان مساحتها مثل مساحة العمود مائة وستّون. والباب فى مساحتها مثل الباب فى مساحة العمود سواء. وانّما اعنى فى مساحة البئر الفارغ منها واعنى بمساحة العمود جسم العمود.

(27) فان اردت مساحة سطح العمود اعنى بسطحه تبييضه ان بيّض او صبغه ان صبغ او لِبسه ان لبس.

فاذا اردت ذلك فخذ استدارته من اربعة جوانبه فتجده ستّة عشر ذراعاً فاضربه فى سمكه وهو عشرة فتكون مائة وستّين ذراعاً فهو مساحة سطحه سوى مساحة ⟨سطح⟩ اسفله وسطح اعلاه.

---

6 فتمسّك] فتسمك. 10 خمسة] خمس. 11 المدوّرة] المدور. 12 يمسحونه] مسحوه. 15 مربّع] مربعاً. 19 عندهم] عليهم. ونقصتُ] ونقصته. عن] من. 20 اليه] عليه. السلطان] السطان. 22 ورأسه] وار *pr. scr.* اربعة فى اربعة] ٤ في ٤. عشرة] ١٠. 24 اربعة فى اربعة] ٤ في ٤. ستّة عشر] ١٦. 25–24 مائة وستّين] ١٦٠. 26 اربعة فى اربعة] ٤ في ٤. عشرة] ١٠. 27 مائة وستّون] ١٦٠. 29 تبييضه] تبيضه. 30 اربعة] اربع. ستّة عشر] ١٦. 31–30 مائة وستّين] ١٦٠ (*add. in marg.*). 31 سوى] سوا.

فان اردت مساحة سطح اسفله فاضرب طوله فى عرضه فتكون ستّة عشر ذراعاً وهو مساحة سطح اسفله وكذلك مساحة سطح اعلاه ستّة عشر ذراعاً.

(28) فان كان العمود مدوّراً متساوى الطرفين وكان قطر قاعدته اربعة وقطر دائرة اعلاه اربعة فاسقط من المائة وستّين ذراعاً سُبعها ونصف سُبعها وهو اربعة وثلاثون وسُبعان فتبقى مائة وخمسة وعشرون وخمسة اسباع وهو مساحة جسم العمود.

(29) وكذلك مساحة سطحه ان تأخذ دوره فيكون اثنى عشر واربعة اسباع لانّ دائرة قطره اربعة اذرع فان دورها ان تضرب القطر فى ثلاثة وسُبع ابداً فتكون اثنى عشر واربعة اسباع فتضربه فى السمك وهو عشرة فتكون مائة وخمسة وعشرين وخمسة اسباع وهو مساحة سطح العمود سوى مساحة سطح قاعدته وسطح اعلاه. فان اردت (101v) مساحة سطح قاعدته فامسحه على نحو ما قد فسّرتُ لك من مساحة الدائرة وكذلك سطح اعلاه.

(30) فان قال عمود مربّع مخروط اسفله خمسة فى خمسة واعلاه اربعة فى اربعة وعموده عشرة اعنى بعموده الخطّ المستقيم الذى يخرج من وسط ⟨سطح⟩ اعلاه الى وسط سطح قاعدته كم مساحة جسمه.

امّا حسابه فانّ مساحة جسمه مائتان وثلاثة وثُلث. وبابه ان تضرب خمسة فى خمسة فتكون خمسة وعشرين وتضرب اربعة فى اربعة فتكون ستّة عشر ثمّ تضرب خمسة فى اربعة فتكون عشرين ⟨فتجمع الخمسة وعشرين والستّة عشر والعشرين فتكون واحداً وستّين فتأخذ ثُلثها فيكون عشرين وثُلثاً⟩ فتضربه فى العمود وهو عشرة فتكون مائتين ⟨وثلاثة وثُلثاً⟩ وهو مساحة جسم العمود.

وان شئت فاجمع الخمسة والاربعة فتكون تسعة فتأخذ نصفها فتكون اربعة ونصفاً فتضربها فى مثلها فتكون عشرين ورُبعاً فاحفظها ثمّ تأخذ فضل ما بين الخمسة والاربعة فتجده واحداً فخذ نصفه فيكون نصفاً فاضربه فى مثله فيكون رُبعاً فخذ ثُلثه فيكون نصف سُدس فزده على العشرين والرُبع الذى حفظتَ فيكون عشرين وثُلثاً فاضربها فى العمق وهو عشرة فتكون مائتين وثلاثة وثُلثاً وهو مساحة جسم العمود.

(31) فان كان العمود مدوّراً فاسقط من مائتين وثلاثة وثُلث سُبعها ونصف سُبعها وهو ثلاثة واربعون واربعة اسباع فيبقى مائة وتسعة وخمسون وخمسة اسباع وثُلث سُبع وهو مساحة جسم العمود.

(32) فان اردت مساحة سطح هذا العمود على ما قلنا من تربيع اسفله وتربيع اعلاه.

فخذ تدوير اسفله فتجده عشرين وتدوير اعلاه فتجده ستّة عشر فاجمعهما فيكونان ستّة وثلاثين فخذ نصفها فتكون ثمانية عشر فاضربها فى سمك العمود على سطحه وهو اكثر من عشرة بشىء يسير فيكون مائة وثمانين ذراعاً وشيئاً وهو مساحة سطحه كما تدور سوى مساحة سطح اسفله وسطح اعلاه [ستّة عشر].

(33) فان كان العمود مدوّراً واردت مساحة سطحه.

---

[1] ستّة عشر] ١٦. [2] ستّة عشر] ١٦. [3] اربعة] ٤. اربعة] ٤. المائة وستّين] ١٦٠. [4] اربعة وثلاثون وسُبعان] ٣٤ ٢/٧. مائة وخمسة وعشرون وخمسة اسباع] ١٢٥ ٥/٧. [6] ان][1] وان شيت فى مساحه سطحه ان. اثنى عشر واربعة اسباع] ١٢ ٤/٧. اربعة] ٤. [7] اثنى عشر واربعة اسباع] ١٢ ٤/٧. [7–8] مائة وخمسة وعشرين وخمسة اسباع] ١٢٥ ٥/٧. [9] فامسحه] فامسح. [10] خمسة فى خمسة] ٥ في ٥. اربعة فى اربعة] ٤ في ٤. [12] مائتان وثلاثة وثُلث] ٢٠٣ ١/٣. خمسة فى خمسة] ٥ في ٥. خمسة وعشرين] ٢٥. [13] اربعة فى اربعة] ٤ في اربعه. ستّة عشر] ١٦. خمسة فى اربعة] ٥ في ٤. عشرين] ٢٠. [14] مائتين] ٢٠٠. [16] تسعة] ٩. فتكون][1] فتكو. اربعة ونصفاً] ٤ ١/٢. [17] عشرين] ٢٠. [19] مائتين وثلاثة وثُلثاً] ٢٠٣ ١/٣. [20] مائتين وثلاثة وثُلث] ٢٠٣ ١/٣. ثلاثة واربعون واربعة اسباع] ٤٣ ٤/٧. [21] مائة وتسعة وخمسون وخمسة اسباع وثُلث سُبع] ١٥٩ ٥/٧ ١/٣. [23] وتدوير] وتدير. ستّة عشر] ١٦. فيكونان] فيكونا. ستّة وثلاثين] ٣٦. [24] ثمانية عشر] ١٨. [25] وشيئاً] وشي. ستّة عشر] ١٦.

فخذ تدوير (102r) اسفله فتجده خمسة عشر وخمسة اسباع لانّ قطر دائرة اسفله خمسة اذرع فتضربه فى ثلاثة وسُبع وخذ تدوير اعلاه فتجده اثنى عشر واربعة اسباع فاجمعهما فيكونان ثمانية وعشرين وسُبعين فخذ نصفها فتكون اربعة عشر وسُبعاً فاضربها فى سمك العمود على سطحه وهو اكثر من عشرة بشىء يسير فتكون مائة وأحداً واربعين وثلاثة اسباع وهو مساحة سطح العمود سوى مساحة سطح قاعدته وسطح اعلاه ومساحة سطح قاعدته مثل مساحة دائرة قطرها خمسة اذرع فمساحته تسعة عشر ونصف وسُبع ومساحة سطح اعلاه اثنا عشر واربعة اسباع.

(34) ومساحة الخشب تجرى على نحو ما فسّرتُ لك.

فان كان مربّعاً واردت مساحة جسمه فعلى ما فسّرتُ لك من مساحة جسم العمود وان اردت مساحة جسم الخشب اذا كان مدوّراً فعلى ما بيّنتُ لك من مساحة العمود ان كان مدوّراً وان كان مربّعاً واردت مساحة سطحه فعلى ما مثّلتُ لك من مساحة سطح العمود المربّع وان كان مدوّراً فعلى ما قدّمتُ لك بيانه من مساحة سطح العمود المدوّر.

ولا نعلم للخشب ولا لشىء من المجسّمات اكثر من مساحة المجسم والسطح وهو ما قد بيّنتُ لك واهل مصر جروا فى مساحة خشبهم على شيء [لم يتقدّمهم فيه احد] لا هو مساحة الجسم ولا هو مساحة السطح وقد فسّرتُ لك فيما تقدّم من كتابى كيف يمسحونه وانّه مضحكة وسخريّة وبيّنتُ ما يدخل عليهم فيه من النقص والضعف وقلّة التمييز. فنسأل اللّه السلامة والعافية فى الديّن والدنيا والآخرة انّه على كلّ شىء قدير.

(35) فان قال عمود اسفله ستّة فى ستّة وسمكه [الذى هو عموده] عشرة اذرع كم مساحة جسمه اذا كان العمود مربّعاً مخروطاً ورأسه مثل الزجّ.

فامّا حسابه فان مساحة جسمه مائة وعشرون ذراعاً. وبابه (102v) ان تضرب ستّة فى ستّة فتكون ستّة وثلاثين فتأخذ ثُلثها ابداً فتكون اثنى عشر فتضربها فى العمود وهو عشرة فتكون مائة وعشرين وهو مساحة جسم العمود.

(36) فان كان العمود مدوّراً فاسقط من المائة والعشرين سُبعها ونصف سُبعها وهو خمسة وعشرون وخمسة اسباع فيبقى اربعة وتسعون وسُبعان وهو مساحة جسم العمود اذا كان مدوّراً.

(37) فان اردت مساحة سطحه وهو مربّع.

فخذ تدوير اسفله من اربعة جوانب فتكون اربعة وعشرين فخذ نصفها فتكون اثنى عشر فاضربها فى سمكه على سطح العمود وهو عشرة واقلّ من نصف فتكون مائة وستّة وعشرين الّا شيئاً وهو مساحة سطحه سوى مساحة سطح قاعدته ومساحة سطح قاعدته ستّة وثلاثون.

(38) فان اردت مساحة سطحه اذا كان مدوّراً.

فخذ تدوير اسفله فتجده ثمانية عشر وستّة اسباع فخذ نصفه فتكون تسعة وثلاثة اسباع فاضربه فى سمكه على سطحه وهو عشرة واقلّ من نصف فتكون تسعة وتسعين الّا شيئاً فذلك مساحة سطحه سوى مساحة سطح قاعدته ومساحة سطح قاعدته ثمانية وعشرون وسُبعان.

---

[1] خمسة عشر وخمسة اسباع] ١٥ ٥/٧. خمسة] ٥. ثلاثة] مثله. [2] اثنى عشر واربعة اسباع] ١٢ ٤/٧. فاجمعهما] فاجمعها. فيكونان] فتكون. ثمانية وعشرين وسُبعين] ٢٨ ٢/٧ (*add. in marg.*). [3] اربعة عشر وسُبعاً] ١٤ ١/٧. [3–4] مائة وأحداً واربعين وثلاثة اسباع] ١٤١ ٣/٧. [5] خمسة] ٥. تسعة عشر] ١٩ (*post quod* ٩ ادرع *add. et del.*). اثنا عشر واربعة اسباع] ١٢ ٤/٧. [9] مثّلتُ] مثلث. [11] اكثر من مساحة] مساحة اكثر من. [13] يمسحونه] يمسحوه. مضحكة وسخريّة] بصحكبة وسحوبة. [15] ستّة فى ستّة] ٦ في ٦. عشرة] ١٠. [17] مائة وعشرون] ١٢٠. ستّة فى ستّة] ٦ في ٦. ستّة وثلاثين] ٣٦. [18] اثنى عشر] ١٢. مائة وعشرين] ١٢٠. [19] خمسة وعشرون وخمسة اسباع] ٢٥ ٥/٧. [20] اربعة وتسعون وسُبعان] ٩٤ ٢/٧. [22] اربعة] اربع. اربعة وعشرين] ٢٤. اثنى عشر] ١٢. [23] مائة وستّة وعشرين] ١٢٦. شيئاً] شي. سوى] سوا. [24] ستّة وثلاثون] ٣٦. [26] ثمانية عشر وستّة اسباع] ١٨ ٦/٧. تسعة وثلاثة اسباع] ٩ ٣/٧. [27] تسعة وتسعين] ٩٩. شيئاً] شي. [28] ثمانية وعشرون وسُبعان] ٢٨ ٢/٧.

(35′) ومساحة الآبار على نحو ما قد بيّنتُ لك من مساحة اجسام الاعمدة مربّعة او مدوّرة او مخروطة مثل الزجّ سواء بسواء ولا نجاوزه الى غيره ولا نتعدّاه الى سواه ان شاء اللّه تعالى.

مساحة الكرة

نبدأ اوّلاً بمساحة سطحها فانّه انّما يعلم مساحة جسمها من قبل سطحها.

(39) فاذا اردت ان تعلم كم مساحة سطحها وقد سأل عن كرة قطرها سبعة اذرع كم مساحة سطحها.

فاعلم كم تدوير منطقتها ودور منطقتها هو بمنزلة دائرة قطرها سبعة اذرع فاضرب سبعة فى ثلاثة وسُبع ابداً فتكون اثنين وعشرين وهو تدوير منطقتها ثمّ اضرب تدوير منطقتها وهو اثنان وعشرون فى قطرها وهو سبعة اذرع فتكون مائة واربعة وخمسين وهو مساحة سطح الكرة. (103^r) وان شئت فاضرب قطر الكرة فى مثله ثمّ فى ثلاثة وسُبع فما كان فهو مساحة سطح الكرة.

(40) واذا اردت ان تعلم كم مساحة جسم هذه الكرة وقد قلنا ان مساحة سطحها مائة واربعة وخمسون.

فاضرب المائة واربعة وخمسين فى سُدس قطر الكرة وهو فى هذه المسئلة واحد وسُدس فتكون مائة وتسعة وسبعين وثُلثين وهو مساحة جسم الكرة.

وكذلك تفعل ابداً اذا اردت ان تعلم مساحة جسم الكرة 〈ويجب〉 ان تعلم اوّلاً مساحة سطحها ومعرفة مساحتها ان تضرب دور منطقتها فى قطرها فما كان فهو مساحة سطحها [ايضاً وهو موافق بعضه لبعض] ثمّ تضرب مساحة سطحها فى سُدس قطرها فما كان فهو مساحة جسمها.

(41) واعلم ان كلّ مخروط لا رأس له اعنى ان رأسه مثل الزجّ فان مساحة جسمه ان تأخذ ثُلث مساحة قاعدته فتضربه فى عموده فما كان فهو مساحة جسمه كيف كانت قاعدته مخمّسة كانت او مسدّسة او مسبّعة او مثمّنة او مثلّثة او مربّعة او مدوّرة كائنةً ما كانت فاعلم ذلك.

(42) ومساحة سطحه ان تأخذ نصف تدوير اسفله فتضربه فى ارتفاعه على سطح من سطوحه اعنى انّك تضع طرف الخيط على 〈نصف〉 ضلع من اضلاع قاعدته مع الأرض ثمّ طوّفه بالسطح القائم على القاعدة عند طرف الخيط وترفعه قليلاً قليلاً الى ان تنتهى الى رأس المخروط ثمّ تذرع الخيط فما كان فهو ارتفاعه فتضربه فى نصف تدوير اسفله فما كان فهو مساحة سطحه.

(43) فامّا مساحة القوس فعلى ما ذكر بطلميوس فى المقالة السادسة من كتاب المجسطى وقد ذكر محمّد بن موسى الخوارزمى فى كتاب الجبر والمقابلة وهو ان قال تأخذ نصف تدوير القوس فتضربه فى نصف قطر (103^v) الدائرة التى منها تلك القوس فما اجتمع فاحفظه ثمّ انقص سهم القوس من نصف قطر الدائرة ان كانت القوس اقلّ من نصف دائرة وان كانت اكثر من نصف دائرة فانقص نصف قطر المدوّرة من سهم القوس واضرب ما بقى فى نصف الوتر وانقصه ممّا حفظتَ ان كان القوس اقلّ من نصف دائرة وزده عليه ان كانت اكثر من نصف دائرة فما 〈 كان〉 بعد الزيادة والنقصان فهو مساحة القوس.

ومعرفة قطر الدائرة التى منها تلك القوس ان تضرب نصف وتر القوس فى مثله وتقسم ما اجتمع على سهم القوس فما خرج فزد عليه سهم القوس فما كان فهو قطر الدائرة التى منها تلك القوس.

---

2 سواء بسواء] سوي بسوي. ولا] الا. نتعدّاه] يتعداه. 5 سبعة] ٧. 6 سبعة] ٧. سبعة فى ثلاثة وسُبع] ٧ في ٣ $\frac{1}{7}$. 7 اثنين وعشرين] ٢٢. اثنان وعشرون] ٢٢. سبعة] ٧. 8 مائة واربعة وخمسين] ١٥٤. 8–9 ثلاثة وسُبع] ٣ $\frac{1}{7}$. 10 مائة واربعة وخمسون] ١٥٤. 11 المائة واربعة وخمسين] ١٥٤. 11–12 مائة وتسعة وسبعين وثُلثين] ١٩٩ $\frac{2}{3}$ (*sic*). 13 تفعل] فافعل. 16 فان] وان. 19 سطحه] سطحها. 20 طوّفه] طرفه. 21 تنتهى] ينتهي. 22 مساحة] ارتفاع. 26–27 وان كانت اكثر ... اقلّ من نصف دائرة] (المدور *et* حفضت *scr.*) *iter.* 27 وزده] وزد.

ومثال ذلك اذا قيل لك قوس وترها ثمانية وسهمها اثنان كم قطر الدائرة التى منها هذه القوس.

فاضرب نصف الوتر وهو اربعة فى مثله فتكون ستّة عشر فاقسمه على السهم وهو اثنان فيخرج ثمانية فزد عليها السهم وهو اثنان فتكون عشرة وهو قطر الدائرة التى منها هذه القوس.

باب معرفة الاوتار

(44) فان قال مدوّرة قطرها عشرة اذرع كم وتر اوسع مربّع متساوى الاضلاع يقع فيها.

امّا حسابه فانّه جذر خمسين. وبابه ان تضرب قطر الدائرة وهو عشرة فى مثله فتكون مائة فتأخذ نصفها فتكون خمسين فتقول جذر خمسين كلّ جانب من المربّع وهو قريب من سبعة ونصف سُبع وذلك بابه وحسابه. وهذه صورته. (104ʳ)

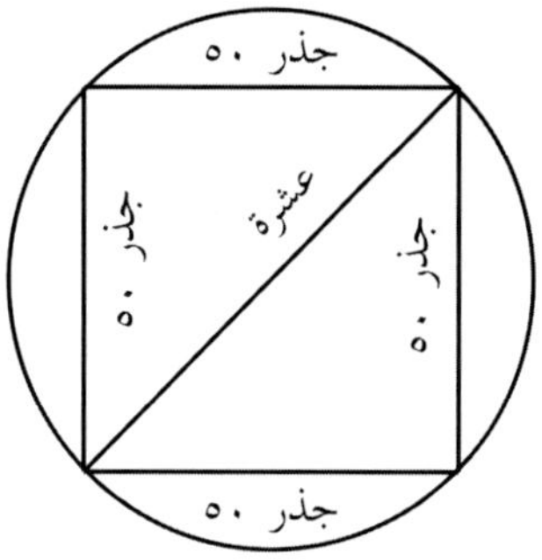

(45) فان قال دائرة قطرها عشرة اذرع كم ⟨وتر⟩ اوسع مربّع متساوى الاضلاع والزوايا يقع عليها من خارج ويكون كلّ جانب من جوانب المربّع يماسّ الخطّ الذى يحيط بالدائرة.

امّا حسابه فانّ كلّ جانب من المربّع مثل قطر الدائرة سواء على ما ترى فى الصورة بيانه فمعلوم انّ كلّ جانب من المربّع عشرة.

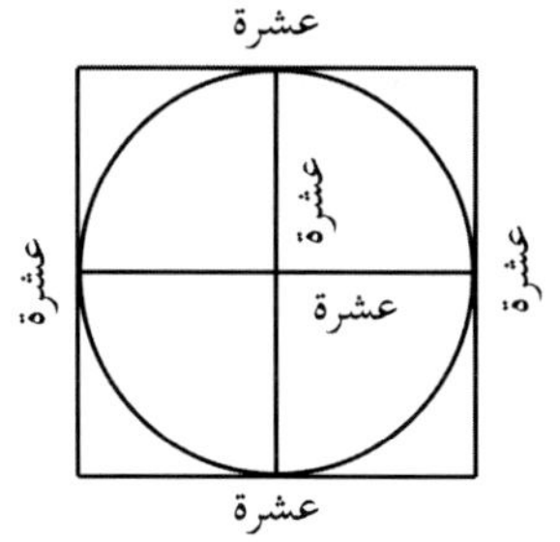

(46) فان قال مربّع كلّ جانب منه عشرة اذرع كم قطر الدائرة التى تقع عليه من خارج.

امّا حسابه فانّه مثل قطر المربّع سواء وهو جذر مائتين وهو قريب من اربعة عشر وسُبع.

(47) فان قال كم قطر الدائرة التى تقع داخل المربّع الذى كلّ جانب منه عشرة اذرع.

امّا حسابه فانّ قطر الدائرة التى تقع داخل هذا المربّع مثل احد جوانب المربّع وهو عشرة.

(48) فان قال مدوّرة قطرها عشرة اذرع كم كلّ جانب من المثلّث المتساوى الاضلاع الذي يقع داخل هذه الدائرة.

---

[1] ثمانية] ٨. اثنان] ٢. [2] اربعة] ٤. ستّة عشر] ١٦. اثنان] ٢. ثمانية] ٨. [3] اثنان] ٢. عشرة] ١٠. [7] سبعة ونصف سُبع] ٧ ١/٢ ١/٧. [11] سواء] سوي. ترى] ترا. [14] المربّع] المربعه. وهو جذر] اوهو جدر. اربعة عشر وسُبع] ١٤ ١/٧.

امّا حسابه فانّه جذر خمسة وسبعين. وبابه ان تضرب قطر الدائرة وهو عشرة فى مثله فتكون مائة فتسقط منها رُبعها وهو خمسة وعشرون فتبقى خمسة وسبعون فتقول جذر خمسة وسبعين كلّ جانب من المثلّث المتساوى الاضلاع الذى يقع فى الدائرة التى قطرها عشرة وهو قريب من ثمانية وثُلثين.

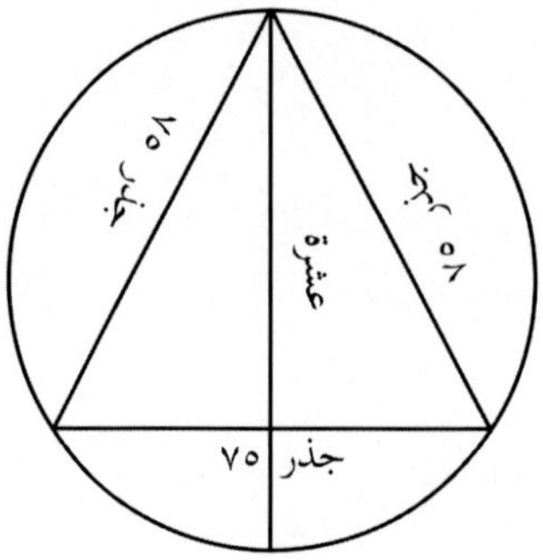

(49) فان قال كم كلّ جانب من المثلّث المتساوى الاضلاع الذى يماسّ الدائرة من خارجها اذا كان قطر الدائرة عشرة.

امّا (104ᵛ) حسابه فان كلّ جانب من المثلّث الذى يقع خارج الدائرة جذر ثلاث مائة. وبابه ان تضرب قطر الدائرة وهو عشرة فى مثله فتكون مائة فتضرب المائة فى ثلاثة ابداً فتكون ثلثمائة فجذر ثلثمائة كلّ جانب من المثلّث الذى يقع خارج الدائرة التى قطرها عشرة وهو قريب من سبعة عشر وثُلث. وهذه صورتها.

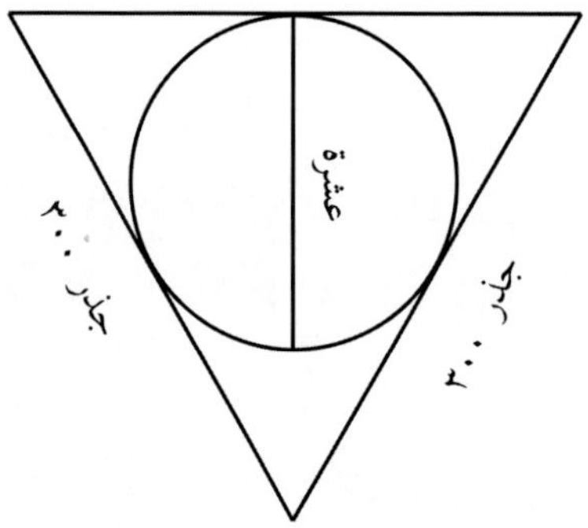

(50) فان قال مثلّث متساوى الاضلاع [والزوايا] كلّ جانب منه عشرة كم قطر اوسع دائرة تقع فيه يماسّ اضلاعه الخطّ المحيط بالدائرة.

امّا حسابه فانّ قطر الدائرة التى تقع داخل المثلّث جذر ثلاثة وثلاثين وثُلث وهو اكثر من خمسة ونصف ورُبع. وبابه ان تضرب العشرة فى مثلها فتكون مائة فتأخذ ثُلثها ابداً وهو ثلاثة وثلاثون وثُلث فتقول جذر ثلاثة وثلاثين وثُلث قطر الدائرة.

(51) فان قال كم قطر الدائرة التى تقع خارج هذا المثلّث اذا كان يماسّ زوايا الخطّ المحيط بالدائرة.

امّا حسابه فانّه جذر ⟨مائة و⟩ثلاثة وثلاثين وثُلث. وبابه ان تضرب العشرة فى مثلها فتكون مائة فتزيد عليها ثُلثها ابداً وهو ثلاثة وثلاثون وثُلث فتكون مائة وثلاثة وثلاثين وثُلثاً ⟨فتقول جذر مائة وثلاثة وثلاثين وثُلث⟩ قطر الدائرة التى تقع خارج المثلّث كلّ جانب منه عشرة وهو اكثر من احد عشر ورُبع وسُدس وثُمن.

[1] خمسة وسبعين] ٧٥. مائة] ١٠٠ (*add. in marg.*). [2] خمسة وعشرون] ٢٥. خمسة وسبعون] ٧٥. خمسة وسبعين] ٧٥.
[3] ثمانية وثُلثين] ٣٨ (*sic*). [6] ثلاث مائة] ٣٠٠. [8] سبعة عشر وثُلث] ١٧$\frac{1}{3}$. [11] ثلاثة وثلاثين وثُلث] ٣٣$\frac{1}{3}$. [12] العشرة] عشره. وثلاثون] وثلاثين. [12–13] ثلاثة وثلاثين وثُلث] ٣٣$\frac{1}{3}$. [15] ثلاثة وثلاثين وثُلث] ٣٣$\frac{1}{3}$. [16] ثلاثة وثلاثون وثُلث] ٣٣$\frac{1}{3}$. مائة وثلاثة وثلاثين وثُلثاً] ١٣٣$\frac{1}{3}$. [17] كلّ] التى كلّ.

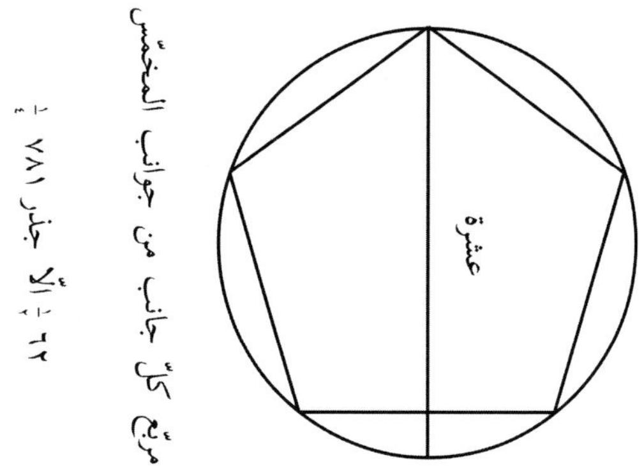

(52) فان قال دائرة قطرها عشرة [فى داخلها مخمّس] كم طول كلّ جانب من جوانب المخمّس المتساوى الجوانب الذى يقع داخل هذه الدائرة ويماسّ زوايا الخطّ المحيط بها.

فاضرب قطر الدائرة فى مثله [فما اجتمع من الضرب قسمة مربّع قطر الدائرة] ثمّ خذ خمسة اثمان مربّع قطر الدائرة فاضربه فى (105r) ثُمن مربّع قطر الدائرة فما اجتمع فخذ جذره فما خرج فانقصه من خمسة اثمان مربّع قطر الدائرة فما بقى فخذ جذره فما كان فهو طول وتر خُمس الدائرة. فاضرب قطر الدائرة وهو عشرة فى مثله فتكون مائة فخذ خمسة اثمانها فتكون اثنين وستّين ونصفاً فاضربها فى ثُمنها وهو اثنا عشر ونصف فتكون سبع مائة وأحداً وثمانين ورُبعاً فخذ جذرها فما كان فانقصه من اثنين وستّين ونصف فما بقى فخذ جذره فما خرج فهو طول وتر خُمس الدائرة. وهذه صورتها.

(53) فان قال دائرة قطرها عشرة يحيط بها مخمّس كم طول كلّ جانب من جوانبه.

فاضرب قطر الدائرة فى مثله ثمّ فى خمسة ابداً واحفظ ما يجتمع ثمّ اضعف قطر الدائرة واضربه فى مثله فما اجتمع فاضربه فى الذى حفظتَ فما بلغ فخذ جذره فما خرج فانقصه من الذى حفظتَ فما بقى فخذ جذره فما كان فهو طول كلّ جانب من جوانب المخمّس. فاضرب عشرة فى مثله ثمّ فى خمسة فتكون خمس مائة فاحفظها ثمّ اضعف العشرة فتكون عشرين فاضربها فى مثلها فتكون اربع مائة فاضربها فى الخمس مائة التى حفظتَ ⟨فتكون مائتى الف فخذ جذرها فما كان فانقصه من الخمس مائة التى حفظتَ⟩ فما بقى فخذ جذره فما كان فهو طول كلّ جانب من جوانب المخمّس.

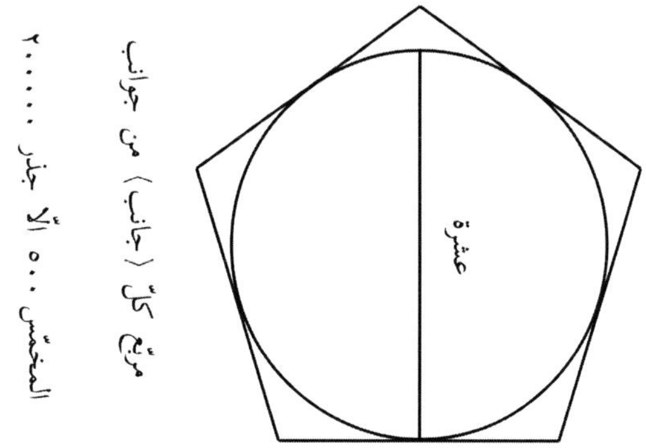

(54) فان قال دائرة قطرها عشرة كم طول كلّ جانب من جوانب المعشّر الذى يقع فيها اذا كانت زواياه تماسّ الخطّ المحيط بالدائرة وكانت جوانبه متساوية.

---

[1] الجوانب] الجانب *corr. ex* [6] اثنين وستّين ونصفاً] ٦٢ ½. فاضربها] فاضربه. اثنا عشر ونصف] ١٢ ½. [6–7] سبع مائة وأحداً وثمانين ورُبعاً] ٧٨١ ¼. [7] اثنين وستّين ونصف] ٦٢ ½. [9] طول] *add. in marg.* [12] خمسة] ٥. خمس مائة] ٥٠٠. [13] اربع مائة] ٤٠٠. الخمس مائة] ٥٠٠. [17] وكانت] وكان.

فاضرب القطر وهو عشرة فى مثله فتكون مائة فخذ رُبعها ونصف ثُمنها ابداً فتكون واحداً وثلاثين ورُبعاً فخذ جذرها فانقص منه رُبع القطر ابداً وهو فى هذه المسئلة اثنان ونصف فما بقى فهو طول كلّ جانب من جوانب المعشّر.

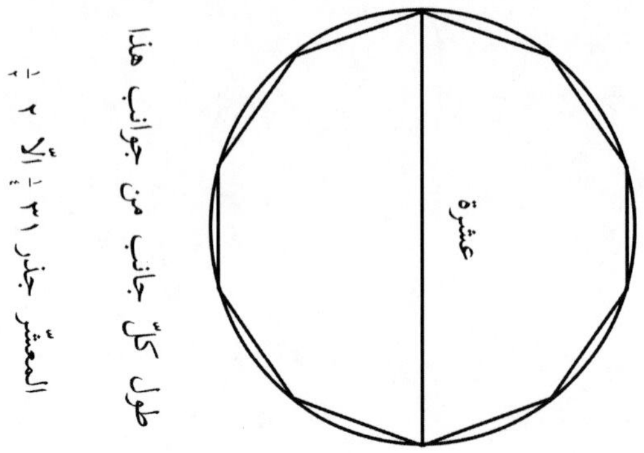

(55) فان قال كم كلّ جانب من جوانب المعشّر المتساوى الجوانب الذى يحيط بالدائرة من خارجها ويماسّ (105^v) جوانبه الخطّ المحيط بالدائرة.

فاضرب قطر الدائرة وهو فى هذه المسئلة عشرة فى مثله فتكون مائة فاحفظها ثمّ اضربها فى اربعة اخماسها وهو ثمانون فتكون ثمانية آلاف فخذ جذرها فما خرج فانقصه من المائة التى حفظتَها فما بقى فخذ جذره فما كان فهو طول كلّ جانب من جوانب المعشّر الذى يقع من خارج هذه الدائرة علي هذه الصورة.

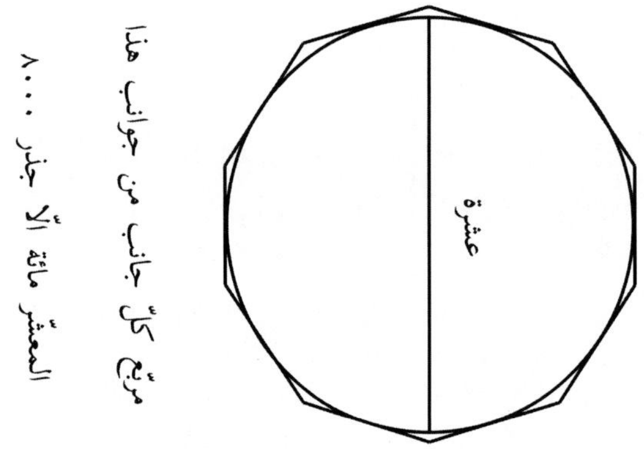

(56) فان قال مخمّس متساوى الاضلاع والزوايا كلّ جانب منه عشرة كم قطر الدائرة التى تحيط به.

امّا حسابه فان مربّع قطر الدائرة التى تحيط به مائتان وجذر ثمانية آلاف. وبابه ان تضرب العشرة فى مثلها [ابداً] فتكون مائة فتضعفها فتكون مائتين ثمّ تضرب العشرة فى اربعة ابداً فتكون اربعين فتضربها فى المائتين فتكون ثمانية آلاف فتأخذ جذر الثمانية آلاف فما خرج زدتَه على المائتين فما اجتمع فخذ جذره فما كان فهو قطر الدائرة.

(57) قال مخمّس متساوى الجوانب والزوايا كلّ جانب منه عشرة كم قطر الدائرة التى فى داخله اذا كان اضلاع المخمّس تماسّ الخطّ المحيط بالدائرة.

امّا حسابه فان مربّع قطر الدائرة التى يحيط بها هذا المخمّس مائة وجذر ثمانية آلاف. وبابه ان تضرب العشرة فى مثلها [ابداً] فتكون مائة ثمّ تضرب العشرة فى ثمانية ابداً فتكون ثمانين فتضربها فى المائة فتكون ثمانية آلاف فتأخذ جذرها فما كان فزده على المائة فما اجتمع فخذ جذره فما كان فهو قطر الدائرة التى تقع داخله.

---

1 واحداً وثلاثين ورُبعاً] ٣١ ¼. 2 اثنان ونصف] ٢ ½. 6 مثله] مثلها. فاحفظها] فاضربها *corr. ex*. 7 ثمانون] ٨٠. ثمانية الآف] ٨٠٠٠. 10 مائتان] مايتين. ثمانية آلاف] ٨٠٠٠. 11 اربعين] ٤٠. 12 ثمانية آلاف] ٨٠٠٠. الثمانية آلاف] ٨٠٠٠. 16 ثمانين] ٨٠. ثمانية آلاف] ٨٠٠٠.

(58) قال دائرة قطرها عشرة كم يكون كلّ جانب من جوانب المسدّس المتساوى الجوانب الذى يقع فيها اذا كانت زواياه تماسّ الخطّ المحيط بالدائرة.

امّا حسابه فان كلّ جانب من جوانب المسدّس (106[r]) مثل نصف قطر الدائرة سواء وهو خمسة. وكذلك كلّ مسدّس يقع فى دائرة فان كلّ جانب منه مثل نصف قطر الدائرة.

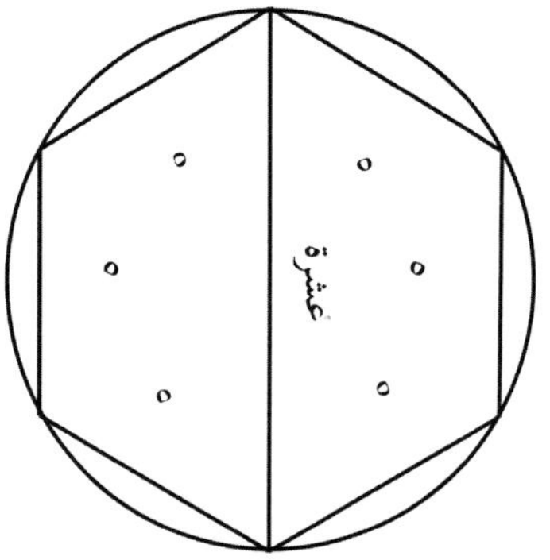

(59) قال دائرة قطرها عشرة كم كلّ جانب من جوانب المسدّس المتساوى الجوانب الذى يحيط بها من خارج وتكون جوانب المسدّس تماسّ الخطّ المحيط بالدائرة.

امّا حسابه فانّ كلّ جانب من المسدّس جذر ثلاثة وثلاثين وثُلث. وبابه ان تضرب العشرة فى مثلها فتكون مائة فتأخذ ثُلثها ابداً وهو ثلاثة وثلاثون وثُلث فتقول جذر ثلاثة وثلاثين وثُلث كلّ جانب من جوانب المسدّس.

(60) قال مسدّس متساوى الجوانب والزوايا كلّ جانب منه عشرة كم قطر الدائرة التى تقع فيه.

امّا حسابه فانّ قطر الدائرة التى تقع فيه جذر ثلاث مائة. وبابه ان تضرب العشرة فى مثلها فتكون مائة فتضربها فى ثلاثة ابداً فتكون ثلاث مائة فتقول قطر الدائرة جذر ثلاث مائة.

(61) قال كم قطر الدائرة التى تقع عليه.

امّا حسابه فانّ قطرها عشرون. وبابه ان تضعف العشرة ابداً فتكون عشرين فتقول قطر الدائرة التى تقع خارج المسدّس اذا كان الخطّ المحيط بها يماسّ زوايا المسدّس عشرون وذلك ما اردناه.

(62) قال دائرة قطرها عشرة كم كلّ جانب من المثمّن المتساوى الجوانب الذى يقع داخل الدائرة وتكون زواياه تماسّ الخطّ المحيط بالدائرة.

امّا حسابه فان مربّع كلّ جانب من جوانب المثمّن ⟨خمسون الّا جذر الف ومائتين و⟩خمسين. وبابه ان تضرب ⟨ عشرة التى هى⟩ قطر الدائرة فى مثلها فيكون مائة فتأخذ نصفها ابداً فتكون خمسين فتحفظها ثمّ تضرب المائة فى ثُمنها ابداً وهو فى هذه المسئلة اثنا عشر ونصف فتكون الفاً ومائتين وخمسين فتأخذ جذرها فما كان فانقصه من الخمسين التى حفظتَ فما بقى فخذ جذره فما خرج فهو كلّ جانب من جوانب المثمّن فمعلوم (106[v]) على ما بيّنّا ان مربّع كلّ جانب من جوانب المثمّن خمسون الّا جذر الف ومائتين وخمسين. وذلك ما اردنا ان نبيّن.

---

[1] المتساوى] المستوى. [7] ثلاثة وثلاثين وثُلث] ٣٣ ١/٣. [8] ثلاثة وثلاثون وثُلث] ٣٣ ١/٣ (*add. in marg.*). ثلاثة وثلاثين وثُلث] ٣٣ ١/٣. [10] ثلاث مائة] ٣٠٠. [11] ثلاثة] ٣. ثلاث مائة] ٣٠٠. ثلاث مائة] ٣٠٠. [13] عشرون] عشرين. [14] عشرون] عشرين. [17] خمسين] ٢/٥ (*sic*). [19] اثنا عشر ونصف] ١٢ ١/٢. الفاً ومائتين وخمسين] ١٢٥٠. [21] خمسون الّا جذر الف ومائتين وخمسين] ٥٠ الا جذر ١٢٥٠ (حمسس الا حدر ٥٢٥٠ ، *sic in figura*).

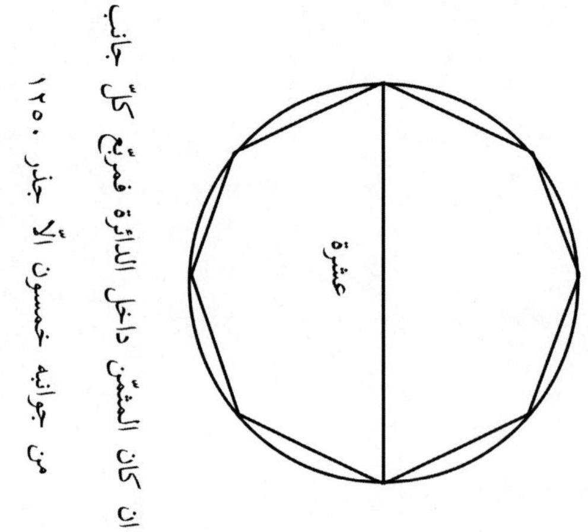

(63) قال دائرة قطرها عشرة كم كلّ جانب من جوانب المثمّن الذى يحيط بالدائرة من خارجها ويماسّ جوانبه الخطّ المحيط بالدائرة.

امّا حسابه فانّ كلّ جانب من جوانب المثمّن جذر مائتين الّا عشرة. وبابه ان تضرب قطر الدائرة فى مثله ثمّ تضعفه فيكون مائتين فتأخذ جذره فما كان نقصتَ منه قطر الدائرة وهو عشرة فما بقى فهو كلّ جانب من جوانب المثمّن وهو جذر مائتين الّا عشرة ويكون بالتقريب اربعة وسُبعاً.

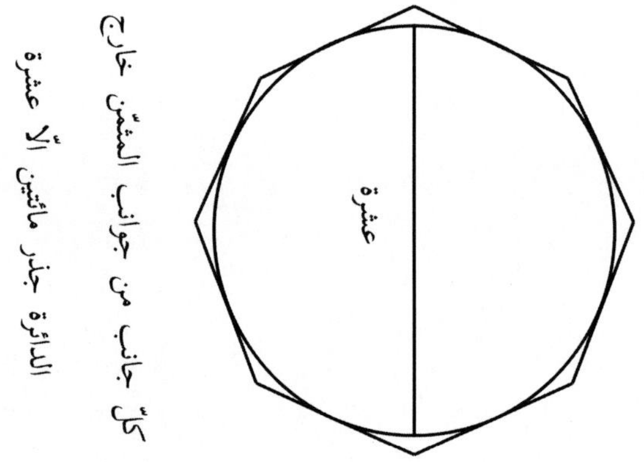

(64) قال كم يكون قطر الدائرة التى تقع فى المثمّن المتساوى الاضلاع والزوايا اذا كان كلّ جانب من المثمّن عشرة وكانت اضلاع المثمّن تماسّ الخطّ المحيط بالدائرة.

امّا حسابه فانّ قطر الدائرة عشرة وجذر مائتين. وبابه ان تضرب العشرة فى مثلها فتكون مائة فتضعفها فتكون مائتين فتأخذ جذرها فما كان زدته على العشرة فما اجتمع فهو قطر الدائرة وهو بالتقريب اربعة وعشرون وسُبع.

(65) قال كم قطر الدائرة التى تقع على المثمّن من خارجه اذا كان كلّ جانب منه عشرة وكان الخطّ ⟨المحيط بالدائرة⟩ يماسّ زوايا المثمّن.

امّا حسابه فانّ مربّع قطر الدائرة اربع مائة وجذر ثمانين الفاً. وبابه ان تضعف العشرة فتكون عشرين فتضربها فى مثلها فتكون اربع مائة ثمّ تضرب العشرة فى مثلها فتكون مائة فتضعفها فتكون مائتين فتضربها فى الاربع مائة

[1] عشرة] ١٠. [3] مائتين] ٢٠٠. [4] مائتين] ٢٠٠. [5] مائتين] ٢٠٠. اربعة وسُبعاً] $٤\frac{1}{7}$. [9] اربعة وعشرون وسُبع] $٢٤\frac{1}{7}$. [12] اربع مائة] ٤٠٠. [13] اربع مائة] ٤٠٠. مائتين] ٢٠٠. الاربع مائة] ٤٠٠.

فتكون ثمانين الفاً فتأخذ جذرها فما كان زدتَه على الاربع مائة فما اجتمع أخذتَ جذره فما خرج فهو قطر الدائرة. [فقد تبيّن انّ مربّع قطر الدائرة جذر ثمانين الفاً واربع مائة].

(66) قال (107[r]) معشّر متساوى الجوانب والزوايا كلّ جانب منه عشرة [من العدد] كم قطر الدائرة التى تقع خارج هذا المعشّر اذا كان زوايا المعشّر تماسّ الخطّ المحيط بالدائرة.

امّا حسابه فانّ قطر الدائرة عشرة وجذر خمس مائة. وبابه ان تضرب العشرة فى مثلها ثمّ فى خمسة فتكون خمس مائة فتأخذ جذرها فما كان زدتَه على العشرة فما بلغ فهو قطر الدائرة وهو عشرة وجذر خمس مائة.

(67) فان اردت ان تعلم قطر الدائرة الداخلة فى هذا المعشّر وتماسّ اضلاعه محيطها من خارج.

امّا حسابه فانّ مربّع قطر الدائرة التى تقع داخل هذا المعشّر خمس مائة وجذر مائتى الف. وبابه ان تضرب العشرة فى مثلها ثمّ فى خمسة فتكون خمس مائة فاحفظها ثمّ اضعف العشرة فتكون عشرين فاضربها فى مثلها فتكون اربع مائة فاضربها فى الخمس مائة التى حفظتَ فتكون مائتى الف فخذ جذرها فما كان فزده على الخمس مائة التى حفظتَ فما بلغ فخذ جذره فما خرج فهو قطر الدائرة. [فتبيّن على ما وصفنا انّ مربّع قطر الدائرة التى تقع فى معشّر متساوى الجوانب وكلّ جانب منه عشرة خمس مائة وجذر مائتى الف.] واللّه اعلم بغيبه.

كمل هذا المختصر
بحمد اللّه وحسن عونه
والصلاة على النبىّ الكريم محمّد وآله وسلّم
وذلك فى يوم الاثنين ثامن ذى القعدة سنة ٧٥٨

---

1 ثمانين الفاً] ٨٠٠٠٠. الاربع مائة] ٤٠٠. 2 مربّع] بمربع. ثمانين الفاً واربع مائة] ٨٠٤٠٠ (*sic*). 3 من العدد] من العشرة *corr. ex*. 4 المعشّر][1] المحيط به *add.* 5 خمس مائة] ٥٠٠. خمسة] ٥. 6 خمس مائة] ٥٠٠. العشرة] ١٠. خمس مائة] ٥٠٠. 8 التى تقع] الدي يقع. خمس مائة] ٥٠٠. مائتى الف] ٢٠٠٠٠٠. 9 خمس مائة] ٥٠٠. فاحفظها ثمّ اضعف العشرة فتكون عشرين] (٢٠ *scr.*) *iter.* 10 اربع مائة] ٤٠٠. الخمس مائة] ٥٠٠. مائتى الف] ٢٠٠٠٠٠. 10–11 الخمس مائة] ٥٠٠. 11 خرج] بلغ *corr. ex*.

Book on Mensuration

by the venerable scholar
Shujāʿ ibn Aslam, known as Abū Kāmil,
may God in His magnanimity have mercy upon him
Amen!

In the name of God the merciful, the compassionate. God bless our lord Muḥammad and his companions.

These are the words of Shujāʿ ibn Aslam, known as Abū Kāmil.

Mensuration of pieces of ground

It would hardly be appropriate, for what I shall write about knowing the mensuration of pieces of ground, that proof be furnished — unlike what I did in many of my treatises.[14] Had I offered proofs for what I shall state about mensuration, I would have just repeated many things already in the treatise of Euclid, and the (present) book would have grown longer and become difficult for the students. Indeed, I have written this book for beginners, as an introduction to that science, and for those anxious to study it. Accordingly, I shall expound my description of mensuration with due order, straight arrangement, agreeable composition and convincing account. Those who adopt the methods I shall expound and follow the way I shall establish will find correct answers, escape mistakes, be safe from errors, and knowledge of proofs will become easy for them.

Should they then want the proof of that and what the calculations might generate as problems involving geometry, mensuration, reckoning or other, they are to look at Euclid's treatise. Thus doing, they will master the subject to the utmost degree and have it established with a firmer foundation; after grasping and understanding correctly this subject so as to leave about it no doubt, no uncertainty, no tarnishing stain, they will know how to set the proof for (the aforesaid problems involving) geometry, mensuration, reckoning and others of all kinds. There is strength only in God the most high and powerful. God is sufficient for us, and the most reliable.

(1) Someone says to you: A quadrilateral piece of ground with equal sides and diagonals has each side ten cubits; what is its area?

According to calculation, its area is 100. The method is (as follows). You multiply the length, namely 10 cubits, which is one of the sides, by the width, which is the other side, namely 10. This gives 100. Such is its area.[15] Here is the figure.

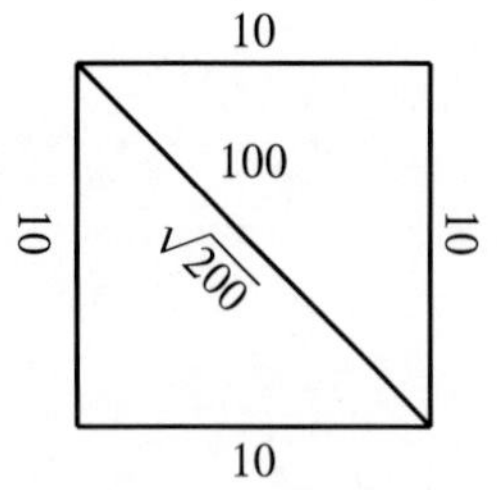

(2) If he says: What is its diagonal?

Multiply one of the sides, namely 10, by itself; this gives 100. Then multiply the adjacent side, namely 10, by itself; this gives 100. Add them both; this gives 200. Take the root of 200; this gives one of its diagonals, the other being the same.

[14] We do not know to which treatises Abū Kāmil is alluding.

[15] Square with side $a = 10$, required the area S; then $S = a^2 = 100$.

The root of a number is a number which when multiplied by itself gives a result equal to the number you want the root of. For example, the root of 4 is 2; for if you multiply the 2 by itself,[16] it gives 4. The root of 9 is 3, for 3 by 3 is 9. The root of 16 is 4, for 4 by 4 is 16.

The root of 200, of which we said that it is the diagonal of the quadrilateral, cannot be expressed and there is no root to the 200; for one cannot find a number[17] which, when multiplied by itself, gives 200. If you want the root of a number close to 200 or to which the 200 is close,[18] such is $14 + \frac{1}{7}$. For if you multiply it by itself, the 200 will be close to this; indeed, multiplying the $14 + \frac{1}{7}$ by itself gives $200 + \frac{1}{7}\frac{1}{7}$. Then if you are asked about the diagonal of a quadrilateral having 10 cubits as each side, say "the root of 200;" if one says to you "close to it," say "$14 + \frac{1}{7}$."[19]

(3) Someone says to you: Its diagonal is ten cubits; what is its area?

Multiply the diagonal, namely 10, by itself; this gives 100. Take its half; this gives 50, and such is its area.[20]

(4) Someone says to you: Its diagonal is ten; what is each of its sides?

According to calculation, each of its sides is the root of 50. The method is (as follows). You multiply the 10 by itself; this gives 100. You take its half; this gives 50. Then you will say: Each of its sides is the root of 50.[21]

(5) Someone says to you: A quadrilateral[22] has its two lengths equal, its two widths equal and its two diagonals equal, with length eight and width six; what is its area?

According to calculation, its area is 48 cubits.[23] The method is (as follows). You multiply one length, namely 8, by one width, namely 6; this gives 48. Such is its area.[24] Here is the figure.

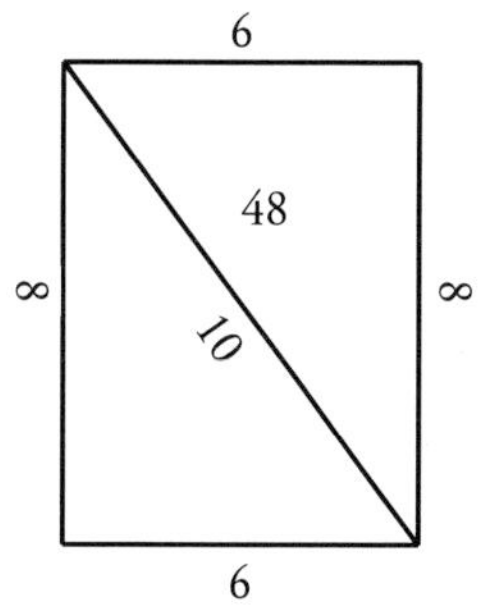

(6) If he says: What is its diagonal?

Multiply one length, namely 8, by itself; this gives 64. Multiply one width, namely 6, by itself; this gives 36. You add them both; this gives 100. You take its root; this gives 10, which is its diagonal.[25]

(7) If he says: A quadrilateral has equal lengths, widths and diagonals, with diagonal ten and the length longer than the width by two; what is its area?

---

[16] "The 2" : Verbal algebra commonly uses the article when a quantity is known, either because it is given at the outset or because it has been calculated. We find the same in Greek mathematical texts.

[17] That is, rational.

[18] That is, a rational quantity $m + \frac{p}{q}$ with $\left(m + \frac{p}{q}\right)^2 < 200$ and $200 < \left(m + \frac{p}{q}\right)^2$, respectively.

[19] Square with side $a = 10$, required the diagonal D; then $D = \sqrt{2a^2} = \sqrt{200} \cong 14 + \frac{1}{7}$. Same example in Didymos [Heiberg 1927, 12–15].

[20] Square with diagonal $D = 10$, required the area S; then $S = \frac{1}{2} D^2 = 50$.

[21] Square with diagonal $D = 10$, required the side a; then $a = \sqrt{\frac{1}{2} D^2} = \sqrt{50}$.

[22] In the feminine in the text: It refers to "piece of ground" (أرض, feminine in Arabic). Same below, Nos 7 and 9 (circle).

[23] Here and in what follows, for "square cubits."

[24] Rectangle with sides $a = 8$, $b = 6$, required the area S; then $S = a \cdot b = 48$.

[25] Rectangle with sides $a = 8$, $b = 6$, required the diagonal D; then $D = \sqrt{a^2 + b^2} = 10$.

The method is (as follows). You multiply the diagonal, namely 10, by itself; this gives 100. You subtract from it the product of the difference between length and width by itself, thus 4, which leaves 96. You take its half; this gives 48, and such is its area.[26]

(8) If he says: What are its length and width?

Halve the 2; this gives 1. Multiply it by itself; this gives 1. Add it to its area, which is 48; this gives 49. Take its root; this gives 7. Add to it the 1; this gives 8, which is one length. Subtract the 1 from the 7; this leaves 6, which is one width.[27] God knows best.

Calculation of circles

(9) If he says: A circle has a diameter of ten cubits; what is its area?

The method is (as follows). You multiply the diameter, namely 10, by itself; this gives 100. You subtract from the 100 its seventh and half a seventh, which is $21 + \frac{3}{7}$; this leaves $78 + \frac{4}{7}$. Such is its area.[28] Here is the figure.

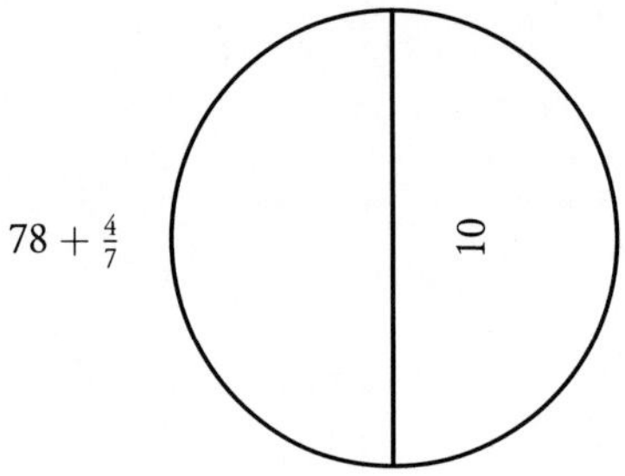

(10) If he says: A circle has a diameter of ten cubits; what is its circumference?

Multiply the diameter by three and a seventh, always;[29] this gives $31 + \frac{3}{7}$, and such is its circumference.[30]

(11) If he says: What is its area from its circumference and diameter?

Multiply half the diameter, thus 5, by half the circumference, thus (by) $15 + \frac{5}{7}$; this gives $78 + \frac{4}{7}$, and such is its area.[31]

You will proceed likewise for each circular piece of ground when you know the value of its diameter.[32]
You may also multiply the diameter by itself and subtract from the result its seventh and half a seventh; the remainder will be the area.[33]
You may also multiply the diameter by three and a seventh, always; the result will be its circumference. Then multiply half the diameter by half the circumference; the result will be its area.[34]

(12) If he says: Its circumference is ten cubits; what is its area?

---

[26] Rectangle with difference between the sides $a - b = 2$ and diagonal $D = 10$, required the area S; then $S = \frac{1}{2}\left[D^2 - (a-b)^2\right] = 48$. Indeed, $S = \frac{1}{2}\left[a^2 + b^2 - (a-b)^2\right] = ab$.

[27] Rectangle with difference between the sides $a - b = 2$ and area $S = 48$, required a, b; then $a,\ b = \sqrt{S + \left(\frac{a-b}{2}\right)^2} \pm \frac{a-b}{2}$ $\left(= \frac{a+b}{2} \pm \frac{a-b}{2}\right) = 7 \pm 1$. This amounts to finding two quantities of which we know product and difference (*Elements* II.6; already solved in this way in Mesopotamian mathematics).

[28] Circle with diameter $D = 10$, required the area S; then $S \cong D^2 - \left(\frac{1}{7} + \frac{1}{2}\frac{1}{7}\right) D^2 = 78 + \frac{4}{7}$. From $S = \frac{\pi}{4} D^2$ with $\pi \cong 3 + \frac{1}{7}$. Same example in Heron's *Metrica* [III, 66].

[29] "always" : This quantity is independent of the values given in the problem.

[30] Circle with diameter $D = 10$, required the circumference P; then $P \cong \left(3 + \frac{1}{7}\right) D = 31 + \frac{3}{7}$.

[31] Archimedes' formula to determine the area from circumference and diameter: $S = \frac{1}{2} D \cdot \frac{1}{2} P$.

[32] Refers to No 10.

[33] See No 9. Probably an interpolation.

[34] See Nos 10 and 11. Probably interpolated as well.

Multiply the 10 by itself; this gives 100. Then subtract from it its eighth, always, thus, in this problem, $12 + \frac{1}{2}$; this leaves $87 + \frac{1}{2}$. Divide it by eleven, always; the result is $7 + \frac{10}{11} + \frac{1}{2}\frac{1}{11}$, and such is its area.[35]

(13) If he says: Its circumference is ten cubits; what is its diameter?

Multiply the 10 by seven, always; this gives 70. Divide it by twenty-two, always; the result is $3 + \frac{4}{22}$, and such is its diameter.[36]

(14) If he says: Its area is a hundred and fifty-four; what is its diameter?

Multiply the 154 by fourteen, always; this gives 2156. Divide it by eleven, always; the result is 196. You take its root; this gives 14, and such is its diameter.[37]

(15) If he says: Its area is a hundred and fifty-four; what is its circumference?

Multiply the 154 by twelve and four sevenths, always; this gives 1936. You take its root; this gives 44, and such is its circumference.[38] God knows best.

Calculation of triangles

(16) If he says: A triangle has respective sides three, four, five; what is its area?

According to calculation, its area is 6. If in a triangle you multiply the two shorter sides each by itself and the sum of the results equals the product of the longest side by itself, then, for its area, you will multiply together the two shorter sides and take half the product; the result will be the area.[39] So multiply the 3 by the 4, which gives 12. Then take half of it; this gives 6, and that is the area of the triangle. Such are the method and the calculation.

You have multiplied together the two shorter sides, namely 3 and 4, and taken half the result because the sum of the products of 3 by itself and 4 by itself is 25, and this equals the product of the longer side, namely 5, by itself. [*Such are the method and the calculation.*][40] Here is the figure.

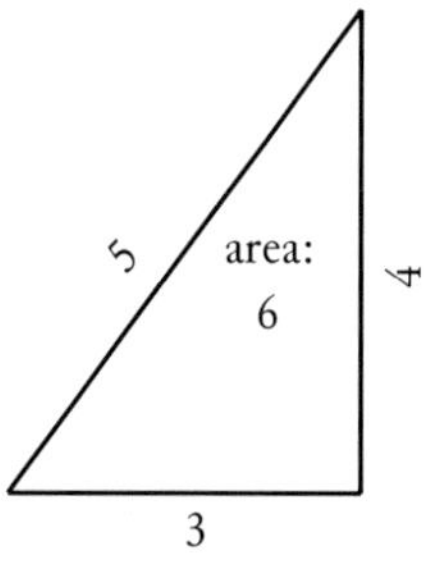

(17) Someone says: A triangle has respective sides fifteen, fourteen, thirteen; what is its area?

According to calculation, its area is 84. If in a triangle you multiply the two shorter sides each by itself and the sum of the results is larger than the product of the longer side by itself, and you want its area by means of the height,[41] then you are to know that such a triangle has three heights each of which falls on one of the sides of this triangle, inside it.[42]

Then, if you want to know one of these three heights (proceed as follows). Look for the height which falls on the side 14 — for an integer comes out, without any fraction. If you want this, multiply

---

[35] Circle with circumference $P = 10$, required the area S; then $S \cong \frac{1}{11}\left(P^2 - \frac{1}{8}P^2\right)$. Indeed, from Nos 11 and 10, $S = \frac{1}{2}D \cdot \frac{1}{2}P \cong \frac{1}{2}\frac{7}{22}P \cdot \frac{1}{2}P = \frac{7}{8}\frac{1}{11}P^2$.

[36] Given the circumference, find the diameter: $D \cong \frac{7}{22}P$. Converse of No 10. Here the fraction is not reduced.

[37] Given the area, find the diameter: $D \cong \sqrt{\frac{14}{11}S}$. Converse of No 9.

[38] Given the area, find the circumference: $P \cong \sqrt{\left(12 + \frac{4}{7}\right)S}$. Converse of No 12, with $12 + \frac{4}{7} = \frac{88}{7}$.

[39] Given the shorter legs a, b of a right-angled triangle, thus with $a^2 + b^2 = c^2$, find the area S; then $S = \frac{1}{2}ab$.

[40] The previous sentence might bc interpolated as well.

[41] For another possibility, see No 19.

[42] Acute-angled triangle: $a^2 + b^2 > c^2$.

the 14 by itself; this gives 196. [*Multiply any of the other sides you wish by itself*]. Then multiply the 13 by itself; this gives 169. Add it to the 196; this gives 365. Subtract from it the product of the other side, namely 15, by itself, which is 225; this leaves 140. You take its half; this gives 70. You divide it by the 14; the result is 5, which is the projection on the side 14 of the adjacent 13.[43] Then you multiply the 5 by itself; this gives 25. You subtract it from the 13 by itself, which is 169; this leaves 144. You take its root; this gives 12, which is the height on the side 14.[44] Multiply this height, thus 12, by half of the 14, which is 7; this gives 84, and such is the area.[45]

If you want to know the projection on the side 14 in another manner (proceed as follows). Multiply the 13 by itself and the 15 by itself, and take the difference between the results; you find 56. You take its half; this gives 28. Then you divide it by the 14; the result is 2. You subtract it from half of the 14; this leaves 5, which is the projection on the 14 of the adjacent 13.[46] If you add the 2 to half of the 14 this makes 9, which is the projection on the 14 of the adjacent 15.[47] If you multiply the 5 by itself and subtract the result from the product of the 13 by itself, this leaves 144; then you take its root; this gives 12, which is the height.[48] If you also multiply the 9 by itself and subtract the result from the product of the 15 by itself, this leaves 144. You take its root; this gives 12, which is the height.[49]

If you come across such a triangle[50] and proceed as I have explained to you, then the sum of the projections in both directions equals the length of the side on which is projected;[51] and if you multiply each of the two projections (on one side) by itself and subtract (each) result from this multiplication from each of the two products of the two (projected) sides of the triangle by itself, (namely) each of the adjacent from that to which it is adjacent, and the remainder from one of them is equal to the remainder from the other, then you did right;[52] but if it is different, then you made a mistake, so perform the computation again. When you know the height, multiply it by half the (corresponding) base; the result will be the area of the triangle.[53] There are for the determination of the height many treatments, (but) what I have explained to you is sufficient.[54] The height of this triangle on the 15 is $11 + \frac{1}{5}$; on the 13, $12 + \frac{12}{13}$.

(18) If he says: A triangle has respective sides four, thirteen, fifteen; what is its area?

According to calculation, its area is 24 (square) cubits. If in a triangle you multiply the two shorter sides each by itself and the sum of the results is less than the product of the longest side by itself, and you want its area by means of the height, then you are to know that such a triangle has three heights with the one falling on the longer side being inside the triangle and the two heights on the shorter sides outside the triangle: one falls on the extension of one of the shorter sides outside the triangle and the other outside the other side, when it is extended outside the triangle.[55]

If you want to know the height falling inside the triangle on the longest side, proceed as I have explained to you in the problem preceding this one, namely as follows. You multiply the 4 by itself and

43 Acute-angled triangle with sides $a = 14$, $b = 13$, $c = 15$; the projection of b on a is: $b_a = \frac{(a^2+b^2)-c^2}{2a} = 5$. Same example and formulae in Greek and Roman sources: Heron, *Metrica* [III, 12–15], *Geometrica* [IV, 234–241]; Nipsus [Blume, Lachmann and Rudorff 1848–1852, I, 299].

44 $h_a = \sqrt{b^2 - b_a^2} = 12$.

45 $S = \frac{1}{2}\, ah_a = 84$.

46 $b_a = \frac{1}{2}\, a - \frac{c^2-b^2}{2a}$. Other form of the previous relation (note 43).

47 $c_a = a - b_a = \frac{1}{2}\, a + \frac{c^2-b^2}{2a}$.

48 Already calculated above (note 44).

49 $h_a = \sqrt{c^2 - c_a^2}$.

50 Acute-angled, thus with each height inside.

51 We must always have: $c_a + b_a = a$, $c_b + a_b = b$, $a_c + b_c = c$.

52 We must always have: $h_a^2 = b^2 - b_a^2 = c^2 - c_a^2$, $h_b^2 = c^2 - c_b^2 = a^2 - a_b^2$, $h_c^2 = b^2 - b_c^2 = a^2 - a_c^2$.

53 $S = \frac{1}{2}\, ah_a = \frac{1}{2}\, bh_b = \frac{1}{2}\, ch_c$.

54 Abū Kāmil considered the projection (and thus the height) on $a = 14$ only (for the above reason: h_a is an integer, unlike h_c and h_b, given thereafter).

55 Obtuse-angled triangle: $a^2 + b^2 < c^2$, with sides $a = 4$, $b = 13$, $c = 15$, thus h_c within the triangle. The formulae below are known from Greek (Heron) and Roman (Nipsus) sources; see Cantor [1875, 41], and Blume, Lachmann and Rudorff [1848–1852, I, 297].

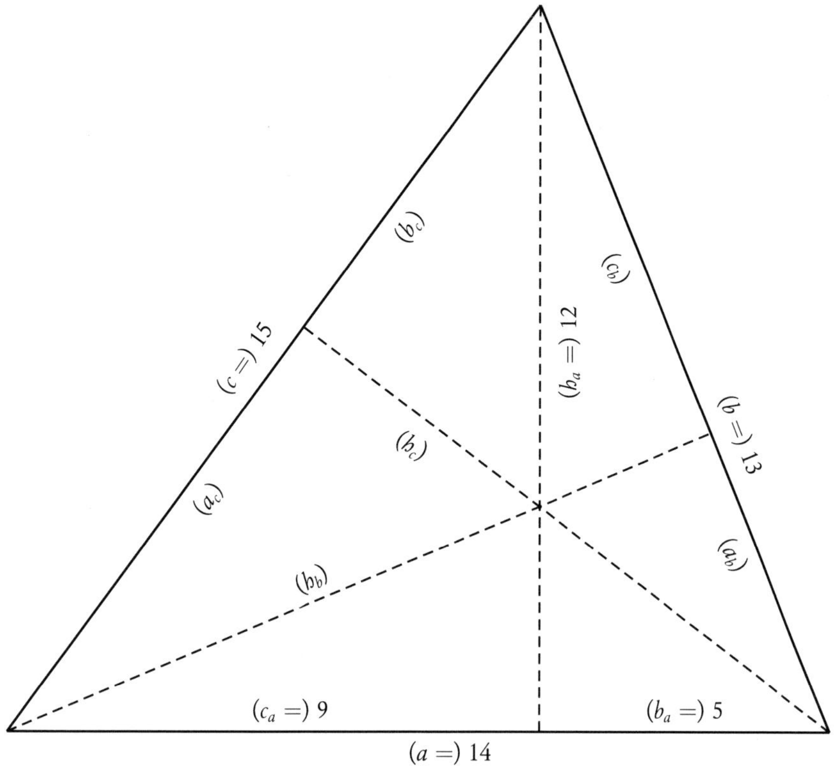

the 15 by itself, and add the results; this gives 241. You subtract from it the 13 by itself; this leaves 72. You take its half; this gives 36. You divide it by the 15; the result is $2 + \frac{2}{5}$, which is the projection on the 15 of the adjacent 4.[56]

If you want to know how much is the projection of the adjacent 13, subtract the $2 + \frac{2}{5}$ from the 15; this leaves $12 + \frac{3}{5}$, which is the projection.[57] If you want that by the way I have taught you,[58] multiply the 13 by itself, the 15 by itself, and add the results; this gives 394. Subtract from it the 4 by itself; this leaves 378. Take its half; this gives 189. Divide it by the 15; the result is $12 + \frac{3}{5}$, and this is the projection on the 15 of the adjacent 13.[59]

If you want to know the length of the height, multiply the $2 + \frac{2}{5}$ by itself and subtract the result from the product of the 4 by itself; this leaves $10 + \frac{1}{5} + \frac{1}{5}\frac{1}{5}$. Take its root; this is $3 + \frac{1}{5}$, and such is the length of the height.[60] Multiply it by half of the 15, namely $7 + \frac{1}{2}$; this gives 24, which is its area.[61] If you multiply the projection of the side 13 (on the 15), namely $12 + \frac{3}{5}$, by itself and subtract the result from the product of the 13 by itself, this leaves $10 + \frac{1}{5} + \frac{1}{5}\frac{1}{5}$, precisely equal to the remainder from the first. Then you take its root; this gives $3 + \frac{1}{5}$, which is the length of the height.[62] If it is different, you have made a mistake, so perform the computation again.

If you want to know by how much each of the two shorter lines is produced to the foot of the corresponding height (proceed as follows).[63] Begin with one of them, say the 4. Place the foot of the

---

56 $a_c = \frac{a^2+c^2-b^2}{2c} = 2 + \frac{2}{5}$. Same, *mutatis mutandis*, as in No 17.

57 $b_c = c - a_c = 12 + \frac{3}{5}$.

58 The same way as for calculating a_c.

59 $b_c = \frac{c^2+b^2-a^2}{2c}$.

60 $h_c = \sqrt{a^2 - a_c^2}$.

61 $S = \frac{1}{2}\, ch_c$.

62 $h_c = \sqrt{b^2 - b_c^2}$.

63 Calculation of b_a (extension of a) and c_a $(= a + b_a)$.

height on its extension, at the place where it falls. The extension is 5 cubits, and the length of this line from its meeting point with the extremity of the 15 to the place of the foot of the height on it is 9 cubits.[64] The method is (as follows). You multiply the 4 by itself, and (the) 13 by itself, and add the results; this gives 185. You subtract it from the 15 by itself, which is 225; this leaves 40. You take half of it; this gives 20. You divide it by the 4; this gives 5, which is the extension of the line of 4 to the foot of the height.[65]

If you want to know the height, multiply the 5 by itself, and subtract the result from the product of the 13 by itself; this leaves 144. Then you take its root; this gives 12, which is the length of the height at the extremity of the (line of) 5 cubits belonging to the line of 4, the whole of which is 9.[66] Then if you want to know the area of this triangle in this way, multiply the height, namely 12, by half of the 4, which is 2; this gives 24, which is the area of the triangle.[67]

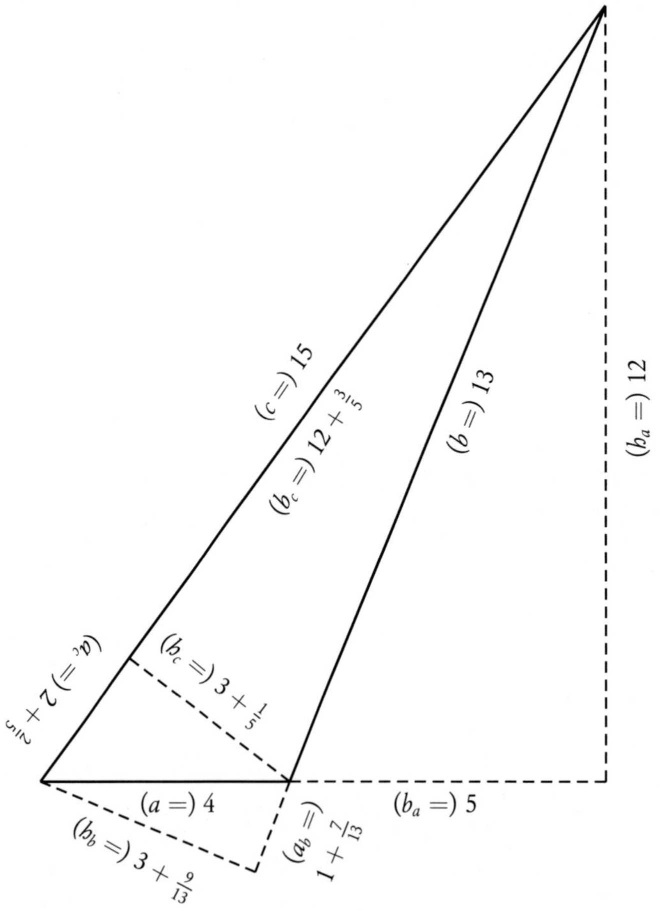

If you want to know by how much the line of 13 is produced to the foot of the height, then the extension is $1 + \frac{7}{13}$;[68] and the length of this same line, from its meeting point with the extremity of the line of 15 to the foot of the height on it, is $14 + \frac{7}{13}$.[69] If you want to know how long is the height drawn

[64] $b_a = 5$ (see below), and thus $a + b_a = 9 = c_a$.

[65] $b_a = \frac{c^2-(a^2+b^2)}{2a}$.

[66] $h_a = \sqrt{b^2 - b_a^2}$.

[67] $S = \frac{1}{2}\, ah_a$.

[68] $a_b = \frac{c^2-(a^2+b^2)}{2b} = \frac{20}{13}$, here simply given.

[69] Adding $a_b = 1 + \frac{7}{13}$ to $b = 13$.

from the meeting point of the line of 4 with the line of 15 to the foot of the height on the extension of the line of 13, which (extension) is $1 + \frac{7}{13}$ (proceed as follows). Multiply $1 + \frac{7}{13}$ by itself, subtract the result from the product of the 4 by itself, and take the root of the remainder; this gives $3 + \frac{9}{13}$, which is the height.[70] If you want to know the area of the triangle according to this way also, multiply the height, namely $3 + \frac{9}{13}$, by half of the 13, which is $6 + \frac{1}{2}$; this gives 24, which is the area of the triangle.[71]

(19) There is for the area of a triangle a single method, applicable to all triangles, which is (as follows). You add the three sides of the triangle; then you take half the sum and keep it in mind. Then you consider the excess of this half over each of the sides of the triangle, and you multiply together these excesses; next you multiply the result by half the (sum of the) sides kept in mind. You take the root of the product, and the result will be the area of the triangle.[72]

Example of that. If he says: A triangle with respective sides fifteen, fourteen, thirteen.[73] If you want to know its area, add its three sides, namely 15, 14, 13; this gives 42. Take half of it; this gives 21. Keep it in mind. Then consider the excess of this over each of the sides of the triangle; you find as the excess of this over the 15, 6, and over the 14, 7, and over the 13, 8. Then you multiply 6 by 7 (and the result) by 8; this gives 336. You multiply it by half the (sum of the) sides of the triangle which you have kept in mind, namely 21; this gives 7056. You take its root; this gives 84, which is the area of the triangle. This method works for any triangle. Retain it, for it is a correct rule.

(20) If he says: An equilateral triangle has each side ten cubits; what is its area?

According to calculation, its area is the root of 1875. If you want its area by means of the height (proceed as follows). It is known that its height falls in the middle of whichever of the sides you want the height to fall on.[74] So you multiply half of one of the sides, thus 5, by itself, and you subtract this from the product of any of the sides you want by itself; this leaves 75. So you will say: The height is the root of 75.[75] Then you multiply it by half the side the height falls on, thus by 5. Then (in order to calculate the result[76]) you multiply 5 by itself; this gives 25. You multiply it by the 75; this gives 1875. Its root is the area (*taksīr*), namely, approximately, $43 + \frac{1}{3}$.[77]

For any equilateral triangle of which you want (to know) the area approximately, multiply one of its sides by itself, take of the result a third plus a tenth; this will give the area of the triangle.[78]

If you want to know its area using the excesses, proceed as I have explained to you;[79] your result for the area will be the root of 1875.

For any equilateral triangle, the determination of its height is (as follows). You multiply one of the sides by itself, then you subtract from the result a fourth of it and you take the root of the remainder; the result will be the height.[80]

---

[70] $h_b = \sqrt{a^2 - a_b^2}$.

[71] $S = \frac{1}{2}\, b h_b$.

[72] If a, b, c, are the sides of the triangle and σ half the perimeter, then $S = \sqrt{\sigma(\sigma - a)(\sigma - b)(\sigma - c)}$. This formula, found in Heron's *Metrica* [III, 18–25] — also in the *Geometrica* [IV, 248–251] — and thus commonly attributed to him, originates most probably with Archimedes. Half the perimeter is greater than any side, for each side is less than the sum of the other two, as stated below.

[73] See No 17.

[74] The reader may infer it from an earlier statement (note 51; the sides being equal, so are the projections on the third side). See also No 22.

[75] For an equilateral triangle with side a, the height is $h = \sqrt{a^2 - \left(\frac{a}{2}\right)^2} = \sqrt{\frac{3}{4}\, a^2}$.

[76] Introducing 5 under the radical.

[77] $S = \frac{1}{2}\, ah = \sqrt{\frac{3}{16}\, a^4} = \sqrt{1875} \cong 43 + \frac{26}{86} = 43 + \frac{13}{43} \cong 43 + \frac{1}{3}$. Same example in Heron's *Metrica* [III, 48].

[78] Since $h = \sqrt{\frac{3}{4}\, a^2} = \frac{1}{2}\sqrt{3}\, a = \frac{1}{20}\sqrt{300}\, a$ while $\sqrt{300} \cong 18 - \frac{24}{36} = 17 + \frac{1}{3}$, then $S = \frac{1}{2}\, ah \cong \frac{1}{2}\, a \cdot \frac{1}{20}\left(17 + \frac{1}{3}\right) a = \frac{13}{30}\, a^2 = \left(\frac{1}{3} + \frac{1}{10}\right) a^2$. Same formula in (Ps.-)Heron's *Geometrica* [IV, 222; Cantor 1875, 40]. Applying this formula gives the previous result.

[79] See No 19. The actual computation is found below (No 20′).

[80] Restating generally the rule first employed.

(21) If he says: An equilateral triangle has a height of ten cubits; what is each side of it?

Multiply the height by itself; this gives 100. Add to it its third, always, thus $33 + \frac{1}{3}$; this gives $133 + \frac{1}{3}$. Each side of the triangle will be the root of $133 + \frac{1}{3}$.[81]

(22) If he says: Each of two sides of a triangle is ten and the other, namely the base, is twelve; what is its area?

Knowing its area by means of its height proceeds (as follows). You draw the height on 12 — this is easier since it falls in the middle. (Indeed,) in any triangle with two equal sides of which you want to draw the height on the other side,[82] it will fall in its middle, be this (other side) equal to one of the two sides[83] or different — but it must be less than the sum of the two other sides: if it is more than the sum, or equal, then a triangle cannot be made from these three lines since in any triangle each pair of sides must be longer than the other side.[84] If you want to find its height on the side 12 — and we have informed you before that it can only fall in its middle — multiply half of the 12, thus 6, by itself, and subtract the result from the product of one of the two sides by itself; this leaves 64. Then you take its root; this gives 8, which is the height. Multiply it by half the base, namely 6; this gives 48, which is the area of the triangle.[85]

If you want to know its area by the excesses, proceed as I explained to you.[86] That is, you add together the three sides of the triangle; this gives 32. You take its half; this is 16. Then you consider its excess over each of the sides of the triangle. You find, as the excesses over the two sides which are 10, 6, and, over the side which is 12, 4. You multiply 6 by 6 (and the result) by 4; this gives 144. You multiply it by 16, which is half the (sum of the) sides of the triangle; this gives 2304. You take its root; this gives 48, which is the area of the triangle.

(20′) Likewise if you want the area of the (above) equilateral triangle, with each side ten, by means of the excesses.

Add together its three sides; this gives 30. Take its half; this gives 15. Consider the excess it has over each of the sides of the triangle; you find 5 for its excess over each side. Then you multiply 5 by 5 (and the result) by 5; this gives 125. You multiply it by half the (sum of the) three sides of the triangle, namely 15; this gives 1875. You take its root; the result will be the area of the triangle. So you will say: The area of the triangle is the root of 1875, which is, approximately, $43 + \frac{1}{3}$.[87] Such are the method and the calculation.

(23) If he says: A quadrilateral with different sides has one side nineteen with, facing it, five and another side fifteen with, facing it, thirteen; what is its area?[88]

According to calculation, its area is 144 (square) cubits. If you want its area, determine its height first.[89]

The determination of its height when the side which we said to be 5 is parallel to the side which we said to be 19 is (as follows). One knows that there is parallelism (in this quadrilateral) when the heights drawn from the extremities of the line of 5 and falling on the line of 19 are equal; if they are not equal, the two sides are not parallel. If you want to know how much is the height drawn from one extremity

---

[81] Given the height h of an equilateral triangle, find its side a; then $a = \sqrt{\frac{4}{3} h^2}$. Converse of No 20.

[82] That is, the unequal side.

[83] As in No 20 (note 74).

[84] Equivalent to saying that a straight line is the shortest distance between two points.

[85] Given the sides $a = b$ and c of an isosceles triangle, find its area S; then $S = \frac{1}{2} c h_c$, with $h_c = \sqrt{a^2 - \left(\frac{c}{2}\right)^2}$ (it falls in the middle of c, as stated). This is also found in the Heronian corpus; see Cantor [1875, 40].

[86] See No 19.

[87] As seen above (No 20).

[88] The sides 19 and 5 are parallel, as stated below.

[89] Area of a trapezium with parallel sides $a = 19$, $c = 5$ and lateral sides $b = 15$, $d = 13$. Find first the projections d_a and b_a, then the height h between the parallel sides, from which the area can be calculated.

of the line of 5 and falling on the line of 19 and where it will fall in it (proceed as follows). Subtract the 5 from the 19; this leaves 14.[90] Remove the surface having as two of its sides the heights issued from the extremities of the line of 5 and falling on the line of 19, and as the two other sides the line of 5 and the equal one opposite within the line of 19. Having removed it, then imagine two triangular pieces (left) from the surface: for the first, one of the sides is 13, the other is the height, the other is the (corresponding) section from the 19; for the other triangle, one side is the other height, of which we said that it is equal to the other height, the other side is 15, the other side is the (corresponding) section from the line of 19. This (last), with the other section from the (line of) 19 belonging to the other triangle, makes 14. So when the two triangles have been associated one with the other and we have joined together the two sides which are the heights, there arises a triangle with respective sides 15, 13, 14.

Determining its height follows what I have explained to you.[91] That is, you multiply, if you want to find the height on the 14, 15 by itself, and 13 by itself, then you take the difference between the results; this gives 56. You take its half; this gives 28. You divide it by the 14; the result is 2. You subtract it from half of the 14, and you add it to half of the 14. This gives, as the location of the foot of the height on 14 in the direction of 13, 5, and in the direction of 15, 9.[92] For the determination of the height, you may multiply the 5 by itself and subtract the result from the product of the 13 by itself, which leaves 144; or you may multiply the 9 by itself and subtract the result from the product of the 15 by itself, which leaves 144 as well. Then you take its root; you find 12, which is the height.[93]

Then you return to the surface of which we said that one of its sides is 19 with, facing it, 5, while the other side is 15 with, facing it, 13. We have already shown that the height drawn from one of the extremities of the line of 5 is 12. Add then the 5 and the 19; this gives 24. Take its half; this gives 12. Multiply it by the height, namely 12; this gives 144, and that is the area.[94] Such are the method and the calculation.

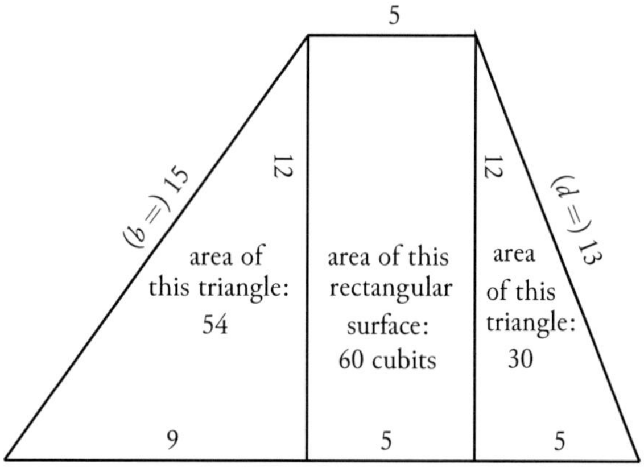

area of this whole surface: 144

(24) The official land surveyors, when they come across such a piece of ground and want to measure it, add the 5 and the 19, then take half of this, thus finding 12. Then they add the 13 and the 15, then take half of this, thus finding 14. They multiply the 14 by the 12; this gives 168. Such is, in their opinion, the area.[95] This is not correct, whereas what we have indicated previously is correct.

[90] As the base of the two lateral triangles.

[91] See No 17. Here the sides are $a - c = 14$, $d = 13$, $b = 15$.

[92] $d_a = d_{a-c} = \frac{1}{2}(a-c) - \frac{b^2-d^2}{2(a-c)} = 5$, $b_a = b_{a-c} = \frac{1}{2}(a-c) + \frac{b^2-d^2}{2(a-c)} = 9$.

[93] $h = \sqrt{d^2 - d^2_{a-c}} = \sqrt{b^2 - b^2_{a-c}} = 12$.

[94] $S = \frac{a+c}{2}\, h = 144$.

[95] Thus, if a and c are the parallel sides and b, d the other two, $S = \frac{a+c}{2} \cdot \frac{b+d}{2}$. This is found as an approximation formula in Egypt [Vogel 1958–1989, I, 66; Cantor 1875, 43], Mesopotamia [Vogel 1958–1989, II, 70; Høyrup 2002, 230], Greece [Heron's *Geometrica*, IV, 208], and Rome [Blume, Lachmann and Rudorff 1848–1852, 355; Cantor 1875, 137, 144]. In the 11th century, al-Baghdādī refers for this wrong formula to *qaul*, or *ṭarīq*, *al-fars*, thus to the saying, or the method, of the Persians [Saidan 1985, 340 & 344]. The result is too large since the height is replaced by the arithmetical mean of

They proceed likewise for any quadrilateral piece of ground having different sides.[96] They add the two sides facing each other and take half the result; then they add the two other sides and take half the result. Then they multiply the two halves, and the result is the area. We have already told you that this is not correct.[97]

They proceed likewise[98] for an equilateral triangle. They take one of its sides and multiply it by half another side; the result is the area.[99] This is incorrect; correct is what I have explained to you.

For the scalene triangle they multiply the longest side in it by half the longer of the two (other) sides. The result is, according to them, its area. When they want to be more precise, in their view, they add the sides, then take a third of the result and multiply it by half the longest side. When they want to ameliorate the corrected result, they multiply a third of the (sum of the) sides ⟨by a sixth of the (sum of the) sides⟩; the result is the area according to them.[100] There is nothing correct in that; what is correct is what I have told you. So adhering to that you will not go astray. There is no power and no strength save in God.

When the official land surveyors arrive at a piece of ground resembling a circle, they take its perimeter, then take a third of it and multiply it by itself. The result is its area according to them. If they want to be more precise, they multiply a third of the perimeter by a fourth of the perimeter; the result is, according to them, its area.[101]

They proceed likewise for a many-sided piece of ground, that is, one having five, six or more sides. They measure it in the way I have presented to you for measuring a circle. They may also divide it for taking out quadrilaterals and triangles; then they measure the quadrilaterals in their aforesaid way for measuring quadrilaterals, and the triangles using their previously explained method for measuring triangles.

(25) The correct way for (calculating) the area of a many-sided piece of ground is to divide it entirely into triangles, measure (*dhara'a*) each one as I have explained to you for the area of triangles, then add all that; the result will be the area of the piece of ground. Should, during the partitioning, a rectangle come out, measure it without dividing it into two triangles [*for its area, you multiply its length by its width; the result is the area of the quadrilateral*];[102] adding it to the area from the remainder of the partitioning will give the area of the piece of ground.

The land owners proceed for their faulty measurements in accordance with the surveyors' way, in their opinion equitable to them. They do not change to anything else and do not know anything else either. Most of them, when I measured for them with the correct way and came to a result less than the area as measured by the official surveyors, were not satisfied with it, did not change the former result and were more favorably disposed to the official land surveyors in spite of their operating wrongly.

Mensuration of bodies and their surfaces

(26) If he says: A quadrilateral pillar, its base four by four, its top four by four, and its height ten; what is its volume?

---

the lateral sides.

96 Thus an irregular quadrilateral (no two sides parallel). For other polygonal figures, see below.

97 But only for the particular case with two parallel sides. For the more general case, see below, No 25.

98 That is, erroneously.

99 The side being a, we have for the area $S = \frac{1}{2}a^2$. See Blume, Lachmann and Rudorff [1848–1852, I, 354] and Cantor [1875, 137]. This relation can be inferred from the (wrong) trapezium formula by putting one side equal to zero. Here too, the result will be too large since one side is taken to be the height.

100 If $c > b > a$, then $S = \frac{1}{2}bc$, or $S = \frac{1}{3}P \cdot \frac{1}{2}c$, or $S = \frac{1}{3}P \cdot \frac{1}{6}P$. In the first formula, the base is taken as the longest side and the height as the second longest side (commonly employed for the case $b = a$, see Cantor [1875, 33, 144]; in the second formula, this height becomes the arithmetical mean of the three sides. Whichever the case, the result will be too large. (The third formula is conjectural since the text is lacunary.)

101 Thus, for a circle, $S = \left(\frac{1}{3}P\right)^2$, or $S = \frac{1}{3}P \cdot \frac{1}{4}P$. The second is found in Mesopotamia [Neugebauer and Sachs 1945, 44; Vogel 1958–1959, II, 74] and in Didymos [Heiberg 1927, 14–15]. The first formula corresponds to $4\pi = 9$, the second to $4\pi = 12$ — not quite so bad but still making the area too large.

102 Clearly a former gloss.

When the body has equal extremities, multiply the length of its base by the width of its base, and multiply the result by its altitude; this will give the volume of the pillar. So multiply 4 by 4, which gives 16, then multiply it by the altitude, which is 10; this gives 160 (cubic) cubits, which is the volume of the pillar.[103]

(26′) Likewise if there is a quadrilateral well, its base four by four, its upper part four by four, and its altitude, which is its depth, ten.

Its volume will be like the volume of the pillar, 160. The method for measuring it is just like the method for measuring the pillar. But for the volume of the well is meant its hollow part, whereas by volume of the pillar is meant its solid content.

(27) If you want to measure the surface of the pillar — where by surface is meant its whiteness if it is white, its dye if it is dyed, or its drape if it is draped.[104]

If you want that, take its perimeter from its four sides; you will find 16 cubits. Multiply it by its altitude, which is 10. This gives 160 (square) cubits, and this is the area of its surface, excepting the area of its base and that of its top.[105]

If you want the area of the surface of its base, multiply its length by its width; this gives 16 cubits, which is the area of the surface of its base. Likewise, the area of the surface of its top is 16 cubits.

(28) If the pillar is circular with identical extremities and the diameter of its base is four and (thus) the diameter of the circle on its top is four, subtract from the 160 cubits a seventh and half a seventh of it, which is $34 + \frac{2}{7}$; this leaves $125 + \frac{5}{7}$, which is the volume of the pillar.[106]

(29) Likewise, the area of its surface is obtained (as follows). You take its circumference; this gives $12 + \frac{4}{7}$. For the circumference of a circle with a diameter of four cubits is obtained by multiplying the diameter by three and a seventh, always, which gives $12 + \frac{4}{7}$.[107] Then you multiply it by the altitude, which is 10; this gives $125 + \frac{5}{7}$, which is the area of the surface of the pillar without the area of the surface of its base and that of its top.[108] If you want the area of the surface of its base, evaluate it in the manner I have explained to you for evaluating a circle; and likewise for the surface of its top.[109]

(30) If he says: A pyramidal pillar[110] has its base five by five, its top four by four, and its height (*ʿamūd*) ten, where under height is meant the straight line extending from the middle of the surface of its top to the middle of the surface of its base;[111] what is its volume?

According to calculation, its volume is $203 + \frac{1}{3}$. The method is (as follows). You multiply 5 by 5, which gives 25, and you multiply 4 by 4, which gives 16. Then you multiply 5 by 4, which gives 20. ⟨You add 25, 16, 20; this makes 61. Then you take a third of it; this gives $20 + \frac{1}{3}$.⟩ You multiply it by the height, which is 10; this gives $203 + \frac{1}{3}$, which is the volume of the pyramid.[112]

If you wish, add the 5 and the 4; this gives 9. Then you take its half; this is $4 + \frac{1}{2}$. You multiply it by itself; this gives $20 + \frac{1}{4}$. Keep it in mind. Then you take the difference between the 5 and the 4; you

---

[103] Given, for a (right) prism, the sides of the base $a = 4 = b$ and the height $h = 10$, the volume is $V = abh = 160$.

[104] This, perhaps an interpolation, means that we are to consider the area of the lateral faces only.

[105] If the perimeter of the base is P and the height h, the area of the lateral faces will be $S = Ph$. The choice of 4 for the side of the base is particularly unfortunate for the reader since it leads to the same result as before (this time in square cubits).

[106] Given for a (right) cylinder the diameter of the base D and the height h, the volume will be $V = D^2h\left(1 - \frac{1}{7} - \frac{1}{2}\frac{1}{7}\right) = 125 + \frac{5}{7}$, according to the approximation for π already encountered (No 9).

[107] This would seem to be an early reader's addition, for the rule is known (No 10); but see No 33.

[108] Given, for a cylinder, the diameter of the base D and the height h, the area of the lateral face is $S = Ph = Dh\left(3 + \frac{1}{7}\right)$.

[109] See No 9.

[110] Frustum of a (right) pyramid.

[111] Thus not the slant height (on the face), to be encountered later on (Nos 32–33, 37–38).

[112] Required the volume of a frustum of a pyramid, with side of the square base a^2, side of the square top b^2 and height h; then $V = \frac{h}{3}\left(a^2 + ab + b^2\right) = 203 + \frac{1}{3}$. Known in ancient Egypt [Vogel 1958–1959, I, 71]. The text has 200, obviously a reader's (or copyist's) correction in view of the previous lacuna.

find 1. Take its half; this is $\frac{1}{2}$. Multiply it by itself; this gives $\frac{1}{4}$. Take a third of it; this gives $\frac{1}{2}\frac{1}{6}$. Add it to the $20 + \frac{1}{4}$ kept in mind. This gives $20 + \frac{1}{3}$. Multiply it by the depth (*sic*), which is 10; this gives $203 + \frac{1}{3}$, which is the volume of the pillar.[113]

(31) If the pillar is circular, subtract from $203 + \frac{1}{3}$ its seventh and half a seventh, thus $43 + \frac{4}{7}$; this leaves $159 + \frac{5}{7} + \frac{1}{3}\frac{1}{7}$, which is the volume of the pillar.[114]

(32) If you want the surface area of this pillar — according to what we have said about the squareness of its base and the squareness of its top — (proceed as follows).

Take the perimeter of its base, for which you find 20, and the perimeter of its top, for which you find 16. Add them; this gives 36. Take its half; this is 18. Multiply it by the altitude of the pillar on its surface, which is greater than 10 by something small; this gives 180 cubits and something, and such is its surface area as you go around, without the surface area of the base and that of the top [*16*].[115]

(33) If the pillar is circular and you want its surface area (proceed as follows).

Take the circumference of its base, for which you find $15 + \frac{5}{7}$; for the diameter of the base circle is 5 cubits and so you multiply it by three and a seventh. Take the circumference of its top, for which you find $12 + \frac{4}{7}$. Add them; this gives $28 + \frac{2}{7}$. Take its half; this is $14 + \frac{1}{7}$. Multiply it by the altitude of the pillar on its surface, which is greater than 10 by something small. This gives $141 + \frac{3}{7}$, which is the surface area of the pillar without the surface area of its base and that of its top.[116] The surface area of its base being like the area of a circle with diameter 5 cubits, its area is $19 + \frac{1}{2} + \frac{1}{7}$, and the surface area of the top is $12 + \frac{4}{7}$.[117]

(34) The measurement of wood is performed in the way I have explained to you.

If (the piece) is quadrilateral[118] and you want its volume, proceed according to what I have explained to you for the volume of a pillar. If you want the volume of a circular piece of wood,[119] proceed as I have shown to you for the volume of a circular pillar. If it is quadrilateral and you want its surface area, proceed according to what I have described to you for the surface area of a quadrilateral pillar. If it is circular, proceed according to the previous explanation about the surface area of a circular pillar.

We do not know, neither for wood nor for any solid, more than measuring volume and surface, which is what I have explained to you. People of Egypt, for measuring their wood, proceed according to something [*for which there is no one preceding them*] which is measurement neither of volume nor of surface.[120] I have explained to you in the preceding part of my book how they measure, how it is ridiculous and laughable, and I have explained what it involves for them in terms of deficiency, weakness, lack of common sense. We ask God for well-being and welfare for the pious, for this world and the other since He has power over everything.

---

[113] $V = h\left[\frac{1}{3}\left(\frac{a-b}{2}\right)^2 + \left(\frac{a+b}{2}\right)^2\right]$. Other form of the previous formula. Known from Heron's *Stereometrica* [Cantor 1875, 61]; see also Vogel [1958–1959, II, 81] and Thureau-Dangin [1938, 37].

[114] Volume of the frustum of a (right) cone: same as before, but the sum of the previous areas a^2, ab and b^2 must be reduced by the factor $\frac{1}{7} - \frac{1}{2}\frac{1}{7}$. It should be stated that what was the side of the square is now taken as the diameter of the circle.

[115] Frustum of a pyramid with $a = 5$ as the side of the base, $b = 4$ as the side of the top, thus $P = 4a$ and $P' = 4b$, and h_s slant height; the lateral surface area of the frustum is then $S = \frac{P+P'}{2} \cdot h_s$. Here $h_s = \sqrt{100 + \frac{1}{4}} \cong h = 10$. The area of the top given at the end must be an interpolation; again (note 105), same value as the perimeter.

[116] Lateral surface area of the frustum of a cone. Same as before (No 32), but taking $P = D(3 + \frac{1}{7})$, $P' = D'(3 + \frac{1}{7})$, and again $h_s \cong h$.

[117] As seen in No 9, $S = D^2\left(1 - \frac{1}{7} - \frac{1}{2}\frac{1}{7}\right)$.

[118] That is, parallelepipedal.

[119] That is, cylindrical.

[120] Since we are told in the next sentence that this has already been seen, whereas Abū Kāmil merely mentioned the erroneous formulae for *plane* surfaces, an early reader might have noted here his fruitless search.

(35) If he says: A pillar (*ʿamūd*) has base six by six and altitude (*sumk*) [*which is its height* (*ʿamūd*)] ten cubits;[121] what is its volume if the pillar is quadrilateral and pyramidal with its top like a spearhead?[122]

According to calculation, its volume is 120 cubits. The method is (as follows). You multiply 6 by 6; this gives 36. You take a third of it, always; this gives 12. You multiply it by the height, which is 10; this gives 120, which is the volume of the pillar.[123]

(36) If the pillar is circular, subtract from the 120 its seventh and half a seventh, which is $25 + \frac{5}{7}$; this leaves $94 + \frac{2}{7}$, which is the volume of the pillar when it is circular.[124]

(37) If you want its surface area, and it is quadrilateral (proceed as follows).

Take the perimeter of its base from the four sides; this gives 24. Take its half; this gives 12. Multiply it by the altitude on the surface of the pillar, which is 10 and less than $\frac{1}{2}$; this gives 126 less something, which is the surface area without the surface area of its base; and the surface area of its base is 36.[125]

(38) If you want its surface area when it is circular (proceed as follows).

Take the circumference of its base; you will find $18 + \frac{6}{7}$. Take its half; this gives $9 + \frac{3}{7}$. Multiply it by the altitude on its surface, which is 10 and less than $\frac{1}{2}$; this gives 99 less something. Such is the surface area without the surface area of its base; and the surface area of its base is $28 + \frac{2}{7}$.[126]

(35′) The volume of wells follows my explanations to you about the volumes of pillars, quadrilateral, circular or pointed like a spearhead, word for word, and we shall neither go beyond it to something else nor extend it to something else. If God the most high wishes.

Mensuration of the sphere

We begin first with its surface area; indeed, its volume is determined by means of its surface.

(39) Then, if you want to know its surface area: It is asked, for a sphere with a diameter of seven cubits, what its surface area is.

You are to know how much is the circumference of its great circle, the circumference of its great circle referring to a circle with diameter seven cubits. So multiply 7 by three and a seventh, always;[127] this gives 22, which is the circumference of its great circle. Then multiply the circumference of the great circle, namely 22, by its diameter, which is 7 cubits; this gives 154, which is the surface area of the sphere. If you wish, multiply the diameter of the sphere by itself, then by three and a seventh; the result will be the surface area of the sphere.[128]

(40) If you want to know the volume of that sphere, of which we have said that its surface area is 154.

Multiply the 154 by a sixth of the diameter of the sphere, which is, in this problem, $1 + \frac{1}{6}$; this gives $179 + \frac{2}{3}$. Such is the volume of the sphere.[129]

You will always do the same if you want to know the volume of the sphere. You are to determine first its surface area; the knowledge of its area is obtained by multiplying the circumference of its great

[121] The interpolator wanted to explain that *sumk* (thickness) is here taken in the sense of altitude (*ʿamūd*). But for solid bodies Abū Kāmil avoids the ambiguous *ʿamūd* (for its other meaning is "pillar").

[122] Square pyramid.

[123] Volume of a square pyramid with side of the base $a = 6$ and height $h = 10$; then $V = \frac{1}{3} a^2 \cdot h$.

[124] Cone with *diameter* of the base $D = 6$ and same height as before; then $V = \frac{1}{3} D^2 \cdot h \left(1 - \frac{1}{7} - \frac{1}{2}\frac{1}{7}\right)$.

[125] Lateral surface area of a square pyramid with side of the base $a = 6$ and height $h = 10$. Then $S = \frac{1}{2} P \cdot h_s$, with the slant height $h_s = \sqrt{10^2 + 3^2} = \sqrt{109} \cong 10 + \frac{9}{20}$.

[126] Lateral surface area of a cone with diameter of the base $D = 6$, thus perimeter $P = D\left(3 + \frac{1}{7}\right) = 18 + \frac{6}{7}$; then $S = \frac{1}{2} P \cdot h_s \cong \frac{1}{2}\left(18 + \frac{6}{7}\right)\left(10 + \frac{9}{20}\right) = 98 + \frac{37}{70}$.

[127] This "always" is superfluous since the procedure is by now well known.

[128] Given the diameter $D = 7$ of a sphere, required its area S; then $S = P \cdot D \cong \left(3 + \frac{1}{7}\right) D^2 = 154$, with $P \cong \left(3 + \frac{1}{7}\right) D = 22$, the circumference of a great circle of the sphere.

[129] Given the surface area $S = 154$ of a sphere, find its volume V; then $V = \frac{1}{6} D \cdot S = 179 + \frac{2}{3}$.

circle by its diameter, the result being its surface area [*as well, and it will correspond to one another*].[130] Then you multiply the surface area by a sixth of its diameter; the result will be its volume.

(41)[131] You are to know that the volume of any pyramid without a top, that is to say that its top is like a spearhead, is obtained by taking a third of the area of its base, which you then multiply by its height; the result will be the volume, be the base pentagonal, hexagonal, heptagonal, octagonal, triangular, quadrilateral, circular — whichever it is. You must know that.[132]

(42) Its surface area is obtained (as follows). You take half the perimeter of its base, which you then multiply by the height on one of its surfaces.[133] That is,[134] you place the end of the string in the middle of some side of the base, on the ground; then go around with it on the surface standing on the base from the end of the string, and you raise it slowly till you reach the top of the pyramid; then you measure the string, and the result will be the (slant) height; then you multiply it by half the perimeter of the base. The result will be the surface area.

(43) The area of a segment follows what Ptolemy has shown in the sixth book of his Almagest, and which Muḥammad ibn Mūsā al-Khwārizmī has shown in his Algebra. He said namely:[135] You take half the length of the arc, then multiply it by half the diameter of the circle corresponding to this segment. Keep in mind the result. Then subtract the arrow of the segment from half the diameter of the circle if the segment is less than a semicircle; if it is greater than a semicircle subtract half the diameter of the circle from the arrow of the segment. Multiply the remainder by half the chord, and subtract the result from what you have kept in mind if the arc is less than a semicircle, and add it to it if it is greater than a semicircle. The result after the addition or subtraction will be the area of the segment.[136]

Knowing the diameter of the circle corresponding to that segment is (as follows). You multiply half the chord of the segment by itself and divide the result by the arrow of the segment. Add to the result the arrow of the segment. It will give the diameter of the circle corresponding to that segment.

Example of this. Someone says to you: A segment has chord eight and arrow two; what is the diameter of the circle corresponding to this segment?

Multiply half of the chord, thus 4, by itself; this gives 16. Divide it by the arrow, which is 2; the result is 8. Add to it the arrow, which is 2; this gives 10, which is the diameter of the circle corresponding to this segment.[137]

Determination of chords[138]

(44) If he says: A circle has a diameter of ten cubits; what is the chord of the largest equilateral (and equiangular) quadrilateral contained in it?

According to calculation, it is the root of 50. The method is (as follows). You multiply the diameter of the circle, namely 10, by itself; this gives 100. You take its half; this gives 50. Then you will say: Each side of the quadrilateral is the root of 50, which is, approximately, $7 + \frac{1}{2}\frac{1}{7}$. Such are the method and the calculation.[139] Here is the figure.

---

130 Perhaps a gloss to the end of No 39.

131 Nos 41–42, possibly 43, are misplaced.

132 $V = \frac{1}{3} s \cdot h$, with s the area of the base and h the height.

133 $S = \frac{1}{2} P \cdot h_s$. See No 37.

134 The genuineness of what follows is dubious.

135 See Rosen [1831, 73, 52–53 for the Arabic (not quoted literally by Abū Kāmil)]. The reference to Ptolemy must be to *Almagest* VI.7, towards the end [Heiberg 1898–1903, I, 513–518; Toomer 1984, 302–305].

136 Given, for the segment, the arc-length a, the chord t, the arrow h, and calculating the diameter D of the circle (see below), then the area of the segment is either $S = \frac{1}{2} a \cdot \frac{1}{2} D - \left(\frac{1}{2} D - h\right) \frac{1}{2} t$ or $S = \frac{1}{2} a \cdot \frac{1}{2} D + \left(h - \frac{1}{2} D\right) \frac{1}{2} t$ according to whether $\frac{1}{2} D > h$ or $\frac{1}{2} D < h$.

137 Given for the segment the chord $t = 8$, the arrow $h = 2$, find the diameter D of the corresponding circle; then $D = \frac{t^2}{4h} + h = 10$.

138 Rather, relation between the diameter of a circle and the side of an inscribed or circumscribed regular polygon.

139 Required the side a_4 of the square inscribed in a circle with given diameter $D = 10$; then $a_4 = \sqrt{\frac{1}{2} D^2} = \sqrt{50} \cong$

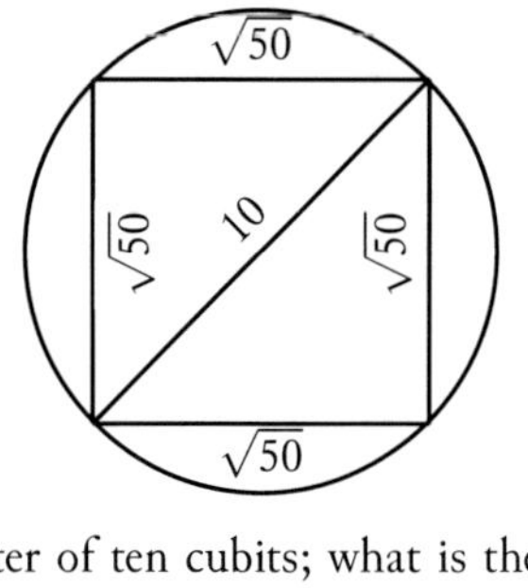

(45) If he says: A circle has a diameter of ten cubits; what is the chord of the largest (*sic*) equilateral and equiangular quadrilateral containing it with each of the sides of the quadrilateral touching the circumference of the circle?

According to calculation, each side of the quadrilateral is precisely equal to the diameter of the circle, as will be shown by a glance at the figure. Thus it is known that each side of the quadrilateral is 10.[140]

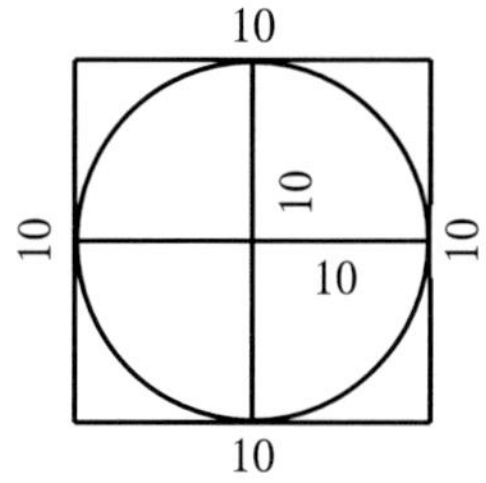

(46) If he says: A quadrilateral has each side ten cubits; what is the diameter of the circle containing it?

According to calculation, it is precisely equal to the diagonal of the quadrilateral, namely the root of 200, which is slightly less than (*qarīb min*) $14 + \frac{1}{7}$.[141]

(47) If he says: What is the diameter of the circle contained in the quadrilateral with each side ten cubits?

According to calculation, the diameter of the circle contained in this quadrilateral is equal to one of the sides of the quadrilateral, thus 10.[142]

(48) If he says: A circle has a diameter of ten cubits; what is each side of the equilateral triangle contained in this circle?

According to calculation, it is the root of 75. The method is (as follows). You multiply the diameter of the circle, namely 10, by itself; this gives 100. You subtract from it its fourth, (always,) thus 25; this leaves 75. Then you will say: Each side of the equilateral triangle contained in the circle with diameter 10 is the root of 75, which is slightly less than $8 + \frac{2}{3}$.[143]

$7 + \frac{1}{2}\,\frac{1}{7}$. See No 2 or No 46 (approximation of $\sqrt{200}$).

[140] Required the side a'_4 of the square circumscribed about a circle with given diameter $D = 10$; then $a'_4 = D = 10$.

[141] Required the diameter of a circle circumscribed about a square with given side $a_4 = 10$; then $D = \sqrt{2\,a_4^2} = \sqrt{200} \cong 14 + \frac{1}{7}$; see No 2.

[142] Required the diameter D of the circle inscribed in a square with given side $a_4 = 10$; then $D' = a_4 = 10$. See No 45. Thus no "calculation" is necessary.

[143] Required the side a_3 of the equilateral triangle inscribed in a circle with given diameter $D = 10$; then $a_3 = \sqrt{\frac{3}{4}\,D^2} = \sqrt{75} \cong 8 + \frac{11}{16} \cong 8 + \frac{2}{3}$. The manuscript has 38 instead of $8 + \frac{2}{3}$ (easily explainable Arabic confusion).

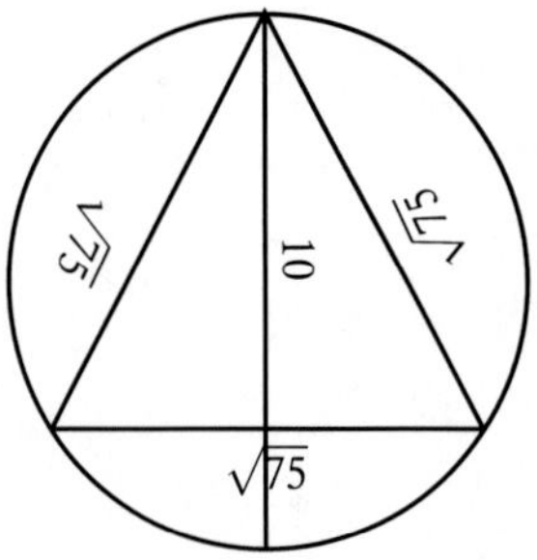

(49) If he says: What is each side of the equilateral triangle touching the circle from outside if the diameter of the circle is ten?

According to calculation, each side of the triangle containing the circle is the root of 300. The method is (as follows). You multiply the diameter of the circle, namely 10, by itself; this gives 100. You multiply the 100 by three, always; this gives 300. Then each side of the triangle containing the circle with diameter 10 will be the root of 300, which is slightly less than $17 + \frac{1}{3}$.[144] Here is the figure.

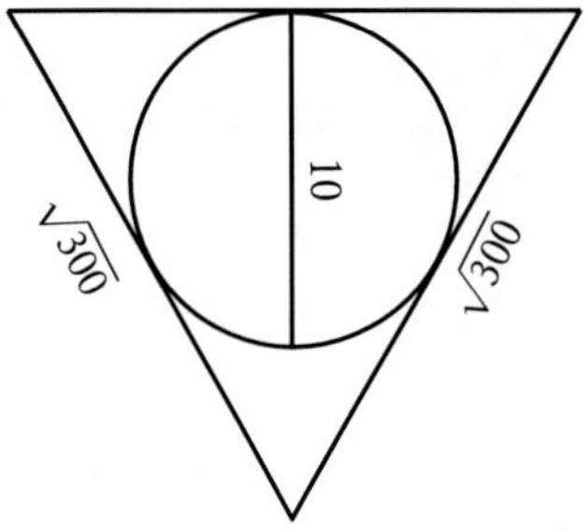

(50) If he says: An equilateral [*and equiangular*][145] triangle has each side ten; what is the diameter of the largest circle contained in it, with its sides touching the circumference of the circle?

According to calculation, the diameter of the circle contained in the triangle is the root of $33 + \frac{1}{3}$, which is larger than $5 + \frac{1}{2} + \frac{1}{4}$. The method is (as follows). You multiply the 10 by itself, which gives 100. You take a third of it, always, which is $33 + \frac{1}{3}$. So you will say: The diameter of the circle is the root of $33 + \frac{1}{3}$.[146]

(51) If he says: What is the diameter of the circle containing this triangle if the angles touch the circumference of the circle?

According to calculation, it is the root of $133 + \frac{1}{3}$. The method is (as follows). You multiply the 10 by itself; this gives 100. You add to it its third, always, which is $33 + \frac{1}{3}$; this gives $133 + \frac{1}{3}$. So you will say: The diameter of the circle containing the triangle with each side 10 is the root of $133 + \frac{1}{3}$, which is larger than $11 + \frac{1}{4} + \frac{1}{6} + \frac{1}{8}$.[147]

(52) If he says: A circle [*containing a pentagon*] has diameter 10; what is the length of each of the sides of the equilateral pentagon contained in this circle with the angles touching the circumference of the circle?

---

[144] Required the side a_3' of the equilateral triangle circumscribed about a circle with given diameter $D = 10$; then $a_3' = \sqrt{3\,D^2} = \sqrt{300} \cong 17 + \frac{11}{34} \cong 17 + \frac{1}{3}$ $(= 2\sqrt{75}$, see previous note).

[145] Superfluous specification, not found before.

[146] Required the diameter D' of the circle inscribed in an equilateral triangle with given side $a_3 = 10$; then, using the results seen in No 49, $D' = \sqrt{\frac{1}{3}\,a_3^2} = \sqrt{\frac{100}{3}} = \frac{1}{3}\sqrt{300} \cong \frac{1}{3}(17 + \frac{1}{3}) = 5 + \frac{2}{3} + \frac{1}{9} = 5 + \frac{21}{27} > 5 + \frac{21}{28} = 5 + \frac{3}{4}$.

[147] Required the diameter D of the circle circumscribed about an equilateral triangle with given side $a_3 = 10$; then, from No 48, $D = \sqrt{\frac{4}{3}\,a_3^2} = \sqrt{133 + \frac{1}{3}} \cong 12 - \frac{1}{24}\left(10 + \frac{2}{3}\right) = 11 + \frac{1}{24}\left(13 + \frac{1}{3}\right) \cong 11 + \frac{13}{24} = 11 + \frac{6+4+3}{24} = 11 + \frac{1}{4} + \frac{1}{6} + \frac{1}{8}$, indeed smaller than the exact value.

Multiply the diameter of the circle by itself. [*The result of the multiplication is a fraction of the square of the diameter of the circle.*][148] Then take five eighths of the square of the diameter of the circle, (always,) and multiply it by an eighth of the square of the diameter of the circle. Take the root of the result. Subtract this from five eighths of the square of the diameter of the circle. Take the root of the remainder. Then this will be the length of the chord of a fifth of the circle. So multiply the diameter of the circle, namely 10, by itself; this gives 100. Take $\frac{5}{8}$ of it; this gives $62 + \frac{1}{2}$. Multiply it by $\frac{1}{8}$ of this (100), thus (by) $12 + \frac{1}{2}$; this gives $781 + \frac{1}{4}$. Take its root, and subtract the result from $62 + \frac{1}{2}$. Take the root of the remainder; the result will be the length of the chord of a fifth of the circle.[149] Here is the figure.

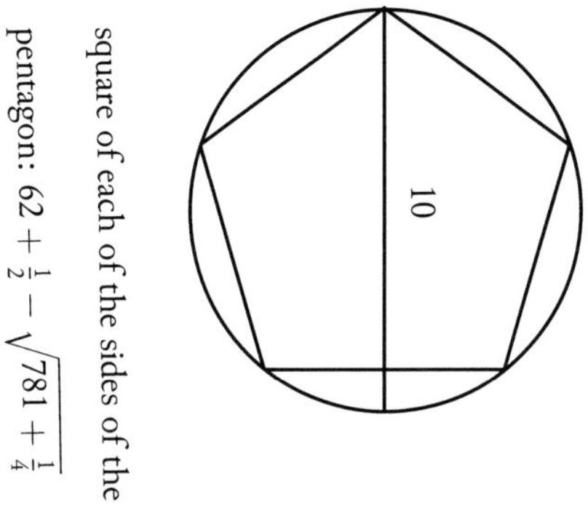

(53) If he says: A circle with diameter ten is surrounded by a pentagon; what is the length of each of its sides?

Multiply the diameter of the circle by itself, then by five, always, and keep in mind the result. Then double the diameter of the circle, and multiply this by itself. Multiply the result by what you have kept in mind. Take the root of the result. Subtract this from what you have kept in mind. Take the root of the remainder. This will give the length of each of the sides of the pentagon. So multiply 10 by itself, then by 5; this gives 500. Keep it in mind. Then double the 10, which gives 20. Multiply it by itself; this gives 400. Multiply it by the 500 which you have kept in mind; ⟨this gives 200000. Take its root, and subtract the result from the 500 kept in mind.⟩ Take the root of the remainder; this will give the length of each of the sides of the pentagon.[150]

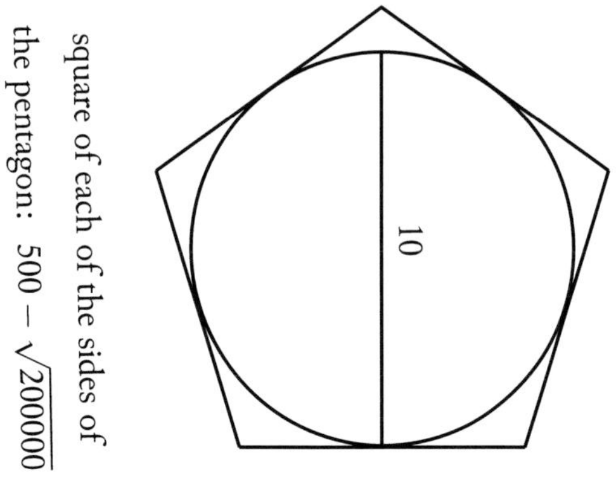

(54) If he says: A circle has diameter ten; what is the length of each of the sides of the decagon contained in it with its angles touching the circumference of the circle and its sides equal?

---

148 Early gloss (faulty) to the next sentence.

149 Required the side a_5 of the regular pentagon inscribed in a circle with given diameter $D = 10$; then $a_5 = \sqrt{\frac{5}{8}D^2 - \sqrt{\frac{5}{8}D^2 \cdot \frac{1}{8}D^2}} = \sqrt{62 + \frac{1}{2} - \sqrt{781 + \frac{1}{4}}}$.

150 Required the side a'_5 of a regular pentagon circumscribed about a circle with given diameter $D = 10$; then $a'_5 = \sqrt{5\,D^2 - \sqrt{(2\,D)^2 \cdot 5\,D^2}} = \sqrt{500 - \sqrt{200000}}$.

Multiply the diameter, namely 10, by itself; this gives 100. Take of it a fourth and half an eighth, always; this gives $31 + \frac{1}{4}$. Take its root. Then subtract from it a fourth of the diameter, always, thus $2 + \frac{1}{2}$ in this problem; the remainder will be the length of each of the sides of the decagon.[151]

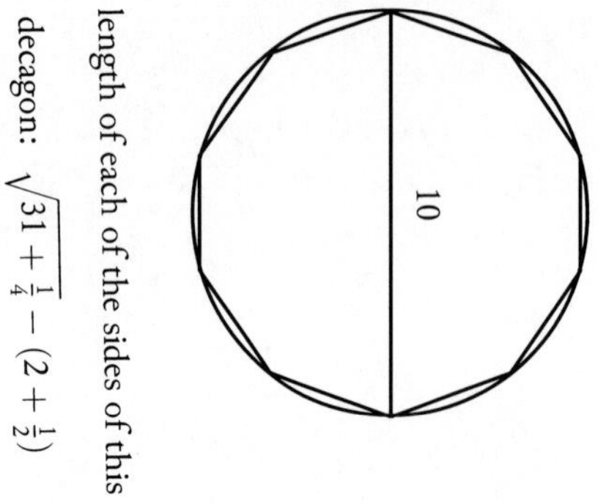

(55) If he says: What is each of the sides of the equilateral decagon containing the (same) circle with its sides touching the circumference of the circle?

Multiply the diameter of the circle, namely 10 in this problem, by itself; this gives 100. Keep it in mind. Then multiply it by its four fifths, namely 80; this gives 8000. Take its root, and subtract the result from the 100 which you have kept in mind. Take the root of the remainder. The result will be the length of each of the sides of the decagon containing this circle, as in this figure.[152]

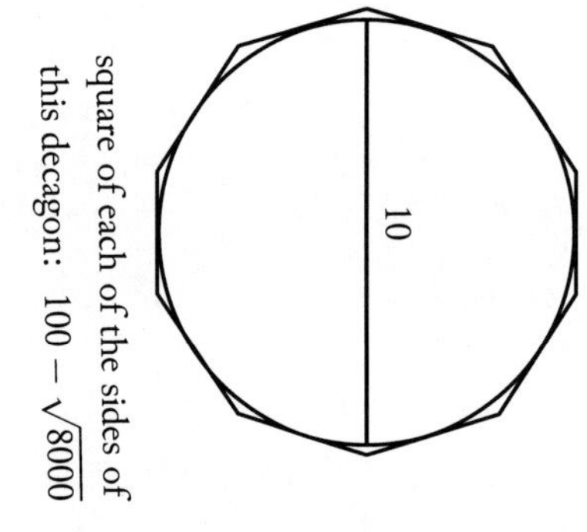

(56) If he says: Each side of an equilateral and equiangular pentagon is ten; what is the diameter of the circle surrounding it?

According to calculation, the square of the diameter of the circle surrounding it is 200 plus the root of 8000. The method is (as follows). You multiply the 10 by itself [*always*];[153] this gives 100. You double it; this gives 200. Then you multiply the 10 by four, always; this gives 40. You multiply it by the 200; this gives 8000. You take the root of the 8000, and you add the result to the 200. Take the root of the sum; this will give the diameter of the circle.[154]

[151] Required the side a_{10} of a regular decagon inscribed in a circle with given diameter $D = 10$; then $a_{10} = \sqrt{\left(\frac{1}{4} + \frac{1}{2}\frac{1}{8}\right) D^2} - \frac{1}{4} D = \sqrt{31 + \frac{1}{4}} - (2 + \frac{1}{2})$.

[152] Required the side a'_{10} of a regular decagon circumscribed about a circle with given diameter $D = 10$; then $a'_{10} = \sqrt{D^2 - \sqrt{D^2 \cdot \frac{4}{5} D^2}} = \sqrt{100 - \sqrt{8000}}$.

[153] See note 29. Same in No 57.

[154] Required the diameter D of the circle circumscribed about a regular pentagon with given side $a_5 = 10$; then $D = \sqrt{2\, a_5^2 + \sqrt{2\, a_5^2 \cdot 4\, a_5}} = \sqrt{200 + \sqrt{8000}}$.

(57) He said: Each side of an equilateral and equiangular pentagon is ten; what is the diameter of the circle contained in it with the sides of the pentagon touching the circumference of the circle?

According to calculation, the square of the diameter of the circle surrounded by this pentagon is 100 plus the root of 8000. The method is (as follows). You multiply the 10 by itself [*always*]; this gives 100. Then you multiply the 10 by eight, always; this gives 80. You multiply it by the 100; this gives 8000. You take its root, and add the result to the 100. Take the root of the sum; this will give the diameter of the circle contained in it.[155]

(58) He said: A circle has diameter ten; what is each of the sides of the equilateral hexagon contained in it with its angles touching the circumference of the circle?

According to calculation, each of the sides of the hexagon is precisely equal to half the diameter of the circle, thus 5. Likewise, each side of the hexagon inscribed in a circle will be equal to half the diameter of the circle.[156]

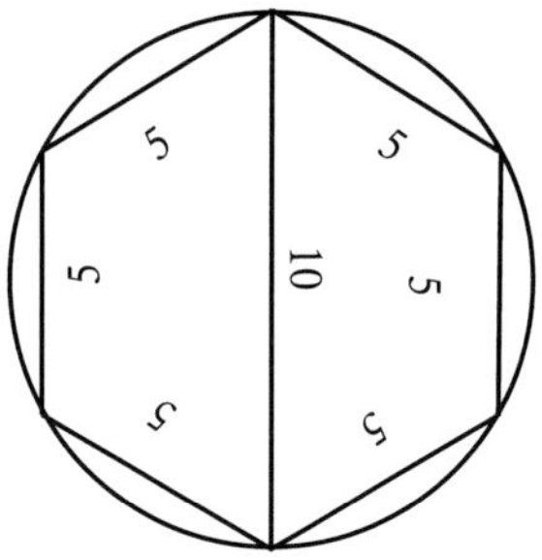

(59) He said: A circle has diameter ten; what is each of the sides of the equilateral hexagon surrounding it with the sides of the hexagon touching the circumference of the circle?

According to calculation, each side of the hexagon is the root of $33 + \frac{1}{3}$. The method is (as follows). You multiply the 10 by itself; this gives 100. You take a third of it, always, which is $33 + \frac{1}{3}$. So you will say: Each of the sides of the hexagon is the root of $33 + \frac{1}{3}$.[157]

(60) He said: Each side of an equilateral and equiangular hexagon is ten; what is the diameter of the circle contained in it?

According to calculation, the diameter of the circle contained in it is the root of 300. The method is (as follows). You multiply the 10 by itself; this gives 100. You multiply it by three, always; this gives 300. So you will say: The diameter of the circle is the root of 300.[158]

(61) He said: What is the diameter of the circle containing it?

According to calculation, its diameter is 20. The method is (as follows). You double the 10, always; this gives 20. So you will say: The diameter of the circle containing the hexagon with its circumference touching the angles of the hexagon is 20, and this is what we wanted.[159]

(62) He said: A circle has diameter ten; what is each side of an equilateral octagon contained in the circle with its angles touching the circumference of the circle?

According to calculation, the square of each of the sides of the octagon is 50 minus the root of 1250. The method is (as follows). You multiply 10, the diameter of the circle, by itself; this gives 100. You take

---

[155] Required the diameter D' of the circle inscribed in a regular pentagon with given side $a_5 = 10$; then $D' = \sqrt{a_5^2 + \sqrt{a_5^2 \cdot 8\, a_5}} = \sqrt{100 + \sqrt{8000}}$.

[156] Required the side a_6 of the regular hexagon inscribed in a circle with given diameter $D = 10$; then $a_6 = \frac{1}{2} D = 5$.

[157] Required the side a_6' of a regular hexagon circumscribed about a circle with given diameter $D = 10$; then $a_6' = \sqrt{\frac{1}{3} D^2} = \sqrt{33 + \frac{1}{3}}$. No figure for this case.

[158] Required the diameter of the circle inscribed in a hexagon with given side $a_6 = 10$; then $D' = \sqrt{3\, a_6^2} = \sqrt{300}$.

[159] Required the diameter of the circle circumscribed about a hexagon with given side $a_6 = 10$; then $D = 2\, a_6 = 20$.

its half, always; this is 50, which you keep in mind. Then you multiply the 100 by its eighth, always, thus in this problem (by) $12 + \frac{1}{2}$; this gives 1250. You take its root and subtract the result from the 50 kept in mind. Take the root of the remainder; the result will be each of the sides of the octagon.[160] Then it is known, according to what we have shown, that the square of each of the sides of the octagon is 50 minus the root of 1250, and this is what we wanted to show.

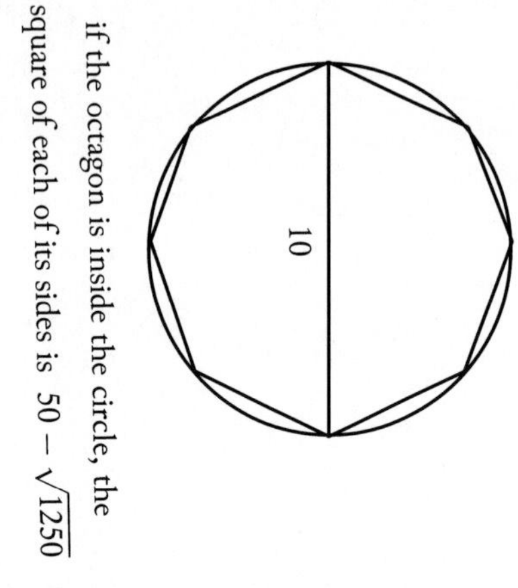

(63) He said: A circle has diameter ten; what is each of the sides of the octagon surrounding the circle with its sides touching the circumference of the circle?

According to calculation, each of the sides of the octagon is the root of 200 minus 10. The method is (as follows). You multiply the diameter of the circle by itself, then you double it; this gives 200. You take its root, and you subtract from the result the diameter of the circle, thus 10. The remainder will be each of the sides of the octagon, which is the root of 200 minus 10, that is, approximately, $4 + \frac{1}{7}$.[161]

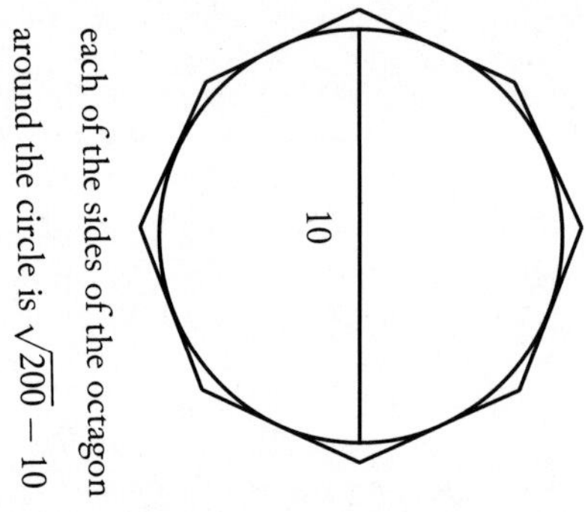

(64) He said: What is the diameter of the circle contained in an equilateral and equiangular octagon with each side ten and with the sides of the octagon touching the circumference of the circle?

According to calculation, the diameter of the circle is 10 plus the root of 200. The method is (as follows). You multiply the 10 by itself; this gives 100. You double it; this gives 200. You take its root and add the result to the 10; the sum will be the diameter of the circle, which is, approximately, $24 + \frac{1}{7}$.[162]

---

[160] Required the side a_8 of a regular octagon inscribed in a circle with a given diameter $D = 10$; then $a_8 = \sqrt{\frac{1}{2}D^2 - \sqrt{D^2 \cdot \frac{1}{8}D^2}} = \sqrt{50 - \sqrt{1250}}$. The next, superfluous sentence might be an addition.

[161] Required the side a'_8 of a regular octagon circumscribed about a circle with a given diameter $D = 10$; then $a'_8 = \sqrt{2D^2} - D = \sqrt{200} - 10 \cong 4 + \frac{1}{7}$ with the approximation seen in No 46. The figure is missing in the manuscript, but not the accompanying text.

[162] Required the diameter of the circle inscribed in a given regular octagon with side $a'_8 = 10$; then $D' = a_8 + \sqrt{2\,a_8^2} = 10 + \sqrt{200} \cong 24 + \frac{1}{7}$ (see previous note).

(65) He said: What is the diameter of the circle containing an octagon with each of its sides ten and the circumference touching the angles of the octagon?

According to calculation, the square of the diameter of the circle is 400 plus the root of 80000. The method is (as follows). You double the 10, which gives 20. You multiply it by itself; this gives 400. Then you multiply the 10 by itself; this gives 100. You double it; this gives 200. You multiply it by the 400; this gives 80000. You take its root, and add the result to the 400. You take the root of the sum; the result will be the diameter of the circle.[163] [*It has thus become clear that the square of the diameter of the circle is the root of 80000 plus 400.*][164]

(66) He said: Each side of an equilateral and equiangular decagon is ten [*in number*];[165] what is the diameter of the circle containing this decagon with the angles of the decagon touching the circumference of the circle?

According to calculation, the diameter of the circle is 10 plus the root of 500. The method is (as follows). You multiply the 10 by itself, then (the result) by five (always); this gives 500. You take its root, and you add the result to the 10; this gives the diameter of the circle, namely 10 plus the root of 500.[166]

(67) If you want to know the diameter of the circle inside this decagon with its sides touching the circumference from outside.

According to calculation, the square of the diameter of the circle contained in this decagon is 500 plus the root of 200000. The method is (as follows). You multiply the 10 by itself, then (the result) by five; this gives 500, which you keep in mind. Then you double the 10; this gives 20. Multiply it by itself; this gives 400. Multiply it by the 500 kept in mind; this gives 200000. Take its root, and add the result to the 500 kept in mind. Take the root of the result. This will give the diameter of the circle.[167] [*It has thus become clear according to our explanations that the square of the diameter of the circle contained in an equilateral decagon with each of its sides 10 is 500 plus the root of 200000.*] God, the invisible Being, knows best.

End of this short work,
with praise to God for the benefit of His help,
and blessings on the noble prophet Muḥammad
and his companions, and salvation be granted.
This was done on Monday eighth of Dhū'l-Qaʿda in the year 758.

---

[163] Required the diameter of the circle circumscribed about a given regular octagon with side $a_8 = 10$; then $D = \sqrt{(2\,a_8)^2 + \sqrt{(2\,a_8)^2 \cdot (2\,a_8^2)}} = \sqrt{400 + \sqrt{80000}}$.

[164] The whole sentence is clearly a gloss. In a verbal mathematical expression involving a root this root must be mentioned in the second place. Indeed, when expressed in words, $\sqrt{80000} + 400$ can be understood as $\sqrt{80400}$, which is precisely what the copyist transcribing that into numerical symbols understood.

[165] This specification, common in the similar problems of the *Algebra*, seems incongruous here.

[166] Required the diameter of the circle circumscribed about a regular decagon with given side $a_{10} = 10$; then $D = a_{10} + \sqrt{5\,a_{10}^2} = 10 + \sqrt{500}$.

[167] Required the diameter D' of the circle inscribed in a regular decagon with given side $a_{10} = 10$; then $D' = \sqrt{5\,a_{10}^2 + \sqrt{5\,a_{10}^2 \cdot (2\,a_{10})^2}} = \sqrt{500 + \sqrt{200000}}$.

References

Manuscript Sources

Tehran, Parliament Library 26. AD 1357 (758 AH).

Modern Scholarship

Anbouba, A., 1963. Un algébriste arabe, Abu Kamil Šuğa' ibn Aslam. Horizons techniques du Moyen-Orient 3, 6–15. (Reprinted in this volume.)

Berggren, J.L., Borwein, J., Borwein, P., 2004. Pi: a source book. Springer-Verlag, New York.

Blume, F., Lachmann, K., Rudorff, A., 1848–1852. Die Schriften der römischen Feldmesser (2 vols). G. Reimer, Berlin. Reprinted: G. Olms, Hildesheim, 1967.

Cantor, M., 1875. Die römischen Agrimensoren und ihre Stellung in der Geschichte der Feldmesskunst. B.G. Teubner, Berlin. Reprinted: M. Sändig, Wiesbaden, 1968.

Ḥājjī Khalīfa, 1835–1858. Lexicon bibliographicum et encyclopædicum a Haji Khalfa compositum, ed. et lat. vert. Flügel, G. (7 vols), Oriental translation fund, London and Leipzig.

Heiberg, J.L., 1927. Mathematici graeci minores. In: Det Kgl. Danske videnskabernes selskab, Historisk-filologiske Meddelelser, vol. XIII, 3. Bianco Lunos Bogtrykkeri, Copenhagen.

Heron, 1899–1914. Heronis Alexandrini Opera quae supersunt omnia, eds. Schmidt, W., et al. B.G. Teubner, Leipzig.

Høyrup, J., 2002. Lengths, widths, surfaces. A portrait of Old Babylonian algebra and its kin. Springer-Verlag, New York.

Ibn al-Nadīm, 1871–1872. Kitâb al-Fihrist, ed. Flügel, G., et al. (2 vols). F.C.W. Vogel, Leipzig.

Neugebauer, O., Sachs, A., 1945. Mathematical Cuneiform Texts. American Oriental Society, New Haven.

Ptolemy, 1898–1903. Claudii Ptolemaei Opera quae exstant omnia, I: Syntaxis mathematica, ed. Heiberg, J., (2 vols). B.G. Teubner, Leipzig.

Rosen, F., 1831. The Algebra of Mohammed ben Musa. Oriental Translation Fund, London. Reprinted: Institut für Geschichte der arabisch-islamischen Wissenschaften, Frankfurt, 1997.

Saidan, S., 1985. التكملة فى الحساب لعبد القاهر بن طاهر البغدادى. Institute of Arab manuscripts, Kuwait.

Sesiano, J., 1996. Le Kitāb al-Misāḥa d'Abū Kāmil. Centaurus 38, 1–21.

Sezgin, F., 1974. Geschichte des arabischen Schrifttums (vol. V). E.J. Brill, Leiden.

Suter, H., 1900. Die Mathematiker und Astronomen der Araber und ihre Werke. B.G. Teubner, Leipzig. Reprinted: Johnson Corp., New York, 1972.

——— 1911. Das Buch der Seltenheiten der Rechenkunst von Abū Kāmil el-Miṣrī. Bibliotheca Mathematica 3. F. 11, 100–120.

Thureau-Dangin, F., 1938. Textes mathématiques babyloniens. E.J. Brill, Leiden.

Toomer, G., 1984. Ptolemy's Almagest. Duckworth, London.

Vogel, K., 1958–1959. Vorgriechische Mathematik. In: Mathematische Studienhefte für den mathematischen Unterricht an Höheren Schulen, Heft 1–2. Schroeder, Hannover / Schöningh, Paderborn.

Hebrew Texts on the Regular Polyhedra

Tzvi Langermann

Abstract I offer here an edition and translation of a Hebrew text on the five regular polyhedra, the so-called Platonic solids, translated from the Arabic and, in my view, descending ultimately from a Greek original. The text survives in a single mansucript, now at Oxford; the translator is Qalonymos ben Qalonymos. Neither the Arabic nor the presumed Greek Vorlagen are extant. There are grounds for speculating that the Arabic original is due to al-Kindī, who reworked many Greek texts, but I find no firm evidence that this is the case. An Arabic text by Muḥyī al-Dīn al-Maghribī draws on the same source; it has been published by J.P. Hogendijk. The author of another Hebrew geometry, extant only in a virtually unstudied Hebrew manuscript now preserved at Mantua, has had access to the text on polyhedra published here. The Hebrew text is collated to the other witnesses; some relevant materials found only in the Mantua manuscript are also edited and translated.

Introduction

The existence of a small group of texts dealing with the five Platonic solids was first announced by Jan Hogendijk and me in a short note in *Historia Mathematica* [Hogendijk and Langermann 1984]. Professor Hogendijk [1994] soon published the relevant chapter by Muḥyī al-Dīn al-Maghribī; his publication has greatly facilitated my own work. My study of the Hebrew texts has been long delayed. With their publication, I am completing a project begun nearly three decades ago. It is a pleasure and honor to publish it in a volume dedicated to Len Berggren, and I hope very much that it answers to his high standards.

Three texts make up this small group. The earliest is clearly the Hebrew text, or rather, the as yet undiscovered Arabic text that was rendered into Hebrew by Qalonymos ben Qalonymos early in the fourteenth century [Lévy 1997, 445–447]. The unique copy, found in MS Oxford Bodley d4 [Neugebauer no. 2773], ff. 181a–187b, 189a–195b, is the text that I present here, in edition and translation. The lost Arabic original may in turn be a translation of a Greek work; that possibility will be discussed below. The second text is that of Muḥyī al-Dīn al-Maghribī, which surely made use of the Arabic from which Oxford Bodley d4 was prepared; an edition, translation, and full discussion of this text is found in Hogendijk's article. A third text is a chapter from a Hebrew geometrical compendium found in a manuscript at Mantua; its author, who may have been called Eliezer, had access to a better copy of Qalonymos' translation than that found in Oxford Bodley d4; he may have had additional sources as well at his disposal. This manuscript is described below in a separate section.[1]

[1] I published a short note in Hebrew on this text, available in David [1995, 74–75].

N. Sidoli and G. Van Brummelen (eds.), *From Alexandria, Through Baghdad: Surveys and Studies in the Ancient Greek and Medieval Islamic Mathematical Sciences in Honor of J.L. Berggren*, DOI 10.1007/978-3-642-36736-6_20,

© Springer-Verlag Berlin Heidelberg 2014

Greek or Arabic Original? Al-Kindī the Author?

Hypsicles tells us in the introduction to his monograph on the regular solids that Apollonius wrote two books on the ratios obtaining between the dodecahedron and icosahedron. The first was found to be faulty, and was revised by Hypsicles' father and his friend, Basilius of Tyre. Later, however, Hypsicles found a second, much better version prepared by Apollonius himself. He further reports that the later version is widely available. His own tract, later incorporated into Euclid's *Elements* as book XIV, is meant as sort of commentary on Apollonius.[2]

Neither a title nor the name of an author is reported in Oxford Bodley d4. However, that manuscript does have an introductory paragraph, as well as a number of authorial remarks scattered through the propositions. These passages offer some rich clues, but, as I see it today, they are not sufficient to decide firmly whether the text was originally written in Greek, or if it is a product of Islamic civilization. The most suggestive, and problematic, piece of evidence is the name found at the beginning of the introduction:

> That which you mentioned from SNQLWNYS,[3] who corrected Apollonius' book, within which [are discussions of] the ratio of the dodecahedron to the icosahedron when they are inscribed in the same circle, and it is Book XIV of the work of Euclid; and his praising the value of knowledge of the ratio which obtains between the two figures, and his saying that there is in this something of the attainment of precious matters.

Ostensibly the author is referring here to the introduction to Book XIV. The name transcribed in Hebrew as SNQLWNYS would then be a warped transcription of Hypsicles. There are a number of problems with this interpretation, beyond the transcription of the name. To begin with, although one certainly gets the impression from the introduction to Book XIV that the ratio between the two figures is praiseworthy and precious, there is no statement to that effect, as one would expect from the passage cited above.

Moreover, our author always refers to Apollonius, rather than Hypsicles, as the author of the materials that now form book XIV of Euclid. The simplest explanation would be that he saw Apollonius' monograph, the revised one that Hypsicles testifies to being "accessible to all;" this treatise is no longer extant. Alternatively, one might maintain, and so our author may have thought, that the real author of book XIV was Apollonius; this is the view of al-Kindī, as we shall soon see. In the very next sentence from the introduction, our author remarks, "And I think that Apollonius' intent in this was comprehensive and all-encompassing." Our author certainly thought that Apollonius intended to cover all five of the Platonic solids. The phrasing of the introduction is not transparent, and the translation into Hebrew does not make it any more so; but it seems clear enough and that he (our author) felt that he was fulfilling Apollonius' original intent by writing his comprehensive treatise. At a number of places in the treatise, our author compares his own proofs with those given by Apollonius, usually commenting that his own proofs are easier for students. (See the ends of Propositions 13, 15, and 16, in sentences 97, 115, and 127.) The first proof in Proposition 13 is said to be due to Apollonius — and it is very similar, though not entirely identical, to XIV, 8, lemma, while the second proof is the author's own. Indeed, at the end of the introduction (sentence I8) we read, "I attributed every proposition that Apollonius mentioned [to him] in order to distinguish between my treatise and his."

Perhaps there is another hint as to the author's identity in the envoi to the treatise, where he announces his wish to make another contribution to the study of the regular polyhedra: "[267.] I think that I shall write another treatise and demonstrate that there are no figures other than these five that can be inscribed in the sphere, now that this treatise [has been finished]." Interestingly enough, the Mantua manuscript adds, after its corresponding theorem: "It is also manifest that no figures other than these five which have been discussed can be inscribed in the sphere." Indeed, the final theorem in

[2] Heath [1921, vol. II, p. 192] reports, without any reference to sources, that the second version of Apollonius' treatise contained the proposition that the surfaces of the dodecahedron and icosahedron are in the same ratio as their volumes, when they are inscribed in the same sphere.

[3] In accordance with standard practice, I transcribe the consonants (the Hebrew alphabet has consonants only; vocalization is indicated by vowel points) in capitals letters. As we shall see, there are different possibilities for vocalization, with significant differences between them.

Euclid's *Elements* (XIII, 18) includes a proof to this effect. We do not know if the author did fulfill his wish, and so this passage in the end cannot settle the question of authorship.

At the end of Proposition 26 (passage 191), our author remarks: "This completes the relationships of the four (sic) figures one to the other. There remain the relationships which Apollonius mentioned between the dodecahedron and the icosahedron. This is the beginning of the explication of his treatise." In the middle of Proposition 31 [corresponding to XIV, 6, first proof], our author says [sentence 215], "Apollonius said: Let us demonstrate this in another way." As we have seen, the propositions that compare Apollonius' proofs with those of our author deal with the two figures discussed by Apollonius; they have apparently been moved forward for organizational purposes. In brief, our treatise is a combination of original materials and an expanded or alternate version of Apollonius.

Let us now turn to the transcription of the name, which in Hebrew is the consonantal string MSNQLWNYS (מסנקלוניס). Our first problem is this: is the first letter, *mem*, the first letter of the name, or is it an attached preposition, a shorted form of *min*, "from?" In order to decide this question, we must first decide whether the word immediately preceding, a verb *cum* pronominal suffix, is in the third person singular or the second person. In ordinary Hebrew usage, one would expect the verb to be in the third person; it is not at all usual to attach the pronominal suffix to a verb in the second person. In this case, the *mem* is part of the name, and the full consonantal string is the subject of the preceding verb. Thus the name looks to be something like Mesenklonis, unattested (even approximately) by any source that I am aware of. Taking the verb to be in the second person would be unusual, but probably would not annoy medieval Hebrew writers as much as it does modern users. This reading is smoother insofar as it continues the direct address to the unnamed person for whom the treatise was written. In this case, the *mem* is a prepositional prefix; our translation above follows this second option.

It would surely make sense if our author were referring to Hypsicles, in which the name was horribly garbled. Even so, the Hebrew forms for Hypsicles, as it appears in copies of Euclid's *Elements*, are quite different: אפסקלאוס or אפסקליאוס.[4] I have found one Greek name transcribed into Hebrew that is very close to the one in our text (without the initial *mem*): סינביליקוס, or SINBILIQUS. It is found in *Meyashsher Aqov*, an extremely interesting monograph and a source of several citations from Hellenistic mathematicians, unparalleled in the medieval literature.[5] In this case, it is clear that the intended name is Simplicius.[6] Now Simplicius has preserved for us some important mathematics, and his commentary on Book I of Euclid's *Elements* was utilized by al-Nayrīzī.[7] A. I. Sabra [1968] has shown that there was more of Simplicius' commentary preserved in Arabic than the portions relayed by al-Nayrīzī. However, there is no evidence that Simplicius wrote on the Platonic solids, and there is thus no point in pursuing this otherwise tantalizing avenue.

If our text is the product of Islamicate culture, then the prime candidate for its author is al-Kindī. He is reported to have written a work, no longer extant, "correcting" (*iṣlāḥ*) books XIV and XV of Euclid [Sezgin 1974, 105, item 4b]. We do possess another tract of his, which explains why the ancients associated the five solids with the five elements (four terrestrial elements and aether).[8] The connection between the elements and the five solids is noted in the introduction to our treatise: "You saw from the works of the ancients, that they likened the five figures to the elements." By way of a citation from al-Kindī's lost treatise *On the Aims of Euclid's Book* we know that the latter believed Apollonius (Ablīnus al-Najjār) to have been the original author of a work in fifteen parts. That work was neglected, until one of the kings of Alexandria took an interest in geometry, and asked Euclid to correct and explain them; ever after, the *Elements* have been linked to Euclid. Two more books were later discovered by

[4] These are the forms recorded by Steinschneider [1893, p. 504, note 24]. The translations of Euclid preceded those of Qalonymos, and one may expect (though of course one cannot prove) that he was aware of the name from the Hebrew versions of Euclid.

[5] For example, *Meyashsher Aqov* has a full and accurate report of Nichomedes' proof that the conchoid and its asymptote never meet; interestingly enough, Eutocius is another source for this information, and his commentary to Archimedes' *On the Sphere and the Cylinder* is found in the same codex under study here, Oxford Bodley d4. See Langermann [1988, 33–39].

[6] Gluskina 1983, f. 106a line 12 and p. 97, n. 21 f. 106a line 12 of the facsimile; see note 21 on p. 97.

[7] Simplicius' contribution is discussed briefly by Heath [1921, II, 538–540].

[8] Available in an English translation by Rescher [1967].

Hypsicles, and these were annexed to the book restored by Euclid.[9] T. L. Heath has suggested that this (seemingly) fanciful story is grounded in al-Kindī's misunderstanding of Hypsicles' preface to Book XIV (likely due in turn to a faulty translation).[10] Be that as it may, there should be little surprise in al-Kindī's (if he is the author of our treatise) referring to the author of Book XIV as Apollonius — even if, elsewhere in the treatise, he refers to the author of the thirteen earlier books as "Euclid."

There are additional connections to Apollonius in the texts preserved in MS Bodley d4. Two of the Hebrew items found uniquely in this codex depend in one way or another on Apollonius' *Conics* (see items 7 and 13 described in the following section). The author of the polyhedra text declares that he is supplementing, or commenting on, something by Apollonius. We thus have a small collection of unique Apollonian texts in Hebrew.[11]

al-Kindī had a great interest in Hellenistic philosophy and science, and he — or he and his "school" of students, assistants, and advisors — were quite free in both adapting and interfering in the works of the ancients. The work done by al-Kindī and/or his associates on the writings of Plotinus has been the subject of the closest scrutiny; however, he and/or his school applied themselves to the writings of Nicomachus, Ptolemy, and others, including, as we have just seen, Euclid. Interestingly enough, the only extant revision of al-Kindī's in the field of mathematics, his thorough-going reworking of Nicomachus' Arithmetic, survives only in Hebrew.[12]

The approach to "Apollonius" in our text is certainly in line with al-Kindī's style, though it remains puzzling that in the titles to his lost work, no mention is made of either Euclid or Hypsicles. The pious remarks made in the course of the introduction are fitting for a Muslim author, but this is no conclusive proof: some works that are definitely translations (such as Pythagorean Golden Verses) were adjusted to the sensibilities of a Muslim audience.

Oxford Bodley d4

This very important codex, which contains several unique items of great interest for the history of mathematics, has not been fully described anywhere.[13] In the entry on Eutocius in his monumental study of Hebrew translations, Steinschneider discusses at length a note found in MS Bodleian Uri 433 [Huntington 96, no. 2008 in Neubauer's catalogue], according to which "Elia ha-Levi" possessed a copy of Eutocius' commentary on the two books of Archimedes' *On the Cylinder and the Circle*. The same codex, which comprised four quires, had a copy of an explication of a doubt besetting the last premise (*haqdama*) of the *Conics*; we shall shortly suggest an explanation for this puzzling reference. Since the notice makes no mention of the language of these items, Steinschneider assumes (correctly)

[9] The citation is found in the bibliography of al-Nadīm, available in English in Dodge (vol. 2, 1970, 635–636); see Sezgin, 1967, p. 105, item 4a. Note that in our text, the name is transcribed Abōlōniyūs (or some mild variant), but never Ablīnus, and never with the moniker al-Najjār, "the carpenter."

[10] See Heath [1925, vol. 1, 5–6]; cf. John Murdoch [1971, 438].

[11] Aptly described as such by Lévy [1997, 438–439]; indeed, they are more deserving of the moniker Apollonian than the Hebrew texts on the asymptote, mentioned there as well, since the latter presuppose that the reader has not seen the *Conics*.

After this study had been completed, I learned from the anonymous reader (whom I take the opportunity to thank) of the new in-depth study of Euclid's Books XIV–XV [Vitrac and Djebbar 2011 and 2012]. One of the co-authors kindly supplied me with a preprint of the second part, which includes a treatment of related texts, such as the one I am publishing here [Vitrac and Djebar 2012, 86–90]; the authors discuss the Hebrew text on the polyhedra on the basis of an early draft of this paper.

[12] For al-Kindī's role in constructing the Arabic Plotinus, see most recently Hansberger [2011], and the literature cited there. The small, extant part of his redaction of Ptolemy was published by Rosenthal [1956]. Al-Kindī's version of Nicomachus, still unpublished, survives in several Hebrew manuscripts; see Freudenthal and Lévy [2004]. Jolivet and Rashed [1997] have published volumes of close studies on al-Kindī.

[13] For now one must still refer to the catalogue of Neubauer and Cowley [1906, columns 187–189].

that they are Hebrew translations, likely from the Arabic, though possibly directly from the Greek.[14] Both of these are found uniquely in MS Oxford Bodley d4. There are several owner's marks in Oxford d4, but none answer to the name Elia. Steinschneider adds, "Es ist aber sonst keine Spur einer solchen Uebersetzungen zu finden." Our manuscript was acquired by the Bodleian Library after Neugebauer had completed the first volume of his catalogue, and after Steinschneider had made his own study of the holdings of that library.[15]

We learn from Steinschneider that someone — his name can now be deciphered as Elia Prigotta — had a copy of our manuscript, or another copy; and this was known to the owner of Uri 433, who was studying the related text by Ibn al-Samḥ.[16] Hence there was a small circle of savants, in Italy I would guess, who were studying advanced treatises in geometry.

I present a fuller account of the contents of Oxford Bodley d4 than any other as yet published; there is still much work to be done. References in parentheses are to Steisnchneider's *Die hebraeischen Uebersetzungen des Mittelalters*:

1. Euclid, *Data*, ff. 3a–23b (p. 510).
2. Theodosios, *Spherics*, ff. 24a-60b (pp. 541–542).
3. Menelaos, *Spherics*, ff. 61b–104a (pp. 515-516).
4. Theorems "from another book," ff. 104b–105a.
5. Archimedes, *Sphere and Cylinder*, ff. 108a–141b (p. 502).
6. "*Sefer Misha'lim bi-Tishboret*," ff. 142a-152. A translation of an as yet unidentified Arabic text; *mish'alim* would be an appropriate (if unusual) rendering of the Arabic *maṭālib* or *masā'il*, meaning in this context "problems."[17]
7. Abu Saʿdān, a small monograph on the triangle which makes use of Apollonius's *Conics*.
8. Thābit bin Qurra, *The Cutting Figure*, ff. 156b–164a (p. 589).
9. Jābir bin Aflah, *The Cutting Figure*, ff. 164a–169a (pp. 544–545).[18]
10. Jābir bin Aflah, comments to Menelaos, ff. 169a–173b.[19]
11. A theorem on the parallelepiped: the square on the its diagonal is equal to the sum of the squares of its three sides. The scribe notes at the end, "I found this proposition in one of the books of the scholars...," f. 174a.
12. "The remarks of one of the ancients on the book of Menelaos," f. 174b–177b.
13. Explication of a "doubt" that has been raised concerning the last premise (*haqdama*) of the *Conics*. Another Apollonian text unknown from other sources, is found in this same codex on ff. 177b–181a. The reference to a premise is puzzling; there are some definitions at the beginning of Apollonius' *Conics*, but no premises. However, at the beginning of MS Oxford Bodley Marsh 667, which contains a copy of the *Conics* in the recension of the Banū Mūsā, there are a series of theorems, beginning on f. 2a, that serve as premises: *al-ashkāl allatī yuḥtāj ilayhā fī tashīl fahm hādhā al-kitāb*, "the theorems required to facilitate the understanding of this book." The last of these has two methods of proving a theorem concerning similar triangles; and indeed, the Hebrew text refers to a second method. Moreover, there is a long, dense marginal note in MS Marsh 667, indicating that perhaps this theorem raised some difficulties. Therefore, we may tentatively identify this short text as a gloss on the last premise in the recension of the Banū Mūsā.
14. The text on the five solids, published here; ff. 181a–187b, 189a–195b.
15. Eutocius' commentary to Archimedes' *Sphere and the Cylinder*, ff. 188a–188b, 195b–206b, incomplete. Steinschneider's remarks, referred to in the discussion of item 13 above, is found in his entry on Eutocius; but he knew of no manuscripts.

---

[14] Steinschneider [1893, 513]. Oxford d4 does not display Elia's name where one would expect an owner's mark to be; but not all those who possessed manuscripts inscribed their own names.

[15] Neubauer and Cowley report [1906, xii] that the codex was acquired in 1886; they provide no additional information.

[16] This text is published and studied by Lévy [1996].

[17] Cf. Lévy [1997, 438–439], regarding items 6, 7.

[18] See Lévy [1997, 435], concerning items 8 and 9, both of which take up materials from Menelaos.

[19] See Lévy [1997, 434 and note 7].

We are concerned here with item 14, the text on the five regular solids. The bulk of the text was translated by Kalonymos ben Kalonymos ben Meir on 29 Shebat 5069 (1309), as we read in the colophon: "The translation of this treatise was completed on 21 Shebat [50]69 [10 February 1309]. I, Kalonymos ben Kalonymos ben Meir of blessed memory, translated it from the Arabic language into the Hebrew language, when I reached my twenty-second year. May God in His mercy give me the privilege of learning and understanding..." However, one leaf containing two propositions was missing from Kalonymous' original translation; fortunately, Kalonymos Todros possessed a copy of the Arabic, and was able to render the missing portion into Hebrew. "Said Kalonymos Todros the scrutinizer [i.e. he who scrutinized the MS]: These two propositions, i.e. Propositions 30 and 31, were missing from the translation of the leader (*nasī'*) and scholar R. Kalonymos ben R. Kalonymos, of blessed memory, of the treatise on the five solids, because a leaf was missing from the Arab text from which he translated. Several years afterwards, I found that leaf and translated it." Immediately afterwards there is note by a certain Miles of Marseilles, who found the missing portion and added it to the text: "Miles of Marseilles wrote this on the third of Ellul [50] 95 [5 August 1334] in Aigues. It was missing from the treatise on the comparison of the five solids, from the middle. That treatise is written in this codex which is before me, and now, after attaining these two propositions by the grace of God, this codex is complete, Thank God."

The propositions are not numbered in the manuscript, but there are internal cross-references by proposition number, and these agree with our own consecutive numbering. However, there are some exceptions; see the notes to sentences 223 and 233.

Bodley d4 is riddled with errors; most are minor, and due to graphic similarities between Hebrew letters, especially *beit* and *kaf*. In sentence 201 the scribe actually wrote out "aleph" (A) and "baf" (sic, instead of "*kāf*" = K), which indicates that the text was dictated to him, and that he was probably in a hurry. The Hebrew consistently reads *mētar* (chord) instead of *mōtar* (excess). This is a trivial orthographical error but a significant error in translation and leads one to believe that our manuscript was copied from another Hebrew text. In most cases, I have been able to produce satisfactory emendations. In the case of a small number of propositions (nos. 12, 17, 26, 27, 33), the text was so corrupt as to make an attempt at emendation useless. In those cases, I have been able to identify the goal of the proposition, and to indicate the corresponding item in Hogendijk's study. Moreover, in some cases I have found a better copy of the same theorem in the Mantua manuscript (to be described in the following section); in that case, the theorem has been copied from Mantua, and placed with the other material from that manuscript.

The Hebrew text has been transcribed and the passages given a sequential numbering in order to facilitate reference to them, as well as comparison between text and translation. Those theorems that I deemed too corrupt to attempt an emendation have been transcribed exactly as they appear, and put in a smaller font. The sequential numbering skips over them. Emendations — correcting the many mislabeled points, or occasional supplements to the text — are indicated by square brackets. In the translation, I have not recorded the mislabelings, but only alerted the reader that the letter is my emendation by using the square brackets. Occasional dittographies in the Hebrew text are indicated by angular brackets. Other symbols used are:

! to indicate that this is exactly how the word appears in the text (sic!),
? to indicate a reading of which I am unsure, and
| to indicate a line break in the manuscript.

There are no figures at all in the Oxford manuscript; the ones accompanying the translation are my reconstructions.

The mathematical contents are similar to those in the text of Muḥyī al-Dīn published by Hogendijk, and are described very ably by the latter in the introduction to his article. I will confine myself to one observation on the relationship between the Hebrew text and that of Muḥyī al-Dīn. I will employ here, and occasionally in the translation and notes as well, the same notation used by Hogendijk:

e_i, f_i, s_i, p_i, v_i the edge, face, surface, perpendicular, and volume of a regular polyhedron with i faces (i=4, 6, 8, 12, 20) inscribed in a fixed sphere

R radius of sphere circumscribing the polyhedra
r_i radius of the circle circumscribing face f_i
a, x_1, x_2 a is an arbitrary line segment divided in mean and extreme ratio into segments x_1 and x_2
d, h side and altitude of a fixed equilateral triangle
$c(n)$ side of a regular n-gon inscribed in a circle
$c(6)$ radius of this circle (= side of inscribed hexagon)
$c(2.5)$ diagonal of the pentagon inscribed in the circle

Proposition 35 in particular is important for establishing the relationship between the Oxford text and the treatise of Muḥyī al-Dīn. In the version of Muḥyī al-Dīn, we have the construction of Proposition 34 (find c such that $a^2 + x_2^2 = c^2$) and, in Proposition 35, the construction of the five (actually six) segments ordered according to the ratios of the surface areas of the five figures. Hogendijk notes two oddities of this presentation: the fact that Proposition 34 is not used in Proposition 35, and the fact that six rather than five segments are constructed.[20] Now we note that in the Bodleian text, there is no independent construction corresponding to Proposition 34 of Muḥyī al-Dīn. That same construction is included as part of this and the succeeding proposition. In my opinion, Muḥyī al-Dīn decided that he could construct the five segments of Proposition 35 in a way simpler than that of the Bodleian text which he had before him; his result is Proposition 35 of the Muḥyī al-Dīn's text. Muḥyī al-Dīn, therefore, took the construction of Proposition 35 of the Bodley d4 text and made it into his own Proposition 34, thus (1) simplifying, in his view, Proposition 35 and (2) simplifying as well his Propositions 36 and 37, which make use of Proposition 34. I checked the version of this proposition in the Mantua MS and found that it is almost word for word identical to that of Oxford Bodley d4. Moreover, it is one of the few propositions in that manuscript which is not accompanied by a figure. Reference is made in this proposition to two earlier propositions by number. In sentence 244, it is stated that $s_{12} : s_{20} = \sqrt{a^2 + x_1^2} : \sqrt{a^2 + x_2^2}$ was shown in Proposition 34. According to our count, this was shown in Proposition 32. In sentence 246, it is stated that $s_6 : s_8 = h : d$ was shown in Proposition 11; this agrees with our count.

Mantua 2

Mantua 2 [Comunita Israelitica MS ebr. 2] is a geometrical encyclopedia of unknown authorship. Part Nine deals with the regular polyhedra, and a subsection entitled to "On the Ratios of the Five Figures, with regard to their Heights, Bases, Surface Areas and Volumes," presents most of the material found in our texts on the polyhedra. In general, both the wording and the ordering of the propositions follow those of Oxford Bodley d4. Some of the differences between Mantua 2 and *Oxford d4* stem from a rearrangement of the material on the part of the editor of Mantua. There are also a few items not found in either Bodley d4 or Muḥyī al-Dīn.

First, let us discuss the editorial adjustments. Two propositions found in Bodley d4 which, strictly speaking, do not deal with polyhedra, have been placed elsewhere in the text. Bodley d4 Proposition 1 (= Muḥyī al-Dīn 1) is found on ff. 34b–35a, as the last proposition of Part III, "On Circles and the Figures [drawn] therein." Bodley Proposition 15 (= Muḥyī al-Dīn 4) is found on f.65a among a group of propositions dealing with the "Golden Ratio," which forms part of Part IV, "On the Ratio[s] of Plane [Figures] and their Sides." On the other hand, *Elements* XIII, 18, in which one constructs the five segments representing the edges of the polyhedra, is included among the material on polyhedra in the subsection of Part IX which interests us.

In the category of "editorial adjustments" one must also include the omission of all the references to Euclid and Apollonius which are found in Bodley d4, as well as the introduction and closing remarks. In fact, Mantua 2, which is entitled "The Book of Euclid," does not mention any mathematician by

[20] These were noted by Hogendijk in the preprint of his article (pp. 48–49) which he kindly shared with me. Cf. Hogendijk [1994, n. 35.1].

name anywhere, with the exception of one of the propositions on spherics, where both Menelaos and Theodosios are cited by name.

Mantua 2 also contains some new material. Proposition 10 demonstrates that $z_{12}^2 + z_6^2 =$ five times the square on the radius of the circle which circumscribes the face of the dodecahedron.[21] The second novelty is a rather obscure remark at the end of Proposition 25 by one Eliezer: "Said Eliezer: it has been shown in a general way that the ratio of the volume of the cube to the volume of the octahedron is equal to the ratio of the surface area of the cube to the surface area of the octahedron, and the ratio of the volume of the dodecahedron to the volume of the icosahedron is equal to the ratio of the surface area of the dodecahedron to the surface area of the icosahedron, by our having made the demonstration general in demonstrating the ratio of all solid figures which lie in the same sphere like the ratio of their surface areas in the preceding manner in demonstrating, with regard to all figures, the surface areas (!), which lie in one sphere."

The last part of this passage is not very clear. Perhaps Eliezer wishes to point out that we arrive at the ratios $s_6 : s_8 = v_6 : v_8$ and $s_{12} : s_{20} = v_{12} : v_{20}$ by assuming that all figures lie in one sphere. This is the only note by Eliezer in our text, that is to say, it is the only place in a rather long text where there is any hint as to author, translator, or scribe. Eliezer could be any one of these three. Perhaps it is noteworthy that only the topic of polyhedra prompted Eliezer to comment.

The third and most important passage containing new material is found in Proposition 26. Two constructions are given for the five segments which represent the edges of the polyhedra. One is taken from *Elements* XIII, 18. The other is a more purely "geometrical" construction, similar to that of Muḥyī al-Dīn 31 but, in my opinion, simpler and more concise.

Where does this new material, especially the construction of Proposition 26, come from? I rule out any direct connection between Mantua and Muḥyī al-Dīn for several reasons. Mantua has no equivalent of Muḥyī al-Dīn 36, the construction of the segments representing the faces of the polyhedra; and the construction of the segments representing the surface areas, heights, and volumes follows Bodley d4, not Muḥyī al-Dīn's different approach. On the other hand, one cannot contend that the material was taken from a better version of Bodley d4, since the introduction of Bodley d4 clearly states that the goal of that text is "to complete the ratios of these five figures one to another, regarding their heights, surface areas, and volumes;" there is no mention of the edges.

The simplest explanation is that the compiler of Mantua realized that Euclid had already given a construction for the edges and, therefore, he included "edges" in the title of this subsection and imported XIII, 18 in order to complete the discussion. The other construction would then seem to be his own contribution to the subject. It is also noteworthy that in this other method — and in the corresponding construction in Muḥyī al-Dīn — one proves as well the ordering of the segments according to length, something which is not done in any of the other constructions. I take this to reflect the influence of XIII, 18, where this ordering is also shown. In this connection, one may also point out that none of the texts know of the proofs of Pappus V, Propositions 52–56, regarding the relative volumes of the polyhedra of equal surface area.

The compendium found in MS Mantua certainly warrants a study of its own. I have utilized it only for the purpose of emending or supplementing MS Bodley d4. Following the translation of the text found in MS Bodley d4, I display from MS Mantua only (1) propositions or parts of propositions which are missing or unintelligible in Bodley d4, and (2) the new material. The proposition numbers refer to their order in Mantua; see the table for corresponding proposition numbers in Bodley d4 or Muḥyī.[22]

Afterword

I have found another short text dealing with the polyhedra in Hebrew. Though this text is unrelated to the three texts with which we have been dealing, a short note may be in order. The text described the

[21] See our translation and notes.

[22] Note that the propositions are not numbered in the text.

construction of the five solids, and it is found in MS Oxford Bodley, Neubauer 2006, ff. 35a–36a. The author is probably Judah b. Solomon ha-Cohen (thirteenth century), the compiler of the encyclopedia of the sciences. In Judah's encyclopedia, there is no discussion of the polyhedra beyond what may be found in Book XIII of Euclid. Even the material of Book XIV is omitted.[23]

Particularly interesting is the historical context within which this treatise was written. The work begins: "Said the author. In my youth, while I was still in Spain, the emperor's philosopher asked me, how we may construct (*naḥōq*) in a given sphere each one of the five solids, and how to construct within each one of them the given sphere (!), and the proofs of each [construction] as well..." The emperor is clearly Frederick II of Sicily, but the identity of his philosopher remains a puzzle [Manekin 2000, 469].

Oxford Bodley d4

Text

[Preface]

[I1] יבאר לך השם יתע' תעלומות הדברים וישירך לחדרי האמת וינקה אותך מהסכלות ויוליכ בכל דרך טוב הכינותי יצליחך הבורא ית' לאהבתו ויטהר לבך לאמונתו [I2] מה שזכרתו מסנקלוניס המתקן לספר אבלוניוס ומתוכה כהתיחס(!) תמונה בעלת שתים עשרה תושבות כל תמונה בעלת עשרים תושבות נקוום בכדור אחד והוא המאמר הי"ד המיוסד לאקלידס ושגבו שעור הידיעה בזה היחס אשר בין שתי אלו התמונות אמרו למה שבזה מהשגת ענינים יקרים [I3] ואחשוב (181b) כוונת אבלוניוס בזה כוללת יקיף קצתם על קצתם [I4] ואתה זכרת שכבר נאמר שכונת ספר אקלידס בלי אות אמנם הוא ידיעת בתמונות החמשה הנזכרות במאמר הי"ג ממנו הנקוום בכדור אחד וראית מפי הראשונים באלו התמונות החמשה לדמות אותם ביסודות [I5] ואתה למה שראית זה חשקה נפשך והשתוקקה לדעת יחסי אלו התמונות החמשה קצתם מקצת ואלו תשלם זה היה בזה מן ההשגה והידיעה כמו מה שיש בהתיחס שתי התמונות אשר זכרם אבלונוס ובאר שהוא כמו שאמרתי ואמת מה שעיונת [? שעיינתי] לפי שנושא אלו התמונות החמשה אצל הראשונים נושא אחד. וגם כן שלא יפול בכדור זולתם [I6] ואני למה שראיתי יקר מחשבתך נכספה נפשי לעזרך ולגורש מושך (!?) וכללתי עשתנותי להשלים התיחסות אלו התמונות החמשה קצתם מקצת מעמודיהם ושטחיהם וגשמיהם אחרי השתחויתי לאל ואליו השלכתי יהבי ושאלתי העזר והסעד ומידו היתה עלי לפתוח לו הכל כיכלתי ישתבח לנצח ויתעלה לעד אמן.

[I7] כתבתי אליך בזה המאמר ובארתי בו כונתך וקבצתי מה שזכרו אבלוניוס בהתיחס שתי התמונות להיות זה המאמר עמוד בזה הענין [I8] ויחסתי כל תמונה שזכרה אבלוניוס אליו להבחין בין מאמרי למאמרו ובאל אעזר ית' וזה ראש המאמר.

[Proposition 1]

[1] נרצה לבאר שמרובע עמוד כל משולש | שוות הצלעות שלשת רבעי מרובע צלעו [2] משל זה אם(?) | נתן המשולש השוה הצלעות משלש אב"ג ועמוד א"ד ואומר שמרובע א"ד | שלשת רבעי מרובע ב"ג [3] מופת זה שא"ב כמו ב"ג וזוית ב' כמו זוית ג"ד | וזוית אד"ב כמו זוית אד"ג לפי שכל אחת מהם נצבת [4] וצלעו א"ד משותף | לשני המשולשים וב"ד כמו ד"ג מרובע ב"ג ארבעת דמיוני מרובע ב"ד | [5] מרובע א"ב אשר הוא כמו מרובע ב"ג כמו שני מרובעי א"ד ד"ב | [6] ומרובע[י] א"ד ד"ב ארבעת דמיוני מרובע ב"ד ונפול מרובע ב"ד המשותף | וישאר מרובע א"ד שלושת דמיוני מרובע ב"ד [7] ומרובע ב"ג ארבעת | דמיוני מרובע ב"ד ומרובע א"ד שלשת רבעי מרובע ב"ג שהוא | צלע משולש וזה מה שרצינו לבאר. |

[Proposition 2]

[8] נרצה לבאר שצלע | בעל שמונה תושבות כמו עמוד משולש בעל | ארבעת תושבות הנקוים בכדור אחד [9] משל זה שניח קטרי הכדור | (182a) הכדור א"ב ונעשה עליו חצי עגלה א"ד ה"ב [10] ונוציא עמוד ז"ה ונבדיק ה"ב | בקו ה"ב וח"ב צלע משולש בעל שמונה תושבות הנקוים בכדור | אשר קטרו א"ב [11] ונתן א"ג שלושה(!) א"ב ונוציא עמוד ג"ד ונבדיק ב"ד בקו | ד"ב וד"ב צלע משולש בעל ארבע תושבות [12] ונעשה

[23] On this encyclopedia see Fontaine [2000], Langermann [2000], and Lévy [2000].

משלוש [שוה] הצלעות יהיה | כל צלע מצלעותיו כמו ד"ב והוא משולש אכ"ט ועמודו ח"ל ואומר | שעמוד ח"ל כמו ה"ב. [13] מופתו אנו נחלק בשני חציים ל"ב על מ' ומ"ט | אחד וחצי כמו ט"ל [14] ומרובע א"ב אחד וחצי כמו מרובע ב"ד כפי מה | שנזכר אקלידס בספרו ומרובע א"ב כפל מרובע ח"ב [15] וב"ט כפל ט"ל | [16] שלושה שעורים ה' [מרובע ה"ב] ומרובע א"ב ומרובע ב"ד ושלשה אחרים על מספרם | ט"מ וט"ל וט"ב כל שנים על יחס שנים והיה המתחלף ביניהם [17] יחס | מרובע ב"ה ממרובע א"ב כיחס ט"ל מט"ב | ויחס מרובע ב"א אל מרובע ב"ד כיחס ט"מ | אל ט"ל וכיחס השווי יחס מרובע ה"ב ממרובע | ב"ד כיחס ט"מ של ט"ב [18] ויחס ט"מ אל ט"ב | שלושה רבעי דמיון כיחס מרובע ח"ל | אל מרובע ט"ב [19] ויחס מרובע ח"ל ממרובע | ט"ב >ביחס מרובע ט"ב< כיחס מרובע ח"ב ממרובע ב"ד ומרובע ד"ב | כמו מרובע ט"ב ויחס מרובע ח"ל לפי שהם שוים וח"ב גם כן כמו ח"ל. |

[Proposition 3]

[20] נרצה לבאר שמדת משולש בעל שמונה תושבות | שלושת רבעי מדת משולש בעל ארבעת תשבות הנקוים | בכדור אחד [21] ומשל זה אנו נתן משולש אב"ג בעל ארבעת תושבות | ועמוד א"ד . ואומר שהמשולש השוה הצלעות אשר על א"ד והוא משולש | בעל שמונה תושבות שלשה רבעי ממשולש אב"ג. [22] מופתו שיחס | המשולש השוה הצלעות שהוא על א"ד אל משולש אב"ג הוא כיחס א"ד ב"ג | נשנה [23] ויחס א"ד מב"ג הוא כיחס מרובע א"ד אל מרובע ב"ג [24] אם כן | יחס מרובע א"ד >ו<ממרובע ב"ג כיחס המשולש | השוה הצלעות אשר על א"ד והוא המשולש בעל | שמונה תושבות אל משולש אב"ג והוא משולש בעל | (182b) ארבעת תושבות [25] ומרובע א"ד שלשה רבעי מרובע ב"ג ומשולש בעל שמונה | תושבות שלשה רביע(!) משולש בעל ארבע תושבות זה מה שרצינו לבאר. |

[Proposition 4]

[26] נרצה לבאר ששטח בעל שמונה תושבות אחד וחצי כמו שטח בעל ארבעת | תושבות הנקוים בכדור אחד [27] מופתו שארבעת דמיוני | משולש בעל ארבעת תושבות הוא שטח בעל ארבעה תושבות ושמונה | דמיוני משולש בעל שמונה תושבות הוא שטח בעל שמונה תושבות [28] וארבע | דמיוני משלש בעל שמונה תושבות הוא חצי שטח בעל שמונה תושבות | ויחס ארבע דמיוני משולש בעל שמונה תושבות מארבע דמיוני [שטח] בעל [ארבע] תושבות | כיחס משלש בעל שמונה תושבות ממשולש בעל ארבעת תושבות [29] ומשלש | בעל שמונה תושבות שלושה רביעי משולש בעל ארבע תושבות וארבע >ד'< | דמיוני משלש בעל שמונה תושבות שלשה רבעי ארבע דמיוני משולש בעל ארבע | תושבות ושטח בעל שמונה תושבות אם כן באלו אחת וחצי כמו שטח בעל | ארבע תושבות וזה מה שרצינו לבאר. |

[Proposition 5]

[30] נרצה לבאר שהמוגשם | הנכחי השטחי' אשר תושבתו שני דמיוני בעל משלש ארבע | תושבות ורומו כמו תשיעית קוטר הכדור משלש בעלי (!) כמו | מוגשם בעל ארבעה תושבות [31] משל זה אנו נניח קוטר הכדור א"ג ונעשה עליו | חצי עגלה אד"ב ונתן א"ג שליש א"ב [32] ונוציא עמוד ג"ד ונדביק ד"ב וד"ב צלע | משלש בעל ארבעה תושבות [33] ואומר שהמוגשם הנכחי השטח' אשר תושבתו | שני דמיוני המשלש שוה הצלעות העשוי אל ד"ב | וגבהו כמו תשיעית הכדור הוא כמו מוגשם | בעל ארבע תושבות [34] מופתו שב"ג הוא העמוד | היוצא מנקרת ראש המחודד אל תושבת המחודד | [35] לפי המגורר אשר תושבתו תושבת המחודד הוא | שלשה דמיוני המחודד וכבר באר אקלידס במאמר השנים עשר מספרו | [36] והמוגשם אשר תושבתו שני דמיוני תושבת המחודד וגבהו כמו שתות הד"ב | גם כן כמו המחודד לפי שהוא כמו המגורר הנזכר [37] אמנם ג"ב הוא ששה | תשיעיות א"ב ושתות א"ב אם כן תשיעית א"ב [ג"ב] והמוגשם אשר תושבותו שני | דמיוני >תושבת המחודד< משלש בעל ארבע תושבות וגבהו כמו תשיעית קוטר | >קוטר< הכדור הוא כמו המחודד וזה מה שרצינו לבאר. |

[Proposition 6]

[38] נרצה לבאר שהמוגשם הנכחי השטחי' אשר תושבתו כמו מרובע צלע בעל | שמונה תושבות ורומו כמו תשיעית קוטר הכדור הוא כמו שליש | מוגשם בעל שמונה תושבות [39] משל זה נניח קוטר הכדור א"ב ונעשה עליו חצי | עגלה אג"ב והמרכז ד' ונוציא עמוד ד"ג ונדביק ד"ג וג"ב משודש בעל שמונה | תושבות [40] ואומר שהמוגשם הנוכחי השטחי' אשר תושבתו כמו מרובע >כ<ג"ב | ורומו כמו >תושבת< תשיעית א"ב הוא כמו שליש מוגשם בעל שמונה תושבות. | [41] מופת זה כשנעשה מרובע יהיה צלעו כמו ג"ב ונוציא ממרכזו עמוד כמו | א"ד [42] ונגיע קשת העמוד בזויות המרובע [43] ונעמוד על המרובע מחודד תושבתו | המרובע

וגבהו כמו חצי קוטר הכדור והוא כמו חצי המוגשם בעל שמונה | תושבות [44] והמוגשם הנכחי השטחי' אשר תושבתו זאת המרובע וגבהו כמו | א"ד שלושה דמיוני המחודד הנזכר [45] לפי שהמחודד בכל אשר תושבתו חצי המרובע | הנזכר כשחולק על קטרו וגבהו כמו א"ג [א"ד] [46] והמוגשם הנזכר חצי דמיון המוגשם | בעל שמונה תושבות [47] והמוגשם הנכחי השטחים אשר תושבתו המרובע הנזכר | וגבהו כמו שליש א"ד הוא שוה לחצי | מוגשם בעל שמונה תושבות [48] והמוגשם | הנכחי השטחי' אשר תושבתו המרובע | וגבהו שני שליש א"ד אשר הוא שליש | א"ב הוא כמו מוגשם בעל שמונה תושבות | >והמוגשם הנכחי השטחי אשר תושבותו | המרובע הנזכר וגבהו כמו שליש א"ד הוא שוה לחצי מוגשם בעל שמונה | תושבות והמוגשם הנכחי השטחי אשר תושבותו המרובע וגבהו שני | שליש א"ד אשר הוא שליש א"ב הוא כמו מוגשם בעל שמונה תושבות< [49] והמוגשם | אם כן אשר תושבותו המרובע וגבהו שליש א"ד [שליש א"ב] וזה תשיעית א"ב [כמו שליש] | המוגשם בעל שמונה תושבות וזה מה שרצינו לבאר. |

[Proposition 7]

[50] נרצה לבאר שיחס מוגשם בעל ארבע תושבות אל מוגשם בעל שמונה | תושבות הנקוים בכדור אחד כיחס צלע כל משולש שוה הצלעות | אל שלושה דמיוני עמדו וזה כמו יחס שליש צלע מצלע [משולש] הוא שוה הצלעות מעמדו | (183b) [51] משל זה אם נתן א"ב צלע משולש בעל ארבע תושבות [52] ויהיה א"ג עמוד עליו ויהיה | [א"ג] עמוד משלש בעל [ארבע] תושבות והוא גם כן כמו צלע משלש בעל שמונה תושבות | [53] ונשלים שטח ב"ג הנכחי הצלעות ויהיה א"ד כמו א"ג [54] ואומר שיחס מוגשם בעל | ארבעת תושבות אל בעל מוגשם שמונה תושבות כיחס א"ב משלשה דמיוני | א"ד [55] מופת זה נוציא ד"ה נכחי לא"ג וא"ה מרובע צתע בעל שמונה תושבות | [56] ושטח ב"ג כמו [שני דמיוני] משולש בעל ארבע תושבות [57] ויחס המוגשם הנכחי השטחים אשר | תושבתו א"ה [ב"ג] וגבהו תשיעית קוטר [הכדור] והוא כמו מוגשם בעל ארבע תושבות אל | המוגשם אשר תושבותו א"ה וגבהו תשיעית קוטר הכדור והוא כמו שליש מוגשם | בעל שמונה תושבות כיחס שטח ב"ג אל מרובע א"ה | [58] ואם כן יחס מוגשם בעל ארבע תושבות אל שליש | מוגשם בעל שמונה תושבות כיחס שטח ב"ג ממרובע | א"ה [59] ואמנם יחס שטח ב"ג ממרובע א"ה כיחס א"ב | מא"ד [60] אם כן יחס מוגשם בעל ארבע תושבות אל | שליש מוגשם בעל שמונה תושבות כיחס א"ב אשר | הוא צלע משולש השוות הצלעות אל א"ד אשר הוא עמוד המשלש השוה הצלעות | [61] אם כן יחס מוגשם בעל ארבע תושבות למוגשם בעל שמונה תושבות כלו | כיחס צלע משלש השוה הצלעות אל שלשה דמיוני עמודה. וזה מה שרצינו לבאר. |

[Proposition 8]

[62] נרצה לבאר שמרובע המעקב ומשלש בעל שמונה תושבות | הנקוים בכדור אחד תקיף >ב"ו ט"ה ואומר שט"ה כמו חייב מופתו | אנו< בהם עגלה אחת [63] משל זה אנו נתן מרובע המעקב א"ב וחצי קוטר העגלה | אשר תקיף בו ח"ב ומשלש בעל שמונה תושבות אד"ה וחצי קוטר העגלה אשר | תקיף בו ט"ה [64] ואומר שט"ה כמו ח"ב [65] מופתו אנו נתן קוטר הכדור קו ז' | ומרובע ז' שלשה דמיוני מרובע כ"ב [66] ומרובע | כ"ב כמו כפל מרובע ח"ב ומרובע ז' אם כן | ששה דמיון מרובע [ח"ב [67] ומרובע] א"ה שלושה דמיוני מרובע | ט"ה [68] ומרובע ז' ששה דמיוני ט"ה [69] וכבר היה גם | כן מרובע ז' ששה דמיוני מרובע ח"ב ומרובע | ח"ב אם כן כמו מרובע ט"ה ו ט"ה כמו ח"ב. זה | (184a) זה מה שרצינו לבאר. [70] והתבאר במה שתארנו שם שהעמוד היוצא ממרכז הכדור | אל ומשלש בעל שמונה התושבות לפי שהמרובע והמשלש כלו בעגלה אחת. |

[Proposition 9]

[71] נרצה לבאר שמרובע צלע משלש בעל שמונה תושבות אחד וחצי כמו מרובע | צלע המעוקב [72] מופתו שמרובע המעוקב ומשלש בעל שמונה תושבות | יפלו בעגלה אחת ויחס מרובע צלע המשלש אל מרובע חצי קוטר העגלה אשר תפל | עליו כיחס שלשה כל [אל] אחד ויחס מרובע חצי קוטר העגלה אל מרובע]צלע[המעוקב | לפי שהיה כמו המרובע אשר יפול באותה העגלה כיחס אחד אל שנים [73] ובשווי יחס | מרובע צלע משלש בעל שמונה תושבות אל מרובע צלע המעוקב כיחס שלשה | אל שנים וזה אחד החצי [וחצי] ומרובע צלע בעל שמונה תושבות אחד וחצי אם | כן כמו מרובע צלע המעוקב וזה מה שרצינו לבאר. |

[Proposition 10]

[74] נרצה לבאר שיחס | משלש בעל שמונה תושבות אל מרובע המעוקב כיחס חצי עמוד | כל משולש שוה הצלעות אל שני שלישי צלע[ו] וזה עם [גם] יחס עמוד המשולש אל | צלעו ושלש צלעו [75] משל זה אנו נתן משלש צלעו בעל שמונה תושבות משלש אב"ג | ונעשה על ב"ג מרובע ג"ה ונוציא זא"ד נכחי | לב"ג וב"ז כמו עמוד משלש אב"ג ונחלק | ד"ב בשני חצאים על ט' ויהיה ב"ח שני שליש | ב"ג [76] ואומר שיחס משלש אב"ג

אל מרובע | המעוקב כיחס ב"ט אל ב"ח [77] מופתו ששטח | ד"ג כמו כפל משלש א"בג ויחס מרובע ה"ג אל חצי שטח ד"ג והוא משלש אב"ג כיחס | ה"ב מב"ט [78] וכבר בארנו במרובע [שמרובע] צלע משלש בעל שמונה תושבות אחד וחצי כמו >בה"ד< | מרובע [צלע] המעוקב [79] וזה כיחס ב"ה מב"ח לפי שהוא דמיוניו [דמיונה] וחצי דמיונה ויחס משלש | אב"ג אל >משלש< מרובע ה"ג כיחס ב"ט מב"ה ויחס מרובע ה"ג אל מרובע צלע | המעוקב כיחס ה"ב מב"ח [80] ובשוה יחס משלש אב"ג אל מרובע המעוקב כיחס | ב"ט מב"ח. וזה מה שרצינו לבאר. |

[Proposition 11]

[81] נרצה לבאר שיחס שטח בעל ששה | תושבות משטח בעל שמונה תושבות הנקוים בכדור אחד | כיחס צלע כל משלש שוה הצלעות מעמודו [82] משל זה נתן מרובע המעוקב | ח' ומשלש בעל השמונה תושבות ב' ויהיה משלש שוה הצלעות עליו גד"ה ועמודו | ג"ז [83] ואומר שיחס שטח בעל ששה תושבות משטח בעל שמונה תושבות כיחס | (184b) ד"ה מג"ז [84] מופתו אנו נתן ג"ט חצי העמוד וב"ח שני שלישי הצלע [85] ומבואר ממה | שהקדמנו שיחס מרובע ח' אל משלש ב' כיחס [ד"ח] מז"ט [86] ויחס ששה דמיוני ח' | הוא שטח בעל ששה תושבות ויחס ששה דמיוני ב' הוא שלשה רבעי שטח בעל | שמונה תושבות [87] ויחס שטח בעל ששה תושבות מהשלשה רבעי שטח בעל | שמונה תושבות כיחס ד"ז [ד"ח] מז"ט [88] ויהיה ז"ב שני שלישי העמוד וז"ט שלשה רבעי | ז"ב [89] ויחס שטח בעל ששה תושבות משלשה רבעי שטח בעל השמונה תושבות | כיחס ד"ח מז"ט ויחס שלשה רבעי שטח בעל | השמונה תושבות משטח בעל שמונה תושבות | כלו כיחס ז"ט מז"ב ובשוה יחס שטח בעל | ששה תושבות משטח בעל שמונה תושבות | כיחס ד"ח מז"ב [90] וזה כמו יחס שני שלישי | הצלע משני שלישי העמוד והוא כמו יחס הצלע | מן העמוד ויחס שטח בעל ששה תושבות משטח בעל שמונה תושבות כיחס | צלע ד"ה מעמוד ג"ז. וזה מה שרצינו לבאר. |

[Proposition 12]

נרצה לבאר שיחס מוגשם בעל ששה תושבות ממוגשם בעל שמונה תושבות כיחס | צלע משלש שוה הצלעות מעמודו. מופתו שמוגשם בעל ששה תושבות הנו | כמו ששה דמיוני המחודד אשר תושבותו מרובע המעוקב ??? ומבואר שהמחודד | אשר תושבותו מרובע כמו שטח המעוקב וגבהו כמו העמוד היוצא ממרכז | הכדור אל מרובע המעוקב שהוא ששה דמיוני המחודד הנזכר לפי ששטח אלו | המחודדים גבהם אחד ויחס התושבת אל התושבת כיחס המחודד אל המחודד | ושטח המעוקב אשר הוא כמו התושבת המחודד ששה דמיוני מרובע המעוקב | אשר הוא תושבת המחודד הראשון והמחודד אם כן ששה דמיוני המחודד | וכבר בארנו שמוגשם המעוקב ששה דמיוני המחודד והמחודד אשר תושבתו | כמו שטח המעוקב ורומו כמו העמוד היוצא ממרכז הכדור אל מרובע | המעוקב שוה למוגשם המעוקב. וכמו זה יתבאר שמחודד אשר תושבתו | כמו שטח בעל שמונה תושבות ורומו כמו העמוד היוצא ממרכז הכדור אל | משלש בעל שמונה תושבות כמו מוגשם בעל השמונה התושבות והעמוד | היוצא ממרכז הכדור אל מרובע המעוקב כמו העמוד היוצא ממרכז הכדור | (185a) הכדור אל משלש בעל השמונה תושבות והמחודד אשר תושבתו כמצו שטח | המעוקב וגבהו העמוד היוצא ממרכז הכדור אל מרובע המעוקב הוא | כמו מוגשם המעוקב ואשר רומו כמו רום המחודד אשר תושבתו כמו שטח | בעל שמונה תושבות וגבהו כמו העמוד היוצא ממרכז הכדור אל משלש בעל | שמונה תושבות ויחס המחודד כמו יחס התושבת אל התושבת הנה יחס מוגשם | בעל ששה תושבות אל מוגשם בעל שמונה תושבות כיחס שטח בעל ששה | תושבות לשטח בעל שמונה תושבות וכבר | בארנו שזה כמו יחס צלע משלש שוה צלעות | מעמודו ויחס אם כן מוגשם בעל ששה | תושבות ממוגשם בעל שמונה תושבות | כיחס צלע משלש שוה הצלעות מעמודו | וזה מה שרצינו לבאר. |

[Proposition 13]

[91] נרצה לבאר שכל קוים מתחלפים וחלק כל אחד מהם על יחס בעל | אמצע ושתי קצוות הנה יחס אחד שני הקוים אל הקו השני כיחס | החלק הגדול מן הקו הראשון אל החלק הקטן [הגדול] מהקו האחר [92] משל זה שקוי | א"ב וד"ה מתחלפים יחלק כל אחד על יחס בעל אמצע שתי קצוות על ג' | ועל ז' ואומר שיחס א"ב מד"ה כיחס א"ג הגדול [מד"ז הגדול] מיחס [וכיחס] ג"ב הקטן ומז"ה הקטן | [93] מופת זה אנו נוציא ב"ז [ב"ח] כמו ג"ב וה"ט כמו ה"ז [94] ושטח א"ב בב"ג כמו [מרובע א"ג] [ו]שטח | ה"ד בה"ז כמו מרובע ח ה"ז [ד"ז] ויחס ארבעה כפלי השטח אשר הוא מא"ב בב"ג כל [אל] | מרובע א"ג כיחס ארבעה כפלי השטח אשר יהיה מהכאת ד"ה בה"ז אל | מרובע ד"ז וכאשר הרכבנו היה יחס ארבעה כפלי שטח א"ב בב"ג עם מרובע | א"ג אל המרובע א"ג כיחס ארבעה כפלי הכאת ד"ה בה"ז עם מרובע ד"ז | אל מרובע ד"ז וארבעה כפלי א"ב בב"ג עם מרובע א"ג הוא כמו מרובע א"ח | וארבעה כפלי ד"ה בה"ז עם מרובע ד"ז הוא כמו מרובע ד"ט וכבר באר זה | אקלידוס במאמר השני [95] הנה יחס מרובע ה"א [ח"א] אל מרובע ה"ג [א"ג] כיחס מרובע ד"ט | אל מרובע> א"ל< ד"ז ויחס ד"א אל א"ג כיחס ט"ד אל ד"ו [ד"ז] וכשהבדלנו יהיה יחס | ח"ג מג"ח [מג"א] כיחס ז"ט מז"ד ויחס ב"ג | מג"ה [מג"ח] והוא חציו כמו יחס ז"ה מז"ט | (185b) >ז"ה מד"ו< לפי שהוא חציו ויחס ג"ב מא"ג

כיחס ז"ה מז"ד וכשהמרנו זה יהיה יחס ג"ב | מז"ה כיחס א"ג מד"ז . ועוד שיחס ג"ב אל מג"א כיחס ז"ה מז"ד
וכשהרכבנו יהיה | יחס א"ב מא"ג כיחס ה"ד מד"ז וכאשר המרינו יהיה יחס א"ב מד"ה כיחס א"ג מד"ז | [96]
וכבר היה יחס א"ג מד"ז כיחס ג"ב מז"ה וזה מה שרצינו לבאר [97] וזאת | התמונה זכרה אבלוניוס בזה הצד
וכבר עשינו אנחנו פנים אחרים | והוא זה [98] וכן שני הקוים א"ב ד"ה ונחלק כל אחד מהם של יחס בעל אמצע
ושתי | קצות על ג'<ד'> [ו]על ז' ואומר שיחס א"ב אל ד"ה כיחס א"ג החלק הגדול אל ד"ז הגדול | [ו]כיחס ג"ב
החלק הקטן אל ז"ה החלק הקטן [99] מופת זה אנו נוציא עמוד ב"ח כמו א"ב | ועמוד ה"ט כמו ד"ה ונדביק
ה"א [ח"א] וא"ג וח"ג] וט"ד וט"ז [100] ומפני שא"ב כמו א"ב וזוית ב' נצבת | תהיה זוית א' חצי נצבת ולו(!)
הזוית ד' חצי נצבת [101] ומרובע א"ב עם מרובע ב"ג שלשה | דמיוני מרובע [א"ג] וא"ב כמו ז"ב [ח"ב] ומרובע
ח"ג כמו שני מרובעי ח"ב וב"ג ומרובע ח"ג | כמו שלשה דמיוני מרובע א"ג [102] וכמו כן נבאר | שמרובע ט"ז
שלשה דמיוני מרובע ד"ז [ו]יחס ח"ג | מג"א כיחס ט"ז מז"ד וזוית ד' כמו זוית א' הנה | זוית אח"ג <אד"ה> כמו
זוית דט"ז ומשלש אז"ג [אח"ג] | דומה למשלש דנ"ע [דט"ז] [103] וזוית גז"ב [גח"ב] כמו זוית זט"ה | ושתי
זויות ב"ה נצבות הנה משלש חכ"ג [חב"ג] דומה | למשלש טז"ה ומשלש אח"ב דומה למשלש דט"ח ומשלש
אח"ג דומה למשלש דט"ז | למשלש [ומשלש] גח"ב דומה למשלש זט"ה [104] הנה יחס א"ח אל ד"ט כיחס
<א"ג אל ז"ט וכיחס > א"ג | אל ד"ז וכיחס ג"ב אל ז"ה. וזה מה שרצינו לבאר. |

[Proposition 14]

[105] נרצה לבאר | שצלע המשושה כשנחלק על יחס בעל אמצע ושתי קצוות הגדול | צלע מוגשם [מעושר]
[106] משל זה נתן צלע משושה [ד"ז] ונחלקהו על יחס בעל אמצע | ושתי קצוות על ה' ותהיה חלוקת ה"ז הגדול
ה"ז ואומר שה"ז צלע המעושר. | [107] מופתו אם נתן ג"ב צלע משושה כמו ד"ז וא"ג צלע מעושר וא"ב כבר
נחלק | על יחס בעל אמצע ושתי קצוות והחלק הגדול ממנה ג"ב ויחס [108] א"ב מד"ז כיחס | ג"ה [ג"ב] מה"ז
כפי מה שבארנו בשאלה אשר קודם(!) [109] וזאת כשנמיר יהיה יחס א"ב | אל ב"ג כיחס ד"ז מז"ה ואמנם יחס
א"ב מב"ג | כיחס ב"ג מג"א ויחס ב"ג מג"א יחס ד"ז מז"ה | וזה מה שרצינו לבאר. |

[Proposition 15]

[110] נרצה לבאר [שמרובע] צלע המחומש ומרובע אשר | יעשה מיתר לשני מחומש העגלה כמו חמש
דמיוני מרובע | קוטר העגלה אשר תקיף במחומש [111] משל זה אנו נתן עגלת | א"ג ז"ב וצלע מחומש ג"ב
ויהיה קשת ג"ז כמו קשת ז"ב ומרכז העגלה | ד' ונדביק זד"א ונדביק א"ג וא"ג מיתר לשני מחומש העגלה ואומר
| שמרובע א"ג עם מרובע ג"ב חמשה דמיוני מרובע ד"ז [112] מופתו אנו | נדביק ג"ז וג"ז צלע מעושר ומרובע
א"ג ו[מרובע] ג"ז כמו מרובע א"ז אשר | הוא ארבע דמיוני מרובע ד"ז לפי שזוית ג' נצבת [113] ונתן מרובע
ד"ז משותף | מרובע א"ג ד"ז וז"ד [וג"ז] חמשה דמיני מרובע | ד"ז [114] אמנם מרובע ג"ז וז"ד כמו מרובע ג"ב
| לפי שג"ז צלע מעשרה(!) וד"ז צלע משותף [משושה] וג"ב | צלע מחומשה(!) [115] ומרובע א"ג וג"ב חמשה
| דמיוני מרובע ד"ה. וזה מה שרצינו | לבאר. וזאת התמונה זכרה אבלוניוס | בזאת המלאכה. |

[Proposition 16]

[116] נרצה לבאר שהמשלש בעל העשרים | תושבות ומחומש בעל השתים עשרה תושבות [תקיף בהם עגלה
אחת [117] משל זה אנו נתן] מחומש | א'ב'ג'ד'ה' ומרכז העגלה אשר תקיף בו ז' חצי קוטר העגלה ז"ד משלש
בעל | העשרים תושבות משלש חט"ב ומרכז העגלה אשר תקיף בו ל' וחצי קט[ר] | ל"ב ואומר של"ב כמו ז"ד
[118] מופת זה אנו נדבק [ב"]ה ומבואר שב"ה | צלע המעוקב [119] ונתן קוטר הכדור מנ"ש [מ"נ] וחצי קוטר
העגלה בעלת העשרים | תושבות אשר כבר באר אוקלידוס | במאמר שלש עשרה שצלע מחמשה | הנה צלע
המשלש בעל העשרים | תושבות צ"ע ויתחלק מ[״נ] על יחס | בעל אמצע ושתי קצוות על ק' | ונתן חלקה הגדול
ק"נ [וית]חלק צ"ע | גם כן על יחס בעל אמצע ושתי קצוות על ס' והחלק הגדול ה"ע [ס"ע] | (186b) וה"ע צלע
מעושר [120] ונחלק גם כן ב"ה על יחס בעל אמצע שתי קצוות | על פ' והחלק הגדול פ"ה ופ"ה כמו א"ה [121]
ויחס מרובע מ"נ אל מרובע צ"ע | כיחס מרובע ק"נ אל מרובע ס"ע ומרובע מ"נ כבר באר אקלידוס במאמר |
השלשה עשר מספרו שהוא חמשה דמיוני מרובע צ"ע ומרובע ק"נ | חמשה דמיוני מרובע ע"ס ומרובע מ"נ נ"ק
חמשה דמיוני צ"ע | וס"ע [122] אמנם מרובע צ"ע וע"ס הוא כמו מרובע ז"ט [ח"ט] לפי שצ"ע צלע | משושה
וס"ע צלע מעושר [123] ומרובע מ"נ נ"ק חמשה דמיוני מרובע ח"ט | ומרובע ח"ט שלשה דמיוני מרובע ב"ל
ומרובע מ"נ נ"ק<ג > חמשה | עשר דמיוני מרובע ל"ד [ל"ב] [124] ועוד שיחס מרובע מ"נ אל מרובע ב"ה |
כיחס מרובע ק"נ אל מרובע פ"ה ומרובע מ"נ שלשה דמיוני מרובע | ב"ה לפי שהוא כמו צלע המעוקב ומרובע
ק"נ שלשה דמיוני מרובע | פ"ה כפי מה שבאר אקלידס במאמר השלשה עשר מספרו ומרובע | מ"נ ונ"ק שלשה
דמיוני מרובע ב"ה וה"פ [125] אמנם מרובע ב"ה וה"פ חמשה | דמיוני מרובע ז"ד ומרובע מ"נ ונ"ק חמשה עשר

דמיוני מרובע ז"ד | [126] וכבר התבאר שהמחובר מ"נ ונ"ק חמשה עשר דמיוני מרובע ל"ד [ל"ב] | ומרובע ל"ב [127] אם כן כמו [מרובע] ז"ד ול"ב כמו ז"ד. וזה מה שרצינו לבאר. | וזאת התמונה זכרה אבלוניוס כפי צד ל"ב זאת המלאכה אלא שזאת אשר | זכרנו יותר קרובה להבין למתלמדים. |

[Proposition 17]

נרצה לבאר | שכל קו יחלק על יחס בעל אמצע ושתי קצוות הנה | יחס הקו המחזיק על הקו ועל חלוקת הגדול אל הקו המחזיק על | הקו ועל חלוקת הקטן כיחס צלע המעוקב אל צלע המשלש בעל | בעל עשרים תושבות הנקוים בכדור אחד. המשל זה אנו נניח קו א"ב | צלע המשושה העגלה אשר תקיף במחומש בעל שתים עשרה תושבות | ובמשלש בעל עשר תושבות הנקוים בכדור אחד ונחלקהו על יחס | בעל אמצע ושתי קצוות על ה' ונוציא מ"ה עמוד על ד"ה והוא ה"ג ויהיה | ה"ג כמו ד"ב ונדביק ג"ב מפני שג"ה צלע משושה וה"ב צלע מעושר | יהיה ב צלע מחומש העגלה ומחומש העגלה הוא מחומש בעל | שתים עשרה תושבות וג"ב צלע מחומש בעל שתים עשרה תושבות | ונדביק ג"ד ויהיה ד"ג צלע המעוקב ויהיה צלע משלש בעל | (187a) עשרים תושבות ומבואר שג' יחזיק על ד"ב וב"ה וג"ד יחזיק על ד"ה | וד"ב ואומר שיחס א"ב אל ז' כיחס ג"ב אל ג"ד. מופת זה שמרובע ג"ד | שלשה דמיוני מרובע ה"ב ומרובע ז' שלשה דמיוני מרובע ה"ב וזה לפי | שז' צלע המשלש השוה הצלעות הנופל בעגלה וד' חצי קוטר העגלה | אשר תקיף בו ואם כן יחס ה"ב מד"ג | כיחס ד"ג מא' ועוד שא"ב צלע המעוקב | וב"ג צלע מחומש בעל שתים עשרה | תושבות וא"ב כבר נחלק על יחס בעל | אמצע ושתי קצוות יתחלק הגדול ב"ג | וד"ב כבר מחולק על יחס בעל אמצע ושתי קצוות על ה' ויחס א"ב | מב"ד כיחס ג"ב מב"ה וכבר התבאר שיחס ד"ב מז' כיחס ב"ה מג"ד | ובשוה יחס א"ב מז' כיחס ג"ב מג"ד. וזה מה שרצינו לבאר. | וזה התמונה זכרה אבלוניוס כיוצא בזאת המלאכה אלא שזה | יותר קרוב להבין למתלמדים. |

[Proposition 18]

[128] נרצה לבאר שיחס שטח המעוקב | משטח בעל העשרים תושבות כיחס מרובע צלע המחומש | אל צלע [שלושה] דמיוני ושליש כמו המשלש השוה הצלעות אשר צלעו | יחזיק <ק> על שלשה דמי[ו]ני [מרובע] צלע מעושר העגלה אשר תקיף במחומש | [129] משל זה אנו נתן א"ב צלע משושה ונחלקהו על יחס בעל אמצע ושתי | קצוות על ג' [130] ויהיה ד' מחזיק א"ב וא"ג וד' צלע מחומש ויהיה ה' מחזיק | על א"ב וב"ג [131] ואם כן ה' יחזיק על שלשה דמיוני מרובע א"ג אשר הוא | צלע מעושר [132] ואומר שיחס שטח המועקב אל שטח בעל העשרים | תושבות כיחס מרובע ד' של שלשה דמיונה(!) ושליש כמו משלש השוה | הצלעות העשוי על ה' [133] מופתו כבר התבאר בתמונה שלפני זאת | שיחס צלע המעוקב אל צלע משלש בעל עשרים תושבות כיחס | ד' אל ה' ויחס מרובע [צלע] המעוקב אל מרובע צלע המשלש בעל העשרים | תושבות כיחס <מרובע ד' בעל העשרים תושבות כיחס> מרובע ד' אל | מרובע ה' [134] ובחלוף יחס מרובע המעוקב אל מרובע ד' כיחס <אל> | מרובע צלע משלש בעל העשרים תושבות אל מרובע ה' [135] ויחס | (187b) מרובע כל קו אל שני דמיוני משלשה [...] ויחס מרובע ד' אל שני דמיוני | משלשו כיחס מרובע ה' אל שני דמיוני משלשו | [136] ואם כן יחס מרובע [צלע] המעוקב אל מרובע ד' | כיחס מרובע צלע משלש בעל העשרים תושבות אל מרובע ה' ויחס | מרובע ד' [המעוקב] אל שני דמיוני משלש ד' אשר הוא שני דמיוני משלש בעל | העשרים תושבות כיחס מרובע א"ז [א"ד] אשר הוא צלע משלש בעל העשרים | תושבות אל שני דמיוני משלש ה' [137] ויחס שני [ששה] דמיוני מרובע המעוקב | אשר הוא כמו שטח המעוקב אל שנים עשר דמיוני משלש בעל העשרים | תושבות אשר הוא שלשה חומשי שטח בעל עשרים תושבות כיחס מרובע | המעוקב אל שני דמיוני משלש בעל העשרים תושבות אשר הוא כיחס | מרובע ד' אל שני דמיוני משלש ה' [138] ויחס שטח המעוקב של שלשה | חומשי שטח בעל העשרים תושבות כיחס מרובע ד' אל שני דמיוני | משלש ה' ויחס שלשה חומשי שטח בעל העשרים תושבות אל שטח | בעל העשרים תושבות כלו כיחס שני דמיוני משלש ה' אל שלשה דמיוני | ושלישו כמו משלש ה' וכל שטח יחס המעוקב אל שטח בעל העשרים | תושבות כיחס מרובע ד' אל שלשה דמיוני השלשה כמו המשלש השוה | הצלעות העשוי על ה' וזה מה שרצינו לבאר [139] והתבאר ממה שתארנו | שיחס שלשה חומשי שטח בעל העשרים תושבות אל שטח המעוקב | כיחס שני דמיוני משלש ה' אל מרובע ד'. וזה מה שרצינו לבאר. |

[Proposition 19]

[140] נרצה לבאר שיחס שטח בעל עשרים תושבות אל שטח בעל | השמונה תושבות כיחס חמשה דמיוני מרובע צלע מעושר | עגלה אל מרובע צלע מחומש [141] ויהיה קו א' צלע חמשה [מחמש] וקו ב' צלע | מעשרה (!) ואומ' שיחס שטח בעל העשרים תושבות אל שטח בעל | השמונה תושבות כיחס חמשה דמיונ' מבובע [מרובע] ב' למרובע א' | [142] מופתו אנו נתן ג' יחזיק על שלשה דמיוני ב' וכבר בארנו שיחס | שלשה חמשי שטח בעל העשרים תושבות אל שטח המעוקב כיחס | שני דמיוני משלש ג' אל מרובע א' [143] וכבר דברנו שיחס שטח המעוקב | אל שטח בעל השמונה (?)תושבות כיחס מרובע א' > וזה צלע היה אל < | (199a) אל שטח שני רשמי משלשו לפי שאנו כבר בארנו במה שקדם שיחס שטח | המעוקב אל שטח בעל השמונה [תושבות]

כיחס צלע כל משלש שוה הצלעות | אל עמודו והוא כיחס מרובע [צלעו] אל שני דמיוני משלשו >ונתוהו(?) מרובע | ה' >(?) [144] ויחס שלשה מחומשי שטח בעל עשרים תושבות אל שטח המעוקב | כיחס שני דמיוני משלש א' [ג'] אל מרובע א' ויחס שטח המעוקב אל שטח | בעל השמונה תושבות כיחס מרובע א' אל שני דמיוני משלש[ו] ובשוה | יחס שלשה חומשי שטח בעל | העשרים תושבות אל שטח בעל | השמונה תושבות כיחס שני דמיוני משלש ג' אל שני דמיוני משלש | א' וזה כיחס משלש ג' למשלש א' [145] ויחס משלש ג' למשלש א' כיחס >משלש | ג'< [שלושה דמיוני מרובע ב'] למרובע א'(?) ויחס שלשה חומשי שטח בעל עשרים תושבות אל שטח | בעל השמונה תושבות כיחס מרובע ב' אשר הוא כמו שלשה דמיוני | מרובע ב' אל מרובע א' ויחס שלשה חומשי שטח בעל העשרים תושבות | אל שטח בעל העשרים תושבות כלו כיחס שלושה דמיוני מרובע ב' | אל חמשה דמיוני >אל< מרובע ב' ובשוה יחס [שטח] בעל העשרים תושבות אל | שטח השמונה תושבות כיחס חמשה דמיוני מרובע ב' אל | מרובע א'. זה מה שרצינו לבאר. |

[Proposition 20]

[146] נרצה לבאר שיחס מיתר [מותר] שטח | העשרים תושבות על שטח בעל השמונה תושבות [אל שטח בעל השמונה תושבות] כיחס | החלק הקטן מכל קו יחלק של יחס בעל אמצע ושתי קצוות אל הקו כלו | ובעבור זה יהיה יחס שטח בעל העשרים תושבות משטח בעל השמונה | תושבות כיחס החלק הקטן מכל קו יחלק על יחס בעל אמצע ושתי קצוות | עם הקו כלו אל הקו כלו וזה כיחס שני דמיוני צלע כל מעושר עם צלע | המשושה מקובצים אל צלע המשושה והמעושר מקובצים [147] משל זה | אנו נתן עגלה (?) עליה הג"ב ומרכזה א' וצלע המחומש הנופל בהג"ב ונחלק | קשת ג"ב בשני חציים על ד' ונדביק א"ד >ד"א< ונוציאהו אל ה' ונדביק ה"ג וה"ב | ונחלק ג' על יחס בעל אמצע ושתי קצוות [148] ואומר שיחס מיתר [מותר] שטח | בעל העשרים תושבות על שטח בעל השמונה תושבות מותר [אל שטח בעל שמונה תושבות] כיחס | ה"ח מה"ג [149] מופתו אנו נדביק א"ג וא"ב וב"ד ומשלש הנ"ג [הג"ב] דומה למשלש | (199b) אב"ד מפני שזוית דא"ב דומה לזוית ה' >וזוית בא"ד כמו זוית ה' ומשלש אד"ב | דומה למשלש הג"ב< ויחס חמש דמיוני מרובע ד"ב אשר הוא צלע מעושר | אל מרובע ג"ב אשר הוא צלע מחומש כיחס חמשה דמיוני מרובע א"ב ממרובע | ה"ג [150] וחמשה דמיוני מרובע ג"א הוא כמו שני מרובע ה"ג וג"ז [וג"ב] לפי שה"ג | עשה מיתר לשני חומשי העגלה וג"ז יעשה מיתר לחמישיתה [151] ויחס חמשה | דמיוני מרובע ד"ב אל מרובע ג"ב כבר בארנו שהוא כיחס שטח בעל | העשרים תושבות משטח בעל השמונה | תושבות ויחס שני מרובעי ה"ג וג"ד [וג"ב] אל | מרובע ה"ג כיחס שטח בעל העשרים | תושבות אל שטח בעל השמונה תושבות | כשהבדלנו היה יחס מיתר [מותר] שטח בעל | העשרים תושבות על שטח בעל השמונה | תושבות אל שטח בעל השמונה תושבות כיחס מרובע ח"ג למרובע ג"ב | [152] ומפני שיחס ה"ז מז"ג [ה"ח מח"ג] כיחס ח"ג מג"ה יהיה יחס מרובע ז"ג [ח"ג] אל מרובע | ג"ה כיחס ח"ג וה"ג [ח"ה מה"ג] ויחס אם כן מיתר [מותר] שטח בעל העשרים תושבות על | שטח בעל השמונה תושבות [אל שטח בעל שמונה תושבות] כיחס ה"ח החלק הקטן [אל] הקו כלו וכשהרכבנו | יחס שטח בעל העשרים תושבות משטח בעל השמונה תושבות כיחס החלק | הקטן מן הקו כלו יחד [עם הקו כלו] אל הקו כלו ולפי שצלע מעושר כשידביק בצלע | משושה הנה הקו כלו יחלק על יחס בעל אמצע ושתי קצוות והחלק הקטן | צלע מעושר יהיה יחס שטח בעל העשרים תושבות משטח בעל השמונה | תושבות כיחס צלע שני דמיוני מעושר עם צלע משושה יחד מצלע | משושה [וצלע ה]מעושר יחד וזה מה שרצינו לבאר. |

[Proposition 21]

[153] נרצה לבאר | שהמוגשם הנכחי השטחים אשר תושבתו כמו מחומש | בעל העשרים תושבות וגבהו כמו שני דמיוני קוטר הכדור המקיף המחומש [המוגשם] | בעל העשרים תושבות כשהוציא [כמו המוגשם בעל העשרים תושבות [154] מופתו כשנוציא] ממרכזו עמוד על שטח [המחומש] בשני | הצדדים יחד וחתך בשני העמודים הצד האחד כמו [חצי] צלע המחומש משושה העגלה | אשר תקיף במחומש ובצד האחר(?) כמו צלע המעושר הנופל בה [155] והגעת | (189a) <והגעת> כל אחת משתי הנקודות בזוית המחומש [צ"ל בעל העשרים תושבות] יחדש על המחומש שני | מחודדים ראש אחד מהם נקודת חתוך צלע המשושה וראש השני נקודת חתוך | צלע המעושר [156] ושני אלו המחודדים שוים למוגשם הנבחר [צ"ל הנכחי] השטחים אשר | תושבתו כמו המחומש וגבהו כמו שליש [צ"ל ששית] צלע המשושה ושליש צלע המעושר | לפי שצלע המעושר ו[חצי צלע] המשושה כמו גובה שני המחודדים וצלע [חצי] המשושה | והמעושר כמו חצי קוטר הכדור וזה יבואר בתמונה בעל העשרים תושבות | [157] והמוגשם הנכחי השטחים אשר תושבתו כמו המחומש וגבהו שליש חצי | קוטר הכדור וזה כמו שתות קוטר הכדור שוה לשני אלו המחודדים ושני אלו | המחודדים כמו החמשה מחודדים אשר תושבתם החמשה משלשים אשר עמדו | מצלע נקדת חתוך צלע המעושר בזויות המחומש [בעל העשרים תושבות] וראשה מרכז הכדור אשר | הוא נקודת צלע [חצי] המשושה [158] ובמוגשם בעל העשרים תושבות עשרים מחודדים | כמו אחד מאלו החמשה שהמוגשם בעל העשרים תושבות ארבע דמיוני | אלו החמשה המחודדים ואלו החמשה

המחודדים כמו שני המחודדים אשר | ראש אחד מהם נקדת חתוך צלע המעושר <והראש השני נקודת חתוך צלע המעושר> והראש השני נקדת חתוך צלע המשושה ותושבתו המחומש ושני | אלו המחודדים כמו מוגשם הנכחי | השטחים אשר תושבתו כמו המחומש | וגבהם כמו שתות קוטר הכדור [159] ומוגשם בעל העשרים תושבות ארבעה דמיוני | המוגשם אשר תושבתו כמו המחומש וגבהו כמו ארבעה שתות קוטר הכדור | והמוגשם אשר תושבתו כמו המחומש וגבהו כמו ארבעה שתויות קוטר | הכדור הוא ארבעה דמיוני המוגשם אשר תושבתו כמו המחומש וגבהו כמו | שתות קוטר הכדור ומוגשם בעל העשרים תושבות אם כן כמו המוגשם | אשר תושבתו כמו המחומש הנופל בעגלה בעלת העשרים תושבות אשר | צלע המחומשה כמו המשלש בעל העשרים תושבות וגבהו כמו שני [שליש] קוטר הכדור | וזה מה שרצינו לבאר. |

[Proposition 22]

[160] נרצה לבאר שהעמוד היוצא ממרכז | העגלה אל צלע המחומש הנופל בה הוא כמו חצי צלע המשושה | והמעושר הנופלים בה יחד [161] משל זה שעגלה אב"ג בה צלע מחומש עליו | ב"ג מרכז העגלה ד' ונוציא מד' אל ג"ב עמוד עליו ד"ח ונוציאהו על יושר אל | נקודת ה' ממקיף העגלה ונדביק ב"ה וב"ה צלע מעושר. ואומר שד"ח כמו חצי | (189b) א"ה [ד"ה] וה"ב יחד [162] מופתו אנו נוציא הד"א ונדביק קו ב"ד ונתן [ח"ז] כמו ה"א [ה"ח] ונדביק ב"ז | [163] וז"ח <וה"ב> כמו ה"ח וח"ב [משותף] ושתי זויות [בח"ז ובח"ה] נצבות וב"ז כמו ה"ב וזוית בה"ז כמו זוית בז"ה | [164] וקשת א"ב ארבעה דמיוני קשת ב"ה וזוית אד"ב | ארבעה דמיוני זוית בד"ה וזוית אד"ב כמו [שני דמיוני] זוית | בה"ד וזוית בה"ז אשר היא כמו זוית בז"ה <דומה | לזוית בד"ה וזוית בז"ה> כמו שתי זויות בד"ז | ודב"ז <ושתי זויות בד"א דב"ז כמו זוית בד"ז> [165] וזוית | ב[ד]"ח כמו זוית זב"ד וקו א"ז [ד"ז] כמו קו ז"ב וז"ב כמו | ב"ה שהוא צלע מעושר. וד"ז כמו [כן] צלע המעושר [166] וז"ה כמו ח"ה <וד"ז חצי ד"ז> וז"ח חצי | ז"ה [167] וד"ח אם כן חצי ד"ז וה"ב וד"ז [וד"ה] > וזה < הוא צלע משושה וה"ב צלע מעושר וד"ז וד"ח] | אם כן חצי צלע המשושה וצלע המעושר וזה מה שרצינו לבאר. |

[Proposition 23]

[168] נרצה לבאר שיחס העמוד היוצא ממרכז הכדור אל משלש בעל העשרים | תושבות אל קוטר הכדור אשר יהיה בבעל העשרים התושבות כיחס חצי צלע | המשושה והמעושר הנופלים בעגלה אל שני דמיוני עמוד המשלש השוה הצלעות | העשוי על צלע המחומש העשוי באותה העגלה [169] משל זה אנו נתן מחומש בעל | עשרים תושבות מחמש א"ב גד"ה ומרכזו ז' ונעשה על ג"ד משלש שוה הצלעות | עליו גד"ל ונדביק ז"ל ונחתוך גד"ל [צ"ל ויחתוך ג"ד] על ח' ומבואר כי זח"ל עמוד על ג"ד ויהיה | קו ט' כמו העמוד היוצא ממרכז הכדור אל משלש בבעל העשרים התושבות | [ויהיה קו כ'] קוטר הכדור [170] ואומר שיחס ט' אל ב' [כ'] כיחס ז"ח אשר כבר התבאר שהוא חצי צלע | המשושה והמעושר אל שני דמיוני ח"ל אשר הוא | כמו עמוד המשלש [171] מופת זה כי ג"ד צלע | המחומש בעל עשרים תושבות ומשלש גל"ד | שוה הצלעות ומבואר שהוא משלש בעל עשרים | תושבות ונגיע ז"ד וז"ג [172] ומבואר שהמוגשם הנכחי השטחים אשר תושבתו | כמו שטח בעל העשרים תושבות וגבהו כמו שליש ט' אשר הוא עמוד היוצא | ממרכז הכדור אל משלש בעל העשרים תושבות כמו מוגשם בעל העשרים | תושבות [173] לפי שזה המוגשם בעל עשרים תושבות כמו המחודד אשר תושבתו כמו | משלש בעל העשרים תושבות וגבהו כמו עמוד ט' ומוגשם בעל העשרים תושבות | (190a) תושבות עשרים דמיונים כמו זה המחודד ומוגשם בעל העשרים תושבות כמו | המוגשם אשר תושבתו כמו שטח בעל העשרים תושבות וגבהו כמו שליש ט' [174] וכבר | בארנו שהמוגשם הנכחי השטחים אשר תושבתו כמו מחמש אב"ג ד"ה וגבהו כמו | שני שלישי ב' [כ'] הוא כמו מוגשם בעל עשרים תושבות [175] והמוגשם אשר תושבתו | כמו שטח בעל העשרים תושבות וזה כמו עשרים כמו משלש גל"ד וגבהו כמו | שליש ט' שוה למוגשם אשר תושבתו כמו מחומש אב"ג ד"ה וגבהו שני שליש ק' [כ']. | [176] והמוגשם הנכחי השטחים השוים תושבותיהם מספיקים לגבהם [177] ויחס מחומש | אב"ג ד"ה אל עשרים כמו משלש גל"ד כיחס משלש גז"ד אשר הוא חומש מחומש אב"ג | ד"ה אל עשרים [ארבעה] כמו משלש גל"ד <וזה ארבעה דמיוני משלש גז"ד> ויחס משלש גז"ד [אל] | ארבעה דמיוני משלש גל"ד כיחס > של[י]ש [ט'] משני שליש ב' [כ'] ויחס אם כן משלש | גז"ד אל משלש גל"ד כיחס שליש ט' אל שתות ב' [כ'] [178] ויחס משלש גז"ד ממשלש גל"ד | כיחס ז"ח מן ח"ל ויחס ז"ח מן ח"ל כיחס ט' מחצי ב' [כ'] אם כן יחס ז"ח משני | דמיוני ח"ל כיחס ט' אל ד' [כ'] [179] וכבר בארנו שקו ז"ח כמו חצי צלע משושה וחצי צלע | מעושר הנופלים בעגלה מחומש אב"ג ד"ה [180] וכבר התבאר שיחס העמוד | היוצא ממרכז הכדור אל משלש בעל העשרים תושבות אל קוטר הכדור אשר | יקיף בבעל העשרים תושבות כיחס חצי צלע המשושה המתעשר (!) [והמעושר] הנופלים | בעגלה אל שני דמיוני עמוד המשלש השוה הצלעות העשוי

על צלע המחומש | העשוי באותה עגלה וזה מה שרצינו לבאר. |

[Proposition 24]

[181] נרצה לבאר שיחס [חצי] קוטר | הכדור אל העמוד היוצא ממרכז הכדור של משלש בעל השמונה | תושבות כיחס שני דמיוני עמוד כל משלש שוה הצלעות אל צלעו [182] מפני | שהעמוד היוצא ממרכז הכדור של מרובע המעוקב הוא כמו חצי צלע המעוקב | וכבר התבאר ממה שזכר אקלידס במאמר השלשה עשר שמרובע קוטר | הכדור שלשה דמיוני מרובע צלע המעוקב ומרובע חצי קוטר הכדור שלשה | דמיוני מרובע חצי צלע המעוקב אשר הוא העמוד היוצא ממרכז הכדור אל | מרובע המעוקב [183] ומרובע עמוד כל משלש שוה הצלעות ממרובע [שלשה דמיוני מרובע] חצי צלע[ו] | <שלשה דמיוני> [184] אם כן יחס מרובע חצי [ה]קוטר של <ה>מרובע [העמוד] היוצא ממרכז | הכדור אל מרובע המעוקב כיחס מרובע עמוד משלש שוה הצלעות ממרובע | חצי צלעו ויחס חצי קוטר הכדור אל העמוד היוצא ממרכז הכדור אל מרובע | (190b) מעוקב כיחס עמוד המשלש שוה הצלעות מחצי צלעו [185] ויחס אם כן חצי קוטר | כדור <כלו> אל העמוד היוצא ממרכז הכדור אל מרובע המעוקב והוא כמו | עמוד היוצא ממרכז הכדור אל משלש בעל השמונה תושבות כיחס שני דמיוני | עמוד משלש שוה הצלעות מחצי צלעו. וזה מה שרצינו לבאר. |

[Proposition 25]

[186] נרצה לבאר [שיחס] <ש>העמוד היוצא ממרכז הכדור של משלש בעל העשרים | תושבות אל העמוד היוצא ממרכז הכדור אל משלש בעל | השמונה תושבות הנקוים בכדור אחד כיחס צלע משושה ומעושר הנופלים | בעגלה אחד אל צלע מחומש אותה העגלה [187] מופת זה אנו כבר בארנו | כיחס [שיחס] העמוד היוצא ממרכז הכדור אל משלש בעל העשרים תושבות אל | קוטר הכדור כיחס חצי צלע המשושה והמעושר אל שני דמיוני עמוד | משלש שוה הצלעות העשוי אל [על] צלע המשלש [המחמש] העשוי באותה העגלה אשר | יפולו בה [188] וכבר בארנו גם כן שיחס קוטר הכדור מן העמוד היוצא | ממרכז הכדור אל משלש בעל השמונה תושבות כיחס שני דמיוני עמוד | כל משלש שוה הצלעות מחצי צלעו [189] ויהיה בשווי [ה]יחס העמוד היוצא | ממרכז הכדור אל משלש בעל העשרים תושבות אל העמוד היוצא ממרכז | הכדור אל משלש בעל השמונה תושבות כיחס חצי המשושה והמעושר | מחצי צלע המחומש [190] ויחס החצי מהחצי כמו הכל מהכל ויחס עמוד היוצא | ממרכז הכדור אל משלש בעל העשרים תושבות [אל העמוד היוצא ממרכז הכדור אל משלש בעל השמונה תושבות] כיחס צלע המשושה והמעושר | אל צלע המחומש הנופלים בעגלה אחת. וזה מה שרצינו לבאר. |

[Proposition 26]

נרצה לבאר שיחס מוגשם בעל העשרים תושבות אל מוגשם בעל | השמונה תושבות כיחס צלע כל מעושר נופל בעגלה אל צלע | המחומש הנופל בה אל צלע המחומש וזה כי יחס החלק הקטן מכל קו | יחלק על בעל שתי קשתות עם הקו המחזיק על הקו כלו ועל החלק | הגדול אל הקו המחזיק של הקו כלו ועל החלק הקטן. משל זה אנחנו נתן | מוגשם הנכחי השטחים תושבות כמו שטח בעל העשרים הנשבות וגבהו | שליש העמוד היוצא ממרכז הכדור אל משלש בעל העשרים תושבות ומבואר | שהוא כמו מוגשם בעל העשרים תושבות ונתן מוגשם הנכחי השטחים | תושבות כמו שטח בעל השמונה תושבות וגבהו שליש העמוד היוצא | (191a) היוצא ממרכז הכדור אל משלש הכדור אל משלש בעל השמונה תושבות ומבואר כי מוגשם ב' | כמו מוגשם בעל השמונה תושבות ויהיה ד' צלע מעושר עגלה צלע מחומשה | ואומר שיחס מוגשם א' אל מוגשם ב' כיחס ד' וב' יחד ח' מופת זה אנו | נתן מוגשם ג' נכחי השטחים ותהיה מושבות כמו יתרון שטח מוגשם בעל | תושבות על ב' שטח בעל השמונה תושבות וגבהו שליש העמוד היוצא ממרכז | הכדור של משלש בעל העשרים תושבות ומבואר שמוגשם ג' כמו יתרון | מוגשם בעל העשרים תושבות על מוגשם בעל השמונה תושבות וכמו כן | בכל שני מוגשמים נכחי השטחים הנה יחס המוגשם מחובר מיחס | התושבת אל תושבת מיחס הגובה של הגובה ויחס מוגשם א' אל מוגשם ב' | מחובר מיחס תושבת ג' אשר הוא מותר שטח בעל העשרים תושבות על | שטח בעל השמונה תושבות ומיחס גובה מוגשם ג' שהוא שליש העמוד | היוצא ממרכז הכדור של משלש בעל העשרים תושבות אל גובה ב' שהוא | שליש העמוד היוצא ממרכז הכדור אל משלש בעל השמונה תושבות | ויחס צלע משושה ומעושר ועגלה אשר ד' מעושרה וה' מחומשה וכבר | בארנו שיחס מוגשם שטח בעל העשרים | תושבות אל שטח מוגשם בעל השמונה תושבות | כיחס צלע מעושר מצלע משושה ומעושר | וכן יחס ד' מן ה' ובארנו גם כן שיחס העמוד | היוצא ממרכז הכדור אל משלש בעל העשרים | תושבות של העמוד היוצא ממרכז הכדור | אל משלש בעל השמונה תושבות כיחס צלע משושה ומעושר של צלע | המחומש הנופל באותה העגלה וזה כיחס ה' אל ח' אם כן יחס מוגשם ב' | מחובר מיחס ד' אל ה' ומיחס ה' אל ח' ואמנם היחס המחובר מז' אל ה' מה' | אל ח' הוא יחס ד' אל ח' ויהיה מוגשם ג' אל מוגשם ה' כיחס ד' אל ח' וכבר | הרכבנו היה יחס ג' וב' אל ב' כיחס ד' וח' אל ח' וג' וב' הוא כמו א' ויחס א' אל | ב' כיחס ד' וח ' אל ח' וזה מה שרצינו לבאר . ומפני צלע המשושה כשנחלק | על יחס בעל אמצעי שתי קצוות ויהיה החלק הגדול צלע מעושר ויהיה | הקו המחזיק של צלע המשושה והמעושר הוא צלע המחומש הנה יחס מוגשם | (191b) בעל העשרים תושבות למוגשם בעל שמונה תושבות כיחס החלק הגדול מכל קו | חלק של יחס בעל ארבעה (!) אמצע שתי

קצוות עם הקו המחזיק על קו | כל ועל החלק הגדול אל הקו המחזיק על הקו כלו ועל החלק הקטן הגדול . וזה | מה שרצינו לבאר. | [191] וכבר נשלם יחס התמונות הארבעה | קצתם מקצתם ונשאר היחס אשר זכרו אבלוניוס | בין תמונות בעלת שתים עשרה תושבות ובעל העשרים תושבות וזה | תחלת באור מאמרו. |

[Proposition 27]

נרצה לבאר שהשטח אשר מהכאת | העמוד היוצא ממרכז העגלה אשר תקיף במחומש | בעל שתים עשרה תושבות. מופת זה אנו נדביק זגוז"ד ושטח בעל | מחומש אב"ג ד"ה ומחומש אב"ג ד"ה חמשה | דמיוני משלש זג"ד ושטח בעל שתים | עשרה תושבות ממשלש זג"ד והשטח | אשר מהכאת ג"ד בט"ז הוא כפל זג"ד | השטח בעל שתים עשרה תושבות שלשים | כמו השטח אשר יהיה מהכאת ג"ד בט"ז וזה מה שרצינו לבאר. |

[Proposition 28]

[192] נרצה לבאר שהשטח אשר מהכאת העמוד היוצא ממרכז העגלה אשר | תקיף במשלש בעל העשרים תושבות <אל צלע המשלש> בצלע | המשלש חלק משלשים בשטח [בעל] העשרים [תושבות] [193] משל זה אב"ג משלש בעל | עשרים תושבות ומרכז העגלה אשר תקיף | בו ד' ד"ה בב"ג וד"ה עמוד על ב"ג ואומר שהשטח אשר | [מהכאת] ד"ה [ב]ב"ג הוא חלק משלשים משטח בעל | העשרים תושבות [194] [מופתו שטח בעל א] עשרים [תושבות] הוא עשרים כמו משלש | אב"ג ומשלש אב"ג שלשים [צ"ל שלשה] דמיוני משלש | דב"ג ושטח בעל העשרים תושבות ששים | כמו משלש דב"ג [195] ושטח אשר מ[הכאת] ד"ה בב"ג כפל משלש דב"ג אם כן שטח | בעל העשרים תושבות כמו [שלושם דמיוני] השטח אשר מד"ה מ[הכאת] ד"ה בב"ג וזה מה | שרצינו לבאר. |

[Proposition 29]

[196] נרצה לבאר שיחס שטח בעל השתים | (192a) עשרה תושבות [אל שטח בעל העשרים תושבות] כיחס השטח אשר מהכאת העמוד היוצא מהעגלה | אשר תקיף במחומש בעל שתים עשרה תושבות אליו בצלע המחומש של השטח | אשר מהכאת העמוד היוצא ממרכז העגלה אשר תקיף במשלש בעל העשרים | תושבות אל הצלע המשלש [197] מופת זה שיחס צלע [צ"ל שטח] בעל השתים עשרה תושבות | אל שטח בעל העשרים תושבות בצלע המשלש כיחס חלק שטח משלשים משטח בעל שתים | עשרה תושבות אל חלק משלשים משטח בעל העשרים תושבות וחלק משלשים | משטח בעל שתים עשרה תושבות הוא שטח מהכאת העמוד בצלע | המחומש והחלק משלשים משטח בעל העשרים הוא השטח אשר מהכאת | העמוד בצלע המשלש [198] ויחס שטח בעל שתים עשרה תושבות משטח בעל | העשרים תושבות כיחס השטח אשר מהכאת העמוד היוצא ממרכז העגלה | אשר תקיף במחומש בעל השתים עשרה תושבות אל צלע המחומש מצלע | המחומש אל השטח אשר מהכאת העמוד היוצא ממרכז העגלה אשר תקיף | משלש בעל העשרים תושבות צלע משלש כצלע [צ"ל בצלע] משלש בעל העשרים | תושבות וזה מה שרצינו לבאר. |

[Proposition 30]

[199] נרצה לבאר שכל מחומש | יפול בעגלה הוא השטח אשר מ[הכאת] שלש רביעי קוטר העגלה | בחמשה שתויות מאשר יהיה מותר [מיתר] לשני חומשי העגלה שוה למחומש [צ"ל שבה המחומש] | [200] דמיון זה עגלה [מחומש] אב"ג ד"ה <וכל משלש שוה הצלעות עליו אב"ג ד"ה> וקוטר | העגלה אה"ז [אח"ד] ונגיע בח"ה ויהיה ב"ל שלש ב"ח הנה היה אם כן [ב"ל] שתות [ב'] ול"ה | חמש שתיות ב"ה ויהיה ט' מרכז העגלה ונגיע ט"ה ונחלק ט"ז בשני חציים | על הכאת [כא"ף] הנה אלף (!) כאף שלשה רביעי | א"ז [201] ואומר שהשטח אשר מהכאת אלף | כאף בלמד היא כמו מחומש אב"ג | ד"ה [202] מופתו שב"ח שלשה דמיוני ב"ל | וח"ה שלשה דמיוני ב"ל וא"ב [וא"כ] שלשה דמיוני | [תצי] ב"ט [א"ט] וב"ט [וכ"ט] הוא חצי א"ט [203] הנה יחס ח"ה מן | ב"ל כיחס א"ב [א"כ] מן חצי א"ט הנה הכאת | ח"ה בחצי א"ט בהכאת [כהכאת] א"ב בז"ל [א"כ בב"ל] והכאת ח"ה בחצי א"ט הוא <משלש א"ט הנה | הכאת א"ב כב"ל> (!) כמו משלש אה"ט הנה אם כן הכאת א"ב [א"כ] בחמשה דמיוני ב"ל | (192b) היא חמשה דמיוני משלש אה"ט [204] וחמשה דמיוני משלש אה"ט הוא מחמש אב"ד ד"ה | וחמשה דמיוני ב"ל הוא ל"ה הנה הכאת א"ב [א"כ] בל"ה הוא מחומש אב"ג ד"ה וזה מה | שרצינו לבאר. |

[Proposition 31]

[205] נרצה לבאר שיחס שטח בעל שתים עשרה תושבות | אל שטח בעל עשרים תושבות הנקוים בכדור

אחד כיחס צלע | המעוקב יקיף בו אותו כדור אל צלע משלש בעל עשרים תושבות [206] דמיונו | אם נשים
העגלה אשר תקיף במחומש בעל שתים עשרה תושבות ובמשלש בעל | העשרים תושבות עגלת אג"ב ויהיה
מרכזה ה' [צ"ל ד'] וצלע המשלש הנופל בה והוא | צלע משלש בעל העשרים תושבות א"ב וצלע המחומש הנופל
בה והוא צלע | מחומש בעל השתים עשרה תושבות א"ג ויהיה [ט'] צלע המעוקב אשר יפול באותו | הכדור.
ואומר שיחס בעל השתים עשרה תושבות לשטח בעל העשרים | תושבות כיחס ט' אל א"ב [207] מופתו אנו
נוציא מן ד' עמוד אל [א"]ג אל ז' [וד"ז] כמו | חצי צלע המשושה וחצי צלע המעושר ונוציא גם כן מן ד' עמוד
אל א"ב והוא | ד"ה הנה ד"ה הוא כמו חצי צלע המשושה [208] [וקו ט'] כאשר נחלק על יחס בעל אמצע ושתי
| קצוות כשהחלק [צ"ל יהיה החלק] היותר גדול א"ג [209] וצלע המשושה ומעושר כאשר נחלקו על | יחס בעל
אמצע ושתי קצוות הוא [צ"ל היה] החלק היותר גדול צלע המשושה וחצי | צלע המעושר [וחצי צלע המשושה]
כאשר נחלק על יחס בעל אמצע ושתי קצוות הוא [צ"ל היה] החלק | היותר גדול חצי צלע המשושה וד"ז כאשר
נחלק על יחס בעל אמצע ושתי | קצוות היה החלק הגדול ד"ה [210] הנה יחס ט' מן א"ג כיחס ד"ז מן ד"ה הנה
| הכאת ה' [צ"ל ט'] בד"ה כהכאת א"ג בד"ז [211] וכבר בארנו | ש[יחס] הכאת א"ג בד"ז >יחס< אל הכאת
א"ב בד"ה | כיחס שטח בעל עשרה תושבות | משטח בעל העשרים תושבות והכאת א"ג | בד"ז כהכאת ט'
בד"ה [212] הנה אם כן יחס הכאת | ט' בד"ה אל הכאת א"ב בד"ה כיחס שטח בעל | השתים עשרה תושבות
משטח בעל העשרים | תושבות >אל< [ו]יחס הכאת ט' בד"ה אל הכאת א"ב בד"ה הוא כיחס ט' מא"ב הנה |
אם כן יחס [ט'] מן א"ב כיחס שטח בעל שתים עשרה תושבות משטח בעל העשרים | תושבות וזה מה שרצינו
לבאר. | (193a) [213] אמר קלונימוס טודרוס המעיין שתי תמונות אלו והן תמונות ל' ול"א | נפלו מהעתקת
הנשיא החכם כ' קלונימוס ב"ר קלונימוס נ"ע | למאמר התמונות החמש המוגשמות לפי שנפל עליה [צ"ל עלה]
מספר הערבי | שהעתיק זה ממנו ולשנות (!) מספר אחר זה מצאתי העלה ההוא והעתקתיו. | [214] כתב זה
מולש מרשילי ג' אלול צ"ה באייגש. והיה זה חסר מהמאמר | בהקשי המוגשמות החמש מהאמצע והמאמר ההוא
כתוב בקובץ הזה | לפי והנו(!) אחר הגעת שתי תמונות אלו לידינו בחסדי השם שלם | בקובץ הזה הודות לאל. |
[215] אמר אבולוניוס ונבאר זה בפנים אחרים נקוה העגלה אשר תקיף | המחומש בעל השתים עשרה תושבות
ובמשלש בעל העשרים | תושבות עגלת אב"ג ד"ה והמחומש העשוי ב[ה] מחומש אב"ג ד"ה והמשלש אשר |
ב[ה] משלש אז"ח ונדביק ב"ה וב"ה צלע המעוקב ויהיה כל [צ"ל ב"ל] חמשה שתויותיו | ונוציא עמוד אב"ט
[צ"ל אכ"ט] והמרכז כ' ומבואר כי כ"ט רביעית קוטר העגולה וא"ט | שלשה רבעי קוטר העגלה ואומר שיחס
[שטח] בעל שתים עשרה תושבות | משטח בעל ההעשרים תושבות כיחס ב"ה מז"ח [216] מופת זה שהכאת
א"ט | בכ"ל [בב"ל] הוא מחמש אב"ג ד"ה אשר הוא מחומש בעל שתים עשרה תושבות [217] והכאת | א"ט
בז"ט הנה [היא] משלש אז"ח אשר משלש בעל העשרים תושבות ויחס | מחומש אב"ג ד"ה אל משלש אז"ח
כיחס כ"ל [ב"ל] מז"ט ויחס שתים עשרה | דמיוני כ"ל [ב"ל] מעשרים דמיוני ז"ט כיחס [שתים עשרה] דמיון
מחומש אב"ג ד"ה וזה | שטח בעל שתים עשרה תושבות אל עשרים דמיוני משלש אז"ח וזה שטח | בעל העשרים
תושבות [218] ואם כן יחס בעל שתים עשרה תושבות אל שטח | בעל העשרים תושבות כיחס שנים | עשרה
דמיוני כ"ל [ב"ל] אל עשרים דמיוני | ז"ט [ו]מפני שכ"ל [שב"ל] חמשה שתיות ב"ה | יהיה שנים עשרה דמיוני
כל [ב"ל] הוא עשרים (?) | דמיוני ב"ה ועשרים דמיוני ז"ט הוא עשרים [עשרה] | דמיוני ז"ח [219] ויחס שטח
בעל העשרים [צ"ל השתים עשרה] תושבות | (193b) משטח בעל העשרים תושבות כיחס עשרה דמיוני ב"ה
מעשרה דמיוני ז"ח וזה | כיחס ב"ה מז"ח ויחס שטח בעל שתים עשרה תושבות מבעל העשרים | תושבות כיחס
צלע המעוקב והוא ב"ה אל צלע משלש בעל העשרים | תושבות והוא קו ז"ה וזה מה שרצינו לבאר. |

[Proposition 32]

[220] נרצה לבאר שיחס שטח שתים עשרה תושבות משטח | בעל העשרים תושבות כיחס הקו אשר יחזיק
על קו יחלק על יחס בעל | אמצע ושתי קצוות כלו ועל חלוקת [החלק] היותר גדול אל הקו אשר יחזיק על קו |
כלו ועל חלוקת [החלק] היותר קטן וזה הצלע המחומש יפול | בעגלה אל הקו אשר יחזיק על >הק"ו< שלשה
דמיוני מרובע צלע מעושר אותה | העגלה [221] משל זה שא"ב חולק על בעל יחס ושתי קצוות על ג' והחלק |
היותר גדול א"ג ויצאו א"ד וב"ה עמודי[ם] על א"ב וב"ג. ואומר שיחס שטח | בעל השתים עשרה תושבות אל
שטח העמוד בעל העשרים תושבות כיחס ד"ג מג"ה | [222] מופת זה אנו כבר בארנו שיחס שטח בעל שתים
עשרה תושבות משטח | בעל העשרים תושבות כיחס צלע המעוקב מצלע המשלש בעל העשרים | תושבות
[223] והתבאר בתמונת י"ח גם כן שיחס צלע המעוקב מצלע משלש | בעל העשרים [תושבות] כיחס ד"ג מג"ה
ואם כן יהיה יחס | שטח בעל שתים עשרה תושבות משטח | בעל העשרים תושבות כיחס ד"ג מג"ה | [224]
ואלו נתן ג"ד [א"ד] צלע משושה וא"ג צלע מעושר | יהיה אמנם ד"ג צלע מחומש לפי שא"ד [צלע] | משושה (!)

וא"ג צלע מעושר [225] ועוד(?) שג"ה יחזיק | על שלשה דמיוני מרובע צלע המעושר וזה מה שרצינו לבאר. |

[Proposition 33]

נרצה לבאר שיחס מוגשם בעל השתים עשרה תושבות אל מוגשם | בעל העשרים תושבות שהם שטח בעל השתים עשרה תושבות | אל שטח בעל העשרה תושבות כיחס הקו המחזיק על כל הקו וחלק על יחס | בעל אמצע ושתי קצוות ועל חלוקת היותר גדול של הקו המחזיק על הקו כלו | ועל חלוקת היותר קטן וזה יחס צלע כל מחומש מן הקו המחזיק על שלשה | דמיוני מרובע המעושר הנופל עמו בעגלה אחת. מופת זה שהעגלה אשר ... | (194a) דמיוני מרובע חצי הצלע לפי שהוא משלש | שוה הצלעות ויחס ב' אל ג' כיחס ד' על ה' | יחס א' אל ב' כיחס ה' אל ז'. ובשווי יחס א' | אל ג' כיחס ד' אל ז' ויחס ד' אל ז' כיחס | שלש ד' אל שלש ז' אשר הוא ה' ויחס ה' ויחס א' אל ג' כיחס שליש ד' אל ה' וזה מה שרצינו | לבאר. |

[Proposition 34]

[226] נרצה למוא הקוים החמשה המסודרים על יחס עמוד | התמונות החמשה היוצאות ממרכז הכדור אל שטחיהם. ונבאר | זה ונבחנהו בתמונה אחת [227] ונניח עגלה עליה <א"ב> וצלע המחומש הנופל בה | א"ב ומרכז העגלה ג' ונעשה על א"ב משלש שוה הצלעות עליו אה"ב [228] ונוציא עמוד | גד"ה ויהיה ד"ז שליש ד"ה ויהיה ח' כמו ד"ז [229] וט' כמו א"ד וב' [כ'] כמו ט' ול' כמו ג"ד ומ' כמו | ל' [230] ואומר שיחס אל ט' כיחס עמוד המחודד אל עמוד המעוקב אשר הוא כמו | עמוד בעל השמונה תושבות ויחס ט' אל כ' כיחס העמוד המעוקב אל עמוד משלש | בעל השמונה תושבות ויחס כ' אל ל' כיחס עמוד בעל השמונה תושבות המעוקב אל עמוד | בעל העשרים תושבות אשר כמו עמוד בעל השתים עשרה תושבות ויחס ל' אל | מ' כיחס עמוד בעל העשרים תושבות אל עמוד בעל השתים עשרה תושבות | [231] מופת זה אנו כבר בארנו שיחס עמוד המחודד אל עמוד המעוקב כיחס שליש | ד"ה והוא ד"ז אל א"ד וד"ז כמו ח' וא"ד כמו ט' ויחס ח' אל ט' כיחס העמוד היוצא | ממרכז הכדור אל משלש המחודד הנופל בה אל העמוד היוצא ממרכז הכדור | הנזכר אל מרובע המעוקב הנופל בכדור הנזכר [232] ויחס ט' אל כ' כיחס עמוד | המעוקב אל עמוד בעל השמונה תושבות לפי שהוא כמוהו לפי שמרובע | המעוקב ומשלש בעל השמונה תושבות יפלו בעגלה [233] וכבר בארנו במה שקדם | בתמונה כ"ז מזה המאמר שיחס עמוד בעל השמונה תושבות אל עמוד בעל | העשרים תושבות כיחס חצי צלע מחומש אל חצי צלע משושה וחצי צלע מעשר | יחס [יחד] וא"ד הוא חצי [...] מעושר יחד והוא כמו ל' [234] ויחס העמוד בעל השמונה תושבות אל | עמוד בעל העשרים תושבות כיחס ב' [כ'] אל ל' | יחס עמוד בעל העשרים תושבות אל עמוד | בעל השתים עשרה תושבות כיחס ל' אל מ' | לפי שהוא כמוהו לפי שמשלש בעל | (194b) העשרים תושבות ומחומש בעל שתים עשרה תושבות יפלו בעגלה אחת [235] וכבר סודרו | קוי ז"ט [ח"ט] כל"מ על יחס עמוד התמונות החמשה וזה מה שרצינו לבאר. |

[Proposition 35]

[236] נרצה שנמצא הקוים החמשה המסודרים על יחס שטחי התמונות החמשה | ונבאר זה ונבחנהו בתמונה אחת [237] ונניח קו עליה א"ב ונחלקהו על יחס בעל | אמצע ושתי קצוות על ג' ונוציא א"ב על יושר אל ד' ויהיה ב"ד כמו ב"ג [238] והיה קו ה"ז | מחזיק על שליש מרובע א"ד ונחלקהו על יחס בעל אמצע שתי קצוות על ח' [239] ונעשה על | א"ד חצי עגלה אט"ד ונפיל בה ד"ט כמו קו ז"ח ונוציא ה"ט [ד"ט] על יושר אל ב' [כ'] ויהיה ט"ב [ט"כ] | כמו ה"ז ונדביק א"ט ונוציא עמוד כ"ל כמו ט"א ונדביק ל"ט [240] ונוציא מב' עמוד והוא ב"מ | ונעמיד זוית בא"ג שליש נצבת ויהיה א"ס שליש א"ב [241] ואומר שיחס א"ט [ל"ט] מא"ד כיחס | שטח בעל העשרים תושבות משטח בעל השמונה תושבות ויחס א"ב מנ"מ | כיחס שטח בעל השמונה תושבות משטח המעוקב ויחס נ"מ מכ"ס [מב"ס] כיחס שטח | המעוקב משטח המחודד [242] מופת זה כי כ"ט הוא כמו ה"ז וט"ד כמו ז"ח וקו ד"ב [ד"כ] | כבר נחלק על יחס בעל אמצע ושתי קצוות על ט' ומרובע ב"ד [כ"ד] ומרובע ז"ט [ד"ט] שלשה | דמיוני מרובע כ"ט לפי שכ"ט כמו ה"ז [243] ומרובע א"ד כמו שני מרובעי א"ט ט"ד | לפי שזוית ט' נצבת ומרובעי א"ט ט"ד כמו שני מרובעי ט"ד כ"ט [כ"ד] וחבור [וחסור] מרובעי ד"ט | המשותף ישאר מרובע א"ט כמו מרובע ב"ד [כ"ד] וא"ט כמו ב"ד [כ"ד] וכמו כן ל"ב [ל"כ] כמו ב"ד [כ"ד] | [244] וקו ל"ט הוא המחזיק על הקו כלו והוא ב"ד [כ"ד] ועל החלק היותר גדול והוא ב"ד [כ"ט] וה"ט [א"ד] | הוא המחזיק על הקו כלו ועל החלק היותר קטן והוא ט"ד אם כן יחס ל"ט מא"ד כיחס | שטח בעל שתים עשרה תושבות משטח בעל | העשרים תושבות כפי מה שהתבאר מל"ד | מזה המאמר [245] ועוד שא"ב נחלק על יחס | בעל אמצע ושתי קצוות על ג"ד [ג'] וב"ד כמו | ב"ג ויחס א"ד אשר הוא החלק הקטן עם הקו | כלו אל א"ב כיחס שטח בעל העשרים | תושבות משטח בעל השמונה תושבות כפי מה שבארנו [246] ועוד שמשלש אנ"מ שוה | הצלעות לפי שזוית א' שליש נצבת וא"ב עמוד ויחס א"ב עמודו מנ"מ צלעו | כיחס שטח בעל שמונה תושבות המעוקב כפי מה שבארנו בי"א מזה | (195a) <מזה> [247] ועוד שאנו כבר בארנו שיחס שטח המחודד מן שטח בעל

שמונה | תושבות כיחס ס"ב מא"ב וזה שני שלישי (!) ויחס א"ב מנ"ה כיחס שטח בעל | שמונה תושבות משטח
המעוקב אם כן יחס שטח המעוקב משטח המחודד | כיחס נ"מ מב"ז [מב"ס] וקוי ל"ט ד"א א"ב נ"מ ד"ה [ב"ס]
החמשה מסודרים על יחס שטחי | תמונה (!) החמשה הנקוות בכדור אחד. וזה מה שרצינו לבאר. |

[Proposition 36]

[250] נרצה לבאר איך נמצא הקוים החמשה המסודרים על יחס גודל | מוגשמות החמשה הנקוות [בכדור
אחד ונבאר זה ונבחנהו בתמונה | אחת [251] ונניח קו א"ב צלע מחומש שוה הצלעות נקוה בעגלה ונוציאהו |
על יושר אל ג' ויהיה ב"ג כמו [צלע המעושר] הנופל באותה עגלה [252] ויהיה קו ה"ד יחזיק | שליש(?) מרובע
א"ג ונחלקהו על יחס בעל אמצע ושתי קצוות על ז' [253] ונקוה | על א"ג חצי עגלה אח"ג ונקוה <ב"ז> ג"ח כמו
ז"ח [ז"ה] ונדביק א"ח ונוציא ג"ח אל ט' ויהיה | ח"ט כמו ד"ה ונוציא עמוד ט"ב [ט"כ] כמו ט"ג ונדביק ב"ח
[כ"ח] [254] ונוציא כ"ל [א"ל] ונעמיד זוית | בח"ל [בא"ל] שליש נצבת ונוציא כ"ז [מב' עמוד] אל מ' ונעמיד
בא"מ שליש נצבת גם כן ויהיה ל"נ (?) | שליש ל"מ [255] ואומר שיחס ב"ח [כ"ח] מא"ג כיחס מוגשם בעל
שתים עשרה תושבות | ממוגשם בעל העשרים תושבות ויחס א"ג אל א"ב כיחס מוגשם בעל העשרים תושבות
אל מוגשם בעל [שמונה תושבות ויחס א"ב אל ל"מ כיחס מוגשם בעל שמונה תושבות אל מוגשם בעל] ששה
תושבות ויחס ל"מ מל"נ כיחס מוגשם בעל | ששה תושבות ממחודד בעל ארבע תושבות [256] מופת זה שד"ה
נחלק על יחס | בעל אמצע ושתי קצוות על ז' וד"ה כמו ט"ח וח"ג כמו ז"ה וג"ט כבר נחלק על | יחס בעל אמצע
ושתי קצוות והחלק היותר גדול ט"ח מרובע[י] ט"ג ג"ח שלשה | דמיוני מרובע ט"ח [257] [ו]לפי שט"ח כמו ד"ה
ומרובע א"ג כמו מרובע[י] א"ח א"ג [ח"ג] מרובעי | א"ח ח"ג כמו שני מרובעי ט"ג ג"ח נפיל מרובע <מרובע>
ח"ג ישאר מרובע א"ח כמו | מרובע ט"ג וא"ח כמו ט"ג וכמו כ"ט [258] וב"ח [וכ"ח] יחזיק על הקו כלו של
החלק היותר | גדול וא"ג יחזיק על הקו כלו ועל החלק היותר קטן ויחס ב"ח [כ"ח] מא"ג כיחס | מוגשם בעל
שתים עשרה תושבות ממוגשם בעל העשרים תושבות כפי | מה שבארנו [259] ועוד שא"ג צלע מחומש וצלע
מעושר נופלים בעגלה <הנה ט"ח> | וא"ב צלע מחומש ויחס ג"א אל א"ב כיחס מוגשם בעל העשרים תושבות
ממוגשם | בעל השמונה [תושבות] [260] ועוד שמשלש אמ"ל שוה הצלעות ועמודו א"ב ויחס א"ב ממ"ל | כיחס
מוגשם בעל שמונה תושבות ממוגשם בעל שמונה [ששה] תושבות כפי מה | (195b) שבארנו <ועוד שיחס מ"ל
מא"ב כיחס מוגשם בעל ששה תושבות ממוגשם | בעל שמונה תושבות כמו מה שבארנו> [261] ויחס א"ב |
העמוד אל שליש הצלע והוא ל"כ [ל"נ] כיחס מוגשם בעל | שמונה תושבות ממוגשם המחודד כמו שבארנו |
[262] וביחס השווי יהיה יחס מ"ל מל"נ (?) כיחס | מוגשם בעל ששה תושבות ממוגשם המחודד בעל | ארבעה
תושבות [263] וקוי ב"ח [כ"ח] א"ג א"ב נ"ל [מ"ל] ל"כ [ל"נ] החמשה | מסודרים על יחס גשמי התמונות
החמשה הנקויות | בכדור אחד . זה מה שרצינו לבאר. | ז |

[Envoi]

[264] כל מה שזכרנו בזה המאמר | מאותו המאמר מבלי(?) [צ"ל מאלו] התמונות החמשה אמנם רצינו בו
מה שהיה | נקויה בכדור אחד [265] וכמו כן מה שזכרנו מחומש אמנם רצינו בו מה שהיה המחומש | השוה
הצלעות והזויות וכמו כן המשושה והמעושר וגם שהם נקוים בעגלה אחת | [266] ואחר שכבר הגענו אל זה
המקום כבר נשלם המכוון אשר רצינו בו והתהלה לאל | יתע' העוזר אל הטוב [267] ואני חושב שאעשה עוד
מאמר לבאר בו שלא | יפול בכדור זולת אלו התמונות החמשה אחר שכבר [נגמר?] זה המאמר אחרי התפללתי
| לשם יתע' שזכינו להשלים ויצליחני (!) לידיעת [ה]אמת ישתבח ויתעלה אמן. |

[Colophon]

[268] נשלמה העתקת זה המאמר בכ"א ס"ט לפרט בעיר ארלד"י | [269] והעתקיו אני קלונימוס ב"ר
קלונימוס זצ"ל ב"ר מאיר ע"ס יש"י מלשון ערבי | ללשון עברי בהגיעי לשתים ועשרים משנותי השם ברחמיו יזכני
להבין וללמד | אל מה שיגיעני לאהבתו אמן. |

Translation

[I1.] May Exalted God clarify for you the hidden meanings of things and set you on the straight path to the chambers of truth. May He cleanse you of foolishness and lead you on every good road which has been prepared.[24] May the Exalted Creator grant you success in loving Him and purify your heart in His belief.

[I2.] That which you mentioned from SNQLWNYS, who corrected Apollonius' book, within which [? are discussions] concerning the ratio of the dodecahedron to the icosahedron when they are inscribed in the same sphere, and it is Book XIV of the work of Euclid;[25] and his praising the value of knowledge of the ratio which obtains between the two figures, and his saying that there is in this something of the attainment of precious matters.

[I3.] And I think that Apollonius' intent in this was comprehensive, encompassing [the ratios?] of each to the other.[26]

[I4.] You mentioned that it has been said that the intent of Euclid's book is the knowledge of the five figures mentioned in Book XIV, [when] they are drawn in the same sphere. You saw from the works of the ancients, that they likened the five figures to the elements.[27]

[I5.] When you saw this, your soul was filled with longing and desire to know the ratios of these figures one to another. If this were to be achieved, there would be the sort of attainment and knowledge as there is in the ratio of the two figures one to another that Apollonius discussed, clarifying it as I said (!),[28] and the truth is as I have reasoned (?), for the subject of these five figures among the ancients is one [and the same], and, furthermore, none others are inscribed in the sphere.

[I6.] When I saw the preciousness of your thought, my soul longed to help you, and drive out [your] darkness.[29] I gathered my strength (?) to complete the ratios of these five figures one to another regarding their altitudes, surface areas, and volumes. But first I bow[30] to the Lord, pinning my hope upon Him, and asking Him for help and support; from His hand was I able to begin it (?), all according to my ability, my He be Praised and Exalted forever and ever.[31] Amen.

[I7.] I wrote to you in this treatise and showed you [according] to your intent. I collected that which Apollonius discussed regarding the ratios of the two figures one to another, since this treatise rests upon that issue.

[I8.] I attributed every proposition that Apollonius mentioned [to him] in order to distinguish between my treatise and his. I ask the Lord — may He be Exalted — for help. This is the beginning of the treatise.

[Proposition 1][32]

[1.] We wish to show that the square on the altitude of any equilateral triangle is three-fourths the

[24] The Hebrew has *hakhīnōtī*, "I prepared." It seems likely that the verb in the Arabic ended in the unvocalized letter *tā*, indicating feminine singular, third person, but Qalonymos, for whatever reason read it as the ending of the first person singular. This seems to allude to a phrase from a traditional text that I have not been able to identify.

[25] The Hebrew has here the very unusual *ha-meyusod li-Uqlidis*. Two possible interpretations may be suggested: (1) the author uses here the Hebrew verb *yasad* to indicate authorship, as in the phrase, *meyosodo shel X*, i.e. a work by X; (2) there is either a pun or some corruption of the text, and the author has in mind *ha-Yesodot*, the Hebrew title of Euclid's *Elements*. I cannot suggest the Arabic form that is being translated here.

[26] I take this to mean that the author our treatise thought that Apollonius intended to cover the ratios of all five regular solids, one to the other.

[27] The connection between the "primary bodies" of the cosmos and the regular solids is an important part of the cosmology of the Plato's *Timaeus* (53C–57C); see, e.g., Cornford, 1935, 210–228. Proclus [1970, 57] averred that Euclid too was a Platonist and "this is why he thought the goal of the *Elements* as a whole to be the construction of the so-called Platonic figures."

[28] According to Hebrew syntax, the subject of "clarifying" must be Apollonius; yet it seems that the remarks immediately following in the first person are those of our author. The precise meaning is unclear, but the thrust seems clear enough: our author finds support for his claim that Apollonius intended to deal with all five solids in his (Apollonius, it must be) praising the knowledge of these ratios.

[29] "Drive out [your] darkness" is my (mild) emendation of the otherwise incomprehensible Hebrew phrase.

[30] Literally: "After first bowing."

[31] Once again, there are some rough edges to the Hebrew syntax, a roughness that surely comes through in my translation; but the pious intent of our author is clear enough.

[32] This is found in MS Mantua, ff. 34b–35a, but not in the section on the regular solids; see the introduction.

square of [one of] its sides.

[2.] For example, if we take as the equilateral triangle, triangle *ABC* and altitude *AD*, I say that $(AD)^2 = 3(BC)^2/4$. [3.] Its proof is: $AB = BC$ and angle B = angle C, and angle ADB = angle ADC, because each of them is right. [4.] Side *AD* is common to the two triangles, $BD = DC$, and $(BC)^2 = 4(BD)^2/4$. [5.] $(AB)^2 = (BC)^2 = (AD)^2 + (DB)^2$. [6.] $(AD)^2 + (DB)^2 = 4(BD)^2$; we subtract $(BD)^2$ which is common, and there remains $(AD)^2 = 3(BD)^2$. [7.] $(BC)^2 = 4(BD)^2$ and $(AD)^2 = 3(BC)^2/4$, which [*BC*] is the side of the triangle. Q.E.D.

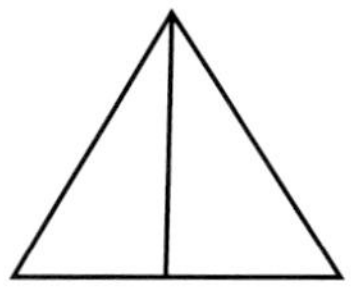

Figure 1: Proposition 1, 3.

[Proposition 2][33]

[8.] We wish to show that the edge of the octahedron is equal to the altitude of the [face of the] tetrahedron, when both are drawn in the same sphere.

[9.] For example, we assume the diameter of the sphere to be *AB*, and we construct on it the semicircle *ADEB*. [10.] We draw the perpendicular *ZE* and connect *E*, *B* by line *EB*, and *EB* is the edge of the octahedron drawn in the sphere whose diameter is *AB*.[34] [11.] We make *AC* [one-third of] *AB* and draw the perpendicular *CD*, and we connect *B*, *D* by line *BD*, and *BD* is the edge of the tetrahedron.[35] [12.] We construct an [equi-]lateral triangle, letting each of its sides be equal to *DB*; it is triangle [*H*]*BT*, and its altitude *HL*. I say that the altitude *HL* equals *EB*. [13.] Its proof is: we bisect *LB* at *M*, and $MT = 3(TL)/2$. [14.] $(AD)^2 = 3(BD)^2/2$ as Euclid noted in his book,[36] and $(AB)^2 = 2([E]B)^2$. [15.] $BT = 2TL$.[37] [16.] Three magnitudes, $(E[B])^2$, $(AB)^2$, $(BD)^2$, and three others in their number, *TM*, *TL*, and *TB*, taken two at a time are in the same ratio as [the corresponding] two [of the second group][38] and they are perturbed in this way.[39]

[17.] $(BE)^2 : (AB)^2 = TL : TB$ and $(BA)^2 : (BD)^2 = TM : TL$, and by the equality of the ratio, $(EB)^2 : (BD)^2 = TM : TB$. [18.] $TM : TB = 3 : 4 = (HL)^2 : (TB)^2$. [19.] $(HL)^2 : (TB)^2 = ([E]B)^2 : (DB)^2$ and $(BD)^2 = (TB)^2$; $(HL)^2 [= (EB)^2]$, because they [*DB* and *TB*] are equal, and [*E*]*B* is also equal to *HL*.

[Proposition 3]

[20.] We wish to show that the measure [i.e. area of the face] of the octahedron is three-fourths the measure of the tetrahedron when both are inscribed in the same sphere.

[21.] For example, we take triangle *ABC* of the tetrahedron and [its] altitude *AD*. I say that the equilateral triangle [drawn] on *AD*, which is the [face of the] octahedron, is three-fourths triangle *ABC*. [22.] Its proof is: the ratio of the equilateral triangle on *AD* to triangle *ABC* is as the ratio of $AD : BC$

[33] A different proof is displayed in MS Mantua.

[34] *Elements* XIII, 14.

[35] *Elements* XIII, 13.

[36] Ibid.

[37] *HL*, the altitude of equilateral triangle *HBT*, bisects *BT*.

[38] *Elements* V, 20.

[39] See definition 18 at the beginning of Euclid. The Hebrew here is very cumbersome.

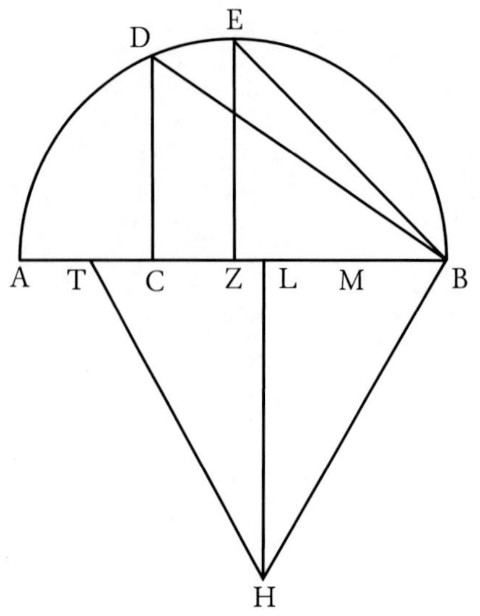

Figure 2: Proposition 2, 5.

in duplicate.[40] [23.] The ratio of $AD : BC$ [in duplicate] is $(AD)^2 : (BC)^2$. [24.] $(AD)^2 : (BC)^2$ is equal to the ratio of the equilateral triangle on *AD*, which is the [face of the] octahedron, to triangle *ABC*, which is [the face of the] tetrahedron. [25.] $(AD)^2 = 3(BC)^2/4$ and the [face of the] octahedron is three-fourths of [face of the] tetrahedron. Q.E.D.

[Proposition 4]

[26.] We wish to show that the area [ie., total surface area] of the octahedron is one and one-half [times] the area of the tetrahedron, when both are inscribed in the same sphere.

[27.] Its proof is that four times [the face of] the tetrahedron is the area of the tetrahedron and eight times the [face of the] octahedron is the area of the octahedron, and four times the [face of the] octahedron is half the area of the octahedron. [28.] The ratio of four times the [face of the] octahedron to four times the [face of] the [tetra-]hedron [i.e., pyramid] is equal to the ratio of [face of the] octahedron to the [face of the] tetrahedron. [29.] The [face of the] octahedron is three-fourths the [face of the] tetrahedron, and the area of the octahedron is thus one and one-half [times] the area of the tetrahedron. Q.E.D.

[Proposition 5]

[30.] We wish to show that the parallelepiped whose base is double the face of the tetrahedron and whose altitude is one-ninth the diameter of the sphere is equal [in volume] to the tetrahedron.

[31.] For example, we take *AB* as the diameter of the sphere and draw on it the semicircle *ADB* and we make $AC = (AB)/3$. [32.] We draw the perpendicular *CD*, and we join *DB*; *DB* is the edge of the tetrahedron. [33.] I say that the parallelepiped whose base is double the equilateral triangle drawn on *DB* and whose altitude is one-ninth the diameter of the sphere is equal to the tetrahedron. [34.] It proof is: *BC* is the perpendicular which is drawn from the apex of the cone to the base of the cone. [35.] For the cylinder whose base is the base of the cone is three times the cone, and Euclid has shown this in the twelfth book of his text.[41] [36.] The solid whose base is double the base of the cone and whose altitude is $[CB]/6$ is also equal to the cone because it is [one-third] the above-mentioned

[40] *Elements* VI, 19.

[41] *Elements* XII, 10.

cylinder.[42] [37.] However, $CB = 6(AB)/9$. Therefore, and the solid whose base is double the face of the tetrahedron and whose altitude is one-ninth the diameter of the sphere is equal to the cone. Q.E.D.

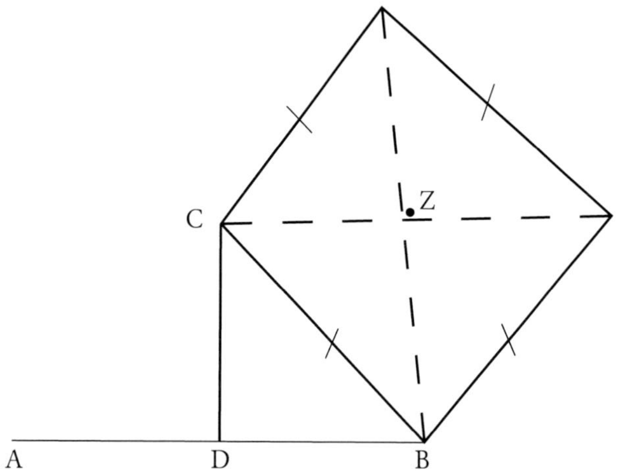

Figure 3: Proposition 6.

[Proposition 6]

[38.] We wish to show that the parallelepiped whose base is equal to the square on the edge of the octahedron and whose altitude is equal to one-ninth the diameter of the sphere is equal [in volume] to one-third of the octahedron.

[39.] For example, we take *AB* as the diameter of the sphere and we construct upon it semicircle *ACB* with center *D*. We draw the perpendicular *CD* and we join *C*, *B*. *CB* is [the edge] of the octahedron. [40.] I say the parallelepiped whose base is the square on *BC* and whose altitude is *AB*/9 is equal to one-third the octahedron. [41.] Its proof is: when we construct a square [whose] side is *CB*, and we draw from its center a perpendicular. [42.] And we extend the arc of the perpendicular to the corners of the square. [43.] And we place upon the square a pyramid whose base is the square and whose altitude is half the diameter of the sphere, [then] it [the "pyramid"] is equal to half the octahedron. [44.] The parallelepiped whose base is this square and whose altitude is *AD* is three times the aforementioned cone. [45.] For every cone, in whose base is half the aforementioned square, when it is divided along its diameter, and its altitude is *A*[*D*] ...[43] [46.] Then this solid is half the octahedron. [47.] The parallelepiped whose base is the aforementioned square and whose altitude is $(AD)/3$ is equal to half the octahedron. [48.] The parallelepiped whose base is the square and whose altitude is is equal to the octahedron. [49.] Therefore, the solid [i.e., the parallelepiped] whose base is the square and whose

[42] Our author takes for granted that the volumes of these solids is strictly a function of the base times the altitude, and thus one can double the base and cut the altitude in half without changing the volume. On this point, see Heath's notes (taken from the work of Legendre) in volume 3 of the *Elements*, pp. 391, 423.

[43] I have left the sentence incomplete, exactly as it appears in the manuscript. *Elements* XII, 7 proves that the pyramid is one-third the prism (parallelepiped) for the prism with a triangular base. Our author is dealing with solids with a square base, and it seems that the intention of this sentence is that if we cut the square along the diameter of the sphere, we will then have two pyramids with triangular bases and can then apply Euclid's theorem, i.e.

$$2(\frac{1}{3}(\frac{BC^2}{2} \cdot AD)) = \frac{1}{3}(BC^2 \cdot AD).$$

altitude is one-third of [one third of] $A[B]$, which is one-ninth of AB is [one-third] the octahedron. Q.E.D.

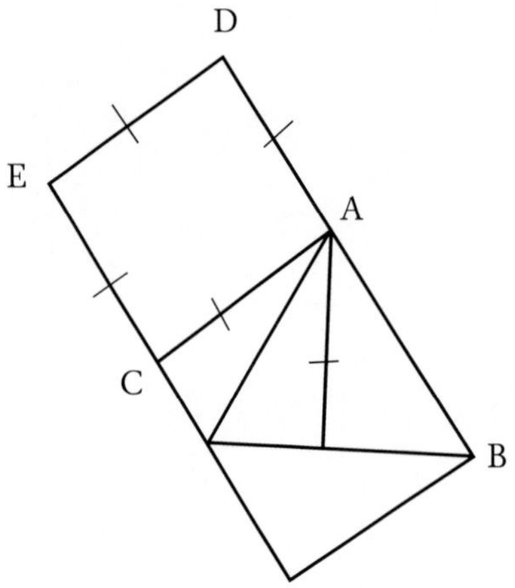

Figure 4: Proposition 7.

[Proposition 7]

[50.] We wish to show that the ratio of the [volume of the] tetrahedron to the [volume of the] octahedron, when both are inscribed in the same sphere, is as the ratio of the side of any equilateral triangle to three times its altitude, which is the same as the ratio of one-third the side of the equilateral triangle to its altitude.

[51.] For example, we take AB as the edge of the tetrahedron. [52.] AC is perpendicular to it and [AC] is the altitude of the triangle of the tetrahedron; it is also equal to the edge of the octahedron. [53.] We complete the parallelogram BC; let $AD = AC$. [54.] I say that that the ratio of the tetrahedron to the octahedron is equal to the ratio of AB to three times AD. [55.] Its proof is: we draw DE parallel to AC; AE is the square of the edge of the octahedron. [56.] The area BC is [twice] the triangle of the tetrahedron. [57.] The ratio of the parallelepiped whose base is [BC] and whose altitude is one-ninth the diameter [of the sphere], which [solid] is equal to the tetrahedron, to the solid whose base is AE and whose altitude is one-ninth the diameter of the sphere, which [solid] is equal to one-third the octahedron, is equal to the ratio of the area BC to the square AE. [58.] Therefore, the ratio of the tetrahedron to one-third of the octahedron is equal to the ratio of the area BC to the square AE.[44] [59.] However, the ratio of the area BC to the square AE is equal to $AB : AD$. [60.] Therefore, the ratio of the tetrahedron to one-third of the octahedron is equal to the ratio of AB, which is the side of the equilateral triangle, to AD, which is the altitude of the equilateral triangle. [61.] Therefore, the ratio of the tetrahedron to the entire octahedron is equal to the ratio of the side of the equilateral triangle to three times its altitude. Q.E.D.

[Proposition 8]

[62.] We wish to show that the square [i.e. the face] of the cube and the triangle of the octahedron, when inscribed in the same sphere, are circumscribed in the same circle. [63.] For example, we take

[44] This is the crucial step, and it makes use of *Elements* XI, 32; hence we have:

$$\frac{\text{tetrahedron}}{\text{octahedron}/3} = \frac{AD \cdot AC}{AB \cdot AC} = \frac{AD}{AB}.$$

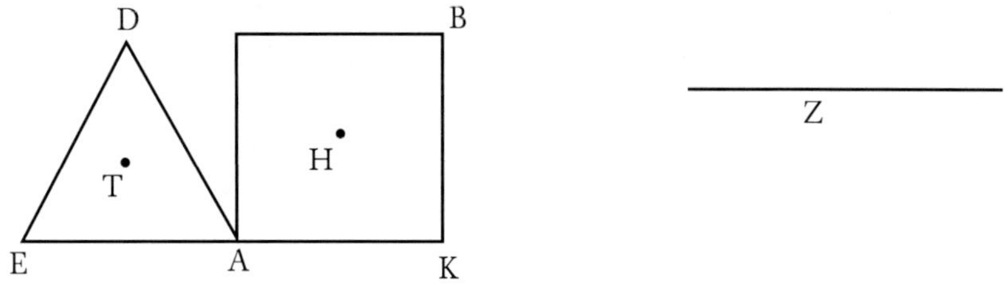

Figure 5: Proposition 8.

AB as the square of the cube, and the radius of the circle which circumscribes it is *HB*. The [face of the] octahedron is *ADE*, and the radius of the circle which circumscribes it is *TE*. [64.] I say that $TE = HB$. [65.] Its proof is: we take line *Z* as the diameter of the sphere. $Z^2 = 3(KB)^2$.[45] [66.] $(KB)^2 = 2(HB)^2$.[46] Z^2 is equal to six times the square [on *HB*] [67.] [The square on] *AE* is equal to $3(TE)^2$.[47] [68.] Z^2 is equal to six times [the square on] *TE*.[48] [69.] Already $Z^2 = 6H(B^2)$, therefore, and $HB = TE$, Q.E.D.

[70.] It is clear from what we have described that the perpendicular which is drawn from the center of the sphere to the [face of the] octahedron [equals the perpendicular drawn to the face of the cube] because the square [of the cube] and the triangle [of the octahedron] fall in the same circle.

[Proposition 9]

[71.] We wish to show that the square on the edge of the octahedron is one and one-half times the square on the edge of the cube.

[72.] Its proof is: the square of the cube and the triangle of the octahedron fall in the same circle. The ratio of the square on the edge of the octahedron to the square on the radius of the circle which circumscribes it is 3 : 1, and the ratio of the square on the radius of the circle to the square [on the edge of the] cube, because it [the square] is like the square which falls in [that] same circle, is 1 : 2.[49] [73.] By the equality of the ratio, the [ratio of the] square on the edge of the octahedron to the square on the edge of the cube is 3 : 2, which is one and one-half. The square on the edge of the octahedron is, therefore, one and one-half times the square on the edge of the cube, Q.E.D.

[Proposition 10]

[74.] We wish to show that the ratio of the face of the octahedron to the face of the cube is as the ratio of half the altitude of any equilateral triangle to two-thirds of one of [its] sides, which is also the ratio of the altitude of the triangle to one and one-third of [one of its] sides.

[75.] For example, we take as the face of the octahedron triangle *ABC*. On *BC* we construct square *CE*. We draw *ZAD* parallel to *BC*, with *BD* equal to the altitude of triangle *ABC*. We divide *BD* in half at *T*, and let $BH = 2(BC)/3$. [76.] I say that the ratio of triangle *ABC* to the face of the cube is like $BT : BH$. [77.] Its proof is: the area *DC* is double the triangle *ABC*. The ratio of the square *EC* to half the area *DC*, which is triangle *ABC*, is equal to $EB : BT$. [78.] We have already shown [that] the square on the edge of the octahedron is 1½ times the square [on the edge] of the cube. [79.] This is equal to $BE : BH$ because it [*BE*] is its [*BH*] equivalent and half [again] its equivalent. The ratio of the triangle *ABC* to the square *EC* is equal to $BT : BE$, and the ratio of the square *EC* to the square

[45] We must assume that KB is the side of the square. By *Elements* XIII, 15, $Z^2 = 3KB^2$.

[46] $HB = \frac{T_2}{Z} KB$, Hence $2HB^2 = KB$.

[47] *Elements* XIII, 12.

[48] *Elements* XIII,14, $Z^2 = 2AE^2$.

[49] This follows from the preceding proposition and from *Elements* XIII, 12.

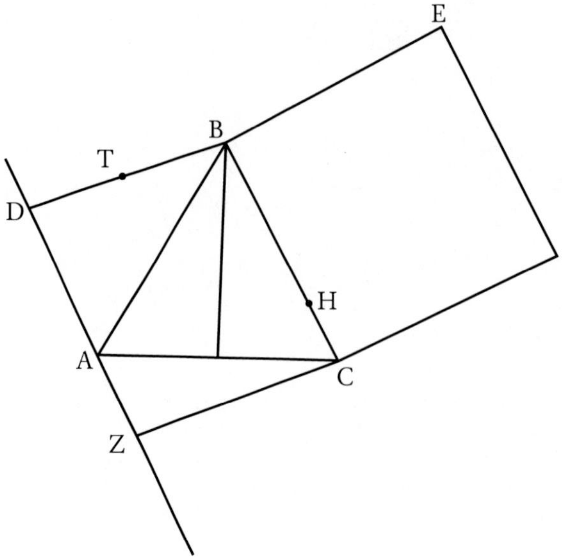

Figure 6: Proposition 10.

on the edge of the cube is equal to $EB : BH$. [80.] By the equality of the ratio, [the ratio of] triangle ABC to the face of the cube is equal to $BT : BH$. Q.E.D.

[Proposition 11]

[81.] We wish to show that the ratio of the surface [i.e. total surface area] of the cube to the surface of the octahedron, when both are inscribed in the same sphere, is like the ratio of the side of the equilateral triangle to its altitude.

[82.] For example, we take H as the face of the cube and B as the face of the octahedron, and let the equilateral triangle be CDE and CZ its altitude. [83.] I say that the ratio of the surface of the cube to the surface of the octahedron is equal to $DE : CZ$. [84.] Its proof is: we take the CT as half the altitude and DH' as two-thirds the side.[50] [85.] It has been shown in the preceding that the ratio of the square on H to the triangle B is like $[DH']$: ZT.[51] [86.] Six times H is the surface of the cube and six times B is three-quarters the surface of the octahedron.[52] [87.] The ratio of the surface of the cube to three-quarters the surface of the octahedron is equal to $D[H']$: ZT. [88.] Let ZB' be two-thirds the altitude; ZT is [thus] $3(ZB')/4$. [89.] The ratio of the surface of the cube to three-quarters the surface of the octahedron is equal to $DH' : ZT$. The ratio of three-quarters the surface of the octahedron to the entire surface of the octahedron is equal to $ZT : ZB'$. By the equality of the ratio, [the ratio] of the surface of the cube to the surface of the octahedron is equal to $DH' : ZB'$. [90.] This is as the ratio of two-thirds the side to two-thirds the altitude, which is like the ratio of the side to the altitude. The ratio of the surface of the cube to the surface of the octahedron is like the ratio of side DE to altitude CZ. Q.E.D.

[50] There is a duplication in the lettering. B and H are used both to indicate faces of polyhedra and points on segments. The second time these letters appear, and, in general, wherever such duplication occurs in the following propositions, we use the "prime" symbol; in this case, B' and H'.

[51] $CT = ZT$.

[52] Literally: "the ratio of six times H ... the ratio of six times B."

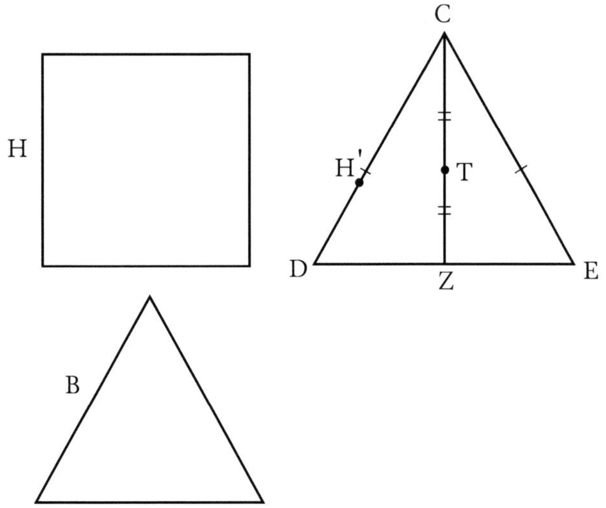

Figure 7: Proposition 11.

[Proposition 12]

[The proof of this proposition is very corrupt, and I have not attempted to reconstruct it. However, the enunciation is clear enough: "the ratio of the volume of the cube to the volume of the octahedron is like the ratio of the side of an equilateral triangle to its altitude."]

[Proposition 13]

[91.] We wish to show that for any [two] different lines, each one of which is cut in mean and extreme ratio, the ratio of one of the lines to the other is like the ratio of the greater segment of the first line to the [greater] segment of the other line.

[92.] For example, lines AB and DE which are not identical are each divided in mean and extreme ratio at C and Z [respectively]. I say that $AB : DE$ is the same as the ratio of AC, the greater [segment to DZ, the greater segment, and the same as the ratio of] CB, the smaller [segment] to ZE, the smaller [segment]. [93.] Its proof is: we extend $B[H] = CB$ and $ET = EZ$. [94.] $AB \cdot BC = ([AC^2 \text{ and}]\ ED) \cdot EZ = [D]Z^2$. $4AB \cdot BC : (AC)^2 = 4DE \cdot EZ : (DZ)^2$. *Componendo*, $4(AB \cdot BC) + (AC)^2\ [: (AC)^2] = 4DE \cdot EZ + (DZ)^2 : (DZ)^2$. $4AB \cdot BC + (AC)^2 = (AH)^2$ and $4DE \cdot EZ + (DZ)^2 = (DT)^2$, as Euclid explained in his second book. [95.] Thus $([H]A)^2 : ([A]C)^2 = (DT)^2 : (DZ)^2$ and $HA : AC = TD : DZ$. After we have separated, $HC : C[A] = ZT : ZD$.[53] The ratio of $BC : C[H]$, which is its half [ie. $BC = \frac{1}{2}HC$], is like $ZE : ZT$, which is [also] its half. Converting, $CB : CA = ZE : ZD$. Compounding, $AB : AC = ED : DZ$, and after converting, $AB : DE = AC : DZ$. [96.] $AC : DZ$ has already [been shown] to be equal to $CB : ZE$. Q.E.D.

[97.] This proposition was noted by Apollonius in this way. We have formulated a different approach, and it is this: [98.] [We take] the two lines AB and DE and divide each one of them in mean and extreme ratio at C and Z [respectively]. I say that $AB : DE$ is as the ratio of AC, the greater segment, to DZ, the greater [segment and] as the ratio of CB, the smaller segment, to ZE, the smaller segment. [99.] Its proof is: we draw the perpendicular BH equal to AB and the perpendicular ET equal to DE, and join $[H]A$, $[H]C$, TD and TZ. [100.] Since $AB = [H]B$ and angle B is right, angle A is half of a right [angle], and (?) angle D is half of a right [angle]. [101.] $(AB)^2 + (BC)^2 = 3[(AC)]^2$. $AB = [H]B$

[53] For the *separando*, recall that $AH = AC + CB + BH$ and $DT = DZ + ZE + ET$.

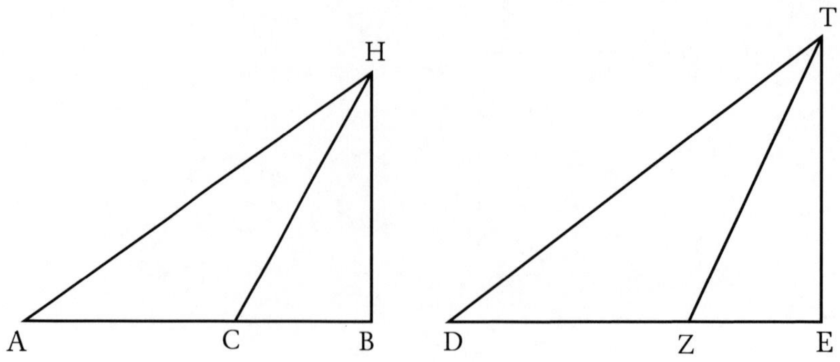

Figure 8: Proposition 13, part 2.

and $(HC)^2 = (HB)^2 + (BC)^2$, $(HC)^2 = 3(AC)^2$.[54] [102.] We show likewise that $(TZ)^2 = 3(DZ)^2$. $HC : CA = TZ : ZD$ and angle D = angle A. Thus angle AHC = angle DTZ and triangle $A[H]C$ is similar to triangle $D[TZ]$. [103.] Angle $C[H]B$ = angle ZTE and the two angles B, E are right; thus triangle HBC is similar to triangle TZE, triangle AHB is similar to triangle $DT[E]$, triangle AHC is similar to triangle DTZ, [and] triangle CHB is similar to triangle ZTE. [104.] Thus $AH : DT = AC : Z[D] = CB : ZE$.[55] Q.E.D.

[Proposition 14]

[105.] We wish to show that when the side of a hexagon is cut in mean and extreme ratio, the greater [segment] is the side of a [decagon].

[106.] For example, we take $[DZ]$, the side of a hexagon, and cut it in mean and extreme ratio at E, and EZ is the greater segment. I say that EZ is the side of a decagon. [107.] Its proof is: we take CB as the side of a hexagon equal to DZ and AC as the side of a decagon. AB has been cut in mean and extreme ratio and its greater segment is CB.[56] [108.] $AB : DZ = C[B] : EZ$, as we demonstrated in the preceding problem. [109.] When we convert this: $AB : BC = DZ : ZE$. However, $AB : BC = BC : CA$[57] and [therefore] $BC : CA = (DZ) : ZE$. $CB = [DZ]$ and [thus] $AC = EZ$. AC is the side of a decagon and EZ is the side of a decagon. Q.E.D.

[Proposition 15]

[110.] We wish to show that the [square on the] side of the pentagon and the square on the chord [subtending] two-[fifths] of the circle is equal to five times the square on the radius of the circle which circumscribes the pentagon.

[111.] For example, we take circle $ACZB$ and CB as the side of the pentagon. Let arc CZ be equal to arc ZB and D be the center of the circle. We join ZDA and we join AC; AC is the chord [subtending] two-fifths of the circle.[58] I say that $(AC)^2 + (CB)^2 = 5(DZ)^2$. [112.] Its proof is: we join CZ; CZ is the side of the decagon. $(AC)^2 + (CZ)^2 = (AZ)^2 = 4(DZ)^2$ since angle C is right.[59] [113.] We add $(ZD)^2$ to both sides. [114.] However, $(CZ)^2 + (ZD)^2 = (CB)^2$ because CZ is the side of the decagon,

[54] Cf. *Elements* XIII, 4.

[55] There is one missing step which we can easily supply, namely $AB : DE = AH : DT$.

[56] *Elements* XIII, 9.

[57] This follows from the definition of cutting a segment in mean and extreme ratio.

[58] That Z, D, and A are collinear is shown in *Elements* XIII, 11.

[59] Since CZ is the side of a decagon, $\angle CDZ = 36°$ and $\angle CAZ = 18°$. Similarly, it can be shown that $\angle CZA = 72°$. Hence, $\angle ZCA = 90°$.

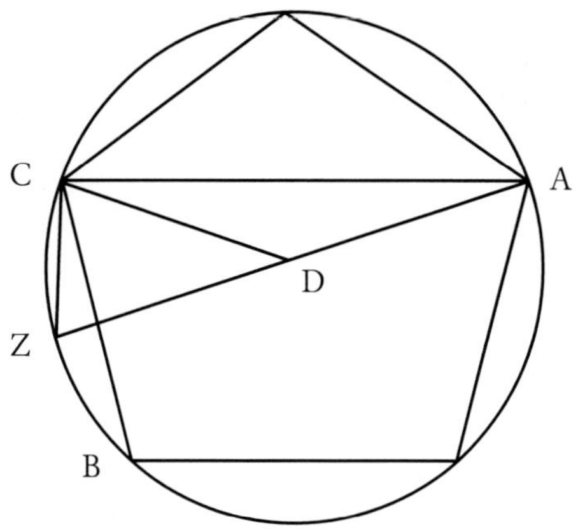

Figure 9: Proposition 15.

DZ is the side of a (hexagon) and CB is the side of a pentagon.[60] [115.] $(AC)^2 + (CB)^2 = 5(D[Z])^2$. Q.E.D. This proof was given by Apollonius in this way.

[Proposition 16]

[116.] We wish to show that the triangle of the icosahedron and the pentagon of the dodecahedron [are circumscribed by the same circle.

[117.] For example, we take] pentagon $ABCDE$. The center of the circle circumscribing it is Z and the radius of the circle is DZ. The triangle of the icosahedron is HTB', the center of circumscribing circle is L, and the radius is LB'. I say that $LB' = ZD$. [118.] Its proof is: we join $[B]E$; it is clear that BE is the edge of a cube.[61] [119.] We take MN as the diameter of the sphere, and the radius of the circle [circumscribing] the triangle of the icosahedron — [after recalling that] Euclid explained in book XIII that the side of the pentagon is the side of the triangle of the icosahedron — is XO.[62] We cut $[MN]$ in mean and extreme ratio at K, taking KN as the greater segment. XO is also (?) divided in mean and extreme ratio at S and $S[O]$ is the greater segment. [120.] We cut BE also in mean and extreme ratio at P. The greater segment is PE and $PE = AE$.[63] [121.] $(MN)^2 : (XO)^2 = (KN)^2 : (SO)^2$. Euclid has shown in Book XIII of his work that $(MN)^2 = 5(XO)^2$.[64] $(KN)^2 = 5(OS)^2$. $(MN)^2 + (KN)^2 = 5((XO)^2 + (SO)^2)$. [122.] However, $(XO)^2 + (SO)^2 = ([H]T)^2$ since XO is the side of the hexagon and $[S]O$ is the side of the decagon. [123.] $(MN)^2 + (NK)^2 = 5(HT)^2$. $(HT)^2 = 3(B'L)^2$. $(MN)^2 + (NK)^2 = 15(L[B'])^2$. [124.] Furthermore, $(MN)^2 : (BE)^2 = (KN)^2 : (PE)^2$ and $(MN)^2 = 3(BE)^2$, since it $[BE]$ is the edge of a cube. $(KN)^2 = 3(PE)^2$ according to what Euclid demonstrated in Book XIII of his work. $(MN)^2 + (NK)^2 = 3((BE)^2 + (EP)^2)$.[65] [125.] However, $(BE)^2 + (EP)^2 = 5(ZD)^2$,[66] and $(MN)^2 + (KN)^2 = 15(ZD)^2$. [126.] It has already been shown that $(MN)^2 + (NK)^2 = 15(LB')^2$. [127.] Therefore, $ZD = LB'$. Q.E.D. This proposition was given Apollonius in a similar way. However, that which I have given is easier for students.

[60] *Elements* XIII, 10.

[61] *Elements* XIII, 17.

[62] In *Elements* XIII, 16 Euclid uses pentagons to construct the triangles of the icosahedron, such that the side of the equilateral triangle of the icosahedron is also the side of a pentagon, and this property is the key to the whole proof.

[63] *Elements* XIII, 8.

[64] $MN^2 = 5XO^2$ follows from the porism to XIII, 16.

[65] $MN^2 = 3BE^2$ follows from XIII, 15.

[66] See preceding proposition.

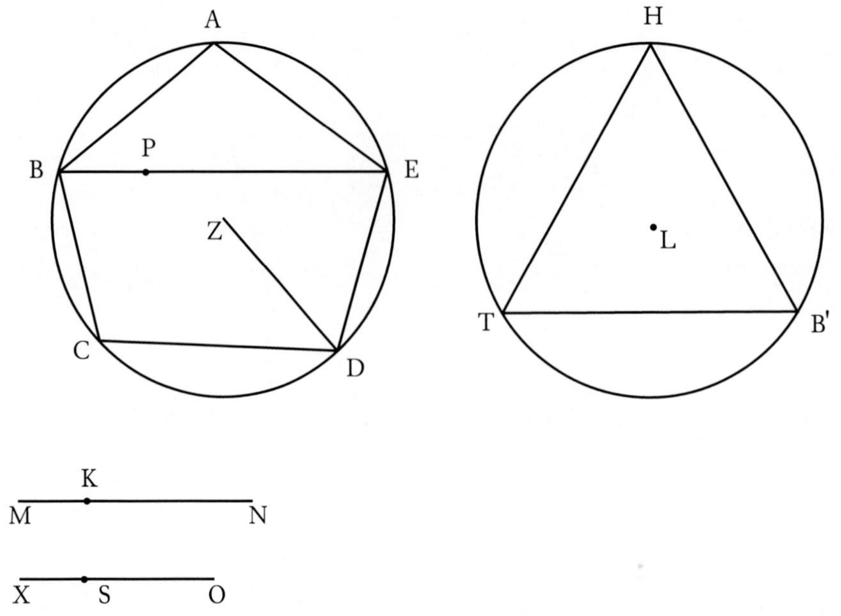

Figure 10: Proposition 16.

[Proposition 17]

[This text seems to be hopelessly corrupt. The proof looks to be similar or identical to that of the Arabic.]

[Proposition 18]

[128.] We wish to show that the ratio of the [surface] area of the cube to the [surface] area of the icosahedron is as the ratio of the square of the side of the pentagon to [three] and a third times the equilateral triangle whose side equals the root of three times the square on a decagon of the circle which circumscribes the pentagon.[67]

[129.] For example, we take AB as the side of the hexagon and divide it in mean and extreme ratio at C. [130.] $D = \sqrt{(AB)^2 + (AC)^2}$ and D is the side of the pentagon.[68] [131.] $E = \sqrt{(AB)^2 + (BC)^2} = \sqrt{3(AC)^2}$ [where AC] is the side of the decagon.[69] [132.] I say that the ratio of the [surface] area of the cube to the [surface] area of the icosahedron equals the ratio of D^2 to $3\frac{1}{3}$ the equilateral triangle on E. [133.] Its proof is: it has already been shown in the preceding proposition that the ratio of the edge of the cube to the edge of the icosahedron $= D : E$, and the ratio of the [square on the] edge of the cube to the square on the edge of the icosahedron $= D^2 : E^2$. [134.] Inverting, the ratio of the square [on the edge] of the cube to $D^2 =$ the ratio of the square on the edge of the icosahedron to E^2. [135.] The ratio of any line to twice its triangle...;[70] D^2 : twice the triangle on $D = E^2$: twice the

[67] The Arabic text specifies (relying on *Elements* XIII 1,6, presumably) that the side of the pentagon is also the edge of the icosahedron. This crucial fact is not mentioned in the Hebrew.

[68] *Elements* XIII, 10.

[69] *Elements* XIII, 4.

[70] This sentence is incomplete in the Hebrew. The Arabic here reads: "For any two lines, the ratio of the square on the one to the square on the other is equal to the triangle on the one to the triangle to the other." (Triangles are to each other as the squares on their sides by VI, 19). The Hebrew has simply multiplied both sides by ½.

triangle on E. [136.] Therefore, the square [on the edge] of the cube : D^2 = the square on the edge of the icosahedron : E^2. And [the square on the edge of the cube] : twice the triangle on D, which is twice the face of the icosahedron $=[D]^2$, which is the edge of the icosahedron : twice the triangle on E. [137.] The ratio of six times the square of the cube, which is the [surface] area of the cube, to twelve times the face of the icosahedron, which is 3/5 the [surface] area of the icosahedron, is equal to the ratio of the face of the cube to twice the face of the icosahedron, which is the same as the ratio of D^2 to twice the triangle on E. [138.] The ratio of the surface area of the cube to three-fifths the surface area of the icosahedron is equal to the ratio to twice the triangle on E. The ratio of three-fifths the [surface] area of the icosahedron to the surface area of the icosahedron is equal to the ratio of D^2 to twice the triangle on E to three and one-third the triangle on E. The complete ratio of the surface area of the cube to the surface area of the icosahedron is equal to the ratio of D^2 to three and one-third the equilateral triangle on E. Q.E.D.

[139.] It is clear from what we have described that the ratio of three-fifths the surface area of the icosahedron to the surface area of the cube is the same as the ratio of twice the triangle on $E : D^2$.

[Note: Mantua Proposition 18 contains a porism demonstrating (after emending the text) the following relationship:

$$\frac{s_6}{s_{20}} = \frac{\sqrt{a^2 + x_1^2}}{\sqrt{a^2 + x_2^2}}.]$$

[Proposition 19]

[140.] We wish to show that the ratio of the [surface] area of the icosahedron to the [surface] area of the octahedron is equal to the ratio of five times the square on the side of the decagon of the circle to the square on the side of the pentagon.

[141.] Let A be the side of the pentagon and B the side of the decagon. I say that the ratio of the [surface] area of the icosahedron to the [surface] area of the octahedron is equal to $5B^2 : A^2$. [142.] Its proof is: we take $C = 3B^2$. We have already shown that the ratio of three-fifths the [surface] area of the icosahedron to the [surface] area of the cube is equal to twice the triangle on $C = A^2$. [143.] We have already said that the ratio of the [surface] area of the cube to the [surface] area of the octahedron is equal to the ratio of A^2 to twice the triangle on it [i.e. on A], for we have already shown in the preceding that the ratio of the [surface] area of the cube to the [surface] area of the octahedron is equal to the ratio of the side of any equilateral triangle to its altitude, which is like the ratio of the square [of its side] to twice its triangle. [144.] The ratio of three-fifths the [surface] area of the icosahedron to the [surface] area of the cube is equal to the ratio of twice the triangle on [C] to A^2. The ratio of the surface area of the cube to the [surface] area of the octahedron is equal to the ratio of A^2 to twice the triangle [on A]. In the equality of the ratio, the ratio of three-fifths the [surface] area of the icosahedron to the [surface] area of the octahedron is equal to the ratio of twice the triangle on C to twice the triangle on A, which in turn is equal to the ratio of the triangle on C to the triangle on A. [145.] The ratio of the triangle on C to the triangle on A is equal to [$3B^2$]: A^2. The ratio of three-fifths the [surface] area of the icosahedron to the [surface] area of the octahedron is $(C)^2$ which is the same as $3B^2$ to A^2. The ratio of three-fifths the [surface] area of the icosahedron to the [surface] area of the entire icosahedron is equal to the ratio, $3B^2 : 5B^2$. In the equality of the ratio, the ratio of the [surface area of the] icosahedron to the [surface] area of the octahedron is $5B^2 : A^2$. Q.E.D.

[Proposition 20]

[146.] We wish to show that the ratio of the [excess] of the [surface] area of the icosahedron over the [surface] area of the octahedron [to the surface area of the octahedron] is equal to the ratio of the smaller segment of any line cut in mean and extreme ratio to the entire line. For this reason, the ratio of the [surface] area of the icosahedron to the [surface] area of the octahedron will be equal to the sum of the smaller segment of any line cut in mean and extreme ratio and the entire line to the entire line, which is the same as the ratio of the sum of two sides of any decagon and the side of the hexagon to

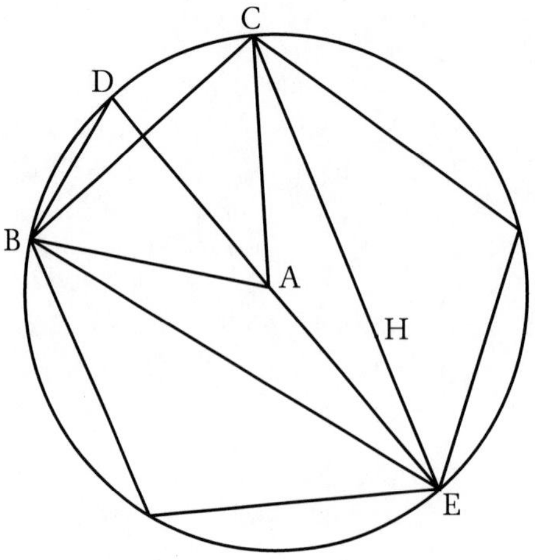

Figure 11: Proposition 20.

the sum of the side of the decagon and the side of the hexagon.

[147.] For example, we take circle *ECB* with center *A*, and the side of the inscribed pentagon is *CB*. We bisect [arc] *CB* at *D*, join *AD*, and extend it to *E*. We join *EC*, *EB*, and divide *EC* in mean and extreme ratio [at H].[71] [148.] I say that the ratio of the [excess] of the [surface] area of the icosahedron over the [surface] area of the octahedron [to the surface area of the octahedron] is equal to $HE : EC$. [149.] Its proof is: we join *AC*, *AB*, and *BD*. Triangle $E[C]B$ is similar to triangle *ABD* because $\angle DAB = \angle E$. The ratio of $5(DB)^2$, which is the side of the decagon, to $(CB)^2$, which is the side of the pentagon, is equal to $5(AB)^2 : (EC)^2$. [150.] $5(AC)^2 = (EC)^2+(C[B])^2$, since *EC* is a chord subtending two-fifths of the circle, and *CB* is a chord subtending its fifth.[72] [151.] $5(DB)^2 : (CB)^2$, as we have already shown, is equal to the ratio of the [surface] area of the icosahedron to the [surface] area of the octahedron. $(EC)^2 + (C[B])^2 : (EC)^2$ is equal to the [surface] area of the icosahedron to the [surface] area of the octahedron. *Separando*, the ratio of the excess of the [surface] area of the icosahedron over the [surface] area of the octahedron to the [surface] area of the octahedron is equal to $(HC)^2 : (C[E])^2$.[73] [152.] Since $E[H]$: $[HC]= HC : CE$, $([H]C)^2 : (CE)^2 = H[E :] EC$. Therefore, the ratio of the [excess] of the [surface] area of the icosaderon over the [surface] area of the octahedron [to the surface area of the octahedron] is equal to the ratio of *EH*, the smaller segment, to the whole line. *Componendo*, the ratio of the [surface] area of the icosahedron to the [surface] area of the octahedron is equal to the ratio of the sum of the smaller segment of the entire line [and the entire line] to the entire line. Since, when the side of the decagon is joined to the side of the hexagon, the whole line has been cut in mean and extreme ratio, and the smaller segment is the side of the decagon, the ratio of the [surface] area of the icosahedron to the [surface] area of the octahedron is equal to the ratio of the sum of two sides of the decagon and the side of the hexagon to the sum of the side of the

[71] *EH* is the smaller segment.

[72] *CA* and *AB* are radii, chord *EC* subtends two-fifths of the circle and *CB* one-fifth, because each arc of the pentagon subtends one-fifth of the circle's circumference.

[73] From Proposition 19, since $CH = CB$ (*Elements* XIII, 8).

decagon and [the side of the] hexagon.[74] Q.E.D.

[Proposition 21][75]

[153.] We wish to show that the parallelepiped whose base is equal to the pentagon of the icosahedron[76] and whose altitude is two-[thirds] the diameter of the sphere which circumscribes the icosahedron [is equal in volume to the icosahedron.

[154.] Its proof is:] we draw from its center a perpendicular to the plane [of the pentagon] together in both directions (!). One side is equal to [half] the side of the hexagon of the circle which circumscribes the pentagon, and the other side is equal to the side of the decagon drawn therein.[77] [155.] Extension of each of the points (!) to the angle of the [icosahedron] produces on the pentagon two pyramids. The vertex of the one is the point of intersection of the side of the hexagon (!), and the vertex of the other is the point of intersection of the side of the decagon (!). [156.] These two pyramids are equal to the [parallele]piped whose base is equal to the pentagon and whose altitude is equal to [one-sixth] the side of the hexagon and one-third the side of the decagon. For the side of the decagon and [half] the side of the hexagon and the decagon are equal to the altitudes of the two pyramids, and [half] the side of the hexagon and the decagon are equal to the radius of the sphere; this is clear from the icosahedron.[78] [157.] The parallelepiped whose base is equal to the pentagon and whose altitude is one-third the radius of sphere, which is one-sixth the diameter of the sphere, is equal to the two pyramids. These two pyramids are equal to the five pyramids whose bases are equal to the triangles which stand from the side of the intersection of the side of the decagon in the angle of the icosahedron (!) and whose vertex is the center of the sphere, which is the point of intersection of [half] the side of the hexagon. [158.] There are in the icosahedron twenty pyramids equal to each one of these five, and the volume of the icosahedron is equal to four times these five pyramids. These five pyramids are equal to the two pyramids, the vertex of the one being the point of intersection of the side of the decagon, and the vertex of the other, the point of intersection of [half] the side of the hexagon; the base is the pentagon. These two pyramids are equal to the parallelepiped whose base is equal to the pentagon and whose altitude is equal to one-sixth the diameter of the sphere. [159.] The volume of the icosahedron is four times the solid whose base is equal to the pentagon and whose altitude is one-sixth the diameter of the sphere. The solid whose base is equal to the pentagon and whose altitude is four-sixth's the diameter of the sphere is four times the solid whose base is equal to the pentagon and whose altitude is one-sixth the diameter of the sphere. Therefore, the icosahedron is equal in volume to the solid whose base is the pentagon drawn on the circle of the icosahedron (!), where the side of the pentagon is equal to [the edge of] the icosahedron, and whose altitude is equal to two-[thirds] the diameter of the sphere. Q.E.D.

[Proposition 22]

[160.] We wish to show that the perpendicular drawn from the center of the circle to the side of the inscribed pentagon is equal to half the sum of the side[s] of the inscribed hexagon and decagon.

[161.] For example: circle ABC, in which BC is the side of the pentagon and D is the center. From D we draw DH perpendicular to CB and extend it rectilinearly to E on the circumference of the circle. We join BE, and BE is the side of the decagon. I say that $DH = ([D]E + EB)/2$. [162.] Its proof is: we draw EDA, join BD and take $[HZ] = E[H]$. We join BZ. [163.] $ZH = EH$, HB [is

[74] This key step is not expressed clearly in either the Hebrew or the Arabic texts. By the definition of mean and extreme ratio, $CH^2 = EC \cdot EH$. Hence, $CH^2 : EC^2 = EC \cdot EH : EC^2 = EH : EC$.

[75] The text of this proposition is also in a bad state. The proof is exactly that of the Arabic, and on that basis I have been able to more or less reconstruct the Hebrew text. It should be pointed out that the Hebrew consistently reads "the side of the hexagon" instead of the correct "half the side of the hexagon". Several puzzling phrases have been translated verbatim, followed by (!).

[76] "The pentagon of the icosahedron" is a "regular pentagon formed by five coplanar angular points of the icosahedron" (Hogendijk, Preprint, p. 45, note to prop. 24). See Hogendijk [1994, 211], which makes essentially the same point.

[77] *Elements* XIII, 16.

[78] "This is clear from the icosahedron" may be a reference to *Elements* XIII, 16.

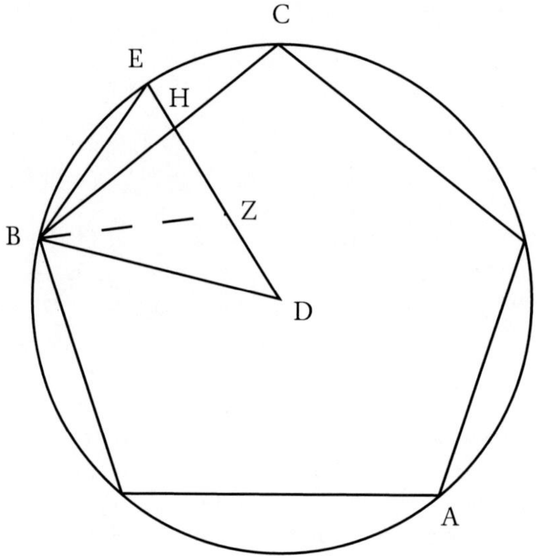

Figure 12: Proposition 22.

common], and the two angles [*BHZ* and *BHE*] are right. $BZ = EB$ and $\angle BEZ = \angle BZE$. [164.] Arc $AB = 4(\text{arc } BE)$ and $\angle ADB = 4\angle BDE$. $\angle ADB = [2]\angle BED$. $\angle BEZ = \angle BZE = \angle BDZ + \angle DBZ$.[79] [165.] $\angle B[D]H = \angle ZBD$ and $[D]Z = ZB$. $ZB = BE$, which is the side of the decagon. *DZ* is [thus] the side of the decagon. [166.] $ZH = HE$ and $ZH = ZE/2$. [167.] *DH*, therefore, is half [the sum of] *DZ* and *EB*. *D*[*E*] is the side of the hexagon and *EB* is the side of the decagon. Therefore, *D*[*H*] is half [the sum of] the side of the hexagon and the side of the decagon.[80] Q.E.D.

[Proposition 23]

[168.] We wish to show that the ratio of the perpendicular drawn from the center of the sphere to the face of the icosahedron to the diameter of the sphere in which the icosahedron is inscribed is equal to [the ratio of] half [the sum of] the side of the hexagon and the side of the decagon inscribed in the circle to twice the altitude of the equilateral triangle drawn on the side of the pentagon which is inscribed in the same circle.

[169.] For example, we take as the pentagon of the icosahedron[81] pentagon *ABCDE* with center *Z*, and we construct on *CD* the equilateral triangle *CDL*. We join *ZL*, intersecting *CD* at *H*; it is evident that *ZHL* is perpendicular to *CD*. Let line *T* be equal to the perpendicular drawn from the center of the sphere to the face of the icosahedron [and $K =$] the diameter of the sphere. [170.] I say that $T : [K] =$ the ratio of *ZH*, which has already been demonstrated to be equal to half [the sum of] the sides of the hexagon and the decagon, to 2*HL*, which is the altitude of the triangle. [171]. Its proof is: *CD* is the side of the pentagon of the icosahedron. Triangle *CLD* is equilateral and is evidently a face of the icosahedron. We join *ZD* and *ZC*. [172.] It is clear that the parallelepiped, whose base is equal to the surface area of the icosahedron and whose altitude is $T/3$, which [i.e., *T*] is the perpendicular drawn from the center of the sphere to the face of the icosahedron, is equal in volume to the icosahedron. [173.] For the icosahedron is like the pyramid[82] whose base is the face of the icosahedron and whose altitude is *T*; the volume of the [entire] icosahedron is twenty times this pyramid. The icosahedron is

---

[79] *Elements* I, 5 and 32.

[80] The text here omits the final steps of the proof. (Recall that $EB = BZ = DZ =$ the side of the decagon.) $DH = DZ + ZH = HE + BE$. $2DH = DH + HE + BE = DE + BE$. $DH = (DE + BE)/2$.

[81] For "pentagon of the icosahedron," see note 76, above.

[82] That is to say, the tetrahedral section of the icosahedron is like the pyramid described.

equal in volume to the solid whose base is equal to the surface area of the icosahedron and whose altitude is *T*/3. [174.] We have already shown that the parallelepiped whose base is equal to pentagon *ABCDE* and whose altitude is 2[*K*]/3 is equal in volume to the icosahedron.[83] [175.] The solid whose base is equal to the surface area of the icosahedron, which is twenty times triangle *CLD*, and whose altitude is *T*/3, is equal to the solid whose base is pentagon *ABCDE* and whose altitude is 2[*K*]/3. [176.] "In equal parallelepiped solids the bases are reciprocally proportionate to the altitudes."[84] [177.] Pentagon *ABCDE*: 20·triangle *CLD* = the ratio of triangle *CZD*, which is 1/5 of pentagon *ABCDE*, to [four] times triangle *CLD*. Triangle *CZD* : 4·triangle *CLD* = *T*/3 : 2(*K*)/3. Therefore, triangle *CZD* : triangle *CLD* = *T*/3 : (*K*)/6. [178.] Triangle *CZD* : triangle *CLD* = *ZH* : *HL*. *ZH* : *HL* = *T* : [*K*]/2. Therefore, *ZH* : 2*HL* = *T* : [*K*]. [179.] We have already shown that *ZH* = half [the sum of] the side of the hexagon and the side of the decagon inscribed in the circle of [i.e. circumscribing] pentagon *ABCDE*. [180.] It has thus been shown that the ratio of the perpendicular drawn from the center of the sphere to the face of the icosahedron to the diameter of the sphere which circumscribes the icosahedron is equal to the ratio of half [the sum of] the sides of the hexagon and the decagon drawn in the circle to twice the altitude of the equilateral triangle constructed on the side of the pentagon which is drawn in the same circle. Q.E.D.

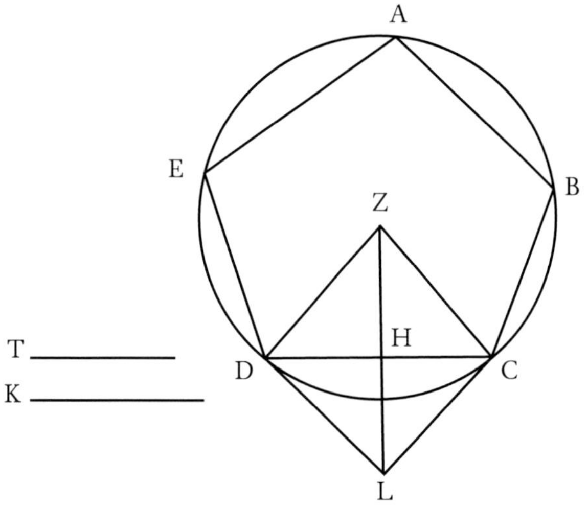

Figure 13: Proposition 23.

[Proposition 24]

[181.] We wish to show that the ratio of the [semi-]diameter of the sphere to the perpendicular drawn from the center of the sphere to the face of the octahedron is equal to the ratio of twice the altitude of any equilateral triangle to its side.

[182.] For the perpendicular drawn from the center of the sphere to the face of the cube is equal to half the edge of the cube. It has already been shown by Euclid in Book XIII that the square on the diameter of the sphere is three times the square on the edge of the cube.[85] The square on the semidiameter of the sphere is three times the square on half the edge of the cube, which is the perpendicular drawn from the center of the sphere to the face of the cube. [183.] The square on the altitude of any

[83] Proposition 21.

[84] A direct quotation from *Elements* XI, 34.

[85] *Elements* XIII, 15.

equilateral triangle is three times the square on half the side.[86] [184.] Therefore, the ratio of the square on the semidiameter to the square on the perpendicular drawn from the center of the sphere to the face of the cube is equal to the ratio of the square on the altitude of the equilateral triangle to the square on half its side. The ratio of the [semi-]diameter of the sphere to the perpendicular drawn from the center of the sphere to the face of the cube is as the ratio of the altitude of the equilateral triangle to half its side. [185.] Therefore, the ratio of the semidiameter of the sphere to the perpendicular drawn from the center of the sphere to the face of the cube, which is equal to the perpendicular drawn from the center of the sphere to the face of the octahedron, is equal to the ratio of twice the altitude of the equilateral triangle to [its side].[87] Q.E.D.

[Proposition 25]

[186.] We wish to show that [the ratio of] the perpendicular drawn from the center of the sphere to the face of the icosahedron to the perpendicular drawn from the center of the sphere to the face of the octahedron, when both are inscribed in the same sphere, is equal to the ratio of the sum of the sides of the hexagon and the decagon inscribed in the circle to the side of the pentagon [inscribed in] the same circle.

[187.] Its proof is: we have already shown that] the ratio of the perpendicular drawn from the center of the sphere to the face of the icosahedron to the diameter of the sphere is equal to the ratio of half the sum of the sides of the hexagon and the decagon to twice the altitude of the equilateral triangle constructed on the side of the pentagon [inscribed] in the same circle in which they [i.e., the hexagon and the decagon] are drawn.[88] [188.] We have already shown that the ratio of the diameter of the sphere to the perpendicular drawn from the center of the sphere to the face of the octahedron is equal to the ratio of twice the altitude of any equilateral triangle to half its side.[89] [189.] By the equality [of the ratio], the ratio of the perpendicular drawn from the center of the sphere to the face of the icosahedron to the perpendicular drawn from the center of the sphere to the face of the octahedron is equal to the ratio of half the sum of the sides of the hexagon and the decagon to half the side of the pentagon.[90] [190.] The ratio of the half to the half is equal to the ratio of the whole to the whole. The ratio of the perpendicular drawn from the center of the sphere to the face of the icosahedron [to the perpendicular drawn from the center of the sphere to the face of the octahedron] is equal to the ratio of the sum of the sides of the hexagon and the decagon to the side of the pentagon when [all] are inscribed in the same circle. Q.E.D.

[Proposition 26]

[The Hebrew text of this proposition in the Oxford Bodley d4 manuscript is in very bad shape; I reproduce the corresponding theorem in the Mantua manuscript below. Though Mantua preserves a far better text, it too is not free of errors, both mathematical (see following paragraph) and orthographical. Clearly Eliezer, the presumed author of the Mantua compendium, had a copy that transmitted some errors from the Ursache; he has, however, changed a few technical terms. The theorem seeks to prove that the ratio of the volume of the icosahedron to the volume of the octahedron is equal to the ratio of the side of an inscribed decagon plus the side of the inscribed pentagon to the side of its pentagon. This is incorrect, as Hogendijk has shown. Moreover, there is another error in the proof. We are told to construct three parallelepipeds: *A*, equal in volume to the icosahedron; *B*, equal in volume to the octahedron; and *G*, whose base is equal to the excess of the surface of the icosahedron over the surface octahedron, and its altitude equal to one-third of the perpendicular of the icosahedron. Then, it is said,

[86] Proposition 1.

[87] That the perpendiculars drawn to the faces of the octahedron and the cube are equal; see porism to Proposition 8 (sentence 70).

[88] Proposition 23.

[89] Proposition 24, doubling both sides of the equation.

[90] Recall that the equilateral triangle in Proposition 23 has been constructed on the side of the pentagon.

$G = A - B$. This is incorrect as well.[91] The Hebrew texts in the Bodley d4 manuscript, like the Arabic, displays the erroneous equation $G = A - B$. This same error appears also in the Mantua manuscript as well; thus all known versions contain the same error.]

[191.] This completes the relationships of the four (!) figures one to the other. There remain the relationships which Apollonius mentioned between the dodecahedron and the icosahedron. This is the beginning of the explication of his treatise.

[Proposition 27]

[The text here is in bad shape; the copyist missed a few lines which belong to the beginning of the proposition. From the end of the proof it appears that the proof is the same as in the Arabic, but the lettering of the figure is different. Fortunately, the thirteenth proposition in the corresponding chapter of the Mantua MS preserves the text: here is our translation:]

[*Mantua* Proposition 13 (part one)]

When the perpendicular is drawn from the center of the circle which circumscribes the pentagon to one of the sides of pentagon, thirty times the product of the side and the perpendicular are equal to the surface area of the dodecahedron. Let the equilateral pentagon in the circle be $ABCDE$, and its center be Z [the text here reads E]. I say that thirty times the product of AB and ZH are equal to 12 times the pentagon $ABCDE$. Its proof is: we join DZ and CZ. Now the surface area of the dodecahedron is twelve times pentagon $ABCDE$, and it is five times triangle ZCD. The surface area of the dodecahedron is sixty times triangle ZCD. $CD \cdot Z[H]$ is twice [triangle] ZCD. Hence the surface area of the dodecahedron is thirty times $CD \cdot ZH$, as we intended. Or we may say: the pentagon is five times triangle $(C)ZD$ and $ZH \cdot CD$ is twice triangle CZD. Hence the pentagon is two and one-half times $CD \cdot ZH$. Therefore twelve pentagons will be [equal to] thirty times $CD \cdot ZH$. Now $DC \cdot HZ$ is one time triangle CDZ, so that five [times] $CD \cdot ZH$ is ten times triangle $C[D]Z$, and ten times triangle CDZ is double pentagon $ABCDE$. When you have multiplied this six times, thirty times $CD \cdot [H]Z$ is equal to twelve times pentagon $ABCDE$. Hence thirty times $CD \cdot HZ$ is equal to the surface area of the dodecahedron...

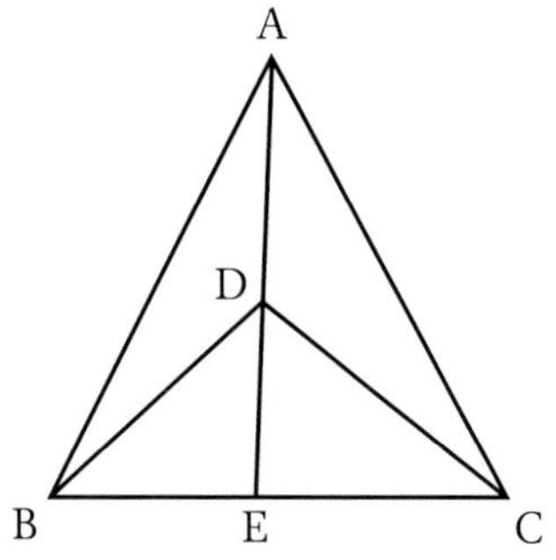

Figure 14: Proposition 28.

[Proposition 28]

[192.] We wish to show that the area which is the product of the perpendicular drawn from the center of the circle which circumscribes the face of the icosahedron and [the edge of the solid] is

[91] See Hogendijk [1994], note 29.2, and note 29.7 for a relatively simple correction to the proof, indicating that the errors may have creeped into the theorem in the process of its transmission.

one-thirtieth the surface area of the icosahedron.

[193.] For example, triangle *ABC* is the face of the icosahedron, and [*D*] is the center of the circle which circumscribes it. *DE* is perpendicular to *BC*. I say that the area which is the product of *DE* and *BC* is 1/30 the surface area of the icosahedron. [194.] [Its proof is: the surface area of the icosahedron is] 20 times triangle *ABC*, and triangle *ABC* is [three] times triangle *DBC*. The surface area of the icosahedron is 60 times triangle *DBC*. [195.] The area which [is the product of] *DE* and *BC* is two times triangle *DBC*. Therefore the surface area of the icosahedron is [thirty times] the area which is [the product of] *DE* and *BC*. Q.E.D.

[Proposition 29]

[196.] We wish to show that the ratio of the surface area of the dodecahedron [to the surface area of the icosahedron] is equal to the ratio of the area which is the product of the perpendicular drawn from the center of the circle circumscribing the face of the dodecahedron to it [i.e. the pentagon] and the side of the pentagon to the area which is the product of the perpendicular drawn from the center of the circle circumscribing the face of the icosahedron to the side of the triangle and the side of the triangle.

[197.] Its proof is: the ratio of [the surface area] of the dodecahedron to the surface area of the icosahedron is equal to the ratio of one-thirtieth the surface area of the dodecahedron to one-thirtieth of the surface area of the icosahedron. One-thirtieth the surface area of the dodecahedron is equal to the product of the altitude [of the dodecahedron] and the side of the pentagon,[92] and one-thirtieth the surface area of the icosahedron is equal to the product of the altitude and the side of the triangle. [198.] The ratio of the surface area of the dodecahedron to the surface area of the icosahedron is equal to the ratio of the product of the perpendicular drawn from the center of the circle which circumscribes the face of the dodecahedron to the edge and the edge, to the product of the perpendicular drawn from the center of the circle which circumscribes the face of the icosahedron and the edge.

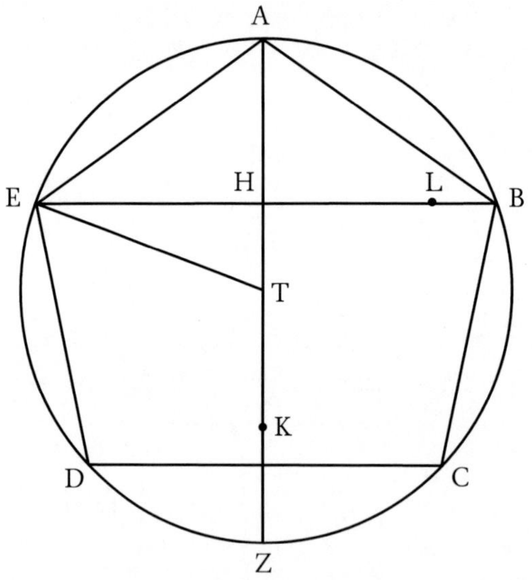

Figure 15: Proposition 30.

[92] The altitude of the dodecahedron is the perpendicular drawn from the center of the circumscribing circle to the edge of the pentagon.

[Proposition 30]

[199.] We wish to show that the area of any inscribed pentagon is equal to the product of three-fourths the diameter of the circle and five-sixths the chord subtending two-fifths of the circle (containing) the pentagon.

[200.] For example, the [pentagon] is *ABCDE*, and the diameter of the circle is *A*[*H*]*Z*. We join *BHE*. Let $BL = BH/3$; hence $[BL]=[B]E/6$ and $LE = 5BE/6$. Let *T* be the center of the circle. We join *TE* and bisect *TZ* at [*K*]. $A[K]= 3AZ/4$. [201.] I say that $AK \cdot LE$ is the area of pentagon *ABCDE*. [202.] Its proof is: $BH = 3[B]L$. $HE = 3BL$. $AK= 3(AT)/2$. $[K]T = AT/2$. [203.] $HE : [B]L = A[K] : \frac{1}{2}AT$. $HE \cdot (AT/2) = A[K] \cdot [B]L$. $HE \cdot (AT/2)$ is [the area of] triangle *AET*. Hence $A[K] \cdot 5[B]L$ is five times triangle *AET*. [204.] Five times triangle *AET* is equal [in area] to pentagon *ABCDE*. $5BL = LE$. Therefore, $A[K] \cdot LE$ is [equal to the area of] pentagon *ABCDE*. Q.E.D.

[205.] We wish to show that the ratio of the surface area of the dodecahedron to the surface area of the icosahedron, when both are inscribed in the same sphere, is equal to the ratio of the edge of the cube circumscribed by the same sphere to the edge of the icosahedron.

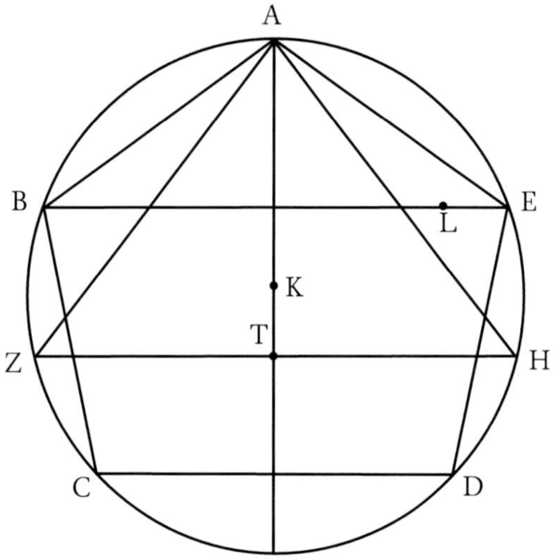

Figure 16: Proposition 31, diagram 1.

[Proposition 31]

[206.] For example, we take circle *ACB* as the circle which circumscribes the pentagon of the dodecahedron and the triangle of the icosahedron. Let its center be [*D*], and the side of the triangle inscribed in it, which is the edge of the icosahedron, *AB*, and the side of the pentagon inscribed in it, which is the edge of the dodecahedron, *AC*. Let [*T*] be the edge of the cube which is inscribed in that same sphere. I say that the ratio of the surface area of the dodecahedron to the surface area of the icosahedron is equal to $T : AB$. [207.] Its proof is: we draw from *D* a perpendicular to *AC*, [intersecting it at] *Z*. [*DZ*] is equal to half the side of the hexagon and half the side of the decagon.[93] [XIV, 1; Proposition 22] Similarly, we draw from *D* a perpendicular to *AB*, and it is *DE*. *DE* is half the side of the hexagon. [208.] When [line *T*] is divided in mean and extreme ratio, [the greater segment will be] *AC*.[94] [209.]

[93] $DZ = \frac{1}{2}(c(6) + c(10))$: Proposition 22.

[94] Porism to *Elements* XIII, 17.

When the sides of the hexagon and decagon [joined together] are divided in mean and extreme ratio, the greater segment will be the side of the hexagon. When half the side of the hexagon [and half the side of the decagon joined together] are divided in mean and extreme ratio, the greater segment will be half the side of the hexagon. When [DZ] is divided in mean and extreme ratio, the greater segment is [DE]. [210.] $T : AC = DZ : DE$. [T] $\cdot DE = AC \cdot DZ$. [211.] We have already shown that [the ratio] of $AC \cdot DZ$ to $AB \cdot DE$ is equal to the ratio of the surface area of the dodecahedron to the surface area of the icosahedron, and $AC \cdot DZ = T \cdot DE$.[95] [212.] Therefore, $T \cdot DE : AB \cdot DE =$ the ratio of the surface area of the dodecahedron to the surface area of the icosahedron. Therefore, $T : AB =$ the ratio of the surface area of the dodecahedron to the surface area of the icosahedron.

[213.] Said Kalonymos Todros the scrutinizer [i.e. he who scrutinized the MS] : These two propositions, i.e. Propositions 30 and 31, were missing from the translation of the leader (*nasī'*) and scholar R. Kalonymos ben R. Kalonymos, of blessed memory, of the treatise on the five solids, because a leaf was missing from the Arab text from which he translated. Several years afterwards, I found that leaf and translated it. [214.] Miles of Marseilles wrote this on the third of Ellul [50] 95 in Aigues. It was missing from the treatise on the comparison of the five solids, from the middle. That treatise is written in this codex which is before me, and now, after attaining these two propositions by the grace of God, this codex is complete, Thank God.

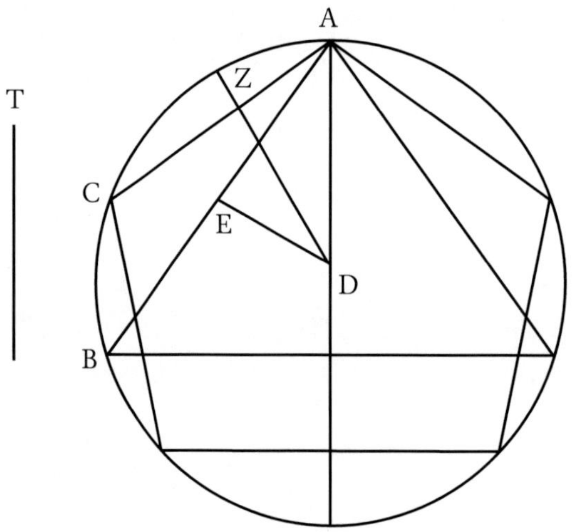

Figure 17: Proposition 31, diagram 2.

[215.] Apollonius said: Let us demonstrate this in another way. We draw a circle which circumscribes the face of the dodecahedron and the face of the icosahedron, circle $ABCDE$. The pentagon constructed therein is pentagon $ABCDE$, and the triangle therein is triangle AZH. We join BE, and BE is the edge of the cube.[96] Let BL be five-sixths of it [i.e. BE]. We draw the perpendicular $A[K]T$, and the center is K. It is clear that KT is one-fourth the diameter of the circle, and AT is three-fourths the diameter of the circle. I say that the ratio of [the surface area of] the dodecahedron to [the surface area] of the icosahedron is equal to $BE : ZH$. [216.] Its proof is: $AT \cdot (B)L$ is [the area of] pentagon $ABCDE$, which is the face of the dodecahedron. [217.] $AT[\cdot]ZT$ is [the area of] triangle AZH, which is the face of the icosahedron. The ratio of pentagon $ABCDE$ to triangle $AZH = [B]L : ZT$. $12([B]L) : 20(ZT) =$ the ratio of [twelve times] pentagon $ABCDE$, which is the surface area of the dodecahedron, to twenty times triangle AZH, which is the surface area of the icosahedron. [218.] Therefore, the ratio of the surface area of the dodecahedron to the surface area of the icosahedron $= 12[BL]: 20(ZT)$. Since

[95] See Proposition 29.

[96] *Elements* XIII, 17.

$[B]L = 5(BE)/6$, $12([B]L) = 10(BE)$. $20(ZT) = [10](ZH)$. [219.] The ratio of the surface area of the [dodecahedron] to the surface area of the icosahedron is equal to $10BE : 10ZH = BE : ZH$. The ratio of the surface area of the dodecahedron to the surface area of the icosahedron is equal to the ratio of the edge of the cube, which is *BE*, to the edge of the icosahedron, which is the line *ZH*. Q.E.D.

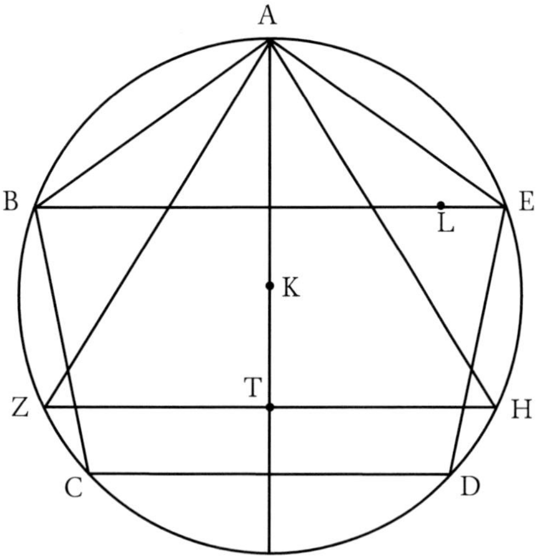

Figure 18: Proposition 31, part 2.

[Proposition 32]

[220.] We wish to show that the ratio of the surface area of the dodecahedron to the surface area of the icosahedron is equal to the ratio of the line that is the sum of the squares (*yaḥaziq ʿal*) of a line divided in mean and extreme ratio and its greater part, to the the line that is the sum of the squares of the line divided in mean and extreme ratio and its smaller part,[97] which is [in turn] equal to the ratio of the side of an inscribed pentagon to the square root of three times the square on the side of the decagon which is inscribed in the same circle.

[221.] For example, let *AB* be divided in mean and extreme ratio at *C*, and *AC* is the greater segment. Let *AD* and *BE* [both be equal to *AB* and] perpendicular to *AB* and *BC*.[98] I say that the ratio of the surface area of the dodecahedron to the surface area of the icosahedron is equal to $DC : CE$. [222.] Its proof is: we have already shown that the ratio of the surface area of the dodecahedron to the surface area of the icosahedron is equal to the ratio of the edge of the cube to the edge of the icosahedron.[99] [223.] It has also been shown in Proposition 18[100] that the ratio of the edge of the cube to the edge of the icosahedron is equal to $DC : CE$. Therefore, the ratio of the surface area of the dodecahedron to the surface area of the icosahedron $= DC : CE$. [224.] If we take [*AD*] to be the side of the hexagon [then] *AC* is the side of the decagon. Then *DC* is the side of the pentagon, since *AD* is the side of a hexagon and *AC* the side of the decagon.[101] [225.] Moreover, *CE* is equal to the root of three times

97 Using Hogendijk's notation: $\sqrt{a^2 + x_1^2} : \sqrt{a^2 + x_2^2}$.

98 The proposition takes $BE = AD = AB$.

99 Proposition 31.

100 The correct reference according to our count would be Proposition 17 (= Proposition 6, Hogendijk [1994]). Our count is verified by the numbering in the margin of a few of the propositions towards the end of the text. Nevertheless, given the textual problems of MS Bodley d4, there could be some other explanation.

101 $AC = c(10)$. Proposition 14. $DC = c(5)$. *Elements* XIII, 10.

the square on the side of the decagon.[102] Q.E.D.

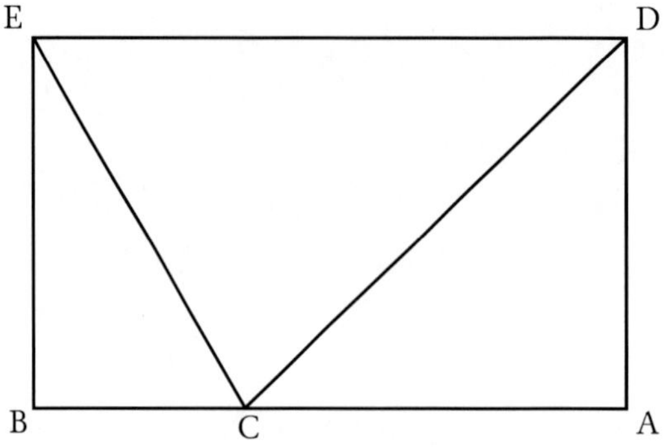

Figure 19: Proposition 32.

[Proposition 33]

[This proposition is the only one to display its numbering in the text; in the margin it is listed as thirty-three, and indeed it is the thirty-third in our count. It begins on f. 193b, l. 23, as follows:]

We wish to show that the ratio of the volume of the dodecahedron to the volume of the icosahedron, which (is the ratio) of the surface area of the dodecahedron to the surface area of the icosahedron, is equal to $\sqrt{a^2 + x_1^2} : \sqrt{a^2 + x_2^2}$, as this is the ratio of the side of the pentagon to $\sqrt{3c(10)^2}$, when both are inscribed in the same circle. Its proof is: the circle which ...

[This concludes f. 193b. The text at the top of f. 194a reads:]

... times the square on half the side, since it is an equilateral triangle. $B : C = D : E$. $A : B = E : Z$. By the equality of the ratio, $A : C = D : Z$ (!). $D : Z = D/3 : Z/3$ which [i.e. $Z/3$]$= E$. $A : C = D/3 : E$. Q.E.D.

[Clearly, there is a lacuna in the Oxford manuscript. In fact, our Proposition 33 follows easily from the preceding propositions (cf. Hogendijk [1994, notes to Proposition 30]). The lines at the top of f. 194a bear some resemblance to the end of Proposition 32 in the Arabic. That proposition is missing from our text, but is in fact assumed in the next proposition (cf. sentence 231). Perhaps the missing page contained as well Proposition 31 of the Arabic, also missing from the Bodleian text, as well as the beginning of Proposition 32 of the Arabic.

The first proposition in the section of interest in the Mantua MS, displays a full Hebrew version of Hogendijk's Proposition 32; see below. In addition, the seventeenth proposition in the Mantua MS exhibits a better version of Bodley d4 Proposition 33; this too is published below.]

[Proposition 34]

[226.] We wish to find the five lines ordered in the ratio of the perpendiculars of the five figures drawn from the center of the sphere to their faces. We will show this and examine it in one figure.

[227.] Assume a circle; the side of the inscribed pentagon is AB, and the center of the circle is C. On AB construct the equilateral triangle AEB. [228.] We draw the perpendicular CDE (!). $DZ = DE/3$. $H = DZ$. [229.] $T = AD$. $K = T$. $L = CD$. $M = L$. [230.] I say that $H : T = p_4 : p_6$ and $[p_6] = p_8$. [See Hogendijk 1994, n. 31.1.] $T : K = p_6 : p_8$. $K : L = p_8 = p_{20}$, which $[p_{20}] = p_{12}$. $L : M = p_{20} : p_{12}$. [231.] Its proof is: we have already shown that $p_4 : p_6 =$ the ratio of $D/3 = DZ$ to

[102] $CE = \sqrt{3c(10)^2}$. From *Elements* XIII, 4, we have $a^2 + x_2^2 = 3x_1^2$. Here $CE = \sqrt{BE^2 + BC^2} = \sqrt{a^2 + x_2^2}$, since $a = BE = c(6)$, and $c_1 = c(10)$ by Proposition 14.

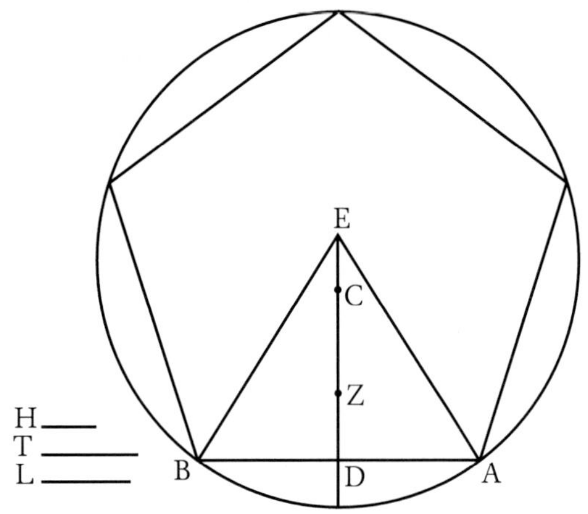

Figure 20: Proposition 34.

AD. $DZ = H$. $AD = T$. $H : T = p_4 : p_6$.[103] [232.] $T : K = p_6 : p_8$, since they are the same, for the square of the cube and the triangle of the octahedron are inscribed in [the same] circle.[104] [233.] We have already shown in Proposition 27 of this treatise that $p_8 : p_{20} = c(5)/2 : (c(6)/2 + c(10)/2)$ and AD is half ... of the decagon together, and it is like L.[105] [234.] $p_8 : p_{20} = L : M$, since they are the same, for the triangle of the icosahedron and the pentagon of the dodecahedron fall in the same circle. [235.] Lines [H], T, K, L, and M have thus been arranged according to the ratio of perpendiculars of the five figures. Q.E.D.

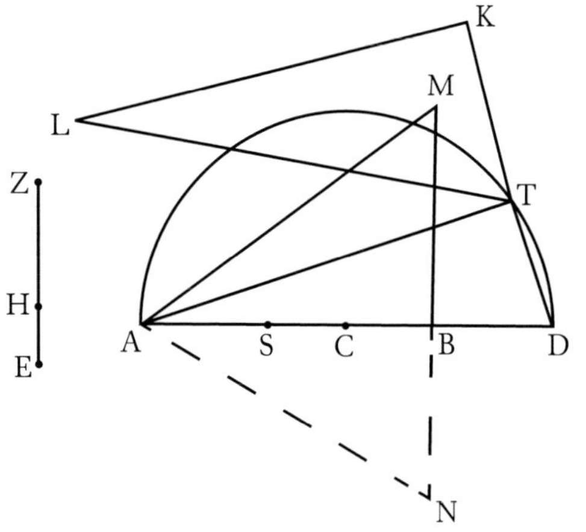

Figure 21: Proposition 35.

[103] $p_4 : p_6 = DE/3 : AD$. This was proven in Proposition 32 of the Arabic, but not in any proposition in the Bodleian text; cf. our notes to the preceding proposition.

[104] Proposition 8, lemma (sentence 70).

[105] By our count, the correct reference is to Proposition 25. In the Arabic text, this is in fact Proposition 27. There are obviously some words missing in the text here: it appears that the scribe skipped a line. The missing part should read, "... the side of the pentagon, and CD is half the side of the hexagon and the side..."

[Proposition 35][106]

[236.] We wish to find the five lines which are ordered in the ratio of the surface areas of the five figures. We shall demonstrate and examine this in one figure.

[237.] Take line AB and divide it in mean and extreme ratio at C. Extend AB rectilinearly to D, such that $BD = BC$. [238.] Let line $EZ = \sqrt{(AD)^2/3}$, and divide it in mean and extreme ratio at H. [239.] Construct on AD semicircle ATD, and draw in it $DT = ZH$. Extend $[D]T$ rectilinearly to $[K]$ such that $T[K] = EZ$. Join AT, and draw the perpendicular KL [to KD and equal to] AT. Join LT. [240.] Draw from B the perpendicular BM, and construct $\angle BAM = 30°$. Let $AS = AB/3$. [241.] I say that $[L]T : AD = s_{12} : s_{20}$. $AD : AB = s_{20} : s_8$. $AB : NM = s_8 : s_6$. $NM : [B]S = s_6 : s_4$. [242.] Its proof is: $KT = EZ$ and $TD = ZH$. DK has been divided in mean and extreme ratio at T. $([K]D)^2 + ([D]T)^2 = 3(KT)^2$, since $KT = EZ$.[107] [243.] $(AD)^2 = (AT)^2 + (TD)^2$, since angle T is a right angle. $(AT)^2 + (TD)^2 = (TD)^2 + ([D])^2$. [Subtracting] the common $(DT)^2$, $(AT)^2 = ([K]D)^2$. $AT = KD$. $[L] = KD$. [244.] $LT = \sqrt{a^2 + x_1^2} = \sqrt{(KD)^2 + (KT)^2}$. $[AD] = \sqrt{a^2 + x_2^2} = \sqrt{KD^2 + TD^2}$. Therefore, $LT : AD = s_{12} : s_{20}$, according to what was shown in [proposition] thirty-four of this treatise. [245.] Furthermore, AB is divided in mean and extreme ratio at C, and $BD = BC$. Thus the ratio of $AD(= a + x_2) : AB = s_{20} : s_8$, as we have demonstrated. [246.] Furthermore, triangle ANM is equilateral, since angle $A = 30°$, and AB is an altitude. The ratio of AB, the altitude to NM, the side, is the same as $s_8 : s_6$, according to what we demonstrated in [proposition number] eleven of this [treatise].[108] [247.] Furthermore, we have already shown that $s_4 : s_8 = SB : AB = 2 : 3$. [248.] $AB = NM = S_8 : s_6$. Hence, $s_6 : s_4 = NM : [B]S$. [249.] The five lines LT, DA, AB, NM, $[BS]$ are ordered in the ratios of the surface areas of the five figures when they are drawn in the same sphere. Q.E.D.

[Proposition 36][109]

[250.] We wish to demonstrate how to find the five lines ordered according to the ratios of the volumes of the five solids when they are drawn in one [sphere].

[251.] Take line AB as the side of an equilateral pentagon inscribed in a circle and extend it rectilinearly to C; let C be [the side of the decagon] which falls in the same circle. [252.] Let $ED = \sqrt{(AC)^2/3}$, and divide it in mean and extreme ratio at Z. [253.] Draw on AC the semicircle AHC, and draw $CH = Z[E]$. Join AH. Extend CH to T, and let $HT = DE$. Draw the perpendicular $TK = TC$ and join $[K]H$. [254]. Draw $[A]L$, making $\angle B[A]L = 30°$. Draw [from B a perpendicular] to M and make $\angle BAM = 30°$ also. Let $LN = LM/3$. [255.] I say that $[K]H : v_{12} : v_{20}$, $AC : AV = v_{20} : v[_8, AB : LM = v_8 : v]_6$, and $LM : LN = v_6 : v_4$. [256.] Its proof is: DE is divided in mean and extreme ratio at Z, $DE = TH$, $HC = ZE$; CT has been divided in mean and extreme ratio, $X_1 = TH$. $(TC)^2 + CH)^2 = (TH)^2$. [257.] Since $TH = DE$ and $(AC)^2 - (AH)^2 + ([H]C)^2$, $(AH)^2 + (HC)^2 = (TC)^2 + (CH)^2$. Subtracting $(HC)^2$, $(AH)^2 = (TC)^2$, and $AH = TC = KT$. [258.] $[K]H = \sqrt{a^2 + x_1^2}$ and $AC = \sqrt{a^2 + x_2^2}$; $[K]H : HC = v_{12} : V_{20}$, as we have shown. [259.] Furthermore, $AC = c(5) + c(10)$, when inscribed in a circle, and $AB = c(5)$. $CA : AB = v_{12} : v_8$. [260.] Triangle AML is equilateral, and its altitude is AB. $AB : ML = v_8 : v_{[6]}$, as we have shown. [261.] The ratio of AB, the altitude to one-third the side, which is $L[N] = v_8 : v_4$, as we have shown. [262.] By the equality of the ratio, $ML : LN = v_6 : v_4$. [263.] The five lines $[K]H$, AC, AB, $[M]L$, $L[N]$ are ordered in the ratios of the volumes of the five figures which are drawn in one sphere. Q.E.D.

---

[106] See the Introduction, p. 415, for further discussion of this proposition.

[107] *Elements* XIII, 4.

[108] Our author never tells us to construct $\triangle ANM$, but this is trivial; simply fold over $\triangle AMB$, as in our figure.

[109] As noted by Hogendijk [1994, 229, n. 37.2], this proposition makes use of the incorrect relationship $v_{20} : V_8 = c(10) + c(5) : c(5)$. The Hebrew text can be emended in the same way as the Arabic to produce the correct relationship.

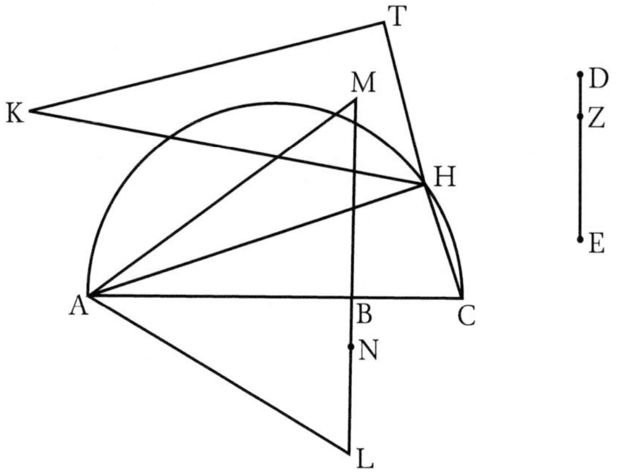

Figure 22: Proposition 36.

[Envoi]

[264.] Every time we mentioned in this section of the treatise (*bi-ze ha-ma'amar me-oto ha-amar*) the five figures, we intended that they be drawn in one sphere. [265.] Similarly, when we said "pentagon," we intended an equilateral and equiangular pentagon, and so also for the hexagon and decagon, and, additionally, that they be drawn in one circle. [266.] Now that we have reached this place, the task which we have intended upon has been completed. Praise to God, may He be Exalted, Who helps [to achieve] the good. [267.] I think that I shall write another treatise and demonstrate that there are no figures other than these five that can be inscribed in the sphere,[110] now that this treatise [has been finished], after I pray to God, may He be Exalted, who privileged me to complete (?) [this treatise], that (?) He grant me success in the moment of truth, may He be Praised and Exalted. Amen.

[Colophon]

[268.] The translation of this treatise was completed on 12 Shebat 5069 [=2.2.1309] in the city of Arldy. [269.] I Kalonymos b. R. Kalonymos of blessed memory, b.R. Meir c.s.y.sh.y. (?), translated it from Arabic to Hebrew, when I reached my twenty-second year. May God in His Mercy privilege me to understand and to learn that which will enable me to arrive at his love.[111] Amen.

Mantua 2

Text

[From Book IX in Mantua 2, which bears the title:]

ביחסי הה' תמונות אשר בין עמודיהם ותושבתי ה'
ושטחיה וגשמיה'

[110] Mantua 29 ends with "It is also manifest that no figures other than these five which have been discussed can be inscribed in the sphere."

[111] That is, to reach the level of knowledge and understanding at which one can truly love God.

[M1] יחס העמוד היוצא ממרכז הכדור אל משלש המחודד אל העמוד היוצא | ממרכז הכדור אל מרובע המעוקב כיחס שליש עמוד כל | משולש שוה הצלעות מחצי צלעות (!) ויהיה עמוד המחודד א' וחצי קוטר | הכדור ב' ועמוד המעוקב ג' ויהיה עמוד המשולש השוה הצלעו' קו ד'| וה' חצי צלעו וז' אחד וחצי כמו צלעו וז' שלשה דמיוני ה' אומ' שיחס | עמוד א' לעמוד ג' כיחס שליש ד' לה'. מופתו שהעמוד היוצא | ממרכז הכדור אל משולש המחודד הוא שתות קוטר הכדור שהוא || שליש חצי הקוטר כמו שקדם ולזה (?) וא' <חצי> שליש ב'. וכזה ה' היה שליש | ז' וכבר התבאר שמרובע קוטר הכדור ג' דמיוני מרובע צלע המעוקב| ומרובע חצי הקוטר ג' דמיוני חצי מרובע צלע המעוקב והוא כמו | העמוד היוצא ממרכז הכדור אל מרובע המעוקב ולזה מרובע ב' | שלשה דמיוני מרובע ג'. ומרובע ד' שלשה דמיוני מרובע ה' לפי| שמרובע עמוד המשולש שלשה *דמיוני מרובע חצי הצלע כי הוא משלש | שוה הצלעות ויחס ב' לג' כיחס ד' לה'. ויחס א' לב' כיחס ה' לז' ועל| השווי יחס א' לג' כיחס ד' לז'. ויחס ד' לז' כיחס שליש ד' לשליש ז' שהוא | ה' הנה יחס א' לג' כיחס שליש ד' לה'. והוא המכוון.*[112]

[Alternative proof to Proposition 2][113]

[M2] (הנה ה"ב כמו ח"ל). ולו פנים יותר קרובים שלפי| שמרובע קוטר הכדור כפל מרובע צלע בעל הח' תושבות והוא כמו || מרובע צלע המחודד לבעל הד' תושבות וחציו יהיה מרובע צלע בעל | הח' תושבות חצי מרובע קוטר הכדור ומרובע צלע המחודד שני שלישי | מרובע קוטר הכדור והוא כמו חצי מרובע קוטר הכדור וששיתו [ומרובע] | העמוד הנופל במשלש תושבות המחודד ג' רביעי מרובע צלע המחודד| הנה מרובע המרובע חצי קוטר הכדור הנה הוא שוה לצלע בעל ח' | תושבות.

[This text is missing from sentence 70 in Bodley d4:]

[M7] והתבאר מזה שהעמוד היוצא | ממרכז הכדור של מרובע הוא תושבת המעוקב כמו העמוד היוצא | ממרכז הכדור אל משלש בעל השמונה תושבות וזה לפי שהמרובע| והמשלש יפלו בעגולה אחת ותהיה העגולה ההיא שוות (!) המרחק | ממרכז הכדור בהם יחד.

[The following proposition is missing from both Bodley d4 and Muḥyī al-Dīn:]

[M10] ביחס הבעלת (!) יב תושבות מחומש ו' והבעל כ' תושבות משולשות | מרובע צלע מחומש בעל י"ב תושבות ומרובע צלע מעוקב אשר | בכדור אחד חמשה דמיוני מרובע חצי קוטר העגולה | שתקיף במחומש בעל י"ב תושבות וזה כי כבר התבאר שצלע המעוקב | הוא מיתר לזוית שיקיפו בו שני צלעי המחומש והמיתר ההוא עם צלע | המחומש הם שוים בכח לחמשה דמיוני חצי קוטר העגולה.

[M13] כשהוצא ממרכז העגולה תקיף במחומש עמוד אל אחד מצלעות | המחומש הזה ל' פעמים כמו השטח שיקיפו בו צלעו | המחומש והעמוד שוה לשטח י"ב תושבות. ויהיה המחומש השוה | הצלעות בעגול א'ב'ג'ד'ה' ומרכז ה' (!) ואומ' כי ל' פעם כמו השטח שיקיפו | בו שני קוי א"ב ז"ח שוה לי"ב פעם כמו מחומש א'ב'ג'ד'ה'. מופתו שנגדיע [צ"ל שנגיע]| קוי ד"ז ז"ג הנה שטח בעל י"ב תושבות הוא י"ב כמו מחומש א'ב'ג'ד'ה' והוא | חמשה דמיוני משלש זג"ד ושטח בעל י"ב תושבות ששי' כמו משלש| זג"ד והשטח שמהכאת ג"ד ב"ז הוא כפול זג"ד הנה שטח בעל הי"ב| תושבות שלושי' כמו שטח ג"ד בז"ח המכוון או נאמ' שהמחומש ה' כפלים משולש בז"ד (!) אם כן כפל המחומש הוא עשרה כפלי משלש | בז"ד (!) [צ"ל גז"ד] ושטח ז"ח בג"ד כפל משולש גז"ד הנה המחומש שנים כפלים | וחצי שטח ג"ד בז"ח. ולכן יהיו י"ב מחומשים שלשים כפלי שטח ג"ד| בז"ח. הנה השטח שיקיפו בו ד"ג ח"ז <ב"י> אחד כפל משלש גד"ז | ולזה חמשה דמיוני השטח שיקיפו בו קוי ג"ד ז"ח כמו עשרה דמיוני |משלש גז"[ד] ועשרה דמיוני משלש גד"ז הם כפל משלש א'ב'ג'ד'ה' וששׁ [צ"ל וכאשר] | כפלת זה ששה פעמים היה ל' פעמים כמו השטח שיקיפו בו שני | קוי ג"ד ד"ז שוה לי"ב מחומשים והי"ב מחומשים הם שטח התמונה | שיקיפו בה י"ב תושבות הנה ל' פעם כמו השטח שיקיפו בו קוי ג"ד | ז"ח שוה לשטח הי"ב תושבות ובזה נתבאר.

[M17] ... מופתו שהעגולה שתקיף | במחומש בעל הי"ב תושבות ובמשולש בעל העשרים תושבות הוא | עגולה אחת כמו שהתבאר. והעמוד היוצא ממנו [צ"ל ממרכז] הכדור אל מחומש | בעל הי"ב תושבות כמו העמוד היוצא ממרכז הכדור אל משולש בעל הח' [צ"ל הכ'] תושבות. והמחודד אשר תושבתו מחומש בעל י"ב תושבות

[112] The section marked off by * is the same as the fragment of this proposition exhibited in Bodley d4, 194a, 1–6.

[113] In Bodley d4, the proposition does not end with the standard Q.E.D., hence we may surmise that this proof somehow fell out of Bodley d4 and is not an addition of the compiler of Mantua 2.

וראשו | מרכז הכדור גבהו גבה המחודד אשר תושבתו משולש בעל כ' תושבות| וראשו מרכז הכדור. והמחודדים השווים אשר בגובה אחר יחסם | כיחס תושבתיהם. ויחס המחודד אשר תושבתו מחומש בעל י"ב | תושבות אל המחודד שתושבתו משולש בעל כ' תושבות . וכן יחס | בעל י"ב (!) כמו מחומש בעל י"ב תושבות וראשו מרכז הכדור אל עשרי' | פנים (!) כמו המחודד שתושבתו משולש בעל העשרים תושבות וראשו | מרכז הכדור הנה הוא כיחס שנים עשר דמיונים כמו המחומש | בעל בי"ב תושבות אל עשרים דמיוני משולש בעל העשרים תושבות| אבל שנים עשר כמו המחודד שתושבתו המחומש וראשו מרכז | הכדור הוא מוגשם בעל הי"ב תושבות. ושנים עשר כמו מחומש | בעל י"ב תושבות הוא שטח בעל ה"ב תושבות ועשרים כמו | המחודד שתושבתו משולש בעל העשרים תושבות וראשו מרכז | הכדור הוא מוגשם בעל הכ' תושבות ועשרים כמו המשולש בעל | העשרים תושבות <וראשו מרכז הכדור> הוא שטח בעל העשרים תושבו' | ואם כן יחס מוגשם בעל י"ב תושבות ממוגשם בעל העשרים תושבות | כיחס שטח בעל י"ב תושבות אל שטח בעל העשרים תושבות | והתבאר גם כן שיחס שטח בעל י"ב תושבות אל שטח בעל הכ' | תושבות כיחס הקו המחזיק על הקו שיחלק על יחס בעל אמצע ושני קצוות ועל החלק היותר גדול אל הקו שיחזיק על הקו כלו | והחלק הקטן וזה שיחס [צ"ל כיחס] צלע המחומש אל הקו שיחזיק על ג' דמיוני | מרובע צלע המעושר הנופל עמו בעגולה אחת הנה אם כן יחס | מוגשם בעל י"ב תושבות אל מוגשם בעל כ' כיחס הקו הנכחי [צ"ל הכחי?=המחזיק] על | כל הקו שנחלק על יחס באו"ק והחלק הגדול אל הקו שיחזיק על כל הקו| אל החלק הקטן | וזה כי יחס [צ"ל כיחס] צלע בעל המחומש הנופל בעגולה אל הקו | המחזיק על ג' דמיוני מרובע צלע המעושר הנופל בה.

[A porism to Bodley d4, Proposition 19:]

[M19] ויתבאר שיחס צלע מעוקב צלע (!) בעל עשרים תושבות כיחס הקו המחזיק על המרובע ההוא [צ"ל ההוה] מן כל הקו יחלק על יחס באו"ק עם מרובע ההוא מהחלק הקטן [צ"ל הגדול] אל הקו המחזיק על כל [ה]קו עם המרובע ההוה.

[Remarks of Eliezer, compiler (?) of Mantua 2, at the end of M25:]

אמר אליעזר יתבאר בביאור כולל שיחס מוגשם בעל ששה תושבות למוגשם בעל ח' תושבות כיחס שטח בעל ששה תושבות לשטח בעל ח' תושבות ושיחס מוגשם בעל י"ב תושבות למוגשם בעל כ' תושבות כיחס שטח בעל י"ב תושבות לשטח בעל כ' תושבות בעשותינו הביאור כולל בביאור יחס מוגשמו כל התמונות יפלו בכדור אחד כיחס שטחיהם על הדרך שקדם מאתנו בביאור בכל תמונ' שטחות (!) יפול בכדור אחד.

[Remarks of Eliezer, compiler (?) of Mantua 2, at the end of M25:]

"בבחינת צלעי התמונות החמש ועמודיהם"

[M26] למצוא הקוים החמשה המסודרים על יחס צלעי התמונות | החמשה נניח עגול בגד"ה ויהיה צלע המחומש הנופל | בו ג"ד וצלע המשלש הנופל בו א"ב [צ"ל ד"ב] | והקו העושה מיתר לשני | מחומשי העגולה ההיא ה"ד ונוציא על קו ה"ד מנקודת ד' עמוד | ד"ז ונניחהו שוה לה"ד ונוציא קו ה"ז ועל הקו ההוא נקיף עגול | הד"ז קטרו ז"ה. ויהיה צלע המשלש הנופל בו ל"א [צ"ל ל"מ] ואומר | שיחס צלע המחודד אל צלע בעל השמונה תושבות כיחס ה"ז ול"מ || ויחס צלע בעל [שמונה] תושבות לצלע המעוקב כיחס ל"מ לה"ד ויחס צלע המעוקב לצלע בעל העשרים תושבות כיחס ה"ד לב"ד ויחס צלע | בעל העשרים תושבות לצלע בעל הי"ב תושבות כיחס ב"ד לג"ד | מופתו שמרובע קוטר הכדור אחד וחצי כמו [מרובע] צלע המחודד| ומרובע קוטר הכדור שלשה דמיוני מרובע צלע המעוקב. ומרובע| צלע המחודד כמו שני דמיוני מרובע צלע המעוקב. הנה יחס צלע | המחודד לצלע המעוקב כיחס ז"ה לה"ד. וכבר התבאר שמרובע | <צלע> המעוקב ומשלש בעל השמונה תושבות יפלו בעגולה אחת | וה"ד ול"א [צ"ל ול"מ] יפלו בעגול לד"מ ולזה יהיה יחס ה"ד ול"מ כיחס צלע | המעוקב לצלע בעל הח' תושבות ולזה יהיה יחס צלע המחודד | לצלע בעל הח' התושבות לצלע המעוקב (!) כיחס [ז"ה ל]ל"מ. ועוד| לפי שצלע המעוקב כשנחלק על יחס באו"ק היה החלק היותר | גדול ממנו צלע בעל י"ב תושבות וה"ד כשנחלק על יחס באו"ק| היה החלק היותר ארוך כמו ג"ד כי ה"ד שני חומשי עגול כד"ג יהיה | יחס ה"ד לג"ד כיחס צלע המעוקב לצלע בעל י"ב תושבות ולפי | שהתבאר שמשולש בעל העשרים תושבות ומחומש בעל י"ב| תושבות יפלו בעגולה אחת ו[לזה יהיה] יחס צלע בעל י"ב תושבות לצלע | העשרים תושבות כיחס ג"ד לב"ד ויחס ה"ד לב"ד כיחס צלע | מעוקב לצלע בעל עשרים תושבות ויחס ב"ד לג"ד כיחס צלע | בעל עשרים תושבות לצלע בעל י"ב תושבות וכבר התבאר | שיחס צלע המחודד לצלע בעל הח' תושבות כיחס זמ"ה [צ"ל ז"ה] לל"מ | ויחס צלע בעל ח' תושבות לצלע המעוקב כיחס ל"מ לה"ד| ויחס צלע מעוקב לצלע משולש בעל עשרים תושבות כיחס | ה"ד לד"ב. ויחס בעל עשרים תושבות לבעל י"ב תושבות כיחס | ב"ג [צ"ל ב"ד] לג"ד הנה כבר מצאנו

חמשה קוים מסודרים על יחס צלעות || התמונה החמש' והתבאר שז"ה ארוך מן ל"מ. ול"מ ארוך מן ה"ד| וא"ד [צ"ל ה"ד] ארוך מן ב"ד וב"ד ארוך מן <א"ג> ג"ד ולזה יהיה [צלע] המחודד ארוך מצלע בעל הח' תושבות וצלע בעל הח' תושבות ארוך מצלע | המעוקב וצלע המעוקב ארוך מצלע בעל העשרים תושבות וצלע | בעל העשרים תושבות ארוך מצלע בעל שתים עשרה תושבות | והוא המכוון ...

...ולו דרך אחר (!) רצוני הנחת צלעות התמונות | החמש נקוים בכדור אחד בתמונה אחת והוא שנניח | קוטר הכדור ב"א ונחלקהו על ג' יהיה א"ג כפל ג"ב ונקוה | על א"ב חצי עגול אד"ב ונוציא עמוד ג"ד ונדביק קוי ב"ד ד"א ויהיה | המרכז ה' ונוציא עמוד ה"ז ונגיע ז"ה ונוציא מן א' עמוד א"ט על | א"ב ונניחהו כמו ב"א ונוציא ט"ה ויהיה כ' במקום שיחתוך ט"ה| המקיף ונוציא עמוד כ"ל ונניח ה"מ כמו ה"ל. ונוציא עמוד מ"נ | ונדביק כ"ג (?) [צ"ל ב"נ] ונחלק ב"ד על יחס האו"ק ויהיה היותר ארוך. ב"ס | ואומר שא"ד כמו צלע המחודד בעל ד' תושבות. וב"ד צלע המעוקב| וז"ב צלע בעל שמונה תושבות. ונ"ב צלע בעל עשרים תושבות| וב"ס צלע הי"ב תושבות הנקוים כלם בכדור שקוטרו א"ב| מופתו שלפי שיחס א"ב לא"ג שהוא היחס אחד וחצי לאחד הוא | כיחס מרובע ב"א למרובע א"ד יהיה מרובע ב"א כמו מרובע א"ד וחציו וא"ב קוטר הכדור הנה מרובע קוטר הכדור דמיון וחצי מרובע | א"ד הנה א"ד צלע המחודד בעל ד' תושבות . וגם כן לפי שא"ב| שלשה דמיוני ב"ג יהיה מרובע א"ב שלשה דמיוני מרובע | צלע המעוקב. הנה ב"ד צלע המעוקב| וגם כן לפי שא"ב | כפל ב"ה ומרובע א"ב כפל מרובע כ"ז [צ"ל ב"ז] ומרובע | קוטר הכדור כפל מרובע צלע בעל הח' תושבות חהחה ז"ב בעל || השמונה תושבות. וגם כן לפי שט"א כמו א"ב וב"א כפל א"ה ויחסו | לא"ה כיחס כ"ל לל"ה יהיה כ"ל כפל ל"ה. ומרובע כ"ל | ד' דמיוני מרובע ל"ה ולזה יהיה מרובע ב"ה [צ"ל כ"ה] וכן מרובע ב"ה השוה | לו חמשה דמיוני מרובע ה"ל ולפי שב"ג חצי ג"א יהיה ב"ג כפל ג"ה| וכ"ה [צ"ל וב"ה] שלשה דמיוני א"ה [צ"ל ג"ה] ולזה יהיה מרובע ב"ה חמשה דמיוני מרובע | ג"ה וכבר היה חמשה דמיוני מרובע ה"ל הנה ה"ל ארוך מן ג"ה ולזה | תפול מ' בין ג' וב'. הנה מרובע כ"ה חמשה דמיוני מרובע ל"ה וא"ב כפל | ב"ה ול"א כפל ל"ה. הנה מרובע א"ב חמשה דמיוני מרובע ל"מ. ומרובע| קוטר הכדור חמשה דמיוני מרובע חצי קוטר עגולה בעל הכ' תושבות | הנה ל"מ כמו חצי קוטר עגולת בעלת הכ' (?) תושבות. וקוטר הכדור| כמו חצי קוטר עגולת בעלת העשרים תושבות עם שני צלעי המעושר | וא"ל כמו מ"ב כי ה"ג כמו ה"ל אם כן כל אחד מן א"ל מ"ב צלע המעושר | ומ"נ כמו כ"ל וכ"ל כפל ה"ל והוא שוה למ"ל הנה מ"נ שוה למ"ל ומ"ל | צלע משושה ואם כן מ"נ צלע משושה וב"מ (?) צלע מעושר. וזוית מ' | נצבת הנה ג"ב [צ"ל נ"ב] המחזיק עליהם צלע המחומש וצלע המחומש | שוה לצלע בעל עשרים תושבות הנה כ"ב [צ"ל נ"ב] צלע בעל עשרים תושבות | וגם כן לפי שא"ד ארוך מן ז"ב וז"ד ארוך מן ב"ד וב"ד ארוך מן | ב"נ ונחלק ב"ד על יחס באו"ק על נקודת ס' והיה החלק | היותר ארוך כ"ב [צ"ל ס"ב] וכבר התבאר כשנחלק צלע המעוקב על יחס | באו"ק שהחלק היותר ארוך הוא צלע בעל הי"ב תושבות. הנה ס"ב | צלע בעל הי"ב תושבות . וא"ג כפל ג"ב ולזה יהיה מרובע א"ג ארבע | דמיוני מרובע ב"ג. ומרובע ד"ב שלשה דמיוני מרובע כ"ג [צ"ל ב"ג] הנה א"ג | יותר ארוך מן ב"ד וא"מ יותר ארוך מן ב"ד וכאשר <ל"מ ארוך> חולק | א"מ על יחס באו"ק היה ל"מ החלק היותר ארוך. ולזה יהיה ל"מ ארוך | מן ס"ב ול"מ שוה למ"נ ומ"נ גדול מן ס"ב ומ"נ קטן מן ב"נ כי זוית || א' נצבת אם כן נ"ב ארוך מן ב"ס והוא המכוון.

[At the end of Proposition 29, the last one in this section of Mantua:]

ויתבאר גם כן שלא יפול בכדור זולת אלה התמונות החמשה הנזכרים.

[Mantua 2, 172a–173a, better version of Proposition 26:]

יחס מוגשם בעל עשרים תושבות אל מוגשם בעל השמונה תושבות כיחס צלע כל מעושר נופל בעגלה עם צלע המחומש הנופל בה אל צלע המחומש. וזה כיחס החלק הקטן מכל קו נחלק על יחס באו"ק עם הקו המחזיק על הקו כלו ועל החלק הגדול אל הקקו המחזיק על הקו כלו והחלק הקטן.

דמיונו יהיה מוגשם א' נכחי השטחים תושבתו כמו שטח בעל העשרים תושבות ומבואר שהוא כמו מוגשם בעל העשרים תושבות [172b] ונניח מוגשם ב' נכחי השטחים תושבתו כמו שטח בעל הח' תושבות וגבהו שליש העמוד היוצא ממרכז הכדור אל משלש בעל הח' תושבות. ומבואר שמוגשם ב' כמו מוגשם בעל השמונה תושבות. ונניח ד' צלע מעושר וח' צלע מחומש מעגלה אחת. ואמור שיחס מוגשם א' למוגשם ב' כיחס ד' ב' [ח'] יחד אל ח'.

מופת זה אנו נתן מוגשם ג' נכחי השטחים ותהיה תושבתו כמו יתרון שטח מוגשם בעל שמונה תושבות על שטח בעל שמוה תושבות וגבהו שליש העמוד היוצא ממרכז הכדור אל משולש בעל העשרים תושבות. ומבוא שמוגשם ג' כמו יתרון <ג' כמו מוגשם בעל העשרים תושבות על> על מוגשם בעל הח' תושבות. וכל שני מוגשמים נכחיי השטחים הנה יחסם מחובר מיחס התושבות לתושבת ומיחס הגובה לגובה.

הנה יחס מוגשם א' למוגשם ב' מחור מיחס תושבת ג' שהוא מותר שטח בעל בעשרים תושבות על שטח

בעל השמונה תושבות ומיחס גובה מוגשם ג' שהוא שליש העמוד היוצא ממרכז הכדור אל משולש בעל העשרים תושבות אל גובה ב' שהוא שליש העמוד היוצא ממרכז הכדור אל משולש בעל השמונה תושבות.

ויהיה [ה'] צלע משושה ומשולש [צ"ל וצלע מעושר] לעגולה אשר ד' מעושרה וח' מחומשה.וכבר התבאר שיחס מוגשם [צ"ל יתרון] שטח בעל העשרים תושבות [על שטח בעל שמונה התושבות] לשטח בעל השמונה תושבות כיחס צלע מעושר לצלע משושה ומעושר. וזה כיחס ד' לה'. והתבאר גם כן שיחס העמוד היותא ממרכז הכדור על [צ"ל אל] משולש בעל העשרים תושבות אל העמוד היוצא ממרכז הכדור על [צ"ל אל] משולש בעל שמונה תושבות כיחס צלע משושה ומעושר אל צלע המחומש הנופלים בעגול אחד.וזה כיחס ה' לח'. אם כן יחס מוגשם ג' אל מוגשם ב' מחובר מיחס ד' לה' ומיחס ה' לח'. וזה הוא כיחס ד' לח'.

הנה יחס מוגשם ג' למוגשם [173a] ב' כיחס ד' לה'. וכשהרכבנו היה יחס ג' וב' אל ב' כיחס ד' וח' אל ח' וג' וב' הוא כמו א'.הנה יחס א' לב' כיחס ד' וח' אל ח' והוא המכוון.

ולפי שצלע המשושה כשמחלק יחס באו"ק יהיה החלק הגדול צלע המעושר ויהיה הקו אשר יחזיק על צלע המשושה והמעושר הוא צלע המחומש. הנה יחס מוגשם בעל העשרים תושבות למוגשם בעל שמונה תושבות כיחס החלק הגדול אבל [צ"ל אל] [מ]קו יחלק על יחס בעל אמצעי ושתי קצוות עם הקו המחזיק על הקו כלו ועל החלק הגדול אל הקו המחזיק על הקו כלו ועל החלק הקטן [צ"ל הגדול].

ונשלם ביאורו עם ביאורי יחס התמונות הד' קצתם לקצת ונשאר ביאור יחס בעל הי"ב תושבות ליחס [צ"ל לבעל] הכ' תושבות.

Translation

[Mantua 1]

$p_4 : p_6 = h/3 : d/2$. Let p_4 be A, the radius of the sphere B, p_6 is C, $h = D$, $d/2 = E$. $3d/2 = Z$. $Z = 3E$. I say that $A : C = D/3 : E$. Its proof is: p_4 is equal to one-sixth the diameter of the sphere, which is one-third the radius of the sphere, as we saw earlier. Hence $A = B/3$. Similarly, $E = Z/3$. It has already been shown that the square on the diameter is equal to $3z_6^2$.[114] The square on the radius $= 3(z_6/2)^2$ which [i.e. $z_6/2$] is equal to p_6. [See sentence 182 of Bodley d4.] Therefore, $B^2 = 3C^2$. $D^2 = 3E^2$, for $h^2 = 3\{(d/2)^2$, since it is an equilateral triangle. $B : C = D : E$. $A : B = E : Z$. By the equality of the ratio, $A : C = D : Z$. $D : Z = D/3 : Z/3$, which [i.e. $Z/3$] is equal to E. Therefore $A : C = D/3 : E$. Q.E.D.}[115]

[Mantua 2 (additional proof)]

It has another easier method. Because the square on the diameter of the sphere is double the square on the edge of the octahedron, and it is equal to the square on the edge of the tetrahedron and its half, the square on the edge of the octahedron is half the square on the diameter of the sphere, and the square on the edge of the tetrahedron is two-thirds the square on the diameter of the sphere, viz. half the square on the diameter of the sphere and its sixth. The [square on the] height of the triangle of the tetrahedron is ¾ the square on the edge of the tetrahedron. Therefore, the square [on the height of the tetrahedron is equal to ¾(2/3) the square on the diameter of the sphere)] is equal to half the square on the diameter, hence it is equal to the edge of the octahedron.[116]

[Mantua 7]

It is clear from this that the perpendicular drawn from the center of the sphere to the square which is the face of the cube is equal to the perpendicular drawn from the center of the sphere to the face of

---

[114] *Elements* XIII, 15.

[115] The section at the end enclosed in curly brackets (with stars in the text) is the same as the fragment in Bodley d4.

[116] $\text{Height}^2 = z_8^2$, hence height $= z_8$. The scribe has skipped a few lines. What is the source of this additional proof? Is it due to the compiler of Mantua, or perhaps to his better version of Bodley d4? If the latter is the case, then one would assume that, as elsewhere in Bodley d4, one of the proofs is due to Apollonius.

the octahedron. This is because the square and the triangle fall in one circle, and that circle will be for both of them the same distance from the center of the sphere.[117]

[Mantua 10]

With regard to the dodecahedron and the icosahedron: the square on the edge of the dodecahedron and the square on the edge of the cube, when [both] are in one sphere, is equal to five times the square on the radius of the circle which circumscribes the face of the dodecahedron. This is so, for it has already been shown that the edge of the cube is the chord subtending the angle surrounded by two sides of the pentagon, and that chord along with the side of the pentagon are equal "in power" [*bi-ko'ah*, i.e. the squares of each are together equal to the square of] 5 times the radius of the circle.[118]

[Mantua 17]

...Its proof is: the same circle circumscribes the pentagon of the dodecahedron and the triangle of the icosahedron, as was shown previously. The perpendicular which is drawn (from the center) of the sphere to the pentagon of the dodecahedron is equal to the perpendicular drawn from the center of the sphere to the triangle of the (icosahedron). The height of the pyramid whose base is the pentagon of the dodecahedron and whose vertex is the center of the sphere is equal to the height of the pyramid whose base is the triangle of the icosahedron and whose vertex is the center of the sphere. Pyramids of equal height are in ratio to their bases. The ratio of the pyramid whose base is the pentagon of the dodecahedron to the pyramid whose base is the triangle of the icosahedron...;[119] similarly (!) the ratio of the dodecahedron [to the icosahedron is equal to the ratio of twelve times the solid whose base is] the face of the dodecahedron and whose vertex is the center of the sphere to twenty times the pyramid whose base is the triangle of the icosahedron and whose vertex is the center of the sphere. Hence it is equal to the ratio of twelve times the pentagon of the dodecahedron to twenty times the triangle of the icosahedron. However, twelve times the pyramid whose base is the pentagon and whose vertex is the center of the sphere is the volume of the dodecahedron and twelve times the pentagon is the surface area of the dodecahedron. Twenty times the pyramid whose base is the triangle of the of the icosahedron and whose vertex is the center of the sphere is the volume of the icosahedron, and twenty times the triangle is the surface area of the icosahedron. Therefore, the ratio of the volume of the dodecahedron to the volume of the icosahedron is equal to the ratio of the surface area of the dodecahedron to the surface area of the icosahedron. It has been shown as well that the ratio of the surface area of the dodecahedron to the surface area of the icosahedron is equal to $\sqrt{a^2 + x_1^2} : \sqrt{a^2 + x_2^2}$, and that this $= c(5) : \sqrt{3c(10)^2}$. Therefore, now, the ratio of the volume of the dodecahedron to the volume of the icosahedron is equal to $\sqrt{a^2 + x_1^2} : \sqrt{a^2 + x_2^2} = c(5) : \sqrt{3c(10)^2}$.[120]

[Mantua 26 (first method)]

To find the five lines ordered according to the ratio of the edges of the five figures. Take circle *BCDE*; let *CD* be the side of the pentagon inscribed therein, [*D*]*B* the side of the inscribed triangle, *ED* the line which subtends two-fifths of that circle. We draw from point *D*, *DZ* perpendicular to *ED*, and make it equal to *ED*. We draw line *EZ* and on it circle *EDZ*, with diameter *ZE*. Let the side of the

---

[117] Cf. sentence 70 in Bodley d4.

[118] This proposition, which is found in neither Bodley d4 nor Muḥyī, follows trivially from two other propositions: 1. $c(5)^2 + c(2\frac{1}{2})^2 = 5c(6)^2$, and 2. $c(2\frac{1}{2}) = z_6$. The latter is assumed by both Bodley d4 and Muḥyī, and by Pappus V, 56, who says this has been shown in Book XIII of Euclid [Ver Eecke 1982, 340, n. 4]. I think it is worth noting that this relationship was taken as a proven fact by Pappus and all the polyhedra texts. It is true, as Ver Eecke shows, that it follows readily from Euclid. Nevertheless, less subtle matters were more than once the subject of full-blown proofs in Hellenistic and medieval mathematical texts.

[119] The lacuna can be supplied with "is equal to the ratio of the pentagon to the triangle."

[120] This completes Bodley d4 33 = Muḥyī 30. It is somewhat verbose, but it does without parallelepipeds *A* and *B* of Muḥyī.

triangle inscribed in it [circle EDZ] be $L[M]$. I say that $z_4 : z_8 = EZ : LM$, $z_{[8]} : z_6 = LM : ED$, $z_6 : z_{20} = ED : BD$, and $z_{20} : z_{12} = BD : CD$. Its proof is: the square on the diameter of the sphere is one and one-half times [the square on] the edge of the tetrahedron. The square on the diameter of the sphere is three times the square on the edge of the cube. $z_4^2 = 2z_6^2$. Thus $z_4 : z_6 = ZE : ED$. It has already been shown that the faces of the cube and the octahedron are inscribed in the same circle, and ED and LM [both] lie in circle LDM. Therefore, $ED : LM = z_6 : z_8$. Thus $z_4 : z_8 : z_6 =[ZE]: LM : ED$. Furthermore, when the edge of the cube is divided in mean and extreme ratio, the greater segment is the edge of the dodecahedron, and when ED is divided in mean and extreme ratio, the greater segment is equal to CD, for ED subtends two-fifths of circle EDC. $ED : CD = z_6 : z_{12}$. And since it has been shown that the faces of the icosahedron and the dodecahedron are inscribed in the same circle, $z_{12} : z_{20} = CD : BD$, $ED : BD = z_6 : z_{20}$, and $BD : CD = z_{20} : z_{12}$ (!). It has already been shown that $z_4 : z_8 = ZE : LM$, $z_8 : z_6 = LM : ED$, $z_6 : z_{20} = ED : BD$, and $z_{20} : z_{12} = B[D]: CD$. Hence we have found five lines ordered according to the ratio of the edges of the five figures. It is clear that $ZH > LM$, $LM > ED$, $ED > BD$, and $BD > CD$. Therefore, $z_4 > z_8$, $z_8 > z_6$, $z_6 > z_{20}$ and $z_{20} > z_{12}$. Q.E.D.

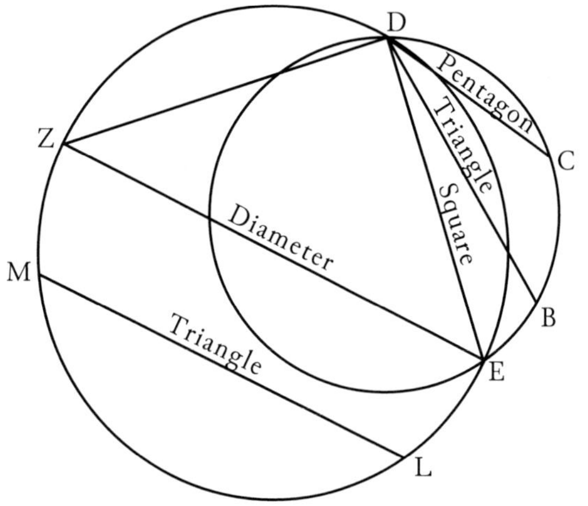

Figure 23: Mantua 26, first method.

[Mantua 26 (second method)]

There is another method for this, i.e., placing the edges of the five figures which are drawn in one sphere, in one figure. It is: we take BA as the diameter of the sphere and divide it at C such that $AC = 2CB$. We draw on AB the semicircle ADB, and draw the perpendicular CD. Connect BD and DA. Let the center be E, draw the perpendicular EZ, and join ZB. From A draw AT perpendicular to AB and let it be equal to AB. Draw TE. Let K be the intersection of TE and the circumference, and draw the perpendicular KL. Let $EM = EL$. Draw the perpendicular MN and connect BN. Divide BD in mean and extreme ratio, and let the greater segment be BS. I say that $AD = z_4$, $BD = z_6$, $ZB = z_8$, $NB = z_{20}$, and $BS = z_{12}$, all of which [polyhedra] are drawn in the sphere whose diameter is AB. Its proof is: since $AB : AC = 1 : 1$, which is like the ratio of BA^2 to AD^2, $BA^2 = 1AD^2$, and AB is the diameter of the sphere. Now the square on the diameter of the sphere is one and one-half times AD^2; hence AD is the edge of the tetrahedron. Moreover, since $AB = 3BC$, $AB^2 = 3z_6^2$; hence $BD = z_6$. Moreover still, since $AV = 2BE$, $AB^2 = 2[B]\,Z^2$ and the square on the diameter of the sphere is double the square on the edge of the octahedron, then $ZB = z_8$. Also, since $TA = AB$, $BA = 2AE$, and $TA : AE = KL : LE$, then $KL = 2LE$, and $KL^2 = 4LE^2$. For this reason, $[K]E^2$, and likewise

BE^2, which is equal to it, is equal to $5EL^2$. Because $BC = CA$, $BC = 2CE$, and $[B]E = 3[C]E$. Therefore, $BE^2 = 9CE^2$, and it $[BE^2]$ was already proven to be equal to $5EL^2$. Hence $EL > CE$, and therefore, M lies between C and B. Now $KE^2 = 5LE^2$, $AB = 2BE$, and $LA = 2LE$; hence $AB^2 = 5LM^2$. The square on the diameter of the sphere is five times the square on the radius of the circle which circumscribes the face of the icosahedron; thus LM is equal to the radius of the circle which circumscribes the face of the icosahedron. The diameter of the sphere is equal to the radius of the circle which circumscribes the face of the icosahedron together with two sides of the decagon, and $AL = MB$, since $EC = EL$. Therefore, both AL and MB are sides of the decagon. $MN = KL$, $KL = 2EL = ML$. Hence $MN = ML$, and ML is the edge of the hexagon; therefore MN is the edge of the hexagon. BM is the side of the decagon, and angle M is a right angle. Thus $[N]B$, which is the root of the sum of their squares, is the side of the pentagon. The side of the pentagon is the edge of the icosahedron, hence $[N]B = z_{20}$. Moreover, since $AD > ZB$, $Z[B] > BD$, $BD > BN$, BD has been divided in mean and extreme ratio at S with $[S]B$ the longer segment, and it has already been shown that when the edge of the cube is divided in mean and extreme ratio, the longer segment is the edge of the dodecahedron; therefore $SB = z_{12}$. $AC = 2CB$, thus $AC^2 = 4BC^2$. $DB^2 = 3BC^2$. Hence $AC > BD$ and $AM > BD$. When AM is divided in mean and extreme ratio, the longer segment is LM. Thus $LM > SB$. $LM = MN$ and $MN > SB$. $MN > BN$, since angle A is a right angle. Therefore $NB > BC$. Q.E.D.

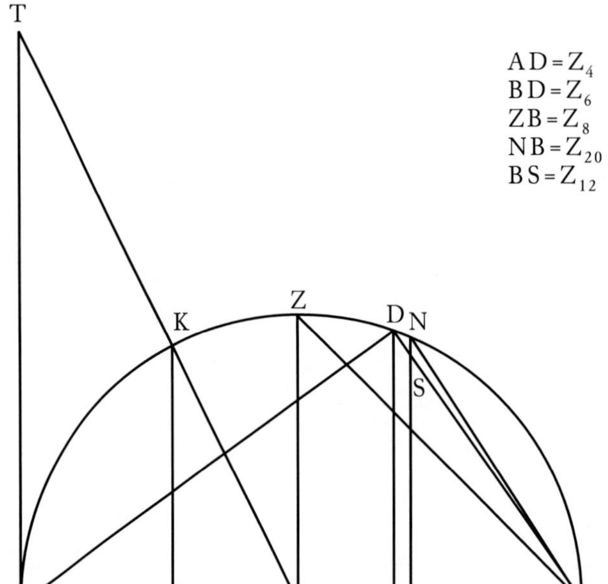

Figure 24: Mantua 26, second method.

[Mantua ff. 172a- (corresponding to Proposition 26 in Bodley d4)]

The ratio of the icosahedral solid to the octahedral solid is equal to the ratio of the sum of the sides of a decagon and pentagon, drawn in the same circle, to the side of the pentagon. This [in turn] is equal to the ratio of the greater part of a line divided in mean and extreme ratio plus the squares on the whole line and on its greater part, to the sum of the squares on the whole line and on its smaller part.[121]

For example, let there be parallelepipedal solid A whose base is equal to the surface area of the icosahedron. Clearly it is equal to the icosahedral solid. Now we take paralellpipedel solid B whose base is equal to the surface area of the octahedron and whose height is one-third the perpendicular dropped from the center of the sphere to the triangle of the octahedron. Clearly, solid B is equal to octahedral solid. Now let us take D to the side of the decagon and H the side of the pentagon drawn in the same circle. I say that $A : B = (D + H) : H$.

It proof is: We take parallelepiped G, whose base is equal to the excess of the surface of the octahedron over the surface of the icosahedron, and whose height is one-third the perpendicular dropped from the center of the sphere to the triangle of the icosahedron. Clearly, solid G is equal to the excess of the icosahedral solid over the octahedral solid. And the ratio between every two paralellpipedal solids is compounded of the ratio of the bases and the ratio of the heights to each other. Now $A : B$ is compounded of the ratio of the base of G, which is the excess of the surface area of the icosahedron over the surface area of the octahedron, and from the ratio of the height of solid G, which is one-third the height of the perpendicular dropped from the center of the sphere to the triangle of the icosahedron, to the height of B, which is one-third of the perpendicular dropped from the center of the sphere to the triangle of the octahedron.

Let [E] be the sum of the side[s] the hexagon and the [decagon of the circle] in which D is the [side of] the decagon and H is [the side of] of the pentagon. It has already been shown that the ratio of the [surface of the] icosahedron [over the surface of the octahedron] to the surface of the octahedron is equal to the ratio of the side of the decagon to the sum of the sides of the hexagon and decagon. And this is like $D : E$.

But it has also been shown that the ratio of the perpendicular dropped from the center of the sphere to the triangle of the icosahedron to the perpendicular dropped from the center of the sphere to the triangle of the octahedron is equal to the ratio of the sum of the sides of the hexagon and decagon to the side of the pentagon, when they [all] lie in the same circle. This is equal to $E : H$. Therefore, $G : B$ is compounded of $D : E$ and $E : H$, which is equal to $D : H$. Thus $G : B = D : H$. Componendo, $(G + B) : B = (B + H) : H$; and $G + B = A$. Thus $A : B = (D + H) : H$. Q.E.D.

Now when the side of the hexagon is divided in mean and extreme ratio, the greater section is the side of the decagon;[122] and the sum of the squares on sides of the hexagon and the decagon is [equal to the square on] the side of the pentagon [in the same circle].[123] Hence the ratio of the icosahedral solid to the octahedral solid is equal to the sum of the greater part of the line divided in mean and extreme ratio and the square on the entire line plus the square on the greater part, to the sum of the squares on the whole line and the [greater] part.

Its proof is now complete, and so are the proofs concerning the ratio of the four figures, one to the other. There remains the proof of the ratio of the dodecahedron to the icosahedron.

Acknowledgements I extend my heartfelt thanks to Michael N. Fried, who graciously read the final draft of this paper and offered some very useful criticisms and pointers. Needless to say, all remaining errors are my own.

---

[121] See Hogendijk [1994, note 29.8].

[122] Bodley d4, Proposition 14.

[123] *Elements*, XIII, 10.

Appendixes

Mantua	Figure in Mantua?	Bodley d4	Hogendijk
1		f. 194a:1–6	32
2	√	2	2
3		3, 4	2 (porism), 17
4		5	18
5	√	6	19
6		7	20
7		8, 9	3
8	√	10, 11	8, 12
9		12	23
10		–	–
11	√	16	5
12	√	17	6
13	√	27, 28	9, 10
14		29	10
15	√	31	16
16	√	32	16 (porism)
17		33	30
18		18	13
19		19	14
20	√	20	15
21		21	24
22	√	23	25
23		24	26
24		25	27
25	√	26	29
26	√	–	31
27	√	34	33
28		35	35
29	√	36	37

Table 1: Correlation between the propositions of Bodley d4, Muḥyī and the commentaries.

Bodley 4d	Hogendijk	Commentary, author of Oxford Bodley d4	Commentary
1	1		
2	2		
3	2 corollary		
4	17		
5	18		
6	19		
7	20		
8	3		Analogous to XIV, 2
9	3		
10	8		
11	12		
12	23		Hebrew text too corrupt to attempt a reconstruction.
13	–	The first proof is that given by Apollonius; the second proof is the author's own.	XIV, 8, lemma; author's proof is geometrical, that of Apollonius is algebraic.
14	–		See XIII, 9
15	4	(115) "The proof was given by Apollonius in this way."	
16	5	(127) "This proposition was given by Apollonius in a similar way. However, that which I have given is easier for students."	
17	6		Hebrew text too corrupt to attempt a reconstruction.
18	13		
19	14		
20	15		
21	24		
22	7		
23	25		
24	26		
25	27		
26	29	(191) "This completes the ratios of the four (!) figures, one to the other... There remain the ratios of which Apollonius mentioned between the dodecahedron and the icosahedron. This is the beginning of the explication of his treatise."	Hebrew text too corrupt to attempt a reconstruction.
27	9		Hebrew text too corrupt to attempt a reconstruction.
28	10		XIV, 5
29	10		XIV, 5
30	–		XIV, 6, alternate proof
31	16	In middle: (1) Note by Kalonymos Todros that he found a page missing from the original and supplied the translation of propositions 30 & 31; (2) numbering given by Kalonymos checks out; (3) a second proof here, (215) " Apollonius said, let us demonstrate this in another way..."	XIV, 6, first proof
32	16		
33	30		Only proposition numbered in the MS; Hebrew text too corrupt to attempt a reconstruction.
34	33		
35	35		
36	37		

Table 2: Correlation between the propositions of Mantua, Bodley d4 and Muḥyī.

References

Manuscript Sources

Mantua 2, Comunita Israelitica MS ebr. 2.

Oxford, Bodleian Library d4, Neugebauer 2776.

Oxford, Bodleian Library, Michael 400, Neubauer 2006.

Oxford, Bodleian Library, Marsh 667.

Oxford, Bodleian Library, Uri 433, Huntington 96, Neugebauer 2008.

Modern Scholarship

Cornford, F.M., 1935. Plato's Cosmology. The Timaeus of Plato, London. Reprinted: Hackett, Indianapolis, 1997.

David, A., (ed.), 1995. From the Collections of the Institute of Microfilmed Hebrew Manuscripts. National Library, Jerusalem.

Dodge, B. (ed. and trans.), 1970. The Fihrist of al-Nadīm, 2 vols., Columbia, New York and London.

Fontaine, R., 2000. Judah ben Solomon ha-Cohen's Midrash ha-Ḥokhmah: Its sources and use of sources. In: Harvey, S. (ed.), The Medieval Hebrew Encyclopedias of Science and Philosophy. Kluwer, Netherlands, pp. 191–210.

Freudenthal, G., and Lévy, T., 2004. De Gérase à Bagdad: Ibn Bahriz, Kindi, et leur recension arabe de l'Introduction arithmétique de Nicomaque, d'après la version hébraique de Qalonymos ben Qalonymos d'Arles. In: Morelon, R., Hasnaoui, A. (eds.), De Zénon d'Elée à Poincaré: Recueil d'études en hommage à Roshdi Rashed. Peeters, Louvain-Paris, pp. 479–544.

Gluskina, G.M., 1983. Alfonso of Valladolid, Meyashsher Aqov. Facsimile edition and annotated Russian translation. Nauka, Moscow.

Hansberger, R., 2011. Plotinus Arabus rides again. Arabic Sciences and Philosophy 21, 57–84.

Heath, T.L., 1921. A History of Greek Mathematics, 2 vols. Clarendon Press, Oxford. Reprinted: Dover, New York, 1981.

——— 1925. The Thirteen Books of Euclid's Elements, second edition, 3 vols. Cambridge University Press, Cambridge. Reprinted: Dover, New York, 1956.

Hogendijk, J.P., Langermann, T., 1984. A hitherto unknown Hellenistic treatise on the regular polyhedra. Historia Mathematica 11, 325–326.

Hogendijk, J.P., 1994. An Arabic text on the comparison of the five regular polyhedra: Book XV of the revision of the Elements by Muḥyī al-Dīn al-Maghribī. Zeitschrift fur Geschichte des arabisch-islamischen Wissenschaften 8, 133–233.

Jolivet, J., Rashed, R., 1997. Oeuvres philosophiques et scientifiques d'al-Kindi. Vol. 1: L'Optique et la catoptrique. Vol. 2: Métaphysique et cosmologie. Brill, Leiden.

Lévy, T., 1996. Fragment d'Ibn al-Samh sur le cylinder et ses sections planes, conserve dans une version hébraique. In: Rashed, R. (ed.), Les mathématiques infinitésimales du IX au XIe siècle, vol. 1, Fondateurs et commentateurs, Routledge, London, pp. 927-973, 1080-1083.

——— 1997. The mathematical bookshelf of the medieval Hebrew scholar. Science in Context 10, 431–451.

——— 2000. Mathematics in the Midrash ha-Ḥokhmah of Judah ben Solomon ha-Cohen. In: Harvey, S. (ed.), The Medieval Hebrew Encyclopedias of Science and Philosophy. Kluwer, Netherlands, pp.-300–312.

Langermann, Y.T., 1988. The scientific writings of Mordekhai Finzi. Italia 7, 7–44. Reprinted: Langermann, Y.T., The Jews and the Sciences in the Middle Ages. Ashgate, Aldershot, 1999, essay IX.

——— 2000. Some remarks on Judah ben Solomon ha-Cohen and his Encyclopedia Midrash ha-Ḥokhmah. In: Harvey, S. (ed.), The Medieval Hebrew Encyclopedias of Science and Philosophy. Kluwer, Netherlands, pp. 371–389.

Manekin, C., 2000. The Hebrew translations of the middle ages: Moritz Steinschneider. In: Harvey, S. (ed.), The Medieval Hebrew Encyclopedias of Science and Philosophy. Kluwer, Netherlands, pp. 468–519.

Murdoch, J., 1971. Euclid: Transmission of the Elements. In: Gillespie, C. (ed.), Dictionary of Scientific Biography, vol. 4. Scribner, New York, pp. 437–459.

Neubauer, A., Cowley, A.E., 1906. Catalogue of the Hebrew Manuscripts in the Bodleian Library, vol. 2, Oxford.

Proclus, 1970. A Commentary on the First Book of Euclid's Elements. Translated, with Introduction and Notes, by Glenn R. Morrow. Princeton University Press, Princeton. Paperback edition, Princeton, 1992.

Rescher, N. 1967. Studies in Arabic Philosophy. University of Pittsburgh, Pittsburgh.

Rosenthal, F., 1956. Al-Kindī and Ptolemy. In: Studi Orientalistici in onore di Giorgio Levi dell Vida, II, Rome, pp. 436–456.

Sabra, A.I., 1968. Simplicius' proof of Euclid's parallels postulate. Journal of the Warburg and Courtauld Institutes 31, 1–24. Reprinted: Sabra, A.I., Optics, Astronomy and Logic, Ashgate, Aldershot, 1994, essay XIII.

Sezgin, F., 1974. Geschichte des arabischen Schrifttums, V, Mathematik. Brill, Leiden.

Steinschneider, M., 1893. Die hebraeischen Uebersetzungen des Mittelaters und die Juden als Dolmetscher. Kommissionsverlag des Bibliographischen bureaus, Berlin. Reprinted: Akademische Druck, Graz, 1956.

Ver Eecke, P., 1982. Pappus d'Alexandrie: La collection mathématique. Albert Blanchard, Paris.

Vitrac, B., Djebbar, A., 2011. Le livre XIV des Éléments d'Euclide: Versions grecques et arabes, première partie. SCIAMVS 12, 29–158.

——— 2012. Le livre XIV des Éléments d'Euclide: Versions grecques et arabes, seconde partie. SCIAMVS 13, 3–158.

A Treatise by al-Bīrūnī on the Rule of Three and its Variations

Takanori Kusuba

Abstract Bīrūnī's treatise on the rule of three and its variations, *Maqāla fī rāshīkāt al-Hind*, is an amalgam of Indian and Greek mathematics. He applies examples of Sanskrit origin and explains the rules based on the Greek theory of ratios. The printed Hyderabad version of the treatise is an uncritical copy of the only known manuscript, which has pages bound in an incorrect order. This is a study of the treatise, preliminary to producing a critical edition.

Introduction

Bīrūnī (A.D. 973–1048) was one of the most prolific scholars of the Middle Ages. In his youth he studied Greek science. Later he accompanied Sultan Mahmud on several expeditions to India. As an interrogator of Indian prisoners, Bīrūnī acquired a knowledge of Indian science. One result of this was his *Kitāb fī taḥqīq mā li'l-hind min maqūlatin* (Book of Verification of What is Said About India), usually referred to simply as Bīrūnī's *India*. It has been recognized that Bīrūnī's knowledge of Sanskrit was not profound. Regarding Bīrūnī's treatment of the astronomy of Brahmagupta, Pingree [1983, 356] writes that "Bīrūnī has a strong tendency to attribute Aristotelian physical theories and Greek geometrical proofs to what his pandit translated, thereby seriously distorting the Indian argument." A study of Bīrūnī's treatment of Indian mathematics is indispensable. However, only one work by Bīrūnī on Indian mathematics, titled *Maqāla fī rāshīkāt al-Hind* (A Treatise on the Rāshīka of India),[1] is extant. It deals with the rule of three and its variations.

The rule of three teaches one how to find a number x that is the fourth proportion to three given numbers a, b, c, such that $a : b = c : x$. With this rule and the rule of false position, one can solve any linear equation in one unknown.[2] Giving a sketch of the history of the rule, Smith [1923–1925, 2, 483] says "the mercantile Rule of Three seems to have originated among the Hindus. It was called by this name by Brahmagupta and Bhāskara II, and the name is also found among the Arabs and medieval Latin writers." Regarding the history of the rule of three in India, Sarma [2002, 134–135] observes that "Āryabhaṭa not only gives the name *Trairāśika* (that which consists of numerical quantities or terms) for the Rule of Three, but mentions as well the technical terms for the four numerical quantities involved (*pramāṇa, phala-rāśi, icchā-rāśi, icchā-phala*) and gives the formula for solving the problem. Subsequent writers, notably Brahmagupta in his *Brāhmasphuṭasiddhānta* (A.D. 628) and Bhāskara I in his commentary (A.D. 629) on the *Āryabhaṭīya* elaborate upon this brief statement by Āryabhaṭa, but employ the same terminology, albeit with slight modification. It is on the basis of the writings of these mathematicians that histories of mathematics generally trace the origin of the Rule of Three to India." In this paper, I examine Bīrūnī's treatment of the rule and its variations in his *Maqāla fī*

[1] The origin of the word *rāshīka*, of which *rāshīkāt* is an Arabicized plural, is explained below; see page 474.

[2] See Chabert, et al. [1999, 106–107].

N. Sidoli and G. Van Brummelen (eds.), *From Alexandria, Through Baghdad: Surveys and Studies in the Ancient Greek and Medieval Islamic Mathematical Sciences in Honor of J.L. Berggren*, DOI 10.1007/978-3-642-36736-6_21,

© Springer-Verlag Berlin Heidelberg 2014

rāshīkāt al-Hind. Although Bīrūnī mentions no Indian mathematician by name, I will first explain the rule of three with reference to the works of Āryabhaṭa and Brahmagupta. In their texts, *Āryabhaṭīya* and *Brāhmasphuṭasiddhānta*, the authors gave only the rules. Hence, I also make use of Indian commentaries to these works.

The Rule of Three and its Variations in India

In Indian sources, the rule of three is first mentioned in the *Āryabhaṭīya* written by Āryabhaṭa, who was 23 years old in 499. The text reads:

> Having multiplied the fruit-quantity in the rule of three quantities by the requisition-quantity, the result is divided by the argument. From it is the required fruit.[3]

The *Āryabhaṭīya* consists of rule-stanzas to be learnt by heart, but provides no examples. In his commentary, Bhāskara I gives an example in order to show how to apply the rule.

> Five palas of sandalwood was bought by me for nine rupas. Then how much sandalwood will be obtained for one rupa? [4]

A pala is a unit of weight, and a rupa a monetary unit. This is the proportion $9 : 5 = 1 : x$. Using the terminology set out by Āryabhaṭa, in this example, 9 is the argument, 5 the fruit, and 1 the requisition. The answer is $x = 5/9$ pala. This is an example of a proportional relationship, such as the price-quantity relationship of a commodity. Commenting on this example, Bhāskara I quotes a verse of unknown origin for the positioning of the quantities on a working surface.

> In order to perform regarding the rule of three quantities[5] the wise should know that the two similar [quantities] among the [three] quantities should be set down at the beginning and the end, and the dissimilar quantity in the middle.[6]

In Sanskrit mathematical texts, the quantities given in an example are arranged in a certain configuration. The three quantities given in this example were presumably aligned horizontally, although the text does not specify the direction.

$$9 \quad 5 \quad 1$$

Given three quantities, the first and the last terms should be of the same genus, or units. In general, the three quantities given in a problem are written down as follows, where a is the argument, b the fruit, and c the requisition:

$$a \quad b \quad c$$

The answer is given by the operation $(c \times b) \div a$.

Bhāskara I and the Rule of Five Quantities

Bhāskara I explains the rule of five quantities as an application of proportions. This is his explanation, in dialogue form.[7]

---

[3] trairāśikaphalarāśiṃ tam athecchārāśinā hataṃ kṛtvā/
labdhaṃ pramāṇabhājitaṃ tasmād icchāphalam idaṃ syāt// [Shukla 1976, 115].

[4] Shukla [1976, 117].

[5] Hereafter, I translate the Sanskrit term *trairāśika* as rule of three quantities, not rule of three.

[6] ādyantayos tu sadṛśau vijñeyau sthāpanāsu rāśīnām/
asadṛśarāśir madhye trairāśikasādhanāya budhiḥ// [Shukla 1976, 117].

[7] [Question] and [Answer] are supplied by me, based on my own interpretation.

> [Question] Only the rule of three quantities is mentioned by the teacher Āryabhaṭa. How should different proportions (*anupāta*), such as the rule of five quantities, and so on, be understood? [Answer] It is replied. Only the seed (*bīja*) of proportion is taught by the teacher. By means of this seed of proportions all beginning with the rule of five quantities has been established. [Question] Why? [Answer] Because the rule of five quantities, and so on, are combinations of the rules of three quantities. [Question] How are the rules of five quantities combined? [Answer] In the rule of five quantities, two rules of three quantities are combined, in the rule of seven quantities three rules of three quantities, in the rule of nine quantities four rules of three quantities, and so on.[8]

Here is one of his examples for the rule of five quantities:

> The interest of one hundred in a month is five. Say how much the interest for twenty invested for six months is, if you undetrstand [Ārya]bhaṭa's mathematics.[9]

After setting down the given quantities Bhāskara explains the procedure as follows:

> Procedure. The first rule of three: 100, 5, 20. What has been obtained is one rupa: 1. The second rule of three: If with one month one rupa [has been obtained], with six [months], how many [will be obtained]? What has been obtained is six rupas.
>
> This very [series of] computation[s] performed simultaneously is the rule of five quantities. And here, of a hundred in a month [gives] the two argument quantities, [a hundred and one], five is the fruit quantity, what is [the interest obtained] with twenty by means of a six month [loan]? Thus, twenty and six are the requisition quantities. Here, exactly as before, the requisition quantity multiplied by the fruit quantity [is] divided by the two argument quantities and the fruit [quantity] is obtained exactly as before. The rule of three quantities is arranged twice.[10]

That is, the first rule of three quantities is the following: 100 rupa (principal): 5 (interest) = 20 rupa (principal): a. So a, the interest for 20 rupa in one month, is 1. The second rule of three quantities is then as follows: 1 month : 1 (interest) = 6 month : b. So b, the interest for 20 rupa in 6 months, is 6.

The explanation that Bhāskara gives is as follows: I is the interest for the principal P in T months, and i is the interest for the principal p in t months. When P, T, I, p, t are known and i is required, i' such that $P : I = p : i'$ is obtained. Then, i such that $T : i' = t : i$ is obtained. That is, $i = \frac{i't}{T} = \frac{t}{T} \times \frac{pI}{P} = \frac{ptI}{PT}$. The quantities given are set down in the following arrangement:

$$\begin{matrix} P & p \\ T & t \\ I & i \end{matrix}$$

In this example:

$$\begin{matrix} 100 & 20 \\ 1 & 6 \\ 5 & \end{matrix}$$

---

[8] atra trairāśikam eva kevalam abhitam ācāryāryabhaṭena pañcarāśikādayo 'nupātaviśeṣāḥ katham avagantavyaḥ/ ucyate/ anupātabījamātram evācāryenopadiṣṭam tenānupātabījena sarvam eva pañcarāśikādikaṃ siddhyati/ kutaḥ/ pañcarāśikādīnāṃ
trairāśikasaṅghātatvāt/ tasmāt pañcarāśyādyas trairāśikasaṃhitāḥ pañcarāśike
trairāśikadvayaṃ saṃhataṃ saptarāśike trairāśikatrayaṃ navarāśike trairāśikacatuṣṭayam/ [Shukla 1976, 116].

[9] śatavṛddhir māse syāt pañca kiyān masaṣaṭprayuktāyā/
vṛddhiṃ vada viṃśatyā yadi bhaṭagaṇitaṃ tvayā buddham// [Shukla 1976, 119].

[10] karaṇam prathamatrairāśikam 100 5 20/ labdhaṃ rūpakaḥ 1/ dvitīyatrairāśikam yadi māsena rūpakaḥ ṣaḍbhiḥ kiyanta iti labdhaṃ rūpakaḥ ṣaṭ/ etad eva gaṇitaṃ yugapat kriyamāṇaṃ pañcarāśikaṃ bhavati/ tatrāpi śatasya māse iti [śataṃ rūpaṃ ca] pramāṇarāśidvayam pañceti phalarāśiḥ viṃśatyā ṣaḍbhir māsaiḥ kim iti viṃśatiḥ ṣaṭ ca icchārāśiḥ/ tatra pūrvavad eva icchārāśiḥ phalarāśinā guṇitaḥ pramāṇarāśibhyāṃ vibhajyate phalaṃ pūrvavad eva/ trairāśikam evaitad dvidhā vyavasthitam/ [Shukla 1976, 120].

The Brāhmasphuṭasiddhānta *by Brahmagupta*

In 628, Brahmagupta completed the *Brāhmasphuṭasiddhānta*.[11] It also consists of stanzas giving only the rules, but no examples. Here is a portion of the material on the rule of three quantities and related operations, from Chapter 12. It first states the inverse rule of three quantities.

> In the rule of three quantities there are argument, fruit and requisition. The first and the last are of the similar quantity. The requisition multiplied by the fruit [and] divided by the argument is the [desired] fruit.
>
> The product of the argument and the fruit, being divided by the requisition, is the fruit in the inverse rule of three quantities.
>
> In the case of uneven terms, from three up to eleven, transposition of the fruit on both sides takes place. The product of the more numerous terms on one side, being divided by that of the fewer terms on the other, is to be known as the result. In all fractions, transposition of the denominators, in a like manner, on both sides takes place.[12]

In the case of the rule of three quantities, Brahmagupta may have written down quantities in two columns. When fractions are given, the denominators are transposed with each other.

Pṛthūdaka's Commentary

Although Chapter 12 of the *Brāhmasphuṭasiddhānta* simply gives rules, the commentator Caturveda Pṛthūdakasvāmin (fl. 864, hereafter Pṛthūdaka) gave a number of examples. His example for the rule of five quantities is for investment interest, similar to the example discussed above. For the rule of seven quantities Pṛthūdaka gives the following example:

> If three pieces of cloth, five cubits long and two wide, cost six panas, and ten have been purchased three wide and six long, tell the price.[13]

The numbers are set down as follows:

2	3
5	6
3	10
6	

The operation is to move the 6 to the empty space, as

2	3
5	6
3	10
	6

Then the product of all the terms in the longer column is taken, that is, $3 \times 6 \times 10 \times 6 = 1080$, and the product of all the terms in the shorter column is taken, that is, $2 \times 5 \times 3 = 30$. Finally, the product of the longer column is divided by that of the shorter column, $1080 \div 30 = 36$. So, the answer is 36 panas.

[11] Pingree [1983] demonstrated that Bīrūnī studied this text from a now lost commentary on *Brāhmasphuṭasiddhānta* composed by Balabhadra, made in the eighth century. The order of chapters listed in Bīrūnī's *India* is different from that of the text I use [Sachau 1910, 1, 154–155].

[12] trairāśike pramāṇaṃ phalam icchādyantayoḥ sadṛśarāśī/
icchā phalena guṇitā pramāṇabhaktā phalaṃ bhavati//
vyastatrairāśikaphalam ichhābhaktaḥ pramāṇaphalaghātaḥ//
trairāśikādiṣu phalaṃ viṣameṣv ekādaśānteṣu/
phalasaṅkramaṇaṃ ubhayato bahurāśivadho 'lpavadhahṛto jñeyam//
sakaleṣv evaṃ bhinneṣūbhayataś chedasaṅkramaṇam// [Dvivedī 1902, 178].

[13] Colebrooke [2005, 285]. Colebrooke referes to Pṛthūdaka as CH.

Pṛthūdaka's example for nine quantities concerns the price of bricks, and that for eleven quantities deals with the amount of grain an elephant consumes.

Following the other rules, Brahmagupta gives a rule for bartering.

> In the barter of commodities, transposition of prices, being first terms, takes place; and the rest of the process is the same as above directed.[14]

Pṛthūdaka supplies the following example.

> If a hundred mangoes be purchased for ten panas; and of pomegranates for eight; how many pomegranates [should be exchanged] for twenty mangoes?[15]

The numbers are set down as follows:

10	8
100	100
20	

The operation is to transpose the prices and move the 20 to the empty space, as

8	10
100	100
	20

Following the same procedure as above, the answer is $(10 \times 100 \times 20) \div (8 \times 100) = 25$ pomegranates.

Bīrūnī's Work on Indian Mathematics

The treatise *Maqāla fī rāshīkāt al-Hind* is included in the *Rasā'il al-Bīrūnī* published by The Osmania Oriental Publications Bureau in Hyderabad.[16] The published text is an uncritical edition of an Arabic manuscript, copied in 631H (1233/1234), Bankipore 2468 (now 2519) in the Khuda Bakhsh Oriental Public Library in Patna, India.[17] The Hyderabad text is edited following the manuscript, the pages of which are bound in an incorrect order.[18] In order to facilitate reading the Hyderabad text, I propose that the text should be read in the following order, which is required by the sense:

p. 13, l. 18. عليه to p. 15, l. 2 الوسط
p. 16, l. 6 واحد to p. 13, l. 18 واحد
p. 15, l. 2 العشرة to p. 16, l. 6 فيجتمع

[14] Colebrooke [2005, 285–286].
prāgmūlyavyatyāso bhāṇḍapratibhāṇḍake 'nyaduktasamam/ [Dvivedī 1902, 181].

[15] Colebrooke [2005, 286].
pādadaśabhir āmrāṇāṃ dāḍimānāṃ tathāṣṭabhiḥ/
yadā śataṃs tadāmrais tu viṃśatyā kati dāḍimāḥ//
I have used the MS Egelling 2769, India Office Library, folio. 53v.

[16] Bīrūnī catalogued his works and those of Rāzī. Adding to the catalogue, Boilot [1955] and Khan [1982] compiled a bibliography of Bīrūnī's work. Among 180 works listed in Boilot, or 183 in Khan, those numbered 34 to 41 seem to be treatises on arithmetic. Out of these 8 treatises, only No. 38, the treatise of concern in this paper, is extant. As well as the number of books, Bīrūnī usually mentioned the number of folios occupied by each work. This treatise occupies 15 folios. Boilot and Khan read No. 39 as *Fī saklab al-a'dād* (On the Saklab of the Numbers). However, the word *saklab* means nothing. According to the Arabic manuscript of the *Fihrist* preserved in the library of University of Leiden, Or. 133.2, *saklab* can be read as *sankalita*. *Sankalita* is the transliteration into Arabic of the Sanskrit word *saṃkalita*, which means "sum of addition" or the sum of a mathematical series.

[17] I thank Professor Michio Yano for color photos of manuscript 2519-XXXVII taken by him.

[18] Hogendijk [1982, 135–137] discusses the ordering of the pages in manuscript.

We will see two of these in the translations, given below, of two of Bīrūnī's examples for the rule of five quantities. Along with the incorrectly bound pages, reading this treatise is complicated by textual corruption in the manuscript, especially towards the end of the treatise.

The unusual word *rāshīka*, in the title, can be explained as follows. The word *rāshi* is an Arabic transliteration of a Sanskrit word, *rāśi*, which means heap. The Sanskrit term means quantity in a mathematical context, or a sign of the zodiac or 30 degrees in astronomy. With the suffix *-ka*, the term *rāśika* is used in Sanskrit mathematical texts as *trairāśika*, the rule of three quantities. Bīrūnī formed an Arabic plural as *rāshīkāt*.

Although Youschkevitch [1976, 45] has briefly enumerated the topics of the treatise, the text has never been critically studied. The manuscript includes tabular arrays of numbers, which appear almost randomly in the Hyderabad version. A critical edition of this text is still needed.

The treatise can be divide into five parts, according to subject matter: (1) A general treatment of the Greek concept of ratio, (2) an introduction to the Indian concept of *trayrāshka* and other rules, (3) a general introduction to composed ratios, (4) the use of composed ratios as a theoretical basis for the Indian rules, and (5) examples of the various rules. In the remainder of the paper, I will discuss the contexts of the treatise, divided into these five sections.

Greek Theory of Ratio and Proportion [Hyderabad 1948, 1–3]

Bīrūnī begins his discussion with a definition of ratio:

> A ratio among homogeneous magnitudes is one of the forms of relations that occur in respect of quantity. Therefore, by means of this [i.e. ratio], one of two is known from the other, if it is unknown.[19]

As claimed by Euclid, in *Elements* Book 5, Definition 3, a ratio must be between two magnitudes of the same genus. Bīrūnī then refers explicitly to Euclid's Definition 8, which states that a proportion must have at least three terms.

> Euclid said: A proportion consists of at least three terms.[20]

In order to explain the rule of three quantities, Bīrūnī uses the Euclidean concept of proportion.

> Euclid demonstrated in the sixteenth [proposition] of the sixth [book] that < if four straight line are proportional > the plane [rectangle contained] by the first of them and the fourth is equal to the plane [rectangle contained] by the second and the third.[21]

That is, where $A : B = C : D$, then $A \times D \div B = C$ and $A \div B \times C = D$. Next, Bīrūnī discusses various terminology related to Greek ratio theory, such as inverse ratio (عكس النسبة), permutation (ابدال), composition (تركيب), separation (تقصيل) and conversion (قلب). These terms are of Greek origin and are defined and treated in *Elements* V.

The Trayrāshīka of India [Hyderabad 1948, 3–4]

Bīrūnī discusses the rule of three quantities as follows:

[19] The Arabic text reads: النسبة فيما بين المقادير المتجانسة هي صورة من صور الاضافات تحصل لها من جهة الكمية فيعرف بها احدهما من الاخهر ان كان غير معلوم [Hyderabad 1948, 1].

[20] The text reads: قال اقليدس ان التناسب اقل ما يكون في ثلاثة حدود [Hyderabad 1948, 1].

[21] The text reads: وقد بين اقليدس في السادس عشر من السادسة ان سطح الاول منها في الرابع مساو لسطح الثاني في الثالث [Hyderabad 1948, 2].

> The Indians called it[22] *trayrāshīka* (تري راشيك), that is, possessor of three positions. *Rāshi* is a sign of the zodiac and *rāshīka* is the position of a constellation. Astrologers call the twelve houses *rāshīka*. They call these three because the known [things] in the data [of a problem] are three.
>
> They are people who take in their arithmetic numerical methods, so that they are skilled in it. They rely on correction by test and investigation of examples, not on being occupied with justification by geometrical demonstration. They draw two intersecting lines such that there result four places in this example.
>
> They say: If five is [produced] by fifteen, then by what is three [produced]? They move fifteen to the empty place and multiply it by what is above it, that is three. Forty-five results. Then they divide it by five and nine results. That is placed necessarily in the empty place so that three is [produced] by nine. This is what we mentioned because the corresponding [member] in multiplication results in this quadrangle in two diagonals.[23]

Bīrūnī uses double columns for setting out the terms of the rule of three quantities. In the manuscript, the numbers are arranged as follows:

15	5
	3

This arrangement, however, does not agree with what he describes in the text. According to Bīrūnī's explanation, given above, the arrangement should be changed to

5	3
15	

According to the text, 15 is moved to the empty space, and the operation is $(15 \times 3) \div 5 = 9$.

If, however, Bīrūnī had followed the procedure described by Brahmagupta the arrangement would have been

15
5 3

Then 5 and 3 would have been transposed, as

15
3 5

Finally, the operation would be $(15 \times 3) \div 5 = 9$.

Composition of Ratios [Hyderabad 1948, 4–6]

Bīrūnī discusses two kinds of composition of ratios. In the first case, the magnitudes form a geometric sequence. His example is 3, 9, 27, 81, 243. Bīrūnī says that the ratio of the first to the third is composed of the ratio of the first to the second and of the ratio of the second to the third, and also that the ratio of the first to the fourth is composed of the ratio of the first to the second and of the ratio of the second to the third and of the ratio of the third to the fourth. In reference to this type of compound ratio, he gives a quotation of *Elements* VI, Definition 5, as follows:

> For this [purpose] Euclid said: The ratio is said to be composed of a number of ratios when [ratios] are multiplied one by the other and a ratio results.[24]

In the second case, the series is not geometric. The example he gives is 5, 10, 30, 120. He says that a ratio composed of two ratios is used for the calculation of the sines of arcs of the sector figure (شكل قطاع), the Arabic name for the so-called Menelaus Theorem.

---

[22] "It" translates the feminine pronoun, *hā*; it could also mean "them." In either case, however, it is not clear what it refers to.

[23] Hyderabad [1948, 3, ll. 9–15].

[24] The text reads: ولهذا قال اقليدس يقال ان النسبة مؤلفة من عدد نسب اذا ضوعفت بعضها ببعض فاحدثت تلك النسبة [Hyderabad 1948, 5].

Theoretical Basis of the Rule of Five Quantities [Hyderabad 1948, 6–13]

When the ratio of A to B is composed of the ratio of G to D and of the ratio of E to W, I use the symbol $\otimes$ to represent the composition, so that a composed ratio can be written as

$$A : B = (G : D) \otimes (E : W). \quad (1)$$

When Bīrūnī wants to talk, in general terms, about the operations of a composed ratio, he uses ordinal numbers, such that A is the first term, B the second term, and so on, as in the discussion above. When one of these six magnitudes, say B, is unknown and other five are known, Bīrūnī explains how to find it. First, let $G : D = A : T$, where T is a magnitude between A and B. Then find B, such that $E : W = T : B$. That is, when only B is unknown, for example, first $(A \times D) \div G = T$ is used to find T, and then $(T \times W) \div E = B$ is used to find B. That is, $((A \times D) \div G) \times W \div E = B$.

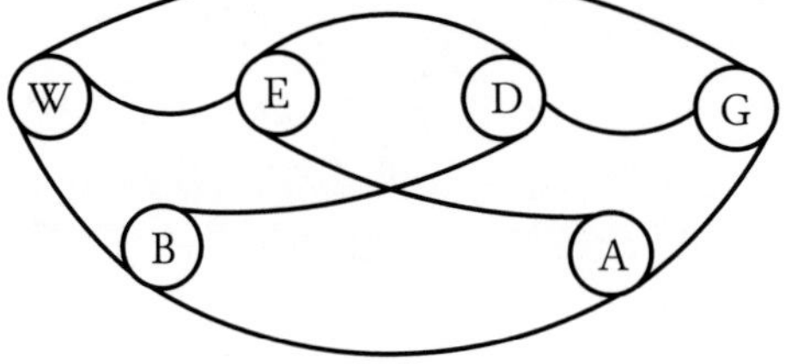

Figure 1: *Maqāla fī rāshīkāt al-Hind*, Diagram 1.

In connection with this discussion on composed ratio, Bīrūnī mentions a number of his predecessors by name, Thābit ibn Qurra, Abū al-Abbās al-Nayrīzī, Ibn al-Baghdādī, Abū Jaʿfar al-Khāzin, Abū Saʿīd al-Sijzī, but refers to none of their works.

In order to discuss various possibilities for relating any two terms among the six terms of the compound ratio, Bīrūnī gives two diagrams. In Figure 1, terms connected by a line have a ratio.

In Figure 2, terms connected by a line cannot have a ratio.

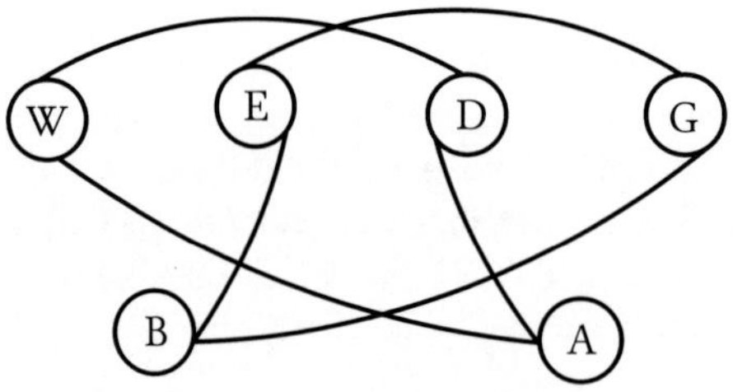

Figure 2: *Maqāla fī rāshīkāt al-Hind*, Diagram 2.

Following the diagrams, Bīrūnī gives three tables. There are nine lines in Figure 1, hence the possibilities for ratios are as follows: A can be in ratio with B, G and E, B in ratio with D and W, G in ratio with D and W, D in ratio with E, and E in ratio with W. Each of these ratios is composed in two ways; therefore there are in total eighteen compound ratios. The first table is as follows:

1	A	B	G	D	E	W
2	A	B	G	W	E	D
3	A	G	B	D	E	W
4	A	G	B	W	E	D
5	A	E	B	D	G	W
6	A	E	B	W	G	D
7	B	D	A	G	W	E
8	B	D	A	E	W	G
9	B	W	A	E	D	G
10	B	W	A	G	D	E
11	G	D	A	B	W	E
12	G	D	A	E	W	B
13	G	W	D	B	A	E
14	G	W	A	B	D	E
15	D	E	G	A	B	W
16	D	E	B	A	G	W
17	E	W	A	B	D	G
18	E	W	A	G	D	B

There are captions at the head of each column. The 1st column reads "number of the composed ratio." The 2nd and 3rd columns read "ratio composed of two ratios." The 4th and 5th columns read "first ratio." And the 6th and 7th columns read "second ratio."

Hence, using symbols, we can understand this table as stating that if

$$A : B = (G : D) \otimes (E : W)$$

then

$$\begin{aligned}
A : B &= (G : W) \otimes (E : D),\\
A : G &= (B : D) \otimes (E : W),\\
A : G &= (B : W) \otimes (E : D),\\
A : E &= (B : D) \otimes (G : W),\\
A : E &= (B : W) \otimes (G : D),\\
B : D &= (A : G) \otimes (W : E),\\
B : D &= (A : E) \otimes (W : G),\\
B : W &= (A : E) \otimes (D : G),\\
B : W &= (A : G) \otimes (D : E),\\
G : D &= (A : B) \otimes (W : E),\\
G : D &= (A : E) \otimes (W : B),\\
G : W &= (D : B) \otimes (A : E),\\
G : W &= (A : B) \otimes (D : E),\\
D : E &= (G : A) \otimes (B : W),\\
D : E &= (B : A) \otimes (G : W),\\
E : W &= (A : B) \otimes (D : G),\\
E : W &= (A : G) \otimes (D : B).
\end{aligned}$$

The second table is as follows:

1	A	B	G	D	W	E
2	A	G	B	D	W	E
3	A	E	B	D	W	G
4	B	D	A	G	E	W
5	B	W	A	G	E	D
6	G	D	A	B	E	W
7	G	W	A	B	E	D
8	D	E	A	B	G	W
9	E	W	A	B	G	D

Once again, there are captions for the columns. The 1st column reads "number of combinations." The 2nd and 3rd columns read "omitted two equal ratios." And the 4th to 7th columns read "remaining proportional magnitudes."

The second table should be read as meaning that

$$A = B \Longrightarrow G : D = W : E,$$
$$A = G \Longrightarrow B : D = W : E,$$
$$A = E \Longrightarrow B : D = W : G,$$
$$B = D \Longrightarrow A : G = E : G,$$
$$B = W \Longrightarrow A : G = E : D,$$
$$G = D \Longrightarrow A : B = E : W,$$
$$G = W \Longrightarrow A : B = E : D,$$
$$D = E \Longrightarrow A : B = G : W,$$
$$E = W \Longrightarrow A : B = G : D.$$

The second table handles the case where two of the six quantities are equal.

The third table is as follows:

1	A	B	G	D	W	E
2	A	B	D	G	E	W
3	A	B	E	W	D	G
4	A	B	W	E	G	D
5	A	G	D	B	E	W
6	A	G	E	W	D	B
7	A	G	W	E	B	D
8	A	D	E	B	G	W
9	A	E	W	B	G	D
10	B	G	D	A	W	E
11	B	G	W	D	A	E
12	B	D	E	A	W	G
13	B	D	W	E	G	A
14	B	E	W	D	A	G
15	G	D	E	W	A	B
16	G	D	W	E	B	A
17	G	E	W	D	A	B
18	D	E	W	G	B	A

Again, the headings are labeled. The 1st column reads "number of combinations." The 2nd to 4th columns read "three equal magnitudes." The 5th column reads "the ratio of [one of] three to this." The 6th column reads "equal to the ratio of this." And the 7th column reads "to this."

The third table means that

$$A = B = G \Longrightarrow A : D = W : E,$$
$$A = B = D \Longrightarrow A : G = E : W,$$
$$A = B = E \Longrightarrow A : W = D : G,$$
$$A = B = W \Longrightarrow A : E = G : D,$$
$$A = G = D \Longrightarrow A : B = E : W,$$
$$A = G = E \Longrightarrow A : W = D : B,$$
$$A = G = W \Longrightarrow A : E = B : D,$$
$$A = D = E \Longrightarrow A : B = G : W,$$
$$A = E = W \Longrightarrow A : B = G : D,$$
$$B = G = D \Longrightarrow B : A = W : E,$$
$$B = G = W \Longrightarrow B : D = A : E,$$
$$B = D = E \Longrightarrow B : A = W : G,$$
$$B = D = W \Longrightarrow B : E = G : A,$$
$$B = E = W \Longrightarrow B : D = A : G,$$
$$G = D = E \Longrightarrow G : W = A : B,$$
$$G = D = W \Longrightarrow G : E = B : A,$$
$$G = E = W \Longrightarrow G : D = A : B,$$
$$D = E = W \Longrightarrow D : G = B : A.$$

Although the captions are somewhat cryptic, the third table must be for the case where three of the six quantities are equal.

I have not seen such tables in Sanskrit mathematical texts; however, similar tables are found in Thābit ibn Qurra's work on the composition of ratios.[25]

Examples of Rules of Various Quantities [Hyderabad 1948, 13–30]

Bīrūnī gives a number of examples for the rules of five and greater quantities. The calculation procedure for all of these is obvious from the text; however, in some cases it is not clear what the terms mean, or if the example provided can really be solved by the methods of these rules.

Examples of the Rule of Five Quantities (پنج راشيك)

For the rule of five quantities, Bīrūnī gives the following example:

> [p. 13, l. 7] When ten dirhams gain in two months five dirhams, then how many [dirhams] do eight [dirhams] gain in three months? They place them as [they did] in this form. Magnitudes of the composed ratio are always at the bottom and both are dirhams produced by the capital in a [certain] period. In order to calculate the unknown, they transfer five to the empty space and multiply it by three, then by eight. So it becomes a hundred and twenty. They keep it. Then they multiply two by ten. So it becomes twenty. They divide the kept [number] by it. Then six results and this is a gain of eight dirhams in three months. This is thus because five is the first, and the desired, the second, and ten, the third, and eight, the fourth, and two, the fifth, and three, the sixth. Therefore, according to its decomposition, which we mentioned, into four proportional numbers twice it is necessary that the first is multiplied by the fourth and the result is divided by the third. Then there results in this case [p. 13, l. 18, to p. 15, l. 2] a mean between the first and the second. We multiply it by the sixth and divide the result by the fifth. Then, the second comes out. It is known that multiplication occurs by each one of the fourth and the sixth and division occurs by each one of the third and the fifth. Therefore when two multiplications are joined because the

[25] See Lorch [2001, 286, 288, 300].

first is multiplied by the fourth and the result [is multiplied] by the sixth or the first is multiplied by the sixth and the result [is multiplied] by the fourth, it is necessary that the two divisions also [are joined] because the result is divided by the third and what results [is divided by] by the fifth. However what they operated is shortened; the division of the result, I mean the kept [number] in their operation, by the product of the third and the fifth, I mean divisor. This is by the shorter [procedure] easier, and by application of fractions more difficult. Because our example [is that] the ratio of five to the desired is composed of the ratio of ten to eight and of the ratio of < two > to three. The two magnitudes for multiplication are on the side in which the empty space is. Therefore, they transfer five to there in order that the multiplicands are joined at one side and the divisor at the other side.

The numbers are set down as follows:[26]

10	8
2	3
5	

Bīrūnī then uses the concept of compound ratio to explain how to find one unknown when the remaining five are given. Using the terms of Equation 1, above, his verbal description can be summarized as

$$B = (\frac{A \times D}{G}) \times \frac{W}{E} = \frac{ADW}{GE}.$$

Following this, Bīrūnī gives another example for the rule of five quantities, and then discusses the case where a term other than B is required.

[p. 14, l. 3] It is possible that the same [problem] is interpreted in the other way. Example: It is said: [When] ten men excavate in two days five cubits, then how many [cubits] do eight men excavate in three days? Then it is possible that the knowns in these places differ such that the intended unknown is days or number of men or the principal. Whenever the unknown, I mean the empty place, is not in the last line, the method to know it is that what is at the bottom line is exchanged, then each one of what is there is placed in other place, [and] then each one of them is multiplied by what is above it and the other and one of them is divided by the result from the side in which the empty place is. Then the desired results.

Its example is that it is said: Ten men excavate five cubits in two days, then in how many days do eight men excavate six cubits? Its way is that we place [p. 15] five at the place of six and six at the place of five. Then we multiply six by two and then by ten then [p. 15, l. 2, to p. 16, l. 6] one hundred and twenty results. We keep it. Then we multiply five by eight. Then forty results. Then we divide the kept [number] by it. Then three results and this is the desired days. This is because the unknown is the sixth of the magnitudes of the composed ratio. Therefore, it is necessary that we multiply the first by the fourth and divide the result by the third and we divide the product of the second by the fifth by the result from this division, then the sixth results. But the first is in other side of the fourth in the diagram of the *rāshīka*. Therefore, if the bottom is brought over, it is in one [and the same] side there. Similarly, the second is with the fifth and third in side. It is necessary that the product of the second by the fifth is divided by the mean in order to derive the sixth. The product of the first by the fourth is the product of the mean by the third. Therefore the dividend by it is a multiplicand by a number. It is necessary that the dividend is also a multiplicand by it in order that there results from them both, what results from the other, a multiplicand by it. And the mean, which is a dividend for it, is a multiplicand by the third, for its result is from the multiplication of the first by the fourth. Therefore, it is necessary that the dividend, which is the product of the second by the fifth, is multiplied by it, in order to derive from it what was derived first.

The numbers are set down in the following tabular arrangement:[27]

10	8
2	
5	6

Again using the terms of Equation 1, Bīrūnī gives a verbal description of the operations, which can be summarized as

$$BE \div \frac{AD}{G} = W = \frac{BEG}{AD}.$$

Following this example, another case is explained, namely:

[26] In Hyderabad [1948] text, the table appears on page 14.

[27] In the Hyderabad [1948] text the table appears on page 15.

	8
2	3
5	6

Generally, using the terms in Equation 1, the tabular arrangement of terms is

G	D
E	W
A	B

When five out of six magnitudes are given and the remaining one is required, it is given by means of the operations implied by the composed ratio of Equation 1. Hence, A and B (called the fruits in Indian mathematics) are exchanged and the product of the terms in the column with more terms is divided by the product of the terms in the column with fewer terms.

Example of the Rule of Barter

Brahmagupta gives a separate rule for barter, which in Sanskrit is called *bhāṇḍapratibhāṇḍa*, but Bīrūnī does not mention any special name for this rule.

Bīrūnī's discussion of this rule is as follows:

> [p. 19, l. 2] One hundred quinces are said to be [sold] for ten dirhams and one hundred pomegranates [are sold] for eight dirhams. We want to know how many pomegranates are for twenty quinces. First we place those [numbers] as was stated. Then, we exchange the two places at the top and return to the rule of five quantities. So, we operate by what proceeded from the movement of the twenty to the empty place, and we multiply it by one hundred, and then by ten. One thousand and ten[28] is divided by eight hundreds. Then there results twenty-five, the number of pomegranates in quinces.

The tabular arrangement is:

10	8
100	100
20	

As Bīrūnī explains, we exchange 10 and 8 in the top row

8	10
100	100
20	

and then move 20 to the empty space

8	10
100	100
	20

The operation is $(20 \times 100 \times 10) \div (100 \times 8) = 25$. This example is same as that given by Pṛthūdaka.

Example of the Inverse Rule of Three Quantities

Bīrūnī gives a peculiar example for the inverse of the rule of three quantities. The text is as follows:

> [p. 19, l. 10] If a girl of twenty costs eight dirhams, then how much is a girl of forty? Its method is that eight is not moved to the empty place but [p. 20] it is multiplied by what is above it. One hundred and sixty results. We divide [it] by forty. Then, four is the cost of the girl of forty. The Indians call this [kind of operation] the inverse rule of three quantities [بست تري راشيك, *vyastatrairāśika* in Sanskrit] ...

[28] This should be twenty thousand; that is $20 \times 100 \times 10$.

The tabular arrangement is as follows:[29]

20	40
8	

The operation is $(8 \times 20) \div 40 = 4$.[30]

Examples of Seven and More Quantities

Bīrūnī's example for the rule of seven quantities is stated as follows:

> [p. 20, l. 7] Its example: A piece of sandalwood the length of which is five digits and the width is three digits and the thickness is four digits is [sold] for thirty dirhams, [then] how much is its piece with eight digits length, six digits width, and two digits thickness? They place those [numbers] according to the previous illustration, with each kind face to face. Then they move three[31] to the [empty] space. They follow the method mentioned in the rule of five quantities. Then the kept [number] is two thousand and eight hundred and eighty. The divisor is sixty. The price of the desired piece is forty-eight [dirhams].

The tabular layout for this example should be as follows:[32]

5	8
3	6
4	2
30	

Following the text, the procedure is to move 30 to the empty space

5	8
3	6
4	2
	30

Take the products of the terms in the two columns, 2880 and 60, and divide the product of the longer column by that of the shorter column. The answer is 48 dirhams.

Bīrūnī explains this calculation with reference to the rule of three quantities: $5 \times 3 \times 4 = 60$ is volume of one piece of sandalwood, $8 \times 6 \times 2 = 96$ the volume of the other.

In the remainder of this section, I will briefly survey the examples and tables that Bīrūnī gives for the rules of nine to seventeen quantities. His rationale for the operations in these rules is based on a generalization of the concept of composed ratio, such that where the tabular arrangement is

a	b
c	d
e	f
g	h
i	x

and, where x is the required number, the composed ratio is

$$i : x = (a : b) \otimes (c : d) \otimes (e : f) \otimes (g : h).$$

[29] In the table given in the Hyderabad [1948, 20] printing, the number 8 is missing, but it is present in the manuscript.

[30] In his *India*, Bīrūnī refers to the *vyastatrairāśika* and gives a similar example: "If the price of a harlot of 15 years be, for example, 10 denars, how much will it be when she is 40 years old?" [Sachau 1910, 1, 313].

[31] The mathematical sense indicates that this should be thirty.

[32] In Hyderabad [1948, 21] we find

3	8
3	6
5	2
30	

, while the manuscript has

4	8
3	6
5	2
30	

.

The example for the rule of nine quantities deals with the price of bricks.[33] The rows of the table are length, breadth, thickness, number of workers and dirhams. The tabular layout is as follows:[34]

5	8
4	6
3	2
30	20
60	

The solution to the problem follows the pattern set out above.

The example of the rule of eleven quantities deals with constructing walls.[35] The rows are length, breadth, thickness, number, number of workers, and dirhams. The tabular arrangement is as follows:

10	15
3	4
8	7
2	3
6	9
40	

Again, the solution to the problem follows the pattern.

Bīrūnī goes on to give examples for the rules of thirteen, fifteen and seventeen quantities; however, these are plagued with textual and other difficulties. I have not seen any examples of rules of thirteen or higher quantities in Sanskrit mathematical texts.

The rule of thirteen quantities deals with the distance of stones thrown by a catapult (منجنيق).[36] I do not understand this example, because I do not know details of the catapult construction, and the names of objects in the rows are sometimes unclear, due to textual corruption. The rows are side, thickness, unclear*, weight of stone, unclear*, number of people, and distance. The tabular arrangement is as follows:[37]

7	10
15	20
13*	17
5	4
300	260
20	25
100	

Presumably, the operation follows the same pattern, but the text seems to be corrupted.

The example of the rule of fifteen quantities deals with the exchange prices of fruit.[38] I give a literal translation of this example; however, the problem set in this example does not seem to be well handled by the rule of fifteen quantities. The text reads:

> [p. 24, l. 7] When they say, for example: Twelve manas[39] of raisins (زبيب) is five manas of dried dates (تمر) and eight manas of dried dates is two manas of sugar candy (فانيد), and six manas of sugar candy is three manas of

[33] Hyderabad [1948, 22.4–2, table on 22].

[34] In both the Hyderabad [1948, 22] text and the manuscript it appears as

4	8
3	6
5	2
30	20
60	

.

[35] Hyderabad [1948, 23.8–14, table on 24].

[36] Hyderabad [1948, 23.19–24.4, table on 26].

[37] In the third row, both the manuscript and the Hyderabad [1948, 22] have "13" but there is some problem with this, and it is not supported by the text.

[38] Hyderabad [1948, 24.7–11, table on 25].

[39] Mana is a unit of weight.

sugar (سكر) and fifteen manas of sugar is twenty dirhams. The exchange rate changes; eight manas of raisins is seven manas of dried datess and four manas of dried dates is nine manas of sugar candy and seven manas of sugar candy is two manas of sugar. Then by how many dirhams is five manas of sugar purchased?

The tabular arrangement is as follows:

12	8
5	7
8	4
2	9
6	7
3	2
15	5
20	

The operation follows the same pattern as the others.

The example of seventeen quantities, which is obscure, seems to be on the irrigation of arable land.[40]

Conclusion

The treatise by Bīrūnī is a strange mixture of Indian and Greek mathematics. In Sanskrit texts, the rule of three quantities is explained using proportional relationships such as the price-quantity relationship of a commodity. Bīrūnī's explanation of the rule, however, is not associated with commercial problems because he wants to derive it from Greek concepts of ratio, in which the two terms of the ratio must be of the same kind. Under such a conception of ratio, the Indian rule of three quantities is troublesome, because we perform arithmetical operations on things of different kinds. For the rule of five quantities and its variations, Bīrūnī argues for a theoretical basis using the Greek theory of composition of ratios, as developed by Thābit ibn Qurra and others.[41]

Perhaps because these rules had no natural place within the framework of theoretical Greco-Arabic mathematics, he simply transliterated the Sanskrit terms *vyastatrairāśika*, *pañcarāśika*, and so on, and did not coin Arabic terminology for the rule of three quantities and its variations. The examples that Bīrūnī gives for the rules of five quantities, and so on, are of Sanskrit origin.

The sequence of the treatment of the rules and examples given by Brahmagupta and Pṛthūdaka was followed by later Indian mathematicians such as Bhāskara II. The order presented by Bīrūnī, however, is somewhat different. This can be summarized as follows:

Brahmagupta	Bhāskara II	Bīrūnī
Rule of three	Rule of three	Rule of three
Inverse rule of three	Inverse rule of three	Rule of five
Rule of five etc.	Rule of five etc.	Barter
Barter	Barter	Inverse rule of three
		Rule of seven quantities etc.

[40] See Hyderabad [1948, 25, table on 27].

[41] In fact, Greek although mathematicians used composed ratios, they never developed a successful theory of their usage. This theory was left to Arabic mathematicians to complete.

References

Primary Sources

Āryabhaṭīya with commentaries of Bhāsakara I and Someśvara: Shukla [1976].
Brāhmasphuṭasiddhānta: Dvivedī [1902].
Maqāla fī rāshīkāt al-Hind: Hyderabad [1948].

Secondary Sources

Boilot, D.J., 1955. L'Oeuvre d'al-Beruni, Essai Bibliographique. Mélanges de l'Institute Dominican d'Études Orientales du Caire 2, 161–256.

Chabert, J.-L., et al. 1999. A History of Algorithms. Weeks, C. (trans.). Springer, New York.

Colebrooke, H.T., 2005. Classics of Indian Mathematics: Algebra, with Arithmetic and Mensuration, from the Sanskrit of Brahmagupta and Bhāskara (with a Foreword by S. R. Sarma). Sharada Publishing House, Delhi. Originally published 1817.

Dvivedī, S. (ed.), 1902. Brāhmasphuṭasiddhānta, with the Editor's Commentary in Sanskrit. Medical Hall Press, Benares.

Hogendijk, J.P., 1982. Rearranging the Arabic mathematical and astronomical manuscript Bankipore 2468. Journal for the History of Arabic Science 6, 133–159.

Hyderabad = *Rasá'ilul-Bírúní*, 1948. The Osmania Oriental Publications Bureau, Hyderabad.

Khan, A.S., 1982. A Bibliography of the Works of Abu'l-Raihan al-Biruni. Indian National Science Academy, New Delhi.

Lorch, R., 2001. Thābit ibn Qurra On the Sector Figure and Related Texts. Institut für Geschichte der arabisch-islamischen Wissenschaften, Frankfurt am Main. Reprinted: Dr. Erwin Rauner, Ausburg, 2008.

Pingree, D.E., 1983. Brahmagupta, Balabhadra, Pṛthūdaka and al-Bīrūnī. Journal of American Oriental Society 103, 353–360.

Sachau, C.E., 1910. Alberuni's India (2 vols.). Kegan Paul, Trench, Tübner & Co., London.

Sarma, S.R., 2002. Rule of three and its variations in India. In: Dold-Samplonius, Y., Dauben, J.W., Folkerts, M., van Dalen, B. (eds.), From China to Paris: 2000 Years Transmission of Mathematical Ideas, pp. 133–156.

Shukla, K.S. (ed.), 1976. Āryabhaṭīya of Āryabhaṭa with the commentary of Bhāskara I and Someśvara. Indian National Science Academy, New Delhi.

Smith, D.E., 1923–1925. History of Mathematics. 2 vols. Ginn and Company, Boston. Reprinted: Dover, New York, 1958

Youschkevitch, A.P., 1976. Les mathématiques Arabes: VIII-XV siècles. Cazenave, M., Jaouiche, K. (trad.). J. Vrin, Paris.

Safavid Art, Science, and Courtly Education in the Seventeenth Century

Sonja Brentjes

Abstract My paper studies several manuscripts of ʿAbd al-Raḥmān Ṣūfī's *Kitāb al-kawākib al-thābita*, which were produced at the Safavid court, a provincial court at Mashhad or by artists related to either of them. The purpose is to contribute in a small manner to a cultural history of science in a specific context of one of the major post-classical societies. Its main claim is that the Safavid elite paid considerable attention to, and invested substantial resources in, reproducing famous illustrated scientific manuscripts. Content and aesthetic point to Timurid inspirations. In a limited sense, one could speak of a Safavid engagement with translating scientific, medical and geographical texts from Arabic into Persian. The integration of art, science and translation could be described as a specific feature of courtly interest in scholarly knowledge under the Safavids.

Introduction

The Safavid period is usually seen as one with substantial innovation in the arts and a vivacious engagement in philosophical inquiries of the divine, but with little interest as well as sophistication in the mathematical sciences and natural philosophy. There is, however, reason to believe that this view may be inadequate.

In the first decades of Safavid rule, the mutakallim Shams al-Dīn Khafrī (d. c. 1550) studied planetary models in the framework of *ʿilm al-hayʾa* and proposed new solutions for modeling planetary orbs. These models stand in the tradition of Naṣīr al-Dīn Ṭūsī (1201–74) and Qutb al-Dīn Shīrāzī (1236–1311). According to Saliba, they represent the most sophisticated and technically complex achievements known so far [Saliba 1994, 2004].

In the late phase of Safavid rule, astrolabe makers in Isfahan fabricated astrolabes with compasses and sundials that show on one of their plates a Mecca-centered map of the Islamic world.[1] The grid of the map is curved in its horizontal parallels and has been differently interpreted as either representing arcs of ellipses or arcs of circles [King 2004, 843]. A ruler fixed in Mecca links any of the marked places on the map to the religious center of the Islamic world, showing its *qibla*. The mathematical knowledge needed for this map goes back to the ninth or the tenth centuries and was, in all likelihood, developed by scholars working in Baghdad and other cities of Buyid Iran [King 2004, 842].

The more than one hundred years between these two highlights of science and instrumentation in the Safavid Empire are characterized by texts on arithmetic, geometry, algebra, and elementary astronomy that only rarely rise above the level of the introductory. The most often taught, translated, or copied texts in mathematics and astronomy of the period were writings by Bahāʾ al-Dīn ʿĀmilī (1547–1622) such as *Khulāṣat al-ḥisāb* (Essence of Arithmetic) and *Tashrīḥ al-aflāk* (Explanation of

[1] See, for images of these instruments, King, D.A., *Safavid World-Maps Centred on Mecca: A third example and some new insights on their original inspiration* in King [2004].

N. Sidoli and G. Van Brummelen (eds.), *From Alexandria, Through Baghdad: Surveys and Studies in the Ancient Greek and Medieval Islamic Mathematical Sciences in Honor of J.L. Berggren*, DOI 10.1007/978-3-642-36736-6_22,
© Springer-Verlag Berlin Heidelberg 2014

the Orbs). Practicing astronomers or astrologers commented on or applied Ulugh Beg's (1394–1449) *Zīj-i jadīd* (The New Astronomical Handbook). Hence, while the practice in the mathematical and astronomical disciplines continued, it did so apparently on an elementary level and without outstanding representatives. In contrast to the mathematical sciences, it is not very clear whether Safavid scholars had any interest in problems of natural philosophy.

Reports by travelers from various European countries support the assumption that the rather bleak picture of the sciences in Safavid courtly culture may be flawed or at least incomplete. Although these reports pose their own methodological problems in regard to their description of Safavid activities, knowledge, and beliefs, because their authors created them within the canonical rules of seeing, thinking, representing, arguing, and demonstrating of their own cultures, Safavid manuscripts and paintings leave no doubt that the visitors and their hosts looked at each other with curiosity and attentiveness.[2] The least we can infer from the numerous travel accounts published in the seventeenth and early eighteenth centuries in Europe about the Safavid Empire is that the visitors thought that inhabitants of Safavid Iran were eager to learn and study [Texeira 1902, 251; Richard 1993, vol. 1, 124–5; vol. 2, 127, 270, 304]. They are said to have privileged mathematics, astronomy, astrology, medicine, and philosophy in the writings of ancient Greek scholars such as Euclid, Theodosius, or Ptolemy and in the books of medieval Islamic scholars such as Ibn Sīnā (d. 1036), Naṣīr al-Dīn Ṭūsī, and a number of other scholars whose names the visitors either knew from their own times at university at home such as Aḥmad b. Muḥammad b. Kathir Farghānī (9th c.) or added from sources newly accessible to them through translation [Brentjes 2004, 395–404, 413–414]. Visitors also told of the interests of high-ranking Safavid court officials in new technologies, instruments, and machines brought to Iran by foreign craftsmen and "adventurers" [Richard 1993, 220–232]. Safavid paintings lend credit to such reports in regard to military technology and fortress planning [Vesel 2001, 267].

In this paper, I survey several manuscripts produced at the Safavid court, or in its vicinity, which suggest that we need to find fresh ways of looking at the sciences in the seventeenth century within Safavid culture. A few years ago, Živa Vesel pointed to the importance of illustrated scientific manuscripts produced during the rule of Shāh ʿAbbās I (r. 1588–1629) and Shāh ʿAbbās II (1642–1666) characterizing them as "of exceptional quality" [Vesel 2001, 267]. She suggested studying the history of the sciences in Safavid Iran as an aspect of courtly patronage for the arts. Related views had been formulated earlier from the perspective of art history by Anthony Welch and Barbara Schmitz. Welch emphasized the value that certain Safavid scientific manuscripts have for the study of Safavid art history, while Schmitz raised the opposite point by claiming that a group of Persian scientific manuscripts in The New York Public Library "suggest that a revived interest in classic Arabic and Persian scientific texts began during the reign of Shāh ʿAbbās I" [Schmitz 1992, 61; Welch 1972, 69]. Reflecting the prevalent views on the history of science in post-classical Islamic societies according to which there was no worthwhile scientific activity in these societies, Schmitz speculated cautiously that this apparent revival could have been caused by the growing number of European visitors to Iran and the new scientific knowledge they brought with them [Schmitz 1992, 61].

A second possibility for casting a fresh look on history of science in Safavid Iran is a comparative investigation of different social spaces of education. This approach opposes the verdict of George Maqdisi, and others, that the ancient sciences were excluded from regular teaching at the madrasa, as well as from courtly patronage after 1200. Empirical experience with manuscripts has indicated repeatedly that such views on the sciences in post-classical Islamic societies are misconstrued. But little systematic work has been done so far to create more nuanced pictures about where the sciences were practiced in different Islamic societies after 1200 and by whom, with the exception of Ilkhanid Iran. Hence this paper situates itself as a small contribution to a cultural history of science in Islamic societies that aims to integrate the ancient sciences in their contemporary cultural contexts.

The following survey draws on the results of several art historians which are brought together and

[2] See, for such problems, for instance the report given by the Capuchin Raphaël du Mans in his first report written in 1660 where he enumerates authors and disciplines studied and practiced in Isfahan in his time while evaluating their relevance, meaning and 'correctness' according to French beliefs and practices [Richard 1993, vol. 1, 122–131, 133–137, 171–173].

compared with additional manuscripts produced in the late sixteenth and during the seventeenth centuries. It indicates that the Safavid educated elite paid considerable attention to illustrated classical scientific books, commissioned their translations into Persian, and financed their continuous reproduction by calligraphers and painters. Content and illustrations point to Timurid ancestors as sources of inspiration. The variations in detail in regard to the illustrations between the extant manuscripts show that different manuscripts were taken as models for the Safavid reproductions. Hence, while it was undoubtedly good tone to pay for one's own copy of an illustrated version of ʿAbd al-Raḥmān Ṣūfī's (903–986) *Kitāb ṣuwar al-kawākib al-thābita* (Book of the Constellations of the Fixed Stars) or Dioskorides' *Materia medica*, the Safavid elite did not insist on privileging one particular set of representations of the constellations, zodiacal signs, plants, or animals. And while this elite sponsored Persian translations and summaries, it also paid for illustrated copies of scientific texts in Arabic. Such texts were appreciated for their educational and artistic value, if not for their scholarly content. This is documented by a collection ordered by the Great Vizier Khalife Soltan for Shāh ʿAbbās II and copied at the command of the Great Vizier ʿAlījān Muḥammad Khān for Shāh Sulaymān (r. 1666–1694).[3] This collection presents, at its end, the images from Ṣūfī's *Star Catalog*, after bringing together shorter and longer texts on various mathematical sciences and natural philosophy with other disciplines of the rational, traditional, and occult sciences.

In the course of the seventeenth century, a number of manuscripts on astronomy, astrology, medicine, and technology were illustrated with paintings that show a close relation to the styles developed in Isfahan. These manuscripts include an undated copy of Quṭb al-Dīn Shīrāzī's *al-Tuḥfa al-shāhiyya* (The Royal Gift); a group of three Persian copies of ʿAbd al-Raḥmān Ṣūfī's *Star Catalog*, two of which were made between 1630 and 1634 in Mashhad, while one is undated, but placed by Anthony Welch in the middle of the seventeenth century; at least two copies of the Arabic text of this work, one produced possibly around 1630, the other being undated; a new Persian translation of Dioskorides' *Materia medica* ordered by Shāh ʿAbbās I; at least three Persian copies of the *Materia medica* made in 1645 and 1657; and a work about the astrological meanings of each of the 360° degrees of the sky made in Isfahan in 1663 [Schmitz 1992, 60–1, 122–3; Vesel 2001, 293–7; Welch 1972, vol. 2, 69–70].[4] In addition to these copies, there are other illustrated scientific manuscripts produced in seventeenth-century Safavid Iran that may or may not be related to the art of the court.[5] I will first discuss the illustrations of Shīrāzī's *al-Tuḥfa al-shāhiyya*. In the second section, I turn to the three Persian versions of Ṣūfī's *Star Catalog*. In the final section, I will survey the decorative features of the collection made for Shāh ʿAbbās II and Shāh Sulaymān.

An Illustrated Copy of Quṭb al-Dīn Shīrāzī's *al-Tuḥfa al-shāhiyya*

Illustrating a work on planetary theory with drawings by a student of a master painter was a rare enterprise in an Islamic society. In the Safavid period, this is what happened. An unknown patron ordered a copy of one of Quṭb al-Dīn Shīrāzī's two major works on planetary theory to be singled out by adorning it with images taken from the tradition of Ṣūfī's *Star Catalog*. The chosen images represent *Andromeda*, *Andromeda* with *Cetus*, *Auriga*, *Boötes*, and another male figure identified by Vesel as *Hercules* [Vesel 2001, 285].[6] These figurative representations of four constellations are joined to a model of Mercury, a visualization of the relative positions of the poles of various orbs, a parallax diagram,

[3] MS Boston, Arthur M. Sackler Museum, 1984.463.

[4] Welch wrote that the text of the manuscript is Arabic. The text on the two folios published in the Catalog is however Persian. That is why I have included this manuscript in the group of Persian copies of Ṣūfī's *Star Catalog*.

[5] See, for instance, Schmitz [1992, 61].

[6] This figure deviates from the standard representation of the constellation in Arabic manuscripts. It stands upright and holds a small, short object in his left hand. The standard representation shows a kneeling figure with a sickle in one hand.

Figure 1: Andromeda with Cetus. Photo courtesy of Riḍā–i Abbāsī Museum.

and a diagram for a specific spherical triangle.[7] The attractive figures are drawn with black ink. Their bodies are portrayed with an elegantly swinging movement. Female and male figures have the same kind of gently rounded faces with almond-shaped eyes, thick brows, short noses, and thin, perhaps slightly smiling, not fully executed lips. They also share, in principle, the same hairstyle. A long lock dangles in front of one ear while the rest of the hair flows freely behind the ears unto the shoulders and down the back. The male figures wear a characteristic Safavid turban and a long shirt under a half-long jacket which is girded by a sash. They hold insignia in their left or right hands, which helps to identify them. *Andromeda* wears a long, pleasantly floating robe that is girded by a sash, similar to that of the male figures. The robe opens from her midriff swinging over long trousers or naked legs. The arms of her robe cover her to her elbows. Her upper body is fully covered. She shows a slightly protruding belly, but no bosom. She wears a diadem on her head and bracelets around her arms. According to Vesel and Richard the figures are drawn in the style of Riẓā–i ʿAbbāsī.[8] This should not come as much of a surprise since Riẓā–i ʿAbbāsī set the norm for many of the court painters from the 1620s [Schmitz 1992, 60, 122]. A comparison with images discussed by Anthony Welch in his paper about the visualization of love in Safavid painting confirms the close relationship in style to Safavid court painters and in particular Riẓā–i ʿAbbāsī. It also shows that the images in Shīrāzī's *Tuḥfa* lack the maturity of the paintings made by this master and the other painters documented by Welch [2001, 302–317, 305–315]. The comparison brings to light that the postures, gestures, and expressions, while following in principle the standard visual codes used for illustrating Ṣūfī's *Star Catalog* for centuries, reverberate clearly with Safavid images of "graceful dandies (who) make smaller claims on heroism, ... embark on no adventures, exterminate no dragons, achieve no notable exploits" [Welch 2001, 301–317]. Such images, produced for albums, were, however, not meant to function within the context of manuscripts as the illustrations of Shīrāzī's text. Welch interpreted them as representations of ideals of love and Sufi worldview. Like the images discussed by Welch, the pictures in Shīrāzī's *Tuḥfa* show idealized, idyllic, beautiful youth [Welch 2001, 303]. But they lack any symbolism that would relate them to Sufi themes. The ease with which the new Safavid painting style was integrated into a manuscript on planetary theory like Shīrāzī's *Tuḥfa* probably results from certain features that characterize the figures in Ṣūfī's *Star Catalog*. The tradition of illustrating this work with individual, gently moving figures clad in playfully swinging dresses may have induced the Safavid painters to treat them in the manner of images destined for an album.

A comparison with the Persian and Arabic manuscripts of Ṣūfī's *Star Catalog*, and of the art historical literature and catalogs that discuss them, brings to light that the illustrations found in Shīrāzī's *Tuḥfa* are closely related to illustrations in MS Geneva, Prince Sadruddin Aga Khan Collection, 9, and a single folio published in 1962 by Ernst Grube, that then belonged to the private collection of Rudolf M. Riefstahl in Toledo, USA. The relationship between these illustrations is particularly strong in the swing of the skirt of *Boötes* in Shīrāzī's *Tuḥfa* and the single folio from the Riefstahl Collection, in one of the three representations of *Andromeda* in Shīrāzī's *Tuḥfa* and the figure of *Andromeda* in the manuscript of the Sadruddin Aga Khan Collection as published by Welch, and of *Cassiopeia* in MS Tehran, Majlis Shūrā 196 as published by Nasr [Grube 1962, pl. 123; Welch 1972, vol. 2, MS 9, f 7v; Nasr 1976, 100, fig. 39]. What I have called here the swing of the skirt is a deviation of the windblown folds of the dresses of Buyid or Seljuq provenance as found in much earlier Arabic manuscripts such as MS Oxford, Bodleian Library, March 144. Wellesz already pointed to the loss of vitality of these folds in some later manuscripts and published an image of *Virgo* from a thirteenth-century Latin codex illustrated in Sicily that shows a schematized rendering of these folds indicating the possible type of variations that led to the swing in the Safavid manuscripts [Wellesz 1964, 85–92, 87, 88, fig. 9, 89, fig. 10]. Grube's suggestion that the figure of *Boötes* from the Riefstahl Collection is quite similar to the drawing of the same figure in MS St. Petersburg, Public Library, Ar. 119 and that "it must be from such a model that the Riefstahl drawing has been copied" cannot be maintained, however.[9] Not only are the illustrations in later Safavid copies of Ṣūfī's *Star Catalog* and in Shīrāzī's *Tuḥfa* much closer to

[7] See, for a description of the diagrams in English, Ragep [1993, vol. 1, 164–174, 194–200, 258–260, 314–316].

[8] Oral information from Vesel. Vesel [2001, 283] dated the manuscript to the sixteenth or seventeenth century and characterized the paintings as done in Isfahani style.

[9] Grube [1962, 138].

the image displayed on the single folio, but also the pictures in the manuscript copied in Naʾin are differently dressed, much more stereotyped, and follow in several cases a different illustrative tradition [Schjellerup 1874, pl. 1–3].

While it is indubitable that the illustrations in the manuscripts discussed here are all very closely related among themselves and, at the same time, distinct from certain of the Arabic copies produced in seventeenth-century Safavid Iran, they differ nonetheless in many of their details and thus were apparently made by different painters who used different models. Since most of them do not have inscriptions, the identification of their painters is fraught with difficulties, as the diverging opinions among art historians show. Grube dated the image of *Boötes* towards the end of the seventeenth century without any specific argument, while linking its style to the early seventeenth century [Grube 1962, 128]. Welch dated the manuscript of the Sadruddin Collection to the middle of the seventeenth century by ascribing its style to a painter strongly influenced by Muḥammad Yūsuf. He considered the artist who drew the sixty-five pen and ink illustrations as "very competent and sometimes gifted" and "obviously trained in an Isfahan atelier, or one closely related to it " [Welch 1972, vol. 2, 70]. In addition, there are persuasive stylistic connections between some of the illustrations in the copy of the Persian translation of Ṣūfī's *Star Catalog* extant in The New York Public Library as Spencer, Pers. 6 (called from now on: Spencer manuscript) and the single image of *Boötes* in the Riefstahl Collection. While *Boötes*' dress and the object he holds in one hand are closely related to the dress and object of *Boötes* in Shīrāzī's *Tuḥfa*, the face of the Riefstahl *Boötes* is very similar to the face of the unusual representation of *Cetus* in the Spencer manuscript [Schmitz 1992, fig. 127]. This then links the images in the Safavid copy of Shīrāzī's *Tuḥfa* to specific copies of Ṣūfī's *Star Catalog* and their painters.

Illustrated Copies of a New Safavid Translation of Ṣūfī's Star Catalog into Persian

In 1630/31, a new Persian translation of Ṣūfī's *Star Catalog* was made at the command of ʿAbū l-Fatḥ Manūchihr Khān (d. 1636). Manūchihr was a son of the general of Georgian origin Qarajaghāy Khān (d. 1623), who served Shāh ʿAbbās I. Like his father, he was a general and governor of Mashhad. He served under ʿAbbās I and Ṣafī (r. 1629-1642) [Schmitz 1992, 55, 123]. The translator of the text was Ḥasan b. Saʿd Qāʾinī, an astrologer at Manūchihr's court. The name of the scribe of the Spencer manuscript is given repeatedly as Muḥammad Bāqir al-Ḥāfiẓ whom Schmitz tentatively identified with the calligrapher Muḥammad Bāqir b. Mullā Maḥmūd Gīlānī [Schmitz 1992, 123]. In a preface written by the translator, the paintings of the Spencer manuscript, produced between 1630 and 1633, are attributed to Malik Ḥusayn Iṣfahānī. Schmitz argues that it was rather his son, Muḥammad ʿAlī, who painted most, if not all of the images [Schmitz 1992, 123–124]. She allows, however, that the second copy of this work produced in 1633–34 and extant in Cairo may include one painting executed by the father, that of *Virgo*.[10] She also suggests that possibly more than these two painters were involved in the execution of the extant manuscripts of Qāʾinī's translation [Schmitz 1992, 123]. In contrast, Farhad claimed that nothing in the images of the Spencer manuscript contradicted their ascription to the father [Babaie, Babayan, Baghdiantz-McCabe and Farhad 2004, 128]. While this is certainly true, the pictorial basis for making claims of a general character seems to be rather small. The only point of the divergent evaluations of Schmitz and Farhad, which I believe can be settled at the moment in a satisfactory manner, is the relationship between the images of *Virgo* in the two manuscripts. A comparison of these two images indicates that *Virgo* in the Cairo manuscript is a slightly unfinished and less skilled version of *Virgo* in the Spencer manuscript. The colors chosen for *Virgo* in the Spencer manuscript are finely harmonized with each other. In the Cairo manuscript, however, the color of the over-coat contrasts with the color range chosen for the chemise and trousers. The wheat ears in *Virgo*'s left hand in the Cairo manuscript show a coarser form, less in agreement with the ears' natural appearance than in the Spencer manuscript. Certain parts of *Virgo*'s body in the Cairo manuscript, such as her right ear or her chin, are less proportionate and worked out. Even *Virgo*'s wings in the Cairo manuscript taken by

[10] MS Cairo, Dār al-Kutub, MMF9; King [1986, pl. III]; Edwards and Signell [1982, 13]; Schmitz [1992, 123].

Schmitz as the indicator for Malik Ḥusayn's authorship of the image fall short of the elegance and completion they show in the Spencer manuscript [Schmitz 1992, 123; Edwards and Signell 1982, 13]. Thus, I propose considering the images of *Virgo* in the Spencer manuscript as the work of Malik Ḥusayn and those in the Cairo manuscript as the work of another, possibly junior painter. The miniatures of a third manuscript, MS Tehran, Malik Library, 6037, considered by Schmitz "stylistically similar" to those in the Spencer and the Cairo manuscripts are the work of another painter or perhaps group of painters [Schmitz 1992, 123].[11] The manuscript is dated 1598, which was unknown to Schmitz.[12] Hence, it is much earlier than the artistic undertakings related to Qāʾinī's Persian translation.

The text of Qāʾinī's Persian translation and its illustrations in the Spencer and the Cairo manuscripts show a number of interesting features, as Farhad has already pointed out. Her observations on the art historical novelties of these illustrations are of particular importance. She stressed the new pictorial style in a scientific manuscript, the unusually large format of the illustrations, and their refined drawing and painting techniques [Babaie, Babayan, Baghdiantz-McCabe and Farhad 2004, 128]. Some of her ideas about the scholarly relevance of the translation are less persuasive, since she took claims made by the translator at face value. In his preface, Qāʾinī reports that he discovered deviations between the values in the Arabic text of the *Star Catalog* and what he could observe in the heavens in his own time. That is why he decided to correct the Arabic text following values found in Ṭūsī's *Zīj-i Īlkhānī* and Ulugh Beg's *Zīj-i jadīd*.[13] For Qāʾinī, the authority for settling scientific disputes rested obviously with acknowledged authors, not with nature or himself in his capacity as an observer. A comparison with other manuscripts of Ṣūfī's *Star Catalog* suggests, however, to take any such claim of observational acuity and revisionist project with some caution. Such claims did not necessarily reflect scientific practices and aims. They highlight rather other matters of cultural relevance. Such an interpretation is seconded by the observation that the rhetoric of cosmic change, scientific dispute, and necessary corrections was an important element of representation that already characterized ʿAbd al-Raḥmān Ṣūfī's own preface to his work. Furthermore, after having surveyed the copy of Ṣūfī's *Star Catalog* extant in the Bibliothèque nationale in Paris, which was produced for Ulugh Beg, Blochet stated that the coordinates in this manuscript were taken from the *Ilkhanid Tables* of Naṣīr al-Dīn Ṭūsī.[14] Since Ulugh Beg and his astronomers worked themselves on new astronomical tables, they could have easily replaced the outdated values by their own observations. Their choice of the *Ilkhanid Tables* may have been inspired not by science alone, but by the well-known Timurid cultural politics of representing themselves as legitimate heirs of the Mongols. Such an interpretation gains additional plausibility by the fact that Ulugh Beg's copy of Ṣūfī's *Star Catalog* derived from a copy owned by Naṣīr al-Dīn Ṭūsī [Richard 1997, 78]. Since the Safavids on their part represented themselves as legitimate political and cultural heirs of the Timurids, Qāʾinī's choice of the tables of Naṣīr al-Dīn Ṭūsī and Ulugh Beg for making up for nature's changes may reflect this imitation as much as it represents his and his patron's interests in the heavens and their accurate portrayal. The possibility of using such scientific works effectively for political ends increased their attractiveness for rulers and courtiers as did their artistic appearance. Taking up one aspect of Farhad's arguments about the meaning of such a major project executed by a ghulām, it may perhaps even be considered that the scientific content of the work provided a less contentious entrance into patronage of the arts and knowledge and the claims to status and position among the elites connected with such an act than the choice of a major work of literature due to the latter's long use by ruling families and members of the military aristocracy for expressing their claims to power and interpretation of the world [Babaie, Babayan, Baghdiantz-McCabe and Farhad 2004, 129, 134, 137].

[11] I have not seen the manuscript itself, but only two more miniatures than the ones available to Schmitz. I thank Živa Vesel for her help in this and other aspects relevant to my discussion in this paper. See Vesel, Tourkin and Porter, with Richard and Ghasemloo [2009, 85–89]. Farhad repeated Schmitz's position and even suggested that the manuscript was of a slightly later date than the Spencer and the Cairo manuscripts [Babaie, Babayan, Baghdiantz-McCabe and Farhad 2004, 130].

[12] I owe this information to Živa Vesel.

[13] See, for instance, MSS Paris, BnF, Supplément Persan 1551 (16th c.), Arabe 4670 (17th c.).

[14] MS Paris, BnF, Arabe 5036; Blochet [1912, 48–49].

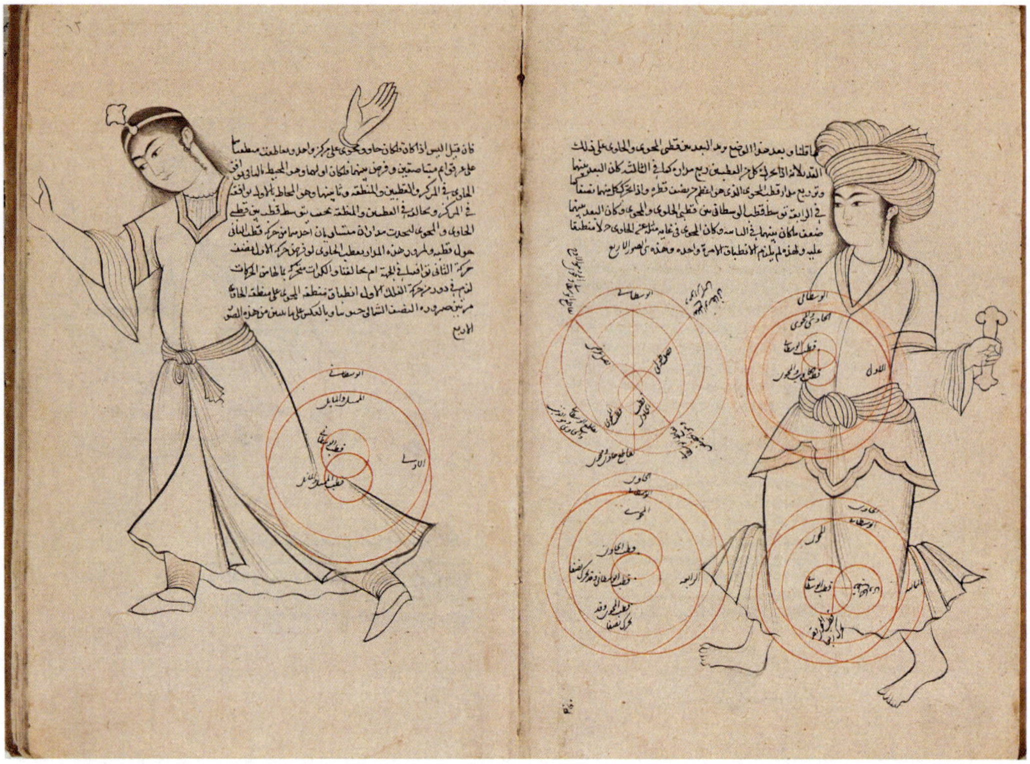

Figure 2: Andromeda and Boötes. Photo courtesy of Riżā–i ʿAbbāsī Museum, Tehran.

Figure 3: Andromeda with Cetus (left) and Virgo (right), BnF Arabe 4670, 5036. Photo courtesy of Bibliothèque nationale, France.

Art historians have long recognized that the copies of Qāʾinī's Persian translation of Ṣūfī's *Star Catalog* and their paintings were situated within the life of the Safavid courtly elite. In 1962, Grube wrote that the paintings of the New York manuscript "are the largest and most magnificent of the Isfahan school known, disregarding the wall-paintings of the ʿAlī Qāpū and the Chahīl Sūtūn in Isfahan. They are unquestionably not all by the same hand, but they are all of first quality. The entire aspect of the magnificently written and produced volume would suggest an origin in the court school of Isfahan. Riżā-i Abbāsī was then still alive and it must have been under this immediate direction that the paintings were executed" [Grube 1962, 129–130]. Grube's view that the paintings in the Spencer manuscript belong to the most magnificent specimen of the Safavid period has been seconded by Anthony Welch who considered them as belonging to the most beautiful illustrations found in scientific manuscripts [Welch 1973, 76–77, 82]. Farhad proposed a new perspective from which to evaluate these images. She declines see them as an extension of, or competition with, the painting style en vogue at court in Isfahan. She considers it more appropriate to interpret the activities in Mashhad as creating an independent space for the arts of the book destined to reinforce the ghulāms' claims to a Persian identity [Babaie, Babayan, Baghdiantz-McCabe and Farhad 2004, 134–137].

Two miniatures in particular have drawn the attention of art historians – the painting of *Sagittarius* and the painting of a camel. The former is important since it seems to portray Manūchihr, an assumption based on specific features of the turban that *Sagittarius* wears [Schmitz 1992, 123]. This interpretation wins credit due to the individualized features of *Sagittarius*' face. Farhad took this inscription of the governor's face into the pictorial cycle as a repetition of what she calls "a well-established Persian practice" of emphasizing the fact of patronage [Babaie, Babayan, Baghdiantz-McCabe and Farhad 2004, 129]. It happened, however, very seldomly that a scientific manuscript was appropriated to such an end. The only other scientific manuscript known to me that shows in one of its illustrations features of a patron is the copy of the *Star Catalog* produced for Ulugh Beg [Richard 1997, 78]. This parallelism is probably insufficient for claiming a conscious analogy between the Timurid prince and the Georgian convert. Manūchihr seems rather to have aimed to invoke a parallel between himself and ʿAbbās I, as Farhad suggests based on Jalāl al-Dīn Munajjim's history of ʿAbbās I. The transformation of Isfahan into the shah's new capital apparently was linked with *Sagittarius* [Babaie, Babayan, Baghdiantz-McCabe and Farhad 2004, 129]. Farhad also considers the possibility that the governor's birthday fell in the same sign [Babaie, Babayan, Baghdiantz-McCabe and Farhad 2004, 129]. Such a multiplicity of meaning can only have increased the sign's attraction to the governor.

The image of the camel is an addition to the constellations usually portrayed in Ṣūfī's *Star Catalog*. It shows a naturalistic rendering of a seated camel. On its back lies a sumptuous saddle cloth. A few stars cover its body, but no names are added for identifying them. This additional visualization of a group of stars is backed by Ṣūfī's text as Schmitz has pointed out since Ṣūfī emphasized that certain stars in the constellation of *Cassiopeia* look like the hump of an Arabic camel, while others resemble its neck [Schmitz 1992, 125]. The saddle cloth of the camel is closely related to such representations on two Safavid paintings extant in the Sadruddin Collection (IR.M.31 and IR.M.36). IR.M.31 shows a single camel dated by Anthony Welch and Stuart C. Welch to the late sixteenth century. Its saddle cloth contains the same type of floral and ornamental motifs used by the painter of the camel in the Spencer manuscript [Welch and Welch 1982, 178]. IR.M.36 depicts two fighting camels the right of which carries a saddle cloth with some motifs found in the Spencer manuscript. Welch and Welch [1982, 184] ascribe this miniature to Isfahan and date it to 1630, i.e. close to the production of the Spencer manuscript. Neither of the two miniatures bears an inscription, however, and Welch and Welch did not try to ascribe them to any particular Safavid artist. The painting of the camel in the Spencer manuscript is, however, not the first effort to translate the textual comments into visual experience. In an illustration of *Cassiopeia* in a twelfth-century Arabic manuscript, a camel is shown behind the female figure.[15] In a sixteenth-century Arabic manuscript ascribed to Egypt, *Cassiopeia* sitting on her throne holds the leash of a lying camel with a saddle cloth and a saddle in her right hand.[16] Hence, the inclusion of a camel in the Spencer manuscript may not reflect artistic innovation on the side of the

[15] MS Oxford, Bodleian Library, Hunt. 212; Wellesz [1964, 89–90].

[16] MS Paris, BnF, Arabe 2490.

Figure 4: Auriga, Spencer Pers. 6. Photo courtesy of the New York Public Library.

Safavid painter, but the model of the *Star Catalog* he used.

Two other illustrations in the Spencer manuscript, *Cetus* and *Andromeda*, show further remarkable alterations of the two major palettes of visual codes set, according to Wellesz, by early manuscripts of the work [Wellesz 1964, 86–89].[17] *Cetus*, the monster sent by Neptune to devour Andromeda chained to a rock and slain by Perseus with the cut off head of Medusa, is here represented as a young man with hands and arms too big for his body and interpreted by Schmitz as a giant [Schmitz 1992, 127 fig. 127]. The form of his face comes very close to the face of *Boötes* in the Riefstahl Collection, while Schmitz pointed to the similarities of his skullcap with that of a groom in a miniature held by the Arthur M. Sackler Museum, Harvard University and signed by Muḥammad ʿAlī b. Malik Ḥusayn [Schmitz 1992, 123]. *Andromeda* is shown from the back and only clad in a diaphanous, red colored, striped skirt. In one image, *Cetus*, now shown in his traditional iconography as a sea monster, has swum under the skirt and bites her in one of her thighs. In the other image, an additional, bigger fish covers *Andromeda*'s back trying to get under her skirt from above. In both images, *Andromeda* holds a torch and is chained by a long chain attached to her wrists. This motif is quite interesting since it differs from what is known about a chained *Andromeda* in Arabic manuscripts. According to Wellesz, "it is only her feet which are fettered, an indication that the connexion with the original myth no longer exists" [Wellesz 1964, 89]. It may be that the cultural background of Manūchihr's family explains this return to ancient Greek mythology in *Andromeda*'s visualization.

In contrast to many other extant manuscripts of Ṣūfī's *Star Catalog*, the paintings of a number of the constellations and zodiacal signs in the Spencer manuscript are fully colored. The painters used a comparatively broad scale of colors ranging from paler (yellow, orange, purple, green, grey) to darker kinds (red, mauve, blue, green, brown, black) [Schmitz 1992, 124–127]. The figures show volume and compactness, a result of the coloring and the application of new painting techniques as Farhad has argued [Babaie, Babayan, Baghdiantz-McCabe and Farhad 2004, 128]. They also show details of what a particular painter privileged when painting a woman or a man in color. *Virgo*, for instance, is not only fully dressed in a robe, underskirt, and trousers and wears two jeweled necklaces and earrings as traditionally prescribed in the visual standards of the *Star Catalog*. She also has fine, long, black hairs arranged carefully, a full bosom, a slightly rounded belly, a full face with a visible double chin, a gracious posture, and red colored finger and toe nails matching her red lips. The clothes, headgears, hairstyles, and faces of the various figures, while documenting the shared visual practices, differ clearly in their details when compared to Shīrāzī's *Tuḥfa*. The single lock before an ear is missing in some of the male figures, while taking a different form in other male figures. The noses, fingers, and toes are substantially longer and drawn, as other parts too, with visibly more care and capability. In addition, *Hercules* wears an Indian turban not found in Shīrāzī's *Tuḥfa*, differs in his posture, and holds no object in his hand [Schmitz 1992, 124].

Another sumptuously illustrated Safavid copy of Ṣūfī's *Star Catalog* is preserved in the Majlis Library in Tehran which has been dated by Vesel to around 1630 [Vesel 2001, 296]. It contains an Arabic text. Its images, while done by a different painter than those in the copies of Qāʾinī's Persian translation, show a clear relationship to those in the Spencer manuscript. *Virgo*, for instance, is largely identical with *Virgo* in the Spencer manuscript, although she differs in a number of smaller, and a few substantial, details, such as the fact that her free hand is placed before her body, not pointing to the side away from it. Since this type of gesture can be found in earlier manuscripts too, the painter(s) of this manuscript apparently worked from a different model than those in Mashhad [Wellesz 1964, 88, fig. 9; 89, fig. 10]. All in all, *Virgo* in this Arabic copy of the *Star Catalog* is executed with much less finesse and care. As a result, she is less lively and more stereotyped. The color palette and the background may suggest an Indian inspiration of a Safavid painter or perhaps even an Indian, possibly Deccani, origin of the picture the painter of which reproduced a Safavid model.[18]

The quantity and quality of illustrated copies of Ṣūfī's *Star Catalog* in Arabic and Persian produced during the seventeenth century in Iran, and perhaps outside its borders, in styles reminiscent of Riẓā–i

[17] For images from other manuscripts compare Upton [1933, 179–197, 184, 188, 190–191, 193], Tourkin and Porter, with Richard and Ghasemloo [2009, 114, 116–133], and Caiozzo [1999, 11–51,41–42].

[18] For other images from this manuscript see Vesel, Tourkin and Porter, with Richard and Ghasemloo [2009, 131].

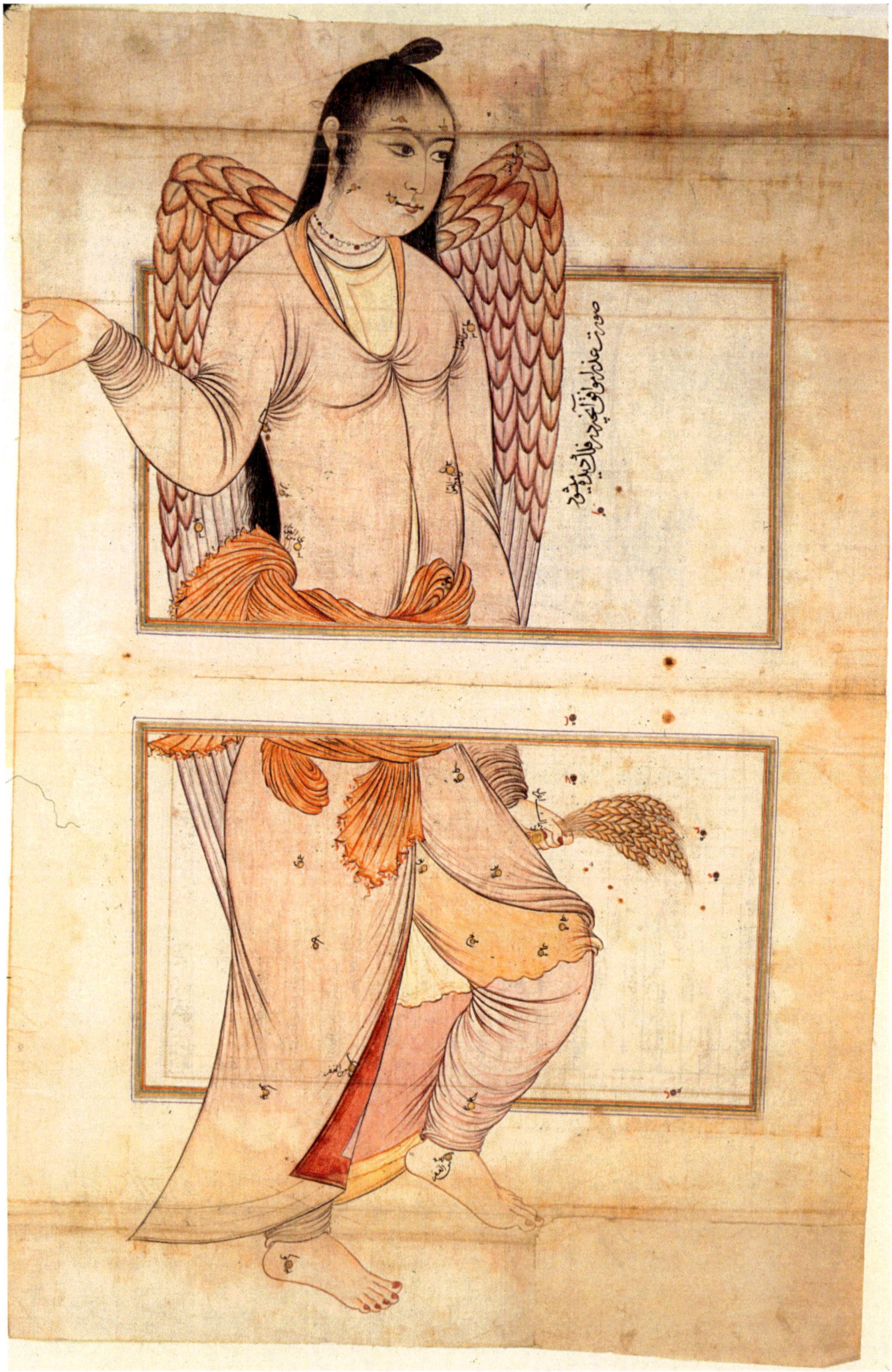

Figure 5: Virgo, Spencer Pers. 6. Photo courtesy of the New York Public Library.

ʿAbbāsī should encourage us to question the prevalent mode of thinking about the sciences in Safavid Iran. Safavid patrons obviously invested considerable sums, if not fortunes, into the production of these manuscripts. Painters considered the text and its illustrations not beneath their skills and capabilities. They invested in adapting the models available to them to contemporary tastes and styles. Some of them treated the painting of star constellations with the same professional standard they applied to other themes and turned them into art objects of high quality. The knowledge of the stars and their distribution in the night sky was apparently valued highly when it came in the format of a lavishly illustrated book. Courtly education, art, pleasure, and political rhetoric achieved a new symbiosis in these Safavid scientific manuscripts.

On the Images of Ṣūfī's *Star Catalog* in a Collection Composed for Shāh ʿAbbās II and Copied for Shāh Sulaymān

The collection preserved at the Arthur M. Sackler Museum, Harvard University is a rare document of courtly efforts to ensure royal education in a tasteful setting. In contrast to the Timurid collections of literature, science, and art, this collection was not produced to show off the prince's encyclopedic knowledge and appreciation of refined, artful style. The collection rather was ordered by two successive great viziers for their successive masters. A broad array of fields of knowledge from religion to science has been brought together in more than one hundred, mostly short pieces of texts in Arabic and Persian. Many of them are extracts from famous, often much older writings such as Euclid's *Elements*, Ibn Sīnā's comments on philosophical issues such as matter and form, Naṣīr al-Dīn Ṭūsī's commentary on Fakhr al-Dīn Rāzī's (1149–1210) commentary on Ibn Sīnā's *Ishārāt wa-tanbihāt*, or Quṭb al-Dīn Shīrāzī's commentary on the theoretical part of Ibn Sīnā's medical encyclopedia. Other texts summarize briefly components of a particular domain of knowledge such as algebra, surveying, arithmetic, number theory, music, astrology, numerology, optics, geography, natural philosophy, metaphysics, logic, Shiʿī law, ḥadīth, or ʿirfān. In addition to these short texts, which fill the margins of the pages, the center of the pages are covered by either two longer texts of different lengths and written in different sizes running in form of parallel horizontal rows or by texts running in form of one or more parallel vertical columns of subsequent series of lines posited in an acute angle. These longer texts include the dedication to, as well as the titles of, the Safavid rulers, writings by the two great viziers on elementary issues of arithmetic and mathematical riddles, debates on various subject matters such as unity and multiplicity, the supra- and sublunar world, the rules of grammar or poetic compositions, or the structure of the universe. Occasionally, other formats structure a page. Examples are a circular table in the center of the page surrounded by texts written in six different directions, or pages divided in halves filled with lines that begin in opposite corners and run in opposite directions. Tables destined for learning numbers, the names of God, the four- or five-fold structures of the natural world, the distances between various towns of Iran, geographical coordinates, astrological categories, astronomical divisions, and the names of the months are interspersed between the texts either filling an entire page or parts of it in various sizes. Diagrams are placed between the lines of the first central text or are imbedded between the pieces of text in the margins. They show a hand giving the numerical values of different parts of the fingers, a quote from Ptolemy's *Astrology*, a *qibla numa* for Tabriz, a plan of the Kaʿba, an allusion to Jerusalem, geometrical figures from Euclid's *Elements*, the circular orbs of the components of the universe, a planetary model, eclipses, grammatical units arranged in circles, a model of the eyes, and numerological arrangements. Further visual forms of presenting knowledge are a world map taken from a version of Zakariyāʾ Qazwīnī's (1203–1283) *ʿAjāʾib al-makhlūqāt* (The Wonders of the Created (Things)) and the star constellations and zodiacal signs from Ṣūfī's *Star Catalog*.

In addition to the complex arrangements of the textual units, several forms of artistic decoration have been applied. Different kinds of colored illuminations emphasize the titles of short pieces of text, function as separators, or replace the central texts. Several types of calligraphy and the use of ten colors for writing text, tables, and diagrams, often applying four, five, or six of these colors on one folio, are a further level of illumination that contributes to rendering the manuscript into a piece of art. Floral and ornamental bands and stripes provide yet another layer of decoration.

The last four folios of the manuscript are devoted to the 48 constellations from Ṣūfī's *Star Catalog*. The choice of the representations follows no clear scheme. Some of them represent Ṣūfī's category of the figure according to "what is seen in the heavens," while others reflect his other category of "what is seen on the globe." The images do not follow strictly the order of the illustrations in the *Star Catalog*, but start with the zodiacal signs, which in themselves proceed according to Ṣūfī's order. Then the star constellations begin, the first being *Ursa Minor* followed by *Ursa Maior* and *Draco*. With one exception, the exchange between *Pegasus* and *Equuleus*, the order followed in the Sackler manuscript agrees with the standard version of the *Star Catalog*. There are a few peculiarities in the naming of the constellations in the Sackler manuscript. *Andromeda with the Northern Fish* (Andromeda with Cetus) is called *al-Rishā'* (the Rope). *Andromeda* does not carry her Arabic name, but a transliteration of the Greek form. *Auriga*'s Arabic name is reduced to *al-ʿInān* (the Rein). And the name of *Equus Maior* is missing altogether.

The visual representation of the figures is much simpler than those painted in styles related to the Isfahan school. The artist used black ink, red, and gold. The figures are slim and in many cases modestly dressed. Most of the figures show crowns, hats, and dresses that can be found in miniatures painted long before the late seventeenth century. This applies in particular to the short skirts of *Boötes*, *Hercules*, and *Cassiopeia* and to the headgears of *Boötes*, *Cassiopeia*, *Andromeda*, and *al-Rishā'*.[19] Only *Auriga*'s and *Orion*'s turbans and some of the sashes resemble Safavid styles as seen in the previously discussed copies of the *Star Catalog*. Certain images also show some resemblance to representations of the constellations in copies of Qazwīnī's *ʿAjā'ib al-makhlūqāt*. This applies for instance to the headgear of *Sagittarius* and the cross in *Centaurus*' hand. The slimness of the figures as a whole, however, agrees with trends in Safavid painting in the second half of the seventeenth century. It also fits the overall aesthetic appearance of the collection with its fine ornamental and floral illuminations, neatly drawn diagrams, and use of thin qalams for writing all but the main text in the center of a folio. This renunciation of any explicit pictorial reference to the Safavid court and the style it patronized in the more sumptuous paintings of the period in a manuscript explicitly produced for two Safavid rulers by two Safavid top ranking administrators raises interesting questions as to the meaning and function of the star constellations within the overall plan of the collection. In the current state of investigation, such questions cannot be answered in a well-founded and satisfying manner. The only feature that can be emphasized is the almost complete isolation of the figures from their usual scientific context. Neither do they show any of the stars they are meant to represent. Nor is there any table giving the stars' coordinates. All that remains from the *Star Catalog*, in addition to the figures, are their Arabic and Arabicised Greek names. In combination with some other parts of the collection this feature may emphasize a didactic and mnemonic purpose, and hence underscores the possible educational meaning of the entire collection.

Some Reflections in Place of Conclusions

This selective survey of some of the illustrated scientific manuscripts produced in the seventeenth century in the environment of the Safavid court and in relationship to the art styles developed in Isfahan points to the need to revisit the history of the sciences in this period and space. The material basis for such a revision is much broader than outlined in this paper. A systematic collection of illuminated, as well as dedicated, scientific manuscripts and an analysis of their relationships to the courts and other social and cultural spaces in Safavid Iran is a necessary preliminary step. It has the potential to open our eyes to new questions about the attraction that scientific texts exercised on Safavid patrons, readers, artists, scholars, and collectors, the impact such knowledge had in various circles, the degree to which such texts were available for carrying messages others than their scientific content and for opening up

[19] Compare, for instance, Lentz and Lowry [1989, 146–148, 171, 175, 186–187, 194].

to new forms of layout and design, or the processes of standardization and differentiation to which such illuminated manuscripts can bear witness.

References

Manuscript Sources

Boston, Arthur M. Sackler Museum, 1984.463.
Cairo, Dār al-Kutub, MMF9.
New York, The New York Public Library, Spencer, Pers. 6.
Oxford, Bodleian Library, Hunt. 212.
Paris, BnF, Arabe 4670.
Paris, BnF, Arabe 5036.
Paris, BnF, Supplément Persan 1551.
Tehran, Majlis Shūrā, 197.
Tehran, Malik Library, 6037.

Modern Scholarship

Babaie, S., Babayan, K., Baghdiantz-McCabe, I., Farhad, M., 2004. Slaves to the Shah. New Elites of Safavid Iran. I.B. Tauris, London and New York.
Blochet, E., 1912. Catalog des manuscrits persans de la Bibliothèque Nationale. Tome Deuxième. Imprimerie Nationale, Paris.
Brentjes, S., 2004. Early Modern Western European travellers in the Middle East and their reports about the sciences. In: Pourjavady, N., Vesel, Ž. (eds.), Sciences, techniques et instruments dans le monde iranien (X^{e}-XIXe siècle). Tehran, Presses Universitaires d'Iran, Institut Français de Recherche en Iran, pp. 379–420.
Caiozzo, A., 1999. Le ciel de l'astronome, le ciel de l'astrologue et celui du sorcier, trois conceptions des cieux dans les manuscrits enluminés de l'orient médiéval. In: Gyselen, R. (ed.), La Science des cieux. Sages, mages, astrologues. Peeters, Paris, pp. 11–51.
Fogg Art Museum, 1973. Fogg Art Museum, Harvard University, Cambridge, Mass. 19 January-24 February, 1974. Fogg Art Museum, New York.
Grube, E., 1962. Muslim Paintings from the XIII to the XIX Century from Collections in the United States and Canada. Catalog of the Exhibition. Neri Pizza Editore, Venezia.
King, D.A., 1986. A Survey of the Scientific Manuscripts in the Egyptian National Library. American Research Center in Egypt, Cairo.
—— 2004. Safavid world-maps centred on Mecca: A third example and some new insights on their original inspiration. In: King, D.A., In Synchrony with the Heavens: Studies in Astronomical Timekeeping and Instrumentation in Islamic Civilization, 2 vols. Brill, Leiden.
Lentz, T.W., Lowry, G.D., 1989. Timur and the Princely Vision. Persian Art and Culture in the Fifteenth Century. Los Angeles County Museum of Art, Arthur M. Sackler Gallery, Los Angeles.
Nasr, S.H., 1976. Islamic Science: An Illustrated Study. World of Islam Festival Publishing Co. Ltd, London.
Edwards, H., Signell, K., 1982. Patterns and Precision. The Arts and Sciences of Islam. National Committee to Honor the Fourteenth Centennial of Islam, Washington, D.C.
Ragep, F.J., 1993. Naṣīr al-Dīn Ṭūsī's *Memoir on Astronomy* (al-Tahdkira fī ʿilm al-hay'a), 2 vols. Springer-Verlag, New York.
Richard, F., 1993. Raphaël du Mans missionaire en Perse au XVIIe siècle, 2 vols. Editions L'Harmattan, Paris.
—— 1997. Splendeurs persanes: Manuscrits du XIIe au XVIIe siècle. Bibliothèque nationale de France, Paris.

Saliba, G., 1994. A sixteenth-century Arabic critique of Ptolemaic astronomy: The work of Shams al Dīn al-Khafrī. Journal for the History of Astronomy 25, 15–38.

—— 2004. Šams al-Dīn al-Ḫafrī's (d. 1550) last work on theoretical astronomy. In: Pourjavady, N., Vesel, Ž. (eds.), Sciences, techniques et instruments dans le monde iranien (X^e^–XIXe siècle). Presses Universitaires d'Iran / Institut Français de Recherche en Iran, Tehran, pp. 55–66.

Schjellerup, H.C.F.C., 1874. Descriptions des étoiles fixes composée au milieu du dixième siècle de notre ère par l'astronome persan Abd-al-Rahman Al-Sûfi. L'Académie Impériale des sciences, St.-Pétersbourg.

Schmitz, B., 1992. Islamic Manuscripts in The New York Public Library. With contributions by Latif Khayyat, Svat Soucek, Massoud Pourfarrokh. Oxford University Press and The New York Public Library, New York and Oxford.

Teixeira, P., 1902. The Travels of Pedro Teixeira. Sinclair, W.F., Ferguson, D. (eds.), The Hakluyt Society, London.

Upton, J.M., 1933. A manuscript of "The Book of the Fixed Stars" by ʿAbd ar-Raḥmān aṣ-Ṣūfī. The Metropolitan Museum of Art Studies 4, 179–197.

Vesel, Ž., 2001. Science and scientific instruments. In: Pourjavady, N., Parham, M. (eds.),The Splendour of Iran. vol. 3: Islamic Period. Applied and Decorative Arts. The Cultural Continuum, Booth-Clibborn Editions, London, pp. 260–311.

Vesel, Ž., Tourkin, S., Porter, Y., in cooperation with Richard, F. and Ghasemloo F. (eds.), 2009. Images of Islamic Science. Illustrated Manuscripts from the Iranian World. Bibliothèque Iranienne 67, Institut français de recherche en Iran, published with the aid of The Soudavar Foundation, Université Sorbonne Nouvelle-Paris III Monde Iranien (CNRS-Paris III-EPHE-INALCO), Fondation van Berchem, Islamic Azad University, Tehran.

Welch, A., 1972. Collection of Islamic Art. Prince Sadruddin Aga Khan. 4 vols. Chateau de Bellerive, Geneva.

—— 1973. Shah ʿAbbas and the Arts of Isfahan. An Exhibition at Asia House Gallery, New York, 11 October-2 December 1973. The Asia Society, New York.

—— 2001. Worldly and otherworldy love in Safavi painting. In: Hillenbrand, R., (ed.), Persian Painting. From the Mongols to the Qajars. I.B. Tauris & Co. Ltd., London, pp. 302–317.

Welch, A., Welch, S.C., 1982. Arts of the Islamic Book. Cornell University Press, Ithaca.

Wellesz, E., 1964. Islamic astronomical imagery. Classical and Bedouin tradition. Oriental Art 10, 85–92.

Translating Playfair's *Geometry* into Arabic: Mathematics and Missions

Gregg De Young

Abstract The paper describes the Arabic translation of John Playfair's *Elements of Geometry* (first edition, London, 1795) by Cornelius Van Dyck. Playfair's treatise was popular in nineteenth century Britain and even more popular in America, where it appeared in dozens of editions, condensations, and extracts. Van Dyck had gone to Beirut as a medical missionary in 1840 and spent more than fifty years in Beirut devoted to education, especially in the sciences. Perceiving a need for modern science textbooks in Arabic, he either translated or wrote more than a dozen textbooks. His translation of Playfair produced generally a literal rendition of the English text, including its "algebraic notation." It also carefully preserved the basic architecture of Playfair's treatise. At the same time, there appears to be a deliberate attempt to adopt a non-traditional technical terminology and a non-standard transliteration system for diagram labels. These choices probably reflect the mindset of the American missionary community — that education, especially in modern mathematics and the sciences, might be a tool for religious conversion as well as social reform. Apparently Van Dyck wished to accentuate the discontinuity between traditional Arabic Euclidean discourse and that of modern European mathematics. Van Dyck's translation was part of a larger movement to import European learning in the sciences into the Ottoman Empire, so it faced potential competition from other Arabic and Turkish translations that were being printed in Istanbul and Cairo under the auspices of the Ottoman administration. Nevertheless, there is some bibliographic evidence that Van Dyck's translation was reprinted at least once in Beirut.

Introduction

In the nineteenth century, editions of several traditional Arabic geometry textbooks were printed.[1] These printed texts bear witness to a continuing (or renewed) interest in mathematics education in the Arabic-speaking world. The printed editions were of the traditional texts that had been taught in madrasas for more than five centuries.[2] Although the basic characteristics of Euclidean mathematics had not changed dramatically, new interpretations of fundamental mathematics and new approaches to mathematics pedagogy had been developing in Europe since the mid-eighteenth century.[3] These newer approaches had penetrated the Islamic world through Europeans who taught them in the new academies founded in the territories of the aging Ottoman Empire or in colonial schools. These instructors, finding

[1] Brentjes [2002, 320] summarizes the history of early Arabic printed mathematical works. I have discussed these printed editions in more detail in a forthcoming study, "Nineteenth Century Geometry Textbooks Printed in Arabic."

[2] The most important of these was the *Taḥrīr Kitāb Uqlīdis*, composed by Naṣīr al-Dīn al-Ṭūsī in 646 AH / AD 1248. De Young [2012] provides a more detailed survey.

[3] This development, in part connected to the Scottish Enlightenment, saw a new emphasis on Euclidean geometry as a tool for right thinking, rather than merely a problem-solving technique. Thus there was an increasing interest in teaching Euclid to students, including an understanding of proof structure and argumentation. It was this time period also that saw an increased interest in use of algebraic symbolism for understanding of geometry [Goldstein 2000].

N. Sidoli and G. Van Brummelen (eds.), *From Alexandria, Through Baghdad: Surveys and Studies in the Ancient Greek and Medieval Islamic Mathematical Sciences in Honor of J.L. Berggren*, DOI 10.1007/978-3-642-36736-6_23,

© Springer-Verlag Berlin Heidelberg 2014

no suitable modern textbooks to use in their classrooms, began to translate already existing European treatises into local vernaculars. This paper examines the Arabic translation of John Playfair's *Elements of Geometry* made by Cornelius Van Alen Van Dyck and situates his translation in the educational landscapes of the nineteenth century Ottoman era.

The Author: John Playfair (1748–1819)

Playfair's mathematical reputation is currently in eclipse. He is better known to historians today for his defense of Hutton's geological theory of the earth and his reformulation of Hutton's often convoluted arguments.[4] To historians of mathematics, he is perhaps better known for his eponymous "Playfair's Axiom" which he substituted for the parallel lines postulate of Euclid.[5]

Although Playfair was early recognized for the breadth and depth of his talent in mathematics, it was only in 1785 that he received an academic appointment, becoming Joint Professor of Mathematics (with Adam Ferguson) at Edinburgh University. He relinquished his post in 1805 to accept an appointment to the chair in Natural Philosophy. One of the original Fellows of the Royal Society of Edinburgh, he regularly published in its *Transactions*, as well as in the *Edinburgh Review*. It was at the Royal Society meetings that he regularly met many of the leading scientific figures of the Scottish Enlightenment.

Playfair's first major publication was *Elements of Geometry* (Edinburgh, 1795). The work was one of the most widely used geometry textbooks in English during the nineteenth century [Ackerberg-Hastings 2002, 70]. Although he published no further mathematical monographs, Playfair continued to review mathematical publications, many of them produced by European mathematicians, in the *Edinburgh Review*.[6] His last major work was *Dissertation: Exhibiting the Progress of the Mathematical and Natural Sciences since the Revival of Letters* which appeared as a supplement to the *Encyclopaedia Britannica* [Playfair 1816].[7]

The Textbook: Playfair's *Elements of Geometry*

Playfair's *Elements of Geometry* (first edition Edinburgh, 1795) was popular enough to go through thirteen editions or reprintings in Britain, the last in 1813. The work was even more popular in North America, where it was printed and reprinted at least thirty-three times during the nineteenth century [Ackerberg-Hastings 2002, 43, n. 1]. The text treated plane geometry (using books I–VI of Euclid's *Elements*), along with supplementary materials on plane and spherical trigonometry, which Playfair described as "the application of arithmetic to geometry," and some material on solid geometry in which Playfair rewrote portions of books XI and XII of the *Elements*. The material on plane and spherical trigonometry appears to be Playfair's reworking of standard textbook material (he explicitly mentions Maskelyne, Taylor, and Napier as sources) [Playfair 1819, xiii–xiv].

Playfair, in his introduction to the second edition, explicitly claimed the work of Robert Simson [1781] as the greatest influence on his edition of Euclid. It was Simson (1687–1768), that meticulous mathematical scholar, whose new edition of Euclid did so much to place the *Elements* on a firm logical

[4] Challinor [1975], for example, identifies Playfair's fields of scientific activity as "mathematics, physics, geology." Nevertheless, only the first two brief paragraphs summarize his work in physics and mathematics. The remaining seven paragraphs expound the main points of his *Illustrations of the Huttonian Theory of the Earth* (Edinburgh, 1802). Ackerberg-Hastings [2002; 2008] has given the most recent elaboration of Playfair's role in the mathematics of his day.

[5] This axiom stated that "Two straight lines that intersect one another cannot both be parallel to the same straight line [Playfair 1806, 7]." The axiom does not originate with Playfair — it had been known since at least the time of Proclus [Heath 1956, 1, 220; Proclus 1970, 273] — nor did he ever claim it as his own creation.

[6] Ackerberg-Hastings [2008, 94] has compiled a basic bibliographic listing of these reviews.

[7] The first 54 pages of the treatise deal with advances in mathematics, leaving some 390 pages devoted to a discussion of natural philosophy.

foundation (a foundation even more solid than that which Euclid himself had provided) by applying Greek analytic methods. Simson's Euclid was a brilliant scholarly *tour de force*, but its complex mathematical and logical argumentation made it unsuitable for use by beginning students. His efforts were directed solely toward the scholarly and mathematical community, as indicated by the small print runs for both the original Latin and his later English translation.[8] His publisher, however, thought that the treatise might also find a role as a textbook, coinciding as it did with the advent of a new interest in teaching geometry in the schools of England. By the end of the eighteenth century, Simson's work had gone through some twenty six editions in Britain and had been translated into Portuguese, Spanish, French and German [Ackerberg-Hastings 2002, 48].

Playfair, who spent much of his active career in education, having perceived that the results of Simson's erudition needed to be presented in a form more easily accessible to the student, immediately set himself the task of writing a textbook to fill the gap. He took most of his material from Simson, of course. But he also made two important (at least in his eyes) innovations [Playfair 1846, preface]. First, he employed an algebraic notation in the discussion of ratios and proportion in book V, rather than the purely verbal quasi-geometrical formulations of the traditional Euclidean approach.[9] One byproduct of this use of algebraic notation was the removal of the traditional geometric diagrams from book V, so that this part of the treatise looks quite different from the standard Euclidean textbook. Second, he introduced a different statement (now known as "Playfair's Axiom") in place of the perennially problematic parallel postulate of Euclid.[10] Playfair did not intend his *Geometry* to open new paths or reshape the field. His only goal was a textbook both appropriate to and useful for beginning students of his day. In this aim he was remarkably successful, if the numerous nineteenth century editions and reprintings (especially in America) may be taken as an indication of popularity.

The Translator: Cornelius Van Alen Van Dyck (1818–1895)

Cornelius Van Alen Van Dyck,[11] certified in medicine from Jefferson Medical College of Philadelphia (1837), was dispatched in 1840 as a medical missionary to the Ottoman province of Syria (encompassing most of what is now Jordan, Lebanon, Syria, and parts of Palestine/Israel) under the auspices of the American Board of Commissioners for Foreign Missions.[12] There is no evidence that Van Dyck had any theological training prior to his arrival in Syria. He was ordained to the ministry at the mission in 1846 [Recent Intelligence 1846, 211–212]. Convinced that education was integral to the success of any missionary work, Van Dyck devoted himself primarily to teaching and organizing schools while continuing to practice his medical skills. But schools need textbooks, and no textbooks appropriate to modern liberal education were available in Arabic. So Van Dyck, a gifted linguist known today primarily for his translation of the Bible into Arabic (completed 1864), set himself the task of preparing Arabic textbooks in a variety of scientific subjects: geography (1852), algebra (1853), geometry (1857),

[8] The first print run Latin was only 543, while the English was only 803 [Ackerberg-Hastings 2002, 47]. A typical print run for a mathematics textbook in the nineteenth century would have been at least 1000 copies [Ackerberg-Hastings 2002, p. 43, n. 1].

[9] Algebraic symbolism was also used in book II beginning with the second (1804) edition. Playfair implies that the use of algebraic symbolism is his own innovation, but similar use of symbols had been introduced more than a century earlier in Isaac Barrow's 1660 edition of Euclid [Barrow-Green 2006, 19, fig. 12; Ackerberg-Hastings 2002, 57, n. 65]. Whether Playfair was aware of earlier uses of "algebraic" notation is unknown. Another possible inspiration for Playfair's "algebraic" approach was Thomas Reid (1710–1796), who had taught students in Aberdeen that algebraic reasoning could be applied in geometry [Wood 2004, 56].

[10] "There being two intersecting lines, it is impossible for both of them to be parallel to one and the same given line" [Playfair 1846, 11].

[11] Saʿdi [1937] provides almost the only English source for Van Dyck's biography and I have relied on it extensively in this section. For a more complete listing of Van Dyck's Arabic publications, see Sarkīs [1928–1930, col. 1462–1465].

[12] Probably the ABCFM intended Van Dyck to use his medical skills primarily to benefit his fellow missionaries [Williams 1982, 274]. But Saʿdi [1937, 26–27] informs us that Van Dyck also worked among the native Syrian Druze population.

chemistry (1869), logarithms (1873) and logarithmic tables (1873), astronomy (1874) and history of astronomy (1874), as well as a medical pathology textbook (1878). All were translated from or based on European scientific works.

About 1860, the ABCFM missionary community in Beirut began to develop plans for a college for boys — an institution of higher learning in which the sciences would hold pride of place.[13] The ABCFM, however, declared that they were unwilling to support this undertaking, so the newly organized Syrian Protestant College was funded through private donors in the USA and UK.

With the opening of the medical department (1867), Van Dyck was appointed professor of internal medicine and general pathology, although he remained on the staff of the Protestant Mission as well. He also agreed to teach chemistry until another instructor could be found. (Edwin Lewis was appointed professor of chemistry in 1870.) In 1870, the College acquired a plot of land and began construction of a permanent campus. When construction of the Lee Astrophysical Observatory was completed in 1874, Van Dyck served as its director.[14] The equipment, including a 10-inch Newtonian reflector and a meridian transit instrument, was largely funded by Van Dyck from the income of his medical practice. Assisted by Fāris Nimr, he made regular observations of sunspot activity and collected daily meteorological data which were telegraphed to the Imperial Observatory in Istanbul twice daily [Saʿdi 1937, 33].

In 1882, Van Dyck resigned from the medical faculty to protest the College administration's dismissal of Professor Lewis over remarks about Darwin's theory that he had made in his 1881 commencement address. These remarks had been picked up and circulated by the Arabic journal *al-Muqtaṭaf*, edited by two Arab Christian converts closely associated with the College and personal proteges of Van Dyck.[15] With no teaching duties, Van Dyck devoted himself to his medical work and to preparation of a series of Arabic introductory surveys of several scientific fields: Introduction to Natural Science, Physics, Chemistry, Physical Geography, Geology, Astronomy, each in a small compact volume.[16]

Basic Print Architecture of Van Dyck's Translation

Van Dyck's Arabic translation opens with a title page (Figure 1). The title is a literal translation into Arabic from the title page of the sixth edition of Playfair's textbook. Playfair's name is not mentioned, however. Instead, we find "Translated by Cornelius Van Dyck" where we would normally expect the author's name. All the text on the title page is set in large font and centered, with considerable white space above and below. The page is bounded by a solid double line, inside which is a generic woodcut border of foliage motifs. The same style of decorated border is repeated on the following pages containing the translator's preface and the historical excursus (to be discussed below). Foliage motifs are also placed at the top of the first page of text, framing the heading of the initial section ("Fundamentals of Geometry" and "Book One" each centered in a type line, one above the other, and separated by a "double-dagger" motif) almost as though marking the boundary of an illuminated *ʿunwān* or title

[13] According to the original plan, there were to be six faculties: Arabic language and literature; mathematics, astronomy and engineering; chemistry, botany, geology and natural science; modern languages; medicine; and law and jurisprudence. "Prospectus and Programme of the Syrian Protestant Collegiate Institute Beirut," p. 2 as quoted by Elshakry [2007, 192.]

[14] Substantial funds for the construction were provided by the Turkish Mission Aid Society of London [Tibawi 1967, 202]. The AUB history of the observatory, published online at http://www.aub.edu.lb/tour/nojava/b12.html, reports that the building was named for Henry Lee, a merchant from Manchester, who donated 150 pounds sterling toward its construction.

[15] The history of the episode has been told in Jeha [1991] and Kaya [2005]. Khūrī [1991] focuses on the role Van Dyck played in the episode. Elshakry [2008, 707–710] notes that Lewis's Arabic speech about the relations between knowledge and wisdom had completely ignored the historical meanings of Arabic terms such as *ʿilm* and *maʿrifa*. Van Dyck had similarly ignored historical terminological usage in his translation of Playfair, as described below.

[16] Their influence may be estimated from the fact that copies of many of these books found their way into the Majlis Shura Library in Tehran.

panel.[17] These statements serve as running headers on right and left pages of the treatise respectively. A similar enclosed title panel begins each book in the treatise. What is distinctively missing is the "bismillah" or pious invocation of God's name — a feature very rarely omitted from traditional Arabic manuscripts. Its omission can scarcely have been an oversight.

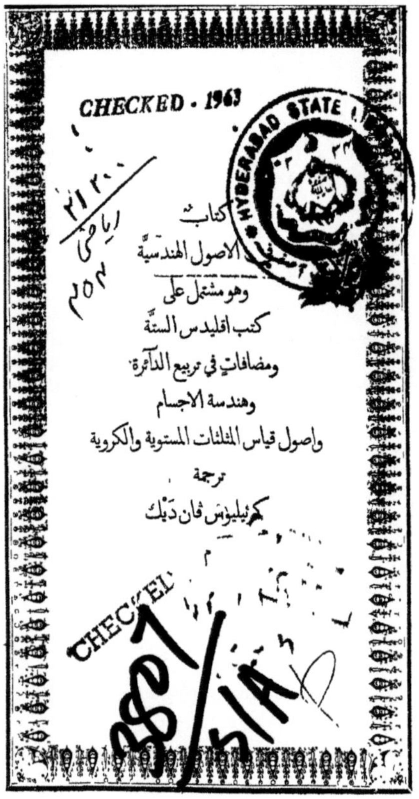

Figure 1: The title page of Cornelius Van Dyck's translation of Playfair's *Geometry*. It does not name Playfair, but places Van Dyck's name where we would expect to find the author's name, although Van Dyck is clearly identified as translator.

Following this title page and a blank page is a short *muqaddima* (introduction or preface) [Van Dyck 1857, 3]. In style, it is similar to that of a traditional Arabic manuscript:

> Thanks [be] to God (Allah) — who is such that [human] imaginations cannot circumscribe the circle of His working and who is far beyond magnitudes of figures and measurements of bodies. Now, the American Cornelius Van Dyck, who is in need of his Omnipotent Lord, declares: Since I perceive the need of teachers in this country for geometry books through which the aimed-for utility may be realized, I have devoted myself to the translation of this instructive treatise. It comprises six books of Euclid and other additions concerning the squaring of the circle, and the geometry of solids, and the basics of plane and spherical trigonometry. And may God (Allah) grant that students may be helped by it and that those [who are] desirous [of learning] may profit from it. And may He, namely the Most Merciful, make it beneficial for the sake of its noble intention.[18]

---

[17] For an example of a classic *ʿunwān* panel used in contemporary lithograph, see Scheglova [1999, 16].

[18] الحمد لله الذي لا تحيط بدائرة عمله الأوهام. وهو المنزّه عن مقادير الأشكال ومساحة الأجسام. أمّا بعد فيقول العبد الفقير إلى ربه القدير كرنيلس ڤان ديك الأمريكاني إنني لما رأيت افتقار المدارس في هذه البلاد إلى الكتب الهندسية الّتي بها نتمّ القائدة المقصودة منها

This introduction follows many of the long-established rhetorical features found in Arabic manuscripts. It begins, for example, with pious apostrophes. Van Dyck, describing himself in the third person as would any classical Arabic author, outlines the basic contents of his book and explains why he has composed it. He concludes with a pious prayer that God will make his efforts useful to the intended audience. It is interesting that in the midst of this very traditional introductory statement, Van Dyck explicitly identifies himself as American. Since he went to considerable lengths to identify with local culture and customs, even adopting native dress [Saʿdi 1937, 26], this explicit national or cultural identity appears oddly out of character.

On the next page Van Dyck has given a brief historical introduction to the *Elements* (*nubdha tārīkhiyya*) [Van Dyck 1857, 4]:

> The philosopher Euclid, author of the treatise *Elements of Geometry*, lived in Egypt about 270 BCE in the time of Ptolemy Lagus. It is said that he was born in Alexandria and it is said that his parents are unknown. He became a teacher of mathematical sciences in the school of Alexandria and had many pupils, among whom was the king, Ptolemy, himself. It is said that the king asked him one day if there were not an easier way to reach an understanding of mathematics. He replied that there is no royal road to that [subject]. He had writings on cosmography and optics, but his most famous writing was the *Elements of Geometry*. This treatise is still the best that has been written on this topic until our day. Nevertheless, there entered into it various alterations and deficiencies over the passing centuries. The Scottish professor Simson returned it to its original [pristine form]. Then some teachers added to it several propositions in order thereby that it become more related to the state of mathematics in this [our] age. And the best and most useful of these treatises is the text written by the Scottish teacher Playfair. And it is this [text of Playfair] upon which I have relied in this translation. And in God (Allah) alone is success.[19]

Following these introductory statements, we come to the translation itself. Each page of the text is surrounded with a double-ruled border to delimit the print area (Figure 2). Centered above and outside this border is a running title. Above the right-hand pages we find "Uṣūl al-Handasah" (Fundamentals of Geometry). Above the left-hand pages, we find "Al-kitāb al-ʾawwal" (Book One, or whatever ordinal number denotes the current section of the treatise).[20] Outside the ruled boundary at the outer corner at the top of each page is a page number in Arabic script numerals.

Not every book of the *Elements* contains definitions, but whenever there are definitions in the text, they precede the propositions of the book. Playfair's definitions are numbered, and each definition begins on a new line, so it is easy to locate any definition.[21] Van Dyck follows the same procedure. There are diagrams accompanying several of Playfair's definitions apparently to give an illustration of the concept being defined. Van Dyck reproduces these diagrams in his translation. These diagrams are rarely labeled and are placed immediately to the left of the text of the definition they are intended to illustrate.[22]

اعتنيت بترجمة هذا الكتاب المقيد وهو مشتمل على كتب أقليدس الستة ومضافات أخرى في تربيع الدائرة وهندسة الأجسام وأصول قياس المثلثات المستوية والكرّوية. والله المسئول أن ينفع به الطالبين ويفيد الراغبين ويجعله مخلصاً لوجهه الكريم وهو أرحم الراحمين.

19 إن الفيلسوف أقليدس صاحب كتاب الأصول الهندسية عاش في بلاد مصر ق م نحو ٢٨٠ في عصر الملك بطلميوس لاغوس. قيل ولد في الإسكندرية وقيل مولده مجهول وصار معلم العلوم التعليمية في مدرسة الإسكندرية وكثر تلاميذه ومنهم الملك بطلميوس نفسه. قيل سأله الملك يوما ألا يوجد سبيل أسهل لمعرفة التعاليم فقال لا توجد سكة سلطانية لذلك. وله مؤلفات في علم الهيئة والبصريات وأشهر مؤلفاته الأصول الهندسية ولم تزل إلى أيامنا هذه أفضل ما صنف في هذا الفن. غير أنه قد دخل عليها بعض التغيرات والنقائص على تمادي الأجيال. وقد رجعها إلى أصلها المعلم سمسون الاسكوتسي ثمّ أضاف إليها بعض المعلمين عدة قضايا لكي تصير بذلك أكثر مناسبة لحال التعليم في هذا العصر. وأحسن نس.خها وأكثرها فائدة النسخة التي اعتني بها المعلم بلايفار السكوتسي وهي المعوّل عليها في هذه الترجمة وبالله التوفيق.

20 Euclid's text is traditionally divided into thirteen books or sections. In Byzantine Greek manuscripts, these sections are denominated only through alphanumeric labels, either the letter alone or else the ordinal numeral spelled out. The Arabic Euclidean transmission had designated each of these sections with the term "maqālah". This term can refer either to a stand-alone treatise or to a division of a larger treatise. Van Dyck chose the term "kitāb", a term referring more often to an independent volume or treatise, in his translation. I will discuss the choice of vocabulary and its significance later.

21 In Arabic manuscripts, definitions are not numbered and are typically presented in a single block of text, although the beginning of each definition may be indicated by some kind of delimiter symbol or by over-lining the initial word.

22 Diagrams illustrating definitions are not included in Arabic primary transmission manuscripts, although they are some-

Each proposition in Van Dyck's translation is separated from the next by a horizontal double-pointed device centered in a blank line. These devices are of several slightly varying designs and are apparently used in random sequence. Individual propositions are introduced by a heading (that is, the proposition number and its classification as theorem or problem), paralleling that found in Playfair's treatise. Following this heading, the enunciation of the proposition begins on a new line. Both the heading and the enunciation are centered on the page and are set in a slightly larger type face than the text of the demonstration. This use of larger type for the enunciation corresponds to the use of italic type to set off the enunciation from the remainder of the proposition in Playfair's treatise.

Van Dyck generally follows the lead of Playfair's text for the paragraphing within the demonstration.[23] The diagram is placed near the beginning of the demonstration and aligned on the left margin of the text area.[24] (In Playfair's treatise, diagrams are aligned with the right-hand margin of the text area.) The diagram is separated from the text of the demonstration by generous white space. The area reserved for the diagram is either square or rectangular, depending on the characteristics of the diagram. Punctuation is not used in the Arabic text although some vowel indicators have been included, often to distinguish active from passive verbs.

Playfair included references to earlier propositions at key points in his treatise. Although in the first edition these references were placed in the margins, in later editions they are placed in parentheses within the text.[25] Thus, (46.1) refers to proposition forty six of book I. Many, but not all, of Playfair's references are included in the text of Van Dyck's translation. The presence or absence of a particular reference in the Arabic seems to reflect a translator choice. Following the example of Playfair's later editions, these references are set off from the text using parentheses. Thus we find, for example, (ق ٤٦ ك ١) pointing back to proposition 46 of book I.

Playfair included additional theorems / problems in book I, II, III, V, and VI, labeling them "Additions to Book..." to which is added the appropriate ordinal numeral.[26] These added propositions appear also in Van Dyck's translation, labeled "muḍāfat ilā al-Kitāb..."[27] Playfair labels these added propositions alphabetically, beginning from A. Van Dyck uses the letters of the Arabic alphabet in *abjad* ordering, rather than simply transliterating Playfair's alphabetic labels according to his own scheme, as described below. Although the bulk of the added propositions are placed at the end of these books, this is not uniformly the case. In books II and VI, some of the added propositions are inserted earlier in the text. In book VI, the additional propositions are divided into two groups. The first group is labeled "Additions to Book VI." These propositions are followed by "Problems Relating to the Sixth Book." Playfair labels these related problems continuously with the added propositions, from L to Y. This structural division into two problem sets consecutively numbered is also present in Van Dyck's Arabic translation; the latter problems are called "ʿAmaliyāt mulḥaqāt bi-l-kitāb al-sādis." Van Dyck also labels both sets of propositions continuously, using traditional *abjad* alphanumerical values.

---

times added in various commentaries or secondary editions. On the other hand, some Byzantine Euclidean manuscripts do contain diagrams accompanying the definitions. These Greek diagrams are usually quite small and appear ancillary to the verbal text. Whether a direct relationship exists between these Byzantine diagrams and those found in the Arabic transmission has yet to be established.

23 Arabic manuscripts are not divided into paragraphs, although sections of the demonstration are sometimes indicated by writing the initial word in bold script or by over-lining the initial word.

24 In many manuscripts, diagrams are placed near the end of the demonstration, and sometimes even follow the demonstration.

25 The tradition of placing references to previous propositions in the margins goes back at least as far as the 1660 English edition of Euclid by Isaac Barrow [Barrow-Green 2006, 19, fig. 12]. It is difficult to determine when the change to parenthetical in-text references was introduced. In both the 1806 and 1826 editions of Playfair, the references are placed in the margin, with superscript lower case Roman letters used to indicate the statement in the text to which the reference referred. By the 1837 American edition, the references have moved into the text.

26 These sets of added propositions had reached their complete form by 1837, but the 1826 edition, for example, still had not added any propositions in book I.

27 Although Van Dyck uses a perfectly acceptable Arabic translation, in traditional Arabic Euclidean discourse one would have expected the term "ziyādāt" for additions.

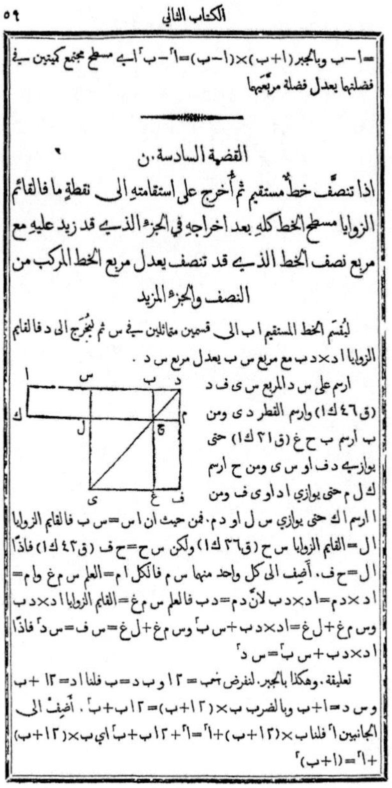
٥٩ الكتاب الثاني

=ا−ب وبالجبر (ا+ب)×(ا−ب)=ا٢−ب٢ اي مسطح مجتمع كميتين في
فضلتهما يعدل فضلة مربعيهما

القضية السادسة. ن

اذا تنصف خطٌ مستقيم ثم أُخرج على استقامتهِ الى نقطةٍ ما فالقائم
الزوايا مسطح الخط كلهِ بعد اخراجهِ في الجزء الذي قد زيد عليهِ مع
مربع نصف الخط الذي قد تنصف يعدل مربع الخط المركب من
النصف والجزء المزيد

لِيُقسَم الخط المستقيم ا ب الى قسمين متماثلين في س ثم ليُخرَج الى د فالقائم
الزوايا ا د×د ب مع مربع س ب يعدل مربع س د .

ارسم على س د المربع س ى ف د
(ق٤٦ ك١) وارسم القطر د ى ومن
ب ارسم ب ح غ (ق٣١ ك١) حتى
يوازي د ف او س ى ومن ح ارسم
ك ل م حتى يوازي ا د او ى ف ومن
ا ارسم ا ك حتى يوازي س ل او د م. فمن حيث ان ا س=س ب فالقائم الزوايا
ا ل=القائم الزوايا س ح (ق٣٦ ك١) ولكن س ح=ح ف (ق٤٣ ك١) فاذًا
ا ل=ح ف. أَضِف الى كل واحد منهما س م فالكل ا م=العلم س م غ وا م=
ا د×د م=ا د×د ب لانَّ د م=د ب فالعلم س م غ=القائم الزوايا ا د×د ب
وس م غ+ل غ=ا د×د ب+س ب٢ وس م غ+ل غ=س ف=س د٢ فاذًا
ا د×د ب+س ب٢=س د٢

تعليقة. وهكذا بالجبر. لنفرض ﺍﺏ=١٢ و ب د=ب فلنا ا د=١٢+ب
و س د=ا+ب وبالضرب ب×(١٢+ب)=١٢ب+ب٢ . أَضِفْ الى
الجانبين ا٢ فلنا ب×(١٢+ب)+ا٢=ا٢+١٢ب+ب٢ اي ب×(١٢+ب)
+ا٢=(ا+ب)٢

Figure 2: A sample page of Van Dyck's Arabic translation, showing *Elements* II, 6. It illustrates many features of the basic "architecture" of the printed translation, including the use of algebraic notation, references to earlier propositions, as well as the placement of diagrams and the labeling of propositions.

The Arabic Translation

I turn now to Van Dyck's translation itself. Through the efforts of Van Dyck, Playfair's *Geometry* was effectively "relocated" or transferred into a different linguistic and cultural matrix.[28] This relocation process refashioned the *Geometry* for readers / consumers whose educational goals and epistemological outlooks were sometimes quite different from those of the original author, Playfair, or even the translator, Van Dyck. During the translation / relocation process, Van Dyck was called upon to make numerous choices about how to express the specific content of Playfair's work. For example, he had to decide how to transliterate the letters designating the points, line segments and angles that appear so frequently both in the text and in the diagrams of Playfair's textbook. Van Dyck also had to consider how closely to reproduce the diagrams used by Playfair and the lettering conventions found in them. In one sense, of course, this question is intertwined with the question of transliteration. But the diagram also forms a distinct visual image which has its own internal logic that allows the diagram to be read and processed at least somewhat independently from the verbal text.

Geometry, from the time of Euclid, has had a recognized technical vocabulary. Each time Euclid's work is carried across a linguistic boundary, a fresh technical vocabulary must be created or an existing

[28] The idea of translation as relocation is developed by Elshakry [2008, 704].

technical vocabulary must be adapted to use in a new context. Although Van Dyck is carrying Playfair's *Geometry* into Arabic for the first time, the introduction of Euclidean geometry into Arabic occurred long before. Thus there already existed a well-established technical Arabic vocabulary for discussing Euclidean geometry. Use of this technical vocabulary carried multiple layers of meaning that had developed over centuries, presupposing a familiarity with that historical discussion. This vocabulary might not always be appropriate to more modern interpretations of geometry imported from other traditions. And apart from matters of technical vocabulary, every translator faces innumerable questions about how to render specific linguistic formulae or phrases effectively into a new linguistic matrix. Since Van Dyck was working not just as translator but also as educator, he was always conscious of how his students would read and understand his Arabic rendition of Playfair's textbook. Again and again, this dual role of translator-educator seems to soften the rigidity of the strictly literal word-for-word approach to translation that Van Dyck often adopted. It was apparently his educator role that also prompted him, at several points, to make more radical changes in his Arabic version of Playfair's treatise, changes in substance, not merely in expression. Such modifications of the mathematical content are infrequent, which only makes them more striking when we encounter them.

Transliteration

I begin my survey with the transliteration scheme used by Van Dyck to render the point labels found in Playfair's treatise. We might expect that this would be a simple problem — essentially a letter for letter substitution. But in practice, Van Dyck adopted some unusual choices of Arabic letters to transliterate Playfair's English labels. Nor was he always consistent in transliteration. Table 1 summarizes the main features of the transliteration used by van Dyck.

ا	A
ب ، س	B
س	C
ص	Ç
د ، ت ، ر ، ذ	D
ى ، ر	E
ف ، ق	F
غ ، ع	G
ح	H
م ، ج ، ح	I
ك	K
ل	L
م	M

ن	N
ر ، ق ، و ، ز	O
ن ، ق ، ل ، غ ، ر ، ٢	P
ق ، ر ، ١	Q
١ ، ٣	R
ص ، ض	S
ف ، ت	T
٣	U
ص ، ٢	V
ت ، ص ، ك ، و ، ر ، ج ، ط	X
ن	Y
٢	Z

Table 1: Tables showing the Arabic letters used by Van Dyck to transliterate Playfair's labels. Because Arabic is read from right to left, the frequency of Arabic letters used decreases from right to left.

Often Van Dyck's chosen transliteration seems based more-or-less on phonemic considerations. Thus C becomes س (whose name is pronounced like the English word "seen") and E becomes ى (which functions like the English letter Y and is often pronounced like the Y in the word "really") and G becomes غ. But this pattern is sometimes ignored. Playfair's letter Z, for example, is never transliterated as ز, its Arabic phonemic equivalent. Other cases of deviation from Playfair's lettering system are more puzzling. Where Playfair uses the letter I, for example, Van Dyck sometimes uses م, the phonemic equivalent of the English M (as, for example, in Scholium C of book I). The rationale for this choice is not clear. Such divergences occur only rarely, however, and in most cases the Arabic transliteration follows the lead of the English letters used by Playfair.

Often there seems to be a confusion between Arabic letters of similar orthography (in many cases the differences are only the presence or absence of a dot). For example, د and ذ are frequently interchanged when the letter expected (based on the manuscript tradition) is د. The use of apparently incorrect letters or the failure to distinguish between orthographically similar letters is puzzling and sometimes confusing. Despite these seeming inconsistencies, though, there is rarely a contradiction between the letter in the text and the label in the diagram.

Diagrams

In general, Van Dyck followed the diagram constructions of Playfair exactly. Since Arabic is read from right to left, one might have expected that he would produce diagrams that were mirror images of Playfair's diagrams (that is, flipped left for right) but this is not the case. One result of this very literal reproduction of the diagrams is that the labeling of Van Dyck's diagrams must typically be read from left to right, backwards from the way Euclidean diagrams were usually read in the Arabic manuscript tradition (Figure 3). In adopting this convention, Van Dyck seems to have consciously distanced himself from traditional Arabic Euclidean scholarship.

As is customary in the Euclidean tradition, the diagrams in Playfair's text typically have their labels assigned in alphabetic order following the order in which the points are mentioned in the demonstration. But this pattern is not universally followed. Sometimes letters are omitted from the usual alphanumeric sequence and sometimes letters are introduced out of the expected sequence.[29] Within the Arabic Euclidean tradition, diagram labels were typically assigned following the order of the alphanumeric *abjad* system (although omitting letters "waw" and "yā"), which mirrored the Greek alphanumeric ordering. Van Dyck abandoned this *abjad* ordering for assigning diagram labels, and instead followed Playfair's labeling patterns. In doing so, he again seems to have deliberately distanced his translation from traditional Arabic Euclidean discourse. As a result, the diagrams become considerably more difficult for Arabic speakers to read without recourse to the accompanying text.

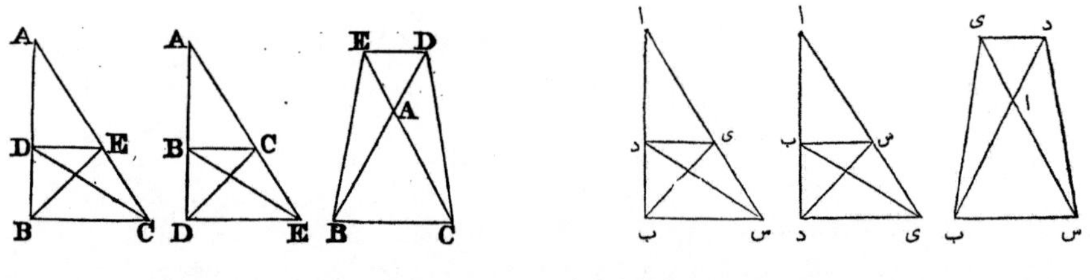

Figure 3: Diagrams for *Elements*, VI, 2. Left, Playfair [1846, 123] diagrams. Right, Van Dyck [1857, 149] diagrams. The Arabic diagrams mirror the English diagrams in both construction and labeling.

The diagram for *Elements* II, 8 (Figure 4) is a rare example where Van Dyck deviates significantly from Playfair's labeling. The rationale for such deviations is unclear. Another example of a diagram modified by Van Dyck is *Elements* III, 8, which has been simplified from that of Playfair (Figure 5). Several labels have been omitted from the Arabic diagram. As a result, points mentioned in the demonstration are

[29] In most cases, Playfair seems simply to be following the lead of Simson. Simson's diagrams, in turn, can be traced back at least as far as the Latin edition of Commandino (1572), on whose work Simson had based his edition. The obvious continuities in diagram construction over this long time period are not in themselves surprising — woodcut blocks were expensive to produce and were frequently re-used. The full history of diagrams in the European print tradition of the *Elements* has still to be written, although there has been some useful preliminary work such as the introductory investigation by Barany [2010]. (Preprint available from http://www.princeton.edu/~mbarany/GhentEuclidPreprint.pdf.)

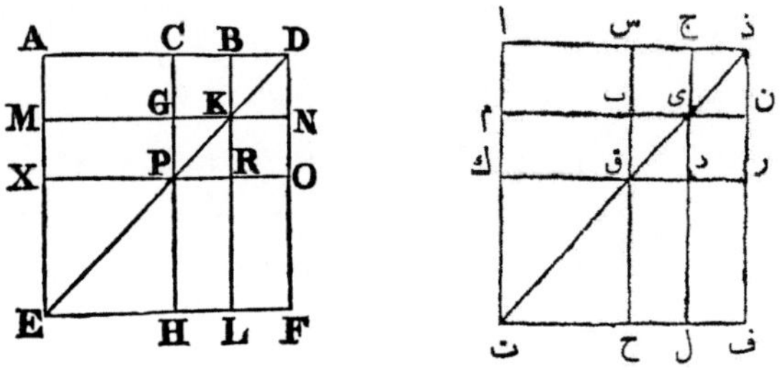

Figure 4: Diagrams for *Elements* II, 8. Left, Playfair's diagram [1846, 53]. Right, Van Dyck's diagram [1857, 61]. Van Dyck deviates from Playfair's labeling and from his own transliteration scheme at several points: G and B have been interchanged, and G is represented here by ج rather than غ; E (ى) has been moved to the position occupied by K and the position of E has been taken by T (ت) while X is represented by the Arabic letter ك which usually transliterates the letter K. Labels P, R, O are represented in Arabic by ق , د , ر respectively.

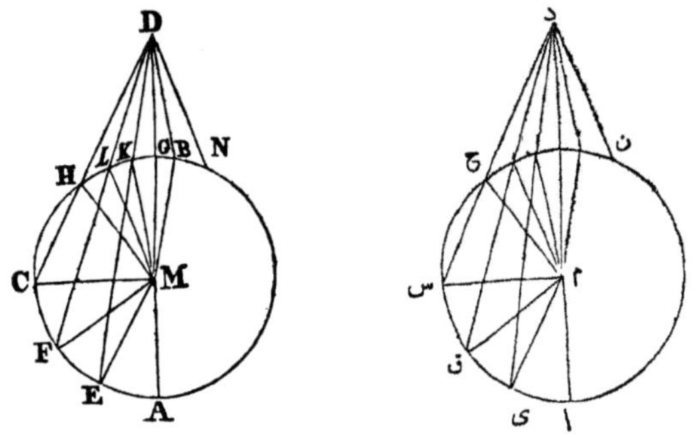

Figure 5: Diagrams for *Elements* III, 8. Left, the diagram from Playfair [1846, 67]. Right, the diagram from Van Dyck [1857, 77].

no longer present in the diagram. It seems unlikely that Van Dyck, with his concern for clarity and his interest in education, would have made such a modification. Perhaps the printer decided not to clutter or crowd the diagram with letters, not considering that these letters are essential for relating the diagram and the text.

Technical Vocabulary

Van Dyck also faced choices when it came to technical vocabulary. He found one set of technical vocabulary in Playfair. There also existed a technical vocabulary that had developed over centuries of Euclidean discourse in Arabic. It would seem easiest, perhaps, to simply substitute existing Arabic terms for Playfair's English terms. But Van Dyck did not choose this path. Instead, he developed an alternative vocabulary which, although it translated Playfair's terms accurately enough, differed in many cases from traditional Arabic terminology. Table 2 summarizes some of these unconventional terminological choices.

Playfair's treatise opens with a section introducing the vocabulary of geometrical discourse. This section is also translated by Van Dyck. The terminology that Van Dyck altered denoted the structure of geometry and geometrical argument, not the names of geometrical entities such as lines, triangles, or hexagons. Apparently, Van Dyck assumed that these entities were the same, no matter what language was used to denominate them, so one could substitute traditional Arabic terminology for English

Playfair's Term	Van Dyck's Term	Traditional Term
Book	كتاب	مقالة
Theorem	نظرية	علمي
Construction Problem	عملية	عملي
Proposition	قضية	شكل
Corollary	فرع	واستبان منه
Lemma	سابقة	مقدّمة
Scholium	تعليقة	ومعناه ، أقول ، حاشية
Axioms	أوليات	أصول الموضوع
Definitions	حدود	حدود
Postulates	مقتضيات ، ممكنات	العلوم المتعارفة
Conversely	بالقلب	بالعكس
"Otherwise"	طريقة أخرى	ببرهان آخر
QED	ما كان علينا أن نبرهن	وذلك ما أردنا أن نبين ، وذلك ما أردناه

Table 2: Table showing Van Dyck's choice if non-traditional terms to transliterate Playfair's English terminology.

terminology without change of meaning. But the more philosophical and logical terminology describing the structure of geometry and its arguments carried many connotations that had arisen over long centuries of use. These Van Dyck did not want to import into his translation. In so doing, he again seems to be distancing himself from the more traditional understanding of geometry. He seems to want to emphasize that the geometry he is translating into Arabic is different from traditional Euclidean discourse found in the Arabic classics and canonical texts.[30]

For example, classifying a proposition as either a theorem — a mathematical result requiring demonstration — or a problem — a construction (whose technique requires adequate demonstration to convince the reader that the procedure does indeed produce the desired result) was not new either to Playfair or to the Arabic Euclidean tradition. In Arabic, the distinction between theorems and problems goes back at least as far as Ibn al-Haytham in his *Kitāb fī ḥall shukūk Kitāb Uqlīdis* (and perhaps back to Proclus in the Greek transmission). In the English transmission, the division into theorem and construction appears already in Recorde's translation (1551) [Easton 1975, 339]. But by choosing to use non-traditional terminology, Van Dyck set his work apart from that of the earlier Arabic tradition, emphasizing that is was something different and new.

Changing Nuances of Diction

In general, we can say that Van Dyck provides a literal rendition of Playfair's text. One example (Figure 6), taken from proposition II, 12, will suffice to illustrate this point. But even though Van Dyck's translation is usually literal, there are minor differences in diction even within this small proof. For example, Playfair cites *Elements* I, 47 twice, but Van Dyck inserts the reference only once, apparently assuming that the second case will be obvious from the repetition of the argument. It is interesting, but probably inconsequential, that in the second edition (Edinburgh, 1804), Playfair cites *Elements* I, 47 only once.[31] Also, Van Dyck states that perpendicular *AD* (*alif dāl*) is constructed after *BC* (*bā sīn*) is

[30] The commencement address of Professor Lewis that eventually led to the resignation of Van Dyck from the faculty of the Syrian Protestant College had similarly disregarded the traditional meanings of Arabic philosophical terminology in order to make a much more European point [Elshakry 2008, 707–709].

[31] In later editions, the reference to proposition I, 47 is given for each statement. Perhaps there is only one citation in the earlier edition because the format of the treatise places the citations in the margin and so they stand out more visibly and

extended to D (*dāl*). This statement implies that point D (*dāl*) must be defined before the side can be extended and before the perpendicular can be constructed. This implication is not present in Playfair's statement. Rather, point D becomes defined only when the perpendicular has been constructed so as to fall on BC extended (indefinitely). And where Playfair says "add AD^2 to both: Then..." we find in Van Dyck's translation "add AD^2 to the two sides; then we have..." Such minor changes in diction only serve to emphasize in general how literally Van Dyck rendered Playfair's text into Arabic. Van Dyck seems to have made a conscious decision to keep the mathematical content and its expression as closely identical to that of Playfair as possible, although he was willing to adapt the diction to a form more congenial to Arabic readers when such adaptation was helpful to his audience and did not alter the presentation of the mathematics itself.

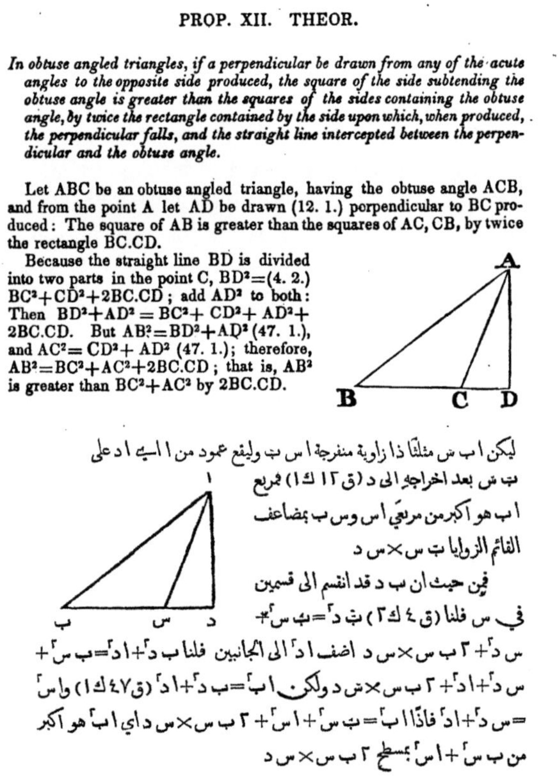
PROP. XII. THEOR.

In obtuse angled triangles, if a perpendicular be drawn from any of the acute angles to the opposite side produced, the square of the side subtending the obtuse angle is greater than the squares of the sides containing the obtuse angle, by twice the rectangle contained by the side upon which, when produced, the perpendicular falls, and the straight line intercepted between the perpendicular and the obtuse angle.

Let ABC be an obtuse angled triangle, having the obtuse angle ACB, and from the point A let AD be drawn (12. 1.) perpendicular to BC produced: The square of AB is greater than the squares of AC, CB, by twice the rectangle BC.CD.

Because the straight line BD is divided into two parts in the point C, $BD^2=$(4. 2.) $BC^2+CD^2+2BC.CD$; add AD^2 to both: Then $BD^2+AD^2=BC^2+CD^2+AD^2+2BC.CD$. But $AB^2=BD^2+AD^2$ (47. 1.), and $AC^2=CD^2+AD^2$ (47. 1.); therefore, $AB^2=BC^2+AC^2+2BC.CD$; that is, AB^2 is greater than BC^2+AC^2 by 2BC.CD.

ليكن ا ب س مثلثًا ذا زاوية منفرجة ا س ب وليقع عمود من ا الى ا د على
ب س بعد اخراجهِ الى د (ق ١٢ ك ١) فمربع
ا ب هو اكبر من مربعَي ا س و س ب بمضاعف
القائم الزوايا ب س × س د
فمن حيث ان ب د قد انقسم الى قسمين
في س فلنا (ق ٤ ك ٢) ب د٢ = ب س٢ +
س د٢ + ٢ ب س × س د اضف ا د٢ الى الجانبين فلنا ب د٢ + ا د٢ = ب س٢ +
س د٢ + ا د٢ + ٢ ب س × س د ولكن ا ب٢ = ب د٢ + ا د٢ (ق ٤٧ ك ١) و ا س٢
= س د٢ + ا د٢ فاذًا ا ب٢ = ب س٢ + ا س٢ + ٢ ب س × س د اي ا ب٢ هو اكبر
من ب س٢ + ا س٢ بسطح ٢ ب س × س د

Figure 6: *Elements* II, 12. Top, the demonstration of Playfair [1837, 57]. The enunciation is included to clarify what is being demonstrated. Bottom, the demonstration in Van Dyck [1857, 65–66]. Its formulation is almost identical to Playfair's text, an indication of the care Van Dyck exercised to make his translation as literal as possible.

Playfair did not see any benefit to be gained from the repetition.

Changes to Playfair's Text Made by Van Dyck

In his translation, Van Dyck occasionally makes more substantive changes to Playfair's text.[32] Some of these changes involve omissions. For example, Playfair had begun his treatise by defining or explaining some general mathematical terms — terms such as geometry, definition, axiom — as well as signs for mathematical operations that are not defined explicitly in Euclid's text (because they are foreign to the way in which Euclid developed his mathematics) such as + and = that are used in his treatise. Van Dyck has taken the basic definitions of terms (excluding the definition for "method") into his treatise. He omits the definitions of algebraic symbols, however, noting that these have already been explained in the volume on algebra. Since he had himself completed an Arabic textbook on algebra [1853], he probably has his own treatise in mind.[33] Apparently Van Dyck assumed that his translation of the *Geometry* would be part of a broader mathematics curriculum in which students would first be introduced to algebra, then proceed to geometry. Another example of omission occurs in the translation of Supplement III, proposition 21: Van Dyck does not include an alternative algebraic demonstration found in Playfair's treatise.

Another kind of change to Playfair's text is illustrated in the Scholium to proposition I, 20. Playfair had used algebraic symbols to express the content of this scholium in his *Geometry* [Playfair 1837, 22]. (The scholium was absent from earlier editions of the *Geometry* [Playfair 1806, 23].) In Van Dyck's translation, however, the scholium is given completely verbally [Van Dyck 1857, 23]. And on a few occasions, Van Dyck replaces material from Playfair with new material, apparently of his own composition. For example, in Supplement III, proposition 10, Van Dyck substitutes an algebraic proof for Playfair's verbal demonstration (Figure 7).

Such drastic alterations to Playfair's work are rare, however. More frequently, as we have already noted, Van Dyck introduces more modest modifications to the text. Presumably many of these lightly edited statements are adopted because Van Dyck felt that they would be more easily understood by his students. To illustrate Van Dyck's editing, consider the first three definitions of book I, which Playfair [1846, 8] stated as:

1. A point is that which has position, but not magnitude.
2. A line is length without breadth.
3. If two lines are such that they cannot coincide in any two points, without coinciding altogether, each of them is called a straight line.

Van Dyck [1857, 4] rendered these definitions into Arabic as follows:

١. النقطة شيء له وضع فقط وليس له طول ولا عرض ولا عمق.
٢. الخط طول بدون عرض أو عمق.
٣. خطان لا يتوفقان فى نقطتين منهما بدون أن يتوافقا بالكلية يسميان مستقيمين. وقيل أيضا الخط المستقيم هو البعد الأقرب بين نقطتين.

Translating Van Dyck's Arabic statements literally into English,[34] we have:

1. The point [is] something that has position only and it has *no length nor breadth nor height.*

[32] I have assumed that any changes from Playfair's last (sixth) edition, the most commonly reprinted version, represent editing by Van Dyck. I have usually consulted the New York reprint by Deans and Homer (1846). Other editions and reprintings have been digitized by Google.

[33] The title page is reproduced in Saʿdi [1937, 41]. I have not been able to examine a copy of this treatise. The American University in Beirut Library has digitized several pages from a handwritten treatise on algebra that belonged to Van Dyck (http://ddc.aub.edu.lb/projects/jafet/manuscripts/ms512.V24kA/index.html), but without access to the printed textbook, it is impossible to correlate the two.

[34] All translations are my own unless otherwise noted.

As the base AC to the base KM, so let the straight line KO be to the straight line S. Then, since AC is to KM as KO to S, and also by hypothesis, AC to KM as KO to AE, KO has the same ratio to S that it has to AE (11. 5.); wherefore AF is equal to S (9. 5.). But the solid AG is

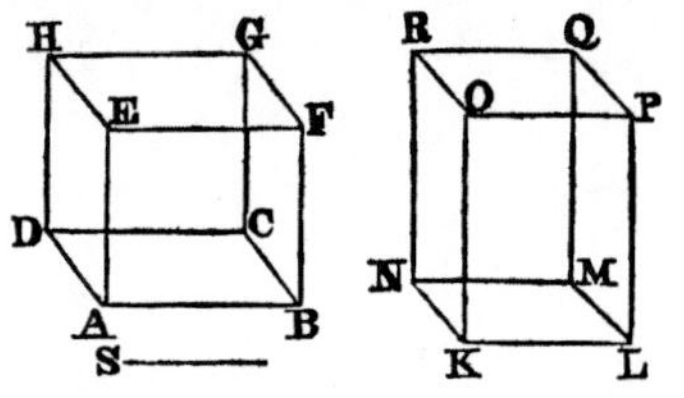

to the solid KQ, in the ratio compounded of the ratios of AE to KO, and of AC to KM (9. 3. Sup.), that is, in the ratio compounded of the ratios of AE to KO, and of KO to S. And the ratio of AE to S is also compounded of the same ratios (def. 10. 5.); therefore, the solid AG has to the solid KQ the same ratio that AE has to S. But AE was proved to be equal to S, therefore AG is equal to KQ.

Again, if the solids AG and KQ be equal, the base AC is to the base KM as the altitude KO to the altitude AE. Take S, so that AC may be to KM as KO to S, and it will be shewn, as was done above, that the solid AG is to the solid KQ as AE to S; now, the solid AG is, by hypothesis, equal to the solid KQ: therefore, AE is equal to S; but, by construction, AC is to KM, as KO is to S; therefore, AC is to KM as KO to AE.

القضية العاشرة·ن

اجسام متوازية السطوح هي متساوية اذا كانت قواعدها وعلوها متناسبة بالتكافؤ· والاجسام المتوازية السطوح المتساوية تكون قواعدها وعلوها متناسبة بالتكافؤ

لتكن نسبة اس : كم :: كو : اى فالجسم اغ = الجسم كق لانهُ بتحويل هذه
النسبة لنا اس×اى = كم
×كو واس×اى = اغ ص
(ق٩ ك٣ م) وكم ×كو
= كق
ثم اذا فُرِض ان اس
×اى = كم×كو لنا
اس : كم :: كو : اى

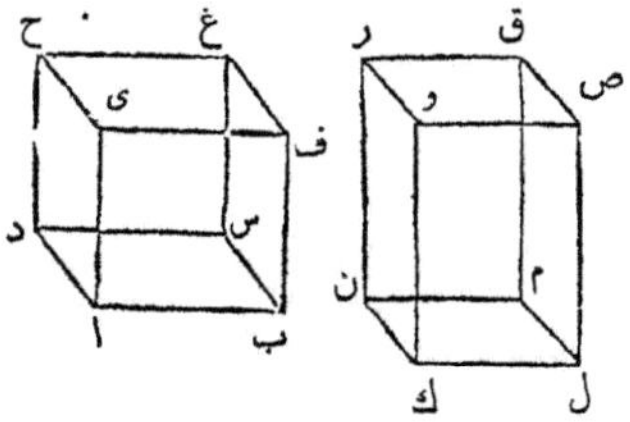

Figure 7: Top, the demonstration of Supplement III, 10 and its diagram according to Playfair [1846, 206]. Bottom, Van Dyck's alternate demonstration [1857, 238]. The demonstration reads (using letter equivalents from Playfair's diagram): "Let ratio $AC : KM :: KO : AE$. Then solid $AG =$ solid KQ because by conversion of this ratio we have $AC \times AE = KM \times KO$ and $AC \times AE = AG$ (III, 9) and $KM \times KO = KQ$. Then since it was specified that $AC \times AE = KM \times KO$ we have $AC : KM :: KO : AE$." Van Dyck's algebraic demonstration no longer requires line S, so it is omitted from the Arabic diagram.

2. The line [is] length without breadth *or height*.
3. Two lines that do not coincide in two of their points without coinciding in totality are called straight. *And one may also say: the straight line is the shortest distance between two points.*

The modifications (explications / additions) to these definitions introduced by Van Dyck are indicated in italic type. Why, we might ask, did Van Dyck introduce these changes into Playfair's text? What might have prompted these specific modifications? In the case of the first two definitions, they do not initially appear to be anything beyond a teacher's explication, making more explicit what is implicit in the English of Playfair. However, the definition of a line undergoes a similar transformation when Legendre's *Éléments de géométrie* was translated into Arabic in Cairo later in the century.[35]

[35] Legendre [1807, 1] formulated the definition as: "La ligne est une longueur sans largeur." This definition is rendered

The addition to the third definition, however, presents a somewhat different case, since it clearly goes beyond simple explanation. Proclus attributed this alternative formulation to Archimedes [Heath 1897, 3; Proclus 1970, 89]. This alternative definition was also used in the Arabic redaction of the *Elements* composed by Muḥyi al-Dīn al-Maghribī (ca.1220–ca. 1283).[36] More recently, the alternative definition was used by Legendre [1807, 1]: "La ligne est le plus court chemin d'un point à un autre." These historical observations, of course, leave open the question of how the statement came to be incorporated into Van Dyck's translation. Could he have been looking at some other textbook that contained this alternative formulation?

Finally, we observe that although Van Dyck's translation is generally a literal rendering of Playfair's text, he was a perceptive translator. He obviously checked the published text of Playfair carefully and corrected occasional printing errors that he found (Figure 8).

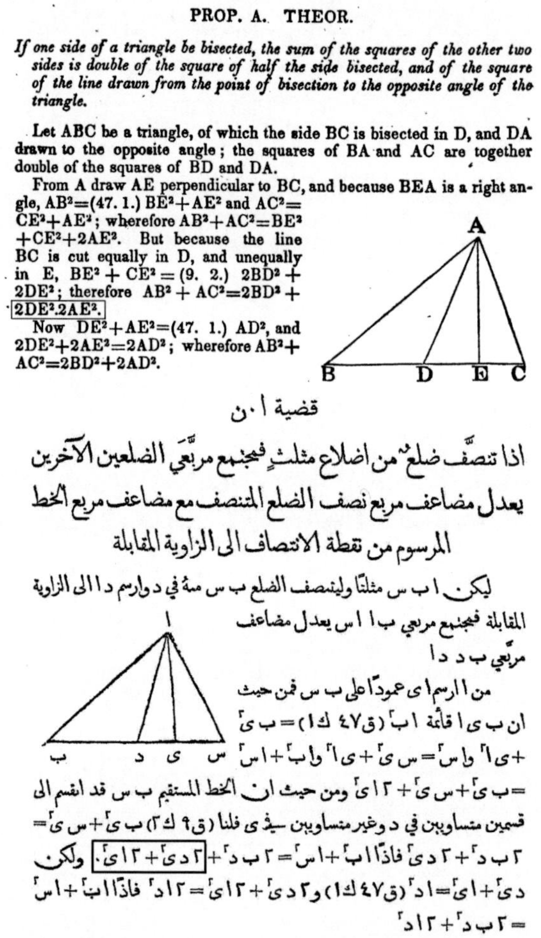
PROP. A. THEOR.

If one side of a triangle be bisected, the sum of the squares of the other two sides is double of the square of half the side bisected, and of the square of the line drawn from the point of bisection to the opposite angle of the triangle.

Let ABC be a triangle, of which the side BC is bisected in D, and DA drawn to the opposite angle; the squares of BA and AC are together double of the squares of BD and DA.

From A draw AE perpendicular to BC, and because BEA is a right angle, AB^2=(47. 1.) BE^2+AE^2 and $AC^2=CE^2+AE^2$; wherefore $AB^2+AC^2=BE^2+CE^2+2AE^2$. But because the line BC is cut equally in D, and unequally in E, BE^2+CE^2 = (9. 2.) $2BD^2+2DE^2$; therefore $AB^2+AC^2=2BD^2+$ $2DE^2.2AE^2$.

Now DE^2+AE^2=(47. 1.) AD^2, and $2DE^2+2AE^2=2AD^2$; wherefore $AB^2+AC^2=2BD^2+2AD^2$.

قضية ا٠ن

اذا تنصّف ضلعٌ من اضلاع مثلثٍ فمجتمع مربّعَي الضلعين الآخرين يعدل مضاعف مربع نصف الضلع المتنصف مع مضاعف مربع الخط المرسوم من نقطة الانتصاف الى الزاوية المقابلة

ليكن ا ب س مثلثًا وليتنصف الضلع ب س منهُ في د وارسم د ا الى الزاوية المقابلة فمجتمع مربعي ب ا ا س يعدل مضاعف مربّعي ب د د ا

من ا ارسم ا ى عمودًا على ب س فمن حيث ان ب ى ا قائمة ا ب٢ (ق٤٧ ك١) = ب ى٢ + ى ا٢ و ا س٢ = س ى٢ + ى ا٢ و ا ب٢ + ا س٢ = ب ى٢ + س ى٢ + ٢ ا ى٢ ومن حيث ان الخط المستقيم ب س قد انقسم الى قسمين متساويين في د وغير متساويين في ى فلنا (ق٩ ك٢) ب ى٢ + س ى٢ = ٢ ب د٢ + ٢ د ى٢ فاذًا ا ب٢ + ا س٢ = ٢ ب د٢ + ٢ د ى٢ + ٢ ا ى٢. ولكن د ى٢ + ا ى٢ = ا د٢ (ق٤٧ ك١) و٢ د ى٢ + ٢ ا ى٢ = ٢ ا د٢ فاذًا ا ب٢ + ا س٢ = ٢ ب د٢ + ٢ ا د٢

Figure 8: Proposition A in the additions to book II. Top, the proposition in Playfair [1837, 59]. Bottom, the proposition in Van Dyck [1857, 68]. A printing error and Van Dyck's correction are highlighted by boxes.

into Arabic [Legendre 1881, 2] as: الخط طول بلا عرض ولا عمق... . Evidently the translator, in this case a native speaker of Arabic, also felt it necessary to make a similar emendation or expanded statement when translating this definition, even though such an explication is not required in either English or French.

[36] See Oxford, Bodleian Library, MS Or 448, fol. 2a.

Which Edition of Playfair's *Geometry* Did Van Dyck Translate?

We do not know how Van Dyck acquired a copy of Playfair's text. Van Dyck traveled to Syria in 1840. If he carried an edition of Playfair's *Geometry* with him on this journey, it would probably have been one of the American reprints. It is possible that Van Dyck was able to request or receive shipments of books during his years in Syria. We also know that Van Dyck returned to America in 1854 for about a year [Saʿdi 1937, 29]. It is therefore possible that, having already published his Arabic treatise on algebra a year earlier and having conceived the idea of complementing it with a treatise on geometry, he took the opportunity to acquire a recent American edition of Playfair during this visit. In any case, editions dated 1856 or later cannot be the source from which Van Dyck is translating since his translation was published in 1857.

Most American reprintings of the *Geometry* were based on the last British edition (1813/1814). A full comparison of the many editions and reprintings during the nineteenth century is beyond the scope of this paper, but I call attention to one important difference between the New York [1846] and Philadelphia [1856] reprint editions as indicative of what still remains unknown about the textual history of Playfair's treatise. The Philadelphia [1856] edition has apparently been typeset anew, since the typeface is slightly larger than that found in New York [1846] edition. More importantly, the diagrams, too, appear to have been redrawn and occasionally modified. For example, the diagram for proposition III, 2 in the Philadelphia [1856] edition contains an additional point not found in the diagram of the New York [1846] edition (Figure 9). This point G is apparently related to a demonstration of the accompanying corollary, but point G is not mentioned in Playfair's text. Van Dyck's diagram for the proposition also does not include point G.

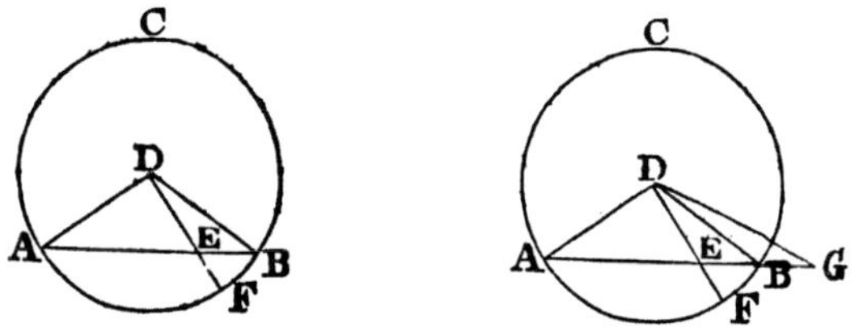

Figure 9: Diagrams of Proposition III, 2 in Playfair's *Geometry* illustrating changes in diagram construction in different editions. Left, the diagram from the New York, 1846 edition, p. 63. Right, the diagram from Philadelphia, 1856 edition, p. 63. Since Van Dyck's translation does not have this altered diagram, he apparently is relying on an earlier edition.

As noted earlier, the placement of cross-references to earlier propositions in earlier editions of the *Geometry* (Philadelphia, 1806; Edinburgh, 1826) were placed in the margins. In later editions, including the American reprints based on the last English edition (New York, 1837 and 1846, etc.) these cross-references were incorporated into the text in parentheses. Van Dyck also includes cross-references parenthetically in his translation, implying that he was using one of these later editions.

Playfair made significant alterations in his text in both the second (1804) and sixth (1813/1814) editions, although these changes were almost all related to the supplementary material, not to the Euclidean books [Ackerberg-Hastings 2002, 52]. The formulation of the "Quadrature of the Circle" (the first of three supplementary sections added by Playfair) in Van Dyck's translation [1857, 195] is the same as that found in several American editions of the *Geomentry* which are based on the last (1814) British edition. This supplement begins with a lemma in the later editions [1837, 163; 1846, 163–164] but in the second edition and the 1826 reprinting of the sixth edition, the section begins with three definitions and an axiom [1804, 207; 1826, 239]. These variants also imply that Van Dyck was using a fairly recent edition (in his day), rather than one of the earlier editions.

Playfair added ten problems at the end of book I, labeled A–K (omitting letter J). These problems were also included by Van Dyck, numbered from "*aliph*" to "*yā*" using the standard *abjad* system [1857,

48–53]. These added propositions were present in both 1837 and 1846 American editions, but absent from the second (Edinburgh, 1804) edition of the *Geometry* as well as from earlier American editions New York (1806), New York (1819), and Philadelphia (1826). Furthermore, the definitions in the second edition were numbered using Roman numerals centered in an otherwise blank line directly above the text of the definition. By the 1837 reprinting, however, the numeration was changed to Arabic numbers and the numbers are placed on the same line as the first line of text of the definition. Van Dyck's Arabic translation follows the numbering pattern found in 1837 and later American editions, which also implies that he was using a fairly recent edition. Perhaps more telling, though, is the number of axioms. In Playfair's second edition (1804) there were only ten numbered axioms, since the statement that became "Playfair's Axiom" was not given a number although it followed axiom ten. In editions following 1837, it has become axiom eleven — the same number assigned to it in Van Dyck's translation [1857, 10]. All these observations also support the hypothesis that Van Dyck was translating from a relatively recent American edition or reprint.

Competition from Other Published Textbooks

Van Dyck's was not the only Arabic geometry textbook to appear in print during the nineteenth century. Editions of the traditional Arabic Euclidean classics were printed in several places throughout the century. The earliest was the *Taḥrīr Kitāb Uqlīdis* by Naṣīr al-Dīn al-Ṭūsī (1176 / 1801) which was typeset in Istanbul.[37] Another thirteenth century classic, the *Ashkāl al-Ta'sīs* of al-Samarqandī, together with its commentary by Qāḍīzāde al-Rūmī, was printed by lithograph in Istanbul (1274 / 1858).[38] The redaction of the *Elements* ascribed to Pseudo-Ṭūsī was printed by lithograph in Fez (1293 / 1876).[39] Finally, al-Ṭūsī's treatise was printed by lithograph in Tehran (1298 / 1881).[40] These works, originally composed centuries earlier, continued to be copied, read, and studied in Islamic education even during the nineteenth century. The existence of printed editions implies that there was a re-awakening of interest in the traditional approaches to mathematics in the Islamic community. Although we have little information about the circulation of these printed editions, they might have provided some competition for Van Dyck's new translation.

Van Dyck's textbook was also not the first translation of European mathematics to appear in the Ottoman Empire. Translation efforts had begun already at the very beginning of the century in Cairo and Istanbul. In Cairo, Muḥammad Surūr translated a treatise on geometry into Arabic and Aḥmad al-Tabbakh translated, *Traité d'arithmétique à l'usage de la marine et de l'artillerie* by E. Bézout (1730–1783), as well as a still-unidentified tract on algebra.[41] These very early translations apparently circulated only in manuscript form. Others, such as the Ottoman Turkish translation of J. Bonnycastle's *Elements of Geometry* (London, 1789) circulated both in manuscript and in print form.[42] Little study has yet been devoted to such translations and the extent to which they may have circulated outside Cairo and Istanbul — the two centers of educational reforms in the Ottoman Empire during the nineteenth century — is unknown.[43]

[37] A digital reproduction is available online: http://nrs.harvard.edu/urn-3:FHCL:2051759.

[38] A digital reproduction is available online: http://pds.lib.harvard.edu/pds/view/11053324.

[39] Two digital reproductions are available online. One has been posted by Harvard University Library: http://pds.lib.harvard.edu/pds/view/11378154. The second (which is not searchable — it can only be read page by page from the beginning) was posted by the Bibliothèque nationale du royaume du Maroc. The link to this online version can be found on this page: http://bnm.bnrm.ma:86/ListeLithographies.aspx?IDC=2. I thank Mahdi Abdeljaouad for calling this online version to my attention.

[40] A digital reproduction is available online: http://www.al-mostafa.com.

[41] These early Arabic translations are currently being studied by Abdeljaouad [2011; 2012; 2013].

[42] Translated by Hüseyin Rıfkı under the title *Usul-i hendese*, it was printed at the Būlāq Press, 1241 / 1825. A digital reproduction is available online: http://reader.digitale-sammlungen.de/resolve/display/bsb10623187.html. At least two manuscript copies exist in Cairo: Dār al-Kutub, handasa turkiyya 42 and Dār al-Kutub, Ṭalʿat, handasa turkiyya 3.

[43] Basic bibliographic information about nineteenth century translations can be collected from İhsanoğlu [1999]. Crozet

Mathematics and Missions

Although Van Dyck is remembered today for his translation of the Bible into Arabic, he also translated and wrote textbooks on the European science and mathematics in Arabic. Although these may initially seem to be very different endeavors, I shall argue that both grew out of the Protestant missionary vision of the ABCFM during the nineteenth century.[44]

The Syrian mission began with a dual vision: to convert non-Christians (mainly Muslims) to the true religion and at the same time to awaken the Eastern churches to a renewed commitment to the Christian gospel, thus thwarting Vatican efforts toward reunification of the Catholic and Orthodox communions. "Prophesy, history, and the present state of the world, seem to unite in declaring, that the great pillars of the Papal and Mahommedan impostures are now tottering to their fall ... Now is the time for the followers of Christ to come forward ... and engage ... in the great work of enlightening and reforming mankind" [Morse, Worcester and Evarts 1811, 28]. And, as the governing committee of the ABCFM instructed a new missionary about to leave for Turkey in 1839, his task was "not to subvert them (the Oriental churches); not to pull down and build anew. It is to reform them; to revive among them ... the knowledge and spirit of the gospel" [Walker 1967, 215]. Lindner [2009, 32-41] summarizes how this sense of mission was born from the confluence of the Great Awakening and its accompanying religious revival among New England Puritan Protestants, the rising American political power and ascendency in the world, together with a strong injection of optimistic and eschatological millenarian thought. These American missionaries dedicated their lives and efforts to bringing "spiritual and worldly reform in foreign lands through an ambitious amalgam of pedagogy, philanthropy and politics" [Elshakry 2007, 174]. Their vision was "a world transformed ... along the lines of an idealized New England community" [Andrew 1978, 331].

But despite the continued efforts of the missionaries there were very few conversions in which individuals officially and openly changed their religious affiliation, nor was there obvious change or development in social or political structures [Peculiar Obstacles 1848, 139–140]. Faced with this disappointing situation, the mission team focused on a new strategy — education.[45] The mission saw the introduction of modern education as essential to any "civilizing" of the Ottoman socio-political system and to preparing the way for the benefits of Christianity to become apparent. Their strategy called for opening boarding schools that would teach students the essentials of modern science and mathematics, something almost completely lacking in traditional schools of the Ottoman Empire, at least according to the missionaries.[46] Religious instruction, of course, would accompany the academic curriculum of the mission schools. As one missionary wrote, the plan was to "bait the hook with arithmetic" [Station Report 1858, 142]. The graduates of these schools "upon returning to their native societies ... could then instruct their own people ..., acting as adjuncts to the American missionaries" [Andrew 1978, 332].

Their strategy of indirect evangelism through education appeared to be effective — at least the number of students in mission schools increased rapidly. The missionaries had been right in their

[2008] gives a comprehensive analysis of the situation in Egypt.

[44] The Protestant Mission to Syria has seen considerable study in recent years. Lindner [2009] suggests that the Protestant Mission and their converts were often isolated, both geographically and culturally, from the Muslim inhabitants of Beirut. Although this may often have been the norm, several missionaries, including Van Dyck, went to considerable lengths to learn Arabic and identified with local culture to the extent that they adopted local dress. Porter's analysis [2004] of the role of Protestant missions vis-à-vis nineteenth century British colonial administration is much broader, although he provides useful insights into the political relations of the ABCFM mission in Syria, since the American missionaries often had to depend on and cooperate with British agents for protection.

[45] It was not a totally new strategy. The ABCFM had earlier attempted to use education to "civilize" native American Indians at the short-lived Foreign Mission School. The episode raised difficulties for missionary efforts among the native Americans and the resulting controversy created deep rifts in the ranks of ABCFM supporters [Andrew 1978]. The lessons of this failed experiment must have been pondered by both the missionaries and the commissioners of the ABCFM as the new strategy unfolded.

[46] Fortna [2002] and Somel [2001] do not paint quite so bleak a picture of Ottoman education. But the missionaries were primarily describing their experience in Beirut, which might not be typical of Ottoman education as a whole.

analysis — there was a considerable hunger for modern education, especially among the growing middle class. But the result, when judged in terms of number of actual conversions, or even socio-political change, remained disappointing. Many students came to be educated, listened to the required religious instruction, attended the prescribed prayers and church services, but made no change in their religious affiliation [Extracts 1837, 445]. The lack of converts deeply troubled Rufus Anderson, general secretary of the ABCFM. As costs of staffing and maintaining the mission in Beirut escalated, he felt it necessary to admonish the missionaries that the "governing object of every mission and of every missionary should not be to liberate, to educate, to enlightened, to polish, but to *convert* men" [Makdisi 1997, 682].

The attempt to evangelize through education was not merely a desperate grasping at straws. Rather, it was part and parcel with the Anglo-American epistemological ideas underlying liberal education in the nineteenth century. One of the distinctive features of these epistemological views was a conviction that mathematics — especially geometry — was essential in training the mind to reach correct conclusions. As Olson [1971, 31] remarks: "Scottish moral philosophers felt bound to study and lecture on the foundations of mathematics for ... they felt that they had only the patterns of mathematics and natural philosophy to indicate how certain knowledge might be obtained." Following a similar vein, John Leslie, who succeeded Playfair in the Chair of Mathematics at Edinburgh, wrote in the preface to his own textbook on geometry [1809, v–vi]: "The study of Mathematics holds forth two capital objects:–While it traces the beautiful relations of figure and quantity, it likewise accustoms the mind to the invaluable exercise of patient attention and accurate reasoning. Of these distinct objects, the last is perhaps the most important in a course of general education. For this purpose, the geometry of the Greeks is the most powerfully recommended..."

Leslie's vision of mathematics as essential to liberal education, a typical nineteenth century pedagogical outlook, was compatible with Scottish Common Sense Philosophy and consistent with views promulgated by such Scottish luminaries as Thomas Reid (1710–1796) and Dugald Stewart (1753–1828). A parallel view of the importance of science and mathematics in liberal education was developing in Britain, associated with the name of Lord Brougham (1778–1868). In his *Discourse on the Objects, Advantages and Pleasures of Science*, which set out many reasons why one should study science, Brougham had stated [1828, 181] that "the highest of all our gratifications in the contemplation of sciences remains: we are raised by them to an understanding of the infinite wisdom and goodness which the Creator has displayed in his works." Brougham was neither a scientist nor a mathematician. Like his illustrious predecessor, Francis Bacon, he was a politician and served as Lord High Chancellor. Convinced that public morality in England had declined enormously, he founded the Society for the Propagation of Useful Knowledge, whose purpose was to promote morality through secular education — especially in mathematics and the sciences. This education was not intended to be so much institutional as individual and thus depended on textbooks designed to be read and studied individually. Brougham combined forces with a young mathematician, Augustus de Morgan (1806–1871), to carry out the production of mathematics textbooks appropriate to his project. Phillips [2005] succinctly summarizes some basic features of mathematics pedagogy in nineteenth century Britain. At the beginning of the century, the focus was primarily rote memorization, limiting the amount of mathematics that was taught. Whewell at Cambridge and De Morgan at University of London spearheaded the drive to introduce more mathematics into the curriculum. But this would only be possible if there were properly trained teachers and appropriate textbooks. Thus De Morgan devoted himself to creating mathematics textbooks geared to the independent student. His program included detailed advise on how the learner should build up a balanced understanding of mathematics, with the ultimate goal of developing the student's rational abilities.

A belief that mathematics was essential in liberal education also spread to the United States in the nineteenth century and became a dominant pedagogical ideology by mid-century.[47] In fact, during the nineteenth century, many American educational institutions shaped their curricula around moral philosophy. As Dorrien [2009, 11] remarks: "moral philosophy was a remarkably uniform enterprise in American schools because one school of philosophy dominated American education: Scottish com-

[47] Redekop [2004, 327–335] outlines the influence of Thomas Reid and Scottish Realism on American religion, moral philosophy, and education.

monsense realism." The Scottish influence in American education was not just in terms of curricula and textbooks, but also through a steady stream of teachers who emigrated from Edinburgh to America to hold key positions in the new universities and schools. They became the leading voices of American education. No doubt the dominance of Scottish realism in American education also contributed to the prodigious popularity of Playfair's *Geometry* in America during the middle decades of the nineteenth century.

It is in the context of the Protestant missionary use of education as a tool for evangelism and against the backdrop of Anglo-American views of mathematics in liberal education, I believe, that we must seek the stimulus for Van Dyck's translation efforts. His lifelong career in Syria seems to have had two foci: to use his medical skills to relieve suffering and to teach Syrian students the essential advances in scientific knowledge they would need in order to take their place as citizens of the modern world. Van Dyck early seems to have realized that the most effective path was to teach students in their own language, rather than force them to learn a new language. So he himself learned Arabic and taught his students in Arabic. But without suitable textbooks, his students found it difficult to study effectively or advance rapidly. Hence Van Dyck's life-long effort to provide science and mathematics textbooks in Arabic for use in the schools.

Although Van Dyck does not reveal why he chose Playfair's *Geometry* for his translation, we can make some guesses. Van Dyck had himself been educated in New England, where the influences of Scottish Realism were at their strongest. New England Protestants were, overall, sympathetic to the Scottish ideals of liberal and "democratic" education, practical education that emphasized pragmatic or utilitarian components rather than abstract theory. These views not only resonated with the pedagogical ideals of the Scottish Common Sense movement, but they were also considerations in the forefront of Playfair's mind as he worked on his *Geometry*. It is quite possible that Van Dyck himself may have studied geometry as a tool for right thinking from an early American edition of Playfair's text. The experience, together with his practical medical training and his personal commitment to missionary ministry in Syria may all have contributed to his choice.

Van Dyck's translation of Playfair's *Geometry* was part of a larger educational endeavor that saw him produce more than a dozen textbooks in Arabic. The first of these textbooks were in mathematics — algebra and geometry.[48] Over the course of his career, Van Dyck produced more than a dozen textbooks in Arabic (it is not always clear whether they were translations or original compositions) introducing many aspects of modern science. The parallels between the Arabic writings of Van Dyck and the publication program of the Society for the Diffusion of Useful Knowledge in nineteenth century Britain are worth noting. Both endeavors grew from the same soil — a desire to reform society through education, and especially an education that will inculcate in the student an ability to reason rightly to reach sound moral decisions. And in both cases, mathematics is foundational because it provides one of the clearest examples of the practice of reasoning and argumentation. Thus we should see Van Dyck's production of Arabic textbooks not merely as a local or individual phenomenon, but as informed and shaped by broader changes in mission strategy and in pedagogical practices throughout the English-speaking world.

Van Dyck's Translation: Influence

We can as yet say little about the direct or indirect influence of Van Dyck's Arabic version of Playfair's *Geometry*. It is impossible to say how many copies were originally printed or how often the translation was reprinted. In addition to the 1857 edition, the catalog of Harvard University Library lists another edition dated 1889. The textbook, according to a later report of dubious veracity, continued to be printed

[48] I have been unable to locate a copy of Van Dyck's algebra textbook, published 1853. Some hints of its contents may be obtained from a manuscript of Van Dyck on algebra. Only a few digital images have been posted online: http://ddc.aub.edu.lb/projects/jafet/manuscripts/ms512.V24kA/index.html. The relation of this manuscript to Van Dyck's printed textbook is still unclear.

and used in the Ottoman Empire and perhaps elsewhere in the Arabic-speaking world long after Arabic instruction had been abandoned at the Syrian Protestant College. If correct, this reported publication information would indicate a continuing popularity for Van Dyck's translation.[49]

In 1886, the Protestant Mission in Beirut claimed to have the largest publishing house in the region. In addition to religious materials, there was a steady outpouring of Arabic textbooks on the sciences and mathematics [Grabill 1971, 55]. Demand for textbooks in the sciences is reported to have been high throughout the eastern Mediterranean and from as far away as China.[50] Although it is difficult to estimate accurately the circulation of these textbooks, it is intriguing that the digital copy of Van Dyck's translation of Playfair's *Geometry* currently available online was owned by the Andhra Pradesh State Central Library in India (Figure 1).

This science publishing, although no doubt lucrative, had originally developed as an outgrowth of American missionaries' use of modern mathematics and science to draw students into mission schools where they could at the same time be exposed to religious instruction. As a proselytizing strategy, the plan proved ineffective, but it did promote production of Arabic textbooks on modern sciences which were published by the mission press. These textbooks, including Van Dyck's translation of Playfair, although produced in response to the burgeoning interest in the new sciences and the desire to create converts to American Protestant Christianity, played an unanticipated role in stimulating interest in and receptivity toward the liberal ideas sweeping across the Muslim world during the nineteenth century.[51]

References

Abdeljaouad, M., 2011. The first Egyptian modern mathematics textbook. International Journal for History of Mathematics Education 6(ii), 1–11.

—— 2012. Teaching European mathematics in the Ottoman Empire during the eighteenth and nineteenth centuries: between admiration and rejection. Zentralblatt für Didaktik der Mathematik 44, 483–498.

—— 2013. L'importance des manuels de Bézout dans le transfert des mathématiques européennes en Turquie et en Égypte au XIXe siècle. In: Barbin, E., Moyon, M. (eds.), Les ouvrages de mathématiques dans l'histoire – entre recherche, enseignement et culture. Presses Universitaires de Limoges, Limoges, pp. 149–160.

Ackerberg-Hastings, A., 2002. Analysis and synthesis in John Playfair's Elements of Geometry. British Journal for History of Science 38, 43–72.

—— 2008. John Playfair on British decline in mathematics. British Society for History of Mathematics Bulletin 23, 81–95.

Andrew, J., 1978. Educating the heathen: The Foreign Mission School controversy and American ideals. Journal of American Studies 12, 331–342.

Barany, M., 2010. Translating Euclid's diagrams into English, 1551–1571. In: Heeffer, A., Van Dyck,

[49] The picture accompanying the report, however, raises doubts concerning the report's veracity (the image can be seen online: http://math.arizona.edu/~dido/usul-euclid.jpg — accessed 4 February 2012). The image appears to have been altered using digital technology since the page length is exactly half that of the original printing [Van Dyck 1857, p. 37]. It would be unusual to find a printed treatise with the dimensions implied by the boundary lines surrounding the text in this illustration. The running header has also been significantly altered from that in the original edition (Figure 2). Because of the odd construction (in terms of both grammar and physical arrangement) of this header, we should probably presume that it, too, has been created using digital technology, perhaps from parts of the title page of the original treatise.

[50] Elshakry [2007, 198] quotes the dramatic statement from the 1868 annual report of the ABCFM (p. 47): "The cry comes from Egypt and Palestine, from Assyria and Northern Africa, and even from Peking, the capital of China, 'Give us Arabic books!'"

[51] My interpretation follows Elshakry [2007]. Makdisi [2008, 15–28] argues that the Protestant Mission in some ways epitomized the confrontation between two colonial powers, the missionary "determination to refashion the world on evangelical terms at a time of ascendant Anglo-American power" and the Ottoman "imperial tradition based on the accommodation of religious difference" that developed in the pre-Tanzimat era.

M. (eds.), Philosophical Aspects of Symbolic Reasoning in Early Modern Mathematics. College Publications, London.
Barrow-Green, J., 2006. Much necessary for all sortes of men: 450 years of Euclid's Elements in English. British Society for History of Mathematics Bulletin 21, 2–25.
Brentjes, S., 2002. Arab countries, Turkey, and Iran. In: Dauben, J., Scriba, C. (eds.), Writing the History of Mathematics: Its Historical Development. Birkhäuser, Basel, pp. 317–328.
Brougham, H., 1828. A Discourse on the Objects, Advantages, and Pleasures of Science. Baldwin and Craddock, Londom
Challinor, J., 1975. Playfair, John. In: C.C. Gillispie (ed.), Dictionary of Scientific Biography, vol. 11. Scribner's, New York, pp. 34–36.
Crozet, P., 2008. Les sciences modernes en Egypte: Transfert et appropriation. Geuthner, Paris.
De Young, G., 2012. Nineteenth century traditional Arabic geometry textbooks. International Journal for History of Mathematics Education 7(ii), 1–34.
Dorrien, G., 2009. Social Ethics in the Making: Interpreting an American Tradition. Blackwell, Chichester.
Easton, J., 1975. Recorde, Robert. In: Gillispie, C.G. (ed.), Dictionary of Scientific Biography, vol. 11. Scribner's, New York, pp. 338–340.
Elshakry, M., 2007. The gospel of science and American evangelism in late Ottoman Beirut. Past and Present 196, 173–214.
——— 2008. Knowledge in motion: The cultural politics of modern science translation in Arabic. Isis 99, 701–730.
Extracts = Anonymous, 1837. Extracts from a joint letter of the missionaries at Beyroot. Missionary Herald 33, 443–447
Fortna, B.C., 2002. Imperial Classroom: Islam, State, and Education in the Late Ottoman Empire. Oxford University Press, Oxford.
Goldstein, J., 2000. A matter of great magnitude: The conflict over arithmetization in 16th and 17th century English editions of Euclid's Elements books I–VI (1561–1795). Historia Mathematica 27, 36–53.
Grabill, J.L., 1971. Protestant Diplomacy and the Near East: Missionary Influence on American Policy, 1820–1927. University of Minnesota Press, Minneapolis.
Heath, T.L., 1956. The Thirteen Books of Euclid's Elements. 2nd edition. Dover, New York. 3 volumes.
——— 1897. The Works of Archimedes. Cambridge University Press, Cambridge. Reprinted: Dover, New York, 2002.
İhsanoğlu, E., et al., 1999. Osmanli Mathematik Literatürü Tarihi. IRCICA, Istanbul.
Jeha, S., 1991. Dārwīn wa-azmat 1882 bi-l-Dāʾira al-Ṭibbiyya wa-l-awwal thawrah ṭullābiyyah fī-l-ʿālam al-ʿarabī bi-l-Kulliyya al-Ṣūriyya al-Injīlliyya. American University of Beirut Press, Beirut.
Kaya, S, 2005. Darwin and the Crisis of 1882 in the Medical Department and the First Student Protest in the Arabic World in the Syrian Protestant College. American University of Beirut, Beirut.
Khūrī, Y., 1991. Al-duktūr Kurnīlyūs Van Dayk wa-naḍa al-dayār al-shāmīya al-ʿalamīya al-qarn al-tāsiʿ ʿashr. Al-farāt li-l-nashr wa-l-tawzīʿ, Beyrut.
Legendre, A.-M., 1807. Éléments de géométrie, septieme édition. Didot, Paris.
——— 1288 [1881]. Uṣūl al-Handasah. ʿAlī Effendī ʿEzzat, (trans.). Revised: Ibrahīm ʿAbdel Ghafār, Taḥdhīb. Būlāq Press, Cairo.
Leslie, J., 1809. Elements of Geometry, Geometrical Analysis, and Plain Trigonometry. J. Ballantyne, Edinburgh.
Lindner, C., 2009. Negotiating the Field: American Protestant Missionaries in Ottoman Syria, 1823–1860. Edinburgh University PhD dissertation, Edinburgh.
Makdisi, U., 1997. Reclaiming the land of the Bible: Missionaries, secularism, and evangelical modernity. American Historical Review 102, 680–713.
——— 2008. The question of American liberalism and the origins of the American Board Mission to the Levant and its historiography. In: Schumann, C. (ed.), Liberal Thought in the Eastern Mediterranean, Late Nineteenth Century until the 1960s. Brill, Leiden, pp. 13–28.

Morse, J., Worcester, S., Evarts, J., 1811. Address to the Christian Public. In: First Ten Annual Reports of the American Board of Commissioners for Foreign Missions. Crocker and Brewster, Boston, 1834, pp. 25–30.
Olson, R., 1971. Scottish philosophy and mathematics, 1750–1830. Journal of the History of Ideas 32, 29–44.
Peculiar Obstacles = Anonymous, 1848. Peculiar obstacles in Syria. Missionary Herald 54, 139–140.
Phillips, C., 2005. Augustus De Morgan and the propagation of moral mathematics. Studies in History and Philosophy of Science 36, 105–133.
Playfair, J., 1804. Elements of Geometry, Containing the First Six Books of Euclid With a Supplement on The Quadrature of the Circle and The Geometry of Solids. Bell & Bradfute, Edinburgh.
——— 1806. Elements of Geometry, Containing the First Six Books of Euclid With a Supplement on The Quadrature of the Circle and The Geometry of Solids. F. Nichols, Philadelphia.
——— 1816. Dissertation Second: Exhibiting the Progress of the Mathematical and Natural Sciences since the Revival of Letters in Europe. In: Playfair, J. G. (ed.), 1822. The Works of John Playfair. Four volumes. A. Constable, Edinburgh. Reprint by: Arno Press, New York, 1975. First published in: Supplement Volume of Encyclopaedia Britannica, 4th, 5th, 6th editions.
——— 1819. Elements of Geometry, Containing the First Six Books of Euclid With a Supplement on The Quadrature of the Circle and The Geometry of Solids. G. Long, New York.
——— 1826. Elements of Geometry, Containing the First Six Books of Euclid With a Supplement on The Quadrature of the Circle and The Geometry of Solids, To Which are Added Elements of Plane and Spherical Trigonometry. Marot & Walter, Philadelphia.
——— 1837. Elements of Geometry, Containing the First Six Books of Euclid With a Supplement on The Quadrature of the Circle and The Geometry of Solids. W. E. Dean, New York.
——— 1846. Elements of Geometry, Containing the First Six Books of Euclid With a Supplement on The Quadrature of the Circle and The Geometry of Solids. W. E. Dean, New York.
——— 1856. Elements of Geometry, Containing the First Six Books of Euclid With a Supplement on The Quadrature of the Circle and The Geometry of Solids. J. B. Lippencott, Philadelphia.
Porter, A., 2004. Religion versus Empire: British Protestant Missionaries and Overseas Expansion, 1700–1914. Manchester University Press, Manchester.
Proclus, 1970. Commentary on the First Book of Euclid's Elements. Morrow, G., (trans.). Princeton University Press, Princeton.
Rafeq, A., 2008. The Syrian university and the French mandate. In: Schuman, C. (ed.), Liberal Thought in the Eastern Mediterranean: Late 19th Century until the 1960s. Brill, Leiden, pp. 75–98.
Recent Intelligence = Anonymous, 1846. Recent intelligence – Syria. Missionary Herald 42, 211–212.
Recorde, R., 1551. The Pathway to Knowledge, containing the First Principles of Geometry. London.
Redekop, B., 2004. Reid's influence in Britain, Germany, France, and America. In: Cuneo, T., van Woudenberg, R. (eds.), The Cambridge Companion to Thomas Reid. Cambridge, Cambridge University Press, pp. 313–340.
Saʿdi, L., 1937. Al-Hakîm Cornelius Van Alen Van Dyck (1818–1895). Isis 27, 20–45.
Sarkīs, Y., 1928–1930. Muʿjam al-Maṭbūʿāt al-ʿarabiyya wa al-muʿarabba. Cairo.
Scheglova, O., 1999. Lithograph versions of Persian manuscripts of Indian manufacture in the nineteenth century. Manuscripta Orientalia 5, 12–22.
Simson, R., 1781. The Elements of Euclid, viz. The First Six Books, together with the Eleventh and Twelfth. J. Balfour, Edinburgh. (First edition, 1756, Sixth edition, 1781.)
Somel, S.A., 2001 Modernization of Public Education in the Ottoman Empire 1839–1908: Islamization, Autocracy, and Discipline. Brill, Leiden.
Station Reports = Anonymous, 1858. Station reports – Syrian Mission – Turkey. Missionary Herald 54, 140–143.
Tibawi, A.L., 1967. The genesis and early history of the Syrian Protestant College, Part II. Middle East Journal 21, 199–212.
Van Dyck, C., 1857. Kitāb fī uṣūl al-handasah. Bayrūt.
——— 1853. Kitāb al-rauḍat al-zahriyyah fī al-uṣūl al-jabariyyah. Bayrūt.

Walker, M., 1967. The American Board and the Oriental churches: A brief survey of policy based on official documents. International Review of Missions 56, 214–223.

Williams, C.P., 1982. Healing and evangelism: The place of medicine in late Victorian Protestant missionary thinking. In: Shiels, W. (ed.), 1982, The Church and Healing. Blackwell, London, pp. 271–285.

Wood, P., 2004. Thomas Reid and the culture of science. In: Cuneo, T., van Woudenberg, R. (eds.), The Cambridge Companion to Thomas Reid. Cambridge University Press, Cambridge, pp. 53–76.

Part III
The Story of π

The Life of π: From Archimedes to ENIAC and Beyond

Jonathan M. Borwein

Abstract The desire to understand π, the challenge, and originally the need, to calculate ever more accurate values of π, the ratio of the circumference of a circle to its diameter, has captured mathematicians — great and less great — for many centuries. And, especially recently, π has provided compelling examples of computational mathematics. π, uniquely in mathematics, is pervasive in popular culture and the popular imagination. In this paper, I intersperse a largely chronological account of π's mathematical and numerical status with examples of its ubiquity.

Preamble: π and Popular Culture

The desire to understand π, the challenge, and originally the need, to calculate ever more accurate values of π, the ratio of the circumference of a circle to its diameter, has challenged mathematicians — great and less great — for many centuries and, especially recently, π has provided compelling examples of computational mathematics. π, uniquely in mathematics, is pervasive in popular culture and the popular imagination.[1]

I shall intersperse this largely chronological account of π's mathematical status with examples of its ubiquity. More details will be found in the selected references at the end of the chapter — especially in *Pi: A Source Book* [Berggren, Borwein and Borwein 2004]. In Berggren, Borwein and Borwein [2004] all material not otherwise referenced may be followed up, as may much other material, both serious and fanciful. Other interesting material is to be found in Eymard and Lafon [2003], which includes attractive discussions of topics such as continued fractions and elliptic integrals.

Fascination with π is evidenced by the many recent popular books, television shows, and movies — even perfume — that have mentioned π. In the 1967 *Star Trek* episode "Wolf in the Fold," Kirk asks, "Aren't there some mathematical problems that simply can't be solved?" And Spock "fries the brains" of a rogue computer by telling it, "Compute to the last digit the value of π." The May 6, 1993 episode of The Simpsons has the character Apu boast, "I can recite π to 40,000 places. The last digit is one." (See Figure 1.)

In November 1996, MSNBC aired a Thanksgiving Day segment about π, including that scene from Star Trek and interviews with the present author and several other mathematicians at Simon Fraser University. The 1997 movie *Contact*, starring Jodie Foster, was based on the 1986 novel by noted astronomer Carl Sagan. In the book, the lead character searched for patterns in the digits of π, and after her mysterious experience found sound confirmation in the base-11 expansion of π. The 1997 book *The Joy of Pi* [Blatner 1997] has sold many thousands of copies and continues to sell well. The

This paper is an updated and revised version of Borwein [2008], and is made with permission of the editor.

[1] The *MacTutor* website, http://turnbull.mcs.st-and.ac.uk/history/, at the University of St. Andrews — my home town in Scotland — is rather a good accessible source for informal mathematical history.

N. Sidoli and G. Van Brummelen (eds.), *From Alexandria, Through Baghdad: Surveys and Studies in the Ancient Greek and Medieval Islamic Mathematical Sciences in Honor of J.L. Berggren*, DOI 10.1007/978-3-642-36736-6_24,
© Springer-Verlag Berlin Heidelberg 2014

Figure 1: Around 250 BCE, Archimedes of Syracuse (287–212 BCE) was the first to show that the "two possible π's" are the same. Clearly for a circle of radius r and diameter d, **Area**$= \pi_1\, r^2$ while **Perimeter** $= \pi_2\, d$, but that $\pi_1 = \pi_2$ is not obvious, and is often overlooked (see Fig. 4). Courtesy of Guilio Einaudi Editori.

1998 movie entitled *Pi* began with decimal digits of π displayed on the screen. And in the 2003 movie *Matrix Reloaded*, the Key Maker warns that a door will be accessible for exactly 314 seconds, a number that *Time* speculated was a reference to π.

As a forceable example, imagine the following excerpt from Eli Mandel's 2002 Booker Prize winning novel *Life of Pi* being written about another transcendental number:

```
My name is
Piscine Molitor Patel
known to all as Pi Patel.
```

For good measure I added

$$\pi = 3.14$$

and I then drew a large circle which I sliced in two with a diameter, to evoke that basic lesson of geometry.

Equally, National Public Radio reported on April 12, 2003 that novelty automatic teller machine withdrawal slips, showing a balance of $314, 159.26, were hot in New York City. One could jot a note on the back and, apparently innocently, let the intended target be impressed by one's healthy savings account. Scott Simon, the host, noted the close resemblance to π. Correspondingly, according to the *New York Times* of August 18 2005, Google offered exactly "14, 159, 265 New Slices of Rich Technology" as the number of shares in its then new stock offering. Likewise, March 14 in North America has become π *Day*, since in the USA the month is written before the day (314). In schools throughout North America, it has become a reason for mathematics projects, especially focussing on π.

As another sign of true legitimacy, on March 14, 2007 the *New York Times* published a crossword in which to solve the puzzle, one had first to note that the clue for 28 DOWN was "March 14, to Mathematicians," to which the answer is PIDAY. Moreover, roughly a dozen other characters in the puzzle are PI — for example, the clue for 5 DOWN was "More pleased" with the six character answer HAPπER. The puzzle is reproduced in Borwein and Bailey [2008].

It is hard to imagine e, γ or $\log 2$ playing the same role. A corresponding scientific example [von Baeyer 2003, 11] is

> A coded message, for example, might represent gibberish to one person and valuable information to another. Consider the number 14159265... Depending on your prior knowledge, or lack thereof, it is either a meaningless random sequence of digits, or else the fractional part of π, an important piece of scientific information.

For those who know *The Hitchhiker's Guide to the Galaxy*, it is amusing that 042 occurs at the digits ending at the fifty billionth decimal place in each of π and $1/\pi$ — thereby providing an excellent answer to the ultimate question, "What is forty-two?" A more intellectual offering is "The Deconstruction of Pi" given by Umberto Eco on page three of his 1988 book *Foucault's Pendulum* [Berggren, Borwein and Borwein 2004, 658].

π

Our central character

$$\pi = 3.14159265358979323...$$

is traditionally defined in terms of the area or perimeter of a unit circle; see Figure 1. The notation of π itself was introduced by William Jones in 1737, replacing "p" and the like, and was popularized by Leonhard Euler who is responsible for much modern nomenclature. A more formal modern definition of π uses the first positive zero of sin defined as a power series. The first thousand and one decimal digits of π are recorded in Figure 2.

3 . 1415926535897932384626433832795028841971693993751058209749445923078164062862089986280348253421170679
8214808651328230664709384460955058223172535940812848111745028410270193852110555964462294895493038196
4428810975665933446128475648233786783165271201909145648566923460348610454326648213393607260249141273
7245870066063155881748815209209628292540917153643678925903600113305305488204665213841469519415116094
3305727036575959195309218611738193261179310511854807446237996274956735188575272489122793818301194912
9833673362440656643086021394946395224737190702179860943702770539217176293176752384674818467669405132
0005681271452635608277857713427577896091736371787214684409012249534301465495853710507922796892589235
4201995611212902196086403441815981362977477130996051870721134999999837297804995105973173281609631859
5024459455346908302642522308253344685035261931188171010003137838752886587533208381420617177669147303
59825349042875546873115956286388235378759375195778185778053217122680661300192787661119590921642019893

Figure 2: 1,001 Decimal Digits of π.

Despite continuing rumors to the contrary, π is not equal to 22/7 (see End Note 1). Of course 22/7 is one of the early continued fraction approximations to π. The first six convergents are

$$3, \frac{22}{7}, \frac{333}{106}, \frac{355}{113}, \frac{103993}{33102}, \frac{104348}{33215}.$$

The convergents are necessarily good rational approximations to π. The sixth differs from π by only $3.31\ 10^{-10}$. The corresponding simple continued fraction starts

$$\pi = [3, 7, 15, 1, 292, 1, 1, 1, 2, 1, 3, 1, 14, 2, 1, 1, 2, 2, 2, 2, 1, 84, 2, 1, 1, ...],$$

using the standard concise notation. This continued fraction is still very poorly understood. Compare that for e which starts

$$e = [2, 1, 2, 1, 1, 4, 1, 1, 6, 1, 1, 8, 1, 1, 10, 1, 1, 12, 1, 1, 14, 1, 1, 16, 1, 1, 18, ...].$$

A proof of this observation shows that e is not a quadratic irrational since such numbers have eventually periodic continued fractions.

Archimedes' famous computation discussed below is:

$$3\frac{10}{71} < \pi < 3\frac{10}{70}. \quad (1)$$

Figure 3 shows this estimate graphically, with the digits shaded modulo ten; one sees structure in 22/7, less obviously in 223/71, and not in π.

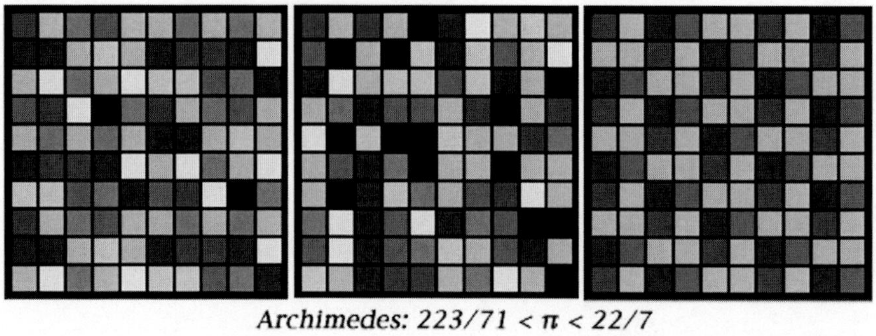

Figure 3: A pictorial proof of Archimedes.

The Childhood of π

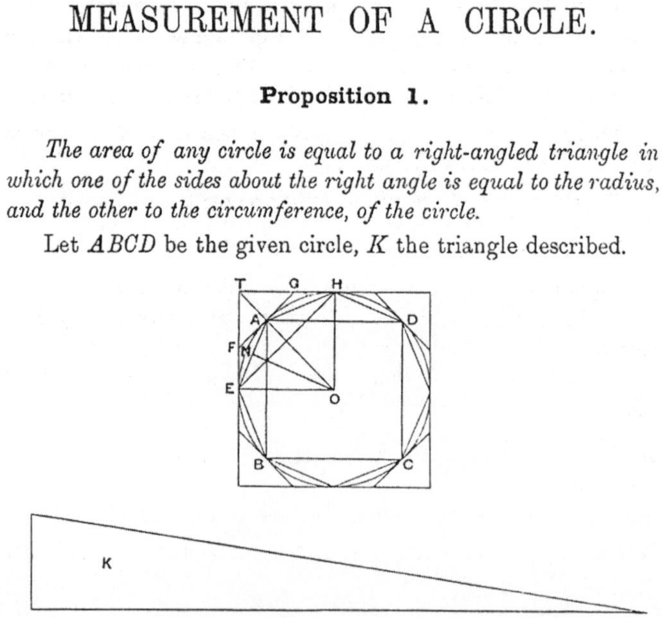
MEASUREMENT OF A CIRCLE.

Proposition 1.

The area of any circle is equal to a right-angled triangle in which one of the sides about the right angle is equal to the radius, and the other to the circumference, of the circle.

Let $ABCD$ be the given circle, K the triangle described.

Figure 4: Construction showing the uniqueness of π, taken from Archimedes' *Measurement of a Circle*. [Heath 1912, 91].

Four thousand years ago, the Babylonians used the approximation $3\frac{1}{8} = 3.125$. Then, or earlier, according to ancient papyri, Egyptians assumed a circle with diameter nine has the same area as a square of side eight, which implies $\pi = 256/81 = 3.1604\ldots$ Some have argued that the ancient Hebrews were satisfied with $\pi = 3$:

> Also, he made a molten sea of ten cubits from brim to brim, round in compass, and five cubits the height thereof; and a line of thirty cubits did compass it round about. (I Kings 7:23; see also II Chronicles 4:2)

One should know that the cubit was a personal not universal measurement. In Judaism's further defense, several millennia later, the great Rabbi Moses ben Maimon Maimonedes (1135–1204) is translated by Langermann, in *The True Perplexity* [Berggren, Borwein and Borwein 2004, 753] as fairly clearly asserting π's irrationality.

> You ought to know that the ratio of the diameter of the circle to its circumference is unknown, nor will it ever be possible to express it precisely. This is not due to any shortcoming of knowledge on our part, as the ignorant think. Rather, this matter is unknown due to its nature, and its discovery will never be attained. (Maimonedes)

In each of these three cases the interest of the civilization in π was primarily in the practical needs of engineering, astronomy, water management and the like. With the Greeks, as with the Hindus, interest was centrally metaphysical and geometric.

Archimedes' Method

The first rigorous mathematical calculation of π was due to Archimedes, who used a brilliant scheme based on **doubling inscribed and circumscribed polygons**

$$6 \mapsto 12 \mapsto 24 \mapsto 48 \mapsto 96$$

and computing the perimeters to obtain the bounds $3\frac{10}{71} < \pi < 3\frac{1}{7}$, that we have recaptured above. The case of 6-gons and 12-gons is shown in Figure 5; for $n = 48$ one already "sees" near-circles. Arguably no mathematics approached this level of rigour again until the 19th century.

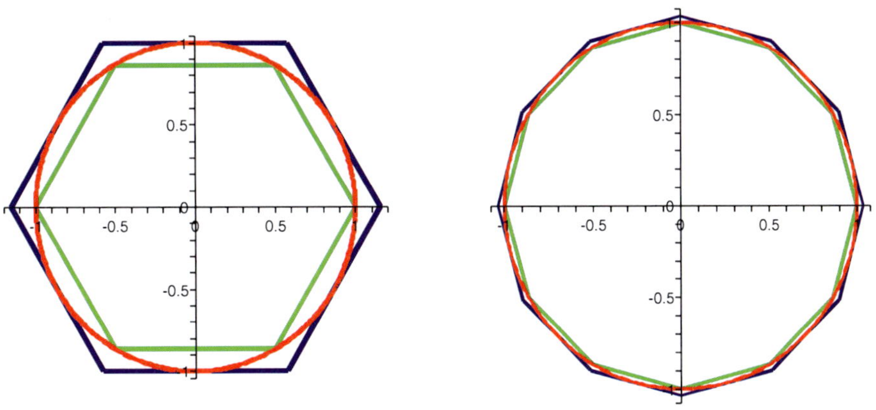

Figure 5: Archimedes' method of computing π with 6- and 12-gons.

Archimedes' scheme constitutes the first true algorithm for π, in that it is capable of producing an arbitrarily accurate value for π. It also represents the birth of numerical and error analysis — all without positional notation or modern trigonometry. As discovered severally in the 19th century, this scheme can be stated as a simple, numerically stable, recursion, as follows [Borwein and Borwein 1987].

Archimedean Mean Iteration (Pfaff-Borchardt-Schwab)

Set $a_0 = 2\sqrt{3}$ and $b_0 = 3$ — the values for circumscribed and inscribed 6-gons. Then define

$$a_{n+1} = \frac{2a_n b_n}{a_n + b_n} \quad (H) \qquad b_{n+1} = \sqrt{a_{n+1} b_n} \quad (G). \tag{2}$$

This converges to π, with the error decreasing by a factor of four with each iteration. In this case, the error is easy to estimate, the limit somewhat less accessible but still reasonably easy [Borwein and Bailey 2008; Borwein and Borwein 1987].

Variations of Archimedes' geometrical scheme were the basis for all high-accuracy calculations of π for the next 1800 years — well beyond its "best before" date. For example, in fifth century CE China, Tsu Ch'ung Chih used a variation of this method to get π correct to seven digits. A millennium later, al-Kāshī in Samarkand "who could calculate as eagles can fly" obtained 2π in *sexagesimal*:

$$2\pi \approx 6 + \frac{16}{60^1} + \frac{59}{60^2} + \frac{28}{60^3} + \frac{01}{60^4} + \frac{34}{60^5} + \frac{51}{60^6} + \frac{46}{60^7} + \frac{14}{60^8} + \frac{50}{60^9},$$

good to 16 decimal places (using $3 \cdot 2^{28}$-gons). This is a personal favorite; reentering it in my computer centuries later and getting the predicted answer gave me goosebumps.

Pre-Calculus Era π Calculations

In Figures 6, 8, and 11 we chronicle the main computational records during the indicated period, only commenting on signal entries.

Little progress was made in Europe during the "dark ages," but a significant advance arose in India (450 CE): *modern positional, zero-based decimal arithmetic* — the "Indo-Arabic" system. This greatly enhanced arithmetic in general, and computing π in particular. The Indo-Arabic system arrived with the Moors in Europe around 1000 CE. Resistance ranged from accountants who feared losing their livelihood to clerics who saw the system as "diabolical" — they incorrectly assumed its origin was Islamic. European commerce resisted into the 18th century, and even in scientific circles usage was limited until the 17th century.

The prior difficulty of doing arithmetic is indicated by college placement advice given a wealthy German merchant in the 16th century:

> A wealthy (15th Century) German merchant, seeking to provide his son with a good business education, consulted a learned man as to which European institution offered the best training. "If you only want him to be able to cope with addition and subtraction," the expert replied, "then any French or German university will do. But if you are intent on your son going on to multiplication and division — assuming that he has sufficient gifts — then you will have to send him to Italy." (George Ifrah, [Borwein and Bailey 2008])

Claude Shannon (1916–2001) had a mechanical calculator wryly called *Throback 1* built to compute in Roman, at Bell Labs in 1953, to show that it was practicable to compute in Roman!

Name	Year	Digits
Babylonians	2000? BCE	1
Egyptians	2000? BCE	1
Hebrews (1 Kings 7:23)	550? BCE	1
Archimedes	250? BCE	3
Ptolemy	150	3
Liu Hui	263	5
Tsu Ch'ung Chi	480?	7
al-Kāshī	1429	14
Romanus	1593	15
van Ceulen (**Ludolph's number**)	1615	35

Figure 6: Pre-calculus π Calculations.

Ludolph van Ceulen (1540–1610)

The last great Archimedean calculation, performed by van Ceulen using 2^{62}-gons — to 39 places with 35 correct — was published posthumously. The number is still called Ludolph's number in parts of Europe and was inscribed on his head-stone. This head-stone disappeared centuries ago but was rebuilt, in part from surviving descriptions, recently as shown in Figure 7. It was reconsecrated on July 5th 2000 with Dutch royalty in attendance. Ludolph van Ceulen, a serious mathematician, was also the discoverer of the cosine formula.

Figure 7: Ludolph's rebuilt tombstone in Leiden. Courtesy of Guilio Einaudi Editori.

π's Adolescence

The dawn of modern mathematics appears in *Viète's* or *Vièta's product* (1579)

$$\frac{2}{\pi} = \frac{\sqrt{2}}{2}\frac{\sqrt{2+\sqrt{2}}}{2}\frac{\sqrt{2+\sqrt{2+\sqrt{2}}}}{2}\cdots$$

considered to be the first truly infinite product; and in the *first infinite continued fraction* for $2/\pi$ given by Lord Brouncker (1620–1684), first President of the Royal Society of London:

$$\frac{2}{\pi} = \cfrac{1}{1+\cfrac{9}{2+\cfrac{25}{2+\cfrac{49}{2+\cdots}}}}.$$

This was based on the following brilliantly "interpolated" product of John Wallis[2] (1616–1703),

$$\prod_{k=1}^{\infty}\frac{4k^2-1}{4k^2} = \frac{2}{\pi}, \tag{3}$$

which led to the discovery of the Gamma function; see End Note 2, and a great deal more.

François Viète (1540–1603)

A flavour of Viète's writings can be gleaned in this quote from his work, first given in English in Berggren, Borwein and Borwein [2004, 759].

> Arithmetic is absolutely as much science as geometry [is]. Rational magnitudes are conveniently designated by rational numbers, and irrational [magnitudes] by irrational [numbers]. If someone measures magnitudes with numbers and by his calculation gets them different from what they really are, it is not the reckoning's fault but the reckoner's.
>
> Rather, says Proclus, "arithmetic is more exact then geometry."[3] To an accurate calculator, if the diameter is set to one unit, the circumference of the inscribed dodecagon will be the side of the binomial [i.e. square root of the difference] $72 - \sqrt{3888}$. Whosoever declares any other result will be mistaken, either the geometer in his measurements or the calculator in his numbers. (François Viète)

This fluent rendition is due to Marinus Taisbak, and the full text is worth reading. It certainly underlines how influential an algebraist and geometer Viète was. Viète, who was the first to introduce literals ("x" and "y") into algebra, nonetheless rejected the use of negative numbers.

Equation (3) may be derived from Leonard Euler's (1707–1783) product formula for π, given below in (4), with $x = 1/2$, or by repeatedly integrating $\int_0^{\pi/2} \sin^{2n}(t)\,dt$ by parts. One may divine (4) as Euler did by *considering* $\sin(\pi x)$ *as an "infinite" polynomial* and obtaining a product in terms of the roots — $0, \{1/n^2 : n = \pm 1, \pm 2, \cdots\}$. It is thus plausible that

$$\frac{\sin(\pi x)}{x} = c\prod_{n=1}^{\infty}\left(1 - \frac{x^2}{n^2}\right). \tag{4}$$

Euler, full well knowing that the whole argument was heuristic, argued that, as with a polynomial, c was the value at zero, and the coefficient of x^2 in the Taylor series must be the sum of the roots. Hence, he was able to pick off coefficients to evaluate the *zeta-function* at two:

$$\zeta(2) := \sum_n \frac{1}{n^2} = \frac{\pi^2}{6}.$$

[2] One of the few mathematicians whom Newton admitted respecting, and also a calculating prodigy!

[3] This phrase was written in Greek.

This also leads to the evaluation of $\zeta(2n) := \sum_{k=1}^{\infty} 1/k^{2n}$ as a rational multiple of π^{2n}:

$$\zeta(2) = \frac{\pi^2}{6}, \zeta(4) = \frac{\pi^4}{90}, \zeta(6) = \frac{\pi^6}{945}, \zeta(8) = \frac{\pi^8}{9450}, \dots$$

in terms of the *Bernoulli numbers*, B_n, where $t/(\exp(t)-1) = \sum_{n\geqslant 0} B_n t^n/n!$, gives a generating function for the B_n which are perforce rational. The explicit formula which solved the so-called *Basel problem*, posed by Pietro Mengoli, is

$$\zeta(2m) = (-1)^{m-1} \frac{(2\pi)^{2m}}{2\,(2m)!} B_{2m};$$

see also Tsumura [2004].

Much less is known about odd integer values of ζ, though they are almost certainly not rational multiples of powers of π. More than two centuries later, in 1976 Roger Apéry, [Berggren, Borwein and Borwein 2004, 439; Borwein and Borwein 1987], showed $\zeta(3)$ to be irrational, and we now also can prove that *at least one of* $\zeta(5), \zeta(7), \zeta(9)$ or $\zeta(11)$ is irrational, but we cannot guarantee which one. All positive integer values are strongly believed to be irrational. Though it is not relevant to our story, Euler's work on the zeta-function also led to the celebrated Riemann hypothesis [Borwein and Bailey 2008].

π's Adult Life with Calculus

In the later 17th century, Newton and Leibniz founded the calculus, and this powerful tool was quickly exploited to find new formulae for π. One early calculus-based formula comes from the integral:

$$\tan^{-1} x = \int_0^x \frac{dt}{1+t^2} = \int_0^x (1 - t^2 + t^4 - t^6 + \cdots)\, dt$$
$$= x - \frac{x^3}{3} + \frac{x^5}{5} - \frac{x^7}{7} + \frac{x^9}{9} - \cdots$$

Substituting $x = 1$ *formally* proves the well-known *Gregory-Leibniz formula* (1671–74)

$$\frac{\pi}{4} = 1 - \frac{1}{3} + \frac{1}{5} - \frac{1}{7} + \frac{1}{9} - \frac{1}{11} + \cdots \tag{5}$$

James Gregory (1638–75) was the greatest of a large Scottish mathematical family. The point $x = 1$, however, is on the boundary of the interval of convergence of the series. Justifying substitution requires a careful error estimate for the remainder or Lebesgue's monotone convergence theorem, etc., but most introductory texts ignore the issue.

A Curious Anomaly in the Gregory Series

In 1988, it was observed that Gregory's series for π,

$$\pi = 4 \sum_{k=1}^{\infty} \frac{(-1)^{k+1}}{2k-1} = 4\left(1 - \frac{1}{3} + \frac{1}{5} - \frac{1}{7} + \frac{1}{9} - \frac{1}{11} + \cdots\right) \tag{6}$$

when truncated to 5,000,000 terms, differs strangely from the true value of π:

```
3.141592453589793238464643383279502784197169399387305820974941822 30...
```

Values differ as expected from truncating an alternating series, in the seventh place — a "4" which should be a "6." But the next 13 digits are correct, and after another blip, for 12 digits. Of the first 46 digits, only four differ from the corresponding digits of π. Further, the "error" digits seemingly occur with a period of 14, as shown above. Such anomalous behavior begs explanation. A great place to start is by using Neil Sloane's Internet-based integer sequence recognition tool, available at http://oeis.org/. This tool has no difficulty recognizing the sequence of errors as twice *Euler numbers*. Even Euler numbers are generated by

$$\sec x = \sum_{k=0}^{\infty} \frac{(-1)^k E_{2k} x^{2k}}{(2k)!}.$$

The first few are $1, -1, 5, -61, 1385, -50521, 2702765$. This discovery led to the following asymptotic expansion:

$$\frac{\pi}{2} - 2\sum_{k=1}^{N/2} \frac{(-1)^{k+1}}{2k-1} \approx \sum_{m=0}^{\infty} \frac{E_{2m}}{N^{2m+1}}. \tag{7}$$

Now the genesis of the anomaly is clear: by chance the series had been truncated at 5,000,000 terms — exactly one-half of a fairly large power of ten. Indeed, setting $N = 10,000,000$ in Equation (7) shows that the first hundred or so digits of the truncated series value are small perturbations of the correct decimal expansion for π. And the asymptotic expansions show up on the computer screen, as we observed above. On a hexadecimal computer with $N = 16^7$ the corresponding strings and hex-errors are:

```
3.243F6A8885A308D313198A2E03707344A4093822299F31D0082EFA98EC4E6C894...
3.243F6A6885A308D31319AA2E03707344A3693822299F31D7A82EFA98EC4DBF694...
          2              -2              A             -7A           2AD2
```

with the first being the correct value of π. In hexadecimal or *hex* one uses "A, B, ..., F" to write 10 through 15 as single "hex-digits." Similar phenomena occur for other constants. (See Berggren, Borwein and Borwein [2004].) Also, knowing the errors means we can correct them and use (7) to make Gregory's formula computationally tractable, notwithstanding the following discussion of complexity!

Calculus Era π Calculations

Used naively, the beautiful formula (5) is computationally useless — so slow that hundreds of terms are needed to compute two digits. Sharp, under the direction of Halley[4] (see Figure 8), actually used $\tan^{-1}(1/\sqrt{3})$ which is geometrically convergent.

Moreover, Euler's (1738) trigonometric identity

$$\tan^{-1}(1) = \tan^{-1}\left(\frac{1}{2}\right) + \tan^{-1}\left(\frac{1}{3}\right) \tag{8}$$

produces a geometrically convergent rational series

$$\frac{\pi}{4} = \frac{1}{2} - \frac{1}{3 \cdot 2^3} + \frac{1}{5 \cdot 2^5} - \frac{1}{7 \cdot 2^7} + \cdots + \frac{1}{3} - \frac{1}{3 \cdot 3^3} + \frac{1}{5 \cdot 3^5} - \frac{1}{7 \cdot 3^7} + \cdots \tag{9}$$

[4] The astronomer and mathematician who was the second Astronomer Royal and worked to develop Greenwich Observatory, and after whom the comet is named.

Name	Year	Correct Digits
Sharp (and Halley)	1699	71
Machin	1706	100
Strassnitzky and Dase	1844	200
Rutherford	1853	440
Shanks	1874	(707) 527
Ferguson (**Calculator**)	1947	808
Reitwiesner et al. (**ENIAC**)	1949	2,037
Genuys	1958	10,000
Shanks and Wrench	1961	100,265
Guilloud and Boyer	1973	1,001,250

Figure 8: Calculus π Calculations.

An even faster formula, found earlier by John Machin, lies similarly in the identity

$$\frac{\pi}{4} = 4\tan^{-1}\left(\frac{1}{5}\right) - \tan^{-1}\left(\frac{1}{239}\right). \tag{10}$$

This was used in numerous computations of π, given in Figure 8, starting in 1706 and culminating with Shanks' famous computation of π to 707 decimal digits accuracy in 1873 (although it was *found in 1945 to be wrong* after the 527-th decimal place, by Ferguson, during the last adding machine-assisted pre-computer computations).[5]

Newton's Arcsin Computation

Newton discovered a different more effective — actually a disguised arcsin — formula. He considering the area A of the left-most region shown in Figure 9. Now, A is the integral

$$A = \int_0^{1/4} \sqrt{x - x^2}\, dx. \tag{11}$$

Also, A is the area of the circular sector, $\pi/24$, less the area of the triangle, $\sqrt{3}/32$. Newton used his newly developed *binomial theorem* in (11):

$$\begin{aligned} A &= \int_0^{\frac{1}{4}} x^{1/2}(1-x)^{1/2}\, dx = \int_0^{\frac{1}{4}} x^{1/2}\left(1 - \frac{x}{2} - \frac{x^2}{8} - \frac{x^3}{16} - \frac{5x^4}{128} - \cdots\right) dx \\ &= \int_0^{\frac{1}{4}} \left(x^{1/2} - \frac{x^{3/2}}{2} - \frac{x^{5/2}}{8} - \frac{x^{7/2}}{16} - \frac{5x^{9/2}}{128} \cdots\right) dx \end{aligned}$$

Integrating term-by-term and combining the above produces

$$\pi = \frac{3\sqrt{3}}{4} + 24\left(\frac{1}{3\cdot 8} - \frac{1}{5\cdot 32} - \frac{1}{7\cdot 128} - \frac{1}{9\cdot 512} \cdots\right).$$

Newton used this formula to compute 15 digits of π. As noted, he later "apologized" for "having no other business at the time." (This was the year of the great plague which closed Cambridge, and of the great fire of London of September 1666.) A standard chronology says "Newton significantly never gave

[5] This must be some sort of record for the length of time needed to detect a mathematical error.

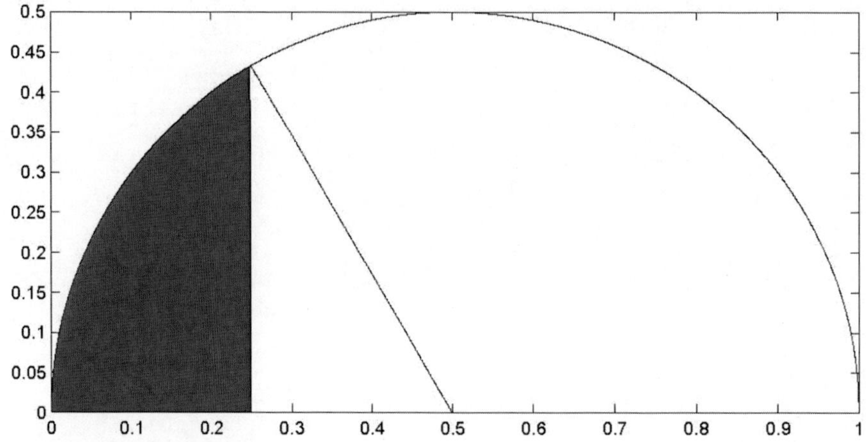

Figure 9: Newton's method of computing π: "I am ashamed to tell you to how many figures I carried these computations, having no other business at the time." Issac Newton, 1666.

a value for π" [Berggren, Borwein and Borwein 2004, 294; Schloper 1950]. *Caveat emptor*, all users of secondary sources.

The Viennese *Computer*

Until quite recently — around 1950 — a computer was a person. Hence the name of ENIAC discussed later. This computer, one Johan Zacharias Dase (1824–1861), would demonstrate his extraordinary computational skill by, for example, multiplying

$$79532853 \times 93758479 = 7456879327810587$$

in 54 seconds; two 20-digit numbers in six minutes; two 40-digit numbers in 40 minutes; two 100-digit numbers in 8 hours and 45 minutes. In 1844, after being shown

$$\frac{\pi}{4} = \tan^{-1}\left(\frac{1}{2}\right) + \tan^{-1}\left(\frac{1}{5}\right) + \tan^{-1}\left(\frac{1}{8}\right)$$

he calculated π to 200 places *in his head* in two months, completing correctly — to my mind — the greatest mental computation ever. Dase later calculated a seven-digit logarithm table, and extended a table of integer factorizations to 10,000,000. Gauss requested that Dase be permitted to assist this project, but Dase died not long afterwards in 1861 by which time Gauss himself already was dead.

An amusing Machin-type identity, that is expressing π as a linear combination of arctan's, due to the Oxford logician Charles Dodgson is

$$\tan^{-1}\left(\frac{1}{p}\right) = \tan^{-1}\left(\frac{1}{p+q}\right) + \tan^{-1}\left(\frac{1}{p+r}\right),$$

valid whenever $1 + p^2$ factors as qr. Dodgson is much better known as Lewis Carroll, the author of *Alice in Wonderland.*

The Irrationality and Transcendence of π

One motivation for computations of π was very much in the spirit of modern experimental mathematics: to see if the decimal expansion of π repeats, which would mean that π is the ratio of two integers (i.e., rational), or to recognize π as *algebraic* — the root of a polynomial with integer coefficients — and later to look at digit distribution. The question of the *rationality of* π was settled in the late 1700s, when Lambert and Legendre proved (using continued fractions) that the constant is irrational.

The question of whether π was algebraic was settled in 1882, when Lindemann proved that π *is transcendental*. Lindemann's proof also settled, once and for all, the ancient Greek question of whether the circle could be squared with straight-edge and compass. It cannot be, because numbers that are the lengths of lines that can be constructed using ruler and compasses (often called *constructible numbers*) are necessarily algebraic, and squaring the circle is equivalent to constructing the value π. The classical Athenian playwright Aristophanes already "knew" this and perhaps derided those who attempted to square the circle in his play *The Birds* of 414 BCE. Likewise, the French Academy had stopped accepting proofs of the three great constructions of antiquity — squaring the circle, doubling the cube, and trisecting the angle — centuries earlier.

We next give, *in extenso*, Ivan Niven's 1947 short proof of the irrationality of π. It well illustrates the ingredients of more difficult later proofs of irrationality of other constants, and indeed of Lindemann's proof of the transcendence of π building on Hermite's 1873 proof of the transcendence of e.

A Proof that π is Irrational

Proof. Let $\pi = a/b$, the quotient of positive integers. We define the polynomials

$$f(x) = \frac{x^n(a - bx)^n}{n!}$$

$$F(x) = f(x) - f^{(2)}(x) + f^{(4)}(x) - \cdots + (-1)^n f^{(2n)}(x)$$

the positive integer being specified later. Since $n!f(x)$ has integral coefficients and terms in x of degree not less than n, $f(x)$ and its derivatives $f^{(j)}(x)$ have integral values for $x = 0$; also for $x = \pi = a/b$, since $f(x) = f(a/b - x)$. By elementary calculus we have

$$\frac{d}{dx}\{F'(x)\sin x - F(x)\cos x\} = F''(x)\sin x + F(x)\sin x = f(x)\sin x$$

and

$$\int_0^{\pi} f(x)\sin x dx = [F'(x)\sin x - F(x)\cos x]_0^{\pi}$$

$$= F(\pi) + F(0). \tag{12}$$

Now $F(\pi) + F(0)$ is an *integer*, since $f^{(j)}(0)$ and $f^{(j)}(\pi)$ are integers. But for $0 < x < \pi$,

$$0 < f(x)\sin x < \frac{\pi^n a^n}{n!},$$

so that the integral in (12) is *positive but arbitrarily small* for n sufficiently large. Thus (12) is false, and so is our assumption that π is rational. **QED**

Irrationality Measures

We end this section by touching on the matter of *measures of irrationality*. The infimum $\mu(\alpha)$ of those $\mu > 0$ for which

$$\left|\alpha - \frac{p}{q}\right| \geqslant \frac{1}{q^{\mu}}$$

for all integers p, q with sufficiently large q, is called the *Liouville-Roth constant* for α and we say that we have an irrationality measure for α if $\mu(\alpha) < \infty$.

Irrationality measures are difficult. Roth's theorem [Borwein and Borwein 1987] implies that $\mu(\alpha) = 2$ for all algebraic irrationals, as is the case for almost all reals. Clearly, $\mu(\alpha) = 1$ for rational α and $\mu(\alpha) = \infty$ if and only if α is a Liouville number such as $\sum 1/10^{n!}$. It is known that $\mu(e) = 2$ while in 1993 Hata showed that $\mu(\pi) \leqslant 8.02$. Similarly, it is known that $\mu(\zeta(2)) \leqslant 5.45, \mu(\zeta(3)) \leqslant 4.8$ and $\mu(\log 2) \leqslant 3.9$.

A consequence of the existence of an irrationality measure μ for π is the ability to estimate quantities such as $\limsup |\sin(n)|^{1/n} = 1$ for integer n, since for large integer m and n with $m/n \to \pi$, we have eventually

$$|\sin(n)| = |\sin(m\pi) - \sin(n)| \geqslant \frac{1}{2}|m\pi - n| \geqslant \frac{1}{2\, m^{\mu-1}}.$$

Related matters are discussed at more length in Amoroso and Viola [2008].

π in the Digital Age

With the substantial development of computer technology in the 1950s, π was computed to thousands and then millions of digits. These computations were greatly facilitated by the discovery soon after of advanced algorithms for the underlying high-precision arithmetic operations. For example, in 1965 it was found that the newly-discovered *fast Fourier transform* (FFT) [Borwein and Borwein 1987; Borwein and Bailey 2008] could be used to perform high-precision multiplications much more rapidly than conventional schemes. Such methods (e.g., for $\div, \sqrt{x}$ see Borwein and Borwein [1987, 1988]; Borwein and Bailey [2008]) dramatically lowered the time required for computing π and other constants to high precision. We are now able to compute algebraic values of algebraic functions essentially as fast as we can multiply, $O_B(M(N))$, where $M(N)$ is the cost of multiplication and O_B counts "bits" or "flops." To convert this into practice: a state-of-the-art processor in 2010, such as the latest AMD Opteron, which runs at 2.4 GHz and has four floating-point cores, each of which can do two 64-bit floating-point operations per second, can produce a total of 9.6 billion floating-point operations per second.

In spite of these advances, into the 1970s all computer evaluations of π still employed classical formulae, usually of Machin-type; see Figure 8. We will see below methods that compute N digits of π with time complexity $O_B(M(N)) \log O_B(M(N))$. Showing that the log term is unavoidable, as seems likely, would provide an algorithmic proof that π is not algebraic.

Electronic Numerical Integrator and Calculator

The first computer calculation of π was performed on ENIAC — a behemoth with a tiny brain from today's vantage point. The ENIAC was built in Aberdeen Maryland by the US Army:

Size/weight. ENIAC had 18,000 vacuum tubes, 6,000 switches, 10,000 capacitors, 70,000 resistors, 1,500 relays, was 10 feet tall, occupied 1,800 square feet and weighed 30 tons.
Speed/memory. A now slow 1.5GHz Pentium does 3 million adds/sec. ENIAC did 5,000, three orders faster than any earlier machine. The first stored-memory computer, ENIAC could hold 200 digits.
Input/output. Data flowed from one accumulator to the next, and after each accumulator finished a calculation, it communicated its results to the next in line. The accumulators were connected to each other manually. The 1949 computation of π to 2,037 places on ENIAC took 70 hours in which output had to be constantly reintroduced as input.

A fascinating description of the ENIAC's technological and commercial travails is to be found in McCartney [1999].

Ballantine's (1939) Series for π

Another formula of Euler for arccot is

$$x \sum_{n=0}^{\infty} \frac{(n!)^2 \, 4^n}{(2\,n+1)!\,(x^2+1)^{n+1}} = \arctan\left(\frac{1}{x}\right).$$

This, intriguingly and usefully, allowed Guilloud and Boyer to reexpress the formula, used by them in 1973 to compute a million digits of Pi, viz, $\pi/4 = 12 \arctan(1/18) + 8 \arctan(1/57) - 5 \arctan(1/239)$ in the efficient form

$$\pi = 864 \sum_{n=0}^{\infty} \frac{(n!)^2\,4^n}{(2\,n+1)!\,325^{n+1}} + 1824 \sum_{n=0}^{\infty} \frac{(n!)^2\,4^n}{(2\,n+1)!\,3250^{n+1}} - 20 \arctan\left(\frac{1}{239}\right),$$

where the terms of the second series are now just decimal shifts of the first.

Ramanujan-Type Elliptic Series.

Truly new types of infinite series formulae, based on elliptic integral approximations, were discovered by Srinivasa Ramanujan (1887–1920), shown in Figure 10, around 1910, but were not well known (nor fully proven) until quite recently when his writings were widely published. They are based on elliptic functions and are described at length in Berggren, Borwein and Borwein [2004]; Borwein and Borwein [1987]; Borwein and Bailey [2008].

Figure 10: Ramanujan's seventy-fifth birthday stamp.

G.N. Watson elegantly describes his feelings on viewing formulae of Ramanujan, such as (13):

> ...a thrill which is indistinguishable from the thrill which I feel when I enter the Sagrestia Nuova of the Cappella Medici and see before me the austere beauty of the four statues representing "Day," "Night," "Evening," and "Dawn" which Michelangelo has set over the tomb of Giuliano de'Medici and Lorenzo de'Medici.

One of these series is the remarkable

$$\frac{1}{\pi} = \frac{2\sqrt{2}}{9801} \sum_{k=0}^{\infty} \frac{(4k)!\,(1103 + 26390k)}{(k!)^4 396^{4k}}. \tag{13}$$

Each term of this series produces an additional *eight* correct digits in the result. When Gosper used this formula to compute 17 million digits of π in 1985, and it agreed to many millions of places with the prior estimates; *this concluded the first proof* of (13), as described in Borwein, Borwein and Bailey [1989]! Actually, Gosper first computed the simple continued fraction for π, hoping to discover some new things in its expansion, but found none.

At about the same time, David and Gregory Chudnovsky found the following rational variation of Ramanujan's formula. It exists because $\sqrt{-163}$ corresponds to an imaginary quadratic field with class number one:

$$\frac{1}{\pi} = 12 \sum_{k=0}^{\infty} \frac{(-1)^k\,(6k)!\,(13591409 + 545140134k)}{(3k)!\,(k!)^3\,640320^{3k+3/2}} \tag{14}$$

Each term of this series produces an additional 14 correct digits. The Chudnovskys implemented this formula using a clever scheme that enabled them to use the results of an initial level of precision to extend the calculation to even higher precision. They used this in several large calculations of π, culminating with a then record computation to over four billion decimal digits in 1994. Their remarkable story was compellingly told by Richard Preston in a prizewinning *New Yorker* article "The Mountains of Pi" (March 2, 1992).

While the Ramanujan and Chudnovsky series are in practice considerably more efficient than classical formulae, they share the property that the number of terms needed increases linearly with the number of digits desired: *if you want to compute twice as many digits of* π, *you must evaluate twice as many terms* of the series.

Relatedly, the Ramanujan-type series

$$\frac{1}{\pi} = \sum_{n=0}^{\infty} \left(\frac{\binom{2n}{n}}{16^n} \right)^3 \frac{42\,n + 5}{16} \tag{15}$$

allows one to compute the billionth binary digit of $1/\pi$, or the like, *without computing the first half* of the series, and is a foretaste of our later discussion of Borwein-Bailey-Plouffe (or BBP) formulae.

Reduced Operational Complexity Algorithms

In 1976, Eugene Salamin and Richard Brent independently discovered a *reduced complexity* algorithm for π. It is based on the **arithmetic-geometric mean iteration** (AGM) and some other ideas due to Gauss and Legendre around 1800, although Gauss, nor many after him, never directly saw the connection to effectively computing π.

Quadratic Algorithm (Salamin-Brent)

Set $a_0 = 1, b_0 = 1/\sqrt{2}$ and $s_0 = 1/2$. Calculate

$$a_k = \frac{a_{k-1} + b_{k-1}}{2} \quad (A) \qquad b_k = \sqrt{a_{k-1} b_{k-1}} \quad (G) \tag{16}$$

$$c_k = a_k^2 - b_k^2, \qquad s_k = s_{k-1} - 2^k c_k \qquad \text{and compute} \quad p_k = \frac{2a_k^2}{s_k}. \tag{17}$$

Name	Year	Correct Digits
Miyoshi and Kanada	1981	2,000,036
Kanada-Yoshino-Tamura	1982	16,777,206
Gosper	1985	17,526,200
Bailey	Jan. 1986	29,360,111
Kanada and Tamura	Sep. 1986	33,554,414
Kanada and Tamura	Oct. 1986	67,108,839
Kanada et. al	Jan. 1987	134,217,700
Kanada and Tamura	Jan. 1988	201,326,551
Chudnovskys	May 1989	480,000,000
Kanada and Tamura	Jul. 1989	536,870,898
Kanada and Tamura	Nov. 1989	1,073,741,799
Chudnovskys	Aug. 1991	2,260,000,000
Chudnovskys	May 1994	4,044,000,000
Kanada and Takahashi	Oct. 1995	6,442,450,938
Kanada and Takahashi	Jul. 1997	51,539,600,000
Kanada and Takahashi	Sep. 1999	206,158,430,000
Kanada-Ushiro-Kuroda	Dec. 2002	1,241,100,000,000
Takahashi	Jan. 2009	1,649,000,000,000
Takahashi	April. 2009	2,576,980,377,524
Bellard	Dec. 2009	2,699,999,990,000

Figure 11: Post-calculus π calculations.

Then p_k converges *quadratically* to π. Note the similarity between the arithmetic-geometric mean iteration (16) (which for general initial values converges fast to a non-elementary limit), and the out-of-kilter harmonic-geometric mean iteration (2) (which in general converges slowly to an elementary limit), and which is an arithmetic-geometric iteration in the reciprocals (see Borwein and Borwein [1987]).

Each iteration of the algorithm *doubles* the correct digits. Successive iterations produce 1, 4, 9, 20, 42, 85, 173, 347 and 697 good decimal digits of π, and takes $\log N$ operations for N digits. Twenty-five iterations computes π to over 45 million decimal digit accuracy. A disadvantage is that each of these iterations must be performed to the precision of the final result. In 1985, my brother Peter and I discovered families of algorithms of this type. For example, here is a genuinely third-order iteration:

Cubic Algorithm

Set $a_0 = 1/3$ and $s_0 = (\sqrt{3}-1)/2$. Iterate

$$r_{k+1} = \frac{3}{1+2(1-s_k^3)^{1/3}}, \qquad s_{k+1} = \frac{r_{k+1}-1}{2} \quad \text{and}$$

$$a_{k+1} = r_{k+1}^2 a_k - 3^k(r_{k+1}^2 - 1).$$

Then $1/a_k$ *converges cubically* to π. Each iteration *triples* the number of correct digits.

Quartic Algorithm

Set $a_0 = 6 - 4\sqrt{2}$ and $y_0 = \sqrt{2} - 1$. Iterate

$$y_{k+1} = \frac{1-(1-y_k^4)^{1/4}}{1+(1-y_k^4)^{1/4}} \qquad \text{and}$$

$$a_{k+1} = a_k(1+y_{k+1})^4 - 2^{2k+3}y_{k+1}(1+y_{k+1}+y_{k+1}^2).$$

Then $1/a_k$ *converges quartically* to π. Note that only the power of 2 or 3 used in a_k depends on k.

Let us take an interlude and discuss:

'Piems' or π-mnemonics. *Piems* are mnemonics in which the length of each word is the corresponding digit of π. Punctuation is ignored. A better piem is both longer and better poetry.

There are many more and longer mnemonics than the samples given below — see Berggren, Borwein and Borwein [2004, 405, 560, 659] for a fine selection.

Now I, even I, would celebrate
In rhyme inapt, the great
Immortal Syracusan, rivaled nevermore,
Who in his wondrous lore,
Passed on before
Left men for guidance
How to circles mensurate. (30)

How I want a drink, alcoholic of course, after the heavy lectures involving quantum mechanics. (15)

See I have a rhyme assisting my feeble brain its tasks ofttimes resisting. (13)

Philosophy of mathematics. In 1997 the first occurrence of the sequence 0123456789 was found (later than expected heuristically) in the decimal expansion of π starting at the $17,387,594,880$th digit after the decimal point. In consequence, the status of several famous *intuitionistic examples* due to Brouwer and Heyting has changed. These challenge the *principle of the excluded middle* — either a predicate holds or it does not — and involve classically well-defined objects that for an intuitionist are ill-founded until one can determine when or if the sequence occurred [Borwein 1998].

For example, consider the sequence which is "0" except for a "1" in the first place where 0123456789 first begins to appear in order if it ever occurs. Did it converge when first used by Brouwer as an example? Does it now? Was it then and is it now well defined? Classically it always was and converged to "0." Intuitionistically it converges now. What if we redefine the sequence to have its "1" in the first place that 0123456789101112 first begins?

Back to the Future

In December 2002, Kanada computed π to over 1.24 trillion decimal digits. His team first computed π in hexadecimal (base 16) to 1,030,700,000,000 places, using the following two arctangent relations:

$$\pi = 48\tan^{-1}\frac{1}{49} + 128\tan^{-1}\frac{1}{57} - 20\tan^{-1}\frac{1}{239} + 48\tan^{-1}\frac{1}{110443}$$

$$\pi = 176\tan^{-1}\frac{1}{57} + 28\tan^{-1}\frac{1}{239} - 48\tan^{-1}\frac{1}{682} + 96\tan^{-1}\frac{1}{12943}.$$

The first formula was found in 1982 by K. Takano, a high school teacher and song writer. The second formula was found by F.C.W. Störmer in 1896. Kanada verified the results of these two computations agreed, and then converted the hex digit sequence to decimal. The resulting decimal expansion was checked by converting it back to hex. These conversions are themselves non-trivial, requiring massive computation.

This process is quite different from those of the previous quarter century. One reason is that reduced operational complexity algorithms require full-scale multiply, divide and square root operations. These in turn require large-scale FFT operations, which demand huge amounts of memory and massive all-to-all communication between nodes of a large parallel system. For this latest computation, even the very large system available in Tokyo did not have sufficient memory and network bandwidth to perform these operations at reasonable efficiency levels — at least not for trillion-digit computations. Utilizing arctans again meant using many more arithmetic operations, but no system-scale FFTs, and it can be implemented using $\times, \div$ by smallish integer values — additionally, hex is somewhat more efficient!

Kanada and his team evaluated these two formulae using a scheme analogous to that employed by Gosper and by the Chudnovskys in their series computations, in that they were able to avoid explicitly storing the multiprecision numbers involved. This resulted in a scheme that is roughly competitive in *numerical* efficiency with the Salamin-Brent and Borwein quartic algorithms they had previously used, but with a significantly lower total memory requirement. Kanada used a 1 Tbyte main memory system, as with the previous computation, yet got six times as many digits. Hex and decimal evaluations included, it ran 600 hours on a 64-node Hitachi, with the main segment of the program running at a sustained rate of nearly 1 Tflop/sec.

手にしているのは π の値が入ったカートリッジテープ

Figure 12: Yasumasa Kanada in his Tokyo office. Courtesy of Guilio Einaudi Editori.

Why π?

What possible motivation lies behind modern computations of π, given that questions such as the irrationality and transcendence of π were settled more than 100 years ago? One motivation is the raw challenge of harnessing the stupendous power of modern computer systems. Programming such calculations are definitely not trivial, especially on large, distributed memory computer systems.

There have been substantial practical spin-offs. For example, some new techniques for performing the fast Fourier transform (FFT), heavily used in modern science and engineering computing, had their roots in attempts to accelerate computations of π. And always the computations help in road-testing computers — often uncovering subtle hardware and software errors.

Beyond practical considerations lies the abiding interest in the fundamental question of the *normality (digit randomness)* of π. Kanada, for example, has performed detailed statistical analysis of his results to see if there are any statistical abnormalities that suggest π is not normal. So far the answer is "no;" see Figure 13 and 14. Indeed the first computer computation of π and e on ENIAC, discussed above, was so motivated by John von Neumann. The digits of π have been studied more than any other single constant, in part because of the widespread fascination with and recognition of π. Kanada reports that the 10 decimal digits ending in position one trillion are 6680122702, while the 10 hexadecimal digits ending in position one trillion are 3F89341CD5.

Decimal Digit	Occurrences
0	99999485134
1	99999945664
2	100000480057
3	99999787805
4	100000357857
5	99999671008
6	99999807503
7	99999818723
8	100000791469
9	99999854780
Total	**1000000000000**

Hex Digit	Occurrences
0	62499881108
1	62500212206
2	62499924780
3	62500188844
4	62499807368
5	62500007205
6	62499925426
7	62499878794
8	62500216752
9	62500120671
A	62500266095
B	62499955595
C	62500188610
D	62499613666
E	62499875079
F	62499937801
Total	**1000000000000**

Figure 13: Seemingly random behavior of π base 10 and 16

Changing world views. In retrospect, we may wonder why in antiquity π was not *measured* to an accuracy in excess of 22/7? Perhaps it reflects not an inability to do so but a very different mind set to a modern experimental — Baconian or Popperian — one. In the same vein, one reason that Gauss and Ramanujan did not further develop the ideas in their identities for π is that an iterative algorithm, as opposed to explicit results, was not as satisfactory for them (especially Ramanujan). Ramanujan much preferred formulae like

$$\pi \approx \frac{3}{\sqrt{67}} \log(5280), \qquad \frac{3}{\sqrt{163}} \log(640320) \approx \pi$$

correct to 9 and 15 decimal places, both of which rely on deep number theory. Contrastingly, Ramanujan in his famous 1914 paper *Modular Equations and Approximations to Pi* found

$$\left(9^2 + \frac{19^2}{22}\right)^{1/4} = 3.14159265\bar{2}58\cdots$$

"empirically, and it has no connection with the preceding theory" [Berggren, Borwein and Borwein 2004, 253]. Only the marked digit is wrong.

Discovering the π Iterations. The genesis of the π algorithms and related material is an illustrative example of experimental mathematics. My brother and I in the early eighties had a family of quadratic algorithms for π [Borwein and Borwein 1987], call them $\mathcal{P}_N$, of the kind we saw above. For $N = 1, 2, 3, 4$ we could prove they were correct but were only conjectured for $N = 5, 7$. In each case the

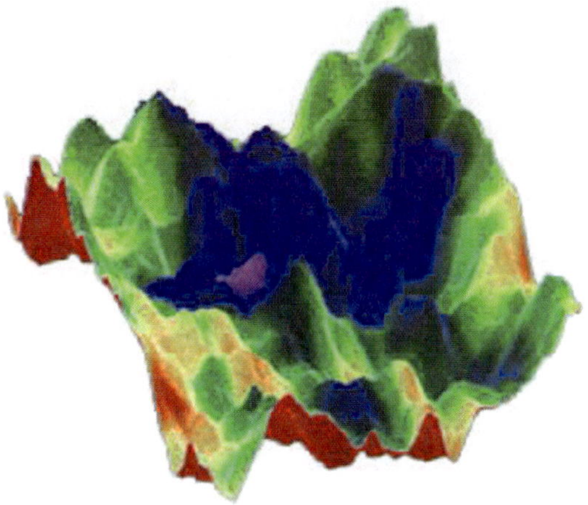

Figure 14: A 'random walk' on the first one million digits of π. Courtesy D. and G. Chudnovsky.

algorithm *appeared* to converge quadratically to π. On closer inspection while the provable cases were correct to 5,000 digits, the empirical versions agreed with π to roughly 100 places only. Now, in many ways, to have discovered a "natural" number that agreed with π to that level — and no more — would have been more interesting than the alternative. That seemed unlikely but recoding and rerunning the iterations kept producing identical results.

Two decades ago even moderately high precision calculation was less accessible, and the code was being run remotely over a phone-line in a Berkeley Unix integer package. After about six weeks, it transpired that the package's square root algorithm was badly flawed, but *only if run with an odd precision of more than sixty digits*! And for idiosyncratic reasons that had only been the case in the two unproven cases. Needless to say, tracing the bug was a salutary and somewhat chastening experience. And it highlights why one checks computations using different sub-routines and methods.

How to Compute the N-th Digits of π

One might be forgiven for thinking that essentially everything of interest with regard to π has been dealt with. This is suggested in the closing chapters of Beckmann's 1971 book *A History of* π. Ironically, the Salamin-Brent quadratically convergent iteration was discovered only five years later, and the higher-order convergent algorithms followed in the 1980s. Then in 1990, Rabinowitz and Wagon discovered a "spigot" algorithm for π — the digits "drip out" one by one. This permits successive digits of π (in any desired base) to be computed by a relatively simple recursive algorithm based on the *all previously* generated digits.

Even insiders are sometimes surprised by a new discovery. Prior to 1996, most folks thought if you want to determine the d-th digit of π, you had to generate the (order of) the entire first d digits. This is not true, at least for hex (base 16) or binary (base 2) digits of π. In 1996, Peter Borwein, Plouffe, and Bailey found an algorithm for computing individual hex digits of π. It (1) yields a modest-length hex or binary digit string for π, from an arbitrary position, using no prior bits; (2) is implementable on any modern computer; (3) requires no multiple precision software; (4) requires very little memory; and (5) has a computational cost growing only slightly faster than the digit position. For example, the millionth hexadecimal digit (four millionth binary digit) of π could be found in four seconds on a 2005 Apple computer.

This new algorithm is not fundamentally faster than the best known schemes if used for computing *all* digits of π up to some position, but its elegance and simplicity are of considerable interest, and is easy to parallelize. It is based on the following at-the-time new formula for π:

$$\pi = \sum_{i=0}^{\infty} \frac{1}{16^i}\left(\frac{4}{8i+1} - \frac{2}{8i+4} - \frac{1}{8i+5} - \frac{1}{8i+6}\right) \tag{18}$$

which was discovered using *integer relation methods* (see Borwein and Bailey [2008]), with a computer search that ran for several months and then produced the (equivalent) relation

$$\pi = 4 \cdot {}_2F_1\left(\begin{matrix}1, \frac{1}{4} \\ \frac{5}{4}\end{matrix}\,\middle|\, -\frac{1}{4}\right) + 2\arctan\left(\frac{1}{2}\right) - \log 5, \tag{19}$$

where the first term is a generalized Gaussian hypergeometric function evaluation.

Maple and *Mathematica* can both now prove (18). A human proof may be found in Borwein and Bailey [2008].

The algorithm in action. In 1997, Fabrice Bellard at INRIA — whom we shall meet again in the section "π in the Third Millennium" — computed 152 binary digits of π starting at the trillionth position. The computation took 12 days on 20 workstations working in parallel over the Internet. Bellard's scheme is based on the following variant of (18):

$$\pi = 4\sum_{k=0}^{\infty} \frac{(-1)^k}{4^k(2k+1)} - \frac{1}{64}\sum_{k=0}^{\infty} \frac{(-1)^k}{1024^k}\left(\frac{32}{4k+1} + \frac{8}{4k+2} + \frac{1}{4k+3}\right),$$

which permits hex or binary digits of π to be calculated somewhat faster than (18) depending on the implementation.

In 1998 Colin Percival, then a 17-year-old student at Simon Fraser University, utilized 25 machines to calculate first the five trillionth hexadecimal digit, and then the ten trillionth hex digit. In September, 2000, he found the quadrillionth binary digit is **0**, a computation that required 250 CPU-years, using 1734 machines in 56 countries. We record some of Percival's computational results in Figure 15. Nor have matters stopped there. As described by Bailey [2011] and Bailey, et al. [2013] in the most recent computation of π using the BBP formula, Tse-Wo Zse of Yahoo! Cloud Computing calculated 256 binary digits of π starting at the *two quadrillionth* bit. He then checked his result using Bellard's variant. In this case, both computations verified that the 24 hex digits beginning immediately after the 500 trillionth hex digit (i.e., after the two quadrillionth binary bit) are E6C1294A ED40403F 56D2D764.

A last comment for this section is that Kanada was able to confirm his 2002 computation in only 21 hours by computing a 20 hex digit string starting at the trillionth digit, and comparing this string to the hex string he had initially obtained in over 600 hours. Their agreement provided enormously strong confirmation. We shall see this use of BBP for verification again when we discuss the most recent record computations of π.

Further BBP Digit Formulae

Motivated as above, constants α of the form

$$\alpha = \sum_{k=0}^{\infty} \frac{p(k)}{q(k)2^k}, \tag{20}$$

where $p(k)$ and $q(k)$ are integer polynomials, are said to be in the class of *binary (Borwein-Bailey-Plouffe) BBP numbers*. I illustrate for $\log 2$ why this permits one to calculate isolated digits in the binary

Position	Hex strings starting at this Position
10^6	26C65E52CB4593
10^7	17AF5863EFED8D
10^8	ECB840E21926EC
10^9	85895585A0428B
10^{10}	921C73C6838FB2
10^{11}	9C381872D27596
1.25×10^{12}	07E45733CC790B
2.5×10^{14}	E6216B069CB6C1

Borweins and Plouffe (MSNBC, 1996)

Figure 15: Percival's hexadecimal findings. Courtesy of Guilio Einaudi Editori.

expansion:

$$\log 2 = \sum_{k=0}^{\infty} \frac{1}{k2^k}. \tag{21}$$

We wish to compute a few binary digits beginning at position $d+1$. This is equivalent to calculating $\{2^d \log 2\}$, where $\{\cdot\}$ denotes fractional part. We can write

$$\{2^d \log 2\} = \left\{ \left\{ \sum_{k=0}^{d} \frac{2^{d-k}}{k} \right\} + \left\{ \sum_{k=d+1}^{\infty} \frac{2^{d-k}}{k} \right\} \right\} \tag{22}$$

$$= \left\{ \left\{ \sum_{k=0}^{d} \frac{\mathbf{2^{d-k} \bmod k}}{k} \right\} + \left\{ \sum_{k=d+1}^{\infty} \frac{\mathbf{2^{d-k}}}{\mathbf{k}} \right\} \right\}. \tag{23}$$

The key observation is that the numerator of the first sum in (23), $2^{d-k} \bmod k$, can be calculated rapidly by *binary exponentiation*, performed modulo k. That is, it is economically performed by a factorization based on the binary expansion of the exponent. For example,

$$3^{17} = ((((3^2)^2)^2)^2) \cdot 3$$

uses only five multiplications, not the usual 16. It is important to reduce each product modulo k. Thus, $3^{17} \bmod 10$ is done as

$$3^2 = 9; 9^2 = 1; 1^2 = 1; 1^2 = 1; 1 \times 3 = 3.$$

A natural question in light of (18) is whether there is a formula of this type and an associated computational strategy to compute individual *decimal* digits of π. Searches conducted by numerous researchers have been unfruitful and recently D. Borwein (my father), Gallway and I have shown that there are no BBP formulae of the *Machin-type* (as defined in Borwein and Bailey [2008]) of (18) for π unless the base is a power of two [Borwein and Bailey 2008].

Ternary BBP formulae. Yet, BBP formulae exist in other bases for some constants. For example, for π^2 we have both binary and ternary formulas (discovered by Broadhurst):

$$\pi^2 = \frac{9}{8}\sum_{k=0}^{\infty}\frac{1}{64^k}\left(\frac{16}{(6k+1)^2} - \frac{24}{(6k+2)^2} - \frac{8}{(6k+3)^2} - \frac{6}{(6k+4)^2} + \frac{1}{(6k+5)^2}\right). \tag{24}$$

$$\pi^2 = \frac{2}{27}\sum_{k=0}^{\infty}\frac{1}{729^k}\left(\frac{243}{(12k+1)^2} - \frac{405}{(12k+2)^2} - \frac{81}{(12k+4)^2} - \frac{27}{(12k+5)^2} - \frac{72}{(12k+6)^2} - \frac{9}{(12k+7)^2} - \frac{9}{(12k+8)^2} - \frac{5}{(12k+10)^2} + \frac{1}{(12k+11)^2}\right). \tag{25}$$

These two formulae have recently been used for record digit computations performed in conjunction with IBM Australia [Bailey, et al. 2013].

Also, the volume V_8 in *hyperbolic space* of the *figure-eight knot complement* is well known to be

$$V_8 = 2\sqrt{3}\sum_{n=1}^{\infty}\frac{1}{n\binom{2n}{n}}\sum_{k=n}^{2n-1}\frac{1}{k} = 2.029883212819307250042405108549\ldots$$

Surprisingly, it is also expressible as

$$V_8 = \frac{\sqrt{3}}{9}\sum_{n=0}^{\infty}\frac{(-1)^n}{27^n}\left\{\frac{18}{(6n+1)^2} - \frac{18}{(6n+2)^2} - \frac{24}{(6n+3)^2} - \frac{6}{(6n+4)^2} + \frac{2}{(6n+5)^2}\right\},$$

again discovered numerically by Broadhurst, and proved in Borwein and Bailey [2008]. A beautiful representation by Helaman Ferguson the mathematical sculptor is given in Figure 16. Ferguson produces art inspired by deep mathematics, but not by a formulaic approach.

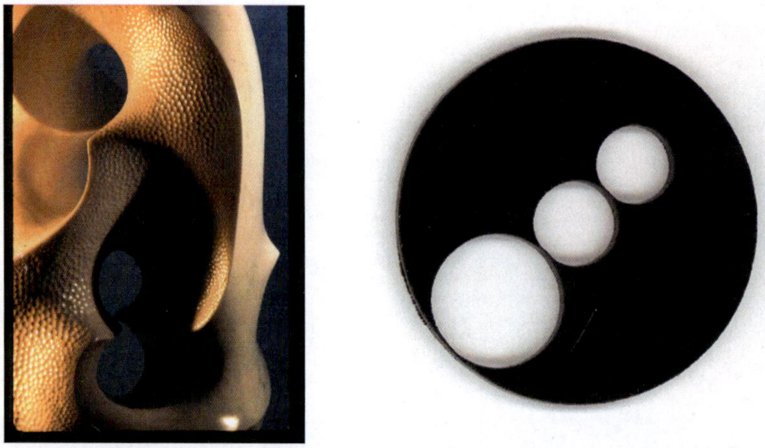

Figure 16: Ferguson's "Eight-Fold Way" and his BBP acrylic circles. These three "subtractive" acrylic circles (white) and the black circle represent the weights $[4, -2, -2, -1]$ in Equation (18). Courtesy of Guilio Einaudi Editori.

Normality and dynamics. Finally, Bailey and Crandall in 2001 made exciting connections between the existence of a b-ary BBP formula for α and its *normality* base b (uniform distribution of base-b digits). They make a reasonable, hence very hard, conjecture about the *uniform distribution of a related chaotic dynamical system.* This conjecture implies: *Existence of a "BBP" formula base b for α ensures the normality base b of α.* For $\log 2$, illustratively,[6] the dynamical system, base 2, is to set $x_0 = 0$ and compute

[6] In this case it is easy to use Weyl's criterion for equidistribution to establish this equivalence without mention of BBP numbers.

$$x_{n+1} \hookleftarrow 2\left(x_n + \frac{1}{n}\right) \mod 1.$$

π in the Third Millennium

Reciprocal Series

A few years ago Jesús Guillera found various Ramanujan-like identities for π using integer relation methods. The three most basic are:

$$\frac{4}{\pi^2} = \sum_{n=0}^{\infty} (-1)^n r(n)^5 (13 + 180n + 820n^2) \left(\frac{1}{32}\right)^{2n+1} \tag{26}$$

$$\frac{2}{\pi^2} = \sum_{n=0}^{\infty} (-1)^n r(n)^5 (1 + 8n + 20n^2) \left(\frac{1}{2}\right)^{2n+1} \tag{27}$$

$$\frac{4}{\pi^3} \stackrel{?}{=} \sum_{n=0}^{\infty} r(n)^7 (1 + 14n + 76n^2 + 168n^3) \left(\frac{1}{8}\right)^{2n+1}, \tag{28}$$

where $r(n) := (1/2 \cdot 3/2 \cdot \cdots \cdot (2n-1)/2)/n!$. Guillera proved (26) and (27) in tandem, using the *Wilf-Zeilberger algorithm* for formally proving hypergeometric-like identities very ingeniously [Borwein and Bailey 2008; Bailey, et al. 2007; Zudilin 2008]. No other proof is known and there seem to be no like formulae for $1/\pi^d$ with $d \geqslant 4$. The third (28) is certainly true,[7] but has no proof, nor does anyone have an inkling of how to prove it, especially as experiment suggests that it has no "mate" unlike (26) and (27) [Bailey, et al. 2007]. My intuition is that if a proof exists it is more a verification than an explication and so I stopped looking. I am happy just to know the beautiful identity is true. A very nice account of the current state of knowledge for Ramanujan-type series for $1/\pi$ is to be found in Baruah, Berndt and Chan [2009].

Guillera [2008a] produced another lovely pair of third millennium identities — discovered with integer relation methods and proved with creative telescoping — this time for π^2 rather than its reciprocal. They are

$$\sum_{n=0}^{\infty} \frac{1}{2^{2n}} \frac{\left(x+\frac{1}{2}\right)_n^3}{(x+1)_n^3} (6(n+x)+1) = 8x \sum_{n=0}^{\infty} \frac{\left(\frac{1}{2}\right)_n^2}{(x+1)_n^2}, \tag{29}$$

and

$$\sum_{n=0}^{\infty} \frac{1}{2^{6n}} \frac{\left(x+\frac{1}{2}\right)_n^3}{(x+1)_n^3} (42(n+x)+5) = 32x \sum_{n=0}^{\infty} \frac{\left(x+\frac{1}{2}\right)_n^2}{(2x+1)_n^2}. \tag{30}$$

Here $(a)_n = a(a+1)\cdots(a+n-1)$ is the rising factorial. Substituting $x = 1/2$ in (29) and (30), he obtained respectively the formulae

$$\sum_{n=0}^{\infty} \frac{1}{2^{2n}} \frac{(1)_n^3}{\left(\frac{3}{2}\right)_n^3} (3n+2) = \frac{\pi^2}{4} \qquad \sum_{n=0}^{\infty} \frac{1}{2^{6n}} \frac{(1)_n^3}{\left(\frac{3}{2}\right)_n^3} (21n+13) = 4\frac{\pi^2}{3}.$$

[7] Guillera ascribes (28) to Gourevich, who used integer relation methods. I've "rediscovered" (28) using integer relation methods with 30 digits. I then checked it to 500 places in 10 seconds, 1200 in 6.25 minutes, and 1500 in 25 minutes, with a naive command-line instruction in *Maple* on a light laptop.

Computational Records

The last decade has seen the record for computation of π broken in some very interesting ways. We have already described Kanada's 2002 computation in the section "Back to the Future," above, and noted that he also took advantage of the BBP formula in the section "How to Compute the N-th Digits of π," above. This stood as a record until 2009 when it was broken three times — twice spectacularly.

Daisuke Takahashi. The record for computation of π went from under 29.37 million decimal digits, by Bailey in 1986, to over 2.649 trillion places by Takahashi in January 2009. Since the same algorithms were used for each computation, it is interesting to review the performance in each case. In 1986 it took 28 hours to compute 29.36 million digits on 1 cpu of the then new CRAY-2 at NASA Ames using (18). Confirmation using the quadratic algorithm 16 took 40 hours. (The computation uncovered hardware and software errors on the CRAY. Success required developing a speedup of the underlying FFT [Borwein and Bailey 2008].) In comparison, on 1024 cores of a 2592 core *Appro Xtreme-X3* system 2.649 trillion digits via (16) took 64 hours 14 minutes with 6732 GB of main memory, and (18) took 73 hours 28 minutes with 6348 GB of main memory. (The two computations differed only in the last 139 places.) In April, Takahashi upped his record to an amazing 2,576,980,377,524 places.

Fabrice Bellard. Near the end of 2009, Bellard magnificently computed nearly 2.7 trillion decimal digits of π (first in binary) using the Chudnovsky series (14). This took 131 days but he only used a single 4-core workstation with a lot of storage and even more human intelligence!

Shiguro Kondo and Alexander Yee. In August 2010, they announced that they had used the Chudnovsky formula to compute 5 trillion digits of π over a 90-day period, mostly on a two-core Intel Xeon system with 96 Gbyte of memory. They confirmed the result in two ways, using the BBP formula (see below), which required 66 hours, and a variant of the BBP formula due to Bellard, which required 64 hours. Changing from binary to decimal required 8 days.

Full details are available at http://www.numberworld.org/misc_runs/pi-5t/details.html. As of October 2011, their record now stands at ten trillion digits [Bailey, et al. 2013].

... Life of Pi

Paul Churchland writing about the sorry creationist battles of the Kansas school board [Churchland 2007, Kindle ed, loc 1589] observes that:

> Even mathematics would not be entirely safe. (Apparently, in the early 1900's, one legislator in a southern state proposed a bill to redefine the value of pi as 3.3 exactly, just to tidy things up.)

As we have seen the life of π captures a great deal of mathematics — algebraic, geometric and analytic, both pure and applied — along with some history and philosophy. It engages many of the greatest mathematicians and some quite interesting characters along the way. Among the saddest and least-well understood episodes was an abortive 1896 attempt in Indiana to legislate the value(s) of π. The bill, reproduced in Berggren, Borwein and Borwein [2004, 231–235], is accurately described by David Singmaster [1985]. Much life remains in this most central of numbers.

At the end of the novel, Piscine (Pi) Molitor writes

> I am a person who believes in form, in harmony of order. Where we can, we must give things a meaningful shape. For example — I wonder — could you tell my jumbled story in exactly one hundred chapters, not one more, not one less? I'll tell you, that's one thing I hate about my nickname, the way that number runs on forever. It's important in life to conclude things properly. Only then can you let go.

We may well not share the sentiment, but we should celebrate that Pi knows π to be irrational.

End Notes

1. Why π is not 22/7. Today, even the computer algebra systems *Maple* or *Mathematica* "know" this since

$$0 < \int_0^1 \frac{(1-x)^4 x^4}{1+x^2}\,dx = \frac{22}{7} - \pi, \tag{31}$$

though it would be prudent to ask why each can perform the integral and whether to trust it. Assuming we do trust it, then the integrand is strictly positive on $(0, 1)$, and the answer in (31) is an area and so strictly positive, despite millennia of claims that π is 22/7. In this case, requesting the indefinite integral provides immediate reassurance. We obtain

$$\int_0^t \frac{x^4 (1-x)^4}{1+x^2}\,dx = \frac{1}{7}t^7 - \frac{2}{3}t^6 + t^5 - \frac{4}{3}t^3 + 4t - 4\arctan(t),$$

as differentiation easily confirms, and so the Newtonian fundamental theorem of calculus proves (31).

One can take the idea in Equation (31) a bit further, as in Borwein and Bailey [2008]. Note that

$$\int_0^1 x^4 (1-x)^4\,dx = \frac{1}{630}, \tag{32}$$

and we observe that

$$\frac{1}{2}\int_0^1 x^4 (1-x)^4\,dx < \int_0^1 \frac{(1-x)^4 x^4}{1+x^2}\,dx < \int_0^1 x^4 (1-x)^4\,dx. \tag{33}$$

Combine this with (31) and (32) to derive: $223/71 < 22/7 - 1/630 < \pi < 22/7 - 1/1260 < 22/7$ and so re-obtain Archimedes' famous computation

$$3\frac{10}{71} < \pi < 3\frac{10}{70}. \tag{34}$$

The derivation above was first popularized in *Eureka*, a Cambridge student journal in 1971.[8] A recent study of related approximations is Lucas [2009]. (See also Borwein and Bailey [2008].)

2. More about Gamma. One may define

$$\Gamma(x) = \int_0^\infty t^{x-1} e^{-t}\,dt$$

for $\mathrm{Re}\, x > 0$. The starting point is that

$$x\Gamma(x) = \Gamma(x+1), \qquad \Gamma(1) = 1. \tag{35}$$

In particular, for integer n, $\Gamma(n+1) = n!$. Also for $0 < x < 1$

$$\Gamma(x)\,\Gamma(1-x) = \frac{\pi}{\sin(\pi x)},$$

since for $x > 0$ we have

$$\Gamma(x) = \lim_{n\to\infty} \frac{n!\, n^x}{\prod_{k=0}^n (x+k)}.$$

[8] Equation (31) was on a Sydney University examination paper in the early sixties and the earliest source I know of dates from the forties [Borwein and Bailey 2008].

This is a nice consequence of the *Bohr-Mollerup theorem* [Borwein and Borwein 1987; Borwein and Bailey 2008], which shows that Γ is the unique log-convex function on the positive half line satisfying (35). Hence, $\Gamma(1/2) = \sqrt{\pi}$ and equivalently we evaluate the *Gaussian integral*

$$\int_{-\infty}^{\infty} e^{-x^2}\, dx = \sqrt{\pi},$$

so central to probability theory. In the same vein, the improper *sinc* function integral evaluates as

$$\int_{-\infty}^{\infty} \frac{\sin(x)}{x}\, dx = \pi.$$

Considerable information about the relationship between Γ and π is to be found in Borwein and Bailey [2008] and Eymard and Lafon [2003].

The Gamma function is as ubiquitous as π. For example, it is shown in Borwein, et al. [2011] that the *expected length*, W_3, of a three-step unit-length random walk in the plane is given by

$$W_3 = \frac{3}{16}\frac{2^{1/3}}{\pi^4}\Gamma^6\left(\frac{1}{3}\right) + \frac{27}{4}\frac{2^{2/3}}{\pi^4}\Gamma^6\left(\frac{2}{3}\right). \tag{36}$$

We recall that $\Gamma(1/2)^2 = \pi$ and that similar algorithms exist for $\Gamma(1/3), \Gamma(1/4)$, and $\Gamma(1/6)$ [Borwein and Borwein 1987; Borwein and Bailey 2008].

3. More about Complexity Reduction. To illustrate the stunning complexity reduction in the elliptic algorithms for π, let us write a *complete set of algebraic equations* approximating π to well over a trillion digits. The number π is transcendental and the number $1/a_{20}$ computed next is algebraic. *Nonetheless they coincide for over 1.5 trillion places.*

Set $a_0 = 6 - 4\sqrt{2}$, $y_0 = \sqrt{2} - 1$ and then solve the system in Figure 17.

$$y_1 = \frac{1-\sqrt[4]{1-{y_0}^4}}{1+\sqrt[4]{1-{y_0}^4}}, a_1 = a_0(1+y_1)^4 - 2^3 y_1(1+y_1+{y_1}^2)$$
$$y_2 = \frac{1-\sqrt[4]{1-{y_1}^4}}{1+\sqrt[4]{1-{y_1}^4}}, a_2 = a_1(1+y_2)^4 - 2^5 y_2(1+y_2+{y_2}^2)$$
$$y_3 = \frac{1-\sqrt[4]{1-{y_2}^4}}{1+\sqrt[4]{1-{y_2}^4}}, a_3 = a_2(1+y_3)^4 - 2^7 y_3(1+y_3+{y_3}^2)$$
$$y_4 = \frac{1-\sqrt[4]{1-{y_3}^4}}{1+\sqrt[4]{1-{y_3}^4}}, a_4 = a_3(1+y_4)^4 - 2^9 y_4(1+y_4+{y_4}^2)$$
$$y_5 = \frac{1-\sqrt[4]{1-{y_4}^4}}{1+\sqrt[4]{1-{y_4}^4}}, a_5 = a_4(1+y_5)^4 - 2^{11} y_5(1+y_5+{y_5}^2)$$
$$y_6 = \frac{1-\sqrt[4]{1-{y_5}^4}}{1+\sqrt[4]{1-{y_5}^4}}, a_6 = a_5(1+y_6)^4 - 2^{13} y_6(1+y_6+{y_6}^2)$$
$$y_7 = \frac{1-\sqrt[4]{1-{y_6}^4}}{1+\sqrt[4]{1-{y_6}^4}}, a_7 = a_6(1+y_7)^4 - 2^{15} y_7(1+y_7+{y_7}^2)$$
$$y_8 = \frac{1-\sqrt[4]{1-{y_7}^4}}{1+\sqrt[4]{1-{y_7}^4}}, a_8 = a_7(1+y_8)^4 - 2^{17} y_8(1+y_8+{y_8}^2)$$
$$y_9 = \frac{1-\sqrt[4]{1-{y_8}^4}}{1+\sqrt[4]{1-{y_8}^4}}, a_9 = a_8(1+y_9)^4 - 2^{19} y_9(1+y_9+{y_9}^2)$$
$$y_{10} = \frac{1-\sqrt[4]{1-{y_9}^4}}{1+\sqrt[4]{1-{y_9}^4}}, a_{10} = a_9(1+y_{10})^4 - 2^{21} y_{10}(1+y_{10}+{y_{10}}^2)$$
$$y_{11} = \frac{1-\sqrt[4]{1-{y_{10}}^4}}{1+\sqrt[4]{1-{y_{10}}^4}}, a_{11} = a_{10}(1+y_{11})^4 - 2^{23} y_{11}(1+y_{11}+{y_{11}}^2)$$
$$y_{12} = \frac{1-\sqrt[4]{1-{y_{11}}^4}}{1+\sqrt[4]{1-{y_{11}}^4}}, a_{12} = a_{11}(1+y_{12})^4 - 2^{25} y_{12}(1+y_{12}+{y_{12}}^2)$$
$$y_{13} = \frac{1-\sqrt[4]{1-{y_{12}}^4}}{1+\sqrt[4]{1-{y_{12}}^4}}, a_{13} = a_{12}(1+y_{13})^4 - 2^{27} y_{13}(1+y_{13}+{y_{13}}^2)$$
$$y_{14} = \frac{1-\sqrt[4]{1-{y_{13}}^4}}{1+\sqrt[4]{1-{y_{13}}^4}}, a_{14} = a_{13}(1+y_{14})^4 - 2^{29} y_{14}(1+y_{14}+{y_{14}}^2)$$
$$y_{15} = \frac{1-\sqrt[4]{1-{y_{14}}^4}}{1+\sqrt[4]{1-{y_{14}}^4}}, a_{15} = a_{14}(1+y_{15})^4 - 2^{31} y_{15}(1+y_{15}+{y_{15}}^2)$$
$$y_{16} = \frac{1-\sqrt[4]{1-{y_{15}}^4}}{1+\sqrt[4]{1-{y_{15}}^4}}, a_{16} = a_{15}(1+y_{16})^4 - 2^{33} y_{16}(1+y_{16}+{y_{16}}^2)$$
$$y_{17} = \frac{1-\sqrt[4]{1-{y_{16}}^4}}{1+\sqrt[4]{1-{y_{16}}^4}}, a_{17} = a_{16}(1+y_{17})^4 - 2^{35} y_{17}(1+y_{17}+{y_{17}}^2)$$
$$y_{18} = \frac{1-\sqrt[4]{1-{y_{17}}^4}}{1+\sqrt[4]{1-{y_{17}}^4}}, a_{18} = a_{17}(1+y_{18})^4 - 2^{37} y_{18}(1+y_{18}+{y_{18}}^2)$$
$$y_{19} = \frac{1-\sqrt[4]{1-{y_{18}}^4}}{1+\sqrt[4]{1-{y_{18}}^4}}, a_{19} = a_{18}(1+y_{19})^4 - 2^{39} y_{19}(1+y_{19}+{y_{19}}^2)$$
$$y_{20} = \frac{1-\sqrt[4]{1-{y_{19}}^4}}{1+\sqrt[4]{1-{y_{19}}^4}}, \mathbf{a_{20}} = a_{19}(1+y_{20})^4 - 2^{41} y_{20}(1+y_{20}+{y_{20}}^2)$$

Figure 17: π to 1.5 trillion places

This quartic algorithm, with the Salamin-Brent scheme, was first used by Borwein, Borwein and Bailey [1989] and was used repeatedly by Yasumasa Kanada (see Figure 12) in Tokyo in computations of π over 15 years or so, culminating in a 200 billion decimal digit computation in 1999. As recorded in Figure 11, it has been used twice very recently by Takahashi. Only thirty-five years earlier in 1963, Dan Shanks — a very knowledgeable participant — was confident that computing a billion digits was forever impossible. Today it is reasonably easy on a modest laptop. A fine self-contained study of this quartic algorithm — along with its cubic confrere also described in section on "Reduced Operational Complexity Algorithms," above — can be read in Guillera [2008b]. The proofs are nicely refined specializations of those in Borwein and Borwein [1988].

4. The Difficulty of Popularizing Accurately. Churchland [2007] offers a fascinating set of essays full of interesting anecdotes — which I have no particular reason to doubt — but the brief quote on page 556, above, contains four inaccuracies. As noted above: (i) the event took place in 1896/7 and (ii) in Indiana (a northern state); (iii) the prospective bill, #246, offered a geometric construction with inconsistent conclusions and certainly offers no one exact value. Finally, (iv) the intent seems to have been pecuniary not hygienic [Singmaster 1985]. As often, this makes me wonder whether mathematics popularization is especially prone to error or if the other disciplines just seem better described because of my relative ignorance.

On April 1, 2009, an article entitled "The Changing Value of Pi" appeared in the *New Scientist* with an analysis of how the value of π has been increasing over time. I hope, but am not confident, that all readers noted that April 1st is April Fool's Day. (See entry seven of http://www.museumofhoaxes.com/hoax/aprilfool/.)

Following π on the Web. One can now follow π on the web through *Wikipedia*, *MathWorld* or elsewhere, and indeed one may check the performance of π by looking up "Pi" at http://www.google.com/trends. It shows very clear seasonal trends, with a large spike around Pi Day.

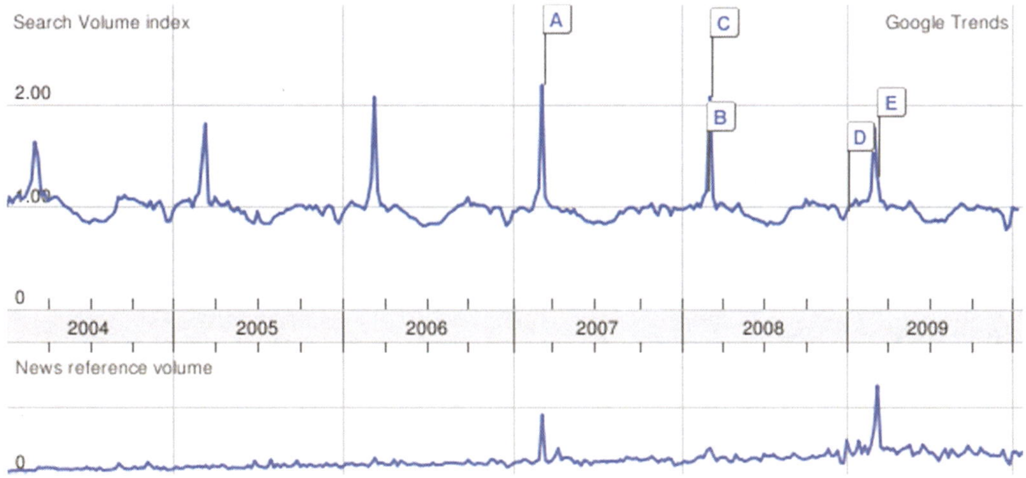

Figure 18: Google's trend line for "Pi."

Acknowledgements. Thanks are due to many, especially my close collaborators P. Borwein and D. Bailey.

References

Amoroso, F., Viola, C., 2008. Irrational and transcendental numbers. In: Bartocci, C., Odifreddi, P. (eds.), Mathematics and Culture, Volume II. La matematica: Problemi e teoremi. Turino: Guilio Einaudi Editori.

Arndt, J., Haenel, C., 2001. Pi Unleashed. Springer-Verlag, New York.

Bailey, D.H., Borwein, J.M., 2011. Exploratory experimentation and computation. Notices of the American Mathematical Society 58, 1410–1419.

Bailey, D., Borwein, J., Calkin, N., Girgensohn, R., Luke, R., Moll, V., 2007. Experimental Mathematics in Action. A.K. Peters, Wellesley, MA.

Bailey, D.H., Borwein, J.M., Mattingly, A., Wightwick, G., 2013. The computation of previously inaccessible digits of π^2 and Catalan's Constant. Notices of the American Mathematical Society 50, 844–854.

Berggren, J.L., Borwein, J.M., Borwein, P.B., 2004. Pi: A Source Book, 3rd edition. New York, Springer.

Baruah, N.D., Berndt, B.C., Chan, H.H., 2009. Ramanujan's series for $1/\pi$: A survey. American Mathematical Monthly 116, 567–587.

Blatner, D., 1997. The Joy of Pi. Walker and Co., New York.

Borwein, J.M., 1998. Brouwer-Heyting sequences converge. Mathematical Intelligencer 20, 14–15.

Borwein, J.M., 2008. La vita di pi greco: from Archimedes to ENIAC and beyond, in Bartocci, C., Odifreddi, P. (eds.), Mathematics and Culture, Volume II. La matematica: Problemi e teoremi, Guilio Einaudi Editori, Turino, pp. 249–285.

Borwein, J.M., Bailey, D.H., 2008. Mathematics by Experiment: Plausible Reasoning in the 21st Century, 2nd expanded edition. A.K. Peters, Wellesley, MA.

Borwein, J.M., Borwein, P.B., 1987. Pi and the AGM. Wiley, New York.

——— 1988. Ramanujan and Pi. Scientific American 256, 112–117. Reprinted: Berndt, B.C., Rankin, R.A. (eds.), 2001, Ramanujan: Essays and Surveys. American Mathematical Society, Providence, pp. 187–199. (Also in: Berggren, Borwein and Borwein [2004].)

Borwein, J.M., Borwein, P.B., Bailey, D.H., 1989. Ramanujan, modular equations and approximations to pi or how to compute one billion digits of pi. American Mathematical Monthly 96, 201–219. Reprinted in: Organic Mathematics Proceedings, http://www.cecm.sfu.ca/organics, 1996. (Also in: Berggren, Borwein and Borwein [2004].)

Borwein, J., Nuyens, D., Straub, A., Wan, J., 2011. Some arithmetic properties of short random walk integrals. The Ramanujan Journal 26, 109–132.

Churchland, P., 2007. Neurophilosophy at Work. Cambridge University Press, Cambridge.

Eymard, P., Lafon, J.-P., 2003. The Number π. American Mathematical Society, Providence.

Guillera, J., 2008a. Hypergeometric identities for 10 extended Ramanujan-type series. Ramanujan Journal 15, 219–234.

——— 2008b. Easy proofs of some Borwein algorithms for π. American Mathematical Monthly 115, 850–854.

Heath, T.L., 1912. The Works of Archimedes, Cambridge University Press, Cambridge.

Lucas, S.K., 2009. Integral approximations to *pi* with nonnegative integrands. American Mathematical Monthly 116, 166–172.

McCartney, S., 1999. ENIAC: The Triumphs and Tragedies of the World's First Computer. Walker and Co., New York.

Schloper, H.C., The chronology of pi. Mathematics Magazine 23, 165–170, 216–228, 279–283. (Also in: Berggren, Borwein and Borwein [2004].)

Singmaster, D., 1985. The legal values of Pi. Mathematical Intelligencer 7, 69–72. (Also in: Berggren, Borwein and Borwein [2004].)

Tsumura, H., 2004. An elementary proof of Euler's formula for $\zeta(2n)$. American Mathematical Monthly, 430–431.

von Baeyer, H.C., 2003. Information: The New Language of Science. Harvard University Press, Cambridge, MA.
Zudilin, W., 2008. Ramanujan-type formulae for $1/\pi$: A second wind. ArXiv:0712.1332v2.

There are many other Internet resources on π; a reliable selection is kept at http://www.experimentalmath.info.

Index of Personal Names

Aaboe, A., 329, 341
Abī, *see* next part of the name
Abū, *see* next part of the name
Aballagh, M., 85, 280, 281, 283–285, 287
ʿAbbās I (Shāh), 488, 489, 492, 495
ʿAbbās II (Shāh), 488, 489, 499
ʿAbdallāh al-Mursī, 286
Abdeljaouad, M., 122, 280–282, 354, 520
ʿAbd al-Ḥamīd, 346, 350, 354
ʿAbd al-Laṭīf, ʿA.I., 244
ʿAbd al-Raḥmān, M., 298, 300, 303, 304, 310, 324, 325
Ibn ʿAbdūn, 112, 286
Abgrall, P., 81, 110–111, 117
Acerbi, F., 26–30, 35, 36, 38–41
Ackerberg-Hastings, A., 504, 505, 519
Adelard of Bath, 103–104
Adıvar, A., 243
Ağargün, A., 79, 123
Agathon, 240
Aghānis, 105
al-Aḥdab, Isaac ibn, 283
Ahlwardt, W., 351
Ahmad, S., 54, 91, 352
Aḥmad ibn Yūsuf, Abū Jaʿfar, 347
Aḥmad ibn Thabāt, 112
Aḥmad ibn Yūsuf, 89
Ahmedov, A., 115
al-Ahwāzī, 63
Ibn al-Akfānī, Muḥammad ibn Ibrāhīm, 351, 354
al-, *see* alphabetically by name
Albertus Magnus, 104
Alhazen, *see* Ibn al-Haytham
ʿAlī (Caliph), 74
Ali, J., 65
ʿAlījān Muḥammad Khān (Vizier), 489
Alkhateeb, H.M., 122
Allard, A., 74
Alvarez, C., 34
Ibn Amājūr, 354
Bahāʾ al-Dīn al-ʿĀmilī, 487
Amīn, A., 347
Amoroso, F., 544
al-Āmulī, 103, 112
Anagnostikas, C., 82
Anbouba, A., 53, 54, 58, 61, 343, 344, 347, 353–355, 361
Anderson, R., 522
Andrew, J., 521
al-Anṭākī, 122
Antiochus IV Epiphanes (King), 150, 151
Apéry, R., 539
Apollonius, 11, 17, 19, 20, 27, 29–110, 113, 149–151, 170, 171, 200, 240, 244, 267, 278, 284, 285, 410–413, 415, 430, 437, 439, 447, 450, 459, 465
Apollonius the Carpenter, 170, 171
Aqāṭun, 11
Aratus, 152
Archimedes, 3, 7, 9–12, 17, 19, 20, 28, 30–32, 35, 36, 39–41, 61, 63, 65, 80–81, 84, 87, 106, 107, 109, 110, 113, 145–149, 151–153, 166, 170, 171, 189, 194, 195, 197, 199–223, 228, 239–246, 259–264, 267, 268, 388, 393, 412, 413, 518, 531, 533–536, 557
Archytas, 20–21, 30, 38, 42, 43
Aristarchus, 27, 32, 43, 107
Ariston, 148
Aristotle, 6, 18, 20, 29, 30, 42, 85, 111–114, 145, 146, 153, 160, 240
Arnaldi, M., 180
Arnzen, R., 104

N. Sidoli and G. Van Brummelen (eds.), *From Alexandria, Through Baghdad: Surveys and Studies in the Ancient Greek and Medieval Islamic Mathematical Sciences in Honor of J.L. Berggren*, DOI 10.1007/978-3-642-36736-6,
© Springer-Verlag Berlin Heidelberg 2014

Arsinoe (Princess), 147
Artmann, B., 19
Āryabhaṭa, 469–471
al-Asad, Mohammad, 87
Ibn ʿAsākir, 348
Asper, M., 33, 38
Assali, S.-A., 288
Athenaeus the Mechanic, 146
Athenaeus of Naucratis, 147, 148
Ibn al-Athīr, 346
Attalus, 149
Attalus I (King), 149
Aujac, G., 29, 145, 153
Autolycus, 27, 63, 153, 229
Avicenna, 113
Avigad, J., 34
Ayyūbī, Aḥmad, 329
Azarian, M., 113, 121
Ibn ʿAzzūz al-Qusanṭīnī, 288, 310

Bülow-Jacosen, A., 27
Babaie, S., 492, 493, 495, 497
Babayan, K., 492, 493, 495, 497
Bachet, C., 52
Bacon, F., 37
Bacon, Francis, 522
Badr al-Dīn Muḥammad, 244, 245
al-Baghdādī, 52, 54, 58, 63, 121, 123, 395, 476
al-Baghdādī, Abd al-Laṭīf, 113
Baghdiantz-McCabe, I., 492, 493, 495, 497
Bagheri, M., 108, 109, 120, 121
Bailey, D.H., 532, 536, 539, 545, 546, 551–559
Ibn Bājja, 278, 283, 285
Abū Bakr, 286, 287
Balabhadra, 472
al-Balawī, Abū Muḥammad, 347, 353
al-Balawī, Ibrāhīm ibn Muḥammad, 329
Ibn Bāmshād al-Qāʾinī, 352, 353
Bancel, F., 110
Ibn al-Bannāʾ, 85, 89, 119, 121, 275–278, 280, 281, 283–286, 288, 297, 298, 302, 306, 307, 312, 313, 315, 317, 360
Banū Mūsā, Muḥammad, Aḥmad, and al-Ḥasan, 76, 81, 107, 111, 244, 245, 285, 348–350, 354, 413
al-Baqqār, Abū ʿAbd Allāh, 117
Bar, *see* next part of the name
Barany, M., 512
Barceló, C., 283
Barker, A., 33, 43
Barrow, Isaac, 260, 505, 509
Baruah, N.D., 555
Abū Barza, 346, 350, 354
Bashmakova, I.G., 10
Basilius of Tyre, 410
Ibn Bāṣo, 83, 311, 338, 339
al-Baṣrī, Ayyūb, 75
al-Battānī, 75, 116, 120, 303, 304, 306–311, 317–320, 322, 354
Bāyazīd II (Sultan), 243
Bebbouchi, R., 78
Becker, O., 8, 18, 191
Beckmann, F., 19
Beckmann, P., 551
Beeson, M., 34
Bekker, I., 149
Bellard, F., 546, 552, 556
Bellosta, H., 29, 31, 79, 105, 107, 109, 112, 114, 115, 119, 123
Bellver, J., 299, 309
Berggren, J.L., 10, 20, 21, 25, 29, 30, 41, 42, 52, 61, 63–65, 76, 80, 82, 84, 88, 90, 101–102, 104, 107, 109–111, 117, 120, 121, 145, 153, 154, 167, 170, 171, 180, 232, 239, 243, 244, 259, 260, 262, 263, 265, 275, 284, 329, 336, 341, 409, 531, 533, 535, 538–540, 542, 545, 548, 550, 556
Bernard, A., 37, 39, 40
Berndt, B.C., 555
Berryman, S., 43, 146, 147, 153
Bézout, E., 520
Besthorn, R.O., 257
Bhāskara I, 469–471
Bhāskara II, 469, 484
Bickerman, E.J., 156
Bilge, K., 346
Bir, A., 244
al-Birjandī, ʿAbd al-ʿAlī, 241, 244
al-Bīrūnī, 51, 63, 65, 76, 77, 79, 82–84, 87, 90, 91, 106–108, 112, 113, 115, 117, 189, 243, 245, 305, 306, 308, 310, 311, 469, 470, 472–476, 479–484
al-Bisṭāmī, ʿAbd al-Raḥmān, 240, 243
Bitsakis, Y., 33, 185
El-Bizri, N., 113
Blatner, D., 531
Blochet, E., 352, 493
Blume, F., 390, 395, 396
Blythe, P.H., 146
Bodnár, I., 146
Boethius, 21
Boilot, D.J., 473
Bonnycastle, J., 520

Borelli, G.A., 259–261, 263–267, 269–272
Borwein, D., 553
Borwein, J.M., 531–533, 535, 536, 538–540, 542, 545–548, 550, 552–559
Borwein, P.B., 531, 533, 535, 536, 538–540, 542, 545–548, 550, 551, 556, 558, 559
Borzacchini, L., 38, 41
Bottecchia Dehò, M.E., 146
Boulahia, N., 115
Bouzari, A., 280, 284, 285, 287
Bouzoubaâ Fennane, K., 104
Bowen, A.C., 21, 30
Boyer, M., 540, 545
Brahmagupta, 54, 76, 469, 470, 472, 473, 475, 481, 484
Breccia, E., 178
Brent, R., 546
Brentjes, S., 30, 77, 88, 91, 103–104, 107, 108, 488, 503
Brethren of Purity, 78
Broadhurst, D.J., 553, 554
Brockelmann, C., 347, 353
Brougham, H.P. (Lord), 522
Brouncker, W., 538
Brouwer, L.E.J., 548
Bryce, S., 124
Buchner, E., 178
Bulgakov, P.G, 65
al-Būnī, 59
Burkert, W., 18, 21
Burnyeat, M.F., 9
al-Būzjānī, *see* Abū al-Wafā
Busard, H.L.L., 63, 103, 286, 287
al-Bustānī, B., 347
Bâltâc, A., 181

Caiozzo, A., 497
Cajori, F., 344
al-Čalabī, D., 356
Calkin, N., 555
Callippus, 145, 167, 180
Calvo, E., 83, 116, 288, 298, 307, 311
Campanus, 76
Camus, A., 361
Cantor, M., 390, 393–396, 398
Carandell, J., 77, 308
Carman, C.C., 33, 43, 158, 160, 162, 163, 168–170
Carpos of Antioch, 149
Carra de Vaux, B., 147, 148, 170
Carroll, L., *see* Dodgson, C.L.
Casiri M., 346, 349
Casulleras, J., 120, 288, 298, 307, 310, 311, 324, 333
Cavaleieri, B.F., 203
Caveing, M., 19
Centrone, B., 18, 21
van Ceulen, Ludolph, 113, 536, 537
Chabás, J., 89, 116, 306, 321, 322
Chabert, J.-L., 469
Chalhoub, S., 123, 345
Challinor, J., 504
Chan, H.H., 555
Charette, F., 115, 119
Chase, W.G., 228
Chemla, K., 39, 42, 74, 89
Cherbonneau, M.A., 275
Cherniss, H., 5
Christianidis, J., 40, 41
Chudnovsky, D., 546, 549, 550
Chudnovsky, G., 546, 549, 550
Churchland, P., 556, 559
Cicero, 145, 148, 166
Clagett, M., 63, 146
Clark, R., 228
Cleomedes, 30
Cleopatra II (Queen), 147
Colebrooke, H.T., 472, 473
Colin, G.S., 283
Comes, M., 81, 90, 116–118, 287, 298, 299, 303, 325
Comes, R., 283
Copernicus, N., 87
Cowley, A.E., 412, 413
Crandall, R., 554
Crates of Mallus, 153
Creese, D., 33, 43
Crozet, P., 81, 104, 105, 110, 112, 115, 520
Crönert, W., 150
Ctesibius, 147
Cultrera, G., 211
Cuomo, S., 26, 32, 34, 37, 39
Curtze, M., 286
Czinczenheim, C., 27, 28, 255, 256

van Dalen, B., 87, 89, 118, 119, 300, 310
d'Alessandro, P., 28, 199
Dallal, A., 77, 83, 84, 87, 102, 108, 116, 118, 308
Dalley, S., 148
Darwīsh ibn Muḥammad ibn Luṭfī, 242
Dase, J.Z., 540, 542
David, A., 409
Davidian, M.-L., 306
Davidson Weinberg, G., 154

De Morgan, A., 522
De Young, G., 30, 32, 63, 77, 78, 103–106, 503
Dean, E., 34
Debarnot, M.-T., 63, 65, 115, 305, 306, 310
Decorps-Foulquier, M., 27, 29, 31, 32, 39, 107
Deleuze, G., 37
della Francesca, P., 209
Demetrius I Soter (King), 150, 151
Democritus, 180
Dhanani, A., 78
Díaz-Fajardo, M., 117, 299, 302, 312
Dicks, D.R., 17
Didymus, 387, 396
Diels, H., 178
Dijksterhuis, E.J., 9, 189, 195
al-Dīnawarī, 351, 353
Diocles, 11, 106, 261, 262, 268
Diodorus Siculus, 147
Diogenes Laertius, 146
Dionysodorus, 150, 261, 262, 268
Diophantus, 11, 17, 29, 35, 40, 43, 58, 277, 345–347, 358
Dioscordes, 489
Dizer, M., 338
Djebbar, A., 28–30, 51, 52, 56, 59, 77, 80, 85, 86, 89–91, 104–106, 108, 111–112, 123, 240, 275, 276, 279–288, 364, 412
Dodge, B., 148, 412
Dodgson, C.L., 542
Dold-Samplonius, Y., 11, 75, 87, 109, 112
Domingues, J.C., 235
Dorce, C., 119, 311
Dorrien, G., 522
Drachmann, A.G., 9, 10, 147, 148
Duke, D., 165, 167
Dupuis, J., 154
Ibn Durayd, 278
Dvivedī, S., 472, 473

Easton, J., 514
Eastwood, B., 149
Ecchellensis, A., 259–261, 264–269, 271, 272
Eco, U., 533
Edmunds, M., 171
Edwards, H., 492, 493
Ehrenkreutz, A.S., 53
Ehrhardt, N., 175
Elia, *see* Prigotta, Elia
Eliezer, 409, 416, 446, 457
Elshakry, M., 506, 510, 514, 521, 524
Endress, G., 113
Eneström, G., 275
Engeler, E., 228
Engroff, J.W., 63, 77
Epicurus, 150
Eratosthenes, 5, 61, 152, 199, 202, 211
Euclid, 5, 7, 9, 17, 19–21, 28–30, 32, 34–36, 38, 39, 41, 42, 52, 63, 74, 75, 77–81, 85, 102–106, 108–112, 114–115, 123, 146, 149, 152, 153, 200, 210, 229, 232, 234, 235, 240, 244, 245, 257, 277, 278, 284, 286, 287, 348, 354, 361, 410–413, 415, 416, 430–433, 439, 445, 460, 474, 475, 488, 499, 504, 505, 508, 510, 516
Euctemon, 180
Eudemus, 17, 38, 149–151
Eudoxus, 6–8, 18, 21, 41, 86, 105, 145, 152, 154, 167, 180, 234
Euler, L., 533, 538–540, 545
Eutocius, 32, 39, 149, 151, 152, 259, 261–264, 266–269, 411–413
Evans, J., 30, 32, 33, 42, 43, 145, 153, 154, 158, 160, 162, 167–169, 180
Evarts, J., 521
Eymard, P., 531, 558
Ibn ʿEzra, Abraham, 279, 303

Fakhr al-Mulk, 344
al-Fanārī, Muḥammad, 239–241
al-Fanārī, Muḥammad Shāh, 240, 241, 243
al-Fārābī, 64, 114
Abū al-Faraj, 245
Farès, N., 75
al-Farghānī, 82, 117, 333, 349, 488
Farhad, M., 492, 493, 495, 497
al-Fārisī, Kamāl al-Dīn, 52, 79, 85, 109, 114, 123, 242, 243, 245
Fatḥallāh al-Shirwānī, 241, 244
Fazlıoğlu, İ., 240, 241, 243, 244
Federspiel, M., 29, 31, 107
Feke, J., 37
Feraru, R.M., 181
Ferguson, A., 504
Ferguson, D.F., 540, 541
Ferguson, H., 554
Fibonacci, Leonardo of Pisa, 343, 355, 360
Finzi, M., 345
Fischler, R., 9
Fletcher, C., 79
Flügel, G., 345, 346, 350, 359
Folkerts, M., 74, 121
Fontaine, R., 417
Forcada, M., 90
Fortna, B.C., 521
Fournarakis, P., 41

Fowler, D., 7–8, 18, 20, 26, 27
Fraser, P.M., 150
Frederick II of Sicily, 417
Freeth, T., 33, 154, 157, 158, 160, 162, 163, 167, 171, 185
Frege, G., 228
Freudenthal, G., 107, 412
Freudenthal, H., 6, 18
Friberg, J., 27, 38, 40
Fried, M., 31, 36, 37, 463
Friederich, S., 228
von Fritz, K., 18

Gabrieli, G., 346
Abū Ǧaᶜfar Aḥmad ibn Yūsuf, *see* Aḥmad ibn Yūsuf, Abū Jaʿfar
Galen, 43
Gallo, I., 150, 151
Gallus, Gaius Sulpicius, 148
Gallway, W.F., 553
Gamba, E., 209
Gandz, S., 54
Gardies, J.-L., 19
Gauss, C.F., 542, 546, 550
Geminus, 30, 35, 41, 42, 153, 154, 167, 180, 186, 187, 233
Gera, D., 151
Gerard of Cremona, 103–104, 116, 255, 256
Ghasemloo, F., 493, 497
al-Ghurbī, 280, 282, 284
Gibbs, S., 180, 181, 183
Gīlānī, Muḥammad Bāqir ibn Mullā Maḥmūd, 492
Girgensohn, R., 555
Gluskina, G.M., 411
de Goeje, M.J., 346
Goldstein, B., 9, 12, 89, 116, 118, 302, 303, 306, 308, 309, 320–322
Goldziher, I., 88
Gosper, W., 546, 549
Gourevich, D., 555
Grabill, J.L., 524
Grant, H., 91
Grasshoff, G., 29
Grattan-Guinness, I., 20
Greaves, J., 260
Gregory, J., 539
Grube, E., 491, 492, 495
Guergour, Y., 105, 111, 121, 283–285
Guesdon, M.-G., 283
Guillelmus, 345
Guillera, J., 555, 559
Guilloud, J., 540, 545
Gunther, R.T., 333, 337, 338
Gutas, D., 102

Ḥabash al-Ḥāsib, 64, 76, 82, 83, 115, 117, 119, 306, 334
al-Habbāk, 299, 305
Habsieger, L., 39
al-Ḥāfiẓ, Muḥammad Bāqir, 492
Hāǧǧī Ḫalīfa, *see* Ḥājjī Khalīfa
Ibn al-Hāʾim, 116, 122, 280, 281, 287, 298, 302, 303, 316, 324, 354
al-Ḥajarī, 324
al-Ḥajjāj ibn Yūsuf, 77–78, 102–104, 350, 356
Ḥājjī Khalīfa, 346, 349–351, 354, 356, 359
Ibrāhim al-Ḥalabī, 113
Halley, E., 29, 540
Hamadanizadeh, J., 64
Hamedani, H.M., 114
Hamm, E.A., 160, 165
al-Ḥanbalī, Abū Yaʿlā, 353
Hansberger, R., 412
Harari, O., 42, 235
Harbili, A., 280, 281, 283, 284
Harley, J., 91
Hartner, W., 54
Harūn al-Rashīd (Caliph), 77, 348
al-Ḥasan al-Murrākushī, 288
al-Ḥaṣṣār, 85, 86, 121, 275, 277, 280, 282, 284, 289
Hata, M., 544
Hauser, F., 148, 343
al-Hawārī, Muḥammad ibn Aḥmad, 299, 300, 312
Hayashi, E., 201, 207
Ibn Haydūr, 280, 282–285
Abū al-Ḫayr, 103
Ibn al-Haytham, 52, 61, 63, 64, 77, 80, 82, 84–87, 103, 104, 106, 107, 109–114, 116, 119–121, 240, 244, 245, 278, 308, 514
al-Ḫāzin, *see* al-Khāzin
Heath, T.L., 9, 17, 43, 146, 149–152, 203, 208, 217, 229, 230, 232, 234, 235, 259–261, 263, 265, 410–412, 504, 518
Hedylos, 147
Heiberg, J.L., 17, 20, 28, 31, 78, 104, 146, 149, 200, 203, 204, 206, 207, 210, 216, 217, 220–223, 257, 259, 387, 396, 400
Heinen, A., 84
Hellman, G., 227
Henry, R., 149
Heraclides of Pontus, 149
Hermann of Carinthia, 103

Hermarchus, 150
Hermelink, H., 54, 59
Hermes, 311
Hermite, C., 543
Heron, 7, 9, 12, 25, 27, 29, 33, 35, 36, 38, 43, 104, 123, 147, 148, 189–192, 194, 195, 197, 360, 388, 390, 393, 395, 398
Herz-Fischler, R., 78
Heslin, P., 178
Hett, W.S., 146
Heyting, A., 548
Hieron II (King), 147, 211
Hilbert, D., 34
Hill, D.R., 12, 90, 148, 170, 171
Hintikka, J., 5
Hipparchus, 11, 39, 145, 152, 160, 161, 167, 168
Hippasus of Metapontum, 7
Hippocrates of Chios, 7, 38, 151
Bar Ḥiyya, Abraham, 286, 287
Hjelmslev, J., 9
Hochheim, A., 53, 344
Hogendijk, J.P., 28, 41, 52, 61, 75, 76, 78–80, 82, 85, 90, 91, 105, 107, 109–115, 117, 119, 120, 242, 244, 260, 278, 284, 285, 287, 310, 311, 409, 414, 415, 443, 446, 447, 451, 452, 454, 463–465, 473
Homer, 19, 37, 153
Hopkins, J.F.P., 90
al-Houjairi, M., 111, 115
Houzel, C., 75, 105, 111, 123
Høyrup, J., 74, 88, 395
al-Ḥubūbī, 121, 289
Ibn Hūd, Yūsuf al-Muʾtaman, *see* al-Muʾtaman
Huffman, C.A., 21, 38
al-Ḥūfī, 277, 289
Hughes, B., 75
Hūlāgū Khān, 86
Hultsch, F., 149
Hunt, D.W.S., 178
Hutton, J., 504
Huxley, G., 150
Huygens, C., 114
al-H̱wārizmī, *see* al-Khwārizmī
Hypatia, 347
Hypsicles, 29, 410–412

Iamblichus, 18, 52
Ibn, *see* next part of the name
Ibrāhīm ibn al-ʿAjamī, 348
Ibrāhīm ibn Sinān, 79, 107, 109, 112, 116, 119, 120, 245, 278, 285
al-Idrīsī, 91
Ifrah, G., 536
İhsanoğlu, E., 101, 240, 242, 243, 280, 520
al-Ījī, 88
Ibn al-Imām, 278
al-ʿImrānī, Aḥmad, 346, 349, 354
Ionescu-Cârligel, C., 181, 183, 185
Irani, R.A.K., 64
Isḥāq ibn Ḥunayn, 77, 103, 257, 262, 284
Ibn Isḥāq al-Tūnisī, 297–299, 301–303, 306–309, 315, 318, 320, 322
al-Iṣṭakhrī, 346, 349, 354
İzgi, C., 245

Jābir ibn Aflaḥ, 86, 104, 287, 413
Jacob ben Machir, 256
Jacob of San Cassiano, 28
al-Jādirī, 288, 299, 305
al-Jahshiyārī, 348, 353
Jalāl al-Dīn Munajjim, 495
Jaouiche, K., 65, 78, 80, 109, 348
al-Jārim, ʿA., 347
Ibn al-Jauzī, 346
al-Jawharī, al-ʿAbbās ibn Saʿīd, 105, 348
al-Jawharī, Saʿīd, 348
Ibn al-Jayyāb, 286
al-Jayyānī, *see* Ibn Muʿādh
Jensen, C., 64
Johannes de Dumpno, 306
Johannes Hispalensis, *see* John of Seville
Johansson, B.G., 121
John of Seville, 357
Jolivet, J., 412
Jones, A., 21, 27, 29, 30, 32, 37, 42, 151, 154, 157, 160, 162, 163, 169, 171, 185, 336
Jones, W., 533
Abū al-Jūd ibn al-Layth, 110, 111, 245, 354
Judah ben Solomon ha-Cohen, 417
al-Jurjānī, Abū Saʿīd al-Ḍarīr, 112

Kaçar, M., 244
Kalonymos Todros, 414, 450, 465
Kalonymus ben Kalonymus ben Meir, 414, 450, 455
Kaltsas, N., 154
Kamāl al-Dīn ibn Yūnus, 111
Abū Kāmil, 56, 112, 123, 276, 343–347, 349–357, 359–361, 364, 386, 390, 398–400
Ibn al-Kammād, 116–117, 302, 306, 315, 316, 318, 321–323
Kanada, Y., 546, 548–550, 552, 556, 559
Kanani, N., 105

al-Karābīsī, 104, 353, 354
al-Karajī, 53–56, 59, 85, 105, 106, 112, 123, 276, 343–345, 353, 355, 356
al-Karkhī, *see* al-Karajī
Karpinski, L.C., 345, 349, 355
al-Kāshī, 51, 53, 54, 56, 64, 66, 77, 87–88, 91, 108, 112, 113, 115, 116, 118, 120, 121, 242, 243, 305, 536
al-Kāshī, Bābā Afḍal al-Dīn, 120
al-Kāshī, ʿImād al-Dīn, 242
Katz, V., 101
Kazarian, M., 39
Kennedy, E.S., 51, 63, 65, 76, 83, 84, 86, 87, 117–121, 297, 298, 300, 303–307, 310–312, 320, 321, 323, 324, 333, 344
Kennedy, M.H., 84, 333
Kepler, Johannes, 81
Ibn al-Khaṭīb, 298
al-Khafrī, Shams al-Dīn, 116, 487
Ibn Khaldūn, 88
Khalife Soltan (Vizier), 489
al-Khalīl Ibn Aḥmad, 278
al-Khalīlī, 64, 75
Khan, A.S., 473
al-Kharaqī, Jamāl al-Zamān, 79
al-Khaṣīb (Wali), 348
al-Khaṭṭābī, M.L., 286
Ibn al-Khawwām, 242, 243
al-Khayyām, ʿUmar, 39, 56, 61, 75, 80, 105, 106, 110, 123, 276, 351
al-Khāzin, Abū Jaʾfar, 56–58, 63, 80, 343, 344, 476
al-Khāzinī, 53, 61, 65, 118
Kheirandish, E., 30, 84–85, 106, 108, 114
Nāṣer-e Khosrow, 121
al-Khujandī, 56, 334, 338
al-Khwārizmī, 53, 54, 74–77, 90, 111, 118–120, 122–123, 240, 276, 286, 301, 309–311, 321, 322, 325, 346, 349–354, 356–359, 361, 400
al-Kindī, 80, 85, 106, 107, 113, 114, 240, 349, 355, 409–412
King, D.A., 51, 52, 54, 63, 64, 74, 76, 83, 86, 88, 90, 91, 117, 118, 275, 298, 323, 329–331, 333–338, 341, 487, 492
Kirschner, P., 228
al-Kishnāwī, 79
Kitcher, P., 227, 228
Klamroth, M., 28, 78, 104
Kleiner, I., 91
Knobloch, E., 115, 200
Knorr, W.R., 5–10, 18, 20, 26, 28, 41, 61, 78, 80, 104, 146, 150, 151, 235, 244, 256, 257
Koelblen, S., 89
Kondo, S., 556
van Konigsveld, P.S., 63
Krafft, F., 146
Krates, *see* Crates
Krikorian-Preisler, H., 310
al-Kūhī, 61, 65, 78–82, 87, 91, 104, 107, 110–111, 113, 117, 120, 171, 240, 259–263, 265–268
Kuhn, T., 19
Kumar, A., 170
Kunitzsch, P., 29, 76, 80–83, 90, 107, 117, 121, 170, 255, 257, 280, 339
Kurd ʿAlī, M., 347
Kūshyar ibn Labbān, 63, 75, 87, 89, 119
Kusuba, T., 107, 231, 469
Köhler, U., 150, 151

Laabid, E., 121, 279, 289
Labarta, A., 283
Lachmann, K., 390, 395, 396
Lafon, J.-P., 531, 558
Lafrance, Y., 5
Lambert, J.H., 543
Lamrabet, D., 279–281, 283, 284
Lando, S., 39
Langermann, T., 409, 411, 417, 535
al-Lārī, Muṣliḥ al-Dīn, 241
Lee, H., 506
Legendre, A.-M., 517, 518, 543, 546
Lehoux, D., 32, 43
Leibniz, G.W., 539
Lentz, T.W., 500
Leslie, J., 522
Levey, M., 345, 357
Lévy, T., 107, 279, 283–285, 287, 409, 412, 413, 417
Lewis, E., 506, 514
Lewis, M.J.T., 32
von Lindemann, C.L.F., 543
Lindner, C., 521
Lippert, J., 346
Lo Bello, A., 30, 104
Lorch, R.P., 28, 29, 65, 76, 80, 82, 83, 85, 86, 91, 105, 107, 115, 117, 170, 171, 255, 333, 345, 358
Lowry, G.D., 500
Lucas, S.K., 557
Luckey, P., 53, 56, 119, 355
Lucore, S., 211
Lucullus, Lucius Licinius, 166

Luke, R., 555
Ibn Lurra, Muḥammad, 346
Luther, I., 109

Mäenpää, P., 34
Māristūn, 148
Mach, R., 52
Machin, J., 540, 541
al-Maghribī, Muḥyī al-Dīn, 119, 170, 311, 409, 414–416, 460, 518
al-Māhānī, 63, 80, 104, 105, 115
Mahoney, M., 4
Maimonedes, Moses ben, 279, 535
Ibn al-Majdī, 54, 280, 281
al-Majrīṭī, *see* Maslama
Makdisi, U., 522, 524
Malik Ḥusayn Iṣfahānī, 492, 493
Ibn Malik Ḥusayn, Muḥammad ʿAlī, 492, 497
Malpangotto, M., 32, 43
al-Maʾmūn (Caliph), 74, 77, 102, 105, 334, 347–349
Mancha, J.L., 116
Mandel, E., 532
Manders, K., 35
Manekin, C., 417
Mansfeld, J., 37
Manūchihr Khān, ʿAbū l-Fatḥ, 492, 495, 497
Maqdisi, G., 488
Marcellus, Marcus Claudius, 148
Marin, M., 121, 280
Marre, A., 275
Martin-Löf, P., 34
Māshaʾallāh, 355
Abū Maʿshar al-Balkhī, 300, 310, 355
Masià-Fornos, R., 35
Maskelyne, N., 504
Maslama al-Majrīṭī, 82, 115, 117, 118, 301, 311
al-Maṣmūdī, 118
Mastrocinque, A., 166
Mattingly, A., 552, 554
Matvievskaya, G., 63, 106, 280
Mawaldi, M., 109, 123, 243
al-Māwardī, Abū'l-Ḥasan, 353
May, K., 3
McCartney, S., 544
McGinnis, J., 113
de' Medici, Fernando II, 268
de' Medici, Lorenzo, 545
de' Medici, Giuliano, 545
Mehmed II (Sultan), 241, 243, 245
Menaechmus, 5
Mendell, H., 42
Menelaus, 28, 41, 42, 86, 87, 106–107, 111, 115, 278, 284, 413, 416, 475
Mengoli, P., 539
Mercier, R., 29, 91, 116
Merrifield, C.W., 260
Meskens, A., 40
Mestres, A., 298, 300, 306, 307, 312, 315, 320–324
Metrodorus, 150
Mez, A., 348
Michelangelo, 545
Mielgo, H., 81, 118, 308, 312
Ben Miled, M., 105, 106
Miles of Marseilles, 414, 450
Millás Vallicrosa, J.M., 302, 303, 306, 312, 322–324
Millás Vendrell, E., 119, 302, 303, 312, 315–318
Miller, N., 35
Minar, E.L., 152
al-Miṣrī, Najm al-Dīn, 119
al-Miṣṣīṣī, 351, 353, 354
Mogenet, J., 27, 28
Moll, V., 555
Montbelli, V., 209
Morelli, G., 40
Morelon, R., 76, 116
Morrison, L.V., 156
Morse, J., 521
Moschion, 147
Moses ben Tibbon, 255–257
Moussa, A., 115, 118
Moyon, M., 281, 282, 287
Muṣṭafā Ṣidqī, 239, 244, 245
Ibn Muʿādh al-Jayyānī, 63, 76, 78, 115, 120, 308, 310, 311, 316
Muʾayyad-zāde ʿAbd al-Raḥmān ibn ʿAlī, 245
Mueller, I., 6, 7, 19, 20, 38, 150, 191, 234, 235
Ibn al-Muftī, 299
Mugler, C., 5, 7, 29, 151
Muḥammad II (Emir), 298
Müller, A., 345–347
Mumma, J., 34
al-Munajjid, S.D., 348
Ibn Munʿim, 59, 85, 89, 123, 277, 278, 281–285, 289
Murdoch, J., 412
Mūsā ibn Shākir, 348
Muṣliḥ, A., 280, 282
al-Muʿtaḍid (Caliph), 76
al-Muʾtaman, Yūsuf ibn Hūd, 85, 105, 106, 110–111, 114, 115, 240, 277, 278, 280,

282–285
al-Mutawwakil (Caliph), 349
Ibn al-Muthannā, 303, 309
Muwafi, A., 61

Nabonassar (King), 157
Nadal, R., 107
Ibn al-Nadīm, 148, 345, 346, 349, 351, 352, 354–356, 360, 412
Nadir, N., 306
Naghsh, I., 280
Naini, A.D., 52
Nallino, C.A., 303, 304, 306–311, 318, 320, 322, 323, 346
Napier, J., 504
Napolitani, P.D., 28, 31, 152, 203
Ibn Naqīb, Aḥmad al-Ḥalabī, 241, 242
al-Naqqāsh, Aḥmad ibn Muḥammad, 333
al-Nasafī, Abū Maḥmūd, 117
al-Nasawī, Abū al-Ḥasan, 260, 264, 268
al-Nāṣir (Caliph), 59
Abū Naṣr Manṣur ibn ʿIrāq, 63, 64, 106, 117
Nasr, S.H., 491
Nastulus, 119
al-Nayrīzī, 77, 102–105, 117, 411, 476
Necipoğlu, G., 87
Netz, R., 19, 26, 29, 31, 33, 37, 39–42, 151, 206, 207, 209, 216, 219–221, 227, 228, 232, 235, 261
Neubauer, A., 412, 413
Neuenschwander, E., 8, 19
Neugebauer, O., 5, 10, 11, 17, 114, 152, 167, 168, 303, 311, 312, 316, 321, 324, 336, 396, 413
Newton, Isaac, 538, 539, 541
Nichomedes, 152, 411
Nicomachus, 21, 106, 107, 111, 277, 412
Nimr, F, 506
Nipsus, 390
al-Nīsābūrī, 86
Niven, I., 543
Noack, B., 27
Noel, W., 31, 206, 207, 209, 219–221
North, J., 86, 120, 307
Nuʿaim ibn Muḥammad ibn Mūsā, 111
al-Nuʿaymī, 88, 108
Abū Nuwās, 348
Nuyens, D., 558

Oaks, J., 122–123
von Ofenheim, J., 124
Oleson, J.P., 148
Olson, R., 522
Özdural, A., 112

Pṛthūdakasvāmin, C., 472, 473, 481, 484
Pambuccian, V., 34, 228, 233
Panza, M., 35, 111
Pappus, 5, 7, 9, 10, 12, 19, 30, 37, 39, 41, 78, 81, 104, 149–152, 233, 235, 460
Parmenides, 4
Parra, M.J., 324
Pascal, Blaise, 59
Pasch, M., 227
Pāsha, Muḥammad, 242
Pedersen, F.S., 303, 309, 320
Pedersen, O., 10, 309
Percival, C., 552
Phaseis, 152
Philippou, A.N., 61
Philippson, R., 151
Phillips, C., 522
Philo of Byzantium, 148
Philolaus, 18, 20–21
Philonides, 150, 151, 171
Photius (Patriarch), 149
Piccinetti. P., 209
Pinel, P., 107
Pingree, D.E., 11, 90, 335, 469, 472
Plassart, A., 150, 151
Plato, 4, 5, 7–9, 20, 21, 38, 148, 152, 153, 229, 430
Plato of Tivoli, 308–310
von Plato, J., 34
Playfair, J., 503–519, 522, 523
Plofker, K., 102
Plooij, E.B., 63, 78
Plotinus, 412
Plutarch, 12, 39, 152
Pollard, S., 227
Polya, G., 109
Polyaenus, 150
Popescu, E., 181
Porphyry, 21
Porter, A., 521
Porter, Y., 493, 497
Posidonius, 149, 166
Price, D.J.d.S., 12, 155–157, 183
Prigotta, Elia, 412, 413
Proclus, 5, 17, 35, 39, 104, 153, 160, 233, 504, 514, 518, 538
Ptolemaeus Chennus, 149
Ptolemy, 10–12, 21, 25, 27, 29, 30, 32, 33, 35–37, 42, 43, 63, 75, 81–86, 89, 107, 114, 116–119, 149, 153, 154, 157, 160, 161, 165, 167, 168, 171, 245, 297, 310,

311, 325, 336, 338, 348, 361, 400, 412, 488, 499
Ptolemy II Philadelphus (King), 147
Ptolemy IV Philopater (King), 149
Ptolemy Lagus (King), 508
Ptolemy VI Philometor (King), 147
Ptolemy VIII Euergetes (King), 147
Puig, R., 83, 90, 287, 288, 298, 302
Pythagoras, 18, 22, 25, 38, 52, 240

al-Qabīṣī, Abū Ṣaqr, 53, 61
Ibn al-Qāḍī, 286
Qāḍīzāde al-Rūmī, 106, 115, 241, 244, 520
al-Qā'inī, Ḥasan ibn Saʿd, 492, 493, 495, 497
al-Qalaṣādī, 121, 275, 281, 283, 289
Qalonymos ben Qalonymos, 409, 411, 430
al-Qalqashandī, 351
Qarajaghāy Khān, 492
Qāsim ibn Yūsuf Abū Naṣr, 121
al-Qaṭrawānī, 276
al-Qayṣarī, Dāwūd, 239, 240
al-Qazwīnī, Zakariyā', 499, 500
Ibn al-Qifṭī, 346, 348, 353
Quandt, A., 206
al-Qummī, 113
Ibn Qunfudh, 277, 311
al-Qurashī, Abū al-Ḥasan, 121
al-Qurashī, Jamāl al-Dīn, 242
al-Qurashī, Abū al-Qāsim, 276, 282, 289, 290, 354
Qūshjī, ʿAlī, 241, 244
Qusṭā ibn Lūqā, 111, 346, 347

Rabinowitz, S., 551
Rackham, H., 153, 166
Ragep, F.J., 86, 116, 239, 241, 491
Ragep, S.P., 239, 248
Ramanujan, S., 545, 550
Raphael du Mans, 488
Ibn al-Raqqām, Abū ʿAbd Allāh Muḥammad, 77, 286, 297–299, 301–312, 315–317, 320–323, 325
Rashed, R., 11, 29, 31, 52–54, 56, 58, 59, 61, 75, 80–82, 84–85, 105–107, 109–115, 117, 119, 120, 122, 123, 266, 285, 286, 344, 352, 412
al-Rashīd (Caliph), *see* Harūn al-Rashīd
al-Rāzī, 473
al-Rāzī, Fakhr al-Dīn, 499
Rebstock, U., 89, 121
Recorde, R., 514
Redekop, B., 522
Regiomontanus, 76, 120
Rehm, A., 175, 177, 178, 180, 183
Reid, T., 505, 522
Reinach, T., 203, 204, 207, 217
Remes, U., 5
Rescher, N., 411
Rhode, E., 149
Ricci, M., 10
Richard, F., 488, 491, 493, 495, 497
Richter-Bernburg, L., 65
Riddell, R.C., 6
Rifāʿī, A.F., 348
Rıfkı, Hüseyin, 520
Ibn Abī al-Rijāl, 311
Rius, M., 118
Riẓā–i ʿAbbāsī, 489, 491, 493, 495, 497
Robbins, F.E., 311
Robson, E., 38, 231
Rodgers, B., 171
Roediger, J., 345
Rome, A., 75
Rommevaux, S., 28, 104, 284
van Roomen, Adriaan, 113
Rosen, F., 352, 400
Rosenfeld, B.A., 11, 51, 56, 61, 65, 78, 101, 115, 240, 242, 280
Rosenthal, F., 412
Rotman, B., 227, 228
Rozhanskaya, M., 53, 61, 65, 113, 243
Rudorff, A., 390, 395, 396
Rufini, E., 204, 205, 217
Ibn Rustah, 346

al-Ṣābī, Abū Isḥāq, 65, 110
Sabra, A.I., 12, 53, 63, 64, 84–85, 88, 102, 108, 112–114, 124, 411
Sacerdote, G., 345, 355
Sachau, C.E., 472, 482
Sachs, A., 396
Sacrobosco, 153
Abū Saʿdān, 413
Saʿdi, L., 505, 506, 508, 516, 519
al-Ṣafadī, 348
Ibn Ṣafwān, 289
al-Ṣāghānī, 65, 110, 117
Ibn Sahl, Abū Saʿd al-ʿAlā, 81, 84, 110, 114, 117
Sahl ibn Bishr, 351
al-Sahlī, Ibrāhīm ibn Saʿīd, 333, 337
al-Sahlī, Muḥammad ibn, 336
Saidan, A., 52–53, 58, 65, 115, 344, 352, 395
Ṣāʿid al-Andalusī, 170, 278, 285
al-Ṣaidanāni, 351, 353, 354
al-Ṣā'igh, ʿA., 356

Saito, K., 19, 25, 31–33, 36, 41–43, 152, 203, 207, 211, 216, 229, 232, 233
Sakkal, M., 112
Saladin (Sultan), 171
Ibn al-Ṣalāḥ, 117, 240
Salamin, E., 546
al-Salāwī, Ibn Qāsim al-Ṣafāʿī, 300
Salem, S.I., 170
Saliba, G., 54, 64, 86, 90, 102, 116, 299, 487
Ibn Ṣāliḥ, 114
Salīm II (Sultan), 242
Salim, M.E., 243
al-Samarqandī, 106, 520
al-Samawʾal, 53–56, 59, 76, 110, 120, 123, 276, 351, 352
Ibn al-Samḥ, 81, 104, 278, 285, 288, 310, 413
Samsó, J., 81, 83, 90, 108, 118, 119, 288, 298, 299, 301–303, 306–312, 315–318, 320, 324
Sanad ibn ʿAlī, 347–349, 351, 353, 354
al-Ṣardafī, 121
Sarhangi, R., 112
Ibn al-Sarī, 78, 106
Sarkīs, Y., 505
Sarma, S.R., 469
Ibn Sartāq, 85, 111, 240, 244, 284
Sarton, G., 344, 345, 347
Sato, T., 9, 205
Saunderson, N., 78
Savage-Smith, E., 76, 83, 91
Sayılı, A., 346, 348
Ibn Sayyid, 278, 283, 285
Sayyid ʿAlī Raʾīs, 242
Schöne, H., 189, 191–195
Schaldach, K., 178
Schamp, J., 149
Scheglova, O., 507
Schiapparelli, G., 307, 310
Schirmer, O., 333
Schjellerup, H.C.F.C., 492
Schmidl, P., 119, 337, 338
Schmidt, O., 324
Schmitz, B., 488, 489, 491–493, 495, 497
Schneider, I., 9
Schoy, C., 83
Schramm, M., 10
Sedillot, L.P., 56
Sedley, D., 150
Sefrin-Weis, H., 30
Seleucus IV (King), 151
Şeşen, R., 245, 246
Sesiano, J., 11, 27, 40, 51, 53, 56, 58, 61, 65, 74, 78–80, 90, 112, 121–122, 282–284
Sezgin, F., 11, 51, 63, 117, 260, 262, 299, 347, 360, 411, 412
Shanks, D., 540, 559
Shanks, W., 540, 541
al-Shannī, Abū ʿAbdallāh, 111, 245
Shannon, C., 536
Shapiro, A.E., 11
al-Shaqqāq, 121
Sharp, A., 540
Ibn al-Shāṭir, 86
Shehadeh, K., 170, 171
al-Shīrāzī, Quṭb al-Dīn, 86, 103, 244, 487, 489, 491, 492, 497, 499
al-Shirwānī, Muḥammad ibn Ibrāhīm, 245
Shukla, K.S., 470, 471
Shukrzāde Fayḍallāh Sarmad, 245
Sidoli, N., 28, 29, 32, 33, 35, 36, 41–43, 106, 107, 117, 152, 171, 229, 231–233
Signell, K., 492, 493
al-Sijzī, 28, 41, 78, 80, 104, 107, 109–113, 476
Simon, H.A., 228
Simon, S., 532
Simplicius, 12, 38, 104, 411
Simson, R., 504, 505, 508, 512
Ibn Sīnā, 488, 499
Ibn Sinān, Muṣliḥ al-Dīn, 243
Sinān ibn al-Fatḥ, 350, 351, 353, 354
Sinclair, N., 91
Sindī ibn Shāhak, 348
Singmaster, D., 556, 559
Sloane, N., 540
Smart, T., 81
Smith, A.M., 43, 114, 308
Smith, Adam, 227
Smith, D.E., 344, 469
Socrates, 21
Solomon, J., 30
Somel, S.A., 521
Souffrin, P., 10
Souissi, M., 275
Stömer, F.C.W., 548
Stückelberger, A., 29
Stanley, R.P., 39
Steele, J.M., 33, 185
Steinschneider, M., 275, 345, 355, 411–413
Stenius, A., 41
Stephenson, F.R., 156
Stewart, D., 522
Strabo, 150, 166
Strato of Lampsacus, 146, 147

Straub, A., 558
al-Ṣūfī, ʿAbd al-Raḥmān, 310, 487, 489, 491–493, 495, 497, 499, 500
Sulaymān (Shāh), 489, 499
Surūr, M., 520
Suter, H., 65, 275, 303, 321, 322, 343, 345, 355, 359, 360
Sweller, J., 228
Szabó, Á., 4, 6, 9, 19

Ṭabarī, Muḥammad ibn Ayyūb, 121
al-Ṭabarī, 348
al-Tabbakh, Aḥmad, 520
al-Tabrīzī, Alī ibn ʿAbdallah, 360
Taha, A., 107
Ibn Ṭāhir, 283
Ibn Taifūr, Aḥmad ibn abī Ṭāhir, 348
al-Taimī, Abū ʿUbaida, 353
Taisbak, C.M., 27, 30, 189, 538
Takahashi, D., 546, 556, 559
Takahashi, K., 29
Takano, K., 548
Tannery, P., 5, 17, 19
Taqī al-Dīn ibn Maʿrūf, 239, 242, 243, 245, 248
Taqī al-Dīn al-Rāṣid, *see* Taqī al-Dīn ibn Maʿrūf
Ṭāšköprüzāde, Aḥmad ibn Muṣṭafā, 351
Taylor, B., 504
Tchernetska, N., 31, 206, 207, 216, 219–221
Tee, G., 64
Teixeira, Pedro, 488
Tekeli, S., 61
al-Ṭfayyash, M., 282
Thābit ibn Qurra, 52, 65, 76–79, 81, 89, 103, 105, 107, 109, 111, 113, 115, 116, 122, 123, 240, 244, 245, 257, 260, 262, 282, 284, 286, 355, 413, 476, 484
Thales, 17
Tharp, L., 235, 236
Theaetetus, 4, 8, 234
Theodorus, 4, 8
Theodosius, 7, 27–29, 32, 33, 41–43, 63, 85, 107, 111, 115, 152, 170, 227, 229, 231, 232, 234, 235, 255, 413, 416, 488
Theon of Alexandria, 28, 37, 116, 150
Theon of Smyrna, 154, 160
Theophrastus, 43
Thomaides, Y., 40
Thomas, R.S.D., 21, 228
Thorndike, A.S., 33, 158, 160, 162, 168, 169
Thorndike, L., 153
Thureau-Dangin, F., 398
Tibawi, A. L., 506
Tibbetts, G., 91
Tichenor, M., 64
Tihon, A., 28
Todd, R.B., 30
Toomer, G.J., 10, 11, 63, 65, 81, 106, 148–150, 154, 168, 260, 267, 302, 303, 305, 306, 309, 311, 315, 323, 400
Torricelli, E., 10
Tourkin, S., 493, 497
Tropfke, J., 11, 355
Tse-Wo Zse, 552
Tsu Ch'ung Chih (Zu Chongzhi), 536
Tsumura, H., 539
Ṭūlūn, Aḥmad ibn, 347, 353
Turner, A.J., 335
Ṭūqān, Q., 349, 351
al-Ṭūsī, Naṣīr al-Dīn, 54, 84, 86–87, 89, 103, 106, 107, 109, 113, 114, 116, 231, 234, 244, 245, 260, 262–264, 266, 278, 348, 487, 488, 493, 499, 503, 520
al-Ṭūsī, Sharaf al-Dīn, 56, 74–75, 266
Tybjerg, K., 33, 43

Ulugh Beg, 53, 84, 87, 108, 115, 244, 488, 493, 495
ʿUmar (Caliph), 74
ʿUmar al-Khayyām, *see* al-Khayyām
Unguru, S., 6, 20, 31, 36, 37
Upton, J.M., 497
al-ʿUqbānī, 277, 289
al-Uqlīdisī, 53
al-ʿUrḍī, 86, 116
Ibn abī Uṣaibiʿa, 347, 349, 353
ʿUthmānī, A., 281

Vahabzadeh, B., 78, 105, 106, 123
van, *see* next part of the name
Van Dyck, C., 503–521, 523, 524
Van Brummelen, A., 124
Van Brummelen, G., 28, 30, 41, 43, 75, 90, 104, 107, 109–111, 115, 116, 118–120
Van der Waerden, B.L., 6, 8, 17–19
Ver Eecke, P., 149, 151, 152, 170, 203, 259–261, 460
Vernet, J., 90, 298, 306, 307, 315
Vesel, Ž., 488, 489, 491, 493, 497
Vettius Valens, 310
Viète, François, 109, 538
Viladrich, M., 90
Villuendas, M.V., 63
Vinel, N., 29
Viola, C., 544

Vitrac, B., 17, 19, 28–30, 104, 106, 284, 412
Vitruvius, 147, 150
Vlasschaert, A.-M., 283
Vogel, K., 395–398
Vogt, H., 18
von, *see also* next part of the name
von Baeyer, H.C., 532
von Neumann, J., 549
Voogt, C.S., 260

Abū al-Wafā᾽ al-Būzjānī, 53, 61, 63, 87, 112–114, 118, 122, 286, 305, 306, 344
al-Wafā᾽ī, 338
Wagner, E., 348
Wagner, R., 37
Wagon, S., 551
Wālīs al-Miṣrī (= Vettius Valens), 310
Walker, M., 521
Wallis, J., 538
Wan, J., 558
Wartenberg, I., 283
Waterhouse, W.C., 5
Watson, G.N., 545
Weil, A., 6
Weinberg, J., 345, 355
Welch, A., 488, 489, 491, 492, 495
Welch, S.C., 495
Wellesz, E., 491, 495, 497
Wells, D.G., 228
Weyl, H., 554
Whewell, W., 522
Whitehead, D., 146
Wieber, R., 61
Wiedemann, E., 148, 170, 243, 343
Wiegand, T., 175
Wightwick, G., 552, 554
William of Moerbeke, 28, 29
Williams, C. P., 505
Willson, N., 206, 207, 219–221
Wilson, A.I., 171
Wilson, N.G., 31, 40
Winter, T.N., 30, 147
Wittgenstein, L., 227, 228
Woepcke, F., 56, 58, 171, 262, 263, 275, 276, 337, 344, 345
Wood, P., 505
Woodward, D., 91
Worcester, S., 521
Wright, M.T., 33, 157, 163

Xenikakis, K., 155

Yadegari, M., 54
Yaḥyā ibn Abī Manṣūr, 90, 316
Abū Yaḥyā al-Marwazī, 354
Yaltkaya, S., 346
al-Yanyawī, Asʿad Efendī, 80, 243–245, 329
Yaʿqūb ibn Ṭāriq, 76, 89
Yāqūt al-Ḥamawī, 353
Ibn al-Yāsamīn, 85, 276, 277, 281, 282, 284, 289
Yazdī, Muḥammad Bāqir, 52
Yee, A., 556
Youschkevitch, A.P., 51, 53, 474
Ibn Yūnus, 63, 64
Abū Yūsuf, 348
Yūsuf, Muḥammad, 492

Zacut, Abraham, 324
Ibn Zakariyā᾽, 284
al-Zanjānī, 54, 79
Ibn al-Zarqālluh, 81, 83, 116–117, 122, 297, 302, 306, 312, 315, 317, 318, 322–324
Zeki, S., 349, 351
Zeller, E., 18
Zemouli, T., 276, 289
Zeno, 4
Zenodorus, 150
Zerrouki, M., 289
Zeuthen, H.G., 5, 8, 17, 19, 37, 42, 235
Zhmud, L., 18, 38
Zudilin, W., 555

Index of Ancient and Medieval Titles

al-Aʿdād al-mutaḥābba, Risāla fī (Thābit ibn Qurra), 52, 282
Aderameti, Liber (ʿAbd al-Raḥmān), 287
al-Aḥkām al-sulṭāniya (al-Ḥanbalī), 353
al-Aḥkām al-sulṭāniya (al-Māwardī), 353
Aims of Euclid's Book, On the (al-Kindī), 411
ʿAjāʾib al-makhlūqāt (al-Qazwīnī), 499, 500
al-Aʿlāq al-nafīsa (Ibn Rustah), 346
Alfonsine Tables, 316
Algebra (Ibn al-Bannāʾ), 360
Algebra (Abū Kāmil), 56, 123, 344–347, 349–352, 354–356, 359–361, 363, 364, 407
Algebra (al-Khayyām), 75
Algebra (al-Khwārizmī), 74, 111, 122, 349, 352, 400
Algebra (al-Qurashī), 282
Algebra (Sharaf al-Dīn al-Ṭūsī), 56, 74–75
Almagest (Ptolemy), 11, 29, 84–86, 89, 154, 157, 160, 168, 245, 260, 298, 299, 303, 305, 306, 309, 311, 316–319, 323, 324, 400
  Arabic, 81, 348
Almagest (Abū al-Wafāʾ), 118
Almanac (Ibn al-Zarqālluh), 306, 322–324
Almanach perpetuum (Zacut), 324
ʿAmal al-dāʾira al-maqsūma bi-sabʿa aqsām mutasāwiya (Archimedes), 244
ʿAmal al-mīzān al-ṭabīʿī, Risāla fī (Taqī al-Dīn), 243
ʿAmal al-murabbaʿ al-musāwī li-l-dāʾira (Archimedes), 243
ʿAmal sāʿat al-māʾ (Archimedean), 245
Analemma (Ptolemy), 82
Analysis and Synthesis (Ibrāhım ibn Sinān), 79
Anthology of Problems (Ibrāhīm ibn Sinān), 112
Architecture (Vitruvius), 147, 150
Area of the Parabola, Epistle on the (Ibrāhīm Ibn Sinān), 278
Arithmetic (al-Khwārizmī), 120
Arithmetica (Nicomachus), *see Introduction to Arithmetic*
Arithmetic for Government Officials (Abū al-Wafāʾ), 53
Arithmetics (Diophantus), 11, 40, 277
  Arabic, 17, 58, 346, 347, 358
al-Arkān fī al-muʿāmalāt ʿalā ṭarīq al-burhān, Kitāb, 288
Arshimīdis fī taksīr al-dāʾira, Maqālā (Archimedes), 245
De arte mensurandi (Johannes de Muris), 287
Āryabhaṭīya (Āryabhaṭa), 469, 470
Ashkāl al-Taʾsīs (al-Samarqandī), 520
al-Ashkāl al-misāḥiya, Risāla fī (Ibn al-Bannāʾ), 286
De aspectibus (al-Kindī), 85
Assumptions (Archimedean), 11, 259, 260, 262, 264
  Thābit ibn Qurra's, 109
Astrology (Ptolemy), *see Four Books, Mathematical Treatise in*
al-ʿAyn, Kitāb (al-Khalīl Ibn Aḥmad), 278

al-Badīʿ fī al-ḥisāb (al-Karajī), 56, 343, 344, 353, 354
Baghdād, Kitāb (Ibn Taifūr), 348
al-Bāhir (al-Samawʾal), 352
Balance of Wisdom (al-Khāzinī), 53, 65
al-Bayān wa al-tadhkār, Kitāb (al-Ḥaṣṣār), 280, 284, 289
Bibliotheca historica (Diodorus Siculus), 147
Book, see next full word
Brāhmasphuṭasiddhānta (Brahmagupta), 469, 470, 472

N. Sidoli and G. Van Brummelen (eds.), *From Alexandria, Through Baghdad: Surveys and Studies in the Ancient Greek and Medieval Islamic Mathematical Sciences in Honor of J.L. Berggren*, DOI 10.1007/978-3-642-36736-6,
© Springer-Verlag Berlin Heidelberg 2014

Muḥyi al-Dīn al-Maghribī's version, 518
Playfair's version, 504–506, 508–520, 523, 524
Arabic, 503, 504, 506–513, 515–520, 523, 524
Recorde's version, 514
Simson's version, 504, 505, 512
Elements (Menelaus), 41, 107
Embadorum, Liber (Bar Ḥiyya), 285, 287
Epistle, *see* next full word
Epistles (Brethren of Purity), 78
Equilibrium of Planes (Archimedes), 7, 10, 146
Exposure of the Errors of the Astronomers (al-Samawʾal), 120
Extraction of Amicable Numbers (Thābit ibn Qurra), 123
Extraction of the Meridian Lines (al-Jurjānī), 112

al-Fakhrī fī al-jabr wa al-muqābala (al-Karajī), 344, 345, 354, 355
Fallacies (Euclid), 36
Farāʾiḍ al-fawāʾid (al-Kāshī, ʿImād al-Dīn), 242
al-Fawāʾid al-bahāʾiyya fī al-qawāʿid al-ḥisābiyya (Ibn al-Khawwām), 242
Fihrist (al-Nadīm), 148, 343, 345, 346, 348, 349, 351–355, 473
Filling the Gaps in Archimedes' Sphere and Cylinder (al-Kūhī), 259, 262, 266
Fiqh al-ḥisāb (Ibn Munʿim), 59, 277, 278, 281–284, 289
Five Solids, Treatise of, 413, 414
Four Books, Mathematical Treatise in (Ptolemy), 310, 311, 499
Fundamentals of Rules on the Principles of Useful Things (al-Fārisī), 109
Fundamental Theorems (al-Samarqandī), 106
Fundamenta tabularum (Ibn ʿEzra), 303

Geography (Ptolemy), 29, 30
Geography (Strabo), 150, 166
Geometry (Heronian), 390, 393
Geometrical Elements (Menelaus), *see Elements*
Geometric Problems (al-Jurjānī), 112

Habitations (Theodosius), 29
Ḥākimī Zīj, *see Zīj al-Ḥākimī*
Ḥall shukūk Kitāb Uqlīdis, Kitāb fī (Ibn al-Haytham), *see Solution of the Difficulties in Euclid's Elements*
Handy Tables (Ptolemy), 27, 29, 316
Harmonics (Ptolemy), 30, 33, 43
The Harmonious Arrangement of Numbers, 122
Ḥaṭṭ al-niqāb (Ibn Zakariyāʾ), 284
Ḥaṭṭ al-niqāb ʿan wujūh aʿmāl al-ḥisāb (Ibn Qunfudh), 277
Ḥāwī al-lubāb fī sharḥ Talkhīṣ aʿmāl al-ḥisāb (Ibn al-Majdī), 280, 281
Heavens (Aristotle), 146, 160
Heavens (Cleomedes), 30
Ḥisāb al-muʿāmalāt, 283
Fī al-ḥujja al-mansūba ilā Suqrāṭ fī al-murabbaʿ wa quṭrihī (Thābit ibn Qurra), 286

Īḍāḥ al-maqāṣid (al-Fārisī), 242
Ifrād al-maqāl fī amr al-ẓilāl, Kitāb fī (al-Bīrūnī), *see Shadows* (al-Bīrūnī)
Iggeret ha-Mispar (al-Aḥdab), 283
al-Ikmāl fī al-handasa (Ibn Sartāq), 85, 240
Ikhbār al-ʿulamāʾ (Ibn al-Qifṭī), 346
ʿIlal al-jabr wa al-muqābala (al-Karajī), 355
Inbāṭ al-miyāh al-khafīya (al-Karajī), 344
India (Bīrūnī), 469, 482
Instruments, Treatise on (al-Miṣrī), 119
Interlocking Similar and Corresponding Figures, 112
Introduction to Arithmetic (Nicomachus), 106, 107, 111, 277
Introduction to the Phenomena (Geminus), 30, 153, 154, 167
al-Iqnāʿ fī ʿilm al-misāḥa, 241
Irshād al-qāṣid (Ibn al-Akfānī), 351
Ishārāt wa-l-tanbihāt (Ibn Sīnā), 499
Istikhrāj dilʿ al-musabbaʿ... (al-Kūhī), 61
al-Istikmāl, Kitāb (al-Muʾtaman), *see Perfection*

al-Jabr wa al-muqābala, Kitāb, *see Algebra*
Jadwal al-taqwīm (Ḥabash al-Ḥāsib), 64
Jamharat al-ʿArab (Ibn Durayd), 278
Jāmiʿ Zīj, *see Zīj-i jāmiʿ*

al-Kaʿb wa al-māl wa al-aʿdād al-mutanāsiba, Kitāb (Sinān ibn al-Fatḥ), 350
al-Kāfī fī al-ḥisāb (al-Karajī), 53, 344, 353
Kamāl al-jabr wa tamāmuh wa al-ziyāda fī uṣūlih, Kitāb (Abū Kāmil), 349
al-Kāmil, *see Zīj al-kāmil*
al-Kāmil fī al-taʾrīkh (Ibn al-Athīr), 346
al-Kāmil fī ʿilm al-ghubār, Kitāb (al-Ḥaṣṣār), 277, 282
Kashf al-ẓunūn (Ḥājjī Khalīfa), 346, 351, 356
Keskintos Astronomical Inscription, 32
Key of Arithmetic, *see Calculator's Key*

Keys of the Science of Astronomy (al-Bīrūnī), 63, 115, 305, 306, 310
Key to Transactions (Ṭabarī), 121
al-Khail, Kitāb (al-Taimī), 353
Khaṇḍakhādyaka (Brahmagupta), 316
Khāqānī Zīj, *see Zīj-i Khāqānī*
al-Kharāj, Kitāb (Abū Yūsuf), 348
Khulāṣat al-ḥisāb (al-ʿĀmilī), 487
Khulāṣat al-hayʾa (Sayyid ʿAlī Raʾīs), 242
Khuṭūṭ lawlabiyya (Archimedes), *see Spiral Lines* (Archimedes)
Kifāyat al-fāriḍ al-murtāḍ fī al-tanbīh ʿalā mā aghfalahū jumhūr al-furrāḍ (Ibn Ṣafwān), 289
Kitāb, *see* next full word
al-Kura wa al-usṭuwāna (Archimedes), *see Sphere and Cylinder* (Archimedes)

Lemmas (Archimedean), 107, 259–261, 263, 264, 267, 268
  Arabic, 245
Liber, *see* next full word
The Light of the Moon (Ibn al-Haytham), 114
Little Astronomy, 27
Longitudes and Latitudes of the Persians, 84

Mā yaḥtāju ilayhī al-ṣāniʿ min aʿmāl al-handasa, Kitāb fī (Abū al-Wafāʾ), 286
al-Mabādiʾ wa al-ghāyāt fī ʿilm al-mīqāt, Kitāb (al-Ḥasan al-Murrākushī), 288
The Magic Disposition of Numbers in Squares (Abū al-Wafāʾ), 122
Mahameleth, Liber, 79, 283, 288, 351, 357
Majhūlāt qusiy al-kura (Ibn Muʿādh), 63
al-Maʾkhūdhāt, *see Assumptions* or *Lemmas*
al-Manāẓir (Ibn al-Haytham), 64, 85, 106, 111, 114
Maqāla, *see* next full word
Maqālīd ʿilm al-hayʾa (al-Bīrūnī), *see Keys of the Science of Astronomy*
Marcellus (Plutarch), 152
Maʿrifat kammiyyat muḥīṭ al-dāʾira, Risāla fī (al-Qurashī), 242
Masʿūdīc Canon (al-Bīrūnī), *see Qānūn al-masʿūdī*
Materia medica (Dioscordes), 489
Mathematical Collection (Pappus), 10, 30, 37, 78, 149, 151, 152
Measurement of a Circle (Archimedes), 9, 20, 80, 189, 197, 211, 284, 534
  Arabic, 80, 245, 285
Measurements (Heron), 189, 192, 194, 195, 197, 360, 388, 390, 393
Mechanical Problems (Aristotelian), 30, 146, 147, 149
Mechanics (Heron), 147
Memorandum to Colleagues Explaining the Proof of Amicability (al-Fārisī), 123
Mensuration (Abū Kāmil), 112, 356, 359–361, 386
Mensurationum, Liber (Abū Bakr), 286, 287
Meteorology (Aristotle), 29
Method (Archimedes), 9, 17, 31, 109, 152, 153, 199–224
Metrika, *see Measurements* (Heron)
Meyashsher Aqov (Alfonso of Valladolid), 411
Middle Books (various), 245, 260
Miftāḥ al-ḥisāb (al-Kāshi), 242
Miftāḥ al-saʿāda (Ṭāšköprüzāde), 351
Minhāj al-ṭālib fī taʿdīl al-kawākib (Ibn al-Bannāʾ), *see Zīj* (Ibn al-Bannāʾ)
al-Misāḥa wa al-handasa, Kitāb (Abū Kāmil), *see Mensuration* (Abū Kāmil)
Misāḥat al-dāʾira, *see Measurement of a Circle* (Archimedes)
Mishaʾlim bi-Tishboret, Sefer, 413
Motion of Animals (Aristotle), 153
De motu octave sphere, 116
Moving Sphere (Autolycus), 153
al-Muʿāmalāt, Kitāb (Ibn al-Samḥ), 288
al-Muḥīṭiyya, Risāla (al-Kāshī), 242
al-Mukāfaʾa (Aḥmad ibn Yūsuf), 347–349
Mukhtaṣar (al-Ḥūfī), 277
Mukhtasār fī al-misāḥa (Ibn al-Bannāʾ), 278
Mumtaḥan Zīj (Yaḥyā ibn Abī Manṣūr), *see Zīj al-mumtaḥan*
al-Muntaẓam (Ibn al-Jauzī), 346, 347
Mutawassiṭāt, *see Middle Books*

Natāʾij al-afkār fī sharḥ Rawḍat al-azhār, 299, 305
Natural History (Pliny), 153
Nature (Epicurus), 150
Nature of the Gods (Cicero), 166
Neuses (Apollonius), 81
al-Nisba, Risāla fī (Ibn Haydūr), 285

On, *see* next full word
Optics (Euclid), 32, 85, 114
  Arabic, 106
Optics (al-Muʾtaman), 111
Optics (Ptolemy), 43, 84–85, 114

Parapegma (Geminus), 180, 186, 187
Pearls on the Projection of Spheres (al-Bīrūnī), 83

Perfection (al-Muʾtaman), 85, 105, 106, 111, 115, 240, 277, 278, 280, 282, 284, 285
Phaseis (Phaseis), 152
Phenomena (Euclid), 32, 85, 229, 234
Philokalia (Geminus), 153
Philotegni, Liber (Jordanus de Nemore), 287
Physics (Aristotle), 113, 146
Plane Loci (Apollonius), 81
Planetary Hypotheses (Ptolemy), 160, 165
Planisphere (Ptolemy), 29, 82, 107, 117
Polygonal Numbers (Diophantus), 29, 35
Porisms (Euclid), 78
Practica geometriae (Fibonacci), 360
Primer on the Harmonious Arrangement of Numbers, 122
Projection of the Constellations (al-Bīrūnī), 82
Projection of Rays (Ibn Muʿādh), 120

Qānūn al-masʿūdī (al-Bīrūnī), 84, 91, 305, 310, 311
al-Qarasṭūn (Thābit ibn Qurra), 65
Qaṭʿ al-kura wa al-makhrūṭāt (Archimedes), *see Conoids and Spheroids* (Archimedes)
Qaul fī taṣḥīḥ masāʾil al-jabr bi al-barāhīn al-handasīya (Thābit ibn Qurra), 355
Qawl fī samt al-qibla bi al-ḥisāb (Ibn al-Haytham), 308
Qibla, Book of the (al-Maṣmūdī), 118
Quadratorum, Liber (Fibonacci), 343
Quadrature of the Parabola (Archimedes), 200, 209, 211
Quadrature of the Parabola (Ibrāhīm ibn Sinān), 112
al-Qurb fī al-taksīr wa al-taqṭīʿ (ʿAbdallāh al-Mursī), 286

Rafʿ al-ḥijāb ʿan wujūh aʿmāl al-ḥisāb (Ibn al-Bannāʾ), 277, 278, 281, 283, 284
Rashfat al-ruḍāb min thughūr aʿmāl al-ḥisāb (al-Qaṭrawānī), 276
Rāshīkāt al-Hind, Maqāla fī (al-Bīrūnī), 469, 473, 476
Rational-Sided Right Triangles (al-Khāzin), 56–58
The Ratio of the Segments of a Line that Falls on Three Lines (al-Kūhī), 110
Rawḍat al-azhār fī ʿilm waqt al-layl wa al-nahār (al-Jādirī), 299, 305
Republic (Cicero), 148
Republic (Plato), 148
Risāla, see next full word
Risings and Settings (Autolycus), 153

Saklab al-aʿdād (al-Bīrūnī), 473
Sand Reckoner (Archimedes), 11
Saydi Abuothmi, Liber (Saʿīd Abū ʿUthmān), 287
Sectio Canonis (Euclid), 21
Sector Figure (Jābir ibn Aflaḥ), 413
Sector Figure (Thābit ibn Qurra), 105, 115, 413
Seeing the Stars (Ibn al-Haytham), 84
Sefer, see next full word
Selected Problems (Ibrāhīm ibn Sinān), 107
Shadows (al-Bīrūnī), 63, 305, 306
Sharḥ Kalima min kalimāt al-Aflāṭūniyya, 243
Sharḥ al-iksīr fī ʿilm al-taksīr (Ibn al-Qāḍī), 286
Sharḥ al-ishārāt wa al-tanbīhāt (al-Ṭūsī), 278
Sharḥ Mukhtaṣar al-Ḥūfī (al-ʿUqbānī), 277
Sharḥ al-Talkhīṣ (al-ʿUqbānī), 277
Sharḥ tarjamat kitāb al-maʾkhūdhāt li-Arshimīdis (Archimedes, al-Nasawī), 246
Sharh al-Urjūza al-Yasmīnīya (Ibn al-Hāʾim), 354
al-Shafīya ʿan al-shakk fī al-khuṭūṭ al-mutawāzīya, Risāla (al-Ṭūsī), 348
Sindhind Zīj (al-Khwārizmī), *see Zīj al-Sindhind*
Sīrat Aḥmad ibn Ṭūlūn (al-Balawī), 347
On the Sizes and Distances of the Sun and Moon (Aristarchus), 32, 43, 107
Sketch of Astronomical Hypotheses (Proclus), 160
Solid Geometry (Heron), 398
Solution of the Difficulties in Euclid's Elements (Ibn al-Haytham), 103, 104, 514
Solution of Doubts in the Book of the Almagest, Which a Certain Scholar Has Raised (Ibn al-Haytham), 84
Sphaerica, see Spherics
Sphere (Sacrobosco), 153
Sphere and Cylinder (Archimedes), 31, 35, 80, 87, 107, 110, 200, 211, 259, 261, 262, 264, 266, 268, 411, 413
  Arabic, 80, 243, 245, 262, 285
Spherics (Menelaus), 28, 41, 86, 106, 111, 115, 278, 284, 413
Spherics (Theodosius), 7, 27, 28, 32, 33, 42, 43, 85, 115, 152, 170, 227, 229, 231, 233–235, 413
  Arabic, 29, 107, 170, 231, 255
  Muḥyi al-Dīn al-Maghribī's version, 170
Spiral Lines (Archimedes), 10, 206, 211, 221, 222

Arabic, 80, 243
Spirals (Archimedes), *see Spiral Lines*
Star Catalog (Ṣūfī), 487, 489, 491–493, 495, 497, 499, 500
Stereometrica (Heron), *see Solid Geometry*
Stomachion (Archimedes), 40
Subḥ al-aʿshā fī ṣināʿat al-inshā' (al-Qalqashandī), 351
The Suda, 171
Sundials (Ibrāhīm ibn Sinān), 112
De superficierum divisionibus liber, 287
Surface Measuring (Ibn ʿAbdūn), 112
Ṣuwar al-kawākib al-thābita, Kitāb (Ṣūfī), *see Star Catalog* (Ṣūfī)
Symposia (Plutarch), 152

Table of Minutes (Abū Naṣr Manṣur ibn ʿIrāq), 64
Tabulae Jahen (Ibn Muʿādh), 120, 308, 316
al-Tadhkira fī ʿilm al-hay'a (al-Ṭūsī), 86
Tadhkirat al-kuttāb fī ʿilm al-ḥisāb (Ibn Naqīb), 241, 242
Taḥdīd nihāyāt al-amākin li-taṣḥiḥ masāfāt al-masākin (al-Bīrūnī), 65, 306
al-Taḥlīl wa al-tarkīb, Kitāb (Ibn al-Haytham), 277, 278
Taḥqīq mā li al-Hind min maqūlatin, Kitāb fī (al-Bīrūnī), *see India*
Taḥqīq mā qālahu al-ʿAllāma Ghiyāth al-Dīn Jamshīd fī bayān al-nisba bayn al-muḥīṭ wa al-quṭr, Risāla fī (Taqī al-Dīn), 242
Taḥrīr Kitāb Uqlīdis (Naṣīr al-Dīn al-Ṭūsī), 103, 106, 503, 520
Taḥrīr Kitāb al-kura wa al-usṭuwāna (Archimedes, al-Ṭūsī), 245
al-Ṭair, Kitāb (Abū Kāmil), 352, 354–356, 359
Tajrīd akhbār kutub al-handasa ʿalā ikhtilāf maqāṣidihā (Ibn Munʿim), 285
al-Takmila fī al-ḥisāb (al-Baghdādī), 52, 54, 58
al-Taksīr, Risāla fī (Ibn ʿAbdūn), 286
Talkhīṣ aʿmāl al-ḥisāb (Ibn al-Bannā'), 280, 281, 283, 360
Talqīḥ al-afkār fī al-ʿamal bi rushūm al-ghubār (Ibn al-Yāsamīn), 276, 277, 284, 289
Tanbīh al-albāb ʿalā masā'il al-ḥisāb (Ibn al-Bannā'), 278
al-Tanbīh wa al-tabṣīr fī qawānīn al-taksīr (Ibn al-Raqqām), 286
Tangencies (Apollonius), 81
Tangent Circles (Archimedean), 11
al-Taqrīb wa al-taysīr li ifādat al-mubtadi' bi ṣināʿat al-taksīr, Kitāb (Ibn al-Jayyāb), 286
Tarbīʿ al-dā'ira wa-nisba mu'allafa, 244
T'arīkh (Ibn ʿAsākir), 348
T'arīkh (al-Ṭabarī), 348
Tashrīḥ al-aflāk (al-ʿĀmilī), 487
Tasṭīḥ al-kuwar (al-Bīrūnī), 65
Tetrabiblos (Ptolemy), *see Four Books, Mathematical Treatise in*
Theaetetus (Plato), 4, 8, 9, 20
Timaeus (Plato), 153, 430
Toledan Tables, 116, 303, 309, 320
Topics (Aristotle), 18
Tracing the Conic Sections (Ibrāhīm ibn Sinān), 112
Treatise, *see* next full word
The True Perplexity (Maimonedes), 535
al-Tuḥfa al-shāhiyya (Shīrāzī), 489, 491, 492, 497
Tuḥfat al-ṭullāb fī sharḥ mā ashkala min Rafʿ al-ḥijāb (Ibn Haydūr), 283, 284

'Umarā' Dimashq (al-Ṣafadī), 348
Unmūdhaj al-ʿulūm (al-Fanārī, Muḥammad Shāh), 240, 241, 243
al-Uṣūl, Kitāb (Ibn al-Bannā'), 276, 281, 283
ʿUyūn al-anbā' fī ṭabaqāt al-aṭibbā (Uṣaibiʿa), 347

al-Waṣāyā, Kitāb (Abū Kāmil), 346, 349, 350, 355, 356, 359
al-Waṣāyā bi al-judhūr, Kitāb (al-Ḥajjāj), 356
al-Waṣāyā bi al-jabr wa al-muqābala, Kitāb (Abū Kāmil), 356
What the Artisan Requires of Geometrical Constructions (Abū al-Wafā'), 112
What is Seen of Sky and Sea (al-Kūhī), 120
al-Wuzarā' wa al-kuttāb, Kitāb (al-Jahshiyārī), 348

Zīj-i Ashrafī, 90
Zīj (Ibn al-Bannā'), 297, 298, 302, 306, 307, 312–317
Zīj (al-Battānī), 304, 307–311, 320, 322, 354
Zīj (Ḥabash), 306
Zīj (Ibn al-Hā'im), 116
Zīj al-Ḥākimī (Ibn Yūnus), 63, 64
Zīj-i Īlkhānī (al-Ṭūsī), 87, 493
Zīj (Ibn Isḥāq), 297–299, 302, 303
Zīj (unknown from Ibn Isḥāq), 297, 298, 302, 306–308, 312–318, 320–324
Zīj al-jadīd (Ulugh Beg), 488, 493
Zīj-i jāmiʿ (Kūshyar ibn Labbān), 89, 119

Zīj al-kāmil fī al-Taʿālīm (Ibn al-Hāʾim), 298, 303, 316
Zīj (Ibn al-Kammād), 306
Zīj-i Khāqānī (al-Kāshī), 87, 91, 116, 118, 120
Zīj (al-Khwārizmī), 309, 321, 322, 325, 354
Zīj (Maslama from al-Khwārizmī), 301, 303, 311, 321, 322
Zīj al-mumtaḥan (Yaḥyā ibn Abī Manṣūr), 118, 310, 316
Zīj al-muqtabas (Ibn al-Kammād), 306, 315, 316, 318, 321, 322
Latin, 306
Zīj al-mustawfī (Ibn al-Raqqām), 297, 299, 300, 302, 305, 307, 310, 313, 320, 322
Zīj-i Nāsirī, 118
Zīj al-qawīm (Ibn al-Raqqām), 297–299, 301, 302, 312–317, 319–325
Zīj al-shāh (al-Khwārizmī), 316
Zīj al-shāmil (Ibn al-Raqqām), 297–299, 302–304, 307, 310, 312–317, 319–323, 325
Zīj al-Sindhind (al-Khwārizmī), 75, 76, 118